Complete Solutions Guide

Trigonometry

SIXTH EDITION

Larson/Hostetler

Dianna L. Zook
Indiana University – Purdue University
Fort Wayne, Indiana

HOUGHTON MIFFLIN COMPANY Boston New York

Printed in the United States of America
ISBN 0-618-31799-6
23456789-CS-07 06 05 04

CONTENTS

PART I

CHAPTER P
Prerequisites

CHAPTER P
Prerequisites

Section P.1 Review of Real Numbers and Their Properties

- You should know the following sets.
  (a) The set of real numbers includes the rational numbers and the irrational numbers.
  (b) The set of rational numbers includes all real numbers that can be written as the ratio p/q of two integers, where $q \neq 0$.
  (c) The set of irrational numbers includes all real numbers which are not rational.
  (d) The set of integers: $\{. . . , -3, -2, -1, 0, 1, 2, 3, . . .\}$
  (e) The set of whole numbers: $\{0, 1, 2, 3, 4, . . .\}$
  (f) The set of natural numbers: $\{1, 2, 3, 4, . . .\}$
- The real number line is used to represent the real numbers.
- Know the inequality symbols.
  (a) $a < b$ means a is less than b.
  (b) $a \leq b$ means a is less than or equal to b.
  (c) $a > b$ means a is greater than b.
  (d) $a \geq b$ means a is greater than or equal to b.
- You should know that

$$|a| = \begin{cases} a, \text{ if } a \geq 0 \\ -a, \text{ if } a < 0. \end{cases}$$

- Know the properties of absolute value.
  (a) $|a| \geq 0$
  (b) $|-a| = |a|$
  (c) $|ab| = |a|\,|b|$
  (d) $\left|\frac{a}{b}\right| = \frac{|a|}{|b|}, b \neq 0$
- The distance between a and b on the real line is $d(a, b) = |b - a| = |a - b|$.
- You should be able to identify the terms in an algebraic expression.
- You should know and be able to use the basic rules of algebra.
- Commutative Property
  (a) Addition: $a + b = b + a$
  (b) Multiplication: $a \cdot b = b \cdot a$
- Associative Property
  (a) Addition: $(a + b) + c = a + (b + c)$
  (b) Multiplication: $(ab)c = a(bc)$
- Identity Property
  (a) Addition: 0 is the identity; $a + 0 = 0 + a = a$.
  (b) Multiplication: 1 is the identity; $a \cdot 1 = 1 \cdot a = a$.
- Inverse Property
  (a) Addition: $-a$ is the additive inverse of a; $a + (-a) = -a + a = 0$.
  (b) Multiplication: $1/a$ is the multiplicative inverse of a, $a \neq 0$; $a(1/a) = (1/a)a = 1$.
- Distributive Property
  (a) $a(b + c) = ab + ac$
  (b) $(a + b)c = ac + bc$

—CONTINUED—

■ Properties of Negation

(a) $(-1)a = -a$

(b) $-(-a) = a$

(c) $(-a)b = a(-b) = -ab$

(d) $(-a)(-b) = ab$

(e) $-(a + b) = (-a) + (-b) = -a - b$

■ Properties of Equality

(a) If $a = b$, then $a + c = b + c$.

(b) If $a = b$, then $ac = bc$.

(c) If $a + c = b + c$, then $a = b$.

(d) If $ac = bc$ and $c \neq 0$, then $a = b$.

■ Properties of Zero

(a) $a \pm 0 = a$

(b) $a \cdot 0 = 0$

(c) $0 \div a = 0/a = 0, a \neq 0$

(d) $a/0$ is undefined.

(e) If $ab = 0$, then $a = 0$ or $b = 0$.

■ Properties of Fractions ($b \neq 0, d \neq 0$)

(a) Equivalent Fractions: $a/b = c/d$ if and only if $ad = bc$.

(b) Rule of Signs: $-a/b = a/-b = -(a/b)$ and $-a/-b = a/b$

(c) Equivalent Fractions: $a/b = ac/bc, c \neq 0$

(d) Addition and Subtraction

1. Like Denominators: $(a/b) \pm (c/b) = (a \pm c)/b$

2. Unlike Denominators: $(a/b) \pm (c/d) = (ad \pm bc)/bd$

(e) Multiplication: $(a/b) \cdot (c/d) = (ac)/(bd)$

(f) Division: $(a/b) \div (c/d) = (a/b) \cdot (d/c) = (ad)/(bc)$ if $c \neq 0$.

Solutions to Odd-Numbered Exercises

1. $-9, -\frac{7}{2}, 5, \frac{2}{3}, \sqrt{2}, 0, 1, -4, 2, -11$

(a) Natural numbers: 5, 1, 2

(b) Integers: $-9, 5, 0, 1, -4, 2, -11$

(c) Rational numbers: $-9, -\frac{7}{2}, 5, \frac{2}{3}, 0, 1, -4, 2, -11$

(d) Irrational numbers: $\sqrt{2}$

3. $2.01, 0.666\ldots, -13, 0.010110111\ldots, 1, -6$

(a) Natural numbers: 1

(b) Integers: $-13, 1, -6$

(c) Rational numbers: $2.01, 0.666\ldots, -13, 1, -6$

(d) Irrational numbers: $0.010110111\ldots$

5. $-\pi, -\frac{1}{3}, \frac{6}{3}, \frac{1}{2}\sqrt{2}, -7.5, -1, 8, -22$

(a) Natural numbers: $\frac{6}{3}$ (since it equals 2), 8

(b) Integers: $\frac{6}{3}, -1, 8, -22$

(c) Rational numbers: $-\frac{1}{3}, \frac{6}{3}, -7.5, -1, 8, -22$

(d) Irrational numbers: $-\pi, \frac{1}{2}\sqrt{2}$

7. $\frac{5}{8} = 0.625$

9. $\frac{41}{333} = 0.\overline{123}$

11. $-1 < 2.5$

13. $-4 > -8$

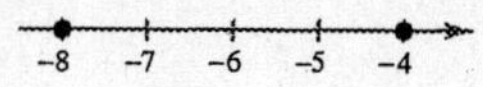

15. $\frac{3}{2} < 7$

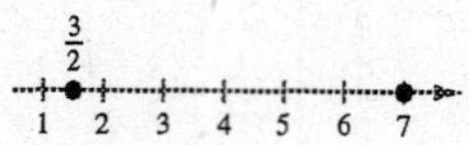

17. $\frac{5}{6} > \frac{2}{3}$

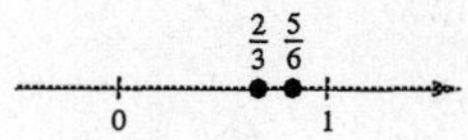

19. The inequality $x \leq 5$ denotes the set of all real numbers less than or equal to 5. The interval is unbounded.

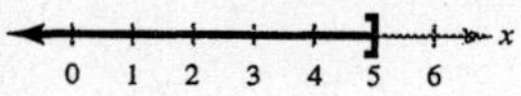

21. The inequality $x < 0$ denotes the set of all negative real numbers. The interval is unbounded.

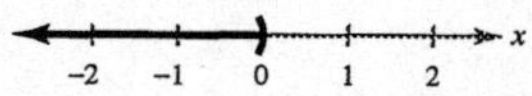

23. The inequality $x \geq 4$ denotes the set of all real numbers greater than or equal to 4. The interval is unbounded.

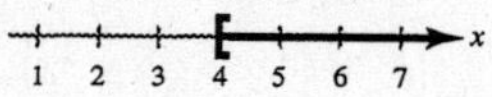

25. The inequality $-2 < x < 2$ denotes the set of all real numbers greater than -2 and less than 2. The interval is bounded.

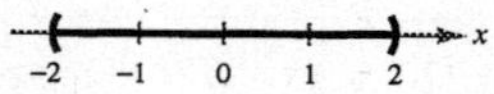

27. The inequality $-1 \leq x < 0$ denotes the set of all negative real numbers greater than or equal to -1. The interval is bounded.

29. $-2 < x \leq 4$

31. $y \geq 0$

33. $10 \leq t \leq 22$

35. $W > 65$

37. This interval consists of all real numbers greater than or equal to zero, but less than 8.

39. This interval consists of all real numbers greater than -6.

41. $|-10| = -(-10) = 10$

43. $|3 - 8| = |-5| = -(-5) = 5$

45. $|-1| - |-2| = 1 - 2 = -1$

47. $\dfrac{-5}{|-5|} = \dfrac{-5}{-(-5)} = \dfrac{-5}{5} = -1$

49. If $x < -2$, then $x + 2$ is negative.

Thus $\dfrac{|x+2|}{x+2} = \dfrac{-(x+2)}{x+2} = -1.$

51. $|-3| > -|-3|$ since $3 > -3$.

53. $-5 = -|5|$ since $-5 = -5$.

55. $-|-2| = -|2|$ since $-2 = -2$.

57. $d(-1, 3) = |3 - (-1)| = |3 + 1| = 4$

59. $d(126, 75) = |75 - 126| = 51$

61. $d\left(-\frac{5}{2}, 0\right) = \left|0 - \left(-\frac{5}{2}\right)\right| = \frac{5}{2}$

63. $d\left(\frac{16}{5}, \frac{112}{75}\right) = \left|\frac{112}{75} - \frac{16}{5}\right| = \frac{128}{75}$

65.

Budgeted Expense, b	*Actual Expense, a*	$\lvert a - b\rvert$	$0.05b$
\$112,700	\$113,356	\$656	\$5635

The actual expense difference is greater than \$500 (but is less than 5% of the budget) so the actual expense does not pass the test.

67. *Budgeted Expense, b* | *Actual Expense, a* | $|a - b|$ | $0.05b$

Budgeted Expense, b	*Actual Expense, a*	$\lvert a - b \rvert$	$0.05b$
\$37,640	\$37,335	\$305	\$1882

Since \$305 < \$500 and \$305 < \$1882, the actual expense passes the "budget variance test."

69. (a)

Year	Expenditures (in billions)	Surplus or Deficit (in billions)
1960	\$92.2	$\lvert 92.5 - 92.2 \rvert$ = \$0.3 surplus
1970	\$195.6	$\lvert 192.8 - 195.6 \rvert$ = \$2.8 deficit
1980	\$590.9	$\lvert 517.1 - 590.9 \rvert$ = \$73.8 deficit
1990	\$1253.2	$\lvert 1032.0 - 1253.2 \rvert$ = \$221.2 deficit
2000	\$1788.8	$\lvert 2025.2 - 1788.8 \rvert$ = \$236.4 surplus

(b)

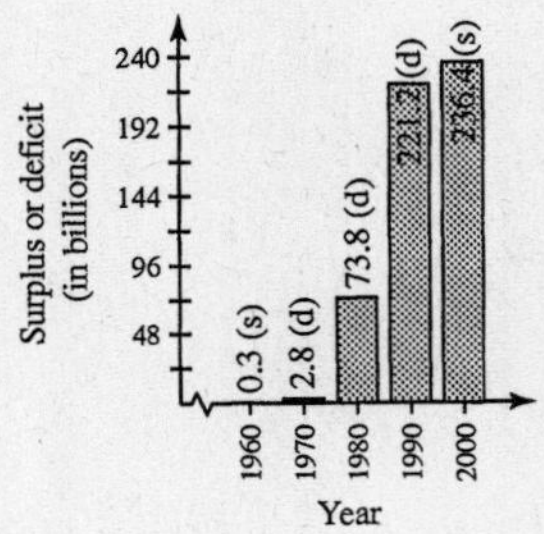

71. $d(57, 236) = |236 - 57| = 179$ miles

73. $d(23°, 60°) = |60° - 23°| = 37°$

75. $d(x, 5) = |x - 5|$ and $d(x, 5) \le 3$, thus $|x - 5| \le 3$.

77. $d(y, 0) = |y - 0| = |y|$ and $d(y, 0) \ge 6$, thus $|y| \ge 6$.

79. $7x + 4$

Terms: $7x, 4$

Coefficient: 7

81. $\sqrt{3}x^2 - 8x - 11$

Terms: $\sqrt{3}x^2, -8x, -11$

Coefficients: $\sqrt{3}, -8$

83. $4x^3 + \frac{x}{2} - 5$

Terms: $4x^3, \frac{x}{2}, -5$

Coefficients: $4, \frac{1}{2}$

85. $4x - 6$

(a) $4(-1) - 6 = -4 - 6 = -10$

(b) $4(0) - 6 = 0 - 6 = -6$

87. $x^2 - 3x + 4$

(a) $(-2)^2 - 3(-2) + 4 = 4 + 6 + 4 = 14$

(b) $(2)^2 - 3(2) + 4 = 4 - 6 + 4 = 2$

89. $\frac{x+1}{x-1}$

(a) $\frac{1+1}{1-1} = \frac{2}{0}$

Division by zero is undefined

(b) $\frac{-1+1}{-1-1} = \frac{0}{-2} = 0$

91. $x + 9 = 9 + x$

Commutative Property of Addition

93. $\frac{1}{(h+6)}(h+6) = 1, h \neq -6$

Multiplicative Inverse Property

95. $2(x+3) = 2x+6$

Distributive Property

97. $1 \cdot (1+x) = 1+x$

Multiplicative Identity Property

99. $x(3y) = (x \cdot 3)y$ Associative Property of Multiplication

$= (3x)y$ Commutative Property of Multiplication

101. $\frac{3}{16} + \frac{5}{16} = \frac{8}{16} = \frac{1}{2}$

103. $\frac{5}{8} - \frac{5}{12} + \frac{1}{6} = \frac{15}{24} - \frac{10}{24} + \frac{4}{24} = \frac{9}{24} = \frac{3}{8}$

105. $12 \div \frac{1}{4} = 12 \cdot \frac{4}{1} = 12 \cdot 4 = 48$

107. $\frac{2x}{3} - \frac{x}{4} = \frac{8x}{12} - \frac{3x}{12} = \frac{5x}{12}$

109. (a)

n	1	0.5	0.01	0.0001	0.000001
$5/n$	5	10	500	50,000	5,000,000

(b) The value of $\frac{5}{n}$ approaches infinity as n approaches 0.

111. False. If $a < b$, then $\frac{1}{a} > \frac{1}{b}$, where $a \neq b \neq 0$.

113. (a) $|u+v| \neq |u| + |v|$ if u is positive and v is negative or vice versa.

(b) $|u+v| \leq |u| + |v|$

They are equal when u and v have the same sign. If they differ in sign, $|u+v|$ is less than $|u| + |v|$.

115. The only even prime number is 2, because its factors are itself and 1.

117. (a) Since $A > 0$, $-A < 0$. The expression is negative.

(b) Since $B < A$, $B - A < 0$. The expression is negative.

119. Yes, if a is a negative number, then $-a$ is positive. Thus, $|a| = -a$ if a is negative.

Section P.2 Solving Equations

- You should know how to solve linear equations.
  $ax + b = 0$
- An identity is an equation whose solution consists of every real number in its domain.
- To solve an equation you can:
  (a) Add or subtract the same quantity from both sides.
  (b) Multiply or divide both sides by the same nonzero quantity.
- To solve an equation that can be simplified to a linear equation:
  (a) Remove all symbols of grouping and all fractions.
  (b) Combine like terms.
  (c) Solve by algebra.
  (d) Check the answer.
- A "solution" that does not satisfy the original equation is called an extraneous solution.
- You should be able to solve a quadratic equation by factoring, if possible.
- You should be able to solve a quadratic equation of the form $u^2 = d$ by extracting square roots.
- You should be able to solve a quadratic equation by completing the square.
- You should know and be able to use the Quadratic Formula: For $ax^2 + bx + c = 0, a \neq 0,$

$$x = \frac{-b \pm \sqrt{b^2 - 4ac}}{2a}.$$

- You should be able to solve polynomials of higher degree by factoring.
- For equations involving radicals or fractional powers, raise both sides to the same power.
- For equations with fractions, multiply both sides by the least common denominator to clear the fractions.
- For equations involving absolute value, remember that the expression inside the absolute value can be positive or negative.

Solutions to Odd-Numbered Exercises

1. $2(x - 1) = 2x - 2$ is an *identity* by the Distributive Property. It is true for all real values of x.

3. $-6(x - 3) + 5 = -2x + 10$ is *conditional.* There are real values of x for which the equation is not true.

5. $4(x + 1) - 2x = 4x + 4 - 2x = 2x + 4 = 2(x + 2)$
This is an *identity* by simplification. It is true for all real values of x.

7. $x^2 - 8x + 5 = (x - 4)^2 - 11$ is an *identity* since $(x - 4)^2 - 11 = x^2 - 8x + 16 - 11 = x^2 - 8x + 5$.

9. $3 + \dfrac{1}{x + 1} = \dfrac{4x}{x + 1}$ is *conditional.* There are real values of x for which the equation is not true.

11.
$$x + 11 = 15$$
$$x + 11 - 11 = 15 - 11$$
$$x = 4$$

13.
$$7 - 2x = 25$$
$$7 - 7 - 2x = 25 - 7$$
$$-2x = 18$$
$$\frac{-2x}{-2} = \frac{18}{-2}$$
$$x = -9$$

15.
$$8x - 5 = 3x + 20$$
$$8x - 3x - 5 = 3x - 3x + 20$$
$$5x - 5 = 20$$
$$5x - 5 + 5 = 20 + 5$$
$$5x = 25$$
$$\frac{5x}{5} = \frac{25}{5}$$
$$x = 5$$

17.
$$2(x + 5) - 7 = 3(x - 2)$$
$$2x + 10 - 7 = 3x - 6$$
$$2x + 3 = 3x - 6$$
$$-x = -9$$
$$x = 9$$

19.
$$x - 3(2x + 3) = 8 - 5x$$
$$x - 6x - 9 = 8 - 5x$$
$$-5x - 9 = 8 - 5x$$
$$-5x + 5x - 9 = 8 - 5x + 5x$$
$$-9 \neq 8$$
No solution

21.
$$\frac{5x}{4} + \frac{1}{2} = x - \frac{1}{2}$$
$$4\left(\frac{5x}{4}\right) + 4\left(\frac{1}{2}\right) = 4(x) - 4\left(\frac{1}{2}\right)$$
$$5x + 2 = 4x - 2$$
$$x = -4$$

23.
$$\tfrac{3}{2}(z + 5) - \tfrac{1}{4}(z + 24) = 0$$
$$4(\tfrac{3}{2})(z + 5) - 4(\tfrac{1}{4})(z + 24) = 4(0)$$
$$6(z + 5) - (z + 24) = 0$$
$$6z + 30 - z - 24 = 0$$
$$5z = -6$$
$$z = -\tfrac{6}{5}$$

25.
$$0.25x + 0.75(10 - x) = 3$$
$$4(0.25x) + 4(0.75)(10 - x) = 4(3)$$
$$x + 3(10 - x) = 12$$
$$x + 30 - 3x = 12$$
$$-2x = -18$$
$$x = 9$$

27.
$$x + 8 = 2(x - 2) - x$$
$$x + 8 = 2x - 4 - x$$
$$x + 8 = x - 4$$
$$8 \neq -4$$
No solution

29.
$$\frac{100 - 4x}{3} = \frac{5x + 6}{4} + 6$$
$$12\left(\frac{100 - 4x}{3}\right) = 12\left(\frac{5x + 6}{4}\right) + 12(6)$$
$$4(100 - 4x) = 3(5x + 6) + 72$$
$$400 - 16x = 15x + 18 + 72$$
$$-31x = -310$$
$$x = 10$$

31.
$$\frac{5x - 4}{5x + 4} = \frac{2}{3}$$
$$3(5x - 4) = 2(5x + 4)$$
$$15x - 12 = 10x + 8$$
$$5x = 20$$
$$x = 4$$

33.
$$10 - \frac{13}{x} = 4 + \frac{5}{x}$$
$$\frac{10x - 13}{x} = \frac{4x + 5}{x}$$
$$10x - 13 = 4x + 5$$
$$6x = 18$$
$$x = 3$$

35.
$$3 = 2 + \frac{2}{z + 2}$$
$$3(z + 2) = \left(2 + \frac{2}{z + 2}\right)(z + 2)$$
$$3z + 6 = 2z + 4 + 2$$
$$z = 0$$

37.
$$\frac{x}{x + 4} + \frac{4}{x + 4} + 2 = 0$$
$$\frac{x + 4}{x + 4} + 2 = 0$$
$$1 + 2 = 0$$
$$3 \neq 0$$
Contradiction : no solution
The variable is divided out.

39. $\dfrac{2}{(x-4)(x-2)} = \dfrac{1}{x-4} + \dfrac{2}{x-2}$ Multiply both sides by $(x-4)(x-2)$.

$$2 = 1(x-2) + 2(x-4)$$
$$2 = x - 2 + 2x - 8$$
$$2 = 3x - 10$$
$$12 = 3x$$
$$4 = x$$

A check reveals that $x = 4$ is an extraneous solution–it makes the denominator zero. There is no real solution.

41. $\dfrac{1}{x-3} + \dfrac{1}{x+3} = \dfrac{10}{x^2-9}$

$$\frac{(x+3)+(x-3)}{x^2-9} = \frac{10}{x^2-9}$$
$$2x = 10$$
$$x = 5$$

43. $\dfrac{3}{x^2-3x} + \dfrac{4}{x} = \dfrac{1}{x-3}$ Multiply both sides by $x(x-3)$.

$$3 + 4(x-3) = x$$
$$3 + 4x - 12 = x$$
$$3x = 9$$
$$x = 3$$

A check reveals that $x = 3$ is an extraneous solution, so there is no solution.

45. $(x+2)^2 + 5 = (x+3)^2$

$$x^2 + 4x + 4 + 5 = x^2 + 6x + 9$$
$$4x + 9 = 6x + 9$$
$$-2x = 0$$
$$x = 0$$

47. $(x+2)^2 - x^2 = 4(x+1)$

$$x^2 + 4x + 4 - x^2 = 4x + 4$$
$$4 = 4$$

The equation is an identity; every real number is a solution.

49. $2x^2 = 3 - 8x$

General form: $2x^2 + 8x - 3 = 0$

51. $(x-3)^2 = 3$

$x^2 - 6x + 9 = 3$

General form: $x^2 - 6x + 6 = 0$

53. $\frac{1}{5}(3x^2 - 10) = 18x$

$3x^2 - 10 = 90x$

General form: $3x^2 - 90x - 10 = 0$

55. $6x^2 + 3x = 0$

$3x(2x+1) = 0$

$3x = 0$ or $2x + 1 = 0$

$x = 0$ or $x = -\frac{1}{2}$

57. $x^2 - 2x - 8 = 0$

$(x-4)(x+2) = 0$

$x - 4 = 0$ or $x + 2 = 0$

$x = 4$ or $x = -2$

59. $x^2 + 10x + 25 = 0$

$(x+5)(x+5) = 0$

$x + 5 = 0$

$x = -5$

61. $3 + 5x - 2x^2 = 0$

$(3-x)(1+2x) = 0$

$3 - x = 0$ or $1 + 2x = 0$

$x = 3$ or $x = -\frac{1}{2}$

63. $x^2 + 4x = 12$

$x^2 + 4x - 12 = 0$

$(x+6)(x-2) = 0$

$x + 6 = 0$ or $x - 2 = 0$

$x = -6$ or $x = 2$

65. $\frac{3}{4}x^2 + 8x + 20 = 0$

$4(\frac{3}{4}x^2 + 8x + 20) = 4(0)$

$3x^2 + 32x + 80 = 0$

$(3x + 20)(x + 4) = 0$

$3x + 20 = 0$ or $x + 4 = 0$

$x = -\frac{20}{3}$ or $x = -4$

67. $x^2 + 2ax + a^2 = 0$

$(x + a)^2 = 0$

$x + a = 0$

$x = -a$

69. $x^2 = 49$

$x = \pm\sqrt{49}$

$= \pm 7$

$= \pm 7.00$

71. $x^2 = 11$

$x = \pm\sqrt{11}$

$x \approx \pm 3.32$

73. $3x^2 = 81$

$x^2 = 27$

$x = \pm\sqrt{27} = \pm 3\sqrt{3}$

$x \approx \pm 5.20$

75. $(x - 12)^2 = 16$

$x - 12 = \pm\sqrt{16}$

$x = 12 \pm 4$

$x = 16$ or $x = 8$

$x = 16.00$ or $x = 8.00$

77. $(x + 2)^2 = 14$

$x + 2 = \pm\sqrt{14}$

$x = -2 \pm \sqrt{14}$

$x \approx 1.74$ or $x \approx -5.74$

79. $(2x - 1)^2 = 18$

$2x - 1 = \pm\sqrt{18}$

$2x = 1 \pm 3\sqrt{2}$

$x = \dfrac{1 \pm 3\sqrt{2}}{2}$

$x \approx 2.62$ or $x \approx -1.62$

81. $(x - 7)^2 = (x + 3)^2$

$x - 7 = \pm(x + 3)$

$x - 7 = x + 3$ or $x - 7 = -x - 3$

$-7 \neq 3$ $\quad 2x = 4$

No solution $\quad x = 2 = 2.00$

83. $x^2 - 2x = 0$

$x^2 - 2x + 1 = 0 + 1$

$(x - 1)^2 = 1$

$x - 1 = \pm\sqrt{1}$

$x = 1 \pm 1$

$x = 0$ or $x = 2$

85. $x^2 + 4x - 32 = 0$

$x^2 + 4x = 32$

$x^2 + 4x + 2^2 = 32 + 2^2$

$(x + 2)^2 = 36$

$x + 2 = \pm\sqrt{36}$

$x = -2 \pm 6$

$x = 4$ or $x = -8$

87. $x^2 + 6x + 2 = 0$

$x^2 + 6x = -2$

$x^2 + 6x + 3^2 = -2 + 3^2$

$(x + 3)^2 = 7$

$x + 3 = \pm\sqrt{7}$

$x = -3 \pm \sqrt{7}$

89. $9x^2 - 18x = -3$

$x^2 - 2x = -\dfrac{1}{3}$

$x^2 - 2x + 1 = -\dfrac{1}{3} + 1$

$(x - 1)^2 = \dfrac{2}{3}$

$x - 1 = \pm\sqrt{\dfrac{2}{3}}$

$x = 1 \pm \sqrt{\dfrac{6}{9}}$

$x = 1 \pm \dfrac{\sqrt{6}}{3}$

91. $8 + 4x - x^2 = 0$

$$-x^2 + 4x + 8 = 0$$
$$x^2 - 4x - 8 = 0$$
$$x^2 - 4x = 8$$
$$x^2 - 4x + 2^2 = 8 + 2^2$$
$$(x-2)^2 = 12$$
$$x - 2 = \pm\sqrt{12}$$
$$x = 2 \pm 2\sqrt{3}$$

93. $2x^2 + x - 1 = 0$

$$x = \frac{-b \pm \sqrt{b^2 - 4ac}}{2a}$$
$$= \frac{-1 \pm \sqrt{1^2 - 4(2)(-1)}}{2(2)}$$
$$= \frac{-1 \pm 3}{4} = \frac{1}{2}, -1$$

95. $16x^2 + 8x - 3 = 0$

$$x = \frac{-b \pm \sqrt{b^2 - 4ac}}{2a}$$
$$= \frac{-8 \pm \sqrt{8^2 - 4(16)(-3)}}{2(16)}$$
$$= \frac{-8 \pm 16}{32} = \frac{1}{4}, -\frac{3}{4}$$

97. $2 + 2x - x^2 = 0$

$$x = \frac{-b \pm \sqrt{b^2 - 4ac}}{2a}$$
$$= \frac{-2 \pm \sqrt{2^2 - 4(-1)(2)}}{2(-1)}$$
$$= \frac{-2 \pm 2\sqrt{3}}{-2} = 1 \pm \sqrt{3}$$

99. $x^2 + 14x + 44 = 0$

$$x = \frac{-b \pm \sqrt{b^2 - 4ac}}{2a}$$
$$= \frac{-14 \pm \sqrt{14^2 - 4(1)(44)}}{2(1)}$$
$$= \frac{-14 \pm 2\sqrt{5}}{2} = -7 \pm \sqrt{5}$$

101. $x^2 + 8x - 4 = 0$

$$x = \frac{-b \pm \sqrt{b^2 - 4ac}}{2a}$$
$$= \frac{-8 \pm \sqrt{8^2 - 4(1)(-4)}}{2(1)}$$
$$= \frac{-8 \pm 4\sqrt{5}}{2}$$
$$= -4 \pm 2\sqrt{5}$$

103. $12x - 9x^2 = -3$

$$-9x^2 + 12x + 3 = 0$$
$$x = \frac{-b \pm \sqrt{b^2 - 4ac}}{2a}$$
$$= \frac{-12 \pm \sqrt{12^2 - 4(-9)(3)}}{2(-9)}$$
$$= \frac{-12 \pm 6\sqrt{7}}{-18} = \frac{2}{3} \pm \frac{\sqrt{7}}{3}$$

105. $9x^2 + 24x + 16 = 0$

$$x = \frac{-b \pm \sqrt{b^2 - 4ac}}{2a}$$
$$= \frac{-24 \pm \sqrt{24^2 - 4(9)(16)}}{2(9)}$$
$$= \frac{-24 \pm 0}{18}$$
$$= -\frac{4}{3}$$

107. $4x^2 + 4x = 7$

$$4x^2 + 4x - 7 = 0$$
$$x = \frac{-b \pm \sqrt{b^2 - 4ac}}{2a}$$
$$= \frac{-4 \pm \sqrt{4^2 - 4(4)(-7)}}{2(4)}$$
$$= \frac{-4 \pm 8\sqrt{2}}{8} = -\frac{1}{2} \pm \sqrt{2}$$

109. $28x - 49x^2 = 4$

$$-49x^2 + 28x - 4 = 0$$
$$x = \frac{-b \pm \sqrt{b^2 - 4ac}}{2a}$$
$$= \frac{-28 \pm \sqrt{28^2 - 4(-49)(-4)}}{2(-49)}$$
$$= \frac{-28 \pm 0}{-98} = \frac{2}{7}$$

111. $8t = 5 + 2t^2$

$$-2t^2 + 8t - 5 = 0$$
$$t = \frac{-b \pm \sqrt{b^2 - 4ac}}{2a}$$
$$= \frac{-8 \pm \sqrt{8^2 - 4(-2)(-5)}}{2(-2)}$$
$$= \frac{-8 \pm 2\sqrt{6}}{-4} = 2 \pm \frac{\sqrt{6}}{2}$$

113. $(y - 5)^2 = 2y$

$$y^2 - 12y + 25 = 0$$

$$x = \frac{-b \pm \sqrt{b^2 - 4ac}}{2a}$$

$$= \frac{-(-12) \pm \sqrt{(-12)^2 - 4(1)(25)}}{2(1)}$$

$$= \frac{12 \pm 2\sqrt{11}}{2} = 6 \pm \sqrt{11}$$

115. $\frac{1}{2}x^2 + \frac{3}{8}x = 2$

$$4x^2 + 3x = 16$$

$$4x^2 + 3x - 16 = 0$$

$$x = \frac{-b \pm \sqrt{b^2 - 4ac}}{2a}$$

$$= \frac{-3 \pm \sqrt{3^2 - 4(4)(-16)}}{2(4)}$$

$$= \frac{-3 \pm \sqrt{265}}{8} = -\frac{3}{8} \pm \frac{\sqrt{265}}{8}$$

117. $5.1x^2 - 1.7x - 3.2 = 0$

$$x = \frac{1.7 \pm \sqrt{(-1.7)^2 - 4(5.1)(-3.2)}}{2(5.1)}$$

$$x \approx 0.976, -0.643$$

119. $-0.067x^2 - 0.852x + 1.277 = 0$

$$x = \frac{-(-0.852) \pm \sqrt{(-0.852)^2 - 4(-0.067)(1.277)}}{2(-0.067)}$$

$$x \approx -14.071, 1.355$$

121. $422x^2 - 506x - 347 = 0$

$$x = \frac{506 \pm \sqrt{(-506)^2 - 4(422)(-347)}}{2(422)}$$

$$x \approx 1.687, -0.488$$

123. $12.67x^2 + 31.55x + 8.09 = 0$

$$x = \frac{-31.55 \pm \sqrt{(31.55)^2 - 4(12.67)(8.09)}}{2(12.67)}$$

$$x \approx -2.200, -0.290$$

125. $x^2 - 2x - 1 = 0$

$$x^2 - 2x = 1$$

$$x^2 - 2x + 1^2 = 1 + 1^2$$

$$(x - 1)^2 = 2$$

$$x - 1 = \pm\sqrt{2}$$

$$x = 1 \pm \sqrt{2}$$

127. $(x + 3)^2 = 81$

$$x + 3 = \pm 9$$

$x + 3 = 9$ or $x + 3 = -9$

$x = 6$ or $x = -12$

129. $x^2 - x - \frac{11}{4} = 0$ Complete the Square

$$x^2 - x = \frac{11}{4}$$

$$x^2 - x + \left(\frac{1}{2}\right)^2 = \frac{11}{4} + \left(\frac{1}{2}\right)^2$$

$$\left(x - \frac{1}{2}\right)^2 = \frac{12}{4}$$

$$x - \frac{1}{2} = \pm\sqrt{\frac{12}{4}}$$

$$x = \frac{1}{2} \pm \sqrt{3}$$

131. $(x + 1)^2 = x^2$ Extract Square Roots

$$x^2 = (x + 1)^2$$

$$x = \pm(x + 1)$$

For $x = +(x + 1)$:

$0 \neq 1$ No solution

For $x = -(x + 1)$:

$$2x = -1$$

$$x = -\frac{1}{2}$$

133. $3x + 4 = 2x^2 - 7$ Quadratic Formula

$$0 = 2x^2 - 3x - 11$$

$$x = \frac{-(-3) \pm \sqrt{(-3)^2 - 4(2)(-11)}}{2(2)}$$

$$= \frac{3 \pm \sqrt{97}}{4} = \frac{3}{4} \pm \frac{\sqrt{97}}{4}$$

135. $4x^4 - 18x^2 = 0$

$$2x^2(2x^2 - 9) = 0$$

$$2x^2 = 0 \Rightarrow x = 0$$

$$2x^2 - 9 = 0 \Rightarrow x = \pm\frac{3\sqrt{2}}{2}$$

137. $x^4 - 81 = 0$

$$(x^2 + 9)(x + 3)(x - 3) = 0$$

$x^2 + 9 = 0 \Rightarrow$ No real solution

$$x + 3 = 0 \Rightarrow x = -3$$

$$x - 3 = 0 \Rightarrow x = 3$$

139. $x^3 + 216 = 0$

$$x^3 + 6^3 = 0$$

$$(x + 6)(x^2 - 6x + 36) = 0$$

$$x + 6 = 0 \Rightarrow x = -6$$

$x^2 - 6x + 36 = 0 \Rightarrow$ No real solution

(By the Quadratic Formula)

141. $5x^3 + 30x^2 + 45x = 0$

$$5x(x^2 + 6x + 9) = 0$$

$$5x(x + 3)^2 = 0$$

$$5x = 0 \Rightarrow x = 0$$

$$x + 3 = 0 \Rightarrow x = -3$$

143. $x^3 - 3x^2 - x + 3 = 0$

$$x^2(x - 3) - (x - 3) = 0$$

$$(x - 3)(x^2 - 1) = 0$$

$$(x - 3)(x + 1)(x - 1) = 0$$

$$x - 3 = 0 \Rightarrow x = 3$$

$$x + 1 = 0 \Rightarrow x = -1$$

$$x - 1 = 0 \Rightarrow x = 1$$

145. $x^4 - x^3 + x - 1 = 0$

$$x^3(x - 1) + (x - 1) = 0$$

$$(x - 1)(x^3 + 1) = 0$$

$$(x - 1)(x + 1)(x^2 - x + 1) = 0$$

$$x - 1 = 0 \Rightarrow x = 1$$

$$x + 1 = 0 \Rightarrow x = -1$$

$x^2 - x + 1 = 0 \Rightarrow$ No real solution (By the Quadratic Formula)

147. $x^4 - 4x^2 + 3 = 0$

$$(x^2 - 3)(x^2 - 1) = 0$$

$$(x + \sqrt{3})(x - \sqrt{3})(x + 1)(x - 1) = 0$$

$$x + \sqrt{3} = 0 \Rightarrow x = -\sqrt{3}$$

$$x - \sqrt{3} = 0 \Rightarrow x = \sqrt{3}$$

$$x + 1 = 0 \Rightarrow x = -1$$

$$x - 1 = 0 \Rightarrow x = 1$$

149. $4x^4 - 65x^2 + 16 = 0$

$$(4x^2 - 1)(x^2 - 16) = 0$$

$$(2x + 1)(2x - 1)(x + 4)(x - 4) = 0$$

$$2x + 1 = 0 \Rightarrow x = -\tfrac{1}{2}$$

$$2x - 1 = 0 \Rightarrow x = \tfrac{1}{2}$$

$$x + 4 = 0 \Rightarrow x = -4$$

$$x - 4 = 0 \Rightarrow x = 4$$

151.
$$x^6 + 7x^3 - 8 = 0$$
$$(x^3 + 8)(x^3 - 1) = 0$$
$$(x + 2)(x^2 - 2x + 4)(x - 1)(x^2 + x + 1) = 0$$
$$x + 2 = 0 \Rightarrow x = -2$$
$x^2 - 2x + 4 = 0 \Rightarrow$ No real solution (By the Quadratic Formula)
$$x - 1 = 0 \Rightarrow x = 1$$
$x^2 + x + 1 = 0 \Rightarrow$ No real solution (By the Quadratic Formula)

153. $\sqrt{2x} - 10 = 0$
$$\sqrt{2x} = 10$$
$$2x = 100$$
$$x = 50$$

155. $\sqrt{x - 10} - 4 = 0$
$$\sqrt{x - 10} = 4$$
$$x - 10 = 16$$
$$x = 26$$

157. $\sqrt[3]{2x + 5} + 3 = 0$
$$\sqrt[3]{2x + 5} = -3$$
$$2x + 5 = -27$$
$$2x = -32$$
$$x = -16$$

159. $-\sqrt{26 - 11x} + 4 = x$
$$4 - x = \sqrt{26 - 11x}$$
$$16 - 8x + x^2 = 26 - 11x$$
$$x^2 + 3x - 10 = 0$$
$$(x + 5)(x - 2) = 0$$
$$x + 5 = 0 \Rightarrow x = -5$$
$$x - 2 = 0 \Rightarrow x = 2$$

161. $\sqrt{x + 1} = \sqrt{3x + 1}$
$$x + 1 = 3x + 1$$
$$-2x = 0$$
$$x = 0$$

163. $(x - 5)^{3/2} = 8$
$$(x - 5)^3 = 8^2$$
$$x - 5 = 8^{2/3}$$
$$x = 5 + 4$$
$$x = 9$$

165. $(x + 3)^{2/3} = 8$
$$(x + 3)^2 = 8^3$$
$$x + 3 = \pm\sqrt{8^3}$$
$$x + 3 = \pm\sqrt{512}$$
$$x = -3 \pm 16\sqrt{2}$$

167. $(x^2 - 5)^{3/2} = 27$
$$x^2 - 5 = 27^{2/3}$$
$$x^2 = 5 + 9$$
$$x^2 = 14$$
$$x = \pm\sqrt{14}$$

169. $3x(x - 1)^{1/2} + 2(x - 1)^{3/2} = 0$
$$(x - 1)^{1/2}[3x + 2(x - 1)] = 0$$
$$(x - 1)^{1/2}(5x - 2) = 0$$
$$(x - 1)^{1/2} = 0 \Rightarrow x - 1 = 0 \Rightarrow x = 1$$
$5x - 2 = 0 \Rightarrow x = \frac{2}{5}$ which is extraneous.

171.

$$x = \frac{3}{x} + \frac{1}{2}$$

$$(2x)(x) = (2x)\left(\frac{3}{x}\right) + (2x)\left(\frac{1}{2}\right)$$

$$2x^2 = 6 + x$$

$$2x^2 - x - 6 = 0$$

$$(2x + 3)(x - 2) = 0$$

$$2x + 3 = 0 \Rightarrow x = -\frac{3}{2}$$

$$x - 2 = 0 \Rightarrow x = 2$$

173.

$$\frac{1}{x} - \frac{1}{x + 1} = 3$$

$$x(x + 1)\frac{1}{x} - x(x + 1)\frac{1}{x + 1} = x(x + 1)(3)$$

$$x + 1 - x = 3x(x + 1)$$

$$1 = 3x^2 + 3x$$

$$0 = 3x^2 + 3x - 1;\ a = 3,\ \ b = 3,\ \ c = -1$$

$$x = \frac{-3 \pm \sqrt{(3)^2 - 4(3)(-1)}}{2(3)} = \frac{-3 \pm \sqrt{21}}{6}$$

175. $\dfrac{20 - x}{x} = x$

$$20 - x = x^2$$

$$0 = x^2 + x - 20$$

$$0 = (x + 5)(x - 4)$$

$$x + 5 = 0 \Rightarrow x = -5$$

$$x - 4 = 0 \Rightarrow x = 4$$

177.

$$\frac{x}{x^2 - 4} + \frac{1}{x + 2} = 3$$

$$(x + 2)(x - 2)\frac{x}{x^2 - 4} + (x + 2)(x - 2)\frac{1}{x + 2} = 3(x + 2)(x - 2)$$

$$x + x - 2 = 3x^2 - 12$$

$$3x^2 - 2x - 10 = 0$$

$a = 3, b = -2, c = -10$

$$x = \frac{-(-2) \pm \sqrt{(-2)^2 - 4(3)(-10)}}{2(3)} = \frac{2 \pm \sqrt{124}}{6} = \frac{2 \pm 2\sqrt{31}}{6} = \frac{1 \pm \sqrt{31}}{3} = \frac{1}{3} \pm \frac{\sqrt{31}}{3}$$

179. $|2x - 1| = 5$

$$2x - 1 = 5 \Rightarrow x = 3$$

$$-(2x - 1) = 5 \Rightarrow x = -2$$

181. $|x| = x^2 + x - 3$

$x = x^2 + x - 3$ OR $-x = x^2 + x - 3$

$x^2 - 3 = 0$ $\quad$ $x^2 + 2x - 3 = 0$

$x = \pm\sqrt{3}$ $\quad$ $(x - 1)(x + 3) = 0$

$x - 1 = 0 \Rightarrow x = 1$

$x + 3 = 0 \Rightarrow x = -3$

Only $x = \sqrt{3}$, and $x = -3$ are solutions to the original equation. $x = -\sqrt{3}$ and $x = 1$ are extraneous.

183. $|x + 1| = x^2 - 5$

$x + 1 = x^2 - 5$ OR $-(x + 1) = x^2 - 5$

$x^2 - x - 6 = 0$ $\quad$ $-x - 1 = x^2 - 5$

$(x - 3)(x + 2) = 0$ $\quad$ $x^2 + x - 4 = 0$

$x - 3 = 0 \Rightarrow x = 3$ $\quad$ $x = \dfrac{-1 \pm \sqrt{17}}{2}$

$x + 2 = 0 \Rightarrow x = -2$

Only $x = 3$ and $x = \dfrac{-1 - \sqrt{17}}{2}$ are solutions to the original equation. $x = -2$ and $x = \dfrac{-1 + \sqrt{17}}{2}$ are extraneous.

185. (a) Female: $y = 0.432x - 10.44$

For $y = 16$: $16 = 0.432x - 10.44$

$26.44 = 0.432x$

$\dfrac{26.44}{0.432} = x$

$x \approx 61.2$ inches

(b) Male: $y = 0.449x - 12.15$

For $y = 19$: $19 = 0.449x - 12.15$

$31.15 = 0.449x$

$69.4 \approx x$

Yes, it is likely that both bones came from the same person because the estimated height of a male with a 19-inch thigh bone is 69.4 inches.

(c)

Height x	Female Femur Length	Male Femur Length
60	15.48	14.79
70	19.80	19.28
80	24.12	23.77
90	28.44	28.26
100	32.76	32.75
110	37.08	37.24

They are approximately equal when $x = 100$ inches.

(d) $0.432x - 10.44 = 0.449x - 12.15$

$1.71 = 0.017x$

$x \approx 100.59$ inches

It is unlikely that a female would be over 8 feet tall, so if a femur of this length was found, it most likely belonged to a very tall male.

187. Let h be the number of feet above flood level after t hours.

$h = -\dfrac{1}{4}t + 8$

$1 = -\dfrac{1}{4}t + 8 \Rightarrow t = 28$ hours

189. $x^2 + x^2 = 5^2$

$2x^2 = 25$

$x^2 = \dfrac{25}{2}$

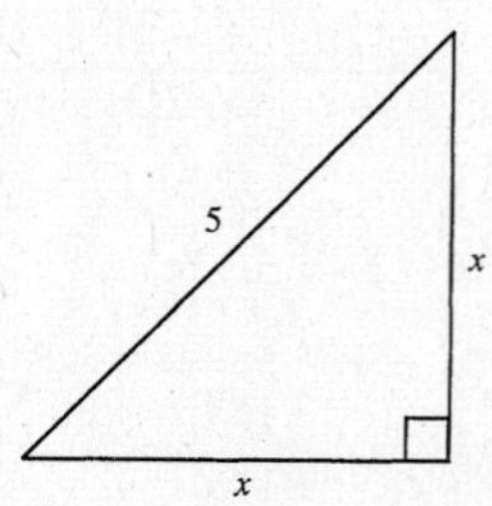

$x = \sqrt{\dfrac{25}{2}} = \dfrac{5}{\sqrt{2}} = \dfrac{5\sqrt{2}}{2} \approx 3.54$ centimeters

191. Let r = speed of the eastbound plane and $r + 50$ = speed of the northbound plane. After 3 hours the eastbound plane has traveled $3r$ miles and the northbound plane has traveled $3(r + 50)$ miles.

$$[3r^2] + [3(r + 50)]^2 = 2440^2$$

$$9r^2 + 9(r^2 + 100r + 2500) = 5{,}953{,}600$$

$$18r^2 + 900r - 5{,}931{,}100 = 0$$

By the Quadratic Formula, $r \approx 550$ (discard the negative value of r as extraneous).

Speed of the eastbound plane: 550 miles per hour

Speed of the northbound plane: 600 miles per hour

193.

$$p = 40 - \sqrt{0.01x + 1}$$

$$37.55 = 40 - \sqrt{0.01x + 1}$$

$$40 - 37.55 = \sqrt{0.01x + 1}$$

$$(2.45)^2 = 0.01x + 1$$

$$x = 500.25$$

The demand is approximately 500 units per day.

195. False. $x(3 - x) = 10$ is a quadratic equation.

197. False. An absolute value equation may have only one solution or no solutions in some cases. For example, $|x| = -7$ has no solution.

199. Answers will vary. For example: Two equations are equivalent if they differ only by algebraic simplification steps and have the same solutions.

$x^2 + 3x + 4 = x + 1$ is equivalent to $x^2 + 2x + 3 = 0$.

201. (a) $3(x + 4)^2 + (x + 4) - 2 = 0$

Let $u = x + 4$.

$$3u^2 + u - 2 = 0$$

$$(3u - 2)(u + 1) = 0$$

$$u = \tfrac{2}{3}, -1$$

$$x = u - 4 = -\tfrac{10}{3}, -5$$

(b) $3(x^2 + 8x + 16) + x + 4 - 2 = 0$

$$3x^2 + 24x + 48 + x + 4 - 2 = 0$$

$$3x^2 + 25x + 50 = 0$$

$$(3x + 10)(x + 5) = 0$$

$$x = -\tfrac{10}{3}, -5$$

(c) Answers will vary. Method (a) is slightly easier since there are less algebraic steps.

203. $x + |x - a| = b$

Solving for x we have:

First equation:

$$x + x - a = b$$

$$x = \frac{a + b}{2}$$

Second equation:

$$x - x + a = b$$

$$a = b$$

Thus, $x = 9$ will be the only solution if $9 = \dfrac{a + b}{2}$ and $a \le b$.

For example,

$a = 0, b = 18$

$a = 2, b = 16$, etc.

Answers will vary.

205. $x + \sqrt{x - a} = b$

When $x = 20$ we have:

$20 + \sqrt{20 - a} = b$

$\sqrt{20 - a} = b - 20 \implies a \le 20$ and $b \ge 20$

Answers will vary.

Possible values for a and b:

$a = 20$ and $b = 20$

$a = 4$ and $b = 24$

$a = 11$ and $b = 23$

$a = 19$ and $b = 21$

Section P.3 The Cartesian Plane and Graphs of Equations

- You should be able to plot points.
- You should know that the distance between (x_1, y_1) and (x_2, y_2) in the plane is

  $d = \sqrt{(x_2 - x_1)^2 + (y_2 - y_1)^2}.$
- You should know that the midpoint of the line segment joining (x_1, y_1) and (x_2, y_2) is

  $\left(\frac{x_1 + x_2}{2}, \frac{y_1 + y_2}{2}\right).$
- You should be able to use the point-plotting method of graphing.
- You should be able to find x- and y-intercepts.
  (a) To find the x-intercepts, let $y = 0$ and solve for x.
  (b) To find the y-intercepts, let $x = 0$ and solve for y.
- You should be able to test for symmetry.
  (a) To test for x-axis symmetry, replace y with $-y$.
  (b) To test for y-axis symmetry, replace x with $-x$.
  (c) To test for origin symmetry, replace x with $-x$ and y with $-y$.
- You should know the standard equation of a circle with center (h, k) and radius r:

  $(x - h)^2 + (y - k)^2 = r^2$

Solutions to Odd-Numbered Exercises

1. A: $(2, 6)$, B: $(-6, -2)$, C: $(4, -4)$, D: $(-3, 2)$

3. $(-3, 4)$

5. $x > 0$ and $y < 0$ in Quadrant IV.

7. $x = -4 \implies x$ is in Quadrant II or III.

$y > 0 \implies y$ is in Quadrant I or II.

Both conditions are met in Quadrant II.

9. $y < -5 \implies y$ is in Quadrant III or IV.

11. $(x, -y)$ is in the second Quadrant means that (x, y) is in Quadrant III.

13. $xy > 0 \Rightarrow$ x and y are either both positive or are both negative. $\Rightarrow$ Quadrant I or III.

15.

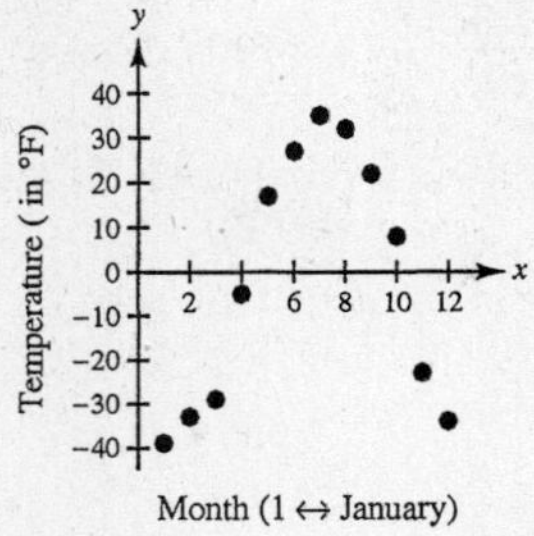

17. (a) The distance between (0, 2) and (4, 2) is 4.

The distance between (4, 2) and (4, 5) is 3.

The distance between (0, 2) and (4, 5) is

$$\sqrt{(4-0)^2 + (5-2)^2} = \sqrt{16+9}$$
$$= \sqrt{25} = 5.$$

(b) $4^2 + 3^2 = 16 + 9 = 25 = 5^2$

19. (a) The distance between $(-1, 1)$ and $(9, 1)$ is 10.

The distance between $(9, 1)$ and $(9, 4)$ is 3.

The distance between $(-1, 1)$ and $(9, 4)$ is $\sqrt{(9-(-1))^2 + (4-1)^2} = \sqrt{100+9} = \sqrt{109}.$

(b) $10^2 + 3^2 = 109 = \left(\sqrt{109}\right)^2$

21. (a)

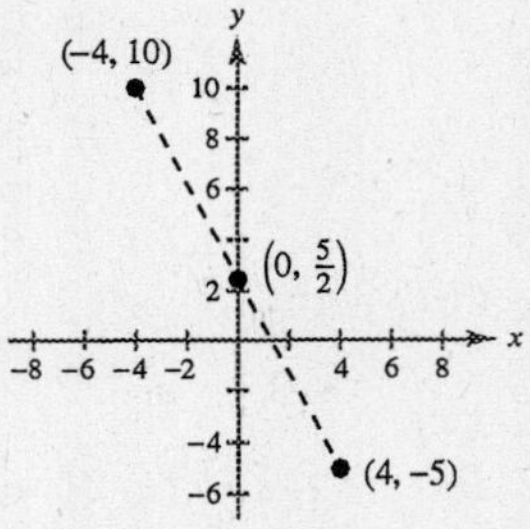

(b) $d = \sqrt{(4+4)^2 + (-5-10)^2}$
$= \sqrt{64 + 225} = 17$

(c) $\left(\dfrac{4-4}{2}, \dfrac{-5+10}{2}\right) = \left(0, \dfrac{5}{2}\right)$

23. (a)

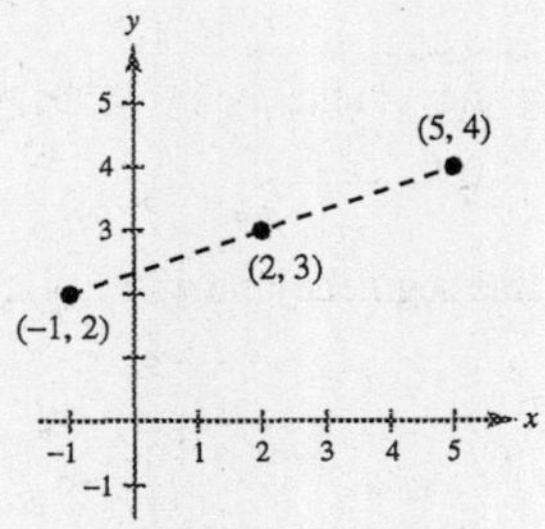

(b) $d = \sqrt{(5+1)^2 + (4-2)^2}$
$= \sqrt{36 + 4} = 2\sqrt{10}$

(c) $\left(\dfrac{-1+5}{2}, \dfrac{2+4}{2}\right) = (2, 3)$

25. (a)

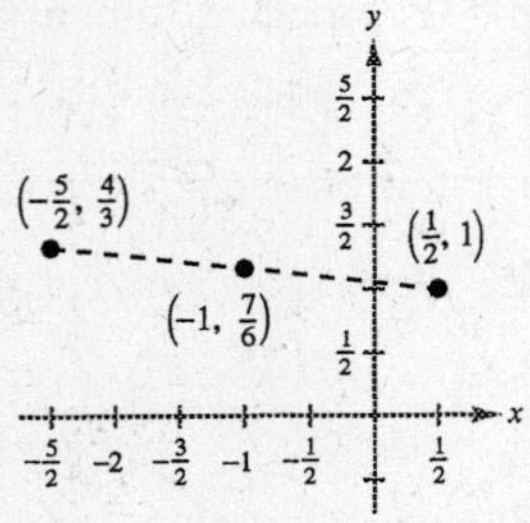

(b) $d = \sqrt{\left(\dfrac{1}{2}+\dfrac{5}{2}\right)^2 + \left(1-\dfrac{4}{3}\right)^2}$

$d = \sqrt{9 + \dfrac{1}{9}} = \dfrac{\sqrt{82}}{3}$

(c) $\left(\dfrac{-\frac{5}{2}+\frac{1}{2}}{2}, \dfrac{\frac{4}{3}+1}{2}\right) = \left(-1, \dfrac{7}{6}\right)$

27. (a)

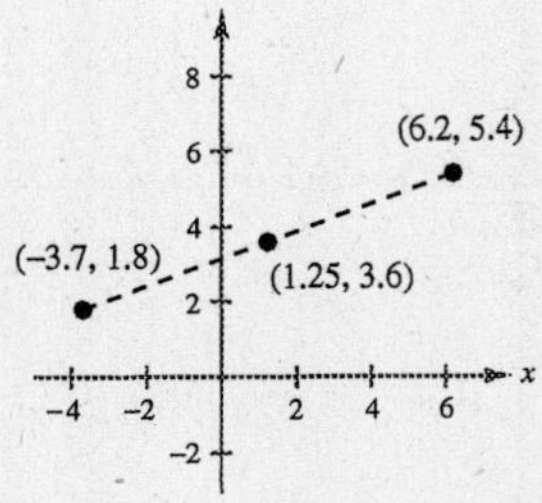

(b) $d = \sqrt{(6.2+3.7)^2 + (5.4-1.8)^2}$
$= \sqrt{98.01 + 12.96}$
$= \sqrt{110.97}$

(c) $\left(\dfrac{6.2-3.7}{2}, \dfrac{5.4+1.8}{2}\right) = (1.25, 3.6)$

29. (a) The number of artists elected each year seems to be nearly steady except for the first few years. Between 6 and 8 artists will be elected in 2004.

(b) The Rock and Roll Hall of Fame began in 1986.

31. $d = \sqrt{(45-10)^2 + (40-15)^2} = \sqrt{35^2 + 25^2} = \sqrt{1850} = 5\sqrt{74} \approx 43$ yards

33. $\left(\frac{1998 + 2000}{2}, \frac{1713.1 + 2225.8}{2}\right) = (1999, 1969.45)$

The annual revenue in 1999 was approximately \$1969.45 million.

35. $y = \sqrt{x + 4}$

(a) $(0, 2)$: $2 \stackrel{?}{=} \sqrt{0 + 4}$

$2 = 2$

Yes, the point *is* on the graph.

(b) $(5, 3)$: $3 \stackrel{?}{=} \sqrt{5 + 4}$

$3 = \sqrt{9}$

Yes, the point *is* on the graph.

37. $y = 4 - |x - 2|$

(a) $(1, 5)$: $5 \stackrel{?}{=} 4 - |1 - 2|$

$5 \neq 4 - 1$

No, the point *is not* on the graph.

(b) $(6, 0)$: $0 \stackrel{?}{=} 4 - |6 - 2|$

$0 = 4 - 4$

Yes, the point *is* on the graph.

39. $y = \frac{3}{4}x - 1$

x	-2	0	1	$\frac{4}{3}$	2
y	$-\frac{5}{2}$	-1	$-\frac{1}{4}$	0	$\frac{1}{2}$
(x, y)	$\left(-2, -\frac{5}{2}\right)$	$(0, -1)$	$\left(1, -\frac{1}{4}\right)$	$\left(\frac{4}{3}, 0\right)$	$\left(2, \frac{1}{2}\right)$

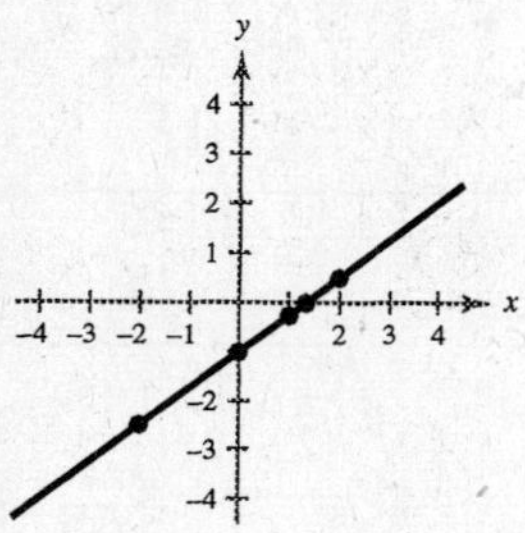

41. $y = 16 - 4x^2$

x-intercepts: $0 = 16 - 4x^2$

$4x^2 = 16$

$x^2 = 4$

$x = \pm 2$

$(-2, 0), (2, 0)$

y-intercept: $y = 16 - 4(0)^2 = 16$

$(0, 16)$

43. $y = 5x - 6$

x-intercept: $0 = 5x - 6$

$6 = 5x$

$\frac{6}{5} = x$

$\left(\frac{6}{5}, 0\right)$

y-intercept: $y = 5(0) - 6 = -6$

$(0, -6)$

45. $y = \sqrt{x + 4}$

x-intercept: $0 = \sqrt{x + 4}$

$0 = x + 4$

$-4 = x$

$(-4, 0)$

y-intercept: $y = \sqrt{0 + 4} = 2$

$(0, 2)$

47. $y = |3x - 7|$

x-intercept: $0 = |3x - 7|$

$0 = 3x - 7$

$\frac{7}{3} = 0$

$\left(\frac{7}{3}, 0\right)$

y-intercept: $y = |3(0) - 7| = 7$

$(0, 7)$

49. $y = 2x^3 - 4x^2$

x-intercepts: $0 = 2x^3 - 4x^2$

$0 = 2x^2(x - 2)$

$x = 0$ or $x = 2$

$(0, 0), (2, 0)$

y-intercept: $y = 2(0)^3 - 4(0)^2 = 0$

$(0, 0)$

51. $y^2 = 6 - x$

x-intercept: $0 = 6 - x$

$x = 6$

$(6, 0)$

y-intercepts: $y^2 = 6 - 0$

$y = \pm\sqrt{6}$

$(0, \sqrt{6}), (0, -\sqrt{6})$

53. $x^2 - y = 0$

$(-x)^2 - y = 0 \Rightarrow x^2 - y = 0 \Rightarrow$ y-axis symmetry

$x^2 - (-y) = 0 \Rightarrow x^2 + y = 0 \Rightarrow$ No x-axis symmetry

$(-x)^2 - (-y) = 0 \Rightarrow x^2 + y = 0 \Rightarrow$ No origin symmetry

55. $y = x^3$

$y = (-x)^3 \Rightarrow y = -x^3 \Rightarrow$ No y-axis symmetry

$-y = x^3 \Rightarrow y = -x^3 \Rightarrow$ No x-axis symmetry

$-y = (-x)^3 \Rightarrow -y = -x^3 \Rightarrow y = x^3 \Rightarrow$ Origin symmetry

57. $y = \dfrac{x}{x^2 + 1}$

$y = \dfrac{-x}{(-x)^2 + 1} \Rightarrow y = \dfrac{-x}{x^2 + 1} \Rightarrow$ No y-axis symmetry

$-y = \dfrac{x}{x^2 + 1} \Rightarrow y = \dfrac{-x}{x^2 + 1} \Rightarrow$ No x-axis symmetry

$-y = \dfrac{-x}{(-x)^2 + 1} \Rightarrow -y = \dfrac{-x}{x^2 + 1} \Rightarrow y = \dfrac{x}{x^2 + 1} \Rightarrow$ Origin symmetry

59. $xy^2 + 10 = 0$

$(-x)y^2 + 10 = 0 \Rightarrow -xy^2 + 10 = 0 \Rightarrow$ No y-axis symmetry

$x(-y)^2 + 10 = 0 \Rightarrow xy^2 + 10 = 0 \Rightarrow$ x-axis symmetry

$(-x)(-y)^2 + 10 = 0 \Rightarrow -xy^2 + 10 = 0 \Rightarrow$ No origin symmetry

61. y-axis symmetry

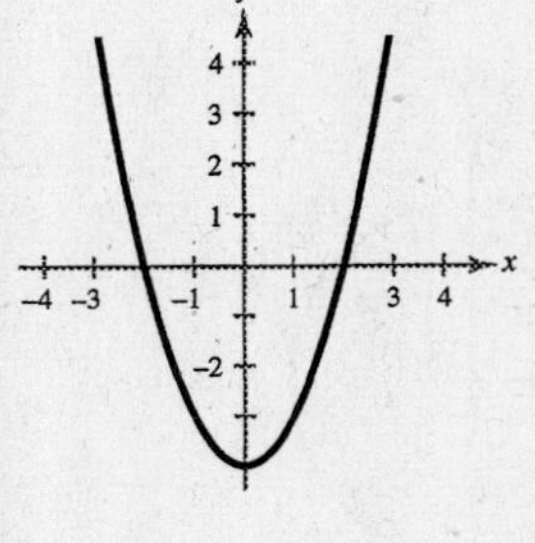

63. Origin symmetry

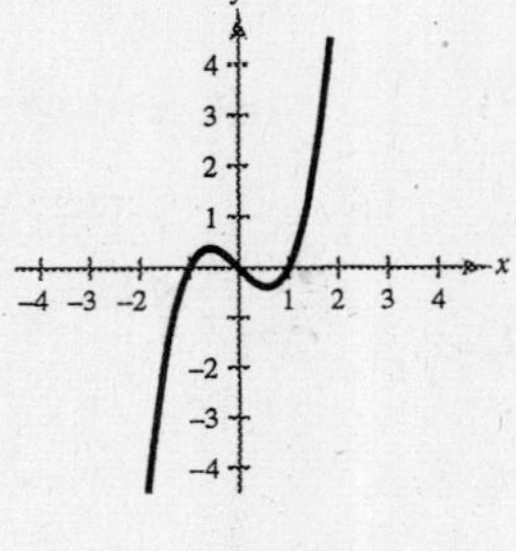

65. $y = -3x + 1$

x-intercept: $\left(\frac{1}{3}, 0\right)$

y-intercept: $(0, 1)$

No axis or origin symmetry

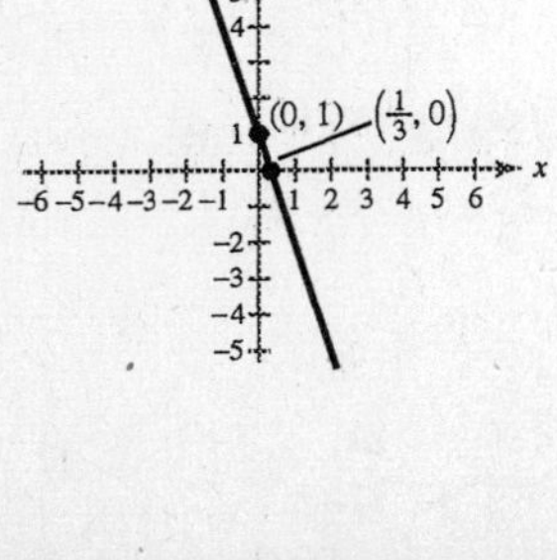

67. $y = x^2 - 2x$

Intercepts: $(0, 0), (2, 0)$

No axis or origin symmetry

x	-1	0	1	2	3
y	3	0	-1	0	3

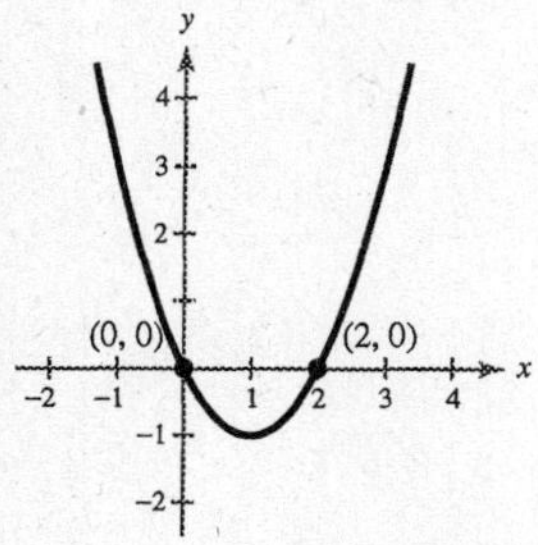

69. $y = x^3 + 3$

Intercepts: $(0, 3), \left(-3\sqrt[3]{-3}, 0\right)$

No axis or origin symmetry

x	-2	-1	0	1	2
y	-5	2	3	4	11

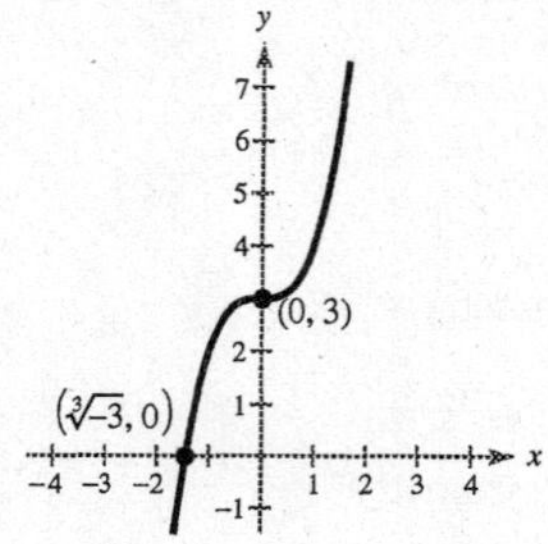

71. $y = \sqrt{x - 3}$

Domain: $[3, \infty)$

Intercept: $(3, 0)$

No axis or origin symmetry

x	3	4	7	12
y	0	1	2	3

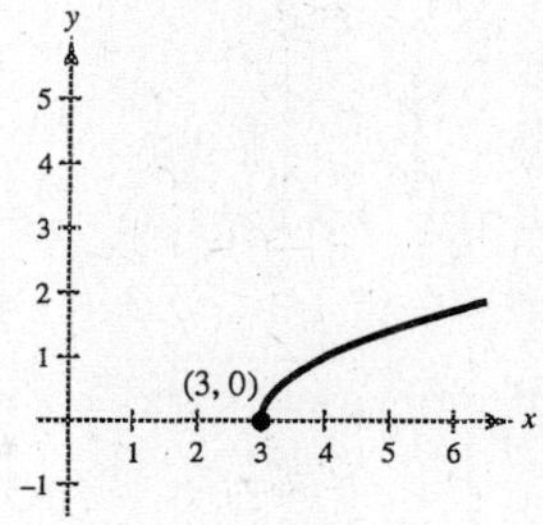

73. $y = |x - 6|$

Intercepts: $(0, 6), (6, 0)$

No axis or origin symmetry

x	-2	0	2	4	6	8	10
y	8	6	4	2	0	2	4

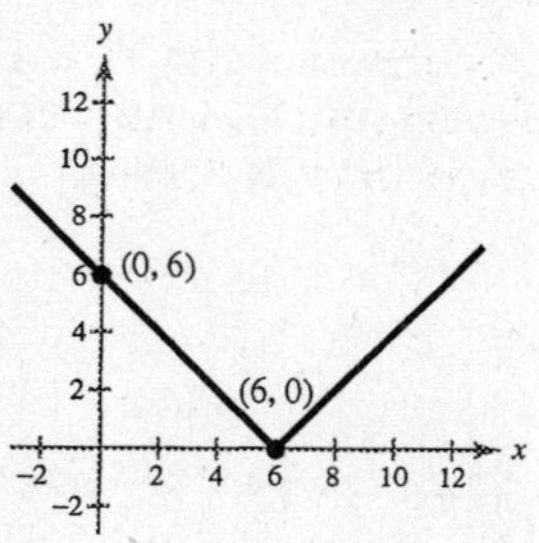

75. $x = y^2 - 1$

Intercepts: $(0, -1), (0, 1), (-1, 0)$

x-axis symmetry

x	-1	0	3
y	0	± 1	± 2

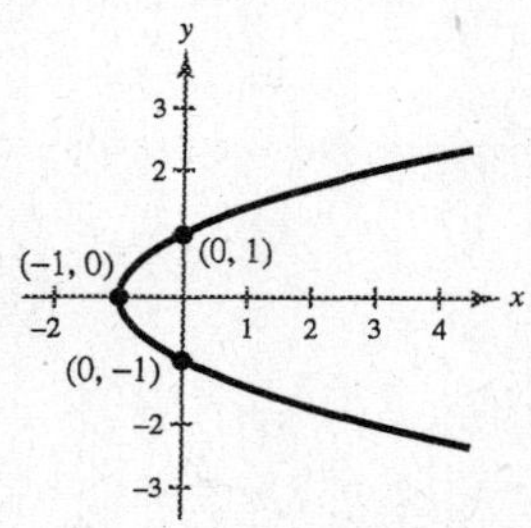

77. Center: $(2, -1)$; radius: 4

Standard form: $(x - 2)^2 + (y - (-1))^2 = 4^2$

$(x - 2)^2 + (y + 1)^2 = 16$

79. Center: $(-1, 2)$; solution point: $(0, 0)$

$(x - (-1))^2 + (y - 2)^2 = r^2$

$(0 + 1)^2 + (0 - 2)^2 = r^2 \Longrightarrow 5 = r^2$

Standard form: $(x + 1)^2 + (y - 2)^2 = 5$

81. Endpoints of a diameter: $(0, 0), (6, 8)$

Center: $\left(\dfrac{0 + 6}{2}, \dfrac{0 + 8}{2}\right) = (3, 4)$

$(x - 3)^2 + (y - 4)^2 = r^2$

$(0 - 3)^2 + (0 - 4)^2 = r^2 \Longrightarrow 25 = r^2$

Standard form: $(x - 3)^2 + (y - 4)^2 = 25$

83. $x^2 + y^2 = 25$

Center: $(0, 0)$

Radius: 5

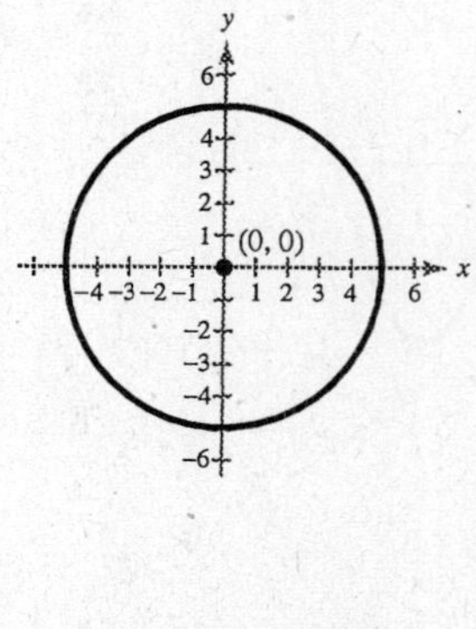

85. $(x - 1)^2 + (y + 3)^2 = 9$

Center: $(1, -3)$

Radius: 3

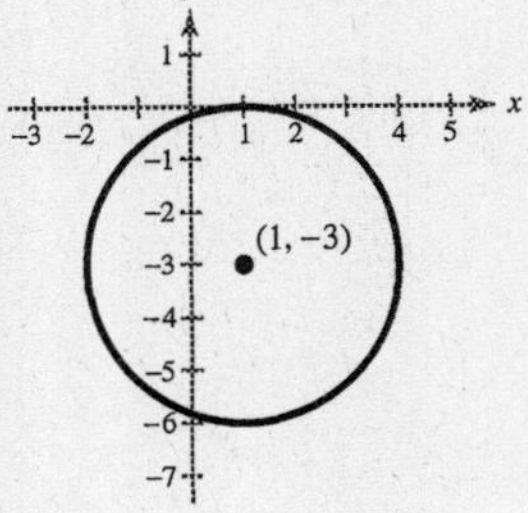

87. $\left(x - \frac{1}{2}\right)^2 + \left(y - \frac{1}{2}\right)^2 = \frac{9}{4}$

Center: $\left(\frac{1}{2}, \frac{1}{2}\right)$

Radius: $\frac{3}{2}$

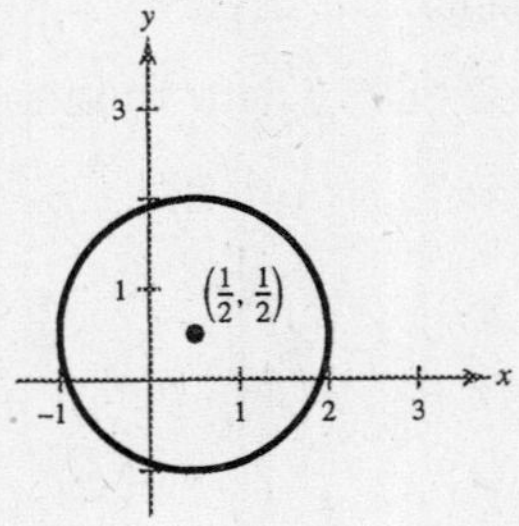

89. $y = 225{,}000 - 20{,}000t,\ 0 \le t \le 8$

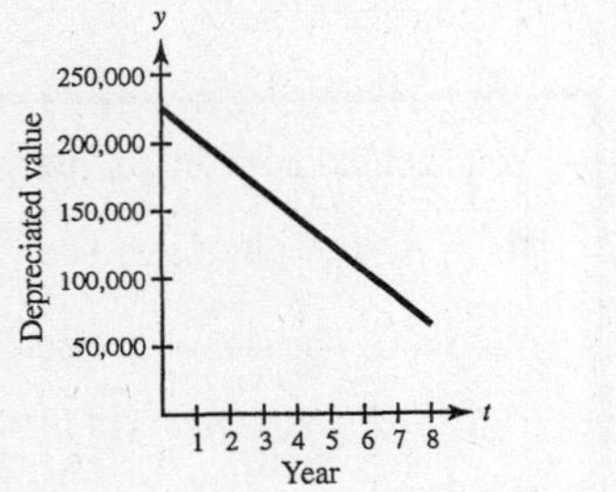

91. (a)

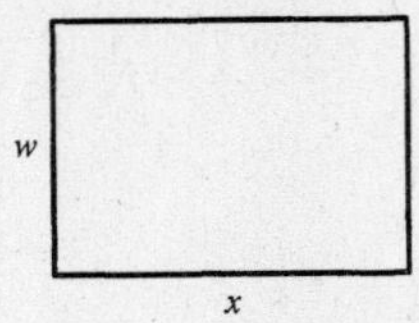

(b) $2x + 2w = 12 \Longrightarrow w = 6 - x$
$A = x \cdot w = x(6 - x)$

(c)

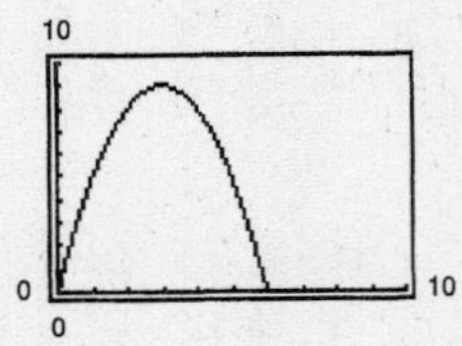

(d) The area is maximum when $x = 3$ and $w = 6 - 3 = 3$.

$x = 3$ meters

$w = 3$ meters

93. $y = -0.0025t^2 + 0.572t + 44.31$

(a) and (b)

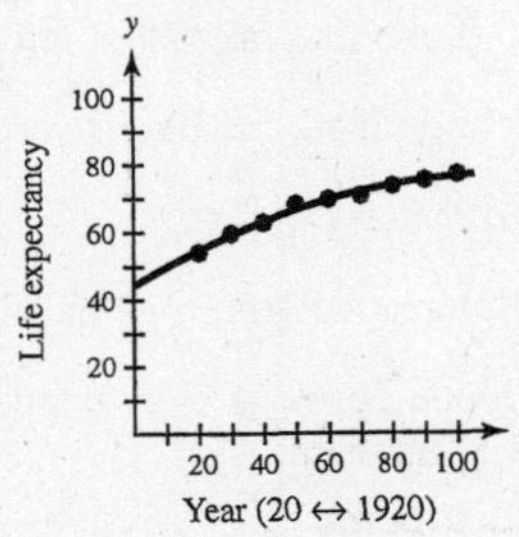

(c) For the year 2005 let $t = 105$: $y \approx 76.8$ years

For the year 2010 let $t = 110$: $y \approx 77.0$ years

(d) No. The graph reaches a maximum of $y \approx 77$ years when $t \approx 114.4$, or during the year 2014. After this time, the model has life expectancy decreasing.

95. False. In order to divide a line segment into 16 equal parts, you would have to use the Midpoint Rule 15 times.

97. False. To find y-intercepts, let x be zero and solve the equation for y.

99. The y-coordinate of any point on the x-axis is zero. Likewise, the x-coordinate of any point on the y-axis is zero.

101. The midpoint of the line segment through $(0, 0)$ and $(a + b, c)$ is $\left(\frac{a+b}{2}, \frac{c}{2}\right)$. The midpoint of the line segment through $(a, 0)$ and (b, c) is also $\left(\frac{a+b}{2}, \frac{c}{2}\right)$. Since the diagonals intersect, the point of intersection must lie on both diagonals. Since $\left(\frac{a+b}{2}, \frac{c}{2}\right)$ lies on both diagonals, it must be their point of intersection. Thus, the diagonals intersect at their midpoints.

103. $y = ax^2 + bx^3$

(a) $a = 1, b = 0 \Longrightarrow$ symmetric with respect to y-axis.

(b) $a = 0, b = 1 \Longrightarrow$ symmetric with respect to the origin.

Section P.4 Linear Equations in Two Variables

You should know the following important facts about lines.

- The graph of $y = mx + b$ is a straight line. It is called a linear equation in two variables.

  (a) The slope (steepness) is m.

  (b) The y-intercept is $(0, b)$.

- The slope of the line through (x_1, y_1) and (x_2, y_2) is

  $$m = \frac{y_2 - y_1}{x_2 - x_1} = \frac{\text{change in } y}{\text{change in } x} = \frac{\text{rise}}{\text{run}}.$$

- (a) If $m > 0$, the line rises from left to right.

  (b) If $m = 0$, the line is horizontal.

  (c) If $m < 0$, the line falls from left to right.

  (d) If m is undefined, the line is vertical.

- Equations of Lines

  (a) Slope-Intercept Form: $y = mx + b$

  (b) Point-Slope Form: $y - y_1 = m(x - x_1)$

  (c) Two-Point Form: $y - y_1 = \frac{y_2 - y_1}{x_2 - x_1}(x - x_1)$

  (d) General Form: $Ax + By + C = 0$

  (e) Vertical Line: $x = a$

  (f) Horizontal Line: $y = b$

- Given two distinct nonvertical lines

  $L_1\colon y = m_1x + b_1$ and $L_2\colon y = m_2x + b_2$

  (a) L_1 is parallel to L_2 if and only if $m_1 = m_2$ and $b_1 \neq b_2$.

  (b) L_1 is perpendicular to L_2 if and only if $m_1 = -1/m_2$.

Solutions to Odd-Numbered Exercises

1. (a) $m = \frac{2}{3}$. Since the slope is positive, the line rises. Matches L_2.

(b) m is undefined. The line is vertical. Matches L_3.

(c) $m = -2$. The line falls. Matches L_1.

3.

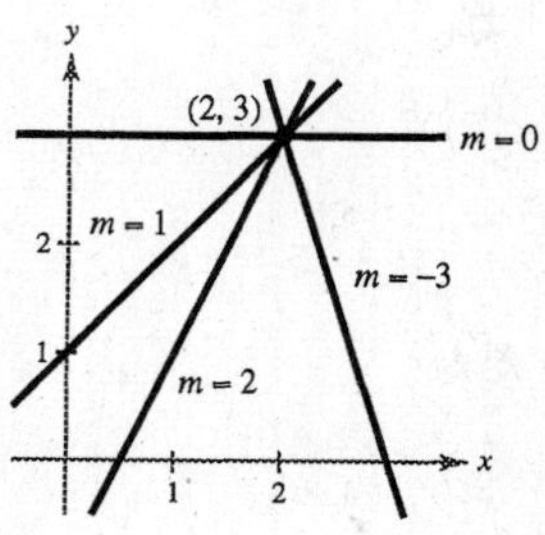

5. Two points on the line: $(0, 0)$ and $(5, 8)$

$$\text{Slope} = \frac{\text{rise}}{\text{run}} = \frac{8}{5}$$

7. Two points on the line: $(0, 8)$ and $(2, 0)$

$$\text{Slope} = \frac{\text{rise}}{\text{run}} = \frac{-8}{2} = -4$$

9. $y = 5x + 3$

Slope: $m = 5$

y-intercept: $(0, 3)$

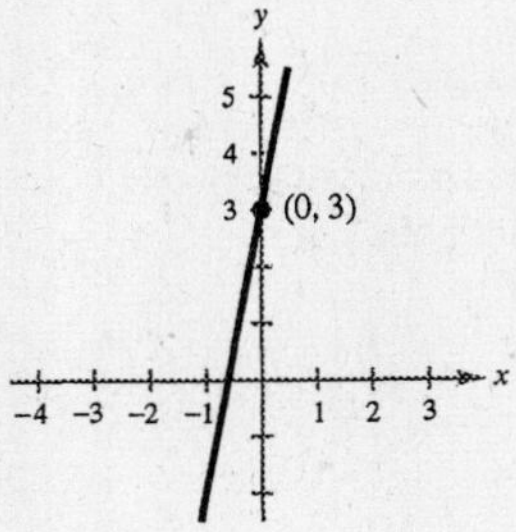

11. $y = -\frac{1}{2}x + 4$

Slope: $m = -\frac{1}{2}$

y-intercept: $(0, 4)$

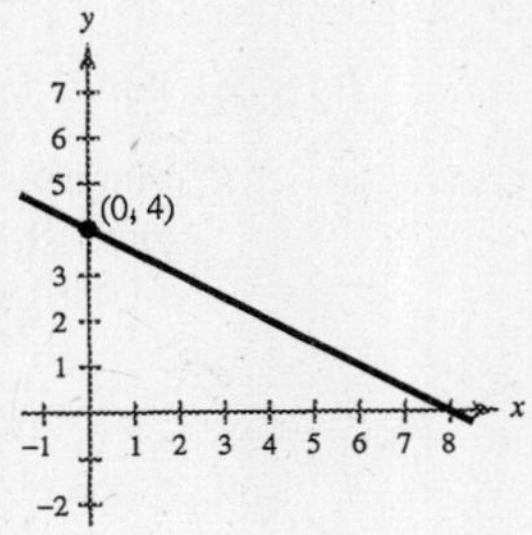

13. $5x - 2 = 0$

$x = \frac{2}{5}$ Vertical line

Slope: undefined

No y-intercept

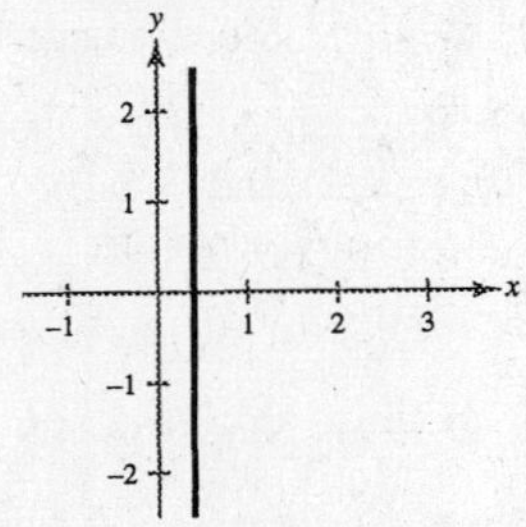

15. $7x + 6y = 30$

$y = -\frac{7}{6}x + 5$

Slope: $m = -\frac{7}{6}$

y-intercept: $(0, 5)$

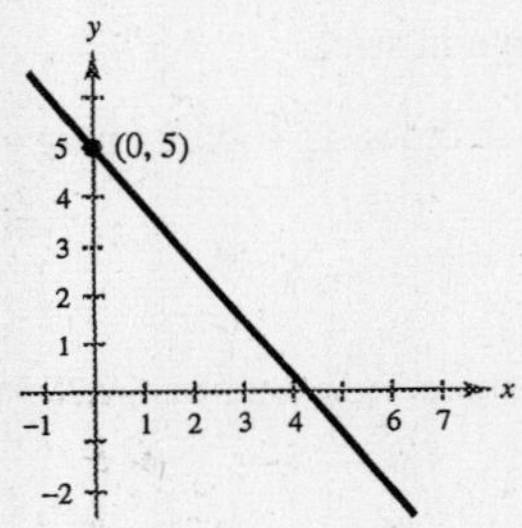

17. $y - 3 = 0$

$y = 3$ Horizontal line

Slope: $m = 0$

y-intercept: $(0, 3)$

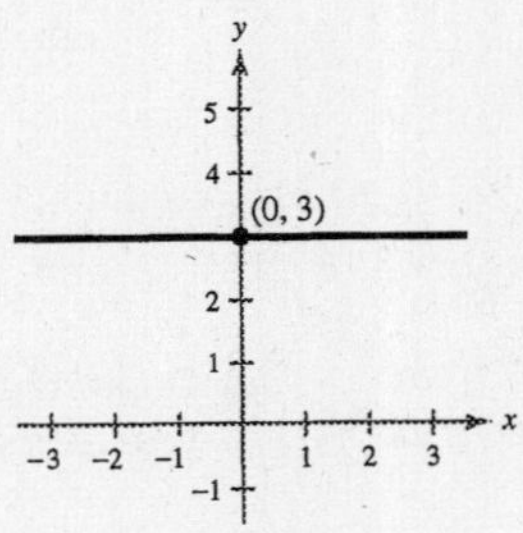

19. $x + 5 = 0$

$x = -5$ Vertical Line

Slope: Undefined

No y-intercept

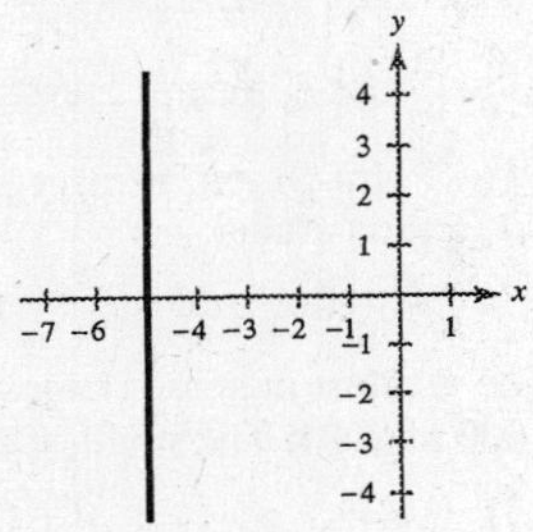

21.

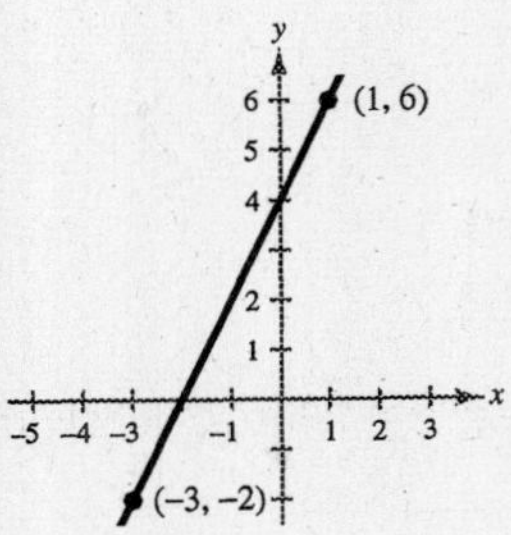

$$m = \frac{6 - (-2)}{1 - (-3)} = \frac{8}{4} = 2$$

23.

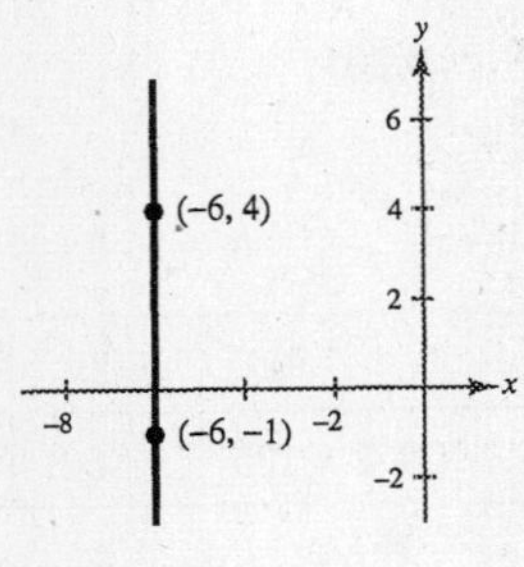

$$m = \frac{4 - (-1)}{-6 - (-6)} = \frac{5}{0}$$

m is undefined.

25.

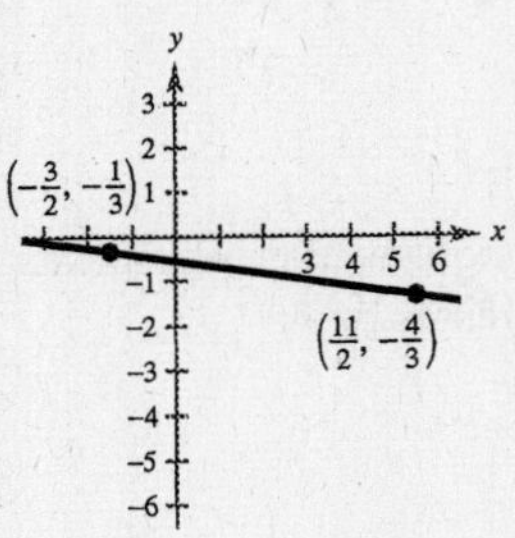

$$m = \frac{-\frac{1}{3} - \left(-\frac{4}{3}\right)}{-\frac{3}{2} - \frac{11}{2}} = -\frac{1}{7}$$

27.

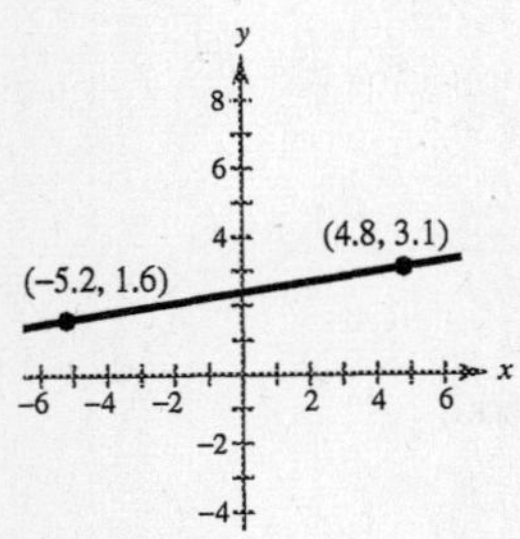

$$m = \frac{1.6 - 3.1}{-5.2 - 4.8} = \frac{-1.5}{-10}$$

$$= 0.15$$

29. Point: $(2, 1)$ Slope: $m = 0$

Since $m = 0$, y does not change. Three points are $(0, 1)$, $(3, 1)$, and $(-1, 1)$.

31. Point: $(5, -6)$ Slope: $m = 1$

Since $m = 1$, y increases by 1 for every one unit increase in x. Three points are $(6, -5)$, $(7, -4)$, and $(8, -3)$.

33. Point: $(-8, 1)$ Slope is undefined.

Since m is undefined, x does not change. Three points are $(-8, 0)$, $(-8, 2)$, and $(-8, 3)$.

35. Point: $(-5, 4)$ Slope: $m = 2$

Since $m = 2 = \frac{2}{1}$, y increases by 2 for every one unit increase in x. Three additional points are $(-4, 6)$, $(-3, 8)$, and $(-2, 10)$.

37. Point: $(7, -2)$ Slope: $m = \frac{1}{2}$

Since $m = \frac{1}{2}$, y increases by 1 unit for every two unit increase in x. Three additional points are $(9, -1)$, $(11, 0)$, and $(13, 1)$.

39. Slope of L_1: $m = \dfrac{9 + 1}{5 - 0} = 2$

Slope of L_2: $m = \dfrac{1 - 3}{4 - 0} = -\dfrac{1}{2}$

L_1 and L_2 are perpendicular.

41. Slope of L_1: $m = \dfrac{0 - 6}{-6 - 3} = \dfrac{2}{3}$

Slope of L_2: $m = \dfrac{\frac{7}{3} + 1}{5 - 0} = \dfrac{2}{3}$

L_1 and L_2 are parallel.

43. (a) $m = 135$. The sales are increasing 135 units per year.

(b) $m = 0$. There is no change in sales.

(c) $m = -40$. The sales are decreasing 40 units per year.

45. (a) The greatest increase (largest positive slope) was from 2000 to 2001. The smallest increase was from 1991 to 1992.

(b) $(1, 0.33)$ and $(11, 2.38)$

$$m = \frac{2.38 - 0.33}{11 - 1} = \frac{2.05}{10} = 0.205$$

(c) On average, the earnings per share increased by $0.205 per year over this 10 year period.

47. (a) and (b)

x	300	600	900	1200	1500	1800	2100
y	-25	-50	-75	-100	-125	-150	-175

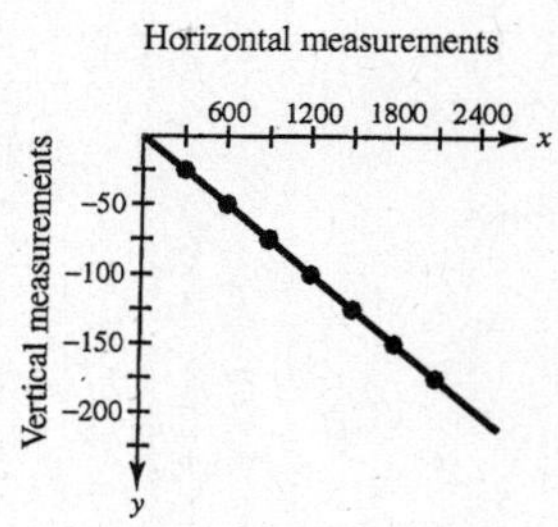

(c) $m = \dfrac{-50 - (-25)}{600 - 300} = \dfrac{-25}{300} = -\dfrac{1}{12}$

$$y - (-50) = -\frac{1}{12}(x - 600)$$

$$y + 50 = -\frac{1}{12}x + 50$$

$$y = -\frac{1}{12}x$$

(d) Since $m = -\frac{1}{12}$, for every 12 horizontal measurements the vertical measurement decreases by 1.

(e) $\dfrac{1}{12} \approx 0.083 = 8.3\%$ grade

49. $(3, 2540), m = 125$

$$V - 2540 = 125(t - 3)$$
$$V - 2540 = 125t - 375$$
$$V = 125t + 2165$$

51. The slope is $m = -20$. This represents the decrease in the amount of the loan each week. Matches graph (b).

53. The slope is $m = 0.32$. This represents the increase in travel cost for each mile driven. Matches graph (a).

55. Point $(0, -2)$; $m = 3$

$$y + 2 = 3(x - 0)$$
$$y = 3x - 2$$

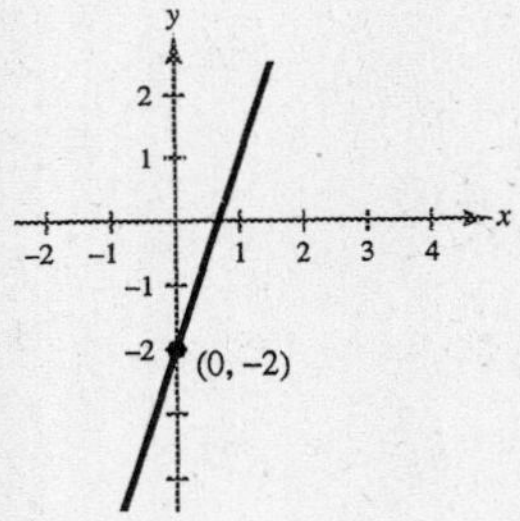

57. Point $(-3, 6)$; $m = -2$

$$y - 6 = -2(x + 3)$$
$$y = -2x$$

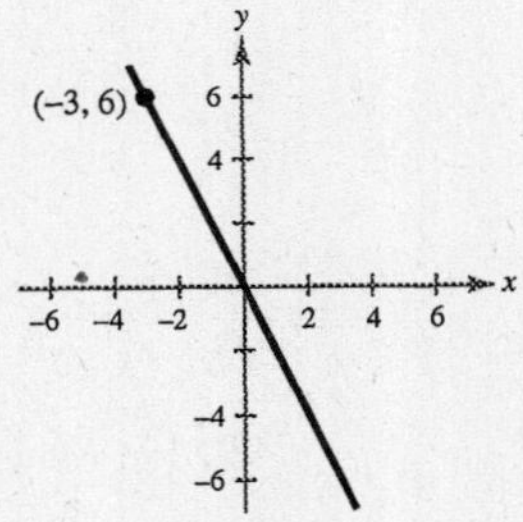

59. $(4, 0)$; $m = -\frac{1}{3}$

$$y - 0 = -\tfrac{1}{3}(x - 4)$$
$$y = -\tfrac{1}{3}x + \tfrac{4}{3}$$

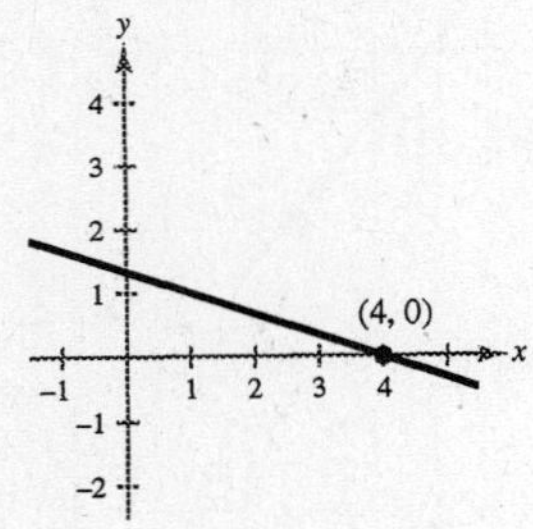

61. Point $(6, -1)$; m is undefined.

The line is vertical.

$x = 6$

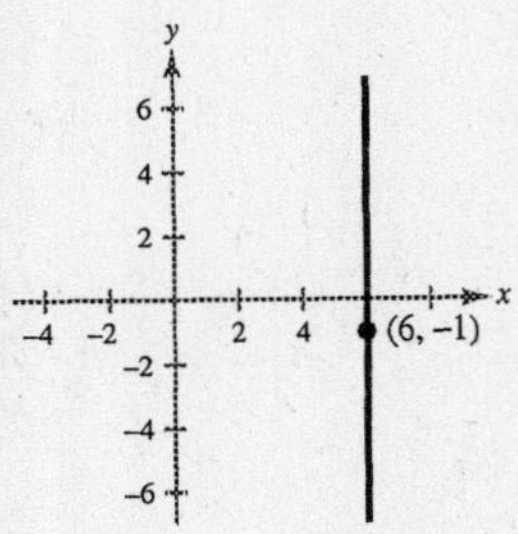

63. Point $\left(4, \frac{5}{2}\right)$; $m = 0$

The line is horizontal.

$y = \frac{5}{2}$

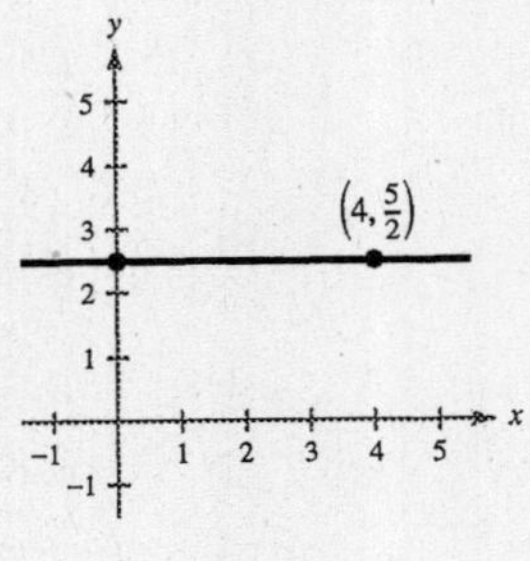

65. Point $(-5.1, 1.8)$; $m = 5$

$$y - 1.8 = 5(x - (-5.1))$$
$$y = 5x + 27.3$$

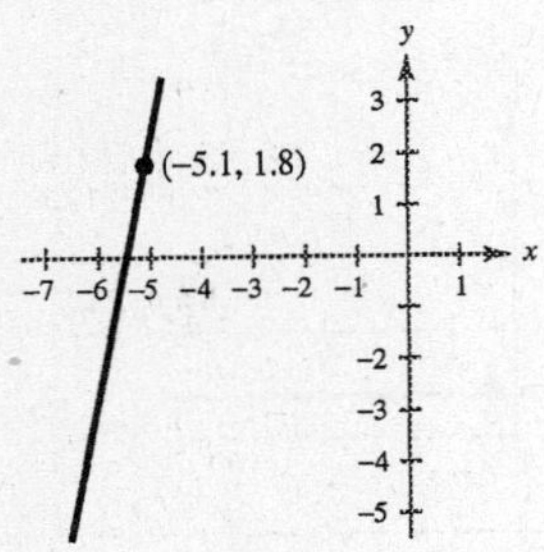

67. $(5, -1)$ and $(-5, 5)$

$$y + 1 = \frac{5 + 1}{-5 - 5}(x - 5)$$
$$y = -\frac{3}{5}(x - 5) - 1$$
$$y = -\frac{3}{5}x + 2$$

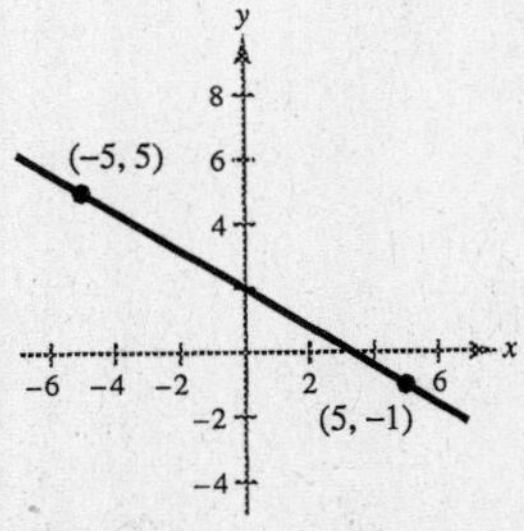

69. $(-8, 1)$ and $(-8, 7)$

Since both points have $x = -8$, the slope is undefined, and the line is vertical.

$x = -8$

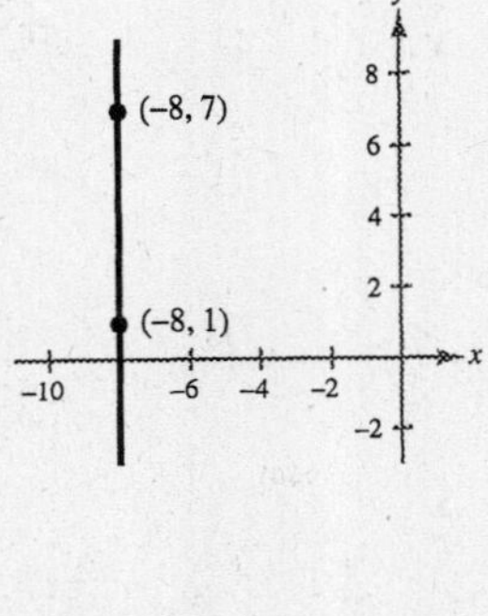

71. $\left(2, \frac{1}{2}\right)$ and $\left(\frac{1}{2}, \frac{5}{4}\right)$

$$y - \frac{1}{2} = \frac{\frac{5}{4} - \frac{1}{2}}{\frac{1}{2} - 2}(x - 2)$$
$$y = -\frac{1}{2}(x - 2) + \frac{1}{2}$$
$$y = -\frac{1}{2}x + \frac{3}{2}$$

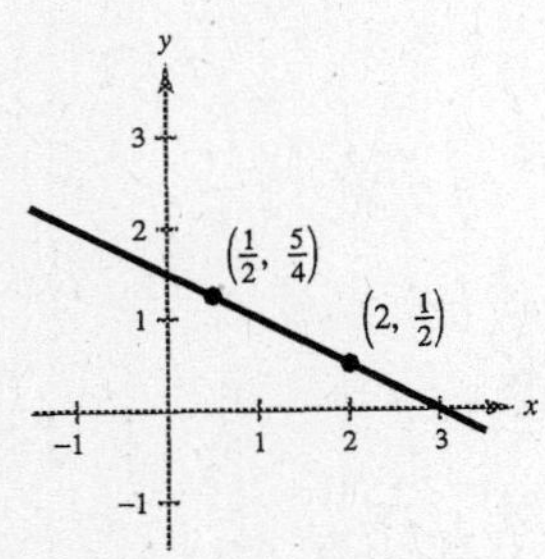

73. $\left(-\frac{1}{10}, -\frac{3}{5}\right)$ and $\left(\frac{9}{10}, -\frac{9}{5}\right)$

$$y - \left(-\frac{3}{5}\right) = \frac{-\frac{9}{5} - \left(-\frac{3}{5}\right)}{\frac{9}{10} - \left(-\frac{1}{10}\right)}\left(x - \left(-\frac{1}{10}\right)\right)$$

$$y = -\frac{6}{5}\left(x + \frac{1}{10}\right) - \frac{3}{5}$$

$$y = -\frac{6}{5}x - \frac{18}{25}$$

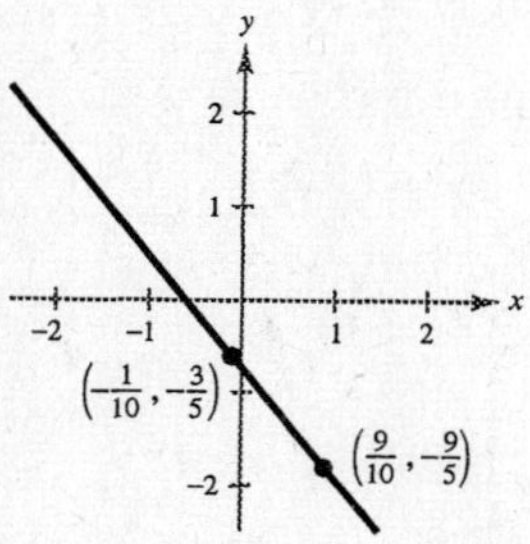

75. $(1, 0.6)$ and $(-2, -0.6)$

$$y - 0.6 = \frac{-0.6 - 0.6}{-2 - 1}(x - 1)$$

$$y = 0.4(x - 1) + 0.6$$

$$y = 0.4x + 0.2$$

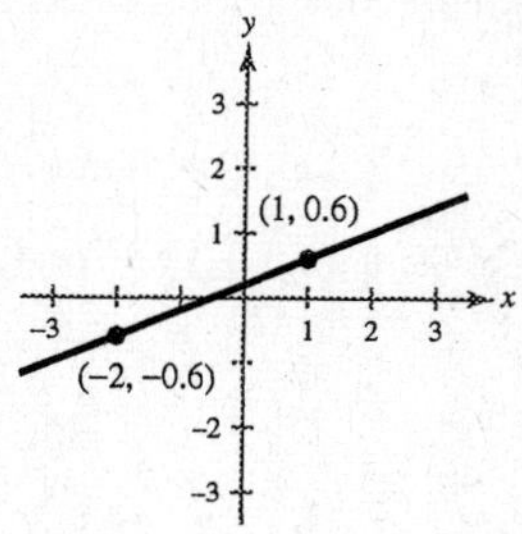

77. $(2, -1)$ and $\left(\frac{1}{3}, -1\right)$

$$y + 1 = \frac{-1 - (-1)}{\frac{1}{3} - 2}(x - 2)$$

$$y + 1 = 0$$

$$y = -1$$

The line is horizontal.

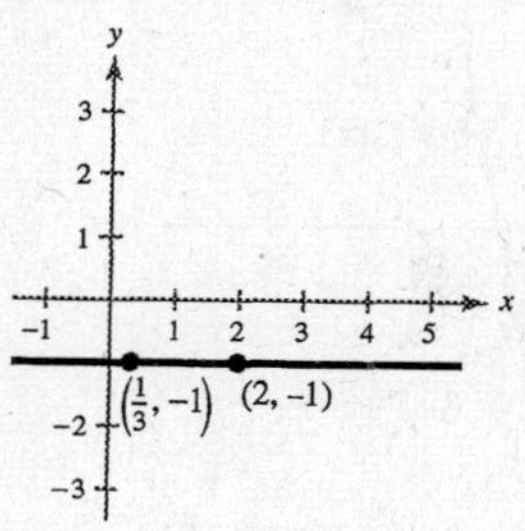

79. $\left(\frac{7}{3}, -8\right)$ and $\left(\frac{7}{3}, 1\right)$

$$m = \frac{1 - (-8)}{\frac{7}{3} - \frac{7}{3}} = \frac{9}{0}$$ and is undefined.

$$x = \frac{7}{3}$$

The line is vertical.

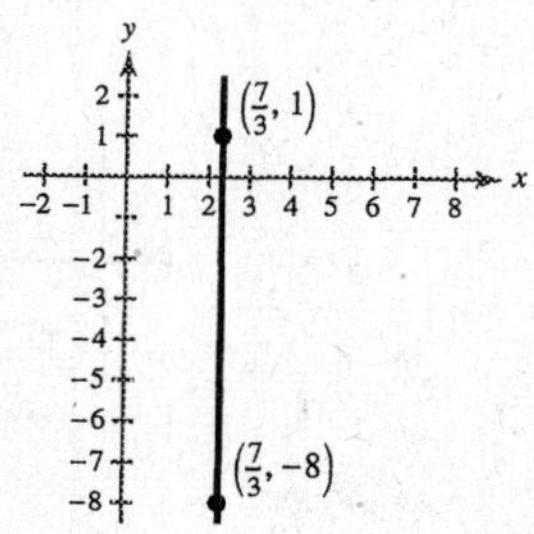

81.
$$\frac{x}{2} + \frac{y}{3} = 1$$
$$3x + 2y - 6 = 0$$

83.
$$\frac{x}{-\frac{1}{6}} + \frac{y}{-\frac{2}{3}} = 1$$
$$6x + \frac{3}{2}y = -1$$
$$12x + 3y + 2 = 0$$

85.
$$\frac{x}{c} + \frac{y}{c} = 1, \; c \neq 0$$
$$x + y = c$$
$$1 + 2 = c$$
$$3 = c$$
$$x + y = 3$$
$$x + y - 3 = 0$$

87. $4x - 2y = 3$

$y = 2x - \frac{3}{2}$

Slope: $m = 2$

(a) $(2, 1), m = 2$

$y - 1 = 2(x - 2)$

$y = 2x - 3$

(b) $(2, 1), m = -\frac{1}{2}$

$y - 1 = -\frac{1}{2}(x - 2)$

$y = -\frac{1}{2}x + 2$

89. $3x + 4y = 7$

$y = -\frac{3}{4}x + \frac{7}{4}$

Slope: $m = -\frac{3}{4}$

(a) $\left(-\frac{2}{3}, \frac{7}{8}\right), m = -\frac{3}{4}$

$y - \frac{7}{8} = -\frac{3}{4}\left(x - \left(-\frac{2}{3}\right)\right)$

$y = -\frac{3}{4}x + \frac{3}{8}$

(b) $\left(-\frac{2}{3}, \frac{7}{8}\right), m = \frac{4}{3}$

$y - \frac{7}{8} = \frac{4}{3}\left(x - \left(-\frac{2}{3}\right)\right)$

$y = \frac{4}{3}x + \frac{127}{72}$

91. $y = -3$

$m = 0$

(a) $(-1, 0)$ and $m = 0$

$y = 0$

(b) $(-1, 0)$, m is undefined.

$x = -1$

93. $x = 4$

m is undefined.

(a) $(2, 5)$, m is undefined.

$x = 2$

(b) $(2, 5), m = 0$

$y = 5$

95. $x - y = 4$

$y = x - 4$

Slope: $m = 1$

(a) $(2.5, 6.8), m = 1$

$y - 6.8 = 1(x - 2.5)$

$y = x + 4.3$

(b) $(2.5, 6.8), m = -1$

$y - 6.8 = (-1)(x - 2.5)$

$y = -x + 9.3$

97. (a) $y = 2x$

(b) $y = -2x$

(c) $y = \frac{1}{2}x$

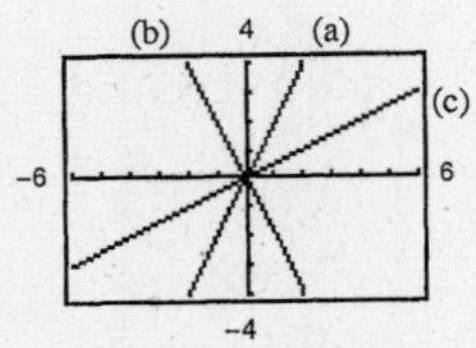

(b) and (c) are perpendicular.

99. (a) $y = -\frac{1}{2}x$

(b) $y = -\frac{1}{2}x + 3$

(c) $y = 2x - 4$

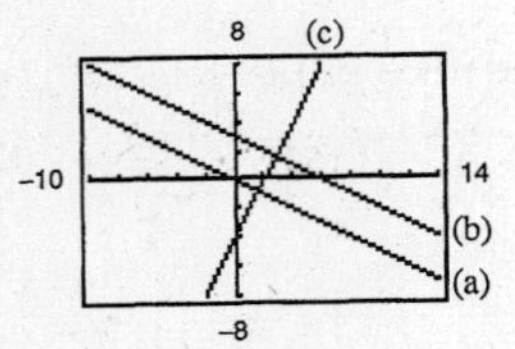

(a) and (b) are parallel.

(c) is perpendicular to (a) and (b).

101. Set the distance between $(4, -1)$ and (x, y) equal to the distance between $(-2, 3)$ and (x, y).

$$\sqrt{(x - 4)^2 + [y - (-1)]^2} = \sqrt{[x - (-2)]^2 + (y - 3)^2}$$

$$(x - 4)^2 + (y + 1)^2 = (x + 2)^2 + (y - 3)^2$$

$$x^2 - 8x + 16 + y^2 + 2y + 1 = x^2 + 4x + 4 + y^2 - 6y + 9$$

$$-8x + 2y + 17 = 4x - 6y + 13$$

$$0 = 12x - 8y - 4$$

$$0 = 4(3x - 2y - 1)$$

$$0 = 3x - 2y - 1$$

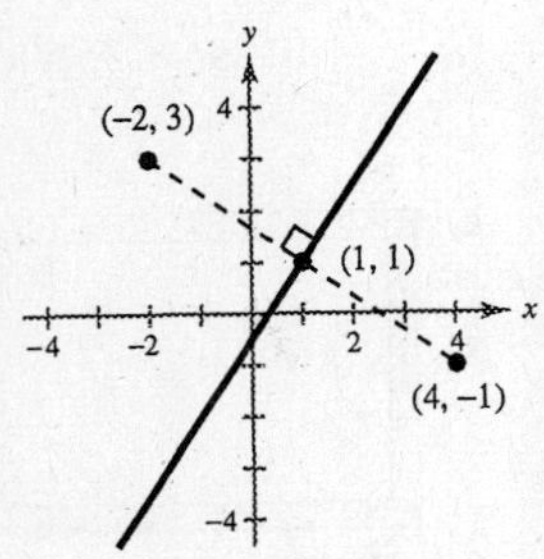

This line is the perpendicular bisector of the line segment connecting $(4, -1)$ and $(-2, 3)$.

103. Set the distance between $\left(3, \frac{5}{2}\right)$ and (x, y) equal to the distance between $(-7, 1)$ and (x, y).

$$\sqrt{(x-3)^2 + \left(y - \tfrac{5}{2}\right)^2} = \sqrt{[x-(-7)]^2 + (y-1)^2}$$
$$(x-3)^2 + \left(y - \tfrac{5}{2}\right)^2 = (x+7)^2 + (y-1)^2$$
$$x^2 - 6x + 9 + y^2 - 5y + \tfrac{25}{4} = x^2 + 14x + 49 + y^2 - 2y + 1$$
$$-6x - 5y + \tfrac{61}{4} = 14x - 2y + 50$$
$$-24x - 20y + 61 = 56x - 8y + 200$$
$$80x + 12y + 139 = 0$$

This line is the perpendicular bisector of the line segment connecting $\left(3, \frac{5}{2}\right)$ and $(-7, 1)$.

105. $(0, 0.18)$ and $(5, 3.65)$

$$y - 0.18 = \frac{3.65 - 0.18}{5 - 0}(t - 0)$$
$$y = 0.694t + 0.18$$

For the year 2005 use $t = 10$: $y = 0.694(10) + 0.18 = \$7.12$

For the year 2010 use $t = 15$: $y = 0.694(15) + 0.18 = \$10.59$

107. Using the points (1998, 28,500) and (2000, 32,900), we have

$$m = \frac{32{,}900 - 28{,}500}{2000 - 1998} = \frac{4400}{2} = 2200$$
$$S - 28{,}500 = 2200(t - 1998)$$
$$S = 2200t - 4{,}367{,}100.$$

When $t = 2005$, we have $S = 2200(2005) - 4{,}367{,}100$, or \$43,900.

109. Using the points $(0, 875)$ and $(5, 0)$, where the first coordinate represents the year t and the second coordinate represents the value V, we have

$$m = \frac{0 - 875}{5 - 0} = -175$$
$$V = -175t + 875,\ 0 \le t \le 5.$$

111. Sale price = List price − 15% of the list price

$$S = L - 0.15L$$
$$S = 0.85L$$

113. (a) $C = 36{,}500 + 5.25t + 11.50t$

$= 16.75t + 36{,}500$

(b) $R = 27t$

(c) $P = R - C$

$= 27t - (16.75t + 36{,}500)$

$= 10.25t - 36{,}500$

(d) $0 = 10.25t - 36{,}500$

$36{,}500 = 10.25t$

$t \approx 3561$ hours

115. (a)

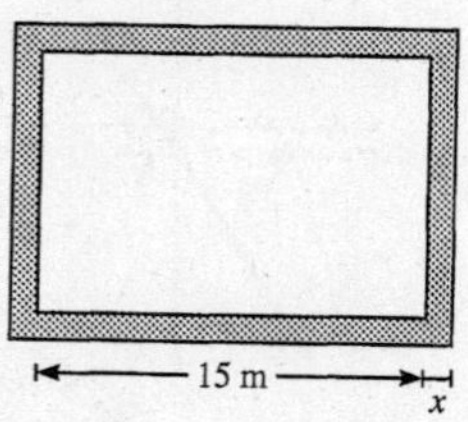

(b) $y = 2(15 + 2x) + 2(10 + 2x)$

$= 8x + 50$

(c)

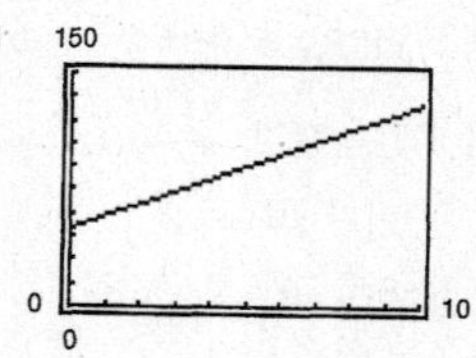

(d) Since $m = 8$, each 1 meter increase in x will increase y by 8 meters.

117. $C = 0.35x + 120$

119. (a) and (b)

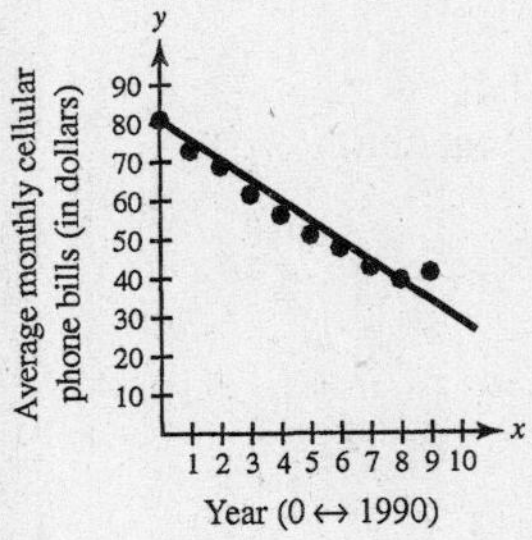

(c) Use two of the data points (0, 80.90) and (8, 39.43)

$$m = \frac{39.43 - 80.90}{8 - 0} \approx -5.2$$

$$y = -5.2x + 80.90$$

(d) The slope approximates the decrease per year in monthly cellular phone bills and the y-intercept approximates the monthly bill in 1990.

(e)

Year	Actual Average Phone Bill	Model Value
1990	\$80.90	\$80.90
1991	\$72.74	\$75.70
1992	\$68.68	\$70.50
1993	\$61.48	\$65.30
1994	\$56.21	\$60.10
1995	\$51.00	\$54.90
1996	\$47.70	\$49.70
1997	\$42.78	\$44.50
1998	\$39.43	\$39.30
1999	\$41.24	\$34.10

The model is a fairly good fit.

(f) For the year 2005, use $x = 15$ in the model.

$y \approx$ \$2.90 per month.

121. False. The slope with the greatest magnitude corresponds to the steepest line.

123. By finding the distance between each pair of points and by using the Pythagorean Theorem, we could determine if A, B, and C are the vertices of a right triangle.

125. No. The slope cannot be determined without knowing the scale on the y-axis. The slopes could be the same.

127. The V-intercept measures the initial cost and the slope measures annual depreciation.

Section P.5 Functions

- Given a set or an equation, you should be able to determine if it represents a function.
- Know that functions can be represented in four ways: verbally, numerically, graphically, and algebraically.
- Given a function, you should be able to do the following.
  (a) Find the domain and range.
  (b) Evaluate it at specific values.
- You should be able to use function notation.

Solutions to Odd-Numbered Exercises

1. Yes, the relationship is a function. Each domain value is matched with only one range value.

3. No, the relationship is not a function. The domain values are each matched with three range values.

5. Yes, it does represent a function. Each input value is matched with only one output value.

7. No, it does not represent a function. The input values of 10 and 7 are each matched with two output values.

9. (a) Each element of A is matched with exactly one element of B, so it does represent a function.

(b) The element 1 in A is matched with two elements, -2 and 1 of B, so it does not represent a function.

(c) Each element of A is matched with exactly one element of B, so it does represent a function.

(d) The element 2 in A is not matched with an element of B, so it does not represent a function.

11. Each is a function. For each year there corresponds one and only one circulation.

13. $x^2 + y^2 = 4 \Longrightarrow y = \pm\sqrt{4 - x^2}$

No, y *is not* a function of x.

15. $x^2 + y = 4 \Longrightarrow y = 4 - x^2$

Yes, y *is* a function of x.

17. $2x + 3y = 4 \Longrightarrow y = \frac{1}{3}(4 - 2x)$

Yes, y *is* a function of x.

19. $y^2 = x^2 - 1 \Longrightarrow y = \pm\sqrt{x^2 - 1}$

No, y *is not* a function of x.

21. $y = |4 - x|$

Yes, y *is* a function of x.

23. $f(x) = 2x - 3$

(a) $f(1) = 2(1) - 3 = -1$

(b) $f(-3) = 2(-3) - 3 = -9$

(c) $f(x - 1) = 2(x - 1) - 3 = 2x - 5$

25. $V(r) = \frac{4}{3}\pi r^3$

(a) $V(3) = \frac{4}{3}\pi(3)^3 = \frac{4}{3}\pi(27) = 36\pi$

(b) $V(\frac{3}{2}) = \frac{4}{3}\pi\left(\frac{3}{2}\right)^3 = \frac{4}{3}\pi\left(\frac{27}{8}\right) = \frac{9}{2}\pi$

(c) $V(2r) = \frac{4}{3}\pi(2r)^3 = \frac{4}{3}\pi(8r^3) = \frac{32}{3}\pi r^3$

27. $f(y) = 3 - \sqrt{y}$

(a) $f(4) = 3 - \sqrt{4} = 1$

(b) $f(0.25) = 3 - \sqrt{0.25} = 2.5$

(c) $f(4x^2) = 3 - \sqrt{4x^2} = 3 - 2|x|$

29. $q(x) = \dfrac{1}{x^2 - 9}$

(a) $q(0) = \dfrac{1}{0^2 - 9} = -\dfrac{1}{9}$

(b) $q(3) = \dfrac{1}{3^2 - 9}$ is undefined.

(c) $q(y + 3) = \dfrac{1}{(y + 3)^2 - 9} = \dfrac{1}{y^2 + 6y}$

31. $f(x) = \dfrac{|x|}{x}$

(a) $f(2) = \dfrac{|2|}{2} = 1$

(b) $f(-2) = \dfrac{|-2|}{-2} = -1$

(c) $f(x - 1) = \dfrac{|x - 1|}{x - 1}$

33. $f(x) = \begin{cases} 2x + 1, & x < 0 \\ 2x + 2, & x \geq 0 \end{cases}$

(a) $f(-1) = 2(-1) + 1 = -1$

(b) $f(0) = 2(0) + 2 = 2$

(c) $f(2) = 2(2) + 2 = 6$

35. $f(x) = \begin{cases} 3x - 1, & x < -1 \\ 4, & -1 \leq x \leq 1 \\ x^2, & x > 1 \end{cases}$

(a) $f(-2) = 3(-2) - 1 = -7$

(b) $f\left(-\frac{1}{2}\right) = 4$

(c) $f(3) = 3^2 = 9$

37. $f(x) = x^2 - 3$

x	-2	-1	0	1	2
$f(x)$	1	-2	-3	-2	1

39. $h(t) = \frac{1}{2}|t + 3|$

t	-5	-4	-3	-2	-1
$h(t)$	1	$\frac{1}{2}$	0	$\frac{1}{2}$	1

41. $f(x) = \begin{cases} -\frac{1}{2}x + 4, & x \le 0 \\ (x-2)^2, & x > 0 \end{cases}$

x	-2	-1	0	1	2
$f(x)$	5	$\frac{9}{2}$	4	1	0

43. $15 - 3x = 0$

$3x = 15$

$x = 5$

45. $\frac{3x - 4}{5} = 0$

$3x - 4 = 0$

$x = \frac{4}{3}$

47. $x^2 - 9 = 0$

$x^2 = 9$

$x = \pm 3$

49. $x^3 - x = 0$

$x(x^2 - 1) = 0$

$x(x + 1)(x - 1) = 0$

$x = 0,\ x = -1, \text{ or } x = 1$

51. $f(x) = g(x)$

$x^2 + 2x + 1 = 3x + 3$

$x^2 - x - 2 = 0$

$(x + 1)(x - 2) = 0$

$x = -1 \text{ or } x = 2$

53. $f(x) = g(x)$

$\sqrt{3x} + 1 = x + 1$

$\sqrt{3x} = x$

$3x = x^2$

$0 = x^2 - 3x$

$0 = x(x - 3)$

$x = 0 \text{ or } x = 3$

55. $f(x) = 5x^2 + 2x - 1$

Since $f(x)$ is a polynomial, the domain is all real numbers x.

57. $h(t) = \frac{4}{t}$

Domain: All real numbers except $t = 0$

59. $g(y) = \sqrt{y - 10}$

Domain: $y - 10 \ge 0$

$y \ge 10$

61. $f(x) = \sqrt[4]{1 - x^2}$

Domain: $1 - x^2 \ge 0$

$-x^2 \ge -1$

$x^2 \le 1$

$x^2 - 1 \le 0$

Critical Numbers: $x = \pm 1$

Test Intervals: $(-\infty, -1), (-1, 1), (1, \infty)$

Test: Is $x^2 - 1 \le 0$?

Solution: $[-1, 1]$ or $-1 \le x \le 1$

63. $g(x) = \frac{1}{x} - \frac{3}{x + 2}$

Domain: All real numbers except $x = 0,\ x = -2$

65. $f(s) = \frac{\sqrt{s - 1}}{s - 4}$

Domain: $s - 1 \ge 0 \Longrightarrow s \ge 1$ and $s \ne 4$

The domain consists of all real numbers s, such that $s \ge 1$ and $s \ne 4$.

67. $f(x) = \frac{x - 4}{\sqrt{x}}$

The domain is all real numbers such that $x > 0$ or $(0, \infty)$.

69. $f(x) = x^2$

$\{(-2, 4), (-1, 1), (0, 0), (1, 1), (2, 4)\}$

71. $f(x) = |x| + 2$

$\{(-2, 4), (-1, 3), (0, 2), (1, 3), (2, 4)\}$

73. By plotting the points, we have a parabola, so $g(x) = cx^2$. Since $(-4, -32)$ is on the graph, we have $-32 = c(-4)^2 \implies c = -2$. Thus, $g(x) = -2x^2$.

75. Since the function is undefined at 0, we have $r(x) = c/x$. Since $(-4, -8)$ is on the graph, we have $-8 = c/-4 \implies c = 32$. Thus, $r(x) = 32/x$.

77.

$$\begin{aligned} f(x) &= x^2 - x + 1 \\ f(2 + h) &= (2 + h)^2 - (2 + h) + 1 \\ &= 4 + 4h + h^2 - 2 - h + 1 \\ &= h^2 + 3h + 3 \\ f(2) &= (2)^2 - 2 + 1 = 3 \\ f(2 + h) - f(2) &= h^2 + 3h \\ \frac{f(2 + h) - f(2)}{h} &= \frac{h^2 + 3h}{h} = h + 3,\ h \neq 0 \end{aligned}$$

79.

$$\begin{aligned} f(x) &= x^3 + 3x \\ f(x + h) &= (x + h)^3 + 3(x + h) \\ &= x^3 + 3x^2h + 3xh^2 + h^3 + 3x + 3h \\ \frac{f(x + h) - f(x)}{h} &= \frac{(x^3 + 3x^2h + 3xh^2 + h^3 + 3x + 3h) - (x^3 + 3x)}{h} \\ &= \frac{h(3x^2 + 3xh + h^2 + 3)}{h} \\ &= 3x^2 + 3xh + h^2 + 3,\ h \neq 0 \end{aligned}$$

81. $g(x) = \dfrac{1}{x^2}$

$$\frac{g(x) - g(3)}{x - 3} = \frac{\frac{1}{x^2} - \frac{1}{9}}{x - 3} = \frac{9 - x^2}{9x^2(x - 3)} = \frac{-(x + 3)(x - 3)}{9x^2(x - 3)} = -\frac{x + 3}{9x^2},\ x \neq 0, 3$$

83. $f(x) = \sqrt{5x}$

$$\frac{f(x) - f(5)}{x - 5} = \frac{\sqrt{5x} - 5}{x - 5}$$

85. $A = s^2$ and $P = 4s \implies \dfrac{P}{4} = s$

$$A = \left(\frac{P}{4}\right)^2 = \frac{P^2}{16}$$

87. (a)

Height, x	Volume, V
1	484
2	800
3	972
4	1024
5	980
6	864

The volume is maximum when $x = 4$ and $V = 1024$ cubic centimeters.

(b)

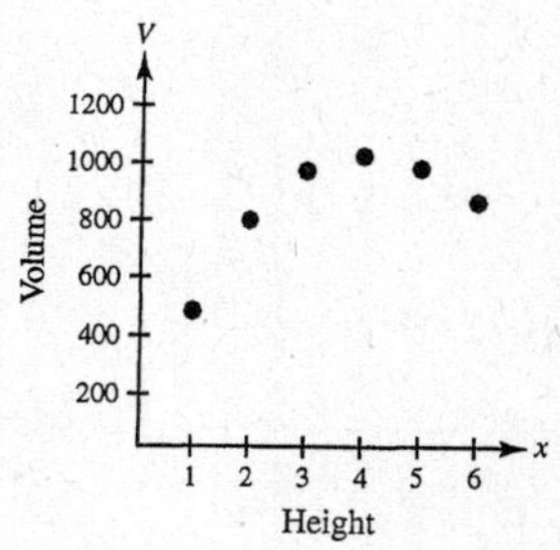

V is a function of x.

(c) $V = x(24 - 2x)^2$

Domain: $0 < x < 12$

89. $A = \frac{1}{2}bh = \frac{1}{2}xy$

Since $(0, y)$, $(2, 1)$, and $(x, 0)$ all lie on the same line, the slopes between any pair are equal.

$$\frac{1-y}{2-0} = \frac{0-1}{x-2}$$

$$\frac{1-y}{2} = \frac{-1}{x-2}$$

$$y = \frac{2}{x-2} + 1$$

$$y = \frac{x}{x-2}$$

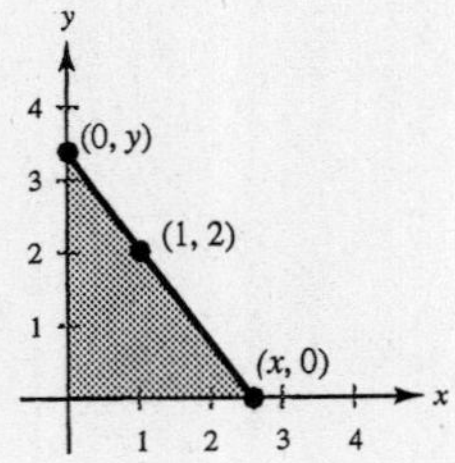

Therefore,

$$A = \frac{1}{2}x\left(\frac{x}{x-2}\right) = \frac{x^2}{2(x-2)}.$$

The domain of A includes x-values such that $x^2/[2(x-2)] > 0$.
Using methods of Section 1.8 we find that the domain is $x > 2$.

91. $p(t) = \begin{cases} 0.543t^2 - 0.75t + 27.8, & 0 \le t \le 4 \\ 1.89t + 27.1 & , \ 5 \le t \le 9 \end{cases}$

where $t = 0$ represents 1990

1990: $t = 0$ and $p(0) = 0.543(0)^2 - 0.75(0) + 27.8$
$= 27.8$ thousand dollars
$= \$27,800$

1994: $t = 4$ and $p(4) = 0.543(4)^2 - 0.75(4) + 27.8$
$= 33.488$ thousand dollars
$= \$33,488$

1996: $t = 6$ and $p(6) = 1.89(6) + 27.1$
$= 38.44$ thousand dollars
$= \$38,440$

1999: $t = 9$ and $p(9) = 1.89(9) + 27.1$
$= 44.11$ thousand dollars
$= \$44,110$

93. (a) Cost = variable costs + fixed costs

$$C = 12.30x + 98,000$$

(b) Revenue = price per unit × number of units

$$R = 17.98x$$

(c) Profit = Revenue − Cost

$$P = 17.98x - (12.30x + 98,000)$$

$$P = 5.68x - 98,000$$

95. (a) $R = n(\text{rate}) = n[8.00 - 0.05(n - 80)], \ n \ge 80$

$$R = 12.00n - 0.05n^2 = 12n - \frac{n^2}{20} = \frac{240n - n^2}{20}, \ n \ge 80$$

(b)

n	90	100	110	120	130	140	150
$R(n)$	\$675	\$700	\$715	\$720	\$715	\$700	\$675

The revenue is maximum when 120 people take the trip.

97. (a)

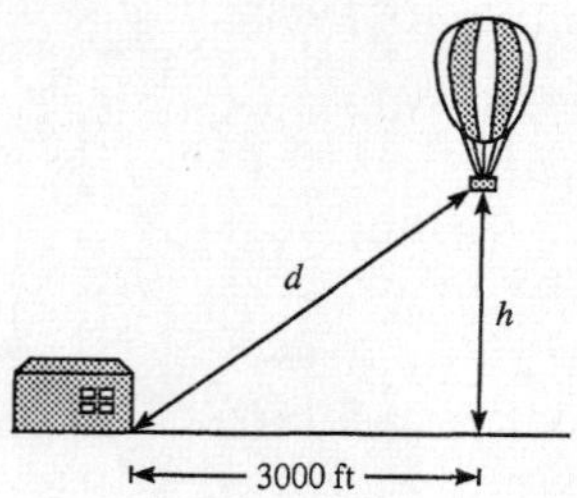

(b) $(3000)^2 + h^2 = d^2$

$$h = \sqrt{d^2 - (3000)^2}$$

Domain: $d \geq 3000$
(since both $d \geq 0$ and $d^2 - (3000)^2 \geq 0$)

99. (a) $\dfrac{f(2001) - f(1996)}{2001 - 1996} = \dfrac{125 - 116}{5} = 1.8$

Approximately 1.8 more species of fish became threatened and/or endangered each year between 1996 and 2001.

(b)–(e) Algebraic Model: $y = \begin{cases} 2x + 104, & 6 \leq x \leq 7 \\ 2x + 103, & 8 \leq x \leq 11 \end{cases}$

Calculator Model: $y \approx 1.7714x + 105.2762$

Year 6↔1996	Actual Number of Fish Species	Number from the Algebraic Model	Number from the Calculator Model
6	116	116	116
7	118	118	118
8	119	119	119
9	121	121	121
10	123	123	123
11	125	125	125

Both models are good fits to the actual data.

101. True, the set represents a function.
Each x-value corresponds to one y-value.

Section P.6 Analyzing Graphs of Functions

- You should be able to determine the domain and range of a function from its graph.
- You should be able to use the vertical line test for functions.
- You should be able to find the zeros of a function.
- You should be able to determine when a function is constant, increasing, or decreasing.
- You should be able to approximate relative minimums and relative maximums from the graph of a function.
- You should know that f is

  (a) odd if $f(-x) = -f(x)$. (b) even if $f(-x) = f(x)$.

Solutions to Odd-Numbered Exercises

1. Domain: $(-\infty, -1] \cup [1, \infty)$

Range: $[0, \infty)$

3. Domain: $[-4, 4]$

Range: $[0, 4]$

5. (a) $f(-2) = 0$ (b) $f(-1) = -1$

(c) $f(\frac{1}{2}) = 0$ (d) $f(1) = -2$

7. (a) $f(-2) = -3$ (b) $f(1) = 0$

(c) $f(0) = 1$ (d) $f(2) = -3$

9. $y = \frac{1}{2}x^2$

A vertical line intersects the graph just once, so y *is* a function of x.

11. $x - y^2 = 1 \Rightarrow y = \pm\sqrt{x - 1}$

y *is not* a function of x.
Some vertical lines cross the graph twice.

13. $x^2 = 2xy - 1$

A vertical line intersects the graph just once, so y *is* a function of x.

15.
$$2x^2 - 7x - 30 = 0$$
$$(2x + 5)(x - 6) = 0$$
$$2x + 5 = 0 \quad \text{or} \quad x - 6 = 0$$
$$x = -\tfrac{5}{2} \quad \text{or} \quad x = 6$$

17. $\dfrac{x}{9x^2 - 4} = 0$

$$x = 0$$

19. $\dfrac{1}{2}x^3 - x = 0$

$$x^3 - 2x = 2(0)$$
$$x(x^2 - 2) = 0$$
$$x = 0 \quad \text{or} \quad x^2 - 2 = 0$$
$$x^2 = 2$$
$$x = \pm\sqrt{2}$$

21.
$$4x^3 - 24x^2 - x + 6 = 0$$
$$4x^2(x - 6) - 1(x - 6) = 0$$
$$(x - 6)(4x^2 - 1) = 0$$
$$(x - 6)(2x + 1)(2x - 1) = 0$$
$$x - 6 = 0, \quad 2x + 1 = 0, \quad 2x - 1 = 0$$
$$x = 6, \quad x = -\tfrac{1}{2}, \quad x = \tfrac{1}{2}$$

23. $\sqrt{2x} - 1 = 0$

$$\sqrt{2x} = 1$$
$$2x = 1$$
$$x = \tfrac{1}{2}$$

25. $3 + \dfrac{5}{x} = 0$

$$3x + 5 = 0$$
$$x = -\frac{5}{3}$$

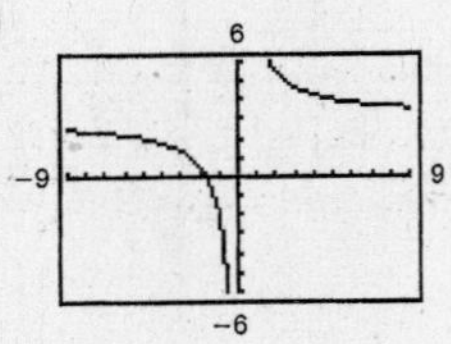

27. $\sqrt{2x + 11} = 0$

$$2x + 11 = 0$$
$$x = -\frac{11}{2}$$

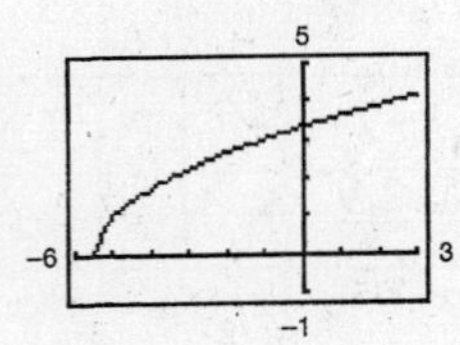

29. $\dfrac{3x - 1}{x - 6} = 0$

$$3x - 1 = 0$$
$$x = \frac{1}{3}$$

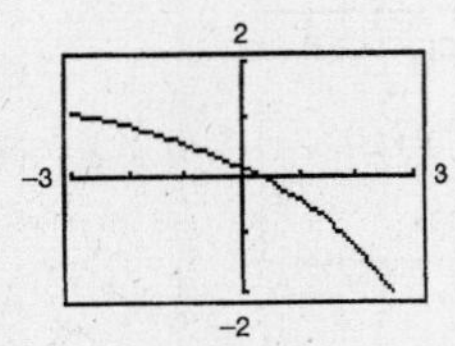

31. $f(x) = \dfrac{3}{2}x$

f is increasing on $(-\infty, \infty)$.

33. $f(x) = x^3 - 3x^2 + 2$

f is increasing on $(-\infty, 0)$ and $(2, \infty)$.

f is decreasing on $(0, 2)$.

35. $f(x) = \begin{cases} x + 3, & x \le 0 \\ 3, & 0 < x \le 2 \\ 2x + 1, & x > 2 \end{cases}$

f is increasing on $(-\infty, 0)$ and $(2, \infty)$.

f is constant on $(0, 2)$.

37. $f(x) = |x + 1| + |x - 1|$

f is increasing on $(1, \infty)$.

f is constant on $(-1, 1)$.

f is decreasing on $(-\infty, -1)$.

39. $f(x) = 3$

(a)

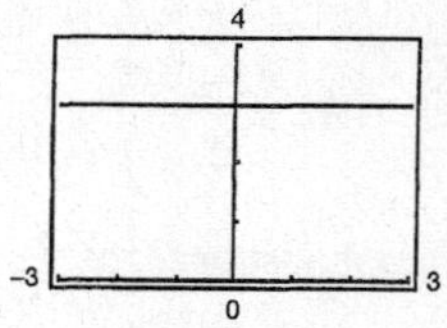

Constant on $(-\infty, \infty)$

(b)

x	-2	-1	0	1	2
$f(x)$	3	3	3	3	3

41. $g(s) = \dfrac{s^2}{4}$

(a)

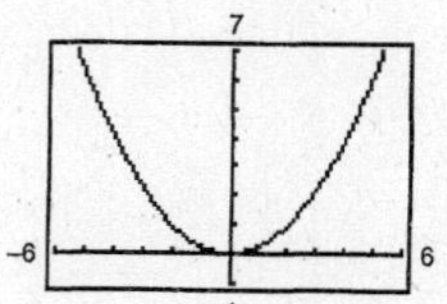

Decreasing on $(-\infty, 0)$

Increasing on $(0, \infty)$

(b)

s	-4	-2	0	2	4
$g(s)$	4	1	0	1	4

43. $f(t) = -t^4$

(a)

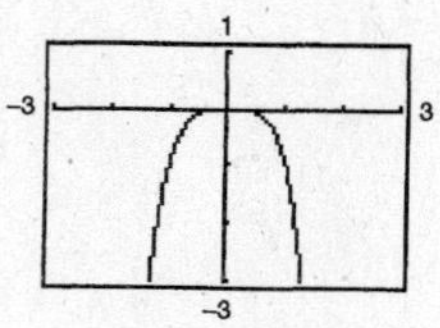

Increasing on $(-\infty, 0)$

Decreasing on $(0, \infty)$

(b)

t	-2	-1	0	1	2
$f(t)$	-16	-1	0	-1	-16

45. $f(x) = \sqrt{1 - x}$

(a)

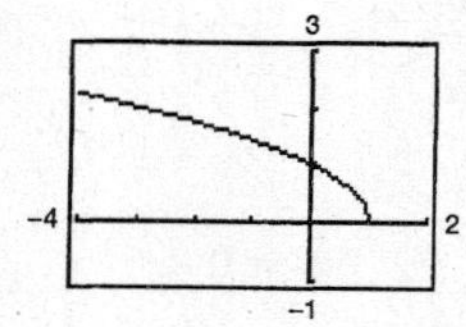

Decreasing on $(-\infty, 1)$

(b)

x	-3	-2	-1	0	1
$f(x)$	2	$\sqrt{3}$	$\sqrt{2}$	1	0

47. $f(x) = x^{3/2}$

(a)

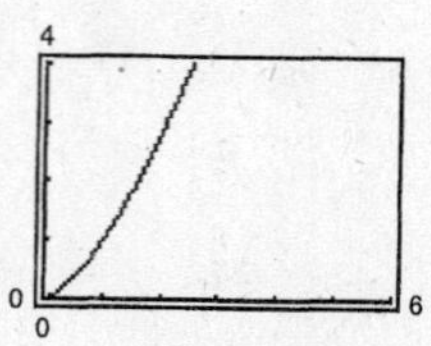

Increasing on $(0, \infty)$

(b)

x	0	1	2	3	4
$f(x)$	0	1	2.8	5.2	8

49. $f(x) = (x - 4)(x + 2)$

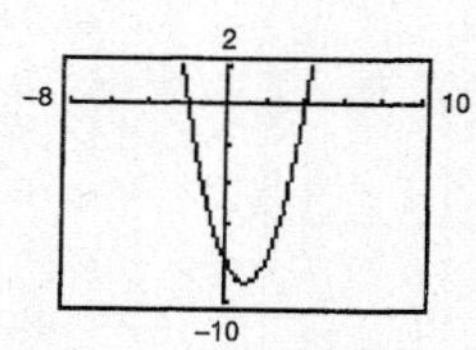

Relative Minimum at $(1, -9)$

51. $f(x) = x(x - 2)(x + 3)$

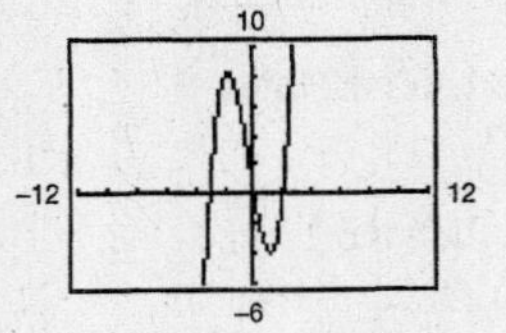

Relative Minimum at $(1.12, -4.06)$

Relative Maximum at $(-1.79, 8.21)$

53. $f(x) = 4 - x$

$f(x) \geq 0$ on $(-\infty, 4]$.

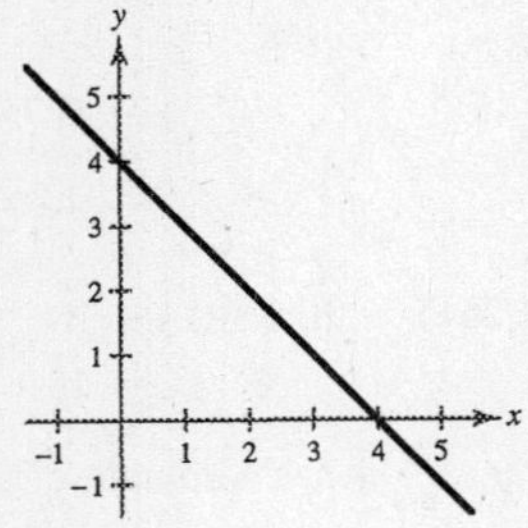

55. $f(x) = x^2 + x$

$f(x) \geq 0$ on $(-\infty, -1]$ and $[0, \infty)$.

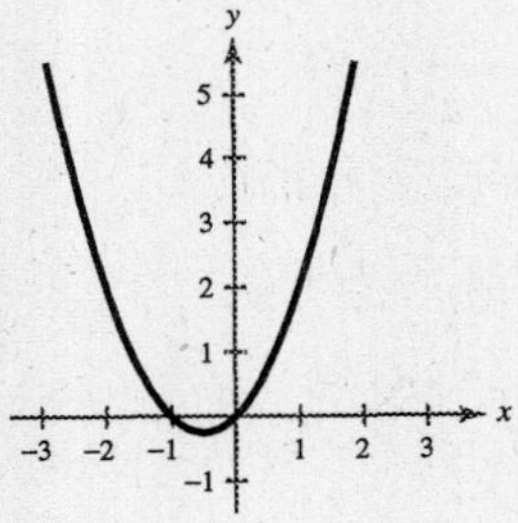

57. $f(x) = \sqrt{x - 1}$

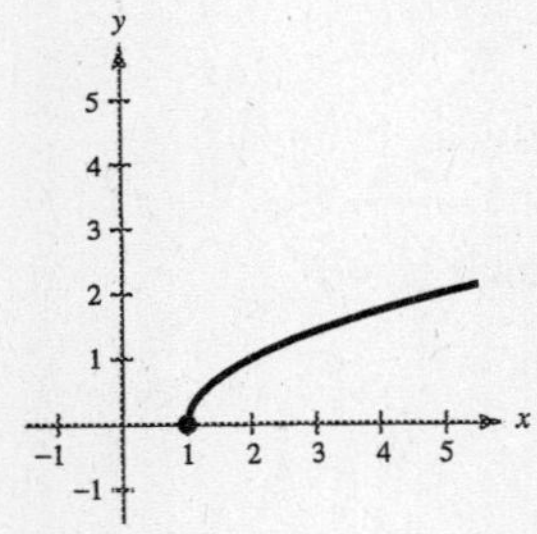

$f(x) \geq 0$ on $[1, \infty)$.

59. $f(x) = -(1 + |x|)$

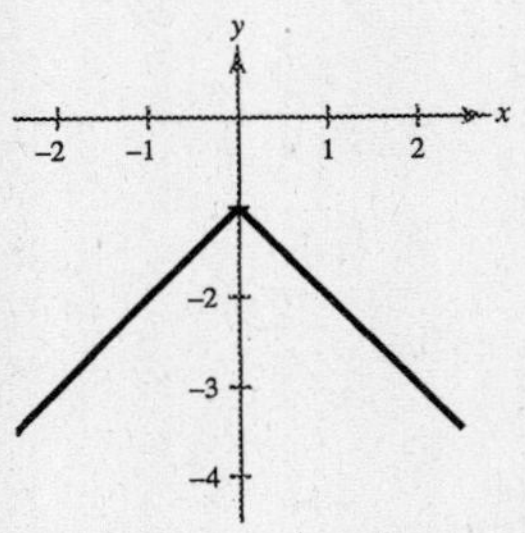

$f(x)$ is never greater than 0. ($f(x) < 0$ for all x.)

61. $f(x) = x^6 - 2x^2 + 3$

$f(-x) = (-x)^6 - 2(-x)^2 + 3$

$= x^6 - 2x^2 + 3$

$= f(x)$

f is even.

63. $g(x) = x^3 - 5x$

$g(-x) = (-x)^3 - 5(-x)$

$= -x^3 + 5x$

$= -g(x)$

g is odd.

65. $f(t) = t^2 + 2t - 3$

$f(-t) = (-t)^2 + 2(-t) - 3$

$= t^2 - 2t - 3$

$\neq f(t), \neq -f(t)$

f is neither even nor odd.

67. $h = \text{top} - \text{bottom}$

$= (-x^2 + 4x - 1) - 2$

$= -x^2 + 4x - 3$

69. $h = \text{top} - \text{bottom}$

$= (4x - x^2) - 2x$

$= 2x - x^2$

71. $L = \text{right} - \text{left}$

$= \frac{1}{2}y^2 - 0$

$= \frac{1}{2}y^2$

73. $L = \text{right} - \text{left}$

$= 4 - y^2$

75. $L = -0.294x^2 + 97.744x - 664.875,\ 20 \le x \le 90$

(a)

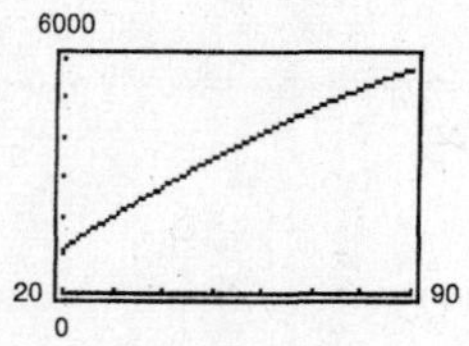

(b) $L = 2000$ when $x \approx 29.9645 \approx 30$ watts

77. (a) For the average salaries of college professors, a scale of \$10,000 would be appropriate.

(b) For the population of the United States, use a scale of 50,000,000.

(c) For the percent of the civilian workforce that is unemployed, use a scale of 1%.

79. False. The function $f(x) = \sqrt{x^2 + 1}$ has a domain of all real numbers.

81. (a) Even. The graph is a reflection in the x-axis.

(b) Even. The graph is a reflection in the y-axis.

(c) Even. The graph is a vertical translation of f.

(d) Neither. The graph is a horizontal translation of f.

83. $\left(-\frac{3}{2}, 4\right)$

(a) If f is even, another point is $\left(\frac{3}{2}, 4\right)$.

(b) If f is odd, another point is $\left(\frac{3}{2}, -4\right)$.

85. $(4, 9)$

(a) If f is even, another point is $(-4, 9)$.

(b) If f is odd, another point is $(-4, -9)$.

87. (a) $y = x$

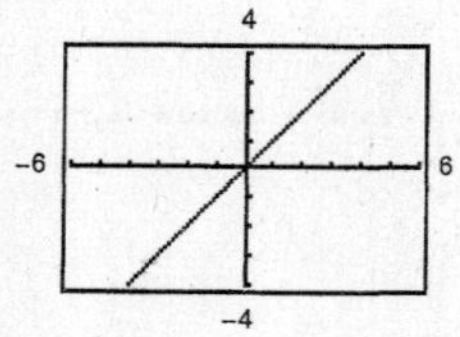

(b) $y = x^2$

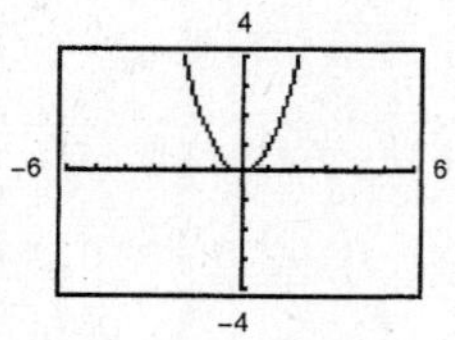

(c) $y = x^3$

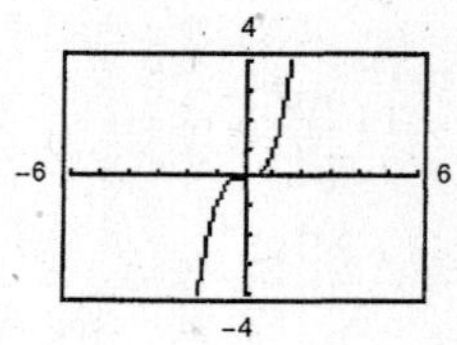

(d) $y = x^4$

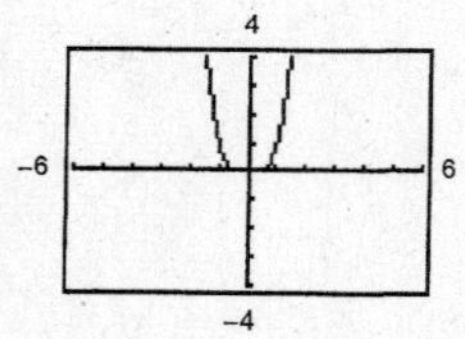

(e) $y = x^5$

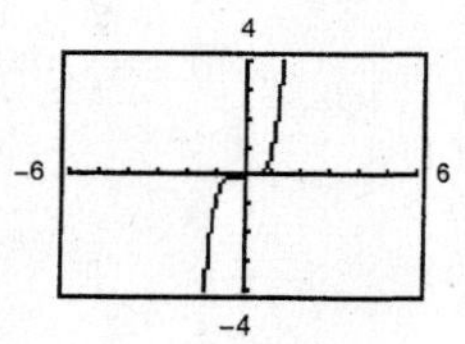

(f) $y = x^6$

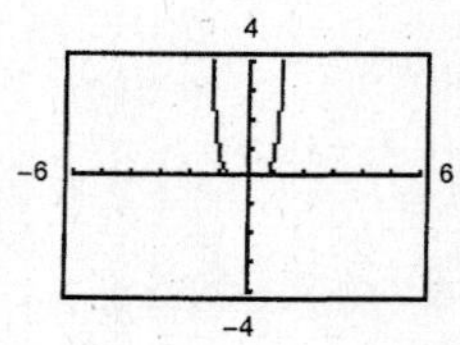

All the graphs pass through the origin. The graphs of the odd powers of x are symmetric with respect to the origin and the graphs of the even powers are symmetric with respect to the y-axis. As the powers increase, the graphs become flatter in the interval $-1 < x < 1$.

Section P.7 A Library of Functions

- You should be able to identify and graph the following types of functions:
  - (a) Linear functions like $f(x) = ax + b$
  - (b) Squaring functions like $f(x) = x^2$
  - (c) Cubic functions like $f(x) = x^3$
  - (d) Square root functions like $f(x) = \sqrt{x}$
  - (e) Reciprocal functions like $f(x) = \frac{1}{x}$
  - (f) Constant functions like $f(x) = c$
  - (g) Absolute value functions like $f(x) = |x|$
  - (h) Step and Piecewise - Defined functions like $f(x) = [\![x]\!]$
- You should be able to determine the following about these common functions:
  - (a) Domain and range
  - (b) x-intercept(s) and y-intercept
  - (c) Symmetries
  - (d) Where it is increasing, decreasing, or constant.
  - (e) If it is odd, even or neither
  - (f) Relative maximums and relative minimums

Solutions to Odd-Numbered Exercises

1. $f(1) = 4,\ f(0) = 6$

$(1, 4)$ and $(0, 6)$

$m = \dfrac{6-4}{0-1} = -2$

$y - 6 = -2(x - 0)$

$y = -2x + 6$

$f(x) = -2x + 6$

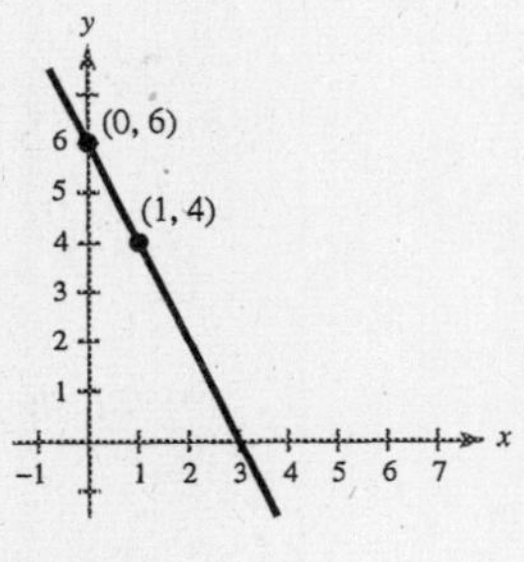

3. $f(5) = -4,\ f(-2) = 17$

$(5, -4)$ and $(-2, 17)$

$m = \dfrac{17-(-4)}{-2-5} = \dfrac{21}{-7} = -3$

$y - (-4) = -3(x - 5)$

$y + 4 = -3x + 15$

$y = -3x + 11$

$f(x) = -3x + 11$

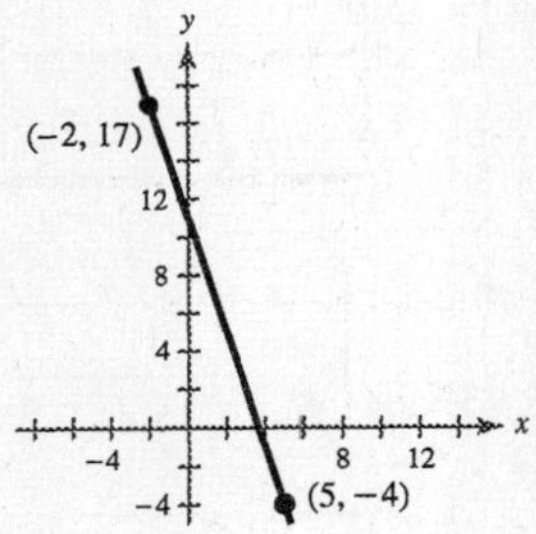

5. $f(-5) = -1,\ f(5) = -1$

$(-5, -1)$ and $(5, -1)$

$m = \dfrac{-1-(-1)}{5-(-5)} = \dfrac{0}{10} = 0$

$y - (-1) = 0(x - (-5))$

$y + 1 = 0$

$y = -1$

$f(x) = -1$

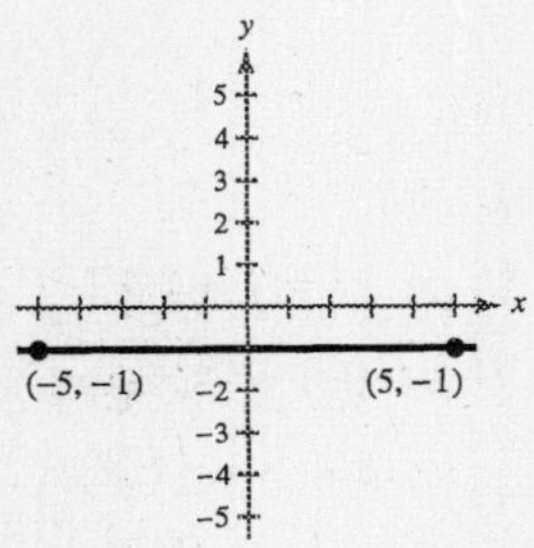

7. $f\left(\frac{1}{2}\right) = -6, f(4) = -3$

$\left(\frac{1}{2}, -6\right)$ and $(4, -3)$

$$m = \frac{-3-(-6)}{4-\frac{1}{2}} = \frac{3}{\frac{7}{2}} = \frac{6}{7}$$

$$y - (-3) = \frac{6}{7}(x - 4)$$

$$y + 3 = \frac{6}{7}x - \frac{24}{7}$$

$$y = \frac{6}{7}x - \frac{45}{7}$$

$$f(x) = \frac{6}{7}x - \frac{45}{7}$$

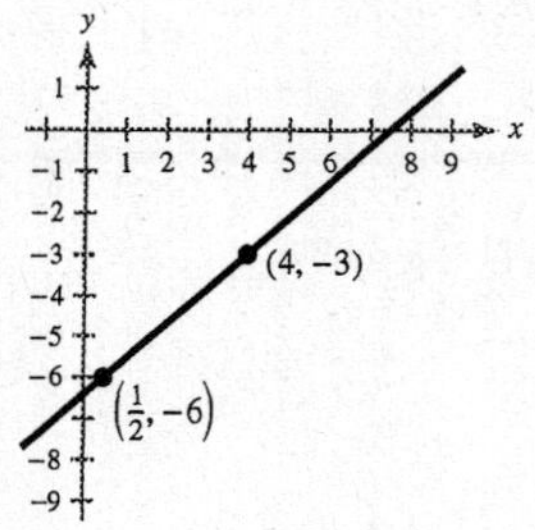

9. $f(x) = -x - \frac{3}{4}$

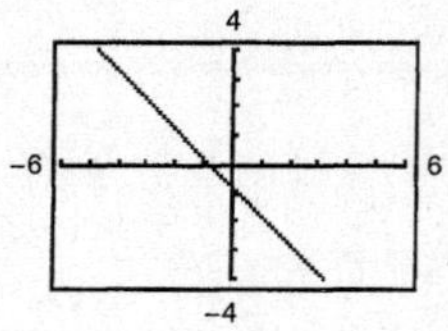

11. $f(x) = -\frac{1}{6}x - \frac{5}{2}$

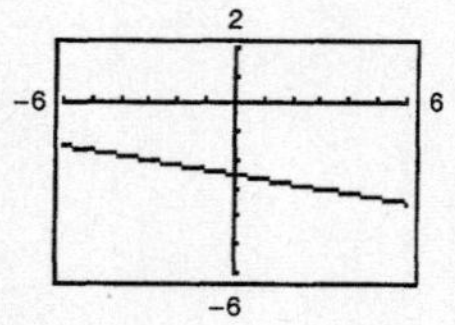

13. $f(x) = x^2 - 2x$

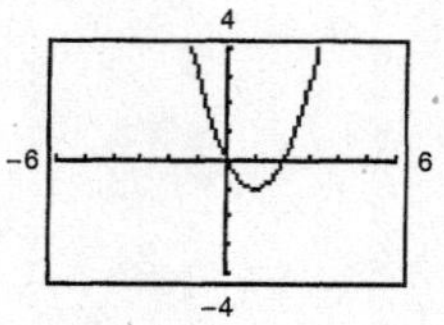

15. $h(x) = -x^2 + 4x + 12$

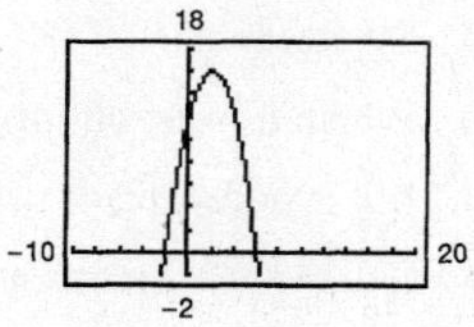

17. $f(x) = x^3 - 1$

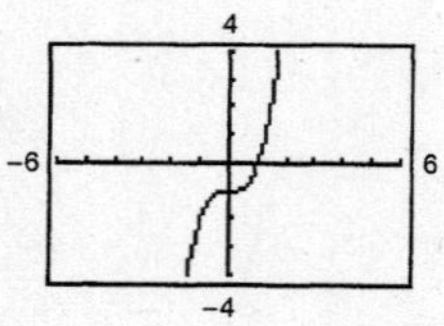

19. $f(x) = (x - 1)^3 + 2$

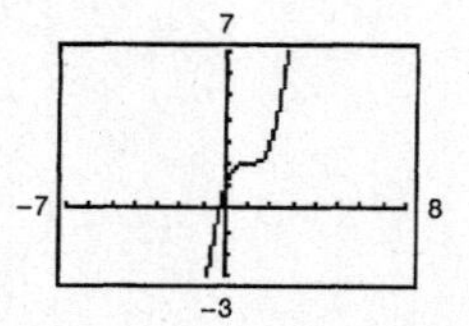

21. $f(x) = 4\sqrt{x}$

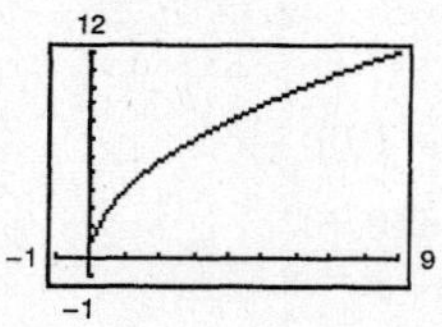

23. $g(x) = 2 - \sqrt{x + 4}$

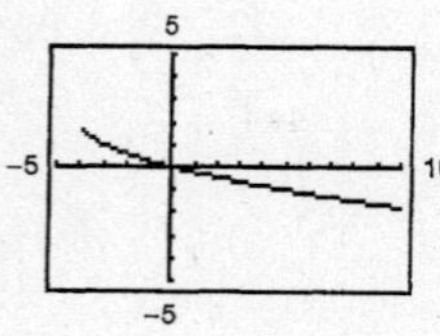

25. $f(x) = -\dfrac{1}{x}$

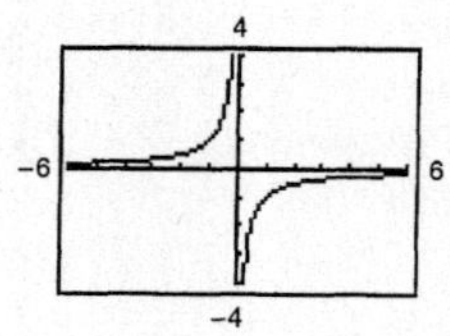

27. $h(x) = \dfrac{1}{x + 2}$

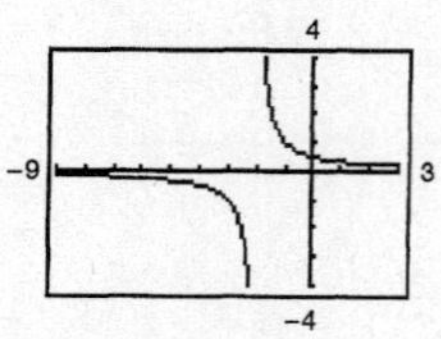

29. $f(x) = [\![x]\!]$

(a) $f(2.1) = 2$

(b) $f(2.9) = 2$

(c) $f(-3.1) = -4$

(d) $f\left(\frac{7}{2}\right) = 3$

31. $h(x) = [\![x + 3]\!]$

(a) $h(-2) = [\![1]\!] = 1$

(b) $h\left(\frac{1}{2}\right) = [\![3.5]\!] = 3$

(c) $h(4.2) = [\![7.2]\!] = 7$

(d) $h(-21.6) = [\![-18.6]\!] = -19$

33. $h(x) = [\![3x - 1]\!]$

(a) $h(2.5) = [\![6.5]\!] = 6$

(b) $h(-3.2) = [\![-10.6]\!] = -11$

(c) $h\left(\frac{7}{3}\right) = [\![6]\!] = 6$

(d) $h\left(-\frac{21}{3}\right) = [\![-22]\!] = -22$

35. $g(x) = 3[\![x - 2]\!] + 5$

(a) $g(-2.7) = 3[\![-4.7]\!] + 5 = 3(-5) + 5 = -10$

(b) $g(-1) = 3[\![-3]\!] + 5 = 3(-3) + 5 = -4$

(c) $g(0.8) = 3[\![-1.2]\!] + 5 = 3(-2) + 5 = -1$

(d) $g(14.5) = 3[\![12.5]\!] + 5 = 3(12) + 5 = 41$

37. $g(x) = -[\![x]\!]$

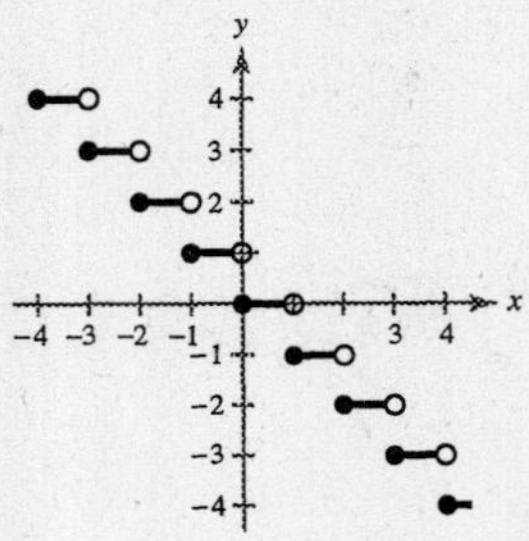

39. $g(x) = [\![x]\!] - 2$

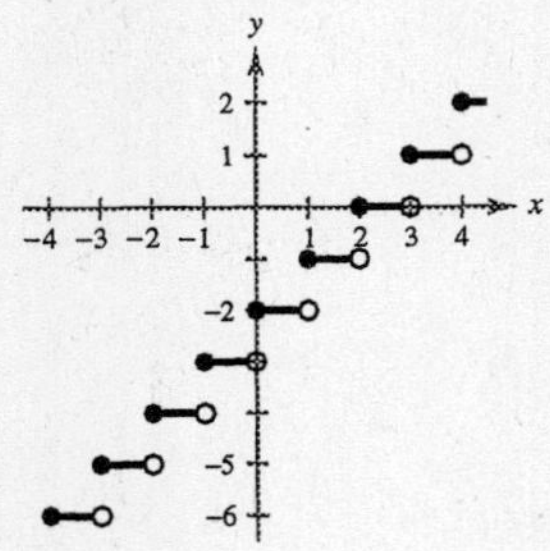

41. $g(x) = [\![x + 1]\!]$

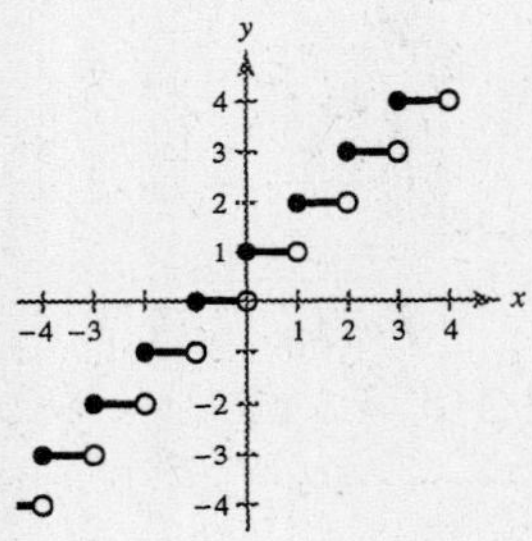

43. $f(x) = \begin{cases} 2x + 3, & x < 0 \\ 3 - x, & x \ge 0 \end{cases}$

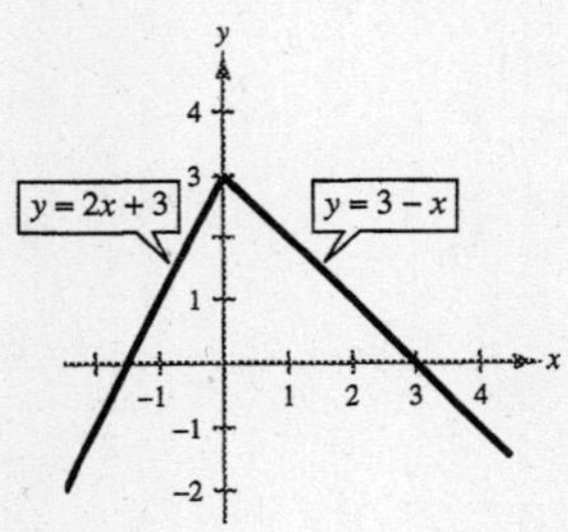

45. $f(x) = \begin{cases} \sqrt{4 + x}, & x < 0 \\ \sqrt{4 - x}, & x \ge 0 \end{cases}$

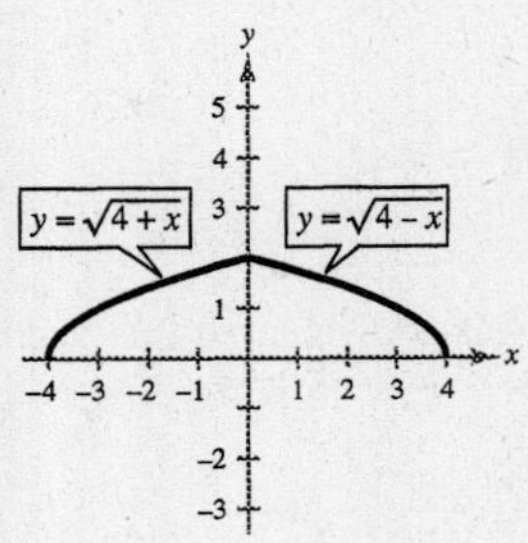

47. $f(x) = \begin{cases} x^2 + 5, & x \le 1 \\ -x^2 + 4x + 3, & x > 1 \end{cases}$

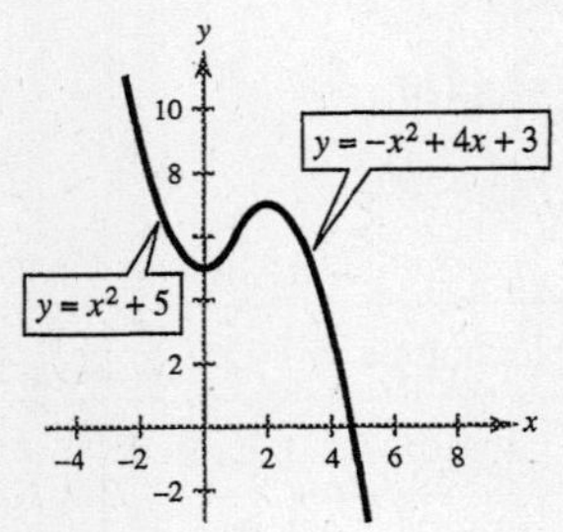

49. $h(x) = \begin{cases} 4 - x^2, & x < -2 \\ 3 + x, & -2 \le x < 0 \\ x^2 + 1, & x \ge 0 \end{cases}$

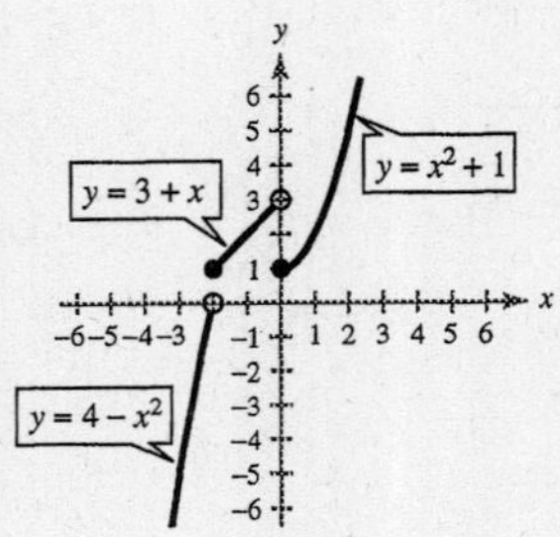

51. $s(x) = 2\left(\frac{1}{4}x - \left[\!\left[\frac{1}{4}x\right]\!\right]\right)$

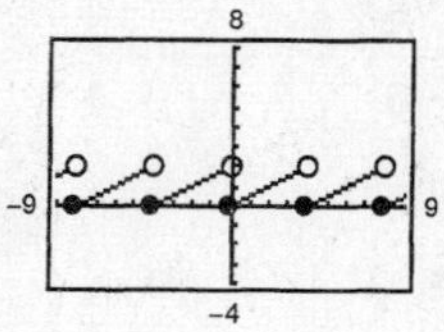

Domain: $(-\infty, \infty)$

Range: $[0, 2)$

Sawtooth pattern

53. Common function: $f(x) = |x|$

55. Common function: $f(x) = x^3$

57. Common function: $f(x) = c$

59. Common function: $f(x) = x$

61. Common function: $f(x) = x^2$

63. $C = 0.60 - 0.42[\![1 - t]\!], t > 0$

(a)

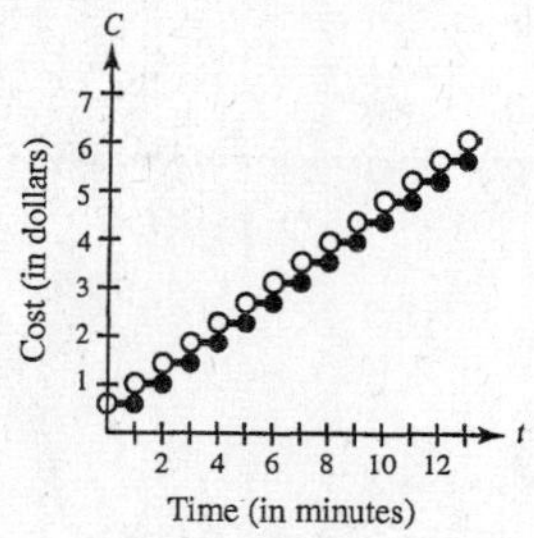

(b) $C(12.5) = \$5.64$

65. $C = 10.75 - 3.95[\![x]\!], x > 0$

(a)

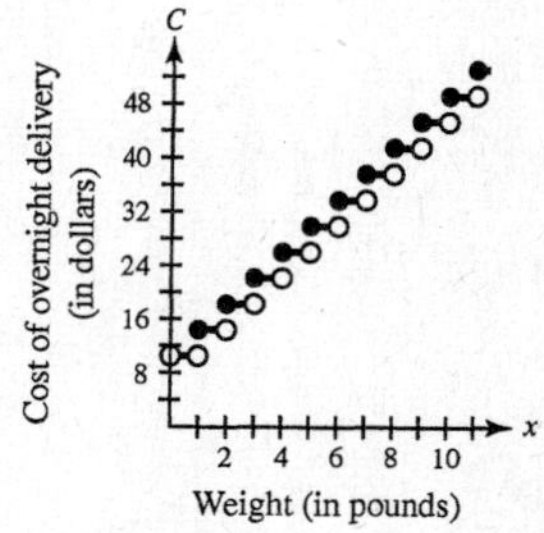

(b) $C(10.33) = 10.75 + 3.95(10) = \$50.25$

67. $W(h) = \begin{cases} 12h, & 0 < h \le 40 \\ 18(h - 40) + 480, & h > 40 \end{cases}$

(a) $W(30) = 12(30) = \$360$

$W(40) = 12(40) = \$480$

$W(45) = 18(5) + 480 = \$570$

$W(50) = 18(10) + 480 = \$660$

(b) $W(h) = \begin{cases} 12h, & 0 < h \le 45 \\ 18(h - 45) + 540, & h > 45 \end{cases}$

69.

Interval	Intake Pipe	Drainpipe 1	Drainpipe 2
$[0, 5]$	Open	Closed	Closed
$[5, 10]$	Open	Open	Closed
$[10, 20]$	Closed	Closed	Closed
$[20, 30]$	Closed	Closed	Open
$[30, 40]$	Open	Open	Open
$[40, 45]$	Open	Closed	Open
$[45, 50]$	Open	Open	Open
$[50, 60]$	Open	Open	Closed

71. True. $f(x) = 2[\![x]\!], 1 \le x < 4$ is equivalent to the given piecewise function.

Section P.8 Shifting, Reflecting, and Stretching Graphs

- You should know the basic types of transformations.

  Let $y = f(x)$ and let c be a positive real number.

1. $h(x) = f(x) + c$	Vertical shift c units upward
2. $h(x) = f(x) - c$	Vertical shift c units downward
3. $h(x) = f(x - c)$	Horizontal shift c units to the right
4. $h(x) = f(x + c)$	Horizontal shift c units to the left
5. $h(x) = -f(x)$	Reflection in the x-axis
6. $h(x) = f(-x)$	Reflection in the y-axis
7. $h(x) = cf(x), c > 1$	Vertical stretch
8. $h(x) = cf(x), 0 < c < 1$	Vertical shrink
9. $h(x) = f(cx), c > 1$	Horizontal shrink
10. $h(x) = f(cx), 0 < c < 1$	Horizontal stretch

Solutions to Odd-Numbered Exercises

1. (a) $f(x) = |x| + c$ Vertical shifts

$c = -1 : f(x) = |x| - 1$ 1 unit down

$c = 1 : f(x) = |x| + 1$ 1 unit up

$c = 3 : f(x) = |x| + 3$ 3 units up

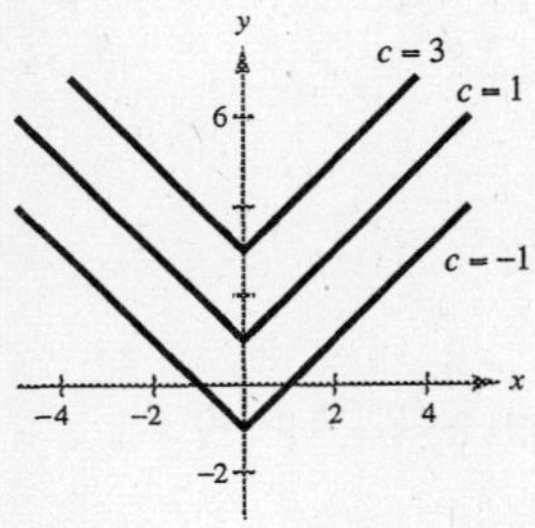

(b) $f(x) = |x - c|$ Horizontal shifts

$c = -1 : f(x) = |x + 1|$ 1 unit left

$c = 1 : f(x) = |x - 1|$ 1 unit right

$c = 3 : f(x) = |x - 3|$ 3 units right

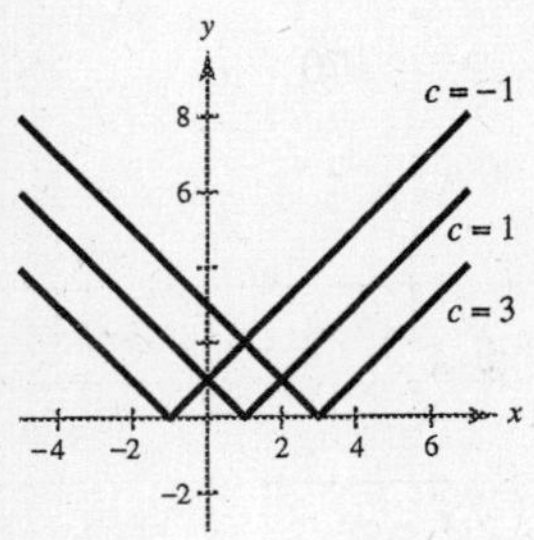

(c) $f(x) = |x + 4| + c$ Horizontal shift 4 units left and a vertical shift

$c = -1 : f(x) = |x + 4| - 1$ 1 unit down

$c = 1 : f(x) = |x + 4| + 1$ 1 unit up

$c = 3 : f(x) = |x + 4| + 3$ 3 units up

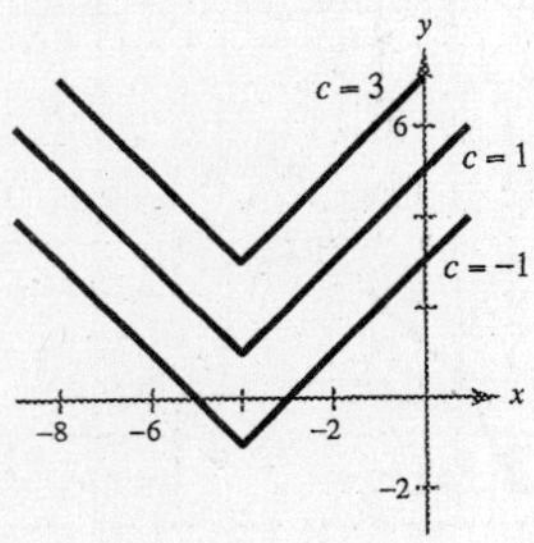

3. (a) $f(x) = [\![x]\!] + c$ — Vertical shifts

$c = -2 : f(x) = [\![x]\!] - 2$ — 2 units down

$c = 0 : f(x) = [\![x]\!]$ — common function

$c = 2 : f(x) = [\![x]\!] + 2$ — 2 units up

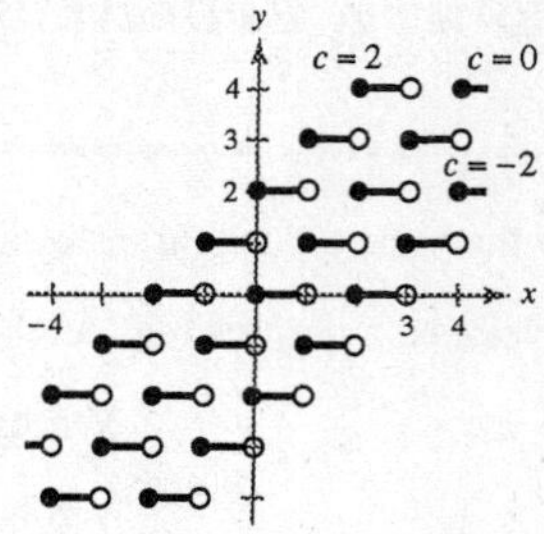

(b) $f(x) = [\![x + c]\!]$ — Horizontal shifts

$c = -2 : f(x) = [\![x - 2]\!]$ — 2 unit right

$c = 0 : f(x) = [\![x]\!]$ — common function

$c = 2 : f(x) = [\![x + 2]\!]$ — 2 units left

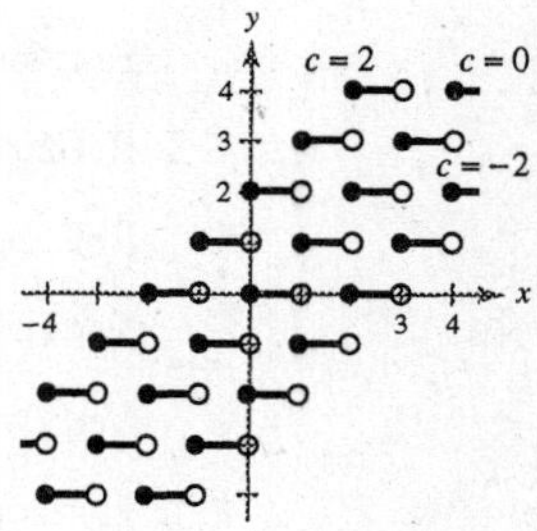

(c) $f(x) = [\![x - 1]\!] + c$ — Horizontal shift 1 unit right and a vertical shift

$c = -2 : f(x) = [\![x - 1]\!] - 2$ — 2 units down

$c = 0 : f(x) = [\![x - 1]\!]$

$c = 2 : f(x) = [\![x - 1]\!] + 2$ — 2 units up

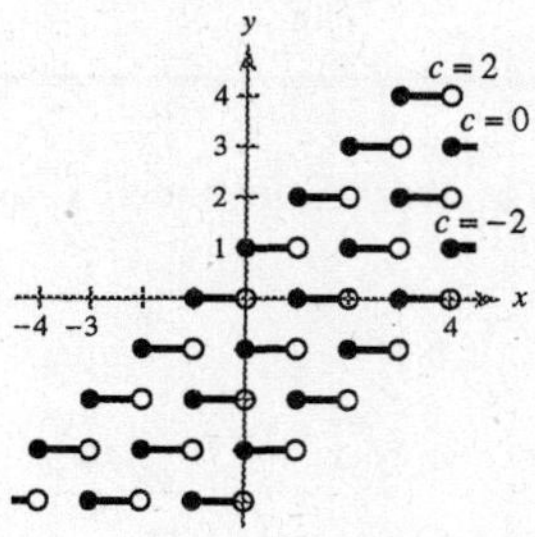

5. (a) $y = f(x) + 2$

Vertical shift 2 units upward

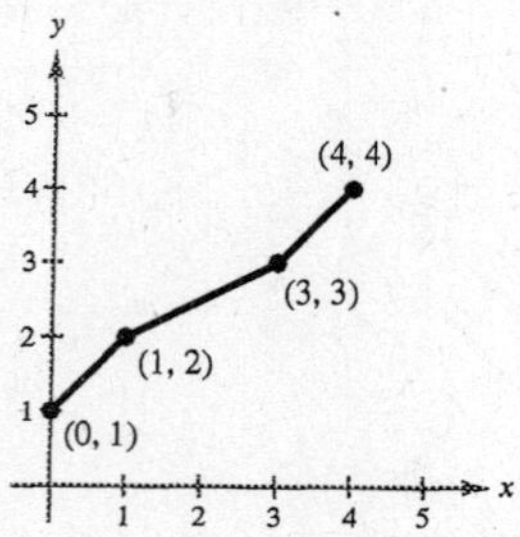

(b) $y = f(x - 2)$

Horizontal shift 2 units to the right

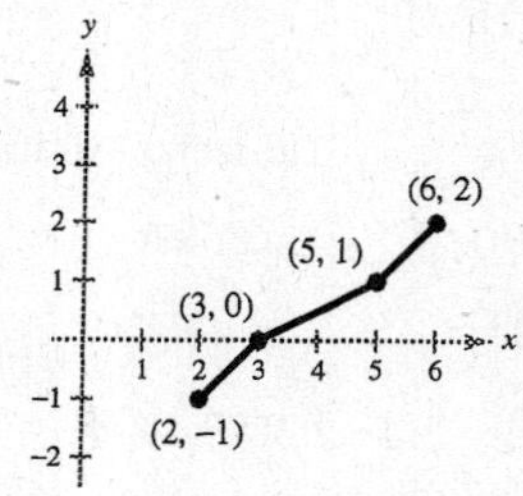

(c) $y = 2f(x)$

Vertical stretch (each y-value is multiplied by 2)

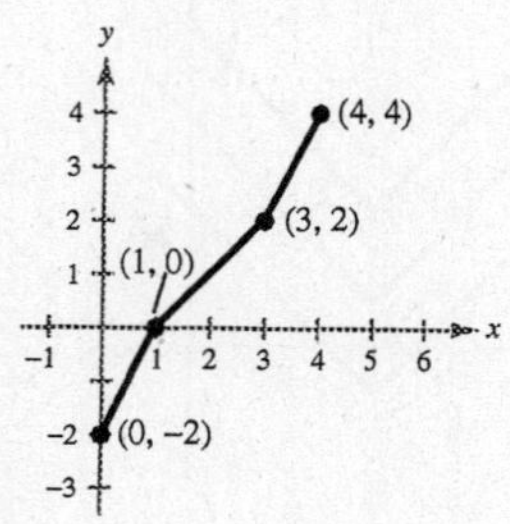

(d) $y = -f(x)$

Reflection in the x-axis

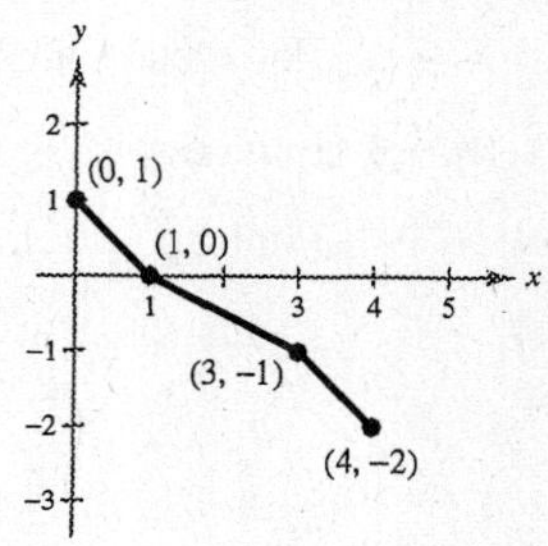

—CONTINUED—

5. —CONTINUED—

(e) $y = f(x + 3)$

Horizontal shift 3 units to the left

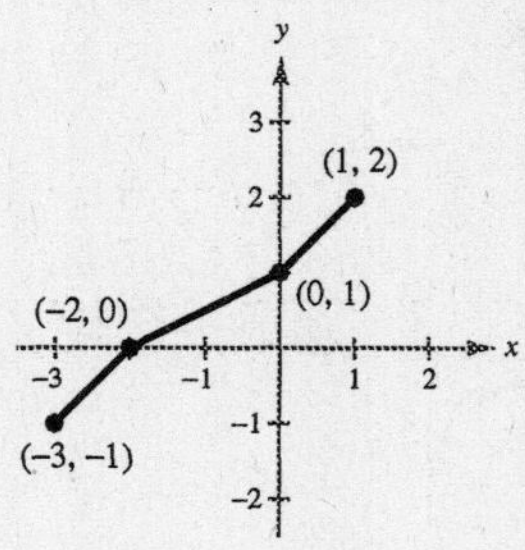

(f) $y = f(-x)$

Reflection in the y-axis

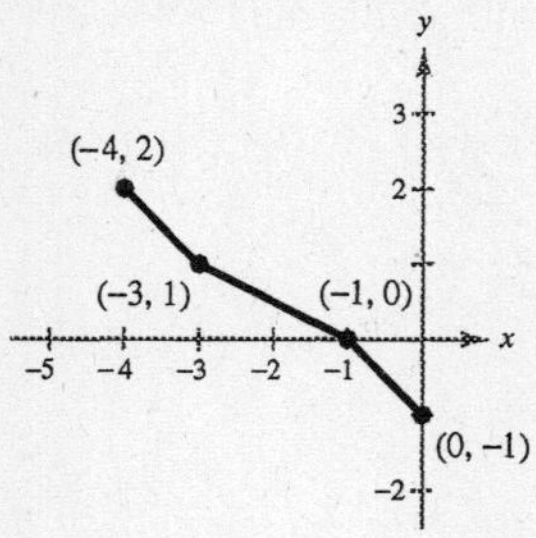

(g) $y = f\left(\frac{1}{2}x\right)$

Horizontal stretch (each x-value is multiplied by 2)

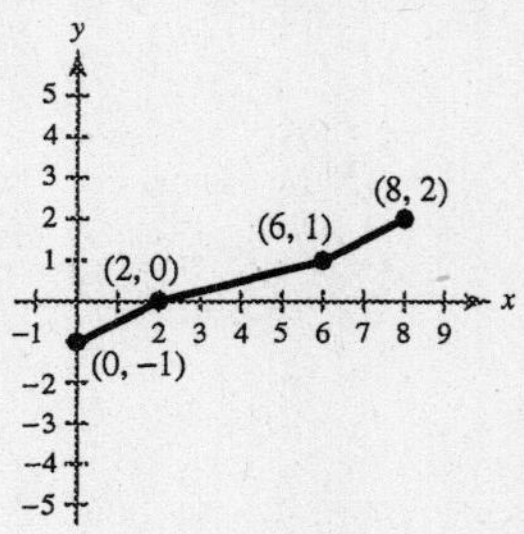

7. (a) $y = f(x) - 1$

Vertical shift 1 unit downward

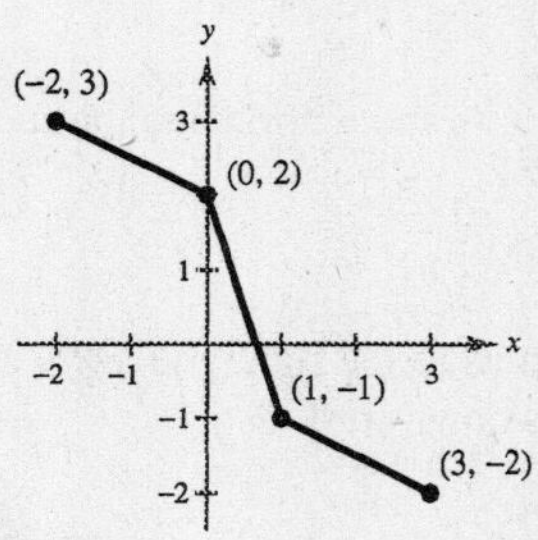

(b) $y = f(x - 1)$

Horizontal shift 1 unit to the right

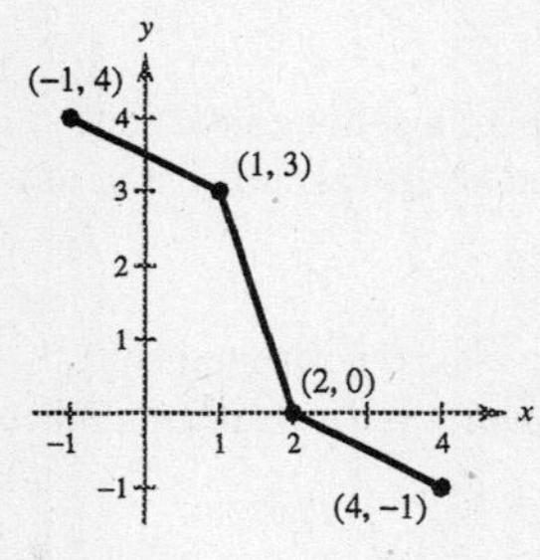

(c) $y = f(-x)$

Reflection about the y-axis

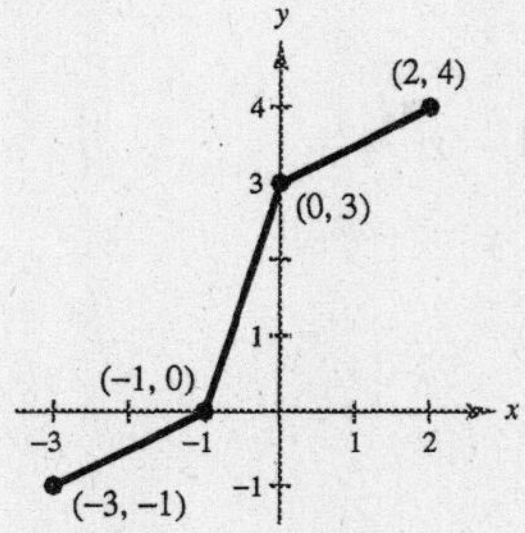

(d) $y = f(x + 1)$

Horizontal shift 1 unit to the left

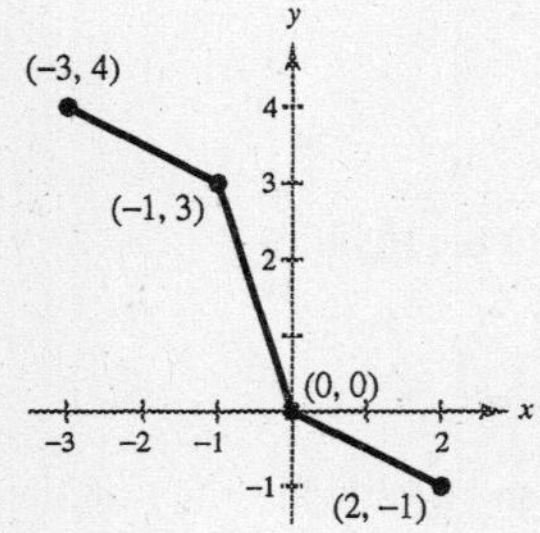

—CONTINUED—

7. —CONTINUED—

(e) $y = -f(x - 2)$

Reflection about the x-axis and a horizontal shift 2 units to the right

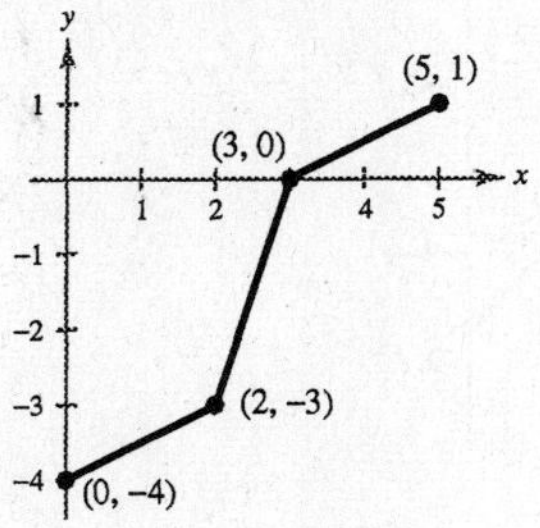

(f) $y = \frac{1}{2}f(x)$

Vertical shrink (each y-value is multiplied by $\frac{1}{2}$)

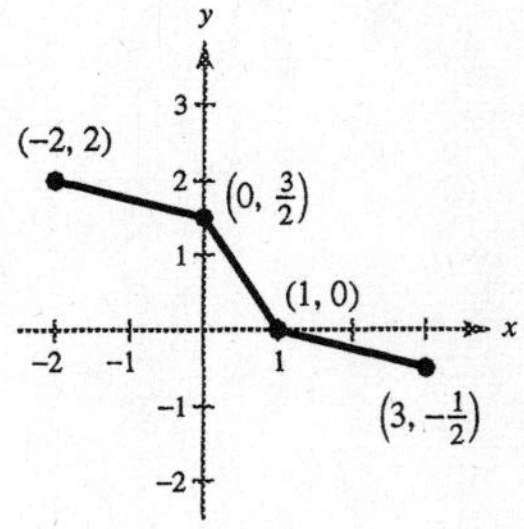

(g) $y = f(2x)$

Horizontal shrink (each x-value is multiplied by $\frac{1}{2}$)

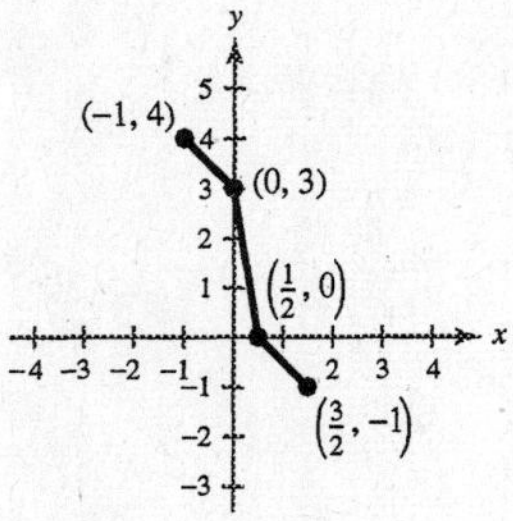

9. (a) Vertical shift 1 unit downward

$f(x) = x^2 - 1$

(b) Reflection about the x-axis, horizontal shift 1 unit to the left, and a vertical shift 1 unit upward

$f(x) = -(x + 1)^2 + 1$

(c) Reflection about the x-axis, horizontal shift 2 units to the right, and a vertical shift 6 units upward

$f(x) = -(x - 2)^2 + 6$

(d) Horizontal shift 5 units to the right and a vertical shift 3 units downward

$f(x) = (x - 5)^2 - 3$

11. (a) Vertical shift 5 units upward

$f(x) = |x| + 5$

(b) Reflection in the x-axis and a horizontal shift 3 units to the left.

$f(x) = -|x + 3|$

(c) Horizontal shift 2 units to the right and a vertical shift 4 units downward

$f(x) = |x - 2| - 4$

(d) Reflection in the x-axis, horizontal shift 6 units to the right, and a vertical shift 1 unit downward

$f(x) = -|x - 6| - 1$

13. Common function: $f(x) = x^3$

Horizontal shift 2 units to the right: $y = (x - 2)^3$

15. Common function: $f(x) = x^2$

Reflection in the x-axis: $y = -x^2$

17. Common function: $f(x) = \sqrt{x}$

Reflection in the x-axis and a vertical shift 1 unit upward: $y = -\sqrt{x} + 1$

19. $f(x) = 12 - x^2$

Common function: $g(x) = x^2$

Reflection in the x-axis and a vertical shift 12 units upward

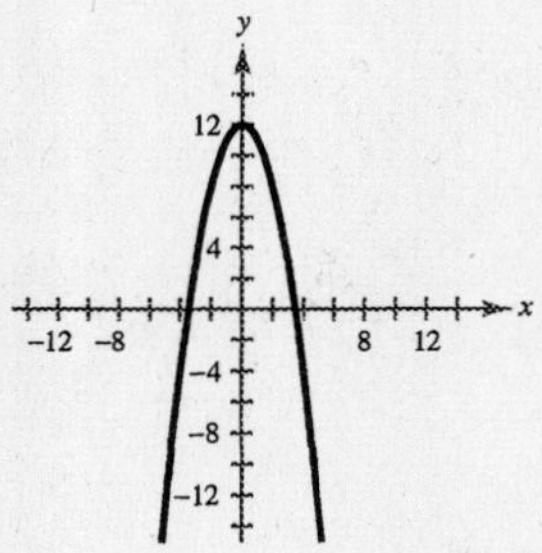

21. $f(x) = x^3 + 7$

Common function: $g(x) = x^3$

Vertical shift 7 units upward

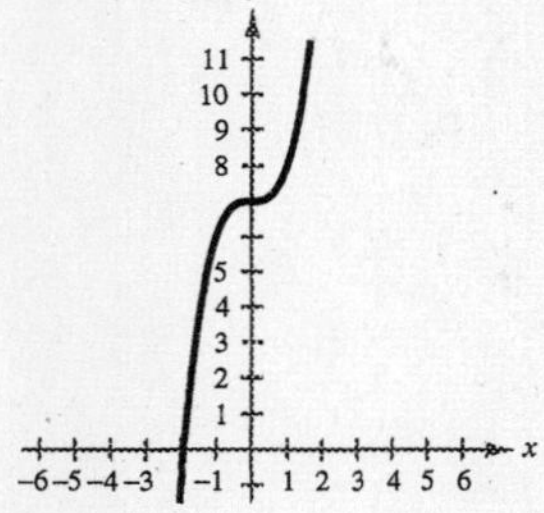

23. $f(x) = 2 - (x + 5)^2$

Common function: $g(x) = x^2$

Reflection in the x-axis, horizontal shift 5 units to the left, and a vertical shift 2 units upward

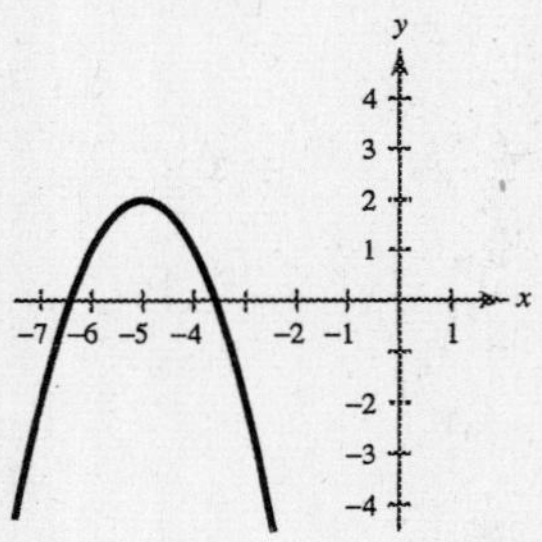

25. $f(x) = (x - 1)^3 + 2$

Common function: $g(x) = x^3$

Horizontal shift 1 unit to the right and a vertical shift 2 units upward

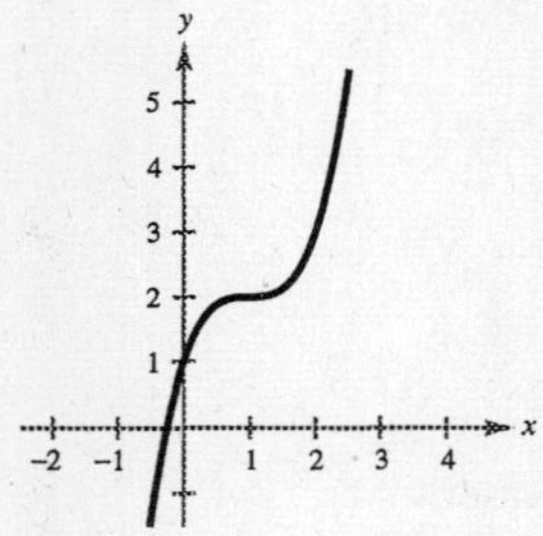

27. $f(x) = -|x| - 2$

Common function: $g(x) = |x|$

Reflection in the x-axis and a vertical shift 2 units downward

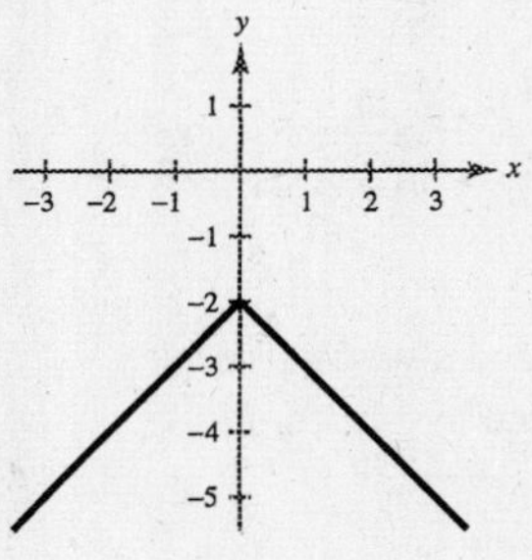

29. $f(x) = -|x + 4| + 8$

Common function: $g(x) = |x|$

Reflection in the x-axis, horizontal shift 4 units to the left, and a vertical shift 8 units upward

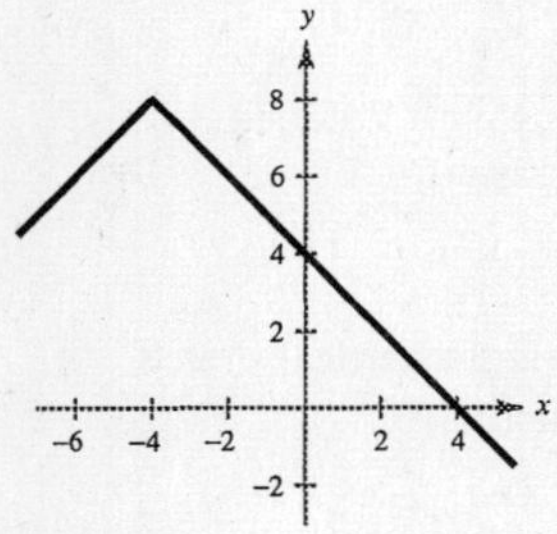

31. $f(x) = 3 - [\![x]\!]$

Common function: $g(x) = [\![x]\!]$

Reflection in the x-axis and a vertical shift 3 units up

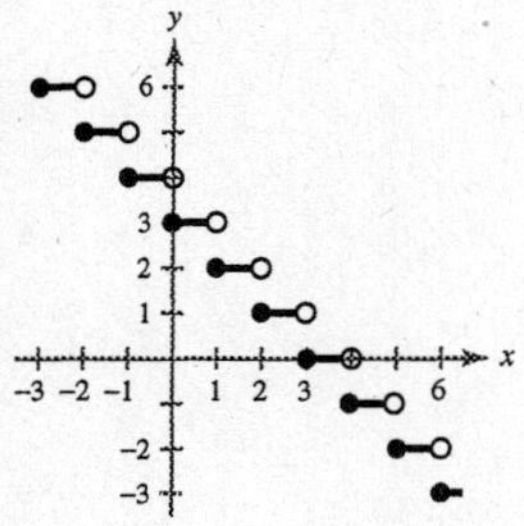

33. $f(x) = \sqrt{x - 9}$

Common function: $g(x) = \sqrt{x}$

Horizontal shift 9 units to the right

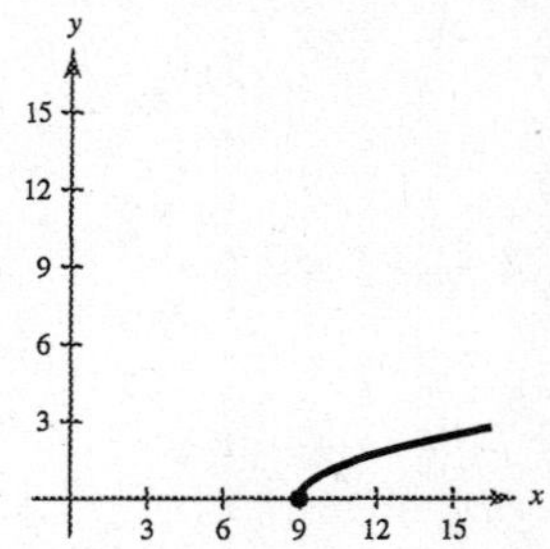

35. $f(x) = \sqrt{7 - x} - 2$ or $f(x) = \sqrt{-(x - 7)} - 2$

Reflection in the y-axis, horizontal shift 7 units to the right, and a vertical shift 2 units downward

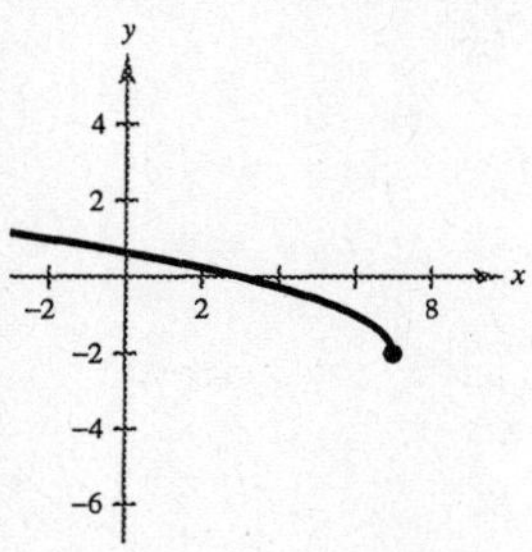

37. $f(x) = \sqrt{\frac{1}{2}x} - 4$

Common function: $g(x) = \sqrt{x}$

Horizontal stretch (each x-value is multiplied by 2) and a vertical shift 4 units down

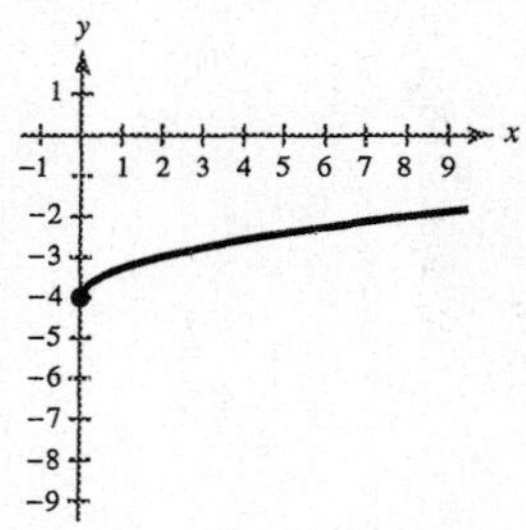

39. $f(x) = x^2$ moved 2 units to the right and 8 units down.

$g(x) = (x - 2)^2 - 8$

41. $f(x) = x^3$ moved 13 units to the right.

$g(x) = (x - 13)^3$

43. $f(x) = |x|$ moved 10 units up and reflected about the x-axis.

$g(x) = -(|x| + 10) = -|x| - 10$

45. $f(x) = \sqrt{x}$ moved 6 units to the left and reflected in both the x and y axes.

$g(x) = -\sqrt{-x + 6}$

47. $f(x) = x^2$

(a) Reflection in the x-axis and a vertical stretch (each y-value is multiplied by 3)

$g(x) = -3x^2$

(b) Vertical shift 3 units upward and a vertical stretch (each y-value is multiplied by 4)

$g(x) = 4x^2 + 3$

49. $f(x) = |x|$

(a) Reflection in the x-axis and a vertical shrink (each y-value is multiplied by $\frac{1}{2}$)

$g(x) = -\frac{1}{2}|x|$

(b) Vertical stretch (each y-value is multiplied by 3) and a vertical shift 3 units downward

$g(x) = 3|x| - 3$

51. Common function: $f(x) = x^3$

Vertical stretch (each y-value is multiplied by 2)
$g(x) = 2x^3$

53. Common function: $f(x) = x^2$

Reflection in the x-axis and a vertical shrink (each y-value is multiplied by $\frac{1}{2}$)

$g(x) = -\frac{1}{2}x^2$

55. Common function: $f(x) = \sqrt{x}$

Reflection in the y-axis and a vertical shrink $\left(\text{each } y\text{-value is multiplied by } \frac{1}{2}\right)$

$g(x) = \frac{1}{2}\sqrt{-x}$

57. Common function: $f(x) = x^3$

Reflection in the x-axis, horizontal shift 2 units to the right and a vertical shift 2 units upward: $g(x) = -(x - 2)^3 + 2$

59. Common function: $f(x) = \sqrt{x}$

Reflection in the x-axis and a vertical shift 3 units downward: $g(x) = -\sqrt{x} - 3$

61. (a) $g(x) = f(x) + 2$

Vertical shift 2 units upward

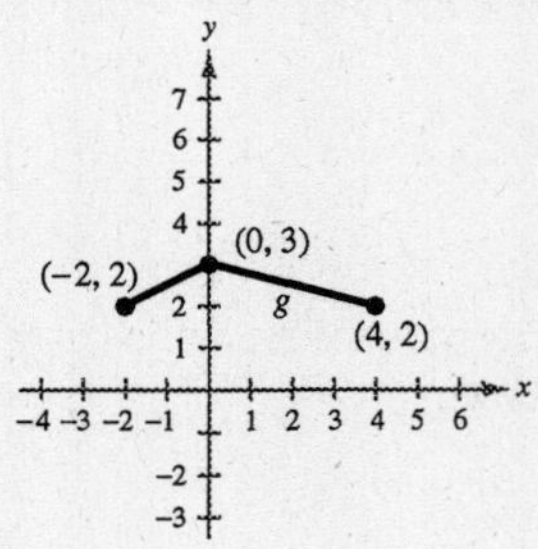

(b) $g(x) = f(x) - 1$

Vertical shift 1 unit downward

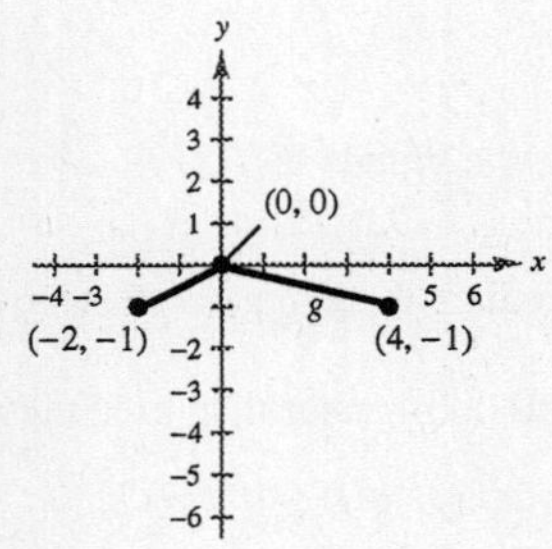

(c) $g(x) = f(-x)$

Reflection in the y-axis

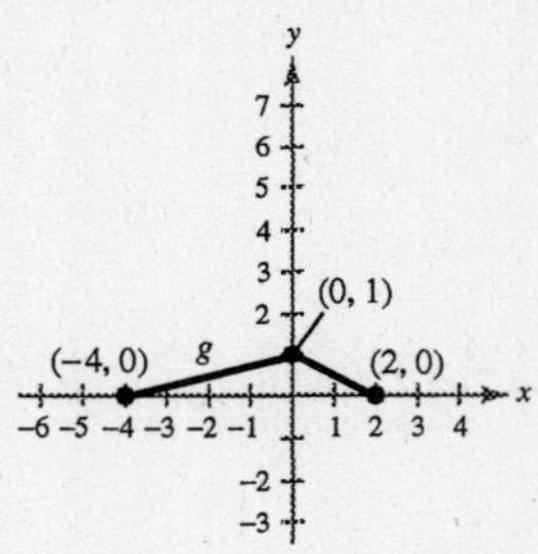

(d) $g(x) = -2f(x)$

Reflection in the x-axis and a vertical stretch (each y-value is multiplied by 2)

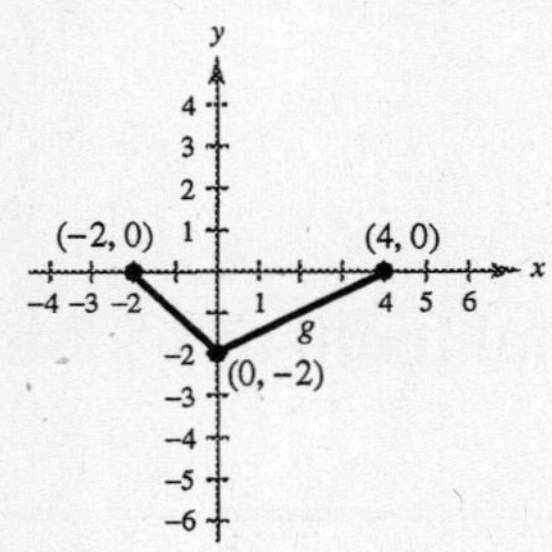

(e) $g(x) = f(4x)$

Horizontal shrink $\left(\text{each } x\text{-value is multiplied by } \frac{1}{4}\right)$

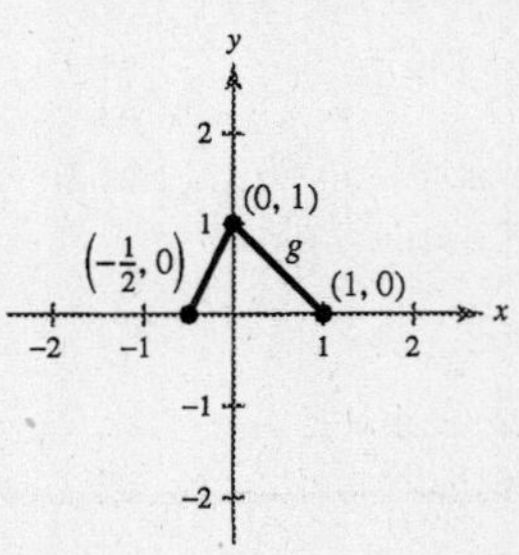

(f) $g(x) = f\left(\frac{1}{2}x\right)$

Horizontal stretch (each x-value is multiplied by 2)

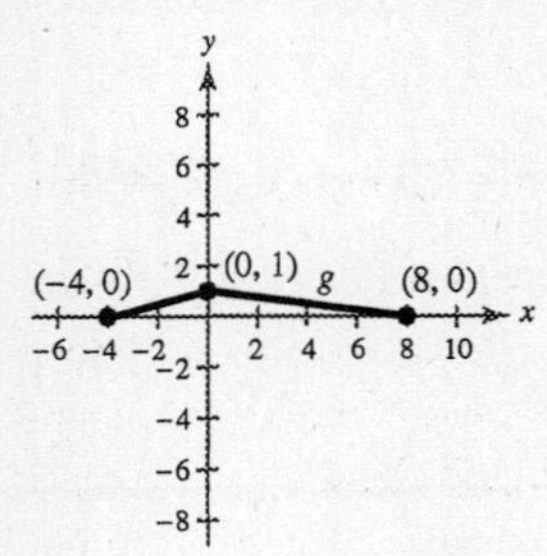

63. $F = f(t) = 20.5 + 0.035t^2$

(a) Common function: $g(x) = x^2$

Vertical shrink by a factor of 0.035 and a vertical shift 20.5 units up

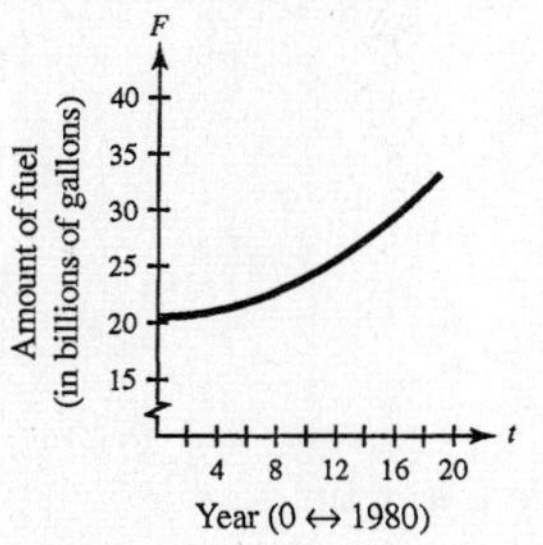

(b) $\dfrac{f(19) - f(0)}{19 - 0} = \dfrac{33.135 - 20.5}{19} = 0.665$

The average increase in fuel used by trucks was 0.665 billion gallons per year between 1980 and 1999.

(c) This represents a horizontal shift 10 units to the left.

$g(t) = f(t + 10) = 20.5 + 0.035(t + 10)^2$

(d) $g(15) = 42.375$ billion gallons

There are many factors involved here. The number of trucks on the road continues to increase but are more fuel efficient. The availability and the cost of overseas and domestic fuel also plays a role in usage.

65. True, since $|x| = |-x|$, the graphs of $f(x) = |x| + 6$ and $f(x) = |-x| + 6$ are identical.

67. (a) The profits were only $\frac{3}{4}$ as large as expected: $g(t) = \frac{3}{4}f(t)$

(b) The profits were \$10,000 greater than predicted: $g(t) = f(t) + 10{,}000$

(c) There was a 2-year delay: $g(t) = f(t - 2)$

69. $y = f(x + 2) - 1$

Horizontal shift 2 units to the left and a vertical shift 1 unit downward.

$(0, 1) \rightarrow (0 - 2, 1 - 1) = (-2, 0)$

$(1, 2) \rightarrow (1 - 2, 2 - 1) = (-1, 1)$

$(2, 3) \rightarrow (2 - 2, 3 - 1) = (0, 2)$

Section P.9 Combinations of Functions

- Given two functions, f and g, you should be able to form the following functions (if defined):
  1. Sum: $(f + g)(x) = f(x) + g(x)$
  2. Difference: $(f - g)(x) = f(x) - g(x)$
  3. Product: $(fg)(x) = f(x)g(x)$
  4. Quotient: $(f/g)(x) = f(x)/g(x)$, $g(x) \neq 0$
  5. Composition of f with g: $(f \circ g)(x) = f(g(x))$
  6. Composition of g with f: $(g \circ f)(x) = g(f(x))$

Solutions to Odd-Numbered Exercises

1.

x	0	1	2	3
f	2	3	1	2
g	-1	0	$\frac{1}{2}$	0
$f+g$	1	3	$\frac{3}{2}$	2

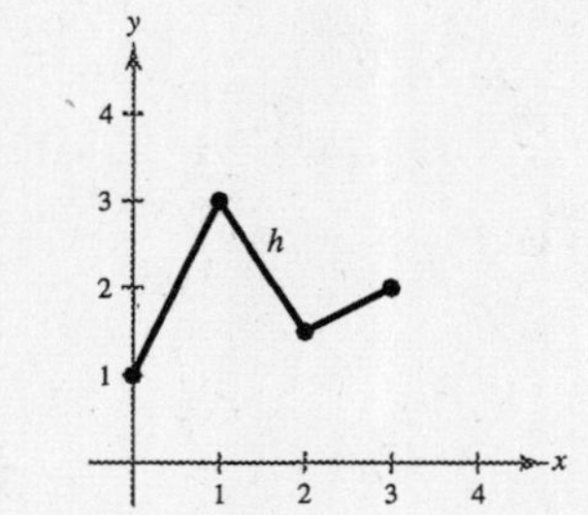

3.

x	-2	0	1	2	4
f	2	0	1	2	4
g	4	2	1	0	2
$f+g$	6	2	2	2	6

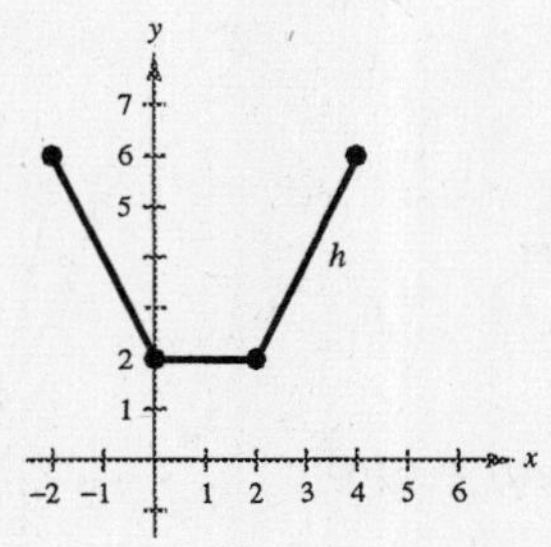

5. $f(x) = x + 2,\ g(x) = x - 2$

(a) $(f + g)(x) = f(x) + g(x) = (x + 2) + (x - 2) = 2x$

(b) $(f - g)(x) = f(x) - g(x) = (x + 2) - (x - 2) = 4$

(c) $(fg)(x) = f(x) \cdot g(x) = (x + 2)(x - 2) = x^2 - 4$

(d) $\left(\frac{f}{g}\right)(x) = \frac{f(x)}{g(x)} = \frac{x + 2}{x - 2}$

Domain: all real numbers except $x = 2$

7. $f(x) = x^2,\ g(x) = 4x - 5$

(a) $(f + g)(x) = f(x) + g(x) = x^2 + (4x - 5) = x^2 + 4x - 5$

(b) $(f - g)(x) = f(x) - g(x) = x^2 - (4x - 5) = x^2 - 4x + 5$

(c) $(fg)(x) = f(x) \cdot g(x) = x^2(4x - 5) = 4x^3 - 5x^2$

(d) $\left(\frac{f}{g}\right)(x) = \frac{f(x)}{g(x)} = \frac{x^2}{4x - 5}$, Domain: all real numbers except $x = \frac{5}{4}$

9. $f(x) = x^2 + 6,\ g(x) = \sqrt{1 - x}$

(a) $(f + g)(x) = f(x) + g(x) = (x^2 + 6) + \sqrt{1 - x}$

(b) $(f - g)(x) = f(x) - g(x) = (x^2 + 6) - \sqrt{1 - x}$

(c) $(fg)(x) = f(x) \cdot g(x) = (x^2 + 6)\sqrt{1 - x}$

(d) $\left(\frac{f}{g}\right)(x) = \frac{f(x)}{g(x)} = \frac{x^2 + 6}{\sqrt{1 - x}} = \frac{(x^2 + 6)\sqrt{1 - x}}{1 - x}$,

Domain: $x < 1$

11. $f(x) = \frac{1}{x},\ g(x) = \frac{1}{x^2}$

(a) $(f + g)(x) = f(x) + g(x) = \frac{1}{x} + \frac{1}{x^2} = \frac{x + 1}{x^2}$

(b) $(f - g)(x) = f(x) - g(x) = \frac{1}{x} - \frac{1}{x^2} = \frac{x - 1}{x^2}$

(c) $(fg)(x) = f(x) \cdot g(x) = \frac{1}{x}\left(\frac{1}{x^2}\right) = \frac{1}{x^3}$

(d) $\left(\frac{f}{g}\right)(x) = \frac{f(x)}{g(x)} = \frac{1/x}{1/x^2} = \frac{x^2}{x} = x,\ x \neq 0$

For Exercises 13–23, $f(x) = x^2 + 1$ and $g(x) = x - 4$.

13. $(f + g)(2) = f(2) + g(2) = (2^2 + 1) + (2 - 4) = 3$

15. $(f - g)(0) = f(0) - g(0) = (0^2 + 1) - (0 - 4) = 5$

17. $(f - g)(3t) = f(3t) - g(3t) = [(3t)^2 + 1] - (3t - 4)$
$= 9t^2 - 3t + 5$

19. $(fg)(6) = f(6)g(6) = (6^2 + 1)(6 - 4) = 74$

21. $\left(\frac{f}{g}\right)(5) = \frac{f(5)}{g(5)} = \frac{5^2 + 1}{5 - 4} = 26$

23. $\left(\frac{f}{g}\right)(-1) - g(3) = \frac{f(-1)}{g(-1)} - g(3)$
$= \frac{(-1)^2 + 1}{-1 - 4} - (3 - 4)$
$= -\frac{2}{5} + 1 = \frac{3}{5}$

25. $f(x) = \frac{1}{2}x,\ g(x) = x - 1,\ (f + g)(x) = \frac{3}{2}x - 1$

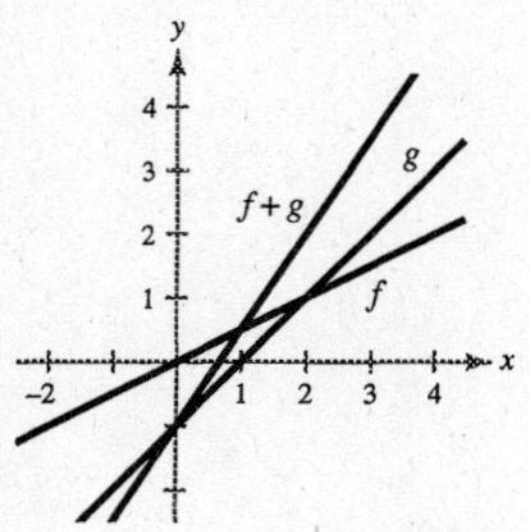

27. $f(x) = x^2,\ g(x) = -2x,\ (f + g)(x) = x^2 - 2x$

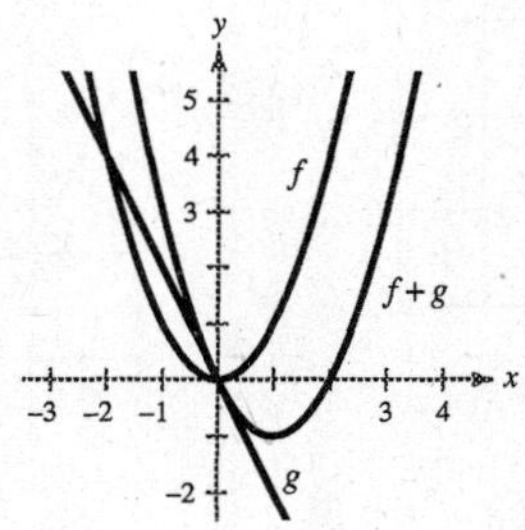

29. $f(x) = 3x,\ g(x) = -\frac{x^3}{10},\ (f + g)(x) = 3x - \frac{x^3}{10}$

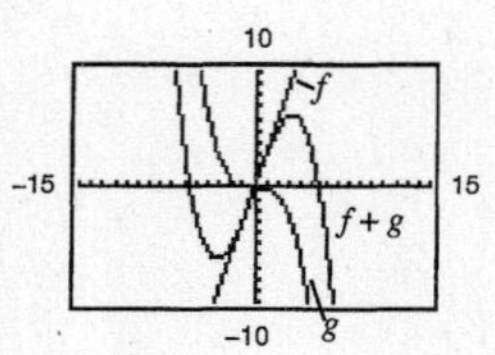

For $0 \le x \le 2$, $f(x)$ contributes most to the magnitude.

For $x > 6$, $g(x)$ contributes most to the magnitude.

31. $T(x) = R(x) + B(x) = \frac{3}{4}x + \frac{1}{15}x^2$

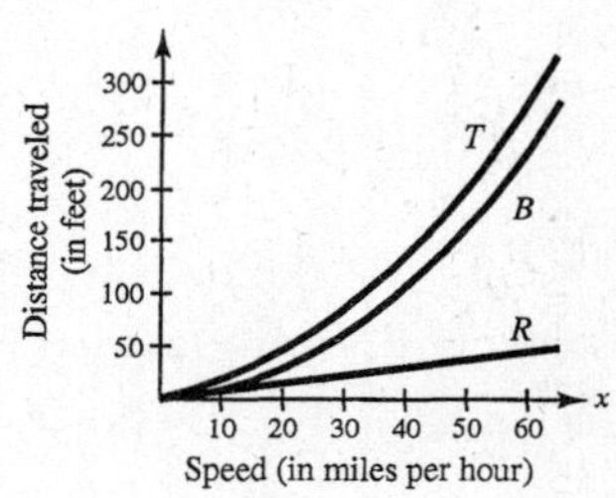

33.

Year	y_1	y_2	y_3
1993	148.9	295.7	39.1
1994	146.2	308.9	40.8
1995	149.2	322.3	44.8
1996	155.0	337.4	47.9
1997	165.5	355.6	52.0
1998	176.1	376.8	54.8
1999	186.5	401.2	58.9

(a) $y_1 \approx 1.344t^2 - 9.38t + 163.6$

$y_2 \approx 17.34t + 238.5$

$y_3 \approx 3.38t + 28.1$

(b) $y_1 + y_2 + y_3 \approx 1.344t^2 + 11.34t + 430.2$

This sum represents the total spent on health services and supplies for the years 1993 through 1999. It includes out-of-pocket payments, insurance premiums, and other types of payments.

(c)

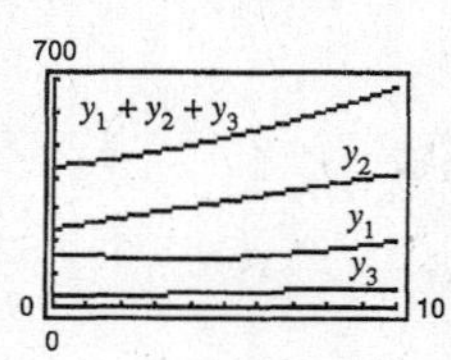

(d) For 2003 use $t = 13$: \$804.76 billion

For 2005 use $t = 15$: \$902.70 billion

35. $f(x) = x^2, g(x) = x - 1$

(a) $(f \circ g)(x) = f(g(x)) = f(x - 1) = (x - 1)^2$

(b) $(g \circ f)(x) = g(f(x)) = g(x^2) = x^2 - 1$

(c) $(f \circ f)(x) = f(f(x)) = f(x^2) = (x^2)^2 = x^4$

37. $f(x) = \sqrt[3]{x - 1},\ g(x) = x^3 + 1$

(a) $(f \circ g)(x) = f(g(x))$

$= f(x^3 + 1)$

$= \sqrt[3]{(x^3 + 1) - 1}$

$= \sqrt[3]{x^3} = x$

(b) $(g \circ f)(x) = g(f(x))$

$= g\left(\sqrt[3]{x - 1}\right)$

$= \left(\sqrt[3]{x - 1}\right)^3 + 1$

$= (x - 1) + 1 = x$

(c) $(f \circ f)(x) = f(f(x))$

$= f\left(\sqrt[3]{x - 1}\right)$

$= \sqrt[3]{\sqrt[3]{x - 1} - 1}$

39. $f(x) = \sqrt{x + 4}$ Domain: $x \geq -4$

$g(x) = x^2$ Domain: all real numbers

(a) $(f \circ g)(x) = f(g(x)) = f(x^2) = \sqrt{x^2 + 4}$

Domain: all real numbers

(b) $(g \circ f)(x) = g(f(x)) = g\left(\sqrt{x + 4}\right) = \left(\sqrt{x + 4}\right)^2 = x + 4$

Domain: $x \geq -4$

41. $f(x) = x^2 + 1$ Domain: all real numbers

$g(x) = \sqrt{x}$ Domain: $x \geq 0$

(a) $(f \circ g)(x) = f(g(x)) = f\left(\sqrt{x}\right) = \left(\sqrt{x}\right)^2 + 1 = x + 1$

Domain: $x \geq 0$

(b) $(g \circ f)(x) = g(f(x)) = g(x^2 + 1) = \sqrt{x^2 + 1}$

Domain: all real numbers

43. $f(x) = |x|$ Domain: all real numbers

$g(x) = x + 6$ Domain: all real numbers

(a) $(f \circ g)(x) = f(g(x)) = f(x + 6) = |x + 6|$ Domain: all real numbers

(b) $(g \circ f)(x) = g(f(x)) = g(|x|) = |x| + 6$ Domain: all real numbers

45. $f(x) = \dfrac{1}{x}$ Domain: all real numbers except $x = 0$

$g(x) = x + 3$ Domain: all real numbers

(a) $(f \circ g)(x) = f(g(x)) = f(x + 3) = \dfrac{1}{x + 3}$

Domain: all real numbers except $x = -3$

(b) $(g \circ f)(x) = g(f(x)) = g\left(\dfrac{1}{x}\right) = \dfrac{1}{x} + 3$

Domain: all real numbers except $x = 0$

47. (a) $(f + g)(3) = f(3) + g(3) = 2 + 1 = 3$

(b) $\left(\dfrac{f}{g}\right)(2) = \dfrac{f(2)}{g(2)} = \dfrac{0}{2} = 0$

49. (a) $(f \circ g)(2) = f(g(2)) = f(2) = 0$

(b) $(g \circ f)(2) = g(f(2)) = g(0) = 4$

51. Let $f(x) = x^2$ and $g(x) = 2x + 1$, then $(f \circ g)(x) = h(x)$.

This is not a unique solution.

53. Let $f(x) = \sqrt[3]{x}$ and $g(x) = x^2 - 4$, then $(f \circ g)(x) = h(x)$.

This answer is not unique.

55. Let $f(x) = 1/x$ and $g(x) = x + 2$, then $(f \circ g)(x) = h(x)$.

This is not a unique solution.

57. Let $f(x) = \dfrac{x+3}{4+x}$ and $g(x) = -x^2$, then $(f \circ g)(x) = h(x)$.

This answer is not unique.

59. (a) $r(x) = \dfrac{x}{2}$

(b) $A(r) = \pi r^2$

(c) $(A \circ r)(x) = A(r(x)) = A\left(\dfrac{x}{2}\right) = \pi\left(\dfrac{x}{2}\right)^2$

$(A \circ r)(x)$ represents the area of the circular base of the tank on the square foundation with side length x.

61. False. $(f \circ g)(x) = 6x + 1$ and $(g \circ f)(x) = 6x + 6$.

63. (a) $f(g(x)) = f(0.03x) = 0.03x - 500{,}000$

(b) $g(f(x)) = g(x - 500{,}000) = 0.03(x - 500{,}000)$

$g(f(x))$ represents 3% of an amount over \$500,000.

65. Let $f(x)$ be an odd function, $g(x)$ be an even function and define $h(x) = f(x)g(x)$. Then

$$\begin{aligned} h(-x) &= f(-x)g(-x) \\ &= [-f(x)]g(x) \qquad \text{Since } f \text{ is odd and g is even.} \\ &= -f(x)g(x) \\ &= -h(x) \end{aligned}$$

Thus, h is odd.

Section P.10 Inverse Functions

- Two functions f and g are inverses of each other if $f(g(x)) = x$ for every x in the domain of g and $g(f(x)) = x$ for every x in the domain of f.
- A function f has an inverse function if and only if no **horizontal** line crosses the graph of f at more than one point.
- The graph of f^{-1} is a reflection of the graph of f about the line $y = x$.
- Be able to find the inverse of a function, if it exists.
  1. Use the Horizontal Line Test to see if f^{-1} exists.
  2. Replace $f(x)$ with y.
  3. Interchange x and y and solve for y.
  4. Replace y with $f^{-1}(x)$.

Solutions to Odd-Numbered Exercises

1. The inverse is a line through $(-1, 0)$.

Matches graph (c).

3. The inverse is half a parabola starting at $(1, 0)$.

Matches graph (a).

5. $f^{-1}(x) = \dfrac{x}{6} = \dfrac{1}{6}x$

$f(f^{-1}(x)) = f\left(\dfrac{x}{6}\right) = 6\left(\dfrac{x}{6}\right) = x$

$f^{-1}(f(x)) = f^{-1}(6x) = \dfrac{6x}{6} = x$

7. $f^{-1}(x) = x - 9$

$f(f^{-1}(x)) = f(x - 9) = (x - 9) + 9 = x$

$f^{-1}(f(x)) = f^{-1}(x + 9) = (x + 9) - 9 = x$

9. $f^{-1}(x) = \dfrac{x - 1}{3}$

$f(f^{-1}(x)) = f\left(\dfrac{x - 1}{3}\right) = 3\left(\dfrac{x - 1}{3}\right) + 1 = x$

$f^{-1}(f(x)) = f^{-1}(3x + 1) = \dfrac{(3x + 1) - 1}{3} = x$

11. $f^{-1}(x) = x^3$

$f(f^{-1}(x)) = f(x^3) = \sqrt[3]{x^3} = x$

$f^{-1}(f(x)) = f^{-1}\left(\sqrt[3]{x}\right) = \left(\sqrt[3]{x}\right)^3 = x$

13. (a) $f(g(x)) = f\left(\dfrac{x}{2}\right) = 2\left(\dfrac{x}{2}\right) = x$

$g(f(x)) = g(2x) = \dfrac{2x}{2} = x$

(b)

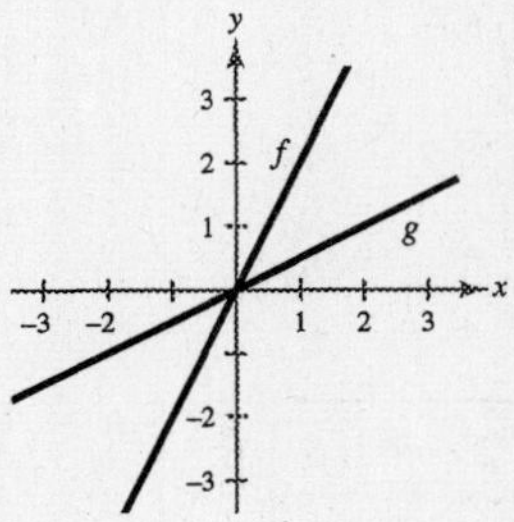

15. (a) $f(g(x)) = f\left(\dfrac{x - 1}{7}\right) = 7\left(\dfrac{x - 1}{7}\right) + 1 = x$

$g(f(x)) = g(7x + 1) = \dfrac{(7x + 1) - 1}{7} = x$

(b)

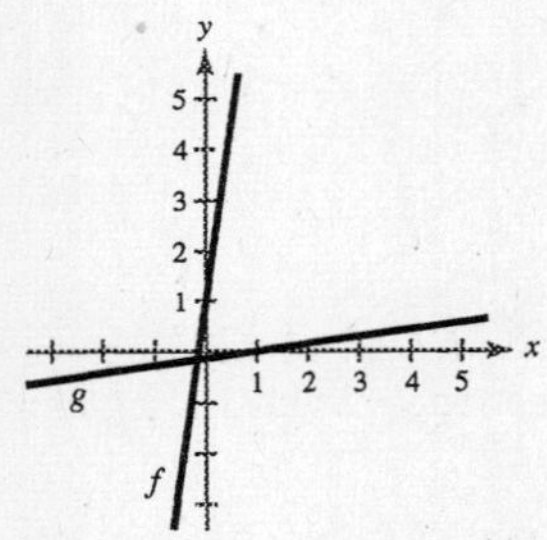

17. (a) $f(g(x)) = f\left(\sqrt[3]{8x}\right) = \dfrac{\left(\sqrt[3]{8x}\right)^3}{8} = \dfrac{8x}{8} = x$

$g(f(x)) = g\left(\dfrac{x^3}{8}\right) = \sqrt[3]{8\left(\dfrac{x^3}{8}\right)} = \sqrt[3]{x^3} = x$

(b)

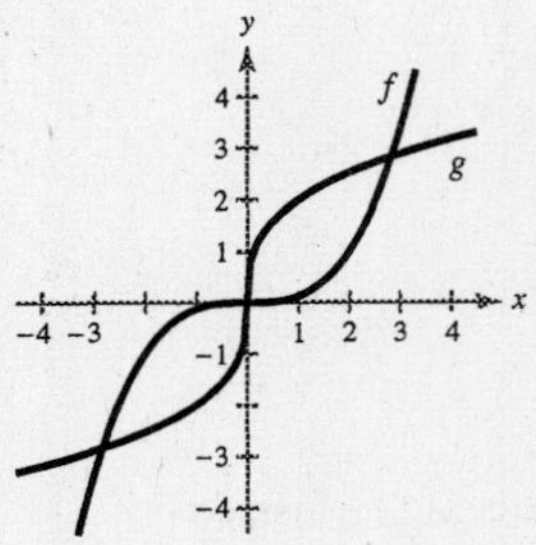

19. (a) $f(g(x)) = f(x^2 + 4),\ x \geq 0$

$= \sqrt{(x^2 + 4) - 4} = x$

$g(f(x)) = g\left(\sqrt{x - 4}\right)$

$= \left(\sqrt{x - 4}\right)^2 + 4 = x$

(b)

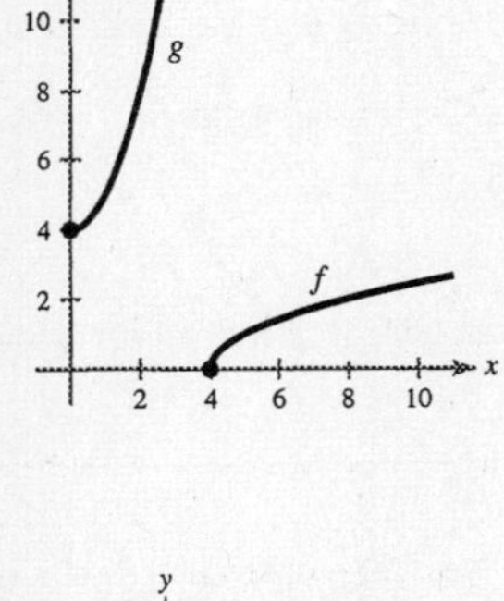

21. (a) $f(g(x)) = f\left(\sqrt{9 - x}\right),\ x \leq 9$

$= 9 - \left(\sqrt{9 - x}\right)^2 = x$

$g(f(x)) = g(9 - x^2),\ x \geq 0$

$= \sqrt{9 - (9 - x^2)} = x$

(b)

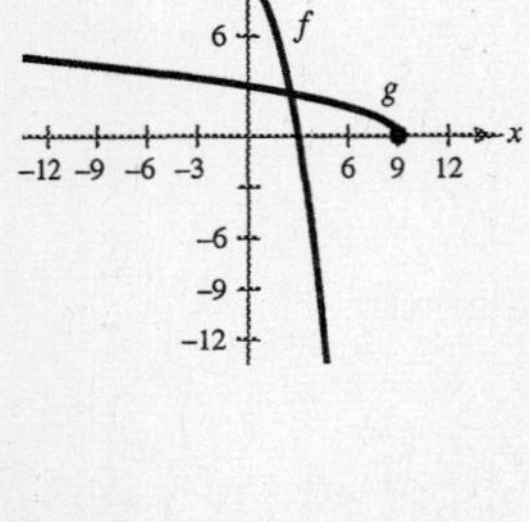

23. (a) $f(g(x)) = f\left(-\dfrac{5x+1}{x-1}\right)$

$$= \frac{\left(-\dfrac{5x+1}{x-1} - 1\right)}{\left(-\dfrac{5x+1}{x-1} + 5\right)} \cdot \frac{x-1}{x-1}$$

$$= \frac{-(5x+1)-(x-1)}{-(5x+1)+5(x-1)}$$

$$= \frac{-6x}{-6}$$

$$= x$$

$$g(f(x)) = g\left(\frac{x-1}{x+5}\right)$$

$$= -\frac{\left[5\left(\dfrac{x-1}{x+5}\right)+1\right]}{\left[\dfrac{x-1}{x+5}-1\right]} \cdot \frac{x+5}{x+5}$$

$$= -\frac{5(x-1)+(x+5)}{(x-1)-(x+5)}$$

$$= -\frac{6x}{-6}$$

$$= x$$

(b)

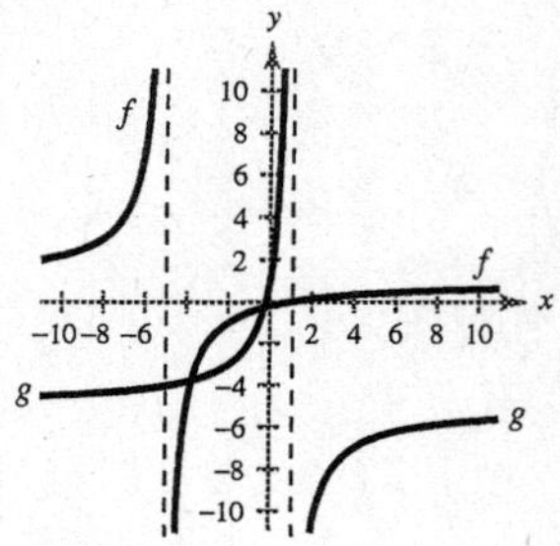

25. No, $\{(-2, -1), (1, 0), (2, 1), (1, 2), (-2, 3), (-6, 4)\}$ does not represent a function. -2 and 1 are paired with two different values.

27.

x	-2	0	2	4	6	8
$f^{-1}(x)$	-2	-1	0	1	2	3

29. Since no horizontal line crosses the graph of f at more than one point, f **has** an inverse.

31. Since some horizontal lines cross the graph of f twice, f does **not** have an inverse.

33. $g(x) = \dfrac{4-x}{6}$

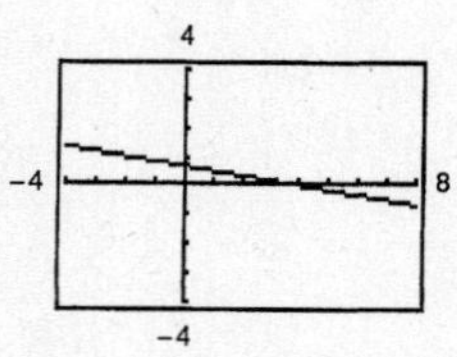

g passes the horizontal line test, so g **has** an inverse.

35. $h(x) = |x+4| - |x-4|$

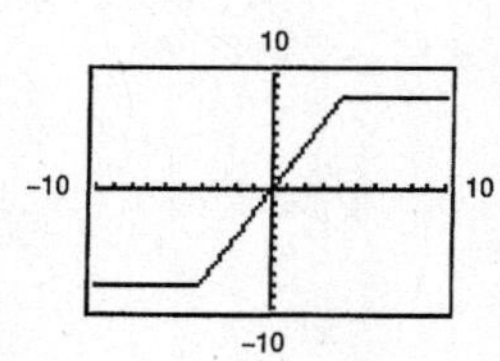

h does not pass the horizontal line test, so h does **not** have an inverse.

37. $f(x) = -2x\sqrt{16-x^2}$

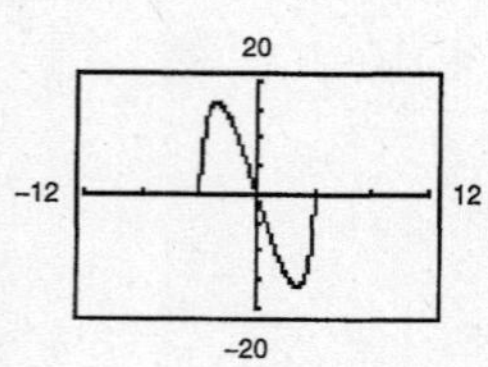

f does not pass the horizontal line test, so f does **not** have an inverse.

39.

$$f(x) = 2x - 3$$
$$y = 2x - 3$$
$$x = 2y - 3$$
$$y = \frac{x+3}{2}$$
$$f^{-1}(x) = \frac{x+3}{2}$$

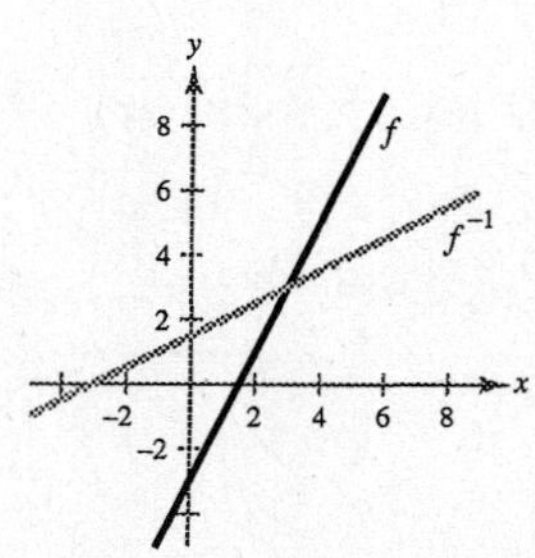

41. $f(x) = x^5 - 2$

$y = x^5 - 2$

$x = y^5 - 2$

$y = \sqrt[5]{x + 2}$

$f^{-1}(x) = \sqrt[5]{x + 2}$

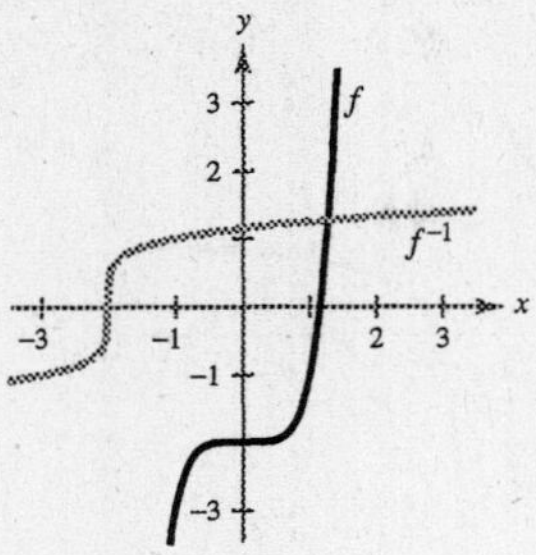

43. $f(x) = \sqrt{x}$

$y = \sqrt{x}$

$x = \sqrt{y}$

$y = x^2$

$f^{-1}(x) = x^2, \ x \geq 0$

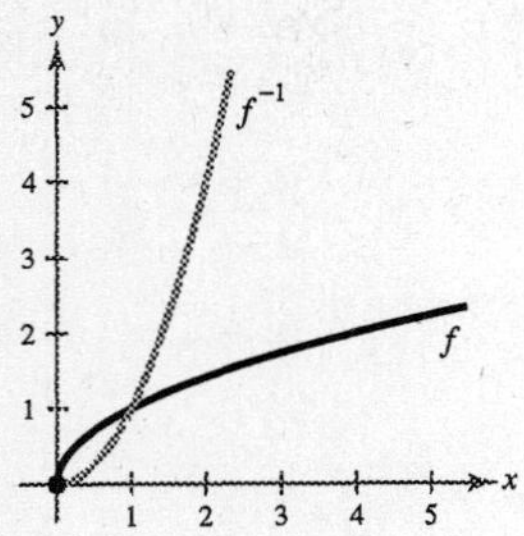

45. $f(x) = \sqrt{4 - x^2}, \ 0 \leq x \leq 2$

$y = \sqrt{4 - x^2}$

$x = \sqrt{4 - y^2}$

$f^{-1}(x) = \sqrt{4 - x^2}, \ 0 \leq x \leq 2$

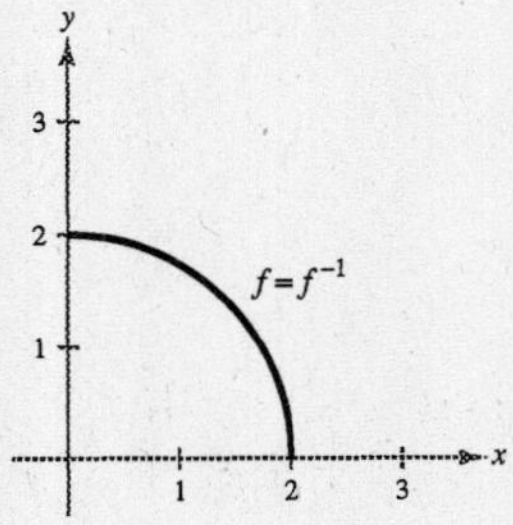

47. $f(x) = \dfrac{4}{x}$

$y = \dfrac{4}{x}$

$x = \dfrac{4}{y}$

$xy = 4$

$y = \dfrac{4}{x}$

$f^{-1}(x) = \dfrac{4}{x}$

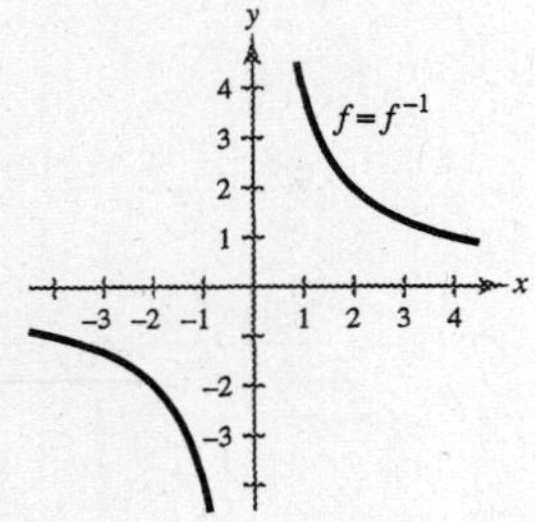

49. $f(x) = \dfrac{x + 1}{x - 2}$

$y = \dfrac{x + 1}{x - 2}$

$x = \dfrac{y + 1}{y - 2}$

$x(y - 2) = y + 1$

$xy - 2x = y + 1$

$xy - y = 2x + 1$

$y(x - 1) = 2x + 1$

$y = \dfrac{2x + 1}{x - 1}$

$f^{-1}(x) = \dfrac{2x + 1}{x - 1}$

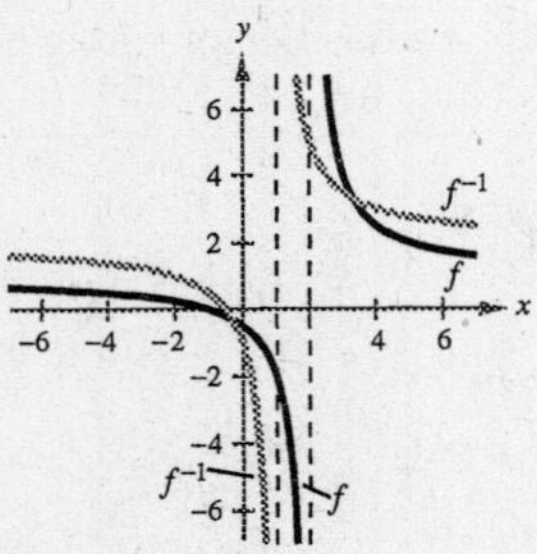

51. $f(x) = \sqrt[3]{x - 1}$

$y = \sqrt[3]{x - 1}$

$x = \sqrt[3]{y - 1}$

$x^3 = y - 1$

$y = x^3 + 1$

$f^{-1}(x) = x^3 + 1$

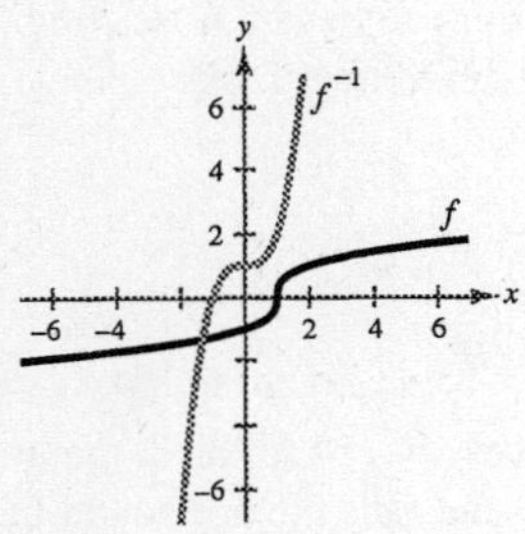

53. $f(x) = \dfrac{6x + 4}{4x + 5}$

$$y = \frac{6x + 4}{4x + 5}$$

$$x = \frac{6y + 4}{4y + 5}$$

$$x(4y + 5) = 6y + 4$$

$$4xy + 5x = 6y + 4$$

$$4xy - 6y = -5x + 4$$

$$y(4x - 6) = -5x + 4$$

$$y = \frac{-5x + 4}{4x - 6}$$

$$f^{-1}(x) = \frac{-5x + 4}{4x - 6} = \frac{5x - 4}{6 - 4x}$$

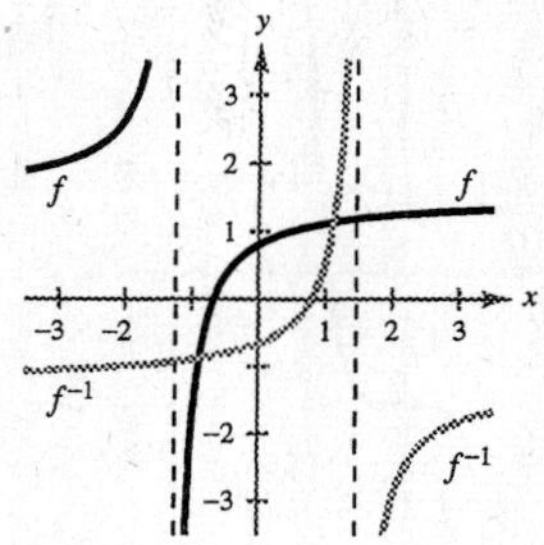

55. $f(x) = x^4$

$$y = x^4$$

$$x = y^4$$

$$y = \pm\sqrt[4]{x}$$

This does not represent y as a function of x.
f does not have an inverse.

57. $g(x) = \dfrac{x}{8}$

$$y = \frac{x}{8}$$

$$x = \frac{y}{8}$$

$$y = 8x$$

This is a function of x, so g has an inverse.
$g^{-1}(x) = 8x$

59. $p(x) = -4$

$$y = -4$$

Since $y = -4$ for all x, the graph is a horizontal line and fails the horizontal line test. p does not have an inverse.

61. $f(x) = (x + 3)^2,\ x \ge -3 \Rightarrow y \ge 0$

$$y = (x + 3)^2,\ x \ge -3,\ y \ge 0$$

$$x = (y + 3)^2,\ y \ge -3,\ x \ge 0$$

$$\sqrt{x} = y + 3,\ y \ge -3,\ x \ge 0$$

$$y = \sqrt{x} - 3,\ x \ge 0,\ y \ge -3$$

This is a function of x, so f has an inverse.

$f^{-1}(x) = \sqrt{x} - 3,\ x \ge 0$

63. $f(x) = \begin{cases} x + 3, & x < 0 \\ 6 - x, & x \ge 0 \end{cases}$

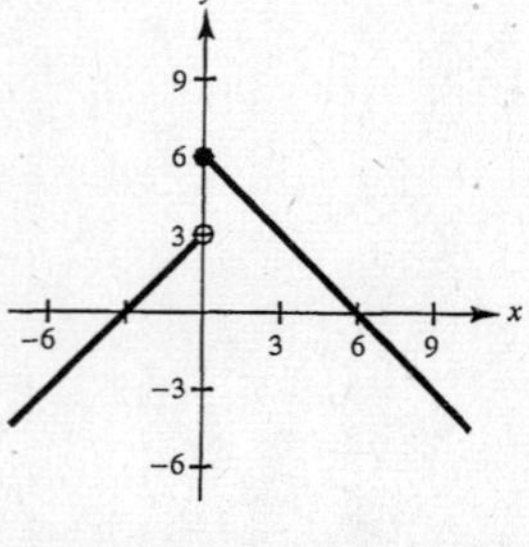

The graph fails the horizontal line test, so $f(x)$ does not have an inverse.

65. $h(x) = -\dfrac{4}{x^2}$

The graph fails the horizontal line test so h does not have an inverse.

67. $f(x) = \sqrt{2x+3} \Rightarrow x \geq -\frac{3}{2}, \ y \geq 0$

$y = \sqrt{2x+3}, \ x \geq -\frac{3}{2}, \ y \geq 0$

$x = \sqrt{2y+3}, \ y \geq -\frac{3}{2}, \ x \geq 0$

$x^2 = 2y+3, \ x \geq 0, \ y \geq -\frac{3}{2}$

$y = \frac{x^2-3}{2}, \ x \geq 0, \ y \geq -\frac{3}{2}$

This is a function of x, so f has an inverse.

$f^{-1}(x) = \frac{x^2-3}{2}, \ x \geq 0$

In Exercises 69, 71, and 73, $f(x) = \frac{1}{8}x - 3, \ f^{-1}(x) = 8(x+3), \ g(x) = x^3, \ g^{-1}(x) = \sqrt[3]{x}$.

69. $(f^{-1} \circ g^{-1})(1) = f^{-1}(g^{-1}(1)) = f^{-1}\left(\sqrt[3]{1}\right) = 8\left(\sqrt[3]{1} + 3\right) = 32$

71. $(f^{-1} \circ f^{-1})(6) = f^{-1}(f^{-1}(6)) = f^{-1}(8[6+3]) = 8[8(6+3)+3] = 600$

73. $(f \circ g)(x) = f(g(x)) = f(x^3) = \frac{1}{8}x^3 - 3$

$y = \frac{1}{8}x^3 - 3$

$x = \frac{1}{8}y^3 - 3$

$x + 3 = \frac{1}{8}y^3$

$8(x+3) = y^3$

$\sqrt[3]{8(x+3)} = y$

$(f \circ g)^{-1}(x) = 2\sqrt[3]{x+3}$

In Exercises 75 and 77, $f(x) = x + 4, \ f^{-1}(x) = x - 4, \ g(x) = 2x - 5, \ g^{-1}(x) = \frac{x+5}{2}$.

75. $(g^{-1} \circ f^{-1})(x) = g^{-1}(f^{-1}(x)) = g^{-1}(x-4) = \frac{(x-4)+5}{2} = \frac{x+1}{2}$

77. $(f \circ g)(x) = f(g(x)) = f(2x-5) = (2x-5) + 4 = 2x - 1$

$(f \circ g)^{-1}(x) = \frac{x+1}{2}$

Note: Comparing Exercises 75 and 77, we see that $(f \circ g)^{-1}(x) = (g^{-1} \circ f^{-1})(x)$.

79. (a) $f^{-1}(103{,}874) = 9$

(b) f^{-1} represents the time in years with 4 corresponding to 1994. The domain of f^{-1} is the number of households (in thousands) in the United States.

(c) $f(t) \approx 1266.54t + 92{,}255.54,\ 4 \le t \le 9$

(d) $y = 1266.54t + 92{,}255.54$

$$t = 1266.54y + 92{,}255.54$$

$$\frac{t - 92{,}255.54}{1266.54} = y$$

$$f^{-1}(t) = \frac{t - 92{,}255.54}{1266.54}$$

(e) $f^{-1}(111{,}254) = \dfrac{111{,}254 - 92{,}255.54}{1266.54} \approx 15$

The model yields 111,254 households (in thousands) in the United States in the year 2005.

81. (a) Yes. Since the values of f increase each year, no two f-values are paired with the same t-value so f does have an inverse.

(b) f^{-1} would represent the year that a given number of miles was traveled by motor vehicles.

(c) Since $f(8) = 2632, f^{-1}(2632) = 8$

(d) No. Since the new value is the same as the value given for 1995, f would not pass the Horizontal Line Test and would not have an inverse.

83. (a)

$$y = 0.03x^2 + 245.50,\ 0 < x < 100$$

$$x = 0.03y^2 + 245.50$$

$$x - 245.50 = 0.03y^2$$

$$\frac{x - 245.50}{0.03} = y^2$$

$$\sqrt{\frac{x - 245.50}{0.03}} = y,\ 245.50 < x < 545.50$$

$$f^{-1}(x) = \sqrt{\frac{x - 245.50}{0.03}}$$

x = temperature in degrees Fahrenheit

y = percent load for a diesel engine

(b)

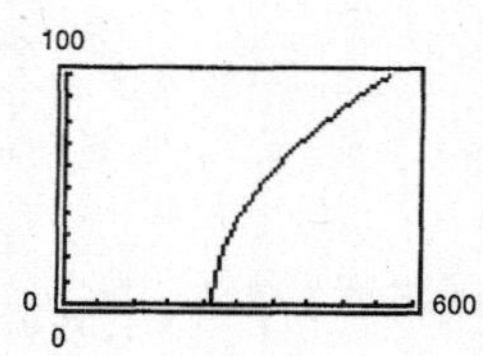

(c) $0.03x^2 + 245.50 \le 500$

$$0.03x^2 \le 254.50$$

$$x^2 \le 8483.33$$

$$x \le 92.10$$

Thus, $0 < x \le 92.10$.

85. False. $f(x) = x^2$ is even and does not have an inverse.

87.

x	1	3	4	6
f	1	2	6	7

x	1	2	6	7
$f^{-1}(x)$	1	3	4	6

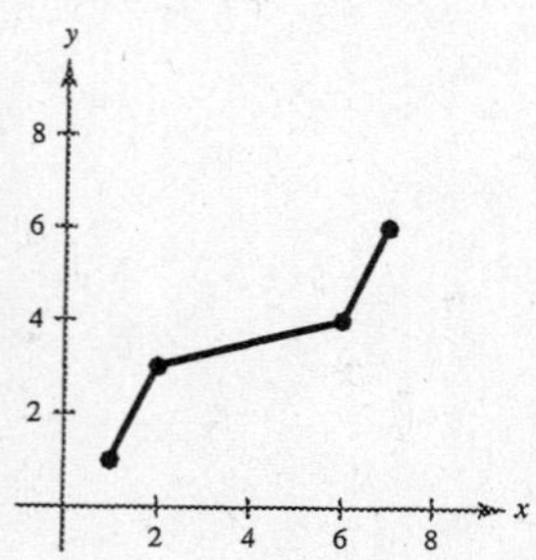

89.

x	-2	-1	3	4
f	6	0	-2	-3

x	-3	-2	0	6
$f^{-1}(x)$	4	3	-1	-2

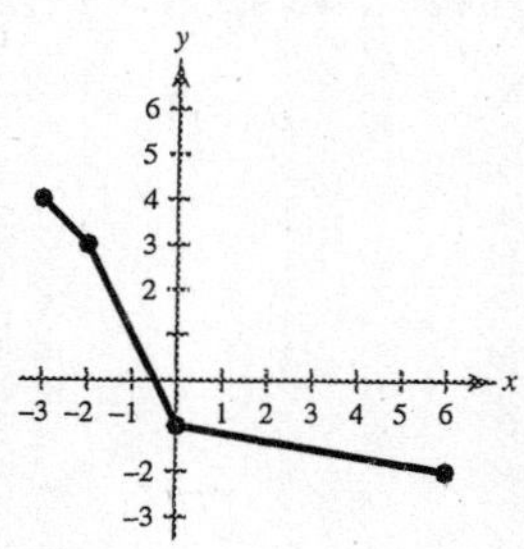

91. If $f(x) = k(2 - x - x^3)$ has an inverse and $f^{-1}(3) = -2$, then $f(-2) = 3$. Thus,

$$f(-2) = k(2 - (-2) - (-2)^3) = 3$$

$$k(2 + 2 + 8) = 3$$

$$12k = 3$$

$$k = \tfrac{3}{12} = \tfrac{1}{4}$$

So, $k = \frac{1}{4}$.

Review Exercises for Chapter P

Solutions to Odd-Numbered Exercises

1. $\{11, -14, -\frac{8}{9}, \frac{5}{2}, \sqrt{6}, 0.4\}$

(a) Natural numbers: 11

(b) Integers: 11, −14

(c) Rational numbers: $11, -14, -\frac{8}{9}, \frac{5}{2}, 0.4$

(d) Irrational numbers: $\sqrt{6}$

3. (a) $\frac{5}{6} = 0.8\overline{3}$

(b) $\frac{7}{8} = 0.875$

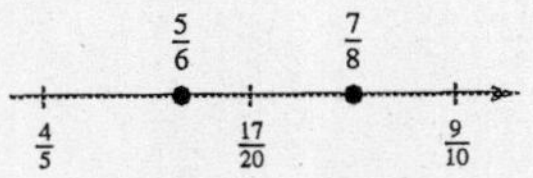

5. $x \le 7$

The set consists of all real numbers less than or equal to 7.

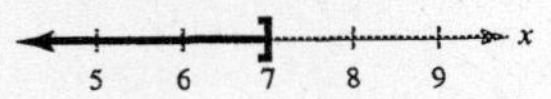

7. $d(-92, 63) = |63 - (-92)| = 155$

9. $d(x, 25) = |x - 25|$ and $d(x, 25) \le 10$, thus, $|x - 25| \le 10$.

11. $-x^2 + x - 1$

(a) $-(1)^2 + 1 - 1 = -1$

(b) $-(-1)^2 + (-1) - 1 = -3$

13. $2x + (3x - 10) = (2x + 3x) - 10$

Illustrates the Associative Property of Addition

15. $0 + (a - 5) = a - 5$ illustrates the Additive Identity property.

17. $|-3| + 4(-2) - 6 = 3 - 8 - 6 = -11$

19. $\frac{5}{18} \div \frac{10}{3} = \frac{\cancel{5}}{\cancelto{6}{18}} \cdot \frac{\cancel{3}}{\cancelto{2}{10}} = \frac{1}{12}$

21. $6[4 - 2(6 + 8)] = 6[4 - 2(14)] = 6[4 - 28] = 6(-24) = -144$

23.

$$6 - (x - 2)^2 = 2 + 4x - x^2$$

$$6 - (x^2 - 4x + 4) = 2 + 4x - x^2$$

$$2 + 4x - x^2 = 2 + 4x - x^2$$

$$0 = 0 \quad \text{Identity}$$

All real numbers are solutions.

25.

$$3x - 2(x + 5) = 10$$

$$3x - 2x - 10 = 10$$

$$x = 20$$

27.

$$4(x + 3) - 3 = 2(4 - 3x) - 4$$

$$4x + 12 - 3 = 8 - 6x - 4$$

$$4x + 9 = -6x + 4$$

$$10x = -5$$

$$x = -\tfrac{1}{2}$$

29. $\frac{x}{5} - 3 = \frac{2x}{3} + 1$

$15\left(\frac{x}{5} - 3\right) = 15\left(\frac{2x}{3} + 1\right)$

$3x - 45 = 10x + 15$

$-7x = 60$

$x = -\frac{60}{7}$

31. $\frac{18}{x} = \frac{10}{x - 4}$

$18(x - 4) = 10x$

$18x - 72 = 10x$

$8x = 72$

$x = 9$

33. $15 + x - 2x^2 = 0$

$2x^2 - x - 15 = 0$

$(2x + 5)(x - 3) = 0$

$x = 3, -\frac{5}{2}$

35. $6 = 3x^2$

$2 = x^2$

$x = \pm\sqrt{2}$

37. $(x + 4)^2 = 18$

$x + 4 = \pm\sqrt{18}$

$x = -4 \pm 3\sqrt{2}$

39. $x^2 - 12x + 30 = 0$

$x = \frac{-(-12) \pm \sqrt{(-12)^2 - 4(1)(30)}}{2(1)} = \frac{12 \pm \sqrt{24}}{2}$

$= \frac{12 \pm 2\sqrt{6}}{2}$

$= 6 \pm \sqrt{6}$

41. $-2x^2 - 5x + 27 = 0$

$2x^2 + 5x - 27 = 0$

$x^2 + \frac{5}{2}x = \frac{27}{2}$

$x^2 + \frac{5}{2}x + \frac{25}{16} = \frac{27}{2} + \frac{25}{16}$

$\left(x + \frac{5}{4}\right)^2 = \frac{241}{16}$

$x + \frac{5}{4} = \pm\sqrt{\frac{241}{16}}$

$x = -\frac{5}{4} \pm \frac{\sqrt{241}}{4}$

43. $5x^4 - 12x^3 = 0$

$x^3(5x - 12) = 0$

$x^3 = 0$ or $5x - 12 = 0$

$x = 0 \qquad x = \frac{12}{5}$

45. $x^4 - 5x^2 + 6 = 0$

$(x^2 - 3)(x^2 - 2) = 0$

$x = \pm\sqrt{3}, \pm\sqrt{2}$

47. $\sqrt{x + 4} = 3$

$x + 4 = 9$

$x = 5$

49. $\sqrt{2x + 3} + \sqrt{x - 2} = 2$

$\sqrt{2x + 3} = 2 - \sqrt{x - 2}$

$2x + 3 = 4 - 4\sqrt{x - 2} + x - 2$

$2x + 3 = x + 2 - 4\sqrt{x - 2}$

$x + 1 = -4\sqrt{x - 2}$

$x^2 + 2x + 1 = 16(x - 2)$

$x^2 - 14x + 33 = 0$

$(x - 3)(x - 11) = 0$

$x = 3$ or $x = 11$

The solutions $x = 3$ and $x = 11$ are extraneous, the equation has no solution.

51. $(x - 1)^{2/3} - 25 = 0$

$(x - 1)^2 = 25^3$

$x - 1 = \pm\sqrt{25^3}$

$x = 1 \pm 125$

$x = 126$ or $x = -124$

53. $|x - 5| = 10$

$x - 5 = 10 \quad \text{or} \quad x - 5 = -10$

$x = 15 \qquad\qquad x = -5$

$x = 15, -5$

55. $x > 0$ and $y = -2$ in Quadrant IV.

57. $(-x, y)$ is in the third quadrant means that (x, y) is in Quadrant IV.

59. (a)

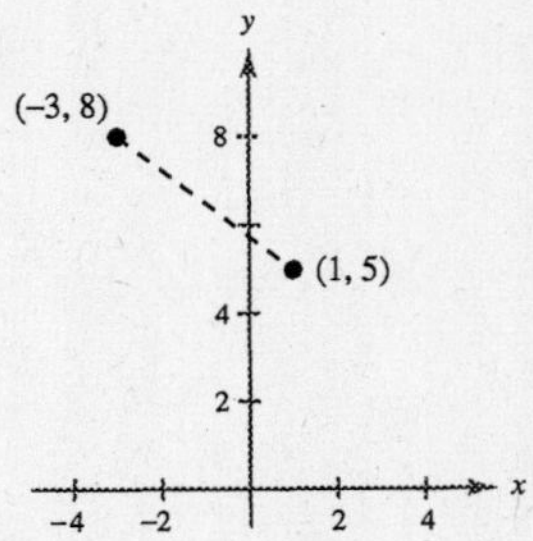

(b) $d = \sqrt{(-3 - 1)^2 + (8 - 5)^2} = \sqrt{16 + 9} = 5$

61.

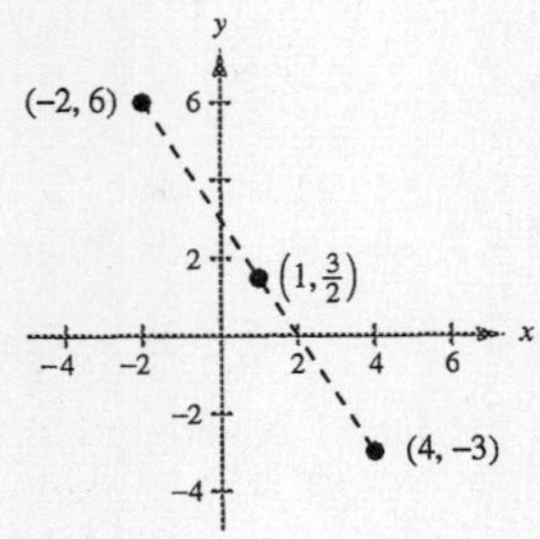

Midpoint: $\left(\dfrac{-2 + 4}{2}, \dfrac{6 + (-3)}{2}\right) = \left(1, \dfrac{3}{2}\right)$

63. $y = 3x - 5$

x	y	(x, y)
-1	-8	$(-1, -8)$
0	-5	$(0, -5)$
1	-2	$(1, -2)$
2	1	$(2, 1)$
3	4	$(3, 4)$

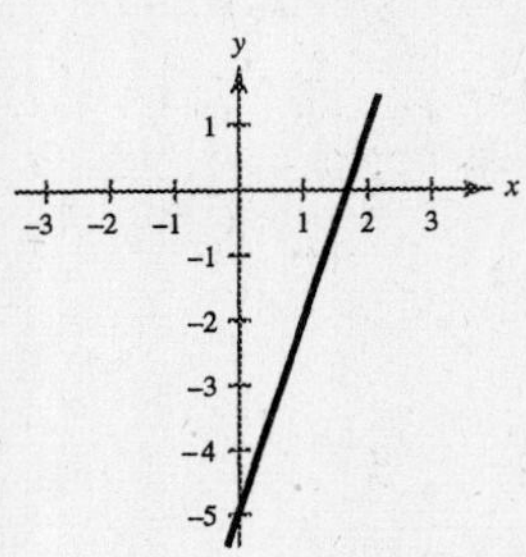

65. $y = x^2 - 3x$

x	y	(x, y)
-1	4	$(-1, 4)$
0	0	$(0, 0)$
1	-2	$(1, -2)$
2	-2	$(2, -2)$
3	0	$(3, 0)$
4	4	$(4, 4)$

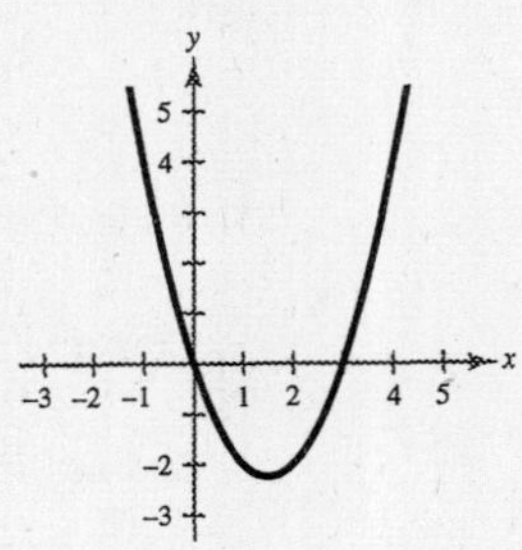

67. $y = \sqrt{5 - x}$

x	y	(x, y)
5	0	$(5, 0)$
4	1	$(4, 1)$
1	2	$(1, 2)$
-4	3	$(-4, 3)$
-11	4	$(-11, 4)$

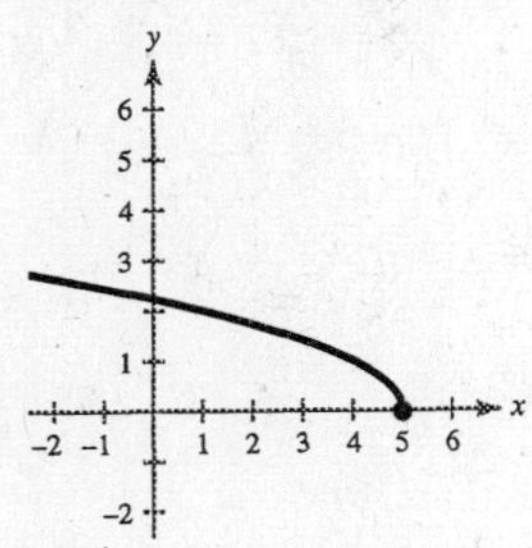

69. $y = 2x - 9$

x-intercept: $0 = 2x - 9 \Rightarrow x = \frac{9}{2}$

$\left(\frac{9}{2}, 0\right)$ is the x-intercept

y-intercept: $y = 2(0) - 9 = -9$

$(0, -9)$ is the y-intercept

71. $y = x\sqrt{9 - x^2}$

x-intercepts: $0 = x\sqrt{9 - x^2} \Rightarrow 0 = x^2(9 - x^2)$

$x = 0 \quad \text{or} \quad x = \pm 3$

The x-intercepts are $(0, 0)$, $(\pm 3, 0)$

y-intercept: $y = 0\sqrt{9 - 0^2} = 0$

The y-intercept is $(0, 0)$

73. $y = 5 - x^2$

$-y = 5 - x^2 \Rightarrow y = -5 + x^2 \Rightarrow$ No x-axis symmetry

$y = 5 - (-x)^2 \Rightarrow y = 5 - x^2 \Rightarrow$ y-axis symmetry

$-y = 5 - (-x)^2 \Rightarrow y = -5 + x^2 \Rightarrow$ No origin symmetry

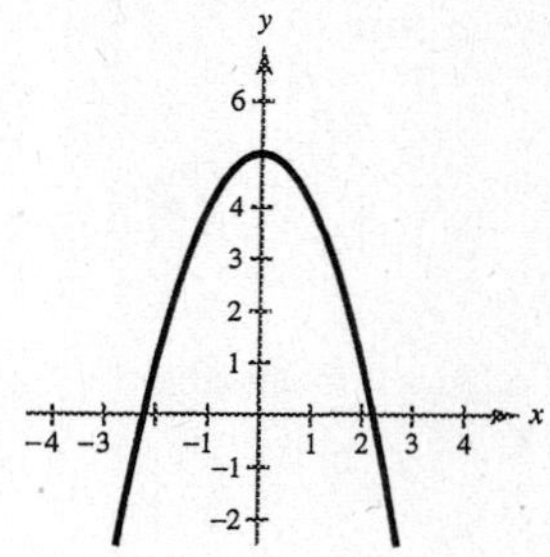

75. $y = x^3 + 3$

$-y = x^3 + 3 \Rightarrow y = -x^3 - 3 \Rightarrow$ No x-axis symmetry

$y = (-x)^3 + 3 \Rightarrow y = -x^3 + 3 \Rightarrow$ No y-axis symmetry

$-y = (-x)^3 + 3 \Rightarrow y = x^3 - 3 \Rightarrow$ No origin symmetry

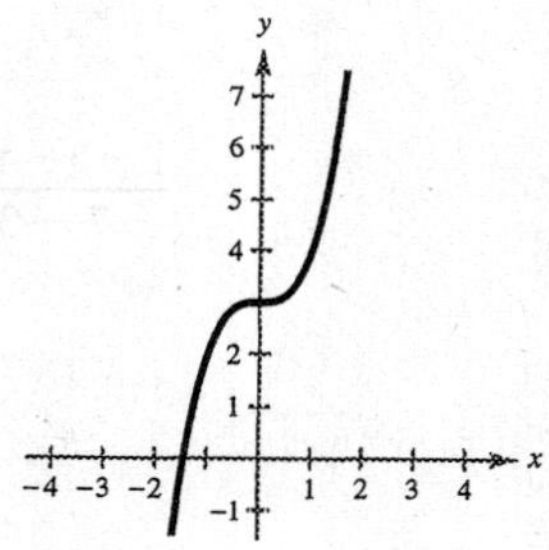

77. $x = -|y|$

$x = -|-y| \Rightarrow x = -|y| \Rightarrow$ x-axis symmetry

$-x = -|y| \Rightarrow x = |y| \Rightarrow$ No y-axis symmetry

$-x = -|-y| \Rightarrow x = |y| \Rightarrow$ No origin symmetry

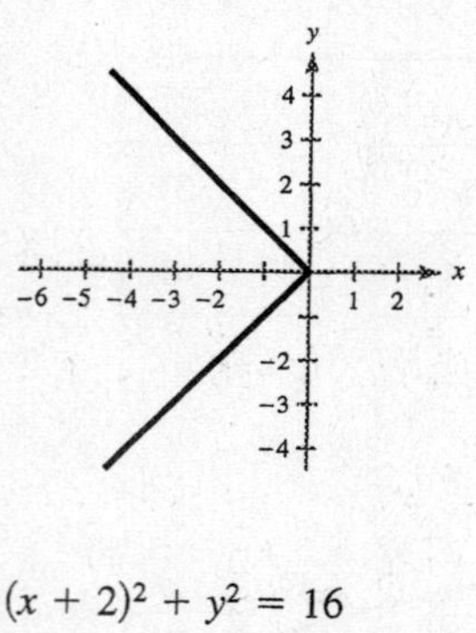

79. $x^2 + y^2 = 9$

Center: $(0, 0)$

Radius: 3

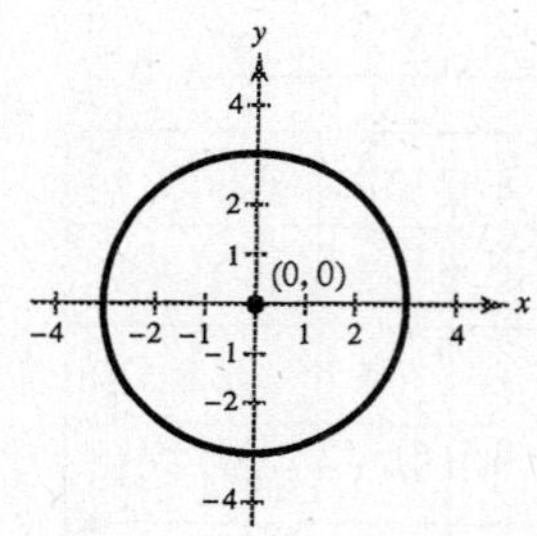

81. $(x + 2)^2 + y^2 = 16$

$(x - (-2))^2 + (y - 0)^2 = 4^2$

Center: $(-2, 0)$

Radius: 4

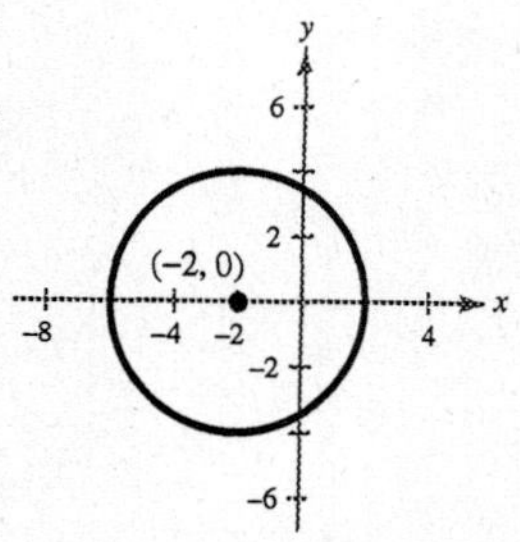

83. Endpoints of a diameter: $(0, 0)$ and $(4, -6)$

Center: $\left(\dfrac{0 + 4}{2}, \dfrac{0 + (-6)}{2}\right) = (2, -3)$

Radius: $r = \sqrt{(2 - 0)^2 + (-3 - 0)^2} = \sqrt{4 + 9} = \sqrt{13}$

Standard form: $(x - 2)^2 + (y - (-3))^2 = \left(\sqrt{13}\right)^2$

$$(x - 2)^2 + (y + 3)^2 = 13$$

85. $y = -2x - 7$

y-intercept: $(0, -7)$

Slope: $m = -2 = -\frac{2}{1}$

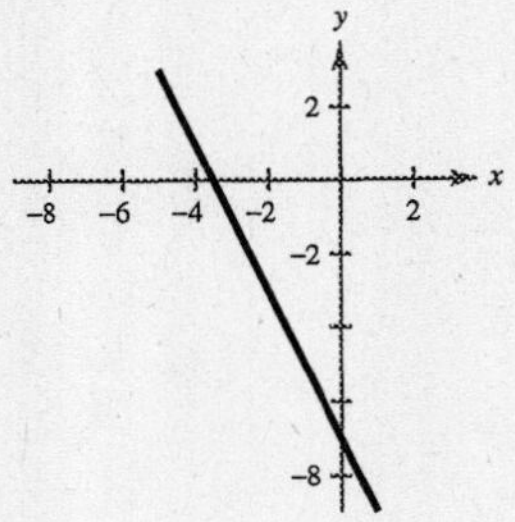

87. $y = 6$

Horizontal line

y-intercept: $(0, 6)$

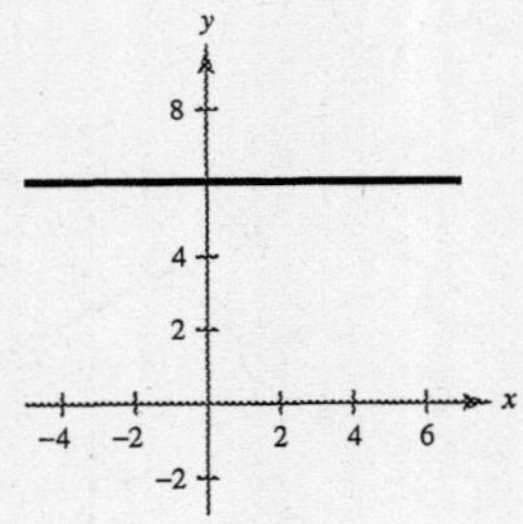

89. $y = 3x + 13$

y-intercept: $(0, 13)$

Slope: $m = 3 = \frac{3}{1}$

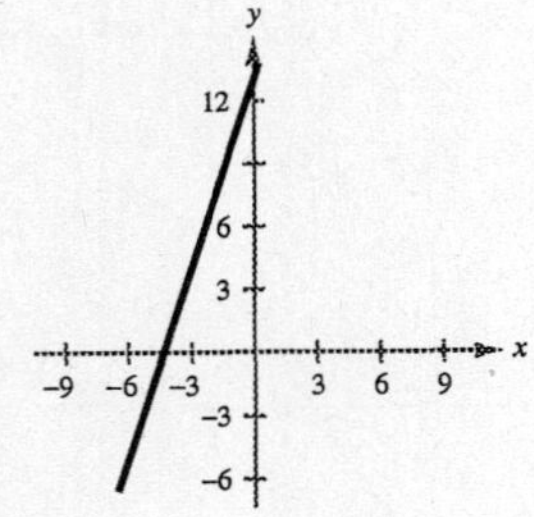

91. $y = -\frac{5}{2}x - 1$

y-intercept: $(0, -1)$

Slope: $m = -\frac{5}{2}$

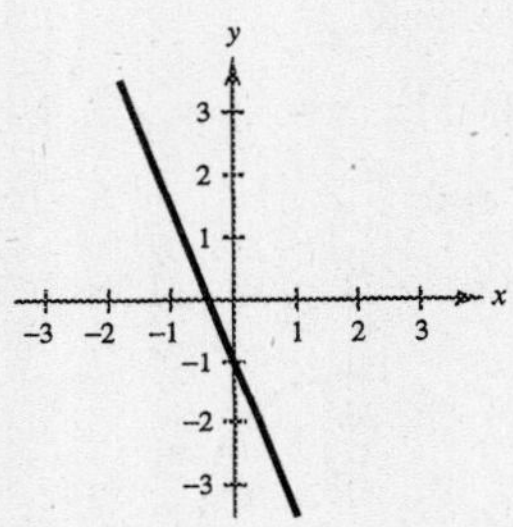

93. $(3, -4)$ and $(-7, 1)$

$$m = \frac{1 - (-4)}{-7 - 3} = \frac{5}{-10} = -\frac{1}{2}$$

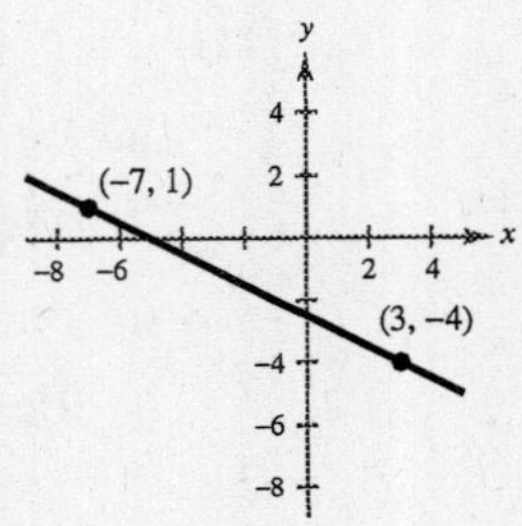

95. $(-2, 5)$ and $(1, 1)$

$$m = \frac{1 - 5}{1 - (-2)} = -\frac{4}{3}$$

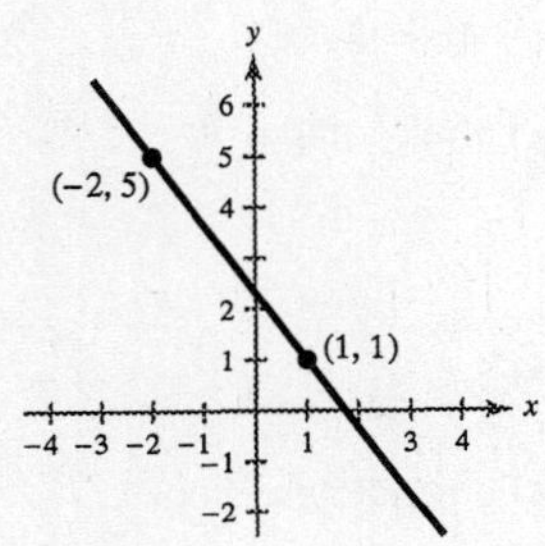

97. $(-4.5, 6)$, $(2.1, 3)$

$$m = \frac{3 - 6}{2.1 - (-4.5)} = \frac{-3}{6.6} = -\frac{30}{66} = -\frac{5}{11}$$

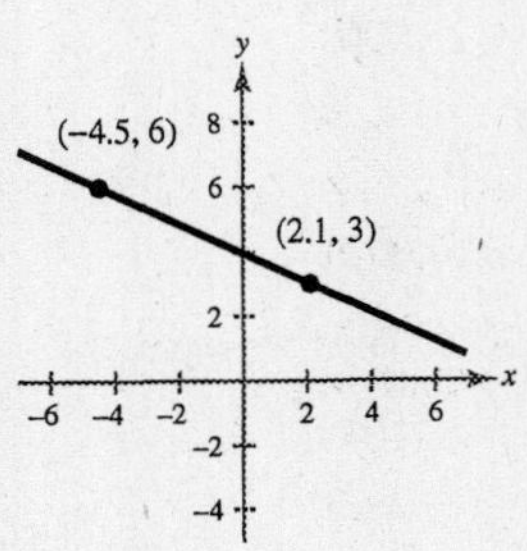

99. $(0, 0)$, $(0, 10)$

$$m = \frac{10 - 0}{0 - 0} = \frac{10}{0} \quad \text{undefined}$$

The line is vertical.

$x = 0$

101. $(-1, 4)$, $(2, 0)$

$$m = \frac{0 - 4}{2 - (-1)} = -\frac{4}{3}$$

$$y - 4 = -\frac{4}{3}(x - (-1)) \quad \text{or}$$

$$y = -\frac{4}{3}x + \frac{8}{3}$$

$$-3y + 12 = 4x + 4$$

$$4x + 3y - 8 = 0$$

103. $(-5, 10)$, $(-1, -6)$

$$m = \frac{-6 - 10}{-1 - (-5)} = \frac{-16}{4} = -4$$

$$y - 10 = -4(x - (-5))$$

$$y - 10 = -4x - 20$$

$$y = -4x - 10$$

$$4x + y + 10 = 0$$

105. $y - (-5) = \frac{3}{2}(x - 0)$

$y + 5 = \frac{3}{2}x$

$y = \frac{3}{2}x - 5$ or

$3x - 2y - 10 = 0$

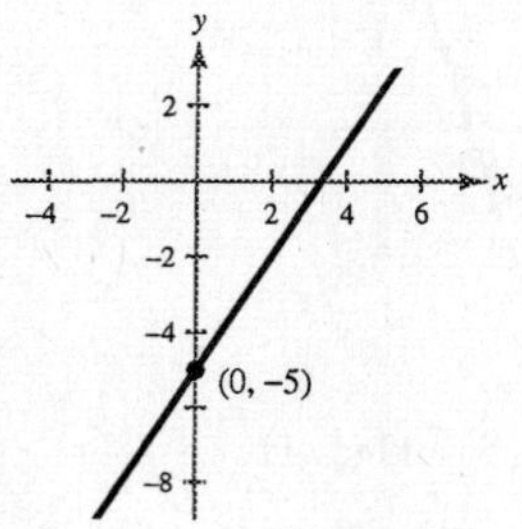

107. $y - (-3) = -\frac{1}{2}(x - 10)$

$y + 3 = -\frac{1}{2}x + 5$

$y = -\frac{1}{2}x + 2$ or

$x + 2y - 4 = 0$

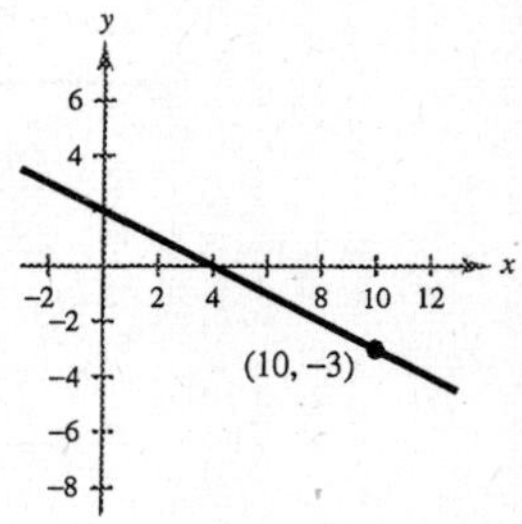

109. $5x - 4y = 8 \Rightarrow y = \frac{5}{4}x - 2$ and $m = \frac{5}{4}$

(a) Parallel slope: $m = \frac{5}{4}$

$y - (-2) = \frac{5}{4}(x - 3)$

$4y + 8 = 5x - 15$

$5x - 4y - 23 = 0$

(b) Perpendicular slope: $m = -\frac{4}{5}$

$y - (-2) = -\frac{4}{5}(x - 3)$

$5y + 10 = -4x + 12$

$4x + 5y - 2 = 0$

111. $x = 3 \Rightarrow m$ is undefined

(a) Parallel slope: m is undefined

Vertical line through $(4, -1)$: $x = 4$ or $x - 4 = 0$

(b) Perpendicular slope: $m = 0$

Horizontal line through $(4, -1)$: $y = -1$ or $y + 1 = 0$

113. $(4, 12{,}500)$ $m = 850$

$V - 12{,}500 = 850(t - 4)$

$V - 12{,}500 = 850t - 3400$

$V = 850t + 9100, \quad 4 \le t \le 9$

115. $(2, 160{,}000), (3, 185{,}000)$

$$m = \frac{185{,}000 - 160{,}000}{3 - 2} = 25{,}000$$

$S - 160{,}000 = 25{,}000(t - 2)$

$S = 25{,}000t + 110{,}000$

For the fourth quarter let $t = 4$. Then we have

$S = 25{,}000(4) + 110{,}000 = \$210{,}000.$

117. $A = \{10, 20, 30, 40\}$ and

$B = \{0, 2, 4, 6\}$

(a) 20 is matched with two elements in the range so it is **not** a function.

(b) function

(c) function

(d) 30 is not matched with any element of B so it is **not** a function.

119. $16x - y^4 = 0$

$y^4 = 16x$

$y = \pm 2\sqrt[4]{x}$

y is **not** a function of x. Some x-values correspond to two y-values.

121. $y = \sqrt{1 - x}$

Each x-value, $x \le 1$, corresponds to only one y-value so y **is** a function of x.

123. $h(x) = \begin{cases} 2x + 1, & x \leq -1 \\ x^2 + 2, & x > -1 \end{cases}$

(a) $h(-2) = 2(-2) + 1 = -3$

(b) $h(-1) = 2(-1) + 1 = -1$

(c) $h(0) = (0)^2 + 2 = 2$

(d) $h(2) = (2)^2 + 2 = 6$

125. $f(x) = \sqrt{25 - x^2}$

Domain: $25 - x^2 \geq 0$

$(5 + x)(5 - x) \geq 0$

Critical Numbers: $x = \pm 5$

Test intervals: $(-\infty, -5)$, $(-5, 5)$, $(5, \infty)$

Solution set: $-5 \leq x \leq 5$

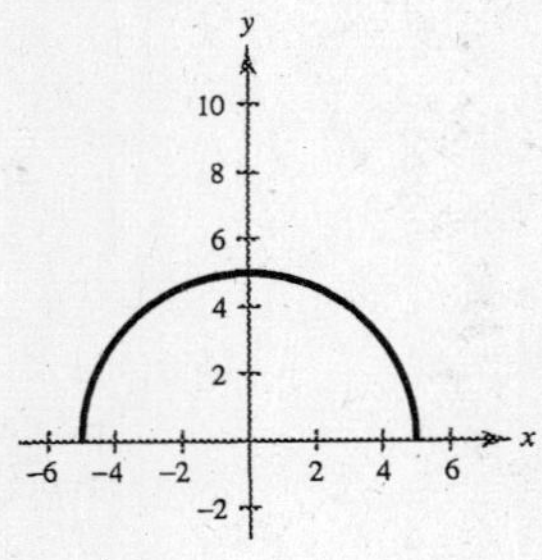

127. $g(s) = \dfrac{5}{3s - 9} = \dfrac{5}{3(s - 3)}$

Domain: All real numbers except $s = 3$.

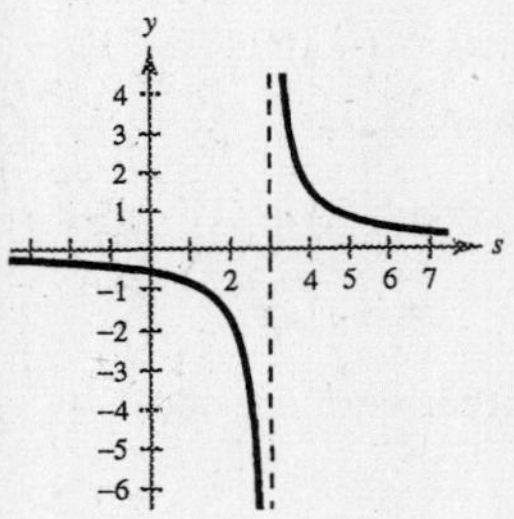

129. $h(x) = \dfrac{x}{x^2 - x - 6} = \dfrac{x}{(x + 2)(x - 3)}$

Domain: All real numbers except $x = -2, 3$.

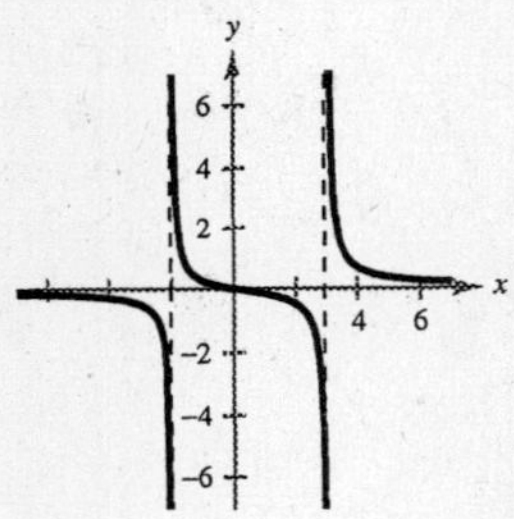

131. $f(x) = 2x^2 + 3x - 1$

$$\frac{f(x + h) - f(x)}{h} = \frac{2(x + h)^2 + 3(x + h) - 1 - (2x^2 + 3x - 1)}{h}$$

$$= \frac{2x^2 + 4xh + 2h^2 + 3x + 3h - 1 - 2x^2 - 3x + 1}{h}$$

$$= \frac{4xh + 2h^2 + 3h}{h}$$

$$= \frac{h(4x + 2h + 3)}{h}$$

$$= 4x + 2h + 3, \; h \neq 0$$

133. $v(t) = -32t + 48$

(a) $v(1) = 16$ ft/sec

(b) $0 = -32t + 48$

$t = \frac{48}{32} = 1.5$ sec

(c) $v(2) = -16$ ft/sec

135.

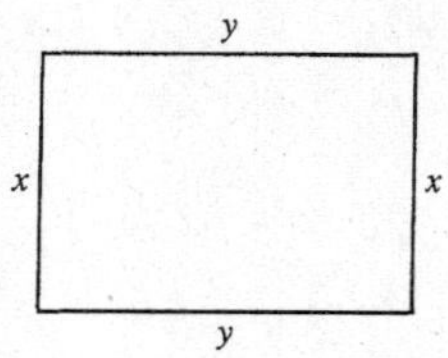

(a) $2x + 2y = 24$

$y = 12 - x$

$A = xy = x(12 - x)$

(b) Since x and y cannot be negative, we have $0 < x < 12$. The domain is $0 < x < 12$.

137. $y = (x - 3)^2$ passes the Vertical Line Test so y is a function of x.

139. $x - 4 = y^2$ does not pass the Vertical Line Test so y is not a function of x.

141. $f(x) = 5x^2 + 4x - 1$

$5x^2 + 4x - 1 = 0$

$(5x - 1)(x + 1) = 0$

Zeros: $x = -1, \frac{1}{5}$

143. $3x^2 - 16x + 21 = 0$

$(3x - 7)(x - 3) = 0$

$3x - 7 = 0$ or $x - 3 = 0$

$x = \frac{7}{3}$ or $x = 3$

145. $f(x) = |x| + |x + 1|$

From the graph in the text, we see that $f(x)$ is increasing on $(0, \infty)$, decreasing on $(-\infty, -1)$, and constant on $[-1, 0]$.

147. $f(x) = x^5 + 4x - 7$

$f(-x) = (-x)^5 + 4(-x) - 7$

$= -x^5 - 4x - 7$

$\neq f(x)$

$\neq -f(x)$

Neither even nor odd

149. $f(x) = 2x\sqrt{x^2 + 3}$

$f(-x) = 2(-x)\sqrt{(-x)^2 + 3}$

$= -2x\sqrt{x^2 + 3}$

$= -f(x)$

f is **odd**

151. $f(2) = -6, f(-1) = 3$

Points: $(2, -6), (-1, 3)$

$m = \dfrac{3 - (-6)}{-1 - 2} = \dfrac{9}{-3} = -3$

$y - (-6) = -3(x - 2)$

$y + 6 = -3x + 6$

$y = -3x$

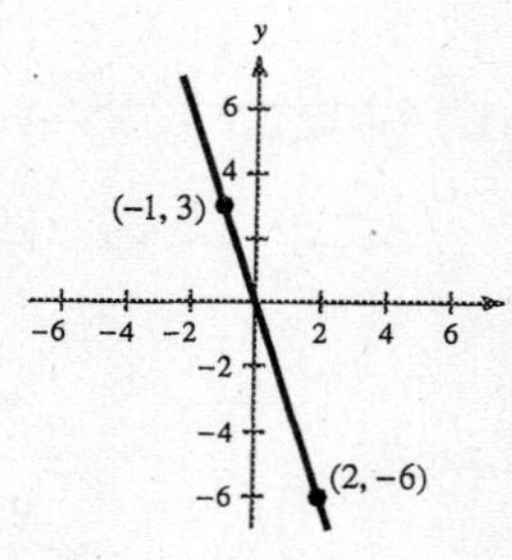

153. $f\left(-\frac{4}{5}\right) = 2, f\left(\frac{11}{5}\right) = 7$

Points: $\left(-\frac{4}{5}, 2\right), \left(\frac{11}{5}, 7\right)$

$m = \dfrac{7 - 2}{\frac{11}{5} - \left(-\frac{4}{5}\right)} = \dfrac{5}{3}$

$y - 2 = \frac{5}{3}\left(x - \left(-\frac{4}{5}\right)\right)$

$y - 2 = \frac{5}{3}x + \frac{4}{3}$

$y = \frac{5}{3}x + \frac{10}{3}$

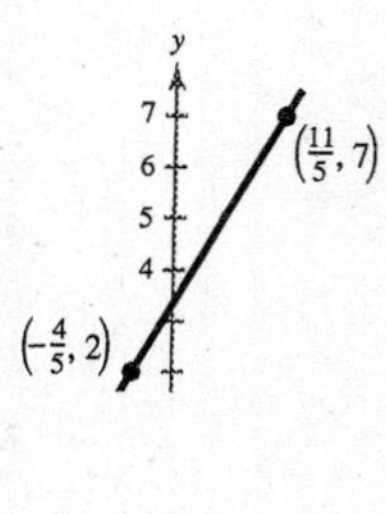

155. $f(x) = 3 - x^2$

Intercepts: $(0, 3), (\pm\sqrt{3}, 0)$

y-axis symmetry

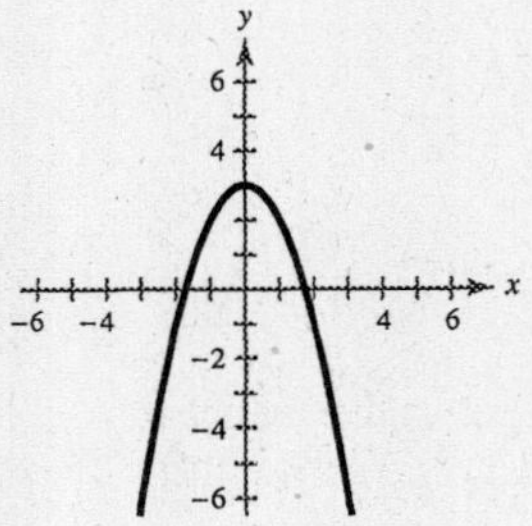

157. $f(x) = -\sqrt{x}$

Domain: $x \geq 0$

Intercept: $(0, 0)$

x	0	1	4	9
y	0	-1	-2	-3

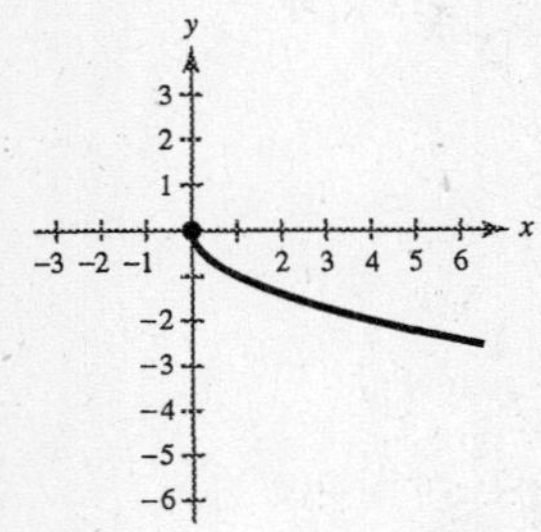

159. $g(x) = \dfrac{3}{x}$

No intercepts

Origin symmetry

x	-3	-1	1	3
y	-1	-3	3	1

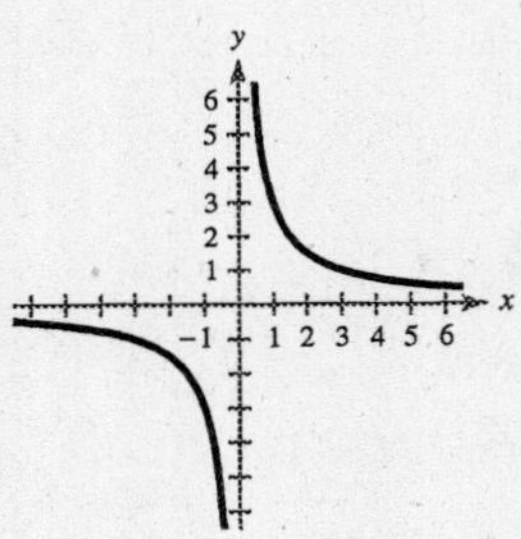

161. $f(x) = [\![x]\!] - 2$

This is the greatest integer function shifted down two units.

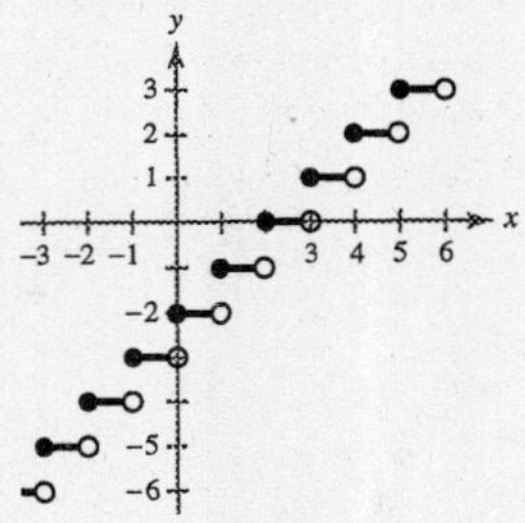

163. $f(x) = \begin{cases} 5x - 3, & x \geq -1 \\ -4x + 5, & x < -1 \end{cases}$

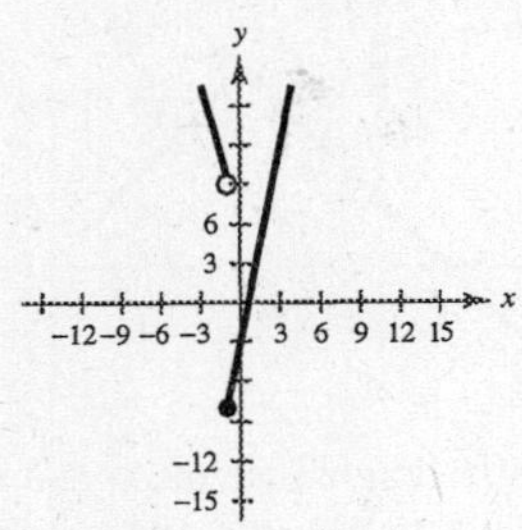

165. Basic function: $f(x) = x^3$

Vertical shift 4 units upward and a horizontal shift 4 units to the left.

167. $f(x) = x^2$

$h(x) = x^2 - 9$

Vertical shift 9 units downward.

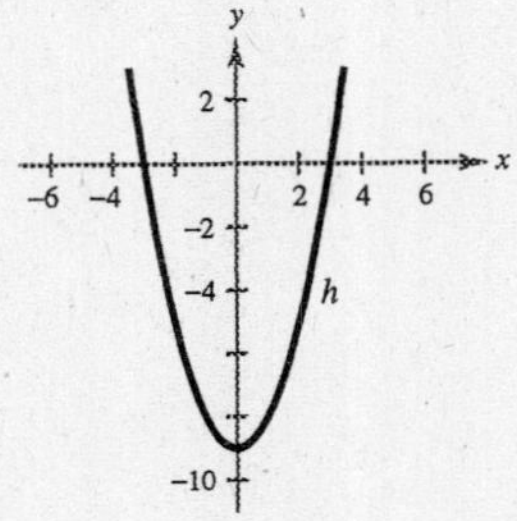

169. $f(x) = \sqrt{x}$

$h(x) = \sqrt{x - 7}$

Horizontal shift 7 units to the right.

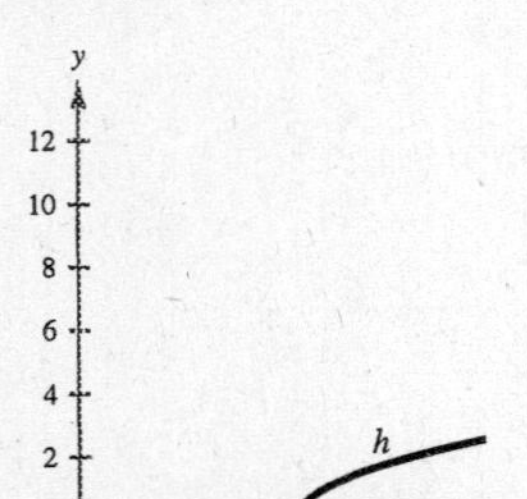

171. $f(x) = x^2$

$h(x) = -(x + 3)^2 + 1$

Reflection in the x-axis, a horizontal shift 3 units to the left, and a vertical shift 1 unit upward.

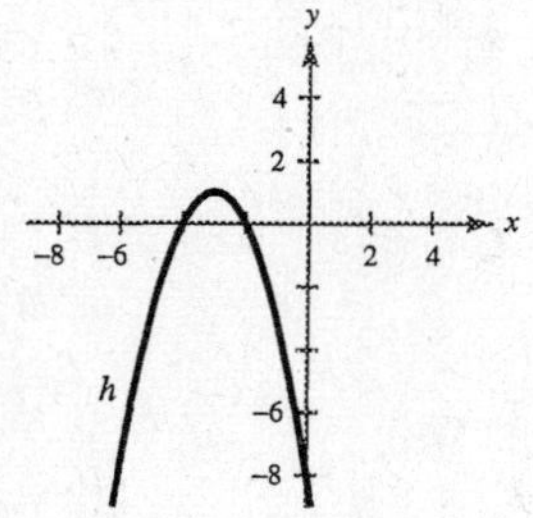

173. $f(x) = [\![x]\!]$

$h(x) = -[\![x]\!] + 6$

Reflection in the x-axis and a vertical shift 6 units upward.

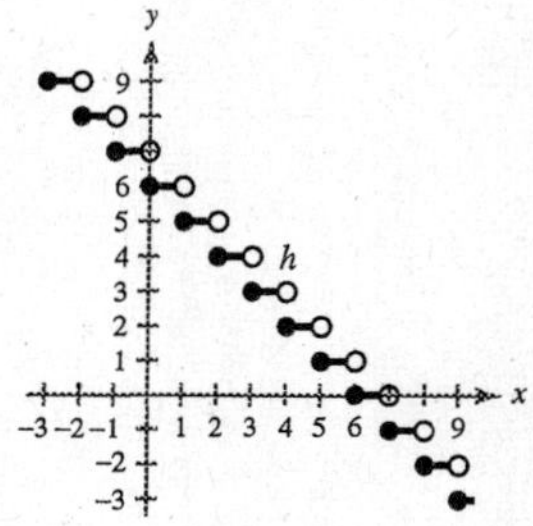

175. $f(x) = |x|,\ h(x) = -|-x + 4| + 6$

$= -|-(x - 4)| + 6$

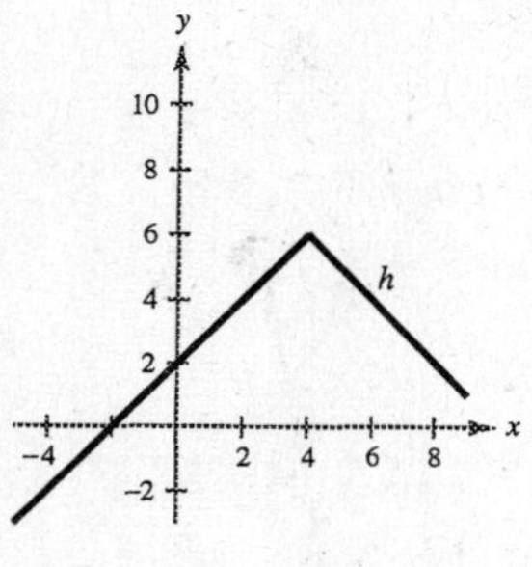

Reflection in both the x- and y-axes; horizontal shift 4 units to the right; vertical shift 6 units upward

177. $f(x) = [\![x]\!]$

$h(x) = 5\,[\![x - 9]\!]$

Vertical stretch by a factor of 5 and a horizontal shift 9 units to the right.

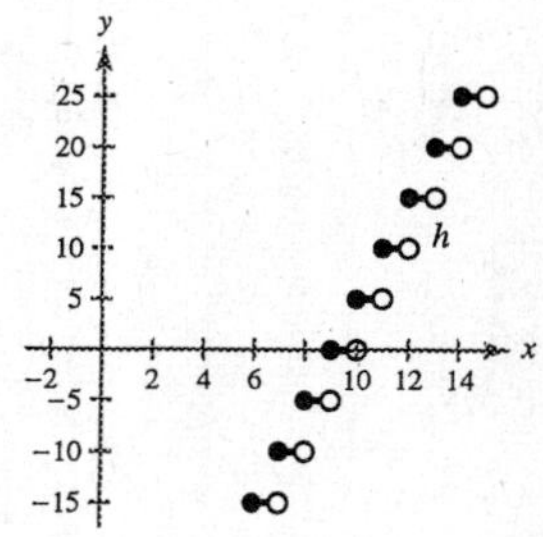

179. $f(x) = \sqrt{x}$

$h(x) = -2\sqrt{x - 4}$

Reflection in the x-axis, a vertical stretch by a factor of 2, and a horizontal shift 4 units to the right.

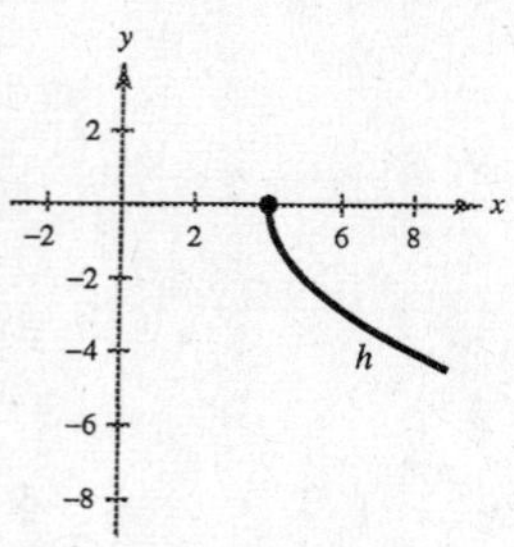

181. $f(x) = x^2 + 3,\ g(x) = 2x - 1$

(a) $(f + g)(x) = f(x) + g(x)$

$= (x^2 + 3) + (2x - 1)$

$= x^2 + 2x + 2$

(b) $(f - g)(x) = f(x) - g(x)$

$= (x^2 + 3) - (2x - 1)$

$= x^2 - 2x + 4$

(c) $(fg)(x) = f(x)g(x)$

$= (x^2 + 3)(2x - 1)$

$= 2x^3 - x^2 + 6x - 3$

(d) $\left(\dfrac{f}{g}\right)(x) = \dfrac{f(x)}{g(x)}$

$= \dfrac{x^2 + 3}{2x - 1}$

Domain: $x \neq \dfrac{1}{2}$

183. $f(x) = \frac{1}{3}x - 3,\ g(x) = 3x + 1$

(a) $(f \circ g)(x) = f(g(x))$

$= f(3x + 1)$

$= \frac{1}{3}(3x + 1) - 3$

$= x + \frac{1}{3} - 3$

$= x - \frac{8}{3}$

(b) $(g \circ f)(x) = g(f(x))$

$= g\left(\frac{1}{3}x - 3\right)$

$= 3\left(\frac{1}{3}x - 3\right) + 1$

$= x - 9 + 1$

$= x - 8$

The domains of f, g, $f \circ g$, and $g \circ f$, are all real numbers.

185. $h(x) = (6x - 5)^3$

Answers will vary.

Possible answers: $f(x) = x^3$ and $g(x) = 6x - 5$

$f(x) = (x - 5)^3$ and $g(x) = 6x$

$f(x) = (x + 10)^3$ and $g(x) = 6x - 15$

187. (5, 62.1) and (5, 144.4)

(6, 74.3) and (6, 155.9)

(7, 85.9) and (7, 168.2)

(8, 94.6) and (8, 173.3)

(9, 109.7) and (9, 198.3)

$y_1 \approx 0.21t^2 + 8.65t + 14.21$

$y_2 \approx 1.41t^2 - 7.28t + 146.85$

189. $f(x) = x - 7$

$f^{-1}(x) = x + 7$

$f(f^{-1}(x)) = f(x + 7) = (x + 7) - 7 = x$

$f^{-1}(f(x)) = f^{-1}(x - 7) = (x - 7) + 7 = x$

191. From the graph we see that the function passes the Horizontal Line Test. It *has* an inverse.

193. $f(x) = 4 - \dfrac{1}{3}x$

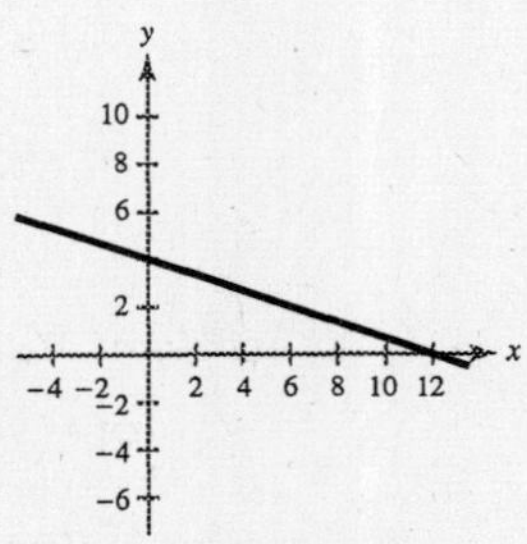

The function passes the Horizontal Line Test. It *has* an inverse.

195. $h(t) = \dfrac{2}{t - 3}$

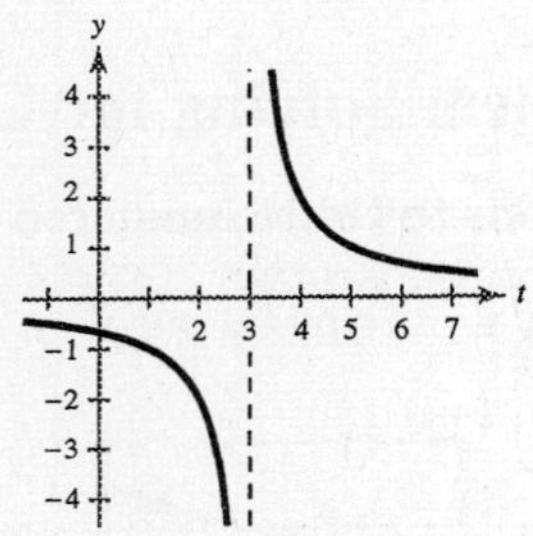

The function passes the Horizontal Line Test. It *has* an inverse.

197. (a) $f(x) = \frac{1}{2}x - 3$

$y = \frac{1}{2}x - 3$

$x = \frac{1}{2}y - 3$

$x + 3 = \frac{1}{2}y$

$2(x + 3) = y$

$f^{-1}(x) = 2x + 6$

(b)

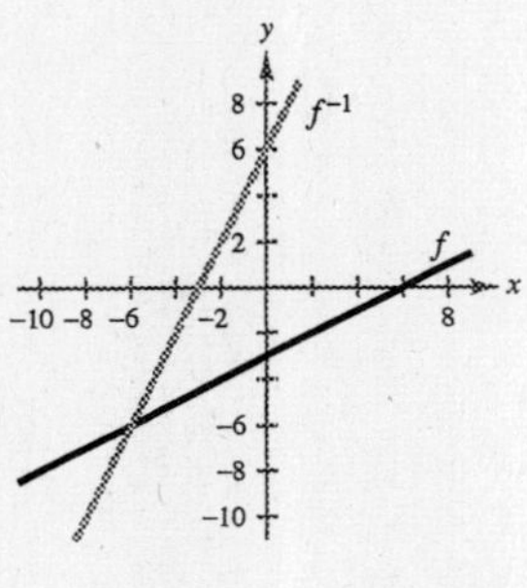

(c) $f^{-1}(f(x)) = f^{-1}\left(\frac{1}{2}x - 3\right)$

$= 2\left(\frac{1}{2}x - 3\right) + 6$

$= x - 6 + 6$

$= x$

$f(f^{-1}(x)) = f(2x + 6)$

$= \frac{1}{2}(2x + 6) - 3$

$= x + 3 - 3$

$= x$

199. (a) $f(x) = \sqrt{x+1}$

$y = \sqrt{x+1}$

$x = \sqrt{y+1}$

$x^2 = y + 1$

$x^2 - 1 = y$

$f^{-1}(x) = x^2 - 1, \ x \geq 0$

Note: The inverse must have a restricted domain.

(b)

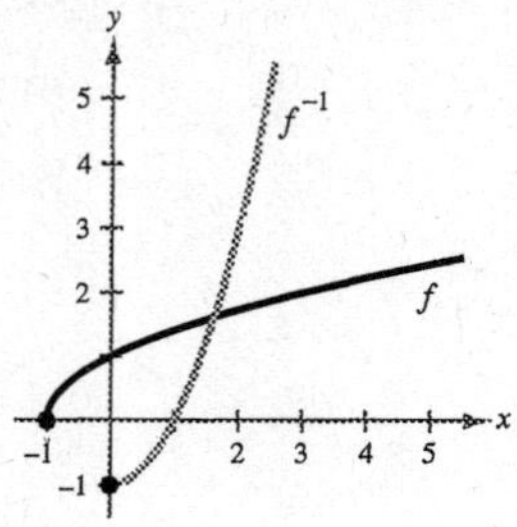

(c) $f^{-1}(f(x)) = f^{-1}\left(\sqrt{x+1}\right)$

$= \left(\sqrt{x+1}\right)^2 - 1$

$= x + 1 - 1$

$= x$

$f(f^{-1}(x)) = f(x^2 - 1)$

$= \sqrt{(x^2-1)+1}$

$= \sqrt{x^2} = x$ for $x \geq 0$.

201. $f(x) = 2(x-4)^2$ is increasing on $[4, \infty)$.

Let $f(x) = 2(x-4)^2, \ x \geq 4$ and $y \geq 0$

$y = 2(x-4)^2$

$x = 2(y-4)^2, \ x \geq 0, \ y \geq 4$

$\frac{x}{2} = (y-4)^2$

$\sqrt{\frac{x}{2}} = y - 4$

$\sqrt{\frac{x}{2}} + 4 = y$

$f^{-1}(x) = \sqrt{\frac{x}{2}} + 4, \ x \geq 0$

203. False. The graph is reflected in the x-axis first, then shifted 9 units to the left, and 13 units down.

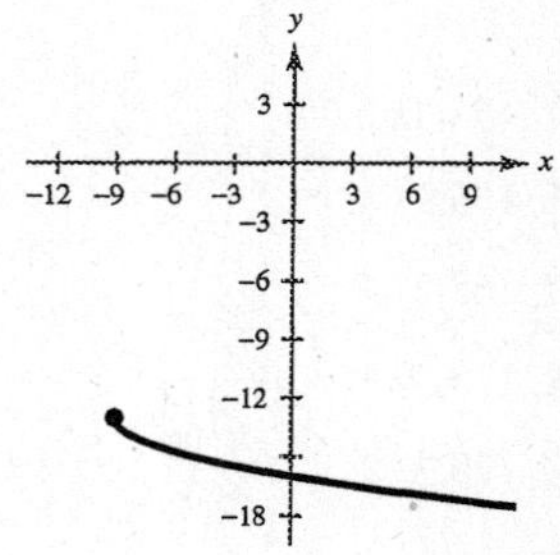

205. A function from set A to set B is a relation that assigns to each element x in the set A exactly one element y in the set B.

Problem Solving for Chapter P

Solutions to Odd-Numbered Exercises

1. (a) $W_1 = 0.07x + 2000$

(b) $W_2 = 0.05x + 2300$

(c)

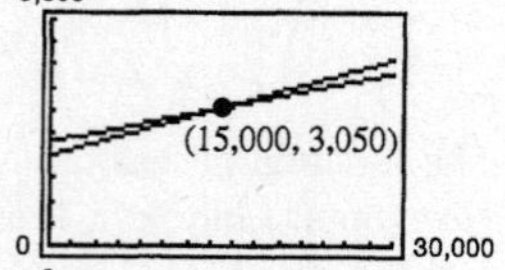

Point of Intersection: (15,000, 3050)

Both jobs pay the same, \$3050, if you sell \$15,000 per month.

(d) If you think you can sell \$20,000 per month, keep your current job with the higher commission rate. For sales over \$15,000 it pays more than the other job.

3. (a) Let $f(x)$ and $g(x)$ be two even functions.

Then define $h(x) = f(x) \pm g(x)$.

$$h(-x) = f(-x) \pm g(-x)$$
$$= f(x) \pm g(x) \text{ since } f \text{ and } g \text{ are even}$$
$$= h(x)$$

So, $h(x)$ is also even.

(b) Let $f(x)$ and $g(x)$ be two odd functions.

Then define $h(x) = f(x) \pm g(x)$.

$$h(-x) = f(-x) \pm g(-x)$$
$$= -f(x) \mp g(x) \text{ since } f \text{ and } g \text{ are odd}$$
$$= -h(x)$$

So, $h(x)$ is also odd. (If $f(x) \neq g(x)$)

(c) Let $f(x)$ be odd and $g(x)$ be even.

Then define $h(x) = f(x) \pm g(x)$.

$$h(-x) = f(-x) \pm g(-x)$$
$$= -f(x) \pm g(x) \text{ since } f \text{ is odd and } g \text{ is even}$$
$$\neq h(x)$$
$$\neq -h(x)$$

So, $h(x)$ is neither odd nor even.

5.
$$f(x) = a_{2n}x^{2n} + a_{2n-2}x^{2n-2} + \cdots + a_2x^2 + a_0$$
$$f(-x) = a_{2n}(-x)^{2n} + a_{2n-2}(-x)^{2n-2} + \cdots + a_2(-x)^2 + a_0$$
$$= a_{2n}x^{2n} + a_{2n-2}x^{2n-2} + \cdots + a_2x^2 + a_0$$
$$= f(x)$$

Therefore, $f(x)$ is even.

7. (a)

April 11:	10 hours
April 12:	24 hours
April 13:	24 hours
April 14:	$23\frac{2}{3}$ hours
Total:	$81\frac{2}{3}$ hours

(b) $\text{Speed} = \dfrac{\text{distance}}{\text{time}} = \dfrac{2100}{81\frac{2}{3}} = \dfrac{180}{7} = 25\frac{5}{7}$ m.p.h.

(c) $D = -\dfrac{180}{7}t + 3400$

Domain: $0 \le t \le \dfrac{1190}{9}$

Range: $0 \le D \le 3400$

(d)

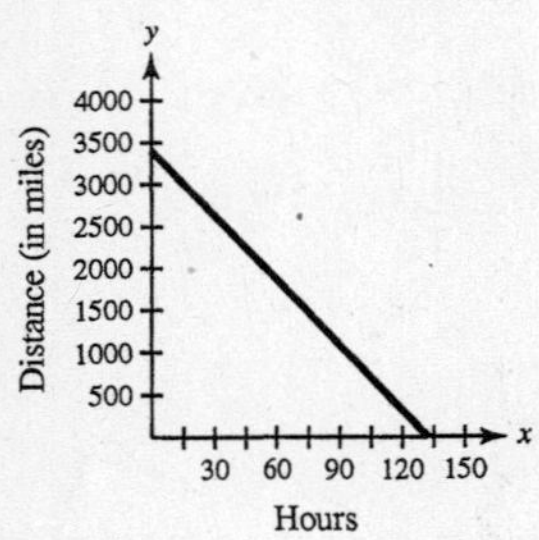

9. (a) The length of the trip in the water is $\sqrt{2^2 + x^2}$, and the length of the trip over land is $\sqrt{1 + (3 - x)^2}$. Hence, the total time is

$$T(x) = \frac{\sqrt{4 + x^2}}{2} + \frac{\sqrt{1 + (3 - x)^2}}{4} \text{ hours.}$$

(b) Domain of $T(x)$: $0 \le x \le 3$

(c)

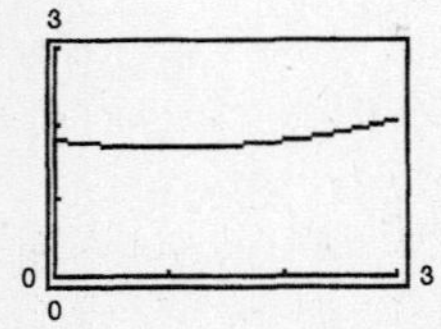

(d) $T(x)$ is a minimum when $x = 1$

(e) To reach point Q in the shortest amount of time, you should row to a point one mile down the coast, and then walk the rest of the way.

11. $f(x) = y = \dfrac{1}{1-x}$

(a) Domain: all $x \neq 1$

Range: all $y \neq 0$

(b) $f(f(x)) = f\left(\dfrac{1}{1-x}\right) = \dfrac{1}{1-\left(\dfrac{1}{1-x}\right)} = \dfrac{1}{\dfrac{1-x-1}{1-x}} = \dfrac{1-x}{-x} = \dfrac{x-1}{x}$

Domain: all $x \neq 0, 1$

(c) $f(f(f(x))) = f\left(\dfrac{x-1}{x}\right) = \dfrac{1}{1-\left(\dfrac{x-1}{x}\right)} = \dfrac{1}{\dfrac{1}{x}} = x$

Domain: all $x \neq 0, 1$

The graph is not a line. It has holes at $(0, 0)$ and $(1, 1)$.

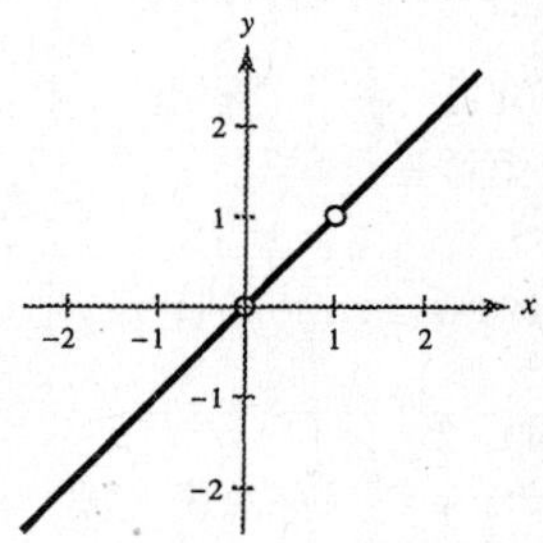

13. (a) $f(x+1)$

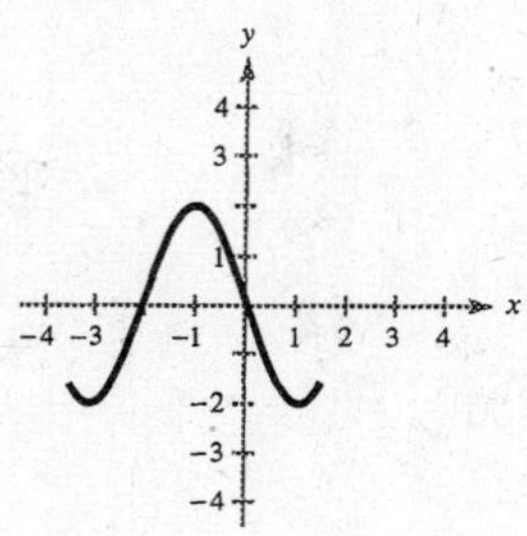

(b) $f(x) + 1$

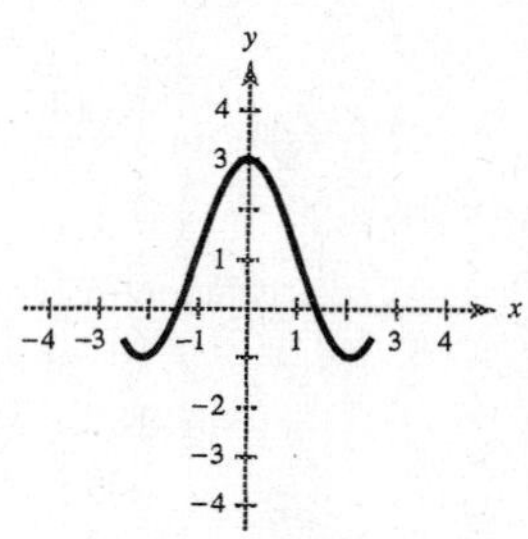

(c) $2f(x)$

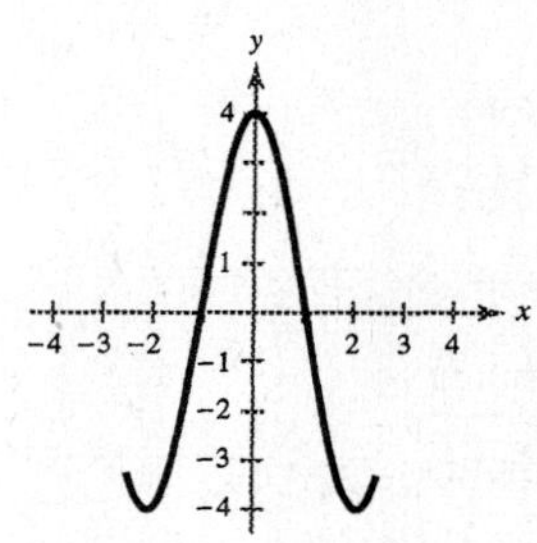

(d) $f(-x)$

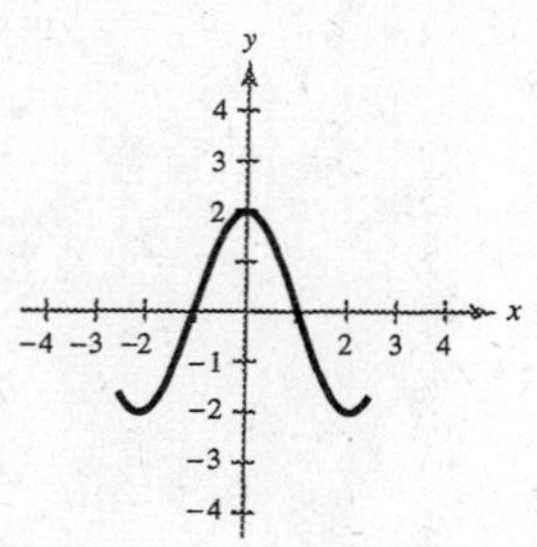

(e) $-f(x)$

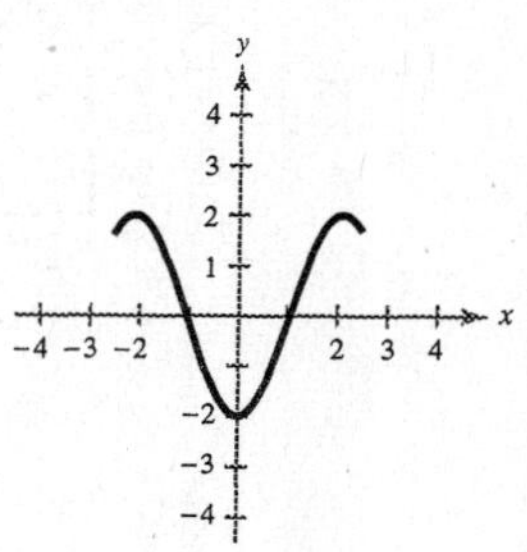

(f) $|f(x)|$

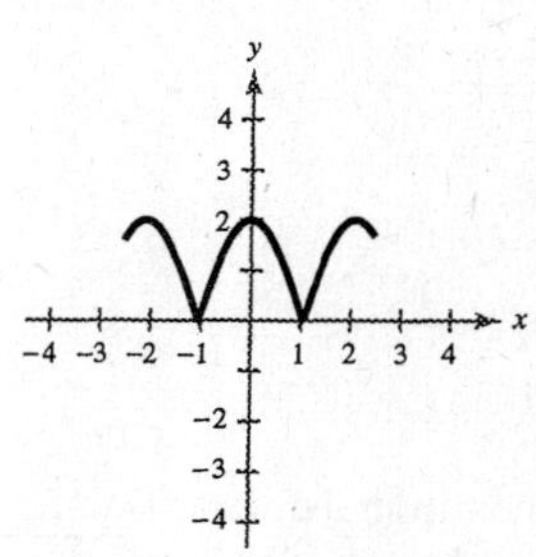

(g) $f(|x|)$

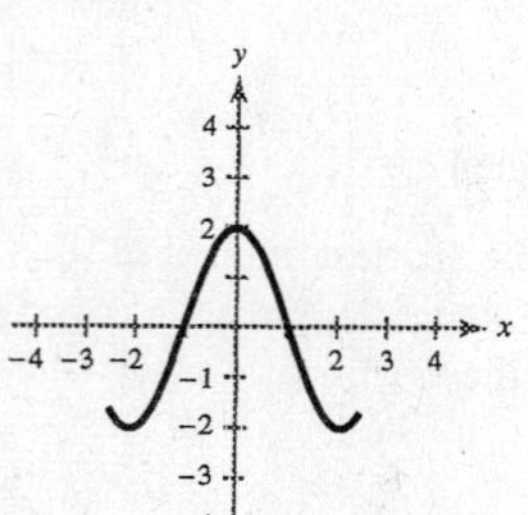

Chapter P Practice Test

1. Solve for x: $5(x - 2) - 4 = 3x + 8$

2. Graph $y = \sqrt{7 - x}$

3. Find the domain $y = \sqrt{25 - x^2}$

4. Write the standard equation of the circle with center $(-3, 5)$ and radius 6.

5. Find the equation of the line through $(2, 4)$ and $(3, -1)$.

6. Find the equation of the line with slope $m = 4/3$ and y-intercept $b = -3$.

7. Find the equation of the line through $(4, 1)$ perpendicular to the line $2x + 3y = 0$.

8. If it costs a company \$32 to produce 5 units of a product and \$44 to produce 9 units, how much does it cost to produce 20 units? (Assume that the cost function is linear.)

9. Given $f(x) = x^2 - 2x + 1$, find $f(x - 3)$.

10. Given $f(x) = 4x - 11$, find $\dfrac{f(x) - f(3)}{x - 3}$

11. Find the domain and range of $f(x) = \sqrt{36 - x^2}$.

12. Which equations determine y as a function of x?

(a) $6x - 5y + 4 = 0$

(b) $x^2 + y^2 = 9$

(c) $y^3 = x^2 + 6$

13. Sketch the graph of $f(x) = x^2 - 5$.

14. Sketch the graph of $f(x) = |x + 3|$.

15. Sketch the graph of $f(x) = \begin{cases} 2x + 1 & \text{if } x \geq 0, \\ x^2 - x & \text{if } x < 0. \end{cases}$

16. Use the graph of $f(x) = |x|$ to graph the following:

(a) $f(x + 2)$

(b) $-f(x) + 2$

17. Given $f(x) = 3x + 7$ and $g(x) = 2x^2 - 5$, find the following:

(a) $(g - f)(x)$

(b) $(fg)(x)$

18. Given $f(x) = x^2 - 2x + 16$ and $g(x) = 2x + 3$, find $f(g(x))$.

19. Given $f(x) = x^3 + 7$, find $f^{-1}(x)$.

20. Which of the following functions have inverses?

(a) $f(x) = |x - 6|$

(b) $f(x) = ax + b,\ a \neq 0$

(c) $f(x) = x^3 - 19$

21. Given $f(x) = \sqrt{\dfrac{3 - x}{x}}$, $0 < x \leq 3$, find $f^{-1}(x)$.

Exercises 22–24, true or false?

22. $y = 3x + 7$ and $y = \frac{1}{3}x - 4$ are perpendicular.

23. $(f \circ g)^{-1} = g^{-1} \circ f^{-1}$

24. If a function has an inverse, then it must pass both the vertical line test and the horizontal line test.

25. Use your calculator to find the least square regression line for the data.

x	-2	-1	0	1	2	3
y	1	2.4	3	3.1	4	4.7

CHAPTER 1
Trigonometry

CHAPTER 1
Trigonometry

Section 1.1 Radian and Degree Measure

You should know the following basic facts about angles, their measurement, and their applications.

- Types of Angles:
  - (a) Acute: Measure between 0° and 90°.
  - (b) Right: Measure 90°.
  - (c) Obtuse: Measure between 90° and 180°.
  - (d) Straight: Measure 180°.
- α and β are complementary if $\alpha + \beta = 90°$. They are supplementary if $\alpha + \beta = 180°$.
- Two angles in standard position that have the same terminal side are called coterminal angles.
- To convert degrees to radians, use $1° = \pi/180$ radians.
- To convert radians to degrees, use 1 radian $= (180/\pi)°$.
- $1' =$ one minute $= 1/60$ of $1°$.
- $1'' =$ one second $= 1/60$ of $1' = 1/3600$ of $1°$.
- The length of a circular arc is $s = r\theta$ where θ is measured in radians.
- Linear speed $= \dfrac{\text{arc length}}{\text{time}} = \dfrac{s}{t}$
- Angular speed $= \theta/t = s/rt$

Solutions to Odd-Numbered Exercises

1.

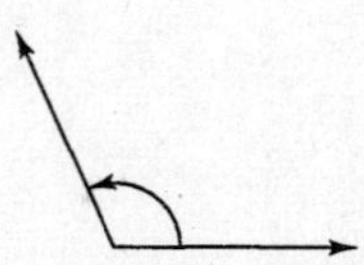

The angle shown is approximately 2 radians.

3.

The angle shown is approximately -3 radians.

5.

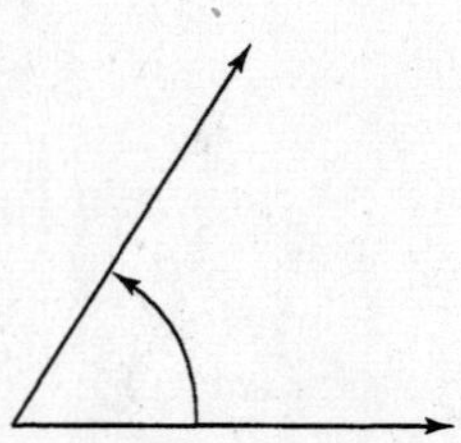

The angle shown is approximately 1 radian.

7. (a) Since $0 < \frac{\pi}{5} < \frac{\pi}{2}$; $\frac{\pi}{5}$ lies in Quadrant I.

(b) Since $\pi < \frac{7\pi}{5} < \frac{3\pi}{2}$; $\frac{7\pi}{5}$ lies in Quadrant III.

9. (a) Since $-\frac{\pi}{2} < -\frac{\pi}{12} < 0$; $-\frac{\pi}{12}$ lies in Quadrant IV.

(b) Since $-\pi < -2 < -\frac{\pi}{2}$; -2 lies in Quadrant III.

11. (a) Since $\pi < 3.5 < \frac{3\pi}{2}$; 3.5 lies in Quadrant III.

(b) Since $\frac{\pi}{2} < 2.25 < \pi$; 2.25 lies in Quadrant II.

13. (a)

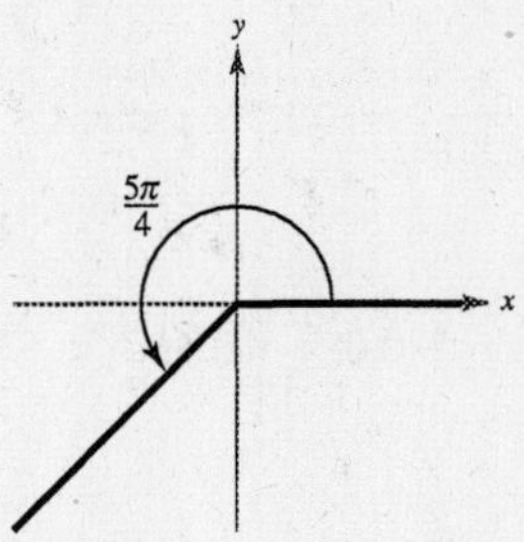

(b) $-\frac{2\pi}{3}$

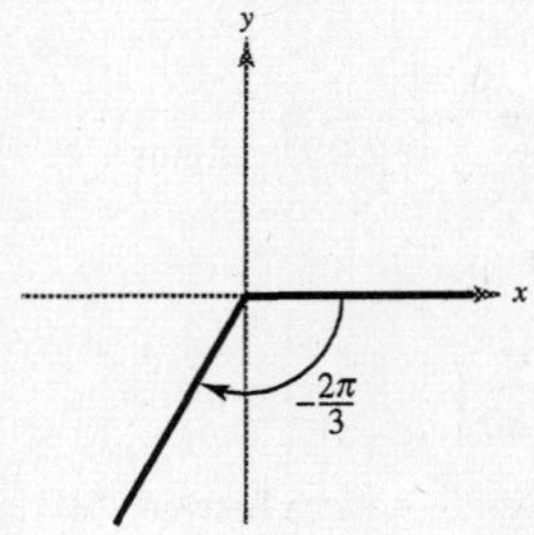

15. (a)

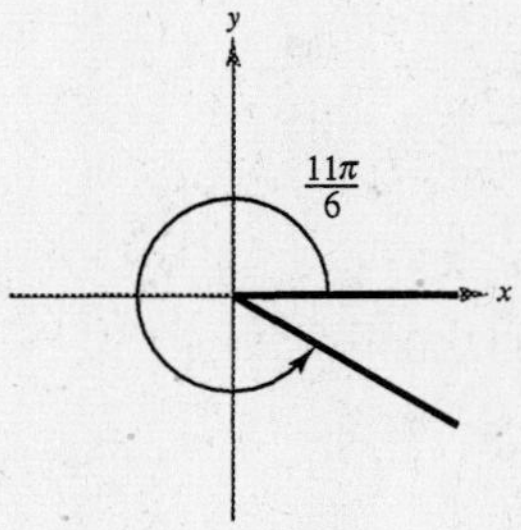

(b) -3

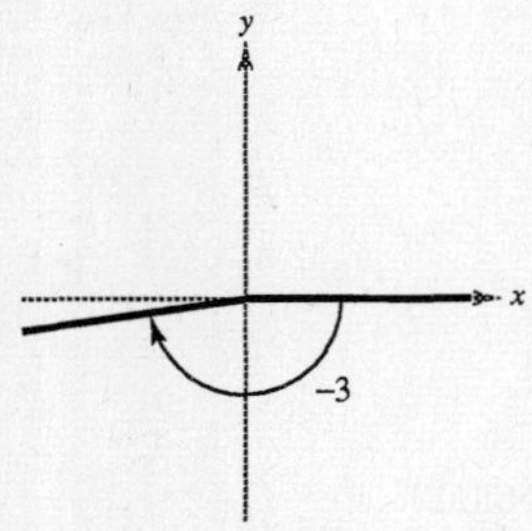

17. (a) Coterminal angles for $\frac{\pi}{6}$

$$\frac{\pi}{6} + 2\pi = \frac{13\pi}{6}$$

$$\frac{\pi}{6} - 2\pi = -\frac{11\pi}{6}$$

(b) Coterminal angles for $\frac{5\pi}{6}$

$$\frac{5\pi}{6} + 2\pi = \frac{17\pi}{6}$$

$$\frac{5\pi}{6} - 2\pi = -\frac{7\pi}{6}$$

19. (a) Coterminal angles for $\frac{2\pi}{3}$

$$\frac{2\pi}{3} + 2\pi = \frac{8\pi}{3}$$

$$\frac{2\pi}{3} - 2\pi = -\frac{4\pi}{3}$$

(b) Coterminal angles for $\frac{\pi}{12}$

$$\frac{\pi}{12} + 2\pi = \frac{25\pi}{12}$$

$$\frac{\pi}{12} - 2\pi = -\frac{23\pi}{12}$$

21. (a) Complement: $\frac{\pi}{2} - \frac{\pi}{3} = \frac{\pi}{6}$

Supplement: $\pi - \frac{\pi}{3} = \frac{2\pi}{3}$

(b) Complement: Not possible; $\frac{3\pi}{4}$ is greater than $\frac{\pi}{2}$.

Supplement: $\pi - \frac{3\pi}{4} = \frac{\pi}{4}$

23. (a) Complement: $\frac{\pi}{2} - 1 \approx 0.57$

Supplement: $\pi - 1 \approx 2.14$

(b) Complement: Not possible. 2 is greater than $\frac{\pi}{2}$.

Supplement: $\pi - 2 \approx 1.14$

25. (a) $30° = 30\left(\frac{\pi}{180}\right) = \frac{\pi}{6}$

(b) $150° = 150\left(\frac{\pi}{180}\right) = \frac{5\pi}{6}$

27. (a) $-20° = -20\left(\frac{\pi}{180}\right) = -\frac{\pi}{9}$

(b) $-240° = -240\left(\frac{\pi}{180}\right) = -\frac{4\pi}{3}$

29. $115° = 115\left(\frac{\pi}{180}\right) \approx 2.007$ radians

31. $-216.35° = -216.35\left(\frac{\pi}{180}\right) \approx -3.776$ radians

33. $532° = 532\left(\frac{\pi}{180}\right) \approx 9.285$ radians

35. $-0.83° = -0.83\left(\frac{\pi}{180}\right) \approx -0.014$ radian

37. (a) $\frac{3\pi}{2} = \frac{3\pi}{2}\left(\frac{180}{\pi}\right)^{\circ} = 270°$

(b) $\frac{7\pi}{6} = \frac{7\pi}{6}\left(\frac{180}{\pi}\right)^{\circ} = 210°$

39. (a) $\frac{7\pi}{3} = \frac{7\pi}{3}\left(\frac{180}{\pi}\right)^{\circ} = 420°$

(b) $-\frac{11\pi}{30} = -\frac{11\pi}{30}\left(\frac{180}{\pi}\right)^{\circ} = -66°$

41. $\frac{\pi}{7} = \frac{\pi}{7}\left(\frac{180}{\pi}\right)^{\circ} \approx 25.714°$

43. $\frac{15\pi}{8} = \frac{15\pi}{8}\left(\frac{180}{\pi}\right)^{\circ} = 337.500°$

45. $-4.2\pi = -4.2\pi\left(\frac{180}{\pi}\right)^{\circ} = -756.000°$

47. $-2 = -2\left(\frac{180}{\pi}\right)^{\circ} \approx -114.592°$

49.

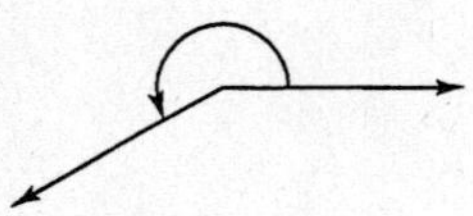

The angle shown is approximately 210°.

51.

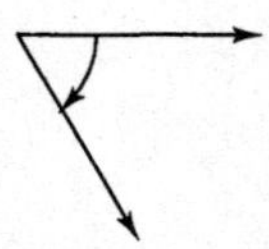

The angle shown is approximately $-60°$.

53.

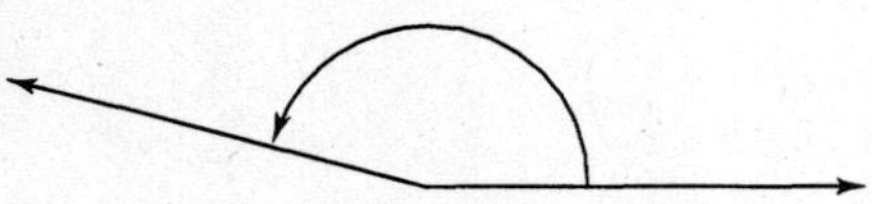

The angle shown is approximately 165°.

55. (a) Since $90° < 130° < 180°$; 130° lies in Quadrant II.

(b) Since $270° < 285° < 360°$; 285° lies in Quadrant IV.

57. (a) Since $-180° < -132°50' < -90°$; $-132° 50'$ lies in Quadrant III.

(b) Since $-360° < -336° < -270°$; $-336°$ lies in Quadrant I.

59. (a)

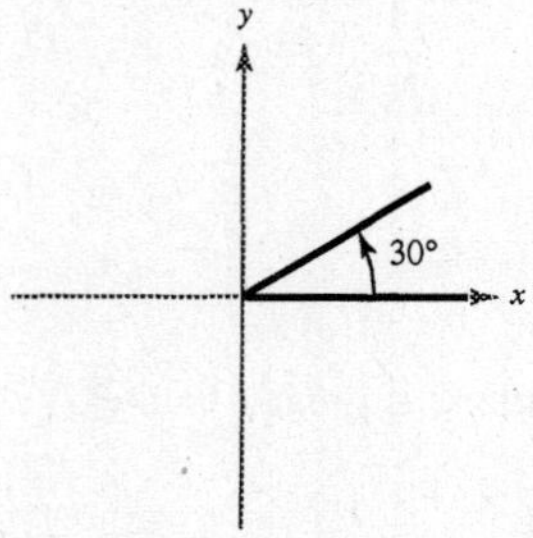

(b)

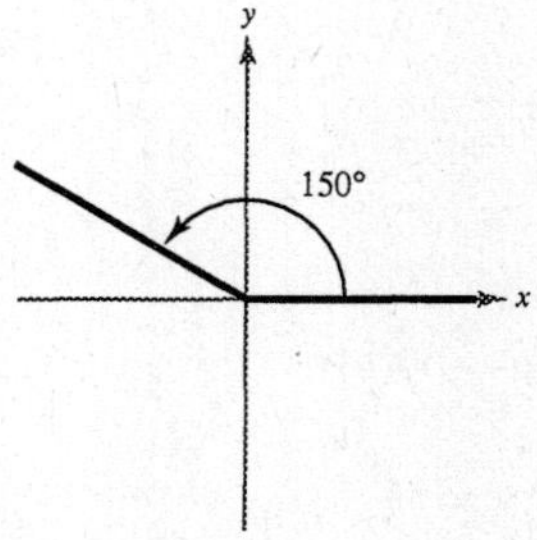

61. (a)

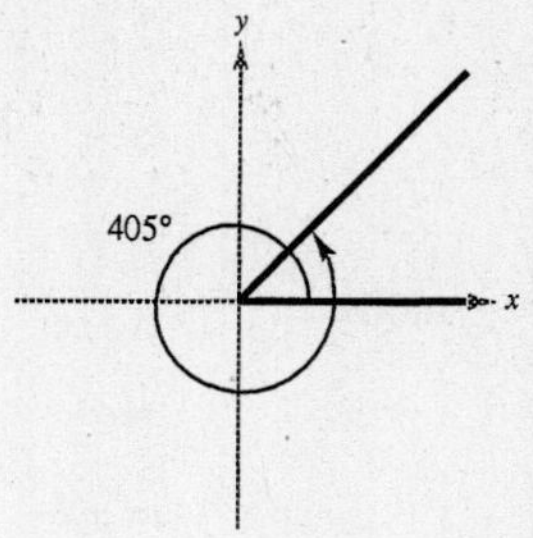

(b)

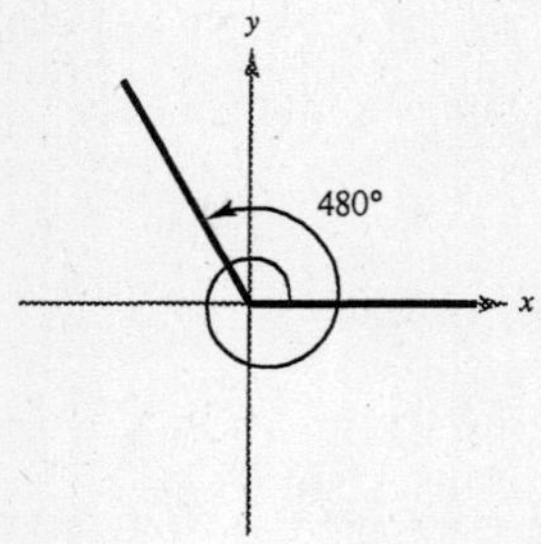

63. (a) Coterminal angles for 45°

$45° + 360° = 405°$

$45° - 360° = -315°$

(b) Coterminal angles for $-36°$

$-36° + 360° = 324°$

$-36° - 360° = -396°$

65. (a) Coterminal angles for 240°

$240° + 360° = 600°$

$240° - 360° = -120°$

(b) Coterminal angles for $-180°$

$-180° + 360° = 180°$

$-180° - 360° = -540°$

67. (a) Complement: $90° - 18° = 72°$

Supplement: $180° - 18° = 162°$

(b) Complement: Not possible; 115° is greater than 90°.

Supplement: $180° - 115° = 65°$

69. (a) Complement: $90° - 79° = 11°$

Supplement: $180° - 79° = 101°$

(b) Complement: Not possible. 150° is greater than 90°.

Supplement: $180° - 150° = 30°$

71. (a) $54° 45' = 54° + \left(\frac{45}{60}\right)° = 54.75°$

(b) $-128° 30' = -128° - \left(\frac{30}{60}\right)° = -128.5°$

73. (a) $85° 18' 30'' = \left(85 + \frac{18}{60} + \frac{30}{3600}\right)° \approx 85.308°$

(b) $330° 25'' = \left(330 + \frac{25}{3600}\right)° \approx 330.007°$

75. (a) $240.6° = 240° + 0.6(60)' = 240° 36'$

(b) $-145.8° = -[145° + 0.8(60')] = -145° 48'$

77. (a) $2.5° = 2° 30'$

(b) $-3.58° = -3° 34' 48''$

79. $s = r\theta$

$6 = 5\theta$

$\theta = \frac{6}{5}$ radians

81. $s = r\theta$

$32 = 7\theta$

$\theta = \frac{32}{7} = 4\frac{4}{7}$ radians

83. $s = r\theta$

$6 = 27\theta$

$\theta = \frac{6}{27} = \frac{2}{9}$ radian

85. $s = r\theta$

$25 = 14.5\theta$

$\theta = \frac{25}{14.5} = \frac{50}{29}$ radians

87. $s = r\theta$, θ in radians

$$s = 15(180)\left(\frac{\pi}{180}\right) = 15\pi \text{ inches}$$

$$\approx 47.12 \text{ inches}$$

89. $s = r\theta$, θ in radians

$s = 3(1) = 3$ meters

91. $\theta = 41° 15' 42'' - 32° 47' 9'' = 8° 28' 33'' \approx 8.47583° \approx 0.14793$ radian

$s = r\theta = 4000(0.14793) \approx 591.72$ miles

93. $\theta = 42° \, 7' \, 15'' - 25° \, 46' \, 37'' = 16° \, 20' \, 38'' \approx 0.285255$ radian

$s = r\theta = 4000(0.285255) \approx 1141.02$ miles

95. $\theta = \dfrac{s}{r} = \dfrac{450}{6378} \approx 0.071 \text{ radian} \approx 4.04°$

97. $\theta = \dfrac{s}{r} = \dfrac{2.5}{6} = \dfrac{25}{60} = \dfrac{5}{12}$ radian

99. (a) 65 miles per hour $= \dfrac{65(5280)}{60} = 5720$ feet per minute

The circumference of the tire is $C = 2.5\pi$ feet.

The number of revolutions per minute is $r = \dfrac{5720}{2.5\pi} \approx 728.3$ revolutions per minute

(b) The angular speed is $\dfrac{\theta}{t}$.

$$\theta = \frac{5720}{2.5\pi}(2\pi) = 4576 \text{ radians}$$

$$\text{Angular speed} = \frac{4576 \text{ radians}}{1 \text{ minute}} = 4576 \text{ radians per minute}$$

101. (a) Arc length of larger sprocket in feet:

$$s = r\theta$$

$$s = \frac{1}{3}(2\pi) = \frac{2\pi}{3} \text{ feet}$$

Therefore, the chain moves $\dfrac{2\pi}{3}$ feet as does the smaller rear sprocket.

Thus, the angle θ of the smaller sprocket is $\left(r = 2 \text{ inches} = \dfrac{2}{12} \text{ feet}\right)$

$$\theta = \frac{s}{r} = \frac{\frac{2\pi}{3} \text{ ft}}{\frac{2}{12} \text{ ft}} = 4\pi$$

and the arc length of the tire in feet is:

$$s = \theta r$$

$$s = (4\pi)\left(\frac{14}{12}\right) = \frac{14\pi}{3} \text{ feet}$$

$$\text{Speed} = \frac{s}{t} = \frac{\frac{14\pi}{3}}{1 \text{ sec}} = \frac{14\pi}{3} \text{ feet per second}$$

$$\frac{14\pi \text{ feet}}{3 \text{ seconds}} \times \frac{3600 \text{ seconds}}{1 \text{ hour}} \times \frac{1 \text{ mile}}{5280 \text{ feet}} \approx 10 \text{ miles per hour}$$

(b) Since the arc length of the tire is $\dfrac{14\pi}{3}$ feet and the cyclist is pedaling at a rate of one revolution per second, we have

$$\text{Distance} = \left(\frac{14\pi}{3}\frac{\text{feet}}{\text{rev}}\right)\left(\frac{1 \text{ mile}}{5280 \text{ feet}}\right)(n \text{ rev.}) = \frac{7\pi}{7920} n \text{ miles}$$

(c) Distance = Rate · Time

$$= \left(\frac{14\pi}{3} \text{ ft/sec}\right)\left(\frac{1 \text{ mile}}{5280 \text{ feet}}\right)(t \text{ sec.}) = \frac{7\pi}{7920} t \text{ miles}$$

(d) The functions are both linear.

103. False. A measurement of 4π radians corresponds to two complete revolutions from the initial to the terminal side of an angle.

105. False. The terminal side of $-1260°$ lies on the negative x-axis.

107. Increases, since the linear speed is proportional to the radius.

109. The arc length is increasing. In order for the angle θ to remain constant as the radius r increases, the arc length s must increase in proportion to r, as can be seen from the formula $s = r\theta$.

111. $f(x) = (x - 2)^5$

Graph of $y = x^5$ shifted to the right by 2 units.

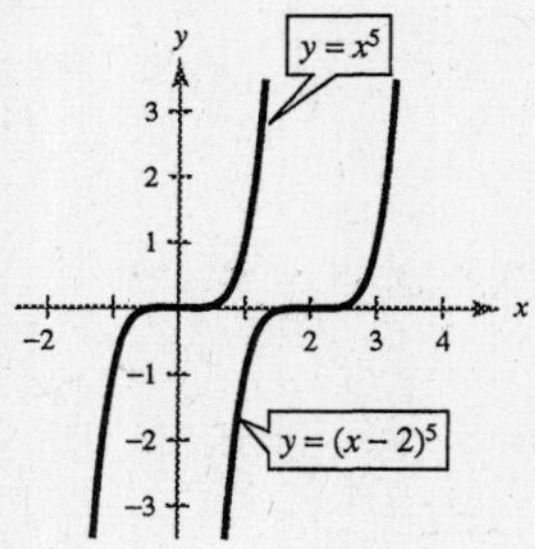

113. $f(x) = 2 - x^5$

Graph of $y = x^5$ reflected in x-axis and shifted upward by 2 units.

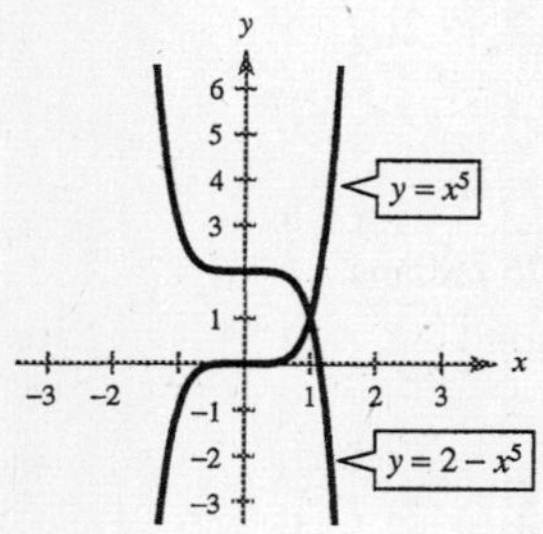

115. $\dfrac{4}{4\sqrt{2}} = \dfrac{4}{4\sqrt{2}} \cdot \dfrac{\sqrt{2}}{\sqrt{2}} = \dfrac{4\sqrt{2}}{8} = \dfrac{\sqrt{2}}{2}$

117. $\dfrac{2\sqrt{3}}{\sqrt{6}} = 2\sqrt{\dfrac{3}{6}} = 2\sqrt{\dfrac{1}{2}} = \dfrac{2}{\sqrt{2}} \cdot \dfrac{\sqrt{2}}{\sqrt{2}} = \sqrt{2}$

119. $\sqrt{2^2 + 6^2} = \sqrt{4 + 36} = \sqrt{40} = \sqrt{4 \cdot 10} = 2\sqrt{10}$

121. $\sqrt{18^2 - 6^2} = \sqrt{324 - 36} = \sqrt{288} = \sqrt{144 \cdot 2} = 12\sqrt{2}$

Section 1.2 Trigonometric Functions: The Unit Circle

- You should know the definition of the trigonometric functions in terms of the unit circle. Let t be a real number and (x, y) the point on the unit circle corresponding to t.

  $\sin t = y \qquad \csc t = \dfrac{1}{y},\quad y \neq 0$

  $\cos t = x \qquad \sec t = \dfrac{1}{x},\quad x \neq 0$

  $\tan t = \dfrac{y}{x},\quad x \neq 0 \qquad \cot t = \dfrac{x}{y},\quad y \neq 0$

- The cosine and secant functions are even.

  $\cos(-t) = \cos t \qquad \sec(-t) = \sec t$

- The other four trigonometric functions are odd.

  $\sin(-t) = -\sin t \qquad \csc(-t) = -\csc t$

  $\tan(-t) = -\tan t \qquad \cot(-t) = -\cot t$

- Be able to evaluate the trigonometric functions with a calculator.

Solutions to Odd-Numbered Exercises

1. $x = -\frac{8}{17}, \quad y = \frac{15}{17}$

$\sin\theta = y = \frac{15}{17}$ $\qquad \csc\theta = \frac{1}{y} = \frac{17}{15}$

$\cos\theta = x = -\frac{8}{17}$ $\qquad \sec\theta = \frac{1}{x} = -\frac{17}{8}$

$\tan\theta = \frac{y}{x} = -\frac{15}{8}$ $\qquad \cot\theta = \frac{x}{y} = -\frac{8}{15}$

3. $x = \frac{12}{13}, \quad y = -\frac{5}{13}$

$\sin\theta = y = -\frac{5}{13}$ $\qquad \csc\theta = \frac{1}{y} = -\frac{13}{5}$

$\cos\theta = x = \frac{12}{13}$ $\qquad \sec\theta = \frac{1}{x} = \frac{13}{12}$

$\tan\theta = \frac{y}{x} = -\frac{5}{12}$ $\qquad \cot\theta = \frac{x}{y} = -\frac{12}{5}$

5. $t = \frac{\pi}{4}$ corresponds to $\left(\frac{\sqrt{2}}{2}, \frac{\sqrt{2}}{2}\right)$.

7. $t = \frac{7\pi}{6}$ corresponds to $\left(-\frac{\sqrt{3}}{2}, -\frac{1}{2}\right)$.

9. $t = \frac{4\pi}{3}$ corresponds to $\left(-\frac{1}{2}, -\frac{\sqrt{3}}{2}\right)$.

11. $t = \frac{3\pi}{2}$ corresponds to $(0, -1)$.

13. $t = \frac{\pi}{4}$ corresponds to $\left(\frac{\sqrt{2}}{2}, \frac{\sqrt{2}}{2}\right)$.

$\sin t = y = \frac{\sqrt{2}}{2}$

$\cos t = x = \frac{\sqrt{2}}{2}$

$\tan t = \frac{y}{x} = 1$

15. $t = -\frac{\pi}{6}$ corresponds to $\left(\frac{\sqrt{3}}{2}, -\frac{1}{2}\right)$.

$\sin t = y = -\frac{1}{2}$

$\cos t = x = \frac{\sqrt{3}}{2}$

$\tan t = \frac{y}{x} = -\frac{1}{\sqrt{3}} = -\frac{\sqrt{3}}{3}$

17. $t = -\frac{7\pi}{4}$ corresponds to $\left(\frac{\sqrt{2}}{2}, \frac{\sqrt{2}}{2}\right)$.

$\sin t = y = \frac{\sqrt{2}}{2}$

$\cos t = x = \frac{\sqrt{2}}{2}$

$\tan t = \frac{y}{x} = 1$

19. $t = \frac{11\pi}{6}$ corresponds to $\left(\frac{\sqrt{3}}{2}, -\frac{1}{2}\right)$.

$\sin t = y = -\frac{1}{2}$

$\cos t = x = \frac{\sqrt{3}}{2}$

$\tan t = \frac{y}{x} = -\frac{1}{\sqrt{3}} = -\frac{\sqrt{3}}{3}$

21. $t = -\frac{3\pi}{2}$ corresponds to $(0, 1)$.

$\sin t = y = 1$

$\cos t = x = 0$

$\tan t = \frac{y}{x}$ is undefined.

23. $t = \frac{3\pi}{4}$ corresponds to $\left(-\frac{\sqrt{2}}{2}, \frac{\sqrt{2}}{2}\right)$.

$\sin t = y = \frac{\sqrt{2}}{2}$ $\qquad \csc t = \frac{1}{y} = \sqrt{2}$

$\cos t = x = -\frac{\sqrt{2}}{2}$ $\qquad \sec t = \frac{1}{x} = -\sqrt{2}$

$\tan t = \frac{y}{x} = -1$ $\qquad \cot t = \frac{x}{y} = -1$

25. $t = -\frac{\pi}{2}$ corresponds to $(0, -1)$.

$\sin t = y = -1$ $\qquad$ $\csc t = \frac{1}{y} = -1$

$\cos t = x = 0$ $\qquad$ $\sec t = \frac{1}{x}$ is undefined.

$\tan t = \frac{y}{x}$ is undefined. $\qquad$ $\cot t = \frac{x}{y} = 0$

27. $t = \frac{4\pi}{3}$ corresponds to $\left(-\frac{1}{2}, -\frac{\sqrt{3}}{2}\right)$.

$\sin t = y = -\frac{\sqrt{3}}{2}$ $\qquad$ $\csc t = \frac{1}{y} = -\frac{2\sqrt{3}}{3}$

$\cos t = x = -\frac{1}{2}$ $\qquad$ $\sec t = \frac{1}{x} = -2$

$\tan t = \frac{y}{x} = \sqrt{3}$ $\qquad$ $\cot t = \frac{x}{y} = \frac{\sqrt{3}}{3}$

29. $\sin 5\pi = \sin \pi = 0$

31. $\cos \frac{8\pi}{3} = \cos \frac{2\pi}{3} = -\frac{1}{2}$

33. $\cos\left(-\frac{15\pi}{2}\right) = \cos\left(\frac{\pi}{2}\right) = 0$

35. $\sin\left(-\frac{9\pi}{4}\right) = \sin\left(\frac{7\pi}{4}\right) = -\frac{\sqrt{2}}{2}$

37. $\sin t = \frac{1}{3}$

(a) $\sin(-t) = -\sin t = -\frac{1}{3}$

(b) $\csc(-t) = -\csc t = -3$

39. $\cos(-t) = -\frac{1}{5}$

(a) $\cos t = \cos(-t) = -\frac{1}{5}$

(b) $\sec(-t) = \frac{1}{\cos(-t)} = -5$

41. $\sin t = \frac{4}{5}$

(a) $\sin(\pi - t) = \sin t = \frac{4}{5}$

(b) $\sin(t + \pi) = -\sin t = -\frac{4}{5}$

43. $\sin \frac{\pi}{4} \approx 0.7071$

45. $\csc 1.3 = \frac{1}{\sin 1.3} \approx 1.0378$

47. $\cos(-1.7) \approx -0.1288$

49. $\csc 0.8 = \frac{1}{\sin 0.8} \approx 1.3940$

51. $\sec 22.8 = \frac{1}{\cos 22.8} \approx -1.4486$

53. (a) $\sin 5 \approx -1$

(b) $\cos 2 \approx -0.4$

55. (a) $\sin t = 0.25$

$t \approx 0.25$ or 2.89

(b) $\cos t = -0.25$

$t \approx 1.82$ or 4.46

57. $y(t) = 3 \sin\left(\frac{\pi}{4}t\right)$

(a)

t	0	$\frac{1}{2}$	1	$\frac{3}{2}$	2
y	0	1.1481	2.1213	2.7716	3

(b) The displacement is maximum when $t = 2 + 8n$, where n is any integer.

(c) $y(t) = 0$ when $t = 4n$, where n is any integer.

59. False. $\sin(-t) = -\sin t$ means the function is odd, not that the sine of a negative angle is a negative number.

For example: $\sin\left(-\frac{3\pi}{2}\right) = -\sin\left(\frac{3\pi}{2}\right) = -(-1) = 1.$

Even though the angle is negative, the sine value is positive.

61. (a) The points have y-axis symmetry.

(b) $\sin t_1 = \sin(\pi - t_1)$ since they have the same y-value.

(c) $\cos(\pi - t_1) = -\cos t_1$ since the x-values have the opposite signs.

63. $\sin 1 = \sin(0.25 + 0.75) \approx 0.8415$

$\sin 0.25 + \sin 0.75 \approx 0.2474 + 0.6816$

≈ 0.9290

65.
$$f(x) = \tfrac{1}{2}(3x - 2)$$
$$y = \tfrac{1}{2}(3x - 2)$$
$$x = \tfrac{1}{2}(3y - 2)$$
$$2x = 3y - 2$$
$$\tfrac{2}{3}x + \tfrac{2}{3} = y$$
$$f^{-1}(x) = \tfrac{2}{3}x + \tfrac{2}{3}$$

67.
$$f(x) = \sqrt{x^2 - 4},\ x \ge 2$$
$$y = \sqrt{x^2 - 4}$$
$$x = \sqrt{y^2 - 4}$$
$$x^2 = y^2 - 4$$
$$\pm\sqrt{x^2 + 4} = y$$
$$f^{-1}(x) = \sqrt{x^2 + 4},\ x \ge 0$$

Section 1.3 Right Triangle Trigonometry

- You should know the right triangle definition of trigonometric functions.

(a) $\sin\theta = \dfrac{\text{opp}}{\text{hyp}}$ (b) $\cos\theta = \dfrac{\text{adj}}{\text{hyp}}$ (c) $\tan\theta = \dfrac{\text{opp}}{\text{adj}}$

(d) $\csc\theta = \dfrac{\text{hyp}}{\text{opp}}$ (e) $\sec\theta = \dfrac{\text{hyp}}{\text{adj}}$ (f) $\cot\theta = \dfrac{\text{adj}}{\text{opp}}$

- You should know the following identities.

(a) $\sin\theta = \dfrac{1}{\csc\theta}$ (b) $\csc\theta = \dfrac{1}{\sin\theta}$ (c) $\cos\theta = \dfrac{1}{\sec\theta}$

(d) $\sec\theta = \dfrac{1}{\cos\theta}$ (e) $\tan\theta = \dfrac{1}{\cot\theta}$ (f) $\cot\theta = \dfrac{1}{\tan\theta}$

(g) $\tan\theta = \dfrac{\sin\theta}{\cos\theta}$ (h) $\cot\theta = \dfrac{\cos\theta}{\sin\theta}$ (i) $\sin^2\theta + \cos^2\theta = 1$

(j) $1 + \tan^2\theta = \sec^2\theta$ (k) $1 + \cot^2\theta = \csc^2\theta$

- You should know that two acute angles α and β are complementary if $\alpha + \beta = 90°$, and that cofunctions of complementary angles are equal.
- You should know the trigonometric function values of 30°, 45°, and 60°, or be able to construct triangles from which you can determine them.

Solutions to Odd-Numbered Exercises

1. $\text{hyp} = \sqrt{6^2 + 8^2} = \sqrt{36 + 64} = \sqrt{100} = 10$

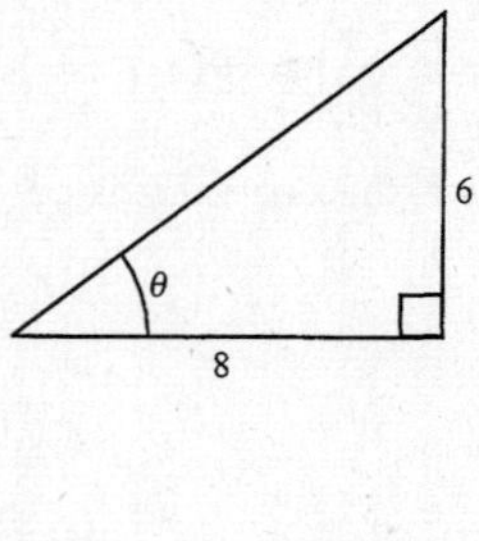

$\sin\theta = \dfrac{\text{opp}}{\text{hyp}} = \dfrac{6}{10} = \dfrac{3}{5}$ $\csc\theta = \dfrac{\text{hyp}}{\text{opp}} = \dfrac{10}{6} = \dfrac{5}{3}$

$\cos\theta = \dfrac{\text{adj}}{\text{hyp}} = \dfrac{8}{10} = \dfrac{4}{5}$ $\sec\theta = \dfrac{\text{hyp}}{\text{adj}} = \dfrac{10}{8} = \dfrac{5}{4}$

$\tan\theta = \dfrac{\text{opp}}{\text{adj}} = \dfrac{6}{8} = \dfrac{3}{4}$ $\cot\theta = \dfrac{\text{adj}}{\text{opp}} = \dfrac{8}{6} = \dfrac{4}{3}$

3. adj $= \sqrt{41^2 - 9^2} = \sqrt{1681 - 81} = \sqrt{1600} = 40$

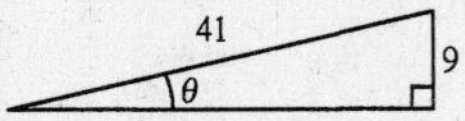

$\sin\theta = \dfrac{\text{opp}}{\text{hyp}} = \dfrac{9}{41}$ $\qquad \csc\theta = \dfrac{\text{hyp}}{\text{opp}} = \dfrac{41}{9}$

$\cos\theta = \dfrac{\text{adj}}{\text{hyp}} = \dfrac{40}{41}$ $\qquad \sec\theta = \dfrac{\text{hyp}}{\text{adj}} = \dfrac{41}{40}$

$\tan\theta = \dfrac{\text{opp}}{\text{adj}} = \dfrac{9}{40}$ $\qquad \cot\theta = \dfrac{\text{adj}}{\text{opp}} = \dfrac{40}{9}$

5. adj $= \sqrt{3^2 - 1^2} = \sqrt{8} = 2\sqrt{2}$

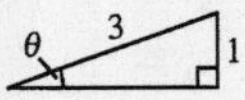

$\sin\theta = \dfrac{\text{opp}}{\text{hyp}} = \dfrac{1}{3}$ $\qquad \csc\theta = \dfrac{\text{hyp}}{\text{opp}} = 3$

$\cos\theta = \dfrac{\text{adj}}{\text{hyp}} = \dfrac{2\sqrt{2}}{3}$ $\qquad \sec\theta = \dfrac{\text{hyp}}{\text{adj}} = \dfrac{3}{2\sqrt{2}} = \dfrac{3\sqrt{2}}{4}$

$\tan\theta = \dfrac{\text{opp}}{\text{adj}} = \dfrac{1}{2\sqrt{2}} = \dfrac{\sqrt{2}}{4}$ $\qquad \cot\theta = \dfrac{\text{adj}}{\text{opp}} = 2\sqrt{2}$

adj $= \sqrt{6^2 - 2^2} = \sqrt{32} = 4\sqrt{2}$

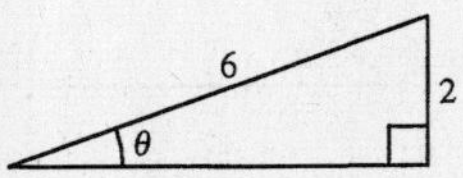

$\sin\theta = \dfrac{\text{opp}}{\text{hyp}} = \dfrac{2}{6} = \dfrac{1}{3}$ $\qquad \csc\theta = \dfrac{\text{hyp}}{\text{opp}} = \dfrac{6}{2} = 3$

$\cos\theta = \dfrac{\text{adj}}{\text{hyp}} = \dfrac{4\sqrt{2}}{6} = \dfrac{2\sqrt{2}}{3}$ $\qquad \sec\theta = \dfrac{\text{hyp}}{\text{adj}} = \dfrac{6}{4\sqrt{2}} = \dfrac{3}{2\sqrt{2}} = \dfrac{3\sqrt{2}}{4}$

$\tan\theta = \dfrac{\text{opp}}{\text{adj}} = \dfrac{2}{4\sqrt{2}} = \dfrac{1}{2\sqrt{2}} = \dfrac{\sqrt{2}}{4}$ $\qquad \cot\theta = \dfrac{\text{adj}}{\text{opp}} = \dfrac{4\sqrt{2}}{2} = 2\sqrt{2}$

The function values are the same since the triangles are similar and the corresponding sides are proportional.

7. opp $= \sqrt{5^2 - 4^2} = 3$

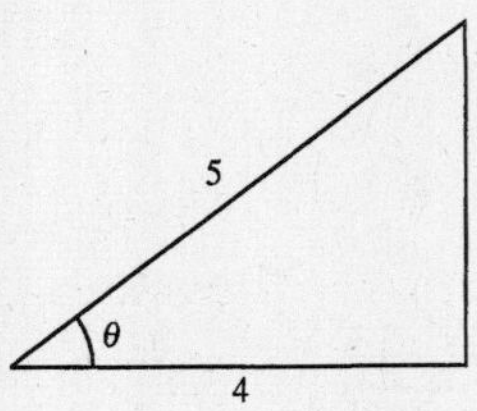

$\sin\theta = \dfrac{\text{opp}}{\text{hyp}} = \dfrac{3}{5}$ $\qquad \csc\theta = \dfrac{\text{hyp}}{\text{opp}} = \dfrac{5}{3}$

$\cos\theta = \dfrac{\text{adj}}{\text{hyp}} = \dfrac{4}{5}$ $\qquad \sec\theta = \dfrac{\text{hyp}}{\text{adj}} = \dfrac{5}{4}$

$\tan\theta = \dfrac{\text{opp}}{\text{adj}} = \dfrac{3}{4}$ $\qquad \cot\theta = \dfrac{\text{adj}}{\text{opp}} = \dfrac{4}{3}$

opp $= \sqrt{1.25^2 - 1^2} = 0.75$

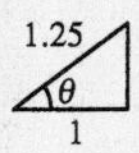

$\sin\theta = \dfrac{\text{opp}}{\text{hyp}} = \dfrac{0.75}{1.25} = \dfrac{3}{5}$ $\qquad \csc\theta = \dfrac{\text{hyp}}{\text{opp}} = \dfrac{1.25}{0.75} = \dfrac{5}{3}$

$\cos\theta = \dfrac{\text{adj}}{\text{hyp}} = \dfrac{1}{1.25} = \dfrac{4}{5}$ $\qquad \sec\theta = \dfrac{\text{hyp}}{\text{adj}} = \dfrac{1.25}{1} = \dfrac{5}{4}$

$\tan\theta = \dfrac{\text{opp}}{\text{adj}} = \dfrac{0.75}{1} = \dfrac{3}{4}$ $\qquad \cot\theta = \dfrac{\text{adj}}{\text{opp}} = \dfrac{1}{0.75} = \dfrac{4}{3}$

The function values are the same since the triangles are similar and the corresponding sides are proportional.

9. Given: $\sin\theta = \frac{3}{4} = \frac{\text{opp}}{\text{hyp}}$

$3^2 + (\text{adj})^2 = 4^2$

$\text{adj} = \sqrt{7}$

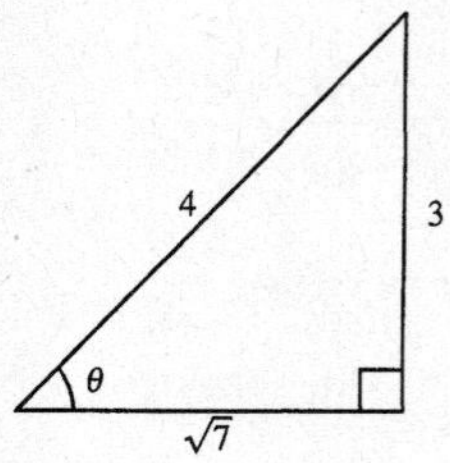

$\cos\theta = \frac{\sqrt{7}}{4}$

$\tan\theta = \frac{3\sqrt{7}}{7}$

$\cot\theta = \frac{\sqrt{7}}{3}$

$\sec\theta = \frac{4\sqrt{7}}{7}$

$\csc\theta = \frac{4}{3}$

11. Given: $\sec\theta = 2 = \frac{2}{1} = \frac{\text{hyp}}{\text{adj}}$

$(\text{opp})^2 + 1^2 = 2^2$

$\text{opp} = \sqrt{3}$

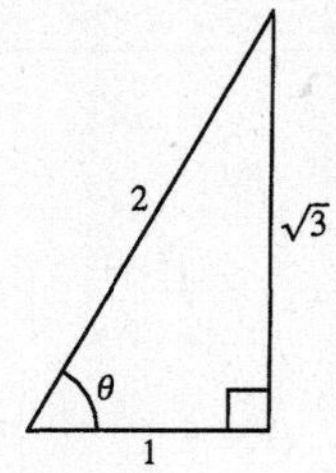

$\sin\theta = \frac{\sqrt{3}}{2}$

$\cos\theta = \frac{1}{2}$

$\tan\theta = \sqrt{3}$

$\cot\theta = \frac{\sqrt{3}}{3}$

$\csc\theta = \frac{2\sqrt{3}}{3}$

13. Given: $\tan\theta = 3 = \frac{3}{1} = \frac{\text{opp}}{\text{adj}}$

$3^2 + 1^2 = (\text{hyp})^2$

$\text{hyp} = \sqrt{10}$

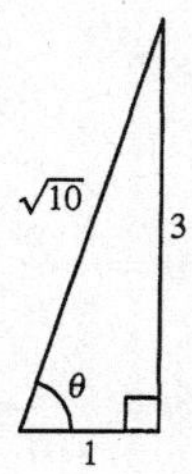

$\sin\theta = \frac{3\sqrt{10}}{10}$

$\cos\theta = \frac{\sqrt{10}}{10}$

$\cot\theta = \frac{1}{3}$

$\sec\theta = \sqrt{10}$

$\csc\theta = \frac{\sqrt{10}}{3}$

15. Given: $\cot\theta = \frac{3}{2} = \frac{\text{adj}}{\text{opp}}$

$2^2 + 3^2 = (\text{hyp})^2$

$\text{hyp} = \sqrt{13}$

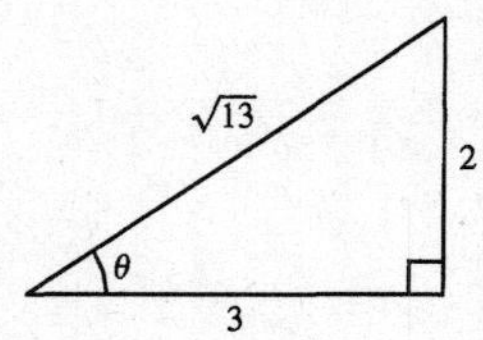

$\sin\theta = \frac{2}{\sqrt{13}} = \frac{2\sqrt{13}}{13}$

$\cos\theta = \frac{3}{\sqrt{13}} = \frac{3\sqrt{13}}{13}$

$\tan\theta = \frac{2}{3}$

$\csc\theta = \frac{\sqrt{13}}{2}$

$\sec\theta = \frac{\sqrt{13}}{3}$

17. $\sin 60° = \frac{\sqrt{3}}{2}$, $\cos 60° = \frac{1}{2}$

(a) $\tan 60° = \frac{\sin 60°}{\cos 60°} = \sqrt{3}$

(b) $\sin 30° = \cos 60° = \frac{1}{2}$

(c) $\cos 30° = \sin 60° = \frac{\sqrt{3}}{2}$

(d) $\cot 60° = \frac{\cos 60°}{\sin 60°} = \frac{1}{\sqrt{3}} = \frac{\sqrt{3}}{3}$

19. $\csc\theta = \frac{\sqrt{13}}{2}$, $\sec\theta = \frac{\sqrt{13}}{3}$

(a) $\sin\theta = \frac{1}{\csc\theta} = \frac{2}{\sqrt{13}} = \frac{2\sqrt{13}}{13}$

(b) $\cos\theta = \frac{1}{\sec\theta} = \frac{3}{\sqrt{13}} = \frac{3\sqrt{13}}{13}$

(c) $\tan\theta = \frac{\sin\theta}{\cos\theta} = \frac{\frac{2\sqrt{13}}{13}}{\frac{3\sqrt{13}}{13}} = \frac{2}{3}$

(d) $\sec(90° - \theta) = \csc\theta = \frac{\sqrt{13}}{2}$

21. $\cos\alpha = \frac{1}{3}$

(a) $\sec\alpha = \frac{1}{\cos\alpha} = 3$

—CONTINUED—

21. —CONTINUED—

(b) $\sin^2\alpha + \cos^2\alpha = 1$

$$\sin^2\alpha + \left(\frac{1}{3}\right)^2 = 1$$

$$\sin^2\alpha = \frac{8}{9}$$

$$\sin\alpha = \frac{2\sqrt{2}}{3}$$

(c) $\cot\alpha = \dfrac{\cos\alpha}{\sin\alpha} = \dfrac{\frac{1}{3}}{\frac{2\sqrt{2}}{3}} = \dfrac{1}{2\sqrt{2}} = \dfrac{\sqrt{2}}{4}$

(d) $\sin(90^\circ - \alpha) = \cos\alpha = \dfrac{1}{3}$

23. (a) $\cos 60^\circ = \dfrac{1}{2}$

(b) $\csc 30^\circ = 2$

(c) $\tan 60^\circ = \sqrt{3}$

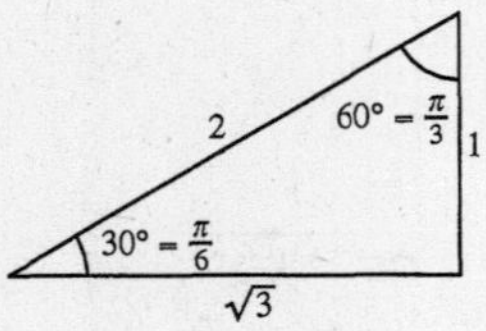

25. (a) $\sin 45^\circ = \dfrac{1}{\sqrt{2}} = \dfrac{\sqrt{2}}{2}$

(b) $\cos 30^\circ = \dfrac{\sqrt{3}}{2}$

(c) $\tan 30^\circ = \dfrac{1}{\sqrt{3}} = \dfrac{\sqrt{3}}{3}$

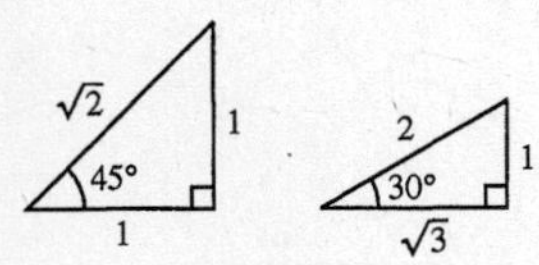

27. (a) $\sin 10^\circ \approx 0.1736$

(b) $\cos 80^\circ \approx 0.1736$

Note: $\cos 80^\circ = \sin(90^\circ - 80^\circ) = \sin 10^\circ$

29. (a) $\sin 16.35^\circ \approx 0.2815$

(b) $\csc 16.35^\circ = \dfrac{1}{\sin 16.35^\circ} \approx 3.5523$

31. (a) $\sec 42^\circ 12' = \sec 42.2^\circ = \dfrac{1}{\cos 42.2^\circ} \approx 1.3499$

(b) $\csc 48^\circ 7' = \dfrac{1}{\sin\left(48 + \frac{7}{60}\right)^\circ} \approx 1.3432$

33. (a) $\cot 11^\circ 15' = \dfrac{1}{\tan 11.25^\circ} \approx 5.0273$

(b) $\tan 11^\circ 15' = \tan 11.25^\circ \approx 0.1989$

35. (a) $\csc 32^\circ 40' 3'' = \dfrac{1}{\sin 32.6675^\circ} \approx 1.8527$

(b) $\tan 44^\circ 28' 16'' \approx \tan 44.4711^\circ \approx 0.9817$

37. (a) $\sin\theta = \dfrac{1}{2} \implies \theta = 30^\circ = \dfrac{\pi}{6}$

(b) $\csc\theta = 2 \implies \theta = 30^\circ = \dfrac{\pi}{6}$

39. (a) $\sec\theta = 2 \implies \theta = 60^\circ = \dfrac{\pi}{3}$

(b) $\cot\theta = 1 \implies \theta = 45^\circ = \dfrac{\pi}{4}$

41. (a) $\csc\theta = \dfrac{2\sqrt{3}}{3} \implies \theta = 60^\circ = \dfrac{\pi}{3}$

(b) $\sin\theta = \dfrac{\sqrt{2}}{2} \implies \theta = 45^\circ = \dfrac{\pi}{4}$

43. (a) $\sin\theta = 0.0145 \implies \theta \approx 0.83^\circ \approx 0.015$ radian

(b) $\sin\theta = 0.4565 \implies \theta \approx 27^\circ \approx 0.474$ radian

45. (a) $\tan\theta = 0.0125 \implies \theta \approx 0.72^\circ \approx 0.012$ radian

(b) $\tan\theta = 2.3545 \implies \theta \approx 67^\circ \approx 1.169$ radians

47. $\tan\theta\cot\theta = \tan\theta\left(\dfrac{1}{\tan\theta}\right) = 1$

49. $\tan\alpha\cos\alpha = \left(\dfrac{\sin\alpha}{\cos\alpha}\right)\cos\alpha = \sin\alpha$

51.
$$\begin{aligned}(1 + \cos\theta)(1 - \cos\theta) &= 1 - \cos^2\theta\\ &= (\sin^2\theta + \cos^2\theta) - \cos^2\theta\\ &= \sin^2\theta\end{aligned}$$

53.
$$\begin{aligned}(\sec\theta + \tan\theta)(\sec\theta - \tan\theta) &= \sec^2\theta - \tan^2\theta\\ &= (1 + \tan^2\theta) - \tan^2\theta\\ &= 1\end{aligned}$$

55. $\dfrac{\sin\theta}{\cos\theta} + \dfrac{\cos\theta}{\sin\theta} = \dfrac{\sin^2\theta + \cos^2\theta}{\sin\theta\cos\theta}$

$= \dfrac{1}{\sin\theta\cos\theta}$

$= \dfrac{1}{\sin\theta} \cdot \dfrac{1}{\cos\theta}$

$= \csc\theta\sec\theta$

57.

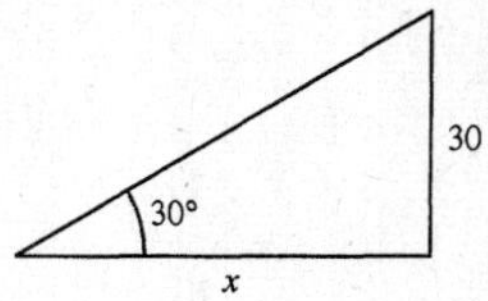

$\tan 30° = \dfrac{30}{x}$

$\dfrac{1}{\sqrt{3}} = \dfrac{30}{x}$

$x = 30\sqrt{3}$

59.

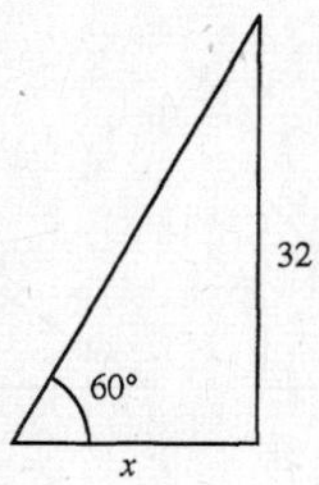

$\tan 60° = \dfrac{32}{x}$

$\sqrt{3} = \dfrac{32}{x}$

$\sqrt{3}x = 32$

$x = \dfrac{32}{\sqrt{3}} = \dfrac{32\sqrt{3}}{3}$

61. (a)

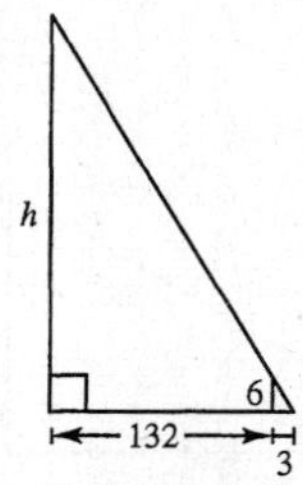

Not drawn to scale

(b) $\cos\theta = \dfrac{6}{3} = \dfrac{h}{135}$

(c) $2(135) = h$

$h = 270$ feet

63. (a)

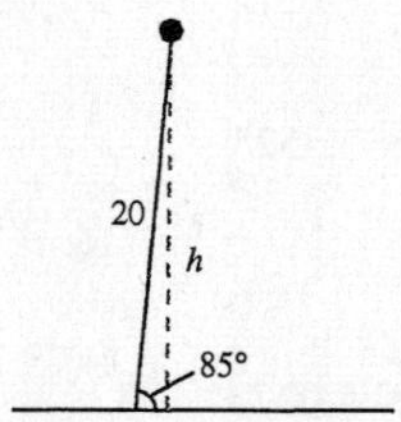

(b) $\sin 85° = \dfrac{h}{20}$

(c) $h = 20\sin 85° \approx 19.9$ meters

(d) The side of the triangle labeled h will become shorter.

(e)

Angle, θ	Height (in meters)
80°	19.7
70°	18.8
60°	17.3
50°	15.3
40°	12.9
30°	10.0
20°	6.8
10°	3.5

(f) The height of the balloon decreases.

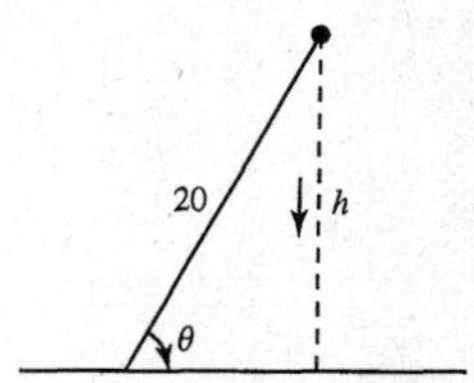

65. $\tan\theta = \dfrac{\text{opp}}{\text{adj}}$

$\tan 54° = \dfrac{w}{100}$

$w = 100\tan 54° \approx 137.6$ feet

67.

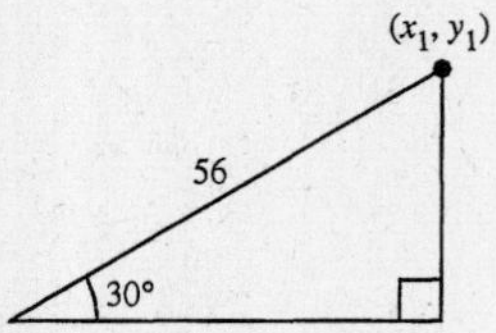

$\sin 30^\circ = \dfrac{y_1}{56}$

$y_1 = (\sin 30^\circ)(56) = \left(\dfrac{1}{2}\right)(56) = 28$

$\cos 30^\circ = \dfrac{x_1}{56}$

$x_1 = \cos 30^\circ(56) = \dfrac{\sqrt{3}}{2}(56) = 28\sqrt{3}$

$(x_1, y_1) = (28\sqrt{3}, 28)$

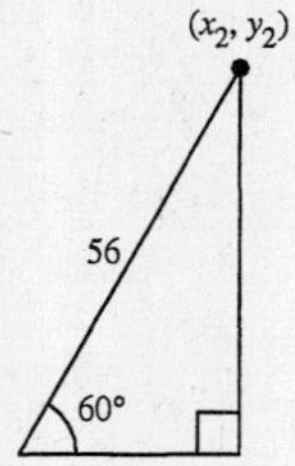

$\sin 60^\circ = \dfrac{y_2}{56}$

$y_2 = \sin 60^\circ(56) = \left(\dfrac{\sqrt{3}}{2}\right)(56) = 28\sqrt{3}$

$\cos 60^\circ = \dfrac{x_2}{56}$

$x_2 = (\cos 60^\circ)(56) = \left(\dfrac{1}{2}\right)(56) = 28$

$(x_2, y_2) = (28, 28\sqrt{3})$

69. $x \approx 9.397,\ y \approx 3.420$

$\sin 20^\circ = \dfrac{y}{10} \approx 0.34$ $\qquad \cot 20^\circ = \dfrac{x}{y} \approx 2.75$

$\cos 20^\circ = \dfrac{x}{10} \approx 0.94$ $\qquad \sec 20^\circ = \dfrac{10}{x} \approx 1.06$

$\tan 20^\circ = \dfrac{y}{x} \approx 0.36$ $\qquad \csc 20^\circ = \dfrac{10}{y} \approx 2.92$

71. True, $\csc x = \dfrac{1}{\sin x} \Rightarrow \sin 60^\circ \csc 60^\circ = \sin 60^\circ\left(\dfrac{1}{\sin 60^\circ}\right) = 1$

73. False, $\dfrac{\sqrt{2}}{2} + \dfrac{\sqrt{2}}{2} = \sqrt{2} \neq 1$

75. False, $\dfrac{\sin 60^\circ}{\sin 30^\circ} = \dfrac{\cos 30^\circ}{\sin 30^\circ} = \cot 30^\circ \approx 1.7321$; $\sin 2^\circ \approx 0.0349$

77. This is true because the corresponding sides of similar triangles are proportional.

79. (a)

θ	0.1	0.2	0.3	0.4	0.5
$\sin\theta$	0.0998	0.1987	0.2955	0.3894	0.4794

(b) As $\theta \to 0$, $\sin\theta \to 0$

81. $\dfrac{x^2 - 6x}{x^2 + 4x - 12} \cdot \dfrac{x^2 + 12x + 36}{x^2 - 36} = \dfrac{x(x - 6)}{(x + 6)(x - 2)} \cdot \dfrac{(x + 6)(x + 6)}{(x + 6)(x - 6)}$

$= \dfrac{x}{x - 2}, x \neq \pm 6$

83. $\dfrac{3}{x+2} - \dfrac{2}{x-2} + \dfrac{x}{x^2+4x+4} = \dfrac{3(x+2)(x-2) - 2(x+2)^2 + x(x-2)}{(x-2)(x+2)^2}$

$= \dfrac{3(x^2-4) - 2(x^2+4x+4) + x^2 - 2x}{(x-2)(x+2)^2}$

$= \dfrac{2x^2 - 10x - 20}{(x-2)(x+2)^2} = \dfrac{2(x^2 - 5x - 10)}{(x-2)(x+2)^2}$

85. $\dfrac{2}{x+3} + \dfrac{4}{x-2} = \dfrac{12}{x^2+x-6}$

$2(x-2) + 4(x+3) = 12$

$2x - 4 + 4x + 12 = 12$

$6x + 8 = 12$

$6x = 4$

$x = \dfrac{2}{3}$

Section 1.4 Trigonometric Functions of Any Angle

- Know the Definitions of Trigonometric Functions of Any Angle.
  If θ is in standard position, (x, y) a point on the terminal side and $r = \sqrt{x^2 + y^2} \neq 0$, then

  $\sin\theta = \dfrac{y}{r}$ $\qquad$ $\csc\theta = \dfrac{r}{y},\ y \neq 0$

  $\cos\theta = \dfrac{x}{r}$ $\qquad$ $\sec\theta = \dfrac{r}{x},\ x \neq 0$

  $\tan\theta = \dfrac{y}{x},\ x \neq 0$ $\qquad$ $\cot\theta = \dfrac{x}{y},\ y \neq 0$
- You should know the signs of the trigonometric functions in each quadrant.
- You should know the trigonometric function values of the quadrant angles 0, $\dfrac{\pi}{2}$, π, and $\dfrac{3\pi}{2}$.
- You should be able to find reference angles.
- You should be able to evaluate trigonometric functions of any angle. (Use reference angles.)
- You should know that the period of sine and cosine is 2π.

Solutions to Odd-Numbered Exercises

1. (a) $(x, y) = (4, 3)$

$r = \sqrt{16 + 9} = 5$

$\sin\theta = \dfrac{y}{r} = \dfrac{3}{5}$ $\qquad$ $\csc\theta = \dfrac{r}{y} = \dfrac{5}{3}$

$\cos\theta = \dfrac{x}{r} = \dfrac{4}{5}$ $\qquad$ $\sec\theta = \dfrac{r}{x} = \dfrac{5}{4}$

$\tan\theta = \dfrac{y}{x} = \dfrac{3}{4}$ $\qquad$ $\cot\theta = \dfrac{x}{y} = \dfrac{4}{3}$

(b) $(x, y) = (8, -15)$

$r = \sqrt{64 + 225} = 17$

$\sin\theta = \dfrac{y}{r} = -\dfrac{15}{17}$ $\qquad$ $\csc\theta = \dfrac{r}{y} = -\dfrac{17}{15}$

$\cos\theta = \dfrac{x}{r} = \dfrac{8}{17}$ $\qquad$ $\sec\theta = \dfrac{r}{x} = \dfrac{17}{8}$

$\tan\theta = \dfrac{y}{x} = -\dfrac{15}{8}$ $\qquad$ $\cot\theta = \dfrac{x}{y} = -\dfrac{8}{15}$

3. (a) $(x, y) = (-\sqrt{3}, -1)$

$r = \sqrt{3 + 1} = 2$

$\sin\theta = \frac{y}{r} = -\frac{1}{2}$ $\csc\theta = \frac{r}{y} = -2$

$\cos\theta = \frac{x}{r} = -\frac{\sqrt{3}}{2}$ $\sec\theta = \frac{r}{x} = -\frac{2\sqrt{3}}{3}$

$\tan\theta = \frac{y}{x} = \frac{\sqrt{3}}{3}$ $\cot\theta = \frac{x}{y} = \sqrt{3}$

(b) $(x, y) = (-4, 1)$

$r = \sqrt{16 + 1} = \sqrt{17}$

$\sin\theta = \frac{y}{r} = \frac{\sqrt{17}}{17}$ $\csc\theta = \frac{r}{y} = \sqrt{17}$

$\cos\theta = \frac{x}{r} = -\frac{4\sqrt{17}}{17}$ $\sec\theta = \frac{r}{x} = -\frac{\sqrt{17}}{4}$

$\tan\theta = \frac{y}{x} = -\frac{1}{4}$ $\cot\theta = \frac{x}{y} = -4$

5. $(x, y) = (7, 24)$

$r = \sqrt{49 + 576} = 25$

$\sin\theta = \frac{y}{r} = \frac{24}{25}$ $\csc\theta = \frac{r}{y} = \frac{25}{24}$

$\cos\theta = \frac{x}{r} = \frac{7}{25}$ $\sec\theta = \frac{r}{x} = \frac{25}{7}$

$\tan\theta = \frac{y}{x} = \frac{24}{7}$ $\cot\theta = \frac{x}{y} = \frac{7}{24}$

7. $(x, y) = (-4, 10)$

$r = \sqrt{16 + 100} = 2\sqrt{29}$

$\sin\theta = \frac{y}{r} = \frac{5\sqrt{29}}{29}$ $\csc\theta = \frac{r}{y} = \frac{\sqrt{29}}{5}$

$\cos\theta = \frac{x}{r} = -\frac{2\sqrt{29}}{29}$ $\sec\theta = \frac{r}{x} = -\frac{\sqrt{29}}{2}$

$\tan\theta = \frac{y}{x} = -\frac{5}{2}$ $\cot\theta = \frac{x}{y} = -\frac{2}{5}$

9. $(x, y) = (-3.5, 6.8)$

$r = \sqrt{12.25 + 46.24} \approx 7.65$

$\sin\theta = \frac{y}{r} = \frac{6.8}{7.65} \approx 0.9$ $\csc\theta = \frac{r}{y} = \frac{7.65}{6.8} \approx 1.1$

$\cos\theta = \frac{x}{r} = -\frac{3.5}{7.65} \approx -0.5$ $\sec\theta = \frac{r}{x} = -\frac{7.65}{3.5} \approx -2.2$

$\tan\theta = \frac{y}{x} = -\frac{6.8}{3.5} \approx -1.9$ $\cot\theta = \frac{x}{y} = -\frac{3.5}{6.8} \approx -0.5$

11. $\sin\theta < 0 \implies \theta$ lies in Quadrant III or in Quadrant IV.

$\cos\theta < 0 \implies \theta$ lies in Quadrant II or in Quadrant III.

$\sin\theta < 0$ *and* $\cos\theta < 0 \implies \theta$ lies in Quadrant III.

13. $\sin\theta > 0 \implies \theta$ lies in Quadrant I or in Quadrant II.

$\tan\theta < 0 \implies \theta$ lies in Quadrant II or in Quadrant IV.

$\sin\theta > 0$ *and* $\tan\theta < 0 \implies \theta$ lies in Quadrant II.

15. $\sin\theta = \frac{y}{r} = \frac{3}{5} \implies x^2 = 25 - 9 = 16$

θ in Quadrant II $\implies x = -4$

$\sin\theta = \frac{y}{r} = \frac{3}{5}$ $\csc\theta = \frac{r}{y} = \frac{5}{3}$

$\cos\theta = \frac{x}{r} = -\frac{4}{5}$ $\sec\theta = \frac{r}{x} = -\frac{5}{4}$

$\tan\theta = \frac{y}{x} = -\frac{3}{4}$ $\cot\theta = \frac{x}{y} = -\frac{4}{3}$

17. $\tan\theta = \frac{y}{x} = \frac{-15}{8}$

$\sin\theta < 0$ and $\tan\theta < 0 \implies \theta$ is in Quadrant IV $\implies$ $y < 0$ and $x > 0$.

$x = 8, y = -15, r = 17$

$\sin\theta = \frac{y}{r} = -\frac{15}{17}$ $\csc\theta = \frac{r}{y} = -\frac{17}{15}$

$\cos\theta = \frac{x}{r} = \frac{8}{17}$ $\sec\theta = \frac{r}{x} = \frac{17}{8}$

$\tan\theta = \frac{y}{x} = -\frac{15}{8}$ $\cot\theta = \frac{x}{y} = -\frac{8}{15}$

19. $\cot\theta = \frac{x}{y} = -\frac{3}{1} = \frac{3}{-1}$

$\cos\theta > 0 \implies \theta$ is in Quadrant IV $\implies x$ is positive;

$x = 3, y = -1, r = \sqrt{10}$

$\sin\theta = \frac{y}{r} = -\frac{\sqrt{10}}{10}$ $\csc\theta = \frac{r}{y} = -\sqrt{10}$

$\cos\theta = \frac{x}{r} = \frac{3\sqrt{10}}{10}$ $\sec\theta = \frac{r}{x} = \frac{\sqrt{10}}{3}$

$\tan\theta = \frac{y}{x} = -\frac{1}{3}$ $\cot\theta = \frac{x}{y} = -3$

21. $\sec\theta = \frac{r}{x} = \frac{2}{-1} \Rightarrow y^2 = 4 - 1 = 3$

$\sin\theta > 0 \Rightarrow \theta$ is in Quadrant II $\Rightarrow y = \sqrt{3}$

$\sin\theta = \frac{y}{r} = \frac{\sqrt{3}}{2}$ $\qquad$ $\csc\theta = \frac{r}{y} = \frac{2\sqrt{3}}{3}$

$\cos\theta = \frac{x}{r} = -\frac{1}{2}$ $\qquad$ $\sec\theta = \frac{r}{x} = -2$

$\tan\theta = \frac{y}{x} = -\sqrt{3}$ $\qquad$ $\cot\theta = \frac{x}{y} = -\frac{\sqrt{3}}{3}$

23. $\cot\theta$ is undefined, $\frac{\pi}{2} \le \theta \le \frac{3\pi}{2} \Rightarrow y = 0 \Rightarrow \theta = \pi$

$\sin\pi = 0$ $\qquad$ $\csc\pi$ is undefined

$\cos\pi = -1$ $\qquad$ $\sec\pi = -1$

$\tan\pi = 0$ $\qquad$ $\cot\pi$ is undefined

25. To find a point on the terminal side of θ, use any point on the line $y = -x$ that lies in Quadrant II. $(-1, 1)$ is one such point.

$x = -1, y = 1, r = \sqrt{2}$

$\sin\theta = \frac{1}{\sqrt{2}} = \frac{\sqrt{2}}{2}$ $\qquad$ $\csc\theta = \sqrt{2}$

$\cos\theta = -\frac{1}{\sqrt{2}} = -\frac{\sqrt{2}}{2}$ $\qquad$ $\sec\theta = -\sqrt{2}$

$\tan\theta = -1$ $\qquad$ $\cot\theta = -1$

27. To find a point on the terminal side of θ, use any point on the line $y = 2x$ that lies in Quadrant III. $(-1, -2)$ is one such point.

$x = -1, y = -2, r = \sqrt{5}$

$\sin\theta = -\frac{2}{\sqrt{5}} = -\frac{2\sqrt{5}}{5}$ $\qquad$ $\csc\theta = \frac{\sqrt{5}}{-2} = -\frac{\sqrt{5}}{2}$

$\cos\theta = -\frac{1}{\sqrt{5}} = -\frac{\sqrt{5}}{5}$ $\qquad$ $\sec\theta = \frac{\sqrt{5}}{-1} = -\sqrt{5}$

$\tan\theta = \frac{-2}{-1} = 2$ $\qquad$ $\cot\theta = \frac{-1}{-2} = \frac{1}{2}$

29. $(x, y) = (-1, 0), r = 1$

$\cos\pi = \frac{x}{r} = \frac{-1}{1} = -1$

31. $\sec\frac{3\pi}{2} = \frac{r}{x} = \frac{1}{0} \Rightarrow$ undefined

since $\frac{3\pi}{2}$ corresponds to $(0, -1)$.

33. $(x, y) = (0, 1), r = 1$

$\tan\frac{\pi}{2} = \frac{y}{x} = \frac{1}{0}$ undefined.

35. $\csc\pi = \frac{r}{y} = \frac{1}{0} \Rightarrow$ undefined

since π corresponds to $(-1, 0)$.

37. $\theta = 203°$

$\theta' = 203° - 180° = 23°$

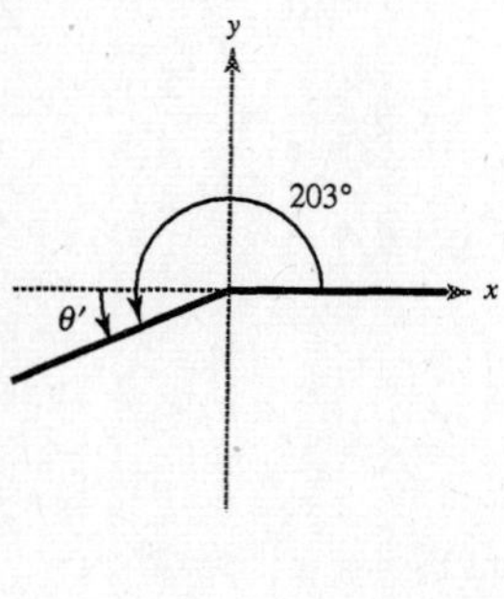

39. $\theta = -245°$

$360° - 245° = 115°$ (coterminal angle)

$\theta' = 180° - 115° = 65°$

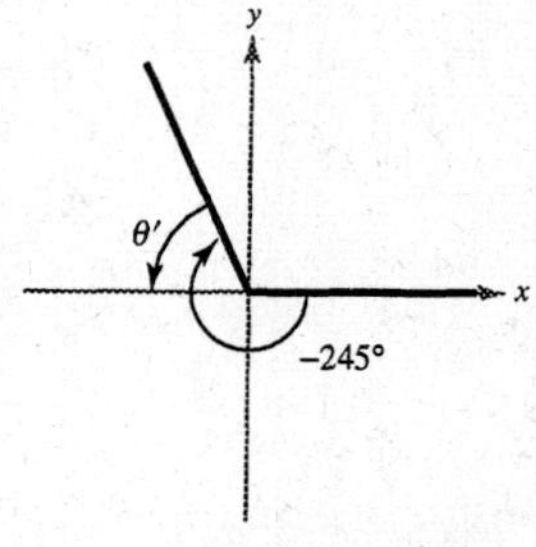

41. $\theta = \dfrac{2\pi}{3}$

$\theta' = \pi - \dfrac{2\pi}{3} = \dfrac{\pi}{3}$

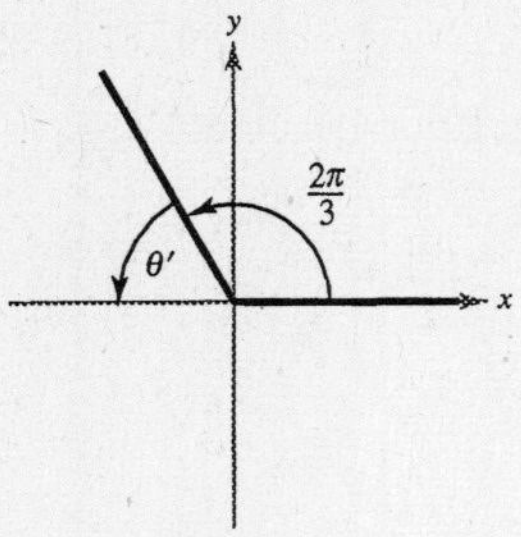

43. $\theta = 3.5$

$\theta' = 3.5 - \pi$

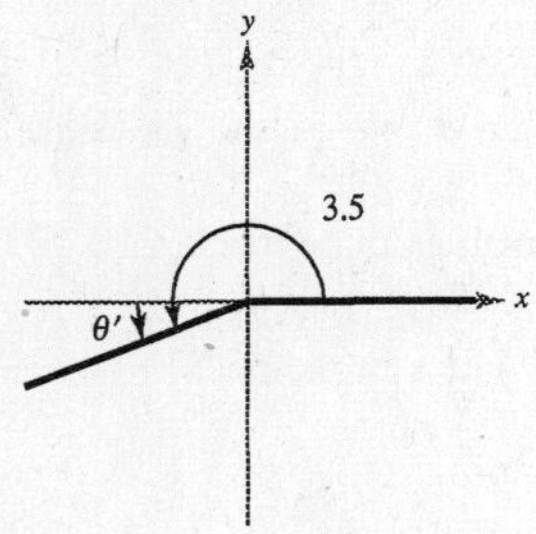

45. $\theta' = 45°$, Quadrant III

$\sin 225° = -\sin 45° = -\dfrac{\sqrt{2}}{2}$

$\cos 225° = -\cos 45° = -\dfrac{\sqrt{2}}{2}$

$\tan 225° = \tan 45° = 1$

47. $\theta' = 30°$, Quadrant I

$\sin 750° = \sin 30° = \dfrac{1}{2}$

$\cos 750° = \cos 30° = \dfrac{\sqrt{3}}{2}$

$\tan 750° = \tan 30° = \dfrac{\sqrt{3}}{3}$

49. $\theta' = 30°$, Quadrant III

$\sin(-150°) = -\sin 30° = -\dfrac{1}{2}$

$\cos(-150°) = -\cos 30° = -\dfrac{\sqrt{3}}{2}$

$\tan(-150°) = \tan 30° = \dfrac{\sqrt{3}}{3}$

51. $\theta' = \dfrac{\pi}{3}$, Quadrant III

$\sin \dfrac{4\pi}{3} = -\sin \dfrac{\pi}{3} = -\dfrac{\sqrt{3}}{2}$

$\cos \dfrac{4\pi}{3} = -\cos \dfrac{\pi}{3} = -\dfrac{1}{2}$

$\tan \dfrac{4\pi}{3} = \tan \dfrac{\pi}{3} = \sqrt{3}$

53. $\theta' = \dfrac{\pi}{6}$, Quadrant IV

$\sin\left(-\dfrac{\pi}{6}\right) = -\sin \dfrac{\pi}{6} = -\dfrac{1}{2}$

$\cos\left(-\dfrac{\pi}{6}\right) = \cos \dfrac{\pi}{6} = \dfrac{\sqrt{3}}{2}$

$\tan\left(-\dfrac{\pi}{6}\right) = -\tan \dfrac{\pi}{6} = -\dfrac{\sqrt{3}}{3}$

55. $\theta' = \dfrac{\pi}{4}$, Quadrant II

$\sin \dfrac{11\pi}{4} = \sin \dfrac{\pi}{4} = \dfrac{\sqrt{2}}{2}$

$\cos \dfrac{11\pi}{4} = -\cos \dfrac{\pi}{4} = -\dfrac{\sqrt{2}}{2}$

$\tan \dfrac{11\pi}{4} = -\tan \dfrac{\pi}{4} = -1$

57. $\theta' = \dfrac{\pi}{2}$

$\sin\left(-\dfrac{3\pi}{2}\right) = \sin \dfrac{\pi}{2} = 1$

$\cos\left(-\dfrac{3\pi}{2}\right) = \cos \dfrac{\pi}{2} = 0$

$\tan\left(-\dfrac{3\pi}{2}\right) = \tan \dfrac{\pi}{2}$ which is undefined

59. $\sin 10° \approx 0.1736$

61. $\cos(-110°) \approx -0.3420$

63. $\tan 4.5 \approx 4.6373$

65. $\tan \dfrac{\pi}{9} \approx 0.3640$

67. $\sin(-0.65) \approx -0.6052$

69. (a) $\sin \theta = \dfrac{1}{2} \Rightarrow$ reference angle is 30° or $\dfrac{\pi}{6}$ and θ is in Quadrant I or Quadrant II.

Values in degrees: 30°, 150°

Values in radians: $\dfrac{\pi}{6}, \dfrac{5\pi}{6}$

—CONTINUED—

69. —CONTINUED—

(b) $\sin\theta = -\frac{1}{2} \Rightarrow$ reference angle is 30° or $\frac{\pi}{6}$ and θ is in Quadrant III or Quadrant IV.

Values in degrees: 210°, 330°

Values in radians: $\frac{7\pi}{6}, \frac{11\pi}{6}$

71. (a) $\csc\theta = \frac{2\sqrt{3}}{3} \Rightarrow$ reference angle is 60° or $\frac{\pi}{3}$ and θ is in Quadrant I or Quadrant II.

Values in degrees: 60°, 120°

Values in radians: $\frac{\pi}{3}, \frac{2\pi}{3}$

(b) $\cot\theta = -1 \Rightarrow$ reference angle is 45° or $\frac{\pi}{4}$ and θ is in Quadrant II or Quadrant IV.

Values in degrees: 135°, 315°

Values in radians: $\frac{3\pi}{4}, \frac{7\pi}{4}$

73. (a) $\tan\theta = 1 \Rightarrow$ reference angle is 45° or $\frac{\pi}{4}$ and θ is in Quadrant I or Quadrant III.

Values in degrees: 45°, 225°

Values in radians: $\frac{\pi}{4}, \frac{5\pi}{4}$

(b) $\cot\theta = -\sqrt{3} \Rightarrow$ reference angle is 30° or $\frac{\pi}{6}$ and θ is in Quadrant II or Quadrant IV.

Values in degrees: 150°, 330°

Values in radians: $\frac{5\pi}{6}, \frac{11\pi}{6}$

75. $\sin\theta = 0.8191$

Quadrant I: $\theta = \sin^{-1} 0.8191 \approx 54.99°$

Quadrant II: $\theta = 180° - \sin^{-1} 0.8191 \approx 125.01°$

77. $\cos\theta = -0.4367 \Rightarrow \theta' \approx 64.11°$

Quadrant II: $\theta \approx 180° - 64.11° = 115.89°$

Quadrant III: $\theta \approx 180° + 64.11° = 244.11°$

79. $\cos\theta = 0.9848 \Rightarrow \theta' \approx 0.175$

Quadrant I: $\theta = \cos^{-1}(0.9848) \approx 0.175$

Quadrant IV: $\theta = 2\pi - \theta' \approx 6.109$

81. $\tan\theta = 1.192 \Rightarrow \theta' \approx 0.873$

Quadrant I: $\theta = \tan^{-1} 1.192 \approx 0.873$

Quadrant III: $\theta = \pi + \theta' \approx 4.014$

83. $\sec\theta = -2.6667 \Rightarrow \theta' = \cos^{-1}\left(\frac{1}{2.6667}\right) \approx 1.1864$

Quadrant II: $\theta = \pi - 1.1864 \approx 1.955$

Quadrant III: $\theta = \pi + 1.1864 \approx 4.328$

85. $\sin\theta = -\frac{3}{5}$

$\sin^2\theta + \cos^2\theta = 1$

$\cos^2\theta = 1 - \sin^2\theta$

$\cos^2\theta = 1 - \left(-\frac{3}{5}\right)^2$

$\cos^2\theta = 1 - \frac{9}{25}$

$\cos^2\theta = \frac{16}{25}$

$\cos\theta > 0$ in Quadrant IV.

$\cos\theta = \frac{4}{5}$

87. $\tan\theta = \frac{3}{2}$

$\sec^2\theta = 1 + \tan^2\theta$

$\sec^2\theta = 1 + \left(\frac{3}{2}\right)^2$

$\sec^2\theta = 1 + \frac{9}{4}$

$\sec^2\theta = \frac{13}{4}$

$\sec\theta < 0$ in Quadrant III.

$\sec\theta = -\frac{\sqrt{13}}{2}$

89. $\cos\theta = \frac{5}{8}$

$\cos\theta = \frac{1}{\sec\theta} \Rightarrow \sec\theta = \frac{1}{\cos\theta}$

$\sec\theta = \frac{1}{\frac{5}{8}} = \frac{8}{5}$

91.

t	N	F
1	32°	−10°
4	53°	31°
7	77°	63°
10	58°	25°
12	37°	−7°

(a) New York City: $N \approx 22.66\sin(0.51t - 2.12) + 54.58$

Fairbanks: $F \approx 37.18\sin(0.51t - 1.91) + 26.17$

(b)

Month	N	F
February	34°	−3°
March	42°	12°
May	64°	48°
June	73°	60°
August	76°	57°
September	69°	43°
November	47°	6°

(c) The periods are the same for both, 12 months.

93. $y(t) = 2\cos 6t$

(a) $y(0) = 2\cos 0 = 2$ centimeters

(b) $y\left(\frac{1}{4}\right) = 2\cos\left(\frac{3}{2}\right) \approx 0.14$ centimeter

(c) $y\left(\frac{1}{2}\right) = 2\cos 3 \approx -1.98$ centimeters

95. False. In each of the four quadrants, the sign of the secant function and the cosine function will be the same since they are reciprocals of each other.

97. As θ increases from 0° to 90°, x decreases from 12 cm to 0 cm and y increases from 0 cm to 12 cm. Therefore, $\sin\theta = \frac{y}{12}$ increases from 0 to 1 and $\cos\theta = \frac{x}{12}$ decreases from 1 to 0. Thus, $\tan\theta = \frac{y}{x}$ increases without bound, and when $\theta = 90°$ the tangent is undefined.

99. $y = x - 8$

Intercepts: $(8, 0), (0, -8)$

Domain: All real numbers or $(-\infty, \infty)$

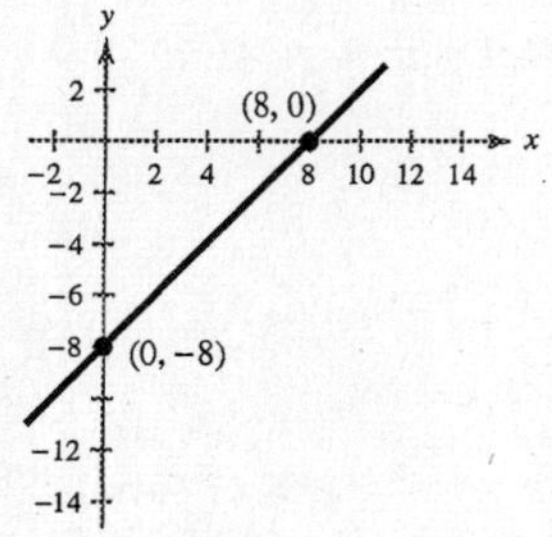

101. $y = x^2 + 3x - 4 = (x + 4)(x - 1)$

Intercepts: $(-4, 0), (1, 0), (0, -4)$

Domain: All real numbers or $(-\infty, \infty)$

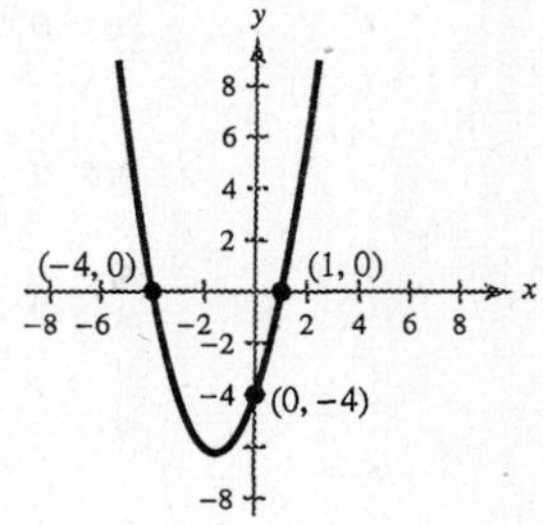

103. $f(x) = x^3 + 8$

Intercepts: $(-2, 0), (0, 8)$

Domain: All real numbers or $(-\infty, \infty)$

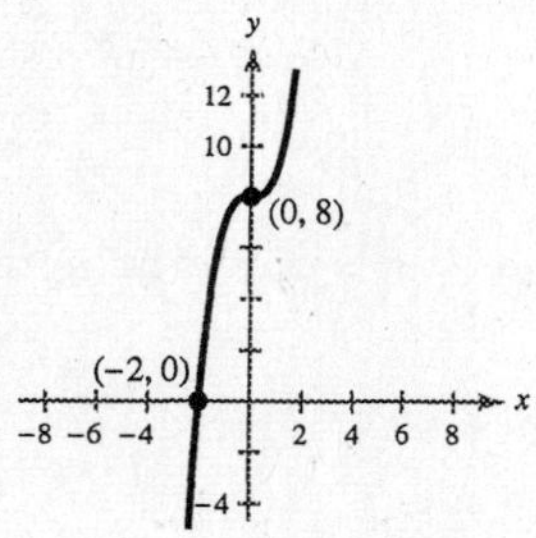

105. $g(x) = \sqrt{x + 5}$

Intercepts: $(-5, 0), (0, \sqrt{5})$

Domain: $x \geq -5$ or $[-5, \infty)$

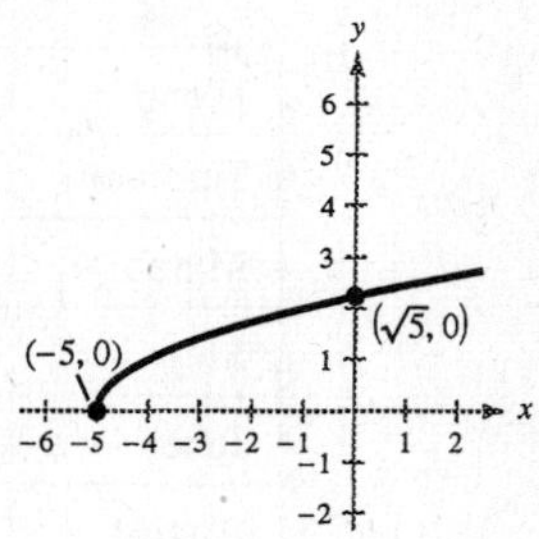

Section 1.5 Graphs of Sine and Cosine Functions

- You should be able to graph $y = a \sin(bx - c)$ and $y = a \cos(bx - c)$. (Assume $b > 0$)
- Amplitude: $|a|$
- Period: $\dfrac{2\pi}{|b|}$
- Shift: Solve $bx - c = 0$ and $bx - c = 2\pi$.
- Key Increments: $\dfrac{1}{4}$ (period)

Solutions to Odd-Numbered Exercises

1. $y = 3 \sin 2x$

Period: $\dfrac{2\pi}{2} = \pi$

Amplitude: $|3| = 3$

3. $y = \dfrac{5}{2} \cos \dfrac{x}{2}$

Period: $\dfrac{2\pi}{\frac{1}{2}} = 4\pi$

Amplitude: $\left|\dfrac{5}{2}\right| = \dfrac{5}{2}$

5. $y = \dfrac{1}{2} \sin \dfrac{\pi x}{3}$

Period: $\dfrac{2\pi}{\frac{\pi}{3}} = 6$

Amplitude: $\left|\dfrac{1}{2}\right| = \dfrac{1}{2}$

7. $y = -2 \sin x$

Period: $\dfrac{2\pi}{1} = 2\pi$

Amplitude: $|-2| = 2$

9. $y = 3 \sin 10x$

Period: $\dfrac{2\pi}{10} = \dfrac{\pi}{5}$

Amplitude: $|3| = 3$

11. $y = \dfrac{1}{2} \cos \dfrac{2x}{3}$

Period: $\dfrac{2\pi}{\frac{2}{3}} = 3\pi$

Amplitude: $\left|\dfrac{1}{2}\right| = \dfrac{1}{2}$

13. $y = \dfrac{1}{4} \sin 2\pi x$

Period: $\dfrac{2\pi}{2\pi} = 1$

Amplitude: $\left|\dfrac{1}{4}\right| = \dfrac{1}{4}$

15. $f(x) = \sin x$

$g(x) = \sin(x - \pi)$

The graph of g is a horizontal shift to the right π units of the graph of f (a phase shift).

17. $f(x) = \cos 2x$

$g(x) = -\cos 2x$

The graph of g is a reflection in the x-axis of the graph of f.

19. $f(x) = \cos x$

$g(x) = \cos 2x$

The period of f is twice that of g.

21. $f(x) = \sin 2x$

$f(x) = 3 + \sin 2x$

The graph of g is a vertical shift 3 units upward of the graph of f.

23. The graph of g has twice the amplitude as the graph of f. The period is the same.

25. The graph of g is a horizontal shift π units to the right of the graph of f.

27. $f(x) = -2 \sin x$

Period: 2π

Amplitude: 2

$g(x) = 4 \sin x$

Period: 2π

Amplitude: 4

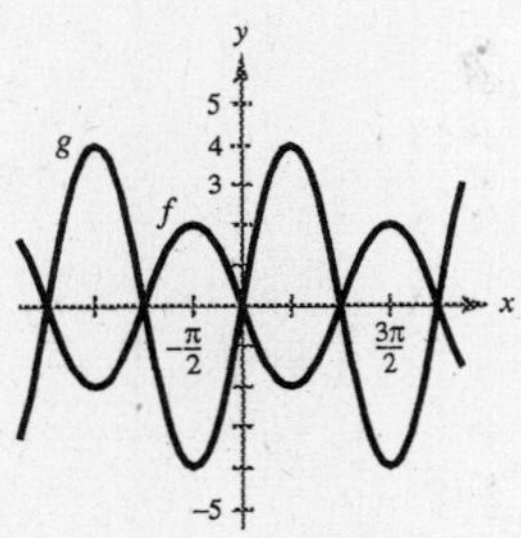

29. $f(x) = \cos x$

Period: 2π

Amplitude: 1

$g(x) = 1 + \cos x$

is a vertical shift of the graph of $f(x)$ one unit upward.

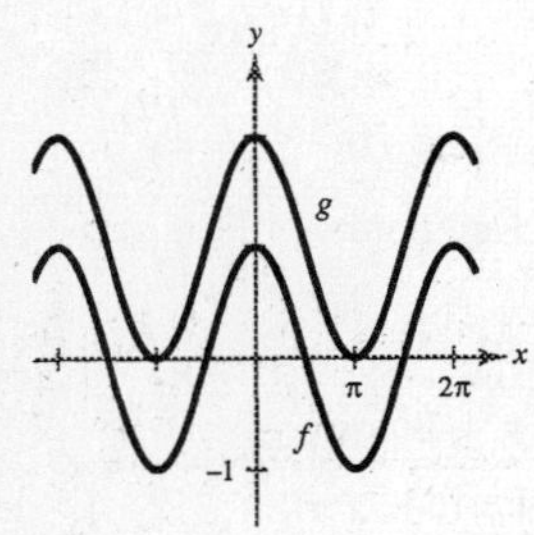

31. $f(x) = -\dfrac{1}{2} \sin \dfrac{x}{2}$

Period: 4π

Amplitude: $\dfrac{1}{2}$

$g(x) = 3 - \dfrac{1}{2} \sin \dfrac{x}{2}$ is the graph of $f(x)$ shifted vertically three units upward.

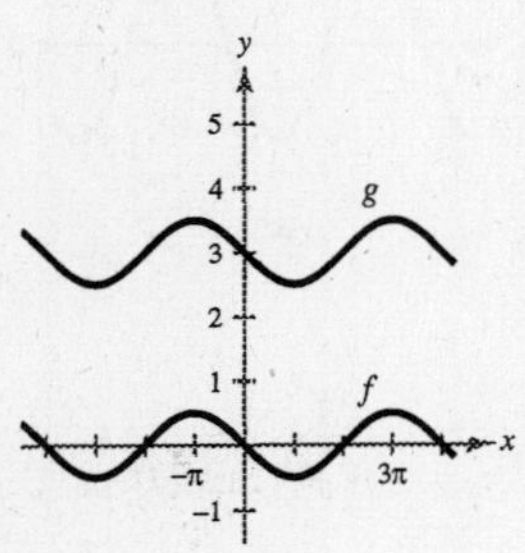

33. $f(x) = 2 \cos x$

Period: 2π

Amplitude: 2

$g(x) = 2 \cos(x + \pi)$ is the graph of $f(x)$ shifted π units to the left.

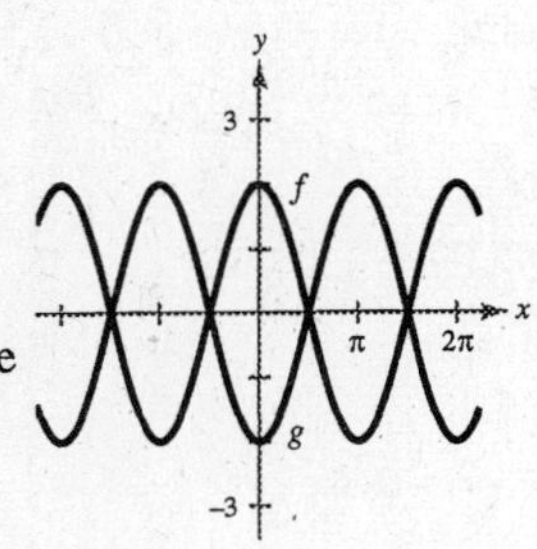

35. $y = -2 \sin 6x;\ a = -2,\ b = 6,\ c = 0$

Period: $\dfrac{2\pi}{6} = \dfrac{\pi}{3}$

Amplitude: $|-2| = 2$

Key points: $(0, 0), \left(\dfrac{\pi}{12}, -2\right), \left(\dfrac{\pi}{6}, 0\right), \left(\dfrac{\pi}{4}, 2\right), \left(\dfrac{\pi}{3}, 0\right)$

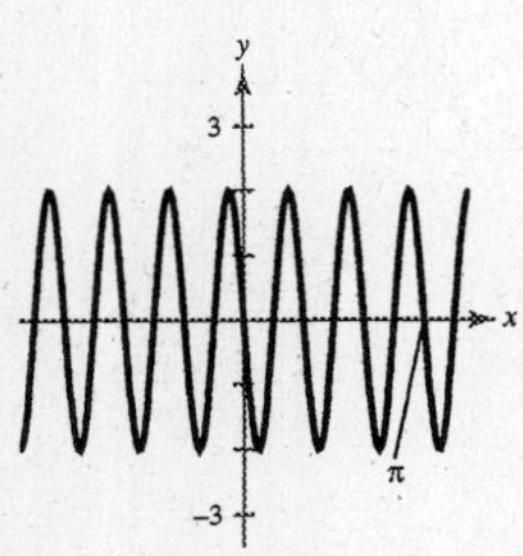

37. $y = \cos 2\pi x$

Period: $\dfrac{2\pi}{2\pi} = 1$

Amplitude: 1

Key points: $(0, 1), \left(\frac{1}{4}, 0\right), \left(\frac{1}{2}, -1\right), \left(\frac{3}{4}, 0\right)$

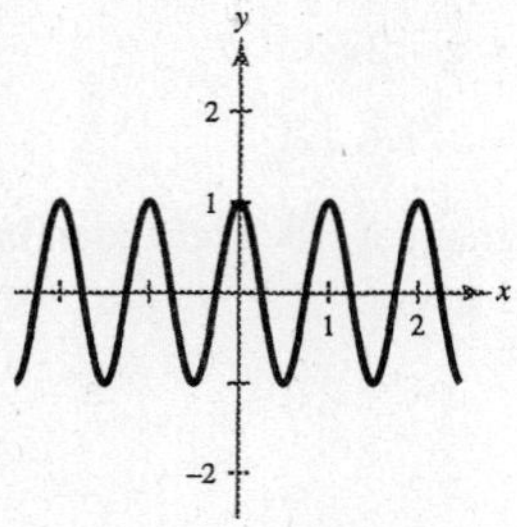

39. $y = -\sin \dfrac{2\pi x}{3}$; $a = -1, b = \dfrac{2\pi}{3}, c = 0$

Period: $\dfrac{2\pi}{\frac{2\pi}{3}} = 3$

Amplitude: 1

Key points: $(0, 0), \left(\frac{3}{4}, -1\right), \left(\frac{3}{2}, 0\right), \left(\frac{9}{4}, 1\right), (3, 0)$

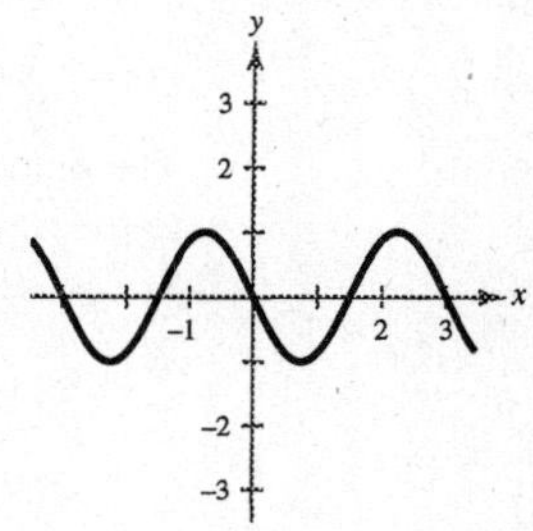

41. $y = \sin\left(x - \dfrac{\pi}{4}\right)$; $a = 1,\ b = 1,\ c = \dfrac{\pi}{4}$

Period: 2π

Amplitude: 1

Shift: Set $x - \dfrac{\pi}{4} = 0$ and $x - \dfrac{\pi}{4} = 2\pi$

$x = \dfrac{\pi}{4}$ $\qquad x = \dfrac{9\pi}{4}$

Key points: $\left(\frac{\pi}{4}, 0\right), \left(\frac{3\pi}{4}, 1\right), \left(\frac{5\pi}{4}, 0\right), \left(\frac{7\pi}{4}, -1\right), \left(\frac{9\pi}{4}, 0\right)$

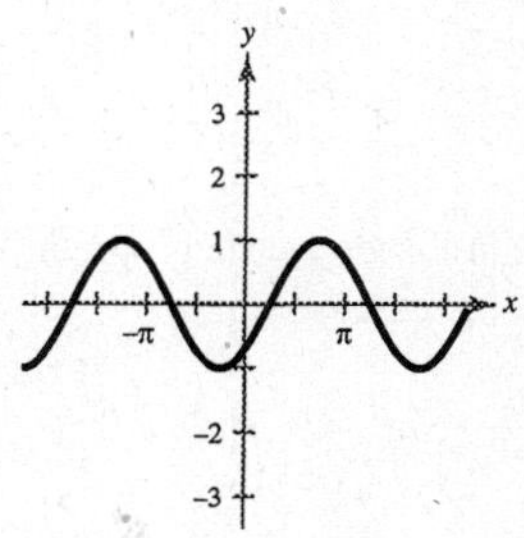

43. $y = 3\cos(x + \pi)$

Period: 2π

Amplitude: 3

Shift: Set $x + \pi = 0$ and $x + \pi = 2\pi$

$x = -\pi$ $\qquad x = \pi$

Key points: $(-\pi, 3), \left(-\frac{\pi}{2}, 0\right), (0, -3), \left(\frac{\pi}{2}, 0\right), (\pi, 3)$

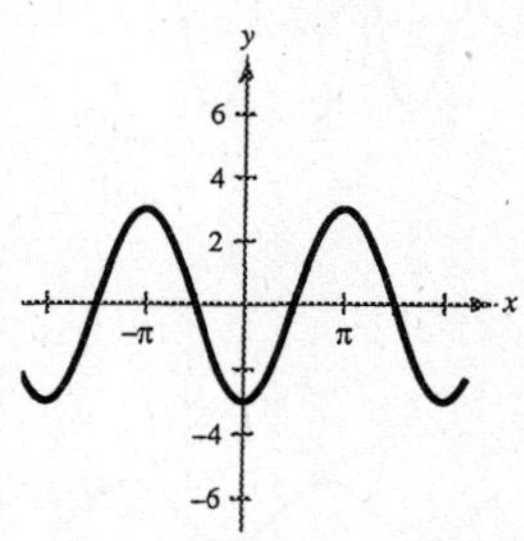

45. $y = 2 - \sin \dfrac{2\pi x}{3}$

Vertical shift 2 units upward of the graph in Exercise 39.

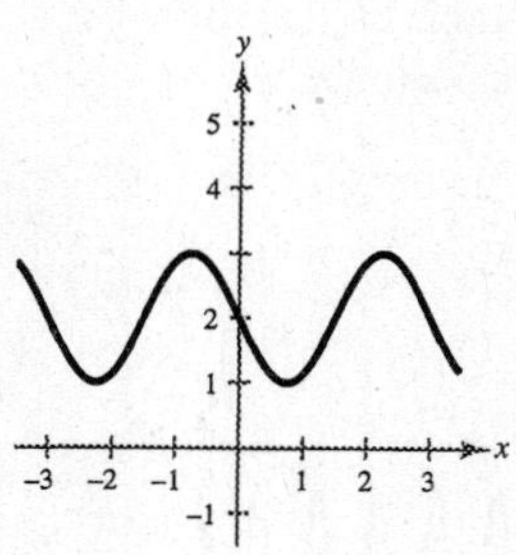

47. $y = 2 + \dfrac{1}{10}\cos 60\pi x$

Period: $\dfrac{2\pi}{60\pi} = \dfrac{1}{30}$

Amplitude: $\dfrac{1}{10}$

Vertical shift 2 units upward

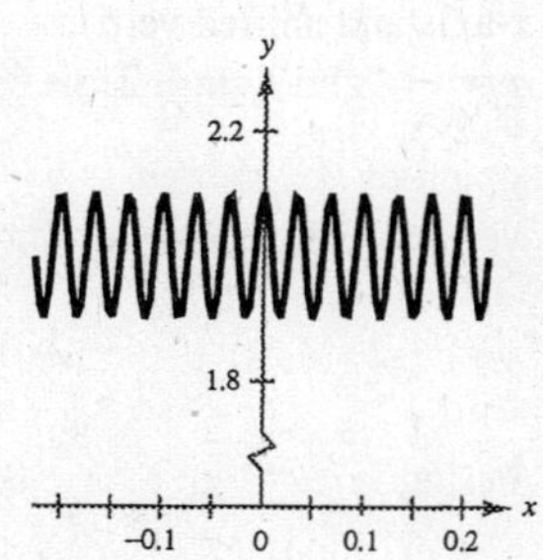

Key points:

$(0, 2.1), \left(\frac{1}{120}, 2\right), \left(\frac{1}{60}, 1.9\right), \left(\frac{1}{40}, 2\right), \left(\frac{1}{30}, 2.1\right)$

49. $y = 3\cos(x + \pi) - 3$

Vertical shift 3 units downward of the graph in Exercise 43.

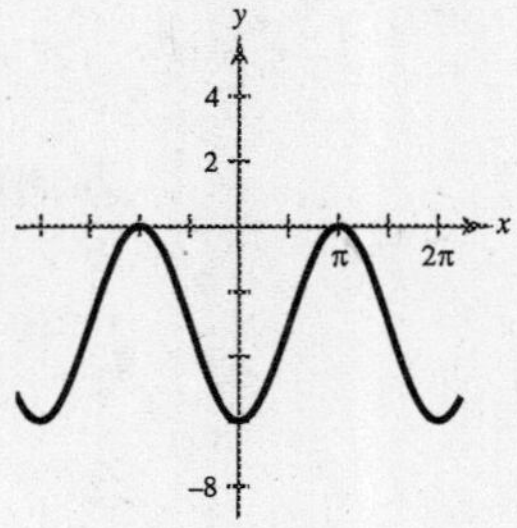

51. $y = \frac{2}{3}\cos\left(\frac{x}{2} - \frac{\pi}{4}\right)$; $a = \frac{2}{3}$, $b = \frac{1}{2}$, $c = \frac{\pi}{4}$

Period: 4π

Amplitude: $\frac{2}{3}$

Shift: $\frac{x}{2} - \frac{\pi}{4} = 0$ and $\frac{x}{2} - \frac{\pi}{4} = 2\pi$

$x = \frac{\pi}{2}$ $\qquad x = \frac{9\pi}{2}$

Key points: $\left(\frac{\pi}{2}, \frac{2}{3}\right)$, $\left(\frac{3\pi}{2}, 0\right)$, $\left(\frac{5\pi}{2}, \frac{-2}{3}\right)$, $\left(\frac{7\pi}{2}, 0\right)$, $\left(\frac{9\pi}{2}, \frac{2}{3}\right)$

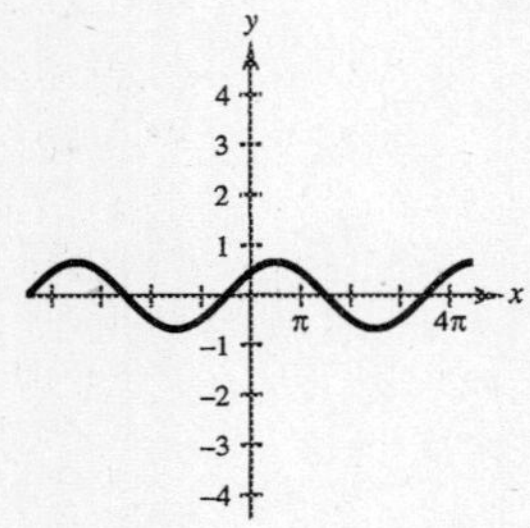

53. $y = -2\sin(4x + \pi)$

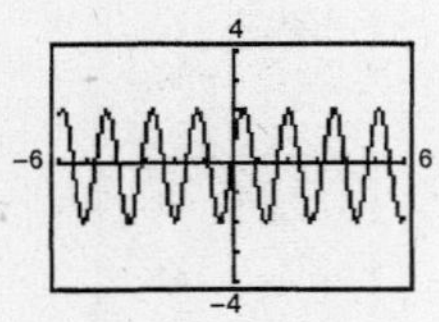

55. $y = \cos\left(2\pi x - \frac{\pi}{2}\right) + 1$

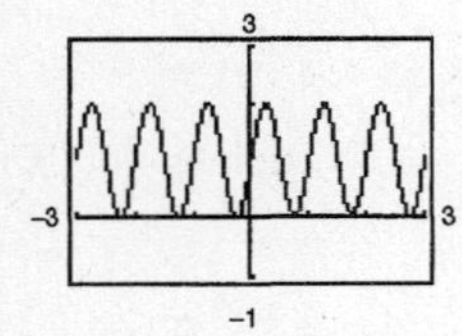

57. $y = -0.1\sin\left(\frac{\pi x}{10} + \pi\right)$

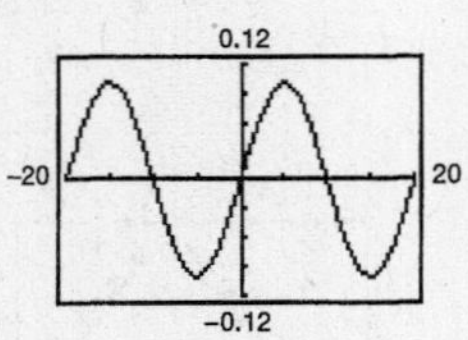

59. $f(x) = a\cos x + d$

Amplitude: $\frac{1}{2}[3 - (-1)] = 2 \Rightarrow a = 2$

Vertical shift 1 unit upward of $g(x) = 2\cos x \Rightarrow d = 1$.
Thus, $f(x) = 2\cos x + 1$.

61. $f(x) = a\cos x + d$

Amplitude: $\frac{1}{2}[8 - 0] = 4$

Since $f(x)$ is the graph of $g(x) = 4\cos x$ reflected in the x-axis and shifted vertically 4 units upward, we have $a = -4$ and $d = 4$. Thus, $f(x) = -4\cos x + 4$.

63. $y = a\sin(bx - c)$

Amplitude: $|a| = |3|$ Since the graph is reflected in the x-axis, we have $a = -3$.

Period: $\frac{2\pi}{b} = \pi \Rightarrow b = 2$

Phase shift: $c = 0$

Thus, $y = -3\sin 2x$.

65. $y = a\sin(bx - c)$

Amplitude: $a = 2$

Period: $2\pi \Rightarrow b = 1$

Phase shift: $bx - c = 0$ when $x = -\dfrac{\pi}{4}$

$$(1)\left(\frac{-\pi}{4}\right) - c = 0 \Rightarrow c = -\frac{\pi}{4}$$

Thus, $y = 2\sin\left(x + \dfrac{\pi}{4}\right)$.

67. $y_1 = \sin x$

$y_2 = -\dfrac{1}{2}$

In the interval $[-2\pi, 2\pi]$, $\sin x = -\dfrac{1}{2}$ when

$x = -\dfrac{5\pi}{6}, -\dfrac{\pi}{6}, \dfrac{7\pi}{6}, \dfrac{11\pi}{6}$.

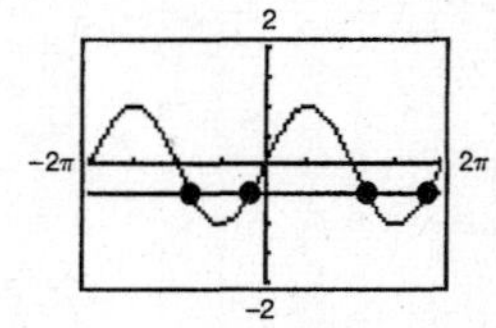

69. $y = 0.85\sin\dfrac{\pi t}{3}$

(a) Time for one cycle $= \dfrac{2\pi}{\dfrac{\pi}{3}} = 6$ sec

(b) Cycles per min $= \dfrac{60}{6} = 10$ cycles per min

(c) Amplitude: 0.85

Period: 6

Key points: $(0, 0), \left(\dfrac{3}{2}, 0.85\right), (3, 0), \left(\dfrac{9}{2}, -0.85\right), (6, 0)$

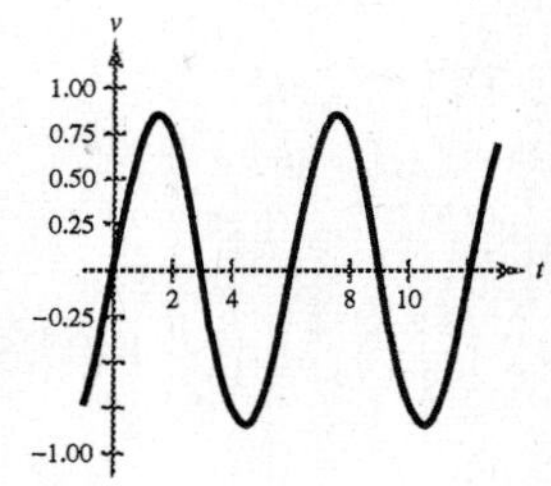

71. $y = 0.001\sin 880\pi t$

(a) Period: $\dfrac{2\pi}{880\pi} = \dfrac{1}{440}$ seconds

(b) $f = \dfrac{1}{p} = 440$ cycles per second

73. (a) $a = \dfrac{1}{2}[\text{high} - \text{low}] = \dfrac{1}{2}[83.7 - 29.0] = 27.35$

$p = 2[\text{high time} - \text{low time}] = 2[7 - 1] = 12$

$b = \dfrac{2\pi}{p} = \dfrac{2\pi}{12} = \dfrac{\pi}{6}$

$\dfrac{c}{b} = 7 \Rightarrow c = 7\left(\dfrac{\pi}{6}\right) \approx 3.67$

$d = \dfrac{1}{2}[\text{high} + \text{low}] = \dfrac{1}{2}[83.7 + 29.0] = 56.35$

$C(t) = 56.35 + 27.35\cos\left(\dfrac{\pi t}{6} - 3.67\right)$

(b)

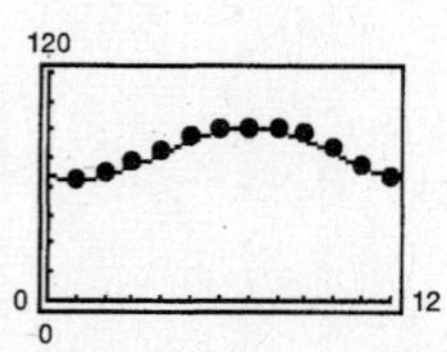

The model is a good fit.

(c)

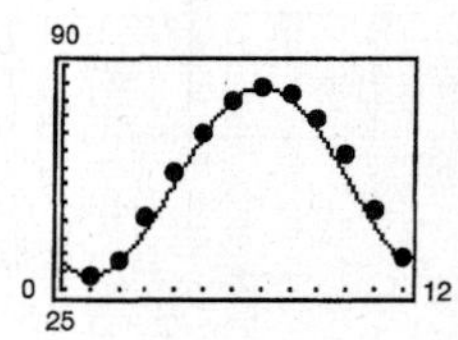

The model is a good fit.

(d) Tallahassee average maximum: 77.60°

Chicago average maximum: 56.35°

Average maximum $= d$, the constant term

(e) The period for both models is $\dfrac{2\pi}{\pi/6} = 12$ months.

This is as we expected since one full period is one year.

(f) Chicago has the greater variability in temperature throughout the year. The amplitude, a, determines this variability since it is $\frac{1}{2}[\text{high temp} - \text{low temp}]$.

75. $C = 30.3 + 21.6\sin\left(\dfrac{2\pi t}{365} + 10.9\right)$

(a) Period $= \dfrac{2\pi}{\dfrac{2\pi}{365}} = 365$

Yes, this is what is expected because there are 365 days in a year.

(b) The average daily fuel consumption is given by the amount of the vertical shift (from 0) which is given by the constant 30.3.

(c)

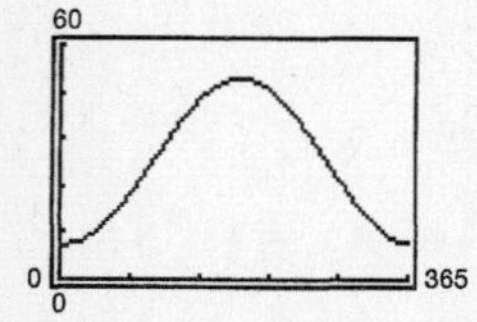

The consumption exceeds 40 gallons per day when $124 < x < 252$.

77. False. $y = \dfrac{1}{2}\cos 2x$ has an amplitude that is **half** that of $y = \cos x$. For $y = a\cos bx$, the amplitude is $|a|$.

79.

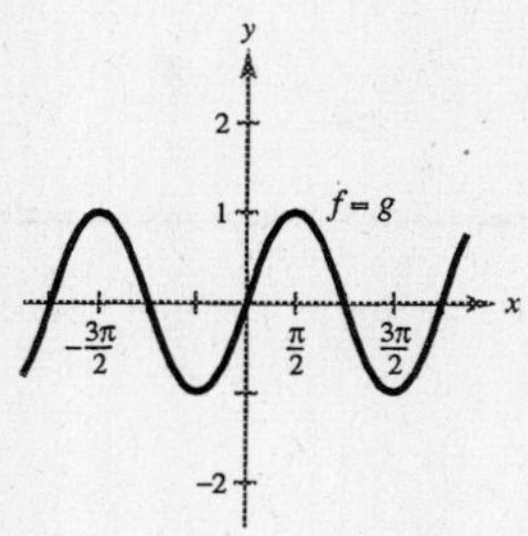

Since the graphs are the same, the conjecture is that $\sin(x) = \cos\left(x - \dfrac{\pi}{2}\right)$.

81.

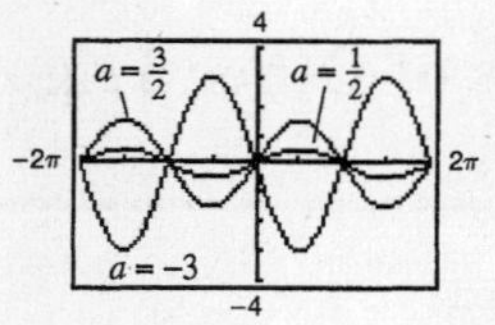

As a changes from $\dfrac{1}{2}$ to $\dfrac{3}{2}$, the amplitude increases. When $a = -3$, the amplitude again increases but the graph is also reflected in the x-axis.

83.

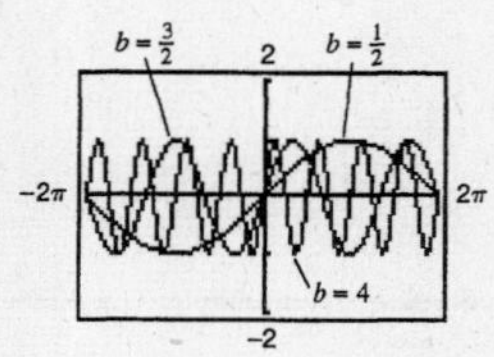

As b increases the period decreases.

85. (a)

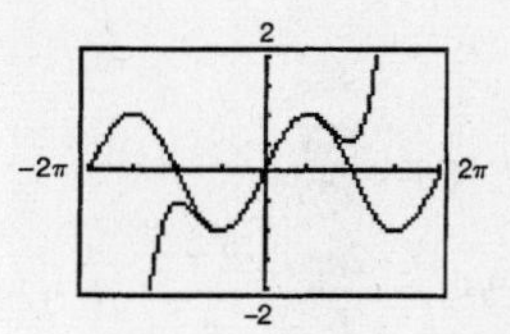

The graphs nearly the same for $-\dfrac{\pi}{2} < x < \dfrac{\pi}{2}$

(b)

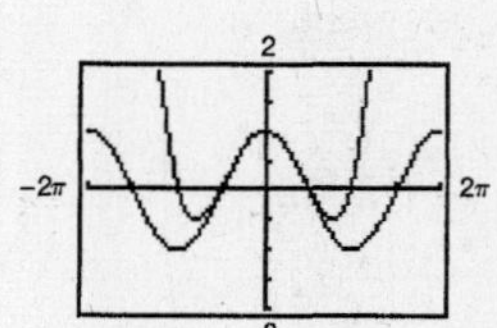

The graphs are nearly the same for $-\dfrac{\pi}{2} < x < \dfrac{\pi}{2}$

(c) $\sin x \approx x - \dfrac{x^3}{3!} + \dfrac{x^5}{5!} - \dfrac{x^7}{7!}$

$\cos x \approx 1 - \dfrac{x^2}{2!} + \dfrac{x^4}{4!} - \dfrac{x^6}{6!}$

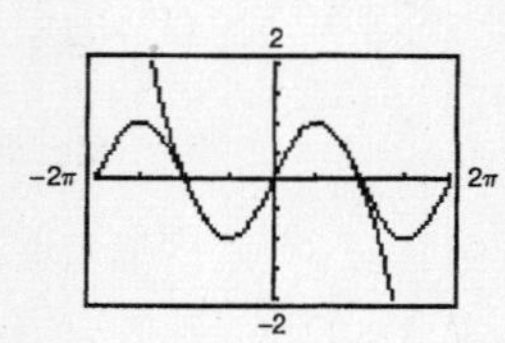

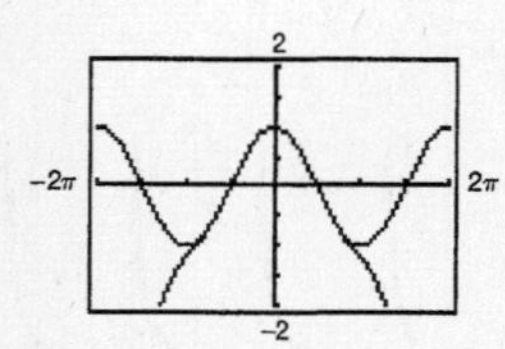

The graphs now agree over a wider range, $-\dfrac{3\pi}{4} < x < \dfrac{3\pi}{4}$

The interval of accuracy increased.

87. $\frac{4}{x} + \frac{4}{1-x} = \frac{4(1-x)+4x}{x(1-x)} = \frac{4-4x+4x}{x(1-x)} = \frac{4}{x(1-x)}$

89. $\frac{3}{x-1} - \frac{2}{x(x-1)} = \frac{3x-2}{x(x-1)}$

91. $f(x) = \frac{2}{11-x}$

Domain: All real numbers except $x = 11$

93. $f(x) = \sqrt{81 - x^2}$

$81 - x^2 \geq 0$

$(9 + x)(9 - x) \geq 0$

Critical Numbers: $x = \pm 9$

Test Intervals: $(-\infty, -9), (-9, 9), (9, \infty)$

Solution: $[-9, 9]$

Domain of $f(x)$: $-9 \leq x \leq 9$

Section 1.6 Graphs of Other Trigonometric Functions

- You should be able to graph

  $y = a \tan(bx - c)$ $y = a \cot(bx - c)$

  $y = a \sec(bx - c)$ $y = a \csc(bx - c)$
- When graphing $y = a \sec(bx - c)$ or $y = a \csc(bx - c)$ you should first graph $y = a \cos(bx - c)$ or $y = a \sin(bx - c)$ because
  (a) The x-intercepts of sine and cosine are the vertical asymptotes of cosecant and secant.
  (b) The maximums of sine and cosine are the local minimums of cosecant and secant.
  (c) The minimums of sine and cosine are the local maximums of cosecant and secant.
- You should be able to graph using a damping factor.

Solutions to Odd-Numbered Exercises

1. $y = \sec 2x$

Period: $\frac{2\pi}{2} = \pi$

Matches graph (e).

3. $y = \frac{1}{2}\cot \pi x$

Period: $\frac{\pi}{\pi} = 1$

Matches graph (a).

5. $y = \frac{1}{2}\sec\frac{\pi x}{2}$

Period $= \frac{2\pi}{b} = \frac{2\pi}{\frac{\pi}{2}} = 4$

Asymptotes: $x = -1, x = 1$

Matches graph (f).

7. $y = \frac{1}{3}\tan x$

Period: π

Two consecutive asymptotes:

$x = -\frac{\pi}{2}$ and $x = \frac{\pi}{2}$

x	$-\frac{\pi}{4}$	0	$\frac{\pi}{4}$
y	$-\frac{1}{3}$	0	$\frac{1}{3}$

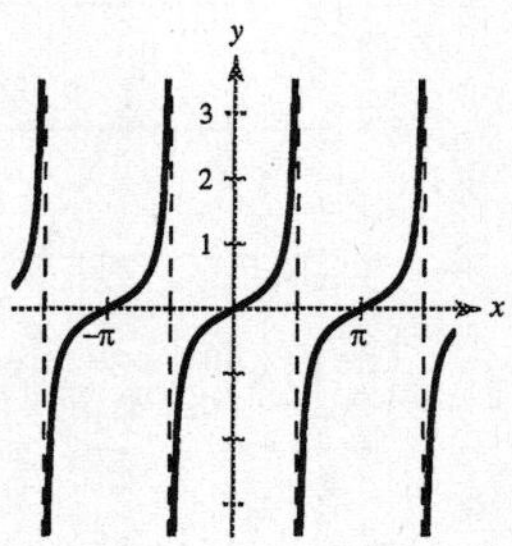

9. $y = \tan 3x$

Period: $\dfrac{\pi}{3}$

Two consecutive asymptotes:

$3x = -\dfrac{\pi}{2} \Rightarrow x = -\dfrac{\pi}{6}$

$3x - \dfrac{\pi}{2} \Rightarrow x = \dfrac{\pi}{6}$

x	$-\dfrac{\pi}{12}$	0	$\dfrac{\pi}{12}$
y	-1	0	1

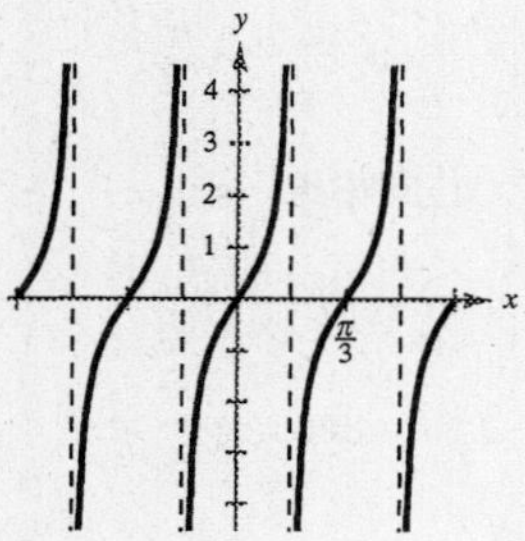

11. $y = -\dfrac{1}{2}\sec x$

Graph $y = -\dfrac{1}{2}\cos x$ first.

Period: 2π

One cycle: 0 to 2π

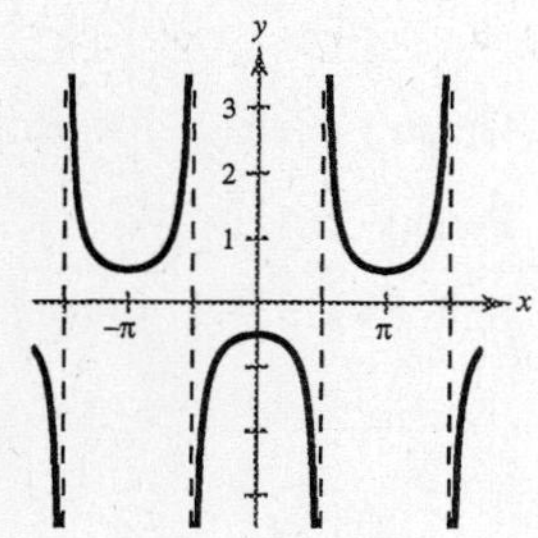

13. $y = \csc \pi x$

Graph $y = \sin \pi x$ first.

Period: $\dfrac{2\pi}{\pi} = 2$

One cycle: 0 to 2

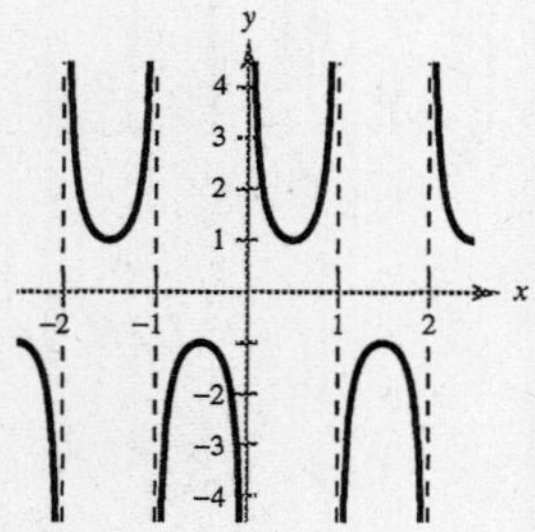

15. $y = \sec \pi x - 1$

Graph $y = \cos \pi x$ first

Period: $\dfrac{2\pi}{\pi} = 2$

One cycle: 0 to 2

Vertical shift 1 unit downward.

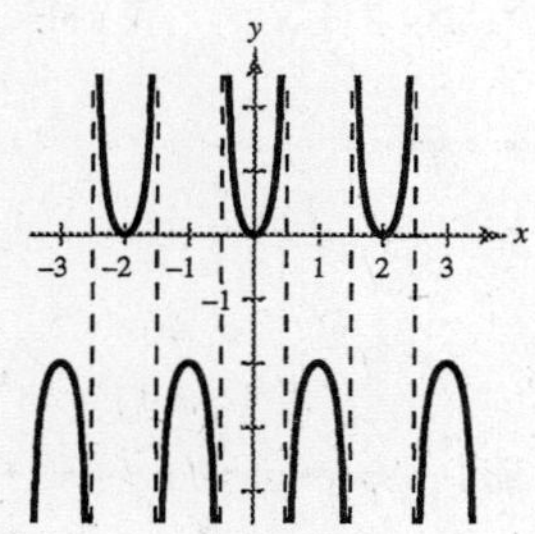

17. $y = \csc \dfrac{x}{2}$

Graph $y = \sin \dfrac{x}{2}$ first.

Period: $\dfrac{2\pi}{\frac{1}{2}} = 4\pi$

One cycle: 0 to 4π

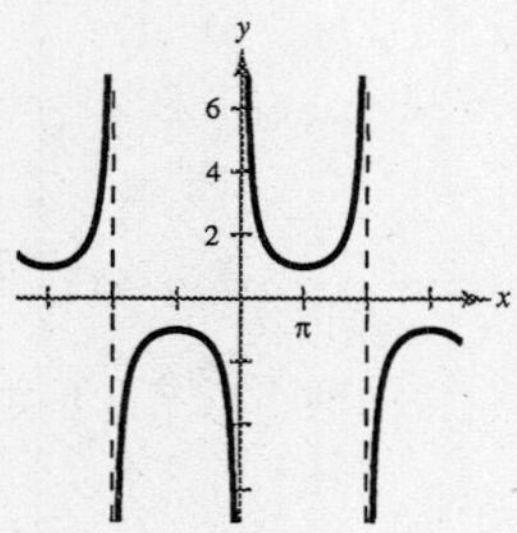

19. $y = \cot \dfrac{x}{2}$

Period: $\dfrac{\pi}{\frac{1}{2}} = 2\pi$

x	$\dfrac{\pi}{2}$	π	$\dfrac{3\pi}{2}$
y	1	0	-1

Two consecutive asymptotes: $\dfrac{x}{2} = 0 \Rightarrow x = 0$

$\dfrac{x}{2} = \pi \Rightarrow x = 2\pi$

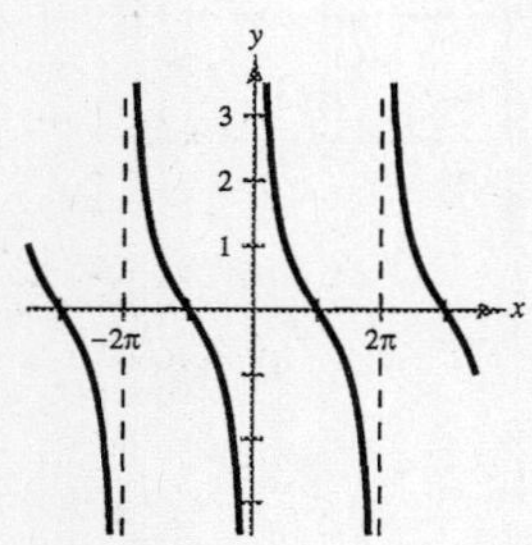

21. $y = \dfrac{1}{2}\sec 2x$

Graph $y = \dfrac{1}{2}\cos 2x$ first.

Period: $\dfrac{2\pi}{2} = \pi$

One cycle: 0 to π

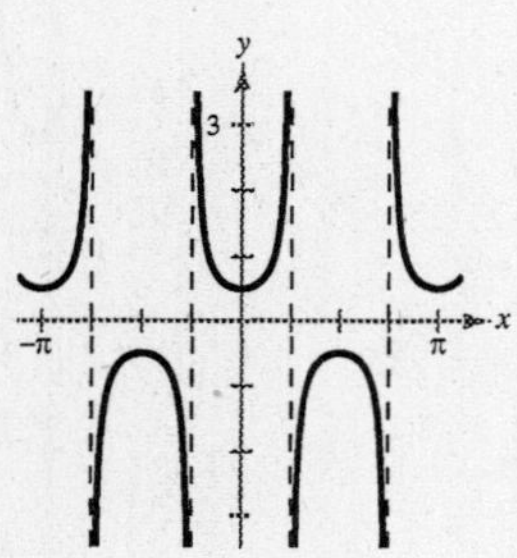

23. $y = \tan \dfrac{\pi x}{4}$

Period: $\dfrac{\pi}{\frac{\pi}{4}} = 4$

Two consecutive asymptotes:

$\dfrac{\pi x}{4} = -\dfrac{\pi}{2} \Rightarrow x = -2$

$\dfrac{\pi x}{4} = \dfrac{\pi}{2} \Rightarrow x = 2$

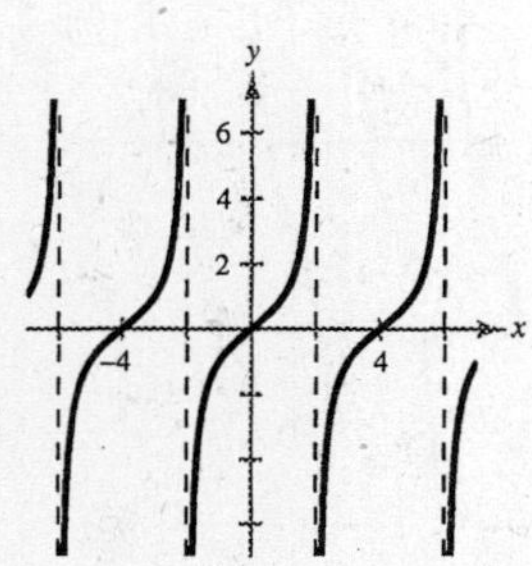

x	-1	0	1
y	-1	0	1

25. $y = \csc(\pi - x)$

Graph $y = \sin(\pi - x)$ first.

Period: 2π

Shift: Set $\pi - x = 0$ and $\pi - x = 2\pi$

$$x = \pi \qquad x = -\pi$$

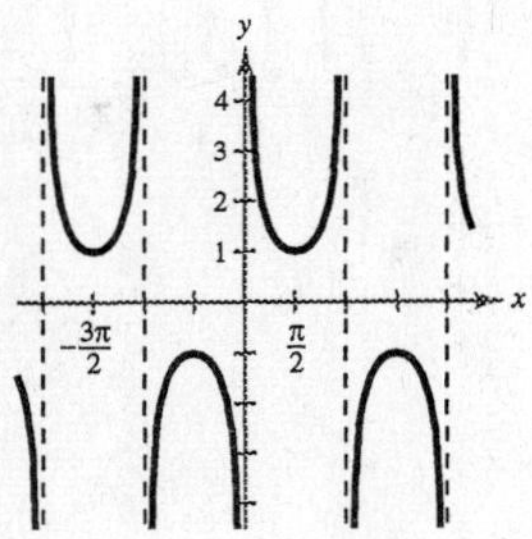

27. $y = \frac{1}{4}\csc\left(x + \frac{\pi}{4}\right)$

Graph $y = \frac{1}{4}\sin\left(x + \frac{\pi}{4}\right)$ first.

Period: 2π

Shift: Set $x + \frac{\pi}{4} = 0$ and $x + \frac{\pi}{4} = 2\pi$

$$x = -\frac{\pi}{4} \quad \text{to} \quad x = \frac{7\pi}{4}$$

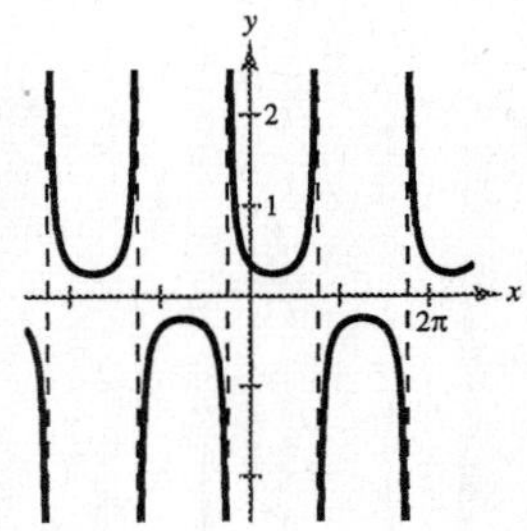

29. $y = \tan\frac{x}{3}$

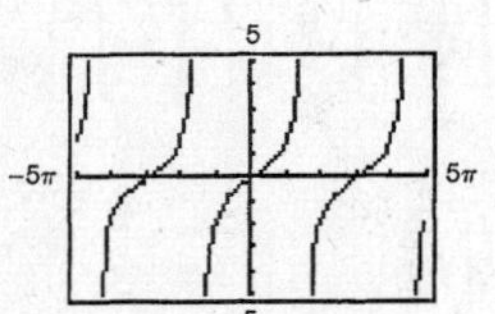

31. $y = -2\sec 4x$

$$= \frac{-2}{\cos 4x}$$

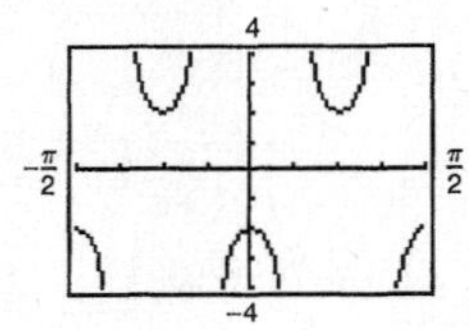

33. $y = \tan\left(x - \frac{\pi}{4}\right)$

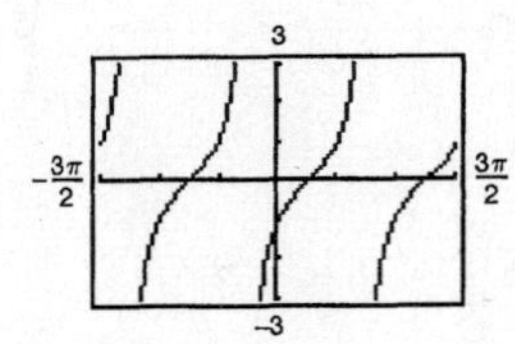

35. $y = -\csc(4x - \pi)$

$$y = \frac{-1}{\sin(4x - \pi)}$$

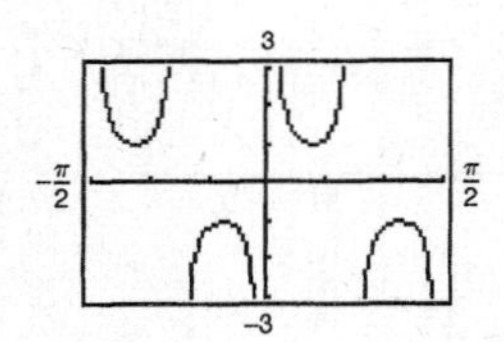

37. $y = 0.1\tan\left(\frac{\pi x}{4} + \frac{\pi}{4}\right)$

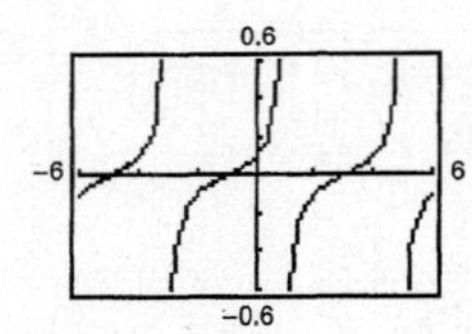

39. $\tan x = 1$

$$x = -\frac{7\pi}{4}, -\frac{3\pi}{4}, \frac{\pi}{4}, \frac{5\pi}{4}$$

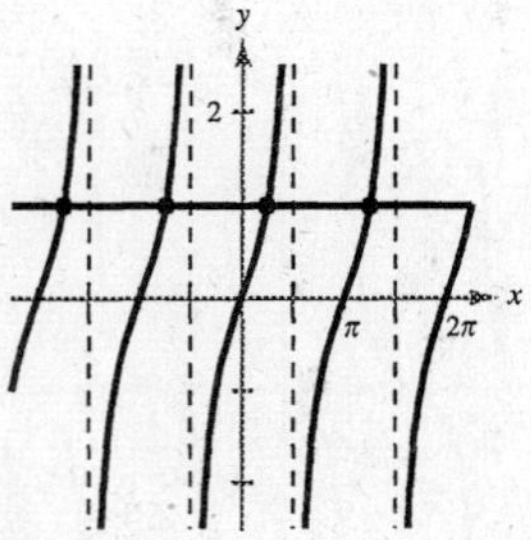

41. $\cot x = -\frac{\sqrt{3}}{3}$

$$x = -\frac{4\pi}{3}, -\frac{\pi}{3}, \frac{2\pi}{3}, \frac{5\pi}{3}$$

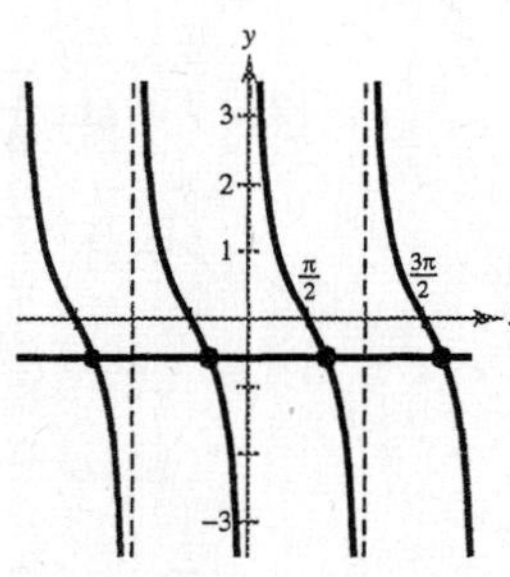

43. $\sec x = -2$

$$x = \pm\frac{2\pi}{3}, \pm\frac{4\pi}{3}$$

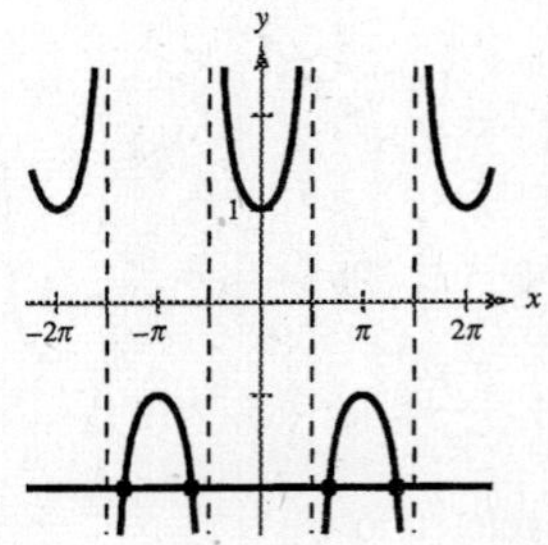

45. $\csc x = \sqrt{2}$

$$x = -\frac{7\pi}{4}, -\frac{5\pi}{4}, \frac{\pi}{4}, \frac{3\pi}{4}$$

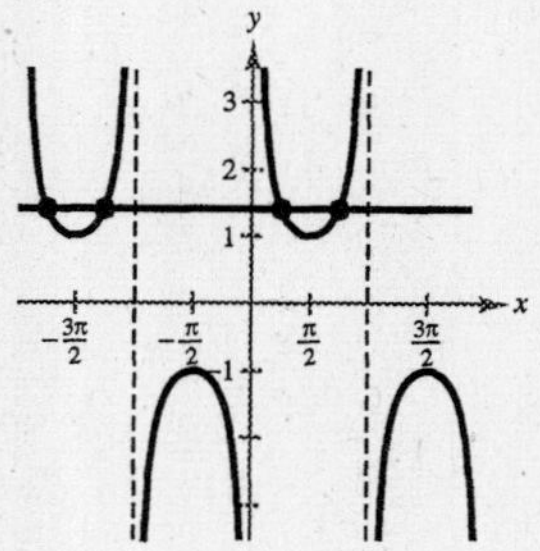

47. The graph of $f(x) = \sec x$ has y-axis symmetry. Thus, the function is even.

49. $f(x) = 2 \sin x$

$g(x) = \frac{1}{2} \csc x$

(a)

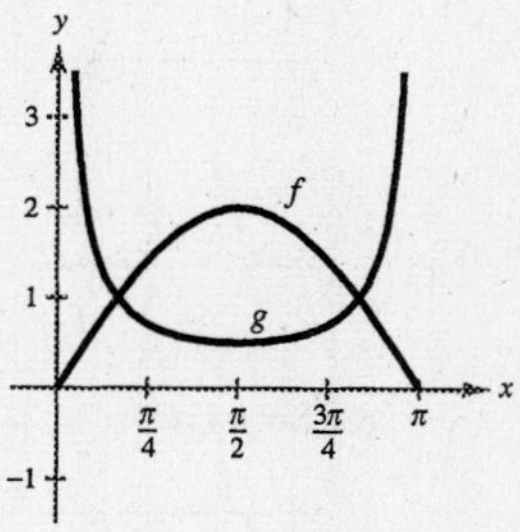

(b) $f > g$ on the interval, $\frac{\pi}{6} < x < \frac{5\pi}{6}$

(c) As $x \to \pi$, $f(x) = 2 \sin x \to 0$ and $g(x) = \frac{1}{2} \csc x \to \pm\infty$ since $g(x)$ is the reciprocal of $f(x)$.

51. $y_1 = \sin x \csc x$ and $y_2 = 1$

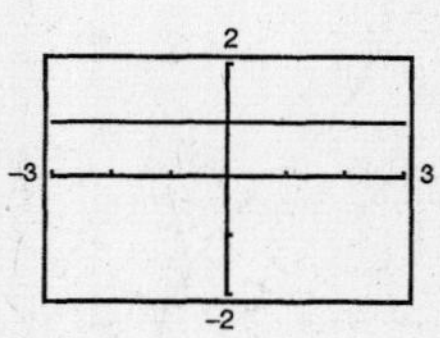

$$\sin x \csc x = \sin x \left(\frac{1}{\sin x}\right) = 1, \sin x \neq 0$$

The expressions are equivalent except when $\sin x = 0$ and y_1 is undefined.

53. $y_1 = \frac{\cos x}{\sin x}$ and $y_2 = \cot x = \frac{1}{\tan x}$

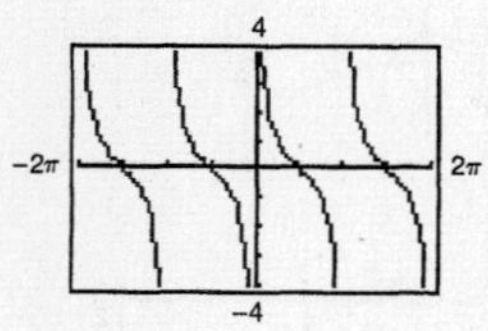

$$\cot x = \frac{\cos x}{\sin x}$$

The expressions are equivalent.

55. $f(x) = |x \cos x|$

As $x \to 0$, $f(x) \to 0$ and $f(x) \geq 0$.

Matches graph (d).

57. $g(x) = |x| \sin x$

As $x \to 0$, $g(x) \to 0$ and $g(x)$ is odd.

Matches graph (b).

59. $f(x) = \sin x + \cos\left(x + \dfrac{\pi}{2}\right),\ g(x) = 0$

$f(x) = g(x)$ The graph is the line $y = 0$.

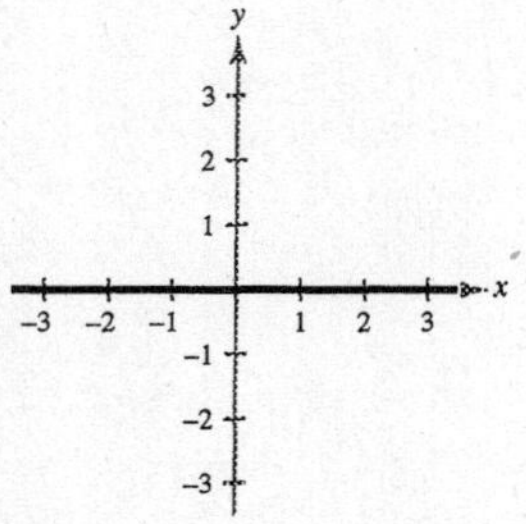

61. $f(x) = \sin^2 x,\ g(x) = \dfrac{1}{2}(1 - \cos 2x)$

$f(x) = g(x)$

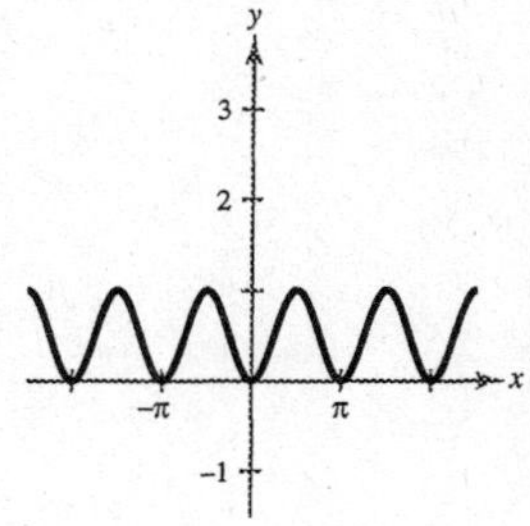

63. $f(x) = x \cos \pi x$

Damping factor: x

As $x \to \infty$, $f(x)$ oscillates and approaches $-\infty$ and ∞.

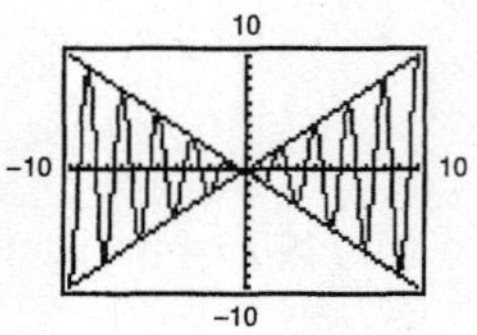

65. $g(x) = x^3 \sin x$

Damping factor: x^3

As $x \to \infty$, $g(x)$ oscillates and approaches $-\infty$ and ∞.

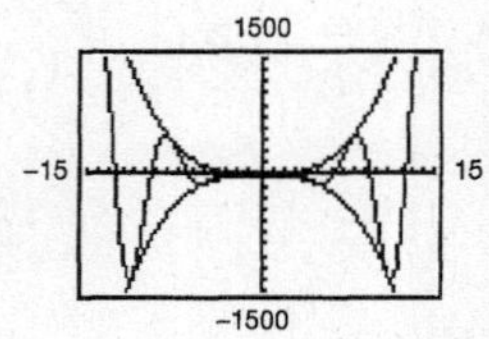

67. $y = \dfrac{6}{x} + \cos x,\ x > 0$

As $x \to 0,\ y \to \infty$.

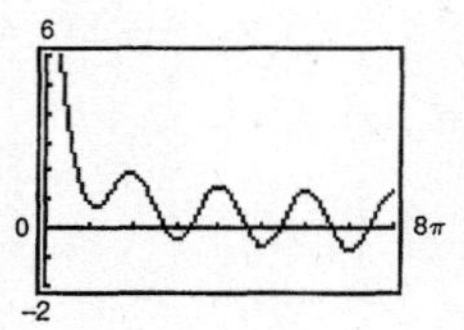

69. $g(x) = \dfrac{\sin x}{x}$

As $x \to 0,\ g(x) \to 1$.

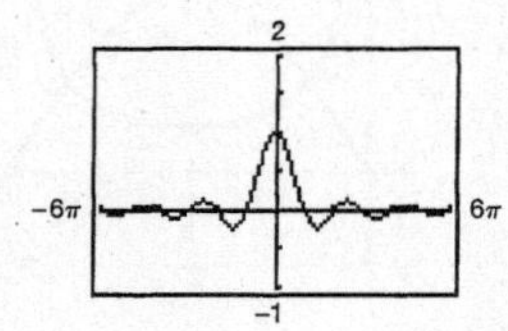

71. $f(x) = \sin \dfrac{1}{x}$

As $x \to 0$, $f(x)$ oscillates between -1 and 1.

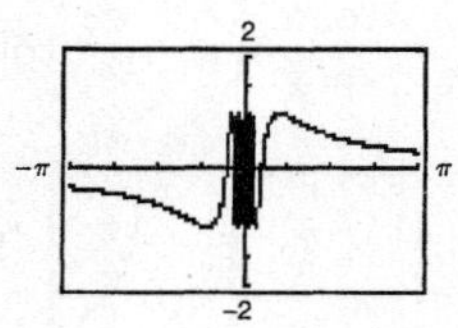

73. $\tan x = \dfrac{7}{d}$

$d = \dfrac{7}{\tan x} = 7 \cot x$

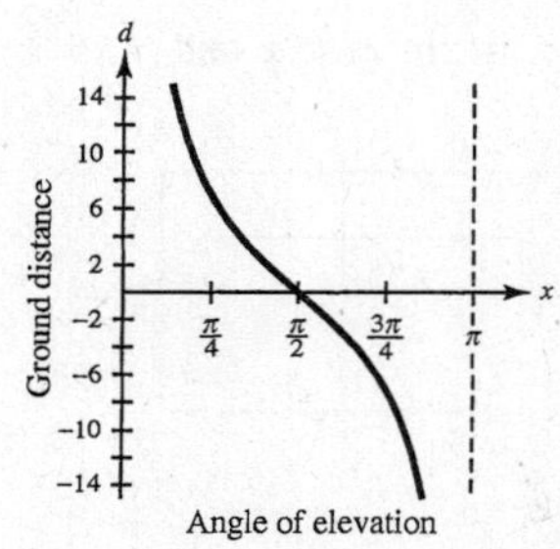

75. $C = 5000 + 2000 \sin \dfrac{\pi t}{12}$

$R = 25{,}000 + 15{,}000 \cos \dfrac{\pi t}{12}$

(a)

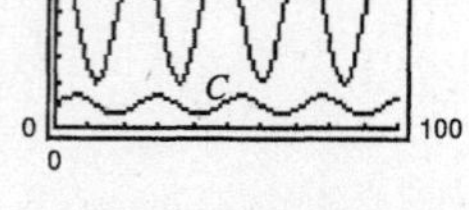

(b) As the predator population increases, the number of prey decreases. When the number of prey is small, the number of predators decreases.

(c) The period for both C and R is:

$$p = \frac{2\pi}{\pi/12} = 24 \text{ months}$$

When the prey population is highest, the predator population is increasing most rapidly.
When the prey population is lowest, the predator population is decreasing most rapidly.
When the predator population is lowest, the prey population is increasing most rapidly.
When the predator population is highest, the prey population is decreasing most rapidly.

In addition; weather, food sources for the prey, hunting, all affect the populations of both the predator and the prey.

77. $H(t) = 54.33 - 20.38 \cos \frac{\pi t}{6} - 15.69 \sin \frac{\pi t}{6}$

$L(t) = 39.36 - 15.70 \cos \frac{\pi t}{6} - 14.16 \sin \frac{\pi t}{6}$

(a) Period of $\cos \frac{\pi t}{6}$: $\frac{2\pi}{\pi/6} = 12$

Period of $\sin \frac{\pi t}{6}$: $\frac{2\pi}{\pi/6} = 12$

Period of $H(t)$: 12

Period of $L(t)$: 12

(b) From the graph, it appears that the greatest difference between high and low temperatures occurs in summer. The smallest difference occurs in winter.

(c) The highest high and low temperatures appear to occur around the middle of July, roughly one month after the time when the sun is northernmost in the sky.

79. True. Since $y = \csc x = \frac{1}{\sin x}$, for a given value of x, the y-coordinate of $\csc x$ is the reciprocal of the y-coordinate of $\sin x$.

81. As $x \rightarrow \frac{\pi}{2}$ from the left, $f(x) = \tan x \rightarrow \infty$.

As $x \rightarrow \frac{\pi}{2}$ from the right, $f(x) = \tan x \rightarrow -\infty$.

83. $f(x) = x - \cos x$

(a)

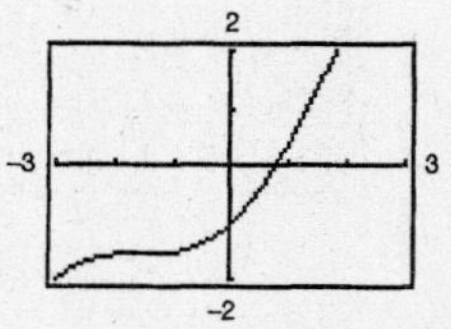

The zero between 0 and 1 appears to occur at $x \approx 0.7391$.

(b) $x_n = \cos(x_{n-1})$

$x_0 = 1$

$x_1 = \cos 1 \approx 0.5403$

$x_2 = \cos 0.5403 \approx 0.8576$

$x_3 = \cos 0.8576 \approx 0.6543$

$x_4 = \cos 0.6543 \approx 0.7935$

$x_5 = \cos 0.7935 \approx 0.7014$

$x_6 = \cos 0.7014 \approx 0.7640$

$x_7 = \cos 0.7640 \approx 0.7221$

$x_8 = \cos 0.7221 \approx 0.7504$

$x_9 = \cos 0.7504 \approx 0.7314$

$\vdots$

This sequence appears to be approaching the zero of f: $x \approx 0.7391$.

85. $y_1 = \sec x$

$y_2 = 1 + \frac{x^2}{2!} + \frac{5x^4}{4!}$

The approximation appears to coincide on the interval $-1.1 \le x \le 1.1$.

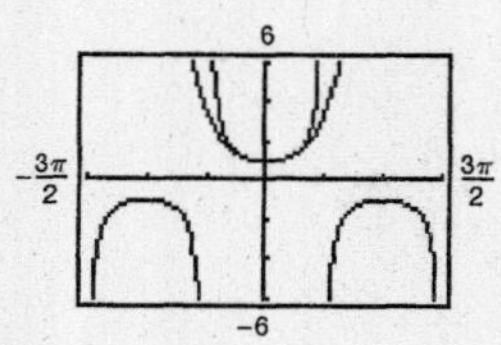

87. $x^2 = 64$

$x = \pm\sqrt{64} = \pm 8$

89. $4x^2 - 12x + 9 = 0$

$(2x - 3)^2 = 0$

$2x - 3 = 0$

$x = \frac{3}{2}$

91. $x^2 - 6x + 4 = 0$ Complete the Square

$x^2 - 6x = -4$

$x^2 - 6x + 9 = -4 + 9$

$(x - 3)^2 = 5$

$x - 3 = \pm\sqrt{5}$

$x = 3 \pm \sqrt{5}$

93. $50 + 5x = 3x^2$

$0 = 3x^2 - 5x - 50$

$0 = (3x + 10)(x - 5)$

$3x + 10 = 0 \Rightarrow x = -\frac{10}{3}$

$x - 5 = 0 \Rightarrow x = 5$

Section 1.7 Inverse Trigonometric Functions

- You should know the definitions, domains, and ranges of $y = \arcsin x$, $y = \arccos x$, and $y = \arctan x$.

Function	Domain	Range
$y = \arcsin x \Rightarrow x = \sin y$	$-1 \le x \le 1$	$-\frac{\pi}{2} \le y \le \frac{\pi}{2}$
$y = \arccos x \Rightarrow x = \cos y$	$-1 \le x \le 1$	$0 \le y \le \pi$
$y = \arctan x \Rightarrow x = \tan y$	$-\infty < x < \infty$	$-\frac{\pi}{2} < x < \frac{\pi}{2}$

- You should know the inverse properties of the inverse trigonometric functions.

$\sin(\arcsin x) = x$ and $\arcsin(\sin y) = y, -\frac{\pi}{2} \le y \le \frac{\pi}{2}$

$\cos(\arccos x) = x$ and $\arccos(\cos y) = y, 0 \le y \le \pi$

$\tan(\arctan x) = x$ and $\arctan(\tan y) = y, -\frac{\pi}{2} < y < \frac{\pi}{2}$

- You should be able to use the triangle technique to convert trigonometric functions of inverse trigonometric functions into algebraic expressions.

Solutions to Odd-Numbered Exercises

1. $y = \arcsin\frac{1}{2} \Rightarrow \sin y = \frac{1}{2}$ for

$-\frac{\pi}{2} \le y \le \frac{\pi}{2} \Rightarrow y = \frac{\pi}{6}$

3. $y = \arccos\frac{1}{2} \Rightarrow \cos y = \frac{1}{2}$ for

$0 \le y \le \pi \Rightarrow y = \frac{\pi}{3}$

5. $y = \arctan\frac{\sqrt{3}}{3} \Rightarrow \tan y = \frac{\sqrt{3}}{3}$ for

$-\frac{\pi}{2} < y < \frac{\pi}{2} \Rightarrow y = \frac{\pi}{6}$

7. $y = \arccos\left(-\frac{\sqrt{3}}{2}\right) \Rightarrow \cos y = -\frac{\sqrt{3}}{2}$ for

$0 \le y \le \pi \Rightarrow y = \frac{5\pi}{6}$

9. $y = \arctan(-\sqrt{3}) \Rightarrow \tan y = -\sqrt{3}$ for

$-\frac{\pi}{2} < y < \frac{\pi}{2} \Rightarrow y = -\frac{\pi}{3}$

11. $y = \arccos\left(-\frac{1}{2}\right) \Rightarrow \cos y = -\frac{1}{2}$ for

$0 \le y \le \pi \Rightarrow y = \frac{2\pi}{3}$

13. $y = \arcsin \dfrac{\sqrt{3}}{2} \Rightarrow \sin y = \dfrac{\sqrt{3}}{2}$ for

$-\dfrac{\pi}{2} \le y \le \dfrac{\pi}{2} \Rightarrow y = \dfrac{\pi}{3}$

15. $y = \arctan 0 \Rightarrow \tan y = 0$ for $-\dfrac{\pi}{2} < y < \dfrac{\pi}{2} \Rightarrow y = 0$

17. $\arccos 0.28 = \cos^{-1} 0.28 \approx 1.29$

19. $\arcsin(-0.75) = \sin^{-1}(-0.75) \approx -0.85$

21. $\arctan(-3) = \tan^{-1}(-3) \approx -1.25$

23. $\arcsin 0.31 = \sin^{-1} 0.31 \approx 0.32$

25. $\arccos(-0.41) = \cos^{-1}(-0.41) \approx 1.99$

27. $\arctan 0.92 = \tan^{-1} 0.92 \approx 0.74$

29. $\arcsin\left(\dfrac{3}{4}\right) = \sin^{-1}(0.75) \approx 0.85$

31. $\arctan\left(\dfrac{7}{2}\right) = \tan^{-1}(3.5) \approx 1.29$

33. This is the graph of $y = \arctan x$. The coordinates are $\left(-\sqrt{3}, -\dfrac{\pi}{3}\right)$, $\left(-\dfrac{\sqrt{3}}{3}, -\dfrac{\pi}{6}\right)$, and $\left(1, \dfrac{\pi}{4}\right)$.

35. $f(x) = \tan x$ and $g(x) = \arctan x$

Graph $y_1 = \tan x$

Graph $y_2 = \tan^{-1} x$

Graph $y_3 = x$

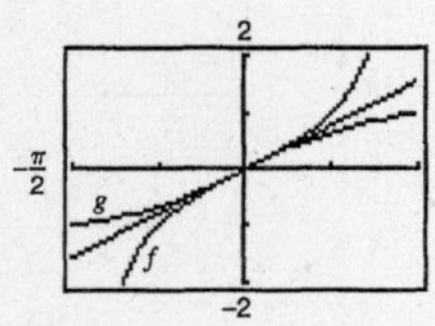

37. $\tan \theta = \dfrac{x}{4}$

$\theta = \arctan \dfrac{x}{4}$

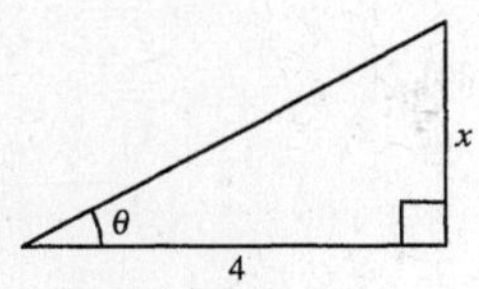

39. $\sin \theta = \dfrac{x+2}{5}$

$\theta = \arcsin\left(\dfrac{x+2}{5}\right)$

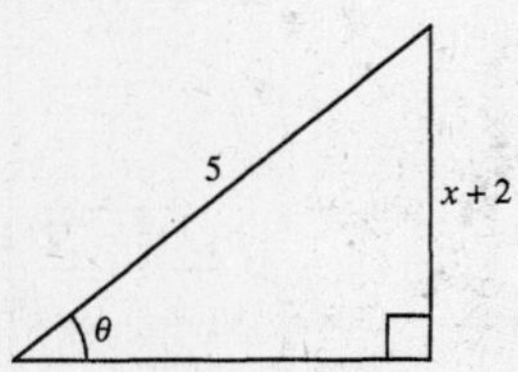

41. $\cos \theta = \dfrac{x+3}{2x}$

$\theta = \arccos\left(\dfrac{x+3}{2x}\right)$

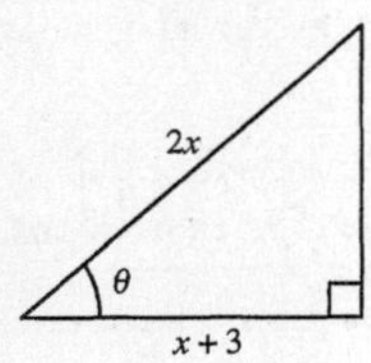

43. $\sin(\arcsin 0.3) = 0.3$

45. $\cos[\arccos(-0.1)] = -0.1$

47. $\arcsin(\sin 3\pi) = \arcsin(0) = 0$

Note: 3π is not in the range of the arcsine function.

49. Let $y = \arctan \dfrac{3}{4}$. Then,

$\tan y = \dfrac{3}{4}$, $0 < y < \dfrac{\pi}{2}$

and $\sin y = \dfrac{3}{5}$.

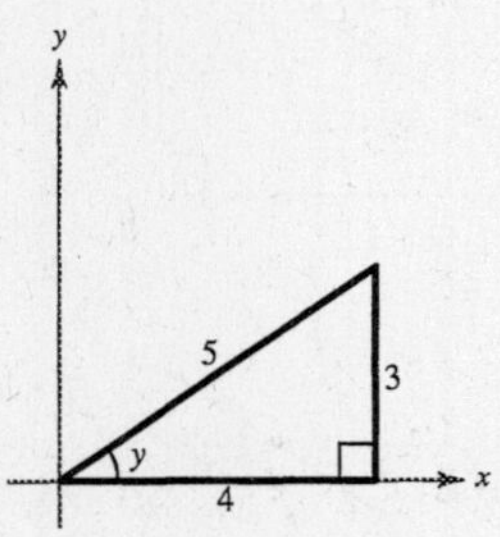

51. Let $y = \arctan 2$. Then,

$\tan y = 2 = \dfrac{2}{1}$, $0 < y < \dfrac{\pi}{2}$

and $\cos y = \dfrac{1}{\sqrt{5}} = \dfrac{\sqrt{5}}{5}$.

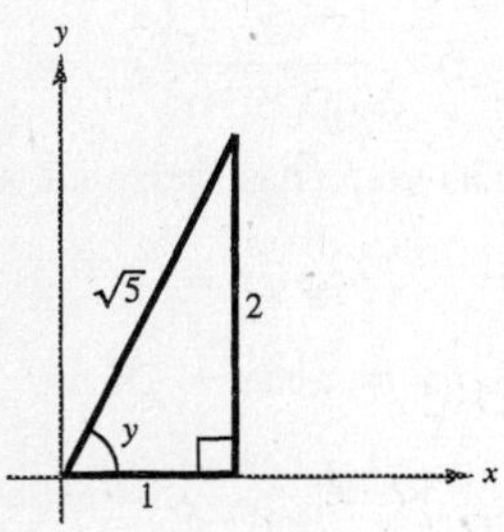

53. Let $y = \arcsin \frac{5}{13}$. Then,

$\sin y = \frac{5}{13}, \; 0 < y < \frac{\pi}{2}$

and $\cos y = \frac{12}{13}$.

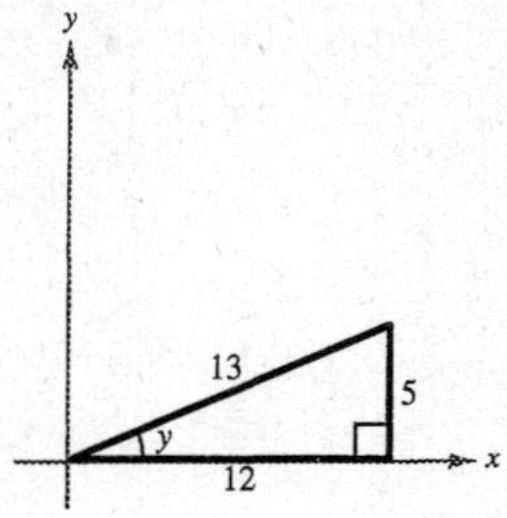

55. Let $y = \arctan\left(-\frac{3}{5}\right)$. Then,

$\tan y = -\frac{3}{5}, \; -\frac{\pi}{2} < y < 0$

and $\sec y = \frac{\sqrt{34}}{5}$.

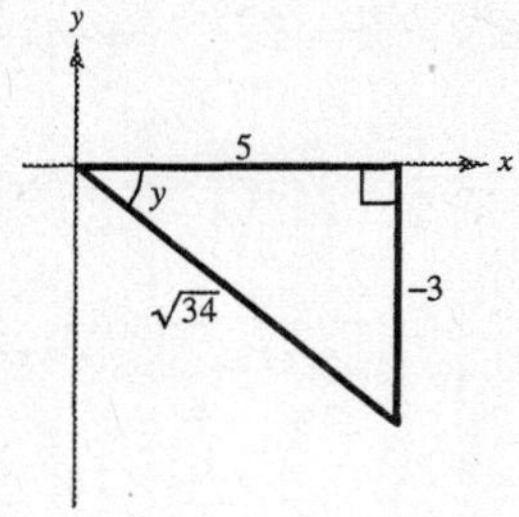

57. Let $y = \arccos\left(-\frac{2}{3}\right)$. Then,

$\cos y = -\frac{2}{3}, \; \frac{\pi}{2} < y < \pi$

and $\sin y = \frac{\sqrt{5}}{3}$.

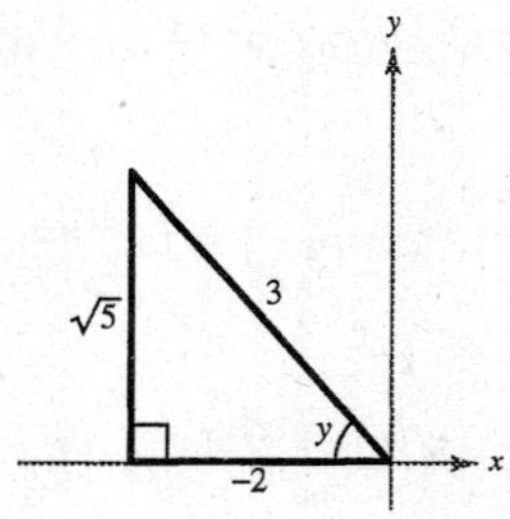

59. Let $y = \arctan x$. Then,

$\tan y = x = \frac{x}{1}$

and $\cot y = \frac{1}{x}$.

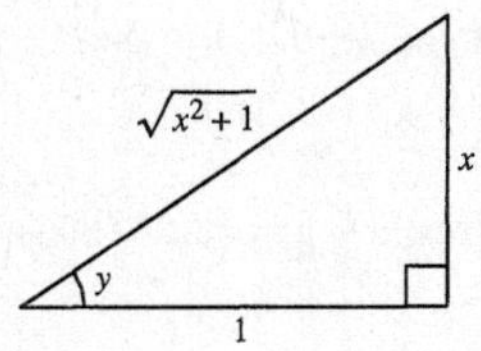

61. Let $y = \arcsin(2x)$. Then,

$\sin y = 2x = \frac{2x}{1}$

and $\cos y = \sqrt{1 - 4x^2}$.

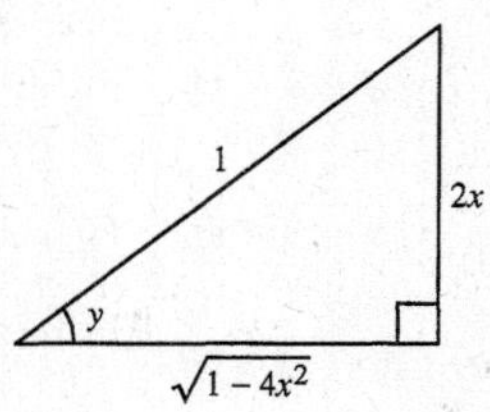

63. Let $y = \arccos x$. Then,

$\cos y = x = \frac{x}{1}$

and $\sin y = \sqrt{1 - x^2}$.

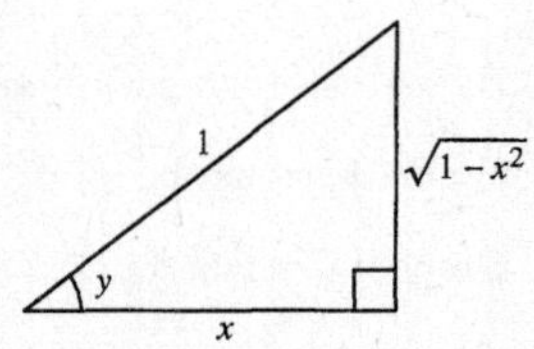

65. Let $y = \arccos\left(\frac{x}{3}\right)$. Then,

$\cos y = \frac{x}{3}$

and $\tan y = \frac{\sqrt{9 - x^2}}{x}$.

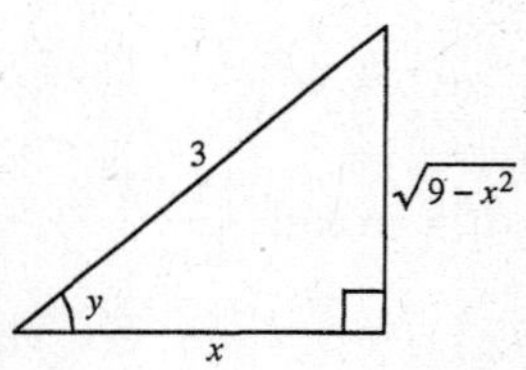

67. Let $y = \arctan \frac{x}{\sqrt{2}}$. Then,

$\tan y = \frac{x}{\sqrt{2}}$

and $\csc y = \frac{\sqrt{x^2 + 2}}{x}$.

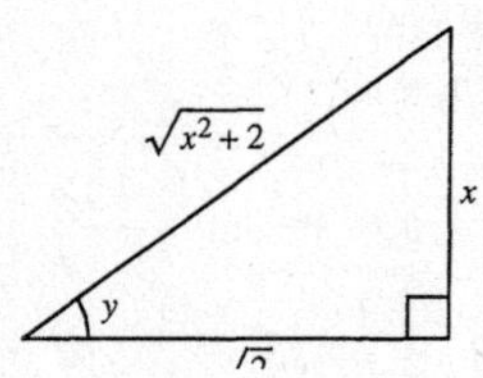

69. $f(x) = \sin(\arctan 2x), \; g(x) = \frac{2x}{\sqrt{1 + 4x^2}}$

Let $y = \arctan 2x$. Then,

$\tan y = 2x = \frac{2x}{1}$

and $\sin y = \frac{2x}{\sqrt{1 + 4x^2}}$.

$g(x) = \frac{2x}{\sqrt{1 + 4x^2}} = f(x)$

The graph has horizontal asymptotes at $y = \pm 1$.

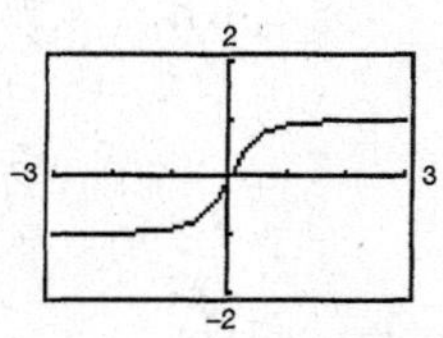

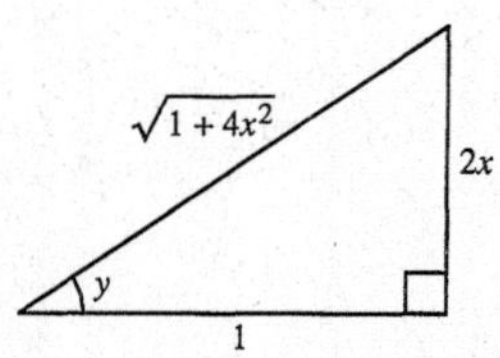

71. Let $y = \arctan \frac{9}{x}$. Then,

$\tan y = \frac{9}{x}$ and $\sin y = \frac{9}{\sqrt{x^2 + 81}}, x > 0; \; \frac{-9}{\sqrt{x^2 + 81}}, x < 0$

Thus, arcsin

$y = \frac{9}{\sqrt{x^2 + 81}}, x > 0; \text{ arcsin } y = \frac{-9}{\sqrt{x^2 + 81}}, x < 0$

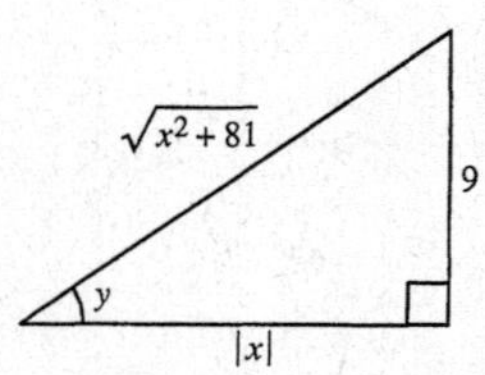

73. Let $y = \arccos \dfrac{3}{\sqrt{x^2 - 2x + 10}}$. Then,

$$\cos y = \frac{3}{\sqrt{x^2 - 2x + 10}} = \frac{3}{\sqrt{(x-1)^2 + 9}}$$

and $\sin y = \dfrac{|x-1|}{\sqrt{(x-1)^2 + 9}}$.

Thus, $y = \arcsin \dfrac{|x-1|}{\sqrt{x^2 - 2x + 10}}$.

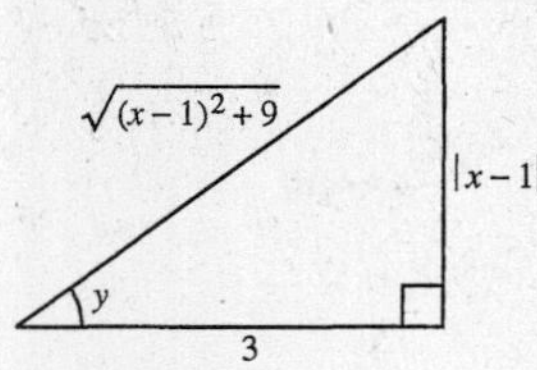

75. $y = 2 \arccos x$

Domain: $-1 \le x \le 1$

Range: $0 \le y \le 2\pi$

Vertical stretch of $f(x) = \arccos x$

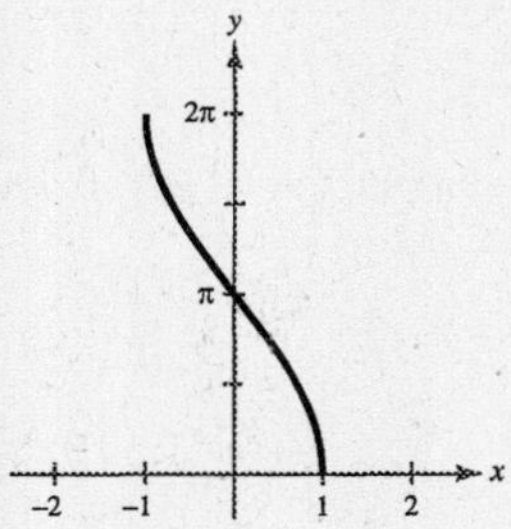

77. The graph of $f(x) = \arcsin(x - 1)$ is a horizontal translation of the graph of $y = \arcsin x$ by one unit.

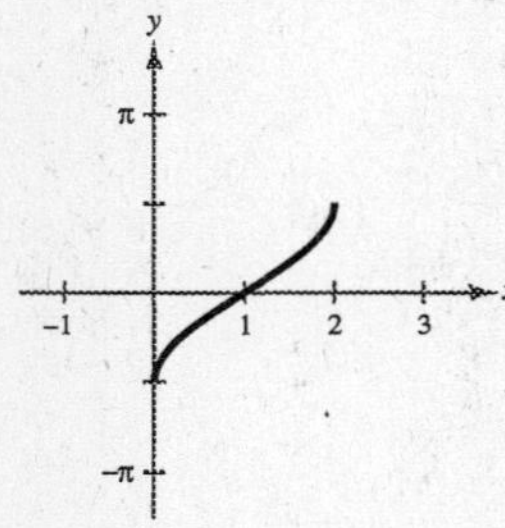

79. $f(x) = \arctan 2x$

Domain: all real numbers

Range: $-\dfrac{\pi}{2} < y < \dfrac{\pi}{2}$

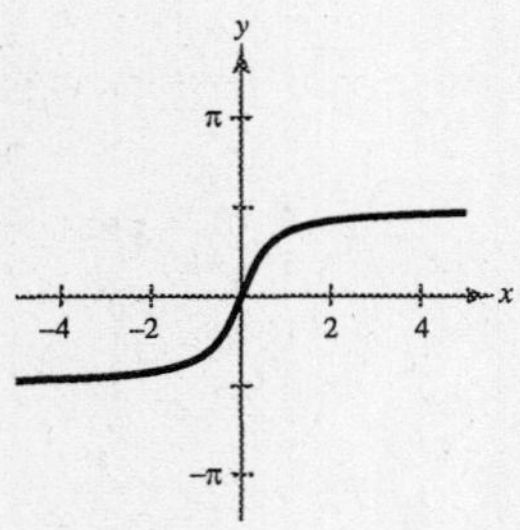

81. $h(v) = \tan(\arccos v) = \dfrac{\sqrt{1 - v^2}}{v}$

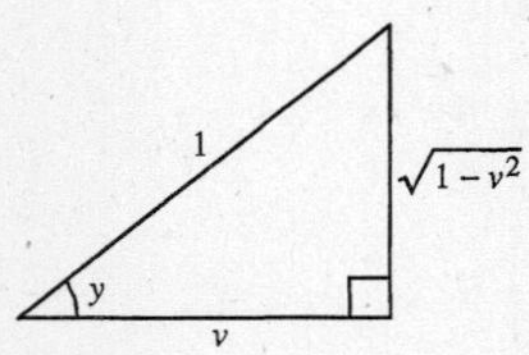

Domain: $-1 \le v \le 1, v \ne 0$

Range: all real numbers

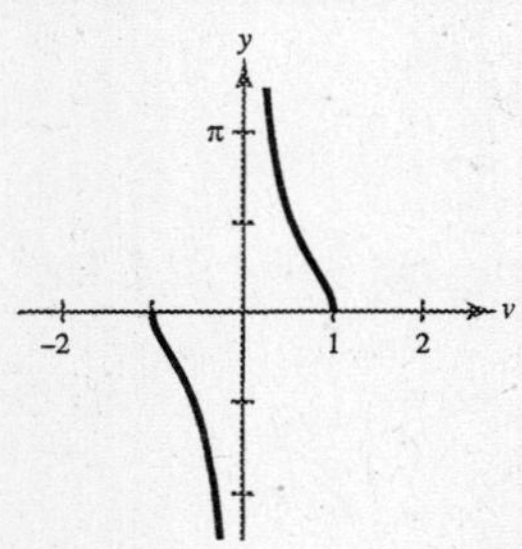

83. $f(x) = 2 \arccos (2x)$

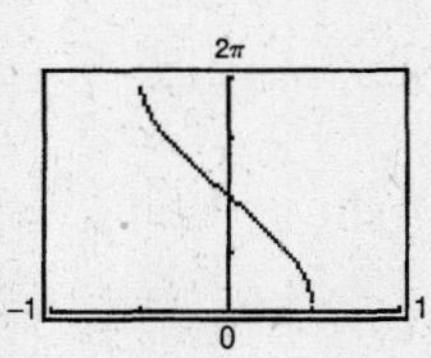

85. $f(x) = \arctan (2x - 3)$

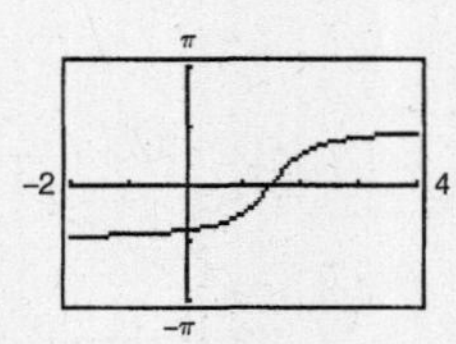

87. $f(x) = \pi - \arcsin \left(\dfrac{2}{3}\right) \approx 2.412$

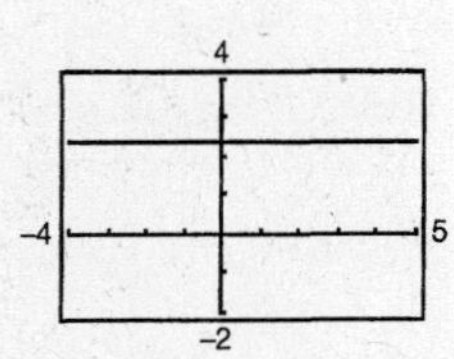

89. $f(t) = 3\cos 2t + 3\sin 2t = \sqrt{3^2 + 3^2}\sin\left(2t + \arctan\dfrac{3}{3}\right)$

$$= 3\sqrt{2}\sin(2t + \arctan 1)$$

$$= 3\sqrt{2}\sin\left(2t + \frac{\pi}{4}\right)$$

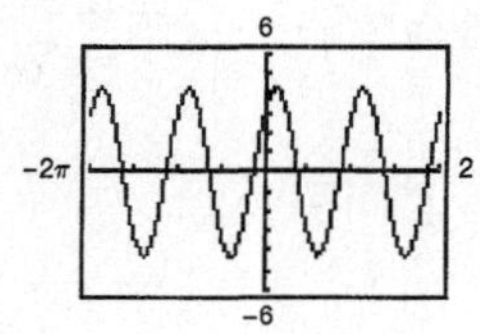

The graphs are the same and implies that the identity is true.

91. (a) $\sin\theta = \dfrac{5}{s}$

$$\theta = \arcsin\frac{5}{s}$$

(b) $s = 40$: $\theta = \arcsin\dfrac{5}{40} \approx 0.13$

$s = 20$: $\theta = \arcsin\dfrac{5}{20} \approx 0.25$

93. $\beta = \arctan\dfrac{3x}{x^2 + 4}$

(a)

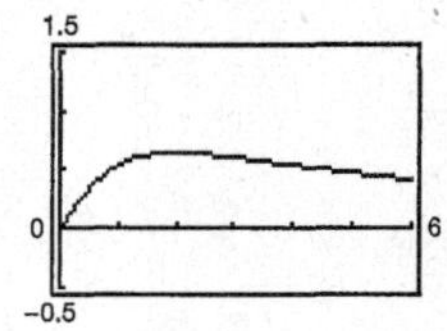

(b) β is maximum when $x = 2$.

(c) The graph has a horizontal asymptote at $\beta = 0$. As x increases, β decreases.

95.

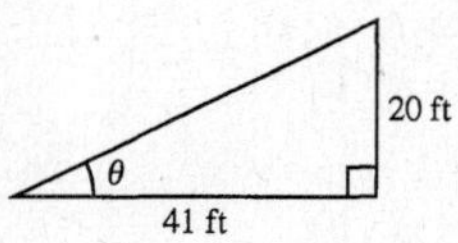

(a) $\tan\theta = \dfrac{20}{41}$

$$\theta = \arctan\left(\frac{20}{41}\right) \approx 26.0°$$

(b) $\tan 26° = \dfrac{h}{50}$

$$h = 50\tan 26° \approx 24.39 \text{ feet}$$

97. (a) $\tan\theta = \dfrac{x}{20}$

$$\theta = \arctan\frac{x}{20}$$

(b) $x = 5$: $\theta = \arctan\dfrac{5}{20} \approx 14.0°$

$x = 12$: $\theta = \arctan\dfrac{12}{20} \approx 31.0°$

99. False; $\arctan 1 = \dfrac{\pi}{4}$. $\dfrac{5\pi}{4}$ is not in the range of the arctangent function.

101. $y = \operatorname{arcsec} x$ if and only if $\sec y = x$ where $x \le -1 \cup x \ge 1$ and $0 \le y < \dfrac{\pi}{2}$ and $\dfrac{\pi}{2} < y \le \pi$. The domain of $y = \operatorname{arcsec} x$ is $(-\infty, -1] \cup [1, \infty)$ and the range is $\left[0, \dfrac{\pi}{2}\right) \cup \left(\dfrac{\pi}{2}, \pi\right]$.

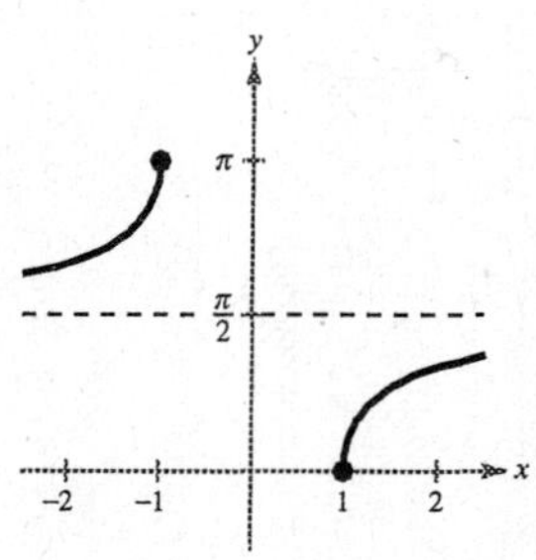

103. (a) $y = \operatorname{arcsec}\sqrt{2} \Rightarrow \sec y = \sqrt{2}$ and $0 \le y < \dfrac{\pi}{2} \cup \dfrac{\pi}{2} < y \le \pi \Rightarrow y = \dfrac{\pi}{4}$

(b) $y = \operatorname{arcsec} 1 \Rightarrow \sec y = 1$ and $0 \le y < \dfrac{\pi}{2} \cup \dfrac{\pi}{2} < y \le \pi \Rightarrow y = 0$

(c) $y = \operatorname{arccot}(-\sqrt{3}) \Rightarrow \cot y = -\sqrt{3}$ and $0 < y < \pi \Rightarrow y = \dfrac{5\pi}{6}$

(d) $y = \operatorname{arccsc} 2 \Rightarrow \csc y = 2$ and $-\dfrac{\pi}{2} \le y < 0 \cup 0 < y \le \dfrac{\pi}{2} \Rightarrow y = \dfrac{\pi}{6}$

105. $f(x) = \sqrt{x}$

$g(x) = 6 \arctan x$

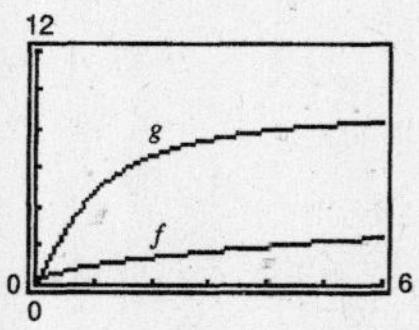

As x increases to infinity, g approaches 3π, but f has no maximum. Using the solve feature of the graphing utility, you find $a \approx 87.54$.

107. Let $y = \arcsin(-x)$. Then,

$$\sin y = -x$$
$$-\sin y = x$$
$$\sin(-y) = x$$
$$-y = \arcsin x$$
$$y = -\arcsin x.$$

Therefore, $\arcsin(-x) = -\arcsin x$.

109. $y = \pi - \arccos x$

$$\cos y = \cos(\pi - \arccos x)$$
$$\cos y = \cos \pi \cos(\arccos x) + \sin \pi \sin(\arccos x)$$
$$\cos y = -x$$
$$y = \arccos(-x)$$

Therefore, $\arccos(-x) = \pi - \arccos x$.

111. Let $\alpha = \arcsin x$ and $\beta = \arccos x$, then $\sin \alpha = x$ and $\cos \beta = x$. Thus, $\sin \alpha = \cos \beta$ which implies that α and β are complementary angles and we have

$$\alpha + \beta = \frac{\pi}{2}$$
$$\arcsin x + \arccos x = \frac{\pi}{2}.$$

113. $\sin \theta = \dfrac{3}{4} = \dfrac{\text{opp}}{\text{hyp}}$

$$(\text{adj})^2 + (3)^2 = (4)^2$$
$$(\text{adj})^2 + 9 = 16$$
$$(\text{adj})^2 = 7$$
$$\text{adj} = \sqrt{7}$$

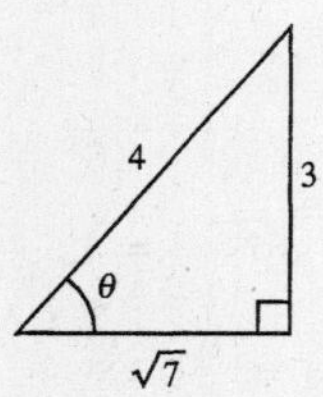

115. $\cos \theta = \dfrac{5}{6} = \dfrac{\text{adj}}{\text{hyp}}$

$$(\text{opp})^2 + (5)^2 = (6)^2$$
$$(\text{opp})^2 + 25 = 36$$
$$(\text{opp})^2 = 11$$
$$\text{opp} = \sqrt{11}$$

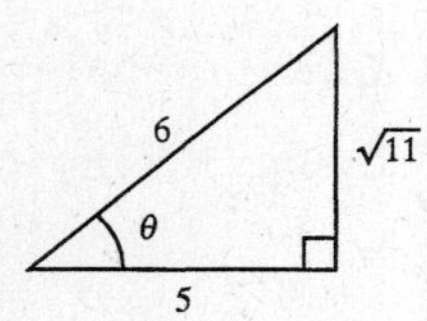

117. Let $x =$ the number of people presently in the group. Each person's share is now $\dfrac{250{,}000}{x}$. If two more join the group, each person's share would then be $\dfrac{250{,}000}{x+2}$.

$$\begin{matrix}\text{Share per person with} \\ \text{two more people}\end{matrix} = \begin{matrix}\text{Original share} \\ \text{per person}\end{matrix} - 6250$$

$$\frac{250{,}000}{x+2} = \frac{250{,}000}{x} - 6250$$
$$250{,}000x = 250{,}000(x+2) - 6250x(x+2)$$
$$250{,}000x = 250{,}000x + 500{,}000 - 6250x^2 - 12500x$$

$$6250x^2 + 12500x - 500{,}000 = 0$$
$$6250(x^2 + 2x - 80) = 0$$
$$6250(x + 10)(x - 8) = 0$$
$$x = -10 \quad \text{or} \quad x = 8$$

$x = -10$ is not possible.

There were 8 people in the original group.

Section 1.8 Applications and Models

- You should be able to solve right triangles.
- You should be able to solve right triangle applications.
- You should be able to solve applications of simple harmonic motion.

Solutions to Odd-Numbered Exercises

1. Given: $A = 20°,\ b = 10$

$$\tan A = \frac{a}{b} \Rightarrow a = b \tan A = 10 \tan 20° \approx 3.64$$

$$\cos A = \frac{b}{c} \Rightarrow c = \frac{b}{\cos A} = \frac{10}{\cos 20°} \approx 10.64$$

$$B = 90° - 20° = 70°$$

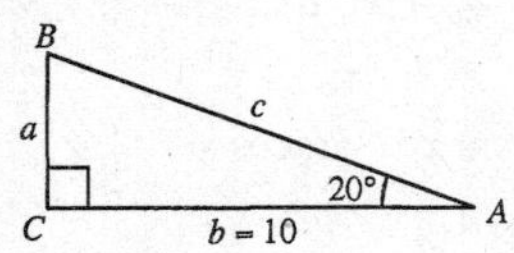

3. Given: $B = 71°,\ b = 24$

$$\tan B = \frac{b}{a} \Rightarrow a = \frac{b}{\tan B} = \frac{24}{\tan 71°} \approx 8.26$$

$$\sin B = \frac{b}{c} \Rightarrow c = \frac{b}{\sin B} = \frac{24}{\sin 71°} \approx 25.38$$

$$A = 90° - 71° = 19°$$

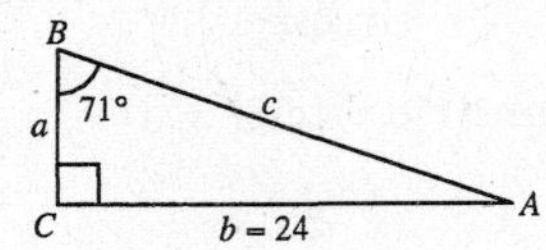

5. Given: $a = 6,\ b = 10$

$$c^2 = a^2 + b^2 \Rightarrow c = \sqrt{36 + 100}$$
$$= 2\sqrt{34} \approx 11.66$$

$$\tan A = \frac{a}{b} = \frac{6}{10} \Rightarrow A = \arctan\frac{3}{5} \approx 30.96°$$

$$B = 90° - 30.96° = 59.04°$$

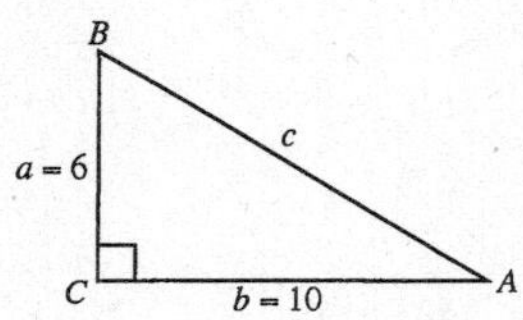

7. $b = 16,\ c = 52$

$$a = \sqrt{52^2 - 16^2}$$
$$= \sqrt{2448} = 12\sqrt{17} \approx 49.48$$

$$\cos A = \frac{16}{52}$$

$$A = \arccos\frac{16}{52} \approx 72.08°$$

$$B = 90° - 72.08° \approx 17.92°$$

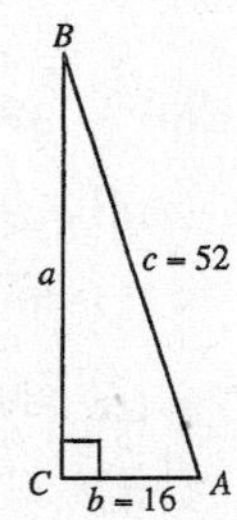

9. $A = 12°15',\ c = 430.5$

$$B = 90° - 12°15' = 77°45'$$

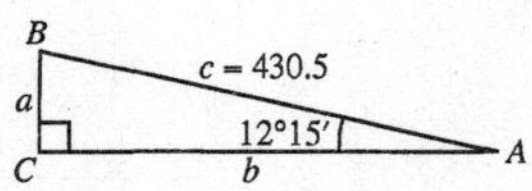

$$\sin 12°15' = \frac{a}{430.5}$$
$$a = 430.5 \sin 12°15' \approx 91.34$$

$$\cos 12°15' = \frac{b}{430.5}$$
$$b = 430.5 \cos 12°15' \approx 420.70$$

11. $\tan\theta = \dfrac{h}{\frac{1}{2}b} \Rightarrow h = \dfrac{1}{2}b\tan\theta$

$$h = \frac{1}{2}(4)\tan 52° \approx 2.56 \text{ inches}$$

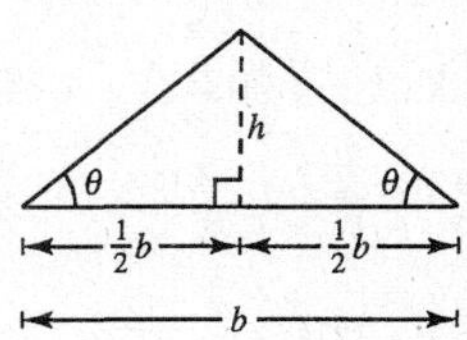

13. $\tan\theta = \dfrac{h}{\frac{1}{2}b} \Rightarrow h = \dfrac{1}{2}b\tan\theta$

$$h = \frac{1}{2}(46)\tan 41° \approx 19.99 \text{ inches}$$

15. $\tan 25° = \dfrac{50}{x}$

$$x = \frac{50}{\tan 25°} \approx 107.2 \text{ feet}$$

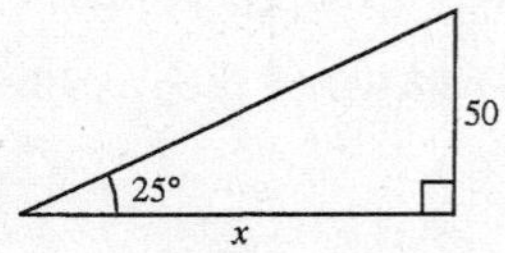

17. $\sin 80° = \dfrac{h}{20}$

$20 \sin 80° = h$

$h \approx 19.7$ feet

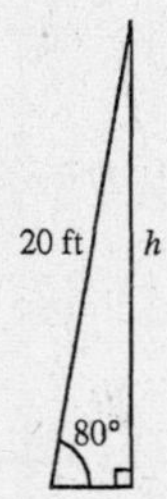

19. (a)

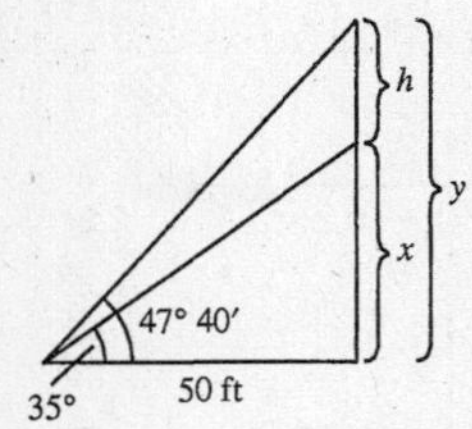

(b) Let the height of the church $= x$ and the height of the church and steeple $= y$. Then,

$\tan 35° = \dfrac{x}{50}$ and $\tan 47°40' = \dfrac{y}{50}$

$x = 50 \tan 35°$ and $y = 50 \tan 47°40'$

$h = y - x = 50(\tan 47°40' - \tan 35°)$.

(c) $h \approx 19.9$ feet

21. $\sin 34° = \dfrac{x}{4000}$

$x = 4000 \sin 34°$

≈ 2236.8 feet

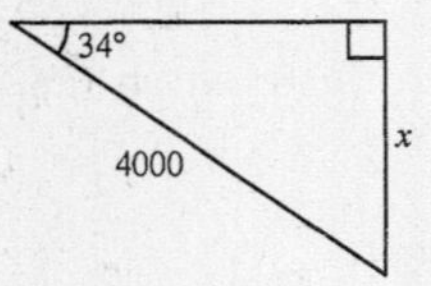

23. (a)

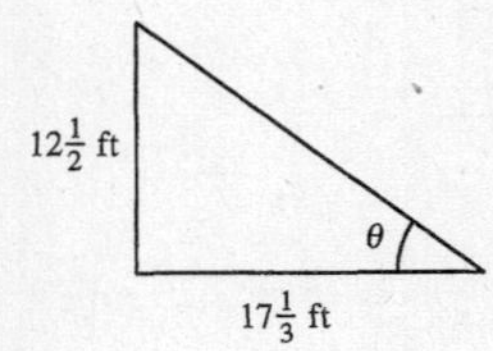

(b) $\tan \theta = \dfrac{12\frac{1}{2}}{17\frac{1}{3}}$

(c) $\theta = \arctan \dfrac{12\frac{1}{2}}{17\frac{1}{3}} \approx 35.8°$

25. 1200 feet + 150 feet − 400 feet = 950 feet

Not drawn to scale

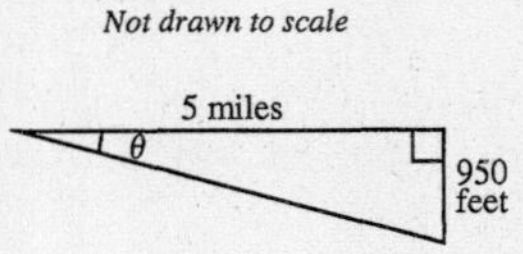

$5 \text{ miles} = 5 \text{ miles}\left(\dfrac{5280 \text{ ft}}{1 \text{ mile}}\right) = 26{,}400 \text{ feet}$

$\tan \theta = \dfrac{950}{26{,}400}$

$\theta = \text{artan}\left(\dfrac{950}{26{,}400}\right) \approx 2.06°$

27. $\sin 10.5° = \dfrac{x}{4}$

$x = 4 \sin 10.5°$

≈ 0.73 mile

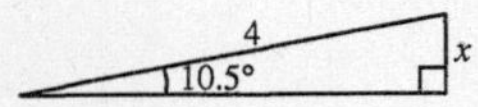

29. The plane has traveled 1.5 (600) = 900 miles.

$\sin 38° = \dfrac{a}{900} \implies a \approx 554$ miles north

$\cos 38° = \dfrac{b}{900} \implies b \approx 709$ miles east

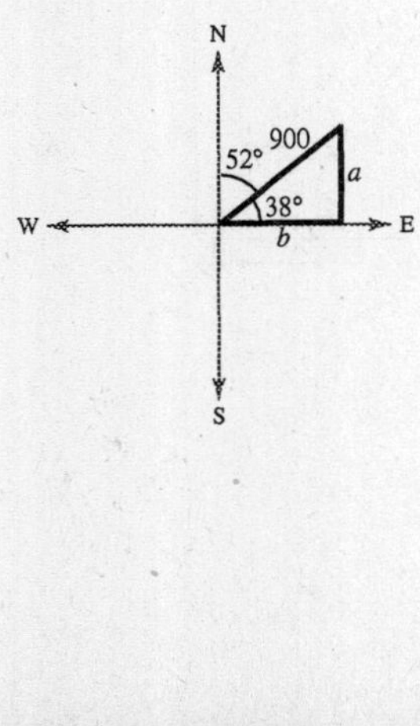

31. $\tan 14° = \dfrac{d}{x} \Rightarrow x = d \cot 14°$

$$\tan 34° = \frac{d}{y} = \frac{d}{30 - x} = \frac{d}{30 - d \cot 14°}$$

$$\cot 34° = \frac{30 - d \cot 14°}{d}$$

$$d \cot 34° = 30 - d \cot 14°$$

$$d = \frac{30}{\cot 34° + \cot 14°}$$

$$\approx 5.46 \text{ kilometers}$$

33. $\tan 6.5° = \dfrac{350}{d} \Rightarrow d \approx 3071.91$ ft

$\tan 4° = \dfrac{350}{D} \Rightarrow D \approx 5005.23$ ft

Distance between ships: $D - d \approx 1933.3$ ft

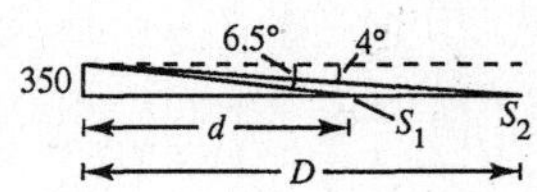

35. $\tan 57° = \dfrac{a}{x} \Rightarrow x = a \cot 57°$

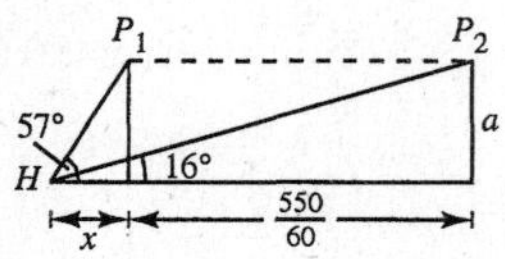

$$\tan 16° = \frac{a}{x + \frac{55}{6}}$$

$$\tan 16° = \frac{a}{a \cot 57° + \frac{55}{6}}$$

$$\cot 16° = \frac{a \cot 57° + \frac{55}{6}}{a}$$

$$a \cot 16° - a \cot 57° = \frac{55}{6} \Rightarrow a \approx 3.23 \text{ miles}$$

$$\approx 17{,}054 \text{ ft}$$

37. L_1: $3x - 2y = 5 \Rightarrow y = \frac{3}{2}x - \frac{5}{2} \Rightarrow m_1 = \frac{3}{2}$

L_2: $x + y = 1 \Rightarrow y = -x + 1 \Rightarrow m_2 = -1$

$$\tan \alpha = \left|\frac{-1 - \frac{3}{2}}{1 + (-1)(\frac{3}{2})}\right| = \left|\frac{-\frac{5}{2}}{-\frac{1}{2}}\right| = 5$$

$$\alpha = \arctan 5 \approx 78.7°$$

39. The diagonal of the base has a length of $\sqrt{a^2 + a^2} = \sqrt{2}a$.

Now, we have $\tan\theta = \dfrac{a}{\sqrt{2}a} = \dfrac{1}{\sqrt{2}}$

$$\theta = \arctan\frac{1}{\sqrt{2}}$$

$$\theta \approx 35.3°.$$

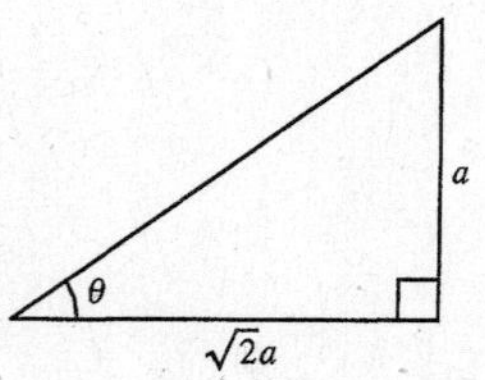

41. $\sin 36° = \dfrac{d}{25} \Rightarrow d \approx 14.69$

Length of side: $2d \approx 29.4$ inches

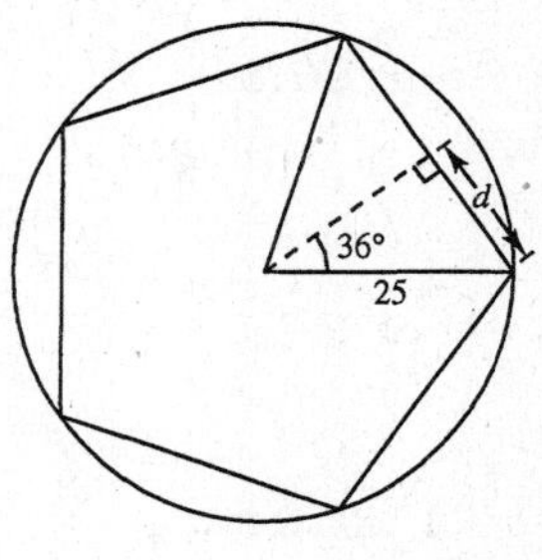

43. $\cos 30° = \dfrac{b}{r}$

$$b = r \cos 30°$$

$$b = \frac{\sqrt{3}r}{2}$$

$$y = 2b = 2\left(\frac{\sqrt{3}r}{2}\right) = \sqrt{3}r$$

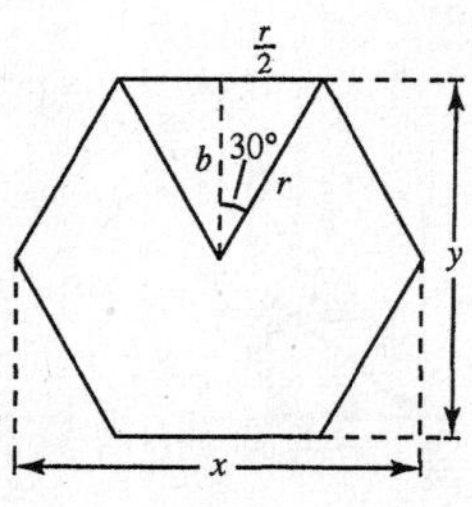

45. $\tan 35° = \dfrac{b}{10}$

$$b = 10 \tan 35° \approx 7$$

$$\cos 35° = \frac{10}{a}$$

$$a = \frac{10}{\cos 35°} \approx 12.2$$

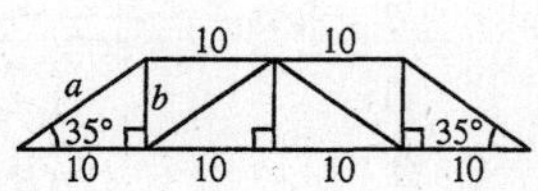

47. $d = 4 \cos 8\pi t$

(a) Maximum displacement $=$ amplitude $= 4$

(b) Frequency $= \dfrac{\omega}{2\pi} = \dfrac{8\pi}{2\pi}$

$= 4$ cycles per unit of time

(c) $8\pi t = \dfrac{\pi}{2} \Rightarrow t = \dfrac{1}{16}$

49. $d = \dfrac{1}{16} \sin 120\pi t$

(a) Maximum displacement $=$ amplitude $= \dfrac{1}{16}$

(b) Frequency $= \dfrac{\omega}{2\pi} = \dfrac{120\pi}{2\pi}$

$= 60$ cycles per unit of time

(c) $120\pi t = \pi \Rightarrow t = \dfrac{1}{120}$

51. $d = 0$ when $t = 0$, $a = 4$, Period $= 2$

Use $d = a \sin \omega t$ since $d = 0$ when $t = 0$.

$\dfrac{2\pi}{\omega} = 2 \Rightarrow \omega = \pi$

Thus, $d = 4 \sin(\pi t)$.

53. $d = 3$ when $t = 0$, $a = 3$, Period $= 1.5$

Use $d = a \cos \omega t$ since $d = 3$ when $t = 0$.

$\dfrac{2\pi}{\omega} = 1.5 \Rightarrow \omega = \dfrac{4\pi}{3}$

Thus, $d = 3 \cos\left(\dfrac{4\pi}{3}t\right) = 3 \cos\left(\dfrac{4\pi t}{3}\right)$.

55. $d = a \sin \omega t$

Period $= \dfrac{2\pi}{\omega} = \dfrac{1}{\text{frequency}}$

$\dfrac{2\pi}{\omega} = \dfrac{1}{264}$

$\omega = 2\pi(264) = 528\pi$

57. $y = \dfrac{1}{4} \cos 16t,\ t > 0$

(a)

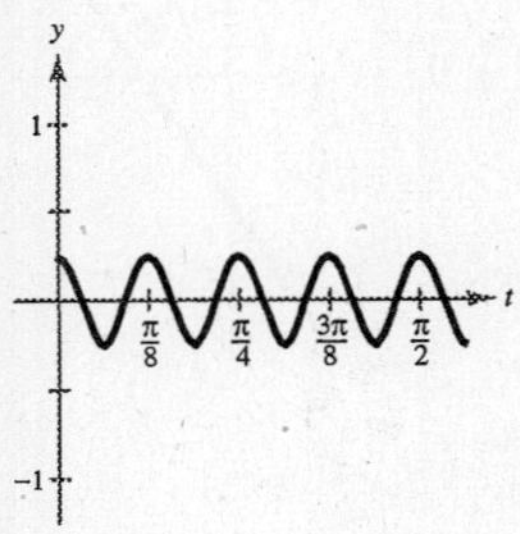

(b) Period: $\dfrac{2\pi}{16} = \dfrac{\pi}{8}$

(c) $\dfrac{1}{4} \cos 16t = 0$ when $16t = \dfrac{\pi}{2} \Rightarrow t = \dfrac{\pi}{32}$

59. False. One period is the time for one complete cycle of the motion.

61. (a) & (b)

Base 1	Base 2	Altitude	Area
8	$8 + 16 \cos 10°$	$8 \sin 10°$	22.1
8	$8 + 16 \cos 20°$	$8 \sin 20°$	42.5
8	$8 + 16 \cos 30°$	$8 \sin 30°$	59.7
8	$8 + 16 \cos 40°$	$8 \sin 40°$	72.7
8	$8 + 16 \cos 50°$	$8 \sin 50°$	80.5
8	$8 + 16 \cos 60°$	$8 \sin 60°$	83.1
8	$8 + 16 \cos 70°$	$8 \sin 70°$	80.7

The maximum occurs when $\theta = 60°$ and is approximately 83.1 square feet.

(c) $A(\theta) = [8 + (8 + 16 \cos \theta)]\left[\dfrac{8 \sin \theta}{2}\right]$

$= (16 + 16 \cos \theta)(4 \sin \theta)$

$= 64\,(1 + \cos \theta)(\sin \theta)$

(d)

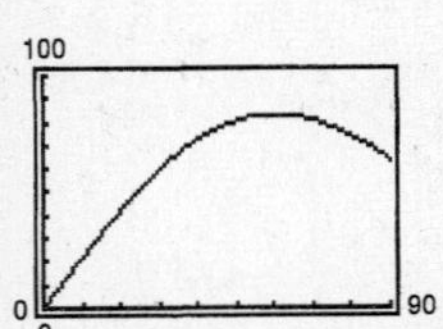

The maximum of 83.1 square feet occurs when $\theta = \dfrac{\pi}{3} = 60°$.

63. (a)

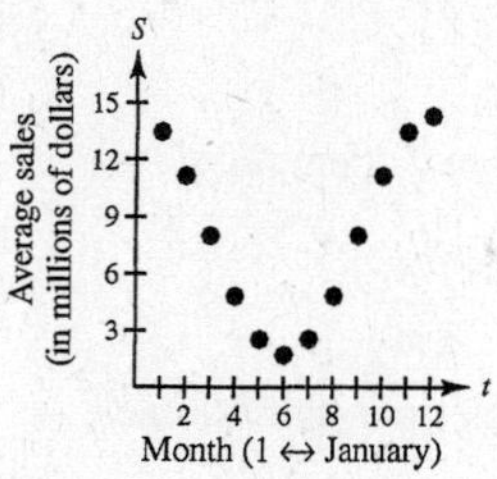

(b) $a = \frac{1}{2}(14.3 - 1.7) = 6.3$

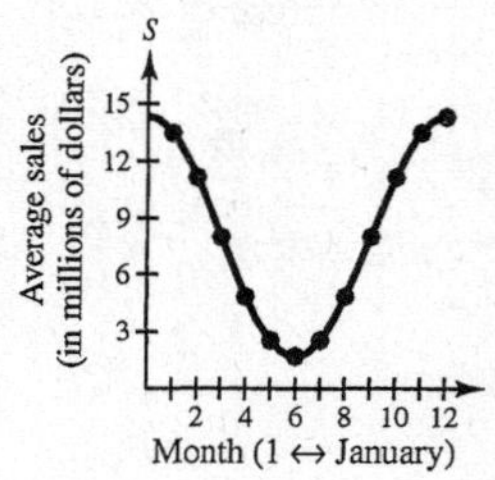

$\frac{2\pi}{b} = 12 \Rightarrow b = \frac{\pi}{6}$

Shift: $d = 14.3 - 6.3 = 8$

$S = d + a \cos bt$

$S = 8 + 6.3 \cos\left(\frac{\pi t}{6}\right)$

Note: Another model is $S = 8 + 6.3 \sin\left(\frac{\pi t}{6} + \frac{\pi}{2}\right)$

The model is a good fit.

(c) Period: $\frac{2\pi}{\pi/6} = 12$

This corresponds to the 12 months in a year. Since the sales of outerwear is seasonal this is reasonable.

(d) The amplitude represents the maximum displacement from average sales of 8 million dollars. Sales are greatest in December (cold weather + Christmas) and least in June.

65. $3x - 2y = 4$

Line

Intercepts: $(0, -2)$ and $\left(\frac{4}{3}, 0\right)$

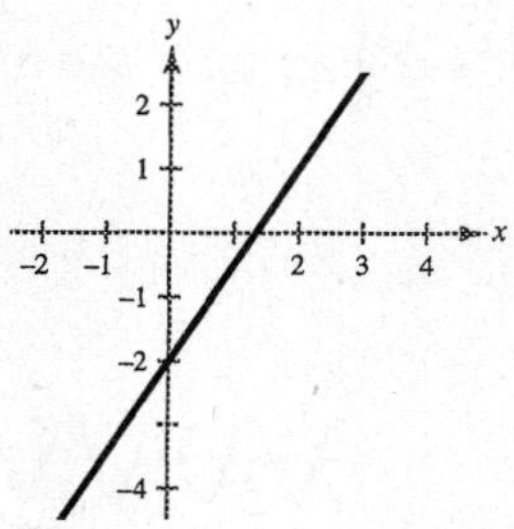

67. $(y - 2)^2 = 8(x + 2)$

Parabola opening to the right

Vertex: $(-2, 2)$

Intercepts: $(0, -2)$ and $(0, 6)$

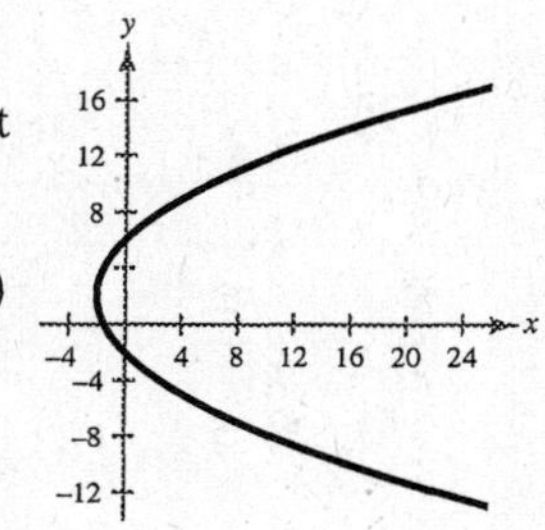

69. $\frac{x^2}{4} + y^2 = 1$

Ellipse with center $(0, 0)$

Horizontal major axis

Vertices: $(\pm 2, 0)$

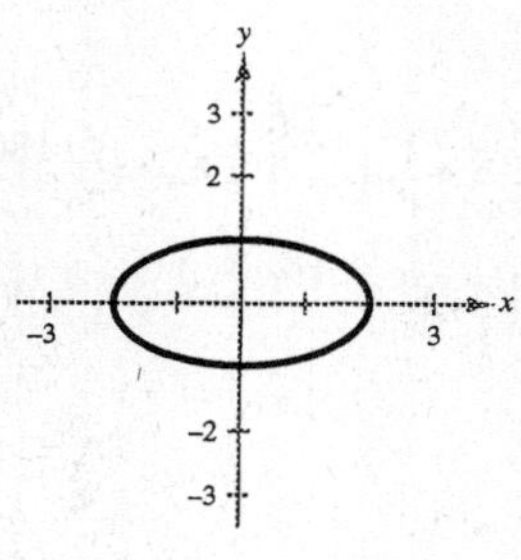

71. $\frac{x^2}{4} + \frac{y^2}{4} = 1$

Circle

Center: $(0, 0)$

Radius: 2

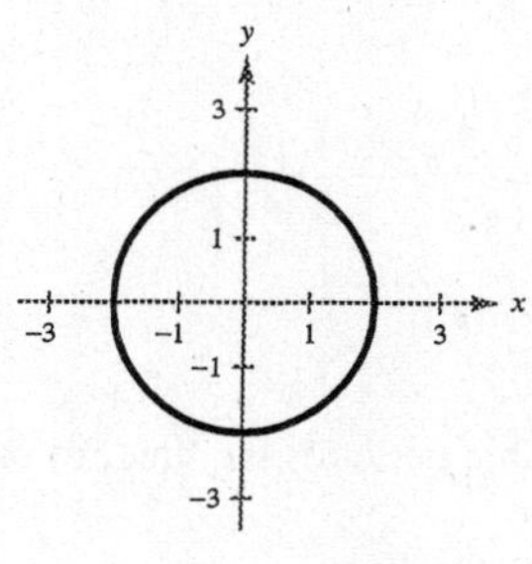

Review Exercises for Chapter 1

Solutions to Odd-Numbered Exercises

1. $\theta \approx 0.5$ radian

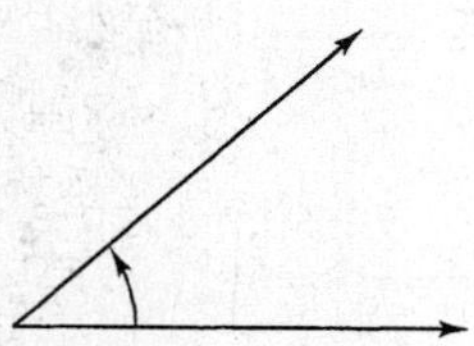

3. $\theta \approx 4.5$ radians

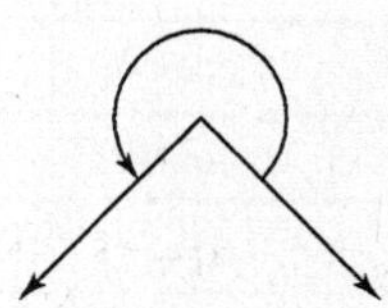

5. $\theta = \dfrac{11\pi}{4}$

Coterminal angles: $\dfrac{11\pi}{4} - 2\pi = \dfrac{3\pi}{4}$

$\dfrac{3\pi}{4} - 2\pi = -\dfrac{5\pi}{4}$

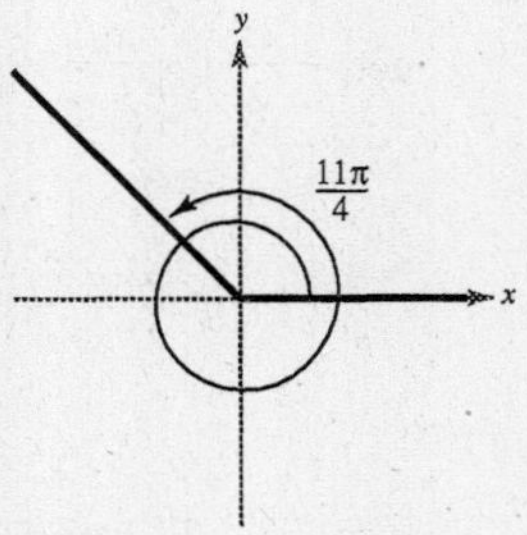

7. $\theta = -\dfrac{4\pi}{3}$

Coterminal angles: $-\dfrac{4\pi}{3} + 2\pi = \dfrac{2\pi}{3}$

$-\dfrac{4\pi}{3} - 2\pi = -\dfrac{10\pi}{3}$

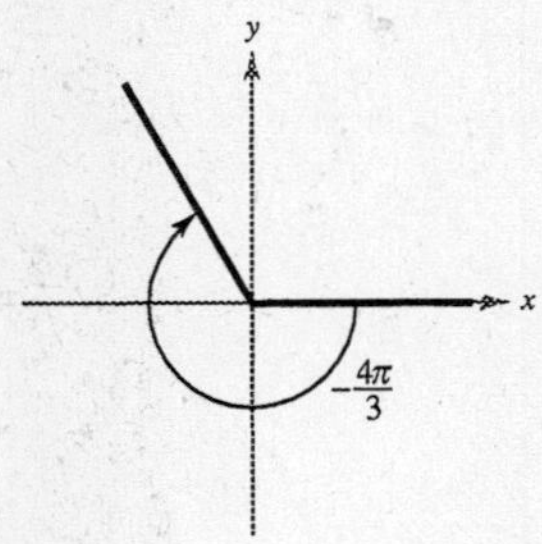

9. $\theta = 70°$

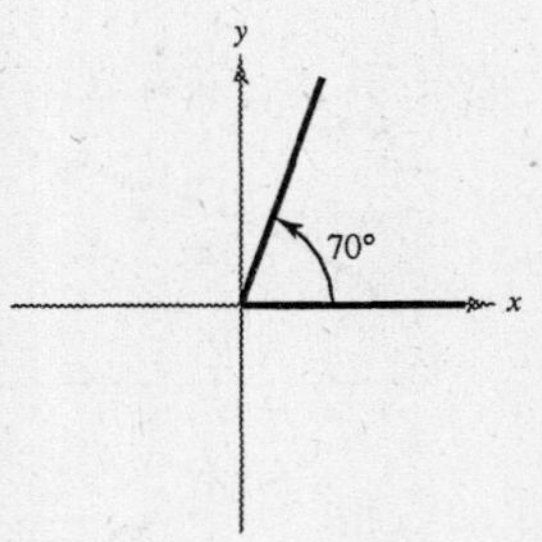

Coterminal angles: $70° + 360° = 430°$

$70° - 360° = -290°$

11. $\theta = -110°$

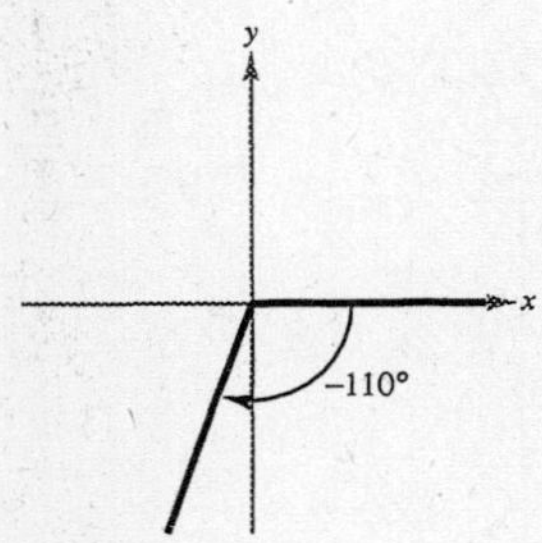

Coterminal angles: $-110° + 360° = 250°$

$-110° - 360° = -470°$

13. $\dfrac{5\pi \text{ rad}}{7} = \dfrac{5\pi \text{ rad}}{7} \cdot \dfrac{180°}{\pi \text{ rad}} \approx 128.57°$

15. $-3.5 \text{ rad} = -3.5 \text{ rad} \cdot \dfrac{180°}{\pi \text{ rad}} \approx -200.54°$

17. $480° = 480° \cdot \dfrac{\pi \text{ rad}}{180°} = \dfrac{8\pi}{3}$ radians ≈ 8.3776 radians

19. $-33°45' = -33.75° = -33.75° \cdot \dfrac{\pi \text{ rad}}{180°} = -\dfrac{3\pi}{16}$ radian ≈ -0.5890 radian

21. (a) Angular speed $= \dfrac{\left(33\frac{1}{3}\right)(2\pi) \text{ radians}}{1 \text{ minute}}$

$= 66\frac{2}{3}\pi$ radians per minute

(b) Linear speed $= \dfrac{6\left(66\frac{2}{3}\pi\right) \text{ inches}}{1 \text{ minute}}$

$= 400\pi$ inches per minute

23. $t = \dfrac{2\pi}{3}$ corresponds to the point $\left(-\dfrac{1}{2}, \dfrac{\sqrt{3}}{2}\right)$.

25. $t = \dfrac{5\pi}{6}$ corresponds to the point $\left(-\dfrac{\sqrt{3}}{2}, \dfrac{1}{2}\right)$.

27. $t = \frac{7\pi}{6}$ corresponds to the point $\left(-\frac{\sqrt{3}}{2}, -\frac{1}{2}\right)$.

$\sin\frac{7\pi}{6} = y = -\frac{1}{2}$ $\qquad \csc\frac{7\pi}{6} = \frac{1}{y} = -2$

$\cos\frac{7\pi}{6} = x = -\frac{\sqrt{3}}{2}$ $\qquad \sec\frac{7\pi}{6} = \frac{1}{x} = -\frac{2\sqrt{3}}{3}$

$\tan\frac{7\pi}{6} = \frac{y}{x} = \frac{1}{\sqrt{3}} = \frac{\sqrt{3}}{3}$ $\qquad \cot\frac{7\pi}{6} = \frac{x}{y} = \sqrt{3}$

29. $t = -\frac{2\pi}{3}$ corresponds to the point $\left(-\frac{1}{2}, -\frac{\sqrt{3}}{2}\right)$.

$\sin\left(-\frac{2\pi}{3}\right) = y = -\frac{\sqrt{3}}{2}$ $\qquad \csc\left(-\frac{2\pi}{3}\right) = \frac{1}{y} = -\frac{2\sqrt{3}}{3}$

$\cos\left(-\frac{2\pi}{3}\right) = x = -\frac{1}{2}$ $\qquad \sec\left(-\frac{2\pi}{3}\right) = \frac{1}{x} = -2$

$\tan\left(-\frac{2\pi}{3}\right) = \frac{y}{x} = \sqrt{3}$ $\qquad \cot\left(-\frac{2\pi}{3}\right) = \frac{x}{y} = \frac{\sqrt{3}}{3}$

31. $\sin\frac{11\pi}{4} = \sin\frac{3\pi}{4} = \frac{\sqrt{2}}{2}$

33. $\sin\left(-\frac{17\pi}{6}\right) = \sin\left(-\frac{5\pi}{6}\right) = -\frac{1}{2}$

35. $\tan 33 \approx -75.31$

37. $\sec\frac{12\pi}{5} = \frac{1}{\cos\left(\frac{12\pi}{5}\right)} \approx 3.24$

39. opp $= 4$, adj $= 5$, hyp $= \sqrt{4^2 + 5^2} = \sqrt{41}$

$\sin\theta = \frac{\text{opp}}{\text{hyp}} = \frac{4}{\sqrt{41}} = \frac{4\sqrt{41}}{41}$ $\qquad \csc\theta = \frac{\text{hyp}}{\text{opp}} = \frac{\sqrt{41}}{4}$

$\cos\theta = \frac{\text{adj}}{\text{hyp}} = \frac{5}{\sqrt{41}} = \frac{5\sqrt{41}}{41}$ $\qquad \sec\theta = \frac{\text{hyp}}{\text{adj}} = \frac{\sqrt{41}}{5}$

$\tan\theta = \frac{\text{opp}}{\text{adj}} = \frac{4}{5}$ $\qquad \cot\theta = \frac{\text{adj}}{\text{opp}} = \frac{5}{4}$

41. adj $= 4$, hyp $= 8$, opp $= \sqrt{8^2 - 4^2} = \sqrt{48} = 4\sqrt{3}$

$\sin\theta = \frac{\text{opp}}{\text{hyp}} = \frac{4\sqrt{3}}{8} = \frac{\sqrt{3}}{2}$ $\qquad \csc\theta = \frac{\text{hyp}}{\text{opp}} = \frac{8}{4\sqrt{3}} = \frac{2\sqrt{3}}{3}$

$\cos\theta = \frac{\text{adj}}{\text{hyp}} = \frac{4}{8} = \frac{1}{2}$ $\qquad \sec\theta = \frac{\text{hyp}}{\text{adj}} = \frac{8}{4} = 2$

$\tan\theta = \frac{\text{opp}}{\text{adj}} = \frac{4\sqrt{3}}{4} = \sqrt{3}$ $\qquad \cot\theta = \frac{\text{adj}}{\text{opp}} = \frac{4}{4\sqrt{3}} = \frac{\sqrt{3}}{3}$

43. $\sin\theta = \frac{1}{3}$

(a) $\csc\theta = \frac{1}{\sin\theta} = 3$

(b) $\sin^2\theta + \cos^2\theta = 1$

$$\left(\frac{1}{3}\right)^2 + \cos^2\theta = 1$$

$$\cos^2\theta = 1 - \frac{1}{9}$$

$$\cos^2\theta = \frac{8}{9}$$

$$\cos\theta = \sqrt{\frac{8}{9}}$$

$$\cos\theta = \frac{2\sqrt{2}}{3}$$

(c) $\sec\theta = \frac{1}{\cos\theta} = \frac{3}{2\sqrt{2}} = \frac{3\sqrt{2}}{4}$

(d) $\tan\theta = \frac{\sin\theta}{\cos\theta} = \frac{\frac{1}{3}}{\frac{2\sqrt{2}}{3}} = \frac{1}{2\sqrt{2}} = \frac{\sqrt{2}}{4}$

45. $\csc\theta = 4$

(a) $\sin\theta = \frac{1}{\csc\theta} = \frac{1}{4}$

(b) $\sin^2\theta + \cos^2\theta = 1$

$$\left(\frac{1}{4}\right)^2 + \cos^2\theta = 1$$

$$\cos^2\theta = 1 - \frac{1}{16}$$

$$\cos^2\theta = \frac{15}{16}$$

$$\cos\theta = \sqrt{\frac{15}{16}}$$

$$\cos\theta = \frac{\sqrt{15}}{4}$$

(c) $\sec\theta = \frac{1}{\cos\theta} = \frac{4}{\sqrt{15}} = \frac{4\sqrt{15}}{15}$

(d) $\tan\theta = \frac{\sin\theta}{\cos\theta} = \frac{\frac{1}{4}}{\frac{\sqrt{15}}{4}} = \frac{1}{\sqrt{15}} = \frac{\sqrt{15}}{15}$

47. $\tan 33° \approx 0.65$

49. $\sin 34.2° \approx 0.56$

51. $\cot 15°14' \approx \cot 15.2333° = \dfrac{1}{\tan 15.2333°} \approx 3.67$

53. $\sin 1°10' = \dfrac{x}{3.5}$

$x = 3.5 \sin 1°10' \approx 0.07$ Kilometer

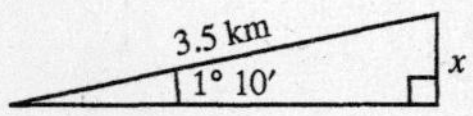

Not drawn to scale

55. $x = 12, y = 16, r = \sqrt{144 + 256} = \sqrt{400} = 20$

$\sin\theta = \dfrac{y}{r} = \dfrac{4}{5}$ $\quad\csc\theta = \dfrac{r}{y} = \dfrac{5}{4}$

$\cos\theta = \dfrac{x}{r} = \dfrac{3}{5}$ $\quad\sec\theta = \dfrac{r}{x} = \dfrac{5}{3}$

$\tan\theta = \dfrac{y}{x} = \dfrac{4}{3}$ $\quad\cot\theta = \dfrac{x}{y} = \dfrac{3}{4}$

57. $x = \dfrac{2}{3}, y = \dfrac{5}{2}$

$$r = \sqrt{\left(\frac{2}{3}\right)^2 + \left(\frac{5}{2}\right)^2} = \frac{\sqrt{241}}{6}$$

$$\sin\theta = \frac{y}{r} = \frac{\frac{5}{2}}{\frac{\sqrt{241}}{6}} = \frac{15}{\sqrt{241}} = \frac{15\sqrt{241}}{241} \qquad \csc\theta = \frac{r}{y} = \frac{\frac{\sqrt{241}}{6}}{\frac{5}{2}} = \frac{2\sqrt{241}}{30} = \frac{\sqrt{241}}{15}$$

$$\cos\theta = \frac{x}{r} = \frac{\frac{2}{3}}{\frac{\sqrt{241}}{6}} = \frac{4}{\sqrt{241}} = \frac{4\sqrt{241}}{241} \qquad \sec\theta = \frac{r}{x} = \frac{\frac{\sqrt{241}}{6}}{\frac{2}{3}} = \frac{\sqrt{241}}{4}$$

$$\tan\theta = \frac{y}{x} = \frac{\frac{5}{2}}{\frac{2}{3}} = \frac{15}{4} \qquad \cot\theta = \frac{x}{y} = \frac{\frac{2}{3}}{\frac{5}{2}} = \frac{4}{15}$$

59. $x = -0.5, y = 4.5$

$r = \sqrt{(-0.5)^2 + 4.5^2} \approx 4.528$

$\sin\theta = \dfrac{y}{r} \approx 1$ $\quad\csc\theta = \dfrac{r}{y} \approx 1$

$\cos\theta = \dfrac{x}{r} \approx -0.1$ $\quad\sec\theta = \dfrac{r}{x} \approx -9$

$\tan\theta = \dfrac{y}{x} \approx -9$ $\quad\cot\theta = \dfrac{x}{y} \approx -0.1$

61. $(x, 4x), x > 0$

$x = x, y = 4x$

$r = \sqrt{x^2 + (4x)^2} = \sqrt{17}x$

$\sin\theta = \dfrac{y}{r} = \dfrac{4x}{\sqrt{17}x} = \dfrac{4\sqrt{17}}{17}$ $\quad\csc\theta = \dfrac{r}{y} = \dfrac{\sqrt{17}x}{4x} = \dfrac{\sqrt{17}}{4}$

$\cos\theta = \dfrac{x}{r} = \dfrac{x}{\sqrt{17}x} = \dfrac{\sqrt{17}}{17}$ $\quad\sec\theta = \dfrac{r}{x} = \dfrac{\sqrt{17}x}{x} = \sqrt{17}$

$\tan\theta = \dfrac{y}{x} = \dfrac{4x}{x} = 4$ $\quad\cot\theta = \dfrac{x}{y} = \dfrac{x}{4x} = \dfrac{1}{4}$

63. $\sec\theta = \frac{6}{5}$, $\tan\theta < 0 \Rightarrow \theta$ is in Quadrant IV.

$r = 6, x = 5, y = -\sqrt{36-25} = -\sqrt{11}$

$\sin\theta = \frac{y}{r} = -\frac{\sqrt{11}}{6}$ $\quad\csc\theta = \frac{r}{y} = -\frac{6\sqrt{11}}{11}$

$\cos\theta = \frac{x}{r} = \frac{5}{6}$ $\quad\sec\theta = \frac{6}{5}$

$\tan\theta = \frac{y}{x} = -\frac{\sqrt{11}}{5}$ $\quad\cot\theta = -\frac{5\sqrt{11}}{11}$

65. $\sin\theta = \frac{3}{8}$, $\cos\theta < 0 \Rightarrow \theta$ is in Quadrant II.

$y = 3, r = 8, x = -\sqrt{55}$

$\sin\theta = \frac{y}{r} = \frac{3}{8}$ $\quad\csc\theta = \frac{8}{3}$

$\cos\theta = \frac{x}{r} = -\frac{\sqrt{55}}{8}$ $\quad\sec\theta = -\frac{8}{\sqrt{55}} = -\frac{8\sqrt{55}}{55}$

$\tan\theta = \frac{y}{x} = -\frac{3}{\sqrt{55}} = -\frac{3\sqrt{55}}{55}$ $\quad\cot\theta = -\frac{\sqrt{55}}{3}$

67. $\cos\theta = \frac{x}{r} = \frac{-2}{5} \Rightarrow y^2 = 21.$

$\sin\theta > 0 \Rightarrow \theta$ is in Quadrant II $\Rightarrow y = \sqrt{21}$

$\sin\theta = \frac{y}{r} = \frac{\sqrt{21}}{5}$

$\tan\theta = \frac{y}{x} = -\frac{\sqrt{21}}{2}$

$\csc\theta = \frac{r}{y} = \frac{5}{\sqrt{21}} = \frac{5\sqrt{21}}{21}$

$\sec\theta = \frac{r}{x} = \frac{5}{-2} = -\frac{5}{2}$

$\cot\theta = \frac{x}{y} = \frac{-2}{\sqrt{21}} = -\frac{2\sqrt{21}}{21}$

69. $\sin\frac{\pi}{3} = \frac{\sqrt{3}}{2}$

$\cos\frac{\pi}{3} = \frac{1}{2}$

$\tan\frac{\pi}{3} = \sqrt{3}$

71. $\sin\left(-\frac{7\pi}{3}\right) = -\sin\frac{\pi}{3} = -\frac{\sqrt{3}}{2}$

$\cos\left(-\frac{7\pi}{3}\right) = \cos\frac{\pi}{3} = \frac{1}{2}$

$\tan\left(-\frac{7\pi}{3}\right) = -\tan\frac{\pi}{3} = -\sqrt{3}$

73. $\sin 495° = \sin 45° = \frac{\sqrt{2}}{2}$

$\cos 495° = -\cos 45° = -\frac{\sqrt{2}}{2}$

$\tan 495° = -\tan 45° = -1$

75. $\sin(-240°) = \sin 60° = \frac{\sqrt{3}}{2}$

$\cos(-240°) = -\cos 60° = -\frac{1}{2}$

$\tan(-240°) = -\tan 60° = -\sqrt{3}$

77. $\sin 4 \approx -0.76$

79. $\sin(-3.2) \approx 0.06$

81. $\sec\frac{12\pi}{5} = \frac{1}{\cos\left(\frac{12\pi}{5}\right)} \approx 3.24$

83. $y = \sin x$

Amplitude: 1

Period: 2π

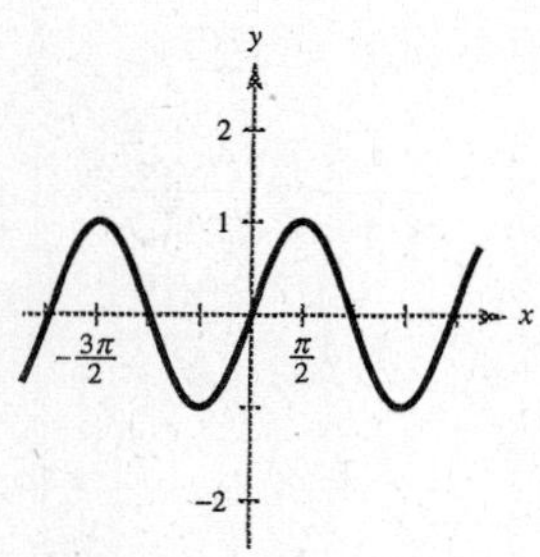

85. $f(x) = 5\sin\frac{2x}{5}$

Amplitude: 5

Period: $\frac{2\pi}{\frac{2}{5}} = 5\pi$

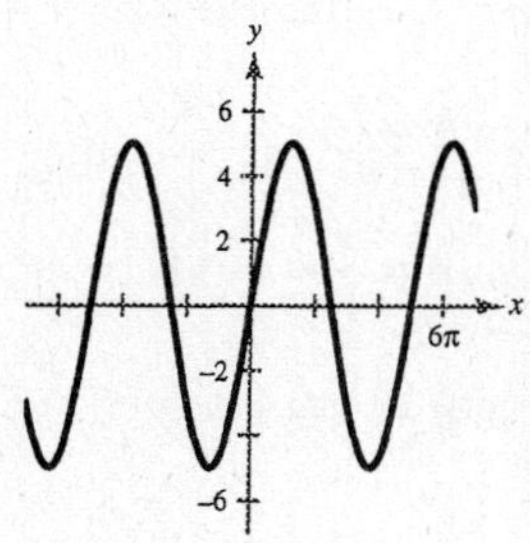

87. $y = 2 + \sin x$

Shift the graph of $y = \sin x$ two units upward.

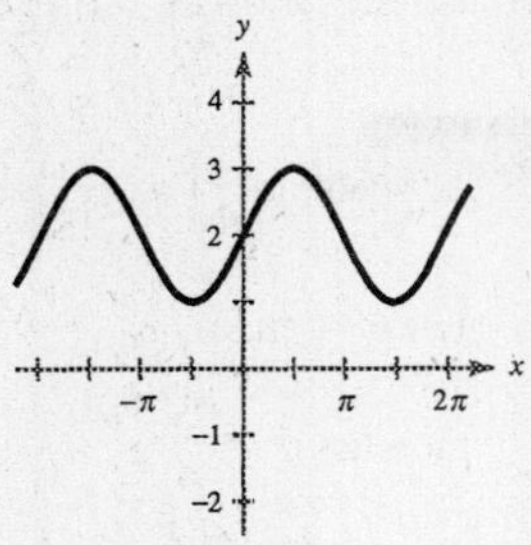

89. $g(t) = \dfrac{5}{2}\sin(t - \pi)$

Amplitude: $\dfrac{5}{2}$

Period: 2π

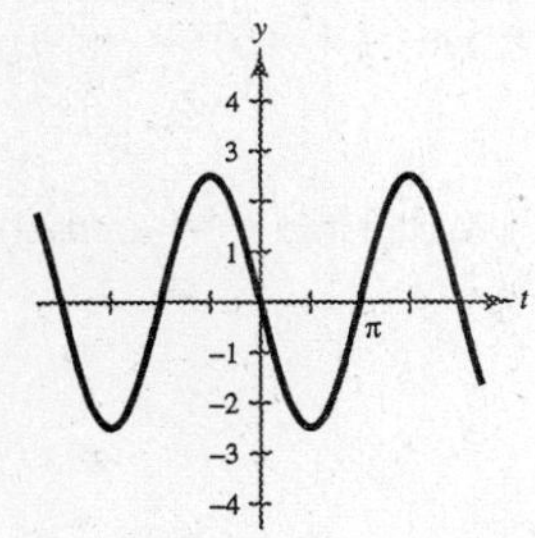

91. $y = a \sin bx$

(a) $a = 2, \dfrac{2\pi}{b} = \dfrac{1}{264} \implies b = 528\pi$

$y = 2\sin(528\pi x)$

(b) $f = \dfrac{1}{\frac{1}{264}} = 264$ cycles per second.

93. $f(x) = \tan x$

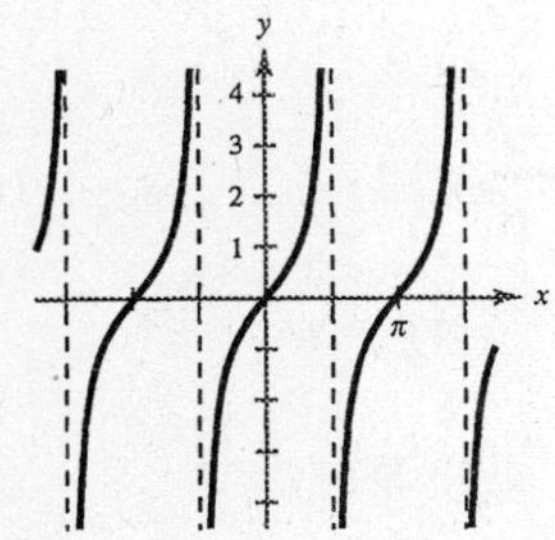

95. $f(x) = \cot x$

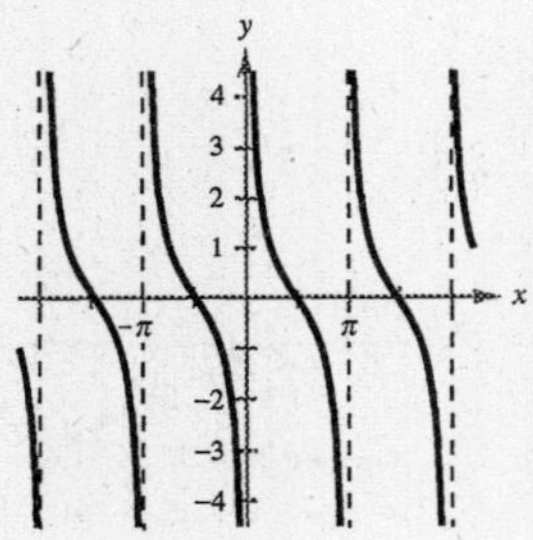

97. $f(x) = \sec x$

Graph $y = \cos x$ first.

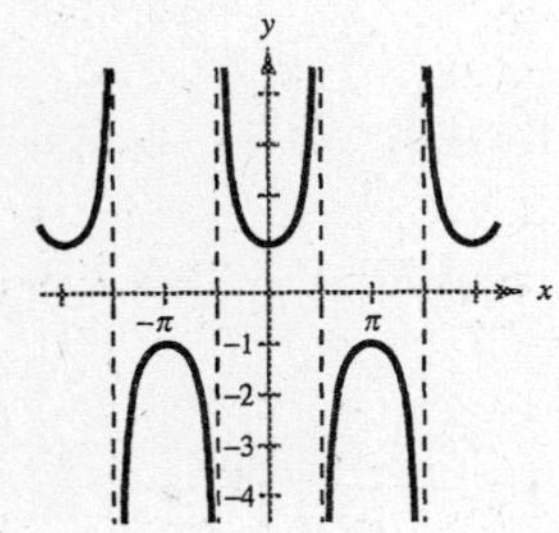

99. $f(x) = \csc x$

Graph $y = \sin x$ first.

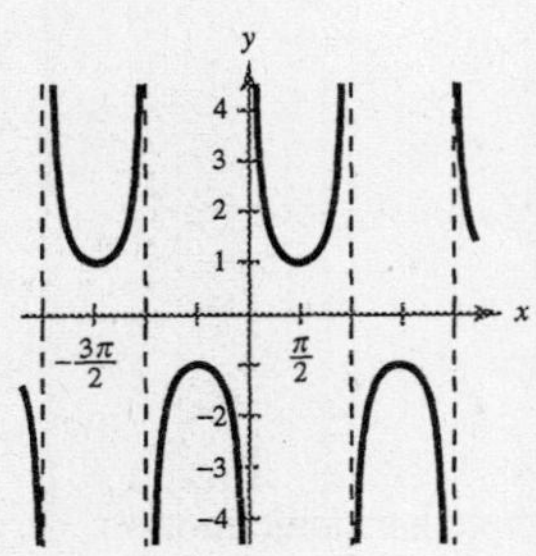

101. $f(x) = x \cos x$

Graph $y = x$ and $y = -x$ first.

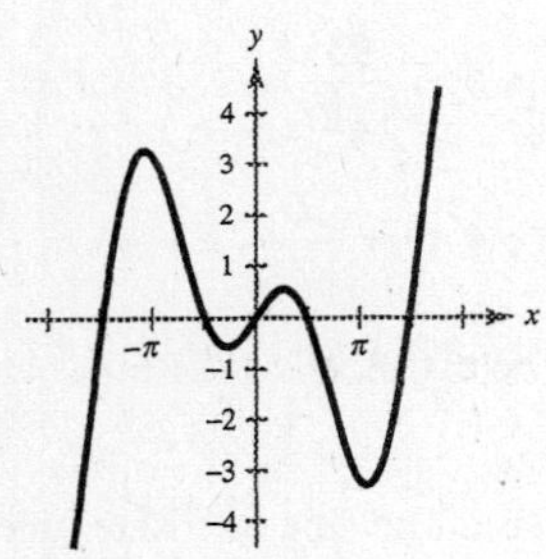

103. $\arcsin\left(-\dfrac{1}{2}\right) = -\arcsin\dfrac{1}{2} = -\dfrac{\pi}{6}$

105. $\arcsin 0.4 \approx 0.41$ radian

107. $\sin^{-1}(-0.44) \approx -0.46$ radian

109. $\arccos\dfrac{\sqrt{3}}{2} = \dfrac{\pi}{6}$

111. $\cos^{-1}(-1) = \pi$

113. $\arccos 0.324 \approx 1.24$ radians

115. $\arctan 0.123 \approx 0.12$ radian

117. $\arctan 5.783 \approx 1.40$ radians

119. $\tan^{-1}(-1.5) \approx -0.98$ radian

121. $\sin(\arcsin 0.72) = 0.72$

123. $\arctan(\tan \pi) = \arctan 0 = 0$

125. $\cos\left(\arctan \frac{3}{4}\right) = \frac{4}{5}$. Use a right triangle.

Let $\theta = \arctan \frac{3}{4}$

then $\tan \theta = \frac{3}{4}$

and $\cos \theta = \frac{4}{5}$.

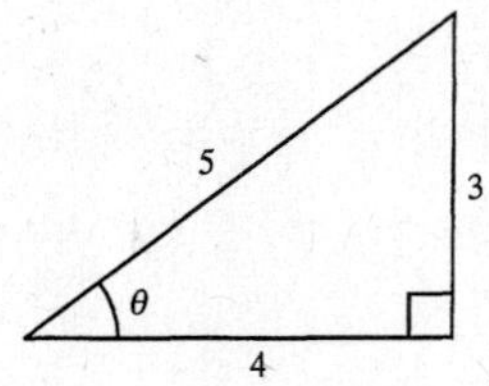

127. $\sec\left(\arctan \frac{12}{5}\right) = \frac{13}{5}$. Use a right triangle.

Let $\theta = \arctan \frac{12}{5}$

then $\tan \theta = \frac{12}{5}$

and $\sec \theta = \frac{13}{5}$.

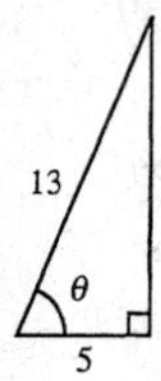

129. $\tan \theta = \frac{70}{30}$

$\theta = \arctan\left(\frac{70}{30}\right) \approx 66.8°$

131. $\sin 48° = \frac{d_1}{650} \Rightarrow d_1 \approx 483$

$\cos 25° = \frac{d_2}{810} \Rightarrow d_2 \approx 734$

$d_1 + d_2 \approx 1217$

$\cos 48° = \frac{d_3}{650} \Rightarrow d_3 \approx 435$

$\sin 25° = \frac{d_4}{810} \Rightarrow d_4 \approx 342$

$d_3 - d_4 \approx 93$

$\tan \theta \approx \frac{93}{1217} \Rightarrow \theta \approx 4.4°$

$\sec 4.4° \approx \frac{D}{1217} \Rightarrow D \approx 1217 \sec 4.4° \approx 1221$

The distance is 1221 miles and the bearing is N 85.6° E.

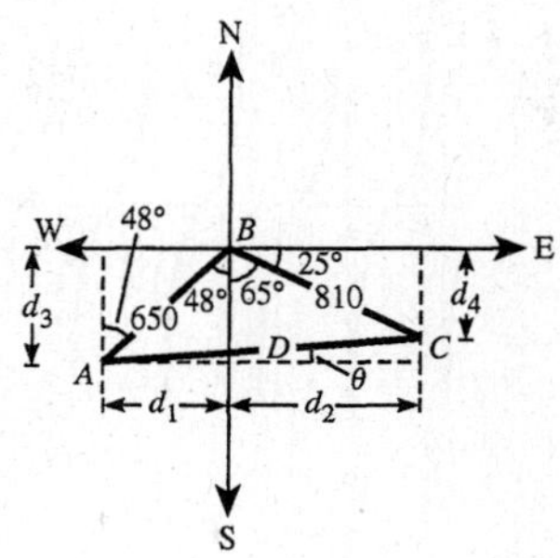

133. False. The sine or cosine functions are often useful for modeling simple harmonic motion.

135. False. For each θ there corresponds exactly one value of y.

137. $y = 3 \sin x$

Amplitude: 3

Period: 2π

Matches graph (d)

139. $y = 2 \sin \pi x$

Amplitude: 2

Period: 2

Matches graph (b)

141. $f(\theta) = \sec \theta$ is undefined at the zeros of $g(\theta) = \cos \theta$ since $\sec \theta = \frac{1}{\cos \theta}$.

143. The ranges for the other four trigonometric functions are not bounded. For $y = \tan x$ and $y = \cot x$, the range is $(-\infty, \infty)$. For $y = \sec x$ and $y = \csc x$, the range is $(-\infty, -1] \cup [1, \infty)$

144. Answers will vary.

Problem Solving for Chapter 1

Solutions to Odd-Numbered Exercises

1. (a) 8:57 − 6:45 = 2 hours 12 minutes = 132 minutes

$$\frac{132}{48} = \frac{11}{4} \text{ revolutions}$$

$$\theta = \left(\frac{11}{4}\right)(2\pi) = \frac{11\pi}{2} \text{ radians or } 990^\circ$$

(b) $s = r\theta = 47.25(5.5\pi) \approx 816.42$ feet

3. (a) $\sin 39^\circ = \dfrac{3000}{d}$

$$d = \frac{3000}{\sin 39^\circ} \approx 4767 \text{ feet}$$

(b) $\tan 39^\circ = \dfrac{3000}{x}$

$$x = \frac{3000}{\tan 39^\circ} \approx 3705 \text{ feet}$$

(c) $\tan 63^\circ = \dfrac{w + 3705}{3000}$

$$3000 \tan 63^\circ = w + 3705$$

$$w = 3000 \tan 63^\circ - 3705 \approx 2183 \text{ feet}$$

5. (a) $h(x) = \cos^2 x$

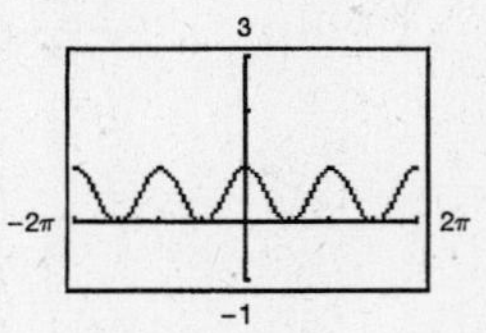

h is even

(b) $h(x) = \sin^2 x$

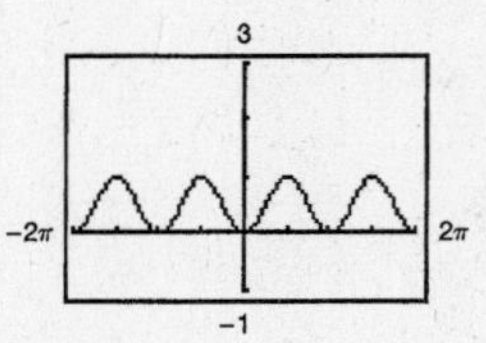

h is even

7. If we alter the model so that $h = 1$ when $t = 0$, we can use either a sine or a cosine model.

$$a = \frac{1}{2}[\max - \min] = \frac{1}{2}[101 - 1] = 50$$

$$d = \frac{1}{2}[\max + \min] = \frac{1}{2}[101 + 1] = 51$$

$$b = 8\pi$$

For the cosine model we have:

$$h = 51 - 50\cos(8\pi t)$$

For the sine model we have:

$$h = 51 - 50\sin\left(8\pi t + \frac{\pi}{2}\right)$$

Notice that we needed the horizontal shift so that the sine value was one when $t = 0$.

Another model would be:

$$h = 51 + 50\sin\left(8\pi t + \frac{3\pi}{2}\right)$$

Here we wanted the sine value to be -1 when $t = 0$.

9. $f(x) = 2\cos 2x + 3\sin 3x$

$g(x) = 2\cos 2x + 3\sin 4x$

(a)

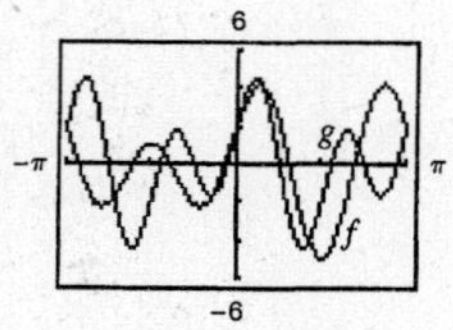

(b) The period of $f(x)$ is 2π.

The period of $g(x)$ is π.

(c) $h(x) = A\cos \alpha x + B\sin \beta x$ is periodic since the sine and cosine functions are periodic.

11. (a) $f(t - 2c) = f(t)$ is true since this is a two period horizontal shift.

(b) $f\left(t + \frac{1}{2}c\right) = f\left(\frac{1}{2}t\right)$ is not true.

$f\left(t + \frac{1}{2}c\right)$ is a horizontal translation of $f(t)$.

$f\left(\frac{1}{2}t\right)$ is a doubling of the period of $f(t)$.

(c) $f\left(\frac{1}{2}(t + c)\right) = f\left(\frac{1}{2}t\right)$ is not true.

$f\left(\frac{1}{2}(t + c)\right) = f\left(\frac{1}{2}t + \frac{1}{2}c\right)$ is a horizontal translation

of $f\left(\frac{1}{2}t\right)$ by half a period.

For example, $\sin\left[\frac{1}{2}(\pi + 2\pi)\right] \neq \sin\left(\frac{1}{2}\pi\right)$.

13. $\arctan x \approx x - \frac{x^3}{3} + \frac{x^5}{5} - \frac{x^7}{7}$

(a)

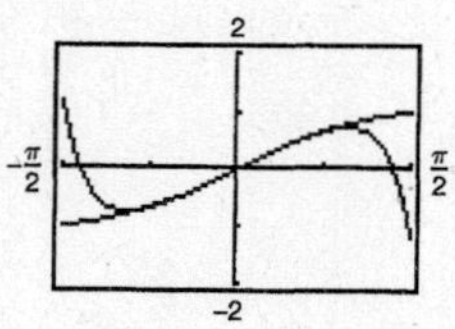

The graphs are nearly the same for $-1 < x < 1$.

(b)

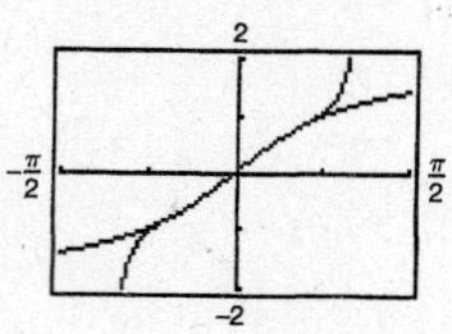

The accuracy of the approximation improved slightly by adding the $\frac{x^9}{9}$ term.

Chapter 1 Practice Test

1. Express 350° in radian measure.

2. Express $(5\pi)/9$ in degree measure.

3. Convert $135°14'12''$ to decimal form.

4. Convert $-22.569°$ to D°M′S″ form.

5. If $\cos\theta = \frac{2}{3}$, use the trigonometric identities to find $\tan\theta$.

6. Find θ given $\sin\theta = 0.9063$.

7. Solve for x in the figure below.

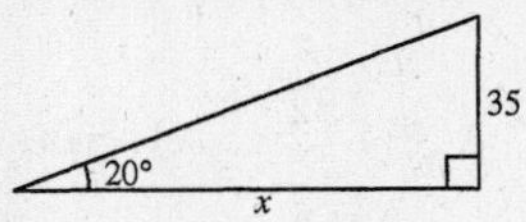

8. Find the magnitude of the reference angle for $\theta = (6\pi)/5$.

9. Evaluate csc 3.92.

10. Find $\sec\theta$ given that θ lies in Quadrant III and $\tan\theta = 6$.

11. Graph $y = 3\sin\dfrac{x}{2}$.

12. Graph $y = -2\cos(x - \pi)$.

13. Graph $y = \tan 2x$.

14. Graph $y = -\csc\left(x + \dfrac{\pi}{4}\right)$.

15. Graph $y = 2x + \sin x$, using a graphing calculator.

16. Graph $y = 3x\cos x$, using a graphing calculator.

17. Evaluate arcsin 1.

18. Evaluate $\arctan(-3)$.

19. Evaluate $\sin\left(\arccos\dfrac{4}{\sqrt{35}}\right)$.

20. Write an algebraic expression for $\cos\left(\arcsin\dfrac{x}{4}\right)$.

For Exercises 21–23, solve the right triangle.

21. $A = 40°,\ c = 12$

22. $B = 6.84°,\ a = 21.3$

23. $a = 5,\ b = 9$

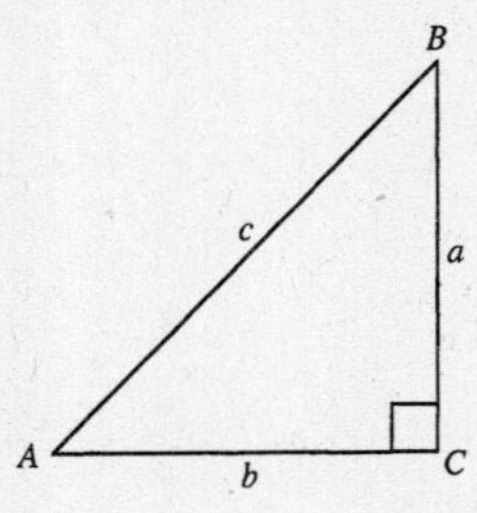

24. A 20-foot ladder leans against the side of a barn. Find the height of the top of the ladder if the angle of elevation of the ladder is 67°.

25. An observer in a lighthouse 250 feet above sea level spots a ship off the shore. If the angle of depression to the ship is 5°, how far out is the ship?

CHAPTER 2
Analytic Trigonometry

CHAPTER 2
Analytic Trigonometry

Section 2.1 Using Fundamental Identities

- You should know the fundamental trigonometric identities.

  (a) Reciprocal Identities

  $$\sin u = \frac{1}{\csc u} \qquad \csc u = \frac{1}{\sin u}$$

  $$\cos u = \frac{1}{\sec u} \qquad \sec u = \frac{1}{\cos u}$$

  $$\tan u = \frac{1}{\cot u} = \frac{\sin u}{\cos u} \qquad \cot u = \frac{1}{\tan u} = \frac{\cos u}{\sin u}$$

  (b) Pythagorean Identities

  $$\sin^2 u + \cos^2 u = 1$$

  $$1 + \tan^2 u = \sec^2 u$$

  $$1 + \cot^2 u = \csc^2 u$$

  (c) Cofunction Identities

  $$\sin\left(\frac{\pi}{2} - u\right) = \cos u \qquad \cos\left(\frac{\pi}{2} - u\right) = \sin u$$

  $$\tan\left(\frac{\pi}{2} - u\right) = \cot u \qquad \cot\left(\frac{\pi}{2} - u\right) = \tan u$$

  $$\sec\left(\frac{\pi}{2} - u\right) = \csc u \qquad \csc\left(\frac{\pi}{2} - u\right) = \sec u$$

  (d) Even/Odd Identities

  $$\sin(-x) = -\sin x \qquad \csc(-x) = -\csc x$$

  $$\cos(-x) = \cos x \qquad \sec(-x) = \sec x$$

  $$\tan(-x) = -\tan x \qquad \cot(-x) = -\cot x$$

- You should be able to use these fundamental identities to find function values.
- You should be able to convert trigonometric expressions to equivalent forms by using the fundamental identities.

Solutions to Odd-Numbered Exercises

1. $\sin x = \frac{\sqrt{3}}{2}$, $\cos x = -\frac{1}{2} \Rightarrow x$ is in Quadrant II.

$$\tan x = \frac{\sin x}{\cos x} = \frac{\sqrt{3}/2}{-1/2} = -\sqrt{3}$$

$$\cot x = \frac{1}{\tan x} = -\frac{1}{\sqrt{3}} = -\frac{\sqrt{3}}{3}$$

$$\sec x = \frac{1}{\cos x} = \frac{1}{-1/2} = -2$$

$$\csc x = \frac{1}{\sin x} = \frac{1}{\sqrt{3}/2} = \frac{2}{\sqrt{3}} = \frac{2\sqrt{3}}{3}$$

3. $\sec \theta = \sqrt{2}$, $\sin \theta = -\frac{\sqrt{2}}{2} \Rightarrow \theta$ is in Quadrant IV.

$$\cos \theta = \frac{1}{\sec \theta} = \frac{1}{\sqrt{2}} = \frac{\sqrt{2}}{2}$$

$$\tan \theta = \frac{\sin \theta}{\cos \theta} = \frac{-\sqrt{2}/2}{\sqrt{2}/2} = -1$$

$$\cot \theta = \frac{1}{\tan \theta} = -1$$

$$\csc \theta = \frac{1}{\sin \theta} = -\sqrt{2}$$

5. $\tan x = \frac{5}{12}$, $\sec x = -\frac{13}{12} \Rightarrow x$ is in Quadrant III.

$$\cos x = \frac{1}{\sec x} = -\frac{12}{13}$$

$$\sin x = -\sqrt{1 - \cos^2 x} = -\sqrt{1 - \frac{144}{169}} = -\frac{5}{13}$$

$$\cot x = \frac{1}{\tan x} = \frac{12}{5}$$

$$\csc x = \frac{1}{\sin x} = -\frac{13}{5}$$

7. $\sec \phi = \frac{3}{2}$, $\csc \phi = -\frac{3\sqrt{5}}{5} \Rightarrow \phi$ is in Quadrant IV.

$$\sin \phi = \frac{1}{\csc \phi} = \frac{1}{-3\sqrt{5}/5} = -\frac{\sqrt{5}}{3}$$

$$\cos \phi = \frac{1}{\sec \phi} = \frac{1}{3/2} = \frac{2}{3}$$

$$\tan \phi = \frac{\sin \phi}{\cos \phi} = \frac{-\sqrt{5}/3}{2/3} = -\frac{\sqrt{5}}{2}$$

$$\cot \phi = \frac{1}{\tan \phi} = \frac{1}{-\sqrt{5}/2} = -\frac{2}{\sqrt{5}} = -\frac{2\sqrt{5}}{5}$$

9. $\sin(-x) = -\frac{1}{3} \Rightarrow \sin x = \frac{1}{3}$, $\tan x = -\frac{\sqrt{2}}{4} \Rightarrow x$ is in Quadrant II.

$$\cos x = -\sqrt{1 - \sin^2 x} = -\sqrt{1 - \frac{1}{9}} = -\frac{2\sqrt{2}}{3}$$

$$\cot x = \frac{1}{\tan x} = \frac{1}{-\sqrt{2}/4} = -2\sqrt{2}$$

$$\sec x = \frac{1}{\cos x} = \frac{1}{-2\sqrt{2}/3} = -\frac{3\sqrt{2}}{4}$$

$$\csc x = \frac{1}{\sin x} = \frac{1}{1/3} = 3$$

11. $\tan \theta = 2$, $\sin \theta < 0 \Rightarrow \theta$ is in Quadrant III.

$$\sec \theta = -\sqrt{\tan^2 \theta + 1} = -\sqrt{4 + 1} = -\sqrt{5}$$

$$\cos \theta = \frac{1}{\sec \theta} = -\frac{1}{\sqrt{5}} = -\frac{\sqrt{5}}{5}$$

$$\sin \theta = -\sqrt{1 - \cos^2 \theta}$$

$$= -\sqrt{1 - \frac{1}{5}} = -\frac{2}{\sqrt{5}} = -\frac{2\sqrt{5}}{5}$$

$$\csc \theta = \frac{1}{\sin \theta} = -\frac{\sqrt{5}}{2}$$

$$\cot \theta = \frac{1}{\tan \theta} = \frac{1}{2}$$

13. $\sin \theta = -1$, $\cot \theta = 0 \Rightarrow \theta = \frac{3\pi}{2}$

$$\cos \theta = \sqrt{1 - \sin^2 \theta} = 0$$

$\sec \theta$ is undefined.

$\tan \theta$ is undefined.

$\csc \theta = -1$

15. $\sec x \cos x = \sec x \cdot \frac{1}{\sec x} = 1$

The expression is matched with (d).

17. $\cot^2 x - \csc^2 x = \cot^2 x - (1 + \cot^2 x) = -1$

The expression is matched with (b).

19. $\dfrac{\sin(-x)}{\cos(-x)} = \dfrac{-\sin x}{\cos x} = -\tan x$

The expression is matched with (e).

21. $\sin x \sec x = \sin x \cdot \dfrac{1}{\cos x} = \tan x$

The expression is matched with (b).

23. $\sec^4 x - \tan^4 x = (\sec^2 x + \tan^2 x)(\sec^2 x - \tan^2 x)$

$= (\sec^2 x + \tan^2 x)(1) = \sec^2 x + \tan^2 x$

The expression is matched with (f).

25. $\dfrac{\sec^2 x - 1}{\sin^2 x} = \dfrac{\tan^2 x}{\sin^2 x} = \dfrac{\sin^2 x}{\cos^2 x} \cdot \dfrac{1}{\sin^2 x} = \sec^2 x$

The expression is matched with (e).

27. $\cot \theta \sec \theta = \dfrac{\cos \theta}{\sin \theta} \cdot \dfrac{1}{\cos \theta} = \dfrac{1}{\sin \theta} = \csc \theta$

29. $\sin \phi (\csc \phi - \sin \phi) = (\sin \phi)\dfrac{1}{\sin \phi} - \sin^2 \phi$

$= 1 - \sin^2 \phi = \cos^2 \phi$

31. $\dfrac{\cot x}{\csc x} = \dfrac{\cos x/\sin x}{1/\sin x}$

$= \dfrac{\cos x}{\sin x} \cdot \dfrac{\sin x}{1} = \cos x$

33. $\dfrac{1 - \sin^2 x}{\csc^2 x - 1} = \dfrac{\cos^2 x}{\cot^2 x} = \cos^2 x \tan^2 x = (\cos^2 x)\dfrac{\sin^2 x}{\cos^2 x}$

$= \sin^2 x$

35. $\sec \alpha \dfrac{\sin \alpha}{\tan \alpha} = \dfrac{1}{\cos \alpha}(\sin \alpha) \cot \alpha$

$= \dfrac{1}{\cos \alpha}(\sin \alpha)\left(\dfrac{\cos \alpha}{\sin \alpha}\right) = 1$

37. $\cos\left(\dfrac{\pi}{2} - x\right)\sec x = (\sin x)(\sec x)$

$= (\sin x)\left(\dfrac{1}{\cos x}\right)$

$= \dfrac{\sin x}{\cos x}$

$= \tan x$

39. $\dfrac{\cos^2 y}{1 - \sin y} = \dfrac{1 - \sin^2 y}{1 - \sin y}$

$= \dfrac{(1 + \sin y)(1 - \sin y)}{1 - \sin y}$

$= 1 + \sin y$

41. $\sin \beta \tan \beta + \cos \beta = (\sin \beta)\dfrac{\sin \beta}{\cos \beta} + \cos \beta$

$= \dfrac{\sin^2 \beta}{\cos \beta} + \dfrac{\cos^2 \beta}{\cos \beta}$

$= \dfrac{\sin^2 \beta + \cos^2 \beta}{\cos \beta}$

$= \dfrac{1}{\cos \beta}$

$= \sec \beta$

43. $\cot u \sin u + \tan u \cos u = \dfrac{\cos u}{\sin u}(\sin u) + \dfrac{\sin u}{\cos u}(\cos u)$

$= \cos u + \sin u$

45. $\tan^2 x - \tan^2 x \sin^2 x = \tan^2 x(1 - \sin^2 x)$

$= \tan^2 x \cos^2 x$

$= \dfrac{\sin^2 x}{\cos^2 x} \cdot \cos^2 x$

$= \sin^2 x$

47. $\sin^2 x \sec^2 x - \sin^2 x = \sin^2 x(\sec^2 x - 1)$

$= \sin^2 x \tan^2 x$

49. $\dfrac{\sec^2 x - 1}{\sec x - 1} = \dfrac{(\sec x + 1)(\sec x - 1)}{\sec x - 1} = \sec x + 1$

51. $\tan^4 x + 2\tan^2 x + 1 = (\tan^2 x + 1)^2$

$= (\sec^2 x)^2$

$= \sec^4 x$

53. $\sin^4 x - \cos^4 x = (\sin^2 x + \cos^2 x)(\sin^2 x - \cos^2 x)$

$= (1)(\sin^2 x - \cos^2 x)$

$= \sin^2 x - \cos^2 x$

55. $\csc^3 x - \csc^2 x - \csc x + 1 = \csc^2 x(\csc x - 1) - 1(\csc x - 1)$

$= (\csc^2 x - 1)(\csc x - 1)$

$= \cot^2 x(\csc x - 1)$

57. $(\sin x + \cos x)^2 = \sin^2 x + 2\sin x \cos x + \cos^2 x$

$= (\sin^2 x + \cos^2 x) + 2\sin x \cos x$

$= 1 + 2\sin x \cos x$

59. $(2\csc x + 2)(2\csc x - 2) = 4\csc^2 x - 4 = 4(\csc^2 x - 1) = 4\cot^2 x$

61. $\dfrac{1}{1 + \cos x} + \dfrac{1}{1 - \cos x} = \dfrac{1 - \cos x + 1 + \cos x}{(1 + \cos x)(1 - \cos x)}$

$= \dfrac{2}{1 - \cos^2 x}$

$= \dfrac{2}{\sin^2 x}$

$= 2\csc^2 x$

63. $\dfrac{\cos x}{1 + \sin x} + \dfrac{1 + \sin x}{\cos x} = \dfrac{\cos^2 x + (1 + \sin x)^2}{\cos x(1 + \sin x)} = \dfrac{\cos^2 x + 1 + 2\sin x + \sin^2 x}{\cos x(1 + \sin x)}$

$= \dfrac{2 + 2\sin x}{\cos x(1 + \sin x)}$

$= \dfrac{2(1 + \sin x)}{\cos x(1 + \sin x)}$

$= \dfrac{2}{\cos x}$

$= 2\sec x$

65. $\dfrac{\sin^2 y}{1 - \cos y} = \dfrac{1 - \cos^2 y}{1 - \cos y}$

$= \dfrac{(1 + \cos y)(1 - \cos y)}{1 - \cos y}$

$= 1 + \cos y$

67. $\dfrac{3}{\sec x - \tan x} \cdot \dfrac{\sec x + \tan x}{\sec x + \tan x} = \dfrac{3(\sec x + \tan x)}{\sec^2 x - \tan^2 x}$

$= \dfrac{3(\sec x + \tan x)}{1}$

$= 3(\sec x + \tan x)$

69. $y_1 = \cos\left(\frac{\pi}{2} - x\right),\ y_2 = \sin x$

x	0.2	0.4	0.6	0.8	1.0	1.2	1.4
y_1	0.1987	0.3894	0.5646	0.7174	0.8415	0.9320	0.9855
y_2	0.1987	0.3894	0.5646	0.7174	0.8415	0.9320	0.9855

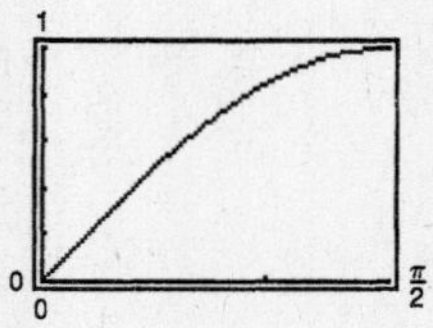

Conclusion: $y_1 = y_2$

71. $y_1 = \frac{\cos x}{1 - \sin x},\ y_2 = \frac{1 + \sin x}{\cos x}$

x	0.2	0.4	0.6	0.8	1.0	1.2	1.4
y_1	1.2230	1.5085	1.8958	2.4650	3.4082	5.3319	11.6814
y_2	1.2230	1.5085	1.8958	2.4650	3.4082	5.3319	11.6814

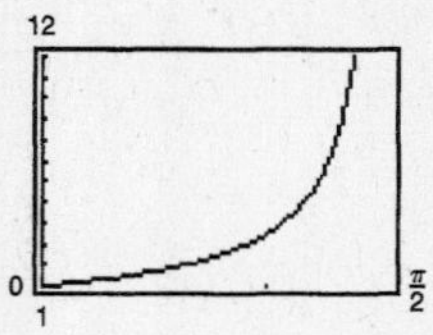

Conclusion: $y_1 = y_2$

73. $y_1 = \cos x \cot x + \sin x = \csc x$

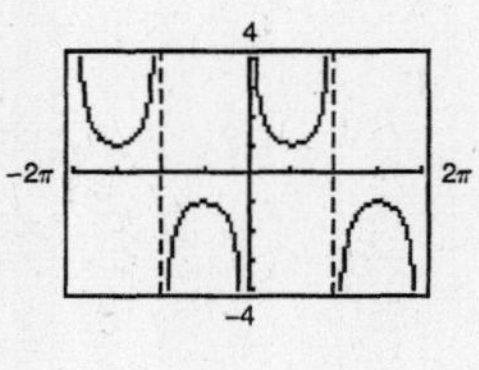

$$\cos x \cot x + \sin x = \cos x\left(\frac{\cos x}{\sin x}\right) + \sin x$$

$$= \frac{\cos^2 x}{\sin x} + \frac{\sin^2 x}{\sin x}$$

$$= \frac{\cos^2 x + \sin^2 x}{\sin x} = \frac{1}{\sin x} = \csc x$$

75. $y_1 = \frac{1}{\sin x}\left(\frac{1}{\cos x} - \cos x\right) = \tan x$

$$\frac{1}{\sin x}\left(\frac{1}{\cos x} - \cos x\right) = \frac{1}{\sin x \cos x} - \frac{\cos x}{\sin x}$$

$$= \frac{1 - \cos^2 x}{\sin x \cos x} = \frac{\sin^2 x}{\sin x \cos x} = \frac{\sin x}{\cos x} = \tan x$$

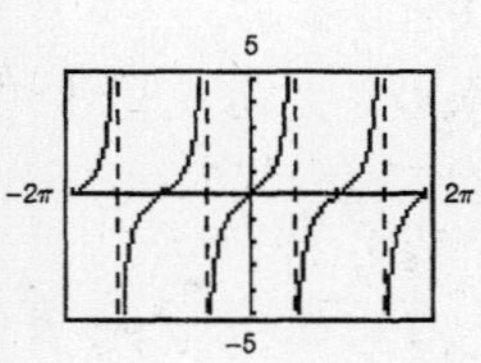

77. Let $x = 3\cos\theta$, then

$$\sqrt{9-x^2} = \sqrt{9-(3\cos\theta)^2} = \sqrt{9-9\cos^2\theta} = \sqrt{9(1-\cos^2\theta)}$$
$$= \sqrt{9\sin^2\theta} = 3\sin\theta$$

79. Let $x = 3\sec\theta$, then

$$\sqrt{x^2-9} = \sqrt{(3\sec\theta)^2-9}$$
$$= \sqrt{9\sec^2\theta - 9}$$
$$= \sqrt{9(\sec^2\theta - 1)}$$
$$= \sqrt{9\tan^2\theta}$$
$$= 3\tan\theta$$

81. Let $x = 5\tan\theta$, then

$$\sqrt{x^2+25} = \sqrt{(5\tan\theta)^2+25}$$
$$= \sqrt{25\tan^2\theta + 25}$$
$$= \sqrt{25(\tan^2\theta + 1)}$$
$$= \sqrt{25\sec^2\theta}$$
$$= 5\sec\theta$$

83. Let $x = 3\sin\theta$, then $\sqrt{9-x^2} = 3$ becomes

$$\sqrt{9-(3\sin\theta)} = 3$$
$$\sqrt{9-9\sin^2\theta} = 3$$
$$\sqrt{9(1-\sin^2\theta)} = 3$$
$$\sqrt{9\cos^2\theta} = 3$$
$$3\cos\theta = 3$$
$$\cos\theta = 1$$
$$\sin\theta = \sqrt{1-\cos^2\theta} = \sqrt{1-(1)^2} = 0$$

85. Let $x = 2\cos\theta$, then $\sqrt{16-4x^2} = 2\sqrt{2}$ becomes

$$\sqrt{16-4(2\cos\theta)^2} = 2\sqrt{2}$$
$$\sqrt{16-16\cos^2\theta} = 2\sqrt{2}$$
$$\sqrt{16(1-\cos^2\theta)} = 2\sqrt{2}$$
$$\sqrt{16\sin^2\theta} = 2\sqrt{2}$$
$$4\sin\theta = 2\sqrt{2}$$
$$\sin\theta = \frac{\sqrt{2}}{2}$$
$$\cos\theta = \sqrt{1-\sin^2\theta} = \sqrt{1-\frac{1}{2}} = \sqrt{\frac{1}{2}} = \frac{\sqrt{2}}{2}$$

87. $\sin\theta = \sqrt{1-\cos^2\theta}$

Let $y_1 = \sin x$ and $y_2 = \sqrt{1-\cos^2 x}$, $0 \le x \le 2\pi$.

$y_1 = y_2$ for $0 \le x \le \pi$, so we have

$\sin\theta = \sqrt{1-\cos^2\theta}$ for $0 \le \theta \le \pi$.

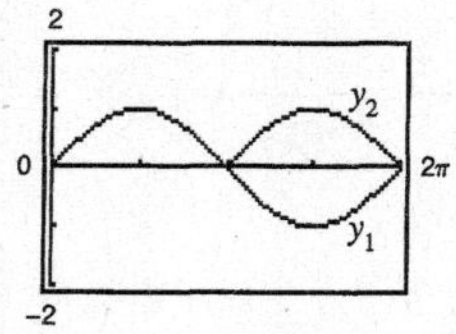

89. $\sec\theta = \sqrt{1+\tan^2\theta}$

Let $y_1 = \dfrac{1}{\cos x}$ and $y_2 = \sqrt{1+\tan^2 x}$, $0 \le x \le 2\pi$.

$y_1 = y_2$ for $0 \le x < \dfrac{\pi}{2}$ and $\dfrac{3\pi}{2} < x \le 2\pi$, so we have

$\sec\theta = \sqrt{1+\tan^2\theta}$ for $0 \le \theta < \dfrac{\pi}{2}$ and $\dfrac{3\pi}{2} < \theta \le 2\pi$.

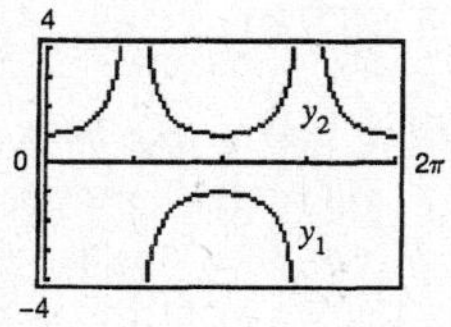

91. (a) $\csc^2 132° - \cot^2 132° \approx 1.8107 - 0.8107 = 1$

(b) $\csc^2 \frac{2\pi}{7} - \cot^2 \frac{2\pi}{7} \approx 1.6360 - 0.6360 = 1$

93. $\cos\left(\frac{\pi}{2} - \theta\right) = \sin \theta$

(a) $\theta = 80°$

$\cos(90° - 80°) = \sin 80°$

$0.9848 = 0.9848$

(b) $\theta = 0.8$

$\cos\left(\frac{\pi}{2} - 0.8\right) = \sin 0.8$

$0.7174 = 0.7174$

95. $\cot \theta = \frac{\cos \theta}{\sin \theta}$

(a) $\theta = 25°$

$\cot \theta = \cot 25° \approx 2.1445$

$\frac{\cos \theta}{\sin \theta} = \frac{\cos 25°}{\sin 25°} \approx 2.1445$

(b) $\theta = \frac{\pi}{8}$

$\cot \theta = \cot \frac{\pi}{8} \approx 2.4142$

$\frac{\cos \theta}{\sin \theta} = \frac{\cos \frac{\pi}{8}}{\sin \frac{\pi}{8}} \approx 2.4142$

97. $\tan\left(\frac{\pi}{2} - \theta\right) = \cot \theta$

(a) $\theta = 5°$

$\tan\left(\frac{\pi}{2} - \theta\right) = \tan(90° - 5°) = \tan(85°) \approx 11.4301$

$\cot \theta = \cot 5° = 11.4301$

(b) $\theta = \frac{11\pi}{12}$

$\tan\left(\frac{\pi}{2} - \theta\right) = \tan\left(\frac{\pi}{2} - \frac{11\pi}{12}\right) = \tan\left(-\frac{5\pi}{12}\right) \approx -3.7321$

$\cot \theta = \cot \frac{11\pi}{12} = -3.7321$

99. $\mu W \cos \theta = W \sin \theta$

$\mu = \frac{W \sin \theta}{W \cos \theta} = \tan \theta$

101. True.
For example, $\sin(-x) = -\sin(x)$ means the graph of $\sin(x)$ is symmetric about the origin.

103. As $x \longrightarrow \frac{\pi^-}{2}$, $\sin x \longrightarrow 1$ and $\csc x \longrightarrow 1$

105. As $x \longrightarrow \frac{\pi^-}{2}$, $\tan x \longrightarrow \infty$ and $\cot x \longrightarrow 0$.

107. $\cos \theta = \sqrt{1 - \sin^2 \theta}$ is not an identity.

$\cos^2 \theta + \sin^2 \theta = 1 \implies \cos \theta = \pm\sqrt{1 - \sin^2 \theta}$

109. $\frac{\sin k\theta}{\cos k\theta} = \tan \theta$ is not an identity.

$\frac{\sin k\theta}{\cos k\theta} = \tan k\theta$

111. $\sin\theta\csc\theta = 1$ is an identity.

$$\sin\theta \cdot \frac{1}{\sin\theta} = 1, \text{ provided } \sin\theta \neq 0.$$

113. Let (x, y) be any point on the terminal side of θ.

Then $r = \sqrt{x^2 + y^2}$ and:

$$\sin^2\theta + \cos^2\theta = \left(\frac{y}{r}\right)^2 + \left(\frac{x}{r}\right)^2$$
$$= \frac{y^2 + x^2}{r^2}$$
$$= \frac{r^2}{r^2}$$
$$= 1$$

115. $(\sqrt{x} + 5)(\sqrt{x} - 5) = (\sqrt{x})^2 - (5)^2 = x - 25$

117. $$\frac{1}{x+5} + \frac{x}{x-8} = \frac{(x-8) + x(x+5)}{(x+5)(x-8)}$$
$$= \frac{x - 8 + x^2 + 5x}{(x+5)(x-8)}$$
$$= \frac{x^2 + 6x - 8}{(x+5)(x-8)}$$

119. $$\frac{2x}{x^2-4} - \frac{7}{x+4} = \frac{2x(x+4) - 7(x^2-4)}{(x^2-4)(x+4)}$$
$$= \frac{2x^2 + 8x - 7x^2 + 28}{(x^2-4)(x+4)}$$
$$= \frac{-5x^2 + 8x + 28}{(x^2-4)(x+4)}$$

Section 2.2 Verifying Trigonometric Identities

- You should know the difference between an expression, a conditional equation, and an identity.
- You should be able to solve trigonometric identities, using the following techniques.
  (a) Work with *one* side at a time. Do not "cross" the equal sign.
  (b) Use algebraic techniques such as combining fractions, factoring expressions, rationalizing denominators, and squaring binomials.
  (c) Use the fundamental identities.
  (d) Convert all the terms into sines and cosines.

Solutions to Odd-Numbered Exercises

1. $\sin t\csc t = \sin t\left(\frac{1}{\sin t}\right) = 1$

3. $(1 + \sin\alpha)(1 - \sin\alpha) = 1 - \sin^2\alpha = \cos^2\alpha$

5. $$\cos^2\beta - \sin^2\beta = (1 - \sin^2\beta) - \sin^2\beta$$
$$= 1 - 2\sin^2\beta$$

7. $$\tan^2\theta + 4 = (\sec^2\theta - 1) + 4$$
$$= \sec^2\theta + 3$$

9. $$\sin^2\alpha - \sin^4\alpha = \sin^2\alpha(1 - \sin^2\alpha)$$
$$= (1 - \cos^2\alpha)(\cos^2\alpha)$$
$$= \cos^2\alpha - \cos^4\alpha$$

11. $\dfrac{\csc^2\theta}{\cot\theta} = \csc^2\theta\left(\dfrac{1}{\cot\theta}\right) = \csc^2\theta\tan\theta$

$= \left(\dfrac{1}{\sin^2\theta}\right)\left(\dfrac{\sin\theta}{\cos\theta}\right) = \left(\dfrac{1}{\sin\theta}\right)\left(\dfrac{1}{\cos\theta}\right)$

$= \csc\theta\sec\theta$

13. $\dfrac{\cot^2 t}{\csc t} = \dfrac{\cos^2 t}{\sin^2 t}\cdot\sin t$

$= \dfrac{\cos^2 t}{\sin t}$

$= \dfrac{1-\sin^2 t}{\sin t} = \dfrac{1}{\sin t} - \dfrac{\sin^2 t}{\sin t}$

$= \csc t - \sin t$

15. $\sin^{1/2} x\cos x - \sin^{5/2} x\cos x = \sin^{1/2} x\cos x(1-\sin^2 x) = \sin^{1/2} x\cos x\cdot\cos^2 x = \cos^3 x\sqrt{\sin x}$

17. $\dfrac{1}{\sec x\tan x} = \cos x\cot x = \cos x\cdot\dfrac{\cos x}{\sin x}$

$= \dfrac{\cos^2 x}{\sin x}$

$= \dfrac{1-\sin^2 x}{\sin x}$

$= \dfrac{1}{\sin x} - \sin x$

$= \csc x - \sin x$

19. $\cot\alpha + \tan\alpha = \dfrac{\cos\alpha}{\sin\alpha} + \dfrac{\sin\alpha}{\cos\alpha}$

$= \dfrac{\cos^2\alpha + \sin^2\alpha}{\sin\alpha\cos\alpha}$

$= \dfrac{1}{\sin\alpha\cos\alpha}$

$= \dfrac{1}{\sin\alpha}\cdot\dfrac{1}{\cos\alpha}$

$= \csc\alpha\sec\alpha$

21. $\dfrac{1}{\tan x} + \dfrac{1}{\cot x} = \dfrac{\cot x + \tan x}{\tan x\cot x}$

$= \dfrac{\cot x + \tan x}{1}$

$= \tan x + \cot x$

23. $\dfrac{\cos\theta\cot\theta}{1-\sin\theta} - 1 = \dfrac{\cos\theta\cot\theta - (1-\sin\theta)}{1-\sin\theta}$

$= \dfrac{\cos\theta\left(\dfrac{\cos\theta}{\sin\theta}\right) - 1 + \sin\theta}{1-\sin\theta}\cdot\dfrac{\sin\theta}{\sin\theta}$

$= \dfrac{\cos^2\theta - \sin\theta + \sin^2\theta}{\sin\theta(1-\sin\theta)}$

$= \dfrac{1-\sin\theta}{\sin\theta(1-\sin\theta)}$

$= \dfrac{1}{\sin\theta}$

$= \csc\theta$

25. $\dfrac{1}{\sin x + 1} + \dfrac{1}{\csc x + 1} = \dfrac{\csc x + 1 + \sin x + 1}{(\sin x + 1)(\csc x + 1)}$

$= \dfrac{\sin x + \csc x + 2}{\sin x\csc x + \sin x + \csc x + 1}$

$= \dfrac{\sin x + \csc x + 2}{1 + \sin x + \csc x + 1}$

$= \dfrac{\sin x + \csc x + 2}{\sin x + \csc x + 2}$

$= 1$

27. $\tan\left(\dfrac{\pi}{2} - \theta\right)\tan\theta = \cot\theta\tan\theta = \left(\dfrac{1}{\tan\theta}\right)\tan\theta = 1$

29. $\dfrac{\csc(-x)}{\sec(-x)} = \dfrac{\dfrac{1}{\sin(-x)}}{\dfrac{1}{\cos(-x)}}$

$= \dfrac{\cos(-x)}{\sin(-x)}$

$= \dfrac{\cos x}{-\sin x}$

$= -\cot x$

31. $\dfrac{\sin x \cos y + \cos x \sin y}{\cos x \cos y - \sin x \sin y} = \dfrac{\dfrac{\sin x \cos y}{\cos x \cos y} + \dfrac{\cos x \sin y}{\cos x \cos y}}{\dfrac{\cos x \cos y}{\cos x \cos y} - \dfrac{\sin x \sin y}{\cos x \cos y}} = \dfrac{\tan x + \tan y}{1 - \tan x \tan y}$

33. $\dfrac{\tan x + \cot y}{\tan x \cot y} = \dfrac{\dfrac{1}{\cot x} + \dfrac{1}{\tan y}}{\dfrac{1}{\cot x} \cdot \dfrac{1}{\tan y}} \cdot \dfrac{\cot x \tan y}{\cot x \tan y} = \tan y + \cot x$

35. $\sqrt{\dfrac{1 + \sin \theta}{1 - \sin \theta}} = \sqrt{\dfrac{1 + \sin \theta}{1 - \sin \theta} \cdot \dfrac{1 + \sin \theta}{1 + \sin \theta}}$

$= \sqrt{\dfrac{(1 + \sin \theta)^2}{1 - \sin^2 \theta}}$

$= \sqrt{\dfrac{(1 + \sin \theta)^2}{\cos^2 \theta}}$

$= \dfrac{1 + \sin \theta}{|\cos \theta|}$

37. $\cos^2 \beta + \cos^2\left(\dfrac{\pi}{2} - \beta\right) = \cos^2 \beta + \sin^2 \beta = 1$

39. $\sin t \csc\left(\dfrac{\pi}{2} - t\right) = \sin t \sec t = \sin t\left(\dfrac{1}{\cos t}\right)$

$= \dfrac{\sin t}{\cos t} = \tan t$

41.

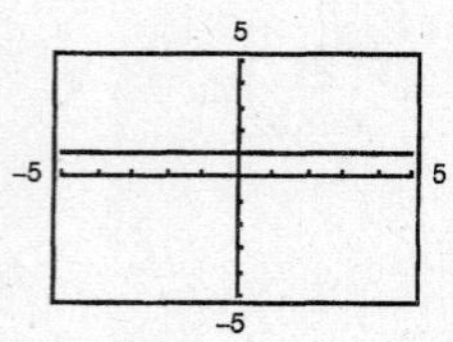

Let $y_1 = \dfrac{2}{(\cos x)^2} - \dfrac{2(\sin x)^2}{(\cos x)^2} - (\sin x)^2 - (\cos x)^2$

and $y_2 = 1$

Is an identity

$2 \sec^2 x - 2 \sec^2 x \sin^2 x - \sin^2 x - \cos^2 x = 2 \sec^2 x(1 - \sin^2 x) - (\sin^2 x + \cos^2 x)$

$= 2 \sec^2 x(\cos^2 x) - 1$

$= 2 \cdot \dfrac{1}{\cos^2 x} \cdot \cos^2 x - 1$

$= 2 - 1$

$= 1$

43. 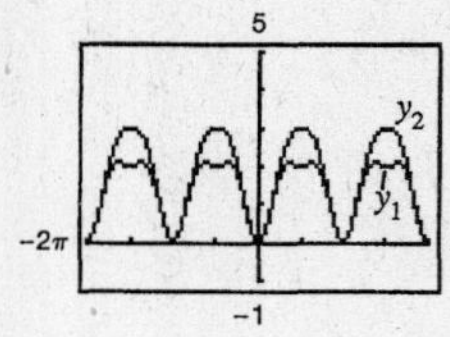

$$2 + \cos^2 x - 3\cos^4 x = (1 - \cos^2 x)(2 + 3\cos^2 x)$$
$$= \sin^2 x(2 + 3\cos^2 x)$$

Let $y_1 = 2 + (\cos x)^2 - 3(\cos x)^4$ and
$y_2 = (\sin x)^2(3 + 2(\cos x)^2)$

Not an identity

45. 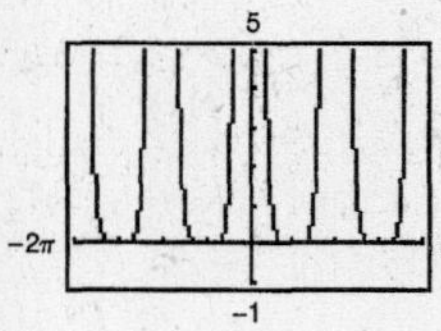

Let $y_1 = \dfrac{1}{(\sin x)^4} - \dfrac{2}{(\sin x)^2} + 1$ and $y_2 = \dfrac{1}{(\tan x)^4}$

Is an identity

$$\csc^4 x - 2\csc^2 x + 1 = (\csc^2 x - 1)^2$$
$$= (\cot^2 x)^2 = \cot^4 x$$

47. 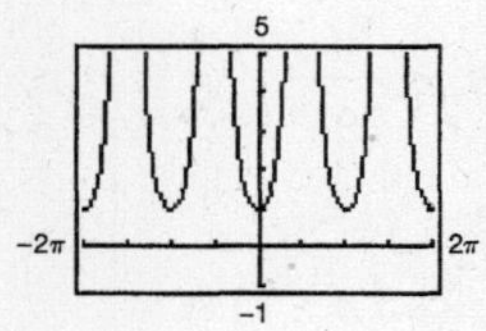

Let $y_1 = \dfrac{1}{(\cos x)^4} - (\tan x)^4$ and $y_2 = 1 + 2(\tan x)^2$

Is an identity

$$\sec^4 \theta - \tan^4 \theta = (\sec^2 \theta + \tan^2 \theta)(\sec^2 \theta - \tan^2 \theta)$$
$$= (1 + \tan^2 \theta + \tan^2 \theta)(1)$$
$$= 1 + 2\tan^2 \theta$$

49. 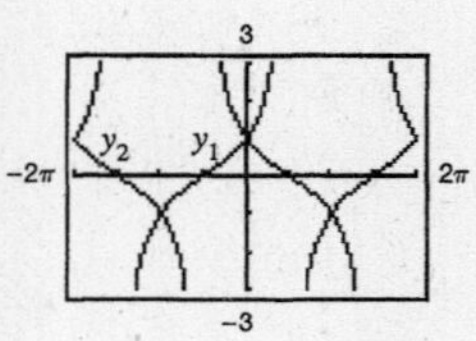

Let $y_1 = \dfrac{\cos x}{(1 - \sin x)}$

and $y_2 = \dfrac{(1 - \sin x)}{\cos x}$

Not an identity

$$\frac{\cos x}{1 - \sin x} = \frac{\cos x}{1 - \sin x} \cdot \frac{1 + \sin x}{1 + \sin x}$$
$$= \frac{\cos x\,(1 + \sin x)}{1 - \sin^2 x}$$
$$= \frac{\cos x\,(1 + \sin x)}{\cos^2 x}$$
$$= \frac{1 + \sin x}{\cos x}$$

51. 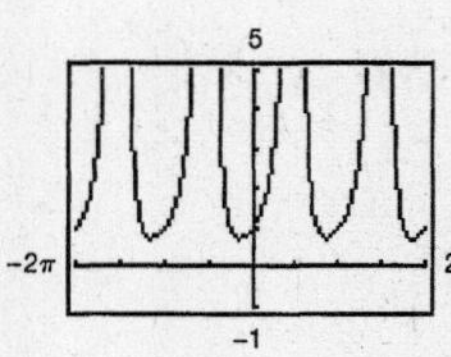

Let $y_1 = \dfrac{((\tan x)^3 - 1)}{(\tan x - 1)}$ and $y_2 = (\tan x)^2 + (\tan x) + 1$

Is an identity

$$\frac{\tan^3 \alpha - 1}{\tan \alpha - 1} = \frac{(\tan \alpha - 1)(\tan^2 \alpha + \tan \alpha + 1)}{\tan \alpha - 1} = \tan^2 \alpha + \tan \alpha + 1$$

53. $\sin^2 25° + \sin^2 65° = \sin^2 25° + \cos^2(90° - 65°) = \sin^2 25° + \cos^2 25° = 1$

55. $\cos^2 20° + \cos^2 52° + \cos^2 38° + \cos^2 70° = \cos^2 20° + \cos^2 52° + \sin^2(90° - 38°) + \sin^2(90° - 70°)$

$$= \cos^2 20° + \cos^2 52° + \sin^2 52° + \sin^2 20°$$
$$= (\cos^2 20° + \sin^2 20°) + (\cos^2 52° + \sin^2 52°)$$
$$= 1 + 1$$
$$= 2$$

57. $\cos x - \csc x \cot x = \cos x - \dfrac{1}{\sin x}\dfrac{\cos x}{\sin x}$

$$= \cos x\left(1 - \frac{1}{\sin^2 x}\right)$$
$$= \cos x(1 - \csc^2 x)$$
$$= -\cos x(\csc^2 x - 1)$$
$$= -\cos x \cot^2 x$$

59. False. For the equation to be an identity, it must be true for all values of θ in the domain.

61. Since $\sin^2 \theta = 1 - \cos^2 \theta$, then $\sin \theta = \pm\sqrt{1 - \cos^2 \theta}$; $\sin \theta \neq \sqrt{1 - \cos \theta}$ if θ lies in Quadrant III or IV.

One such angle is $\theta = \dfrac{7\pi}{4}$.

63. $x^2 - 6x - 12 = 0$

$$x = \frac{-(-6) \pm \sqrt{(-6)^2 - 4(1)(-12)}}{2(1)}$$
$$= \frac{6 \pm \sqrt{36 + 48}}{2}$$
$$= \frac{6 \pm \sqrt{84}}{2}$$
$$= \frac{6 \pm 2\sqrt{21}}{2}$$
$$= 3 \pm \sqrt{21}$$

65. $3x^2 + 6x - 12 = 0$

$$x = \frac{-6 \pm \sqrt{6^2 - 4(3)(-12)}}{2(3)}$$
$$= \frac{-6 \pm \sqrt{36 + 144}}{6}$$
$$= \frac{-6 \pm \sqrt{180}}{6}$$
$$= \frac{-6 \pm 6\sqrt{5}}{6}$$
$$= -1 \pm \sqrt{5}$$

Section 2.3 Solving Trigonometric Equations

- You should be able to identify and solve trigonometric equations.
- A trigonometric equation is a conditional equation. It is true for a specific set of values.
- To solve trigonometric equations, use algebraic techniques such as collecting like terms, taking square roots, factoring, squaring, converting to quadratic form, using formulas, and using inverse functions. Study the examples in this section.

Solutions to Odd-Numbered Exercises

1. $2\cos x - 1 = 0$

(a) $2\cos\dfrac{\pi}{3} - 1 = 2\left(\dfrac{1}{2}\right) - 1 = 0$

(b) $2\cos\dfrac{5\pi}{3} - 1 = 2\left(\dfrac{1}{2}\right) - 1 = 0$

3. $3\tan^2 2x - 1 = 0$

(a) $$\begin{aligned} 3\left[\tan 2\left(\frac{\pi}{12}\right)\right]^2 - 1 &= 3\tan^2\frac{\pi}{6} - 1 \\ &= 3\left(\frac{1}{\sqrt{3}}\right)^2 - 1 \\ &= 0 \end{aligned}$$

(b) $$\begin{aligned} 3\left[\tan 2\left(\frac{5\pi}{12}\right)\right]^2 - 1 &= 3\tan^2\frac{5\pi}{6} - 1 \\ &= 3\left(-\frac{1}{\sqrt{3}}\right)^2 - 1 \\ &= 0 \end{aligned}$$

5. $2\sin^2 x - \sin x - 1 = 0$

(a) $$\begin{aligned} 2\sin^2\frac{\pi}{2} - \sin\frac{\pi}{2} - 1 &= 2(1)^2 - 1 - 1 \\ &= 0 \end{aligned}$$

(b) $$\begin{aligned} 2\sin^2\frac{7\pi}{6} - \sin\frac{7\pi}{6} - 1 &= 2\left(-\frac{1}{2}\right)^2 - \left(-\frac{1}{2}\right) - 1 \\ &= \frac{1}{2} + \frac{1}{2} - 1 \\ &= 0 \end{aligned}$$

7. $2\cos x + 1 = 0$

$$\begin{aligned} 2\cos x &= -1 \\ \cos x &= -\frac{1}{2} \\ x &= \frac{2\pi}{3} + 2n\pi \\ \text{or } x &= \frac{4\pi}{3} + 2n\pi \end{aligned}$$

9. $\sqrt{3}\csc x - 2 = 0$

$$\begin{aligned} \sqrt{3}\csc x &= 2 \\ \csc x &= \frac{2}{\sqrt{3}} \\ x &= \frac{\pi}{3} + 2n\pi \\ \text{or } x &= \frac{2\pi}{3} + 2n\pi \end{aligned}$$

11. $3\sec^2 x - 4 = 0$

$$\begin{aligned} \sec^2 x &= \frac{4}{3} \\ \sec x &= \pm\frac{2}{\sqrt{3}} \\ x &= \frac{\pi}{6} + n\pi \\ \text{or } x &= \frac{5\pi}{6} + n\pi \end{aligned}$$

13. $\sin x(\sin x + 1) = 0$

$\sin x = 0$ or $\sin x = -1$

$x = n\pi$ $\qquad$ $x = \dfrac{3\pi}{2} + 2n\pi$

15. $4\cos^2 x - 1 = 0$

$$\begin{aligned} \cos^2 x &= \frac{1}{4} \\ \cos^2 x &= \pm\frac{1}{2} \end{aligned}$$

$x = \dfrac{\pi}{3} + n\pi$ or $x = \dfrac{2\pi}{3} + n\pi$

17. $2\sin^2 2x = 1$

$$\sin 2x = \pm\frac{1}{\sqrt{2}} = \pm\frac{\sqrt{2}}{2}$$

$$2x = \frac{\pi}{4} + 2n\pi,\ 2x = \frac{3\pi}{4} + 2n\pi,\ 2x = \frac{5\pi}{4} + 2n\pi,\ 2x = \frac{7\pi}{4} + 2n\pi.$$

Thus, $x = \frac{\pi}{8} + n\pi,\ \frac{3\pi}{8} + n\pi,\ \frac{5\pi}{8} + n\pi,\ \frac{7\pi}{8} + n\pi$ or these can be combined as $x = \frac{\pi}{8} + \frac{n\pi}{2},\ x = \frac{3\pi}{8} + \frac{n\pi}{2}$.

19. $\tan 3x(\tan x - 1) = 0$

$\tan 3x = 0$ or $\tan x - 1 = 0$

$3x = n\pi$ $\quad$ $\tan x = 1$

$x = \frac{n\pi}{3}$ $\quad$ $x = \frac{\pi}{4} + n\pi$

21. $\cos^3 x = \cos x$

$\cos^3 x - \cos x = 0$

$\cos x(\cos^2 x - 1) = 0$

$\cos x = 0$ or $\cos^2 x - 1 = 0$

$x = \frac{\pi}{2}, \frac{3\pi}{2}$ $\quad$ $\cos x = \pm 1$

$x = 0,\ \pi$

23. $3\tan^3 x - \tan x = 0$

$\tan x(3\tan^2 x - 1) = 0$

$\tan x = 0$ or $3\tan^2 x - 1 = 0$

$x = 0,\ \pi$ $\quad$ $\tan x = \pm\frac{\sqrt{3}}{3}$

$x = \frac{\pi}{6}, \frac{5\pi}{6}, \frac{7\pi}{6}, \frac{11\pi}{6}$

25. $\sec^2 x - \sec x - 2 = 0$

$(\sec x - 2)(\sec x + 1) = 0$

$\sec x - 2 = 0$ or $\sec x + 1 = 0$

$\sec x = 2$ $\quad$ $\sec x = -1$

$x = \frac{\pi}{3}, \frac{5\pi}{3}$ $\quad$ $x = \pi$

27. $2\sin x + \csc x = 0$

$2\sin x + \frac{1}{\sin x} = 0$

$2\sin^2 x + 1 = 0$

$\sin^2 x = -\frac{1}{2} \Rightarrow$ No solution

29. $2\cos^2 x + \cos x - 1 = 0$

$(2\cos x - 1)(\cos x + 1) = 0$

$2\cos x - 1 = 0$ or $\cos x + 1 = 0$

$\cos x = \frac{1}{2}$ $\quad$ $\cos x = -1$

$x = \frac{\pi}{3}, \frac{5\pi}{3}$ $\quad$ $x = \pi$

31. $2\sec^2 x + \tan^2 x - 3 = 0$

$2(\tan^2 x + 1) + \tan^2 x - 3 = 0$

$3\tan^2 x - 1 = 0$

$\tan x = \pm\frac{\sqrt{3}}{3}$

$x = \frac{\pi}{6}, \frac{5\pi}{6}, \frac{7\pi}{6}, \frac{11\pi}{6}$

33. $\cos 2x = \frac{1}{2}$

$2x = \frac{\pi}{3} + 2n\pi$ or $2x = \frac{5\pi}{3} + 2n\pi$

$x = \frac{\pi}{6} + n\pi$ $\quad$ $x = \frac{5\pi}{6} + n\pi$

35. $\tan 3x = 1$

$$3x = \frac{\pi}{4} + 2n\pi \quad \text{or} \quad 3x = \frac{5\pi}{4} + 2n\pi$$

$$x = \frac{\pi}{12} + \frac{2n\pi}{3} \qquad x = \frac{5\pi}{12} + \frac{2n\pi}{3}$$

These can be combined as $x = \dfrac{\pi}{12} + \dfrac{n\pi}{3}$.

37. $\cos\left(\dfrac{x}{2}\right) = \dfrac{\sqrt{2}}{2}$

$$\frac{x}{2} = \frac{\pi}{4} + 2n\pi \quad \text{or} \quad \frac{x}{2} = \frac{7\pi}{4} + 2n\pi$$

$$x = \frac{\pi}{2} + 4n\pi \qquad x = \frac{7\pi}{2} + 4n\pi$$

39. $y = \sin\dfrac{\pi x}{2} + 1$

From the graph in the textbook we see that the curve has x-intercepts at $x = -1$ and at $x = 3$.

In general, we have: $\sin\left(\dfrac{\pi x}{2}\right) = -1$

$$\frac{\pi x}{2} = \frac{3\pi}{2} + 2n\pi$$

$$x = 3 + 4n$$

41. $y = \tan^2\left(\dfrac{\pi x}{6}\right) - 3$

From the graph in the textbook we see that the curve has x-intercepts at $x = \pm 2$.

In general, we have: $\tan^2\left(\dfrac{\pi x}{6}\right) = 3$

$$\tan\left(\frac{\pi x}{6}\right) = \pm\sqrt{3}$$

$$\frac{\pi x}{6} = \pm\frac{\pi}{3} + n\pi$$

$$x = \pm 2 + 6n$$

43. $6y^2 - 13y + 6 = 0$

$$(3y - 2)(2y - 3) = 0$$

$$3y - 2 = 0 \quad \text{or} \quad 2y - 3 = 0$$

$$y = \frac{2}{3} \qquad y = \frac{3}{2}$$

$$6\cos^2 x - 13\cos x + 6 = 0$$

$$(3\cos x - 2)(2\cos x - 3) = 0$$

$$3\cos x - 2 = 0 \quad \text{or} \quad 2\cos x - 3 = 0$$

$$\cos x = \frac{2}{3} \qquad \cos x = \frac{3}{2} \text{(No solution)}$$

$$x \approx 0.8411 + 2n\pi,\ 5.4421 + 2n\pi$$

45. $2\sin x + \cos x = 0$

$$2\sin x = -\cos x$$

$$2 = -\frac{\cos x}{\sin x}$$

$$2 = -\cot x$$

$$-2 = \cot x$$

$$-\frac{1}{2} = \tan x$$

$$x = \arctan\left(-\frac{1}{2}\right)$$

$$x = \pi - \arctan\left(\frac{1}{2}\right) \approx 2.6779$$

$$\text{or } x = 2\pi - \arctan\left(\frac{1}{2}\right) \approx 5.8195$$

Graph $y_1 = 2\sin x + \cos x$

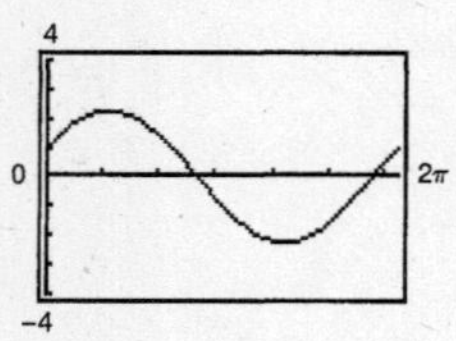

The x-intercepts occur at $x \approx 2.6779$ and $x \approx 5.8195$

47.
$$\frac{1+\sin x}{\cos x}+\frac{\cos x}{1+\sin x}=4$$
$$\frac{(1+\sin x)^2+\cos^2 x}{\cos x(1+\sin x)}=4$$
$$\frac{1+2\sin x+\sin^2 x+\cos^2 x}{\cos x(1+\sin x)}=4$$
$$\frac{2+2\sin x}{\cos x(1+\sin x)}=4$$
$$\frac{2}{\cos x}=4$$
$$\cos x=\frac{1}{2}$$
$$x=\frac{\pi}{3},\ \frac{5\pi}{3}$$

Graph $y_1=\dfrac{1+\sin x}{\cos x}+\dfrac{\cos x}{1+\sin x}-4.$

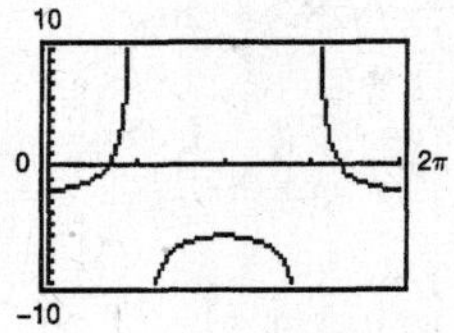

The x-intercepts occur at $x=\frac{\pi}{3}\approx 1.0472$ and $x=\frac{5\pi}{3}\approx 5.2360.$

49. $x\tan x-1=0$

Graph $y_1=x\tan x-1$

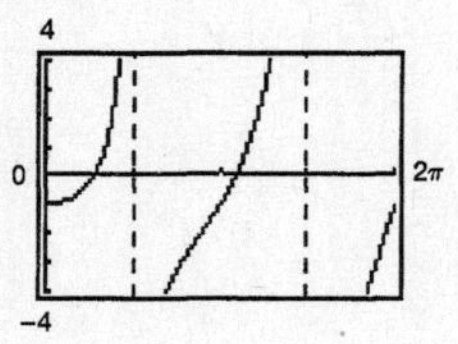

The x-intercepts occur at $x\approx 0.8603$ and $x\approx 3.4256$

51. $\sec^2 x+0.5\tan x-1=0$

Graph $y_1=\dfrac{1}{(\cos x)^2}+0.5\tan x-1.$

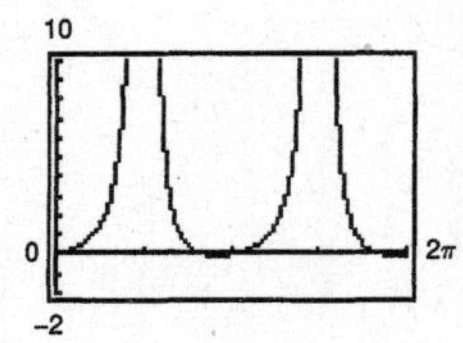

The x-intercepts occur at $x=0$, $x\approx 2.6779$, $x=\pi\approx 3.1416$, and $x\approx 5.8195$.

53. $2\tan^2 x+7\tan x-15=0$

$(2\tan x-3)(\tan x+5)=0$

$2\tan x-3=0$ or $\tan x+5=0$

$\tan x=1.5$ $\qquad\tan x=-5$

$x\approx 0.9828,\ 4.1244$ $\qquad x\approx 1.7682,\ 4.9098$

Graph $y_1=2\tan^2 x+7\tan x-15.$

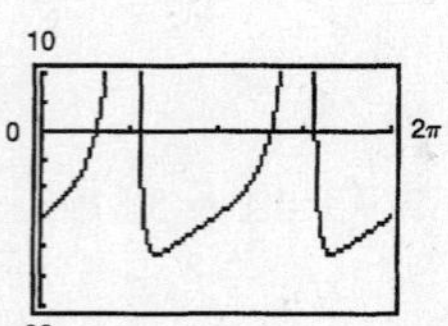

The x-intercepts occur at $x\approx 0.9828$, $x\approx 1.7682$, $x\approx 4.1244$, and $x\approx 4.9098$.

55. $12 \sin^2 x - 13 \sin x + 3 = 0$

$$\sin x = \frac{-(-13) \pm \sqrt{(-13)^2 - 4(12)(3)}}{2(12)}$$

$$= \frac{13 \pm 5}{24}$$

$\sin x = \frac{1}{3}$ or $\sin x = \frac{3}{4}$

$x \approx 0.3398,\ 2.8018$ $\quad x \approx 0.8481,\ 2.2935$

Graph $y_1 = 12 \sin^2 x - 13 \sin x + 3$.

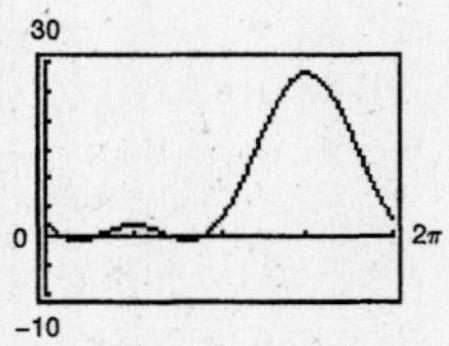

The x-intercepts occur at $x \approx 0.3398$, $x \approx 0.8481$, $x \approx 2.2935$, and $x \approx 2.8018$.

57. $\tan^2 x + 3 \tan x + 1 = 0$

$$\tan x = \frac{-3 \pm \sqrt{3^2 - 4(1)(1)}}{2(1)} = \frac{-3 \pm \sqrt{5}}{2}$$

$\tan x = \dfrac{-3 - \sqrt{5}}{2}$ or $\tan x = \dfrac{-3 + \sqrt{5}}{2}$

$x \approx 1.9357,\ 5.0773$ $\quad x \approx 2.7767,\ 5.9183$

Graph $y_1 = \tan^2 x + 3 \tan x + 1$.

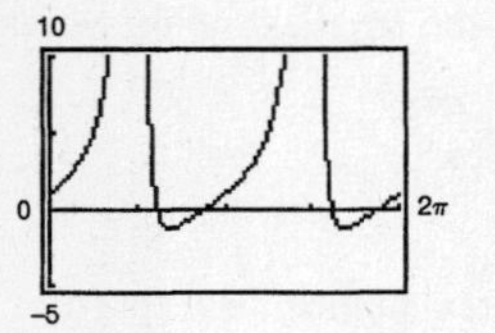

The x-intercepts occur at $x \approx 1.9357$, $x \approx 2.7767$, $x \approx 5.0773$, and $x \approx 5.9183$.

59. $\tan^2 x - 6 \tan x + 5 = 0$

$(\tan x - 1)(\tan x - 5) = 0$

$\tan x - 1 = 0$ or $\tan x - 5 = 0$

$\tan x = 1$ $\quad \tan x = 5$

$x = \dfrac{\pi}{4}, \dfrac{5\pi}{4}$ $\quad x = \arctan 5, \arctan 5 + \pi$

61. $2 \cos^2 x - 5 \cos x + 2 = 0$

$(2 \cos x - 1)(\cos x - 2) = 0$

$2 \cos x - 1 = 0$ or $\cos x - 2 = 0$

$\cos x = \dfrac{1}{2}$ $\quad \cos x = 2$

$x = \dfrac{\pi}{3}, \dfrac{5\pi}{3}$ $\quad$ No solution

63. (a) $f(x) = \sin x + \cos x$

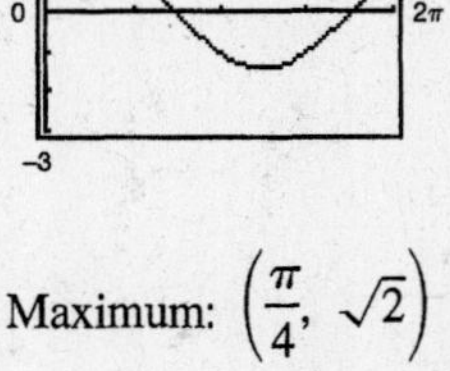

Maximum: $\left(\dfrac{\pi}{4}, \sqrt{2}\right)$

Minimum: $\left(\dfrac{5\pi}{4}, -\sqrt{2}\right)$

—CONTINUED—

63. —CONTINUED—

(b) $\cos x - \sin x = 0$

$$\cos x = \sin x$$

$$1 = \frac{\sin x}{\cos x}$$

$$\tan x = 1$$

$$x = \frac{\pi}{4}, \frac{5\pi}{4}$$

$$f\left(\frac{\pi}{4}\right) = \sin\frac{\pi}{4} + \cos\frac{\pi}{4} = \frac{\sqrt{2}}{2} + \frac{\sqrt{2}}{2} = \sqrt{2}$$

$$f\left(\frac{5\pi}{4}\right) = \sin\frac{5\pi}{4} + \cos\frac{5\pi}{4} = -\sin\frac{\pi}{4} + \left(-\cos\frac{\pi}{4}\right) = -\frac{\sqrt{2}}{2} - \frac{\sqrt{2}}{2} = -\sqrt{2}$$

Therefore, the maximum point in the interval $[0, 2\pi)$ is $(\pi/4, \sqrt{2})$ and the minimum point is $(5\pi/4, -\sqrt{2})$.

65. $f(x) = \tan\frac{\pi x}{4}$

Since $\tan \pi/4 = 1$, $x = 1$ is the smallest nonnegative fixed point.

67. $f(x) = \cos\frac{1}{x}$

(a) The domain of $f(x)$ is all real numbers except 0.

(b) The graph has y-axis symmetry and a horizontal asymptote at $y = 1$.

(c) As $x \longrightarrow 0$, $f(x)$ oscillates between -1 and 1.

(d) There are infinitely many solutions in the interval $[-1, 1]$.

(e) The greatest solution appears to occur at $x \approx 0.6366$.

69. $$y = \frac{1}{12}(\cos 8t - 3\sin 8t)$$

$$\frac{1}{12}(\cos 8t - 3\sin 8t) = 0$$

$$\cos 8t = 3\sin 8t$$

$$\frac{1}{3} = \tan 8t$$

$$8t \approx 0.32175 + n\pi$$

$$t \approx 0.04 + \frac{n\pi}{8}$$

In the interval $0 \le t \le 1$, $t \approx 0.04, 0.43$, and 0.83.

71. $S = 74.50 + 43.75 \sin\frac{\pi t}{6}$

t	1	2	3	4	5	6	7	8	9	10	11	12
S	96.4	112.4	118.3	112.4	96.4	74.5	52.6	36.6	30.8	36.6	52.6	74.5

Sales exceed 100,000 units during February, March, and April.

73. Range $= 1000$ yards $= 3000$ feet

$$v_0 = 1200 \text{ feet per second}$$
$$f = \tfrac{1}{32} v_0^2 \sin 2\theta$$
$$3000 = \tfrac{1}{32}(1200)^2 \sin 2\theta$$
$$\sin 2\theta \approx 0.066667$$
$$2\theta \approx 3.8°$$
$$\theta \approx 1.9°$$

75. (a)

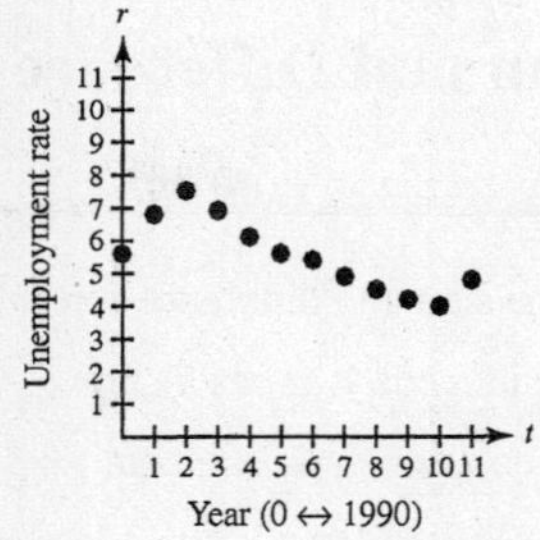

(b) By graphing the curves we see that
$(1) r = 1.39 \sin(0.48t + 0.42) + 5.51$ best fits the data

(c) The constant term gives the average unemployment rate of 5.51%

(d) Period: $\dfrac{2\pi}{0.48} \approx 13$ years

(e) $r \approx 6.5$ when $t \approx 14$ which corresponds to the year 2004

77. True. The period of $2 \sin 4t - 1$ is $\dfrac{\pi}{2}$ and the period of $2 \sin t - 1$ is 2π.

In the interval $[0, 2\pi)$ the first equation has four cycles whereas the second equation has only one cycle, thus the first equation has four times the x-intercepts (solutions) as the second equation.

79. $y_1 = 2\sin x$

$y_2 = 3x + 1$

From the graph we see that there is only one point of intersection.

81.

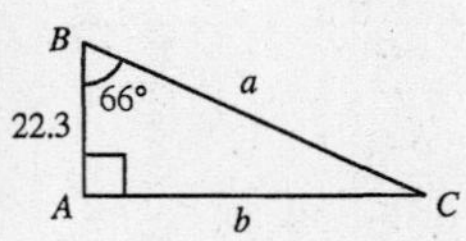

$$C = 90° - 66° = 24°$$
$$\cos 66° = \frac{22.3}{a}$$
$$a \cos 66° = 22.3$$
$$a = \frac{22.3}{\cos 66°} \approx 54.8$$
$$\tan 66° = \frac{b}{22.3}$$
$$b = 22.3 \tan 66° \approx 50.1$$

83. $\theta = 390°$, $\theta' = 390° - 360° = 30°$, θ is in Quadrant I.

$$\sin 390° = \sin 30° = \frac{1}{2}$$
$$\cos 390° = \cos 30° = \frac{\sqrt{3}}{2}$$
$$\tan 390° = \tan 30° = \frac{1}{\sqrt{3}} = \frac{\sqrt{3}}{3}$$

85. $\theta = 495°$, $\theta' = 45°$, θ is in Quadrant II.

$$\sin 495° = \sin 45° = \frac{\sqrt{2}}{2}$$
$$\cos 495° = -\cos 45° = -\frac{\sqrt{2}}{2}$$
$$\tan 495° = -\tan 45° = -1$$

87.

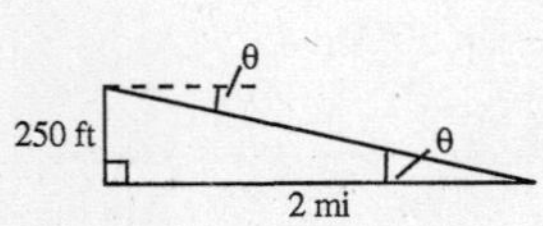

Not drawn to scale

$$\tan \theta = \frac{250 \text{ feet}}{2 \text{ miles}} \times \frac{1 \text{ mile}}{5280 \text{ feet}} \approx 0.02367$$
$$\theta \approx 1.36°$$

Section 2.4 Sum and Difference Formulas

- You should know the sum and difference formulas.

$$\sin(u \pm v) = \sin u \cos v \pm \cos u \sin v$$

$$\cos(u \pm v) = \cos u \cos v \mp \sin u \sin v$$

$$\tan(u \pm v) = \frac{\tan u \pm \tan v}{1 \mp \tan u \tan v}$$

- You should be able to use these formulas to find the values of the trigonometric functions of angles whose sums or differences are special angles.
- You should be able to use these formulas to solve trigonometric equations.

Solutions to Odd-Numbered Exercises

1. (a) $\cos\left(\frac{\pi}{4} + \frac{\pi}{3}\right) = \cos\frac{\pi}{4}\cos\frac{\pi}{3} - \sin\frac{\pi}{4}\sin\frac{\pi}{3}$

$= \frac{\sqrt{2}}{2} \cdot \frac{1}{2} - \frac{\sqrt{2}}{2} \cdot \frac{\sqrt{3}}{2}$

$= \frac{\sqrt{2} - \sqrt{6}}{4}$

(b) $\cos\frac{\pi}{4} + \cos\frac{\pi}{3} = \frac{\sqrt{2}}{2} + \frac{1}{2} = \frac{\sqrt{2} + 1}{2}$

3. (a) $\sin\left(\frac{7\pi}{6} - \frac{\pi}{3}\right) = \sin\frac{5\pi}{6} = \sin\frac{\pi}{6} = \frac{1}{2}$

(b) $\sin\frac{7\pi}{6} - \sin\frac{\pi}{3} = -\frac{1}{2} - \frac{\sqrt{3}}{2} = \frac{-1 - \sqrt{3}}{2}$

5. (a) $\cos(120° + 45°) = \cos 120° \cos 45° - \sin 120° \sin 45°$

$= \left(-\frac{1}{2}\right)\left(\frac{\sqrt{2}}{2}\right) - \left(\frac{\sqrt{3}}{2}\right)\left(\frac{\sqrt{2}}{2}\right)$

$= \frac{-\sqrt{2} - \sqrt{6}}{4}$

(b) $\cos 120° + \cos 45° = -\frac{1}{2} + \frac{\sqrt{2}}{2} = \frac{-1 + \sqrt{2}}{2}$

7. $\sin 105° = \sin(60° + 45°)$

$= \sin 60° \cos 45° + \cos 60° \sin 45°$

$= \frac{\sqrt{3}}{2} \cdot \frac{\sqrt{2}}{2} + \frac{1}{2} \cdot \frac{\sqrt{2}}{2}$

$= \frac{\sqrt{2}}{4}(\sqrt{3} + 1)$

$\cos 105° = \cos(60° + 45°)$

$= \cos 60° \cos 45° - \sin 60° \sin 45°$

$= \frac{1}{2} \cdot \frac{\sqrt{2}}{2} - \frac{\sqrt{3}}{2} \cdot \frac{\sqrt{2}}{2}$

$= \frac{\sqrt{2}}{4}(1 - \sqrt{3})$

$\tan 105° = \tan(60° + 45°)$

$= \frac{\tan 60° + \tan 45°}{1 - \tan 60° \tan 45°}$

$= \frac{\sqrt{3} + 1}{1 - \sqrt{3}} = \frac{\sqrt{3} + 1}{1 - \sqrt{3}} \cdot \frac{1 + \sqrt{3}}{1 + \sqrt{3}}$

$= \frac{4 + 2\sqrt{3}}{-2} = -2 - \sqrt{3}$

9. $\sin 195° = \sin(225° - 30°)$

$= \sin 225° \cos 30° - \cos 225° \sin 30°$

$= -\sin 45° \cos 30° + \cos 45° \sin 30°$

$= -\frac{\sqrt{2}}{2} \cdot \frac{\sqrt{3}}{2} + \frac{\sqrt{2}}{2} \cdot \frac{1}{2}$

$= \frac{\sqrt{2}}{4}\left(1 - \sqrt{3}\right)$

$\cos 195° = \cos(225° - 30°)$

$= \cos 225° \cos 30° + \sin 225° \sin 30°$

$= -\cos 45° \cos 30° - \sin 45° \sin 30°$

$= -\frac{\sqrt{2}}{2} \cdot \frac{\sqrt{3}}{2} - \frac{\sqrt{2}}{2} \cdot \frac{1}{2}$

$= -\frac{\sqrt{2}}{4}\left(\sqrt{3} + 1\right)$

$\tan 195° = \tan(225° - 30°)$

$= \frac{\tan 225° - \tan 30°}{1 + \tan 225° \tan 30°}$

$= \frac{\tan 45° - \tan 30°}{1 + \tan 45° \tan 30°}$

$= \frac{1 - \left(\frac{\sqrt{3}}{3}\right)}{1 + \left(\frac{\sqrt{3}}{3}\right)} = \frac{3 - \sqrt{3}}{3 + \sqrt{3}} \cdot \frac{3 - \sqrt{3}}{3 - \sqrt{3}}$

$= \frac{12 - 6\sqrt{3}}{6} = 2 - \sqrt{3}$

11. $\sin \frac{11\pi}{12} = \sin\left(\frac{3\pi}{4} + \frac{\pi}{6}\right)$

$= \sin \frac{3\pi}{4} \cos \frac{\pi}{6} + \cos \frac{3\pi}{4} \sin \frac{\pi}{6}$

$= \frac{\sqrt{2}}{2} \cdot \frac{\sqrt{3}}{2} + \left(-\frac{\sqrt{2}}{2}\right)\frac{1}{2}$

$= \frac{\sqrt{2}}{4}\left(\sqrt{3} - 1\right)$

$\cos \frac{11\pi}{12} = \cos\left(\frac{3\pi}{4} + \frac{\pi}{6}\right)$

$= \cos \frac{3\pi}{4} \cos \frac{\pi}{6} - \sin \frac{3\pi}{4} \sin \frac{\pi}{6}$

$= -\frac{\sqrt{2}}{2} \cdot \frac{\sqrt{3}}{2} - \frac{\sqrt{2}}{2} \cdot \frac{1}{2} = -\frac{\sqrt{2}}{4}\left(\sqrt{3} + 1\right)$

$\tan \frac{11\pi}{4} = \tan\left(\frac{3\pi}{4} + \frac{\pi}{6}\right)$

$= \frac{\tan \frac{3\pi}{4} + \tan \frac{\pi}{6}}{1 - \tan \frac{3\pi}{4} \tan \frac{\pi}{6}}$

$= \frac{-1 + \frac{\sqrt{3}}{3}}{1 - (-1)\frac{\sqrt{3}}{3}}$

$= \frac{-3 + \sqrt{3}}{3 + \sqrt{3}} \cdot \frac{3 - \sqrt{3}}{3 - \sqrt{3}}$

$= \frac{-12 + 6\sqrt{3}}{6} = -2 + \sqrt{3}$

13. $\sin \frac{17\pi}{12} = \sin\left(\frac{9\pi}{4} - \frac{5\pi}{6}\right)$

$= \sin \frac{9\pi}{4} \cos \frac{5\pi}{6} - \cos \frac{9\pi}{4} \sin \frac{5\pi}{6}$

$= \frac{\sqrt{2}}{2}\left(-\frac{\sqrt{3}}{2}\right) - \left(\frac{\sqrt{2}}{2}\right)\left(\frac{1}{2}\right)$

$= -\frac{\sqrt{2}}{4}\left(\sqrt{3} + 1\right)$

$\cos \frac{17\pi}{12} = \cos\left(\frac{9\pi}{4} - \frac{5\pi}{6}\right)$

$= \cos \frac{9\pi}{4} \cos \frac{5\pi}{6} + \sin \frac{9\pi}{4} \sin \frac{5\pi}{6}$

$= \frac{\sqrt{2}}{2}\left(-\frac{\sqrt{3}}{2}\right) + \frac{\sqrt{2}}{2}\left(\frac{1}{2}\right)$

$= \frac{\sqrt{2}}{4}\left(1 - \sqrt{3}\right)$

$\tan \frac{17\pi}{12} = \tan\left(\frac{9\pi}{4} - \frac{5\pi}{6}\right)$

$= \frac{\tan(9\pi/4) - \tan(5\pi/6)}{1 + \tan(9\pi/4)\tan(5\pi/6)}$

$= \frac{1 - \left(-\sqrt{3}/3\right)}{1 + \left(-\sqrt{3}/3\right)}$

$= \frac{3 + \sqrt{3}}{3 - \sqrt{3}} \cdot \frac{3 + \sqrt{3}}{3 + \sqrt{3}}$

$= \frac{12 + 6\sqrt{3}}{6} = 2 + \sqrt{3}$

15. $285° = 225° + 60°$

$\sin 285° = \sin(225° + 60°)$

$= \sin 225° \cos 60° + \cos 225° \sin 60°$

$= -\frac{\sqrt{2}}{2}\left(\frac{1}{2}\right) - \frac{\sqrt{2}}{2}\left(\frac{\sqrt{3}}{2}\right) = -\frac{\sqrt{2}}{4}(\sqrt{3} + 1)$

$\cos 285° = \cos(225° + 60°)$

$= \cos 225° \cos 60° - \sin 225° \sin 60°$

$= -\frac{\sqrt{2}}{2}\left(\frac{1}{2}\right) - \left(-\frac{\sqrt{2}}{2}\right)\left(\frac{\sqrt{3}}{2}\right) = \frac{\sqrt{2}}{4}(\sqrt{3} - 1)$

$\tan 285° = \tan(225° + 60°)$

$= \frac{\tan 225° + \tan 60°}{1 - \tan 225° \tan 60°} = \frac{1 + \sqrt{3}}{1 - \sqrt{3}} \cdot \frac{1 + \sqrt{3}}{1 + \sqrt{3}}$

$= \frac{4 + 2\sqrt{3}}{-2} = -2 - \sqrt{3} = -(2 + \sqrt{3})$

17. $-165° = -(120° + 45°)$

$\sin(-165°) = \sin[-(120° + 45°)]$

$= -\sin(120° + 45°)$

$= -[\sin 120° \cos 45° + \cos 120° \sin 45°]$

$= -\left[\frac{\sqrt{3}}{2} \cdot \frac{\sqrt{2}}{2} - \frac{1}{2} \cdot \frac{\sqrt{2}}{2}\right]$

$= -\frac{\sqrt{2}}{4}(\sqrt{3} - 1)$

$\cos(-165°) = \cos[-(120° + 45°)]$

$= \cos(120° + 45°)$

$= \cos 120° \cos 45° - \sin 120° \sin 45°$

$= -\frac{1}{2} \cdot \frac{\sqrt{2}}{2} - \frac{\sqrt{3}}{2} \cdot \frac{\sqrt{2}}{2}$

$= -\frac{\sqrt{2}}{4}(1 + \sqrt{3})$

$\tan(-165°) = \tan[-(120° + 45°)]$

$= -\tan(120° + \tan 45°)$

$= -\frac{\tan 120° + \tan 45°}{1 - \tan 120° \tan 45°}$

$= -\frac{-\sqrt{3} + 1}{1 - (-\sqrt{3})(1)}$

$= -\frac{1 - \sqrt{3}}{1 + \sqrt{3}} \cdot \frac{1 - \sqrt{3}}{1 - \sqrt{3}}$

$= -\frac{4 - 2\sqrt{3}}{-2}$

$= 2 - \sqrt{3}$

19. $\frac{13\pi}{12} = \frac{3\pi}{4} + \frac{\pi}{3}$

$\sin \frac{13\pi}{12} = \sin\left(\frac{3\pi}{4} + \frac{\pi}{3}\right)$

$= \sin \frac{3\pi}{4} \cos \frac{\pi}{3} + \cos \frac{3\pi}{4} \sin \frac{\pi}{3}$

$= \frac{\sqrt{2}}{2} \cdot \frac{1}{2} + \left(-\frac{\sqrt{2}}{2}\right)\left(\frac{\sqrt{3}}{2}\right)$

$= \frac{\sqrt{2}}{4}(1 - \sqrt{3})$

$\cos \frac{13\pi}{12} = \cos\left(\frac{3\pi}{4} + \frac{\pi}{3}\right)$

$= \cos \frac{3\pi}{4} \cos \frac{\pi}{3} - \sin \frac{3\pi}{4} \sin \frac{\pi}{3}$

$= -\frac{\sqrt{2}}{2} \cdot \frac{1}{2} - \frac{\sqrt{2}}{2} \cdot \frac{\sqrt{3}}{2} = -\frac{\sqrt{2}}{4}(1 + \sqrt{3})$

$\tan \frac{13\pi}{12} = \tan\left(\frac{3\pi}{4} + \frac{\pi}{3}\right)$

$= \frac{\tan\left(\frac{3\pi}{4}\right) + \tan\left(\frac{\pi}{3}\right)}{1 - \tan\left(\frac{3\pi}{4}\right)\tan\left(\frac{\pi}{3}\right)}$

$= \frac{-1 + \sqrt{3}}{1 - (-1)(\sqrt{3})}$

$= -\frac{1 - \sqrt{3}}{1 + \sqrt{3}} \cdot \frac{1 - \sqrt{3}}{1 - \sqrt{3}}$

$= -\frac{4 - 2\sqrt{3}}{-2}$

$= 2 - \sqrt{3}$

21. $$-\frac{13\pi}{12} = -\left(\frac{3\pi}{4} + \frac{\pi}{3}\right)$$

$$\sin\left[-\left(\frac{3\pi}{4} + \frac{\pi}{3}\right)\right] = -\sin\left(\frac{3\pi}{4} + \frac{\pi}{3}\right)$$

$$= -\left[\sin\frac{3\pi}{4}\cos\frac{\pi}{3} + \cos\frac{3\pi}{4}\sin\frac{\pi}{3}\right]$$

$$= -\left[\frac{\sqrt{2}}{2}\left(\frac{1}{2}\right) + \left(-\frac{\sqrt{2}}{2}\right)\left(\frac{\sqrt{3}}{2}\right)\right]$$

$$= -\frac{\sqrt{2}}{4}(1 - \sqrt{3}) = \frac{\sqrt{2}}{4}(\sqrt{3} - 1)$$

$$\cos\left[-\left(\frac{3\pi}{4} + \frac{\pi}{3}\right)\right] = \cos\left(\frac{3\pi}{4} + \frac{\pi}{3}\right)$$

$$= \cos\frac{3\pi}{4}\cos\frac{\pi}{3} - \sin\frac{3\pi}{4}\sin\frac{\pi}{3}$$

$$= -\frac{\sqrt{2}}{2}\left(\frac{1}{2}\right) - \frac{\sqrt{2}}{2}\left(\frac{\sqrt{3}}{2}\right) = -\frac{\sqrt{2}}{4}(\sqrt{3} + 1)$$

$$\tan\left[-\left(\frac{3\pi}{4} + \frac{\pi}{3}\right)\right] = -\tan\left(\frac{3\pi}{4} + \frac{\pi}{3}\right)$$

$$= -\frac{\tan\frac{3\pi}{4} + \tan\frac{\pi}{3}}{1 - \tan\frac{3\pi}{4}\tan\frac{\pi}{3}} = -\frac{-1 + \sqrt{3}}{1 - (-\sqrt{3})}$$

$$= \frac{1 - \sqrt{3}}{1 + \sqrt{3}} \cdot \frac{1 - \sqrt{3}}{1 - \sqrt{3}} = \frac{4 - 2\sqrt{3}}{-2} = -2 + \sqrt{3}$$

23. $\cos 25° \cos 15° - \sin 25° \sin 15° = \cos(25° + 15°) = \cos 40°$

25. $\dfrac{\tan 325° - \tan 86°}{1 + \tan 325° \tan 86°} = \tan(325° - 86°) = \tan 239°$

27. $\sin 3 \cos 1.2 - \cos 3 \sin 1.2 = \sin(3 - 1.2) = \sin 1.8$

29. $\dfrac{\tan 2x + \tan x}{1 - \tan 2x \tan x} = \tan(2x + x) = \tan 3x$

31. $\sin 330° \cos 30° - \cos 330° \sin 30° = \sin(330° - 30°)$

$$= \sin 300°$$

$$= -\frac{\sqrt{3}}{2}$$

33. $\sin\dfrac{\pi}{12}\cos\dfrac{\pi}{4} + \cos\dfrac{\pi}{12}\sin\dfrac{\pi}{4} = \sin\left(\dfrac{\pi}{12} + \dfrac{\pi}{4}\right)$

$$= \sin\frac{\pi}{3}$$

$$= \frac{\sqrt{3}}{2}$$

35. $\dfrac{\tan 25° + \tan 110°}{1 - \tan 25° \tan 110°} = \tan(25° + 110°)$

$$= \tan 135°$$

$$= -1$$

For Exercises 37–43, we have:

$\sin u = \frac{5}{13}$, u in Quadrant II $\Rightarrow \cos u = -\frac{12}{13}$, $\tan u = -\frac{5}{12}$

$\cos v = -\frac{3}{5}$, v in Quadrant II $\Rightarrow \sin v = \frac{4}{5}$, $\tan v = -\frac{4}{3}$,

37. $\sin(u + v) = \sin u \cos v + \cos u \sin v$

$$= \left(\tfrac{5}{13}\right)\left(-\tfrac{3}{5}\right) + \left(-\tfrac{12}{13}\right)\left(\tfrac{4}{5}\right)$$

$$= -\tfrac{63}{65}$$

39. $\cos(u + v) = \cos u \cos v - \sin u \sin v$

$$= \left(-\tfrac{12}{13}\right)\left(-\tfrac{3}{5}\right) - \left(\tfrac{5}{13}\right)\left(\tfrac{4}{5}\right)$$

$$= \tfrac{16}{65}$$

41. $\tan(u + v) = \dfrac{\tan u + \tan v}{1 - \tan u \tan v} = \dfrac{-\frac{5}{12} + \left(-\frac{4}{3}\right)}{1 - \left(-\frac{5}{12}\right)\left(-\frac{4}{3}\right)} = \dfrac{-\frac{21}{12}}{1 - \frac{5}{9}}$

$$= \left(-\frac{7}{4}\right)\left(\frac{9}{4}\right) = -\frac{63}{16}$$

43. $\sec(v - u) = \dfrac{1}{\cos(v - u)} = \dfrac{1}{\cos v \cos u + \sin v \sin u}$

$$= \frac{1}{\left(-\frac{3}{5}\right)\left(-\frac{12}{13}\right) + \left(\frac{4}{5}\right)\left(\frac{5}{13}\right)} = \frac{1}{\left(\frac{36}{65}\right) + \left(\frac{20}{65}\right)} = \frac{1}{\frac{56}{65}}$$

$$= \frac{65}{56}$$

For Exercises 45–49, we have:

$\sin u = -\frac{7}{25}$, u in Quadrant III $\Longrightarrow \cos u = -\frac{24}{25}$, $\tan u = \frac{7}{24}$

$\cos v = -\frac{4}{5}$, v in Quadrant III $\Longrightarrow \sin v = -\frac{3}{5}$, $\tan v = \frac{3}{4}$

45. $\cos(u + v) = \cos u \cos v - \sin u \sin v$

$$= \left(-\frac{24}{25}\right)\left(-\frac{4}{5}\right) - \left(-\frac{7}{25}\right)\left(-\frac{3}{5}\right)$$

$$= \frac{3}{5}$$

47. $\tan(u - v) = \dfrac{\tan u - \tan v}{1 + \tan u \tan v}$

$$= \frac{\frac{7}{24} - \frac{3}{4}}{1 + \left(\frac{7}{24}\right)\left(\frac{3}{4}\right)} = \frac{-\frac{11}{24}}{\frac{39}{32}} = -\frac{44}{117}$$

49. $\sec(u + v) = \dfrac{1}{\cos(u + v)} = \dfrac{1}{\frac{3}{5}} = \dfrac{5}{3}$

Use Exercise 45 for $\cos(u + v)$.

51. $\sin(\arcsin x + \arccos x) = \sin(\arcsin x)\cos(\arccos x) + \sin(\arccos x)\cos(\arcsin x)$

$$= x \cdot x + \sqrt{1 - x^2} \cdot \sqrt{1 - x^2}$$

$$= x^2 + 1 - x^2$$

$$= 1$$

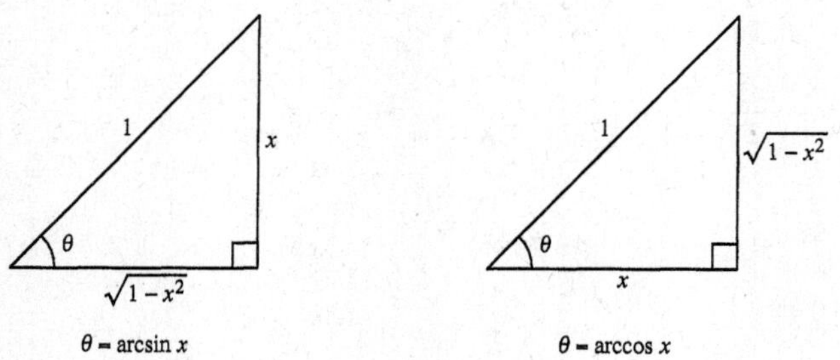

53. $\cos(\arccos x + \arcsin x) = \cos(\arccos x)\cos(\arcsin x) - \sin(\arccos x)\sin(\arcsin x)$

$$= x \cdot \sqrt{1-x^2} - \sqrt{1-x^2} \cdot x$$

$$= 0$$

(Use the triangles in Exercise 51.)

55. $\sin(3\pi - x) = \sin 3\pi \cos x - \sin x \cos 3\pi = (0)(\cos x) - (-1)(\sin x) = \sin x$

57. $\sin\left(\frac{\pi}{6} + x\right) = \sin\frac{\pi}{6}\cos x + \cos\frac{\pi}{6}\sin x = \frac{1}{2}(\cos x + \sqrt{3}\sin x)$

59. $\cos(\pi - \theta) + \sin\left(\frac{\pi}{2} + \theta\right) = \cos\pi\cos\theta + \sin\pi\sin\theta + \sin\frac{\pi}{2}\cos\theta + \cos\frac{\pi}{2}\sin\theta$

$$= (-1)(\cos\theta) + (0)(\sin\theta) + (1)(\cos\theta) + (\sin\theta)(0)$$

$$= -\cos\theta + \cos\theta$$

$$= 0$$

61. $\cos(x + y)\cos(x - y) = (\cos x \cos y - \sin x \sin y)(\cos x \cos y + \sin x \sin y)$

$$= \cos^2 x \cos^2 y - \sin^2 x \sin^2 y$$

$$= \cos^2 x(1 - \sin^2 y) - \sin^2 x \sin^2 y$$

$$= \cos^2 x - \cos^2 x \sin^2 y - \sin^2 x \sin^2 y$$

$$= \cos^2 x - \sin^2 y(\cos^2 x + \sin^2 x)$$

$$= \cos^2 x - \sin^2 y$$

63. $\sin(x + y) + \sin(x - y) = \sin x \cos y + \cos x \sin y + \sin x \cos y - \cos x \sin y$

$$= 2\sin x \cos y$$

65. $\cos\left(\frac{3\pi}{2} - x\right) = \cos\frac{3\pi}{2}\cos x + \sin\frac{3\pi}{2}\sin x$

$$= (0)(\cos x) + (-1)(\sin x)$$

$$= -\sin x$$

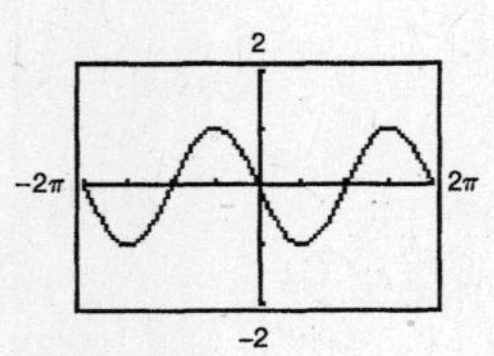

67. $\sin\left(\frac{3\pi}{2} + \theta\right) = \sin\frac{3\pi}{2}\cos\theta + \cos\frac{3\pi}{2}\sin\theta$

$$= (-1)(\cos\theta) + (0)(\sin\theta)$$

$$= -\cos\theta$$

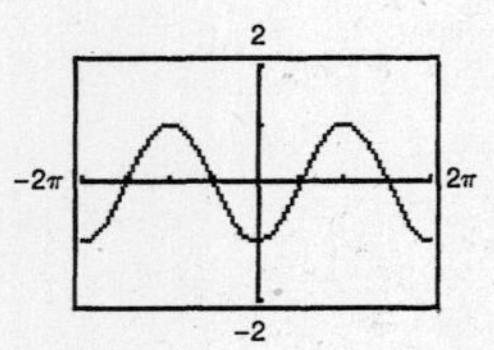

69.

$$\sin\left(x + \frac{\pi}{3}\right) + \sin\left(x - \frac{\pi}{3}\right) = 1$$

$$\sin x \cos\frac{\pi}{3} + \cos x \sin\frac{\pi}{3} + \sin x \cos\frac{\pi}{3} - \cos x \sin\frac{\pi}{3} = 1$$

$$2 \sin x(0.5) = 1$$

$$\sin x = 1$$

$$x = \frac{\pi}{2}$$

71.

$$\cos\left(x + \frac{\pi}{4}\right) - \cos\left(x - \frac{\pi}{4}\right) = 1$$

$$\cos x \cos\frac{\pi}{4} - \sin x \sin\frac{\pi}{4} - \left(\cos x \cos\frac{\pi}{4} + \sin x \sin\frac{\pi}{4}\right) = 1$$

$$-2 \sin x\left(\frac{\sqrt{2}}{2}\right) = 1$$

$$-\sqrt{2} \sin x = 1$$

$$\sin x = -\frac{1}{\sqrt{2}}$$

$$\sin x = -\frac{\sqrt{2}}{2}$$

$$x = \frac{5\pi}{4}, \frac{7\pi}{4}$$

73. Analytically: $\cos\left(x + \frac{\pi}{4}\right) + \cos\left(x - \frac{\pi}{4}\right) = 1$

$$\cos x \cos\frac{\pi}{4} - \sin x \sin\frac{\pi}{4} + \cos x \cos\frac{\pi}{4} + \sin x \sin\frac{\pi}{4} = 1$$

$$2 \cos x\left(\frac{\sqrt{2}}{2}\right) = 1$$

$$\sqrt{2} \cos x = 1$$

$$\cos x = \frac{1}{\sqrt{2}}$$

$$\cos x = \frac{\sqrt{2}}{2}$$

$$x = \frac{\pi}{4}, \frac{7\pi}{4}$$

Graphically: Graph $y_1 = \cos\left(x + \frac{\pi}{4}\right) + \cos\left(x - \frac{\pi}{4}\right)$ and $y_2 = 1$.

The points of intersection occur at $x = \frac{\pi}{4}$ and $x = \frac{7\pi}{4}$.

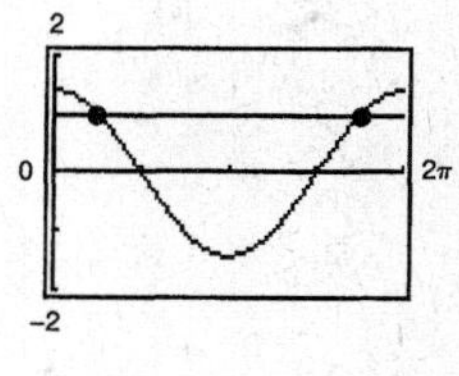

75. $y = \frac{1}{3}\sin 2t + \frac{1}{4}\cos 2t$

(a) $a = \frac{1}{3},\ b = \frac{1}{4},\ B = 2$

$C = \arctan\frac{b}{a} = \arctan\frac{3}{4} \approx 0.6435$

$$y \approx \sqrt{\left(\frac{1}{3}\right)^2 + \left(\frac{1}{4}\right)^2}\sin(2t + 0.6435)$$

$$= \frac{5}{12}\sin(2t + 0.6435)$$

(b) Amplitude: $\frac{5}{12}$ feet

(c) Frequency: $\frac{1}{\text{period}} = \frac{B}{2\pi} = \frac{2}{2\pi} = \frac{1}{\pi}$ cycle per second

77. False. $\sin(u \pm v) = \sin u \cos v \pm \cos u \sin v$.

79. False. $\cos\left(x - \frac{\pi}{2}\right) = \cos x \cos\frac{\pi}{2} + \sin x \sin\frac{\pi}{2}$

$$= (\cos x)(0) + (\sin x)(1)$$

$$= \sin x$$

81. $\cos(n\pi + \theta) = \cos n\pi \cos\theta - \sin n\pi \sin\theta$

$$= (-1)^n(\cos\theta) - (0)(\sin\theta)$$

$= (-1)^n(\cos\theta)$, where n is an integer.

83. $C = \arctan\frac{b}{a} \implies \sin C = \frac{b}{\sqrt{a^2 + b^2}},\ \cos C = \frac{a}{\sqrt{a^2 + b^2}}$

$$\sqrt{a^2 + b^2}\sin(B\theta + C) = \sqrt{a^2 + b^2}\left(\sin B\theta \cdot \frac{a}{\sqrt{a^2 + b^2}} + \frac{b}{\sqrt{a^2 + b^2}} \cdot \cos B\theta\right) = a\sin B\theta + b\cos B\theta$$

85. $\sin\theta + \cos\theta$

$a = 1,\ b = 1,\ B = 1$

(a) $C = \arctan\frac{b}{a} = \arctan 1 = \frac{\pi}{4}$

$$\sin\theta + \cos\theta = \sqrt{a^2 + b^2}\sin(B\theta + C)$$

$$= \sqrt{2}\sin\left(\theta + \frac{\pi}{4}\right)$$

(b) $C = \arctan\frac{a}{b} = \arctan 1 = \frac{\pi}{4}$

$$\sin\theta + \cos\theta = \sqrt{a^2 + b^2}\cos(B\theta - C)$$

$$= \sqrt{2}\cos\left(\theta - \frac{\pi}{4}\right)$$

87. $12\sin 3\theta + 5\cos 3\theta$

$a = 12,\ b = 5,\ B = 3$

(a) $C = \arctan\frac{b}{a} = \arctan\frac{5}{12} \approx 0.3948$

$$12\sin 3\theta + 5\cos 3\theta = \sqrt{a^2 + b^2}\sin(B\theta + C)$$

$$\approx 13\sin(3\theta + 0.3948)$$

(b) $C = \arctan\frac{a}{b} = \arctan\frac{12}{5} \approx 1.1760$

$$12\sin 3\theta + 5\cos 3\theta = \sqrt{a^2 + b^2}\cos(B\theta - C)$$

$$\approx 13\cos(3\theta - 1.1760)$$

89. $C = \arctan\dfrac{b}{a} = \dfrac{\pi}{2} \Rightarrow a = 0$

$\sqrt{a^2 + b^2} = 2 \Rightarrow b = 2$

$B = 1$

$2\sin\left(\theta + \dfrac{\pi}{2}\right) = (0)(\sin\theta) + (2)(\cos\theta) = 2\cos\theta$

91.

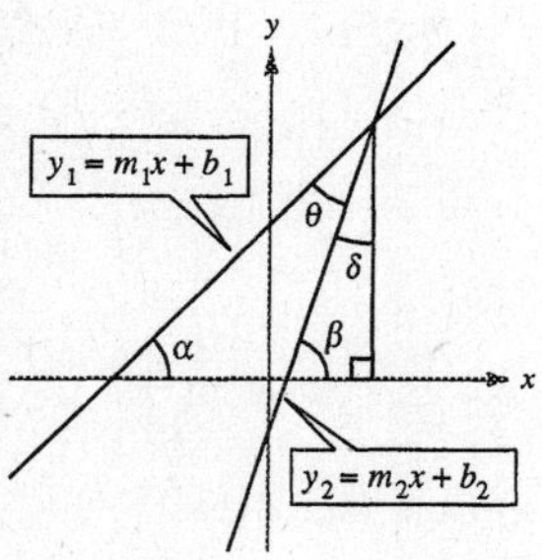

$m_1 = \tan\alpha$ and $m_2 = \tan\beta$

$\beta + \delta = 90° \Rightarrow \delta = 90° - \beta$

$\alpha + \theta + \delta = 90° \Rightarrow \alpha + \theta + (90° - \beta) = 90° \Rightarrow \theta = \beta - \alpha$

Therefore, $\theta = \arctan m_2 - \arctan m_1$

For $y = x$ and $y = \sqrt{3}x$ we have $m_1 = 1$ and $m_2 = \sqrt{3}$

$$\begin{aligned}\theta &= \arctan\sqrt{3} - \arctan 1\\ &= 60° - 45°\\ &= 15°\end{aligned}$$

93.

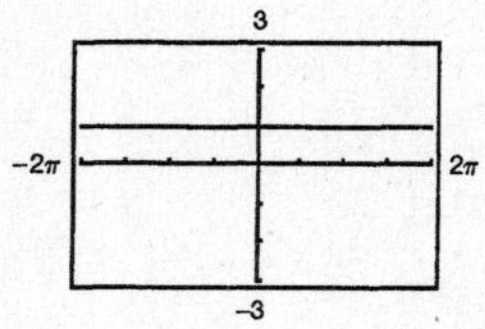

Conjecture: $\sin^2\left(\theta + \dfrac{\pi}{4}\right) + \sin^2\left(\theta - \dfrac{\pi}{4}\right) = 1$

$$\begin{aligned}\sin^2\left(\theta + \frac{\pi}{4}\right) + \sin^2\left(\theta - \frac{\pi}{4}\right) &= \left[\sin\theta\cos\frac{\pi}{4} + \cos\theta\sin\frac{\pi}{4}\right]^2 + \left[\sin\theta\cos\frac{\pi}{4} - \cos\theta\sin\frac{\pi}{4}\right]^2\\ &= \left[\frac{\sin\theta}{\sqrt{2}} + \frac{\cos\theta}{\sqrt{2}}\right]^2 + \left[\frac{\sin\theta}{\sqrt{2}} - \frac{\cos\theta}{\sqrt{2}}\right]^2\\ &= \frac{\sin^2\theta}{2} + \sin\theta\cos\theta + \frac{\cos^2\theta}{2} + \frac{\sin^2\theta}{2} - \sin\theta\cos\theta + \frac{\cos^2\theta}{2}\\ &= \sin^2\theta + \cos^2\theta\\ &= 1\end{aligned}$$

95. To prove the identity for $\sin(u + v)$ we first need to prove the identity for $\cos(u - v)$. Assume $0 < v < u < 2\pi$ and locate u, v, and $u - v$ on the unit circle.

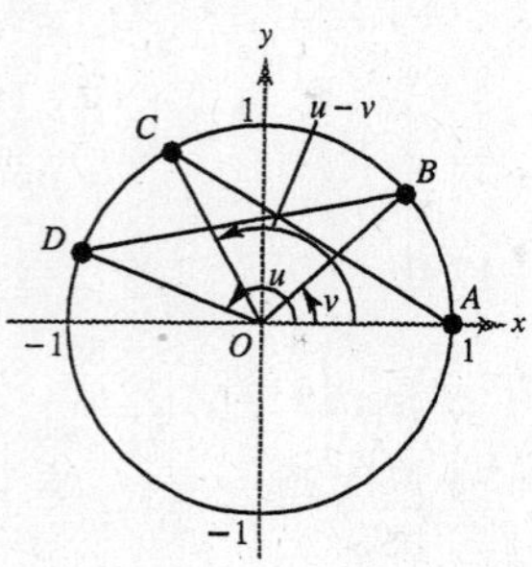

—CONTINUED—

95. —CONTINUED—

The coordinates of the points on the circle are:

$A = (1, 0)$, $B = (\cos v, \sin v)$, $C = (\cos(u - v), \sin(u - v))$, and $D = (\cos u, \sin u)$.

Since ∠DOB = ∠COA, chords AC and BD are equal. By the distance formula we have:

$$\sqrt{[\cos(u - v) - 1]^2 + [\sin(u - v) - 0]^2} = \sqrt{(\cos u - \cos v)^2 + (\sin u - \sin v)^2}$$

$$\cos^2(u - v) - 2\cos(u - v) + 1 + \sin^2(u - v) = \cos^2 u - 2\cos u \cos v + \cos^2 v + \sin^2 u - 2\sin u \sin v + \sin^2 v$$

$$[\cos^2(u - v) + \sin^2(u - v)] + 1 - 2\cos(u - v) = (\cos^2 u + \sin^2 u) + (\cos^2 v + \sin^2 v) - 2\cos u \cos v - 2\sin u \sin v$$

$$2 - 2\cos(u - v) = 2 - 2\cos u \cos v - 2\sin u \sin v$$

$$-2\cos(u - v) = -2(\cos u \cos v + \sin u \sin v)$$

$$\cos(u - v) = \cos u \cos v + \sin u \sin v$$

Now, to prove the identity for $\sin(u + v)$, use cofunction identities.

$$\sin(u + v) = \cos\left[\frac{\pi}{2} - (u + v)\right] = \cos\left[\left(\frac{\pi}{2} - u\right) - v\right]$$

$$= \cos\left(\frac{\pi}{2} - u\right)\cos v + \sin\left(\frac{\pi}{2} - u\right)\sin v$$

$$= \sin u \cos v + \cos u \sin v$$

97. $f(x) = 5(x - 3)$

$$y = 5(x - 3)$$

$$\frac{y}{5} = x - 3$$

$$\frac{y}{5} + 3 = x$$

$$\frac{x}{5} + 3 = y$$

$$f^{-1}(x) = \frac{x + 15}{5}$$

$$f(f^{-1}(x)) = f\left(\frac{x + 15}{5}\right) = 5\left[\frac{x + 15}{5} - 3\right]$$

$$= 5\left(\frac{x + 15}{5}\right) - 5(3)$$

$$= x + 15 - 15$$

$$= x$$

$$f^{-1}(f(x)) = f^{-1}(5(x - 3)) = \frac{5(x - 3) + 15}{5}$$

$$= \frac{5x - 15 + 15}{5}$$

$$= \frac{5x}{5}$$

$$= x$$

99. $f(x) = x^2 - 8$

f is not one-to-one so f^{-1} does not exist.

Section 2.5 Multiple-Angle and Product-to-Sum Formulas

- You should know the following double-angle formulas.
  (a) $\sin 2u = 2 \sin u \cos u$
  (b) $\cos 2u = \cos^2 u - \sin^2 u$
  $= 2\cos^2 u - 1$
  $= 1 - 2\sin^2 u$
  (c) $\tan 2u = \dfrac{2 \tan u}{1 - \tan^2 u}$
- You should be able to reduce the power of a trigonometric function.
  (a) $\sin^2 u = \dfrac{1 - \cos 2u}{2}$
  (b) $\cos^2 u = \dfrac{1 + \cos 2u}{2}$
  (c) $\tan^2 u = \dfrac{1 - \cos 2u}{1 + \cos 2u}$
- You should be able to use the half-angle formulas. The signs of $\sin \frac{u}{2}$ and $\cos \frac{u}{2}$ depend on the quadrant in which $\frac{u}{2}$ lies.
  (a) $\sin \dfrac{u}{2} = \pm\sqrt{\dfrac{1 - \cos u}{2}}$
  (b) $\cos \dfrac{u}{2} = \pm\sqrt{\dfrac{1 + \cos u}{2}}$
  (c) $\tan \dfrac{u}{2} = \dfrac{1 - \cos u}{\sin u} = \dfrac{\sin u}{1 + \cos u}$
- You should be able to use the product-sum formulas.
  (a) $\sin u \sin v = \dfrac{1}{2}[\cos(u - v) - \cos(u + v)]$
  (b) $\cos u \cos v = \dfrac{1}{2}[\cos(u - v) + \cos(u + v)]$
  (c) $\sin u \cos v = \dfrac{1}{2}[\sin(u + v) + \sin(u - v)]$
  (d) $\cos u \sin v = \dfrac{1}{2}[\sin(u + v) - \sin(u - v)]$
- You should be able to use the sum-product formulas.
  (a) $\sin x + \sin y = 2 \sin\left(\dfrac{x + y}{2}\right) \cos\left(\dfrac{x - y}{2}\right)$
  (b) $\sin x - \sin y = 2 \cos\left(\dfrac{x + y}{2}\right) \sin\left(\dfrac{x - y}{2}\right)$
  (c) $\cos x + \cos y = 2 \cos\left(\dfrac{x + y}{2}\right) \cos\left(\dfrac{x - y}{2}\right)$
  (d) $\cos x - \cos y = -2 \sin\left(\dfrac{x + y}{2}\right) \sin\left(\dfrac{x - y}{2}\right)$

Solutions to Odd-Numbered Exercises

Figure for Exercises 1–7

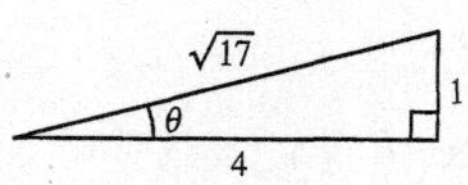

$\sin \theta = \dfrac{\sqrt{17}}{17}$

$\cos \theta = \dfrac{4\sqrt{17}}{17}$

$\tan \theta = \dfrac{1}{4}$

1. $\sin\theta = \dfrac{\sqrt{17}}{17}$

3. $\cos 2\theta = 2\cos^2\theta - 1$

$$= 2\left(\frac{4\sqrt{17}}{17}\right)^2 - 1$$

$$= \frac{32}{17} - 1$$

$$= \frac{15}{17}$$

5. $\tan 2\theta = \dfrac{2\tan\theta}{1 - \tan^2\theta}$

$$= \frac{2\left(\frac{1}{4}\right)}{1 - \left(\frac{1}{4}\right)^2}$$

$$= \frac{\frac{1}{2}}{1 - \frac{1}{16}}$$

$$= \frac{1}{2}\cdot\frac{16}{15}$$

$$= \frac{8}{15}$$

7. $\csc 2\theta = \dfrac{1}{\sin 2\theta}$

$$= \frac{1}{2\sin\theta\cos\theta}$$

$$= \frac{1}{2\left(\frac{\sqrt{17}}{17}\right)\left(\frac{4\sqrt{17}}{17}\right)}$$

$$= \frac{17}{8}$$

9. $\sin 2x - \sin x = 0$

$2\sin x\cos x - \sin x = 0$

$\sin x(2\cos x - 1) = 0$

$\sin x = 0$ or $2\cos x - 1 = 0$

$x = 0,\ \pi$ $\qquad \cos x = \dfrac{1}{2}$

$x = \dfrac{\pi}{3}, \dfrac{5\pi}{3}$

$x = 0, \dfrac{\pi}{3}, \pi, \dfrac{5\pi}{3}$

11. $4\sin x\cos x = 1$

$2\sin 2x = 1$

$\sin 2x = \dfrac{1}{2}$

$2x = \dfrac{\pi}{6} + 2n\pi$ or $2x = \dfrac{5\pi}{6} + 2n\pi$

$x = \dfrac{\pi}{12} + n\pi$ $\qquad x = \dfrac{5\pi}{12} + n\pi$

$x = \dfrac{\pi}{12}, \dfrac{13\pi}{12}$ $\qquad x = \dfrac{5\pi}{12}, \dfrac{17\pi}{12}$

13. $\cos 2x - \cos x = 0$

$\cos 2x = \cos x$

$\cos^2 x - \sin^2 x = \cos x$

$\cos^2 x - (1 - \cos^2 x) - \cos x = 0$

$2\cos^2 x - \cos x - 1 = 0$

$(2\cos x + 1)(\cos x - 1) = 0$

$2\cos x + 1 = 0$ or $\cos x - 1 = 0$

$\cos x = -\dfrac{1}{2}$ $\qquad \cos x = 1$

$x = \dfrac{2\pi}{3}, \dfrac{4\pi}{3}$ $\qquad x = 0$

15.

$$\tan 2x - \cot x = 0$$
$$\frac{2\tan x}{1-\tan^2 x} = \cot x$$
$$2\tan x = \cot x(1-\tan^2 x)$$
$$2\tan x = \cot x - \cot x\tan^2 x$$
$$2\tan x = \cot x - \tan x$$
$$3\tan x = \cot x$$
$$3\tan x - \cot x = 0$$
$$3\tan x - \frac{1}{\tan x} = 0$$
$$\frac{3\tan^2 x - 1}{\tan x} = 0$$
$$\frac{1}{\tan x}(3\tan^2 x - 1) = 0$$
$$\cot x(3\tan^2 x - 1) = 0$$
$$\cot x = 0 \quad \text{or} \quad 3\tan^2 x - 1 = 0$$
$$x = \frac{\pi}{2}, \frac{3\pi}{2} \qquad \tan^2 x = \frac{1}{3}$$
$$\tan x = \pm\frac{\sqrt{3}}{3}$$
$$x = \frac{\pi}{6}, \frac{5\pi}{6}, \frac{7\pi}{6}, \frac{11\pi}{6}$$
$$x = \frac{\pi}{6}, \frac{\pi}{2}, \frac{5\pi}{6}, \frac{7\pi}{6}, \frac{3\pi}{2}, \frac{11\pi}{6}$$

17.

$$\sin 4x = -2\sin 2x$$
$$\sin 4x + 2\sin 2x = 0$$
$$2\sin 2x\cos 2x + 2\sin 2x = 0$$
$$2\sin 2x(\cos 2x + 1) = 0$$
$$2\sin 2x = 0 \quad \text{or} \quad \cos 2x + 1 = 0$$
$$\sin 2x = 0 \qquad \cos 2x = -1$$
$$2x = n\pi \qquad 2x = \pi + 2n\pi$$
$$x = \frac{n}{2}\pi \qquad x = \frac{\pi}{2} + n\pi$$
$$x = 0, \frac{\pi}{2}, \pi, \frac{3\pi}{2} \qquad x = \frac{\pi}{2}, \frac{3\pi}{2}$$

19. $6\sin x\cos x = 3(2\sin x\cos x)$

$= 3\sin 2x$

21. $4 - 8\sin^2 x = 4(1 - 2\sin^2 x)$

$= 4\cos 2x$

23. $\sin u = -\frac{4}{5}, \pi < u < \frac{3\pi}{2} \Rightarrow \cos u = -\frac{3}{5}$

$$\sin 2u = 2\sin u\cos u = 2\left(-\frac{4}{5}\right)\left(-\frac{3}{5}\right) = \frac{24}{25}$$
$$\cos 2u = \cos^2 u - \sin^2 u = \frac{9}{25} - \frac{16}{25} = -\frac{7}{25}$$
$$\tan 2u = \frac{2\tan u}{1-\tan^2 u} = \frac{2\left(\frac{4}{3}\right)}{1-\frac{16}{9}} = \frac{8}{3}\left(-\frac{9}{7}\right) = -\frac{24}{7}$$

25. $\tan u = \frac{3}{4}, 0 < u < \frac{\pi}{2} \Rightarrow \sin u = \frac{3}{5}$ and $\cos u = \frac{4}{5}$

$$\sin 2u = 2\sin u\cos u = 2\left(\frac{3}{5}\right)\left(\frac{4}{5}\right) = \frac{24}{25}$$
$$\cos 2u = \cos^2 u - \sin^2 u = \frac{16}{25} - \frac{9}{25} = \frac{7}{25}$$
$$\tan 2u = \frac{2\tan u}{1-\tan^2 u} = \frac{2\left(\frac{3}{4}\right)}{1-\frac{9}{16}} = \frac{3}{2}\left(\frac{16}{7}\right) = \frac{24}{7}$$

27. $\sec u = -\frac{5}{2}, \frac{\pi}{2} < u < \pi \Rightarrow \sin u = \frac{\sqrt{21}}{5}$ and $\cos u = -\frac{2}{5}$

$$\sin 2u = 2 \sin u \cos u = 2\left(\frac{\sqrt{21}}{5}\right)\left(-\frac{2}{5}\right) = -\frac{4\sqrt{21}}{25}$$

$$\cos 2u = \cos^2 u - \sin^2 u = \left(-\frac{2}{5}\right)^2 - \left(\frac{\sqrt{21}}{5}\right)^2 = -\frac{17}{25}$$

$$\tan 2u = \frac{2 \tan u}{1 - \tan^2 u} = \frac{2\left(-\frac{\sqrt{21}}{2}\right)}{1 - \left(-\frac{\sqrt{21}}{2}\right)^2}$$

$$= \frac{-\sqrt{21}}{1 - \frac{21}{4}} = \frac{4\sqrt{21}}{17}$$

29. $\cos^4 x = (\cos^2 x)(\cos^2 x) = \left(\frac{1 + \cos 2x}{2}\right)\left(\frac{1 + \cos 2x}{2}\right) = \frac{1 + 2 \cos 2x + \cos^2 2x}{4}$

$$= \frac{1 + 2\cos 2x + \frac{1 + \cos 4x}{2}}{4}$$

$$= \frac{2 + 4 \cos 2x + 1 + \cos 4x}{8}$$

$$= \frac{3 + 4 \cos 2x + \cos 4x}{8}$$

$$= \frac{1}{8}(3 + 4 \cos 2x + \cos 4x)$$

31. $(\sin^2 x)(\cos^2 x) = \left(\frac{1 - \cos 2x}{2}\right)\left(\frac{1 + \cos 2x}{2}\right)$

$$= \frac{1 - \cos^2 2x}{4}$$

$$= \frac{1}{4}\left(1 - \frac{1 + \cos 4x}{2}\right)$$

$$= \frac{1}{8}(2 - 1 - \cos 4x)$$

$$= \frac{1}{8}(1 - \cos 4x)$$

33. $\sin^2 x \cos^4 x = \sin^2 x \cos^2 x \cos^2 x = \left(\frac{1 - \cos 2x}{2}\right)\left(\frac{1 + \cos 2x}{2}\right)\left(\frac{1 + \cos 2x}{2}\right)$

$$= \frac{1}{8}(1 - \cos 2x)(1 + \cos 2x)(1 + \cos 2x)$$

$$= \frac{1}{8}(1 - \cos^2 2x)(1 + \cos 2x)$$

$$= \frac{1}{8}(1 + \cos 2x - \cos^2 2x - \cos^3 2x)$$

$$= \frac{1}{8}\left[1 + \cos 2x - \left(\frac{1 + \cos 4x}{2}\right) - \cos 2x\left(\frac{1 + \cos 4x}{2}\right)\right]$$

$$= \frac{1}{16}[2 + 2\cos 2x - 1 - \cos 4x - \cos 2x - \cos 2x \cos 4x]$$

$$= \frac{1}{16}(1 + \cos 2x - \cos 4x - \cos 2x \cos 4x)$$

Figure for Exercises 35–39

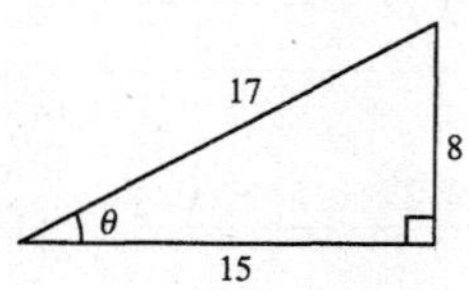

$\sin\theta = \frac{8}{17}$

$\cos\theta = \frac{15}{17}$

35. $\cos\frac{\theta}{2} = \sqrt{\frac{1 + \cos\theta}{2}} = \sqrt{\frac{1 + \frac{15}{17}}{2}} = \sqrt{\frac{32}{34}} = \sqrt{\frac{16}{17}} = \frac{4\sqrt{17}}{17}$

37. $\tan\frac{\theta}{2} = \frac{\sin\theta}{1 + \cos\theta} = \frac{\frac{8}{17}}{1 + \frac{15}{17}} = \frac{8}{17} \cdot \frac{17}{32} = \frac{1}{4}$

39. $\csc\frac{\theta}{2} = \frac{1}{\sin\frac{\theta}{2}} = \frac{1}{\sqrt{\frac{(1 - \cos\theta)}{2}}} = \frac{1}{\sqrt{\frac{1 - \frac{15}{17}}{2}}} = \frac{1}{\sqrt{\frac{1}{17}}} = \sqrt{17}$

41. $\sin 75° = \sin\left(\frac{1}{2} \cdot 150°\right) = \sqrt{\frac{1 - \cos 150°}{2}} = \sqrt{\frac{1 + \frac{\sqrt{3}}{2}}{2}}$

$$= \frac{1}{2}\sqrt{2 + \sqrt{3}}$$

$\cos 75° = \cos\left(\frac{1}{2} \cdot 150°\right) = \sqrt{\frac{1 + \cos 150°}{2}} = \sqrt{\frac{1 - \frac{\sqrt{3}}{2}}{2}}$

$$= \frac{1}{2}\sqrt{2 - \sqrt{3}}$$

$\tan 75° = \tan\left(\frac{1}{2} \cdot 150°\right) = \frac{\sin 150°}{1 + \cos 150°} = \frac{\frac{1}{2}}{1 - \frac{\sqrt{3}}{2}}$

$$= \frac{1}{2 - \sqrt{3}} \cdot \frac{2 + \sqrt{3}}{2 + \sqrt{3}} = \frac{2 + \sqrt{3}}{4 - 3} = 2 + \sqrt{3}$$

43. $\sin 112° 30' = \sin\left(\frac{1}{2} \cdot 225°\right) = \sqrt{\frac{1 - \cos 225°}{2}} = \sqrt{\frac{1 + \frac{\sqrt{2}}{2}}{2}} = \frac{1}{2}\sqrt{2 + \sqrt{2}}$

$\cos 112° 30' = \cos\left(\frac{1}{2} \cdot 225°\right) = -\sqrt{\frac{1 + \cos 225°}{2}} = -\sqrt{\frac{1 - \frac{\sqrt{2}}{2}}{2}} = -\frac{1}{2}\sqrt{2 - \sqrt{2}}$

$\tan 112° 30' = \tan\left(\frac{1}{2} \cdot 225°\right) = \frac{\sin 225°}{1 + \cos 225°} = \frac{-\frac{\sqrt{2}}{2}}{1 - \frac{\sqrt{2}}{2}} = -1 - \sqrt{2}$

45. $\sin\frac{\pi}{8} = \sin\left[\frac{1}{2}\left(\frac{\pi}{4}\right)\right] = \sqrt{\frac{1 - \cos\frac{\pi}{4}}{2}} = \frac{1}{2}\sqrt{2 - \sqrt{2}}$

$\cos\frac{\pi}{8} = \cos\left[\frac{1}{2}\left(\frac{\pi}{4}\right)\right] = \sqrt{\frac{1 + \cos\frac{\pi}{4}}{2}} = \frac{1}{2}\sqrt{2 + \sqrt{2}}$

$\tan\frac{\pi}{8} = \tan\left[\frac{1}{2}\left(\frac{\pi}{4}\right)\right] = \frac{\sin\frac{\pi}{4}}{1 + \cos\frac{\pi}{4}} = \frac{\frac{\sqrt{2}}{2}}{1 + \frac{\sqrt{2}}{2}} = \sqrt{2} - 1$

47. $\sin\frac{3\pi}{8} = \sin\left(\frac{1}{2} \cdot \frac{3\pi}{4}\right) = \sqrt{\frac{1 - \cos\frac{3\pi}{4}}{2}} = \sqrt{\frac{1 + \frac{\sqrt{2}}{2}}{2}} = \frac{1}{2}\sqrt{2 + \sqrt{2}}$

$\cos\frac{3\pi}{8} = \cos\left(\frac{1}{2} \cdot \frac{3\pi}{4}\right) = \sqrt{\frac{1 + \cos\frac{3\pi}{4}}{2}} = \sqrt{\frac{1 - \frac{\sqrt{2}}{2}}{2}} = \frac{1}{2}\sqrt{2 - \sqrt{2}}$

$\tan\frac{3\pi}{8} = \tan\left(\frac{1}{2} \cdot \frac{3\pi}{4}\right) = \frac{\sin\frac{3\pi}{4}}{1 + \cos\frac{3\pi}{4}} = \frac{\frac{\sqrt{2}}{2}}{1 - \frac{\sqrt{2}}{2}} = \frac{\frac{\sqrt{2}}{2}}{\frac{(2 - \sqrt{2})}{2}} = \frac{\sqrt{2}}{2 - \sqrt{2}} = \sqrt{2} + 1$

49. $\sin u = \frac{5}{13}, \frac{\pi}{2} < u < \pi \implies \cos u = -\frac{12}{13}$

$\sin\left(\frac{u}{2}\right) = \sqrt{\frac{1 - \cos u}{2}} = \sqrt{\frac{1 + \frac{12}{13}}{2}} = \frac{5\sqrt{26}}{26}$

$\cos\left(\frac{u}{2}\right) = \sqrt{\frac{1 + \cos u}{2}} = \sqrt{\frac{1 - \frac{12}{13}}{2}} = \frac{\sqrt{26}}{26}$

$\tan\left(\frac{u}{2}\right) = \frac{\sin u}{1 + \cos u} = \frac{\frac{5}{13}}{1 - \frac{12}{13}} = 5$

51. $\tan u = -\frac{5}{8},\ \frac{3\pi}{2} < u < 2\pi \Rightarrow \sin u = -\frac{5}{\sqrt{89}}$ and $\cos u = \frac{8}{\sqrt{89}}$

$$\sin\left(\frac{u}{2}\right) = \sqrt{\frac{1-\cos u}{2}} = \sqrt{\frac{1-\frac{8}{\sqrt{89}}}{2}}\sqrt{\frac{\sqrt{89}-8}{2\sqrt{89}}} = \sqrt{\frac{89-8\sqrt{89}}{178}}$$

$$\cos\left(\frac{u}{2}\right) = -\sqrt{\frac{1+\cos u}{2}} = -\sqrt{\frac{1+\frac{8}{\sqrt{89}}}{2}} = -\sqrt{\frac{\sqrt{89}+8}{2\sqrt{89}}} = -\sqrt{\frac{89+8\sqrt{89}}{178}}$$

$$\tan\left(\frac{u}{2}\right) = \frac{1-\cos u}{\sin u} = \frac{1-\frac{8}{\sqrt{89}}}{-\frac{5}{\sqrt{89}}} = \frac{8-\sqrt{89}}{5}$$

53. $\csc u = -\frac{5}{3},\ \pi < u < \frac{3\pi}{2} \Rightarrow \sin u = -\frac{3}{5}$ and $\cos u = -\frac{4}{5}$

$$\sin\left(\frac{u}{2}\right) = \sqrt{\frac{1-\cos u}{2}} = \sqrt{\frac{1+\frac{4}{5}}{2}} = \frac{3\sqrt{10}}{10}$$

$$\cos\left(\frac{u}{2}\right) = -\sqrt{\frac{1+\cos u}{2}} = -\sqrt{\frac{1-\frac{4}{5}}{2}} = -\frac{\sqrt{10}}{10}$$

$$\tan\left(\frac{u}{2}\right) = \frac{1-\cos u}{\sin u} = \frac{1+\frac{4}{5}}{-\frac{3}{5}} = -3$$

55. $\sqrt{\frac{1-\cos 6x}{2}} = |\sin 3x|$

57. $-\sqrt{\frac{1-\cos 8x}{1+\cos 8x}} = -\frac{\sqrt{\frac{1-\cos 8x}{2}}}{\sqrt{\frac{1+\cos 8x}{2}}}$

$$= -\frac{|\sin 4x|}{|\cos 4x|}$$

$$= -|\tan 4x|$$

59. $\sin\frac{x}{2} + \cos x = 0$

$$\pm\sqrt{\frac{1-\cos x}{2}} = -\cos x$$

$$\frac{1-\cos x}{2} = \cos^2 x$$

$$0 = 2\cos^2 x + \cos x - 1$$

$$= (2\cos x - 1)(\cos x + 1)$$

$\cos x = \frac{1}{2}$ or $\cos x = -1$

$x = \frac{\pi}{3}, \frac{5\pi}{3}$ $\qquad x = \pi$

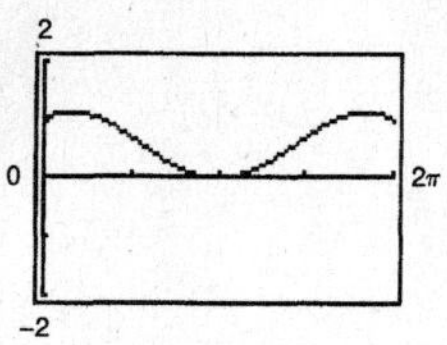

By checking these values in the original equation, we see that $x = \pi/3$ and $x = 5\pi/3$ are extraneous, and $x = \pi$ is the only solution.

61.
$$\cos\frac{x}{2} - \sin x = 0$$
$$\pm\sqrt{\frac{1+\cos x}{2}} = \sin x$$
$$\frac{1+\cos x}{2} = \sin^2 x$$
$$1 + \cos x = 2\sin^2 x$$
$$1 + \cos x = 2 - 2\cos^2 x$$
$$2\cos^2 x + \cos x - 1 = 0$$
$$(2\cos x - 1)(\cos x + 1) = 0$$
$$2\cos x - 1 = 0 \quad \text{or} \quad \cos x + 1 = 0$$
$$\cos x = \frac{1}{2} \qquad \cos x = -1$$
$$x = \frac{\pi}{3}, \frac{5\pi}{3} \qquad x = \pi$$
$$x = \frac{\pi}{3}, \pi, \frac{5\pi}{3}$$

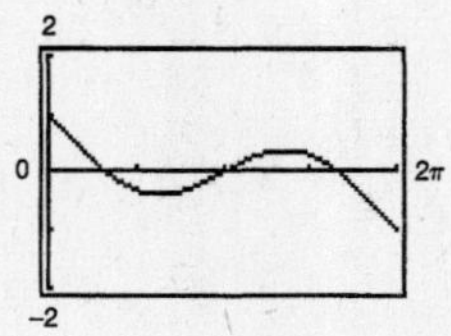

$\pi/3$, π, and $5\pi/3$ are all solutions to the equation.

63. $6\sin\frac{\pi}{4}\cos\frac{\pi}{4} = 6\cdot\frac{1}{2}\left[\sin\left(\frac{\pi}{4}+\frac{\pi}{4}\right) + \sin\left(\frac{\pi}{4}-\frac{\pi}{4}\right)\right] = 3\left(\sin\frac{\pi}{2} + \sin 0\right)$

65. $\cos 4\theta \sin 6\theta = \frac{1}{2}[\sin(4\theta + 6\theta) - \sin(4\theta - 6\theta)] = \frac{1}{2}[\sin 10\theta - \sin(-2\theta)]$
$= \frac{1}{2}(\sin 10\theta + \sin 2\theta)$

67. $5\cos(-5\beta)\cos 3\beta = 5\cdot\frac{1}{2}[\cos(-5\beta - 3\beta) + \cos(-5\beta + 3\beta)] = \frac{5}{2}[\cos(-8\beta) + \cos(-2\beta)]$
$= \frac{5}{2}(\cos 8\beta + \cos 2\beta)$

69. $\sin(x + y)\sin(x - y) = \frac{1}{2}(\cos 2y - \cos 2x)$

71. $\cos(\theta - \pi)\sin(\theta + \pi) = \frac{1}{2}[\sin 2\theta - \sin(-2\pi)]$
$= \frac{1}{2}(\sin 2\theta + \sin 2\pi)$

73. $10\cos 75^\circ \cos 15^\circ = 10\left(\frac{1}{2}\right)[\cos(75^\circ - 15^\circ) + \cos(75^\circ + 15^\circ)] = 5[\cos 60^\circ + \cos 90^\circ]$

75. $\sin 60^\circ + \sin 30^\circ = 2\sin\left(\frac{60^\circ + 30^\circ}{2}\right)\cos\left(\frac{60^\circ - 30^\circ}{2}\right) = 2\sin 45^\circ \cos 15^\circ$

77. $\cos\frac{3\pi}{4} - \cos\frac{\pi}{4} = -2\sin\left(\frac{\frac{3\pi}{4}+\frac{\pi}{4}}{2}\right)\sin\left(\frac{\frac{3\pi}{4}-\frac{\pi}{4}}{2}\right) = -2\sin\frac{\pi}{2}\sin\frac{\pi}{4}$

79. $\sin 5\theta - \sin 3\theta = 2\cos\left(\frac{5\theta + 3\theta}{2}\right)\sin\left(\frac{5\theta - 3\theta}{2}\right) = 2\cos 4\theta \sin\theta$

81. $\cos 6x + \cos 2x = 2\cos\left(\frac{6x + 2x}{2}\right)\cos\left(\frac{6x - 2x}{2}\right) = 2\cos 4x \cos 2x$

83. $\sin(\alpha + \beta) - \sin(\alpha - \beta) = 2\cos\left(\frac{\alpha + \beta + \alpha - \beta}{2}\right)\sin\left(\frac{\alpha + \beta - \alpha + \beta}{2}\right) = 2\cos\alpha\sin\beta$

85. $$\cos\left(\theta + \frac{\pi}{2}\right) - \cos\left(\theta - \frac{\pi}{2}\right) = -2\sin\left[\frac{\left(\theta + \frac{\pi}{2}\right) + \left(\theta - \frac{\pi}{2}\right)}{2}\right]\sin\left[\frac{\left(\theta + \frac{\pi}{2}\right) - \left(\theta - \frac{\pi}{2}\right)}{2}\right]$$

$$= -2\sin\theta\sin\frac{\pi}{2}$$

87. $$\sin 6x + \sin 2x = 0$$

$$2\sin\left(\frac{6x + 2x}{2}\right)\cos\left(\frac{6x - 2x}{2}\right) = 0$$

$$2(\sin 4x)\cos 2x = 0$$

$\sin 4x = 0$ or $\cos 2x = 0$

$4x = n\pi$ $\qquad 2x = \frac{\pi}{2} + n\pi$

$x = \frac{n\pi}{4}$ $\qquad x = \frac{\pi}{4} + \frac{n\pi}{2}$

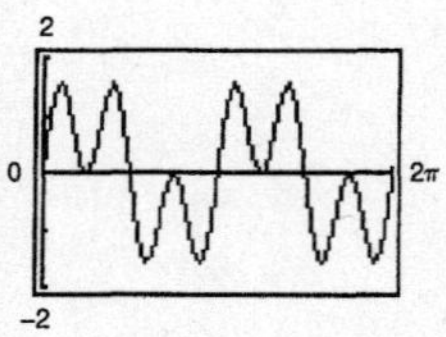

In the interval $[0, 2\pi)$ we have

$x = 0, \frac{\pi}{4}, \frac{\pi}{2}, \frac{3\pi}{4}, \pi, \frac{5\pi}{4}, \frac{3\pi}{2}, \frac{7\pi}{4}$.

89. $$\frac{\cos 2x}{\sin 3x - \sin x} - 1 = 0$$

$$\frac{\cos 2x}{\sin 3x - \sin x} = 1$$

$$\frac{\cos 2x}{2\cos 2x \sin x} = 1$$

$$2\sin x = 1$$

$$\sin x = \frac{1}{2}$$

$$x = \frac{\pi}{6}, \frac{5\pi}{6}$$

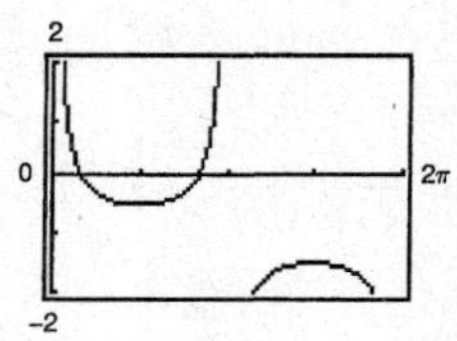

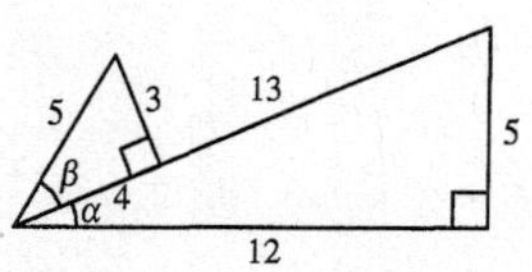

Figure for Exercises 91 and 93

91. $\sin^2\alpha = \left(\frac{5}{13}\right)^2 = \frac{25}{169}$

$\sin^2\alpha = 1 - \cos^2\alpha = 1 - \left(\frac{12}{13}\right)^2$

$= 1 - \frac{144}{169} = \frac{25}{169}$

93. $\sin\alpha\cos\beta = \left(\frac{5}{13}\right)\left(\frac{4}{5}\right) = \frac{4}{13}$

$\sin\alpha\cos\beta = \cos\left(\frac{\pi}{2} - \alpha\right)\sin\left(\frac{\pi}{2} - \beta\right)$

$= \left(\frac{5}{13}\right)\left(\frac{4}{5}\right) = \frac{4}{13}$

95. $\csc 2\theta = \dfrac{1}{\sin 2\theta}$

$= \dfrac{1}{2\sin\theta\cos\theta}$

$= \dfrac{1}{\sin\theta}\cdot\dfrac{1}{2\cos\theta}$

$= \dfrac{\csc\theta}{2\cos\theta}$

97. $\cos^2 2\alpha - \sin^2 2\alpha = \cos[2(2\alpha)]$

$= \cos 4\alpha$

99. $(\sin x + \cos x)^2 = \sin^2 x + 2\sin x\cos x + \cos^2 x$

$= (\sin^2 x + \cos^2 x) + 2\sin x\cos x$

$= 1 + \sin 2x$

101. $1 + \cos 10y = 1 + \cos^2 5y - \sin^2 5y$

$= 1 + \cos^2 5y - (1 - \cos^2 5y)$

$= 2\cos^2 5y$

103. $\sec\dfrac{u}{2} = \dfrac{1}{\cos\dfrac{u}{2}}$

$= \pm\sqrt{\dfrac{2}{1+\cos u}}$

$= \pm\sqrt{\dfrac{2\sin u}{\sin u(1+\cos u)}}$

$= \pm\sqrt{\dfrac{2\sin u}{\sin u + \sin u\cos u}}$

$= \pm\sqrt{\dfrac{\dfrac{2\sin u}{\cos u}}{\dfrac{\sin u}{\cos u} + \dfrac{\sin u\cos u}{\cos u}}}$

$= \pm\sqrt{\dfrac{2\tan u}{\tan u + \sin u}}$

105. $\dfrac{\sin x \pm \sin y}{\cos x + \cos y} = \dfrac{2\sin\left(\dfrac{x\pm y}{2}\right)\cos\left(\dfrac{x\mp y}{2}\right)}{2\cos\left(\dfrac{x+y}{2}\right)\cos\left(\dfrac{x-y}{2}\right)}$

$= \tan\left(\dfrac{x\pm y}{2}\right)$

107. $\dfrac{\cos 4x + \cos 2x}{\sin 4x + \sin 2x} = \dfrac{2\cos\left(\dfrac{4x+2x}{2}\right)\cos\left(\dfrac{4x-2x}{2}\right)}{2\sin\left(\dfrac{4x+2x}{2}\right)\cos\left(\dfrac{4x-2x}{2}\right)}$

$= \dfrac{2\cos 3x\cos x}{2\sin 3x\cos x}$

$= \cot 3x$

109. $\sin\left(\dfrac{\pi}{6} + x\right) + \sin\left(\dfrac{\pi}{6} - x\right) = 2\sin\dfrac{\pi}{6}\cos x$

$= 2\cdot\dfrac{1}{2}\cos x$

$= \cos x$

111.

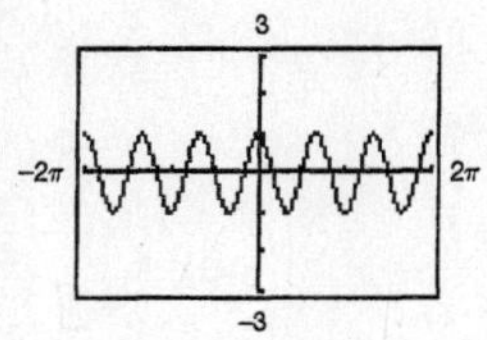

Let $y_1 = \cos(3x)$ and

$y_2 = (\cos x)^3 - 3(\sin x)^2 \cos x$

$$\begin{aligned}\cos 3\beta &= \cos(2\beta + \beta)\\ &= \cos 2\beta \cos\beta - \sin 2\beta \sin \beta\\ &= (\cos^2 \beta - \sin^2 \beta)\cos \beta - 2 \sin \beta \cos \beta \sin \beta\\ &= \cos^3 \beta - \sin^2 \beta \cos \beta - 2 \sin^2 \beta \cos \beta\\ &= \cos^3 \beta - 3 \sin^2 \beta \cos \beta\end{aligned}$$

113.

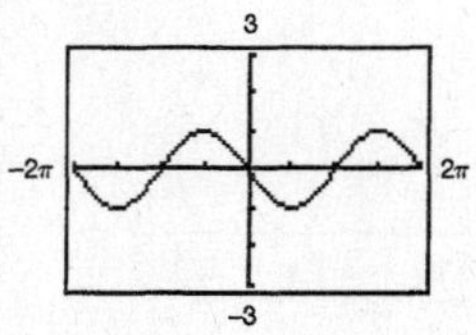

Let $y_1 = \dfrac{(\cos 4x - \cos 2x)}{(2 \sin 3x)}$ and

$y_2 = -\sin x$

$$\begin{aligned}\frac{\cos 4x - \cos 2x}{2 \sin 3x} &= \frac{-2 \sin\left(\frac{4x + 2x}{2}\right) \sin\left(\frac{4x - 2x}{2}\right)}{2 \sin 3x}\\ &= \frac{-2 \sin 3x \sin x}{2 \sin 3x}\\ &= -\sin x\end{aligned}$$

115. $\sin^2 x = \dfrac{1 - \cos 2x}{2} = \dfrac{1}{2} - \dfrac{\cos 2x}{2}$

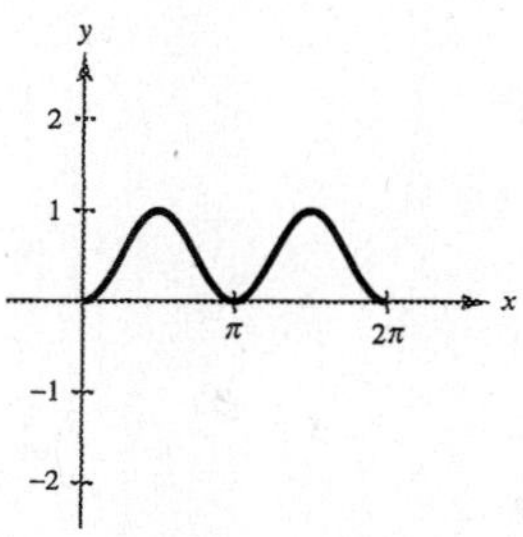

117. $\sin(2 \arcsin x) = 2 \sin(\arcsin x) \cos(\arcsin x) = 2x\sqrt{1 - x^2}$

119. (a) $A = \dfrac{1}{2}bh$

$\cos\dfrac{\theta}{2} = \dfrac{h}{10} \Rightarrow h = 10 \cos\dfrac{\theta}{2}$

$\sin\dfrac{\theta}{2} = \dfrac{(1/2)b}{10} \Rightarrow \dfrac{1}{2}b = 10 \sin\dfrac{\theta}{2}$

$A = 10 \sin\dfrac{\theta}{2} 10 \cos\dfrac{\theta}{2} \Rightarrow A = 100 \sin\dfrac{\theta}{2} \cos\dfrac{\theta}{2}$

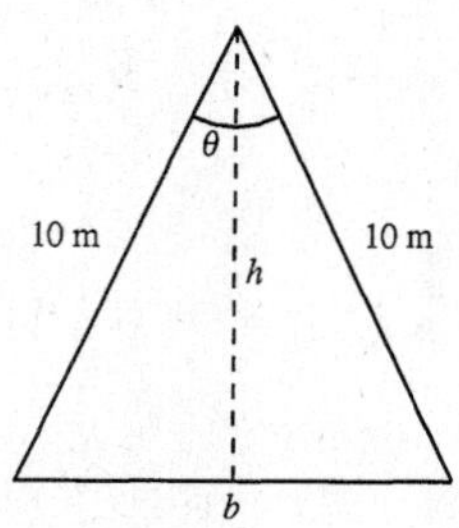

(b) $A = 100 \sin\dfrac{\theta}{2} \cos\dfrac{\theta}{2}$

$A = 50\left(2 \sin\dfrac{\theta}{2} \cos\dfrac{\theta}{2}\right)$

$A = 50 \sin \theta$

When $\theta = \pi/2$, $\sin \theta = 1 \Rightarrow$ the area is a maximum.

$A = 50 \sin\dfrac{\pi}{2} = 50(1) = 50$ square feet

121. $\sin\frac{\theta}{2} = \frac{1}{M}$

(a) $\sin\frac{\theta}{2} = 1$

$\frac{\theta}{2} = \arcsin 1$

$\frac{\theta}{2} = \frac{\pi}{2}$

$\theta = \pi$

(b) $\sin\frac{\theta}{2} = \frac{1}{4.5}$

$\frac{\theta}{2} = \arcsin\left(\frac{1}{4.5}\right)$

$\theta = 2\arcsin\left(\frac{1}{4.5}\right)$

$\theta \approx 0.4482$

(c) $\frac{S}{760} = 1$ $\qquad$ $\frac{S}{760} = 4.5$

$S = 760$ miles per hour $\qquad$ $S = 3420$ miles per hour

(d) $\sin\frac{\theta}{2} = \frac{1}{M}$

$\frac{\theta}{2} = \arcsin\left(\frac{1}{M}\right)$

$\theta = 2\arcsin\left(\frac{1}{M}\right)$

123. False. For $u < 0$,

$$\begin{aligned}\sin 2u &= -\sin(-2u)\\ &= -2\sin(-u)\cos(-u)\\ &= -2(-\sin u)\cos u\\ &= 2\sin u\cos u\end{aligned}$$

125. (a) $y = 4\sin\frac{x}{2} + \cos x$

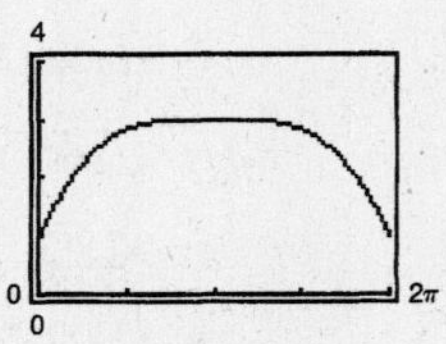

Maximum: $(\pi, 3)$

(b)

$$\begin{aligned}2\cos\frac{x}{2} - \sin x &= 0\\ 2\left(\pm\sqrt{\frac{1+\cos x}{2}}\right) &= \sin x\\ 4\left(\frac{1+\cos x}{2}\right) &= \sin^2 x\\ 2(1+\cos x) &= 1 - \cos^2 x\\ \cos^2 x + 2\cos x + 1 &= 0\\ (\cos x + 1)^2 &= 0\\ \cos x &= -1\\ x &= \pi\end{aligned}$$

127. $f(x) = \sin^4 x + \cos^4 x$

(a) $\sin^4 x + \cos^4 x = (\sin^2 x)^2 + (\cos^2 x)^2$

$$= \left(\frac{1 - \cos 2x}{2}\right)^2 + \left(\frac{1 + \cos 2x}{2}\right)^2$$

$$= \frac{1}{4}[(1 - \cos 2x)^2 + (1 + \cos 2x)^2]$$

$$= \frac{1}{4}(1 - 2\cos 2x + \cos^2 2x + 1 + 2\cos 2x + \cos^2 2x)$$

$$= \frac{1}{4}(2 + 2\cos^2 2x)$$

$$= \frac{1}{4}\left[2 + 2\left(\frac{1 + \cos 2(2x)}{2}\right)\right]$$

$$= \frac{1}{4}(3 + \cos 4x)$$

(b) $\sin^4 x + \cos^4 x = (\sin^2 x)^2 + \cos^4 x$

$$= (1 - \cos^2 x)^2 + \cos^4 x$$

$$= 1 - 2\cos^2 x + \cos^4 x + \cos^4 x$$

$$= 2\cos^4 x - 2\cos^2 x + 1$$

(c) $\sin^4 x + \cos^4 x = \sin^4 x + 2\sin^2 x\cos^2 x + \cos^4 x - 2\sin^2 x\cos^2 x$

$$= (\sin^2 x + \cos^2 x)^2 - 2\sin^2 x\cos^2 x$$

$$= 1 - 2\sin^2 x\cos^2 x$$

(d) $1 - 2\sin^2 x\cos^2 x = 1 - (2\sin x\cos x)(\sin x\cos x)$

$$= 1 - (\sin 2x)\left(\frac{1}{2}\sin 2x\right)$$

$$= 1 - \frac{1}{2}\sin^2 2x$$

(e) No, it does not mean that one of you is wrong. There is often more than one way to rewrite a trigonometric expression.

129. Let x = profit for September, then $x + 0.16x$ = profit for October.

$$x + (x + 0.16x) = 507{,}600$$

$$2.16x = 507{,}600$$

$$x = 235{,}000$$

$$x + 0.16x = 272{,}600$$

Profit for September: \$235,000

Profit for October: \$272,600

131.

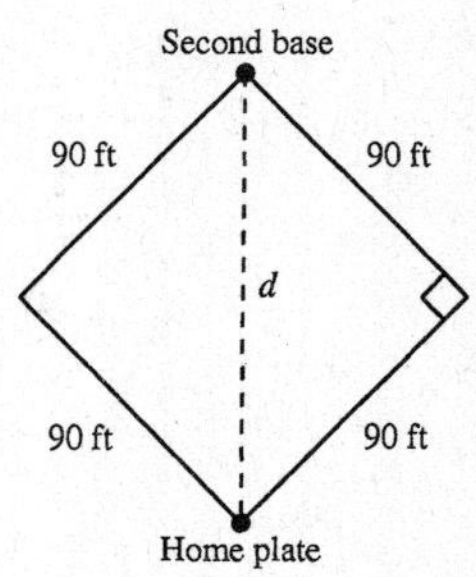

$$d^2 = 90^2 + 90^2$$

$$= 16{,}200$$

$$d = \sqrt{16{,}200}$$

$$= 90\sqrt{2}$$

$$\approx 127 \text{ feet}$$

Review Exercises for Chapter 2

Solutions to Odd-Numbered Exercises

1. $\dfrac{1}{\cos x} = \sec x$

3. $\dfrac{1}{\sec x} = \cos x$

5. $\dfrac{\cos x}{\sin x} = \cot x$

7. $\sin x = \frac{3}{5}, \cos x = \frac{4}{5}$

$$\tan x = \frac{\sin x}{\cos x} = \frac{\frac{3}{5}}{\frac{4}{5}} = \frac{3}{4}$$

$$\cot x = \frac{1}{\tan x} = \frac{4}{3}$$

$$\sec x = \frac{1}{\cos x} = \frac{5}{4}$$

$$\csc x = \frac{1}{\sin x} = \frac{5}{3}$$

9. $\sin\left(\frac{\pi}{2} - x\right) = \frac{\sqrt{2}}{2} \Rightarrow \cos x = \frac{1}{\sqrt{2}} = \frac{\sqrt{2}}{2}$

$$\sin x = -\frac{\sqrt{2}}{2}$$

$$\tan x = \frac{\sin x}{\cos x} = \frac{-\frac{1}{\sqrt{2}}}{\frac{1}{\sqrt{2}}} = -1$$

$$\cot x = \frac{1}{\tan x} = -1$$

$$\sec x = \frac{1}{\cos x} = \sqrt{2}$$

$$\csc x = \frac{1}{\sin x} = -\sqrt{2}$$

11. $\frac{1}{\cot^2 x + 1} = \frac{1}{\csc^2 x} = \sin^2 x$

13. $\tan^2 x(\csc^2 x - 1) = \tan^2 x(\cot^2 x) = \tan^2 x\left(\frac{1}{\tan^2 x}\right) = 1$

15. $\frac{\sin\left(\frac{\pi}{2} - \theta\right)}{\sin \theta} = \frac{\cos \theta}{\sin \theta} = \cot \theta$

17. $\cos^2 x + \cos^2 x \cot^2 x = \cos^2 x(1 + \cot^2 x) = \cos^2 x(\csc^2 x)$

$$= \cos^2 x\left(\frac{1}{\sin^2 x}\right) = \frac{\cos^2 x}{\sin^2 x} = \cot^2 x$$

19. $(\tan x + 1)^2 \cos x = (\tan^2 x + 2\tan x + 1)\cos x$

$$= (\sec^2 x + 2\tan x)\cos x$$

$$= \sec^2 x \cos x + 2\left(\frac{\sin x}{\cos x}\right)\cos x = \sec x + 2\sin x$$

21. $\frac{1}{\csc \theta + 1} - \frac{1}{\csc \theta - 1} = \frac{(\csc \theta - 1) - (\csc \theta + 1)}{(\csc \theta + 1)(\csc \theta - 1)}$

$$= \frac{-2}{\csc^2 \theta - 1}$$

$$= \frac{-2}{\cot^2 \theta}$$

$$= -2\tan^2 \theta$$

23. $\csc^2 x - \csc x \cot x = \frac{1}{\sin^2 x} - \left(\frac{1}{\sin x}\right)\left(\frac{\cos x}{\sin x}\right)$

$$= \frac{1 - \cos x}{\sin^2 x}$$

25. $\cos x(\tan^2 x + 1) = \cos x \sec^2 x$

$$= \frac{1}{\sec x}\sec^2 x$$

$$= \sec x$$

27. $\cos\left(x + \frac{\pi}{2}\right) = \cos x \cos \frac{\pi}{2} - \sin x \sin \frac{\pi}{2}$

$$= (\cos x)(0) - (\sin x)(1)$$

$$= -\sin x$$

29. $\frac{1}{\tan \theta \csc \theta} = \frac{1}{\frac{\sin \theta}{\cos \theta} \cdot \frac{1}{\sin \theta}} = \cos \theta$

31. $\sin^5 x \cos^2 x = \sin^4 x \cos^2 x \sin x$

$$= (1 - \cos^2 x)^2 \cos^2 x \sin x$$

$$= (1 - 2\cos^2 x + \cos^4 x)\cos^2 x \sin x$$

$$= (\cos^2 x - 2\cos^4 x + \cos^6 x)\sin x$$

33. $\sin x = \sqrt{3} - \sin x$

$\sin x = \dfrac{\sqrt{3}}{2}$

$x = \dfrac{\pi}{3} + 2\pi n, \dfrac{2\pi}{3} + 2\pi n$

35. $3\sqrt{3}\tan u = 3$

$\tan u = \dfrac{1}{\sqrt{3}}$

$u = \dfrac{\pi}{6} + n\pi$

37. $3\csc^2 x = 4$

$\csc^2 x = \dfrac{4}{3}$

$\sin x = \pm\dfrac{\sqrt{3}}{2}$

$x = \dfrac{\pi}{3} + 2\pi n, \dfrac{2\pi}{3} + 2\pi n, \dfrac{4\pi}{3} + 2\pi n, \dfrac{5\pi}{3} + 2\pi n$

These can be combined as:

$x = \dfrac{\pi}{3} + n\pi$ or $x = \dfrac{2\pi}{3} + n\pi$

39. $2\cos^2 x - \cos x = 1$

$2\cos^2 x - \cos x - 1 = 0$

$(2\cos x + 1)(\cos x - 1) = 0$

$2\cos x + 1 = 0$ $\qquad$ $\cos x - 1 = 0$

$\cos x = -\dfrac{1}{2}$ $\qquad$ $\cos x = 1$

$x = \dfrac{2\pi}{3}, \dfrac{4\pi}{3}$ $\qquad$ $x = 0$

41. $\cos^2 x + \sin x = 1$

$1 - \sin^2 x + \sin x - 1 = 0$

$-\sin x(\sin x - 1) = 0$

$\sin x = 0$ $\qquad$ $\sin x - 1 = 0$

$x = 0, \pi$ $\qquad$ $\sin x = 1$

$x = \dfrac{\pi}{2}$

43. $2\sin 2x - \sqrt{2} = 0$

$\sin 2x = \dfrac{\sqrt{2}}{2}$

$2x = \dfrac{\pi}{4} + 2\pi n, \dfrac{3\pi}{4} + 2\pi n$

$x = \dfrac{\pi}{8} + \pi n, \dfrac{3\pi}{8} + \pi n$

$x = \dfrac{\pi}{8}, \dfrac{3\pi}{8}, \dfrac{9\pi}{8}, \dfrac{11\pi}{8}$

45. $\cos 4x(\cos x - 1) = 0$

$\cos 4x = 0$ $\qquad$ $\cos x - 1 = 0$

$4x = \dfrac{\pi}{2} + 2\pi n, \dfrac{3\pi}{2} + 2\pi n$ $\qquad$ $\cos x = 1$

$x = \dfrac{\pi}{8} + \dfrac{\pi}{2}n, \dfrac{3\pi}{8} + \dfrac{\pi}{2}n$ $\qquad$ $x = 0$

$x = 0, \dfrac{\pi}{8}, \dfrac{3\pi}{8}, \dfrac{5\pi}{8}, \dfrac{7\pi}{8}, \dfrac{9\pi}{8}, \dfrac{11\pi}{8}, \dfrac{13\pi}{8}, \dfrac{15\pi}{8}$

47. $\sin^2 x - 2\sin x = 0$

$\sin x(\sin x - 2) = 0$

$\sin x = 0$ $\qquad$ $\sin x - 2 = 0$

$x = 0, \pi$ $\qquad$ No solution

49. $\tan^2\theta + \tan\theta - 12 = 0$

$(\tan\theta + 4)(\tan\theta - 3) = 0$

$\tan\theta + 4 = 0$ $\qquad$ $\tan\theta - 3 = 0$

$\theta = \arctan(-4) + n\pi$ $\qquad$ $\theta = \arctan 3 + n\pi$

$\theta = \arctan(-4) + \pi, \arctan(-4) + 2\pi, \arctan 3, \arctan 3 + \pi$

51. $\sin 285° = \sin(315° - 30°)$

$= \sin 315° \cos 30° - \cos 315° \sin 30°$

$= \left(-\frac{\sqrt{2}}{2}\right)\left(\frac{\sqrt{3}}{2}\right) - \left(\frac{\sqrt{2}}{2}\right)\left(\frac{1}{2}\right)$

$= -\frac{\sqrt{2}}{4}(\sqrt{3} + 1)$

$\cos 285° = \cos(315° - 30°)$

$= \cos 315° \cos 30° + \sin 315° \sin 30°$

$= \left(\frac{\sqrt{2}}{2}\right)\left(\frac{\sqrt{3}}{2}\right) + \left(-\frac{\sqrt{2}}{2}\right)\left(\frac{1}{2}\right)$

$= \frac{\sqrt{2}}{4}(\sqrt{3} - 1)$

$\tan 285° = \tan(315° - 30°) = \frac{\tan 315° - \tan 30°}{1 + \tan 315° \tan 30°}$

$= \frac{(-1) - \left(\frac{\sqrt{3}}{3}\right)}{1 + (-1)\left(\frac{\sqrt{3}}{3}\right)} = -2 - \sqrt{3}$

53. $\sin \frac{25\pi}{12} = \sin\left(\frac{11\pi}{6} + \frac{\pi}{4}\right) = \sin \frac{11\pi}{6} \cos \frac{\pi}{4} + \cos \frac{11\pi}{6} \sin \frac{\pi}{4}$

$= \left(-\frac{1}{2}\right)\left(\frac{\sqrt{2}}{2}\right) + \left(\frac{\sqrt{3}}{2}\right)\left(\frac{\sqrt{2}}{2}\right) = \frac{\sqrt{2}}{4}(\sqrt{3} - 1)$

$\cos \frac{25\pi}{12} = \cos\left(\frac{11\pi}{6} + \frac{\pi}{4}\right) = \cos \frac{11\pi}{6} \cos \frac{\pi}{4} - \sin \frac{11\pi}{6} \sin \frac{\pi}{4}$

$= \left(\frac{\sqrt{3}}{2}\right)\left(\frac{\sqrt{2}}{2}\right) - \left(-\frac{1}{2}\right)\left(\frac{\sqrt{2}}{2}\right) = \frac{\sqrt{2}}{4}(\sqrt{3} + 1)$

$\tan \frac{25\pi}{12} = \tan\left(\frac{11\pi}{6} + \frac{\pi}{4}\right) = \frac{\tan \frac{11\pi}{6} + \tan \frac{\pi}{4}}{1 - \tan \frac{11\pi}{6} \tan \frac{\pi}{4}}$

$= \frac{\left(-\frac{\sqrt{3}}{3}\right) + 1}{1 - \left(-\frac{\sqrt{3}}{3}\right)(1)} = 2 - \sqrt{3}$

55. $\sin 60° \cos 45° - \cos 60° \sin 45° = \sin(60° - 45°) = \sin 15°$

57. $\frac{\tan 25° + \tan 10°}{1 - \tan 25° \tan 10°} = \tan(25° + 10°)$

$= \tan 35°$

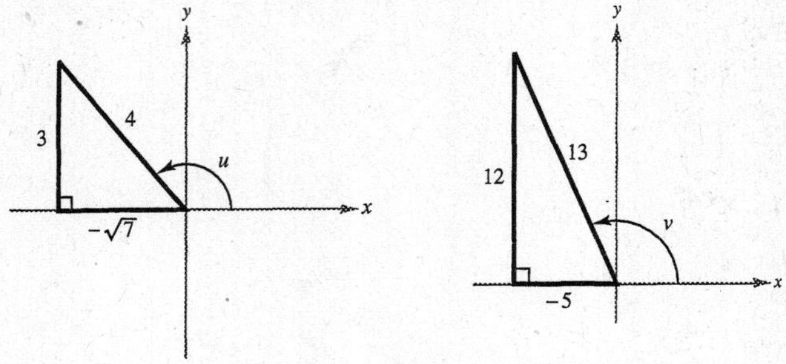

Figures for Exercises 59–63

59. $\sin(u + v) = \sin u \cos v + \cos u \sin v$

$$= \left(\frac{3}{4}\right)\left(-\frac{5}{13}\right) + \left(-\frac{\sqrt{7}}{4}\right)\left(\frac{12}{13}\right)$$

$$= -\frac{3}{52}(5 + 4\sqrt{7})$$

61. $\cos(u - v) = \cos u \cos v + \sin u \sin v$

$$= \left(-\frac{\sqrt{7}}{4}\right)\left(-\frac{5}{13}\right) + \left(\frac{3}{4}\right)\left(\frac{12}{13}\right)$$

$$= \frac{1}{52}(5\sqrt{7} + 36)$$

63. $\cos(u + v) = \cos u \cos v - \sin u \sin v$

$$= \left(-\frac{\sqrt{7}}{4}\right)\left(-\frac{5}{13}\right) - \left(\frac{3}{4}\right)\left(\frac{12}{13}\right)$$

$$= \frac{1}{52}(5\sqrt{7} - 36)$$

65. $\sin\left(x + \frac{\pi}{4}\right) - \sin\left(x - \frac{\pi}{4}\right) = 1$

$$2 \cos x \sin \frac{\pi}{4} = 1$$

$$\cos x = \frac{\sqrt{2}}{2}$$

$$x = \frac{\pi}{4}, \frac{7\pi}{4}$$

67. $\sin\left(x + \frac{\pi}{2}\right) - \sin\left(x - \frac{\pi}{2}\right) = \sqrt{3}$

$$2 \cos x \sin \frac{\pi}{2} = \sqrt{3}$$

$$\cos x = \frac{\sqrt{3}}{2}$$

$$x = \frac{\pi}{6}, \frac{11\pi}{6}$$

69. $\sin 4x = 2 \sin 2x \cos 2x$

$$= 2[2 \sin x \cos x(\cos^2 x - \sin^2 x)]$$

$$= 4 \sin x \cos x(2 \cos^2 x - 1)$$

$$= 8 \cos^3 x \sin x - 4 \cos x \sin x$$

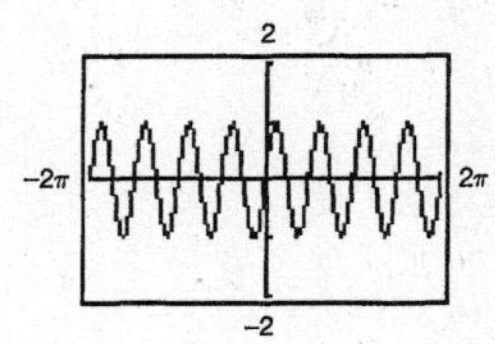

71. $\sin u = -\frac{4}{5}, \pi < u < \frac{3\pi}{2}$

$$\cos u = -\sqrt{1 - \sin^2 u} = \frac{-3}{5}$$

$$\tan u = \frac{\sin u}{\cos u} = \frac{4}{3}$$

$$\sin 2u = 2 \sin u \cos u = 2\left(-\frac{4}{5}\right)\left(-\frac{3}{5}\right) = \frac{24}{25}$$

$$\cos 2u = \cos^2 u - \sin^2 u = \left(-\frac{3}{5}\right)^2 - \left(-\frac{4}{5}\right)^2 = -\frac{7}{25}$$

$$\tan 2u = \frac{2 \tan u}{1 - \tan^2 u} = \frac{2\left(\frac{4}{3}\right)}{1 - \left(\frac{4}{3}\right)^2} = -\frac{24}{7}$$

73. $r = \frac{1}{32} v_0^2 \sin 2\theta$

range = 100 feet

v_0 = 80 feet per second

$$r = \frac{1}{32}(80)^2 \sin 2\theta = 100$$

$$\sin 2\theta = 0.5$$

$$2\theta = 30°$$

$$\theta = 15° \text{ or } \frac{\pi}{12}$$

75. $\tan^2 2x = \frac{\sin^2 2x}{\cos^2 2x} = \frac{\frac{1 - \cos 4x}{2}}{\frac{1 + \cos 4x}{2}} = \frac{1 - \cos 4x}{1 + \cos 4x}$

77. $\sin^2 x \tan^2 x = \sin^2 x\left(\frac{\sin^2 x}{\cos^2 x}\right) = \frac{\sin^4 x}{\cos^2 x}$

$$= \frac{\left(\frac{1 - \cos 2x}{2}\right)^2}{\frac{1 + \cos 2x}{2}} = \frac{\frac{1 - 2\cos 2x + \cos^2 2x}{4}}{\frac{1 + \cos 2x}{2}}$$

$$= \frac{1 - 2\cos 2x + \frac{1 + \cos 4x}{2}}{2(1 + \cos 2x)}$$

$$= \frac{2 - 4\cos 2x + 1 + \cos 4x}{4(1 + \cos 2x)}$$

$$= \frac{3 - 4\cos 2x + \cos 4x}{4(1 + \cos 2x)}$$

79. $\sin(-75°) = -\sqrt{\frac{1 - \cos 150°}{2}} = -\sqrt{\frac{1 - \left(-\frac{\sqrt{3}}{2}\right)}{2}} = -\frac{\sqrt{2 + \sqrt{3}}}{2}$

$$= -\frac{1}{2}\sqrt{2 + \sqrt{3}}$$

$$\cos(-75°) = \sqrt{\frac{1 + \cos 150°}{2}} = \sqrt{\frac{1 + \left(-\frac{\sqrt{3}}{2}\right)}{2}} = \frac{\sqrt{2 - \sqrt{3}}}{2}$$

$$= \frac{1}{2}\sqrt{2 - \sqrt{3}}$$

$$\tan(-75°) = -\left(\frac{1 - \cos 150°}{\sin 150°}\right) = -\left(\frac{1 - \left(-\frac{\sqrt{3}}{2}\right)}{\frac{1}{2}}\right) = -(2 + \sqrt{3})$$

$$= -2 - \sqrt{3}$$

81. $\sin\left(\frac{19\pi}{12}\right) = -\sqrt{\frac{1-\cos\frac{19\pi}{6}}{2}} = -\sqrt{\frac{1-\left(-\frac{\sqrt{3}}{2}\right)}{2}} = -\frac{\sqrt{2+\sqrt{3}}}{2}$

$= -\frac{1}{2}\sqrt{2+\sqrt{3}}$

$\cos\left(\frac{19\pi}{12}\right) = \sqrt{\frac{1+\cos\frac{19\pi}{6}}{2}} = \sqrt{\frac{1+\left(-\frac{\sqrt{3}}{2}\right)}{2}} = \frac{\sqrt{2+\sqrt{3}}}{2}$

$= \frac{1}{2}\sqrt{2-\sqrt{3}}$

$\tan\left(\frac{19\pi}{12}\right) = \frac{1-\cos\frac{19\pi}{6}}{\sin\frac{19\pi}{6}} = \frac{1-\left(-\frac{\sqrt{3}}{2}\right)}{-\frac{1}{2}} = -2-\sqrt{3}$

83. $-\sqrt{\frac{1+\cos 10x}{2}} = -\left|\cos\frac{10x}{2}\right| = -|\cos 5x|$

85. Given $\sin u = \frac{3}{5}, 0 < u < \frac{\pi}{2} \Rightarrow \cos u = \frac{4}{5}$ and $\frac{u}{2}$ is in Quadrant I

$\sin\left(\frac{u}{2}\right) = \sqrt{\frac{1-\cos u}{2}} = \sqrt{\frac{1-4/5}{2}} = \sqrt{\frac{1}{10}} = \frac{\sqrt{10}}{10}$

$\cos\left(\frac{u}{2}\right) = \sqrt{\frac{1+\cos u}{2}} = \sqrt{\frac{1+4/5}{2}} = \sqrt{\frac{9}{10}} = \frac{3\sqrt{10}}{10}$

$\tan\left(\frac{u}{2}\right) = \frac{1-\cos u}{\sin u} = \frac{1-4/5}{3/5} = \frac{1}{3}$

87. $\cos\frac{\pi}{6}\sin\frac{\pi}{6} = \frac{1}{2}\left[\sin\frac{\pi}{3} - \sin 0\right] = \frac{1}{2}\sin\frac{\pi}{3}$

89. $\cos 5\theta\cos 3\theta = \frac{1}{2}[\cos 2\theta + \cos 8\theta]$

91. $\sin 60° + \sin 90° = 2\sin 75°\cos 15°$

93. $\cos\left(x+\frac{\pi}{6}\right) - \cos\left(x-\frac{\pi}{6}\right) = -2\sin x\,\sin\frac{\pi}{6}$

95. (a) $y = 1.5\sin 8t - 0.5\cos 8t = \frac{1}{2}(3\sin 8t - 1\cos 8t)$

Using the identity

$a\sin B\theta + b\cos B\theta = \sqrt{a^2+b^2}\sin(B\theta + C),$

$C = \arctan\frac{b}{a}, a > 0$

(Exercise 83, Section 7.4), we have

$y = \frac{1}{2}\sqrt{(3)^2+(-1)^2}\sin\left(8t + \arctan\left(-\frac{1}{3}\right)\right)$

$= \frac{\sqrt{10}}{2}\sin\left(8t - \arctan\left(\frac{1}{3}\right)\right)$

(b) Amplitude $= \frac{\sqrt{10}}{2}$ feet

(c) Frequency $= \frac{1}{\frac{2\pi}{8}} = \frac{4}{\pi}$ cycles per second

97. False. The correct identity is

$\sin(x + y) = \sin x \cos y + \cos x \sin y.$

99. True. It can be verified using a product-to-sum identity.

$$4 \sin 45° \cos 15° = 4 \cdot \frac{1}{2}[\sin 60° + \sin 30°]$$

$$= 2\left[\frac{\sqrt{3}}{2} + \frac{1}{2}\right] = \sqrt{3} + 1$$

101. No. For an equation to be an identity, the equation must be true for all real numbers. $\sin \theta = \frac{1}{2}$ has an infinite number of solutions but is not an identity.

103. The graph of y_1 is a vertical shift of the graph of y_2 one unit upward so $y_1 = y_2 + 1$.

105. $y = \sqrt{x + 3} + 4 \cos x$

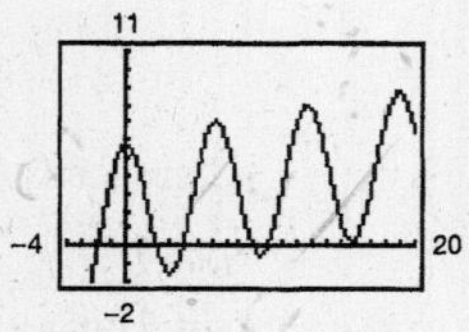

Zeros: $x \approx -1.8431, 2.1758, 3.9903, 8.8935, 9.8820$

Problem Solving for Chapter 2

Solutions to Odd-Numbered Exercises

1. (a) Since $\sin^2 \theta + \cos^2 \theta = 1$ and $\cos^2 \theta = 1 - \sin^2 \theta$:

$$\cos \theta = \pm\sqrt{1 - \sin^2 \theta}$$

$$\tan \theta = \frac{\sin \theta}{\cos \theta} = \pm\frac{\sin \theta}{\sqrt{1 - \sin^2 \theta}}$$

$$\cot \theta = \frac{1}{\tan \theta} = \pm\frac{\sqrt{1 - \sin^2 \theta}}{\sin \theta}$$

$$\sec \theta = \frac{1}{\cos \theta} = \pm\frac{1}{\sqrt{1 - \sin^2 \theta}}$$

$$\cot \theta = \frac{1}{\sin \theta}$$

We also have the following relationships:

$$\cos \theta = \sin\left(\frac{\pi}{2} - \theta\right)$$

$$\tan \theta = \frac{\sin \theta}{\sin\left(\frac{\pi}{2} - \theta\right)}$$

$$\cot \theta = \frac{\sin\left(\frac{\pi}{2} - \theta\right)}{\sin \theta}$$

$$\sec \theta = \frac{1}{\sin\left(\frac{\pi}{2} - \theta\right)}$$

$$\csc \theta = \frac{1}{\sin \theta}$$

(b) $\sin \theta = \pm\sqrt{1 - \cos^2 \theta}$

$$\tan \theta = \frac{\sin \theta}{\cos \theta} = \pm\frac{\sqrt{1 - \cos^2 \theta}}{\cos \theta}$$

$$\csc \theta = \frac{1}{\sin \theta} = \pm\frac{1}{\sqrt{1 - \cos^2 \theta}}$$

$$\sec \theta = \frac{1}{\cos \theta}$$

$$\cot \theta = \frac{1}{\tan \theta} = \pm\frac{\cos \theta}{\sqrt{1 - \cos^2 \theta}}$$

We also have the following relationships:

$$\sin \theta = \cos\left(\frac{\pi}{2} - \theta\right)$$

$$\tan \theta = \frac{\cos\left(\frac{\pi}{2} - \theta\right)}{\cos \theta}$$

$$\csc \theta = \frac{1}{\cos\left(\frac{\pi}{2} - \theta\right)}$$

$$\sec \theta = \frac{1}{\cos \theta}$$

$$\cot \theta = \frac{\cos \theta}{\cos\left(\frac{\pi}{2} - \theta\right)}$$

3. $\sin\left[\frac{(12n+1)\pi}{6}\right] = \sin\left[\frac{1}{6}(12n\pi + \pi)\right]$

$$= \sin\left(2n\pi + \frac{\pi}{6}\right)$$

$$= \sin\frac{\pi}{6} = \frac{1}{2}$$

Thus, $\sin\left[\frac{(12n+1)\pi}{6}\right] = \frac{1}{2}$ for all integers n.

5. $y = -\frac{16}{v^2\cos^2\theta}x^2 + (\tan\theta)x + h_0$

Let $h_0 = 0$ and take half of the horizontal distance: $\frac{1}{2}\left(\frac{1}{32}v^2\sin 2\theta\right) = \frac{1}{64}v^2(2\sin\theta\cos\theta)$

$$= \frac{1}{32}v^2\sin\theta\cos\theta$$

Substitute this expression for x in the model.

$$y = -\frac{16}{v^2\cos^2\theta}\left(\frac{1}{32}v^2\sin\theta\cos\theta\right)^2 + \left(\frac{\sin\theta}{\cos\theta}\right)\left(\frac{1}{32}v^2\sin\theta\cos\theta\right)$$

$$= -\frac{1}{64}v^2\sin^2\theta + \frac{1}{32}v^2\sin^2\theta$$

$$= \frac{1}{64}v^2\sin^2\theta$$

7.

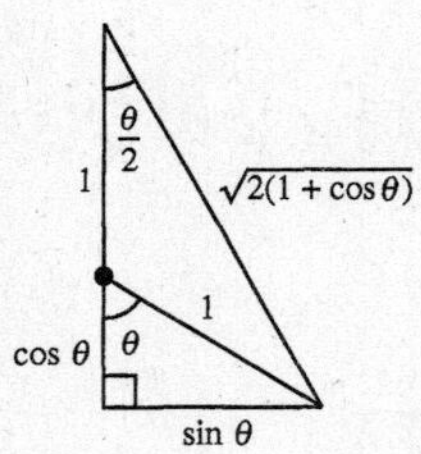

The hypotenuse of the larger right triangle is:

$$\sqrt{\sin^2\theta + (1+\cos\theta)^2} = \sqrt{\sin^2\theta + 1 + 2\cos\theta + \cos^2\theta}$$

$$= \sqrt{2 + 2\cos\theta}$$

$$= \sqrt{2(1+\cos\theta)}$$

$$\sin\left(\frac{\theta}{2}\right) = \frac{\sin\theta}{\sqrt{2(1+\cos\theta)}} = \frac{\sin\theta}{\sqrt{2(1+\cos\theta)}}\cdot\frac{\sqrt{1-\cos\theta}}{\sqrt{1-\cos\theta}}$$

$$= \frac{\sin\theta\sqrt{1-\cos\theta}}{\sqrt{2(1-\cos^2\theta)}} = \frac{\sin\theta\sqrt{1-\cos\theta}}{\sqrt{2}\sin\theta}$$

$$= \sqrt{\frac{1-\cos\theta}{2}}$$

$$\cos\left(\frac{\theta}{2}\right) = \frac{1+\cos\theta}{\sqrt{2(1+\cos\theta)}} = \frac{\sqrt{(1+\cos\theta)^2}}{\sqrt{2(1+\cos\theta)}} = \sqrt{\frac{1+\cos\theta}{2}}$$

$$\tan\left(\frac{\theta}{2}\right) = \frac{\sin\theta}{1+\cos\theta}$$

9. Seward: $D = 12.2 - 6.4\cos\left[\dfrac{\pi(t + 0.2)}{182.6}\right]$

New Orleans: $D = 12.2 - 1.9\cos\left[\dfrac{\pi(t + 0.2)}{182.6}\right]$

(a)

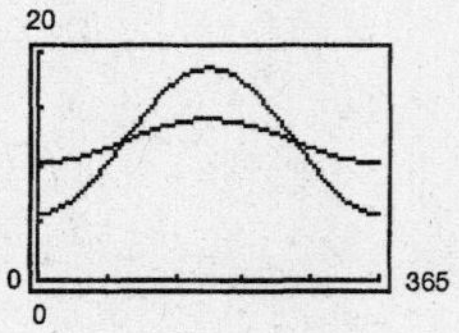

(b) The graphs intersect when $t \approx 91$ and when $t \approx 274$. These values correspond to April 1 and October 1, the spring equinox and the fall equinox.

(c) Seward has the greater variation in the number of daylight hours. This is determined by the amplitudes, 6.4 and 1.9.

(d) Period: $\dfrac{2\pi}{\dfrac{\pi}{182.6}} = 365.2$ days

11. (a) Let $y_1 = \sin x$ and $y_2 = 0.5$

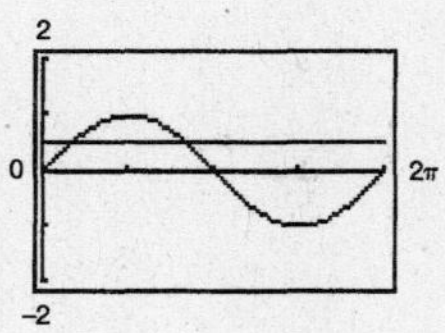

$\sin x \geq 0.5$ on the interval $\left[\dfrac{\pi}{6}, \dfrac{5\pi}{6}\right]$

(b) Let $y_1 = \cos x$ and $y_2 = -0.5$

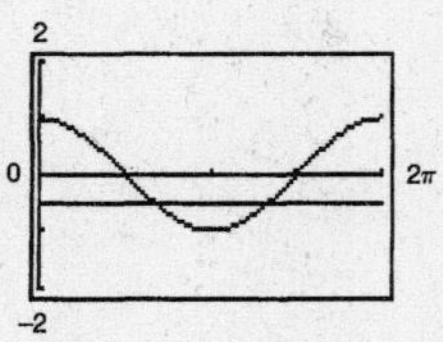

$\cos x \leq -0.5$ on the interval $\left[\dfrac{2\pi}{3}, \dfrac{4\pi}{3}\right]$

(c) Let $y_1 = \tan x$ and $y_2 = \sin x$

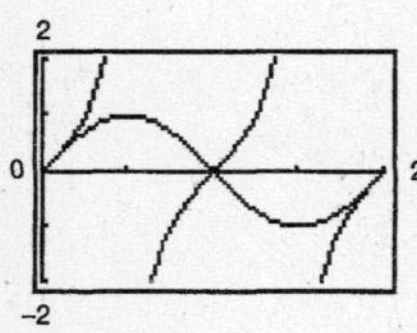

$\tan x < \sin x$ on the intervals $\left(\dfrac{\pi}{2}, \pi\right)$ and $\left(\dfrac{3\pi}{2}, 2\pi\right)$

(d) Let $y_1 = \cos x$ and $y_2 = \sin x$

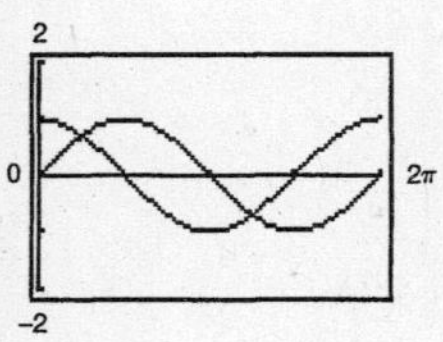

$\cos x \geq \sin x$ on the intervals $\left[0, \dfrac{\pi}{4}\right]$ and $\left[\dfrac{5\pi}{4}, 2\pi\right]$

13. (a) $\sin(u + v + w) = \sin[(u + v) + w]$

$$= \sin(u + v)\cos w + \cos(u + v)\sin w$$

$$= [\sin u \cos v + \cos u \sin v]\cos w + [\cos u \cos v - \sin u \sin v]\sin w$$

$$= \sin u \cos v \cos w + \cos u \sin v \cos w + \cos u \cos v \sin w - \sin u \sin v \sin w$$

(b) $\tan(u + v + w) = \tan[(u + v) + w]$

$$= \frac{\tan(u + v) + \tan w}{1 - \tan(u + v)\tan w}$$

$$= \frac{\left[\dfrac{\tan u + \tan v}{1 - \tan u \tan v}\right] + \tan w}{1 - \left[\dfrac{\tan u + \tan v}{1 - \tan u \tan v}\right]\tan w} \cdot \frac{(1 - \tan u \tan v)}{(1 - \tan u \tan v)}$$

$$= \frac{\tan u + \tan v + (1 - \tan u \tan v)\tan w}{(1 - \tan u \tan v) - (\tan u + \tan v)\tan w}$$

$$= \frac{\tan u + \tan v + \tan w - \tan u \tan v \tan w}{1 - \tan u \tan v - \tan u \tan w - \tan v \tan w}$$

15. $h_1 = 3.75 \sin 733t + 7.5$

$h_2 = 3.75 \sin 733\left(t + \dfrac{4\pi}{3}\right) + 7.5$

(a)

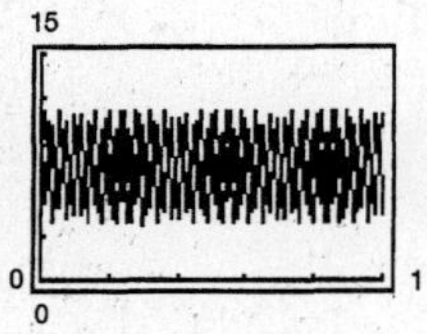

(b) The period for h_1 and h_2 is $\dfrac{2\pi}{733} \approx 0.0086$

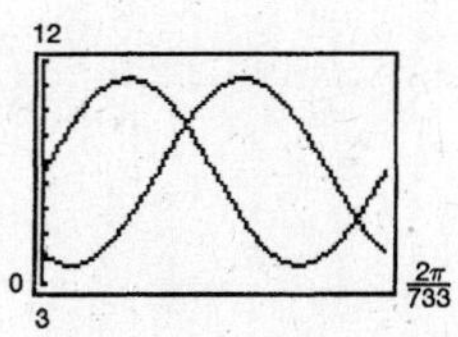

The graphs intersect twice per cycle

There are $\dfrac{1}{\frac{2\pi}{733}} \approx 116.66$ cycles in the interval $[0, 1]$, so the graphs intersect approximately 233.3 times.

Chapter 2 Practice Test

1. Find the value of the other five trigonometric functions, given $\tan x = \frac{4}{11}$, $\sec x < 0$.

2. Simplify $\dfrac{\sec^2 x + \csc^2 x}{\csc^2 x(1 + \tan^2 x)}$.

3. Rewrite as a single logarithm and simplify $\ln|\tan \theta| - \ln|\cot \theta|$.

4. True or false:

$$\cos\left(\frac{\pi}{2} - x\right) = \frac{1}{\csc x}$$

5. Factor and simplify: $\sin^4 x + (\sin^2 x) \cos^2 x$

6. Multiply and simplify: $(\csc x + 1)(\csc x - 1)$

7. Rationalize the denominator and simplify:

$$\frac{\cos^2 x}{1 - \sin x}$$

8. Verify:

$$\frac{1 + \cos \theta}{\sin \theta} + \frac{\sin \theta}{1 + \cos \theta} = 2 \csc \theta$$

9. Verify:

$$\tan^4 x + 2 \tan^2 x + 1 = \sec^4 x$$

10. Use the sum or difference formulas to determine:

(a) $\sin 105°$ (b) $\tan 15°$

11. Simplify: $(\sin 42°) \cos 38° - (\cos 42°) \sin 38°$

12. Verify $\tan\left(\theta + \dfrac{\pi}{4}\right) = \dfrac{1 + \tan \theta}{1 - \tan \theta}$.

13. Write $\sin(\arcsin x - \arccos x)$ as an algebraic expression in x.

14. Use the double-angle formulas to determine:

(a) $\cos 120°$ (b) $\tan 300°$

15. Use the half-angle formulas to determine:

(a) $\sin 22.5°$ (d) $\tan \dfrac{\pi}{12}$

16. Given $\sin = 4/5$, θ lies in Quadrant II, find $\cos(\theta/2)$.

17. Use the power-reducing identities to write $(\sin^2 x) \cos^2 x$ in terms of the first power of cosine.

18. Rewrite as a sum: $6(\sin 5\theta) \cos 2\theta$.

19. Rewrite as a product:

$\sin(x + \pi) + \sin(x - \pi)$.

20. Verify $\dfrac{\sin 9x + \sin 5x}{\cos 9x - \cos 5x} = -\cot 2x$.

21. Verify:

$(\cos u) \sin v = \frac{1}{2}[\sin(u + v) - \sin(u - v)]$.

22. Find all solutions in the interval $[0, 2\pi)$:

$4 \sin^2 x = 1$

23. Find all solutions in the interval $[0, 2\pi)$:

$\tan^2 \theta + \left(\sqrt{3} - 1\right) \tan\theta - \sqrt{3} = 0$

24. Find all solutions in the interval $[0, 2\pi)$:

$\sin 2x = \cos x$

25. Use the quadratic formula to find all solutions in the interval $[0, 2\pi)$:

$\tan^2 x - 6 \tan x + 4 = 0$

CHAPTER 3
Additional Topics in Trigonometry

CHAPTER 3
Additional Topics in Trigonometry

Section 3.1 Law of Sines

- If ABC is any oblique triangle with sides a, b, and c, then

$$\frac{a}{\sin A} = \frac{b}{\sin B} = \frac{c}{\sin C}.$$

- You should be able to use the Law of Sines to solve an oblique triangle for the remaining three parts, given:
  (a) Two angles and any side (AAS or ASA)
  (b) Two sides and an angle opposite one of them (SSA)
    1. If A is acute and $h = b \sin A$:
       (a) $a < h$, no triangle is possible.
       (b) $a = h$ or $a > b$, one triangle is possible.
       (c) $h < a < b$, two triangles are possible.
    2. If A is obtuse and $h = b \sin A$:
       (a) $a \le b$, no triangle is possible.
       (b) $a > b$, one triangle is possible.
- The area of any triangle equals one-half the product of the lengths of two sides times the sine of their included angle.

$$A = \tfrac{1}{2}ab \sin C = \tfrac{1}{2}ac \sin B = \tfrac{1}{2}bc \sin A$$

Solutions to Odd-Numbered Exercises

1. Given: $A = 30°,\ B = 45°,\ a = 20$

$C = 180° - A - B = 105°$

$$b = \frac{a}{\sin A}(\sin B) = \frac{20 \sin 45°}{\sin 30°} = 20\sqrt{2} \approx 28.28$$

$$c = \frac{a}{\sin A}(\sin C) = \frac{20 \sin 105°}{\sin 30°} \approx 38.64$$

3. Given: $A = 25°, B = 35°, a = 3.5$

$C = 180° - A - B = 120°$

$$b = \frac{a}{\sin A}(\sin B) = \frac{3.5}{\sin 25°}(\sin 35°) \approx 4.75$$

$$c = \frac{a}{\sin A}(\sin C) = \frac{3.5}{\sin 25°}(\sin 120°) \approx 7.17$$

5. Given: $A = 36°,\ a = 8,\ b = 5$

$$\sin B = \frac{b \sin A}{a} = \frac{5 \sin 36°}{8} \approx 0.36737 \implies B \approx 21.55°$$

$C = 180° - A - B \approx 180° - 36° - 21.55 = 122.45°$

$$c = \frac{a}{\sin A}(\sin C) = \frac{8}{\sin 36°}(\sin 122.45°) \approx 11.49$$

7. Given: $A = 102.4°, C = 16.7°, a = 21.6$

$B = 180° - A - C = 60.9°$

$$b = \frac{a}{\sin A}(\sin B) = \frac{21.6}{\sin 102.4°}(\sin 60.9°) \approx 19.32$$

$$c = \frac{a}{\sin A}(\sin C) = \frac{21.6}{\sin 102.4°}(\sin 16.7°) \approx 6.36$$

9. Given: $A = 83° 20'$, $C = 54.6°$, $c = 18.1$

$$B = 180° - A - C = 180° - 83° 20' - 54° 36' = 42° 4'$$

$$a = \frac{c}{\sin C}(\sin A) = \frac{18.1}{\sin 54.6°}(\sin 83° 20') \approx 22.05$$

$$b = \frac{c}{\sin C}(\sin B) = \frac{18.1}{\sin 54.6°}(\sin 42° 4') \approx 14.88$$

11. Given: $B = 15° 30'$, $a = 4.5$, $b = 6.8$

$$\sin A = \frac{a \sin B}{b} = \frac{4.5 \sin 15° 30'}{6.8} \approx 0.17685 \implies A \approx 10° 11'$$

$$C = 180° - A - B \approx 180° - 10° 11' - 15° 30' = 154° 19'$$

$$c = \frac{b}{\sin B}(\sin C) = \frac{6.8}{\sin 15° 30'}(\sin 154° 19') \approx 11.03$$

13. Given: $C = 145°$, $b = 4$, $c = 14$

$$\sin B = \frac{b \sin C}{c} = \frac{4 \sin 145°}{14} \approx 0.16388 \implies B \approx 9.43°$$

$$A = 180° - B - C \approx 180° - 9.43° - 145° = 25.57°$$

$$a = \frac{c}{\sin C}(\sin A) \approx \frac{14}{\sin 145°}(\sin 25.57°) \approx 10.53$$

15. Given: $A = 110° 15'$, $a = 48$, $b = 16$

$$\sin B = \frac{b \sin A}{a} = \frac{16 \sin 110° 15'}{48} \approx 0.31273 \implies B \approx 18° 13'$$

$$C = 180° - A - B \approx 180° - 110° 15' - 18° 13' = 51° 32'$$

$$c = \frac{a}{\sin A}(\sin C) = \frac{48}{\sin 110° 15'}(\sin 51° 32') \approx 40.06$$

17. Given: $A = 55°, B = 42°, c = \frac{3}{4}$

$$C = 180° - A - B = 83°$$

$$a = \frac{c}{\sin C}(\sin A) = \frac{0.75}{\sin 83°}(\sin 55°) \approx 0.62$$

$$b = \frac{c}{\sin C}(\sin B) = \frac{0.75}{\sin 83°}(\sin 42°) \approx 0.51$$

19. Given: $a = 4.5$, $b = 12.8$, $A = 58°$

$h = 12.8 \sin 58° \approx 10.86$

Since $a < h$, no triangle is formed.

21. Given: $a = 18, b = 20, A = 76°$

$h = 20 \sin 76° \approx 19.41$

Since $a < h$, no triangle is formed.

23. Given: $a = 125$, $b = 200$, $A = 110°$

No triangle is formed because A is obtuse and $a < b$.

25. Given: $A = 36°$, $a = 5$

(a) One solution if $b \le 5$ or $b = \dfrac{5}{\sin 36°}$

(b) Two solutions if $5 < b < \dfrac{5}{\sin 36°}$

(c) No solution if $b > \dfrac{5}{\sin 36°}$

27. Given: $A = 10°$, $a = 10.8$

(a) One solution if $b \le 10.8$ or $b = \dfrac{10.8}{\sin 10°}$

(b) Two solutions if $10.8 < b < \dfrac{10.8}{\sin 10°}$

(c) No solution if $b > \dfrac{10.8}{\sin 10°}$

29. Area $= \frac{1}{2}ab \sin C = \frac{1}{2}(4)(6) \sin 120° \approx 10.4$

31. Area $= \frac{1}{2}bc \sin A = \frac{1}{2}(57)(85) \sin 43° 45' \approx 1675.2$

33. Area $= \frac{1}{2}ac \sin B = \frac{1}{2}(105)(64)\sin(72°30') \approx 3204.5$

35. $C = 180° - 23° - 94° = 63°$

$$h = \frac{35}{\sin 63°}(\sin 23°) \approx 15.3 \text{ meters}$$

37. $\dfrac{\sin(42° - \theta)}{10} = \dfrac{\sin 48°}{17}$

$\sin(42° - \theta) \approx 0.43714$

$42° - \theta \approx 25.9°$

$\theta \approx 16.1°$

39. Given: $c = 100$

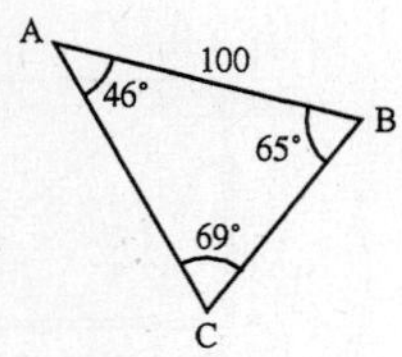

$A = 74° - 28° = 46°$,

$B = 180° - 41° - 74° = 65°$,

$C = 180° - 46° - 65° = 69°$

$$a = \frac{c}{\sin C}(\sin A) = \frac{100}{\sin 69°}(\sin 46°) \approx 77 \text{ meters}$$

41. (a)

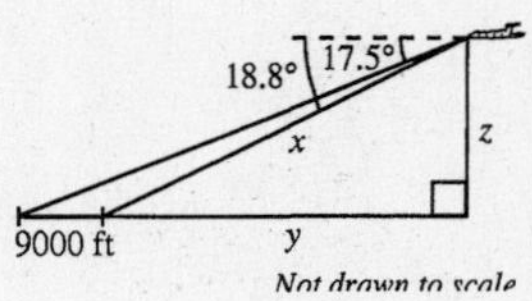

(b) $\dfrac{x}{\sin 17.5°} = \dfrac{9000}{\sin 1.3°}$

$x \approx 119{,}289.1261$ feet ≈ 22.6 miles

(c) $\dfrac{y}{\sin 71.2°} = \dfrac{x}{\sin 90°}$

$y = x \sin 71.2° \approx 119{,}289.1261 \sin 71.2°$

$\approx 112{,}924.963$ feet ≈ 21.4 miles

(d) $z = x \sin 18.8° \approx 119{,}289.1261 \sin 18.8°$

$\approx 38{,}443$ feet ≈ 7.3 miles

43.

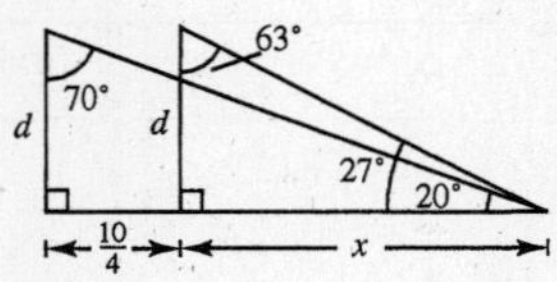

In 15 minutes the boat has traveled

$$(10 \text{ mph})\left(\frac{1}{4}\text{ hr}\right) = \frac{10}{4} \text{ miles}$$

$\tan 63° = \dfrac{x}{d} \quad \Rightarrow d \tan 63° = x$

$\tan 70° = \dfrac{x + (10/4)}{d} \quad \Rightarrow d \tan 70° = x + \dfrac{10}{4}$

$\Rightarrow d \tan 70° - \dfrac{10}{4} = x$

$d \tan 70° - \dfrac{10}{4} = d \tan 63°$

$d \tan 70° - d \tan 63° = \dfrac{10}{4}$

$d(\tan 70° - \tan 63°) = 2.5$

$$d = \frac{2.5}{\tan 70° - \tan 63°} \approx 3.2 \text{ miles}$$

45. True. If one angle of a triangle is obtuse, then there is less than 90° left for the other two angles, so it cannot contain a right angle. It must be oblique.

47. (a) $\dfrac{\sin \alpha}{9} = \dfrac{\sin \beta}{18}$

$\sin \alpha = 0.5 \sin \beta$

$\alpha = \arcsin(0.5 \sin \beta)$

(b) 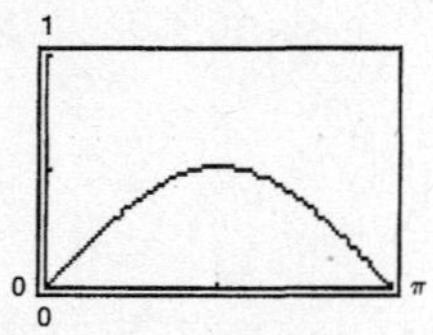

Domain: $0 < \beta < \pi$

Range: $0 < \alpha \le \dfrac{\pi}{6}$

(c) $\gamma = \pi - \alpha - \beta = \pi - \beta - \arcsin(0.5 \sin \beta)$

$\dfrac{c}{\sin \gamma} = \dfrac{18}{\sin \beta}$

$c = \dfrac{18 \sin \gamma}{\sin \beta} = \dfrac{18 \sin[\pi - \beta - \arcsin(0.5 \sin \beta)]}{\sin \beta}$

(d) 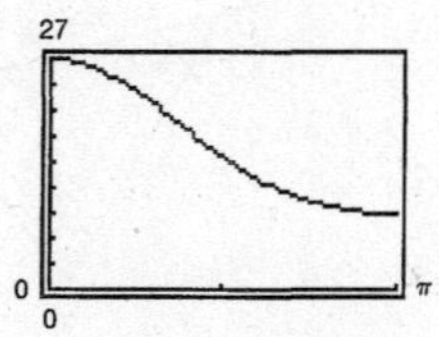

Domain: $0 < \beta < \pi$

Range: $9 < c < 27$

(e)

β	0.4	0.8	1.2	1.6	2.0	2.4	2.8
α	0.1960	0.3669	0.4848	0.5234	0.4720	0.3445	0.1683
c	25.95	23.07	19.19	15.33	12.29	10.31	9.27

As $\beta \to 0$, $c \to 27$

As $\beta \to \pi$, $c \to 9$

49. $\sin x \cot x = \sin x \dfrac{\cos x}{\sin x} = \cos x$

51. $1 - \sin^2\left(\dfrac{\pi}{2} - x\right) = 1 - \cos^2 x = \sin^2 x$

53. $6 \sin 8\theta \cos 3\theta = (6)(\frac{1}{2})[\sin(8\theta + 3\theta) + \sin(8\theta - 3\theta)] = 3(\sin 11\theta + \sin 5\theta)$

Section 3.2 Law of Cosines

- If ABC is any oblique triangle with sides a, b, and c, the following equations are valid.

  (a) $a^2 = b^2 + c^2 - 2bc \cos A$ or $\cos A = \dfrac{b^2 + c^2 - a^2}{2bc}$

  (b) $b^2 = a^2 + c^2 - 2ac \cos B$ or $\cos B = \dfrac{a^2 + c^2 - b^2}{2ac}$

  (c) $c^2 = a^2 + b^2 - 2ab \cos C$ or $\cos C = \dfrac{a^2 + b^2 - c^2}{2ab}$

- You should be able to use the Law of Cosines to solve an oblique triangle for the remaining three parts, given:

  (a) Three sides (SSS)

  (b) Two sides and their included angle (SAS)

- Given any triangle with sides of length a, b, and c, the area of the triangle is

  Area $= \sqrt{s(s-a)(s-b)(s-c)}$, where $s = \dfrac{a+b+c}{2}$. (Heron's Formula)

Solutions to Odd-Numbered Exercises

1. Given: $a = 7, b = 10, c = 15$

$$\cos C = \frac{a^2 + b^2 - c^2}{2ab} = \frac{49 + 100 - 225}{2(7)(10)} \approx -0.5429 \implies C \approx 122.88°$$

$$\sin B = \frac{b \sin C}{c} = \frac{10 \sin 122.88°}{15} \approx 0.5599 \implies B \approx 34.05°$$

$$A \approx 180° - 34.05° - 122.88° \approx 23.07°$$

3. Given: $A = 30°,\ b = 15,\ c = 30$

$$a^2 = b^2 + c^2 - 2bc \cos A$$

$$= 225 + 900 - 2(15)(30) \cos 30° \approx 345.5771$$

$$a \approx 18.59$$

$$\cos B = \frac{a^2 + c^2 - b^2}{2ac} \approx \frac{(18.59)^2 + 900 - 225}{2(18.59)(30)} \approx 0.9150$$

$$B \approx 23.79°$$

$$C \approx 180° - 30° - 23.79° = 126.21°$$

5. $a = 11, b = 14, c = 20$

$$\cos C = \frac{a^2 + b^2 - c^2}{2ab} = \frac{121 + 196 - 400}{2(11)(14)} \approx -0.2695 \implies C \approx 105.63°$$

$$\sin B = \frac{b \sin C}{c} = \frac{14 \sin 105.63°}{20} \approx 0.6741 \implies B \approx 42.38°$$

$$A \approx 180° - 42.38° - 105.63° \approx 31.99°$$

7. Given: $a = 75.4,\ b = 52,\ c = 52$

$$\cos A = \frac{b^2 + c^2 - a^2}{2bc} = \frac{52^2 + 52^2 - 75.4^2}{2(52)(52)} = -0.05125 \implies A \approx 92.94°$$

$$\sin B = \frac{b \sin A}{a} \approx \frac{52(0.9987)}{75.4} \approx 0.68876 \implies B \approx 43.53°$$

$$C = B \approx 43.53°$$

9. Given: $A = 135°, b = 4, c = 9$

$$a^2 = b^2 + c^2 - 2bc \cos A = 16 + 81 - 2(4)(9)\cos 135° \approx 147.9117 \implies a \approx 12.16$$

$$\sin B = \frac{b \sin A}{a} = \frac{4 \sin 135°}{12.16} \approx 0.2326 \implies B \approx 13.45°$$

$$C \approx 180° - 135° - 13.45° \approx 31.55°$$

11. Given: $B = 10° 35', a = 40, c = 30$

$$b^2 = a^2 + c^2 - 2ac \cos B = 1600 + 900 - 2(40)(30)\cos 10° 35' \approx 140.8268 \implies b \approx 11.87$$

$$\sin C = \frac{c \sin B}{b} = \frac{30 \sin 10° 35'}{11.87} \approx 0.4642 \implies C \approx 27.66° \approx 27° 40'$$

$$A \approx 180° - 10° 35' - 27° 40' = 141° 45'$$

13. Given: $B = 125°\,40'$, $a = 32$, $c = 32$

$b^2 = a^2 + c^2 - 2ac\cos B \approx 32^2 + 32^2 - 2(32)(32)(-0.5831) \approx 3242.1 \implies b \approx 56.9$

$A = C \implies 2A = 180° - 125°\,40' = 54°\,20' \implies A = C = 27°\,10'$

15. $C = 43°, a = \frac{4}{9}, b = \frac{7}{9}$

$c^2 = a^2 + b^2 - 2ab\cos C = \left(\frac{4}{9}\right)^2 + \left(\frac{7}{9}\right)^2 - 2\left(\frac{4}{9}\right)\left(\frac{7}{9}\right)\cos 43° \approx 0.2968 \implies c \approx 0.5448$

$\sin A = \frac{a\sin C}{c} = \frac{(4/9)\sin 43°}{0.5448} \approx 0.5564 \implies A \approx 33.80°$

$B \approx 180° - 43° - 33.8° \approx 103.20°$

17. $d^2 = 5^2 + 8^2 - 2(5)(8)\cos 45° \approx 32.4315 \implies d \approx 5.69$

$2\phi = 360° - 2(45°) = 270° \implies \phi = 135°$

$c^2 = 5^2 + 8^2 - 2(5)(8)\cos 135° \approx 145.5685 \implies c \approx 12.07$

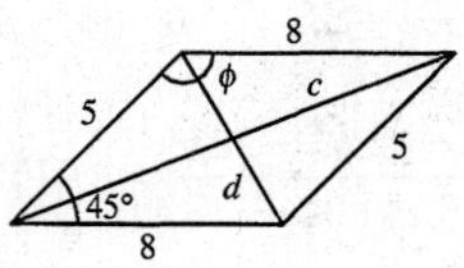

19. $\cos\phi = \frac{10^2 + 14^2 - 20^2}{2(10)(14)}$

$\phi \approx 111.8°$

$2\theta \approx 360° - 2(111.8°)$

$\theta = 68.2°$

$d^2 = 10^2 + 14^2 - 2(10)(14)\cos 68.2°$

$d \approx 13.86$

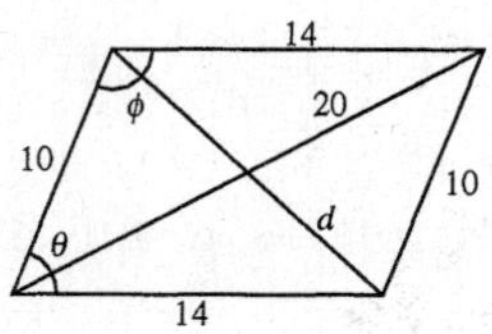

21. $\cos\alpha = \frac{(12.5)^2 + (15)^2 - 10^2}{2(12.5)(15)} = 0.75 \implies \alpha \approx 41.41°$

$\cos\beta = \frac{10^2 + 15^2 - (12.5)^2}{2(10)(15)} = 0.5625 \implies \beta \approx 55.77°$

$z = 180° - \alpha - \beta = 82.82°$

$u = 180° - z = 97.18°$

$b^2 = 12.5^2 + 10^2 - 2(12.5)(10)\cos 97.18° \approx 287.4967 \implies b \approx 16.96$

$\cos\delta = \frac{12.5^2 + 16.96^2 - 10^2}{2(12.5)(16.96)} \approx 0.8111 \implies \delta \approx 35.80°$

$\theta = \alpha + \delta = 41.41° + 35.80° \approx 77.21°$

$2\phi = 360° - 2\theta \implies \phi = \frac{360° - 2(77.21°)}{2} = 102.79°$

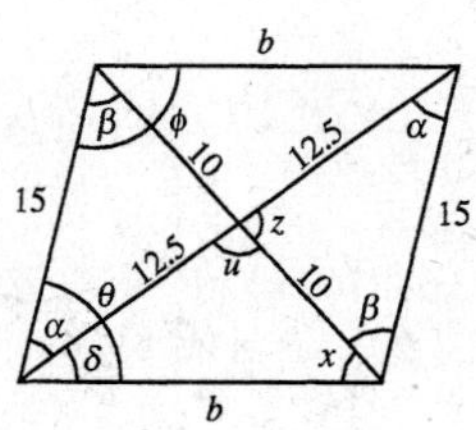

23. $a = 5,\ b = 7,\ c = 10 \implies s = \frac{a + b + c}{2} = 11$

$\text{Area} = \sqrt{s(s-a)(s-b)(s-c)} = \sqrt{11(6)(4)(1)} \approx 16.25$

25. $a = 2.5, b = 10.2, c = 9 \implies s = \frac{a + b + c}{2} = 10.85$

$\text{Area} = \sqrt{s(s-a)(s-b)(s-c)} = \sqrt{10.85(8.35)(0.65)(1.85)} \approx 10.44$

27. $a = 12.32, b = 8.46, c = 15.05 \Rightarrow s = \dfrac{a+b+c}{2} = 17.915$

$\text{Area} = \sqrt{s(s-a)(s-b)(s-c)} = \sqrt{17.915(5.595)(9.455)(2.865)} \approx 52.11$

29. $\cos B = \dfrac{1700^2 + 3700^2 - 3000^2}{2(1700)(3700)} \Rightarrow B \approx 52.9°$

Bearing: $90° - 52.9° = \text{N } 37.1° \text{ E}$

$\cos C = \dfrac{1700^2 + 3000^2 - 3700^2}{2(1700)(3000)} \Rightarrow C \approx 100.2°$

Bearing: $A = 180° - 52.9° - 100.2° = 26.9° \Rightarrow \text{S } 63.1° \text{ E}$

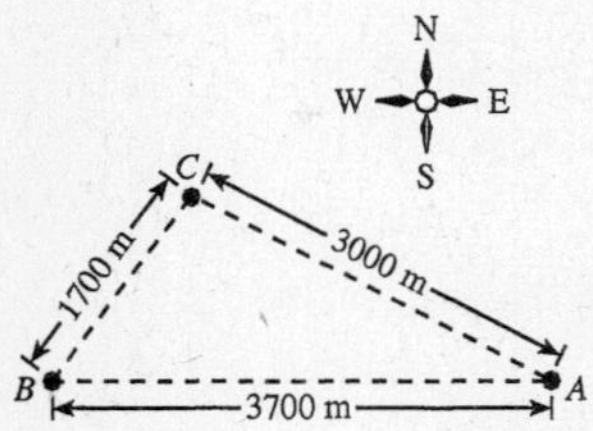

31. $b^2 = 220^2 + 250^2 - 2(220)(250)\cos 105° \Rightarrow b \approx 373.3$ meters

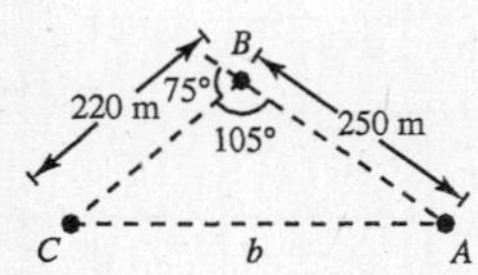

33.

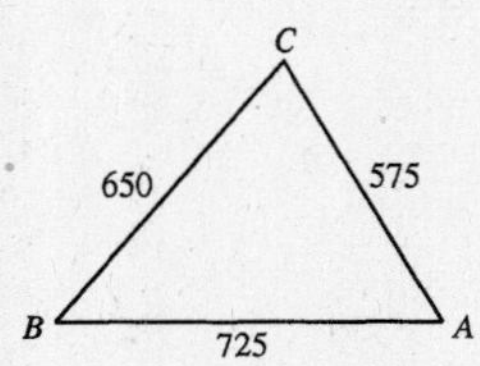

The largest angle is across from the largest side.

$\cos C = \dfrac{650^2 + 575^2 - 725^2}{2(650)(575)}$

$C \approx 72.3°$

35. $C = 180° - 53° - 67° = 60°$

$c^2 = a^2 + b^2 - 2ab\cos C$

$= 36^2 + 48^2 - 2(36)(48)(0.5)$

$= 1872$

$c \approx 43.3$ mi

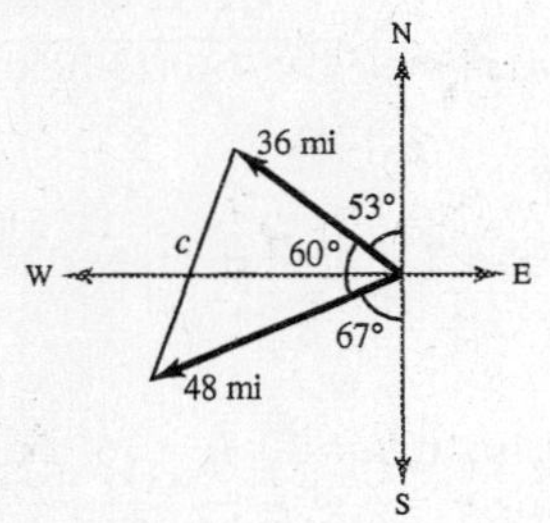

37. (a) $\cos\theta = \dfrac{273^2 + 178^2 - 235^2}{2(273)(178)}$

$\theta \approx 58.4°$

Bearing: N 58.4° W

(b) $\cos\phi = \dfrac{235^2 + 178^2 - 273^2}{2(235)(178)}$

$\phi \approx 81.5°$

Bearing: S 81.5° W

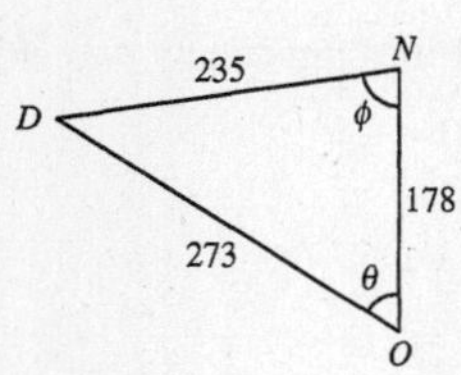

39. $d^2 = 60.5^2 + 90^2 - 2(60.5)(90)\cos 45° \approx 4059.8572 \Rightarrow d \approx 63.7$ ft

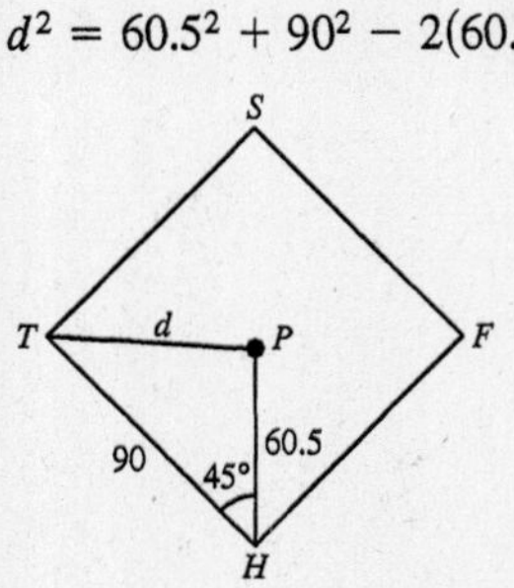

41. $a^2 = 35^2 + 20^2 - 2(35)(20)\cos 42° \implies a \approx 24.2$ miles

43. $\overline{RS} = \sqrt{8^2 + 10^2} = \sqrt{164} = 2\sqrt{41} \approx 12.8$ ft

$\overline{PQ} = \frac{1}{2}\sqrt{16^2 + 10^2} = \frac{1}{2}\sqrt{356} = \sqrt{89} \approx 9.4$ ft

$$\tan P = \frac{10}{16} = \frac{\overline{QS}}{\overline{PS}} = \frac{\overline{QS}}{8} \implies \overline{QS} = 5$$

45. $d^2 = 10^2 + 7^2 - 2(10)(7)\cos\theta$

$$\theta = \arccos\left[\frac{10^2 + 7^2 - d^2}{2(10)(7)}\right]$$

$$s = \frac{360° - \theta}{360°}(2\pi r) = \frac{(360° - \theta)\pi}{45°}$$

d (inches)	9	10	12	13	14	15	16
θ (degrees)	60.9°	69.5°	88.0°	98.2°	109.6°	122.9°	139.8°
s (inches)	20.88	20.28	18.99	18.28	17.48	16.55	15.37

47. $a = 200, b = 500, c = 600 \implies s = \dfrac{200 + 500 + 600}{2} = 650$

Area $= \sqrt{650(450)(150)(50)} \approx 46{,}837.5$ square feet

49. False. The average of the three sides of a triangle is $\dfrac{a + b + c}{3}$, not $\dfrac{a + b + c}{2}$.

51. False. If $a = 10$, $b = 16$, and $c = 5$, then by the Law of Cosines, we would have:

$$\cos A = \frac{16^2 + 5^2 - 10^2}{2(16)(5)} = 1.13125 > 1$$

This is not possible. In general, if the sum of any two sides is less than the third side, then they cannot form a triangle. Here $10 + 5$ is less than 16.

53. $a = 25, b = 55, c = 72$

(a) area of triangle: $s = \frac{1}{2}(25 + 55 + 72) = 76$

area $= \sqrt{76(51)(21)(4)} \approx 570.60$

(b) area of circumscribed circle:

$$\cos C = \frac{25^2 + 55^2 - 72^2}{2(25)(55)} \approx -0.5578 \implies C \approx 123.9°$$

$$R = \frac{1}{2}\left(\frac{c}{\sin C}\right) \approx 43.37$$

area $= \pi R^2 \approx 5909.2$

(c) area of inscribed circle:

$$r = \sqrt{\frac{(s-a)(s-b)(s-c)}{s}} = \sqrt{\frac{(51)(21)(4)}{76}} \approx 7.51$$

area $= \pi r^2 \approx 177.09$

55. $\frac{1}{2}bc(1 + \cos A) = \frac{1}{2}bc\left[1 + \frac{b^2 + c^2 - a^2}{2bc}\right]$

$= \frac{1}{2}bc\left[\frac{2bc + b^2 + c^2 - a^2}{2bc}\right]$

$= \frac{1}{4}[(b + c)^2 - a^2]$

$= \frac{1}{4}[(b + c) + a][(b + c) - a]$

$= \frac{b + c + a}{2} \cdot \frac{b + c - a}{2}$

$= \frac{a + b + c}{2} \cdot \frac{-a + b + c}{2}$

57. $\arcsin(-1) = -\frac{\pi}{2}$

59. $\arctan\sqrt{3} = \frac{\pi}{3}$

61. $\arcsin\left(-\frac{\sqrt{3}}{2}\right) = -\frac{\pi}{3}$

63. Let $\theta = \arcsin 2x$, then

$\sin\theta = 2x = \frac{2x}{1}$ and $\sec\theta = \frac{1}{\sqrt{1 - 4x^2}}$.

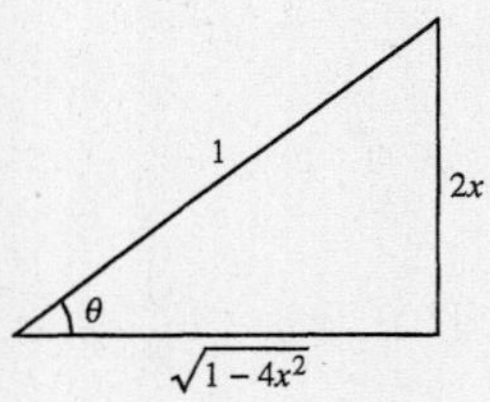

65. Let $\theta = \arctan(x - 2)$, then

$\tan\theta = x - 2 = \frac{x - 2}{1}$ and $\cot\theta = \frac{1}{x - 2}$.

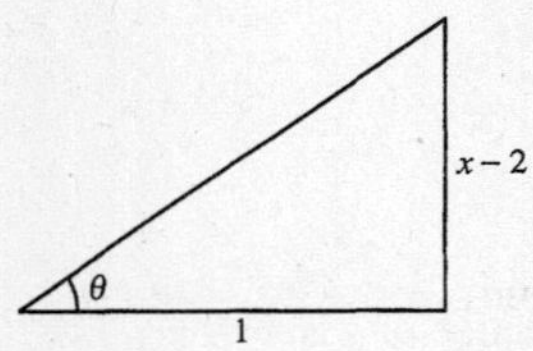

67. $5 = \sqrt{25 - x^2},\ x = 5\sin\theta$

$5 = \sqrt{25 - (5\sin\theta)^2}$

$5 = \sqrt{25(1 - \sin^2\theta)}$

$5 = 5\cos\theta$

$\cos\theta = 1$

$\sec\theta = \frac{1}{\cos\theta} = 1$

$\csc\theta$ is undefined.

69. $-\sqrt{3} = \sqrt{x^2 - 9},\ x = 3\sec\theta$

$-\sqrt{3} = \sqrt{(3\sec\theta)^2 - 9}$

$-\sqrt{3} = \sqrt{9(\sec^2\theta - 1)}$

$-\sqrt{3} = 3\tan\theta$

$\tan\theta = -\frac{\sqrt{3}}{3}$

$\sec\theta = \sqrt{1 + \tan^2\theta} = \sqrt{1 + \left(-\frac{\sqrt{3}}{3}\right)^2} = \frac{2\sqrt{3}}{3}$

$\cot\theta = \frac{1}{\tan\theta} = -\sqrt{3}$

$\csc\theta = -\sqrt{1 + \cot^2\theta} = -\sqrt{1 + (-\sqrt{3})^2} = -2$

71. $\cos\frac{5\pi}{6} - \cos\frac{\pi}{3} = -2\sin\left(\frac{\frac{5\pi}{6} + \frac{\pi}{3}}{2}\right)\sin\left(\frac{\frac{5\pi}{6} - \frac{\pi}{3}}{2}\right)$

$= -2\sin\frac{7\pi}{12}\sin\frac{\pi}{4}$

Section 3.3 Vectors in the Plane

- A vector **v** is the collection of all directed line segments that are equivalent to a given directed line segment $\overrightarrow{PQ}$.
- You should be able to *geometrically* perform the operations of vector addition and scalar multiplication.
- The component form of the vector with initial point $P = (p_1, p_2)$ and terminal point $Q = (q_1, q_2)$ is

  $$\overrightarrow{PQ} = \langle q_1 - p_1, q_2 - p_2 \rangle = \langle v_1, v_2 \rangle = \mathbf{v}.$$
- The magnitude of $\mathbf{v} = \langle v_1, v_2 \rangle$ is given by $\|\mathbf{v}\| = \sqrt{v_1^2 + v_2^2}$.
- If $\|\mathbf{v}\| = 1$, **v** is a unit vector.
- You should be able to perform the operations of scalar multiplication and vector addition in component form.

  (a) $\mathbf{u} + \mathbf{v} = \langle u_1 + v_1, u_2 + v_2 \rangle$ (b) $k\mathbf{u} = \langle ku_1, ku_2 \rangle$
- You should know the following properties of vector addition and scalar multiplication.

  (a) $\mathbf{u} + \mathbf{v} = \mathbf{v} + \mathbf{u}$

  (b) $(\mathbf{u} + \mathbf{v}) + \mathbf{w} = \mathbf{u} + (\mathbf{v} + \mathbf{w})$

  (c) $\mathbf{u} + \mathbf{0} = \mathbf{u}$

  (d) $\mathbf{u} + (-\mathbf{u}) = \mathbf{0}$

  (e) $c(d\mathbf{u}) = (cd)\mathbf{u}$

  (f) $(c + d)\mathbf{u} = c\mathbf{u} + d\mathbf{u}$

  (g) $c(\mathbf{u} + \mathbf{v}) = c\mathbf{u} + c\mathbf{v}$

  (h) $1(\mathbf{u}) = \mathbf{u}, 0\mathbf{u} = \mathbf{0}$

  (i) $\|c\mathbf{v}\| = |c| \, \|\mathbf{v}\|$
- A unit vector in the direction of **v** is $\mathbf{u} = \dfrac{\mathbf{v}}{\|\mathbf{v}\|}$.
- The standard unit vectors are $\mathbf{i} = \langle 1, 0 \rangle$ and $\mathbf{j} = \langle 0, 1 \rangle$. $\mathbf{v} = \langle v_1, v_2 \rangle$ can be written as $\mathbf{v} = v_1\mathbf{i} + v_2\mathbf{j}$.
- A vector **v** with magnitude $\|\mathbf{v}\|$ and direction θ can be written as $\mathbf{v} = a\mathbf{i} + b\mathbf{j} = \|\mathbf{v}\|(\cos\theta)\mathbf{i} + \|\mathbf{v}\|(\sin\theta)\mathbf{j}$ where $\tan\theta = b/a$.

Solutions to Odd-Numbered Exercises

1. Initial point: $(0, 0)$

Terminal point: $(3, 2)$

$\mathbf{v} = \langle 3 - 0, 2 - 0 \rangle = \langle 3, 2 \rangle$

$\|\mathbf{v}\| = \sqrt{3^2 + 2^2} = \sqrt{13}$

3. Initial point: $(2, 2)$

Terminal point: $(-1, 4)$

$\mathbf{v} = \langle -1 - 2, 4 - 2 \rangle = \langle -3, 2 \rangle$

$\|\mathbf{v}\| = \sqrt{(-3)^2 + 2^2} = \sqrt{13}$

5. Initial point: $(3, -2)$

Terminal point: $(3, 3)$

$\mathbf{v} = \langle 3 - 3, 3 - (-2) \rangle = \langle 0, 5 \rangle$

$\|\mathbf{v}\| = \sqrt{0^2 + 5^2} = \sqrt{25} = 5$

7. Initial point: $(-1, 5)$

Terminal point: $(15, 12)$

$\mathbf{v} = \langle 15 - (-1), 12 - 5 \rangle = \langle 16, 7 \rangle$

$\|\mathbf{v}\| = \sqrt{16^2 + 7^2} = \sqrt{305}$

9. Initial point: $(-3, -5)$

Terminal point: $(5, 1)$

$\mathbf{v} = \langle 5 - (-3), 1 - (-5) \rangle = \langle 8, 6 \rangle$

$\|\mathbf{v}\| = \sqrt{8^2 + 6^2} = \sqrt{100} = 10$

11. Initial point: $(1, 3)$

Terminal point: $(-8, -9)$

$\mathbf{v} = \langle -8 - 1, -9 - 3 \rangle = \langle -9, -12 \rangle$

$\|\mathbf{v}\| = \sqrt{(-9)^2 + (-12)^2} = \sqrt{225} = 15$

13.

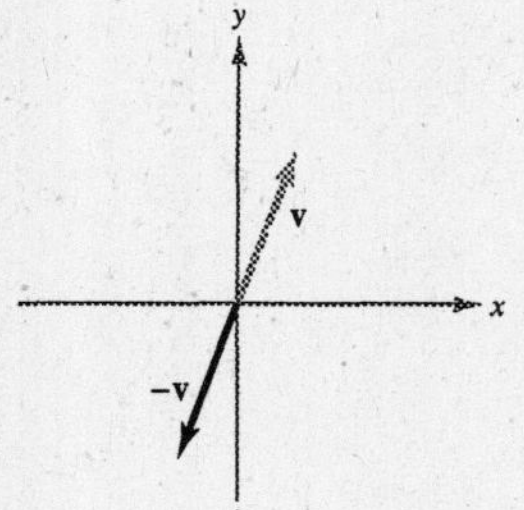

15.

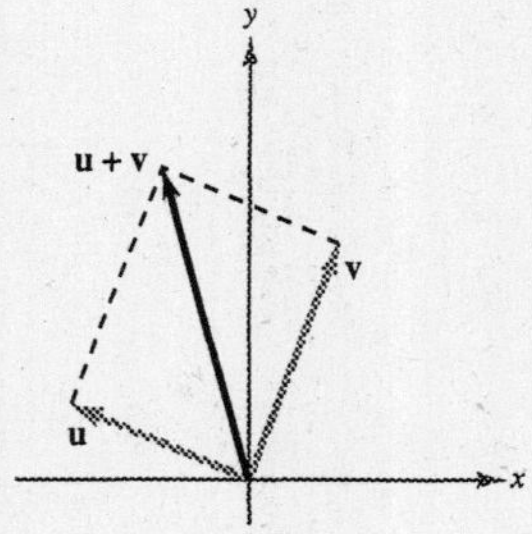

17. $\mathbf{u} + 2\mathbf{v}$

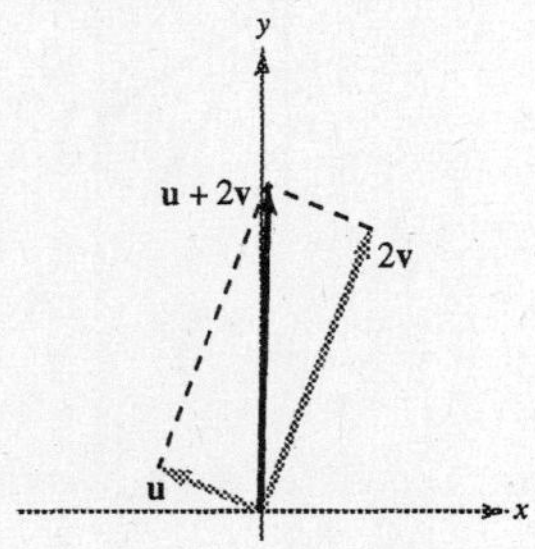

19. $\mathbf{u} = \langle 2, 1 \rangle, \mathbf{v} = \langle 1, 3 \rangle$

(a) $\mathbf{u} + \mathbf{v} = \langle 3, 4 \rangle$

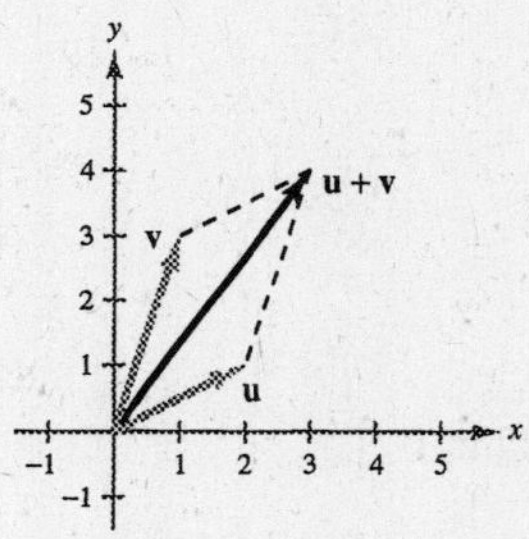

(b) $\mathbf{u} - \mathbf{v} = \langle 1, -2 \rangle$

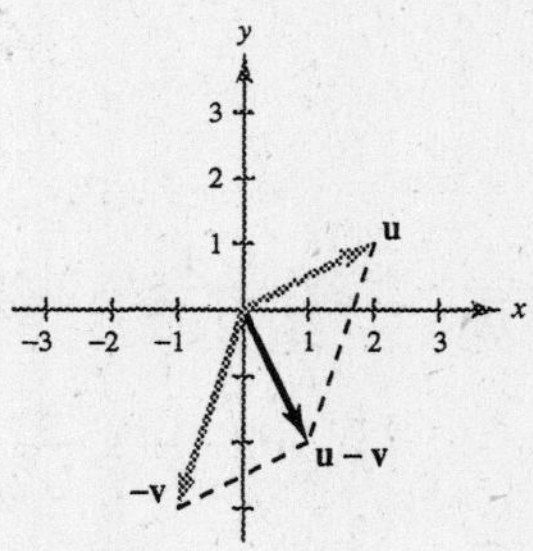

(c) $2\mathbf{u} - 3\mathbf{v} = \langle 4, 2 \rangle - \langle 3, 9 \rangle = \langle 1, -7 \rangle$

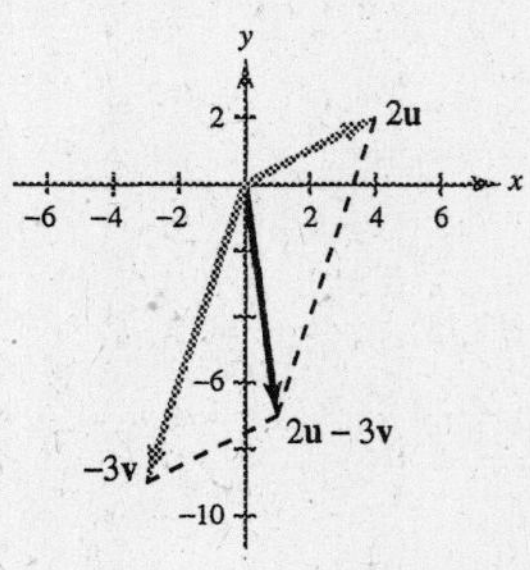

21. $\mathbf{u} = \langle -5, 3 \rangle, \mathbf{v} = \langle 0, 0 \rangle$

(a) $\mathbf{u} + \mathbf{v} = \langle -5, 3 \rangle = \mathbf{u}$

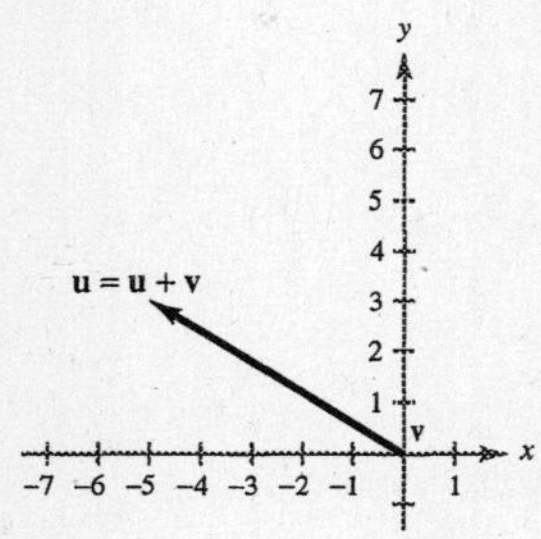

(b) $\mathbf{u} - \mathbf{v} = \langle -5, 3 \rangle = \mathbf{u}$

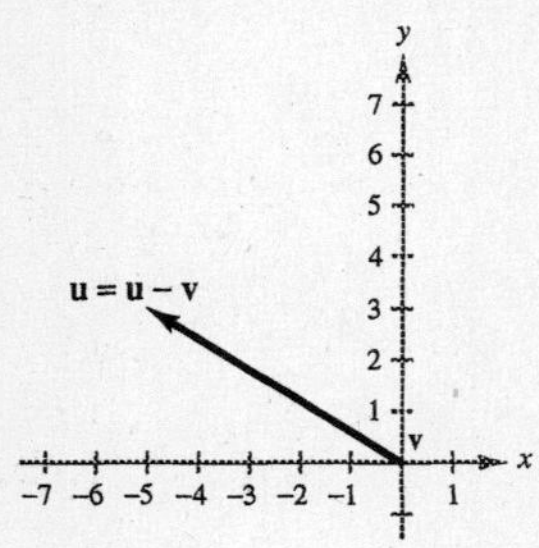

(c) $2\mathbf{u} - 3\mathbf{v} = 2\mathbf{u} = \langle -10, 6 \rangle$

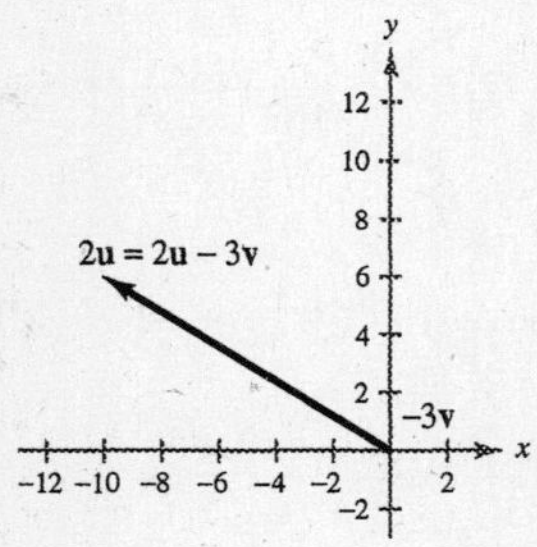

23. $\mathbf{u} = \mathbf{i} + \mathbf{j}, \mathbf{v} = 2\mathbf{i} - 3\mathbf{j}$

(a) $\mathbf{u} + \mathbf{v} = 3\mathbf{i} - 2\mathbf{j}$

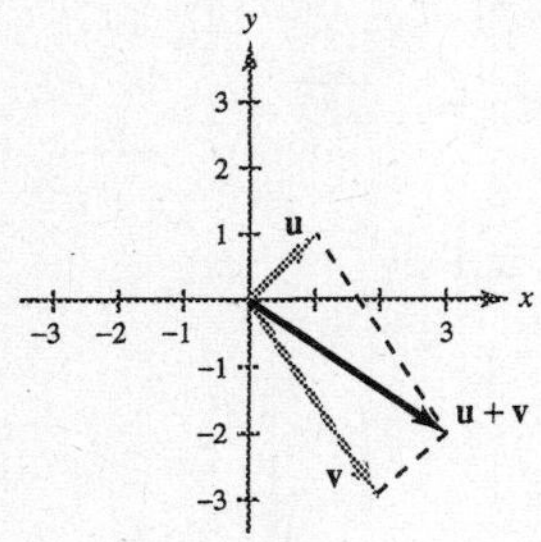

(b) $\mathbf{u} - \mathbf{v} = -\mathbf{i} + 4\mathbf{j}$

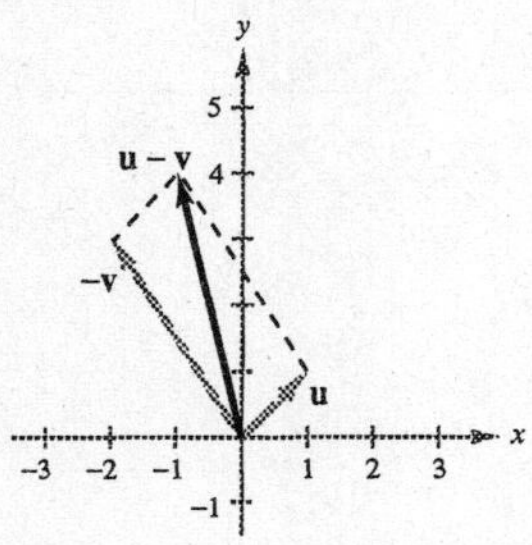

(c) $2\mathbf{u} - 3\mathbf{v} = (2\mathbf{i} + 2\mathbf{j}) - (6\mathbf{i} - 9\mathbf{j}) = -4\mathbf{i} + 11\mathbf{j}$

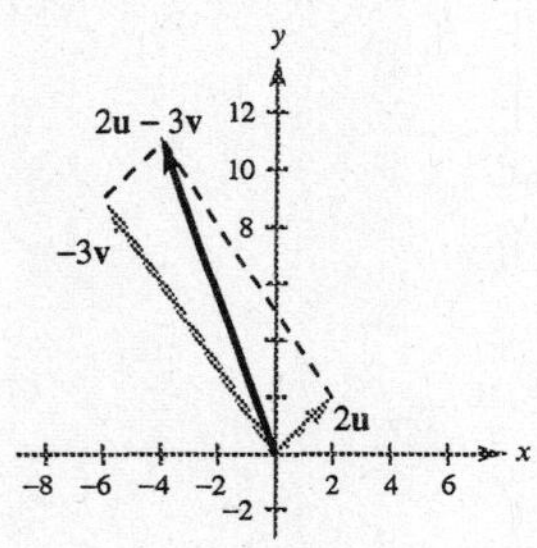

25. $\mathbf{u} = 2\mathbf{i}, \mathbf{v} = \mathbf{j}$

(a) $\mathbf{u} + \mathbf{v} = 2\mathbf{i} + \mathbf{j}$

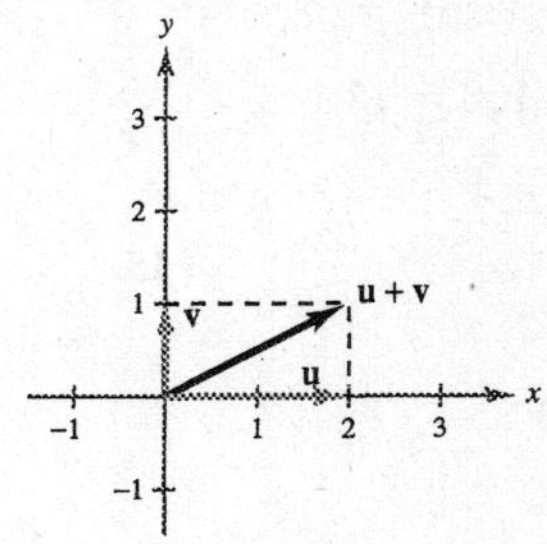

(b) $\mathbf{u} - \mathbf{v} = 2\mathbf{i} - \mathbf{j}$

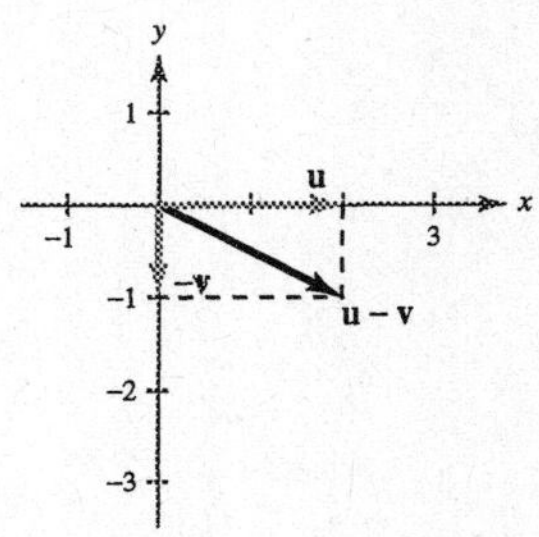

(c) $2\mathbf{u} - 3\mathbf{v} = 4\mathbf{i} - 3\mathbf{j}$

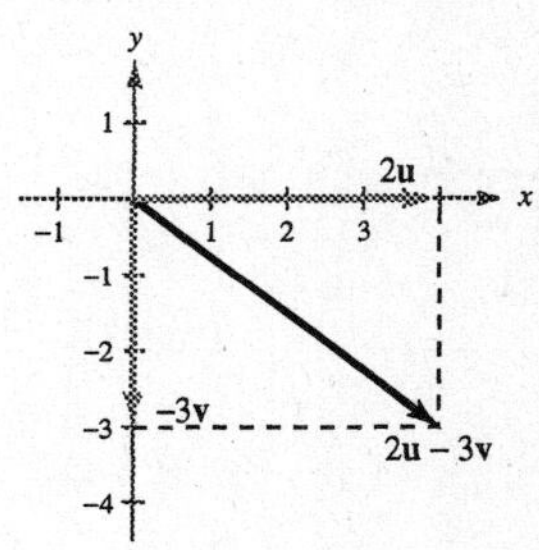

27. $\mathbf{v} = \dfrac{1}{\|\mathbf{u}\|}\mathbf{u} = \dfrac{1}{\sqrt{3^2 + 0^2}}\langle 3, 0\rangle = \dfrac{1}{3}\langle 3, 0\rangle = \langle 1, 0\rangle$

29. $\mathbf{u} = \dfrac{1}{\|\mathbf{v}\|}\mathbf{v} = \dfrac{1}{\sqrt{(-2)^2 + 2^2}}\langle -2, 2\rangle = \dfrac{1}{2\sqrt{2}}\langle -2, 2\rangle$

$= \left\langle -\dfrac{1}{\sqrt{2}}, \dfrac{1}{\sqrt{2}}\right\rangle$

31. $\mathbf{u} = \dfrac{1}{\|\mathbf{v}\|}\mathbf{v} = \dfrac{1}{\sqrt{6^2 + (-2)^2}}(6\mathbf{i} - 2\mathbf{j}) = \dfrac{1}{\sqrt{40}}(6\mathbf{i} - 2\mathbf{j})$

$= \dfrac{1}{2\sqrt{10}}(6\mathbf{i} - 2\mathbf{j}) = \dfrac{3}{\sqrt{10}}\mathbf{i} - \dfrac{1}{\sqrt{10}}\mathbf{j}$

33. $\mathbf{u} = \dfrac{1}{\|\mathbf{w}\|}\mathbf{w} = \dfrac{1}{4}(4\mathbf{j}) = \mathbf{j}$

35. $\mathbf{u} = \dfrac{1}{\|\mathbf{w}\|}\mathbf{w} = \dfrac{1}{\sqrt{1^2 + (-2)^2}}(\mathbf{i} - 2\mathbf{j}) = \dfrac{1}{\sqrt{5}}(\mathbf{i} - 2\mathbf{j})$

$= \dfrac{1}{\sqrt{5}}\mathbf{i} - \dfrac{2}{\sqrt{5}}\mathbf{j} = \dfrac{\sqrt{5}}{5}\mathbf{i} - \dfrac{2\sqrt{5}}{5}\mathbf{j}$

37. $5\left(\dfrac{1}{\|\mathbf{u}\|}\mathbf{u}\right) = 5\left(\dfrac{1}{\sqrt{3^2 + 3^2}}\langle 3, 3\rangle\right) = \dfrac{5}{3\sqrt{2}}\langle 3, 3\rangle$

$= \left\langle \dfrac{5}{\sqrt{2}}, \dfrac{5}{\sqrt{2}}\right\rangle$

39. $9\left(\dfrac{1}{\|\mathbf{u}\|}\mathbf{u}\right) = 9\left(\dfrac{1}{\sqrt{2^2 + 5^2}}\langle 2, 5\rangle\right) = \dfrac{9}{\sqrt{29}}\langle 2, 5\rangle = \left\langle \dfrac{18}{\sqrt{29}}, \dfrac{45}{\sqrt{29}}\right\rangle$

41. $\mathbf{v} = \frac{3}{2}\mathbf{u}$

$= \frac{3}{2}(2\mathbf{i} - \mathbf{j})$

$= 3\mathbf{i} - \frac{3}{2}\mathbf{j} = \langle 3, -\frac{3}{2}\rangle$

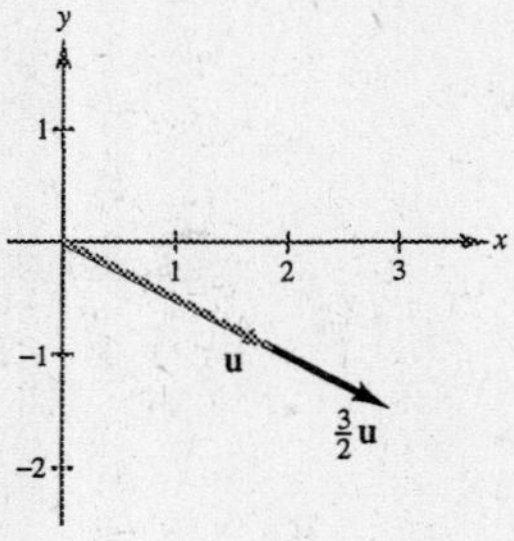

43. $\mathbf{v} = \mathbf{u} + 2\mathbf{w}$

$= (2\mathbf{i} - \mathbf{j}) + 2(\mathbf{i} + 2\mathbf{j})$

$= 4\mathbf{i} + 3\mathbf{j} = \langle 4, 3\rangle$

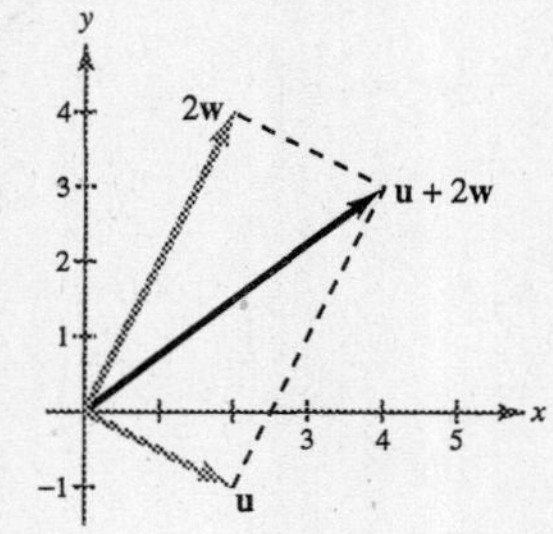

45. $\mathbf{v} = \frac{1}{2}(3\mathbf{u} + \mathbf{w})$

$= \frac{1}{2}(6\mathbf{i} - 3\mathbf{j} + \mathbf{i} + 2\mathbf{j})$

$= \frac{7}{2}\mathbf{i} - \frac{1}{2}\mathbf{j} = \langle \frac{7}{2}, -\frac{1}{2}\rangle$

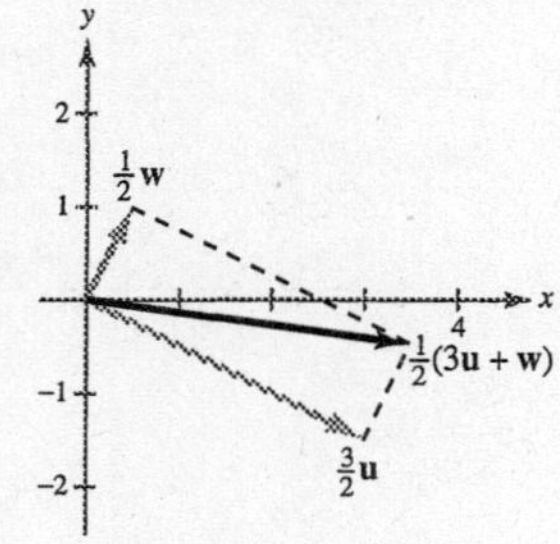

47. $\mathbf{v} = 3(\cos 60°\mathbf{i} + \sin 60°\mathbf{j})$

$\|\mathbf{v}\| = 3, \ \theta = 60°$

49. $\mathbf{v} = 6\mathbf{i} - 6\mathbf{j}$

$\|\mathbf{v}\| = \sqrt{6^2 + (-6)^2} = \sqrt{72} = 6\sqrt{2}$

$\tan \theta = \dfrac{-6}{6} = -1$

Since $\mathbf{v}$ lies in Quadrant IV, $\theta = 315°$.

51. $\mathbf{v} = \langle 3 \cos 0°, 3 \sin 0°\rangle$

$= \langle 3, 0\rangle$

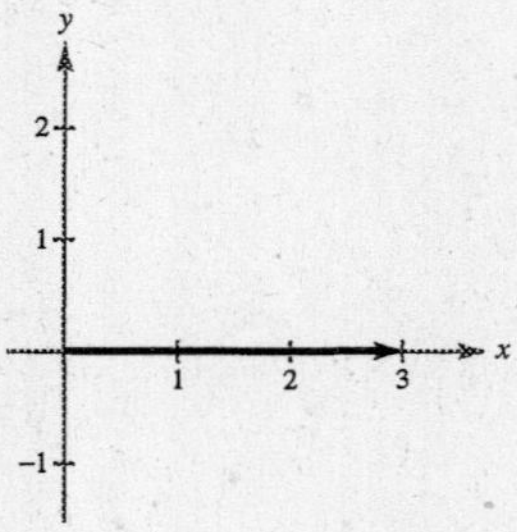

53. $\mathbf{v} = \left\langle \dfrac{7}{2} \cos 150°, \dfrac{7}{2} \sin 150°\right\rangle$

$= \left\langle -\dfrac{7\sqrt{3}}{4}, \dfrac{7}{4}\right\rangle$

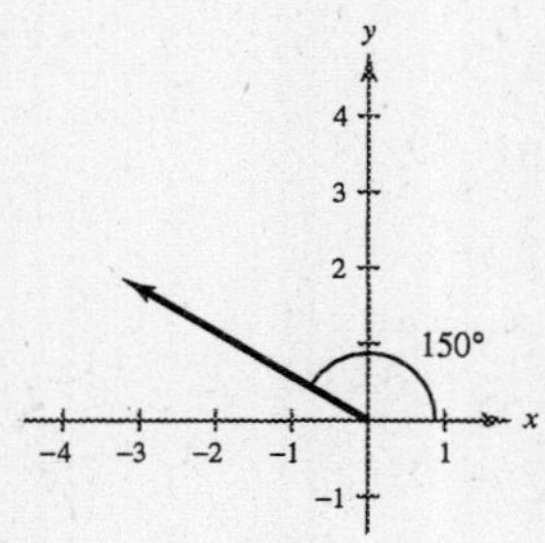

55. $\mathbf{v} = \langle 3\sqrt{2} \cos 150°, 3\sqrt{2} \sin 150°\rangle$

$= \left\langle -\dfrac{3\sqrt{6}}{2}, \dfrac{3\sqrt{2}}{2}\right\rangle$

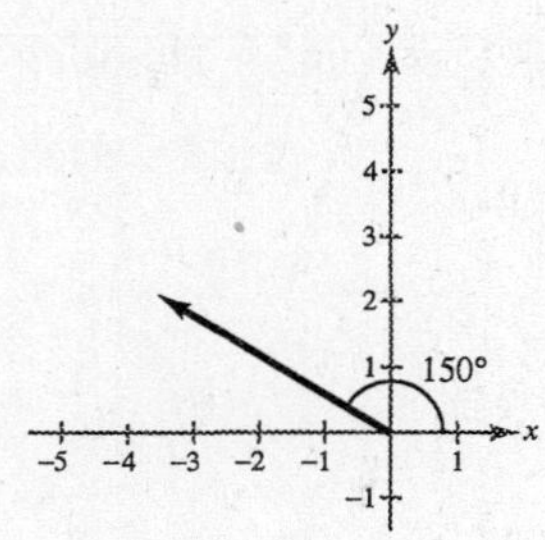

57. $\mathbf{v} = 2\left(\dfrac{1}{\sqrt{1^2 + 3^2}}\right)(\mathbf{i} + 3\mathbf{j})$

$= \dfrac{2}{\sqrt{10}}(\mathbf{i} + 3\mathbf{j})$

$= \dfrac{\sqrt{10}}{5}\mathbf{i} + \dfrac{3\sqrt{10}}{5}\mathbf{j} = \left\langle \dfrac{\sqrt{10}}{5}, \dfrac{3\sqrt{10}}{5}\right\rangle$

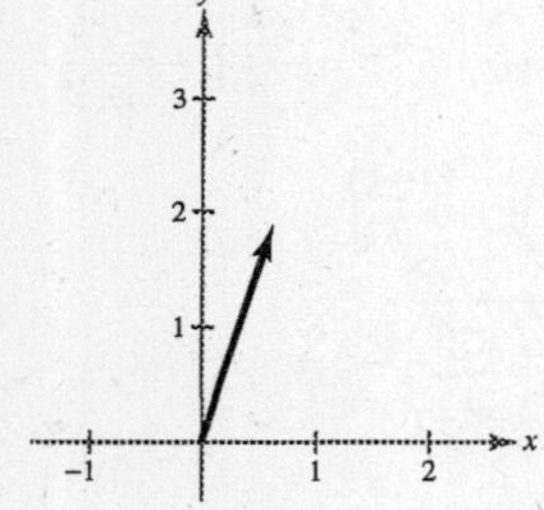

59. $\mathbf{u} = \langle 5 \cos 0°, 5 \sin 0°\rangle = \langle 5, 0\rangle$

$\mathbf{v} = \langle 5 \cos 90°, 5 \sin 90°\rangle = \langle 0, 5\rangle$

$\mathbf{u} + \mathbf{v} = \langle 5, 5\rangle$

61. $\mathbf{u} = \langle 20 \cos 45°, 20 \sin 45°\rangle = \langle 10\sqrt{2}, 10\sqrt{2}\rangle$

$\mathbf{v} = \langle 50 \cos 180°, 50 \sin 180°\rangle = \langle -50, 0\rangle$

$\mathbf{u} + \mathbf{v} = \langle 10\sqrt{2} - 50, 10\sqrt{2}\rangle$

63. $\mathbf{v} = \mathbf{i} + \mathbf{j}$

$\mathbf{w} = 2\mathbf{i} - 2\mathbf{j}$

$\mathbf{u} = \mathbf{v} - \mathbf{w} = -\mathbf{i} + 3\mathbf{j}$

$\|\mathbf{v}\| = \sqrt{2}$

$\|\mathbf{w}\| = 2\sqrt{2}$

$\|\mathbf{v} - \mathbf{w}\| = \sqrt{10}$

$$\cos\alpha = \frac{\|\mathbf{v}\|^2 + \|\mathbf{w}\|^2 - \|\mathbf{v} - \mathbf{w}\|^2}{2\|\mathbf{v}\|\,\|\mathbf{w}\|} = \frac{2 + 8 - 10}{2\sqrt{2} \cdot 2\sqrt{2}} = 0$$

$\alpha = 90°$

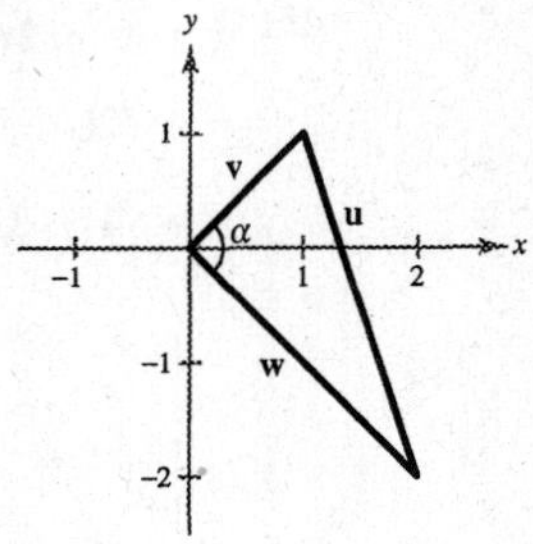

65. $\mathbf{v} = \mathbf{i} + \mathbf{j}$

$\mathbf{w} = 3\mathbf{i} - \mathbf{j}$

$\mathbf{u} = \mathbf{v} - \mathbf{w} = -2\mathbf{i} + 2\mathbf{j}$

$$\cos\alpha = \frac{\|\mathbf{v}\|^2 + \|\mathbf{w}\|^2 - \|\mathbf{v} - \mathbf{w}\|^2}{2\|\mathbf{v}\|\,\|\mathbf{w}\|} = \frac{2 + 10 - 8}{2\sqrt{2}\sqrt{10}} \approx 0.4472$$

$\alpha = 63.4°$

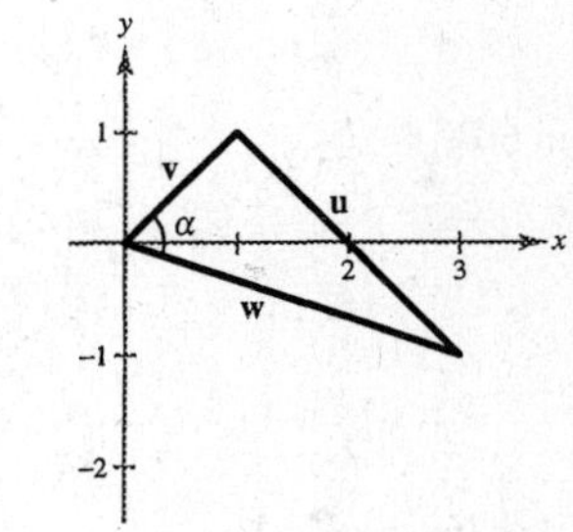

67. Force One: $\mathbf{u} = 45\mathbf{i}$

Force Two: $\mathbf{v} = 60\cos\theta\mathbf{i} + 60\sin\theta\mathbf{j}$

Resultant Force: $\mathbf{u} + \mathbf{v} = (45 + 60\cos\theta)\mathbf{i} + 60\sin\theta\mathbf{j}$

$\|\mathbf{u} + \mathbf{v}\| = \sqrt{(45 + 60\cos\theta)^2 + (60\sin\theta)^2} = 90$

$$2025 + 5400\cos\theta + 3600 = 8100$$

$$5400\cos\theta = 2475$$

$$\cos\theta = \frac{2475}{5400} \approx 0.4583$$

$$\theta \approx 62.7°$$

69. $\mathbf{u} = 300\mathbf{i}$

$$\mathbf{v} = (125\cos 45°)\mathbf{i} + (125\sin 45°)\mathbf{j} = \frac{125}{\sqrt{2}}\mathbf{i} + \frac{125}{\sqrt{2}}\mathbf{j}$$

$$\mathbf{R} = \mathbf{u} + \mathbf{v} = \left(300 + \frac{125}{\sqrt{2}}\right)\mathbf{i} + \frac{125}{\sqrt{2}}\mathbf{j}$$

$$\|\mathbf{R}\| = \sqrt{\left(300 + \frac{125}{\sqrt{2}}\right)^2 + \left(\frac{125}{\sqrt{2}}\right)^2} \approx 398.32 \text{ newtons}$$

$$\tan\theta = \frac{\frac{125}{\sqrt{2}}}{300 + \left(\frac{125}{\sqrt{2}}\right)} \Rightarrow \theta \approx 12.8°$$

71. $\mathbf{u} = (75\cos 30°)\mathbf{i} + (75\sin 30°)\mathbf{j} \approx 64.95\mathbf{i} + 37.5\mathbf{j}$

$\mathbf{v} = (100\cos 45°)\mathbf{i} + (100\sin 45°)\mathbf{j} \approx 70.71\mathbf{i} + 70.71\mathbf{j}$

$\mathbf{w} = (125\cos 120°)\mathbf{i} + (125\sin 120°)\mathbf{j} \approx -62.5\mathbf{i} + 108.3\mathbf{j}$

$\mathbf{u} + \mathbf{v} + \mathbf{w} \approx 73.16\mathbf{i} + 216.5\mathbf{j}$

$\|\mathbf{u} + \mathbf{v} + \mathbf{w}\| \approx 228.5$ pounds

$$\tan\theta \approx \frac{216.5}{73.16} \approx 2.9593$$

$\theta \approx 71.3°$

73. Horizontal component of velocity: $70\cos 35° \approx 57.34$ feet per second

Vertical component of velocity: $70\sin 35° \approx 40.15$ feet per second

75. Cable $\overrightarrow{AC}$: $\mathbf{u} = \|\mathbf{u}\|(\cos 50°\mathbf{i} - \sin 50°\mathbf{j})$

Cable $\overrightarrow{BC}$: $\mathbf{v} = \|\mathbf{v}\|(-\cos 30°\mathbf{i} - \sin 30°\mathbf{j})$

Resultant: $\mathbf{u} + \mathbf{v} = -2000\mathbf{j}$

$\|\mathbf{u}\| \cos 50° - \|\mathbf{v}\| \cos 30° = 0$

$-\|\mathbf{u}\| \sin 50° - \|\mathbf{v}\| \sin 30° = -2000$

Solving this system of equations yields:

$T_{AC} = \|\mathbf{u}\| \approx 1758.8$ pounds

$T_{BC} = \|\mathbf{v}\| \approx 1305.4$ pounds

77. Towline 1: $\mathbf{u} = \|\mathbf{u}\|(\cos 18°\mathbf{i} + \sin 18°\mathbf{j})$

Towline 2: $\mathbf{v} = \|\mathbf{u}\|(\cos 18°\mathbf{i} - \sin 18°\mathbf{j})$

Resultant: $\mathbf{u} + \mathbf{v} = 6000\mathbf{i}$

$\|\mathbf{u}\| \cos 18° + \|\mathbf{u}\| \cos 18° = 6000$

$\|\mathbf{u}\| \approx 3154.4$

Therefore, the tension on each towline is

$\|\mathbf{u}\| \approx 3154.4$ pounds.

79. Airspeed: $\mathbf{u} = (875 \cos 58°)\mathbf{i} - (875 \sin 58°)\mathbf{j}$

Groundspeed: $\mathbf{v} = (800 \cos 50°)\mathbf{i} - (800 \sin 50°)\mathbf{j}$

Wind: $\mathbf{w} = \mathbf{v} - \mathbf{u} = (800 \cos 50° - 875 \cos 58°)\mathbf{i} + (-800 \sin 50° + 875 \sin 58°)\mathbf{j}$

$\approx 50.5507\mathbf{i} + 129.2065\mathbf{j}$

Wind speed: $\|\mathbf{w}\| \approx \sqrt{(50.5507)^2 + (129.2065)^2}$

≈ 138.7 kilometers per hour

Wind direction: $\tan \theta \approx \dfrac{129.2065}{50.5507}$

$\theta \approx 68.6°;\ 90° - \theta = 21.4°$

Bearing: N 21.4° E

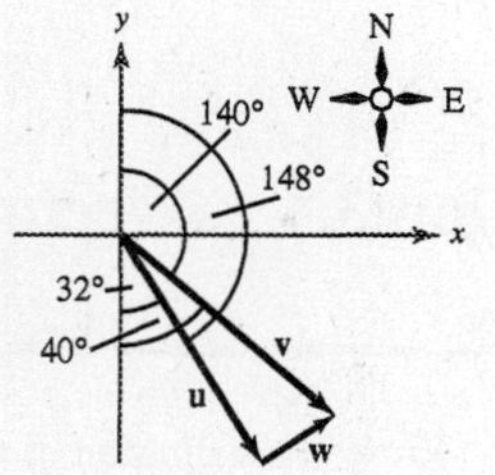

81. $W = FD = (100 \cos 50°)(30) = 1928.4$ foot–pounds

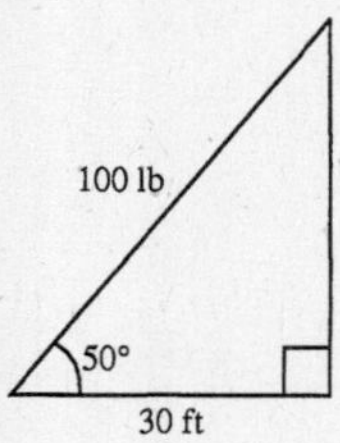

83. True. See Example 1.

85. (a) The angle between them is 0°.

(b) The angle between them is 180°.

(c) No. At most it can be equal to the sum when the angle between them is 0°.

87. Let $\mathbf{v} = (\cos \theta)\mathbf{i} + (\sin \theta)\mathbf{j}$.

$\|\mathbf{v}\| = \sqrt{\cos^2 \theta + \sin^2 \theta} = \sqrt{1} = 1$

Therefore, $\mathbf{v}$ is a unit vector for any value of θ.

89. $\mathbf{u} = \langle 5 - 1, 2 - 6 \rangle = \langle 4, -4 \rangle$

$\mathbf{v} = \langle 9 - 4, 4 - 5 \rangle = \langle 5, -1 \rangle$

$\mathbf{u} - \mathbf{v} = \langle -1, -3 \rangle$ or $\mathbf{v} - \mathbf{u} = \langle 1, 3 \rangle$

91.
$$\begin{aligned} \sqrt{x^2 - 64} &= \sqrt{(8 \sec \theta)^2 - 64} \\ &= \sqrt{64(\sec^2 \theta - 1)} \\ &= 8\sqrt{\tan^2 \theta} \\ &= 8 \tan \theta \quad \text{for } 0 < \theta < \frac{\pi}{2} \end{aligned}$$

93.
$$\begin{aligned} \sqrt{x^2 + 36} &= \sqrt{(6 \tan \theta)^2 + 36} \\ &= \sqrt{36(\tan^2 \theta + 1)} \\ &= 6\sqrt{\sec^2 \theta} \\ &= 6 \sec \theta \quad \text{for } 0 < \theta < \frac{\pi}{2} \end{aligned}$$

95. $\cos x(\cos x + 1) = 0$

$\cos x = 0$ or $\cos x + 1 = 0$

$x = \dfrac{\pi}{2} + n\pi \qquad \cos x = -1$

$x = \pi + 2n\pi$

97. $3 \sec x \sin x - 2\sqrt{3} \sin x = 0$

$$\sin x(3 \sec x - 2\sqrt{3}) = 0$$

$\sin x = 0$ or $3 \sec x - 2\sqrt{3} = 0$

$x = n\pi$ $\quad\quad \sec x = \dfrac{2\sqrt{3}}{3}$

$$\cos x = \frac{3}{2\sqrt{3}} = \frac{\sqrt{3}}{2}$$

$$x = \frac{\pi}{6} + 2n\pi$$

$$x = \frac{11\pi}{6} + 2n\pi$$

Section 3.4 Vectors and Dot Products

- Know the definition of the dot product of $\mathbf{u} = \langle u_1, u_2 \rangle$ and $\mathbf{v} = \langle v_1, v_2 \rangle$.

  $\mathbf{u} \cdot \mathbf{v} = u_1v_1 + u_2v_2$

- Know the following properties of the dot product:
  1. $\mathbf{u} \cdot \mathbf{v} = \mathbf{v} \cdot \mathbf{u}$
  2. $\mathbf{0} \cdot \mathbf{v} = 0$
  3. $\mathbf{u} \cdot (\mathbf{v} + \mathbf{w}) = \mathbf{u} \cdot \mathbf{v} + \mathbf{u} \cdot \mathbf{w}$
  4. $\mathbf{v} \cdot \mathbf{v} = \|\mathbf{v}\|^2$
  5. $c(\mathbf{u} \cdot \mathbf{v}) = c\mathbf{u} \cdot \mathbf{v} = \mathbf{u} \cdot c\mathbf{v}$
- If θ is the angle between two nonzero vectors **u** and **v**, then

  $$\cos \theta = \frac{\mathbf{u} \cdot \mathbf{v}}{\|\mathbf{u}\| \|\mathbf{v}\|}.$$

- The vectors **u** and **v** are orthogonal if $\mathbf{u} \cdot \mathbf{v} = 0$.
- Know the definition of vector components.

  $\mathbf{u} = \mathbf{w}_1 + \mathbf{w}_2$ where $\mathbf{w}_1$ and $\mathbf{w}_2$ are orthogonal, and $\mathbf{w}_1$ is parallel to **v**. $\mathbf{w}_1$ is called the projection of **u** onto **v**

  and is denoted by $\mathbf{w}_1 = \text{proj}_{\mathbf{v}}\mathbf{u} = \left(\dfrac{\mathbf{u} \cdot \mathbf{v}}{\|\mathbf{v}\|^2}\right)\mathbf{v}$. Then we have $\mathbf{w}_2 = \mathbf{u} - \mathbf{w}_1$.

- Know the definition of work.
  1. Projection form: $w = \|\text{proj}_{\overrightarrow{PQ}} \mathbf{F}\| \|\overrightarrow{PQ}\|$
  2. Dot product form: $w = \mathbf{F} \cdot \overrightarrow{PQ}$

Solutions to Odd-Numbered Exercises

1. $\mathbf{u} = \langle 6, 1 \rangle$, $\mathbf{v} = \langle -2, 3 \rangle$

$\mathbf{u} \cdot \mathbf{v} = 6(-2) + 1(3) = -9$

3. $\mathbf{u} = 4\mathbf{i} - 2\mathbf{j}$, $\mathbf{v} = \mathbf{i} - \mathbf{j}$

$\mathbf{u} \cdot \mathbf{v} = 4(1) + (-2)(-1) = 6$

5. $\mathbf{u} = \langle 2, 2 \rangle$

$\mathbf{u} \cdot \mathbf{u} = 2(2) + 2(2) = 8$

The result is a scalar.

7. $\mathbf{u} = \langle 2, 2\rangle,\ \mathbf{v} = \langle -3, 4\rangle$

$$(\mathbf{u}\cdot\mathbf{v})\mathbf{v} = [(2)(-3) + 2(4)]\langle -3, 4\rangle = 2\langle -3, 4\rangle = \langle -6, 8\rangle$$

The result is a vector.

9. $\mathbf{u} = \langle -5, 12\rangle$

$$\|\mathbf{u}\| = \sqrt{\mathbf{u}\cdot\mathbf{u}} = \sqrt{(-5)^2 + 12^2} = 13$$

11. $\mathbf{u} = 20\mathbf{i} + 25\mathbf{j}$

$$\|\mathbf{u}\| = \sqrt{(20)^2 + (25)^2} = \sqrt{1025} = 5\sqrt{41}$$

13. $\mathbf{u} = 6\mathbf{j}$

$$\|\mathbf{u}\| = \sqrt{(0)^2 + (6)^2} = \sqrt{36} = 6$$

15. $\mathbf{u} = \langle 1, 0\rangle,\ \mathbf{v} = \langle 0, -2\rangle$

$$\cos\theta = \frac{\mathbf{u}\cdot\mathbf{v}}{\|\mathbf{u}\|\,\|\mathbf{v}\|} = \frac{0}{(1)(2)} = 0$$

$$\theta = 90^\circ$$

17. $\mathbf{u} = 3\mathbf{i} + 4\mathbf{j},\ \mathbf{v} = -2\mathbf{j}$

$$\cos\theta = \frac{\mathbf{u}\cdot\mathbf{v}}{\|\mathbf{u}\|\,\|\mathbf{v}\|} = -\frac{8}{(5)(2)}$$

$$\theta = \arccos\left(-\frac{4}{5}\right)$$

$$\theta \approx 143.13^\circ$$

19. $\mathbf{u} = 2\mathbf{i} - \mathbf{j},\ \mathbf{v} = 6\mathbf{i} + 4\mathbf{j}$

$$\cos\theta = \frac{\mathbf{u}\cdot\mathbf{v}}{\|\mathbf{u}\|\,\|\mathbf{v}\|} = \frac{8}{\sqrt{5}\sqrt{52}} \Rightarrow \theta \approx 60.26^\circ$$

21. $\mathbf{u} = 5\mathbf{i} + 5\mathbf{j},\ \mathbf{v} = -6\mathbf{i} + 6\mathbf{j}$

$$\cos\theta = \frac{\mathbf{u}\cdot\mathbf{v}}{\|\mathbf{u}\|\,\|\mathbf{v}\|} = 0 \Rightarrow \theta = 90^\circ$$

23. $\mathbf{u} = \left(\cos\frac{\pi}{3}\right)\mathbf{i} + \left(\sin\frac{\pi}{3}\right)\mathbf{j} = \frac{1}{2}\mathbf{i} + \frac{\sqrt{3}}{2}\mathbf{j}$

$$\mathbf{v} = \left(\cos\frac{3\pi}{4}\right)\mathbf{i} + \left(\sin\frac{3\pi}{4}\right)\mathbf{j} = -\frac{\sqrt{2}}{2}\mathbf{i} + \frac{\sqrt{2}}{2}\mathbf{j}$$

$$\|\mathbf{u}\| = \|\mathbf{v}\| = 1$$

$$\cos\theta = \frac{\mathbf{u}\cdot\mathbf{v}}{\|\mathbf{u}\|\,\|\mathbf{v}\|} = \mathbf{u}\cdot\mathbf{v} = \left(\frac{1}{2}\right)\left(-\frac{\sqrt{2}}{2}\right) + \left(\frac{\sqrt{3}}{2}\right)\left(\frac{\sqrt{2}}{2}\right) = \frac{-\sqrt{2} + \sqrt{6}}{4}$$

$$\theta = \arccos\left(\frac{-\sqrt{2} + \sqrt{6}}{4}\right) = 75^\circ = \frac{5\pi}{12}$$

25. $P = (1, 2),\ Q = (3, 4),\ R = (2, 5)$

$$\overrightarrow{PQ} = \langle 2, 2\rangle,\ \overrightarrow{PR} = \langle 1, 3\rangle,\ \overrightarrow{QR} = \langle -1, 1\rangle$$

$$\cos\alpha = \frac{\overrightarrow{PQ}\cdot\overrightarrow{PR}}{\|\overrightarrow{PQ}\|\,\|\overrightarrow{PR}\|} = \frac{8}{(2\sqrt{2})(\sqrt{10})} \Rightarrow \alpha = \arccos\frac{2}{\sqrt{5}} \approx 26.57^\circ$$

$$\cos\beta = \frac{\overrightarrow{PQ}\cdot\overrightarrow{QR}}{\|\overrightarrow{PQ}\|\,\|\overrightarrow{QR}\|} = 0 \Rightarrow \beta = 90^\circ.\ \text{Thus, } \gamma = 180^\circ - 26.57^\circ - 90^\circ = 63.43^\circ.$$

27. $P = (-3, 0), Q = (2, 2), R = (0, 6)$

$$\overrightarrow{QP} = \langle -5, -2\rangle, \overrightarrow{PR} = \langle 3, 6\rangle, \overrightarrow{QR} = \langle -2, 4\rangle$$

$$\cos\alpha = \frac{\overrightarrow{PQ}\cdot\overrightarrow{PR}}{\|\overrightarrow{PQ}\|\,\|\overrightarrow{PR}\|} = \frac{27}{\sqrt{29}\sqrt{45}} \Rightarrow \alpha \approx 41.63^\circ$$

$$\cos\beta = \frac{\overrightarrow{QP}\cdot\overrightarrow{QR}}{\|\overrightarrow{QP}\|\,\|\overrightarrow{PR}\|} = \frac{2}{\sqrt{29}\sqrt{20}} \Rightarrow \beta \approx 85.24^\circ$$

$$\delta = 180^\circ - 41.63^\circ - 85.24^\circ = 53.13^\circ$$

29. $\mathbf{u}\cdot\mathbf{v} = \|\mathbf{u}\|\,\|\mathbf{v}\|\cos\theta$

$$= (4)(10)\cos\frac{2\pi}{3}$$

$$= 40\left(-\frac{1}{2}\right)$$

$$= -20$$

31. $\mathbf{u} = \langle -12, 30\rangle,\ \mathbf{v} = \left\langle \frac{1}{2}, -\frac{5}{4}\right\rangle$

$\mathbf{u} = -24\mathbf{v} \Rightarrow$ **u** and **v** are parallel.

33. $\mathbf{u} = \frac{1}{4}(3\mathbf{i} - \mathbf{j}),\ \mathbf{v} = 5\mathbf{i} + 6\mathbf{j}$

$\mathbf{u} \neq k\mathbf{v} \Rightarrow$ Not parallel

$\mathbf{u} \cdot \mathbf{v} \neq 0 \Rightarrow$ Not orthogonal

Neither

35. $\mathbf{u} = 2\mathbf{i} - 2\mathbf{j},\ \mathbf{v} = -\mathbf{i} - \mathbf{j}$

$\mathbf{u} \cdot \mathbf{v} = 0 \Rightarrow$ **u** and **v** are orthogonal.

37. $\mathbf{u} = \langle 2, 2\rangle,\ \mathbf{v} = \langle 6, 1\rangle$

$$\mathbf{w}_1 = \text{proj}_{\mathbf{v}}\mathbf{u} = \left(\frac{\mathbf{u} \cdot \mathbf{v}}{\|\mathbf{v}\|^2}\right)\mathbf{v} = \frac{14}{37}\mathbf{v} = \frac{14}{37}\langle 6, 1\rangle = \frac{1}{37}\langle 84, 14\rangle$$

$$\mathbf{w}_2 = \mathbf{u} - \mathbf{w}_1 = \langle 2, 2\rangle - \frac{14}{37}\langle 6, 1\rangle = \left\langle -\frac{10}{37}, \frac{60}{37}\right\rangle = \frac{10}{37}\langle -1, 6\rangle = \frac{1}{37}\langle -10, 60\rangle$$

39. $\mathbf{u} = \langle 0, 3\rangle,\ \mathbf{v} = \langle 2, 15\rangle$

$$\mathbf{w}_1 = \text{proj}_{\mathbf{v}}\mathbf{u} = \left(\frac{\mathbf{u} \cdot \mathbf{v}}{\|\mathbf{v}\|^2}\right)\mathbf{v} = \frac{45}{229}\langle 2, 15\rangle$$

$$\mathbf{w}_2 = \mathbf{u} - \mathbf{w}_1 = \langle 0, 3\rangle - \frac{45}{229}\langle 2, 15\rangle = \left\langle -\frac{90}{229}, \frac{12}{229}\right\rangle = \frac{6}{229}\langle -15, 2\rangle$$

41. $\mathbf{u} = \langle 3, 5\rangle$

For **v** to be orthogonal to **u**, $\mathbf{u} \cdot \mathbf{v}$ must equal 0.

Two possibilities: $\langle -5, 3\rangle$ and $\langle 5, -3\rangle$

43. $\mathbf{u} = \frac{1}{2}\mathbf{i} - \frac{2}{3}\mathbf{j}$

For **u** and **v** to be orthogonal, $\mathbf{u} \cdot \mathbf{v}$ must equal 0.

Two possibilities: $\frac{2}{3}\mathbf{i} + \frac{1}{2}\mathbf{j}$ and $-\frac{2}{3}\mathbf{i} - \frac{1}{2}\mathbf{j}$

45. $\mathbf{w} = \|\text{proj}_{\overrightarrow{PQ}}\mathbf{v}\|\,\|\overrightarrow{PQ}\|$ where $\overrightarrow{PQ} = \langle 4, 7\rangle$ and $\mathbf{v} = \langle 1, 4\rangle$.

$$\text{proj}_{\overrightarrow{PQ}}\mathbf{v} = \left(\frac{\mathbf{v} \cdot \overrightarrow{PQ}}{\|\overrightarrow{PQ}\|^2}\right)\overrightarrow{PQ} = \left(\frac{32}{65}\right)\langle 4, 7\rangle$$

$$\mathbf{w} = \|\text{proj}_{\overrightarrow{PQ}}\mathbf{v}\|\,\|\overrightarrow{PQ}\| = \left(\frac{32\sqrt{65}}{65}\right)\left(\sqrt{65}\right) = 32$$

47. (a) $\mathbf{u} = \langle 1650, 3200\rangle,\ \mathbf{v} = \langle 15.25, 10.50\rangle$

$\mathbf{u} \cdot \mathbf{v} = 1650(15.25) + 3200(10.50) = \$58{,}762.50$

This gives the total revenue that can be earned by selling all of the units.

(b) Increase prices by 5%: $1.05\mathbf{v}$ The operation is scalar multiplication.

$$\begin{aligned}\mathbf{u} \cdot 1.05\mathbf{v} &= 1.05\mathbf{u} \cdot \mathbf{v}\\ &= 1.05[1650(15.25) + 3200(10.50)]\\ &= 1.05(58{,}762.50)\\ &= 61{,}700.63\end{aligned}$$

49. $\mathbf{w} = (245)(3) = 735$ newton-meters

51. $\mathbf{w} = (\cos 30°)(45)(20) \approx 779.4$ foot-pounds

53. False. Work is represented by a scalar.

55. (a) $\mathbf{u} \cdot \mathbf{v} = 0 \Rightarrow \mathbf{u}$ and $\mathbf{v}$ are orthogonal and $\theta = \dfrac{\pi}{2}$.

(b) $\mathbf{u} \cdot \mathbf{v} > 0 \Rightarrow \cos\theta > 0 \Rightarrow 0 \le \theta < \dfrac{\pi}{2}$

(c) $\mathbf{u} \cdot \mathbf{v} < 0 \Rightarrow \cos\theta < 0 \Rightarrow \dfrac{\pi}{2} < \theta \le \pi$

57. In a rhombus, $\|\mathbf{u}\| = \|\mathbf{v}\|$. The diagonals are $\mathbf{u} + \mathbf{v}$ and $\mathbf{u} - \mathbf{v}$.

$$\begin{aligned}(\mathbf{u}+\mathbf{v}) \cdot (\mathbf{u}-\mathbf{v}) &= (\mathbf{u}+\mathbf{v}) \cdot \mathbf{u} - (\mathbf{u}+\mathbf{v}) \cdot \mathbf{v} \\ &= \mathbf{u} \cdot \mathbf{u} + \mathbf{v} \cdot \mathbf{u} - \mathbf{u} \cdot \mathbf{v} - \mathbf{v} \cdot \mathbf{v} \\ &= \|\mathbf{u}\|^2 - \|\mathbf{v}\|^2 = 0\end{aligned}$$

Therefore, the diagonals are orthogonal.

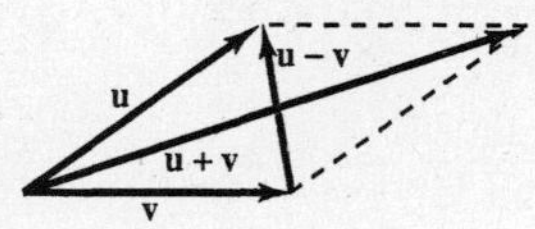

59. $$\begin{aligned}\sqrt{42} \cdot \sqrt{24} &= \sqrt{1008} \\ &= \sqrt{144 \cdot 7} \\ &= 12\sqrt{7}\end{aligned}$$

61. $$\begin{aligned}\sin 2x - \sqrt{3}\sin x &= 0 \\ 2\sin x\cos x - \sqrt{3}\sin x &= 0 \\ \sin x\left(2\cos x - \sqrt{3}\right) &= 0\end{aligned}$$

$\sin x = 0$ or $2\cos x - \sqrt{3} = 0$

$x = 0, \pi$ $\qquad \cos x = \dfrac{\sqrt{3}}{2}$

$x = \dfrac{\pi}{6}, \dfrac{11\pi}{6}$

63. $$\begin{aligned}2\tan x &= \tan 2x \\ 2\tan x &= \frac{2\tan x}{1-\tan^2 x} \\ 2\tan x(1-\tan^2 x) &= 2\tan x \\ 2\tan x(1-\tan^2 x) - 2\tan x &= 0 \\ 2\tan x[(1-\tan^2 x) - 1] &= 0 \\ 2\tan x(-\tan^2 x) &= 0 \\ -2\tan^3 x &= 0 \\ \tan x &= 0 \\ x &= 0, \pi\end{aligned}$$

For Exercises 67 and 69

$\sin u = -\frac{12}{13}$, u **in Quadrant IV** $\Rightarrow$ $\cos u = \frac{5}{13}$

$\cos v = \frac{24}{25}$, v **in Quadrant IV** $\Rightarrow$ $\sin v = -\frac{7}{25}$

65. $$\begin{aligned}\sin(u - v) &= \sin u\cos v - \cos u\sin v \\ &= \left(-\tfrac{12}{13}\right)\left(\tfrac{24}{25}\right) - \left(\tfrac{5}{13}\right)\left(-\tfrac{7}{25}\right) \\ &= -\tfrac{253}{325}\end{aligned}$$

67. $$\begin{aligned}\cos(v - u) &= \cos v\cos u + \sin v\sin u \\ &= \left(\tfrac{24}{25}\right)\left(\tfrac{5}{13}\right) + \left(-\tfrac{7}{25}\right)\left(-\tfrac{12}{13}\right) \\ &= \tfrac{204}{325}\end{aligned}$$

Review Exercises for Chapter 3

Solutions to Odd-Numbered Exercises

1. Given: $A = 35°, B = 71°, a = 8$

$C = 180° - 35° - 71° = 74°$

$$b = \frac{a \sin B}{\sin A} = \frac{8 \sin 71°}{\sin 35°} \approx 13.19$$

$$c = \frac{a \sin C}{\sin A} = \frac{8 \sin 74°}{\sin 35°} \approx 13.41$$

3. Given: $B = 72°, C = 82°, b = 54$

$A = 180° - 72° - 82° = 26°$

$$a = \frac{b \sin A}{\sin B} = \frac{54 \sin 26°}{\sin 72°} \approx 24.89$$

$$c = \frac{b \sin C}{\sin B} = \frac{54 \sin 82°}{\sin 72°} \approx 56.23$$

5. Given: $A = 16°, B = 98°, c = 8.4$

$C = 180° - 16° - 98° = 66°$

$$a = \frac{c \sin A}{\sin C} = \frac{8.4 \sin 16°}{\sin 66°} \approx 2.53$$

$$b = \frac{c \sin B}{\sin C} = \frac{8.4 \sin 98°}{\sin 66°} \approx 9.11$$

7. Given: $A = 24°, C = 48°, b = 27.5$

$B = 180° - 24° - 48° = 108°$

$$a = \frac{b \sin A}{\sin B} = \frac{27.5 \sin 24°}{\sin 108°} \approx 11.76$$

$$c = \frac{b \sin C}{\sin B} = \frac{27.5 \sin 48°}{\sin 108°} \approx 21.49$$

9. Given: $B = 150°, b = 30, c = 10$

$$\sin C = \frac{c \sin B}{b} = \frac{10 \sin 150°}{30} \approx 0.1667 \implies C \approx 9.59°$$

$A \approx 180° - 150° - 9.59° = 20.41°$

$$a = \frac{b \sin A}{\sin B} = \frac{30 \sin 20.41°}{\sin 150°} \approx 20.92$$

11. $A = 75°, a = 51.2, b = 33.7$

$$\sin B = \frac{b \sin A}{a} = \frac{33.7 \sin 75°}{51.2} \approx 0.6358 \implies B \approx 39.48°$$

$C \approx 180° - 75° - 39.48° = 65.52°$

$$c = \frac{a \sin C}{\sin A} = \frac{51.2 \sin 65.52°}{\sin 75°} \approx 48.24$$

13. $\text{Area} = \frac{1}{2}bc \sin A = \frac{1}{2}(5)(7)\sin 27° \approx 7.9$

15. $\text{Area} = \frac{1}{2}ab \sin C = \frac{1}{2}(16)(5)\sin 123° \approx 33.5$

17. $\tan 17° = \dfrac{h}{x + 50} \implies h = (x + 50)\tan 17°$

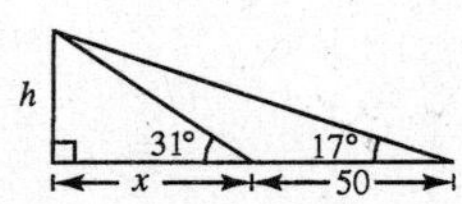

$$h = x \tan 17° + 50 \tan 17°$$

$\tan 31° = \dfrac{h}{x} \implies h = x \tan 31°$

$$x \tan 17° + 50 \tan 17° = x \tan 31°$$

$$50 \tan 17° = x(\tan 31° - \tan 17°)$$

$$\frac{50 \tan 17°}{\tan 31° - \tan 17°} = x$$

$$x \approx 51.7959$$

$$h = x \tan 31°$$

$$\approx 51.7959 \tan 31°$$

$$\approx 31.1 \text{ meters}$$

19.

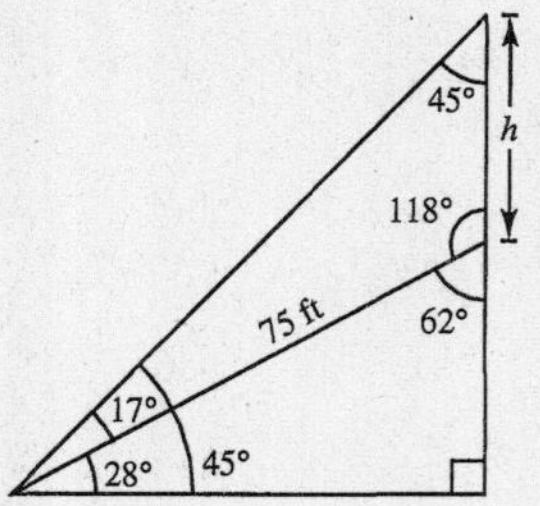

$$\frac{h}{\sin 17°} = \frac{75}{\sin 45°}$$

$$h = \frac{75 \sin 17°}{\sin 45°}$$

$$h \approx 31.01 \text{ feet}$$

21. Given: $a = 5, b = 8, c = 10$

$$\cos C = \frac{a^2 + b^2 - c^2}{2ab} = -0.1375 \implies C \approx 97.90°$$

$$\cos B = \frac{a^2 + c^2 - b^2}{2ac} = 0.61 \implies B \approx 52.41°$$

$$A = 180° - B - C \approx 29.69°$$

23. Given: $a = 2.5, b = 5.0, c = 4.5$

$$\cos B = \frac{a^2 + c^2 - b^2}{2ac} = 0.0667 \implies B \approx 86.18°$$

$$\cos C = \frac{a^2 + b^2 - c^2}{2ab} = 0.44 \implies C \approx 63.90°$$

$$A = 180° - B - C \approx 29.92°$$

25. Given: $B = 110°, a = 4, c = 4$

$$b = \sqrt{a^2 + c^2 - 2ac \cos B} \approx 6.55$$

$$A = C = \tfrac{1}{2}(180° - 110°) = 35°$$

27. Given: $C = 43°, a = 22.5, b = 31.4$

$$c = \sqrt{a^2 + b^2 - 2ab \cos C} \approx 21.42$$

$$\cos B = \frac{a^2 + c^2 - b^2}{2ac} \approx -0.02169 \implies B \approx 91.24°$$

$$A = 180° - B - C \approx 45.76°$$

29. Length of AC $= \sqrt{300^2 + 425^2 - 2(300)(425) \cos 115°}$

≈ 615.1 meters

31. $a = 4,\ b = 5,\ c = 7$

$$s = \frac{a + b + c}{2} = \frac{4 + 5 + 7}{2} = 8$$

$$\text{Area} = \sqrt{s(s - a)(s - b)(s - c)}$$

$$= \sqrt{8(4)(3)(1)} \approx 9.80$$

33. $a = 12.3, b = 15.8, c = 3.7$

$$s = \frac{a + b + c}{2} = \frac{12.3 + 15.8 + 3.7}{2} = 15.9$$

$$\text{Area} = \sqrt{s(s - a)(s - b)(s - c)}$$

$$= \sqrt{15.9(3.6)(0.1)(12.2)}$$

$$= 8.36$$

35.

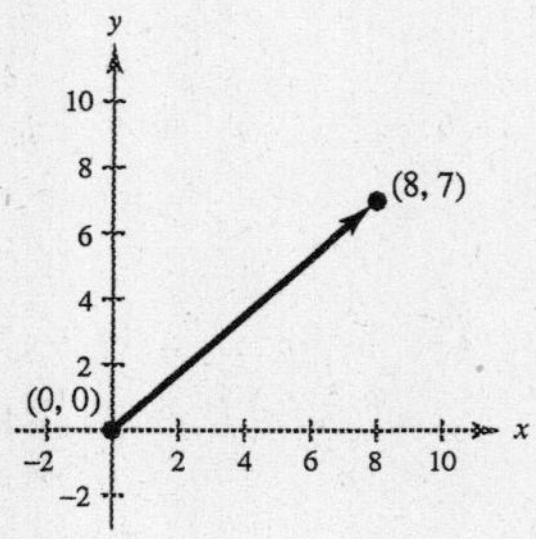

37.

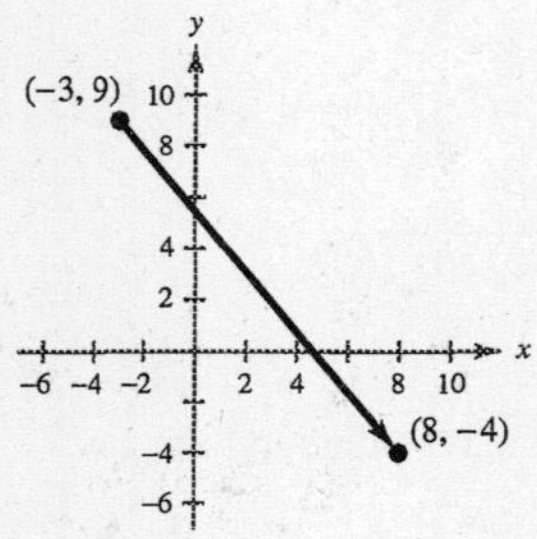

39. Initial point: $(-5, 4)$

Terminal point: $(2, -1)$

$\mathbf{v} = \langle 2 - (-5), -1 - 4 \rangle = \langle 7, -5 \rangle$

41. Initial point: $(0, 10)$

Terminal point: $(7, 3)$

$\mathbf{v} = \langle 7 - 0, 3 - 10 \rangle = \langle 7, -7 \rangle$

43. $\langle 8 \cos 120°, 8 \sin 120° \rangle = \left\langle -4, 4\sqrt{3} \right\rangle$

45. $\mathbf{u} = 6\mathbf{i} - 5\mathbf{j}, \mathbf{v} = 10\mathbf{i} + 3\mathbf{j}$

$$2\mathbf{u} + \mathbf{v} = 2(6\mathbf{i} - 5\mathbf{j}) + (10\mathbf{i} + 3\mathbf{j})$$
$$= 22\mathbf{i} - 7\mathbf{j}$$
$$= \langle 22, -7 \rangle$$

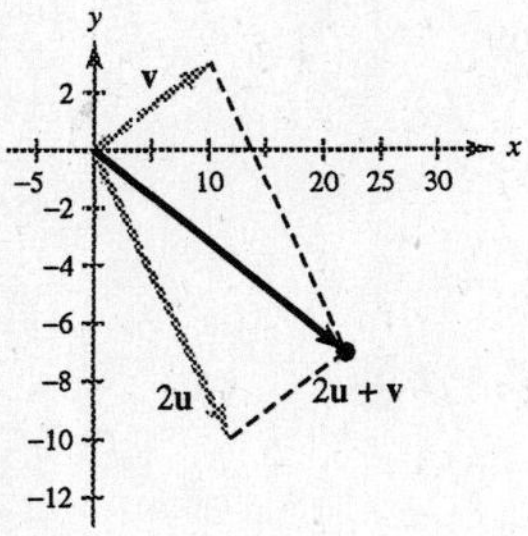

47. $\mathbf{v} = 10\mathbf{i} + 3\mathbf{j}$

$$3\mathbf{v} = 3(10\mathbf{i} + 3\mathbf{j}) = 30\mathbf{i} + 9\mathbf{j}$$
$$= \langle 30, 9 \rangle$$

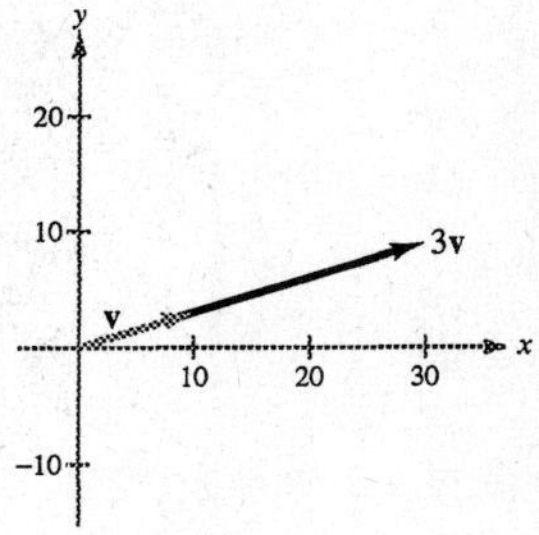

49. $\mathbf{u} = \langle -3, 4 \rangle = -3\mathbf{i} + 4\mathbf{j}$

51. Initial point: $(3, 4)$

Terminal point: $(9, 8)$

$\mathbf{u} = (9 - 3)\mathbf{i} + (8 - 4)\mathbf{j} = 6\mathbf{i} + 4\mathbf{j}$

53. $\mathbf{v} = -10\mathbf{i} + 10\mathbf{j}$

$\|\mathbf{v}\| = \sqrt{(-10)^2 + (10)^2} = \sqrt{200} = 10\sqrt{2}$

$\tan\theta = \dfrac{10}{-10} = -1 \Rightarrow \theta = 135°$ since

$\mathbf{v}$ is in Quadrant II.

$\mathbf{v} = 10\sqrt{2}(\mathbf{i}\cos 135° + \mathbf{j}\sin 135°)$

55. $\mathbf{v} = 7(\cos 60°\,\mathbf{i} + \sin 60°\,\mathbf{j})$

$\|\mathbf{v}\| = 7$

$\theta = 60°$

57. $\mathbf{v} = 5\mathbf{i} + 4\mathbf{j}$

$\|\mathbf{v}\| = \sqrt{5^2 + 4^2} = \sqrt{41}$

$\tan\theta = \dfrac{4}{5} \Rightarrow \theta \approx 38.7°$

59. $\mathbf{v} = -3\mathbf{i} - 3\mathbf{j}$

$\|\mathbf{v}\| = \sqrt{(-3)^2 + (-3)^2} = 3\sqrt{2}$

$\tan\theta = \dfrac{-3}{-3} = 1 \Rightarrow \theta = 225°$

61. Magnitude of resultant:

$$c = \sqrt{85^2 + 50^2 - 2(85)(50)\cos 165°}$$
$$\approx 133.92 \text{ pounds}$$

Let θ be the angle between the resultant and the 85-pound force

$$\cos\theta \approx \frac{(133.92)^2 + 85^2 - 50^2}{2(133.92)(85)}$$
$$\approx 0.9953$$
$$\Rightarrow \theta \approx 5.6°$$

63. Airspeed: $\mathbf{u} = 430(\cos 45°\mathbf{i} - \sin 45°\mathbf{j}) = 215\sqrt{2}(\mathbf{i} - \mathbf{j})$

Wind: $\mathbf{w} = 35(\cos 60° + \sin 60°\,\mathbf{j}) = \frac{35}{2}(\mathbf{i} + \sqrt{3}\mathbf{j})$

Groundspeed: $\mathbf{u} + \mathbf{w} = \left(215\sqrt{2} + \frac{35}{2}\right)\mathbf{i} + \left(\frac{35\sqrt{3}}{2} - 215\sqrt{2}\right)$

$$\|\mathbf{u} + \mathbf{w}\| = \sqrt{\left(215\sqrt{2} + \frac{35}{2}\right)^2 + \left(\frac{35\sqrt{3}}{2} - 215\sqrt{2}\right)^2}$$

≈ 422.30 miles per hour

Bearing: $\tan \theta' = \dfrac{17.5\sqrt{3} - 215\sqrt{2}}{215\sqrt{2} + 17.5}$

$\theta' \approx -40.4°$

$\theta = 90° + |\theta'| = 130.4°$

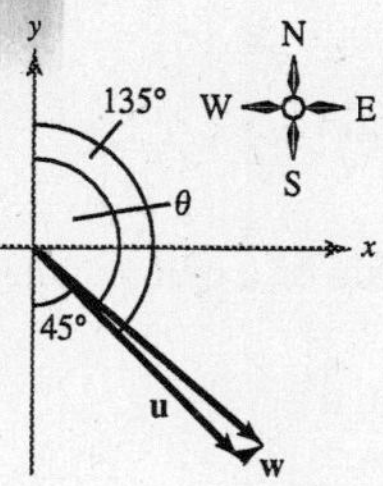

65. $\mathbf{u} = \langle 6, 7\rangle$

$\mathbf{v} = \langle -3, 9\rangle$

$\mathbf{u} \cdot \mathbf{v} = 6(-3) + 7(9) = 45$

67. $\mathbf{u} = 3\mathbf{i} + 7\mathbf{j}$

$\mathbf{v} = 11\mathbf{i} - 5\mathbf{j}$

$\mathbf{u} \cdot \mathbf{v} = 3(11) + 7(-5) = -2$

69. $\mathbf{u} = \langle -3, 4\rangle$

$2\mathbf{u} = \langle -6, 8\rangle$

$2\mathbf{u} \cdot \mathbf{u} = (-6)(-3) + 8(4) = 50$

The result is a scalar.

71. $\mathbf{u} = \langle -3, 4\rangle, \mathbf{v} = \langle 2, 1\rangle$

$\mathbf{u} \cdot \mathbf{v} = (-3)(2) + 4(1) = -2$

$\mathbf{u}(\mathbf{u} \cdot \mathbf{v}) = \mathbf{u}(-2) = -2\mathbf{u} = \langle 6, -8\rangle$

The result is a vector.

73. $\mathbf{u} = \cos\frac{7\pi}{4}\mathbf{i} + \sin\frac{7\pi}{4}\mathbf{j} = \left\langle \frac{1}{\sqrt{2}}, -\frac{1}{\sqrt{2}}\right\rangle$

$\mathbf{v} = \cos\frac{5\pi}{6}\mathbf{i} + \sin\frac{5\pi}{6}\mathbf{j} = \left\langle -\frac{\sqrt{3}}{2}, \frac{1}{2}\right\rangle$

$\cos\theta = \dfrac{\mathbf{u} \cdot \mathbf{v}}{\|\mathbf{u}\|\,\|\mathbf{v}\|} = \dfrac{-\sqrt{3} - 1}{2\sqrt{2}} \implies \theta = \dfrac{11\pi}{12}$

75. $\mathbf{u} = \langle 2\sqrt{2}, -4\rangle, \mathbf{v} = \langle -\sqrt{2}, 1\rangle$

$\cos\theta = \dfrac{\mathbf{u} \cdot \mathbf{v}}{\|\mathbf{u}\|\,\|\mathbf{v}\|} = \dfrac{-8}{(\sqrt{24})(\sqrt{3})} \implies \theta \approx 160.5°$

77. $\mathbf{u} = \langle -3, 8\rangle$

$\mathbf{v} = \langle 8, 3\rangle$

$\mathbf{u} \cdot \mathbf{v} = -3(8) + 8(3) = 0$

$\mathbf{u}$ and $\mathbf{v}$ are orthogonal.

79. $\mathbf{u} = -\mathbf{i}$

$\mathbf{v} = \mathbf{i} + 2\mathbf{j}$

$\mathbf{u} \cdot \mathbf{v} \neq 0 \implies$ Not orthogonal

$\mathbf{v} \neq k\mathbf{u} \implies$ Not parallel

Neither

81. $\mathbf{u} = \langle -4, 3\rangle, \ \mathbf{v} = \langle -8, -2\rangle$

$\mathbf{w}_1 = \text{proj}_{\mathbf{v}}\mathbf{u} = \left(\dfrac{\mathbf{u} \cdot \mathbf{v}}{\|\mathbf{v}\|^2}\right)\mathbf{v} = \left(\dfrac{26}{68}\right)\langle -8, -2\rangle$

$= -\dfrac{13}{17}\langle 4, 1\rangle$

$\mathbf{w}_2 = \mathbf{u} - \mathbf{w}_1 = \langle -4, 3\rangle - \left(-\dfrac{13}{17}\right)\langle 4, 1\rangle$

$= \dfrac{16}{17}\langle -1, 4\rangle$

83. $\mathbf{u} = \langle 2, 7\rangle, \ \mathbf{v} = \langle 1, -1\rangle$

$\mathbf{w}_1 = \text{proj}_{\mathbf{v}}\mathbf{u} = \left(\dfrac{\mathbf{u} \cdot \mathbf{v}}{\|\mathbf{v}\|^2}\right)\mathbf{v} = -\dfrac{5}{2}\langle 1, -1\rangle$

$= \dfrac{5}{2}\langle -1, 1\rangle$

$\mathbf{w}_2 = \mathbf{u} - \mathbf{w}_1 = \langle 2, 7\rangle - \left(\dfrac{5}{2}\right)\langle -1, 1\rangle$

$= \dfrac{9}{2}\langle 1, 1\rangle$

85. $P = (5, 3), Q = (8, 9) \Rightarrow \overrightarrow{PQ} = \langle 3, 6 \rangle$

$W = \mathbf{v} \cdot \overrightarrow{PQ} = \langle 2, 7 \rangle \cdot \langle 3, 6 \rangle = 48$

87. True. sin 90° is defined in the Law of Sines.

89. True, by the definition of a unit vector.

$\mathbf{u} = \dfrac{\mathbf{v}}{\|\mathbf{v}\|}$ so $\mathbf{v} = \|\mathbf{v}\|\mathbf{u}$

91. $\dfrac{a}{\sin A} = \dfrac{b}{\sin B} = \dfrac{c}{\sin C}$ or $\dfrac{\sin A}{a} = \dfrac{\sin B}{b} = \dfrac{\sin C}{c}$

93. A vector in the plane has both a magnitude and a direction.

95. $\|\mathbf{u} + \mathbf{v}\|$ is larger in figure (a) since the angle between **u** and **v** is acute rather than obtuse.

97. The sum of **u** and **v** lies on the diagonal of the parallelogram with **u** and **v** as its adjacent sides.

Problem Solving for Chapter 3

Solutions to Odd-Numbered Exercises

1. $\overrightarrow{PQ}^2 = 4.7^2 + 6^2 - 2(4.7)(6)\cos 25°$

$\overrightarrow{PQ} \approx 2.6409$ feet

$\dfrac{\sin \alpha}{4.7} = \dfrac{\sin 25°}{2.6409} \Rightarrow \alpha \approx 48.78°$

$\theta + \beta = 180° - 25° - 48.78° = 106.22°$

$(\theta + \beta) + \theta = 180° \Rightarrow \theta = 180° - 106.22° = 73.78°$

$\beta = 106.22° - 73.78° = 32.44°$

$\gamma = 180° - \alpha - \beta = 180° - 48.78° - 32.44° = 98.78°$

$\phi = 180° - \gamma = 180° - 98.78° = 81.22°$

$\dfrac{\overrightarrow{PT}}{\sin 25°} = \dfrac{4.7}{\sin 81.22°}$

$\overrightarrow{PT} \approx 2.01$ feet

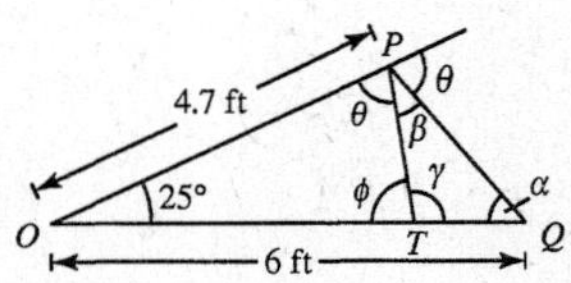

3. (a)

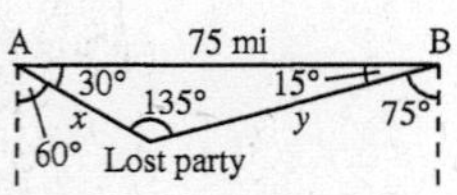

$\dfrac{x}{\sin 15°} = \dfrac{75}{\sin 135°}$ and $\dfrac{y}{\sin 30°} = \dfrac{75}{\sin 135°}$

$x \approx 27.45$ miles $\qquad y \approx 53.03$ miles

(b)

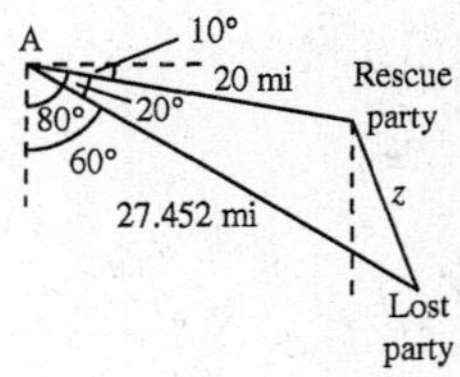

$z^2 = (27.45)^2 + (20)^2 - 2(27.45)(20)\cos 20°$

$z \approx 11.03$ miles

$\dfrac{\sin \theta}{27.45} = \dfrac{\sin 20°}{11.03}$

$\sin \theta \approx 0.8511$

$\theta = 180° - \sin^{-1}(0.8511)$

$\theta \approx 121.7°$

To find the bearing, we have $\theta - 10° - 90° \approx 21.7°$

Bearing: S 21.7° E

5. If $\mathbf{u} \neq 0$, $\mathbf{v} \neq 0$, and $\mathbf{u} + \mathbf{v} \neq 0$, then $\left\|\frac{\mathbf{u}}{\|\mathbf{u}\|}\right\| = \left\|\frac{\mathbf{v}}{\|\mathbf{v}\|}\right\| = \left\|\frac{\mathbf{u}+\mathbf{v}}{\|\mathbf{u}+\mathbf{v}\|}\right\| = 1$ since all of these are magnitudes of **unit** vectors.

(a) $\mathbf{u} = \langle 1, -1\rangle, \quad \mathbf{v} = \langle -1, 2\rangle, \quad \mathbf{u} + \mathbf{v} = \langle 0, 1\rangle$

$\|\mathbf{u}\| = \sqrt{2}, \quad \|\mathbf{v}\| = \sqrt{5}, \quad \|\mathbf{u} + \mathbf{v}\| = 1$

(b) $\mathbf{u} = \langle 0, 1\rangle, \quad \mathbf{v} = \langle 3, -3\rangle, \quad \mathbf{u} + \mathbf{v} = \langle 3, -2\rangle$

$\|\mathbf{u}\| = 1, \quad \|\mathbf{v}\| = \sqrt{18} = 3\sqrt{2}, \|\mathbf{u} + \mathbf{v}\| = \sqrt{13}$

(c) $\mathbf{u} = \left\langle 1, \frac{1}{2}\right\rangle, \quad \mathbf{v} = \langle 2, 3\rangle, \ \mathbf{u} + \mathbf{v} = \left\langle 3, \frac{7}{2}\right\rangle$

$\|\mathbf{u}\| = \frac{\sqrt{5}}{2}, \quad \|\mathbf{v}\| = \sqrt{13}, \|\mathbf{u} + \mathbf{v}\| = \sqrt{9 + \frac{49}{4}} = \frac{\sqrt{85}}{2}$

(d) $\mathbf{u} = \langle 2, -4\rangle, \quad \mathbf{v} = \langle 5, 5\rangle, \quad \mathbf{u} + \mathbf{v} = \langle 7, 1\rangle$

$\|\mathbf{u}\| = \sqrt{20} = 2\sqrt{5}, \|\mathbf{v}\| = \sqrt{50} = 5\sqrt{2}, \|\mathbf{u} + \mathbf{v}\| = \sqrt{50} = 5\sqrt{2}$

7. Initial point: $(0, 0)$

Terminal point: $\left(\frac{\mathbf{u}_1 + \mathbf{v}_1}{2}, \frac{\mathbf{u}_2 + \mathbf{v}_2}{2}\right)$

$$\mathrm{w} = \left\langle \frac{\mathbf{u}_1 + \mathbf{v}_1}{2}, \frac{\mathbf{u}_2 + \mathbf{v}_2}{2}\right\rangle = \frac{1}{2}(\mathbf{u} + \mathbf{v})$$

Initial point: $(\mathbf{u}_1, \mathbf{u}_2)$

Terminal point: $\frac{1}{2}(\mathbf{u}_1 + \mathbf{v}_1, \mathbf{u}_2 + \mathbf{v}_2)$

$$\mathrm{w} = \left\langle \frac{\mathbf{u}_1 + \mathbf{v}_1}{2} - \mathbf{u}_1, \frac{\mathbf{u}_2 + \mathbf{v}_2}{2} - \mathbf{u}_2\right\rangle$$

$$= \left\langle \frac{\mathbf{v}_1 - \mathbf{u}_1}{2}, \frac{\mathbf{v}_2 - \mathbf{u}_2}{2}\right\rangle = \frac{1}{2}(\mathbf{v} - \mathbf{u})$$

9. $W = (\cos\theta)\|F\|\,\|\overrightarrow{PQ}\|$ and $\|F_1\| = \|F_2\|$

(a)

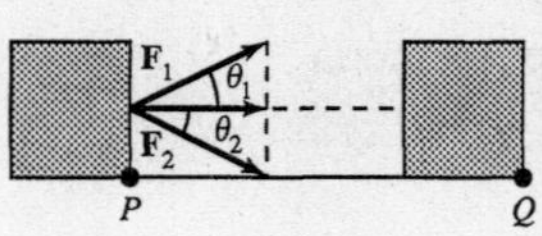

If $\theta_1 = -\theta_2$ then the work is the same since $\cos(-\theta) = \cos\theta$.

(b)

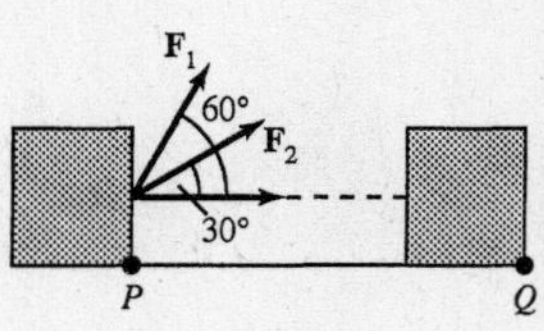

If $\theta_1 = 60°$ then $W_1 = \frac{1}{2}\|F_1\|\,\|\overrightarrow{PQ}\|$

If $\theta_2 = 30°$ then $W_2 = \frac{\sqrt{3}}{2}\|F_2\|\,\|\overrightarrow{PQ}\|$

$W_2 = \sqrt{3}\,W_1$

The amount of work done by F_2 is $\sqrt{3}$ times as great as the amount of work done by F_1.

Chapter 3 Practice Test

For Exercises 1 and 2, use the Law of Sines to find the remaining sides and angles of the triangle.

1. $A = 40°,\ B = 12°,\ b = 100$

2. $C = 150°,\ a = 5,\ c = 20$

3. Find the area of the triangle: $a = 3,\ b = 6,\ C = 130°$.

4. Determine the number of solutions to the triangle: $a = 10,\ b = 35,\ A = 22.5°$.

For Exercises 5 and 6, use the Law of Cosines to find the remaining sides and angles of the triangle.

5. $a = 49,\ b = 53,\ c = 38$

6. $C = 29°,\ a = 100,\ b = 300$

7. Use Heron's Formula to find the area of the triangle: $a = 4.1,\ b = 6.8,\ c = 5.5$.

8. A ship travels 40 miles due east, then adjusts its course 12° southward. After traveling 70 miles in that direction, how far is the ship from its point of departure?

9. $\mathbf{w} = 4\mathbf{u} - 7\mathbf{v}$ where $\mathbf{u} = 3\mathbf{i} + \mathbf{j}$ and $\mathbf{v} = -\mathbf{i} + 2\mathbf{j}$. Find $\mathbf{w}$.

10. Find a unit vector in the direction of $\mathbf{v} = 5\mathbf{i} - 3\mathbf{j}$.

11. Find the dot product and the angle between $\mathbf{u} = 6\mathbf{i} + 5\mathbf{j}$ and $\mathbf{v} = 2\mathbf{i} - 3\mathbf{j}$.

12. $\mathbf{v}$ is a vector of magnitude 4 making an angle of 30° with the positive x-axis. Find $\mathbf{v}$ in component form.

13. Find the projection of $\mathbf{u}$ onto $\mathbf{v}$ given $\mathbf{u} = \langle 3, -1\rangle$ and $\mathbf{v} = \langle -2, 4\rangle$.

14. Given $\|\mathbf{u}\| = 7,\ \theta\mathbf{u} = 35°$

$\|\mathbf{v}\| = 4,\ \theta\mathbf{v} = 123°$

Find the component form of $\mathbf{u} + \mathbf{v}$.

15. Find two vectors orthogonal to $\langle -3, 10\rangle$.

CHAPTER 4
Complex Numbers

CHAPTER 4
Complex Numbers

Section 4.1 Complex Numbers

- Standard form: $a + bi$.

  If $b = 0$, then $a + bi$ is a real number.

  If $a = 0$ and $b \neq 0$, then $a + bi$ is a pure imaginary number.
- Equality of Complex Numbers: $a + bi = c + di$ if and only if $a = c$ and $b = d$
- Operations on complex numbers

  (a) Addition: $(a + bi) + (c + di) = (a + c) + (b + d)i$

  (b) Subtraction: $(a + bi) - (c + di) = (a - c) + (b - d)i$

  (c) Multiplication: $(a + bi)(c + di) = (ac - bd) + (ad + bc)i$

  (d) Division: $\dfrac{a + bi}{c + di} = \dfrac{a + bi}{c + di} \cdot \dfrac{c - di}{c - di} = \dfrac{ac + bd}{c^2 + d^2} + \dfrac{bc - ad}{c^2 + d^2}i$
- The complex conjugate of $a + bi$ is $a - bi$:

  $(a + bi)(a - bi) = a^2 + b^2$
- The additive inverse of $a + bi$ is $-a - bi$.
- $\sqrt{-a} = \sqrt{a}\,i$ for $a > 0$.

Solutions to Odd-Numbered Exercises

1. $a + bi = -10 + 6i$

$a = -10$

$b = 6$

3. $(a - 1) + (b + 3)i = 5 + 8i$

$a - 1 = 5 \Rightarrow a = 6$

$b + 3 = 8 \Rightarrow b = 5$

5. $4 + \sqrt{-9} = 4 + 3i$

7. $2 - \sqrt{-27} = 2 - \sqrt{27}\,i = 2 - 3\sqrt{3}\,i$

9. $\sqrt{-75} = \sqrt{75}\,i = 5\sqrt{3}\,i$

11. $8 = 8 + 0i = 8$

13. $-6i + i^2 = -6i - 1 = -1 - 6i$

15. $\sqrt{-0.09} = \sqrt{0.09}\,i = 0.3i$

17. $(5 + i) + (6 - 2i) = 11 - i$

19. $(8 - i) - (4 - i) = 8 - i - 4 + i = 4$

21. $\left(-2 + \sqrt{-8}\right) + \left(5 - \sqrt{-50}\right) = -2 + 2\sqrt{2}i + 5 - 5\sqrt{2}i = 3 - 3\sqrt{2}i$

23. $13i - (14 - 7i) = 13i - 14 + 7i = -14 + 20i$

25. $-\left(\frac{3}{2}+\frac{5}{2}i\right)+\left(\frac{5}{3}+\frac{11}{3}i\right)=-\frac{3}{2}-\frac{5}{2}i+\frac{5}{3}+\frac{11}{3}i$
$=-\frac{9}{6}-\frac{15}{6}i+\frac{10}{6}+\frac{22}{6}i$
$=\frac{1}{6}+\frac{7}{6}i$

27. $\sqrt{-6}\cdot\sqrt{-2}=\left(\sqrt{6}i\right)\left(\sqrt{2}i\right)=\sqrt{12}i^2=\left(2\sqrt{3}\right)(-1)$
$=-2\sqrt{3}$

29. $\left(\sqrt{-10}\right)^2=\left(\sqrt{10}i\right)^2=10i^2=-10$

31. $(1+i)(3-2i)=3-2i+3i-2i^2=3+i+2=5+i$

33. $6i(5-2i)=30i-12i^2=30i+12=12+30i$

35. $\left(\sqrt{14}+\sqrt{10}i\right)\left(\sqrt{14}-\sqrt{10}i\right)=14-10i^2=14+10=24$

37. $(4+5i)^2=16+40i+25i^2$
$=16+40i-25$
$=-9+40i$

39. $(2+3i)^2+(2-3i)^2=4+12i+9i^2+4-12i+9i^2$
$=4+12i-9+4-12i-9$
$=-10$

41. The complex conjugate of $6+3i$ is $6-3i$.
$(6+3i)(6-3i)=36-(3i)^2=36+9=45$

43. The complex conjugate of $-1-\sqrt{5}i$ is $-1+\sqrt{5}i$.
$\left(-1-\sqrt{5}i\right)\left(-1+\sqrt{5}i\right)=(-1)^2-\left(\sqrt{5}i\right)^2$
$=1+5=6$

45. The complex conjugate of $\sqrt{-20}=2\sqrt{5}i$ is $-2\sqrt{5}i$.
$\left(2\sqrt{5}i\right)\left(-2\sqrt{5}i\right)=-20i^2=20$

47. The complex conjugate of $\sqrt{8}$ is $\sqrt{8}$.
$\left(\sqrt{8}\right)\left(\sqrt{8}\right)=8$

49. $\frac{5}{i}=\frac{5}{i}\cdot\frac{-i}{-i}=\frac{-5i}{1}=-5i$

51. $\frac{2}{4-5i}=\frac{2}{4-5i}\cdot\frac{4+5i}{4+5i}=\frac{2(4+5i)}{16+25}=\frac{8+10i}{41}=\frac{8}{41}+\frac{10}{41}i$

53. $\frac{3+i}{3-i}=\frac{3+i}{3-i}\cdot\frac{3+i}{3+i}$
$=\frac{9+6i+i^2}{9+1}$
$=\frac{8+6i}{10}$
$=\frac{4}{5}+\frac{3}{5}i$

55. $\frac{6-5i}{i}=\frac{6-5i}{i}\cdot\frac{-i}{-i}$
$=\frac{-6i+5i^2}{1}$
$=-5-6i$

57. $\frac{3i}{(4-5i)^2}=\frac{3i}{16-40i+25i^2}=\frac{3i}{-9-40i}\cdot\frac{-9+40i}{-9+40i}$
$=\frac{-27i+120i^2}{81+1600}=\frac{-120-27i}{1681}$
$=-\frac{120}{1681}-\frac{27}{1681}i$

59. $\frac{2}{1+i}-\frac{3}{1-i}=\frac{2(1-i)-3(1+i)}{(1+i)(1-i)}$
$=\frac{2-2i-3-3i}{1+1}$
$=\frac{-1-5i}{2}$
$=-\frac{1}{2}-\frac{5}{2}i$

61. $\dfrac{i}{3-2i}+\dfrac{2i}{3+8i}=\dfrac{i(3+8i)+2i(3-2i)}{(3-2i)(3+8i)}$

$=\dfrac{3i+8i^2+6i-4i^2}{9+24i-6i-16i^2}$

$=\dfrac{4i^2+9i}{9+18i+16}$

$=\dfrac{-4+9i}{25+18i}\cdot\dfrac{25-18i}{25-18i}$

$=\dfrac{-100+72i+225i-162i^2}{625+324}$

$=\dfrac{-100+297i+162}{949}$

$=\dfrac{62+297i}{949}$

$=\dfrac{62}{949}+\dfrac{297}{949}i$

63. $x^2-2x+2=0;\ a=1,\ b=-2,\ c=2$

$x=\dfrac{-(-2)\pm\sqrt{(-2)^2-4(1)(2)}}{2(1)}$

$=\dfrac{2\pm\sqrt{-4}}{2}$

$=\dfrac{2\pm 2i}{2}$

$=1\pm i$

65. $4x^2+16x+17=0;\ a=4,\ b=16,\ c=17$

$x=\dfrac{-16\pm\sqrt{(16)^2-4(4)(17)}}{2(4)}=\dfrac{-16\pm\sqrt{-16}}{8}=\dfrac{-16\pm 4i}{8}=-2\pm\dfrac{1}{2}i$

67. $4x^2+16x+15=0;\ a=4,\ b=16,\ c=15$

$x=\dfrac{-16\pm\sqrt{(16)^2-4(4)(15)}}{2(4)}=\dfrac{-16\pm\sqrt{16}}{8}=\dfrac{-16\pm 4}{8}$

$x=-\dfrac{12}{8}=-\dfrac{3}{2}$ or $x=-\dfrac{20}{8}=-\dfrac{5}{2}$

69. $\dfrac{3}{2}x^2-6x+9=0$ Multiply both sides by 2.

$3x^2-12x+18=0$

$x=\dfrac{-(-12)\pm\sqrt{(-12)^2-4(3)(18)}}{2(3)}=\dfrac{12\pm\sqrt{-72}}{6}=\dfrac{12\pm 6\sqrt{2}i}{6}=2\pm\sqrt{2}i$

71. $1.4x^2-2x-10=0$ Multiply both sides by 5.

$7x^2-10x-50=0$

$x=\dfrac{-(-10)\pm\sqrt{(-10)^2-4(7)(-50)}}{2(7)}=\dfrac{10\pm\sqrt{1500}}{14}=\dfrac{10\pm 10\sqrt{15}}{14}=\dfrac{5\pm 5\sqrt{15}}{7}=\dfrac{5}{7}\pm\dfrac{5\sqrt{15}}{7}$

73. $-6i^3+i^2=-6i^2i+i^2$

$=-6(-1)i+(-1)$

$=6i-1$

$=-1+6i$

75. $-5i^5=-5i^2i^2i$

$=-5(-1)(-1)i$

$=-5i$

77. $\left(\sqrt{-75}\right)^3 = \left(5\sqrt{3}\,i\right)^3$

$= 5^3\left(\sqrt{3}\right)^3 i^3$

$= 125\left(3\sqrt{3}\right)(-1)i$

$= -375\sqrt{3}\,i$

79. $\dfrac{1}{i^3} = \dfrac{1}{-i} = \dfrac{1}{-i}\cdot\dfrac{i}{i} = \dfrac{i}{-i^2} = \dfrac{i}{1} = i$

81. (a) $(2)^3 = 8$

(b) $(-1 + \sqrt{3}i)^3 = (-1)^3 + 3(-1)^2(\sqrt{3}i) + 3(-1)(\sqrt{3}i)^2 + (\sqrt{3}i)^3$

$= -1 + 3\sqrt{3}i - 9i^2 + 3\sqrt{3}\,i^3$

$= -1 + 3\sqrt{3}\,i + 9 - 3\sqrt{3}i$

$= 8$

(c) $\left(-1 - \sqrt{3}i\right)^3 = (-1)^3 + 3(-1)^2\left(-\sqrt{3}\,i\right) + 3(-1)\left(-\sqrt{3}\,i\right)^2 + \left(-\sqrt{3}\,i\right)^3$

$= -1 - 3\sqrt{3}i - 9i^2 - 3\sqrt{3}\,i^3$

$= -1 - 3\sqrt{3}\,i + 9 + 3\sqrt{3}\,i$

$= 8$

83. (a) $i^{40} = i^4 \cdot i^4 \cdot i^4 \cdot i^4 \cdot i^4 \cdot i^4 \cdot i^4 \cdot i^4 \cdot i^4 \cdot i^4$

$= 1 \cdot 1 \cdot 1 \cdot 1 \cdot 1 \cdot 1 \cdot 1 \cdot 1 \cdot 1 \cdot 1$

$= 1$

(b) $i^{25} = i^4 \cdot i^4 \cdot i^4 \cdot i^4 \cdot i^4 \cdot i^4 \cdot i$

$= 1 \cdot 1 \cdot 1 \cdot 1 \cdot 1 \cdot 1 \cdot i$

$= i$

(c) $i^{50} = i^{25} \cdot i^{25} = i \cdot i = i^2 = -1$

(d) $i^{67} = i^{50} \cdot i^{17} = -1 \cdot i^4 \cdot i^4 \cdot i^4 \cdot i^4 \cdot i = -i$

85. False, if $b = 0$ then $a + bi = a - bi = a$.

That is, if the complex number is real, the number equals its conjugate.

87. False.

$i^{44} + i^{150} - i^{74} - i^{109} + i^{61} = (i^4)^{11} + (i^4)^{37}(i^2) - (i^4)^{18}(i^2) - (i^4)^{27}(i) + (i^4)^{15}(i)$

$= (1)^{11} + (1)^{37}(-1) - (1)^{18}(-1) - (1)^{27}(i) + (1)^{15}(i)$

$= 1 + (-1) + 1 - i + i = 1$

89. $(a_1 + b_1i)(a_2 + b_2i) = a_1a_2 + a_1b_2i + a_2b_1i + b_1b_2i^2$

$= (a_1a_2 - b_1b_2) + (a_1b_2 + a_2b_1)i$

The conjugate of this product is $(a_1a_2 - b_1b_2) - (a_1b_2 + a_2b_1)i$.

The product of the conjugates is:

$(a_1 - b_1i)(a_2 - b_2i) = a_1a_2 - a_1b_2i - a_2b_1i + b_1b_2i^2$

$= (a_1a_2 - b_1b_2) - (a_1b_2 + a_2b_1)i$

Thus, the conjugate of the product of two complex numbers is the product of their conjugates.

91. $(4 + 3x) + (8 - 6x - x^2) = -x^2 - 3x + 12$

93. $\left(3x - \frac{1}{2}\right)(x + 4) = 3x^2 + 12x - \frac{1}{2}x - 2 = 3x^2 + \frac{23}{2}x - 2$

95. $-x - 12 = 19$

$-x = 31$

$x = -31$

97. $4(5x - 6) - 3(6x + 1) = 0$

$20x - 24 - 18x - 3 = 0$

$2x - 27 = 0$

$2x = 27$

$x = \frac{27}{2}$

99. $V = \frac{4}{3}\pi a^2 b$

$3V = 4\pi a^2 b$

$\frac{3V}{4\pi b} = a^2$

$\sqrt{\frac{3V}{4\pi b}} = a$

$a = \frac{1}{2}\sqrt{\frac{3V}{\pi b}} = \frac{\sqrt{3V\pi b}}{2\pi b}$

101. Let x = # liters withdrawn and replaced.

$0.50(5 - x) + 1.00x = 0.60(5)$

$2.50 - 0.50x + 1.00x = 3.00$

$0.50x = 0.50$

$x = 1$ liter

Section 4.2 Complex Solutions of Equations

- If f is a polynomial with real coefficients of degree $n > 0$, then f has exactly n solutions (zeros, roots) in the complex number system.
- Given the quadratic equation, $ax^2 + bx + c = 0$, the discriminant can be used to determine the types of solutions.
  - If $b^2 - 4ac < 0$, then both solutions are complex.
  - If $b^2 - 4ac = 0$, then there is one repeating real solution.
  - If $b^2 - 4ac > 0$, then both solutions are real.
- If $a + bi$ is a complex solution to a polynomial with real coefficients, then so is $a - bi$.

Solutions to Odd-Numbered Exercises

1. $2x^3 + 3x + 1 = 0$ has degree three so there are three solutions in the complex number system.

3. $50 - 2x^4 = 0$ has degree four so there are four solutions in the complex number system.

5. $2x^2 - 5x + 5 = 0$

$b^2 - 4ac = (-5)^2 - 4(2)(5) = -15 < 0$

Both solutions are complex. There are no real solutions.

7. $\frac{1}{5}x^2 + \frac{6}{5}x - 8 = 0$

$b^2 - 4ac = \left(\frac{6}{5}\right)^2 - 4\left(\frac{1}{5}\right)(-8) = \frac{196}{25} > 0$

Both solutions are real.

9. $2x^2 - x - 15 = 0$

$b^2 - 4ac = (-1)^2 - 4(2)(-15) = 121 > 0$

Both solutions are real.

11. $x^2 + 2x + 10 = 0$

$b^2 - 4ac = (2)^2 - 4(1)(10) = -36 < 0$

Both solutions are complex. There are no real solutions.

13. $x^2 - 5 = 0$

$x^2 = 5$

$x = \pm\sqrt{5}$

15. $(x + 5)^2 - 6 = 0$

$(x + 5)^2 = 6$

$x + 5 = \pm\sqrt{6}$

$x = -5 \pm \sqrt{6}$

17. $x^2 - 8x + 16 = 0$

$(x - 4)^2 = 0$

$x = 4$

19. $x^2 + 2x + 5 = 0$

$$x = \frac{-2 \pm \sqrt{2^2 - 4(1)(5)}}{2(1)} = \frac{-2 \pm \sqrt{-16}}{2} = \frac{-2 \pm 4i}{2} = -1 \pm 2i$$

21. $4x^2 - 4x + 5 = 0$

$$x = \frac{-(-4) \pm \sqrt{(-4)^2 - 4(4)(5)}}{2(4)} = \frac{4 \pm \sqrt{-64}}{8} = \frac{4 \pm 8i}{8} = \frac{1}{2} \pm i$$

23. $230 + 20x - 0.5x^2 = 0$

$$x = \frac{-20 \pm \sqrt{(20)^2 - 4(-0.5)(230)}}{2(-0.5)} = \frac{-20 \pm \sqrt{860}}{-1} = 20 \pm 2\sqrt{215}$$

25. $8 + (x + 3)^2 = 0$

$(x + 3)^2 = -8$

$x + 3 = \pm\sqrt{-8}$

$x = -3 \pm 2\sqrt{2}i$

27. $f(x) = x^3 - 4x^2 + x - 4$

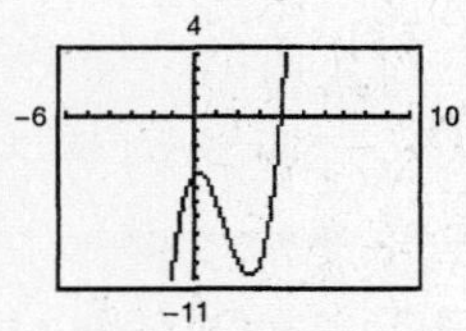

$x^3 - 4x^2 + x - 4 = 0$

$x^2(x - 4) + 1(x - 4) = 0$

$(x^2 + 1)(x - 4) = 0$

$x^2 + 1 = 0 \Rightarrow x = \pm i$

$x - 4 = 0 \Rightarrow x = 4$

Zeros: $x = \pm i, 4$

The graph has one x-intercept and the function has one real zero. The number of real zeros equals the number of x-intercepts. Each x-intercept represents a real solution of the equation $f(x) = 0$.

29. $f(x) = x^4 + 4x^2 + 4$

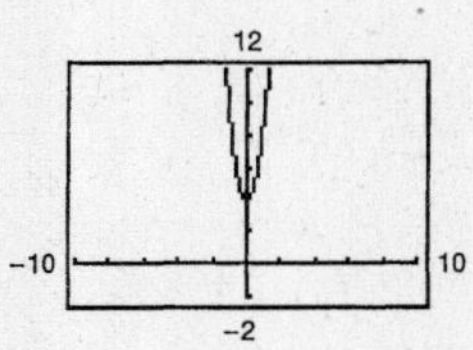

$x^4 + 4x^2 + 4 = 0$

$(x^2 + 2)^2 = 0$

$x^2 + 2 = 0 \Rightarrow x = \pm\sqrt{2}i$

The graph has no x-intercepts and the function has no real zeros. The number of real zeros equals the number of x-intercepts. Each x-intercept represents a real solution of the equation $f(x) = 0$.

31. $f(x) = x^2 + 25$

$= (x + 5i)(x - 5i)$

The zeros of $f(x)$ are $x = \pm 5i$.

33. $h(x) = x^2 - 4x + 1$

h has no rational zeros.

By the Quadratic Formula, the zeros are $x = \dfrac{4 \pm \sqrt{16 - 4}}{2} = 2 \pm \sqrt{3}$.

$h(x) = [x - (2 + \sqrt{3})][x - (2 - \sqrt{3})] = (x - 2 - \sqrt{3})(x - 2 + \sqrt{3})$

35. $f(x) = x^4 - 81$

$= (x^2 - 9)(x^2 + 9)$

$= (x + 3)(x - 3)(x + 3i)(x - 3i)$

The zeros of $f(x)$ are $x = \pm 3$ and $x = \pm 3i$.

37. $f(z) = z^2 - 2z + 2$

$z^2 - 2z + 2 = 0$

$z^2 - 2z = -2$

$z^2 - 2z + 1 = -2 + 1$

$(z - 1)^2 = -1$

$z - 1 = \pm i$

$z = 1 \pm i$

$f(z) = (z - 1 + i)(z - 1 - i)$

39. $g(x) = x^3 + 3x^2 - 3x - 9$

$x^3 + 3x^2 - 3x - 9 = 0$

$x^2(x + 3) - 3(x + 3) = 0$

$(x^2 - 3)(x + 3) = 0$

$x^2 - 3 = 0 \Rightarrow x = \pm\sqrt{3}$

$x + 3 = 0 \Rightarrow x = -3$

$g(x) = (x + 3)(x + \sqrt{3})(x - \sqrt{3})$

41. $h(x) = x^3 - 4x^2 + 16x - 64$

$x^3 - 4x^2 + 16x - 64 = 0$

$x^2(x - 4) + 16(x - 4) = 0$

$(x^2 + 16)(x - 4) = 0$

$x^2 + 16 = 0 \Rightarrow x = \pm 4i$

$x - 4 = 0 \Rightarrow x = 4$

$h(x) = (x - 4)(x + 4i)(x - 4i)$

43. $f(x) = 2x^3 - x^2 + 36x - 18$

$2x^3 - x^2 + 36x - 18 = 0$

$x^2(2x - 1) + 18(2x - 1) = 0$

$(x^2 + 18)(2x - 1) = 0$

$x^2 + 18 = 0 \Rightarrow x = \pm 3\sqrt{2}i$

$2x - 1 = 0 \Rightarrow x = \frac{1}{2}$

$f(x) = 2\left(x - \frac{1}{2}\right)(x + 3\sqrt{2}i)(x - 3\sqrt{2}i)$

$= (2x - 1)(x + 3\sqrt{2}i)(x - 3\sqrt{2}i)$

45. $g(x) = x^4 - 4x^3 + 36x^2 - 144x$

$x^4 - 4x^3 + 36x^2 - 144x = 0$

$x^3(x - 4) + 36x(x - 4) = 0$

$(x^3 + 36x)(x - 4) = 0$

$x(x^2 + 36)(x - 4) = 0$

$x = 0$

$x^2 + 36 = 0 \Rightarrow x = \pm 6i$

$x - 4 = 0 \Rightarrow x = 4$

$g(x) = x(x - 4)(x + 6i)(x - 6i)$

47. $f(x) = x^4 + 10x^2 + 9$

$= (x^2 + 1)(x^2 + 9)$

$= (x + i)(x - i)(x + 3i)(x - 3i)$

The zeros of $f(x)$ are $x = \pm i$ and $x = \pm 3i$.

49. $f(x) = 2x^3 + 3x^2 + 50x + 75$

Since $5i$ is a zero, so is $-5i$.

$$\begin{array}{r|rrrr} 5i & 2 & 3 & 50 & 75 \\ & & 10i & -50+15i & -75 \\ \hline & 2 & 3+10i & 15i & 0 \end{array}$$

$$\begin{array}{r|rrr} -5i & 2 & 3+10i & 15i \\ & & -10i & -15i \\ \hline & 2 & 3 & 0 \end{array}$$

The zero of $2x + 3$ is $x = -\frac{3}{2}$.

The zeros of $f(x)$ are $x = -\frac{3}{2}$ and $x = \pm 5i$.

51. $f(x) = 2x^4 - x^3 + 7x^2 - 4x - 4$

Since $2i$ is a zero, so is $-2i$.

$$\begin{array}{r|rrrrr} 2i & 2 & -1 & 7 & -4 & -4 \\ & & 4i & -8-2i & 4-2i & 4 \\ \hline & 2 & -1+4i & -1-2i & -2i & 0 \end{array}$$

$$\begin{array}{r|rrrr} -2i & 2 & -1+4i & -1-2i & -2i \\ & & -4i & 2i & 2i \\ \hline & 2 & -1 & -1 & 0 \end{array}$$

The zeros of $2x^2 - x - 1 = (2x + 1)(x - 1)$ are $x = -\frac{1}{2}$ and $x = 1$.

The zeros of $f(x)$ are $x = \pm 2i$, $x = -\frac{1}{2}$, and $x = 1$.

Alternate Solution

Since $x = \pm 2i$ are zeros of $f(x)$,
$(x + 2i)(x - 2i) = x^2 + 4$ is a factor of $f(x)$.
By long division we have:

$$\begin{array}{r} 2x^2 - x - 1 \\ x^2 + 0x + 4 \overline{\big) 2x^4 - x^3 + 7x^2 - 4x - 4} \\ \underline{2x^4 + 0x^3 + 8x^2 } \\ -x^3 - x^2 - 4x \\ \underline{-x^3 + 0x^2 - 4x } \\ -x^2 + 0x - 4 \\ \underline{-x^2 + 0x - 4} \\ 0 \end{array}$$

Thus, $f(x) = (x^2 + 4)(2x^2 - x - 1)$
$= (x + 2i)(x - 2i)(2x + 1)(x - 1)$
and the zeros of $f(x)$ are $x = \pm 2i$, $x = -\frac{1}{2}$, and $x = 1$.

Alternate Solution

Since $x = \pm 5i$ are zeros of $f(x)$,
$(x + 5i)(x - 5i) = x^2 + 25$
is a factor of $f(x)$. By long division we have:

$$\begin{array}{r} 2x + 3 \\ x^2 + 0x + 25 \overline{\big) 2x^3 + 3x^2 + 50x + 75} \\ \underline{2x^3 + 0x^2 + 50x } \\ 3x^2 + 0x + 75 \\ \underline{3x^2 + 0x + 75} \\ 0 \end{array}$$

Thus, $f(x) = (x^2 + 25)(2x + 3)$ and the zeros of $f(x)$ are $x = \pm 5i$ and $x = -\frac{3}{2}$.

53. $g(x) = 4x^3 + 23x^2 + 34x - 10$

Since $-3 + i$ is a zero, so is $-3 - i$.

$$\begin{array}{r|rrrr} -3+i & 4 & 23 & 34 & -10 \\ & & -12+4i & -37-i & 10 \\ \hline & 4 & 11+4i & -3-i & 0 \end{array}$$

$$\begin{array}{r|rrr} -3-i & 4 & 11+4i & -3-i \\ & & -12-4i & 3+i \\ \hline & 4 & -1 & 0 \end{array}$$

The zero of $4x - 1$ is $x = \frac{1}{4}$. The zeros of $g(x)$ are $x = -3 \pm i$ and $x = \frac{1}{4}$.

Alternate Solution

Since $-3 \pm i$ are zeros of $g(x)$,

$$\begin{aligned} [x - (-3 + i)][x - (-3 - i)] &= [(x + 3) - i][(x + 3) + i] \\ &= (x + 3)^2 - i^2 \\ &= x^2 + 6x + 10 \end{aligned}$$

is a factor of $g(x)$. By long division we have:

$$\begin{array}{r} 4x - 1 \\ x^2 + 6x + 10 \overline{\big) 4x^3 + 23x^2 + 34x - 10} \\ \underline{4x^3 + 24x^2 + 40x } \\ -x^2 - 6x - 10 \\ \underline{-x^2 - 6x - 10} \\ 0 \end{array}$$

Thus, $g(x) = (x^2 + 6x + 10)(4x - 1)$ and the zeros of $g(x)$ are $x = -3 \pm i$ and $x = \frac{1}{4}$.

55. $f(x) = x^4 + 3x^3 - 5x^2 - 21x + 22$

Since $-3 + \sqrt{2}\,i$ is a zero, so is $-3 - \sqrt{2}\,i$, and

$$\left[x - \left(-3 + \sqrt{2}\,i\right)\right]\left[x - \left(-3 - \sqrt{2}\,i\right)\right]$$
$$= \left[(x + 3) - \sqrt{2}\,i\right]\left[(x + 3) + \sqrt{2}\,i\right]$$
$$= (x + 3)^2 - \left(\sqrt{2}\,i\right)^2$$
$$= x^2 + 6x + 11$$

is a factor of $f(x)$. By long division, we have:

$$\begin{array}{r} x^2 - 3x + 2 \\ x^2 + 6x + 11 \overline{\big)\, x^4 + 3x^3 - 5x^2 - 21x + 22} \\ \underline{x^4 + 6x^3 + 11x^2} \qquad\qquad \\ -3x^3 - 16x^2 - 21x \qquad \\ \underline{-3x^3 - 18x^2 - 33x} \qquad \\ 2x^2 + 12x + 22 \\ \underline{2x^2 + 12x + 22} \\ 0 \end{array}$$

Thus, $f(x) = (x^2 + 6x + 11)(x^2 - 3x + 2)$

$= (x^2 + 6x + 11)(x - 1)(x - 2)$

and the zeros of $f(x)$ are $x = -3 \pm \sqrt{2}\,i$, $x = 1$, and $x = 2$.

57. $f(x) = (x - 1)(x - 5i)(x + 5i)$

$= (x - 1)(x^2 + 25)$

$= x^3 - x^2 + 25x - 25$

Note: $f(x) = a(x^3 - x^2 + 25x - 25)$. where a is any nonzero real number, has the zeros 1 and $\pm 5i$.

59. $f(x) = (x - 6)[x - (-5 + 2i)][x - (-5 - 2i)]$

$= (x - 6)[(x + 5) - 2i][(x + 5) + 2i]$

$= (x - 6)[(x + 5)^2 - (2i)^2]$

$= (x - 6)(x^2 + 10x + 25 + 4)$

$= (x - 6)(x^2 + 10x + 29)$

$= x^3 + 4x^2 - 31x - 174$

Note: $f(x) = a(x^3 + 4x^2 - 31x - 174)$, where a is any nonzero real number, has the zeros 6, and $-5 \pm 2i$.

61. If $3 + \sqrt{2}i$ is a zero, so is its conjugate, $3 - \sqrt{2}i$.

$f(x) = (3x - 2)(x + 1)\left[x - \left(3 + \sqrt{2}i\right)\right]\left[x - \left(3 - \sqrt{2}i\right)\right]$

$= (3x - 2)(x + 1)\left[(x - 3) - \sqrt{2}i\right]\left[(x - 3) + \sqrt{2}i\right]$

$= (3x^2 + x - 2)\left[(x - 3)^2 - \left(\sqrt{2}i\right)^2\right]$

$= (3x^2 + x - 2)(x^2 - 6x + 9 + 2)$

$= (3x^2 + x - 2)(x^2 - 6x + 11)$

$= 3x^4 - 17x^3 + 25x^2 + 23x - 22$

Note: $f(x) = a(3x^4 - 17x^3 + 25x^2 + 23x - 22)$, where a is any nonzero real number, has the zeros $\frac{2}{3}$, -1, and $3 \pm \sqrt{2}i$.

63. $h = -16t^2 + 48t,\ 0 \le t \le 3$

$= -16(t^2 - 3t)$

$= -16\left(t^2 - 3t + \frac{9}{4} - \frac{9}{4}\right)$

$= -16\left[\left(t - \frac{3}{2}\right)^2 - \frac{9}{4}\right]$

$= -16\left(t - \frac{3}{2}\right)^2 + 36$

The maximum height that the baseball reaches is 36 feet when $t = 1.5$ seconds. No, it is not possible for the ball to reach a height of 64 feet.

Alternate Solution

Let $h = 64$ and solve for t.

$$64 = -16t^2 + 48t$$

$16t^2 - 48t + 64 = 0$

$16(t^2 - 3t + 4) = 0$

$t^2 - 3t + 4 = 0$

$$t = \frac{3 \pm \sqrt{9 - 16}}{2} = \frac{3 \pm \sqrt{7}\,i}{2}$$

No, it is not possible since solving this equation yields only imaginary roots.

65. False. The most nonreal complex zeros it can have is two and the Linear Factorization Theorem guarantees that there are 3 linear factors, so one zero must be real.

67. $g(x) = -f(x)$. This function would have the same zeros as $f(x)$ so r_1, r_2, and r_3 are also zeros of $g(x)$.

69. $g(x) = f(x - 5)$. The graph of $g(x)$ is a horizontal shift of the graph of $f(x)$ five units to the right so the zeros of $g(x)$ are $5 + r_1$, $5 + r_2$, and $5 + r_3$.

71. $g(x) = 3 + f(x)$. Since $g(x)$ is a vertical shift of the graph of $f(x)$, the zeros of $g(x)$ cannot be determined.

73. (a) $f(x) = (x - \sqrt{b}i)(x + \sqrt{b}i) = x^2 + b$

(b) $f(x) = [x - (a + bi)][x - (a - bi)]$

$= [(x - a) - bi][(x - a) + bi]$

$= (x - a)^2 - (bi)^2$

$= x^2 - 2ax + a^2 + b^2$

75. $(-3 + 6i) - (8 - 3i) = -3 + 6i - 8 + 3i = -11 + 9i$

77. $(6 - 2i)(1 + 7i) = 6 + 42i - 2i - 14i^2 = 20 + 40i$

79. $g(x) = f(x - 2)$

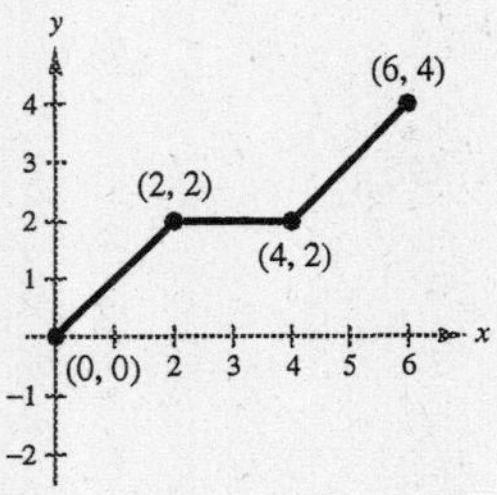

Horizontal shift two units to the right

81. $g(x) = 2f(x)$

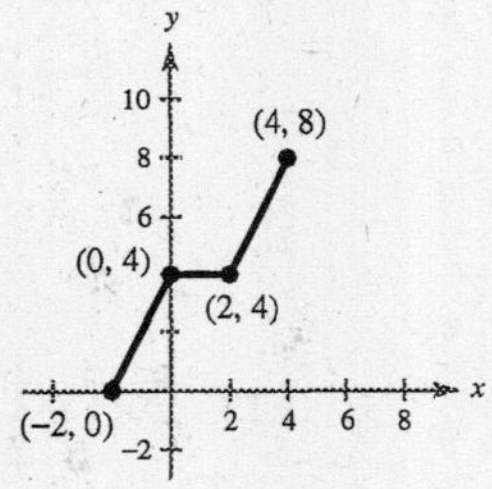

Vertical stretch

83. $g(x) = f(2x)$

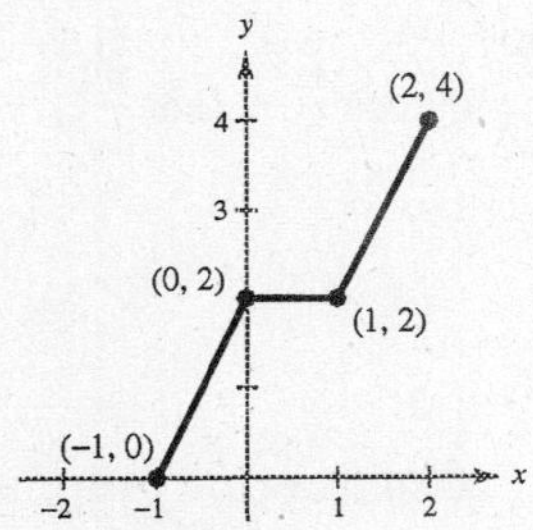

Horizontal shrink

Section 4.3 Trigonometric Form of a Complex Number

- You should be able to graphically represent complex numbers and know the following facts about them.
- The absolute value of the complex numbers $z = a + bi$ is $|z| = \sqrt{a^2 + b^2}$.
- The trigonometric form of the complex number $z = a + bi$ is $z = r(\cos\theta + i\sin\theta)$ where
  (a) $a = r\cos\theta$
  (b) $b = r\sin\theta$
  (c) $r = \sqrt{a^2 + b^2}$; r is called the modulus of z.
  (d) $\tan\theta = \dfrac{b}{a}$; θ is called the argument of z.
- Given $z_1 = r_1(\cos\theta_1 + i\sin\theta_1)$ and $z_2 = r_2(\cos\theta_2 + i\sin\theta_2)$:
  (a) $z_1z_2 = r_1r_2[\cos(\theta_1 + \theta_2) + i\sin(\theta_1 + \theta_2)]$
  (b) $\dfrac{z_1}{z_2} = \dfrac{r_1}{r_2}[\cos(\theta_1 - \theta_2) + i\sin(\theta_1 - \theta_2)]$, $z_2 \neq 0$

Solutions to Odd-Numbered Exercises

1. $|-7i| = \sqrt{0^2 + (-7)^2}$

$= \sqrt{49} = 7$

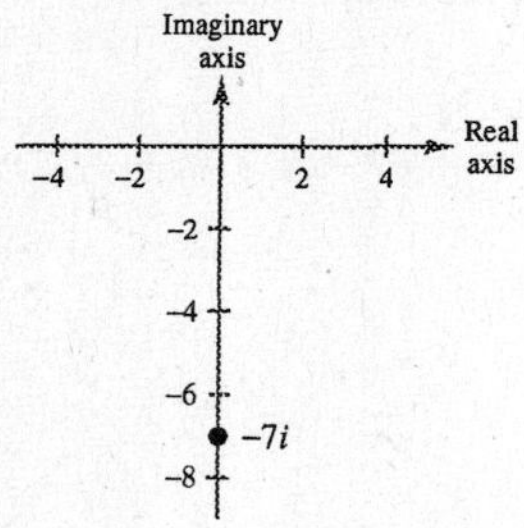

3. $|-4 + 4i| = \sqrt{(-4)^2 + (4)^2}$

$= \sqrt{32} = 4\sqrt{2}$

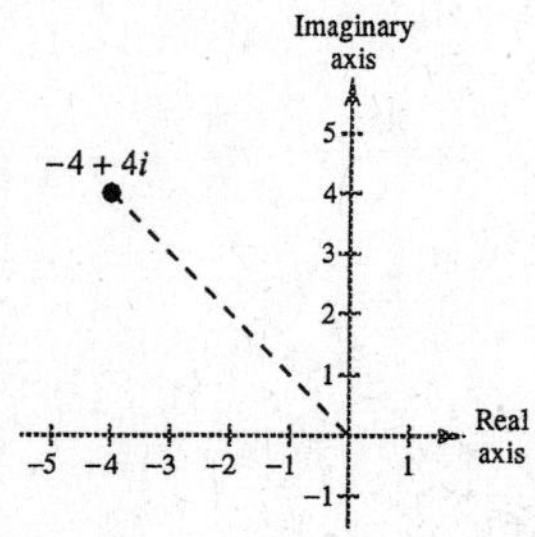

5. $|6 - 7i| = \sqrt{6^2 + (-7)^2}$

$= \sqrt{85}$

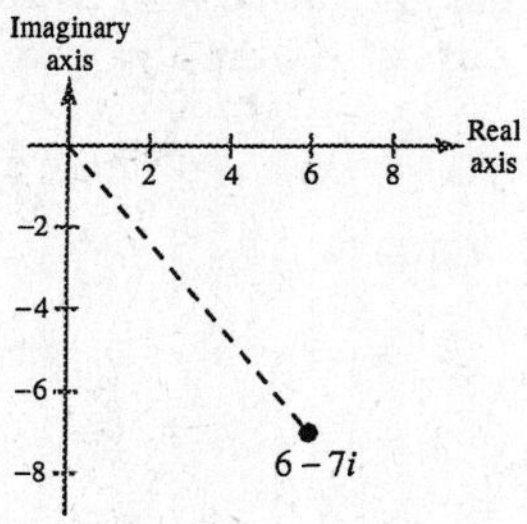

7. $z = 3i$

$r = \sqrt{0^2 + 3^2} = \sqrt{9} = 3$

$\tan\theta = \dfrac{3}{0}$, undefined $\Rightarrow \theta = \dfrac{\pi}{2}$

$z = 3\left(\cos\dfrac{\pi}{2} + i\sin\dfrac{\pi}{2}\right)$

9. $z = 3 - i$

$r = \sqrt{(3)^2 + (-1)^2} = \sqrt{10}$

$\tan\theta = -\dfrac{1}{3}$, θ is in Quadrant IV.

$\theta \approx 5.96$ radians

$z \approx \sqrt{10}(\cos 5.96 + i\sin 5.96)$

11. $z = 3 - 3i$

$r = \sqrt{3^2 + (-3)^2} = \sqrt{18} = 3\sqrt{2}$

$\tan\theta = \dfrac{-3}{3} = -1$, θ is in Quadrant IV $\Rightarrow \theta = \dfrac{7\pi}{4}$.

$z = 3\sqrt{2}\left(\cos\dfrac{7\pi}{4} + i\sin\dfrac{7\pi}{4}\right)$

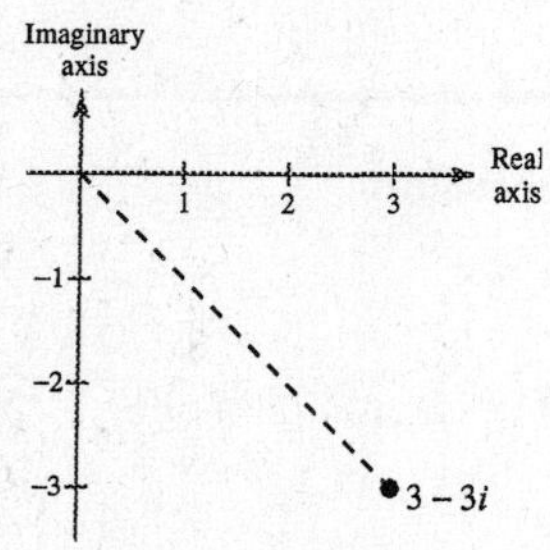

13. $z = \sqrt{3} + i$

$r = \sqrt{\left(\sqrt{3}\right)^2 + 1^2} = \sqrt{4} = 2$

$\tan\theta = \dfrac{1}{\sqrt{3}} = \dfrac{\sqrt{3}}{3} \Rightarrow \theta = \dfrac{\pi}{6}$

$z = 2\left(\cos\dfrac{\pi}{6} + i\sin\dfrac{\pi}{6}\right)$

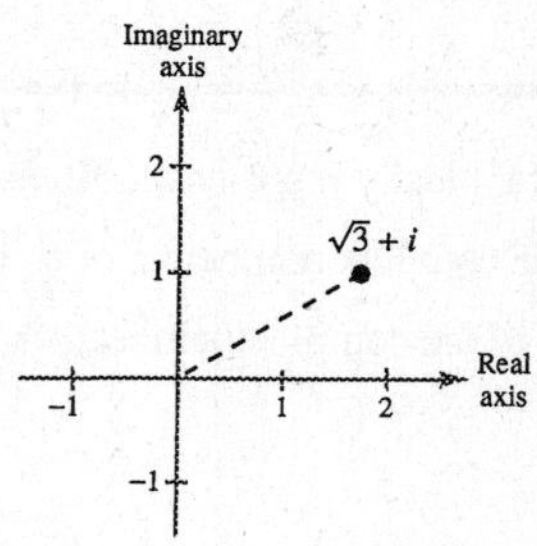

15. $z = -2(1 + \sqrt{3}i)$

$r = \sqrt{(-2)^2 + (-2\sqrt{3})^2} = \sqrt{16} = 4$

$\tan \theta = \dfrac{\sqrt{3}}{1} = \sqrt{3}$, θ is in Quadrant III $\Rightarrow \theta = \dfrac{4\pi}{3}$.

$z = 4\left(\cos \dfrac{4\pi}{3} + i \sin \dfrac{4\pi}{3}\right)$

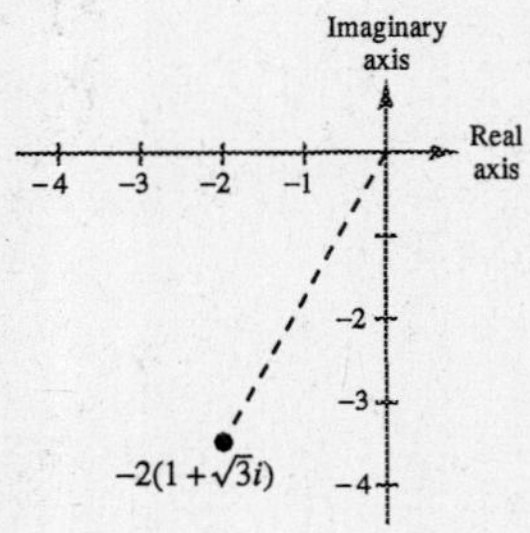

17. $z = -5i$

$r = \sqrt{0^2 + (-5)^2} = \sqrt{25} = 5$

$\tan \theta = \dfrac{-5}{0}$, undefined $\Rightarrow \theta = \dfrac{3\pi}{2}$

$z = 5\left(\cos \dfrac{3\pi}{2} + i \sin \dfrac{3\pi}{2}\right)$

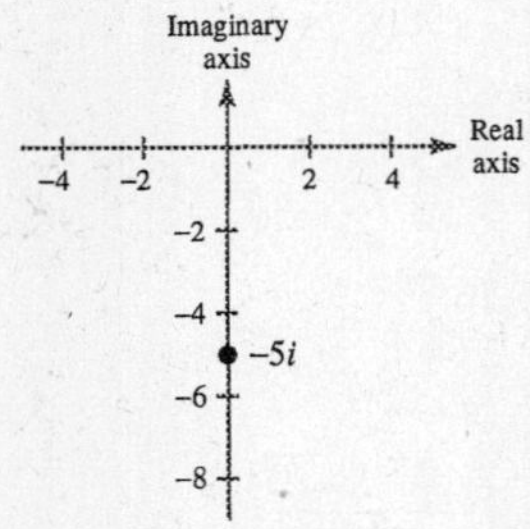

19. $z = -7 + 4i$

$r = \sqrt{(-7)^2 + (4)^2} = \sqrt{65}$

$\tan \theta = \dfrac{4}{-7}$, θ is in Quadrant II $\Rightarrow \theta \approx 2.62$.

$z \approx \sqrt{65}(\cos 2.62 + i \sin 2.62)$

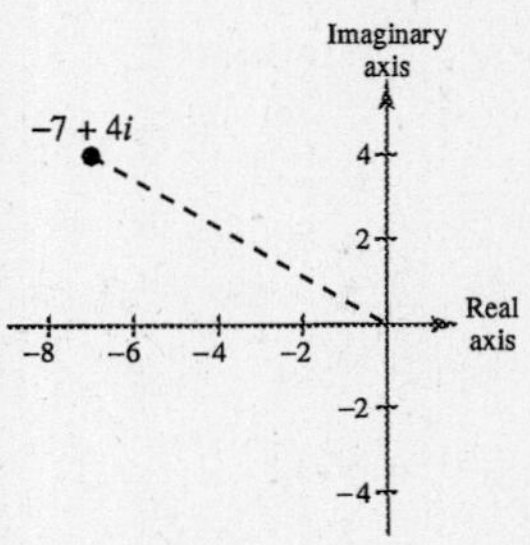

21. $z = 7 + 0i$

$r = \sqrt{(7)^2 + (0)^2} = \sqrt{49} = 7$

$\tan \theta = \dfrac{0}{7} = 0 \Rightarrow \theta = 0$

$z = 7(\cos 0 + i \sin 0)$

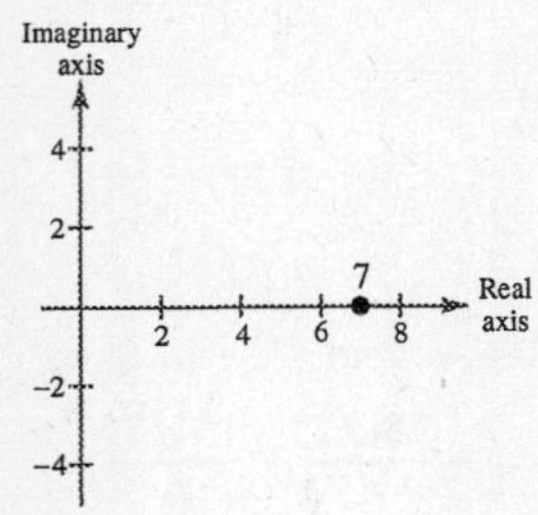

23. $z = 3 + \sqrt{3}i$

$r = \sqrt{(3)^2 + (\sqrt{3})^2} = \sqrt{12}$

$= 2\sqrt{3}$

$\tan \theta = \dfrac{\sqrt{3}}{3} \Rightarrow \theta = \dfrac{\pi}{6}$

$z = 2\sqrt{3}\left(\cos \dfrac{\pi}{6} + i \sin \dfrac{\pi}{6}\right)$

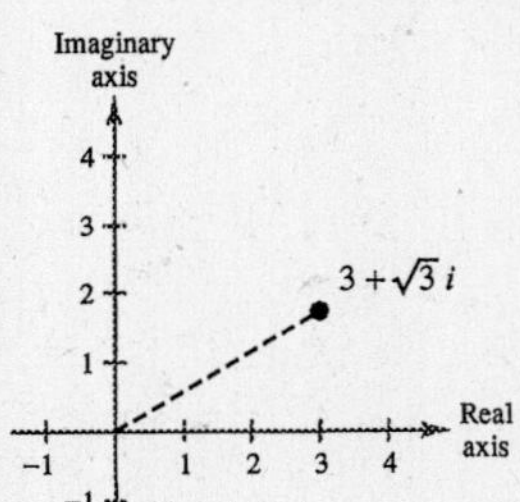

25. $z = -3 - i$

$r = \sqrt{(-3)^2 + (-1)^2} = \sqrt{10}$

$\tan \theta = \dfrac{-1}{-3} = \dfrac{1}{3}$, θ is in Quadrant III $\Rightarrow \theta \approx 3.46$.

$z \approx \sqrt{10}(\cos 3.46 + i \sin 3.46)$

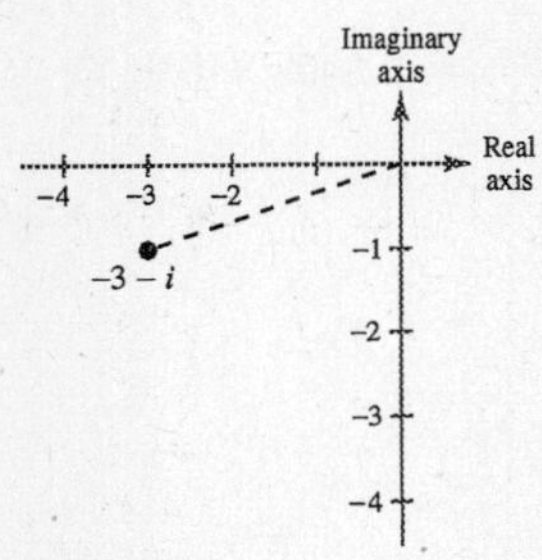

27. $z = 5 + 2i$

$r \approx 5.39$

$\theta \approx 0.38$

$z \approx 5.39(\cos 0.38 + i \sin 0.38)$

29. $z = -3 + i$

$r \approx 3.16$

$\theta \approx 2.82$

$z \approx 3.16(\cos 2.82 + i \sin 2.82)$

31. $z = 3\sqrt{2} - 7i$

$r \approx 8.19$

$\theta = \tan^{-1}\left(-\frac{7}{3\sqrt{2}}\right) + 2\pi \approx 5.26$

$z \approx 8.19(\cos 5.26 + i \sin 5.26)$

33. $z = -8 - 5\sqrt{3}i$

$r \approx 11.79$

$\theta \approx 3.97$

$z \approx 11.79(\cos 3.97 + i \sin 3.97)$

35. $3(\cos 120° + i \sin 120°) = 3\left(-\frac{1}{2} + \frac{\sqrt{3}}{2}i\right)$

$= -\frac{3}{2} + \frac{3\sqrt{3}}{2}i$

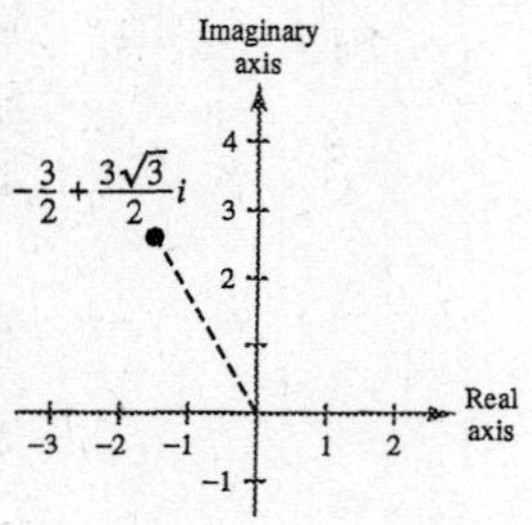

37. $\frac{3}{2}(\cos 300° + i \sin 300°) = \frac{3}{2}\left[\frac{1}{2} + i\left(-\frac{\sqrt{3}}{2}\right)\right]$

$= \frac{3}{4} - \frac{3\sqrt{3}}{4}i$

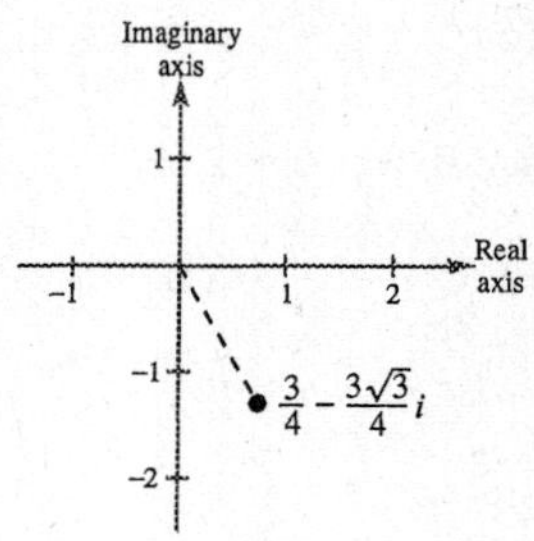

39. $3.75\left(\cos \frac{3\pi}{4} + i \sin \frac{3\pi}{4}\right) = -\frac{15\sqrt{2}}{8} + \frac{15\sqrt{2}}{8}i$

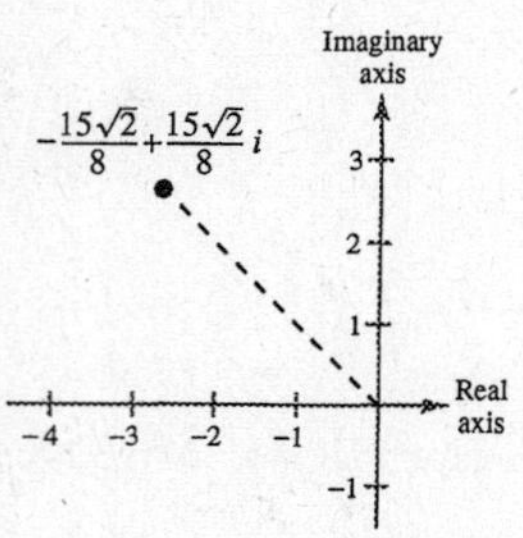

41. $8\left(\cos \frac{\pi}{2} + i \sin \frac{\pi}{2}\right) = 8(0 + i) = 8i$

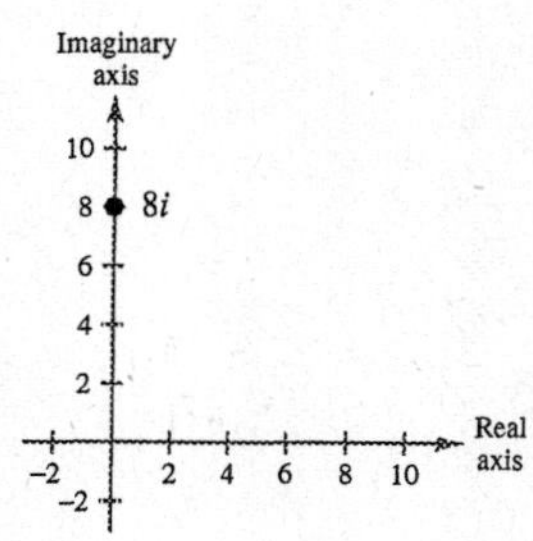

43. $3[\cos(18°45') + i \sin(18°45')] \approx 2.8408 + 0.9643i$

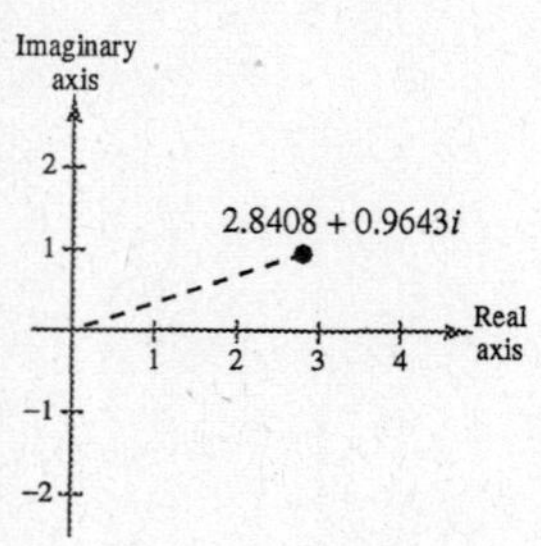

45. $5\left(\cos \frac{\pi}{9} + i \sin \frac{\pi}{9}\right) \approx 4.70 + 1.71i$

47. $3(\cos 165.5° + i \sin 165.5°) \approx -2.90 + 0.75i$

49. $$\left[2\left(\cos\frac{\pi}{4} + i\sin\frac{\pi}{4}\right)\right]\left[6\left(\cos\frac{\pi}{12} + i\sin\frac{\pi}{12}\right)\right] = (2)(6)\left[\cos\left(\frac{\pi}{4} + \frac{\pi}{12}\right) + i\sin\left(\frac{\pi}{4} + \frac{\pi}{12}\right)\right]$$
$$= 12\left(\cos\frac{\pi}{3} + i\sin\frac{\pi}{3}\right)$$

51. $$\left[\tfrac{5}{3}(\cos 140° + i\sin 140°)\right]\left[\tfrac{2}{3}(\cos 60° + i\sin 60°)\right] = \left(\tfrac{5}{3}\right)\left(\tfrac{2}{3}\right)[\cos(140° + 60°) + i\sin(140° + 60°)]$$
$$= \tfrac{10}{9}(\cos 200° + i\sin 200°)$$

53. $$[0.45(\cos 310° + i\sin 310°)][0.60(\cos 200° + i\sin 200°)] = (0.45)(0.60)[\cos(310° + 200°) + i\sin(310° + 200°)]$$
$$= 0.27(\cos 510° + i\sin 510°)$$
$$= 0.27(\cos 150° + i\sin 150°)$$

55. $$\frac{\cos 50° + i\sin 50°}{\cos 20° + i\sin 20°} = \cos(50° - 20°) + i\sin(50° - 20°)$$
$$= \cos 30° + i\sin 30°$$

57. $$\frac{\cos\frac{5\pi}{3} + i\sin\frac{5\pi}{3}}{\cos\pi + i\sin\pi} = \cos\left(\frac{5\pi}{3} - \pi\right) + i\sin\left(\frac{5\pi}{3} - \pi\right) = \cos\left(\frac{2\pi}{3}\right) + i\sin\left(\frac{2\pi}{3}\right)$$

59. $$\frac{12(\cos 52° + i\sin 52°)}{3(\cos 110° + i\sin 110°)} = \frac{12}{3}[\cos(52° - 110°) + i\sin(52° - 110°)]$$
$$= 4[\cos(-58°) + i\sin(-58°)]$$

61. (a) $2 + 2i = 2\sqrt{2}\left(\cos\frac{\pi}{4} + i\sin\frac{\pi}{4}\right)$

$1 - i = \sqrt{2}\left[\cos\left(\frac{7\pi}{4}\right) + i\sin\left(\frac{7\pi}{4}\right)\right]$

(b) $$(2 + 2i)(1 - i) = \left[2\sqrt{2}\left(\cos\frac{\pi}{4} + i\sin\frac{\pi}{4}\right)\right]\left[\sqrt{2}\left(\cos\left(\frac{7\pi}{4}\right) + i\sin\left(\frac{7\pi}{4}\right)\right)\right]$$
$$= 4(\cos 2\pi + i\sin 2\pi) = 4(\cos 0 + i\sin 0) = 4$$

(c) $(2 + 2i)(1 - i) = 2 - 2i + 2i - 2i^2 = 2 + 2 = 4$

63. (a) $-2i = 2\left[\cos\left(\frac{3\pi}{2}\right) + i\sin\left(\frac{3\pi}{2}\right)\right]$

$1 + i = \sqrt{2}\left(\cos\frac{\pi}{4} + i\sin\frac{\pi}{4}\right)$

(b) $-2i(1+i) = 2\left[\cos\left(\frac{3\pi}{2}\right) + i\sin\left(\frac{3\pi}{2}\right)\right]\left[\sqrt{2}\left(\cos\frac{\pi}{4} + i\sin\frac{\pi}{4}\right)\right]$

$= 2\sqrt{2}\left[\cos\left(\frac{7\pi}{4}\right) + i\sin\left(\frac{7\pi}{4}\right)\right]$

$= 2\sqrt{2}\left[\frac{1}{\sqrt{2}} - \frac{1}{\sqrt{2}}i\right] = 2 - 2i$

(c) $-2i(1+i) = -2i - 2i^2 = -2i + 2 = 2 - 2i$

65. (a) $3 + 4i = 5(\cos 0.93 + i\sin 0.93)$

$1 - \sqrt{3}i = 2\left(\cos\frac{5\pi}{3} + i\sin\frac{5\pi}{3}\right)$

(b) $\frac{3+4i}{1-\sqrt{3}i} = \frac{5(\cos 0.93 + i\sin 0.93)}{2\left(\cos\frac{5\pi}{3} + i\sin\frac{5\pi}{3}\right)}$

$\approx 2.5[\cos(-4.31) + i\sin(-4.31)]$

$\approx 2.5(\cos 1.97 + i\sin 1.97)$

$\approx -0.982 + 2.299i$

(c) $\frac{3+4i}{1-\sqrt{3}i} = \frac{3+4i}{1-\sqrt{3}i}\cdot\frac{1+\sqrt{3}i}{1+\sqrt{3}i}$

$= \frac{3 + (4 + 3\sqrt{3})i + 4\sqrt{3}i^2}{1+3}$

$= \frac{3 - 4\sqrt{3}}{4} + \frac{4 + 3\sqrt{3}}{4}i$

$\approx -0.982 + 2.299i$

67. (a) $5 = 5(\cos 0 + i\sin 0)$

$2 + 3i \approx \sqrt{13}(\cos 0.98 + i\sin 0.98)$

(b) $\frac{5}{2+3i} \approx \frac{5(\cos 0 + i\sin 0)}{\sqrt{13}(\cos 0.98 + i\sin 0.98)} = \frac{5}{\sqrt{13}}[\cos(-0.98) + i\sin(-0.98)]$

$\approx \frac{5}{\sqrt{13}}(\cos 5.30 + i\sin 5.30) \approx 0.769 - 1.154i$

(c) $\frac{5}{2+3i} = \frac{5}{2+3i}\cdot\frac{2-3i}{2-3i} = \frac{10-15i}{13} = \frac{10}{13} - \frac{15}{13}i \approx 0.769 - 1.154i$

69. Let $z = x + iy$ such that:

$|z| = 2 \Rightarrow 2 = \sqrt{x^2 + y^2}$

$\Rightarrow 4 = x^2 + y^2$: circle with radius of 2

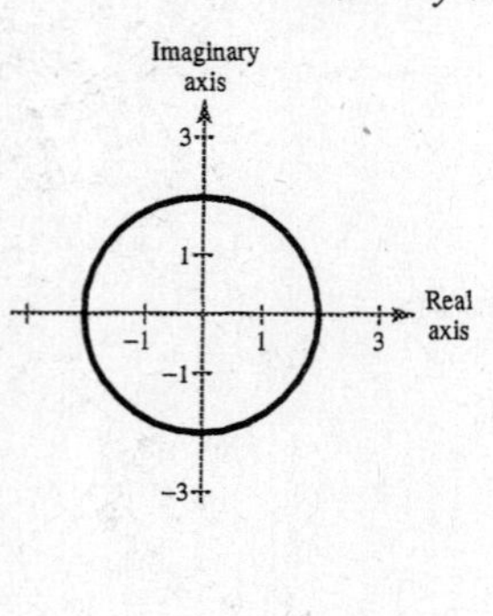

71. Let $\theta = \frac{\pi}{6}$.

Let $z = x + iy$ such that:

$\tan\frac{\pi}{6} = \frac{y}{x}$

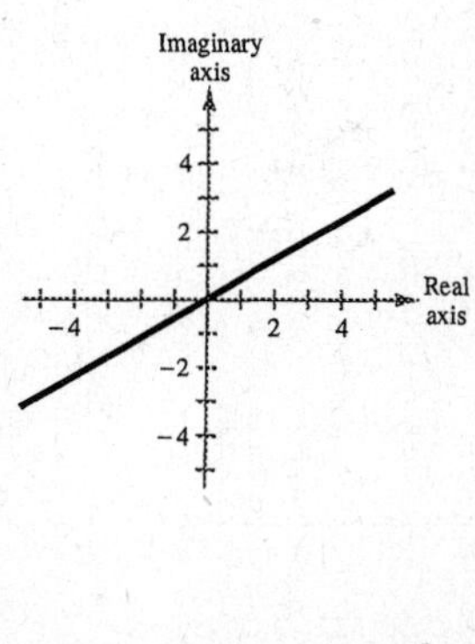

73. True, by the definition of the absolute value of a complex number

75. $\dfrac{z_1}{z_2} = \dfrac{r_1(\cos\theta_1 + i\sin\theta_1)}{r_2(\cos\theta_2 + i\sin\theta_2)} \cdot \dfrac{(\cos\theta_2 + i\sin\theta_2)}{(\cos\theta_2 + i\sin\theta_2)}$

$= \dfrac{r_1}{r_2(\cos^2\theta_2 + \sin^2\theta_2)}[\cos\theta_1\cos\theta_2 + \sin\theta_1\sin\theta_2 + i(\sin\theta_1\cos\theta_2 - \sin\theta_2\cos\theta_1)]$

$= \dfrac{r_1}{r_2}[\cos(\theta_1 - \theta_2) + i\sin(\theta_1 - \theta_2)]$

77. (a) $z\bar{z} = [r(\cos\theta + i\sin\theta)][r(\cos(-\theta) + i\sin(-\theta))]$

$= r^2[\cos(\theta - \theta) + i\sin(\theta - \theta)]$

$= r^2[\cos 0 + i\sin 0]$

$= r^2$

(b) $\dfrac{z}{\bar{z}} = \dfrac{r(\cos\theta + i\sin\theta)}{r[\cos(-\theta) + i\sin(-\theta)]}$

$= \dfrac{r}{r}[\cos(\theta - (-\theta)) + i\sin(\theta - (-\theta))]$

$= \cos 2\theta + i\sin 2\theta$

79. $A = 22°, a = 8$

$B = 90° - A = 68°$

$\tan 22° = \dfrac{8}{b} \Rightarrow b = \dfrac{8}{\tan 22°} \approx 19.80$

$\sin 22° = \dfrac{8}{c} \Rightarrow c = \dfrac{8}{\sin 22°} \approx 21.36$

81. $A = 30°, b = 112.6$

$B = 90° - A = 60°$

$\tan 30° = \dfrac{a}{112.6} \Rightarrow a = 112.6\tan 30° \approx 65.01$

$\cos 30° = \dfrac{112.6}{c} \Rightarrow c = \dfrac{112.6}{\cos 30°} \approx 130.02$

83. $A = 42°15' = 42.25°, c = 11.2$

$B = 90° - A = 47°45'$

$\sin 42.25° = \dfrac{a}{11.2} \Rightarrow a = 11.2\sin 42.25° \approx 7.53$

$\cos 42.25° = \dfrac{b}{11.2} \Rightarrow b = 11.2\cos 42.25° \approx 8.29$

85. $d = 16\cos\dfrac{\pi}{4}t$

Maximum displacement: $|16| = 16$

$16\cos\dfrac{\pi}{4}t = 0 \Rightarrow \dfrac{\pi}{4}t = \dfrac{\pi}{2} \Rightarrow t = 2$

87. $d = \dfrac{1}{16}\sin\dfrac{5}{4}\pi t$

Maximum displacement: $\left|\dfrac{1}{16}\right| = \dfrac{1}{16}$

$\dfrac{1}{16}\sin\dfrac{5}{4}\pi t = 0 \Rightarrow \dfrac{5}{4}\pi t = 0 \Rightarrow t = 0$

Section 4.4 DeMoivre's Theorem

- You should know DeMoivre's Theorem: If $z = r(\cos\theta + i\sin\theta)$, then for any positive integer n,

  $z^n = r^n(\cos n\theta + i\sin n\theta)$.

- You should know that for any positive integer n, $z = r(\cos\theta + i\sin\theta)$ has n distinct nth roots given by

  $$\sqrt[n]{r}\left[\cos\left(\frac{\theta + 2\pi k}{n}\right) + i\sin\left(\frac{\theta + 2\pi k}{n}\right)\right]$$

  where $k = 0, 1, 2, \ldots, n - 1$.

Solutions to Odd-Numbered Exercises

1. $(1+i)^5 = \left[\sqrt{2}\left(\cos\frac{\pi}{4} + i\sin\frac{\pi}{4}\right)\right]^5$

$= (\sqrt{2})^5\left(\cos\frac{5\pi}{4} + i\sin\frac{5\pi}{4}\right)$

$= 4\sqrt{2}\left(-\frac{\sqrt{2}}{2} - \frac{\sqrt{2}}{2}i\right)$

$= -4 - 4i$

3. $(-1+i)^{10} = \left[\sqrt{2}\left(\cos\frac{3\pi}{4} + i\sin\frac{3\pi}{4}\right)\right]^{10}$

$= (\sqrt{2})^{10}\left(\cos\frac{30\pi}{4} + i\sin\frac{30\pi}{4}\right)$

$= 32\left[\cos\left(\frac{3\pi}{2} + 6\pi\right) + i\sin\left(\frac{3\pi}{2} + 6\pi\right)\right]$

$= 32\left(\cos\frac{3\pi}{2} + i\sin\frac{3\pi}{2}\right)$

$= 32[0 + i(-1)]$

$= -32i$

5. $2(\sqrt{3}+i)^7 = 2\left[2\left(\cos\frac{\pi}{6} + i\sin\frac{\pi}{6}\right)\right]^7$

$= 2\left[2^7\left(\cos\frac{7\pi}{6} + i\sin\frac{7\pi}{6}\right)\right]$

$= 256\left(-\frac{\sqrt{3}}{2} - \frac{1}{2}i\right)$

$= -128\sqrt{3} - 128i$

7. $[5(\cos 20° + i\sin 20°)]^3 = 5^3(\cos 60° + i\sin 60°)$

$= \frac{125}{2} + \frac{125\sqrt{3}}{2}i$

9. $\left(\cos\frac{\pi}{4} + i\sin\frac{\pi}{4}\right)^{12} = \cos\frac{12\pi}{4} + i\sin\frac{12\pi}{4}$

$= \cos 3\pi + i\sin 3\pi$

$= -1$

11. $[5(\cos 3.2 + i\sin 3.2)]^4 = 5^4(\cos 12.8 + i\sin 12.8)$

$\approx 608.02 + 144.69i$

13. $(3-2i)^5 \approx \{3.6056[\cos(5.695) + i\sin(5.695)]\}^5$

$\approx (3.6056)^5[\cos(28.475) + i\sin(28.475)]$

$\approx -597 - 122i$

15. $(\sqrt{5} - 4i)^3 \approx \{4.5826[\cos(5.222) + i\sin(5.222)]\}^3$

$\approx (4.5826)^3[\cos(15.666) + i\sin(15.666)]$

$\approx -96.15 + 4.04i$

The exact answer is $-43\sqrt{5} + 4i$

17. $[3(\cos 15° + i\sin 15°)]^4 = 81(\cos 60° + i\sin 60°)$

$= \frac{81}{2} + \frac{8\sqrt{3}}{2}i$

19. $[5(\cos 95° + i\sin 95°)]^3 = 125(\cos 285° + i\sin 285°)$

$\approx 32.3524 - 120.7407i$

21. $\left[2\left(\cos\frac{\pi}{10} + i\sin\frac{\pi}{10}\right)\right]^5 = 2^5\left(\cos\frac{\pi}{2} + i\sin\frac{\pi}{2}\right)$

$= 32i$

23. $\left[3\left(\cos\frac{2\pi}{3} + i\sin\frac{2\pi}{3}\right)\right]^3 = 27(\cos 2\pi + i\sin 2\pi)$

$= 27$

25. (a) Square roots of $5(\cos 120° + i \sin 120°)$:

$$\sqrt{5}\left[\cos\left(\frac{120° + 360°k}{2}\right) + i\sin\left(\frac{120° + 360°k}{2}\right)\right],\ k = 0,\ 1$$

$k = 0$: $\sqrt{5}(\cos 60° + i \sin 60°)$

$k = 1$: $\sqrt{5}(\cos 240° + i \sin 240°)$

(b)

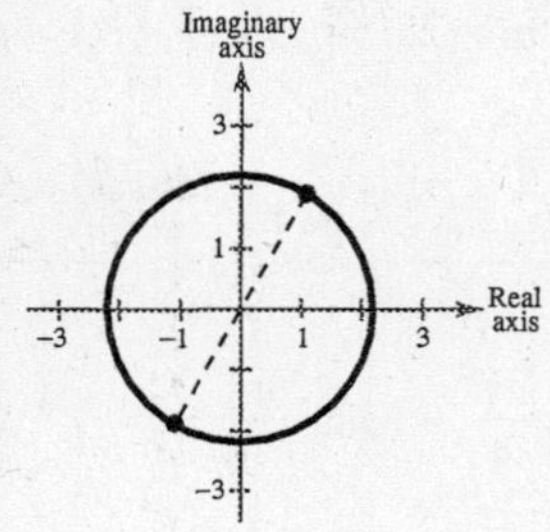

(c) $\dfrac{\sqrt{5}}{2} + \dfrac{\sqrt{15}}{2}i,\ -\dfrac{\sqrt{5}}{2} - \dfrac{\sqrt{15}}{2}i$

27. (a) Cube roots of $8\left(\cos\dfrac{2\pi}{3} + i\sin\dfrac{2\pi}{3}\right)$:

$$\sqrt[3]{8}\left[\cos\left(\frac{\frac{2\pi}{3} + 2\pi k}{3}\right) + i\sin\left(\frac{\frac{2\pi}{3} + 2\pi k}{3}\right)\right],\ k = 0, 1, 2$$

$k = 0$: $2\left(\cos\dfrac{2\pi}{9} + i\sin\dfrac{2\pi}{9}\right)$

$k = 1$: $2\left(\cos\dfrac{8\pi}{9} + i\sin\dfrac{8\pi}{9}\right)$

$k = 2$: $2\left(\cos\dfrac{14\pi}{9} + i\sin\dfrac{14\pi}{9}\right)$

(b)

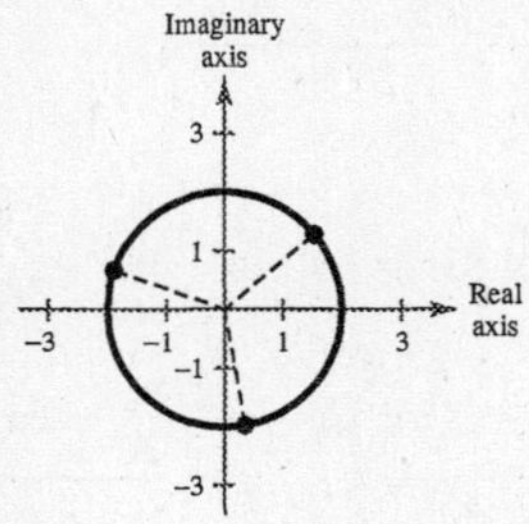

(c) $1.5321 + 1.2856i$

$-1.8794 + 0.6840i$

$0.3473 - 1.9696i$

29. (a) Fifth roots of $243\left(\cos\dfrac{\pi}{6} + i\sin\dfrac{\pi}{6}\right)$

$$\sqrt[5]{243}\left[\cos\left(\frac{\frac{\pi}{6} + 2\pi k}{5}\right) + i\sin\left(\frac{\frac{\pi}{6} + 2\pi k}{5}\right)\right],$$

$k = 0, 1, 2, 3, 4$

$k = 0$: $3\left(\cos\dfrac{\pi}{30} + i\sin\dfrac{\pi}{30}\right)$

$k = 1$: $3\left(\cos\dfrac{13\pi}{30} + i\sin\dfrac{13\pi}{30}\right)$

$k = 2$: $3\left(\cos\dfrac{5\pi}{6} + i\sin\dfrac{5\pi}{6}\right)$

$k = 3$: $3\left(\cos\dfrac{37\pi}{30} + i\sin\dfrac{37\pi}{30}\right)$

$k = 4$: $3\left(\cos\dfrac{49\pi}{30} + i\sin\dfrac{49\pi}{30}\right)$

(b)

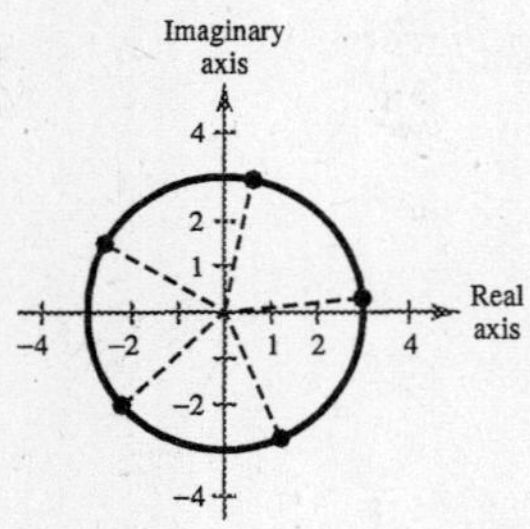

(c) $2.9836 + 0.3136i$

$0.6237 + 2.9344i$

$-2.5981 + 1.5i$

$-2.2294 - 2.0074i$

$1.2202 - 2.7406i$

31. (a) Square roots of $-25i = 25\left(\cos\frac{3\pi}{2} + i\sin\frac{3\pi}{2}\right)$:

$$\sqrt{25}\left[\cos\left(\frac{\frac{3\pi}{2} + 2k\pi}{2}\right) + i\sin\left(\frac{\frac{3\pi}{2} + 2k\pi}{2}\right)\right],\ k = 0, 1$$

$k = 0$: $5\left(\cos\frac{3\pi}{4} + i\sin\frac{3\pi}{4}\right)$

$k = 1$: $5\left(\cos\frac{7\pi}{4} + i\sin\frac{7\pi}{4}\right)$

(b)

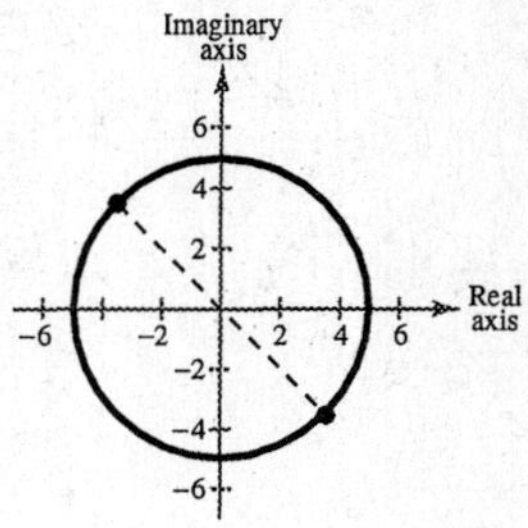

(c) $-\frac{5\sqrt{2}}{2} + \frac{5\sqrt{2}}{2}i,\ \frac{5\sqrt{2}}{2} - \frac{5\sqrt{2}}{2}i$

33. (a) Fourth roots of $81i = 81\left(\cos\frac{\pi}{2} + i\sin\frac{\pi}{2}\right)$

$$\sqrt[4]{81}\left[\cos\left(\frac{\frac{\pi}{2} + 2\pi k}{4}\right) + i\sin\left(\frac{\frac{\pi}{2} + 2\pi k}{4}\right)\right]$$

$k = 0, 1, 2, 3$

$k = 0$: $3\left(\cos\frac{\pi}{8} + i\sin\frac{\pi}{8}\right)$

$k = 1$: $3\left(\cos\frac{5\pi}{8} + i\sin\frac{5\pi}{8}\right)$

$k = 2$: $3\left(\cos\frac{9\pi}{8} + i\sin\frac{9\pi}{8}\right)$

$k = 3$: $3\left(\cos\frac{13\pi}{8} + i\sin\frac{13\pi}{8}\right)$

(b)

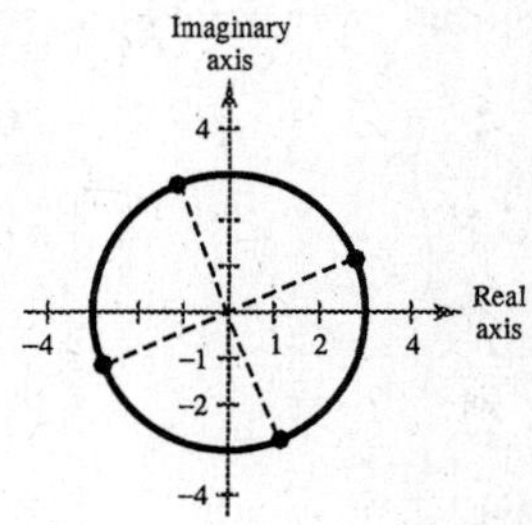

(c) $2.7716 + 1.1481i$

$-1.1481 + 2.7716i$

$-2.7716 - 1.1481i$

$1.1481 - 2.7716i$

35. (a) Cube roots of $-\frac{125}{2}(1 + \sqrt{3}i) = 125\left(\cos\frac{4\pi}{3} + i\sin\frac{4\pi}{3}\right)$:

$$\sqrt[3]{125}\left[\cos\left(\frac{\frac{4\pi}{3} + 2k\pi}{3}\right) + i\sin\left(\frac{\frac{4\pi}{3} + 2k\pi}{3}\right)\right],\ k = 0, 1, 2$$

$k = 0$: $5\left(\cos\frac{4\pi}{9} + i\sin\frac{4\pi}{9}\right)$

$k = 1$: $5\left(\cos\frac{10\pi}{9} + i\sin\frac{10\pi}{9}\right)$

$k = 2$: $5\left(\cos\frac{16\pi}{9} + i\sin\frac{16\pi}{9}\right)$

(b)

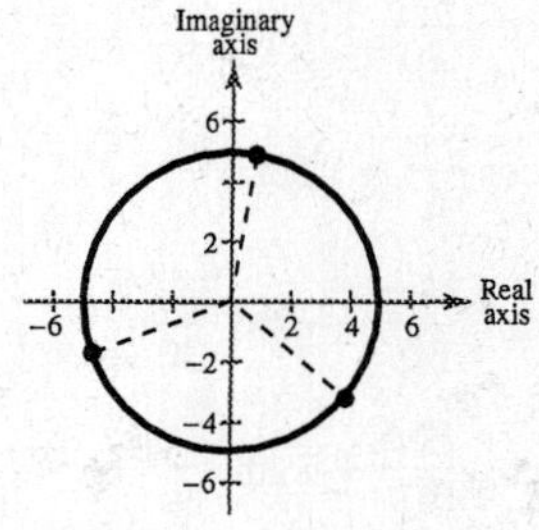

(c) $0.8682 + 4.924i,\ -4.6985 - 1.7101i,\ 3.8302 - 3.2139i$

37. (a) Fourth roots of $16 = 16(\cos 0 + i \sin 0)$

$$\sqrt[4]{16}\left[\cos\left(\frac{0 + 2\pi k}{4}\right) + i \sin\left(\frac{0 + 2\pi k}{4}\right)\right], k = 0, 1, 2, 3$$

$k = 0$: $2(\cos 0 + i \sin 0)$

$k = 1$: $2\left(\cos \frac{\pi}{2} + i \sin \frac{\pi}{2}\right)$

$k = 2$: $2(\cos \pi + i \sin \pi)$

$k = 3$: $2\left(\cos \frac{3\pi}{2} + i \sin \frac{3\pi}{2}\right)$

(b)

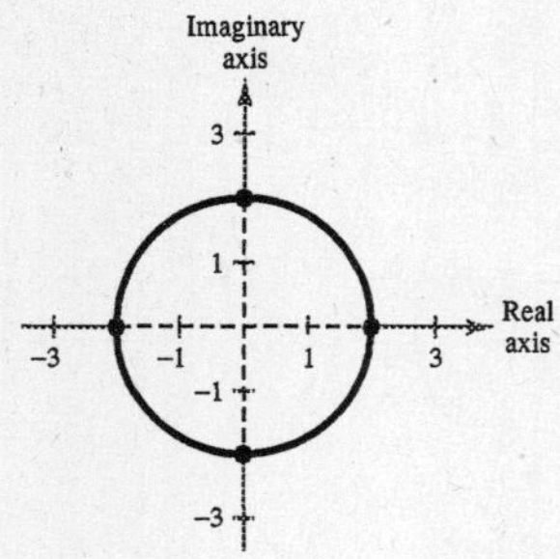

(c) $2, 2i, -2, -2i$

39. (a) Fifth roots of $1 = \cos 0 + i \sin 0$:

$$\cos\left(\frac{2k\pi}{5}\right) + i \sin\left(\frac{2k\pi}{5}\right), k = 0, 1, 2, 3, 4$$

$k = 0$: $\cos 0 + i \sin 0$

$k = 1$: $\cos \frac{2\pi}{5} + i \sin \frac{2\pi}{5}$

$k = 2$: $\cos \frac{4\pi}{5} + i \sin \frac{4\pi}{5}$

$k = 3$: $\cos \frac{6\pi}{5} + i \sin \frac{6\pi}{5}$

$k = 4$: $\cos \frac{8\pi}{5} + i \sin \frac{8\pi}{5}$

(b)

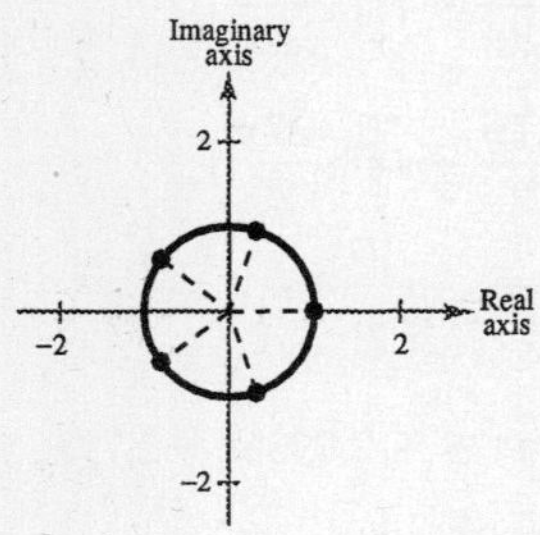

(c) $1, 0.3090 + 0.9511i, -0.8090 + 0.5878i, -0.8090 - 0.5878i, 0.3090 - 0.9511i$

41. (a) The cube roots of $-125 = 125(\cos \pi + i \sin \pi)$ are:

$$\sqrt[3]{125}\left[\cos\left(\frac{\pi + 2\pi k}{3}\right) + i \sin\left(\frac{\pi + 2\pi k}{3}\right)\right], k = 0, 1, 2$$

$k = 0$: $5\left(\cos \frac{\pi}{3} + i \sin \frac{\pi}{3}\right)$

$k = 1$: $5(\cos \pi + i \sin \pi)$

$k = 2$: $5\left(\cos \frac{5\pi}{3} + i \sin \frac{5\pi}{3}\right)$

(b)

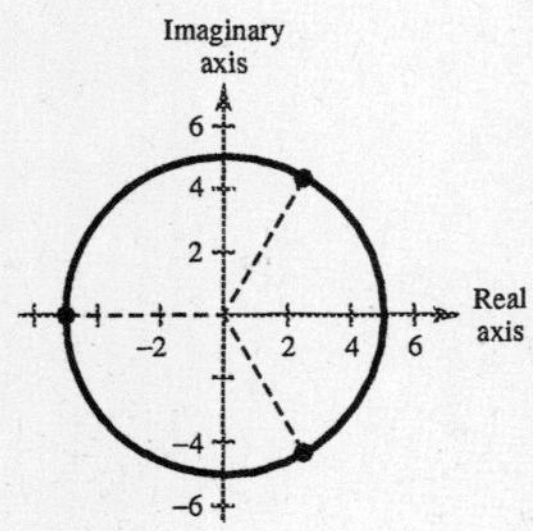

(c) $\frac{5}{2} + \frac{5\sqrt{3}}{2}i, -5, \frac{5}{2} - \frac{5\sqrt{3}}{2}i$

43. (a) The fifth roots of

$128(-1 + i) = 128\sqrt{2}\left(\cos\frac{3\pi}{4} + i\sin\frac{3\pi}{4}\right)$ are:

$$\sqrt[5]{128\sqrt{2}}\left[\cos\left(\frac{\frac{3\pi}{4} + 2\pi k}{5}\right) + i\sin\left(\frac{\frac{3\pi}{4} + 2\pi k}{5}\right)\right],$$

$k = 0, 1, 2, 3, 4$

$k = 0$: $2\sqrt[5]{4\sqrt{2}}\left(\cos\frac{3\pi}{20} + i\sin\frac{3\pi}{20}\right)$

$k = 1$: $2\sqrt[5]{4\sqrt{2}}\left(\cos\frac{11\pi}{20} + i\sin\frac{11\pi}{20}\right)$

$k = 2$: $2\sqrt[5]{4\sqrt{2}}\left(\cos\frac{19\pi}{20} + i\sin\frac{19\pi}{20}\right)$

$k = 3$: $2\sqrt[5]{4\sqrt{2}}\left(\cos\frac{27\pi}{20} + i\sin\frac{27\pi}{20}\right)$

$k = 4$: $2\sqrt[5]{4\sqrt{2}}\left(\cos\frac{7\pi}{4} + i\sin\frac{7\pi}{4}\right)$

(b)

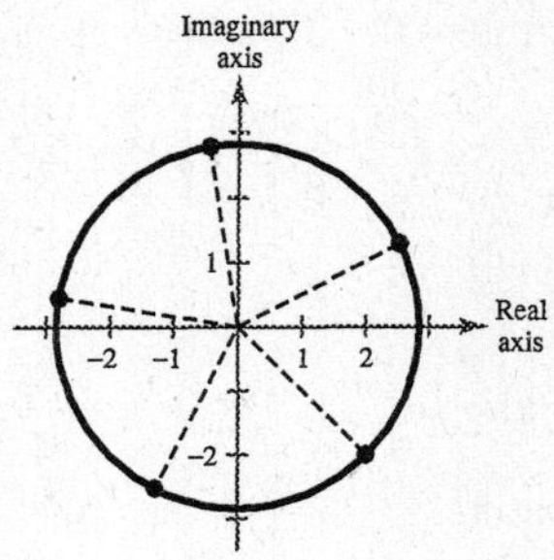

(c) $2.5201 + 1.2841i$, $-0.4425 + 2.7936i$, $-2.7936 + 0.4425i$, $-1.2841 - 2.5201i$, $2 - 2i$

45. $x^4 + i = 0$

$x^4 = -i$

The solutions are the fourth roots of $-i = \cos\frac{3\pi}{2} + i\sin\frac{3\pi}{2}$:

$$\sqrt[4]{1}\left[\cos\left(\frac{\frac{3\pi}{2} + 2k\pi}{4}\right) + i\sin\left(\frac{\frac{3\pi}{2} + 2k\pi}{4}\right)\right],\ k = 0, 1, 2, 3$$

$k = 0$: $\cos\frac{3\pi}{8} + i\sin\frac{3\pi}{8}$

$k = 1$: $\cos\frac{7\pi}{8} + i\sin\frac{7\pi}{8}$

$k = 2$: $\cos\frac{11\pi}{8} + i\sin\frac{11\pi}{8}$

$k = 3$: $\cos\frac{15\pi}{8} + i\sin\frac{15\pi}{8}$

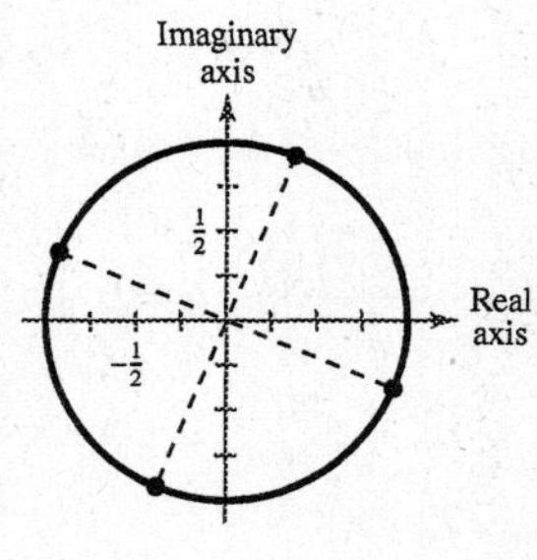

47. $x^6 + 1 = 0$

$x^6 = -1$

The solutions are the sixth roots of $-1 = \cos \pi + i \sin \pi$:

$\sqrt[6]{1}\left[\cos\left(\frac{\pi + 2\pi k}{6}\right) + i \sin\left(\frac{\pi + 2\pi k}{6}\right)\right]$,

$k = 0, 1, 2, 3, 4, 5$

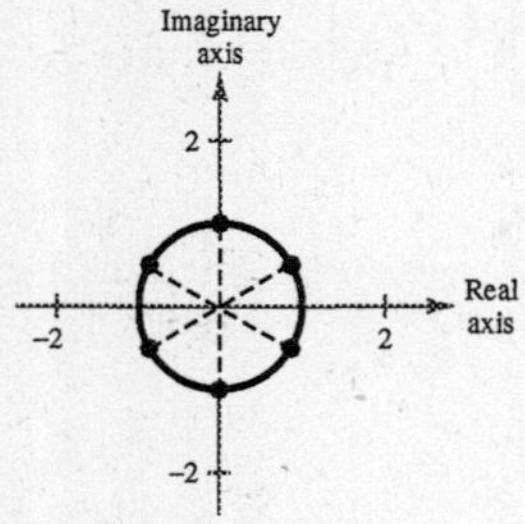

$k = 0$: $\cos \frac{\pi}{6} + i \sin \frac{\pi}{6}$

$k = 1$: $\cos \frac{\pi}{2} + i \sin \frac{\pi}{2}$

$k = 2$: $\cos \frac{5\pi}{6} + i \sin \frac{5\pi}{6}$

$k = 3$: $\cos \frac{7\pi}{6} + i \sin \frac{7\pi}{6}$

$k = 4$: $\cos \frac{3\pi}{2} + i \sin \frac{3\pi}{2}$

$k = 5$: $\cos \frac{11\pi}{6} + i \sin \frac{11\pi}{6}$

49. $x^5 + 243 = 0$

$x^5 = -243$

The solutions are the fifth roots of $-243 = 243(\cos \pi + i \sin \pi)$:

$\sqrt[5]{243}\left[\cos\left(\frac{\pi + 2k\pi}{5}\right) + i \sin\left(\frac{\pi + 2k\pi}{5}\right)\right]$, $k = 0, 1, 2, 3, 4$

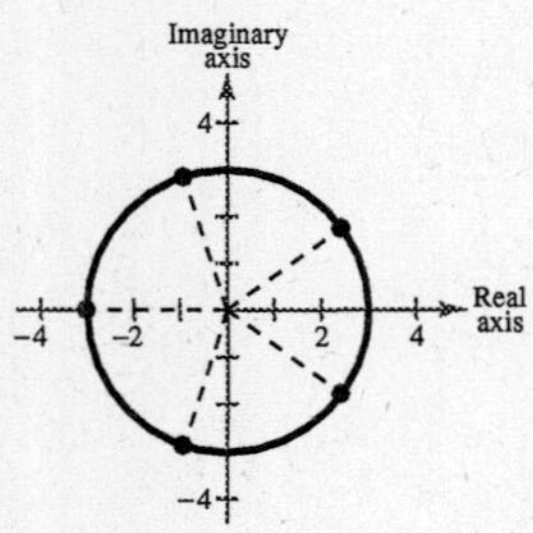

$k = 0$: $3\left(\cos \frac{\pi}{5} + i \sin \frac{\pi}{5}\right)$

$k = 1$: $3\left(\cos \frac{3\pi}{5} + i \sin \frac{3\pi}{5}\right)$

$k = 2$: $3(\cos \pi + i \sin \pi) = -3$

$k = 3$: $3\left(\cos \frac{7\pi}{5} + i \sin \frac{7\pi}{5}\right)$

$k = 4$: $3\left(\cos \frac{9\pi}{5} + i \sin \frac{9\pi}{5}\right)$

51. $x^5 - 32 = 0$

$x^5 = 32$

The solutions are the fifth roots of $32 = 32(\cos 0 + i \sin 0)$:

$$\sqrt[5]{32}\left[\cos\left(\frac{2\pi k}{5}\right) + i \sin\left(\frac{2\pi k}{5}\right)\right], k = 0, 1, 2, 3, 4$$

$k = 0$: $2(\cos 0 + i \sin 0)$

$k = 1$: $2\left(\cos \frac{2\pi}{5} + i \sin \frac{2\pi}{5}\right)$

$k = 2$: $2\left(\cos \frac{4\pi}{5} + i \sin \frac{4\pi}{5}\right)$

$k = 3$: $2\left(\cos \frac{6\pi}{5} + i \sin \frac{6\pi}{5}\right)$

$k = 4$: $2\left(\cos \frac{8\pi}{5} + i \sin \frac{8\pi}{5}\right)$

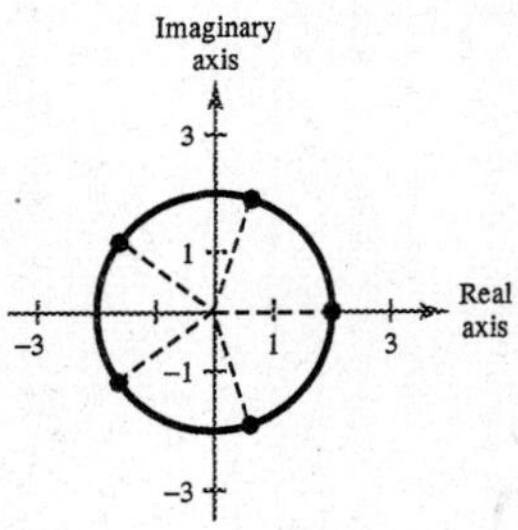

53. $x^4 + 16i = 0$

$x^4 = -16i$

The solutions are the fourth roots of $-16i = 16\left(\cos \frac{3\pi}{2} + i \sin \frac{3\pi}{2}\right)$:

$$\sqrt[4]{16}\left[\cos\left(\frac{\frac{3\pi}{2} + 2\pi k}{4}\right) + i \sin\left(\frac{\frac{3\pi}{2} + 2\pi k}{4}\right)\right], k = 0, 1, 2, 3$$

$k = 0$: $2\left(\cos \frac{3\pi}{8} + i \sin \frac{3\pi}{8}\right)$

$k = 1$: $2\left(\cos \frac{7\pi}{8} + i \sin \frac{7\pi}{8}\right)$

$k = 2$: $2\left(\cos \frac{11\pi}{8} + i \sin \frac{11\pi}{8}\right)$

$k = 3$: $2\left(\cos \frac{15\pi}{8} + i \sin \frac{15\pi}{8}\right)$

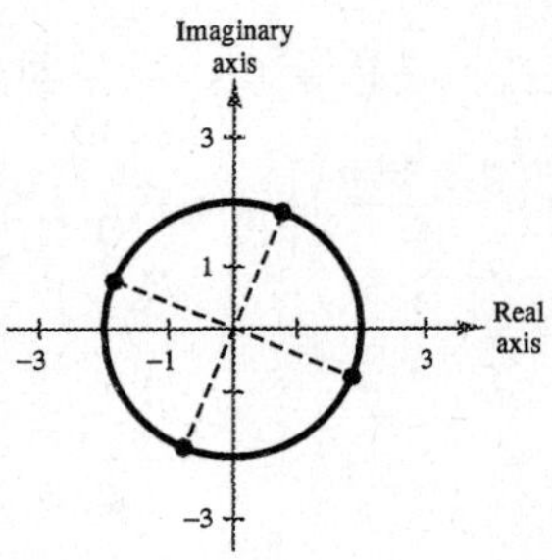

55. $x^4 - 16i = 0$

$x^4 = 16i$

The solutions are the fourth roots of $16i = 16\left(\cos \frac{\pi}{2} + i \sin \frac{\pi}{2}\right)$:

$$\sqrt[4]{16}\left[\cos\left(\frac{\frac{\pi}{2} + 2\pi k}{4}\right) + i \sin\left(\frac{\frac{\pi}{2} + 2\pi k}{4}\right)\right], k = 0, 1, 2, 3$$

$k = 0$: $2\left(\cos \frac{\pi}{8} + i \sin \frac{\pi}{8}\right)$

$k = 1$: $2\left(\cos \frac{5\pi}{8} + i \sin \frac{5\pi}{8}\right)$

$k = 2$: $2\left(\cos \frac{9\pi}{8} + i \sin \frac{9\pi}{8}\right)$

$k = 3$: $2\left(\cos \frac{13\pi}{8} + i \sin \frac{13\pi}{8}\right)$

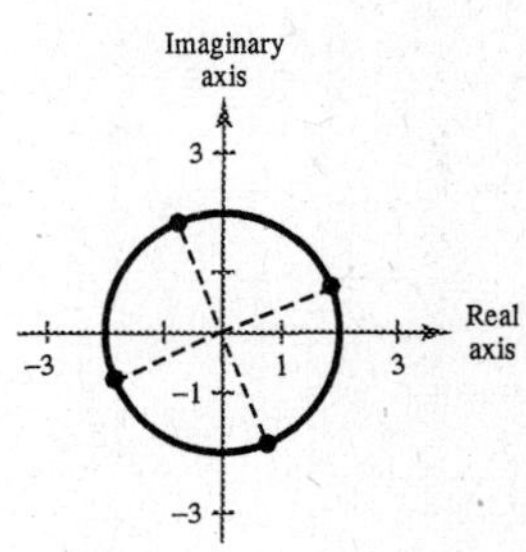

57. $x^3 - (1 - i) = 0$

$$x^3 = 1 - i = \sqrt{2}\left(\cos\frac{7\pi}{4} + i\sin\frac{7\pi}{4}\right)$$

The solutions are the cube roots of $1 - i$:

$$\sqrt[3]{\sqrt{2}}\left[\cos\left(\frac{\frac{7\pi}{4} + 2\pi k}{3}\right) + i\sin\left(\frac{\frac{7\pi}{4} + 2\pi k}{3}\right)\right],\ k = 0, 1, 2$$

$k = 0$: $\sqrt[6]{2}\left(\cos\frac{7\pi}{12} + i\sin\frac{7\pi}{12}\right)$

$k = 1$: $\sqrt[6]{2}\left(\cos\frac{5\pi}{4} + i\sin\frac{5\pi}{4}\right)$

$k = 2$: $\sqrt[6]{2}\left(\cos\frac{23\pi}{12} + i\sin\frac{23\pi}{12}\right)$

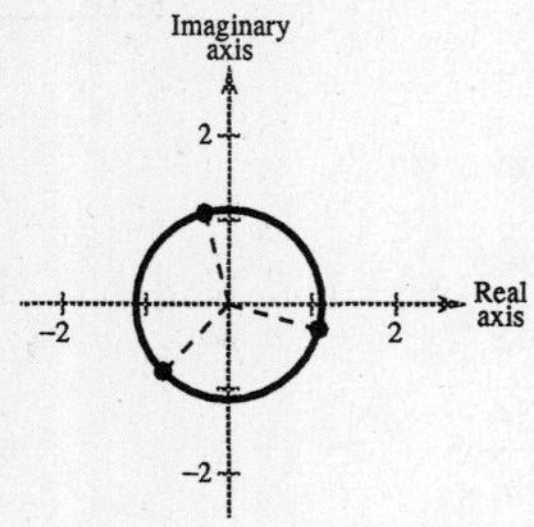

59. $x^6 + (1 + i) = 0$

$$x^6 = -(1 + i) = -1 - i$$

The solutions are the sixth roots of $-1 - i = \sqrt{2}\left(\cos\frac{5\pi}{4} + i\sin\frac{5\pi}{4}\right)$

$$\sqrt[6]{\sqrt{2}}\left[\cos\left(\frac{\frac{5\pi}{4} + 2\pi k}{6}\right) + i\sin\left(\frac{\frac{5\pi}{4} + 2\pi k}{6}\right)\right],\ k = 0, 1, 2, 3, 4, 5$$

$k = 0$: $\sqrt[12]{2}\left(\cos\frac{5\pi}{24} + i\sin\frac{5\pi}{24}\right)$

$k = 1$: $\sqrt[12]{2}\left(\cos\frac{13\pi}{24} + i\sin\frac{13\pi}{24}\right)$

$k = 2$: $\sqrt[12]{2}\left(\cos\frac{7\pi}{8} + i\sin\frac{7\pi}{8}\right)$

$k = 3$: $\sqrt[12]{2}\left(\cos\frac{29\pi}{24} + i\sin\frac{29\pi}{24}\right)$

$k = 4$: $\sqrt[12]{2}\left(\cos\frac{37\pi}{24} + i\sin\frac{37\pi}{24}\right)$

$k = 5$: $\sqrt[12]{2}\left(\cos\frac{15\pi}{8} + i\sin\frac{15\pi}{8}\right)$

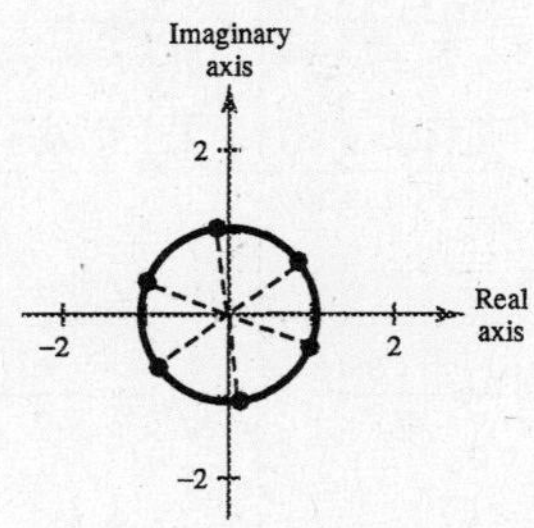

61. False. They are equally spaced along the circle centered at the origin with radius $\sqrt[n]{r}$.

63. $-\frac{1}{2}(1 + \sqrt{3}i) = \cos\frac{4\pi}{3} + i\sin\frac{4\pi}{3}$

$$\left[-\frac{1}{2}(1 + \sqrt{3}i)\right]^6 = \left[\cos\frac{4\pi}{3} + i\sin\frac{4\pi}{3}\right]^6$$

$$= \cos 8\pi + i\sin 8\pi$$

$$= 1$$

65. (a) In trigonometric form we have:

$2(\cos 30° + i\sin 30°)$

$2(\cos 150° + i\sin 150°)$

$2(\cos 270° + i\sin 270°)$

(b) There are three roots evenly spaced around a circle of radius 2. Therefore, they represent the cube roots of some number of modulus 8. Cubing them shows that they are all cube roots of $8i$.

(c) $[2(\cos 30° + i\sin 30°)]^3 = 8i$

$[2(\cos 150° + i\sin 150°)]^3 = 8i$

$[2(\cos 270° + i\sin 270°)]^3 = 8i$

For 67–75, use the following figure.

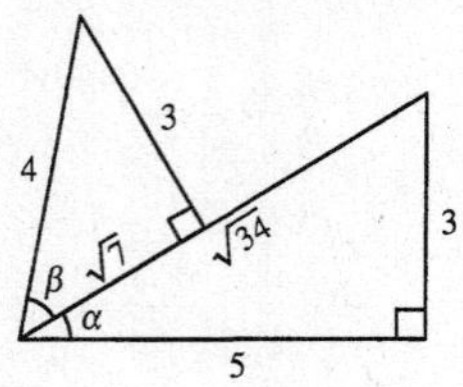

67. $\cos(\alpha + \beta) = \cos\alpha\cos\beta - \sin\alpha\sin\beta$

$$= \frac{5}{\sqrt{34}} \cdot \frac{\sqrt{7}}{4} - \frac{3}{\sqrt{34}} \cdot \frac{3}{4}$$

$$= \frac{5\sqrt{7} - 9}{4\sqrt{34}} = \frac{\sqrt{34}}{136}(5\sqrt{7} - 9)$$

69. $\sin(\alpha - \beta) = \sin\alpha\cos\beta - \cos\alpha\sin\beta$

$$= \frac{3}{\sqrt{34}} \cdot \frac{\sqrt{7}}{4} - \frac{5}{\sqrt{34}} \cdot \frac{3}{4}$$

$$= \frac{3\sqrt{7} - 15}{4\sqrt{34}} = \frac{3\sqrt{34}}{136}(\sqrt{7} - 5)$$

71. $\tan(\alpha + \beta) = \dfrac{\tan\alpha + \tan\beta}{1 - \tan\alpha\tan\beta}$

$$= \frac{\frac{3}{5} + \frac{3}{\sqrt{7}}}{1 - \left(\frac{3}{5}\right)\left(\frac{3}{\sqrt{7}}\right)} \cdot \frac{5\sqrt{7}}{5\sqrt{7}}$$

$$= \frac{3\sqrt{7} + 15}{5\sqrt{7} - 9} \cdot \frac{5\sqrt{7} + 9}{5\sqrt{7} + 9}$$

$$= \frac{240 + 102\sqrt{7}}{94} = \frac{120 + 51\sqrt{7}}{47}$$

73. $\tan 2\alpha = \dfrac{2\tan\alpha}{1 - \tan^2\alpha}$

$$= \frac{2(3/5)}{1 - (3/5)^2}$$

$$= \frac{6}{5} \cdot \frac{25}{16} = \frac{15}{8}$$

75. $\sin\dfrac{\beta}{2} = \sqrt{\dfrac{1 - \cos\beta}{2}} = \sqrt{\dfrac{1 - \sqrt{7}/4}{2}} = \sqrt{\dfrac{4 - \sqrt{7}}{8}} = \dfrac{1}{2}\sqrt{\dfrac{4 - \sqrt{7}}{2}}$

77. $\mathbf{u} = \langle 10, 0 \rangle$

$\|\mathbf{u}\| = \sqrt{10^2 + 0^2} = \sqrt{100} = 10$

$\dfrac{\mathbf{u}}{\|\mathbf{u}\|} = \dfrac{1}{10}\langle 10, 0 \rangle = \langle 1, 0 \rangle$

79. $\mathbf{v} = 12\mathbf{i} - 5\mathbf{j}$

$\|\mathbf{v}\| = \sqrt{12^2 + (-5)^2} = \sqrt{144 + 25} = \sqrt{169} = 13$

$\dfrac{\mathbf{v}}{\|\mathbf{v}\|} = \dfrac{1}{13}(12\mathbf{i} - 5\mathbf{j}) = \dfrac{12}{13}\mathbf{i} - \dfrac{5}{13}\mathbf{j}$

81. $\|\mathbf{u}\|^2 = \mathbf{u} \cdot \mathbf{u} = \langle -3, 4 \rangle \cdot \langle -3, 4 \rangle = 25$

$\|\mathbf{u}\| = \sqrt{25} = 5$

83. $\|\mathbf{u}\|^2 = \mathbf{u} \cdot \mathbf{u} = \langle -9, 40 \rangle \cdot \langle -9, 40 \rangle = 1681$

$\|\mathbf{u}\| = \sqrt{1681} = 41$

85. $\|\mathbf{u}\|^2 = \mathbf{u} \cdot \mathbf{u} = 22^2 + 3^2 = 493$

$\|\mathbf{u}\| = \sqrt{493}$

87. $\mathbf{u} = 13\mathbf{i} + 6\mathbf{j}$

$\|\mathbf{u}\|^2 = \mathbf{u} \cdot \mathbf{u} = 13^2 + 6^2 = 169 + 36 = 205$

$\|\mathbf{u}\| = \sqrt{205}$

Review Exercises for Chapter 4

Solutions to Odd-Numbered Exercises

1. $6 + \sqrt{-4} = 6 + 2i$

3. $i^2 + 3i = -1 + 3i$

5. $(7 + 5i) + (-4 + 2i) = (7 - 4) + (5i + 2i) = 3 + 7i$

7. $5i(13 - 8i) = 65i - 40i^2 = 40 + 65i$

9. $(10 - 8i)(2 - 3i) = 20 - 30i - 16i + 24i^2 = -4 - 46i$

11. $\frac{6+i}{4-i} = \frac{6+i}{4-i} \cdot \frac{4+i}{4+i} = \frac{24 + 10i + i^2}{16+1} = \frac{23+10i}{17}$

$= \frac{23}{17} + \frac{10}{17}i$

13. $\frac{4}{2-3i} + \frac{2}{1+i} = \frac{4}{2-3i} \cdot \frac{2+3i}{2+3i} + \frac{2}{1+i} \cdot \frac{1-i}{1-i}$

$= \frac{8+12i}{4+9} + \frac{2-2i}{1+1}$

$= \frac{8}{13} + \frac{12}{13}i + 1 - i$

$= \left(\frac{8}{13} + 1\right) + \left(\frac{12}{13}i - i\right)$

$= \frac{21}{13} - \frac{1}{13}i$

15. $3x^2 + 1 = 0$

$3x^2 = -1$

$x^2 = -\frac{1}{3}$

$x = \pm\sqrt{-\frac{1}{3}}$

$= \pm\sqrt{\frac{1}{3}}i$

17. $x^2 - 2x + 10 = 0$

$x^2 - 2x + 1 = -10 + 1$

$(x-1)^2 = -9$

$x - 1 = \pm\sqrt{-9}$

$x = 1 \pm 3i$

19. $x^5 - 2x^4 + 3x^2 - 5 = 0$ is a fifth degree polynomial equation, so it has **five** zeros in the complex number system.

21. $\frac{1}{2}x^4 + \frac{2}{3}x^3 - x^2 + \frac{3}{10} = 0$ is a fourth degree polynomial equation, so it has **four** zeros in the complex number system.

23. $6x^2 + x - 2 = 0$

$b^2 - 4ac = 1^2 - 4(6)(-2) = 49 > 0$

Two real solutions

25. $0.13x^2 - 0.45x + 0.65 = 0$

$b^2 - 4ac = (-0.45)^2 - 4(0.13)(0.65) = -0.1355 < 0$

No real solutions

27. $x^2 - 2x = 0$

$x(x-2) = 0$

$x = 0, x = 2$

29. $x^2 + 8x + 10 = 0$

$x^2 + 8x = -10$

$x^2 + 8x + 16 = -10 + 16$

$(x+4)^2 = 6$

$x + 4 = \pm\sqrt{6}$

$x = -4 \pm \sqrt{6}$

31. $2x^2 + 2x + 3 = 0$

$x = \frac{-2 \pm \sqrt{2^2 - 4(2)(3)}}{2(2)}$

$= \frac{-2 \pm \sqrt{-20}}{4}$

$= \frac{-2 \pm 2\sqrt{5}i}{4}$

$= -\frac{1}{2} \pm \frac{\sqrt{5}}{2}i$

33. $f(x) = 2x^3 - 3x^2 + 50x - 75$

$2x^3 - 3x^2 + 50x - 75 = 0$

$x^2(2x - 3) + 25(2x - 3) = 0$

$(2x - 3)(x^2 + 25) = 0$

$2x - 3 = 0 \Rightarrow x = \frac{3}{2}$

$x^2 + 25 = 0 \Rightarrow x = \pm 5i$

Zeros: $x = \frac{3}{2}, \pm 5i$

35. $f(x) = 4x^4 + 3x^2 - 10$

$4x^4 + 3x^2 - 10 = 0$

$(4x^2 - 5)(x^2 + 2) = 0$

$4x^2 - 5 = 0 \Rightarrow x = \pm\frac{\sqrt{5}}{2}$

$x^2 + 2 = 0 \Rightarrow x = \pm\sqrt{2}i$

Zeros: $x = \pm\frac{\sqrt{5}}{2}, \pm\sqrt{2}i$

37. $f(x) = x^3 + 3x^2 - 24x + 28$

Zero: $2 \Rightarrow x - 2$ is a factor of $f(x)$.

$$
\begin{array}{r}
x^2 + 5x - 14 \\
x - 2 \,\overline{)\, x^3 + 3x^2 - 24x + 28} \\
\underline{x^3 - 2x^2} \qquad\qquad \\
5x^2 - 24x \qquad \\
\underline{5x^2 - 10x} \qquad \\
-14x + 28 \\
\underline{-14x + 28} \\
0
\end{array}
$$

Thus, $f(x) = (x - 2)(x^2 + 5x - 14)$

$= (x - 2)(x - 2)(x + 7)$

$= (x - 2)^2(x + 7)$

The zeros of $f(x)$ are $x = 2$ and $x = -7$.

39. $f(x) = x^3 + 3x^2 - 5x + 25$

Zero: $-5 \Rightarrow x + 5$ is a factor of $f(x)$.

$$
\begin{array}{r}
x^2 - 2x + 5 \\
x + 5 \,\overline{)\, x^3 + 3x^2 - 5x + 25} \\
\underline{x^3 + 5x^2} \qquad\qquad \\
-2x^2 - 5x \qquad \\
\underline{-2x^2 - 10x} \qquad \\
5x + 25 \\
\underline{5x + 25} \\
0
\end{array}
$$

Thus, $f(x) = (x + 5)(x^2 - 2x + 5)$ and by the Quadratic Formula, the zeros of $x^2 - 2x + 5$ are $1 \pm 2i$. The zeros of $f(x)$ are $x = -5$ and $x = 1 \pm 2i$.

$f(x) = (x + 5)[x - (1 + 2i)][x - (1 - 2i)]$

$= (x + 5)(x - 1 - 2i)(x - 1 + 2i)$

41. $h(x) = 2x^3 - 19x^2 + 58x + 34$

Zero: $5 + 3i \Rightarrow 5 - 3i$ is also a zero and $(x - 5 - 3i)(x - 5 + 3i) = x^2 - 10x + 34$ is a factor of $h(x)$.

$$
\begin{array}{r}
2x + 1 \\
x^2 - 10x + 34 \,\overline{)\, 2x^3 - 19x^2 + 58x + 34} \\
\underline{2x^3 - 20x^2 + 68x} \qquad \\
x^2 - 10x + 34 \\
\underline{x^2 - 10x + 34} \\
0
\end{array}
$$

Thus, $h(x) = (2x + 1)(x^2 - 10x + 34)$

$= (2x + 1)(x - 5 - 3i)(x - 5 + 3i)$

The zeros of $h(x)$ are $x = -\frac{1}{2}$ and $x = 5 \pm 3i$.

43. $f(x) = x^4 + 5x^3 + 2x^2 - 50x - 84$

Zero: $-3 + \sqrt{5}i \Rightarrow -3 - \sqrt{5}i$ is also a zero and $(x + 3 - \sqrt{5}i)(x + 3 + \sqrt{5}i) = x^2 + 6x + 14$ is a factor of $f(x)$.

$$
\begin{array}{r}
x^2 - x - 6 \\
x^2 + 6x + 14 \,\overline{)\, x^4 + 5x^3 + 2x^2 - 50x - 84} \\
\underline{x^4 + 6x^3 + 14x^2} \qquad\qquad \\
-x^3 - 12x^2 - 50x \qquad \\
\underline{-x^3 - 6x^2 - 14x} \qquad \\
-6x^2 - 36x - 84 \\
\underline{-6x^2 - 36x - 84} \\
0
\end{array}
$$

Thus, $f(x) = (x^2 - x - 6)(x^2 + 6x + 14)$

$= (x + 2)(x - 3)(x + 3 - \sqrt{5}i)(x + 3 + \sqrt{5}i)$

The zeros of $f(x)$ are $x = -2$, $x = 3$, and $x = -3 \pm \sqrt{5}i$.

45. Zeros: $1, 1, \frac{1}{4}, -\frac{2}{3}$

$$f(x) = (x-1)(x-1)(4x-1)(3x+2)$$
$$= (x^2-2x+1)(12x^2+5x-2)$$
$$= 12x^4 - 19x^3 + 9x - 2$$

47. Zeros: $3, 2-\sqrt{3}, 2+\sqrt{3}$

$$f(x) = (x-3)\left(x-2+\sqrt{3}\right)\left(x-2-\sqrt{3}\right)$$
$$= (x-3)(x^2-4x+1)$$
$$= x^3 - 7x^2 + 13x - 3$$

49. Zeros: $\frac{2}{3}, 4, \sqrt{3}i, -\sqrt{3}i$

$$f(x) = (3x-2)(x-4)\left(x-\sqrt{3}i\right)\left(x+\sqrt{3}i\right)$$
$$= (3x^2-14x+8)(x^2+3)$$
$$= 3x^4 - 14x^3 + 17x^2 - 42x + 24$$

51. Zeros: $-\sqrt{2}i, \sqrt{2}i, -5i, 5i$

$$f(x) = \left(x+\sqrt{2}i\right)\left(x-\sqrt{2}i\right)(x+5i)(x-5i)$$
$$= (x^2+2)(x^2+25)$$
$$= x^4 + 27x^2 + 50$$

53. $P = xp - C$

$$9{,}000{,}000 = x(140 - 0.0001x) - (75x + 100{,}000)$$
$$9{,}000{,}000 = 140x - 0.0001x^2 - 75x - 100{,}000$$
$$0.0001x^2 - 65x + 9{,}100{,}000 = 0$$
$$b^2 - 4ac = (-65)^2 - 4(0.0001)(9{,}100{,}000) = 585 > 0$$

There are two real solutions. By the Quadratic Formula we have:

$x \approx 445{,}934$ units or $x \approx 204{,}066$ units

$p \approx \$95.41$ $\quad$ $p \approx \$119.59$

There are two possible values for p that will yield a profit of approximately 9 million.

55.

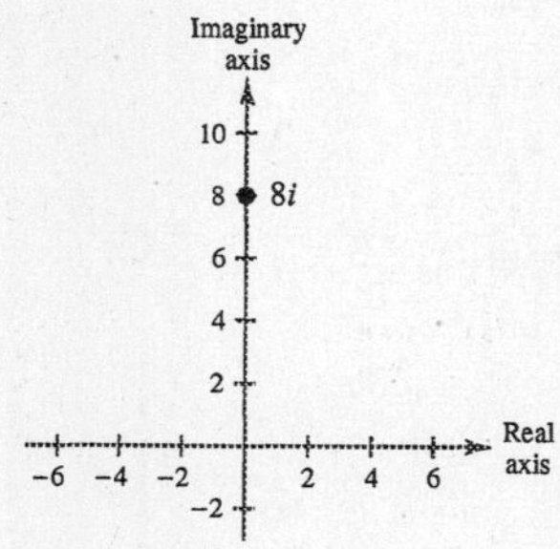

$$|8i| = \sqrt{0^2+8^2} = \sqrt{64} = 8$$

57. $z = 5 + 3i$

$$|z| = \sqrt{5^2+3^2} = \sqrt{34}$$

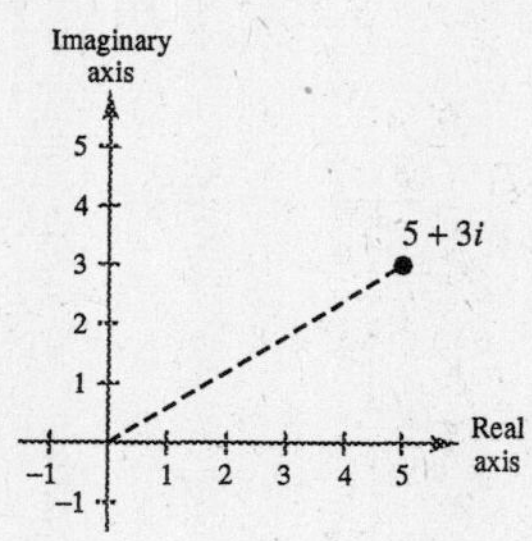

59. $z = 5 - 5i$

$$r = \sqrt{5^2+(-5)^2} = 5\sqrt{2}$$
$$\tan\theta = \frac{-5}{5} = -1 \Rightarrow \theta = \frac{7\pi}{4}$$
$$z = 5\sqrt{2}\left(\cos\frac{7\pi}{4} + i\sin\frac{7\pi}{4}\right)$$

61. $z = -3\sqrt{3} + 3i$

$$r = \sqrt{\left(-3\sqrt{3}\right)^2 + 3^2} = 6$$
$$\tan\theta = \frac{3}{-3\sqrt{3}} = -\frac{\sqrt{3}}{3} \Rightarrow \theta = \frac{5\pi}{6}$$
$$z = 6\left(\cos\frac{5\pi}{6} + i\sin\frac{5\pi}{6}\right)$$

63. $z_1 = 2\sqrt{3} - 2i,\ z_2 = -10i$

(a) $$r_1 = \sqrt{\left(2\sqrt{3}\right)^2 + (-2)^2} = 4$$
$$\tan\theta_1 = \frac{-2}{2\sqrt{3}} = -\frac{\sqrt{3}}{3} \Rightarrow \theta_1 = \frac{11\pi}{6}$$
$$z_1 = 4\left(\cos\frac{11\pi}{6} + i\sin\frac{11\pi}{6}\right)$$
$$r_2 = \sqrt{0^2+(-10)^2} = 10$$
$$\theta_2 = \frac{3\pi}{2}$$
$$z_2 = 10\left(\cos\frac{3\pi}{2} + i\sin\frac{3\pi}{2}\right)$$

(b) $$z_1z_2 = 4 \cdot 10\left(\cos\left(\frac{11\pi}{6} + \frac{3\pi}{2}\right) + i\sin\left(\frac{11\pi}{6} + \frac{3\pi}{2}\right)\right)$$
$$= 40\left(\cos\frac{10\pi}{3} + i\sin\frac{10\pi}{3}\right)$$
$$= 40\left(\cos\frac{4\pi}{3} + i\sin\frac{4\pi}{3}\right)$$
$$\frac{z_1}{z_2} = \frac{4}{10}\left(\cos\left(\frac{11\pi}{6} - \frac{3\pi}{2}\right) + \sin\left(\frac{11\pi}{6} - \frac{3\pi}{2}\right)\right)$$
$$= \frac{2}{5}\left(\cos\frac{\pi}{3} + i\sin\frac{\pi}{3}\right)$$

65. $\left[5\left(\cos\frac{\pi}{12} + i\sin\frac{\pi}{12}\right)\right]^4 = 5^4\left[\cos\left(\frac{4\pi}{12}\right) + i\sin\left(\frac{4\pi}{12}\right)\right]$

$$= 625\left(\cos\frac{\pi}{3} + i\sin\frac{\pi}{3}\right)$$

$$= \frac{625}{2} + \frac{625\sqrt{3}}{2}i$$

67. $(2 + 3i)^6 = \left[\sqrt{13}\left(\cos\left(\arctan\left(\frac{3}{2}\right)\right) + i\sin\left(\arctan\left(\frac{3}{2}\right)\right)\right)\right]^6$

$$= (\sqrt{13})^6\left[\cos\left(6\arctan\left(\frac{3}{2}\right)\right) + i\sin\left(6\arctan\left(\frac{3}{2}\right)\right)\right]$$

$$= 2035 - 828i$$

69. Sixth roots of $-729i = 729\left(\cos\frac{3\pi}{2} + i\sin\frac{3\pi}{2}\right)$:

$$\sqrt[6]{729}\left[\cos\left(\frac{\frac{3\pi}{2} + 2k\pi}{6}\right) + i\sin\left(\frac{\frac{3\pi}{2} + 2k\pi}{6}\right)\right]$$

$k = 0, 1, 2, 3, 4, 5$

$k = 0$: $3\left(\cos\frac{\pi}{4} + i\sin\frac{\pi}{4}\right)$

$k = 1$: $3\left(\cos\frac{7\pi}{12} + i\sin\frac{7\pi}{12}\right)$

$k = 2$: $3\left(\cos\frac{11\pi}{12} + i\sin\frac{11\pi}{12}\right)$

$k = 3$: $3\left(\cos\frac{5\pi}{4} + i\sin\frac{5\pi}{4}\right)$

$k = 4$: $3\left(\cos\frac{19\pi}{12} + i\sin\frac{19\pi}{12}\right)$

$k = 5$: $3\left(\cos\frac{23\pi}{12} + i\sin\frac{23\pi}{12}\right)$

71. $x^4 + 81 = 0$

$x^4 = -81$

The solutions are the fourth roots of $-81 = 81(\cos\pi + i\sin\pi)$:

$$\sqrt[4]{81}\left[\cos\left(\frac{\pi + 2k\pi}{4}\right) + i\sin\left(\frac{\pi + 2k\pi}{4}\right)\right]$$

$k = 0, 1, 2, 3$

$k = 0$: $3\left(\cos\frac{\pi}{4} + i\sin\frac{\pi}{4}\right) = \frac{3\sqrt{2}}{2} + \frac{3\sqrt{2}}{2}i$

$k = 1$: $3\left(\cos\frac{3\pi}{4} + i\sin\frac{3\pi}{4}\right) = -\frac{3\sqrt{2}}{2} + \frac{3\sqrt{2}}{2}i$

$k = 2$: $3\left(\cos\frac{5\pi}{4} + i\sin\frac{5\pi}{4}\right) = -\frac{3\sqrt{2}}{2} - \frac{3\sqrt{2}}{2}i$

$k = 3$: $3\left(\cos\frac{7\pi}{4} + i\sin\frac{7\pi}{4}\right) = \frac{3\sqrt{2}}{2} - \frac{3\sqrt{2}}{2}i$

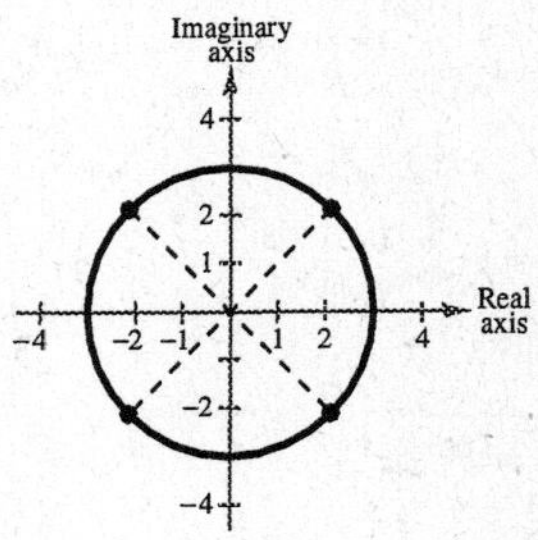

73. $x^3 + 8i = 0$

$x^3 = -8i$

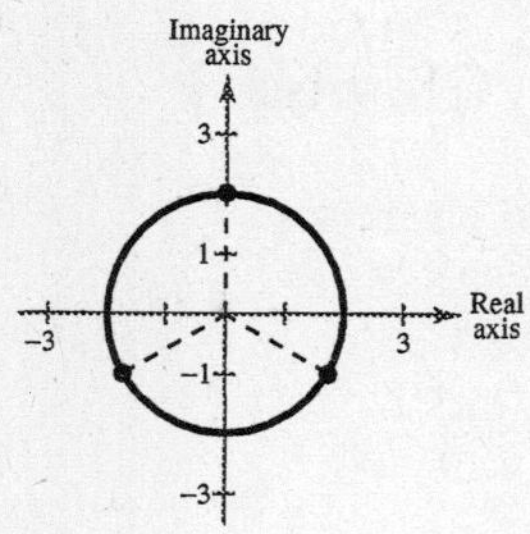

The solutions are the cube roots of

$-8i = 8\left(\cos\frac{3\pi}{2} + i\sin\frac{3\pi}{2}\right)$:

$$\sqrt[3]{8}\left[\cos\left(\frac{\frac{3\pi}{2} + 2k\pi}{3}\right) + i\sin\left(\frac{\frac{3\pi}{2} + 2k\pi}{3}\right)\right]$$

$k = 0, 1, 2$

$k = 0$: $2\left(\cos\frac{\pi}{2} + i\sin\frac{\pi}{2}\right) = 2i$

$k = 1$: $2\left(\cos\frac{7\pi}{6} + i\sin\frac{7\pi}{6}\right) = -\sqrt{3} - i$

$k = 2$: $2\left(\cos\frac{11\pi}{6} + i\sin\frac{11\pi}{6}\right) = \sqrt{3} - i$

75. False.

$\sqrt{-18}\sqrt{-2} = (3\sqrt{2}i)(\sqrt{2}i) = 6i^2 = -6$, whereas

$\sqrt{(-18)(-2)} = \sqrt{36} = 6$.

77. False. A fourth degree polynomial can have at most four zeros, and complex zeros occur in conjugate pairs.

79. (a) From the graph, the 3 roots are

$4(\cos 60° + i\sin 60°)$

$4(\cos 180° + i\sin 180°)$

$4(\cos 300° + i\sin 300°)$

(b) Since there are three evenly spaced roots on the circle of a radius 4, they are cube roots of a complex number of modulus $4^3 = 64$. Cubing them yields -64.

$[4(\cos 60° + i\sin 60°)]^3 = -64$

$[4(\cos 180° + i\sin 180°)]^3 = -64$

$[4(\cos 300° + i\sin 300°)]^3 = -64$

81. From the graph,

$z_1 = 2(\cos\theta + i\sin\theta)$ and

$z_2 = 2(\cos(\pi - \theta) + i\sin(\pi - \theta))$.

$z_1z_2 = 2 \cdot 2(\cos(\theta + \pi - \theta) + i\sin(\theta + \pi - \theta))$

$= 4(\cos\pi + i\sin\pi) = -4$

$\frac{z_1}{z_2} = \frac{2}{2}(\cos(\theta - (\pi - \theta)) + i\sin(\theta - (\pi - \theta)))$

$= \cos(2\theta - \pi) + i\sin(2\theta - \pi)$

$= \cos 2\theta\cos\pi + \sin 2\theta\sin\pi + i(\sin 2\theta\cos\pi - \cos 2\theta\sin\pi)$

$= -\cos 2\theta - i\sin 2\theta$

Problem Solving for Chapter 4

Solutions to Odd-Numbered Exercises

1. (a) $\left(\dfrac{-2 + 2\sqrt{3}i}{2}\right)^3 = 8$

$\left(\dfrac{-2 - 2\sqrt{3}i}{2}\right)^3 = 8$

$(2)^3 = 8$

(b) $\left(\dfrac{-3 + 3\sqrt{3}i}{2}\right)^3 = 27$

$\left(\dfrac{-3 - 3\sqrt{3}i}{2}\right)^3 = 27$

$(3)^3 = 27$

(c) The cube roots of a positive real number "a" are:

(i) $\sqrt[3]{a}$

(ii) $\dfrac{-\sqrt[3]{a} + \sqrt[3]{a}\sqrt{3}i}{2}$

(iii) $\dfrac{-\sqrt[3]{a} - \sqrt[3]{a}\sqrt{3}i}{2}$

3. $(a + bi)(a - bi) = a^2 - abi + abi - b^2i^2 = a^2 = b^2$

Since a and b are real numbers, $a^2 + b^2$ is also a real number.

5. $x^2 - 2kx + k = 0$

$$x^2 - 2kx = -k$$

$$x^2 - 2kx + k^2 = k^2 - k$$

$$(x - k)^2 = k(k - 1)$$

$$x = k \pm \sqrt{k(k - 1)}$$

(a) If the equation has two real solutions, then $k(k - 1) > 0$. This means that $k < 0$ or $k > 1$.

(b) If the equation has two complex solutions, then $k(k - 1) < 0$. This means that $0 < k < 1$.

7. (a) $g(x) = f(x - 2)$

No. This function is a horizontal shift of $f(x)$. Note that x is a zero of g if and only if $x - 2$ is a zero of f; the number of real and complex zeros is not affected by a horizontal shift.

(b) $g(x) = f(2x)$

No. Since x is a zero of g if and only if $2x$ is a zero of f, the number of real and complex zeros of g is the same as the number of real and complex zeros of f.

9.

Interval:	$(-\infty, -2)$,	$(-2, 1)$,	$(1, 4)$,	$(4, \infty)$
Value of $f(x)$:	Positive	Negative	Negative	Positive

(a) Zeros of $f(x)$: $x = -2, x = 1, x = 4$.

(b) The graph touches the x-axis at $x = 1$.

(c) The least possible degree of the function is 4 because there are at least four real zeros (1 is repeated) and a function can have at most the number of real zeros equal to the degree of the function. The degree cannot be odd by the definition of multiplicity.

(d) The leading coefficient of f is positive. From the information in the table, you can conclude that the graph will eventually rise to the left and to the right.

(e) $f(x) = (x + 2)(x - 1)^2(x - 4) = x^4 - 4x^3 - 3x^2 + 14x - 8$

(This answer is not unique.)

(f)

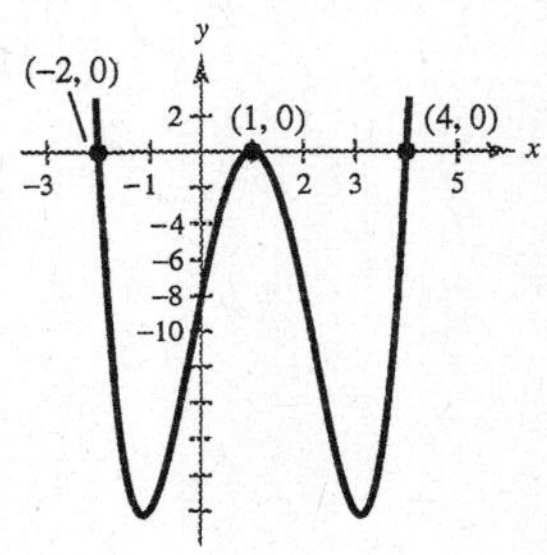

11. (a)

Function	Zeros	Sum of Zeros	Product of Zeros
$f_1(x) = x^2 - 5x + 6$	$2, 3$	5	6
$f_2(x) = x^3 - 7x + 6$	$-3, 1, 2$	0	-6
$f_3(x) = x^4 + 2x^3 + x^2 + 8x - 12$	$-3, 1, \pm 2i$	-2	-12
$f_4(x) = x^5 - 3x^4 - 9x^3 + 25x^2 - 6x$	$-3, 0, 2, 2 \pm \sqrt{3}$	3	0

(b) Conjecture: Sum of Zeros $= -a_{n-1}$

(c) Conjecture: Product of Zeros $= \begin{cases} a_0, \text{ if } n \text{ is even} \\ -a_0, \text{ if } n \text{ is odd} \end{cases}$

13. Let $z = x + yi$ and $\bar{z} = x - yi$

$$|z - 1| \cdot |\bar{z} - 1| = 1$$

$$\sqrt{(x-1)^2 + y^2}\sqrt{(x-1)^2 + (-y)^2} = 1$$

$$(x-1)^2 + y^2 = 1$$

The graph of the solution set is a circle centered at $(1, 0)$ of radius 1.

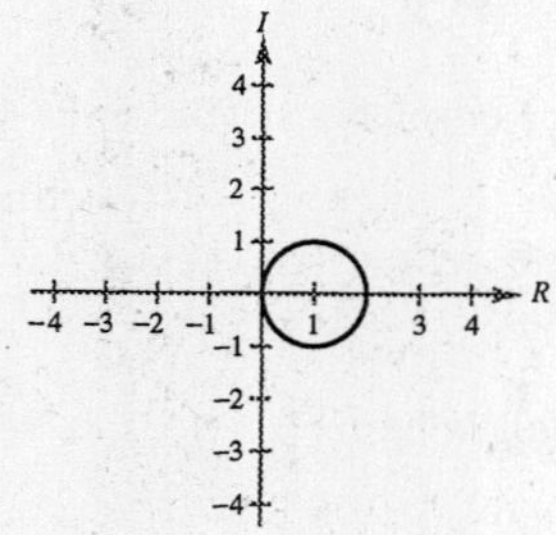

Chapter 4 Practice Test

1. Write $4 + \sqrt{-81} - 3i^2$ in standard form.

2. Write the result in standard form: $\dfrac{3 + i}{5 - 4i}$

3. Use the Quadratic Formula to solve $x^2 - 4x + 7 = 0$.

4. True or false: $\sqrt{-6}\sqrt{-6} = \sqrt{36} = 6$

5. Use the discriminant to determine the type of solutions of $3x^2 - 8x + 7 = 0$.

6. Find all the zeros of $f(x) = x^4 + 13x^2 + 36$.

7. Find a polynomial function that has the following zeros: $3, -1 \pm 4i$

8. Use the zero $x = 4 + i$ to find all the zeros of $f(x) = x^3 - 10x^2 + 33x - 34$.

9. Give the trigonometric form of $z = 5 - 5i$.

10. Give the standard form of $z = 6(\cos 225° + i \sin 225°)$.

11. Multiply $[7(\cos 23° + i \sin 23°)][4(\cos 7° + i \sin 7°)]$.

12. Divide $\dfrac{9\left(\cos \frac{5\pi}{4} + i \sin \frac{5\pi}{4}\right)}{3(\cos \pi + i \sin \pi)}$

13. Find $(2 + 2i)^8$.

14. Find the cube roots of $8\left(\cos \frac{\pi}{3} + i \sin \frac{\pi}{3}\right)$.

15. Find all the solutions to $x^4 + i = 0$.

CHAPTER 5
Exponential and Logarithmic Functions

CHAPTER 5
Exponential and Logarithmic Functions

Section 5.1 Exponential Functions and Their Graphs

- You should know that a function of the form $f(x) = a^x$, where $a > 0$, $a \neq 1$, is called an exponential function with base a.
- You should be able to graph exponential functions.
- You should know formulas for compound interest.

  (a) For n compoundings per year: $A = P\left(1 + \frac{r}{n}\right)^{nt}$.

  (b) For continuous compoundings: $A = Pe^{rt}$.

Solutions to Odd-Numbered Exercises

1. $f(5.6) = (3.4)^{5.6} \approx 946.852$

3. $f(-\pi) = 5^{-\pi} \approx 0.006$

5. $f(\frac{3}{4}) = e^{-3/4} \approx 0.472$

7. $f(x) = 2^x$

Increasing

Asymptote: $y = 0$

Intercept: $(0, 1)$

Matches graph (d).

9. $f(x) = 2^{-x}$

Decreasing

Asymptote: $y = 0$

Intercept: $(0, 1)$

Matches graph (a).

11. $f(x) = 3^x$

$g(x) = 3^{x-4}$

Because $g(x) = f(x - 4)$, the graph of g can be obtained by shifting the graph of f four units to the right.

13. $f(x) = -2^x$

$g(x) = 5 - 2^x$

Because $g(x) = 5 + f(x)$, the graph of g can be obtained by shifting the graph of f five units upward.

15. $f(x) = \left(\frac{3}{5}\right)^x$

$g(x) = -\left(\frac{3}{5}\right)^{x+4}$

Because $g(x) = -f(x + 4)$, the graph of g can be obtained by reflecting the graph of f in the x-axis and shifting the resulting graph four units to the left.

17. $f(x) = 0.3^x$

$g(x) = -0.3^x + 5$

Because $g(x) = -f(x) + 5$, the graph of g can be obtained by reflecting the graph of f in the x-axis and shifting the resulting graph five units upward.

19. $f(x) = \left(\frac{1}{2}\right)^x$

x	-2	-1	0	1	2
$f(x)$	4	2	1	0.5	0.25

Asymptote: $y = 0$

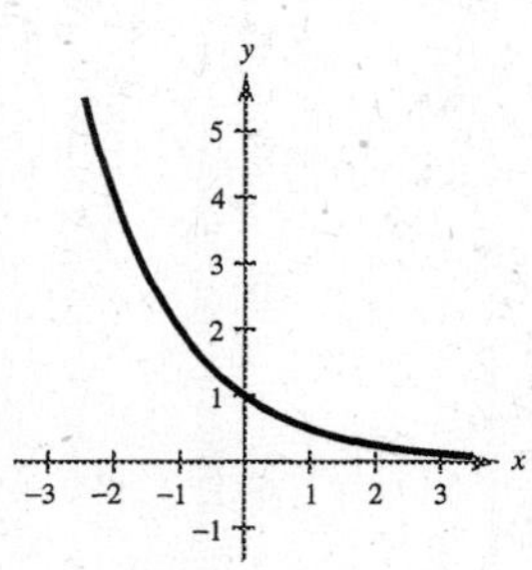

21. $f(x) = 6^{-x}$

x	-2	-1	0	1	2
$f(x)$	36	6	1	0.167	0.028

Asymptote: $y = 0$

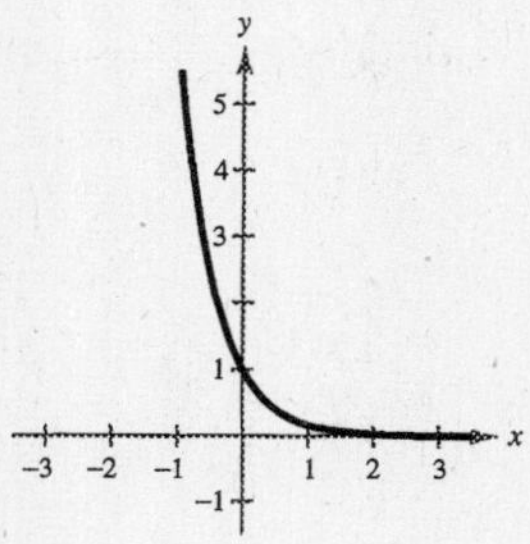

23. $f(x) = 2^{x-1}$

x	-2	-1	0	1	2
$f(x)$	0.125	0.25	0.5	1	2

Asymptote: $y = 0$

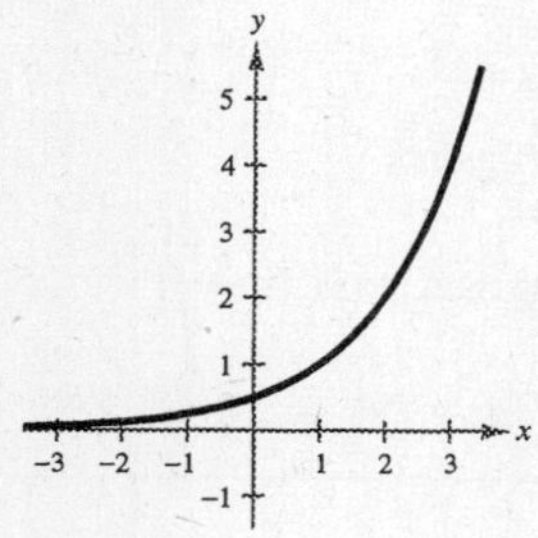

25. $f(x) = e^{x}$

x	-2	-1	0	1	2
$f(x)$	0.135	0.368	1	2.718	7.389

Asymptote: $y = 0$

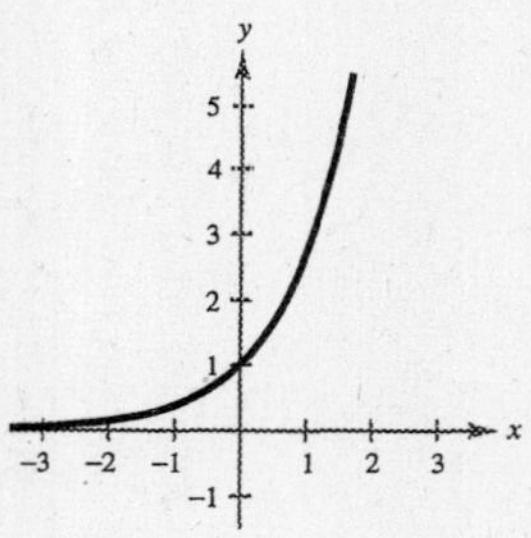

27. $f(x) = 3e^{x+4}$

x	-8	-7	-6	-5	-4
$f(x)$	0.055	0.149	0.406	1.104	3

Asymptote: $y = 0$

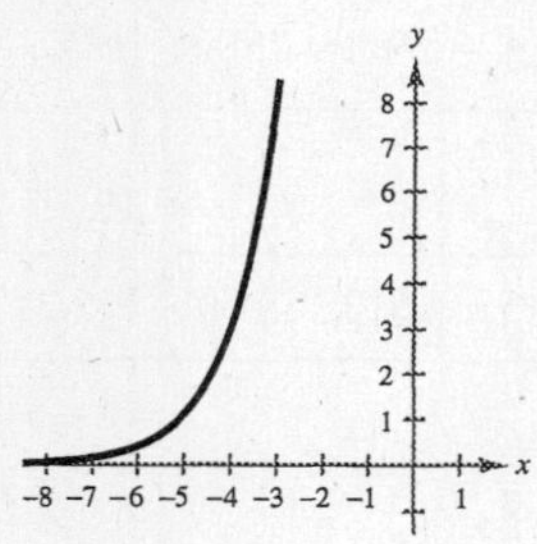

29. $f(x) = 2e^{x-2} + 4$

x	-2	-1	0	1	2
$f(x)$	4.037	4.100	4.271	4.736	6

Asymptote: $y = 4$

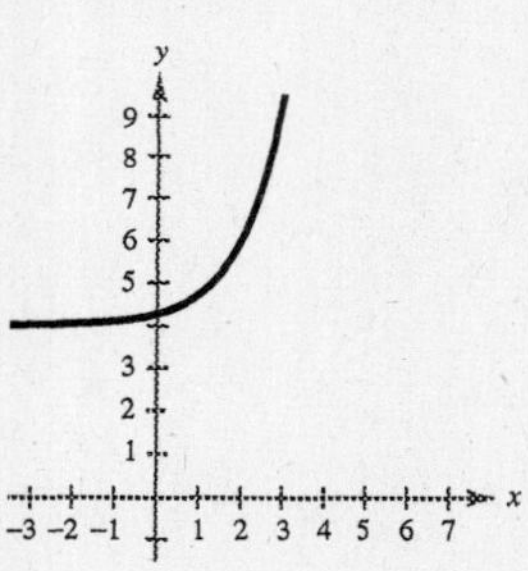

31. $f(x) = 4^{x-3} + 3$

x	-1	0	1	2	3
$f(x)$	3.004	3.016	3.063	3.25	4

Asymptote: $y = 3$

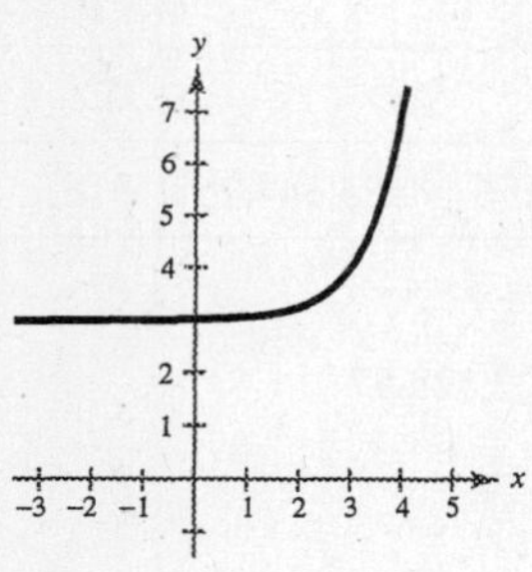

33. $y = 2^{-x^2}$

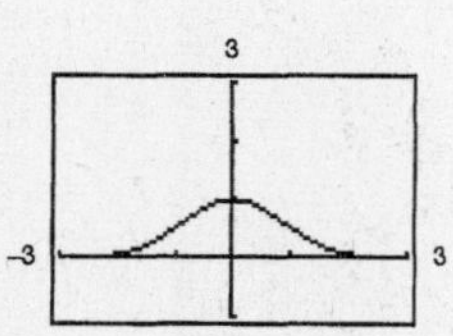

35. $f(x) = 3^{x-2} + 1$

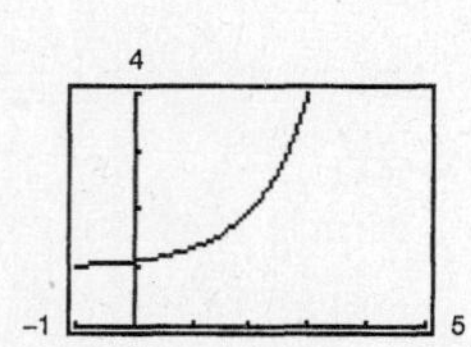

37. $y = 1.08^{-5x}$

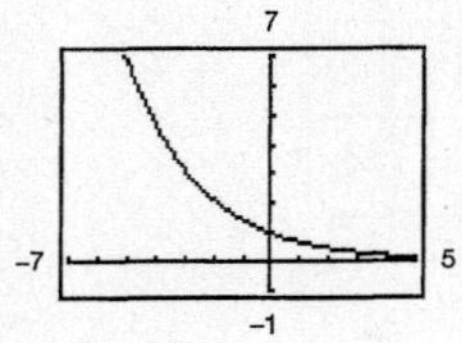

39. $s(t) = 2e^{0.12t}$

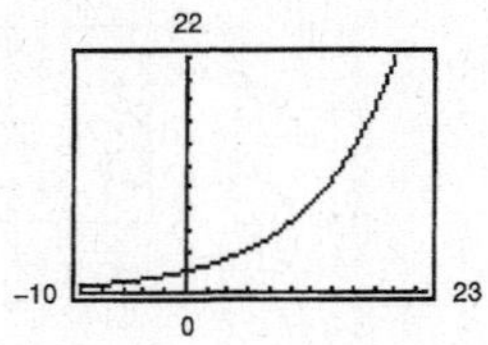

41. $g(x) = 1 + e^{-x}$

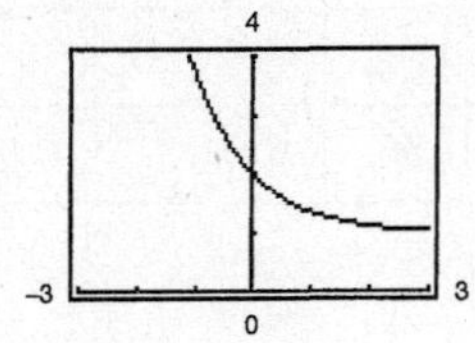

43. $P = \$2500,\ r = 8\%,\ t = 10$ years

Compounded n times per year: $A = P\left(1 + \frac{r}{n}\right)^{nt} = 2500\left(1 + \frac{0.08}{n}\right)^{10n}$

Compounded continuously: $A = Pe^{rt} = 2500e^{0.08(10)}$

n	1	2	4	12	365	Continuous Compounding
A	\$5397.31	\$5477.81	\$5520.10	\$5549.10	\$5563.36	\$5563.85

45. $P = \$2500,\ r = 8\%,\ t = 20$ years

Compounded n times per year: $A = P\left(1 + \frac{r}{n}\right)^{nt} = 2500\left(1 + \frac{0.08}{n}\right)^{20n}$

Compounded continuously: $A = Pe^{rt} = 2500e^{0.08(20)}$

n	1	2	4	12	365	Continuous Compounding
A	\$11,652.39	\$12,002.55	\$12,188.60	\$12,317.01	\$12,380.41	\$12,382.58

47. $A = Pe^{rt}$

$A = 12000e^{0.08t}$

t	10	20	30	40	50
A	\$26,706.49	\$59,436.39	\$132,278.12	\$294,390.36	\$655,177.80

49. $A = Pe^{rt}$

$A = 12000e^{0.065t}$

t	10	20	30	40	50
A	\$22,986.49	\$44,031.56	\$84,344.25	\$161,564.86	\$309,484.08

51. $A = 25{,}000e^{(0.0875)(25)} \approx \$222{,}822.57$

53. $C(10) = 23.95(1.04)^{10} \approx \$35.45$

55. $P(t) = 100e^{0.2197t}$

(a) $P(0) = 100$

(b) $P(5) \approx 300$

(c) $P(10) \approx 900$

57. $Q = 25\left(\frac{1}{2}\right)^{t/1620}$

(a) When $t = 0$, $Q = 25\left(\frac{1}{2}\right)^{0/1620} = 25(1) = 25$ grams.

(b) When $t = 1000$, $Q = 25\left(\frac{1}{2}\right)^{1000/1620} \approx 16.30$ grams.

(c)

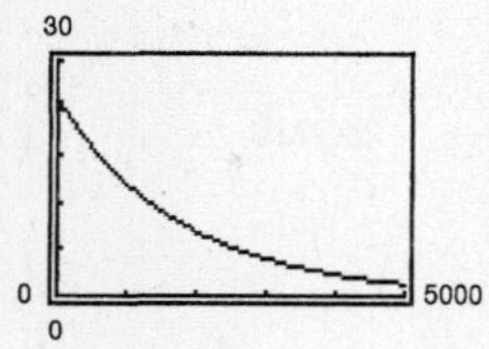

59. $y = \dfrac{100}{1 + 7e^{-0.069x}}$

(a)

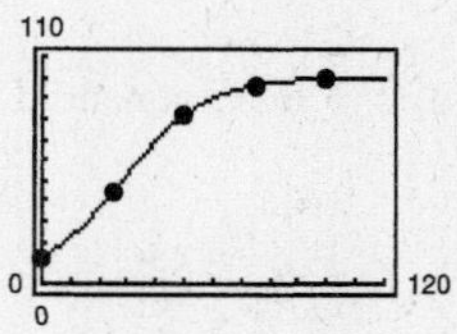

(b)

x	Sample Data	Model
0	12	13
25	44	45
50	81	82
75	96	96
100	99	99

(c) When $x = 36$: $y = \dfrac{100}{1 + 7e^{-0.069(36)}} \approx 63.1\%$.

(d) $\dfrac{2}{3}(100) = \dfrac{100}{1 + 7e^{-0.069x}}$ when $x \approx 38.2$.

61. True. The line $y = -2$ is a horizontal asymptote for the graph of $f(x) = 10^x - 2$.

63.
$$
\begin{aligned}
f(x) &= 3^{x-2} \\
&= 3^x 3^{-2} \\
&= 3^x\left(\frac{1}{3^2}\right) \\
&= \frac{1}{9}(3^x) \\
&= h(x)
\end{aligned}
$$

Thus, $f(x) \neq g(x)$, but $f(x) = h(x)$.

65.
$$
\begin{aligned}
f(x) &= 16(4^{-x}) \\
&= 4^2(4^{-x}) \\
&= 4^{2-x} \\
&= \left(\frac{1}{4}\right)^{-(2-x)} \\
&= \left(\frac{1}{4}\right)^{x-2} \\
&= g(x)
\end{aligned}
$$

and

$$
\begin{aligned}
f(x) &= 16(4^{-x}) \\
&= 16(2^2)^{-x} \\
&= 16(2^{-2x}) \\
&= h(x)
\end{aligned}
$$

and

Thus, $f(x) = g(x) = h(x)$.

67. $y = 3^x$ and $y = 4^x$

x	-2	-1	0	1	2
3^x	$\frac{1}{9}$	$\frac{1}{3}$	1	3	9
4^x	$\frac{1}{16}$	$\frac{1}{4}$	1	4	16

(a) $4^x < 3^x$ when $x < 0$.

(b) $4^x > 3^x$ when $x > 0$.

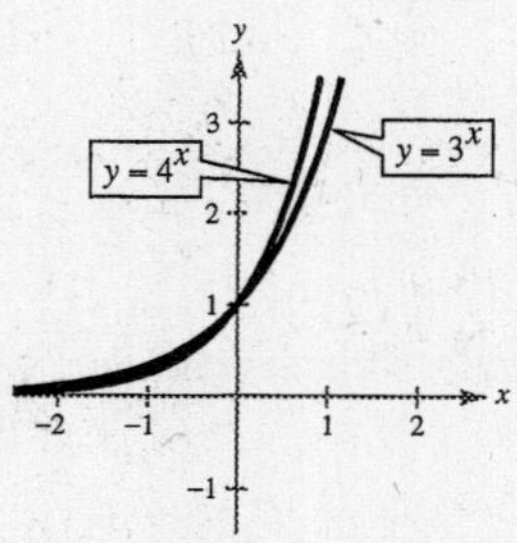

69. (a) $f(x) = \dfrac{8}{1 + e^{-0.5x}}$

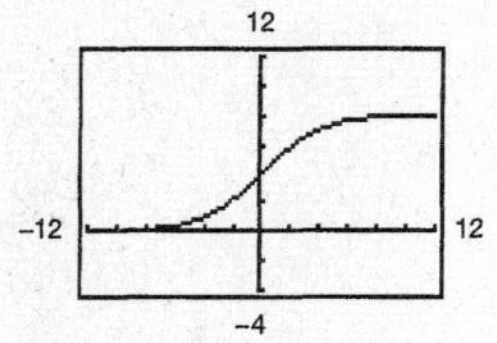

Horizontal asymptotes: $y = 0$ and $y = 8$

(b) $g(x) = \dfrac{8}{1 + e^{-0.5/x}}$

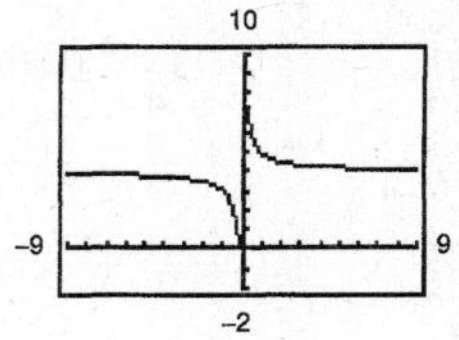

Horizontal asymptote: $y = 4$

Vertical asymptote: $x = 0$

71. $f(x) = \left(1 + \dfrac{0.5}{x}\right)^x$ and $g(x) = e^{0.5}$

(Horizontal line)

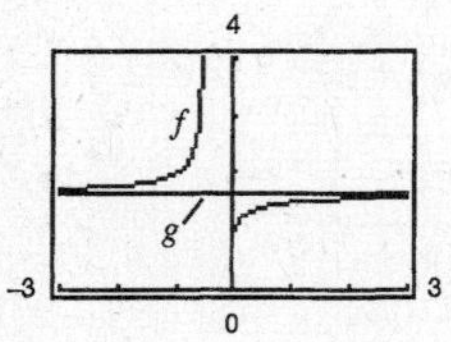

As $x \to \infty$, $f(x) \to g(x)$.

As $x \to -\infty$, $f(x) \to g(x)$.

73. The functions (c) 3^x and (d) 2^{-x} are exponential.

75. $2x - 7y + 14 = 0$

$2x + 14 = 7y$

$\frac{1}{7}(2x + 14) = y$

77. $x^2 + y^2 = 25$

$y^2 = 25 - x^2$

$y = \pm\sqrt{25 - x^2}$

Section 5.2 Logarithmic Functions and Their Graphs

- You should know that a function of the form $y = \log_a x$, where $a > 0$, $a \neq 1$, and $x > 0$, is called a logarithm of x to base a.
- You should be able to convert from logarithmic form to exponential form and vice versa.

  $y = \log_a x \iff a^y = x$
- You should know the following properties of logarithms.
  (a) $\log_a 1 = 0$ since $a^0 = 1$.
  (b) $\log_a a = 1$ since $a^1 = a$.
  (c) $\log_a a^x = x$ since $a^x = a^x$.
  (d) $a^{\log_a x} = x$ Inverse Property
  (e) If $\log_a x = \log_a y$, then $x = y$.
- You should know the definition of the natural logarithmic function.

  $\log_e x = \ln x, x > 0$
- You should know the properties of the natural logarithmic function.
  (a) $\ln 1 = 0$ since $e^0 = 1$.
  (b) $\ln e = 1$ since $e^1 = e$.
  (c) $\ln e^x = x$ since $e^x = e^x$.
  (d) $e^{\ln x} = x$ Inverse Property
  (e) If $\ln x = \ln y$, then $x = y$.
- You should be able to graph logarithmic functions.

Solutions to Odd-Numbered Exercises

1. $\log_4 64 = 3 \implies 4^3 = 64$

3. $\log_7 \frac{1}{49} = -2 \implies 7^{-2} = \frac{1}{49}$

5. $\log_{32} 4 = \frac{2}{5} \implies 32^{2/5} = 4$

7. $\ln \frac{1}{2} = -0.693\ldots \implies e^{-0.693\ldots} = \frac{1}{2}$

9. $5^3 = 125 \implies \log_5 125 = 3$

11. $81^{1/4} = 3 \implies \log_{81} 3 = \frac{1}{4}$

13. $6^{-2} = \frac{1}{36} \implies \log_6 \frac{1}{36} = -2$

15. $e^3 = 20.0855\ldots \implies \ln 20.0855\ldots = 3$

17. $e^x = 4 \implies \ln 4 = x$

19. $f(x) = \log_2 x$

$f(16) = \log_2 16 = 4$ since $2^4 = 16$

21. $f(x) = \log_7 x$

$f(1) = \log_7 1 = 0$ since $7^0 = 1$

23. $g(x) = \ln x$

$g(e^3) = \ln e^3 = 3$ by the Inverse Property

25. $g(x) = \log_a x$

$g(a^2) = \log_a a^2 = 2$ by the Inverse Property

27. $f(x) = \log_{10} x$

$f\left(\frac{4}{5}\right) = \log_{10}\left(\frac{4}{5}\right) \approx -0.097$

29. $f(x) = \ln x$

$f(18.42) = \ln 18.42 \approx 2.913$

31. $g(x) = 2 \ln x$

$g(0.75) = 2 \ln 0.75 \approx -0.575$

33. $f(x) = \log_3 x + 2$

Asymptote: $x = 0$

Point on graph: $(1, 2)$

Matches graph (c).

35. $f(x) = -\log_3(x + 2)$

Asymptote: $x = -2$

Point on graph: $(-1, 0)$

Matches graph (d).

37. $f(x) = \log_3(1 - x)$

Asymptote: $x = 1$

Point on graph: $(0, 0)$

Matches graph (b).

39. $f(x) = \log_4 x$

Domain: $x > 0 \implies$ The domain is $(0, \infty)$.

x-intercept: $(1, 0)$

Vertical asymptote: $x = 0$

$y = \log_4 x \implies 4^y = x$

x	$\frac{1}{4}$	1	4	2
$f(x)$	-1	0	1	$\frac{1}{2}$

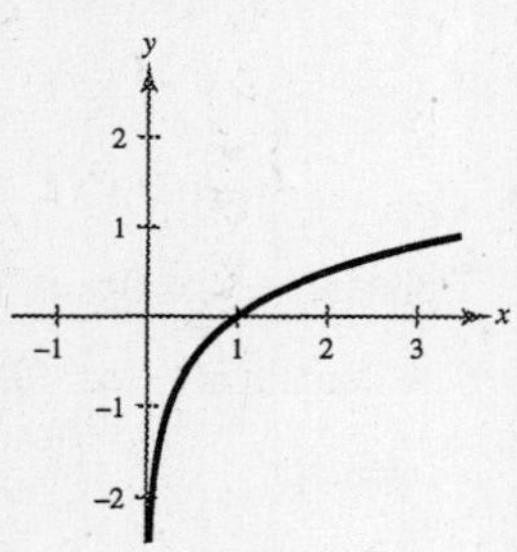

41. $y = -\log_3 x + 2$

Domain: $(0, \infty)$

x-intercept:

$$-\log_3 x + 2 = 0$$
$$2 = \log_3 x$$
$$3^2 = x$$
$$9 = x$$

The x-intercept is $(9, 0)$.

Vertical asymptote: $x = 0$

$y = -\log_3 x + 2$

$\log_3 x = 2 - y \implies 3^{2-y} = x$

x	27	9	3	1	$\frac{1}{3}$
y	-1	0	1	2	3

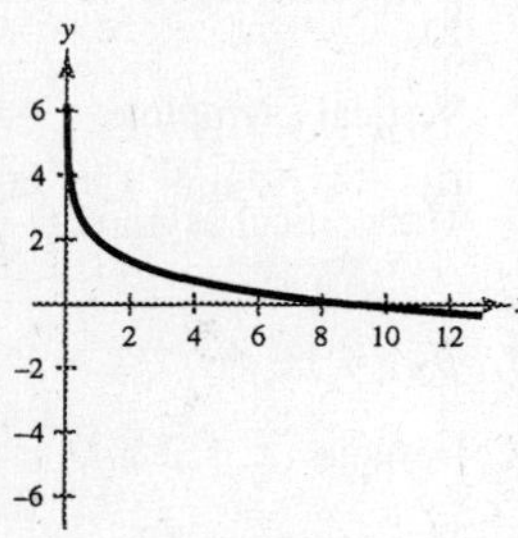

43. $f(x) = -\log_6(x + 2)$

Domain: $x + 2 > 0 \implies x > -2$

The domain is $(-2, \infty)$.

x-intercept:

$$0 = -\log_6(x + 2)$$
$$0 = \log_6(x + 2)$$
$$6^0 = x + 2$$
$$1 = x + 2$$
$$-1 = x$$

The x-intercept is $(-1, 0)$.

Vertical asymptote: $x + 2 = 0 \implies x = -2$

$$y = -\log_6(x + 2)$$
$$-y = \log_6(x + 2)$$
$$6^{-y} - 2 = x$$

x	4	-1	$-1\frac{5}{6}$	$-1\frac{35}{36}$
$f(x)$	-1	0	1	2

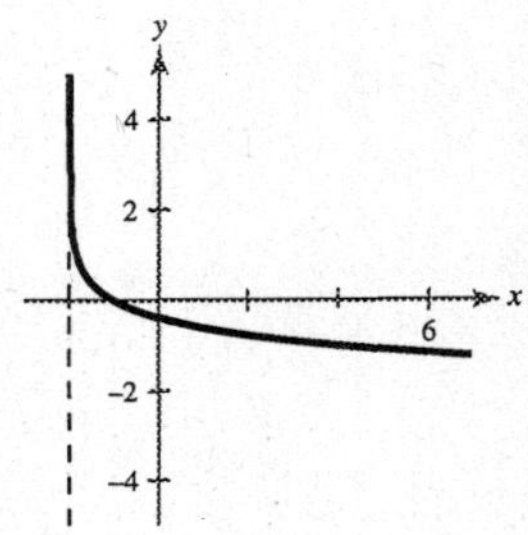

45. $y = \log_{10}\left(\frac{x}{5}\right)$

Domain: $\frac{x}{5} > 0 \implies x > 0$

The domain is $(0, \infty)$.

x-intercept:

$$\log_{10}\left(\frac{x}{5}\right) = 0$$
$$\frac{x}{5} = 10^0$$
$$\frac{x}{5} = 1 \implies x = 5$$

The x-intercept is $(5, 0)$.

Vertical asymptote: $\frac{x}{5} = 0 \implies x = 0$

The vertical asymptote is the y-axis.

x	1	2	3	4	5	6	7
y	-0.70	-0.40	-0.22	-0.10	0	0.08	0.15

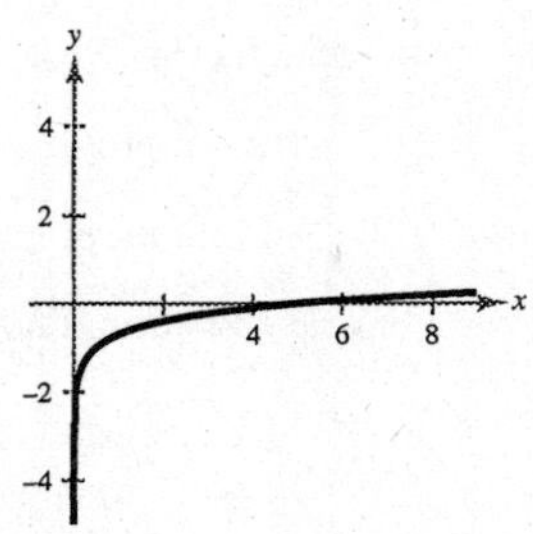

47. $f(x) = \ln(x - 2)$

Domain: $x - 2 > 0 \implies x > 2$

The domain is $(2, \infty)$.

x-intercept:

$$0 = \ln(x - 2)$$
$$e^0 = x - 2$$
$$3 = x$$

The x-intercept is $(3, 0)$.

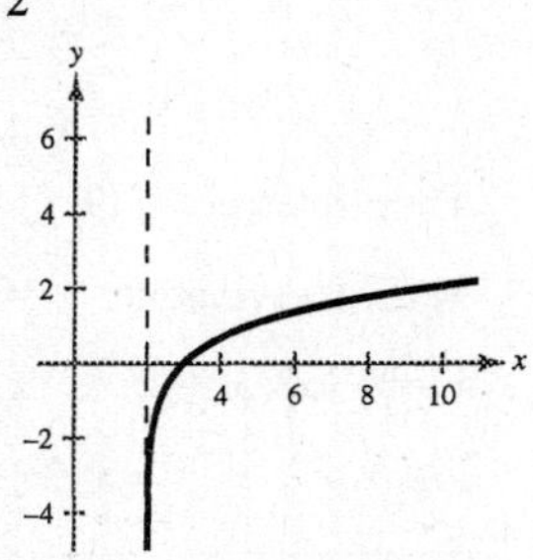

Vertical asymptote: $x - 2 = 0 \implies x = 2$

x	2.5	3	4	5
$f(x)$	-0.69	0	0.69	1.10

49. $g(x) = \ln(-x)$

Domain: $-x > 0 \implies x < 0$

The domain is $(-\infty, 0)$.

x-intercept:

$$0 = \ln(-x)$$
$$e^0 = -x$$
$$-1 = x$$

The x-intercept is $(-1, 0)$.

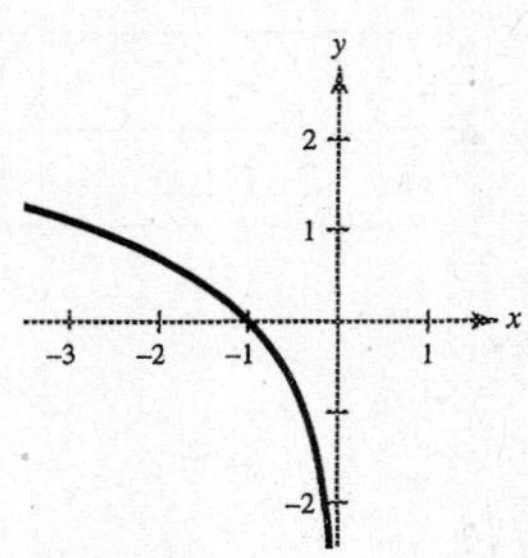

Vertical asymptote: $-x = 0 \implies x = 0$

x	-0.5	-1	-2	-3
$g(x)$	-0.69	0	0.69	1.10

51. $y_1 = \log_{10}(x + 1)$

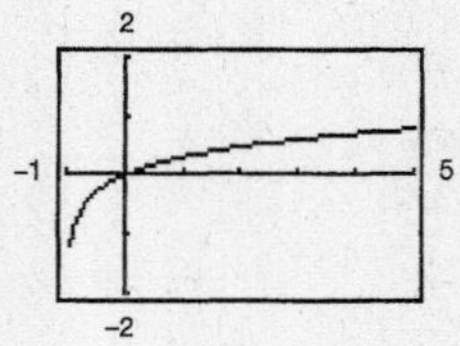

53. $y_1 = \ln(x - 1)$

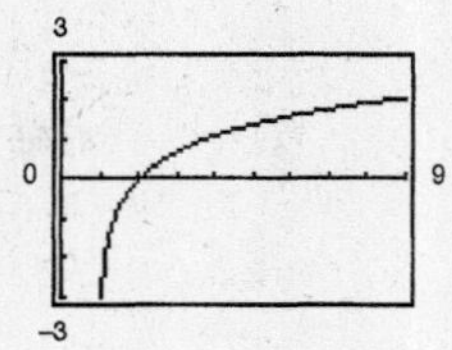

55. $y = \ln x + 2$

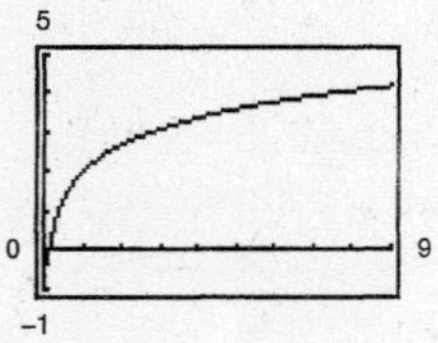

57. $t = 12.542 \ln\left(\dfrac{x}{x - 1000}\right),\ x > 1000$

(a) When $x = \$1100.65$: $t = 12.542 \ln\left(\dfrac{1100.65}{1100.65 - 1000}\right)$

≈ 30 years

When $x = \$1254.68$: $t = 12.542 \ln\left(\dfrac{1254.68}{1254.68 - 1000}\right)$

≈ 20 years

(c) Interest charges: $396{,}234 - 150{,}000 = \$246{,}234$

$301{,}123.20 - 150{,}000 = \$151{,}123.20$

(b) Total amounts: $(1100.65)(12)(30) = \$396{,}234.00$

$(1254.68)(12)(20) = \$301{,}123.20$

(d) The vertical asymptote is $x = 1000$.

The closer the payment is to \$1000 per month, the longer the length of the mortgage will be. Also, the monthly payment must be greater than \$1000.

59. $t = \dfrac{\ln 2}{r}$

(a)

r	0.005	0.01	0.015	0.02	0.025	0.03
t	138.6	69.3	46.2	34.7	27.7	23.1

(b)

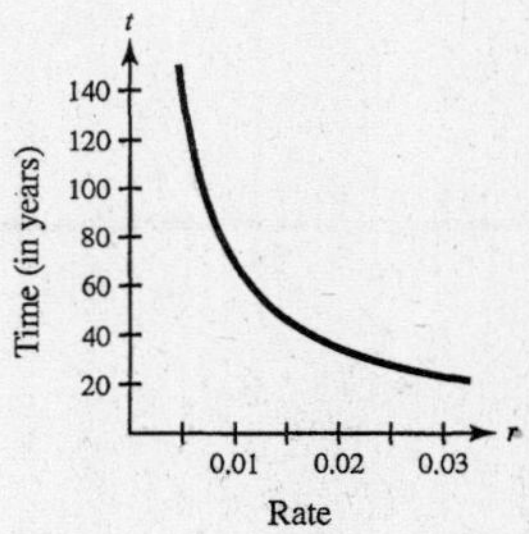

(c) Answers will vary.

61. $f(t) = 80 - 17 \log_{10}(t + 1),\ 0 \le t \le 12$

(a)

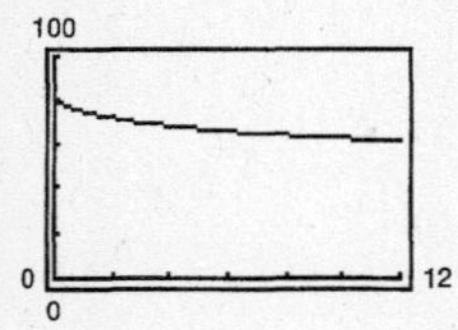

(b) $f(0) = 80 - 17 \log_{10} 1 = 80.0$

(c) $f(4) = 80 - 17 \log_{10} 5 \approx 68.1$

(d) $f(10) = 80 - 17 \log_{10} 11 \approx 62.3$

63. False. Reflecting $g(x)$ about the line $y = x$ will determine the graph of $f(x)$.

65. $f(x) = 3^x,\ g(x) = \log_3 x$

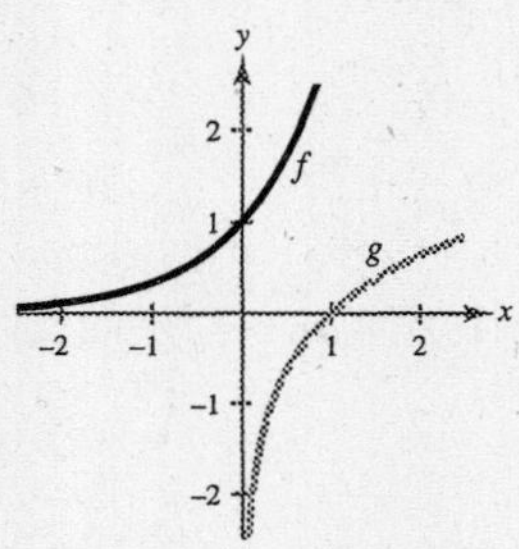

f and g are inverses. Their graphs are reflected about the line $y = x$.

67. $f(x) = e^x,\ g(x) = \ln x$

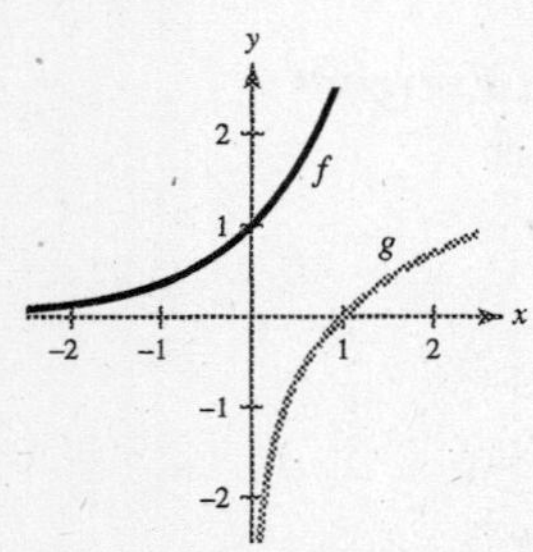

f and g are inverses. Their graphs are reflected about the line $y = x$.

69. (a) $f(x) = \ln x$

$g(x) = \sqrt{x}$

The natural log function grows at a slower rate than the square root function.

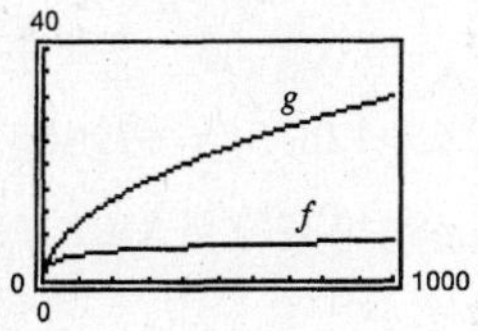

(b) $f(x) = \ln x$

$g(x) = \sqrt[4]{x}$

The natural log function grows at a slower rate than the fourth root function.

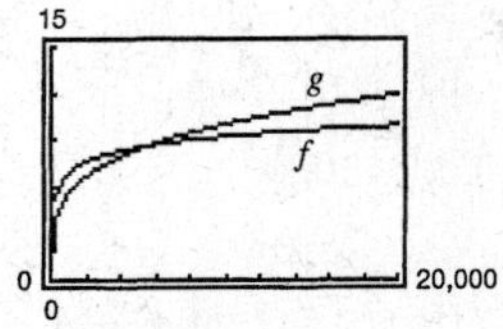

71. The domain of $f(x) = \log_{10} x$ is $x > 0$ or the interval $(0, \infty)$.

73. If x is between 1000 and 10,000 then $f(x)$ is between 3 and 4 since $10^3 = 1000$ and $10^4 = 10,000$.

75. $f(x) = |\ln x|$

(a)

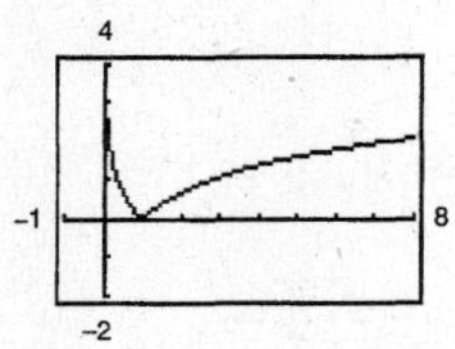

(b) Increasing on $(1, \infty)$

Decreasing on $(0, 1)$

(c) Relative minimum: $(1, 0)$

77. Total Cost $= 83.95 + 37.50t$ Parts and labor

Section 5.3 Properties of Logarithms

- You should know the following properties of logarithms.

  (a) $\log_a x = \dfrac{\log_b x}{\log_b a}$ $\qquad \log_a x = \dfrac{\log_{10} x}{\log_{10} a}$ $\qquad \log_a x = \dfrac{\ln x}{\ln a}$

  (b) $\log_a(uv) = \log_a u + \log_a v$ $\qquad \ln(uv) = \ln u + \ln v$

  (c) $\log_a(u/v) = \log_a u - \log_a v$ $\qquad \ln(u/v) = \ln u - \ln v$

  (d) $\log_a u^n = n \log_a u$ $\qquad \ln u^n = n \ln u$

- You should be able to rewrite logarithmic expressions using these properties.

Solutions to Odd-Numbered Exercises

1. (a) $\log_5 x = \dfrac{\log_{10} x}{\log_{10} 5}$

(b) $\log_5 x = \dfrac{\ln x}{\ln 5}$

3. (a) $\log_{\frac{1}{5}} x = \dfrac{\log_{10} x}{\log_{10}\left(\frac{1}{5}\right)}$

(b) $\log_{\frac{1}{5}} x = \dfrac{\ln x}{\ln\left(\frac{1}{5}\right)}$

5. (a) $\log_x \dfrac{3}{10} = \dfrac{\log_{10}\left(\frac{3}{10}\right)}{\log_{10} x}$

(b) $\log_x \dfrac{3}{10} = \dfrac{\ln\left(\frac{3}{10}\right)}{\ln x}$

7. (a) $\log_{2.6} x = \dfrac{\log_{10} x}{\log_{10} 2.6}$

(b) $\log_{2.6} x = \dfrac{\ln x}{\ln 2.6}$

9. $\log_3 7 = \dfrac{\log_{10} 7}{\log_{10} 3} = \dfrac{\ln 7}{\ln 3} \approx 1.771$

11. $\log_{1/2} 4 = \dfrac{\log_{10} 4}{\log_{10}(1/2)} = \dfrac{\ln 4}{\ln(1/2)} = -2.000$

13. $\log_9(0.4) = \dfrac{\log_{10} 0.4}{\log_{10} 9} = \dfrac{\ln 0.4}{\ln 9} \approx -0.417$

15. $\log_{15} 1250 = \dfrac{\log_{10} 1250}{\log_{10} 15} = \dfrac{\ln 1250}{\ln 15} \approx 2.633$

17. $\log_4 5x = \log_4 5 + \log_4 x$

19. $\log_8 x^4 = 4 \log_8 x$

21. $\log_5 \dfrac{5}{x} = \log_5 5 - \log_5 x = 1 - \log_5 x$

23. $\ln\sqrt{z} = \ln z^{1/2} = \frac{1}{2}\ln z$

25. $\ln xyz^2 = \ln x + \ln y + \ln z^2 = \ln x + \ln y + 2\ln z$

27. $\ln z(z-1)^2 = \ln z + \ln(z-1)^2$

$= \ln z + 2\ln(z-1),\ z > 1$

29. $\log_2 \dfrac{\sqrt{a-1}}{9} = \log_2\sqrt{a-1} - \log_2 9 = \dfrac{1}{2}\log_2(a-1) - \log_2 3^2 = \dfrac{1}{2}\log_2(a-1) - 2\log_2 3,\ a > 1$

31. $\ln\sqrt[3]{\dfrac{x}{y}} = \dfrac{1}{3}\ln\dfrac{x}{y}$

$= \dfrac{1}{3}[\ln x - \ln y]$

$= \dfrac{1}{3}\ln x - \dfrac{1}{3}\ln y$

33. $\ln\left(\dfrac{x^4\sqrt{y}}{z^5}\right) = \ln x^4\sqrt{y} - \ln z^5$

$= \ln x^4 + \ln\sqrt{y} - \ln z^5$

$= 4\ln x + \dfrac{1}{2}\ln y - 5\ln z$

35. $\log_5\left(\dfrac{x^2}{y^2z^3}\right) = \log_5 x^2 - \log_5 y^2z^3$

$= \log_5 x^2 - (\log_5 y^2 + \log_5 z^3)$

$= 2\log_5 x - 2\log_5 y - 3\log_5 z$

37. $\ln\sqrt[4]{x^3(x^2+3)} = \frac{1}{4}\ln x^3(x^2+3)$

$= \frac{1}{4}[\ln x^3 + \ln(x^2+3)]$

$= \frac{1}{4}[3\ln x + \ln(x^2+3)]$

$= \frac{3}{4}\ln x + \frac{1}{4}\ln(x^2+3)$

39. $\ln x + \ln 3 = \ln 3x$

41. $\log_4 z - \log_4 y = \log_4 \dfrac{z}{y}$

43. $2\log_2(x+4) = \log_2(x+4)^2$

45. $\frac{1}{4}\log_3 5x = \log_3(5x)^{1/4} = \log_3\sqrt[4]{5x}$

47. $\ln x - 3\ln(x+1) = \ln x - \ln(x+1)^3$

$= \ln\dfrac{x}{(x+1)^3}$

49. $\log_{10} x - 2\log_{10} y + 3\log_{10} z = \log_{10} x - \log_{10} y^2 + \log_{10} z^3$

$= \log_{10}\dfrac{x}{y^2} + \log_{10} z^3$

$= \log_{10}\dfrac{xz^3}{y^2}$

51. $\ln x - 4[\ln(x+2) + \ln(x-2)] = \ln x - 4\ln(x+2)(x-2)$

$= \ln x - 4\ln(x^2-4)$

$= \ln x - \ln(x^2-4)^4$

$= \ln\dfrac{x}{(x^2-4)^4}$

53. $\frac{1}{3}[2\ln(x+3) + \ln x - \ln(x^2-1)] = \frac{1}{3}[\ln(x+3)^2 + \ln x - \ln(x^2-1)]$

$= \frac{1}{3}[\ln x(x+3)^2 - \ln(x^2-1)]$

$= \frac{1}{3}\ln\frac{x(x+3)^2}{x^2-1}$

$= \ln\sqrt[3]{\frac{x(x+3)^2}{x^2-1}}$

55. $\frac{1}{3}[\log_8 y + 2\log_8(y+4)] - \log_8(y-1) = \frac{1}{3}[\log_8 y + \log_8(y+4)^2] - \log_8(y-1)$

$= \frac{1}{3}\log_8 y(y+4)^2 - \log_8(y-1)$

$= \log_8 \sqrt[3]{y(y+4)^2} - \log_8(y-1)$

$= \log_8\left(\frac{\sqrt[3]{y(y+4)^2}}{y-1}\right)$

57. $\log_2\frac{32}{4} = \log_2 32 - \log_2 4 \neq \frac{\log_2 32}{\log_2 4}$

The second and third expressions are equal by Property 2.

59. $\log_3 9 = 2\log_3 3 = 2$

61. $\log_2 \sqrt[4]{8} = \frac{1}{4}\log_2 2^3 = \frac{3}{4}\log_2 2 = \frac{3}{4}(1) = \frac{3}{4}$

63. $\log_4 16^{1.2} = 1.2(\log_4 16) = 1.2\log_4 4^2 = 1.2(2) = 2.4$

65. $\log_3(-9)$ is undefined. -9 is not in the domain of $\log_3 x$.

67. $\ln e^{4.5} = 4.5$

69. $\ln\frac{1}{\sqrt{e}} = \ln 1 - \ln\sqrt{e} = 0 - \frac{1}{2}\ln e = 0 - \frac{1}{2}(1) = -\frac{1}{2}$

71. $\ln e^2 + \ln e^5 = 2 + 5 = 7$

73. $\log_5 75 - \log_5 3 = \log_5 \frac{75}{3} = \log_5 25 = \log_5 5^2 = 2\log_5 5 = 2$

75. $\log_4 8 = \frac{\log_2 8}{\log_2 4} = \frac{\log_2 2^3}{\log_2 2^2} = \frac{3}{2}$

77. $\log_5\frac{1}{250} = \log_5\left(\frac{1}{125}\cdot\frac{1}{2}\right) = \log_5\frac{1}{125} + \log_5\frac{1}{2}$

$= \log_5 5^{-3} + \log_5 2^{-1}$

$= -3 - \log_5 2$

79. $\ln(5e^6) = \ln 5 + \ln e^6 = \ln 5 + 6 = 6 + \ln 5$

81. $f(t) = 90 - 15\log_{10}(t+1),\ 0 \le t \le 12$

(a) $f(0) = 90$

(b) $f(6) \approx 77$

(c) $f(12) \approx 73$

(d) $75 = 90 - 15\log_{10}(t+1)$

$-15 = -15\log_{10}(t+1)$

$1 = \log_{10}(t+1)$

$10^1 = t+1$

$t = 9$ months

(e) $f(t) = 90 - \log_{10}(t+1)^{15}$

(f)

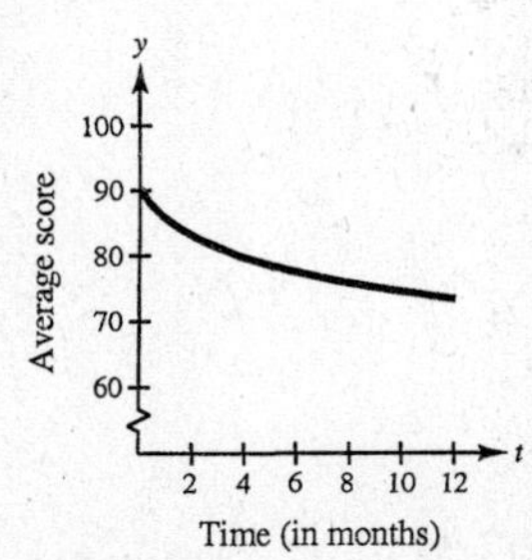

83. $f(x) = \ln x$

False, $f(0) \neq 0$ since 0 is not in the domain of $f(x)$. $f(1) = \ln 1 = 0$

85. False. $f(x) - f(2) = \ln x - \ln 2 = \ln\frac{x}{2} \neq \ln(x - 2)$

87. False. $f(u) = 2f(v) \Rightarrow \ln u = 2\ln v \Rightarrow \ln u = \ln v^2 \Rightarrow u = v^2$

89. Let $x = \log_b u$ and $y = \log_b v$, then $b^x = u$ and $b^y = v$.

$$\frac{u}{v} = \frac{b^x}{b^y} = b^{x-y}$$

Then $\log_b\left(\frac{u}{v}\right) = \log_b(b^{x-y}) = x - y = \log_b u - \log_b v$

91. $f(x) = \log_2 x = \dfrac{\log_{10} x}{\log_{10} 2} = \dfrac{\ln x}{\ln 2}$

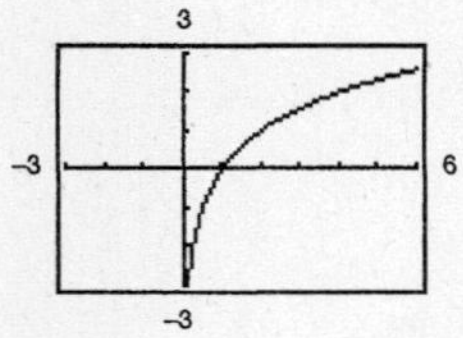

93. $f(x) = \log_{\frac{1}{2}} x = \dfrac{\log_{10} x}{\log_{10} \frac{1}{2}} = \dfrac{\ln x}{\ln\left(\frac{1}{2}\right)}$

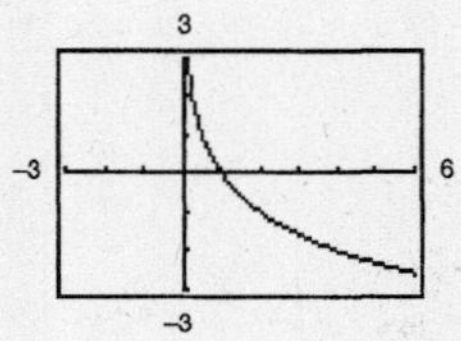

95. $f(x) = \log_{11.8} x = \dfrac{\log_{10} x}{\log_{10} 11.8} = \dfrac{\ln x}{\ln 11.8}$

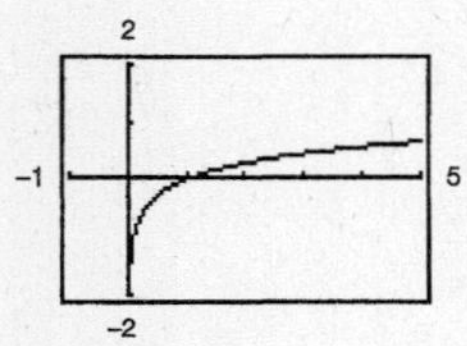

97. $f(x) = \ln\frac{x}{2}$, $g(x) = \dfrac{\ln x}{\ln 2}$, $h(x) = \ln x - \ln 2$

$f(x) = h(x)$ by Property 2.

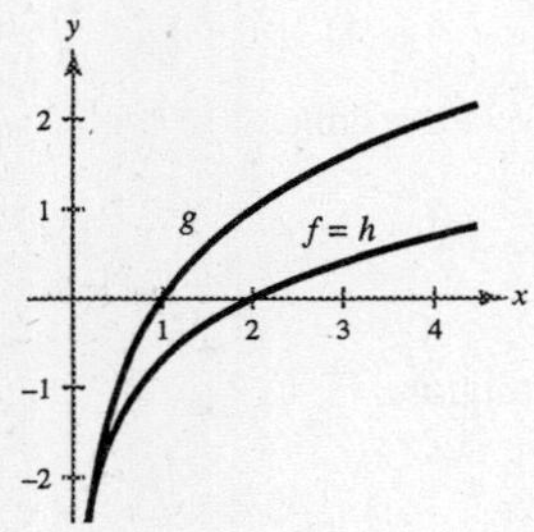

99. $\dfrac{24xy^{-2}}{16x^{-3}y} = \dfrac{24xx^3}{16yy^2} = \dfrac{3x^4}{2y^3}, x \neq 0$

101. $(18x^3y^4)^{-3}(18x^3y^4)^3 = \dfrac{(18x^3y^4)^3}{(18x^3y^4)^3} = 1$ if $x \neq 0, y \neq 0$.

103. $3x^2 + 2x - 1 = 0$

$(3x - 1)(x + 1) = 0$

$3x - 1 = 0 \Rightarrow x = \dfrac{1}{3}$

$x + 1 = 0 \Rightarrow x = -1$

105. $\dfrac{2}{3x + 1} = \dfrac{x}{4}$

$(3x + 1)(x) = (2)(4)$

$3x^2 + x - 8 = 0$

$$x = \frac{-1 \pm \sqrt{1^2 - 4(3)(-8)}}{2(3)} = \frac{-1 \pm \sqrt{97}}{6}$$

Section 5.4 Exponential and Logarithmic Equations

- To solve an exponential equation, isolate the exponential expression, then take the logarithm of both sides. Then solve for the variable.
  1. $\log_a a^x = x$
  2. $\ln e^x = x$
- To solve a logarithmic equation, rewrite it in exponential form. Then solve for the variable.
  1. $a^{\log_a x} = x$
  2. $e^{\ln x} = x$
- If $a > 0$ and $a \neq 1$ we have the following:
  1. $\log_a x = \log_a y \Leftrightarrow x = y$
  2. $a^x = a^y \Leftrightarrow x = y$
- Check for extraneous solutions.

Solutions to Odd-Numbered Exercises

1. $4^{2x-7} = 64$

(a) $x = 5$

$4^{2(5)-7} = 4^3 = 64$

Yes, $x = 5$ *is* a solution.

(b) $x = 2$

$4^{2(2)-7} = 4^{-3} = \frac{1}{64} \neq 64$

No, $x = 2$ *is not* a solution.

3. $3e^{x+2} = 75$

(a) $x = -2 + e^{25}$

$3e^{(-2+e^{25})+2} = 3e^{e^{25}} \neq 75$

No, $x = -2 + e^{25}$ *is not* a solution.

(b) $x = -2 + \ln 25$

$3e^{(-2+\ln 25)+2} = 3e^{\ln 25} = 3(25) = 75$

Yes, $x = -2 + \ln 25$ *is* a solution.

(c) $x \approx 1.219$

$3e^{1.219+2} = 3e^{3.129} \approx 75$

Yes, $x \approx 1.129$ *is* a solution.

5. $\log_4(3x) = 3 \Rightarrow 3x = 4^3 \Rightarrow 3x = 64$

(a) $\log_4(3x) = \dfrac{\ln 3x}{\ln 4}$

$\log_4[3(20.356)] = \dfrac{\ln 61.068}{\ln 4} \approx 2.966 \approx 3$

Yes, 20.356 *is* an approximate solution.

(b) $x = -4$

$3(-4) = -12 \neq 64$

No, $x = -4$ *is not* a solution.

(c) $x = \dfrac{64}{3}$

$3\left(\dfrac{64}{3}\right) = 64$

Yes, $x = \frac{64}{3}$ *is* a solution.

7. $4^x = 16$

$4^x = 4^2$

$x = 2$

9. $5^x = 625$

$5^x = 5^4$

$x = 4$

11. $7^x = \frac{1}{49}$

$7^x = 7^{-2}$

$x = -2$

13. $\left(\frac{1}{2}\right)^x = 32$

$2^{-x} = 2^5$

$-x = 5$

$x = -5$

15. $\left(\frac{3}{4}\right)^x = \frac{27}{64}$

$\left(\frac{3}{4}\right)^x = \left(\frac{3}{4}\right)^3$

$x = 3$

17. $\ln x - \ln 2 = 0$

$\ln x = \ln 2$

$x = 2$

19. $e^x = 2$

$\ln e^x = \ln 2$

$x = \ln 2$

$x \approx 0.693$

21. $\ln x = -1$

$e^{\ln x} = e^{-1}$

$x = e^{-1}$

$x \approx 0.368$

23. $\log_4 x = 3$

$4^{\log_4 x} = 4^3$

$x = 4^3$

$x = 64$

25. $\log_{10} x = -1$

$10^{\log_{10} x} = 10^{-1}$

$x = 10^{-1}$

$x = \frac{1}{10}$

27. $f(x) = g(x)$

$2^x = 8$

$2^x = 2^3$

$x = 3$

Point of intersection: $(3, 8)$

29. $f(x) = g(x)$

$\log_3 x = 2$

$x = 3^2$

$x = 9$

Point of intersection: $(9, 2)$

31. $4(3^x) = 20$

$3^x = 5$

$\log_3 3^x = \log_3 5$

$x = \log_3 5 = \dfrac{\log 5}{\log 3}$ or $\dfrac{\ln 5}{\ln 3}$

$x \approx 1.465$

33. $2e^x = 10$

$e^x = 5$

$\ln e^x = \ln 5$

$x = \ln 5 \approx 1.609$

35. $e^x - 9 = 19$

$e^x = 28$

$\ln e^x = \ln 28$

$x = \ln 28 \approx 3.332$

37. $3^{2x} = 80$

$\ln 3^{2x} = \ln 80$

$2x \ln 3 = \ln 80$

$x = \dfrac{\ln 80}{2 \ln 3} \approx 1.994$

39. $5^{-t/2} = 0.20$

$5^{-t/2} = \dfrac{1}{5}$

$5^{-t/2} = 5^{-1}$

$-\dfrac{t}{2} = -1$

$t = 2$

41. $3^{x-1} = 27$

$3^{x-1} = 3^3$

$x - 1 = 3$

$x = 4$

43. $2^{3-x} = 565$

$\ln 2^{3-x} = \ln 565$

$(3 - x) \ln 2 = \ln 565$

$3 \ln 2 - x \ln 2 = \ln 565$

$-x \ln 2 = \ln 565 - 3 \ln 2$

$x \ln 2 = 3 \ln 2 - \ln 565$

$x = \dfrac{3 \ln 2 - \ln 565}{\ln 2} = 3 - \dfrac{\ln 565}{\ln 2} \approx -6.142$

45. $8(10^{3x}) = 12$

$10^{3x} = \dfrac{12}{8}$

$\log_{10} 10^{3x} = \log_{10}\left(\dfrac{3}{2}\right)$

$3x = \log_{10}\left(\dfrac{3}{2}\right)$

$x = \dfrac{1}{3}\log_{10}\left(\dfrac{3}{2}\right) \approx 0.059$

47. $3(5^{x-1}) = 21$

$$5^{x-1} = 7$$

$$\ln 5^{x-1} = \ln 7$$

$$(x-1)\ln 5 = \ln 7$$

$$x - 1 = \frac{\ln 7}{\ln 5}$$

$$x = 1 + \frac{\ln 7}{\ln 5} \approx 2.209$$

49. $e^{3x} = 12$

$$3x = \ln 12$$

$$x = \frac{\ln 12}{3} \approx 0.828$$

51. $500e^{-x} = 300$

$$e^{-x} = \frac{3}{5}$$

$$-x = \ln\frac{3}{5}$$

$$x = -\ln\frac{3}{5} = \ln\frac{5}{3} \approx 0.511$$

53. $7 - 2e^x = 5$

$$-2e^x = -2$$

$$e^x = 1$$

$$x = \ln 1 = 0$$

55. $6(2^{3x-1}) - 7 = 9$

$$6(2^{3x-1}) = 16$$

$$2^{3x-1} = \frac{8}{3}$$

$$\log_2 2^{3x-1} = \log_2\left(\frac{8}{3}\right)$$

$$3x - 1 = \log_2\left(\frac{8}{3}\right) = \frac{\log(8/3)}{\log 2} \quad \text{or} \quad \frac{\ln(8/3)}{\ln 2}$$

$$x = \frac{1}{3}\left[\frac{\log(8/3)}{\log 2} + 1\right] \approx 0.805$$

57. $e^{2x} - 4e^x - 5 = 0$

$$(e^x + 1)(e^x - 5) = 0$$

$e^x = -1$ (No solution) or $e^x = 5$, $x = \ln 5 \approx 1.609$

59. $e^{2x} - 3e^x - 4 = 0$

$$(e^x + 1)(e^x - 4) = 0$$

$e^x + 1 = 0 \implies e^x = -1$

Not possible since $e^x > 0$ for all x.

$e^x - 4 = 0 \implies e^x = 4 \implies x = \ln 4 \approx 1.386$

61. $\dfrac{500}{100 - e^{x/2}} = 20$

$$500 = 20(100 - e^{x/2})$$

$$25 = 100 - e^{x/2}$$

$$e^{x/2} = 75$$

$$\frac{x}{2} = \ln 75$$

$$x = 2\ln 75 \approx 8.635$$

63. $\dfrac{3000}{2 + e^{2x}} = 2$

$$3000 = 2(2 + e^{2x})$$

$$1500 = 2 + e^{2x}$$

$$1498 = e^{2x}$$

$$\ln 1498 = 2x$$

$$x = \frac{\ln 1498}{2} \approx 3.656$$

65. $\left(1 + \dfrac{0.065}{365}\right)^{365t} = 4$

$$\ln\left(1 + \frac{0.065}{365}\right)^{365t} = \ln 4$$

$$365t\ln\left(1 + \frac{0.065}{365}\right) = \ln 4$$

$$t = \frac{\ln 4}{365\ln\left(1 + \frac{0.065}{365}\right)} \approx 21.330$$

67. $\left(1 + \frac{0.10}{12}\right)^{12t} = 2$

$\ln\left(1 + \frac{0.10}{12}\right)^{12t} = \ln 2$

$12t \ln\left(1 + \frac{0.10}{12}\right) = \ln 2$

$t = \frac{\ln 2}{12 \ln\left(1 + \frac{0.10}{12}\right)} \approx 6.960$

69. $g(x) = 6e^{1-x} - 25$

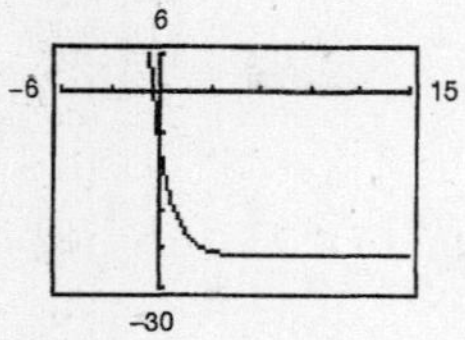

The zero is $x \approx -0.427$.

Algebraically,

$6e^{1-x} = 25$

$e^{1-x} = \frac{25}{6}$

$1 - x = \ln\left(\frac{25}{6}\right)$

$x = 1 - \ln\left(\frac{25}{6}\right)$

$x \approx -0.427$

71. $f(x) = 3e^{3x/2} - 962$

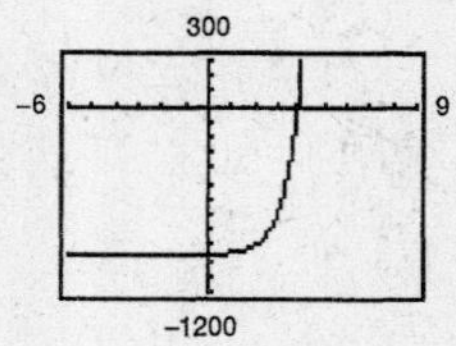

The zero is $x \approx 3.847$.

Algebraically,

$3e^{3x/2} = 962$

$e^{3x/2} = \frac{962}{3}$

$\frac{3x}{2} = \ln\left(\frac{962}{3}\right)$

$x = \frac{2}{3}\ln\left(\frac{962}{3}\right)$

$x \approx 3.847$

73. $g(t) = e^{0.09t} - 3$

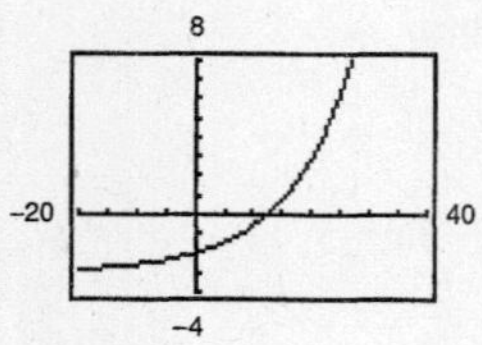

The zero is $t \approx 12.207$.

Algebraically,

$e^{0.09t} = 3$

$0.09t = \ln 3$

$t = \frac{\ln 3}{0.09}$

$t \approx 12.207$

75. $h(t) = e^{0.125t} - 8$

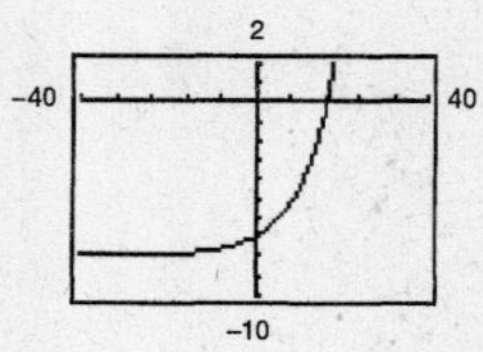

The zero is $t \approx 16.636$.

Algebraically,

$e^{0.125t} - 8 = 0$

$e^{0.125t} = 8$

$0.125t = \ln 8$

$t = \frac{\ln 8}{0.125}$

$t \approx 16.636$

77. $\ln x = -3$

$x = e^{-3} \approx 0.050$

79. $\ln 2x = 2.4$

$2x = e^{2.4}$

$x = \frac{e^{2.4}}{2} \approx 5.512$

81. $\log_{10} x = 6$

$x = 10^6 = 1{,}000{,}000.000$

83. $6 \log_3(0.5x) = 11$

$\log_3(0.5x) = \frac{11}{6}$

$3^{\log_3(0.5x)} = 3^{11/6}$

$0.5x = 3^{11/6}$

$x = 2(3^{11/6}) \approx 14.988$

85. $3 \ln 5x = 10$

$\ln 5x = \frac{10}{3}$

$5x = e^{10/3}$

$x = \frac{e^{10/3}}{5} \approx 5.606$

87. $\ln \sqrt{x + 2} = 1$

$\sqrt{x + 2} = e^1$

$x + 2 = e^2$

$x = e^2 - 2 \approx 5.389$

89. $7 + 3 \ln x = 5$

$3 \ln x = -2$

$\ln x = -\frac{2}{3}$

$x = e^{-2/3} \approx 0.513$

91. $\ln x - \ln(x+1) = 2$

$$\ln\left(\frac{x}{x+1}\right) = 2$$

$$\frac{x}{x+1} = e^2$$

$$x = e^2(x+1)$$

$$x = e^2x + e^2$$

$$x - e^2x = e^2$$

$$x(1 - e^2) = e^2$$

$$x = \frac{e^2}{1 - e^2} \approx -1.157$$

This negative value is extraneous.
The equation has no solution.

93. $\ln x + \ln(x-2) = 1$

$$\ln[x(x-2)] = 1$$

$$x(x-2) = e^1$$

$$x^2 - 2x - e = 0$$

$$x = \frac{2 \pm \sqrt{4 + 4e}}{2}$$

$$= \frac{2 \pm 2\sqrt{1 + e}}{2}$$

$$= 1 \pm \sqrt{1 + e}$$

The negative value is extraneous. The only solution is $x = 1 + \sqrt{1 + e} \approx 2.928$.

95. $\ln(x+5) = \ln(x-1) - \ln(x+1)$

$$\ln(x+5) = \ln\left(\frac{x-1}{x+1}\right)$$

$$x + 5 = \frac{x-1}{x+1}$$

$$(x+5)(x+1) = x - 1$$

$$x^2 + 6x + 5 = x - 1$$

$$x^2 + 5x + 6 = 0$$

$$(x+2)(x+3) = 0$$

$$x = -2 \text{ or } x = -3$$

Both of these solutions are extraneous,
so the equation has no solution.

97. $\log_2(2x - 3) = \log_2(x + 4)$

$$2x - 3 = x + 4$$

$$x = 7$$

99. $\log_{10}(x+4) - \log_{10} x = \log_{10}(x+2)$

$$\log_{10}\left(\frac{x+4}{x}\right) = \log_{10}(x+2)$$

$$\frac{x+4}{x} = x + 2$$

$$x + 4 = x^2 + 2x$$

$$0 = x^2 + x - 4$$

$$x = \frac{-1 \pm \sqrt{17}}{2} \quad \text{Quadratic Formula}$$

Choosing the positive value of x (the negative value is extraneous), we have

$$x = \frac{-1 + \sqrt{17}}{2} \approx 1.562.$$

101. $\log_4 x - \log_4(x-1) = \frac{1}{2}$

$$\log_4\left(\frac{x}{x-1}\right) = \frac{1}{2}$$

$$4^{\log_4\left(\frac{x}{x-1}\right)} = 4^{1/2}$$

$$\frac{x}{x-1} = 4^{1/2}$$

$$x = 2(x-1)$$

$$x = 2x - 2$$

$$-x = -2$$

$$x = 2$$

103. $\log_{10} 8x - \log_{10}(1 + \sqrt{x}) = 2$

$$\log_{10} \frac{8x}{1 + \sqrt{x}} = 2$$

$$\frac{8x}{1 + \sqrt{x}} = 10^2$$

$$8x = 100(1 + \sqrt{x})$$

$$2x = 25(1 + \sqrt{x})$$

$$2x = 25 + 25\sqrt{x}$$

$$2x - 25 = 25\sqrt{x}$$

$$(2x - 25)^2 = (25\sqrt{x})^2$$

$$4x^2 - 100x + 625 = 625x$$

$$4x^2 - 725x + 625 = 0$$

$$x = \frac{725 \pm \sqrt{725^2 - 4(4)(625)}}{2(4)}$$

$$x = \frac{725 \pm \sqrt{515625}}{8}$$

$$x = \frac{25(29 \pm 5\sqrt{33})}{8}$$

$x \approx 0.866$ (extraneous) or $x \approx 180.384$

The only solution is $x = \dfrac{25(29 + 5\sqrt{33})}{8} \approx 180.384$

105. $y_1 = 7$

$y_2 = 2^x$

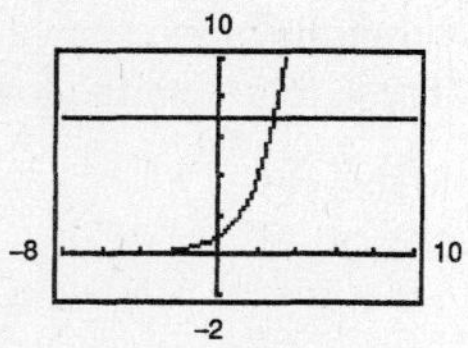

From the graph we have $x \approx 2.807$ when $y = 7$.

Algebraically:

$$2^x = 7$$

$$\ln 2^x = \ln 7$$

$$x \ln 2 = \ln 7$$

$$x = \frac{\ln 7}{\ln 2} \approx 2.807$$

107. $y_1 = 3$

$y_2 = \ln x$

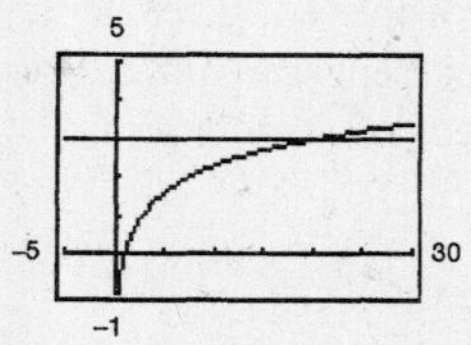

From the graph we have $x \approx 20.086$ when $y = 3$.

Algebraically:

$$3 - \ln x = 0$$

$$\ln x = 3$$

$$x = e^3 \approx 20.086$$

109.

$$A = Pe^{rt}$$

$$2000 = 1000e^{0.085t}$$

$$2 = e^{0.085t}$$

$$\ln 2 = 0.085t$$

$$\frac{\ln 2}{0.085} = t$$

$t \approx 8.2$ years

111.

$$A = Pe^{rt}$$

$$3000 = 1000e^{0.085t}$$

$$3 = e^{0.085t}$$

$$\ln 3 = 0.085t$$

$$\frac{\ln 3}{0.085} = t$$

$t \approx 12.9$ years

113. $p = 500 - 0.5(e^{0.004x})$

(a)

$$p = 350$$

$$350 = 500 - 0.5(e^{0.004x})$$

$$300 = e^{0.004x}$$

$$0.004x = \ln 300$$

$x \approx 1426$ units

(b)

$$p = 300$$

$$300 = 500 - 0.5(e^{0.004x})$$

$$400 = e^{0.004x}$$

$$0.004x = \ln 400$$

$x \approx 1498$ units

115. $V = 6.7e^{-48.1/t}, \ t \geq 0$

(a)

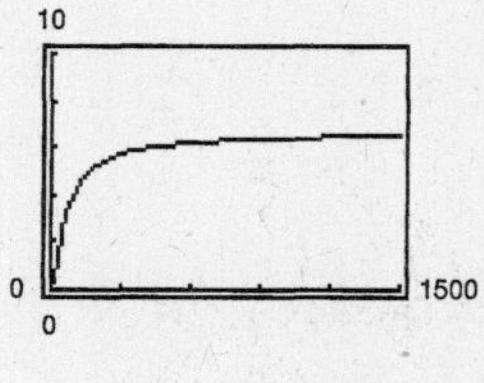

(b) As $t \to \infty$, $V \to 6.7$.

Horizontal asymptote: $V = 6.7$

The yield will approach 6.7 million cubic feet per acre.

(c)

$$1.3 = 6.7e^{-48.1/t}$$

$$\frac{1.3}{6.7} = e^{-48.1/t}$$

$$\ln\left(\frac{13}{67}\right) = \frac{-48.1}{t}$$

$$t = \frac{-48.1}{\ln(13/67)} \approx 29.3 \text{ years}$$

117. (a) From the graph shown in the textbook, we see horizontal asymptotes at $y = 0$ and $y = 100$. These represent the lower and upper percent bounds; the range falls between 0% and 100%.

(b) Males

$$50 = \frac{100}{1 + e^{-0.6114(x-69.71)}}$$

$$1 + e^{-0.6114(x-69.71)} = 2$$

$$e^{-0.6114(x-69.71)} = 1$$

$$-0.6114(x - 69.71) = \ln 1$$

$$-0.6114(x - 69.71) = 0$$

$$x = 69.71 \text{ inches}$$

Females

$$50 = \frac{100}{1 + e^{-0.66607(x-64.51)}}$$

$$1 + e^{-0.66607(x-64.51)} = 2$$

$$e^{-0.6667(x-64.51)} = 1$$

$$-0.66607(x - 64.51) = \ln 1$$

$$-0.66607(x - 64.51) = 0$$

$$x = 64.51 \text{ inches}$$

119. $y = -3.00 + 11.88 \ln x + \dfrac{36.94}{x}$

(a)

x	0.2	0.4	0.6	0.8	1.0
y	162.6	78.5	52.5	40.5	33.9

(b)

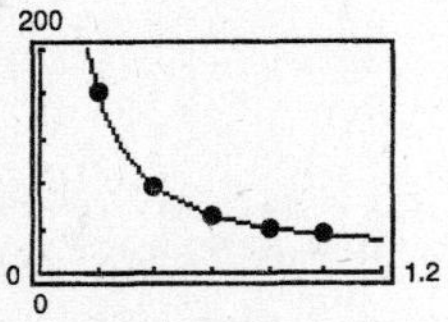

The model seems to fit the data well.

(c) When $y = 30$:

$$30 = -3.00 + 11.88 \ln x + \frac{36.94}{x}$$

Add the graph of $y = 30$ to the graph in part (a) and estimate the point of intersection of the two graphs. We find that $x \approx 1.20$ meters.

(d) No, it is probably not practical to lower the number of g's experienced during impact to less than 23 because the required distance traveled at $y = 23$ is $x \approx 2.27$ meters. It is probably not practical to design a car allowing a passenger to move forward 2.27 meters (or 7.45 feet) during an impact.

121. $\log_a(uv) = \log_a u + \log_a v$

True by Property 1 in Section 5.3.

123. $\log_a(u - v) = \log_a u - \log_a v$

False.

$1.95 \approx \log_{10}(100 - 10) \neq \log_{10} 100 - \log_{10} 10 = 1$

125. Yes, a logarithmic equation can have more than one extraneous solution. See Exercise 95.

127. Yes.

Time to Double

$$2P = Pe^{rt}$$

$$2 = e^{rt}$$

$$\ln 2 = rt$$

$$\frac{\ln 2}{r} = t$$

Time to Quadruple

$$4P = Pe^{rt}$$

$$4 = e^{rt}$$

$$\ln 4 = rt$$

$$\frac{2 \ln 2}{r} = t$$

Thus, the time to quadruple is twice as long as the time to double.

129. $\sqrt{48x^2y^5} = \sqrt{16x^2y^4 3y} = 4|x|y^2\sqrt{3y}$

131. $\sqrt[3]{25}\sqrt[3]{15} = \sqrt[3]{375} = \sqrt[3]{125 \cdot 3} = 5\sqrt[3]{3}$

133. $M = kp^3$

135. $d = kab$

137. $\log_6 9 = \dfrac{\log_{10} 9}{\log_{10} 6} = \dfrac{\ln 9}{\ln 6} \approx 1.226$

139. $\log_{3/4} 5 = \dfrac{\log_{10} 5}{\log_{10}\left(\frac{3}{4}\right)} = \dfrac{\ln 5}{\ln\left(\frac{3}{4}\right)} \approx -5.595$

Section 5.5 Exponential and Logarithmic Models

- You should be able to solve growth and decay problems.
  (a) Exponential growth if $b > 0$ and $y = ae^{bx}$.
  (b) Exponential decay if $b > 0$ and $y = ae^{-bx}$.
- You should be able to use the Gaussian model
  $y = ae^{-(x-b)^2/c}$.
- You should be able to use the logistic growth model
  $$y = \frac{a}{1 + be^{-rx}}.$$
- You should be able to use the logarithmic models
  $y = a + b \ln x, y = a + b \log_{10} x$.

Solutions to Odd-Numbered Exercises

1. $y = 2e^{x/4}$

This is an exponential growth model. Matches graph (c).

3. $y = 6 + \log_{10}(x + 2)$

This is a logarithmic function shifted up 6 units and left 2 units. Matches graph (b).

5. $y = \ln(x + 1)$

This is a logarithmic model shifted left one unit. Matches graph (d).

7. Since $A = 1000e^{0.12t}$, the time to double is given by $2000 = 1000e^{0.12t}$ and we have

$$2000 = 1000e^{0.12t}$$
$$2 = e^{0.12t}$$
$$\ln 2 = \ln e^{0.12t}$$
$$\ln 2 = 0.12t$$
$$t = \frac{\ln 2}{0.12} \approx 5.78 \text{ years.}$$

Amount after 10 years: $A = 1000e^{1.2} \approx \$3320.12$

9. Since $A = 750e^{rt}$ and $A = 1500$ when $t = 7.75$, we have the following.

$$1500 = 750e^{7.75r}$$
$$2 = e^{7.75r}$$
$$\ln 2 = \ln e^{7.75r}$$
$$\ln 2 = 7.75r$$
$$r = \frac{\ln 2}{7.75} \approx 0.089438 = 8.9438\%$$

Amount after 10 years: $A = 750e^{0.089438(10)} \approx \$1834.37$

11. Since $A = 500e^{rt}$ and $A = \$1505.00$ when $t = 10$, we have the following.

$$1505.00 = 500e^{10r}$$

$$r = \frac{\ln(1505.00/500)}{10} \approx 0.110 = 11.0\%$$

The time to double is given by

$$1000 = 500e^{0.110t}$$

$$t = \frac{\ln 2}{0.110} \approx 6.3 \text{ years.}$$

13. Since $A = Pe^{0.045t}$ and $A = 10{,}000.00$ when $t = 10$, we have the following.

$$10{,}000.00 = Pe^{0.045(10)}$$

$$\frac{10{,}000.00}{e^{0.045(10)}} = P \approx \$6376.28$$

The time to double is given by

$$t = \frac{\ln 2}{0.045} \approx 15.40 \text{ years.}$$

15. $500{,}000 = P\left(1 + \dfrac{0.075}{12}\right)^{12(20)}$

$$P = \frac{500{,}000}{\left(1 + \frac{0.075}{12}\right)^{12(20)}} = \frac{500{,}000}{1.00625^{240}} \approx \$112{,}087.09$$

17. $P = 1000$, $r = 11\%$

(a) $n = 1$

$$(1 + 0.11)^t = 2$$

$$t \ln 1.11 = \ln 2$$

$$t = \frac{\ln 2}{\ln 1.11} \approx 6.642 \text{ years}$$

(b) $n = 12$

$$\left(1 + \frac{0.11}{12}\right)^{12t} = 2$$

$$12t \ln\left(1 + \frac{0.11}{12}\right) = \ln 2$$

$$t = \frac{\ln 2}{12 \ln\left(1 + \frac{0.11}{12}\right)} \approx 6.330 \text{ years}$$

(c) $n = 365$

$$\left(1 + \frac{0.11}{365}\right)^{365t} = 2$$

$$365t \ln\left(1 + \frac{0.11}{365}\right) = \ln 2$$

$$t = \frac{\ln 2}{365 \ln\left(1 + \frac{0.11}{365}\right)} \approx 6.302 \text{ years}$$

(d) Continuously

$$e^{0.11t} = 2$$

$$0.11t = \ln 2$$

$$t = \frac{\ln 2}{0.11} \approx 6.301 \text{ years}$$

19. $3P = Pe^{rt}$

$$3 = e^{rt}$$

$$\ln 3 = rt$$

$$\frac{\ln 3}{r} = t$$

r	2%	4%	6%	8%	10%	12%
$t = \frac{\ln 3}{r}$ (years)	54.93	27.47	18.31	13.73	10.99	9.16

21. $3P = P(1 + r)^t$

$$3 = (1 + r)^t$$

$$\ln 3 = \ln(1 + r)^t$$

$$\ln 3 = t \ln(1 + r)$$

$$\frac{\ln 3}{\ln(1 + r)} = t$$

r	2%	4%	6%	8%	10%	12%
$t = \frac{\ln 3}{\ln(1 + r)}$ (years)	55.48	28.01	18.85	14.27	11.53	9.69

23. Continuous compounding results in faster growth.

$A = 1 + 0.075[\![t]\!]$ and $A = e^{0.07t}$

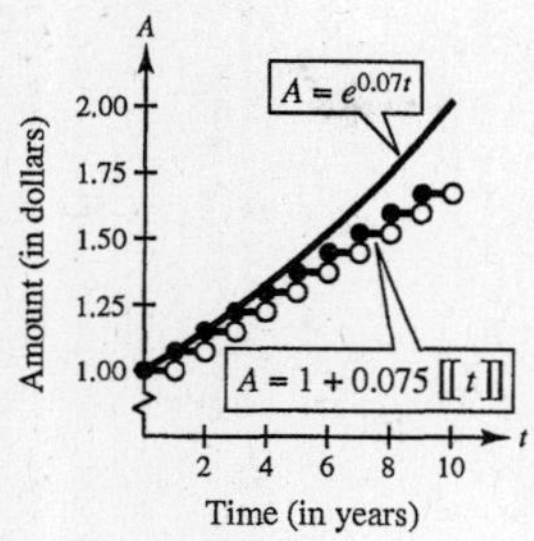

25.

$$\frac{1}{2}C = Ce^{k(1620)}$$
$$0.5 = e^{k(1620)}$$
$$\ln 0.5 = \ln e^{k(1620)}$$
$$\ln 0.5 = k(1620)$$
$$k = \frac{\ln 0.5}{1620}$$

Given $C = 10$ grams after 1000, years we have

$$y = 10e^{[(\ln 0.5)/1620](1000)}$$
$$\approx 6.52 \text{ grams.}$$

27.

$$\frac{1}{2}C = Ce^{k(5730)}$$
$$0.5 = e^{k(5730)}$$
$$\ln 0.5 = \ln e^{k(5730)}$$
$$\ln 0.5 = k(5730)$$
$$k = \frac{\ln 0.5}{5730}$$

Given $y = 2$ grams after 1000 years, we have

$$2 = Ce^{[(\ln 0.5)/5730](1000)}$$
$$C \approx 2.26 \text{ grams.}$$

29.

$$\frac{1}{2}C = Ce^{k(24{,}360)}$$
$$0.5 = e^{k(24{,}360)}$$
$$\ln 0.5 = \ln e^{k(24{,}360)}$$
$$\ln 0.5 = k(24{,}360)$$
$$k = \frac{\ln 0.5}{24{,}360}$$

Given $y = 2.1$ grams after 1000 years, we have

$$2.1 = Ce^{[(\ln 0.5)/24{,}360](1000)}$$
$$C \approx 2.16 \text{ grams.}$$

31.

$$y = ae^{bx}$$
$$1 = ae^{b(0)} \implies 1 = a$$
$$10 = e^{b(3)}$$
$$\ln 10 = 3b$$
$$\frac{\ln 10}{3} = b \implies b \approx 0.7675$$

Thus, $y = e^{0.7675x}$.

33.

$$y = ae^{bx}$$
$$5 = ae^{b(0)} \implies 5 = a$$
$$1 = 5e^{b(4)}$$
$$\frac{1}{5} = e^{4b}$$
$$\ln\left(\frac{1}{5}\right) = 4b$$
$$\frac{\ln\left(\frac{1}{5}\right)}{4} = b \implies b \approx -0.4024$$

Thus, $y = 5e^{-0.4024x}$.

35. $P = 16{,}968e^{0.019t}$

In thousands, 22 million is represented by 22,000.

$$22{,}000 = 16{,}968e^{0.019t}$$
$$\frac{22{,}000}{16{,}968} = e^{0.019t}$$
$$0.019t = \ln\left(\frac{22{,}000}{16{,}968}\right)$$
$$t = \frac{\ln\left(\frac{22{,}000}{16{,}968}\right)}{0.019}$$
$$t \approx 13.7$$

The population will reach 22 million (according to the model) during the later half of the year 2003.

37. $P = 548e^{kt}$

When $t = -30, P = 241$: $241 = 548e^{k(-30)}$

$$\frac{241}{548} = e^{-30k}$$
$$-30k = \ln\left(\frac{241}{548}\right)$$
$$k = \frac{\ln\left(\frac{241}{548}\right)}{-30}$$
$$k \approx 0.0274$$

When $t = 10$: $P = 548e^{0.0274(10)} \approx 720{,}738$

39. $N = 100e^{kt}$

$300 = 100e^{5k}$

$3 = e^{5k}$

$\ln 3 = \ln e^{5k}$

$\ln 3 = 5k$

$k = \dfrac{\ln 3}{5} \approx 0.2197$

$N = 100e^{0.2197t}$

$200 = 100e^{0.2197t}$

$t = \dfrac{\ln 2}{0.2197} \approx 3.15$ hours

41. $y = Ce^{kt}$

$\dfrac{1}{2}C = Ce^{(1620)k}$

$\dfrac{1}{2} = e^{(1620)k}$

$\ln \dfrac{1}{2} = \ln e^{(1620)k}$

$\ln \dfrac{1}{2} = 1620k$

$k = \dfrac{\ln(1/2)}{1620}$

When $t = 100$, we have

$y = Ce^{[\ln(1/2)/1620](100)} \approx 0.958C = 95.8\%C.$

After 100 years, approximately 95.8% of the radioactive radium will remain.

43. (0, 22,000), (2, 13,000)

(a) $m = \dfrac{13,000 - 22,000}{2 - 0} = -4500$

$b = 22,000$

Thus, $V = -4500t + 22,000$.

(b) $a = 22,000$

$13,000 = 22,000e^{k(2)}$

$\dfrac{13}{22} = e^{2k}$

$\ln\left(\dfrac{13}{22}\right) = \ln e^{2k}$

$\ln\left(\dfrac{13}{22}\right) = 2k \Longrightarrow k \approx -0.263$

Thus, $V = 22,000e^{-0.263t}$.

(c) The exponential model depreciates faster in the first two years.

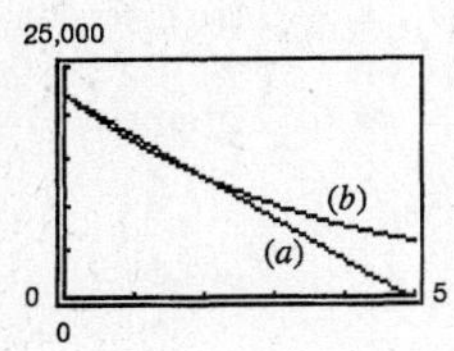

(d)

t	1	3
$V = -4500t + 22,000$	\$17,500	\$8500
$V = 22,000e^{-0.263t}$	\$16,912	\$9995

(e) The slope of the linear model means that the car depreciates \$4500 per year.

45. $S(t) = 100(1 - e^{kt})$

(a) $15 = 100(1 - e^{k(1)})$

$-85 = -100e^{k}$

$\frac{85}{100} = e^{k}$

$0.85 = e^{k}$

$\ln 0.85 = \ln e^{k}$

$k = \ln 0.85$

$k \approx -0.1625$

$S(t) = 100(1 - e^{-0.1625t})$

(b)

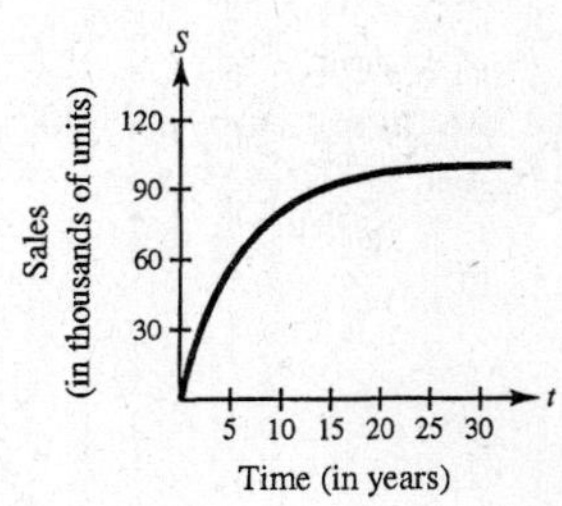

(c) $S(5) = 100(1 - e^{-0.1625(5)})$

$\approx 55.625 = 55,625$ units

47. $S = 10(1 - e^{kx})$

$x = 5$ (in hundreds), $S = 2.5$ (in thousands)

(a) $$\begin{aligned} 2.5 &= 10(1 - e^{k(5)}) \\ 0.25 &= 1 - e^{5k} \\ e^{5k} &= 0.75 \\ \ln e^{5k} &= \ln 0.75 \\ 5k &= \ln 0.75 \\ k &\approx -0.0575 \\ S &= 10(1 - e^{-0.0575x}) \end{aligned}$$

(b) When $x = 7$,
$S = 10(1 - e^{-0.0575(7)}) \approx 3.314$
which corresponds to 3314 units.

49. $N = 30(1 - e^{kt})$

(a) $$\begin{aligned} N &= 19,\ t = 20 \\ 19 &= 30(1 - e^{20k}) \\ 30e^{20k} &= 11 \\ e^{20k} &= \frac{11}{30} \\ \ln e^{20k} &= \ln\left(\frac{11}{30}\right) \\ 20k &= \ln\frac{11}{30} \\ k &\approx -0.050 \\ N &= 30(1 - e^{-0.050t}) \end{aligned}$$

(b) $$\begin{aligned} N &= 25 \\ 25 &= 30(1 - e^{-0.050t}) \\ \frac{5}{30} &= e^{-0.050t} \\ \ln\left(\frac{5}{30}\right) &= \ln e^{-0.050t} \\ \ln\left(\frac{5}{30}\right) &= -0.050t \\ t &= \frac{\ln(5/30)}{-0.050} \approx 36 \text{ days} \end{aligned}$$

51. $R = \log_{10}\frac{I}{I_0} = \log_{10} I$ since $I_0 = 1$.

(a) $R = \log_{10} 80{,}500{,}000 \approx 7.91$

(b) $R = \log_{10} 48{,}275{,}000 \approx 7.68$

(c) $R = \log_{10} 251{,}200 \approx 5.40$

53. $\beta = 10\log_{10}\frac{I}{I_0}$ where $I_0 = 10^{-12}$ watt/m^2.

(a) $\beta = 10\log_{10}\frac{10^{-10}}{10^{-12}} = 10\log_{10} 10^2 = 20$ decibels

(b) $\beta = 10\log_{10}\frac{10^{-5}}{10^{-12}} = 10\log_{10} 10^7 = 70$ decibels

(c) $\beta = 10\log_{10}\frac{10^{-2.5}}{10^{-12}} = 10\log_{10} 10^{9.5} = 95$ decibels

(d) $\beta = 10\log_{10}\frac{1}{10^{-12}} = 10\log_{10} 10^{12} = 120$ decibels

55. $$\begin{aligned} \beta &= 10\log_{10}\frac{I}{I_0} \\ \frac{\beta}{10} &= \log_{10}\frac{I}{I_0} \\ 10^{\beta/10} &= 10^{\log_{10} I/I_0} \\ 10^{\beta/10} &= \frac{I}{I_0} \\ I &= I_0 10^{\beta/10} \end{aligned}$$

$$\%\text{ decrease} = \frac{I_0 10^{9.3} - I_0 10^{8.0}}{I_0 10^{9.3}} \times 100 \approx 95\%$$

57. $\text{pH} = -\log_{10}[\text{H}^+] = -\log_{10}(2.3 \times 10^{-5}) \approx 4.64$

59. $5.8 = -\log_{10}[H^+]$

$-5.8 = \log_{10}[H^+]$

$10^{-5.8} = 10^{\log_{10}[H^+]}$

$10^{-5.8} = [H^+]$

$[H^+] \approx 1.58 \times 10^{-6}$ moles per liter

61. $2.5 = -\log_{10}[H^+]$

$-2.5 = \log_{10}[H^+]$

$10^{-2.5} = 10^{\log_{10}[H^+]}$

$10^{-2.5} = [H^+]$ for the fruit.

$9.5 = -\log_{10}[H^+]$

$10^{-9.5} = [H^+]$ for the antacid tablet

$$\frac{10^{-2.5}}{10^{-9.5}} = 10^7$$

63. $t = -10 \ln \dfrac{T - 70}{98.6 - 70}$

At 9:00 A.M. we have:

$$t = -10 \ln \frac{85.7 - 70}{98.6 - 70} \approx 6 \text{ hours}$$

From this you can conclude that the person died at 3:00 A.M.

65. $u = 120{,}000\left[\dfrac{0.075t}{1 - \left(\dfrac{1}{1 + 0.075/12}\right)^{12t}} - 1\right]$

(a)

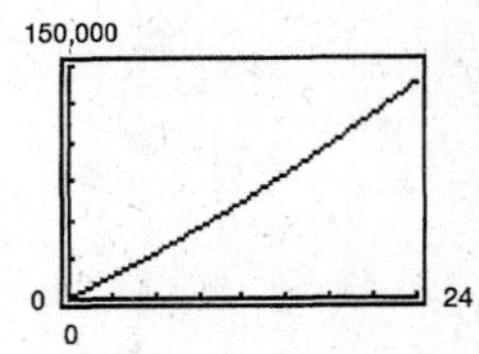

(b) From the graph, $u = \$120{,}000$ when $t \approx 21$ years. It would take approximately 37.6 years to pay \$240,000 in interest. Yes, it is possible to pay twice as much in interest charges as the size of the mortgage. It is especially likely when the interest rates are higher.

67. False. The domain can be the set of real numbers for a logistic growth function.

69. False. The graph of $f(x)$ is the graph of $g(x)$ shifted upward five units.

71. (a) Logarithmic

(b) Logistic

(c) Exponential (decay)

(d) Linear

(e) None of the above (appears to be a combination of a linear and a quadratic)

(f) Exponential (growth)

73. $y = 10 - 3x$

Line

Slope: $m = -3$

y-intercept: $(0, 10)$

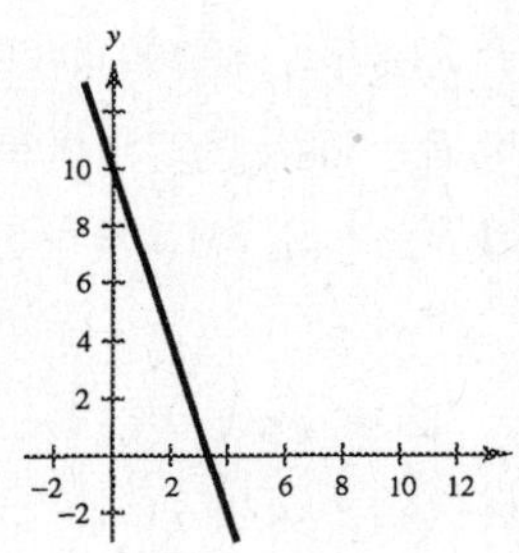

75. $y = -2x^2 - 3$

$y = -2(x - 0)^2 - 3$

Parabola

Vertex: $(0, -3)$

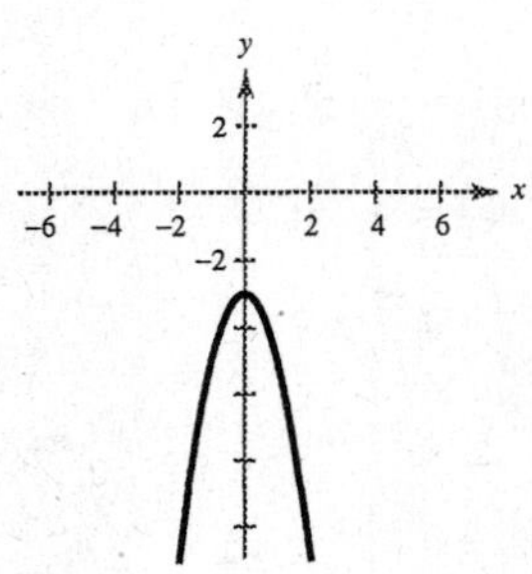

77. $3x^2 - 4y = 0$

$3x^2 = 4y$

$x^2 = \frac{4}{3}y$

Parabola

Vertex: $(0, 0)$

Focus: $\left(0, \frac{1}{3}\right)$

Directrix: $y = -\frac{1}{3}$

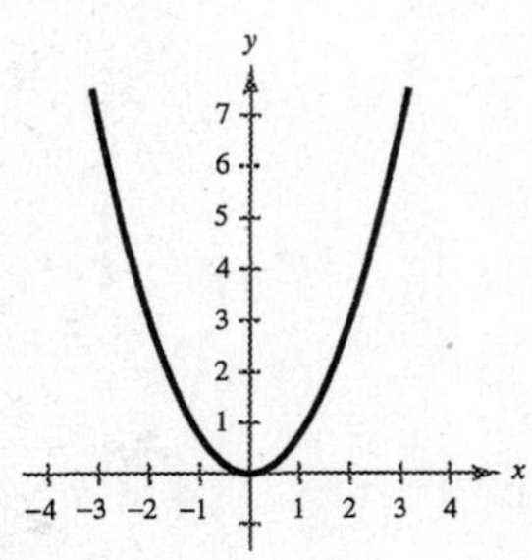

79. $y = \dfrac{4}{1 - 3x}$

Vertical Asymptote: $x = \dfrac{1}{3}$

Horizontal Asymptote: $y = 0$

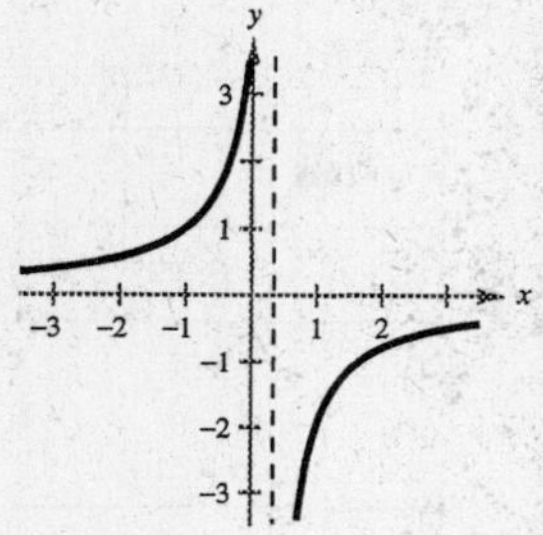

81. $x^2 + (y - 8)^2 = 25$

Circle

Center: $(0, 8)$

Radius: 5

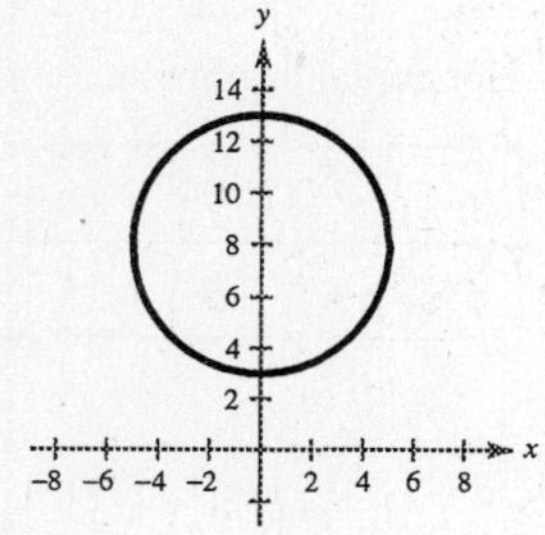

83. $f(x) = 2^{x-1} + 5$

Horizontal Asymptote: $y = 5$

x	-5	-3	-1	0	1	3	5
$f(x)$	5.02	5.06	5.3	5.5	6	9	21

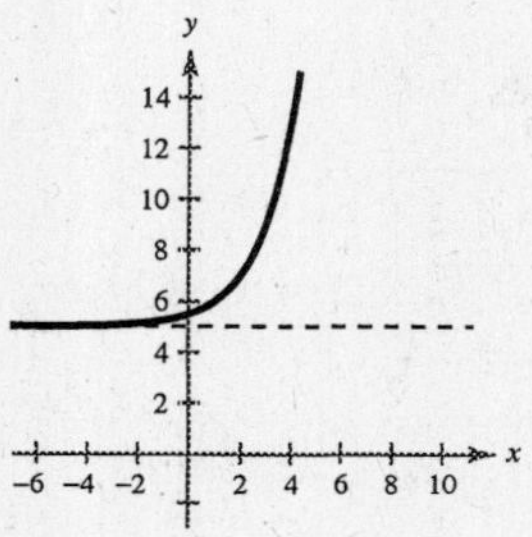

85. $f(x) = 3^x - 4$

Horizontal Asymptote: $y = -4$

x	-4	-2	-1	0	1	2
$f(x)$	-3.99	-3.89	-3.67	-3	-1	5

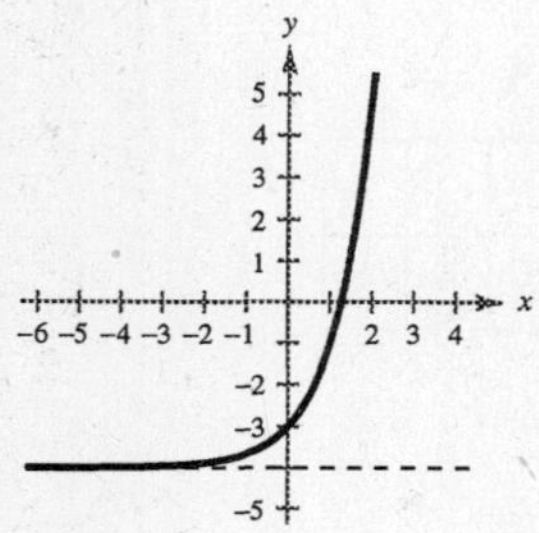

Review Exercises for Chapter 5

Solutions to Odd-Numbered Exercises

1. $(6.1)^{2.4} \approx 76.699$

3. $2^{-0.5\pi} \approx 0.337$

5. $60^{\sqrt{3}} \approx 1201.845$

7. $f(x) = 4^x$

Intercept: $(0, 1)$

Horizontal asymptote: x-axis

Increasing on: $(-\infty, \infty)$

Matches graph (c).

9. $f(x) = -4^x$

Intercept: $(0, -1)$

Horizontal asymptote: x-axis

Decreasing on: $(-\infty, \infty)$

Matches graph (a).

11. $f(x) = 5^x$

$g(x) = 5^{x-1}$

Since $g(x) = f(x - 1)$, the graph of g can be obtained by shifting the graph of f one unit to the right.

13. $f(x) = \left(\frac{1}{2}\right)^x$

$g(x) = -\left(\frac{1}{2}\right)^{x+2}$

Since $g(x) = -f(x + 2)$, the graph of g can be obtained by reflecting the graph of f about the x-axis and shifting $-f$ two units to the left.

15. $f(x) = 4^{-x} + 4$

Horizontal asymptote: $y = 4$

x	-1	0	1	2	3
$f(x)$	8	5	4.25	4.063	4.016

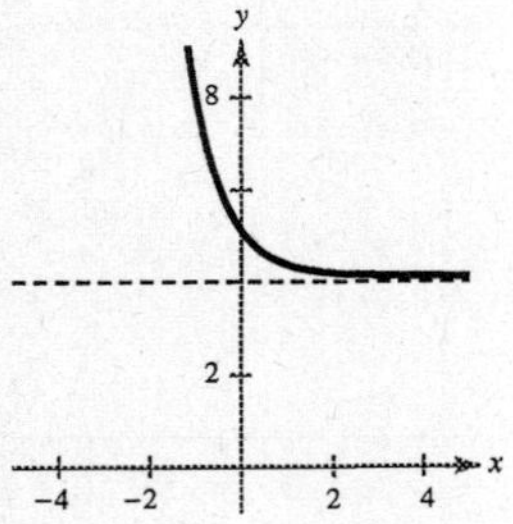

17. $f(x) = -2.65^{x+1}$

Horizontal asymptote: $y = 0$

x	-2	-1	0	1	2
$f(x)$	-0.377	-1	-2.65	-7.023	-18.61

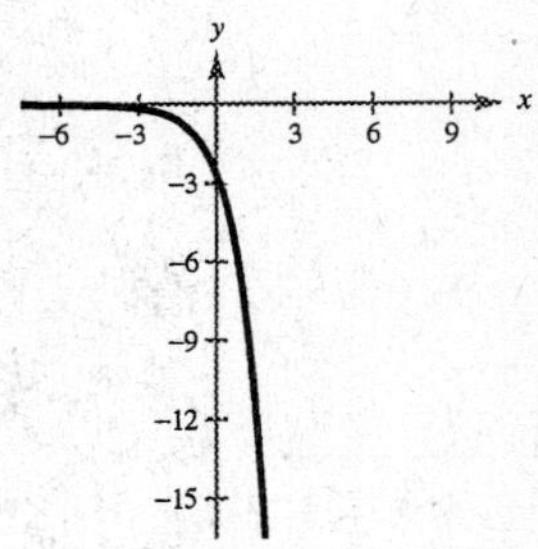

19. $f(x) = 5^{x-2} + 4$

Horizontal asymptote: $y = 4$

x	-1	0	1	2	3
$f(x)$	4.008	4.04	4.2	5	9

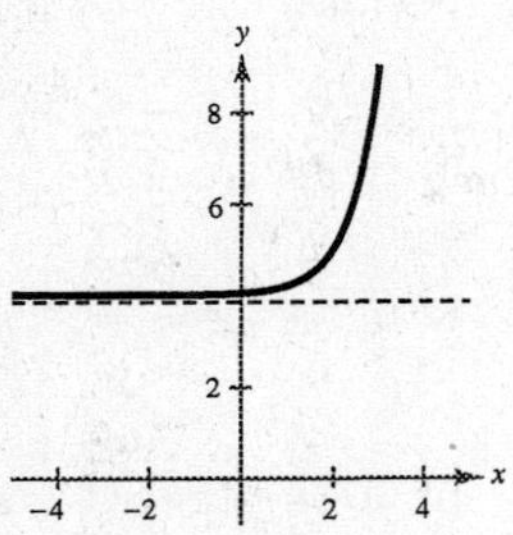

21. $f(x) = \left(\frac{1}{2}\right)^{-x} + 3 = 2^x + 3$

Horizontal asymptote: $y = 3$

x	-2	-1	0	1	2
$f(x)$	3.25	3.5	4	5	7

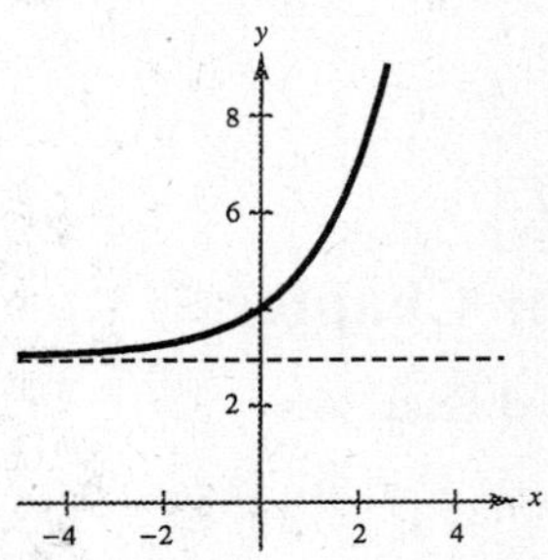

23. $e^8 \approx 2980.958$

25. $e^{-1.7} \approx 0.183$

27. $h(x) = e^{-x/2}$

x	-2	-1	0	1	2
$h(x)$	2.72	1.65	1	0.61	0.37

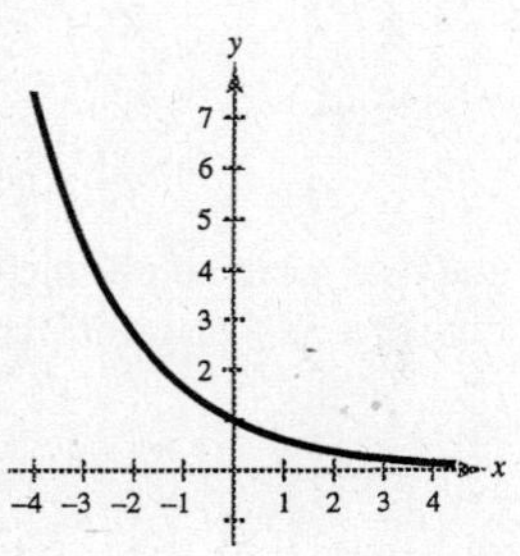

29. $f(x) = e^{x+2}$

x	-3	-2	-1	0	1
$f(x)$	0.37	1	2.72	7.39	20.09

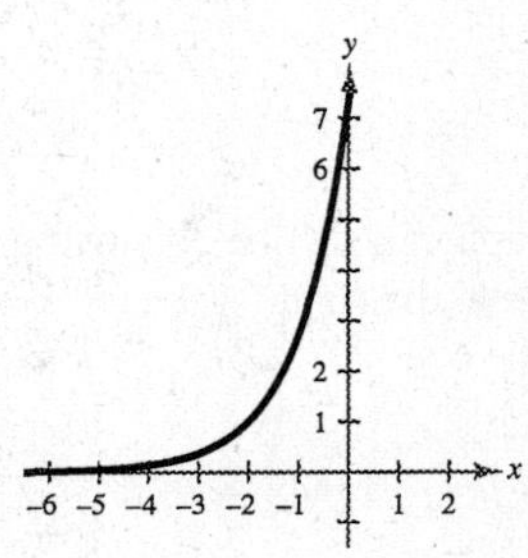

31. $A = 3500\left(1 + \frac{0.065}{n}\right)^{10n}$ or $A = 3500e^{(0.065)(10)}$

n	1	2	4	12	365	Continuous Compounding
A	\$6569.98	\$6635.43	\$6669.46	\$6692.64	\$6704.00	\$6704.39

33. $F(t) = 1 - e^{-t/3}$

(a) $F\left(\frac{1}{2}\right) \approx 0.154$

(b) $F(2) \approx 0.487$

(c) $F(5) \approx 0.811$

35. (a) $A = 50{,}000e^{(0.0875)(35)} \approx \$1{,}069{,}047.14$

(b) The doubling time is

$$\frac{\ln 2}{0.0875} \approx 7.9 \text{ years.}$$

37. $4^3 = 64$

$\log_4 64 = 3$

39. $\log_{10} 1000 = \log_{10} 10^3 = 3$

41. $\log_2 \frac{1}{8} = \log_2 2^{-3} = -3$

43. $g(x) = \log_7 x \Rightarrow x = 7^y$

Domain: $(0, \infty)$

x-intercept: $(1, 0)$

Vertical asymptote: $x = 0$

x	$\frac{1}{7}$	1	7	49
$g(x)$	-1	0	1	2

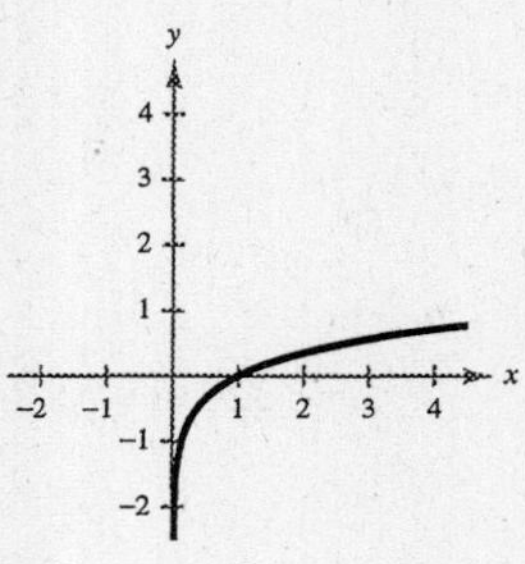

45. $f(x) = \log_{10}\left(\frac{x}{3}\right) \Rightarrow \frac{x}{3} = 10^y \Rightarrow x = 3(10^y)$

Domain: $(0, \infty)$

x-intercept: $(3, 0)$

Vertical asymptote: $x = 0$

x	0.03	0.3	3	30
$f(x)$	-2	-1	0	1

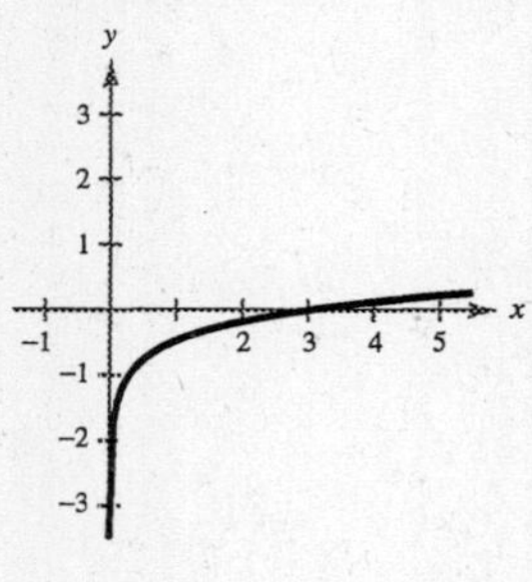

47. $f(x) = 4 - \log_{10}(x + 5)$

Domain: $(-5, \infty)$

x-intercept: $(9995, 0)$

$$\text{Since } 4 - \log_{10}(x + 5) = 0 \Rightarrow \log_{10}(x + 5) = 4$$
$$x + 5 = 10^4$$
$$x = 10^4 - 5 = 9995$$

Vertical asymptote: $x = -5$

x	-4	-3	-2	-1	0	1
$f(x)$	4	3.70	3.52	3.40	3.30	3.22

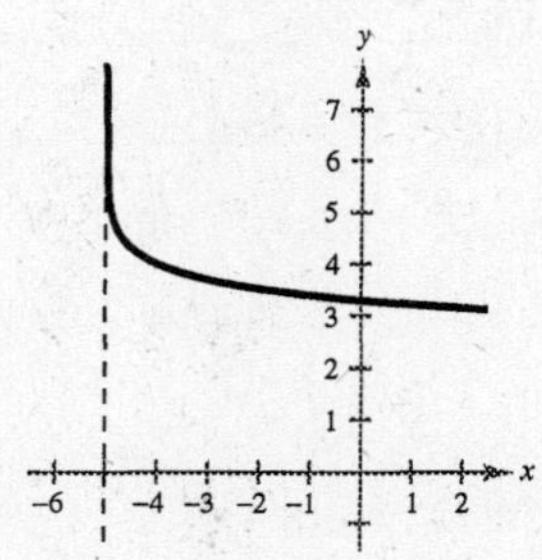

49. $\ln 22.6 \approx 3.118$

51. $\ln e^{-12} = -12$

53. $\ln(\sqrt{7} + 5) \approx 2.034$

55. $f(x) = \ln x + 3$

Domain: $(0, \infty)$

x-intercept: $\ln x + 3 = 0$

$$\ln x = -3$$
$$x = e^{-3}$$
$$(e^{-3}, 0)$$

Vertical asymptote: $x = 0$

x	1	2	3	$\frac{1}{2}$	$\frac{1}{4}$
$f(x)$	3	3.69	4.10	2.31	1.61

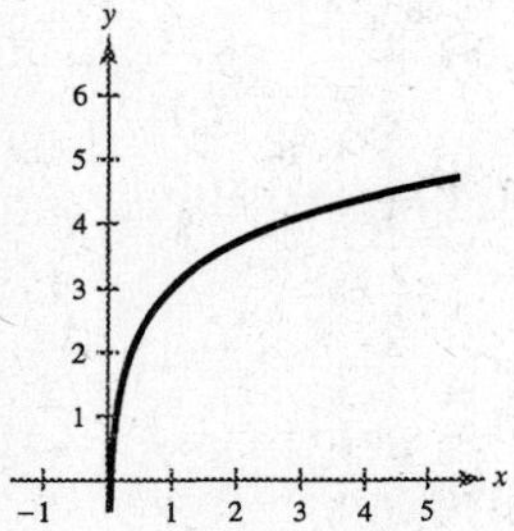

57. $h(x) = \ln(x^2) = 2 \ln|x|$

Domain: $(-\infty, 0) \cup (0, \infty)$

x-intercepts: $(\pm 1, 0)$

Vertical asymptote: $x = 0$

x	± 0.5	± 1	± 2	± 3	± 4
y	-1.39	0	1.39	2.20	2.77

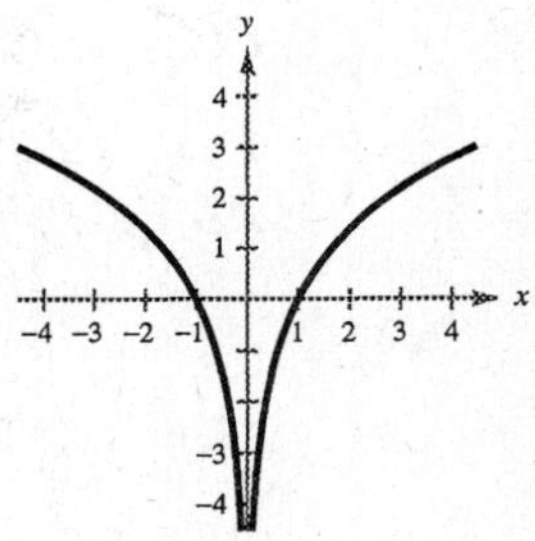

59. $h = 116 \log_{10}(a + 40) - 176$

$h(55) = 116 \log_{10}(55 + 40) - 176 \approx 53.4$ inches

61. $\log_4 9 = \dfrac{\log_{10} 9}{\log_{10} 4} \approx 1.585$

$\log_4 9 = \dfrac{\ln 9}{\ln 4} \approx 1.585$

63. $\log_{1/2} 5 = \dfrac{\log_{10} 5}{\log_{10}(1/2)} \approx -2.322$

$\log_{1/2} 5 = \dfrac{\ln 5}{\ln(1/2)} \approx -2.322$

65. $\log_5 5x^2 = \log_5 5 + \log_5 x^2$

$= 1 + 2 \log_5|x|$

67. $\log_3 \dfrac{6}{\sqrt[3]{x}} = \log_3 6 - \log_3 \sqrt[3]{x}$

$= \log_3(3 \cdot 2) - \log_3 x^{1/3}$

$= \log_3 3 + \log_3 2 - \frac{1}{3}\log_3 x$

$= 1 + \log_3 2 - \frac{1}{3}\log_3 x$

69. $\ln x^2y^2z = \ln x^2 + \ln y^2 + \ln z$

$= 2 \ln|x| + 2 \ln|y| + \ln z$

71. $\ln\left(\dfrac{x+3}{xy}\right) = \ln(x + 3) - \ln xy$

$= \ln(x + 3) - [\ln x + \ln y]$

$= \ln(x + 3) - \ln x - \ln y$

73. $\log_2 5 + \log_2 x = \log_2 5x$

75. $\ln x - \dfrac{1}{4}\ln y = \ln x - \ln \sqrt[4]{y} = \ln\left(\dfrac{x}{\sqrt[4]{y}}\right)$

77. $\frac{1}{3}\log_8(x+4) + 7\log_8 y = \log_8 \sqrt[3]{x+4} + \ln y^7$

$= \log_8\left(y^7 \sqrt[3]{x+4}\right)$

79. $\frac{1}{2}\ln(2x-1) - 2\ln(x+1) = \ln\sqrt{2x-1} - \ln(x+1)^2$

$= \ln\frac{\sqrt{2x-1}}{(x+1)^2}$

81. $t = 50\log_{10}\frac{18{,}000}{18{,}000-h}$

(a) Domain: $0 \le h < 18{,}000$

(b)

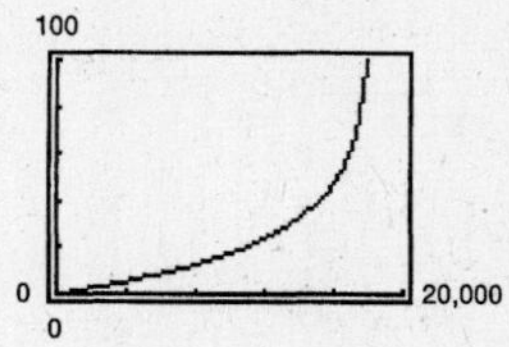

Vertical asymptote: $h = 18{,}000$

(c) As the plane approaches its absolute ceiling, it climbs at a slower rate, so the time required increases.

(d) $50\log_{10}\frac{18{,}000}{18{,}000-4000} \approx 5.46$ minutes

83. $8^x = 512$

$8^x = 8^3$

$x = 3$

85. $6^x = \frac{1}{216}$

$6^x = 6^{-3}$

$x = -3$

87. $e^x = 3$

$x = \ln 3$

89. $\log_4 x = 2$

$x = 4^2 = 16$

91. $\ln x = 4$

$x = e^4$

93. $e^x = 12$

$\ln e^x = \ln 12$

$x = \ln 12 \approx 2.485$

95. $3e^{-5x} = 132$

$e^{-5x} = 44$

$\ln e^{-5x} = \ln 44$

$-5x = \ln 44$

$x = \frac{\ln 44}{-5} \approx -0.757$

97. $2^x + 13 = 35$

$2^x = 22$

$x = \log_2 22 = \frac{\log_{10} 22}{\log_{10} 2}$ or $\frac{\ln 22}{\ln 2}$

$x \approx 4.459$

99. $-4(5^x) = -68$

$5^x = 17$

$\ln 5^x = \ln 17$

$x\ln 5 = \ln 17$

$x = \frac{\ln 17}{\ln 5} \approx 1.760$

101. $e^{2x} - 7e^x + 10 = 0$

$(e^x - 2)(e^x - 5) = 0$

$e^x = 2$ or $e^x = 5$

$\ln e^x = \ln 2$ $\qquad$ $\ln e^x = \ln 5$

$x = \ln 2 \approx 0.693$ $\qquad$ $x = \ln 5 \approx 1.609$

103. $2^{0.6x} - 3x = 0$

Graph $y_1 = 2^{0.6x} - 3x$.

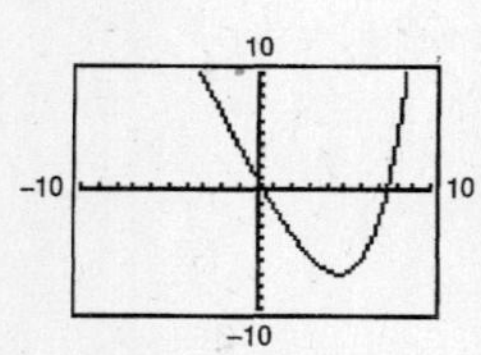

The x-intercepts are at $x \approx 0.39$ and at $x \approx 7.48$.

105. $25e^{-0.3x} = 12$

Graph $y_1 = 25e^{-0.3x}$ and $y_2 = 12$.

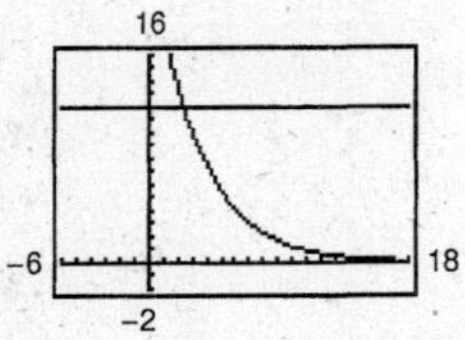

The graphs intersect at $x \approx 2.45$.

107. $\ln 3x = 8.2$

$$e^{\ln 3x} = e^{8.2}$$
$$3x = e^{8.2}$$
$$x = \frac{e^{8.2}}{3} \approx 1213.650$$

109. $2 \ln 4x = 15$

$$\ln 4x = \tfrac{15}{2}$$
$$e^{\ln 4x} = e^{7.5}$$
$$4x = e^{7.5}$$
$$x = \tfrac{1}{4}e^{7.5} \approx 452.011$$

111. $\ln x - \ln 3 = 2$

$$\ln \frac{x}{3} = 2$$
$$e^{\ln \frac{x}{3}} = e^2$$
$$\frac{x}{3} = e^2$$
$$x = 3e^2 \approx 22.167$$

113. $\ln \sqrt{x+1} = 2$

$$\tfrac{1}{2}\ln(x+1) = 2$$
$$\ln(x+1) = 4$$
$$e^{\ln(x+1)} = e^4$$
$$x + 1 = e^4$$
$$x = e^4 - 1 \approx 53.598$$

115. $\log_{10}(x-1) = \log_{10}(x-2) - \log_{10}(x+2)$

$$\log_{10}(x-1) = \log_{10}\left(\frac{x-2}{x+2}\right)$$
$$10^{\log_{10}(x-1)} = 10^{\log_{10}(x-2)/(x+2)}$$
$$x - 1 = \frac{x-2}{x+2}$$
$$(x-1)(x+2) = x - 2$$
$$x^2 + x - 2 = x - 2$$
$$x^2 = 0$$
$$x = 0$$

Since $x = 0$ is not in the domain of $\ln(x-1)$ or of $\ln(x-2)$, it is an extraneous solution. The equation has no solution.

117. $\log_{10}(1-x) = -1$

$$10^{\log_{10}(1-x)} = 10^{-1}$$
$$1 - x = 10^{-1}$$
$$1 - \tfrac{1}{10} = x$$
$$x = 0.900$$

119. $2\ln(x+3) + 3x = 8$

Graph $y_1 = 2\ln(x+3) + 3x$ and $y_2 = 8$.

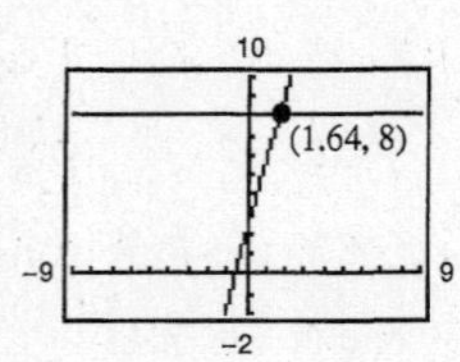

The graphs intersect at approximately (1.64, 8). The solution of the equation is $x \approx 1.64$.

121. $4\ln(x+5) - x = 10$

Graph $y_1 = 4\ln(x+5) - x$ and $y_2 = 10$.

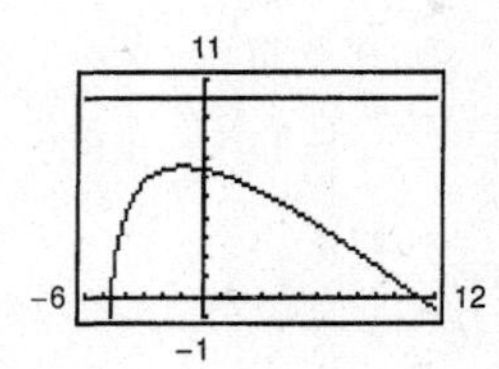

The graphs do not intersect. The equation has no solution.

123. $3(7550) = 7550e^{0.0725t}$

$3 = e^{0.0725t}$

$\ln 3 = \ln e^{0.0725t}$

$\ln 3 = 0.0725t$

$t = \dfrac{\ln 3}{0.0725} \approx 15.2$ years

125. $y = 3e^{-2x/3}$

Exponential decay model

Matches graph (e).

127. $y = \ln(x + 3)$

Logarithmic model

Vertical asymptote: $x = -3$

Graph includes $(-2, 0)$

Matches graph (f).

129. $y = 2e^{-(x+4)^2/3}$

Gaussian model

Matches graph (a).

131. $P = 590e^{0.027t}$

1.5 million = 1500 thousand

$1500 = 590e^{0.027t}$

$\dfrac{150}{59} = e^{0.027t}$

$0.027t = \ln\left(\dfrac{150}{59}\right)$

$t = \dfrac{\ln\left(\dfrac{150}{59}\right)}{0.027} \approx 34.6$ years

According to this model, the population of Phoenix will reach 1.5 million during the year 2004.

133. (a) $20{,}000 = 10{,}000e^{r(5)}$

$2 = e^{5r}$

$\ln 2 = 5r$

$\dfrac{\ln 2}{5} = r$

$r \approx 0.138629 = 13.8629\%$

(b) $A = 10{,}000e^{0.138629}$

$\approx \$11{,}486.98$

135. $y = ae^{bx}$

$2 = ae^{b(0)} \implies a = 2$

$3 = 2e^{b(4)}$

$1.5 = e^{4b}$

$\ln 1.5 = 4b \implies b \approx 0.1014$

Thus, $y \approx 2e^{0.1014x}$.

137. $y = 0.0499e^{-(x-71)^2/128}, 40 \le x \le 100$

(a) Graph $y_1 = 0.0499e^{-(x-71)^2/128}$.

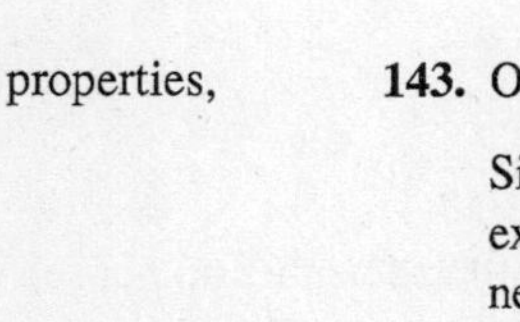

(b) The average test score is 71.

139. $\beta = 10 \log_{10}\left(\dfrac{I}{10^{-16}}\right)$

$125 = 10 \log_{10}\left(\dfrac{I}{10^{-16}}\right)$

$12.5 = \log_{10}\left(\dfrac{I}{10^{-16}}\right)$

$10^{12.5} = \dfrac{I}{10^{-16}}$

$I = 10^{-3.5}$ watt/cm^2

141. True. By the inverse properties, $\log_b b^{2x} = 2x$.

143. Order: $b < d < a < c$

Since graphs (b) and (d) represent exponential decay, b and d are negative.

Since graph (a) and (c) represent exponential growth, a and c are positive.

Problem Solving for Chapter 5

Solutions to Odd-Numbered Exercises

1. $y = a^x$

$y_1 = 0.5^x$

$y_2 = 1.2^x$

$y_3 = 2.0^x$

$y_4 = x$

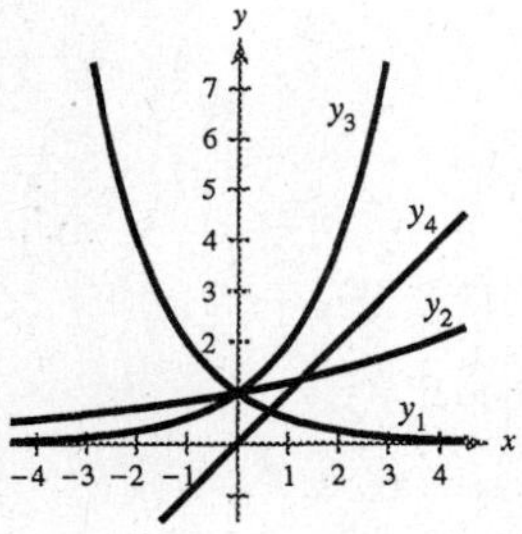

The curves $y = 0.5^x$ and $y = 1.2^x$ cross the line $y = x$.

From checking the graphs it appears that $y = x$ will cross $y = a^x$ for $0 \le a \le 1.44$.

3. The exponential function, $y = e^x$, increases at a faster rate than the polynomial function $y = x^n$.

5. (a) $f(u + v) = a^{u+v}$

$= a^u \cdot a^v$

$= f(u) \cdot f(v)$

(b) $f(2x) = a^{2x}$

$= (a^x)^2$

$= [f(x)]^2$

7. (a)

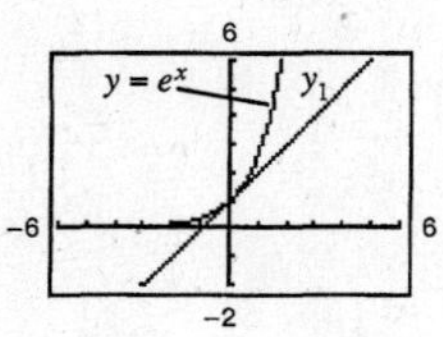

(b)

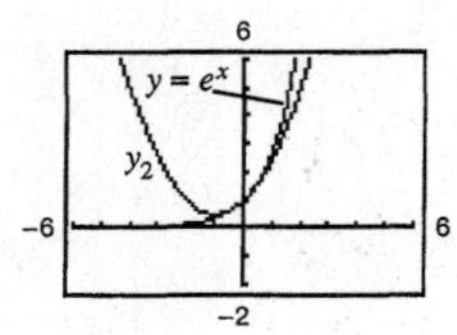

(c)

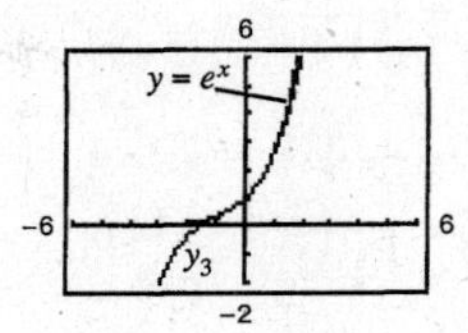

9. $f(x) = e^x - e^{-x}$

$y = e^x - e^{-x}$

$x = e^y - e^{-y}$

$x = \dfrac{e^{2y} - 1}{e^y}$

$xe^y = e^{2y} - 1$

$e^{2y} - xe^y - 1 = 0$

$e^y = \dfrac{x \pm \sqrt{x^2 + 4}}{2}$ Quadratic Formula

Choosing the positive quantity for e^y we have $y = \ln\left(\dfrac{x + \sqrt{x^2 + 4}}{2}\right)$. Thus, $f^{-1}(x) = \ln\left(\dfrac{x + \sqrt{x^2 + 4}}{2}\right)$.

11. Answer (c) $y = 6(1 - e^{-x^2/2})$

The graph passes through (0, 0) and neither (a) nor (b) pass through the origin. Also, the graph has y-axis symmetry and a horizontal asymptote at $y = 6$.

13. $y_1 = c_1\left(\frac{1}{2}\right)^{t/k_1}$ and $y_2 = c_2\left(\frac{1}{2}\right)^{t/k_2}$

$$c_1\left(\frac{1}{2}\right)^{t/k_1} = c_2\left(\frac{1}{2}\right)^{t/k_2}$$

$$\frac{c_1}{c_2} = \left(\frac{1}{2}\right)^{(t/k_2 - t/k_1)}$$

$$\ln\left(\frac{c_1}{c_2}\right) = \left(\frac{t}{k_2} - \frac{t}{k_1}\right)\ln\left(\frac{1}{2}\right)$$

$$\ln c_1 - \ln c_2 = t\left(\frac{1}{k_2} - \frac{1}{k_1}\right)\ln\left(\frac{1}{2}\right)$$

$$t = \frac{\ln c_1 - \ln c_2}{\left(\frac{1}{k_2} - \frac{1}{k_1}\right)\ln\left(\frac{1}{2}\right)}$$

15. (a) $y \approx 252.606(1.0310)^x$

(b) $y \approx 400.88x^2 - 1464.6x + 291{,}782$

(c)

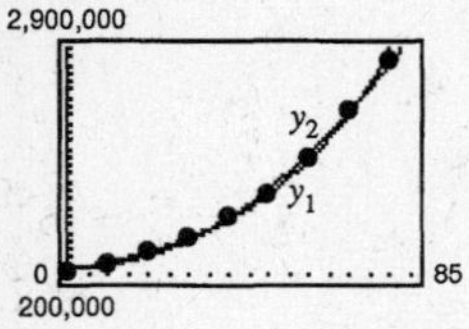

(d) Both models appear to be "good fits" for the data, but neither would be reliable to predict the population of the United States in 2010. The exponential model approaches infinity rapidly.

17. $(\ln x)^2 = \ln x^2$

$(\ln x)^2 - 2\ln x = 0$

$\ln x(\ln x - 2) = 0$

$\ln x = 0$ or $\ln x = 2$

$x = 1$ or $x = e^2$

19. $y_4 = (x - 1) - \frac{1}{2}(x - 1)^2 + \frac{1}{3}(x - 1)^3 - \frac{1}{4}(x - 1)^4$

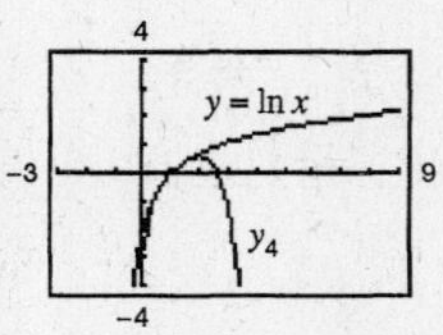

The pattern implies that $\ln x = (x - 1) - \frac{1}{2}(x - 1)^2 + \frac{1}{3}(x - 1)^3 - \frac{1}{4}(x - 1)^4 + \cdots$

21. $y = 80.4 - 11\ln x$

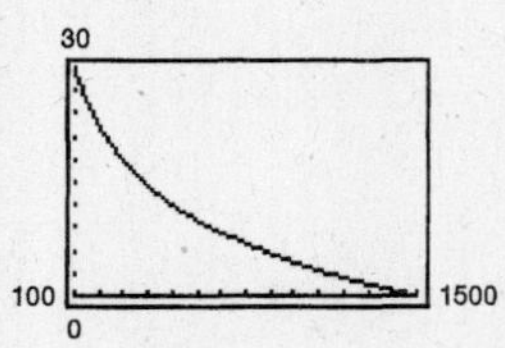

$y(300) = 80.4 - 11\ln 300 \approx 17.7$ ft³/min

Chapter 5 Practice Test

1. Solve for x: $x^{3/5} = 8$.

2. Solve for x: $3^{x-1} = \frac{1}{81}$.

3. Graph $f(x) = 2^{-x}$.

4. Graph $g(x) = e^x + 1$.

5. If \$5000 is invested at 9% interest, find the amount after three years if the interest is compounded

   (a) monthly. (b) quarterly. (c) continuously.

6. Write the equation in logarithmic form: $7^{-2} = \frac{1}{49}$.

7. Solve for x: $x - 4 = \log_2 \frac{1}{64}$.

8. Given $\log_b 2 = 0.3562$ and $\log_b 5 = 0.8271$, evaluate $\log_b \sqrt[4]{8/25}$.

9. Write $5 \ln x - \frac{1}{2} \ln y + 6 \ln z$ as a single logarithm.

10. Using your calculator and the change of base formula, evaluate $\log_9 28$.

11. Use your calculator to solve for N: $\log_{10} N = 0.6646$

12. Graph $y = \log_4 x$.

13. Determine the domain of $f(x) = \log_3(x^2 - 9)$.

14. Graph $y = \ln(x - 2)$.

15. True or false: $\dfrac{\ln x}{\ln y} = \ln(x - y)$

16. Solve for x: $5^x = 41$

17. Solve for x: $x - x^2 = \log_5 \frac{1}{25}$

18. Solve for x: $\log_2 x + \log_2(x - 3) = 2$

19. Solve for x: $\dfrac{e^x + e^{-x}}{3} = 4$

20. Six thousand dollars is deposited into a fund at an annual interest rate of 13%. Find the time required for the investment to double if the interest is compounded continuously.

CHAPTER 6
Topics in Analytic Geometry

CHAPTER 6
Topics in Analytic Geometry

Section 6.1 Lines

- The **inclination** of a nonhorizontal line is the positive angle θ ($\theta < 180°$) measured counterclockwise from the x-axis to the line. A horizontal line has an inclination of zero.
- If a nonvertical line has inclination of θ and slope m, then $m = \tan\theta$.
- If two nonperpendicular lines have slopes m_1 and m_2, then the angle between the lines is given by

$$\tan\theta = \left|\frac{m_2 - m_1}{1 + m_1 m_2}\right|.$$

- The distance between a point (x_1, y_1) and a line $Ax + By + C = 0$ is given by

$$d = \frac{|Ax_1 + By_1 + C|}{\sqrt{A^2 + B^2}}.$$

Solutions to Odd-Numbered Exercises

1. $m = \tan\frac{\pi}{6} = \frac{\sqrt{3}}{3}$

3. $m = \tan\frac{3\pi}{4} = -1$

5. $m = \tan\frac{\pi}{3} = \sqrt{3}$

7. $m = \tan 1.27 \approx 3.2236$

9. $m = -1$

$-1 = \tan\theta$

$\theta = 180° + \arctan(-1) = \frac{3\pi}{4}$ radians $= 135°$

11. $m = 1$

$1 = \tan\theta$

$\theta = \frac{\pi}{4}$ radian $= 45°$

13. $m = \frac{3}{4}$

$\frac{3}{4} = \tan\theta$

$\theta = \arctan\left(\frac{3}{4}\right) \approx 0.6435$ radian $\approx 36.9°$

15. $(6, 1), (10, 8)$

$m = \frac{8 - 1}{10 - 6} = \frac{7}{4}$

$\frac{7}{4} = \tan\theta$

$\theta = \arctan\left(\frac{7}{4}\right) \approx 1.0517$ radians $\approx 60.3°$

17. $(-2, 20), (10, 0)$

$m = \frac{0 - 20}{10 - (-2)} = -\frac{20}{12} = -\frac{5}{3}$

$-\frac{5}{3} = \tan\theta$

$\theta = \pi + \arctan\left(-\frac{5}{3}\right) \approx 2.1112$ radians $\approx 121.0°$

19. $6x - 2y + 8 = 0$

$y = 3x + 4 \Rightarrow m = 3$

$3 = \tan\theta$

$\theta = \arctan 3 \approx 1.2490$ radians $\approx 71.6°$

21. $5x + 3y = 0$

$$y = -\frac{5}{3}x \Rightarrow m = -\frac{5}{3}$$

$$-\frac{5}{3} = \tan\theta$$

$$\theta = \pi + \arctan\left(-\frac{5}{3}\right) \approx 2.1112 \text{ radians} \approx 121.0^\circ$$

23. $3x + y = 3 \Rightarrow y = -3x + 3 \Rightarrow m_1 = -3$

$$x - y = 2 \Rightarrow y = x - 2 \quad \Rightarrow m_2 = 1$$

$$\tan\theta = \left|\frac{1-(-3)}{1+(-3)(1)}\right| = 2$$

$$\theta = \arctan 2 \approx 1.1071 \text{ radians} \approx 63.4^\circ$$

25. $x - y = 0 \quad \Rightarrow y = x \quad \Rightarrow m_1 = 1$

$$3x - 2y = -1 \Rightarrow y = \frac{3}{2}x + \frac{1}{2} \Rightarrow m_2 = \frac{3}{2}$$

$$\tan\theta = \left|\frac{\frac{3}{2} - 1}{1 + \left(\frac{3}{2}\right)(1)}\right| = \frac{1}{5}$$

$$\theta = \arctan\frac{1}{5} \approx 0.1974 \text{ radian} \approx 11.3^\circ$$

27. $x - 2y = 7 \Rightarrow y = \frac{1}{2}x - \frac{7}{2} \quad \Rightarrow m_1 = \frac{1}{2}$

$$6x + 2y = 5 \Rightarrow y = -3x + \frac{5}{2} \Rightarrow m_2 = -3$$

$$\tan\theta = \left|\frac{-3 - \frac{1}{2}}{1 + \left(\frac{1}{2}\right)(-3)}\right| = 7$$

$$\theta = \arctan 7 \approx 1.4289 \text{ radians} \approx 81.9^\circ$$

29. $x + 2y = 8 \Rightarrow y = -\frac{1}{2}x + 4 \Rightarrow m_1 = -\frac{1}{2}$

$$x - 2y = 2 \Rightarrow y = \frac{1}{2}x - 1 \quad \Rightarrow m_2 = \frac{1}{2}$$

$$\tan\theta = \left|\frac{\frac{1}{2} - \left(-\frac{1}{2}\right)}{1 + \left(-\frac{1}{2}\right)\left(\frac{1}{2}\right)}\right| = \frac{4}{3}$$

$$\theta = \arctan\left(\frac{4}{3}\right) \approx 0.9273 \text{ radian} \approx 53.1^\circ$$

31. $0.05x - 0.03y = 0.21 \Rightarrow y = \frac{5}{3}x - 7 \quad \Rightarrow m_1 = \frac{5}{3}$

$$0.07x + 0.02y = 0.16 \Rightarrow y = -\frac{7}{2}x + 8 \Rightarrow m_2 = -\frac{7}{2}$$

$$\tan\theta = \left|\frac{\left(-\frac{7}{2}\right) - \left(\frac{5}{3}\right)}{1 + \left(\frac{5}{3}\right)\left(-\frac{7}{2}\right)}\right| = \frac{31}{29}$$

$$\theta = \arctan\left(\frac{31}{29}\right) \approx 0.8187 \text{ radian} \approx 46.9^\circ$$

33. Let $A = (2, 1)$, $B = (4, 4)$, and $C = (6, 2)$.

Slope of AB: $\quad m_1 = \frac{1-4}{2-4} = \frac{3}{2}$

Slope of BC: $\quad m_2 = \frac{4-2}{4-6} = -1$

Slope of AC: $\quad m_3 = \frac{1-2}{2-6} = \frac{1}{4}$

$$\tan A = \left|\frac{\frac{1}{4} - \frac{3}{2}}{1 + \left(\frac{3}{2}\right)\left(\frac{1}{4}\right)}\right| = \frac{\frac{5}{4}}{\frac{11}{8}} = \frac{10}{11}$$

$$A = \arctan\left(\frac{10}{11}\right) \approx 42.3^\circ$$

$$\tan B = \left|\frac{\frac{3}{2} - (-1)}{1 + (-1)\left(\frac{3}{2}\right)}\right| = \frac{\frac{5}{2}}{\frac{1}{2}} = 5$$

$$B = \arctan 5 \approx 78.7^\circ$$

$$\tan C = \left|\frac{-1 - \frac{1}{4}}{1 + \left(\frac{1}{4}\right)(-1)}\right| = \frac{\frac{5}{4}}{\frac{3}{4}} = \frac{5}{3}$$

$$C = \arctan\left(\frac{5}{3}\right) \approx 59.0^\circ$$

35. Let $A = (-4, -1)$, $B = (3, 2)$, and $C = (1, 0)$.

Slope of AB: $\quad m_1 = \frac{-1-2}{-4-3} = \frac{3}{7}$

Slope of BC: $\quad m_2 = \frac{2-0}{3-1} = 1$

Slope of AC: $\quad m_3 = \frac{-1-0}{-4-1} = \frac{1}{5}$

$$\tan A = \left|\frac{\frac{1}{5} - \frac{3}{7}}{1 + \left(\frac{3}{7}\right)\left(\frac{1}{5}\right)}\right| = \frac{\frac{8}{35}}{\frac{38}{35}} = \frac{4}{19}$$

$$A = \arctan\left(\frac{4}{19}\right) \approx 11.9^\circ$$

$$\tan B = \left|\frac{1 - \frac{3}{7}}{1 + \left(\frac{3}{7}\right)(1)}\right| = \frac{\frac{4}{7}}{\frac{10}{7}} = \frac{2}{5}$$

$$B = \arctan\left(\frac{2}{5}\right) \approx 21.8^\circ$$

$$C = 180^\circ - A - B$$

$$\approx 180^\circ - 11.9^\circ - 21.8^\circ = 146.3^\circ$$

37. $(0, 0) \Rightarrow x_1 = 0$ and $y_1 = 0$

$4x + 3y = 0 \Rightarrow A = 4, B = 3$, and $C = 0$

$$d = \frac{|4(0) + 3(0) + 0|}{\sqrt{4^2 + 3^2}} = \frac{0}{5} = 0$$

Note: The point is *on* the line.

39. $(2, 3) \Rightarrow x_1 = 2$ and $y_1 = 3$

$4x + 3y - 10 = 0 \Rightarrow A = 4, B = 3$, and $C = -10$

$$d = \frac{|4(2) + 3(3) + (-10)|}{\sqrt{4^2 + 3^2}} = \frac{7}{5}$$

41. $(6, 2) \Rightarrow x_1 = 6$ and $y_1 = 2$

$x + 1 = 0 \Rightarrow A = 1, B = 0$, and $C = 1$

$$d = \frac{|1(6) + 0(2) + 1|}{\sqrt{1^2 + 0^2}} = 7$$

43. $(0, 8) \Rightarrow x_1 = 0$ and $y_1 = 8$

$6x - y = 0 \Rightarrow A = 6, B = -1$, and $C = 0$

$$d = \frac{|6(0) + (-1)(8) + 0|}{\sqrt{6^2 + (-1)^2}}$$

$$= \frac{8}{\sqrt{37}} = \frac{8\sqrt{37}}{37} \approx 1.3152$$

45. $A = (0, 0), B = (1, 4), C = (4, 0)$

(a) The slope the line through AC is $m = \dfrac{0 - 0}{4 - 0} = 0$.

The equation of the line through AC is $y = 0$.

The distance between the line and $B = (1, 4)$ is $d = \dfrac{|0(1) + (1)(4) + 0|}{\sqrt{0^2 + 1^2}} = 4$.

(b) The distance between A and C is 4.

$A = \dfrac{1}{2}(4)(4) = 8$ square units.

47. $A = \left(-\dfrac{1}{2}, \dfrac{1}{2}\right), B = (2, 3), C = \left(\dfrac{5}{2}, 0\right)$

(a) The slope of the line through AC is $m = \dfrac{\frac{1}{2} - 0}{\left(-\frac{1}{2}\right) - \frac{5}{2}} = -\dfrac{1}{6}$.

The equation of the line through AC is $y - 0 = -\dfrac{1}{6}\left(x - \dfrac{5}{2}\right) \Rightarrow 2x + 12y - 5 = 0$.

The distance between the line and $B = (2, 3)$ is $d = \dfrac{|2(2) + 12(3) + (-5)|}{\sqrt{2^2 + 12^2}} = \dfrac{35}{\sqrt{148}} = \dfrac{35\sqrt{37}}{74}$.

(b) The distance between A and C is $d = \sqrt{\left[\left(-\dfrac{1}{2}\right) - \left(\dfrac{5}{2}\right)\right]^2 + \left[\left(\dfrac{1}{2}\right) - 0\right]^2} = \dfrac{\sqrt{37}}{2}$.

$A = \dfrac{1}{2}\left(\dfrac{\sqrt{37}}{2}\right)\left(\dfrac{35\sqrt{37}}{74}\right) = \dfrac{35}{8}$ square units

49. $x + y = 1 \Rightarrow (0, 1)$ is a point on the line $\Rightarrow x_1 = 0$ and $y_1 = 1$

$x + y = 5 \Rightarrow A = 1, B = 1$, and $C = -5$

$$d = \frac{|1(0) + 1(1) + (-5)|}{\sqrt{1^2 + 1^2}} = \frac{4}{\sqrt{2}} = 2\sqrt{2}$$

51. Slope: $m = \tan 0.1 \approx 0.1003$

Change in elevation: $\sin 0.1 = \dfrac{x}{2(5280)}$

$x \approx 1054$ feet

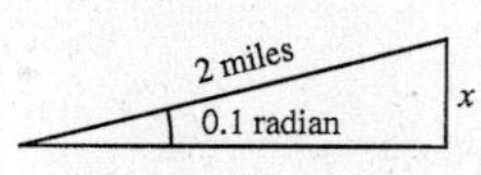

Not drawn to scale

53. (a) $\tan\theta = \frac{1}{3}$

$\theta = \arctan\left(\frac{1}{3}\right) \approx 18.4°$

(b) $\frac{5}{x} = \sin 18.4°$

$x = \frac{5}{\sin 18.4°}$

$x \approx 15.8$ meters

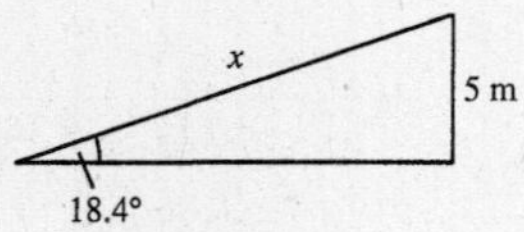

55. $\tan\gamma = \frac{6}{9}$

$\gamma = \arctan\left(\frac{2}{3}\right) \approx 33.69°$

$\beta = 90 - \gamma \approx 56.31°$

Also, since the right triangles containing α and β are equal, $\alpha = \gamma \approx 33.69°$

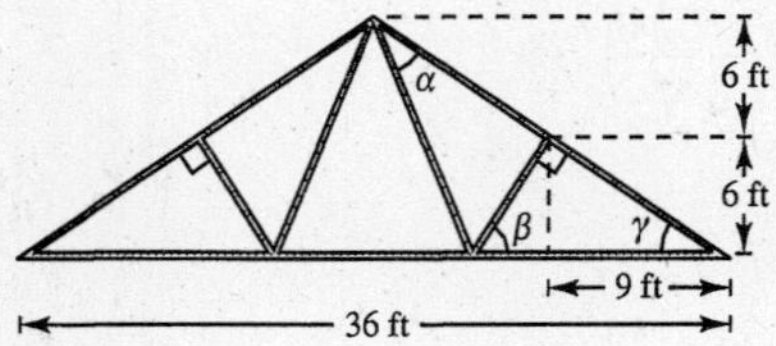

57. True. The inclination of a line is related to its slope by $m = \tan\theta$.

If the angle is greater than $\frac{\pi}{2}$ but less than π, then the angle is in the second quadrant where the tangent function is negative.

59. (a) $(0, 0) \implies x_1 = 0$ and $y_1 = 0$

$y = mx + 4 \implies 0 = mx - y + 4$

$$d = \frac{|m(0) + (-1)(0) + 4|}{\sqrt{m^2 + (-1)^2}} = \frac{4}{\sqrt{m^2 + 1}}$$

(b)

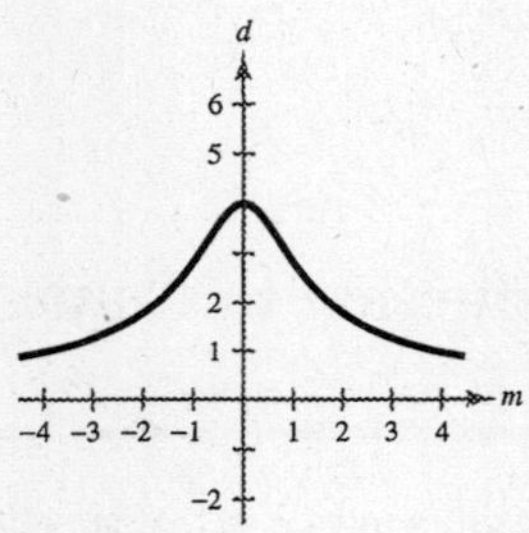

(c) The maximum distance of 4 occurs when the slope m is 0 and the line through (0, 4) is horizontal.

(d) The graph has a horizontal asymptote at $d = 0$. As the slope becomes larger, the distance between the origin and the line, $y = mx + 4$, becomes smaller and approaches 0.

61. $f(x) = (x - 7)^2$

x-intercept: $0 = (x - 7)^2 \implies x = 7$

$(7, 0)$

y-intercept: $y = (0 - 7)^2 = 49$

$(0, 49)$

63. $f(x) = (x - 5)^2 - 5$

x-intercepts: $0 = (x - 5)^2 - 5$

$5 = (x - 5)^2$

$\pm\sqrt{5} = x - 5$

$5 \pm \sqrt{5} = x$

$\left(5 \pm \sqrt{5}, 0\right)$

y-intercept: $y = (0 - 5)^2 - 5 = 20$

$(0, 20)$

65. $f(x) = x^2 - 7x - 1$

x-intercepts: $0 = x^2 - 7x - 1$

$x = \frac{7 \pm \sqrt{53}}{2}$ by the Quadratic Formula

$\left(\frac{7 \pm \sqrt{53}}{2}, 0\right)$

y-intercept: $y = 0^2 - 7(0) - 1 = -1$

$(0, -1)$

67. $f(x) = 3x^2 + 2x - 16$

$= 3\left(x^2 + \frac{2}{3}x\right) - 16$

$= 3\left(x^2 + \frac{2}{3}x + \frac{1}{9}\right) - \frac{1}{3} - 16$

$= 3\left(x + \frac{1}{3}\right)^2 - \frac{49}{3}$

Vertex: $\left(-\frac{1}{3}, -\frac{49}{3}\right)$

69. $f(x) = 5x^2 + 34x - 7$

$$= 5\left(x^2 + \frac{34}{5}x\right) - 7$$

$$= 5\left(x^2 + \frac{34}{5}x + \frac{289}{25}\right) - \frac{289}{5} - 7$$

$$= 5\left(x + \frac{17}{5}\right)^2 - \frac{324}{5}$$

Vertex: $\left(-\frac{17}{5}, -\frac{324}{5}\right)$

71. $f(x) = 6x^2 - x - 12$

$$= 6\left(x^2 - \frac{1}{6}x\right) - 12$$

$$= 6\left(x^2 - \frac{1}{6}x + \frac{1}{144}\right) - \frac{1}{24} - 12$$

$$= 6\left(x - \frac{1}{12}\right)^2 - \frac{289}{24}$$

Vertex: $\left(\frac{1}{12}, -\frac{289}{24}\right)$

73. $f(x) = (x - 4)^2 + 3$

Vertex: $(4, 3)$

y-intercept: $(0, 19)$

x-intercept: None

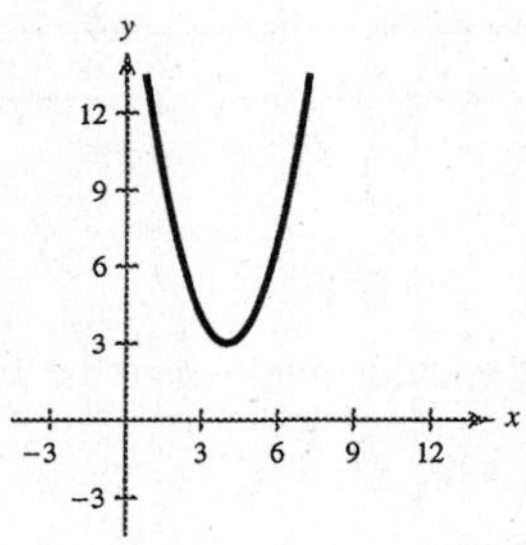

75. $g(x) = 2x^2 - 3x + 1$

$= 2\left(x^2 - \frac{3}{2}x + \frac{9}{16}\right) - \frac{9}{8} + 1$

$= 2\left(x - \frac{3}{4}\right)^2 - \frac{1}{8}$

Vertex: $\left(\frac{3}{4}, -\frac{1}{8}\right)$

y-intercept: $(0, 1)$

x-intercept: $\left(\frac{1}{2}, 0\right), (1, 0)$

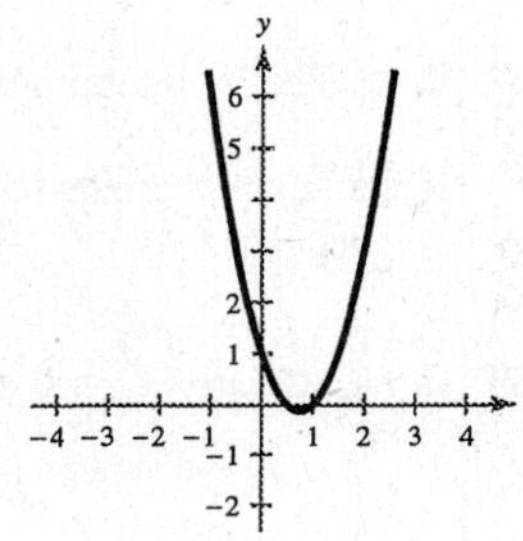

Section 6.2 Introduction to Conics: Parabolas

- A **parabola** is the set of all points (x, y) that are equidistant from a fixed line **(directrix)** and a fixed point **(focus)** not on the line.
- The standard equation of a parabola with vertex (h, k) and:
  (a) Vertical axis $x = h$ and directrix $y = k - p$ is:
  $(x - h)^2 = 4p(y - k), p \neq 0$
  (b) Horizontal axis $y = k$ and directrix $x = h - p$ is:
  $(y - k)^2 = 4p(x - h), p \neq 0$
- The tangent line to a parabola at a point P makes **equal angles** with:
  (a) the line through P and the focus.
  (b) the axis of the parabola.

Solutions to Odd-Numbered Exercises

1. A circle is formed when a plane intersects the top or bottom half of a double-napped cone and is perpendicular to the axis of the cone.

3. A parabola is formed when a plane intersects the top or bottom half of a double-napped cone, is parallel to the side of the cone, and does not intersect the vertex.

5. $y^2 = -4x$
Vertex: $(0, 0)$
Opens to the left since p is negative.
Matches graph (e).

7. $x^2 = -8y$
Vertex: $(0, 0)$
Opens downward since p is negative.
Matches graph (d).

9. $(y - 1)^2 = 4(x - 3)$
Vertex: $(3, 1)$
Opens to the right since p is positive.
Matches graph (a).

11. $y = \frac{1}{2}x^2$

$x^2 = 2y$

$x^2 = 4\left(\frac{1}{2}\right)y \Rightarrow h = 0, k = 0, p = \frac{1}{2}$

Vertex: $(0, 0)$

Focus: $\left(0, \frac{1}{2}\right)$

Directrix: $y = -\frac{1}{2}$

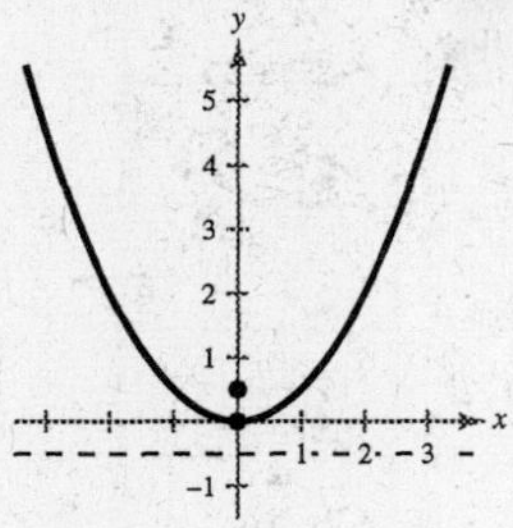

13. $y^2 = -6x$

$y^2 = 4\left(-\frac{3}{2}\right)x \Rightarrow h = 0, k = 0, p = -\frac{3}{2}$

Vertex: $(0, 0)$

Focus: $\left(-\frac{3}{2}, 0\right)$

Directrix: $x = \frac{3}{2}$

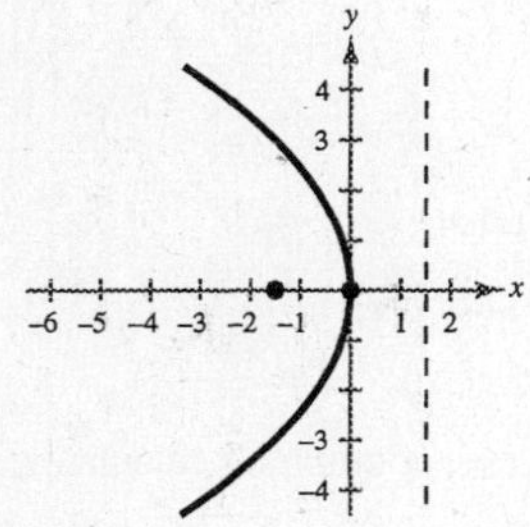

15. $x^2 + 6y = 0$

$x^2 = -6y = 4\left(-\frac{3}{2}\right)y \Rightarrow h = 0, k = 0, p = -\frac{3}{2}$

Vertex: $(0, 0)$

Focus: $\left(0, -\frac{3}{2}\right)$

Directrix: $y = \frac{3}{2}$

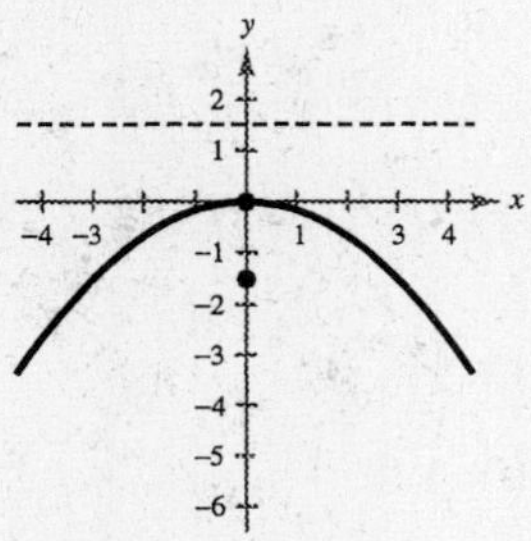

17. $(x - 1)^2 + 8(y + 2) = 0$

$(x - 1)^2 = 4(-2)(y + 2)$

$h = 1, k = -2, p = -2$

Vertex: $(1, -2)$

Focus: $(1, -4)$

Directrix: $y = 0$

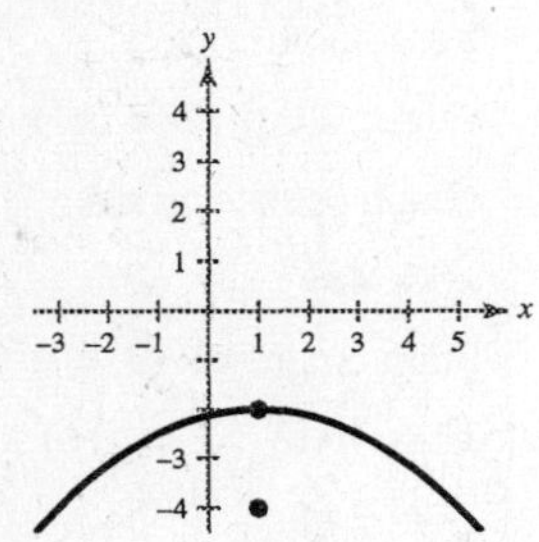

19. $\left(x + \frac{3}{2}\right)^2 = 4(y - 2)$

$\left(x + \frac{3}{2}\right)^2 = 4(1)(y - 2)$

$h = -\frac{3}{2}, k = 2, p = 1$

Vertex: $\left(-\frac{3}{2}, 2\right)$

Focus: $\left(-\frac{3}{2}, 3\right)$

Directrix: $y = 1$

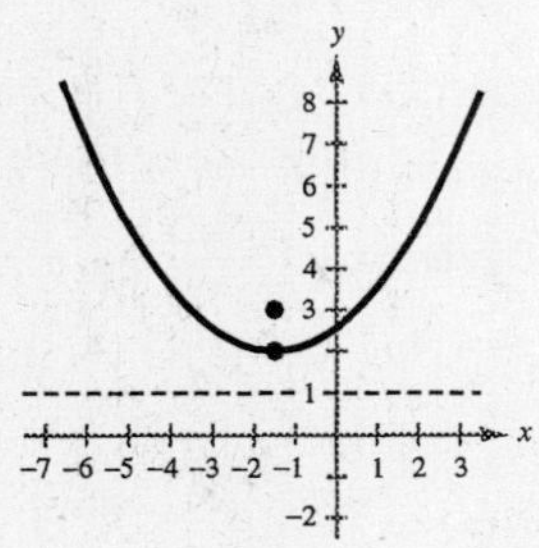

21. $y = \frac{1}{4}(x^2 - 2x + 5)$

$4y = x^2 - 2x + 5$

$4y - 5 + 1 = x^2 - 2x + 1$

$4y - 4 = (x - 1)^2$

$(x - 1)^2 = 4(1)(y - 1)$

$h = 1, k = 1, p = 1$

Vertex: $(1, 1)$

Focus: $(1, 2)$

Directrix: $y = 0$

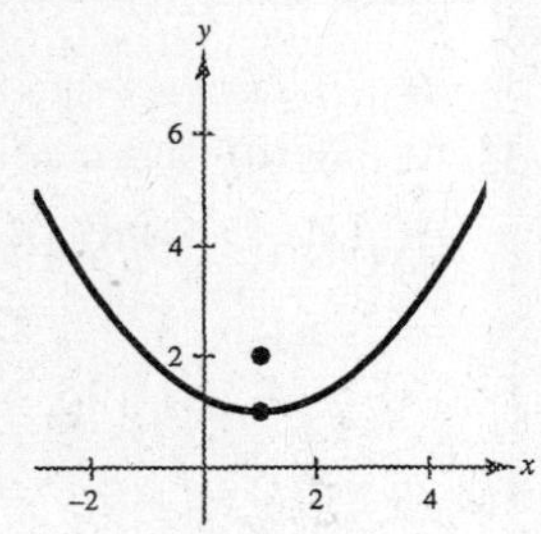

23. $y^2 + 6y + 8x + 25 = 0$

$y^2 + 6y + 9 = -8x - 25 + 9$

$(y + 3)^2 = 4(-2)(x + 2)$

$h = -2, k = -3, p = -2$

Vertex: $(-2, -3)$

Focus: $(-4, -3)$

Directrix: $x = 0$

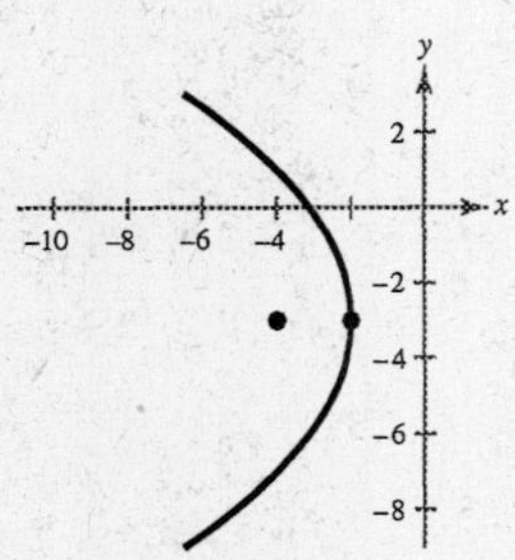

25. $x^2 + 4x + 6y - 2 = 0$

$x^2 + 4x = -6y + 2$

$x^2 + 4x + 4 = -6y + 2 + 4$

$(x + 2)^2 = -6(y - 1)$

$(x + 2)^2 = 4\left(-\frac{3}{2}\right)(y - 1)$

$h = -2, k = 1, p = -\frac{3}{2}$

Vertex: $(-2, 1)$

Focus: $\left(-2, -\frac{1}{2}\right)$

Directrix: $y = \frac{5}{2}$

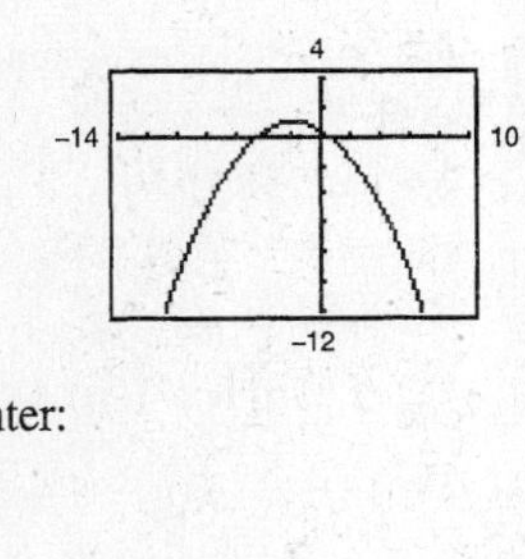

On the graphing calculator, enter:

$y_1 = -\frac{1}{6}(x^2 + 4x - 2)$

27. $y^2 + x + y = 0$

$y^2 + y + \frac{1}{4} = -x + \frac{1}{4}$

$\left(y + \frac{1}{2}\right)^2 = 4\left(-\frac{1}{4}\right)\left(x - \frac{1}{4}\right)$

$h = \frac{1}{4}, k = -\frac{1}{2}, p = -\frac{1}{4}$

Vertex: $\left(\frac{1}{4}, -\frac{1}{2}\right)$

Focus: $\left(0, -\frac{1}{2}\right)$

Directrix: $x = \frac{1}{2}$

To use a graphing calculator, enter:

$y_1 = -\frac{1}{2} + \sqrt{\frac{1}{4} - x}$

$y_2 = -\frac{1}{2} - \sqrt{\frac{1}{4} - x}$

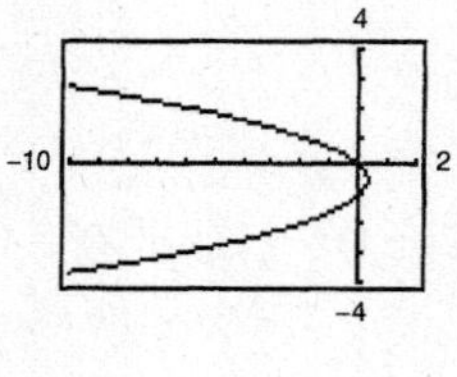

29. $y^2 - 8x = 0 \Rightarrow y = \pm\sqrt{8x}$

$x - y + 2 = 0 \Rightarrow y = x + 2$

The point of tangency is (2, 4).

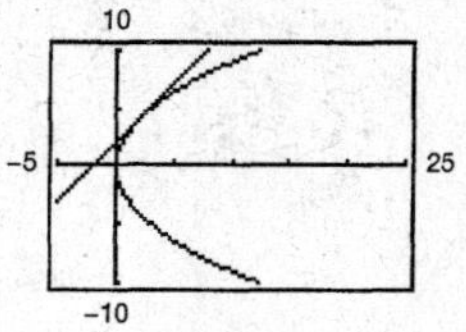

31. Vertex: $(0, 0) \Rightarrow h = 0, k = 0$

Graph opens upward.

$x^2 = 4py$

Point on graph: (3, 6)

$3^2 = 4p(6)$

$9 = 24p$

$\frac{3}{8} = p$

Thus, $x^2 = 4\left(\frac{3}{8}\right)y \Rightarrow x^2 = \frac{3}{2}y$

33. Vertex: $(0, 0) \Rightarrow h = 0, k = 0$

Focus: $\left(0, -\frac{3}{2}\right) \Rightarrow p = -\frac{3}{2}$

$x^2 = 4py$

$x^2 = 4\left(-\frac{3}{2}\right)y$

$x^2 = -6y$

35. Vertex: $(0, 0) \Rightarrow h = 0, k = 0$

Focus: $(-2, 0) \Rightarrow p = -2$

$y^2 = 4px$

$y^2 = 4(-2)x$

$y^2 = -8x$

37. Vertex: $(0, 0) \Rightarrow h = 0, k = 0$

Directrix: $y = -1 \Rightarrow p = 1$

$x^2 = 4py$

$x^2 = 4(1)y$

$x^2 = 4y$

39. Vertex: $(0, 0) \Rightarrow h = 0, k = 0$

Directrix: $x = 2 \Rightarrow p = -2$

$y^2 = 4px$

$y^2 = 4(-2)x$

$y^2 = -8x$

41. Vertex: $(0, 0) \Rightarrow h = 0, k = 0$

Horizontal axis and passes through the point (4, 6)

$y^2 = 4px$

$6^2 = 4p(4)$

$36 = 16p \Rightarrow p = \frac{9}{4}$

$y^2 = 4\left(\frac{9}{4}\right)x$

$y^2 = 9x$

43. Vertex: (3, 1) and opens downward. Passes through (2, 0) and (4, 0).

$y = -(x - 2)(x - 4)$

$= -x^2 + 6x - 8$

$= -(x - 3)^2 + 1$

$(x - 3)^2 = -(y - 1)$

45. Vertex: $(-4, 0)$ and opens to the right. Passes through (0, 4).

$(y - 0)^2 = 4p(x + 4)$

$4^2 = 4p(0 + 4)$

$16 = 16p$

$1 = p$

$y^2 = 4(x + 4)$

47. Vertex: (5, 2)

Focus: (3, 2)

Horizontal axis

$p = 3 - 5 = -2$

$(y - 2)^2 = 4(-2)(x - 5)$

$(y - 2)^2 = -8(x - 5)$

49. Vertex: (0, 4)
Directrix: $y = 2$
Vertical axis

$p = 4 - 2 = 2$

$(x - 0)^2 = 4(2)(y - 4)$

$x^2 = 8(y - 4)$

51. Focus: (2, 2)
Directrix: $x = -2$
Horizontal axis
Vertex: (0, 2)

$p = 2 - 0 = 2$

$(y - 2)^2 = 4(2)(x - 0)$

$(y - 2)^2 = 8x$

53. $(y - 3)^2 = 6(x + 1)$

For the upper half of the parabola:

$y - 3 = +\sqrt{6(x + 1)}$

$y = \sqrt{6(x + 1)} + 3$

55. $x^2 = 2y \Rightarrow p = \frac{1}{2}$

Point: (4, 8)

Focus: $\left(0, \frac{1}{2}\right)$

$d_1 = \frac{1}{2} - b$

$d_2 = \sqrt{(4 - 0)^2 + \left(8 - \frac{1}{2}\right)^2}$

$= \frac{17}{2}$

$d_1 = d_2 \Rightarrow b = -8$

Slope: $m = \dfrac{8 - (-8)}{4 - 0} = 4$

$y = 4x - 8 \Rightarrow 0 = 4x - y - 8$

x-intercept: (2, 0)

57. $y = -2x^2 \Rightarrow x^2 = -\frac{1}{2}y \Rightarrow p = -\frac{1}{8}$

Point: $(-1, -2)$

Focus: $\left(0, -\frac{1}{8}\right)$

$d_1 = b - \left(-\frac{1}{8}\right) = b + \frac{1}{8}$

$d_2 = \sqrt{(-1 - 0)^2 + \left(-2 - \left(-\frac{1}{8}\right)\right)^2}$

$= \frac{17}{8}$

$d_1 = d_2 \Rightarrow b = 2$

slope: $m = \dfrac{-2 - 2}{-1 - 0} = 4$

$y = 4x + 2 \Rightarrow 0 = 4x - y + 2$

x-intercept: $\left(-\frac{1}{2}, 0\right)$

59. $(x - 106)^2 = -\frac{4}{5}(R - 14{,}045)$

$x^2 - 212x + 11{,}236 = -\frac{4}{5}R + 11{,}236$

$R = 265x - \frac{5}{4}x^2$

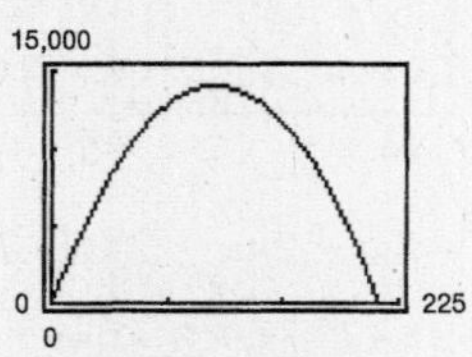

The revenue is maximum when $x = 106$ units.

61. Vertex: (0, 0) $\Rightarrow h = 0, k = 0$

Focus: (0, 4.5) $\Rightarrow p = 4.5$

$(x - h)^2 = 4p(y - k)$

$(x - 0)^2 = 4(4.5)(y - 0)$

$x^2 = 18y$ or $y = \frac{1}{18}x^2$

63. (a) Vertex: (0, 0) $\Rightarrow h = 0, k = 0$

Points on the parabola: $(\pm 16, -0.4)$

$x^2 = 4py$

$(\pm 16)^2 = 4p(-0.4)$

$256 = -1.6p$

$-160 = p$

$x^2 = 4(-160y)$

$x^2 = -640y$

$y = -\frac{1}{640}x^2$

(b) When $y = -0.1$ we have, $-0.1 = -\frac{1}{640}x^2$

$64 = x^2$

$\pm 8 = x$

Thus, 8 feet away from the center of the road, the road surface is 0.1 foot lower than in the middle.

65. (a) $V = 17{,}500\sqrt{2}$ mi/hr

$\approx 24{,}750$ mi/hr

(b) $p = -4100, (h, k) = (0, 4100)$

$(x - 0)^2 = 4(-4100)(y - 4100)$

$x^2 = -16{,}400(y - 4100)$

67. (a) $x^2 = -\dfrac{(32)^2}{16}(y - 75)$

$x^2 = -64(y - 75)$

(b) When $y = 0$, $x^2 = -64(-75) = 4800$.

Thus, $x = \sqrt{4800} = 40\sqrt{3} \approx 69.3$ feet

69. False. It is not possible for a parabola to intersect its directrix. If the graph crossed the directrix there would exist points nearer the directrix than the focus.

71. (a)

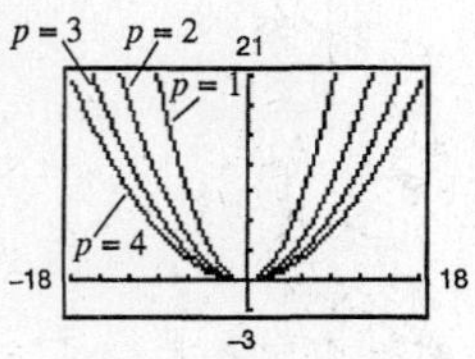

As p increases, the graph becomes wider.

(b) (0, 1), (0, 2), (0, 3), (0, 4)

(c) 4, 8, 12, 16. The chord passing through the focus and parallel to the directrix has length $|4p|$.

(d) This provides an easy way to determine two additional points on the graph, each of which is $|2p|$ units away from the focus on the chord.

73. $y - y_1 = \dfrac{x_1}{2p}(x - x_1)$

Slope: $m = \dfrac{x_1}{2p}$

75. $f(x) = (x - 3)[x - (2 + i)][x - (2 - i)]$

$= (x - 3)[(x - 2) - i][(x - 2) + i]$

$= (x - 3)(x^2 - 4x + 5)$

$= x^3 - 7x^2 + 17x - 15$

77. $g(x) = 6x^4 + 7x^3 - 29x^2 - 28x + 20$

Possible rational roots: $\pm 1, \pm 2, \pm 4, \pm 5, \pm 10, \pm 20,$
$\pm \frac{1}{2}, \pm \frac{5}{2}, \pm \frac{1}{3}, \pm \frac{2}{3}, \pm \frac{4}{3}, \pm \frac{5}{3}, \pm \frac{10}{3}, \pm \frac{20}{3}, \pm \frac{1}{6}, \pm \frac{5}{6}$

$x = \pm 2$ are both solutions.

$$\begin{array}{r|rrrrr} 2 & 6 & 7 & -29 & -28 & 20 \\ & & 12 & 38 & 18 & -20 \\ \hline & 6 & 19 & 9 & -10 & 0 \end{array}$$

$$\begin{array}{r|rrrr} -2 & 6 & 19 & 9 & -10 \\ & & -12 & -14 & 10 \\ \hline & 6 & 7 & -5 & 0 \end{array}$$

$g(x) = (x - 2)(x + 2)(6x^2 + 7x - 5)$

$= (x - 2)(x + 2)(2x - 1)(3x + 5)$

The zeros of $g(x)$ are $x = \pm 2$, $x = \frac{1}{2}$, $x = -\frac{5}{3}$.

79. $A = 35°$, $a = 10$, $b = 7$

$\dfrac{\sin B}{7} = \dfrac{\sin 35°}{10} \Rightarrow \sin B \approx 0.4015 \Rightarrow B \approx 23.67°$

$C \approx 180° - 35° - 23.67° = 121.33°$

$\dfrac{c}{\sin 121.33°} = \dfrac{10}{\sin 35°} \Rightarrow c \approx 14.89$

81. $A = 40°$, $B = 51°$, $c = 3$

$C = 180° - 40° - 51° = 89°$

$\dfrac{a}{\sin 40°} = \dfrac{3}{\sin 89°} \Rightarrow a \approx 1.93$

$\dfrac{b}{\sin 51°} = \dfrac{3}{\sin 89°} \Rightarrow b \approx 2.33$

83. $a = 7$, $b = 10$, $c = 16$

$\cos C = \dfrac{7^2 + 10^2 - 16^2}{2(7)(10)} \approx -0.7643 \Rightarrow C \approx 139.84°$

$\dfrac{\sin B}{10} = \dfrac{\sin 139.84°}{16} \Rightarrow \sin B \approx 0.4031 \Rightarrow B \approx 23.77°$

$A = 180° - B - C \Rightarrow A \approx 16.39°$

85. $A = 65°$, $b = 5$, $c = 12$

$a^2 = 5^2 + 12^2 - 2(5)(12)\cos 65° \Rightarrow a \approx 10.8759$

$\dfrac{\sin B}{5} = \dfrac{\sin 65°}{10.8759} \Rightarrow \sin B \approx 0.4167 \Rightarrow B \approx 24.62°$

$C = 180° - A - B \Rightarrow C \approx 90.38°$

Section 6.3 Ellipses

- An **ellipse** is the set of all points (x, y) the sum of whose distances from two distinct fixed points (**foci**) is constant.
- The standard equation of an ellipse with center (h, k) and major and minor axes of lengths $2a$ and $2b$ is:
  (a) $\frac{(x-h)^2}{a^2} + \frac{(y-k)^2}{b^2} = 1$ if the major axis is horizontal.
  (b) $\frac{(x-h)^2}{b^2} + \frac{(y-k)^2}{a^2} = 1$ if the major axis is vertical.
- $c^2 = a^2 - b^2$ where c is the distance from the center to a focus.
- The eccentricity of an ellipse is $e = \frac{c}{a}$.

Solutions to Odd-Numbered Exercises

1. $\frac{x^2}{4} + \frac{y^2}{9} = 1$

Center: $(0, 0)$

$a = 3, b = 2$

Vertical major axis

Matches graph (b).

3. $\frac{x^2}{4} + \frac{y^2}{25} = 1$

Center: $(0, 0)$

$a = 5, b = 2$

Vertical major axis

Matches graph (d).

5. $\frac{(x-2)^2}{16} + (y+1)^2 = 1$

Center: $(2, -1)$

$a = 4, b = 1$

Horizontal major axis

Matches graph (a).

7. $\frac{x^2}{25} + \frac{y^2}{16} = 1$

Center: $(0, 0)$

$a = 5, b = 4, c = 3$

Foci: $(\pm 3, 0)$

Vertices: $(\pm 5, 0)$

$e = \frac{3}{5}$

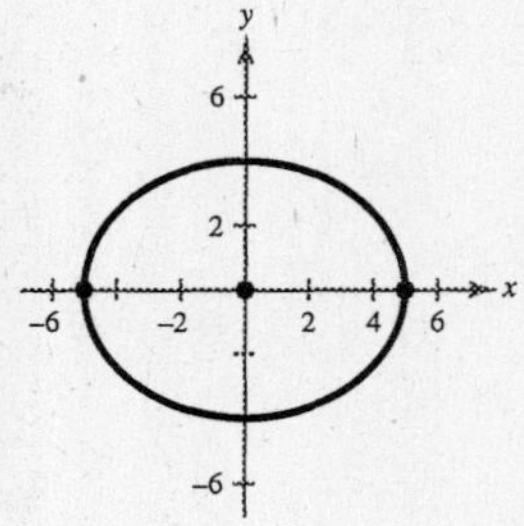

9. $\frac{x^2}{5} + \frac{y^2}{9} = 1$

$a = 3, b = \sqrt{5}, c = 2$

Center: $(0, 0)$

Foci: $(0, \pm 2)$

Vertices: $(0, \pm 3)$

$e = \frac{\sqrt{5}}{3}$

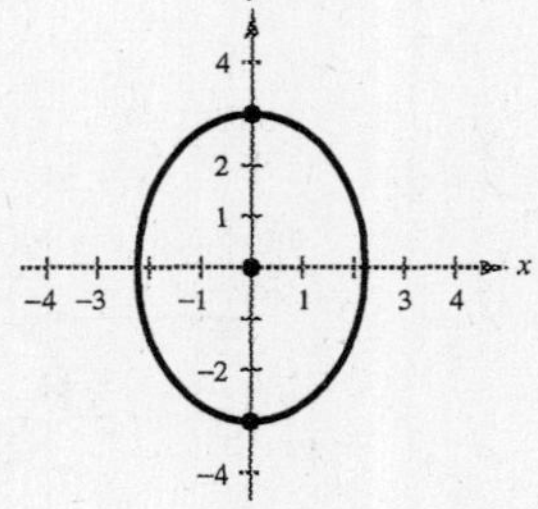

11. $\frac{(x+3)^2}{16} + \frac{(y-5)^2}{25} = 1$

Center: $(-3, 5)$

$a = 5, b = 4, c = 3$

Foci: $(-3, 8)(-3, 2)$

Vertices: $(-3, 10)(-3, 0)$

$e = \frac{3}{5}$

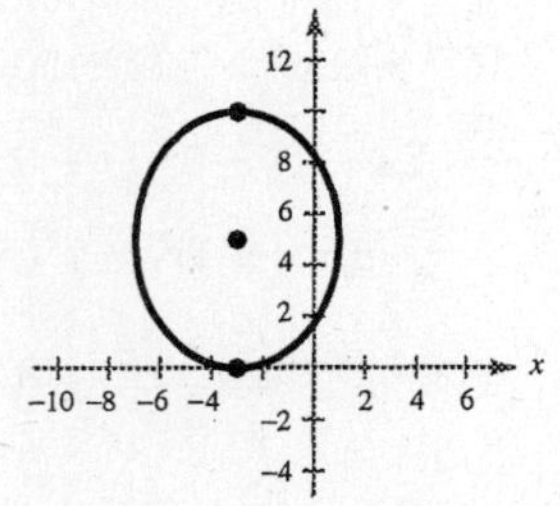

13. $\dfrac{(x+5)^2}{\frac{9}{4}} + (y-1)^2 = 1$

Center: $(-5, 1)$

$a = \dfrac{3}{2}, b = 1, c = \dfrac{\sqrt{5}}{2}$

Foci: $\left(-5 + \dfrac{\sqrt{5}}{2}, 1\right), \left(-5 - \dfrac{\sqrt{5}}{2}, 1\right)$

Vertices: $\left(-\dfrac{7}{2}, 1\right), \left(-\dfrac{13}{2}, 1\right)$

$e = \dfrac{\frac{\sqrt{5}}{2}}{\frac{3}{2}} = \dfrac{\sqrt{5}}{3}$

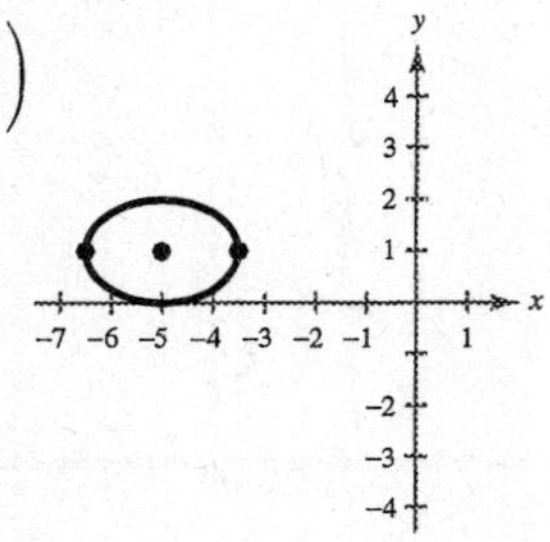

15. $9x^2 + 4y^2 + 36x - 24y + 36 = 0$

$$9(x^2 + 4x + 4) + 4(y^2 - 6y + 9) = -36 + 36 + 36$$

$$9(x+2)^2 + 4(y-3) = 36$$

$$\frac{(x+2)^2}{4} + \frac{(y-3)^2}{9} = 1$$

$a = 3, b = 2, c = \sqrt{5}$

Center: $(-2, 3)$

Foci: $\left(-2, 3 \pm \sqrt{5}\right)$

Vertices: $(-2, 6), (-2, 0)$

$e = \dfrac{\sqrt{5}}{3}$

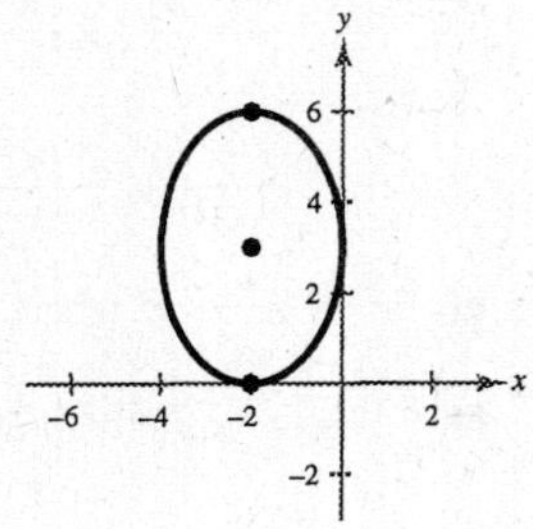

17. $x^2 + 5y^2 - 8x - 30y - 39 = 0$

$$(x^2 - 8x + 16) + 5(y^2 - 6y + 9) = 39 + 16 + 45$$

$$(x-4)^2 + 5(y-3)^2 = 100$$

$$\frac{(x-4)^2}{100} + \frac{(y-3)^2}{20} = 1$$

Center: $(4, 3)$

$a = 10, b = 2\sqrt{5}, c = 4\sqrt{5}$

Foci: $\left(4 \pm 4\sqrt{5}, 3\right)$

Vertices: $(14, 3), (-6, 3)$

$e = \dfrac{4\sqrt{5}}{10} = \dfrac{2\sqrt{5}}{5}$

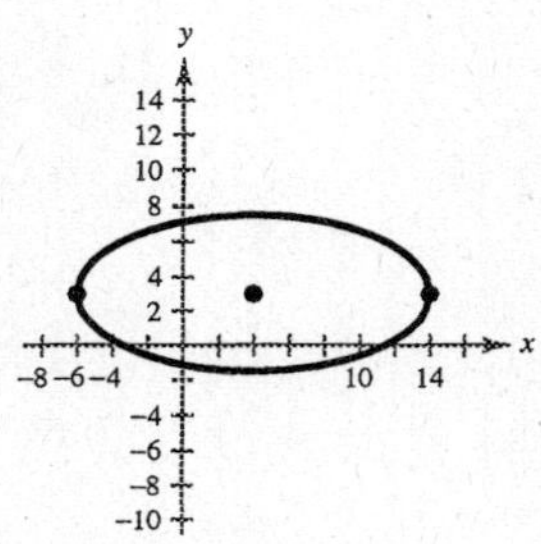

19. $6x^2 + 2y^2 + 18x - 10y + 2 = 0$

$$6\left(x^2 + 3x + \frac{9}{4}\right) + 2\left(y^2 - 5y + \frac{25}{4}\right) = -2 + \frac{27}{2} + \frac{25}{2}$$

$$6\left(x + \frac{3}{2}\right)^2 + 2\left(y - \frac{5}{2}\right)^2 = 24$$

$$\frac{\left(x + \frac{3}{2}\right)^2}{4} + \frac{\left(y - \frac{5}{2}\right)^2}{12} = 1$$

Center: $\left(-\dfrac{3}{2}, \dfrac{5}{2}\right)$

$a = 2\sqrt{3}, b = 2, c = 2\sqrt{2}$

Foci: $\left(-\dfrac{3}{2}, \dfrac{5}{2} \pm 2\sqrt{2}\right)$

Vertices: $\left(-\dfrac{3}{2}, \dfrac{5}{2} \pm 2\sqrt{3}\right)$

$e = \dfrac{2\sqrt{2}}{2\sqrt{3}} = \dfrac{\sqrt{6}}{3}$

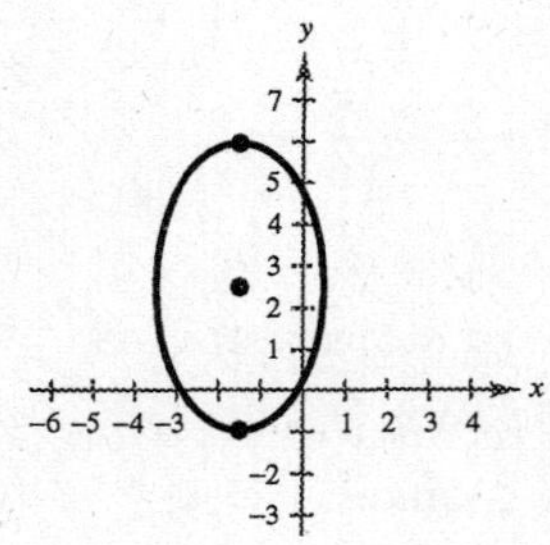

21. $16x^2 + 25y^2 - 32x + 50y + 16 = 0$

$$16(x^2 - 2x + 1) + 25(y^2 + 2y + 1) = -16 + 16 + 25$$

$$16(x-1)^2 + 25(y+1)^2 = 25$$

$$\frac{(x-1)^2}{\frac{25}{16}} + (y+1)^2 = 1$$

$a = \dfrac{5}{4}, b = 1, c = \dfrac{3}{4}$

Center: $(1, -1)$

Foci: $\left(\dfrac{7}{4}, -1\right), \left(\dfrac{1}{4}, -1\right)$

Vertices: $\left(\dfrac{9}{4}, -1\right), \left(-\dfrac{1}{4}, -1\right)$

$e = \dfrac{3}{5}$

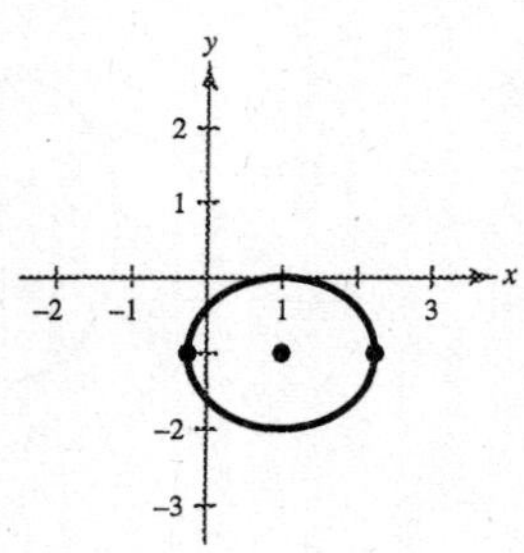

23. $5x^2 + 3y^2 = 15$

$$\frac{x^2}{3} + \frac{y^2}{5} = 1$$

Center: $(0, 0)$

$a = \sqrt{5}, b = \sqrt{3}, c = \sqrt{2}$

Foci: $\left(0, \pm\sqrt{2}\right)$

Vertices: $\left(0, \pm\sqrt{5}\right)$

To graph, solve for y.

$$y^2 = \frac{15 - 5x^2}{3}$$

$$y_1 = \sqrt{\frac{15 - 5x^2}{3}}$$

$$y_2 = -\sqrt{\frac{15 - 5x^2}{3}}$$

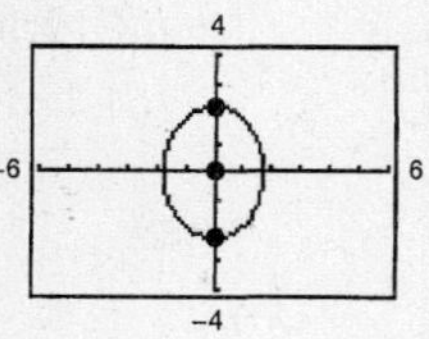

25. $12x^2 + 20y^2 - 12x + 40y - 37 = 0$

$$12\left(x^2 - x + \frac{1}{4}\right) + 20(y^2 + 2y + 1) = 37 + 3 + 20$$

$$12\left(x - \frac{1}{2}\right)^2 + 20(y + 1)^2 = 60$$

$$\frac{\left(x - \frac{1}{2}\right)^2}{5} + \frac{(y + 1)^2}{3} = 1$$

$a = \sqrt{5}, b = \sqrt{3}, c = \sqrt{2}$

Center: $\left(\frac{1}{2}, -1\right)$

Foci: $\left(\frac{1}{2} \pm \sqrt{2}, -1\right)$

Vertices: $\left(\frac{1}{2} \pm \sqrt{5}, -1\right)$

$$e = \frac{\sqrt{10}}{5}$$

To graph, solve for y.

$$(y + 1)^2 = 3\left[1 - \frac{(x - 0.5)^2}{5}\right]$$

$$y_1 = -1 + \sqrt{3\left[1 - \frac{(x - 0.5)^2}{5}\right]}$$

$$y_2 = -1 - \sqrt{3\left[1 - \frac{(x - 0.5)^2}{5}\right]}$$

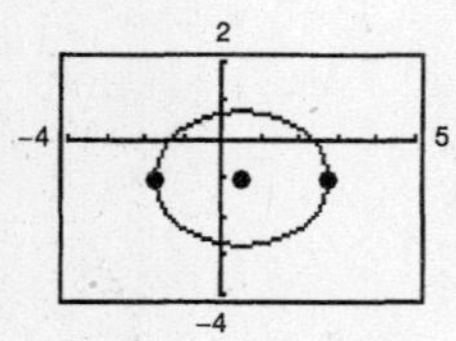

27. Center: $(0, 0)$

$a = 4, b = 2$

Vertical major axis

$$\frac{(x - h)^2}{b^2} + \frac{(y - k)^2}{a^2} = 1$$

$$\frac{x^2}{4} + \frac{y^2}{16} = 1$$

29. Vertices: $(\pm 6, 0)$

$a = 6, c = 2 \Rightarrow b = \sqrt{32} = 4\sqrt{2}$

Foci: $(\pm 2, 0)$

Horizontal major axis

Center: $(0, 0)$

$$\frac{(x - h)^2}{a^2} + \frac{(y - k)^2}{b^2} = 1$$

$$\frac{x^2}{36} + \frac{y^2}{32} = 1$$

31. Foci: $(\pm 5, 0) \Rightarrow c = 5$

Center: $(0, 0)$

Horizontal major axis

Major axis of length $12 \Rightarrow 2a = 12$

$$a = 6$$

$6^2 - b^2 = 5^2 \Rightarrow b^2 = 11$

$$\frac{(x - h)^2}{a^2} + \frac{(y - k)^2}{b^2} = 1$$

$$\frac{x^2}{36} + \frac{y^2}{11} = 1$$

33. Vertices: $(0, \pm 5) \implies a = 5$

Center: $(0, 0)$

Vertical major axis

$$\frac{(x-h)^2}{b^2} + \frac{(y-k)^2}{a^2} = 1$$

$$\frac{x^2}{b^2} + \frac{y^2}{25} = 1$$

Point: $(4, 2)$

$$\frac{4^2}{b^2} + \frac{2^2}{25} = 1$$

$$\frac{16}{b^2} = 1 - \frac{4}{25} = \frac{21}{25}$$

$$400 = 21b^2$$

$$\frac{400}{21} = b^2$$

$$\frac{x^2}{\frac{400}{21}} + \frac{y^2}{25} = 1$$

$$\frac{21x^2}{400} + \frac{y^2}{25} = 1$$

35. Center: $(2, 3)$

$a = 3, \quad b = 1$

Vertical major axis

$$\frac{(x-h)^2}{b^2} + \frac{(y-k)^2}{a^2} = 1$$

$$\frac{(x-2)^2}{1} + \frac{(y-3)^2}{9} = 1$$

37. Center: $(-2, 3)$

$a = 4, \quad b = 3$

Horizontal major axis

$$\frac{(x-h)^2}{a^2} + \frac{(y-k)^2}{b^2} = 1$$

$$\frac{(x+2)^2}{16} + \frac{(y-3)^2}{9} = 1$$

39. Vertices: $(0, 4), (4, 4) \implies a = 2$

Minor axis of length 2 $\implies b = 1$

Center: $(2, 4) = (h, k)$

$$\frac{(x-h)^2}{a^2} + \frac{(y-k)^2}{b^2} = 1$$

$$\frac{(x-2)^2}{4} + \frac{(y-4)^2}{1} = 1$$

41. Foci: $(0, 0), (0, 8) \implies c = 4$

Major axis of length 16 $\implies a = 8$

$b^2 = a^2 - c^2 = 64 - 16 = 48$

Center: $(0, 4) = (h, k)$

$$\frac{(x-h)^2}{b^2} + \frac{(y-k)^2}{a^2} = 1$$

$$\frac{x^2}{48} + \frac{(y-4)^2}{64} = 1$$

43. Center: $(0, 4)$

Vertices: $(-4, 4), (4, 4) \implies a = 4$

$a = 2c \implies 4 = 2c \implies c = 2$

$2^2 = 4^2 - b^2 \implies b^2 = 12$

Horizontal major axis

$$\frac{(x-h)^2}{a^2} + \frac{(y-k)^2}{b^2} = 1$$

$$\frac{x^2}{16} + \frac{(y-4)^2}{12} = 1$$

45. Vertices: $(0, 2), (4, 2) \implies a = 2$

Center: $(2, 2)$

Endpoints of the minor axis: $(2, 3), (2, 1) \implies b = 1$

Horizontal major axis

$$\frac{(x-h)^2}{a^2} + \frac{(y-k)^2}{b^2} = 1$$

$$\frac{(x-2)^2}{4} + \frac{(y-2)^2}{1} = 1$$

47. Vertices: $(\pm 5, 0) \implies a = 5$

Eccentricity: $\frac{3}{5} \implies c = \frac{3}{5}a = 3$

$b^2 = a^2 - c^2 = 25 - 9 = 16$

Center: $(0, 0) = (h, k)$

$$\frac{(x-h)^2}{a^2} + \frac{(y-k)^2}{b^2} = 1$$

$$\frac{x^2}{25} + \frac{y^2}{16} = 1$$

49. (a)

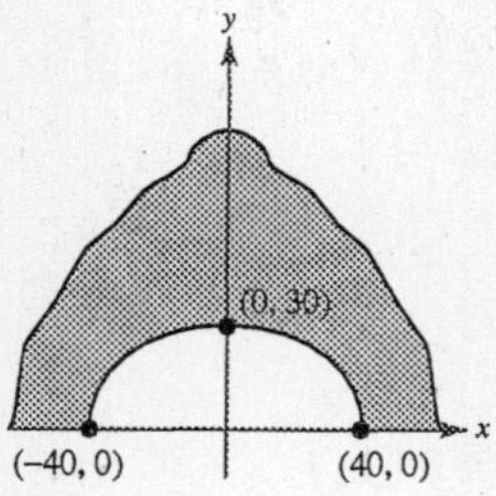

(b) $a = 40, b = 30$

$$\frac{x^2}{a^2} + \frac{y^2}{b^2} = 1$$

$$\frac{x^2}{1600} + \frac{y^2}{900} = 1$$

(c) When $x = \pm 35$, you are five feet from the edge of the tunnel.

$$\frac{35^2}{1600} + \frac{y^2}{900} = 1$$

$$y^2 = 900 - 900\left(\frac{1225}{1600}\right) = \frac{3375}{16}$$

$$y = \pm\sqrt{\frac{3375}{16}} \approx \pm 14.5$$

The height of the tunnel five feet from its edge is 14.5 feet.

51. Area of circle: $\pi r^2 = 100\pi$

Area of ellipse: $\pi(a)(10)$

$10a\pi = 2(100\pi)$

$10a\pi = 200\pi$

$a = 20$

Length of major axis: $2a = 40$

53. $a + c = 6378 + 938 = 7316$

$a - c = 6378 + 212 = 6590$

Solving this system for a and c gives

$a = 6953, c = 363$

$$e = \frac{c}{a} = \frac{363}{6953} \approx 0.052$$

55. $\frac{x^2}{9} + \frac{y^2}{16} = 1$

$a = 4, b = 3, c = \sqrt{7}$

Points on the ellipse: $(\pm 3, 0), (0, \pm 4)$

Length of latus recta: $\frac{2b^2}{a} = \frac{2(3)^2}{4} = \frac{9}{2}$

Additional points: $\left(\pm\frac{9}{4}, -\sqrt{7}\right), \left(\pm\frac{9}{4}, \sqrt{7}\right)$

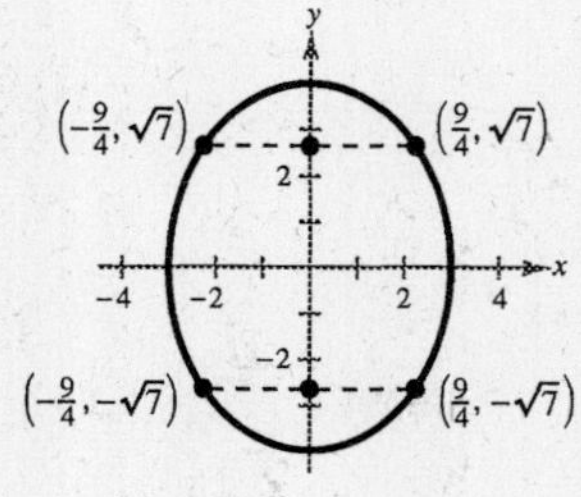

57. $9x^2 + 4y^2 = 36$

$$\frac{x^2}{4} + \frac{y^2}{9} = 1$$

$a = 3, b = 2, c = \sqrt{5}$

Points on the ellipse: $(\pm 2, 0), (0, \pm 3)$

Length of latera recta: $\frac{2b^2}{a} = \frac{2 \cdot 2^2}{3} = \frac{8}{3}$

Additional points: $\left(\pm\frac{4}{3}, -\sqrt{5}\right), \left(\pm\frac{4}{3}, \sqrt{5}\right)$

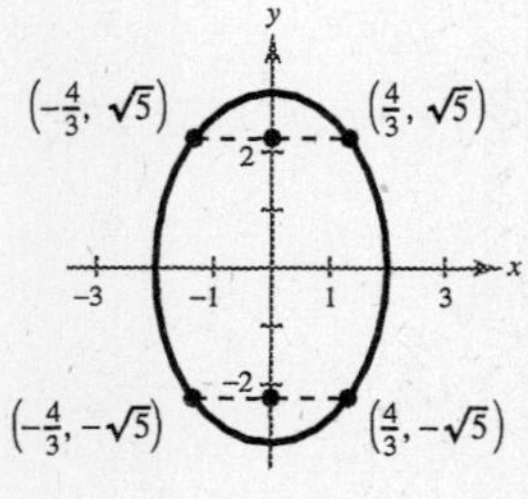

59. False. The graph of $\frac{x^2}{4} + y^4 = 1$ is not an ellipse.

The degree on y is 4, not 2.

61. $\frac{x^2}{a^2} + \frac{y^2}{b^2} = 1$

(a) $a + b = 20 \Rightarrow b = 20 - a$

$A = \pi ab = \pi a(20 - a)$

(b) $264 = \pi a(20 - a)$

$0 = -\pi a^2 + 20\pi a - 264$

$0 = \pi a^2 - 20\pi a + 264$

By the Quadratic Formula: $a \approx 14$ or $a \approx 6$.
Choosing the larger value of a, we have $a \approx 14$ and $b \approx 6$.

The equation of an ellipse with an area of 264 is $\frac{x^2}{196} + \frac{y^2}{36} = 1$.

(c)

a	8	9	10	11	12	13
A	301.6	311.0	314.2	311.0	301.6	285.9

The area is maximum when $a = 10$ and the ellipse is a circle.

(d)

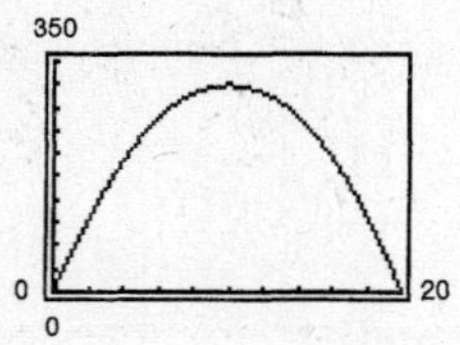

The area is maximum (314.16) when $a = b = 10$ and the ellipse is a circle.

63. $x^2 - 10x = 0$

$x(x - 10) = 0$

$x = 0$

$x - 10 = 0 \Rightarrow x = 10$

65. $x^3 - x = 0$

$x(x^2 - 1) = 0$

$x = 0$

$x^2 - 1 = 0 \Rightarrow x = \pm 1$

67. $f(x) = 5x - 8$

(a) $f(9) = 5(9) - 8 = 37$

(b) $f(-4) = 5(-4) - 8 = -28$

(c) $f(x - 7) = 5(x - 7) - 8 = 5x - 35 - 8 = 5x - 43$

69. $f(x) = \sqrt{x - 12} - 9$

(a) $f(12) = \sqrt{12 - 12} - 9 = 0 - 9 = -9$

(b) $f(40) = \sqrt{40 - 12} - 9 = \sqrt{28} - 9 = 2\sqrt{7} - 9$

(c) $f(-\sqrt{36}) = \sqrt{-\sqrt{36} - 12} - 9 = \sqrt{-6 - 12} - 9 = \sqrt{-18} - 9$

$= 3\sqrt{2}i - 9$

$= -9 + 3\sqrt{2}i$

Section 6.4 Hyperbolas

- A **hyperbola** is the set of all points (x, y) the difference of whose distances from two distinct fixed points **(foci)** is constant.
- The standard equation of a hyperbola with center (h, k) and transverse and conjugate axes of lengths $2a$ and $2b$ is:

  (a) $\dfrac{(x-h)^2}{a^2} - \dfrac{(y-k)^2}{b^2} = 1$ if the traverse axis is horizontal.

  (b) $\dfrac{(y-k)^2}{a^2} - \dfrac{(x-h)^2}{b^2} = 1$ if the traverse axis is vertical.
- $c^2 = a^2 + b^2$ where c is the distance from the center to a focus.
- The asymptotes of a hyperbola are:

  (a) $y = k \pm \dfrac{b}{a}(x - h)$ if the transverse axis is horizontal.

  (b) $y = k \pm \dfrac{a}{b}(x - h)$ if the transverse axis is vertical.
- The eccentricity of a hyperbola is $e = \dfrac{c}{a}$.
- To classify a nondegenerate conic from its general equation $Ax^2 + Cy^2 + Dx + Ey + F = 0$:

  (a) If $A = C$ $(A \neq 0, C \neq 0)$, then it is a circle.

  (b) If $AC = 0$ $(A = 0$ or $C = 0$, but not both), then it is a parabola.

  (c) If $AC > 0$, then it is an ellipse.

  (d) If $AC < 0$, then it is a hyperbola.

Solutions to Odd-Numbered Exercises

1. $\dfrac{y^2}{9} - \dfrac{x^2}{25} = 1$

Center: $(0, 0)$

$a = 3, b = 5$

Vertical transverse axis

Matches graph (b).

3. $\dfrac{(x-1)^2}{16} - \dfrac{y^2}{4} = 1$

Center: $(1, 0)$

$a = 4, b = 2$

Horizontal transverse axis

Matches graph (a).

5. $x^2 - y^2 = 1$

$a = 1, b = 1, c = \sqrt{2}$

Center: $(0, 0)$

Vertices: $(\pm 1, 0)$

Foci: $(\pm\sqrt{2}, 0)$

Asymptotes: $y = \pm x$

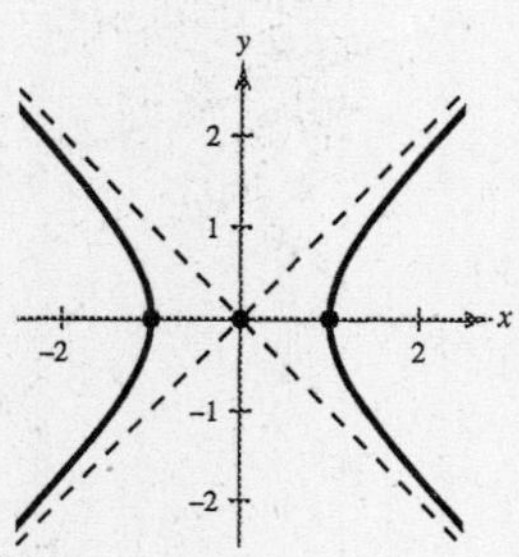

7. $\dfrac{y^2}{25} - \dfrac{x^2}{81} = 1$

$a = 5, b = 9, c = \sqrt{106}$

Center: $(0, 0)$

Vertices: $(0, \pm 5)$

Foci: $(0, \pm\sqrt{106})$

Asymptotes: $y = \pm\dfrac{5}{9}x$

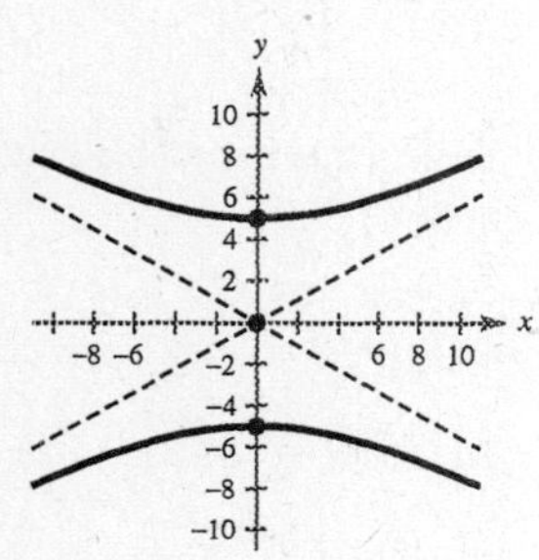

9. $\dfrac{(x-1)^2}{4} - \dfrac{(y+2)^2}{1} = 1$

$a = 2, b = 1, c = \sqrt{5}$

Center: $(1, -2)$

Vertices: $(-1, -2), (3, -2)$

Foci: $\left(1 \pm \sqrt{5}, -2\right)$

Asymptotes: $y = -2 \pm \dfrac{1}{2}(x - 1)$

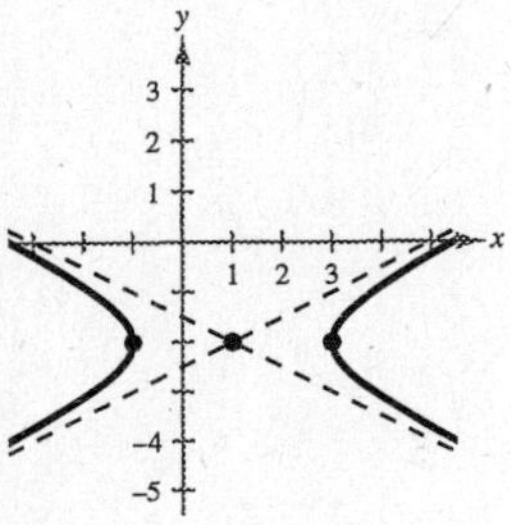

11. $\dfrac{(y+6)^2}{\frac{1}{9}} - \dfrac{(x-2)^2}{\frac{1}{4}} = 1$

$a = \dfrac{1}{3}, b = \dfrac{1}{2}, c = \dfrac{\sqrt{13}}{6}$

Center: $(2, -6)$

Vertices: $\left(2, -\dfrac{17}{3}\right), \left(2, -\dfrac{19}{3}\right)$

Foci: $\left(2, -6 \pm \dfrac{\sqrt{13}}{6}\right)$

Asymptotes: $y = -6 \pm \dfrac{2}{3}(x - 2)$

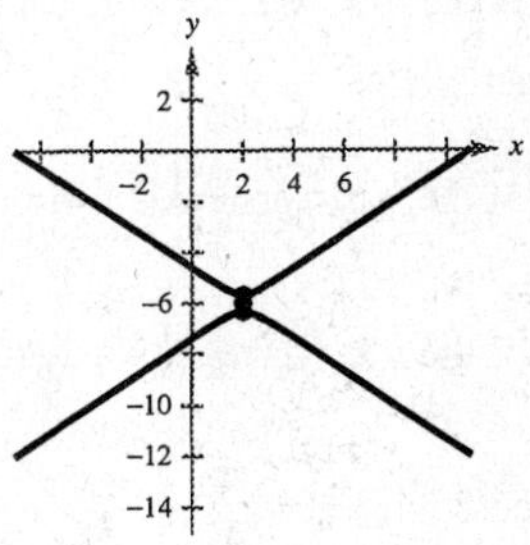

13.

$$
\begin{aligned}
9x^2 - y^2 - 36x - 6y + 18 &= 0 \\
9\left(x^2 - 4x + 4\right) - \left(y^2 + 6y + 9\right) &= -18 + 36 - 9 \\
9(x-2)^2 - (y+3)^2 &= 9 \\
\frac{(x-2)^2}{1} - \frac{(y+3)^2}{9} &= 1
\end{aligned}
$$

$a = 1, b = 3, c = \sqrt{10}$

Center: $(2, -3)$

Vertices: $(1, -3), (3, -3)$

Foci: $\left(2 \pm \sqrt{10}, -3\right)$

Asymptotes: $y = -3 \pm 3(x - 2)$

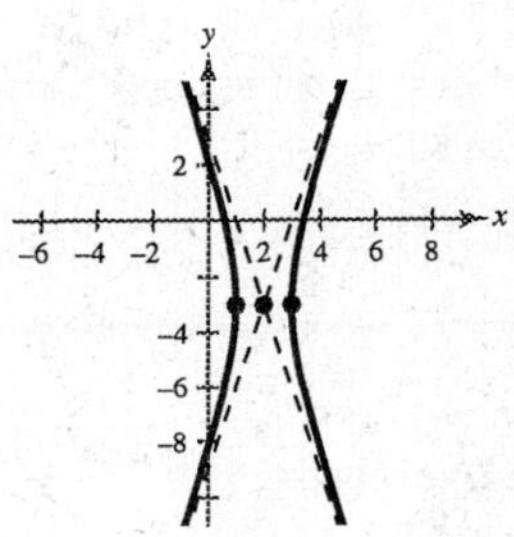

15.

$$
\begin{aligned}
x^2 - 9y^2 + 2x - 54y - 80 &= 0 \\
\left(x^2 + 2x + 1\right) - 9\left(y^2 + 6y + 9\right) &= 80 + 1 - 81 \\
(x+1)^2 - 9(y+3)^2 &= 0 \\
y + 3 &= \pm\frac{1}{3}(x + 1)
\end{aligned}
$$

Degenerate hyperbola is two lines intersecting at $(-1, -3)$.

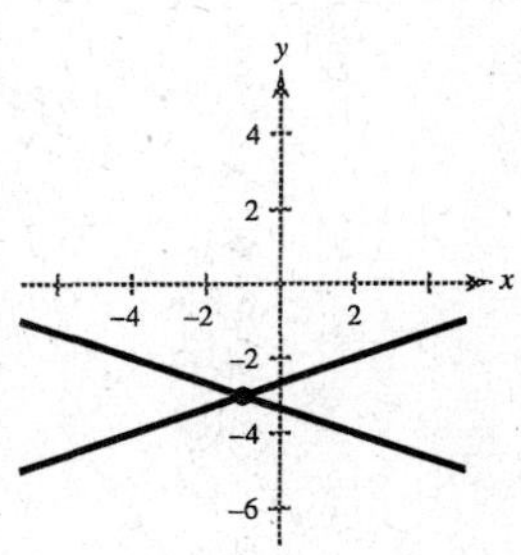

17. $2x^2 - 3y^2 = 6$

$$\frac{x^2}{3} - \frac{y^2}{2} = 1$$

$a = \sqrt{3}, b = \sqrt{2}, c = \sqrt{5}$

Center: $(0, 0)$

Vertices: $\left(\pm\sqrt{3}, 0\right)$

Foci: $\left(\pm\sqrt{5}, 0\right)$

Asymptotes: $y = \pm\sqrt{\frac{2}{3}}x = \pm\frac{\sqrt{6}}{3}x$

To use a graphing calculator, solve for y first.

$$y^2 = \frac{2x^2 - 6}{3}$$

$$\left.\begin{aligned} y_1 &= \sqrt{\frac{2x^2-6}{3}} \\ y_2 &= -\sqrt{\frac{2x^2-6}{3}} \end{aligned}\right\} \text{Hyperbola}$$

$$\left.\begin{aligned} y_3 &= \frac{\sqrt{6}}{3}x \\ y_4 &= -\frac{\sqrt{6}}{3}x \end{aligned}\right\} \text{Asymptotes}$$

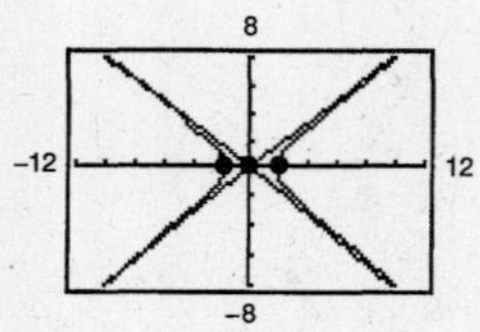

19.

$$9y^2 - x^2 + 2x + 54y + 62 = 0$$

$$9\left(y^2 + 6y + 9\right) - \left(x^2 - 2x + 1\right) = -62 - 1 + 81$$

$$9(y + 3)^2 - (x - 1)^2 = 18$$

$$\frac{(y + 3)^2}{2} - \frac{(x - 1)^2}{18} = 1$$

$a = \sqrt{2}, b = 3\sqrt{2}, c = 2\sqrt{5}$

Center: $(1, -3)$

Vertices: $\left(1, -3 \pm \sqrt{2}\right)$

Foci: $\left(1, -3 \pm 2\sqrt{5}\right)$

Asymptotes: $y = -3 \pm \frac{1}{3}(x - 1)$

To use a graphing calculator, solve for y first.

$$9(y + 3)^2 = 18 + (x - 1)^2$$

$$y = -3 \pm \sqrt{\frac{18 + (x-1)^2}{9}}$$

$$\left.\begin{aligned} y_1 &= -3 + \frac{1}{3}\sqrt{18 + (x-1)^2} \\ y_2 &= -3 - \frac{1}{3}\sqrt{18 + (x-1)^2} \end{aligned}\right\} \text{Hyperbola}$$

$$\left.\begin{aligned} y_3 &= -3 + \frac{1}{3}(x - 1) \\ y_4 &= -3 - \frac{1}{3}(x - 1) \end{aligned}\right\} \text{Asymptotes}$$

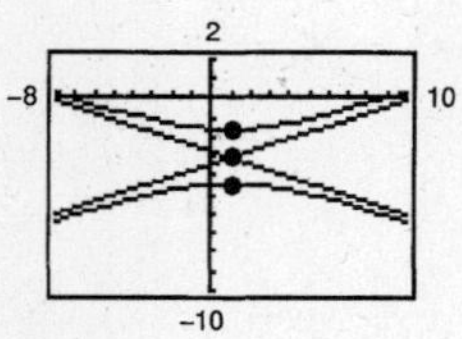

21. Vertices: $(0, \pm 2) \Rightarrow a = 2$

Foci: $(0, \pm 4) \Rightarrow c = 4$

$b^2 = c^2 - a^2 = 16 - 4 = 12$

Center: $(0, 0) = (h, k)$

$$\frac{(y - k)^2}{a^2} - \frac{(x - h)^2}{b^2} = 1$$

$$\frac{y^2}{4} - \frac{x^2}{12} = 1$$

23. Vertices: $(\pm 1, 0) \Rightarrow a = 1$

Asymptotes: $y = \pm 5x \Rightarrow \frac{b}{a} = 5, b = 5$

Center: $(0, 0) = (h, k)$

$$\frac{(x - h)^2}{a^2} - \frac{(y - k)^2}{b^2} = 1$$

$$\frac{x^2}{1} - \frac{y^2}{25} = 1$$

25. Foci: $(0, \pm 8) \Rightarrow c = 8$

Asymptotes: $y = \pm 4x \Rightarrow \frac{a}{b} = 4 \Rightarrow a = 4b$

Center: $(0, 0) = (h, k)$

$c^2 = a^2 + b^2 \Rightarrow 64 = 16b^2 + b^2$

$$\frac{64}{17} = b^2 \Rightarrow a^2 = \frac{1024}{17}$$

$$\frac{(y - k)^2}{a^2} - \frac{(x - h)^2}{b^2} = 1$$

$$\frac{y^2}{\frac{1024}{17}} - \frac{x^2}{\frac{64}{17}} = 1$$

$$\frac{17y^2}{1024} - \frac{17x^2}{64} = 1$$

27. Vertices: $(2, 0), (6, 0) \Rightarrow a = 2$

Foci: $(0, 0), (8, 0) \Rightarrow c = 4$

$b^2 = c^2 - a^2 = 16 - 4 = 12$

Center: $(4, 0) = (h, k)$

$$\frac{(x-h)^2}{a^2} - \frac{(y-k)^2}{b^2} = 1$$

$$\frac{(x-4)^2}{4} - \frac{y^2}{12} = 1$$

29. Vertices: $(4, 1), (4, 9) \Rightarrow a = 4$

Foci: $(4, 0), (4, 10) \Rightarrow c = 5$

$b^2 = c^2 - a^2 = 25 - 16 = 9$

Center: $(4, 5) = (h, k)$

$$\frac{(y-k)^2}{a^2} - \frac{(x-h)^2}{b^2} = 1$$

$$\frac{(y-5)^2}{16} - \frac{(x-4)^2}{9} = 1$$

31. Vertices: $(2, 3), (2, -3) \Rightarrow a = 3$

Passes through the point: $(0, 5)$

Center: $(2, 0) = (h, k)$

$$\frac{(y-k)^2}{a^2} - \frac{(x-h)^2}{b^2} = 1$$

$$\frac{y^2}{9} - \frac{(x-2)^2}{b^2} = 1 \Rightarrow \frac{(x-2)^2}{b^2} = \frac{y^2}{9} - 1 = \frac{y^2-9}{9} \Rightarrow b^2 = \frac{9(x-2)^2}{y^2-9} = \frac{9(-2)^2}{25-9} = \frac{36}{16} = \frac{9}{4}$$

$$\frac{y^2}{9} - \frac{(x-2)^2}{9/4} = 1$$

$$\frac{y^2}{9} - \frac{4(x-2)^2}{9} = 1$$

33. Vertices: $(0, 4), (0, 0) \Rightarrow a = 2$

Passes through the point $\left(\sqrt{5}, -1\right)$

Center: $(0, 2) = (h, k)$

$$\frac{(y-k)^2}{a^2} - \frac{(x-h)^2}{b^2} = 1$$

$$\frac{(y-2)^2}{4} - \frac{x^2}{b^2} = 1 \Rightarrow \frac{x^2}{b^2} = \frac{(y-2)^2}{4} - 1 = \frac{(y-2)^2-4}{4}$$

$$\Rightarrow b^2 = \frac{4x^2}{(y-2)^2-4} = \frac{4\left(\sqrt{5}\right)^2}{(-1-2)^2-4} = \frac{20}{5} = 4$$

$$\frac{(y-2)^2}{4} - \frac{x^2}{4} = 1$$

35. Vertices: $(1, 2), (3, 2) \Rightarrow a = 1$

Asymptotes: $y = x, y = 4 - x$

$\frac{b}{a} = 1 \Rightarrow \frac{b}{1} = 1 \Rightarrow b = 1$

Center: $(2, 2) = (h, k)$

$$\frac{(x-h)^2}{a^2} - \frac{(y-k)^2}{b^2} = 1$$

$$\frac{(x-2)^2}{1} - \frac{(y-2)^2}{1} = 1$$

37. Vertices: $(0, 2), (6, 2) \Rightarrow a = 3$

Asymptotes: $y = \frac{2}{3}x, y = 4 - \frac{2}{3}x$

$\frac{b}{a} = \frac{2}{3} \Rightarrow b = 2$

Center: $(3, 2) = (h, k)$

$$\frac{(x-h)^2}{a^2} - \frac{(y-k)^2}{b^2} = 1$$

$$\frac{(x-3)^2}{9} - \frac{(y-2)^2}{4} = 1$$

39. (a) Foci: $(\pm 150, 0) \implies c = 150$

Center: $(0, 0) = (h, k)$

$$\frac{d_2}{186{,}000} - \frac{d_1}{186{,}000} = 0.001 \implies 2a = 186, a = 93$$

$$b^2 = c^2 - a^2 = 150^2 - 93^2 = 13{,}851$$

$$\frac{x^2}{93^2} - \frac{y^2}{13{,}851} = 1$$

$$x^2 = 93^2\left(1 + \frac{75^2}{13{,}851}\right) \approx 12{,}161$$

$$x \approx 110.3 \text{ miles}$$

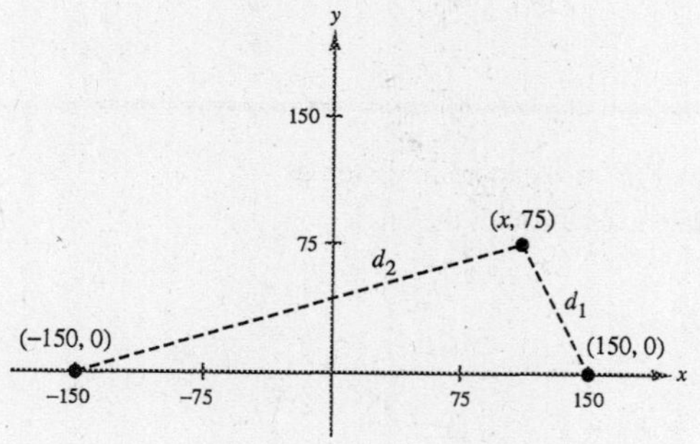

(b) $c - a = 150 - 93 = 57$ miles

(c) $\dfrac{270}{186{,}000} - \dfrac{30}{186{,}000} \approx 0.00129$ second

(d) $\dfrac{d_2}{186{,}000} - \dfrac{d_1}{186{,}000} = 0.00129$

$$2a \approx 239.94$$

$$a \approx 119.97$$

$$b^2 = c^2 - a^2 = 150^2 - 119.97^2 = 8107.1991$$

$$\frac{x^2}{119.97^2} - \frac{y^2}{8107.1991} = 1$$

$$x^2 = 119.97^2\left(1 + \frac{60^2}{8107.1991}\right)$$

$$x \approx 144.2 \text{ miles}$$

Position: $(144.2, 60)$

41. $x^2 + y^2 - 6x + 4y + 9 = 0$

$A = 1, C = 1$

$A = C \implies$ Circle

43. $4x^2 - y^2 - 4x - 3 = 0$

$A = 4, C = -1$

$AC = (4)(-1) = -4 < 0 \implies$ Hyperbola

45. $4x^2 + 3y^2 + 8x - 24y + 51 = 0$

$A = 4, C = 3$

$AC = 4(3) = 12 > 0 \implies$ Ellipse

47. $25x^2 - 10x - 200y - 119 = 0$

$A = 25, C = 0$

$AC = 25(0) = 0 \implies$ Parabola

49. True. For a hyperbola, $c^2 = a^2 + b^2$ or $e^2 = \dfrac{c^2}{a^2} = 1 + \dfrac{b^2}{a^2}$.

The larger the ratio of b to a, the larger the eccentricity $e = \dfrac{c}{a}$ of the hyperbola.

51. Let (x, y) be such that the difference of the distances from $(c, 0)$ and $(-c, 0)$ is $2a$ (again only deriving one of the forms).

$$2a = \left|\sqrt{(x + c)^2 + y^2} - \sqrt{(x - c)^2 + y^2}\right|$$

$$2a + \sqrt{(x - c)^2 + y^2} = \sqrt{(x + c)^2 + y^2}$$

$$4a^2 + 4a\sqrt{(x - c)^2 + y^2} + (x - c)^2 + y^2 = (x + c)^2 + y^2$$

$$4a\sqrt{(x - c)^2 + y^2} = 4cx - 4a^2$$

$$a\sqrt{(x - c)^2 + y^2} = cx - a^2$$

$$a^2(x^2 - 2cx + c^2 + y^2) = c^2x^2 - 2a^2cx + a^4$$

$$a^2(c^2 - a^2) = (c^2 - a^2)x^2 - a^2y^2$$

Let $b^2 = c^2 - a^2$. Then $a^2b^2 = b^2x^2 - a^2y^2 \implies 1 = \dfrac{x^2}{a^2} - \dfrac{y^2}{b^2}$.

53. $x^3 - 16x = x(x^2 - 16) = x(x + 4)(x - 4)$

55. $2x^3 - 24x^2 + 72x = 2x(x^2 - 12x + 36) = 2x(x - 6)^2$

57. $y = 2 \cos x + 1$

Amplitude: 2

Period: 2π

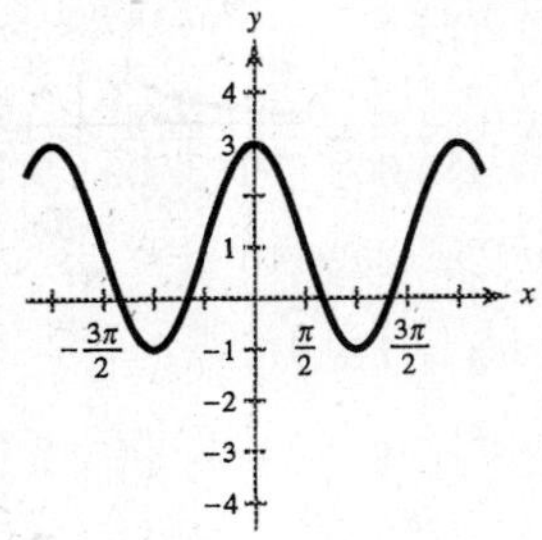

59. $y = \tan 2x$

Period: $\frac{\pi}{2}$

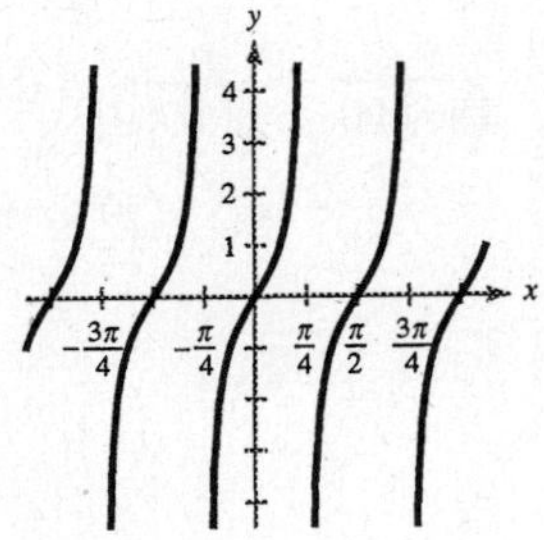

Section 6.5 Rotation of Conics

- The general second-degree equation $Ax^2 + Bxy + Cy^2 + Dx + Ey + F = 0$ can be rewritten as $A'(x')^2 + C'(y')^2 + D'x' + E'y' + F' = 0$ by rotating the coordinate axes through the angle θ, where $\cot 2\theta = (A - C)/B$.
- $x = x' \cos \theta - y' \sin \theta$
  $y = x' \sin \theta + y' \cos \theta$
- The graph of the nondegenerate equation $Ax^2 + Bxy + Cy^2 + Dx + Ey + F = 0$ is:
  (a) An ellipse or circle if $B^2 - 4AC < 0$.
  (b) A parabola if $B^2 - 4AC = 0$.
  (c) A hyperbola if $B^2 - 4AC > 0$.

Solutions to Odd-Numbered Exercises

1. $\theta = 90°$; Point: $(0, 3)$

$x = x' \cos \theta - y' \sin \theta$ $\qquad$ $y = x' \sin \theta + y' \cos \theta$

$0 = x' \cos 90° - y' \sin 90°$ $\qquad$ $3 = x' \sin 90° + y' \cos 90°$

$0 = y'$ $\qquad$ $3 = x°$

So, $(x', y') = (3, 0)$.

3. $\theta = 30°$; Point: $(1, 3)$

$$\begin{aligned} x &= x' \cos \theta - y' \sin \theta \\ y &= x' \sin \theta + y' \cos \theta \end{aligned} \Rightarrow \begin{cases} 1 = x' \cos 30° - y' \sin 30° \\ 3 = x' \sin 30° + y' \cos 30° \end{cases}$$

Solving the system yields $(x', y') = \left(\frac{3 + \sqrt{3}}{2}, \frac{3\sqrt{3} - 1}{2}\right)$.

5. $\theta = 45°$; Point $(2, 1)$

$$\begin{aligned} x &= x' \cos \theta - y' \sin \theta \\ y &= x' \sin \theta + y' \cos \theta \end{aligned} \Rightarrow \begin{cases} 2 = x' \cos 45° - y' \sin 45° \\ 1 = x' \sin 45° + y' \cos 45° \end{cases}$$

Solving the system yields $(x', y') = \left(\frac{3\sqrt{2}}{2}, -\frac{\sqrt{2}}{2}\right)$.

7. $xy + 1 = 0$

$A = 0, B = 1, C = 0$

$$\cot 2\theta = \frac{A - C}{B} = 0 \Rightarrow 2\theta = \frac{\pi}{2} \Rightarrow \theta = \frac{\pi}{4}$$

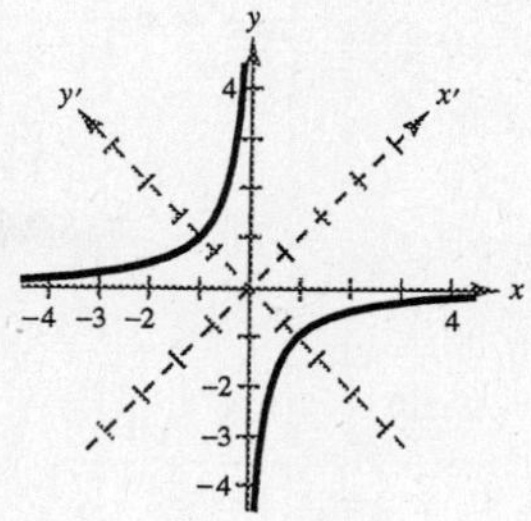

$$x = x'\cos\frac{\pi}{4} - y'\sin\frac{\pi}{4} = x'\left(\frac{\sqrt{2}}{2}\right) - y'\left(\frac{\sqrt{2}}{2}\right) = \frac{x' - y'}{\sqrt{2}}$$

$$y = x'\sin\frac{\pi}{4} + y'\cos\frac{\pi}{4} = x'\left(\frac{\sqrt{2}}{2}\right) + y'\left(\frac{\sqrt{2}}{2}\right) = \frac{x' + y'}{\sqrt{2}}$$

$$xy + 1 = 0$$

$$\left(\frac{x' - y'}{\sqrt{2}}\right)\left(\frac{x' + y'}{\sqrt{2}}\right) + 1 = 0$$

$$\frac{(y')^2}{2} - \frac{(x')^2}{2} = 1$$

9. $x^2 - 2xy + y^2 - 1 = 0$

$A = 1, B = -2, C = 1$

$$\cot 2\theta = \frac{A - C}{B} = 0 \Rightarrow 2\theta = \frac{\pi}{2} \Rightarrow \theta = \frac{\pi}{4}$$

$$x = x'\cos\frac{\pi}{4} - y'\sin\frac{\pi}{4} = x'\left(\frac{\sqrt{2}}{2}\right) - y'\left(\frac{\sqrt{2}}{2}\right) = \frac{x' - y'}{\sqrt{2}}$$

$$y = x'\sin\frac{\pi}{4} + y'\cos\frac{\pi}{4} = x'\left(\frac{\sqrt{2}}{2}\right) + y'\left(\frac{\sqrt{2}}{2}\right) = \frac{x' + y'}{\sqrt{2}}$$

$$x^2 - 2xy + y^2 - 1 = 0$$

$$\left(\frac{x' - y'}{\sqrt{2}}\right)^2 - 2\left(\frac{x' - y'}{\sqrt{2}}\right)\left(\frac{x' + y'}{\sqrt{2}}\right) + \left(\frac{x' + y'}{\sqrt{2}}\right) - 1 = 0$$

$$\frac{(x')^2 - 2(x')(y') + (y')^2}{2} - \frac{2((x')^2 - (y')^2)}{2} + \frac{(x')^2 + 2(x')(y') + (y')^2}{2} - 1 = 0$$

$$2(y')^2 - 1 = 0$$

$$(y')^2 = \frac{1}{2}$$

$$y' = \pm\sqrt{\frac{1}{2}} = \pm\frac{\sqrt{2}}{2}$$

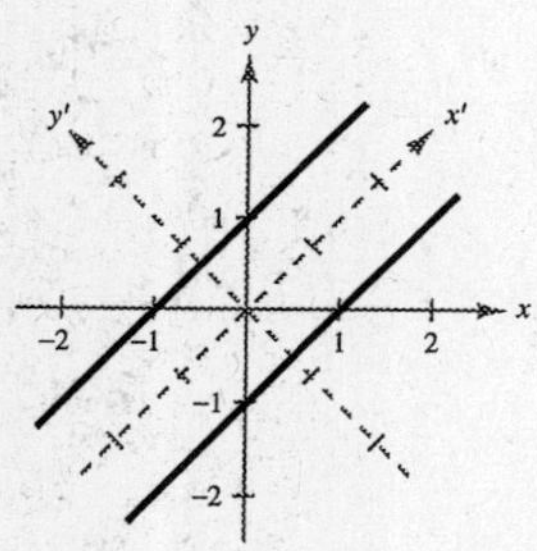

The graph is two parallel lines. Alternate solution.

$$x^2 - 2xy + y^2 - 1 = 0$$

$$(x - y)^2 = 1$$

$$x - y = \pm 1$$

$$y = x \pm 1$$

11. $xy - 2y - 4x = 0$

$A = 0, B = 1, C = 0$

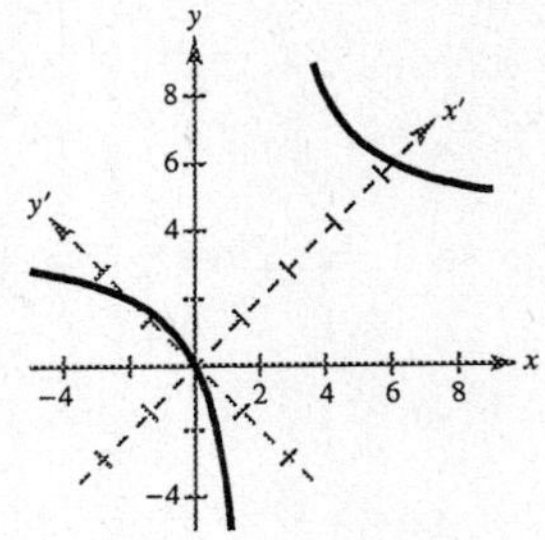

$$\cot 2\theta = \frac{A - C}{B} = 0 \Rightarrow 2\theta = \frac{\pi}{2} \Rightarrow \theta = \frac{\pi}{4}$$

$$x = x'\cos\frac{\pi}{4} - y'\sin\frac{\pi}{4} = x'\left(\frac{\sqrt{2}}{2}\right) - y'\left(\frac{\sqrt{2}}{2}\right) = \frac{x' - y'}{\sqrt{2}}$$

$$y = x'\sin\frac{\pi}{4} + y'\cos\frac{\pi}{4} = x'\left(\frac{\sqrt{2}}{2}\right) + y'\left(\frac{\sqrt{2}}{2}\right) = \frac{x' + y'}{\sqrt{2}}$$

$$xy - 2y - 4x = 0$$

$$\left(\frac{x' - y'}{\sqrt{2}}\right)\left(\frac{x' + y'}{\sqrt{2}}\right) - 2\left(\frac{x' + y'}{\sqrt{2}}\right) - 4\left(\frac{x' - y'}{\sqrt{2}}\right) = 0$$

$$\frac{(x')^2}{2} - \frac{(y')^2}{2} - \sqrt{2}x' - \sqrt{2}y' - 2\sqrt{2}x' + 2\sqrt{2}y' = 0$$

$$\left[(x')^2 - 6\sqrt{2}x' + (3\sqrt{2})^2\right] - \left[(y')^2 - 2\sqrt{2}y' + (\sqrt{2})^2\right] = 0 + (3\sqrt{2})^2 - (\sqrt{2})^2$$

$$(x' - 3\sqrt{2})^2 - (y' - \sqrt{2})^2 = 16$$

$$\frac{(x' - 3\sqrt{2})^2}{16} - \frac{(y' - \sqrt{2})^2}{16} = 1$$

13. $5x^2 - 6xy + 5y^2 - 12 = 0$

$A = 5, B = -6, C = 5$

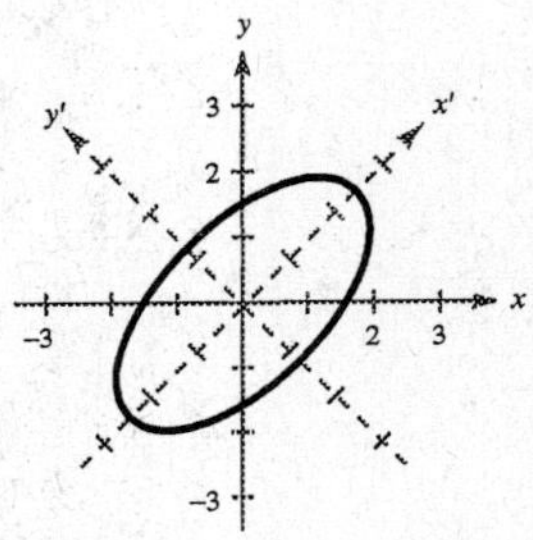

$$\cot 2\theta = \frac{A - C}{B} = 0 \Rightarrow 2\theta = \frac{\pi}{2} \Rightarrow \theta = \frac{\pi}{4}$$

$$x = x'\cos\frac{\pi}{4} - y'\sin\frac{\pi}{4} = x'\left(\frac{\sqrt{2}}{2}\right) - y'\left(\frac{\sqrt{2}}{2}\right) = \frac{x' - y'}{\sqrt{2}}$$

$$y = x'\sin\frac{\pi}{4} + y'\cos\frac{\pi}{4} = x'\left(\frac{\sqrt{2}}{2}\right) + y'\left(\frac{\sqrt{2}}{2}\right) = \frac{x' + y'}{\sqrt{2}}$$

$$5x^2 - 6xy + 5y^2 - 12 = 0$$

$$5\left(\frac{x' - y'}{\sqrt{2}}\right)^2 - 6\left(\frac{x' - y'}{\sqrt{2}}\right)\left(\frac{x' + y'}{\sqrt{2}}\right) + 5\left(\frac{x' + y'}{\sqrt{2}}\right)^2 - 12 = 0$$

$$\frac{5(x')^2}{2} - 5x'y' + \frac{5(y')^2}{2} - 3(x')^2 + 3(y')^2 + \frac{5(x')^2}{2} + 5x'y' + \frac{5(y')^2}{2} - 12 = 0$$

$$2(x')^2 + 8(y')^2 = 12$$

$$\frac{(x')^2}{6} + \frac{(y')^2}{\frac{3}{2}} = 1$$

15. $3x^2 - 2\sqrt{3}xy + y^2 + 2x + 2\sqrt{3}y = 0$

$A = 3, B = -2\sqrt{3}, C = 1$

$\cot 2\theta = \dfrac{A - C}{B} = -\dfrac{1}{\sqrt{3}} \Rightarrow \theta = 60°$

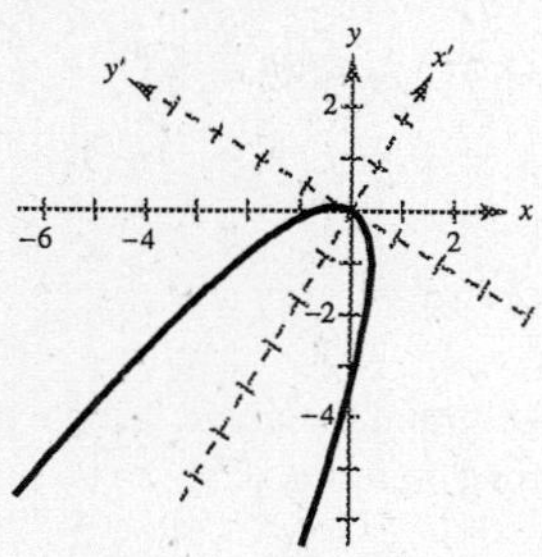

$x = x'\cos 60° - y'\sin 60°$

$= x'\left(\dfrac{1}{2}\right) - y'\left(\dfrac{\sqrt{3}}{2}\right) = \dfrac{x' - \sqrt{3}y'}{2}$

$y = x'\sin 60° + y'\cos 60°$

$= x'\left(\dfrac{\sqrt{3}}{2}\right) + y'\left(\dfrac{1}{2}\right) = \dfrac{\sqrt{3}x' + y'}{2}$

$3x^2 - 2\sqrt{3}xy + y^2 + 2x + 2\sqrt{3}y = 0$

$$3\left(\frac{x' - \sqrt{3}y'}{2}\right)^2 - 2\sqrt{3}\left(\frac{x' - \sqrt{3}y'}{2}\right)\left(\frac{\sqrt{3}x' + y'}{2}\right) + \left(\frac{\sqrt{3}x' + y'}{2}\right)^2 + 2\left(\frac{x' - \sqrt{3}y'}{2}\right) + 2\sqrt{3}\left(\frac{\sqrt{3}x' + y'}{2}\right) = 0$$

$$\frac{3(x')^2}{4} - \frac{6\sqrt{3}x'y'}{4} + \frac{9(y')^2}{4} - \frac{6(x')^2}{4} + \frac{4\sqrt{3}x'y'}{4} + \frac{6(y')^2}{4} + \frac{3(x')^2}{4} + \frac{2\sqrt{3}x'y'}{4} + \frac{(y')^2}{4}$$

$$+ x' - \sqrt{3}y' + 3x' + \sqrt{3}y' = 0$$

$$4(y')^2 + 4x' = 0$$

$$x' = -(y')^2$$

17. $9x^2 + 24xy + 16y^2 + 90x - 130y = 0$

$A = 9, B = 24, C = 16$

$\cot 2\theta = \dfrac{A - C}{B} = -\dfrac{7}{24} \Rightarrow \theta \approx 53.13°$

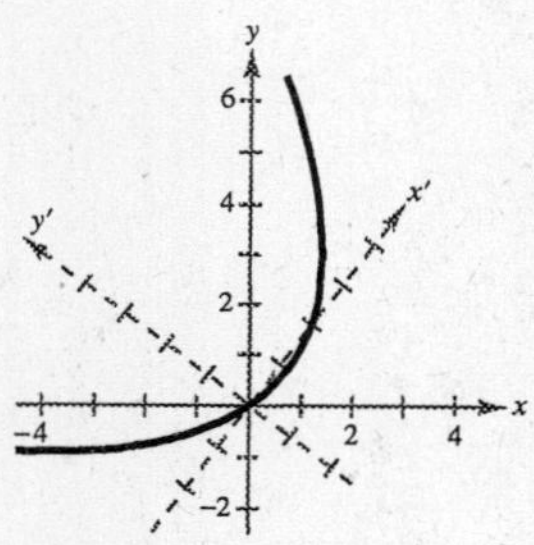

$\cos 2\theta = -\dfrac{7}{25}$

$\sin \theta = \sqrt{\dfrac{1 - \cos 2\theta}{2}} = \sqrt{\dfrac{1 - \left(-\frac{7}{25}\right)}{2}} = \dfrac{4}{5}$

$\cos \theta = \sqrt{\dfrac{1 + \cos 2\theta}{2}} = \sqrt{\dfrac{1 + \left(-\frac{7}{25}\right)}{2}} = \dfrac{3}{5}$

$x = x'\cos\theta - y'\sin\theta$

$= x'\left(\dfrac{3}{5}\right) - y'\left(\dfrac{4}{5}\right) = \dfrac{3x' - 4y'}{5}$

$y = x'\sin\theta + y'\cos\theta$

$= x'\left(\dfrac{4}{5}\right) + y'\left(\dfrac{3}{5}\right)$

$= \dfrac{4x' + 3y'}{5}$

$9x^2 + 24xy + 16y^2 + 90x - 130y = 0$

$$9\left(\frac{3x' - 4y'}{5}\right)^2 + 24\left(\frac{3x' - 4y'}{5}\right)\left(\frac{4x' + 3y'}{5}\right) + 16\left(\frac{4x' + 3y'}{5}\right)^2 + 90\left(\frac{3x' - 4y'}{5}\right) - 130\left(\frac{4x' + 3y'}{5}\right) = 0$$

$$\frac{81(x')^2}{25} - \frac{216x'y'}{25} + \frac{144(y')^2}{25} + \frac{288(x')^2}{25} - \frac{168x'y'}{25} - \frac{288(y')^2}{25} + \frac{256(x')^2}{25} + \frac{384x'y'}{25}$$

$$+ \frac{144(y')^2}{25} + 54x' - 72y' - 104x' - 78y' = 0$$

$$25(x')^2 - 50x' - 150y' = 0$$

$$(x')^2 - 2x' + 1 = 6y' + 1$$

$$y' = \frac{(x')^2}{6} - \frac{x'}{3}$$

19. $x^2 + 2xy + y^2 = 20$

$A = 1, B = 2, C = 1$

$$\cot 2\theta = \frac{A - C}{B} = \frac{1 - 1}{2} = 0 \Rightarrow \theta = \frac{\pi}{4} \text{ or } 45^\circ$$

To graph the conic using a graphing calculator, we need to solve for y in terms of x.

$$(x + y)^2 = 20$$

$$x + y = \pm\sqrt{20}$$

$$y = -x \pm \sqrt{20}$$

Use $y_1 = -x + \sqrt{20}$

and $y_2 = -x - \sqrt{20}$

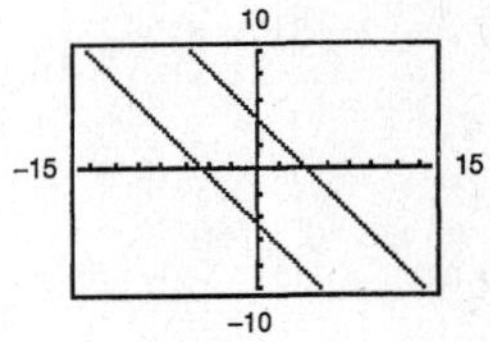

21. $17x^2 + 32xy - 7y^2 = 75$

$$\cot 2\theta = \frac{A - C}{B} = \frac{17 + 7}{32} = \frac{24}{32} = \frac{3}{4} \Rightarrow \theta \approx 26.57^\circ$$

Solve for y in terms of x by completing the square.

$$-7y^2 + 32xy = -17x^2 + 75$$

$$y^2 - \frac{32}{7}xy = \frac{17}{7}x^2 - \frac{75}{7}$$

$$y^2 - \frac{32}{7}xy + \frac{256}{49}x^2 = \frac{119}{49}x^2 - \frac{525}{49} + \frac{256}{49}x^2$$

$$\left(y - \frac{16}{7}x\right)^2 = \frac{375x^2 - 525}{49}$$

$$y = \frac{16}{7}x \pm \sqrt{\frac{375x^2 - 525}{49}}$$

$$y = \frac{16x \pm 5\sqrt{15x^2 - 21}}{7}$$

Use $y_1 = \dfrac{16x + 5\sqrt{15x^2 - 21}}{7}$

and $y_2 = \dfrac{16x - 5\sqrt{15x^2 - 21}}{7}$.

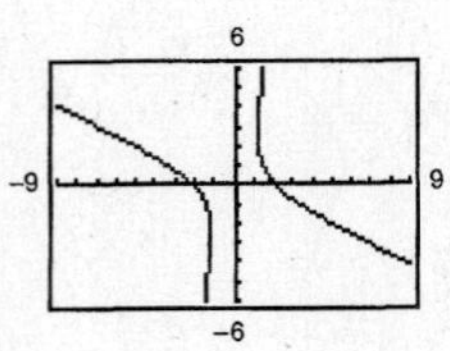

23. $32x^2 + 48xy + 8y^2 = 50$

$$\cot 2\theta = \frac{A - C}{B} = \frac{24}{48} = \frac{1}{2} \Rightarrow \theta \approx 31.72^\circ$$

Solve for y in terms of x by completing the square.

$$8y^2 + 48xy = -32x^2 + 50$$

$$y^2 + 6xy = -4x^2 + \frac{25}{4}$$

$$y^2 + 6xy + 9x^2 = -4x^2 + \frac{25}{4} + 9x^2$$

$$(y + 3x)^2 = 5x^2 + \frac{25}{4}$$

$$y + 3x = \pm\sqrt{5x^2 + \tfrac{25}{4}}$$

$$y = -3x \pm \sqrt{5x^2 + \tfrac{25}{4}}$$

Use $y_1 = -3x + \sqrt{5x^2 + \tfrac{25}{4}}$

and $y_2 = -3x - \sqrt{5x^2 + \tfrac{25}{4}}$

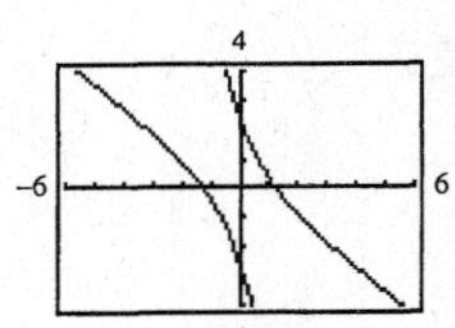

25. $4x^2 - 12xy + 9y^2 + (4\sqrt{13} - 12)x - (6\sqrt{13} + 8)y = 91$

$A = 4, B = -12, C = 9$

$$\cot 2\theta = \frac{A - C}{B} = \frac{4 - 9}{-12} = \frac{5}{12}$$

$$\frac{1}{\tan 2\theta} = \frac{5}{12}$$

$$\tan 2\theta = \frac{12}{5}$$

$$2\theta \approx 67.38°$$

$$\theta \approx 33.69°$$

Solve for y in terms of x with the quadratic formula:

$$4x^2 - 12xy + 9y^2 + (4\sqrt{13} - 12)x - (6\sqrt{13} + 8)y = 91$$

$$9y^2 - (12x + 6\sqrt{13} + 8)y + (4x^2 + 4\sqrt{13}x - 12x - 91) = 0$$

$$a = 9, b = -(12x + 6\sqrt{13} + 8), c = 4x^2 + 4\sqrt{13}x - 12x - 91$$

$$y = \frac{-b \pm \sqrt{b^2 - 4ac}}{2a}$$

$$y = \frac{(12x + 6\sqrt{13} + 8) \pm \sqrt{(12x + 6\sqrt{13} + 8)^2 - 4(9)(4x^2 + 4\sqrt{13}x - 12x - 91)}}{18}$$

$$= \frac{(12x + 6\sqrt{13} + 8) \pm \sqrt{624x + 3808 + 96\sqrt{13}}}{18}$$

Enter $y_1 = \dfrac{12x + 6\sqrt{13} + 8 + \sqrt{624x + 3808 + 96\sqrt{13}}}{18}$

and $y_2 = \dfrac{12x + 6\sqrt{13} + 8 - \sqrt{624x + 3808 + 96\sqrt{13}}}{18}$.

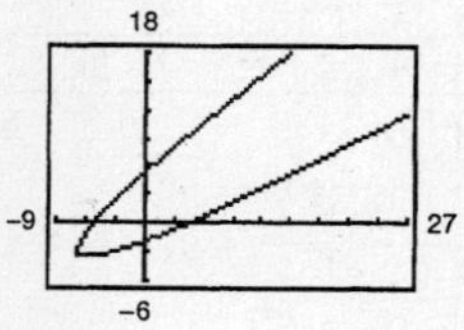

27. $xy + 2 = 0$

$B^2 - 4AC = 1 \implies$ The graph is a hyperbola.

$$\cot 2\theta = \frac{A - C}{B} = 0 \implies \theta = 45°$$

Matches graph (e).

29. $-2x^2 + 3xy + 2y^2 + 3 = 0$

$B^2 - 4AC = (3)^2 - 4(-2)(2) = 25 \implies$ The graph is a hyperbola.

$$\cot 2\theta = \frac{A - C}{B} = -\frac{4}{3} \implies \theta \approx -18.43°$$

Matches graph (b).

31. $3x^2 + 2xy + y^2 - 10 = 0$

$B^2 - 4AC = (2)^2 - 4(3)(1) = -8 \implies$ The graph is an ellipse or circle.

$$\cot 2\theta = \frac{A - C}{B} = 1 \implies \theta = 22.5°$$

Matches graph (d).

33. $16x^2 - 8xy + y^2 - 10x + 5y = 0$

$B^2 - 4AC = (-8)^2 - 4(16)(1) = 0$

The graph is a parabola.

$y^2 + (-8x + 5)y + (16x^2 - 10x) = 0$

$$y = \frac{-(-8x + 5) \pm \sqrt{(-8x + 5)^2 - 4(1)(16x^2 - 10x)}}{2(1)}$$

$$= \frac{8x - 5 \pm \sqrt{(8x - 5)^2 - 4(16x^2 - 10x)}}{2}$$

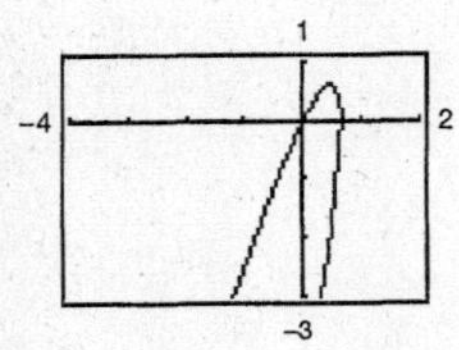

35. $12x^2 - 6xy + 7y^2 - 45 = 0$

$B^2 - 4AC = (-6)^2 - 4(12)(7) = -300 < 0$

The graph is an ellipse.

$7y^2 + (-6x)y + (12x^2 - 45) = 0$

$$y = \frac{-(-6x) \pm \sqrt{(-6x)^2 - 4(7)(12x^2 - 45)}}{2(7)}$$

$$= \frac{6x \pm \sqrt{36x^2 - 28(12x^2 - 45)}}{14}$$

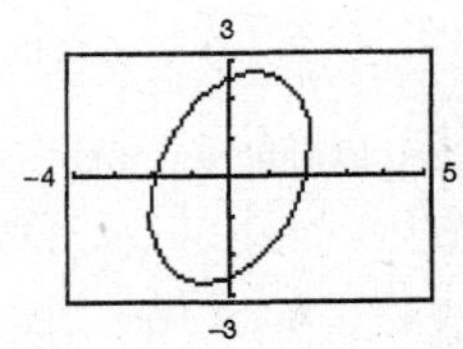

37. $x^2 - 6xy - 5y^2 + 4x - 22 = 0$

$B^2 - 4AC = (-6)^2 - 4(1)(-5) = 56 > 0$

The graph is a hyperbola.

$-5y^2 + (-6x)y + (x^2 + 4x - 22) = 0$

$$y = \frac{-(-6x) \pm \sqrt{(-6x)^2 - 4(-5)(x^2 + 4x - 22)}}{2(-5)}$$

$$= \frac{6x \pm \sqrt{36x^2 + 20(x^2 + 4x - 22)}}{-10}$$

$$= \frac{-6x \pm \sqrt{36x^2 + 20(x^2 + 4x - 22)}}{10}$$

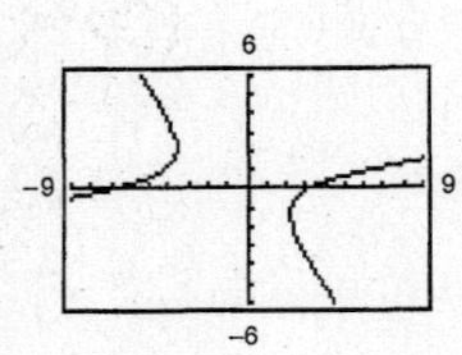

39. $x^2 + 4xy + 4y^2 - 5x - y - 3 = 0$

$B^2 - 4AC = (4)^2 - 4(1)(4) = 0$

The graph is a parabola.

$4y^2 + (4x - 1)y + (x^2 - 5x - 3) = 0$

$$y = \frac{-(4x - 1) \pm \sqrt{(4x - 1)^2 - 4(4)(x^2 - 5x - 3)}}{2(4)}$$

$$= \frac{-(4x - 1) \pm \sqrt{(4x - 1)^2 - 16(x^2 - 5x - 3)}}{8}$$

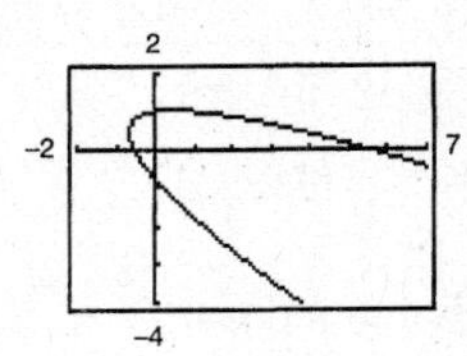

41. $y^2 - 9x^2 = 0$

$y^2 = 9x^2$

$y = \pm 3x$

Two intersecting lines

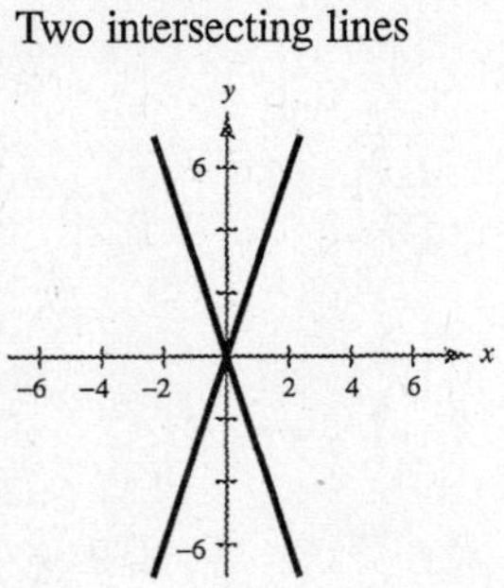

43. $x^2 + 2xy + y^2 - 1 = 0$

$(x + y)^2 - 1 = 0$

$(x + y)^2 = 1$

$x + y = \pm 1$

$y = -x \pm 1$

Two parallel lines

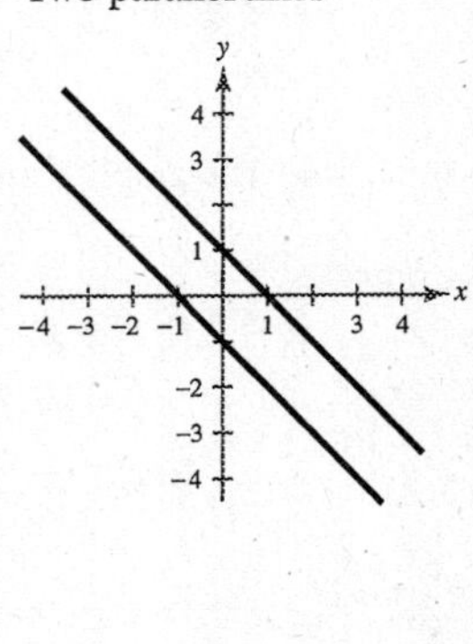

45.

$$\begin{aligned}
-x^2 + y^2 + 4x - 6y + 4 = 0 &\Rightarrow (y-3)^2 - (x-2)^2 = 1\\
x^2 + y^2 - 4x - 6y + 12 = 0 &\Rightarrow (x-2)^2 + (y-3)^2 = 1\\
\hline
2y^2 - 12y + 16 &= 0\\
2(y-2)(y-4) &= 0\\
y = 2 \text{ or } y &= 4
\end{aligned}$$

For $y = 2$: $x^2 + 2^2 - 4x - 6(2) + 12 = 0$

$$x^2 - 4x + 4 = 0$$
$$(x-2)^2 = 0$$
$$x = 2$$

For $y = 4$: $x^2 + 4^2 - 4x - 6(4) + 12 = 0$

$$x^2 - 4x + 4 = 0$$
$$(x-2)^2 = 0$$
$$x = 2$$

The points of intersection are (2, 2) and (2, 4).

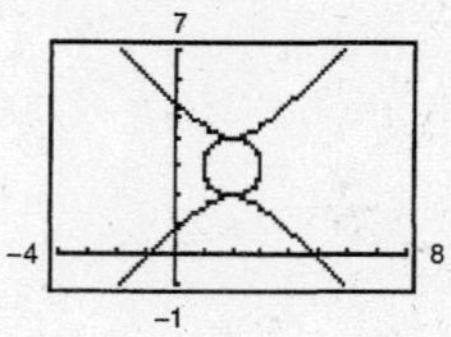

47.

$$\begin{aligned}
-4x^2 - y^2 - 16x + 24y - 16 &= 0\\
4x^2 + y^2 + 40x - 24y + 208 &= 0\\
\hline
24x + 192 &= 0\\
x &= -8
\end{aligned}$$

When $x = -8$: $4(-8)^2 + y^2 + 40(-8) - 24y + 208 = 0$

$$y^2 - 24y + 144 = 0$$
$$(y-12)^2 = 0$$
$$y = 12$$

The point of intersection is $(-8, 12)$.

In standard form the equations are:

$$\frac{(x+2)^2}{36} + \frac{(y-12)^2}{144} = 1$$

$$\frac{(x+5)^2}{9} + \frac{(y-12)^2}{36} = 1$$

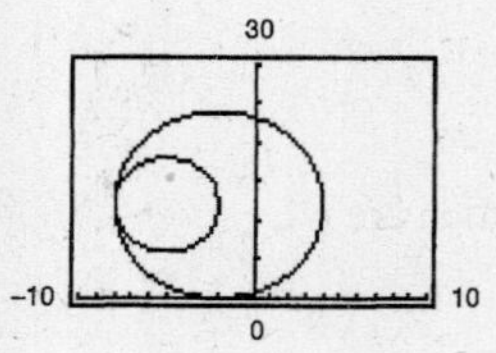

49.

$$\begin{aligned}
x^2 - y^2 - 12x + 16y - 64 &= 0\\
x^2 + y^2 - 12x - 16y + 64 &= 0\\
\hline
2x^2 - 24x &= 0\\
2x(x-12) &= 0\\
x = 0 \text{ or } x &= 12
\end{aligned}$$

When $x = 0$: $0^2 + y^2 - 12(0) - 16y + 64 = 0$

$$y^2 - 16y + 64 = 0$$
$$(y-8)^2 = 0$$
$$y = 8$$

When $x = 12$: $12^2 + y^2 - 12(12) - 16y + 64 = 0$

$$y^2 - 16y + 64 = 0$$
$$(y-8)^2 = 0$$
$$y = 8$$

The points of intersection are (0, 8) and (12, 8).

The standard forms of the equations are:

$$\frac{(x-6)^2}{36} - \frac{(y-8)^2}{36} = 1$$

$$(x-6)^2 + (y-8)^2 = 36$$

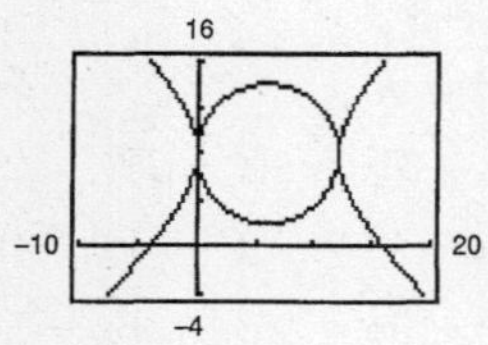

51.
$$\begin{aligned} -16x^2 - y^2 + 24y - 80 &= 0 \\ 16x^2 + 25y^2 - 400 &= 0 \\ \hline 24y^2 + 24y - 480 &= 0 \\ 24(y + 5)(y - 4) &= 0 \\ y = -5 \text{ or } y &= 4 \end{aligned}$$

When $y = -5$: $16x^2 + 25(-5)^2 - 400 = 0$

$16x^2 = -225$

No real solution

When $y = 4$: $16x^2 + 25(4)^2 - 400 = 0$

$16x^2 = 0$

$x = 0$

The point of intersection is $(0, 4)$.

In standard form the equations are:

$$\frac{x^2}{4} + \frac{(y - 12)^2}{64} = 1$$

$$\frac{x^2}{25} + \frac{y^2}{16} = 1$$

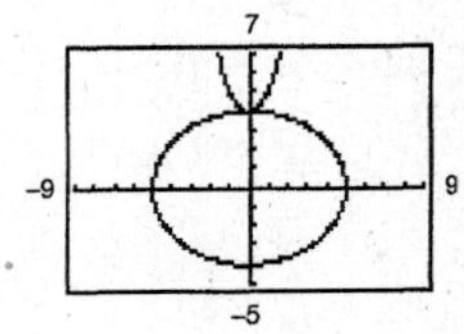

53.
$$\begin{aligned} x^2 + y^2 - 4 &= 0 \\ 3x - y^2 &= 0 \\ x^2 + 3x - 4 &= 0 \\ (x + 4)(x - 1) &= 0 \\ x = -4 \text{ or } x &= 1 \end{aligned}$$

When $x = -4$: $3(-4) - y^2 = 0$

$y^2 = -12$ No real solution

When $x = 1$: $3(1) - y^2 = 0$

$y^2 = 3$

$y = \pm\sqrt{3}$

The points of intersection are $(1, \sqrt{3})$ and $(1, -\sqrt{3})$.

The standard forms of the equations are:

$x^2 + y^2 = 4$

$y^2 = 3x$

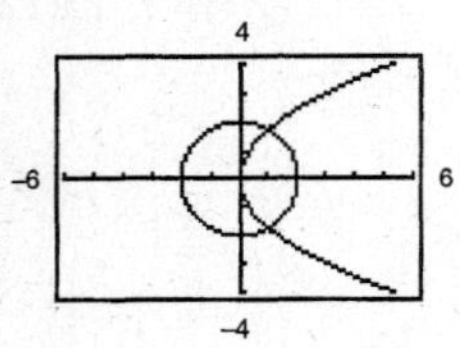

55. $x^2 + 2y^2 - 4x + 6y - 5 = 0$

$-x + y - 4 = 0 \Rightarrow y = x + 4$

$$\begin{aligned} x^2 + 2(x + 4)^2 - 4x + 6(x + 4) - 5 &= 0 \\ x^2 + 2(x^2 + 8x + 16) - 4x + 6x + 24 - 5 &= 0 \\ 3x^2 + 18x + 51 &= 0 \\ 3(x^2 + 6x + 17) &= 0 \\ x^2 + 6x + 17 &= 0 \\ x^2 + 6x + 9 &= -17 + 9 \\ (x + 3)^2 &= -8 \end{aligned}$$

No real solution

No points of intersection

The standard forms of the equations are:

$$\frac{(x - 2)^2}{\frac{27}{2}} + \frac{\left(y + \frac{3}{2}\right)^2}{\frac{27}{4}} = 1$$

$x - y = -4$

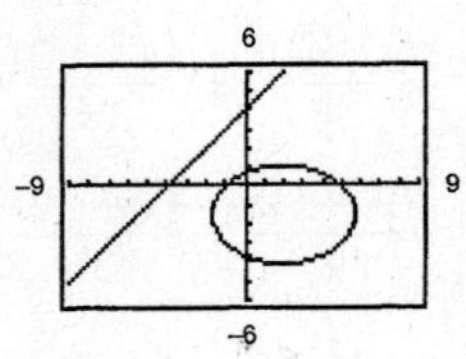

57. $xy + x - 2y + 3 = 0 \Rightarrow y = \dfrac{-x - 3}{x - 2}$

$$\begin{aligned} x^2 + 4y^2 - 9 &= 0 \\ x^2 + 4\left(\frac{-x - 3}{x - 2}\right)^2 &= 9 \\ x^2(x - 2)^2 + 4(-x - 3)^2 &= 9(x - 2)^2 \\ x^2(x^2 - 4x + 4) + 4(x^2 + 6x + 9) &= 9(x^2 - 4x + 4) \\ x^4 - 4x^3 + 4x^2 + 4x^2 + 24x + 36 &= 9x^2 - 36x + 36 \\ x^4 - 4x^3 - x^2 + 60x &= 0 \\ x(x + 3)(x^2 - 7x + 20) &= 0 \\ x = 0 \text{ or } x &= -3 \end{aligned}$$

Note: $x^2 - 7x + 20 = 0$ has no real solution.

When $x = 0$: $y = \dfrac{-0 - 3}{0 - 2} = \dfrac{3}{2}$

When $x = -3$: $y = \dfrac{-(-3) - 3}{-3 - 2} = 0$

The points of intersection are $\left(0, \dfrac{3}{2}\right)$, $(-3, 0)$.

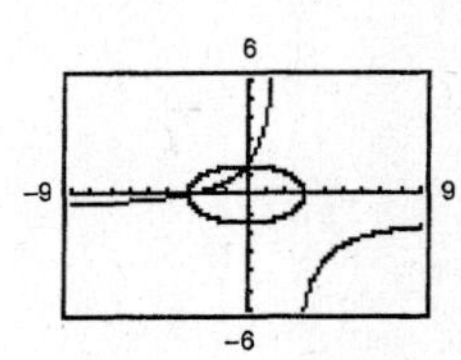

59. $x^2 + xy + ky^2 + 6x + 10 = 0$

$B^2 - 4AC = 1^2 - 4(1)(k) = 1 - 4k > 0 \Rightarrow -4k > -1 \Rightarrow k < \frac{1}{4}$

True. For the graph to be a hyperbola, the discriminant must be greater than zero.

61.
$$\begin{aligned}(x')^2 + (y')^2 &= (x\cos\theta - y\sin\theta)^2 + (y\cos\theta + x\sin\theta)^2\\ &= x^2\cos^2\theta - 2xy\cos\theta\sin\theta + y^2\sin^2\theta + y^2\cos^2\theta + 2xy\cos\theta\sin\theta + x^2\sin^2\theta\\ &= x^2(\cos^2\theta + \sin^2\theta) + y^2(\sin^2\theta + \cos^2\theta) = x^2 + y^2 = r^2\end{aligned}$$

63. $f(x) = |x - 4|$

Intercepts: $(0, 4), (4, 0)$

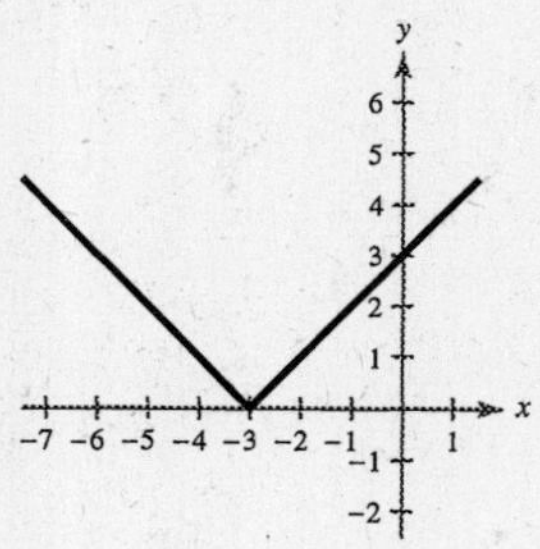

65. $g(x) = \sqrt{4 - x^2}$

Intercepts: $(\pm 2, 0), (0, 2)$

Domain: $-2 \le x \le 2$

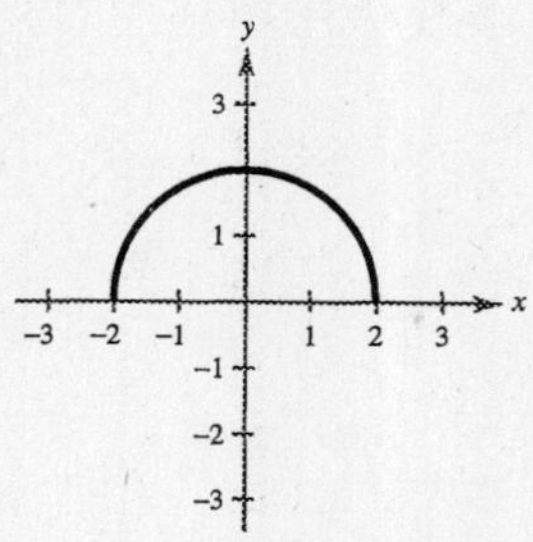

67. $f(t) = -(t - 2)^3 + 3$

x	-1	0	1	2	3
$f(x)$	30	11	4	3	2

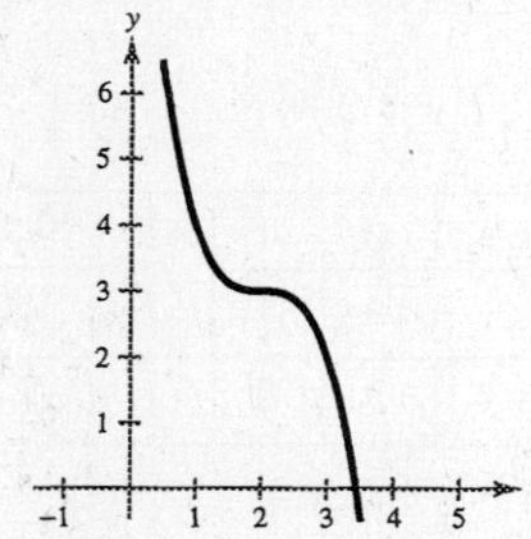

69. $f(t) = [\![t - 5]\!] + 1$

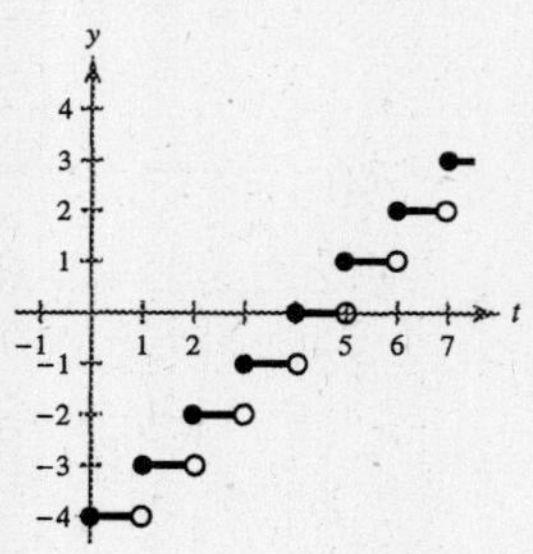

71. $C = 110°, a = 8, b = 12$

Area $= \frac{1}{2}ab\sin C = \frac{1}{2}(8)(12)\sin 110° \approx 45.1052$ square units

73. $a = 11, b = 18, c = 10$

$$s = \frac{11 + 18 + 10}{2} = 19.5$$

Area $= \sqrt{19.5(19.5 - 11)(19.5 - 18)(19.5 - 10)} \approx 48.5998$ square units

75.

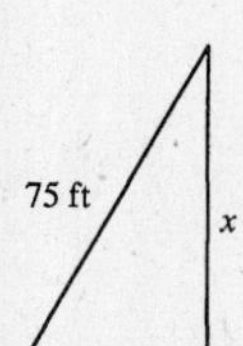

$\sin 60° = \dfrac{x}{75}$

$x = 75 \sin 60°$

$x \approx 64.9519$ feet

Section 6.6 Parametric Equations

- If f and g are continuous functions of t on an interval I, then the set of ordered pairs $(f(t), g(t))$ is a *plane curve* C. The equations $x = f(t)$ and $y = g(t)$ are *parametric equations* for C and t is the *parameter.*
- To eliminate the parameter:
  (a) Solve for t in one equation and substitute into the second equation.
  (b) Use trigonometric identities.
- You should be able to find the parametric equations for a graph.

Solutions to Odd-Numbered Exercises

1. $x = \sqrt{t}, y = 3 - t$

(a)

t	0	1	2	3	4
x	0	1	$\sqrt{2}$	$\sqrt{3}$	2
y	3	2	1	0	-1

(b)

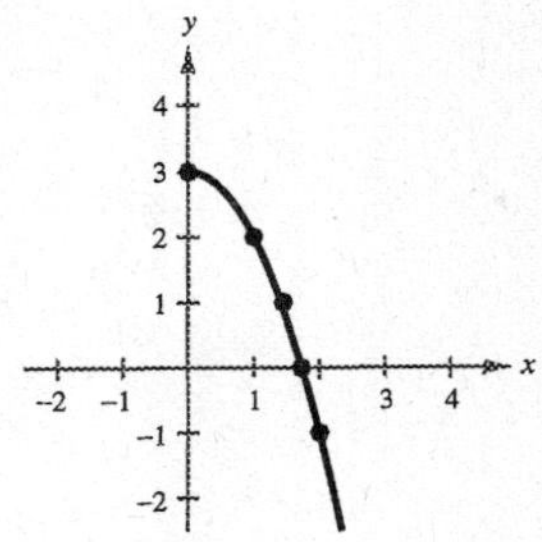

(c) $x = \sqrt{t} \implies x^2 = t$

$y = 3 - t \implies y = 3 - x^2$

The graph of the parametric equations only shows the right half of the parabola, whereas the rectangular equation yields the entire parabola.

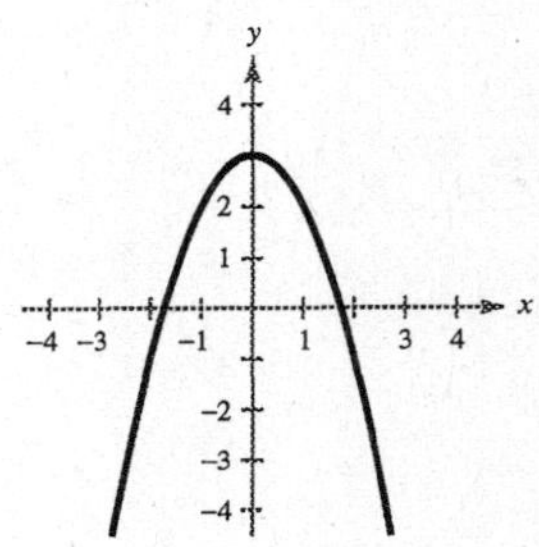

3. $x = 3t - 3 \implies t = \dfrac{x + 3}{3}$

$y = 2t + 1 \implies y = \dfrac{2}{3}(x + 3) + 1 = \dfrac{2}{3}x + 3$

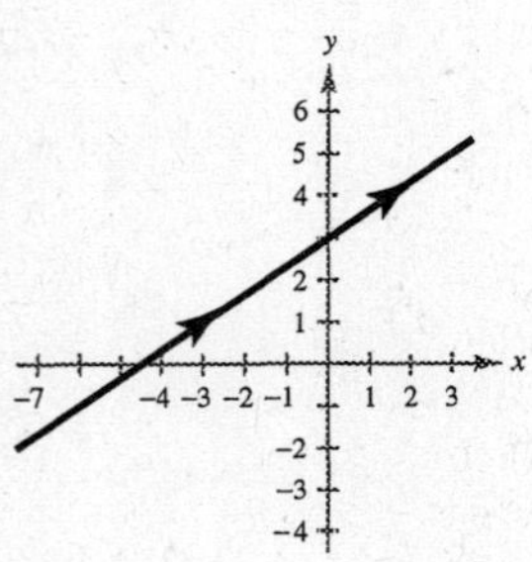

5. $x = \dfrac{1}{4}t \implies t = 4x$

$y = t^2 \implies y = 16x^2$

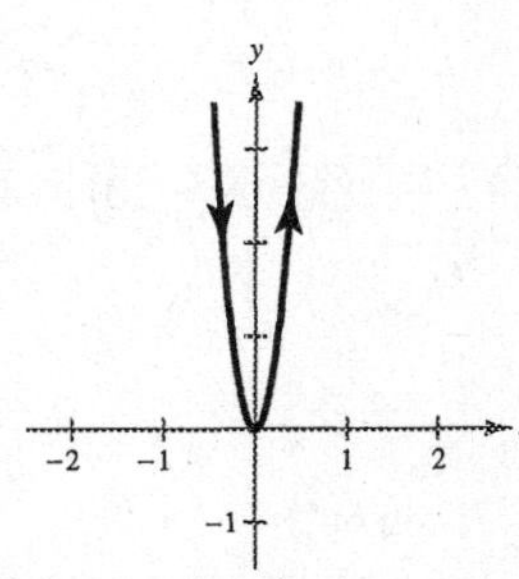

7. $x = t + 2 \implies t = x - 2$

$y = t^2 \implies y = (x - 2)^2$

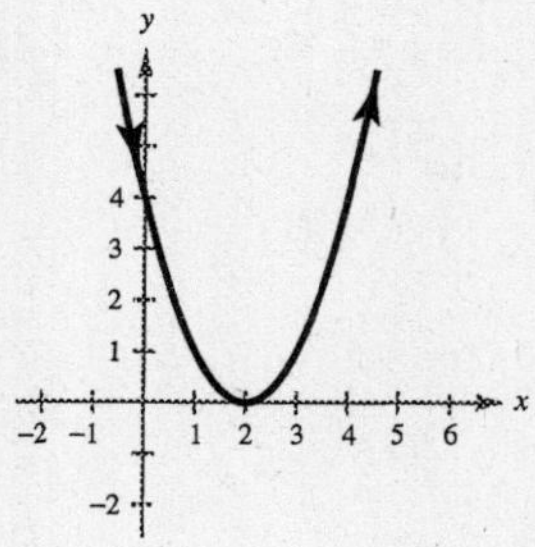

9. $x = t + 1 \implies t = x - 1$

$y = \dfrac{t}{t + 1} \implies y = \dfrac{x - 1}{x} = 1 - \dfrac{1}{x}$

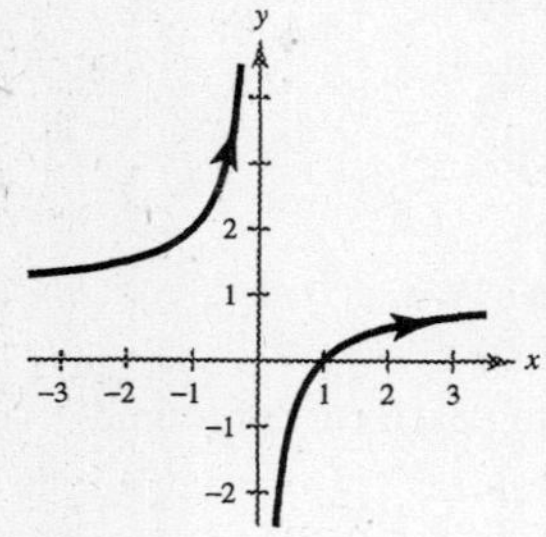

11. $x = 2(t + 1) \implies \dfrac{x}{2} - 1 = t$ or $t = \dfrac{x - 2}{2}$

$y = |t - 2| \implies \quad y = \left|\dfrac{x}{2} - 1 - 2\right| = \left|\dfrac{x}{2} - 3\right| = \left|\dfrac{x - 6}{2}\right|$

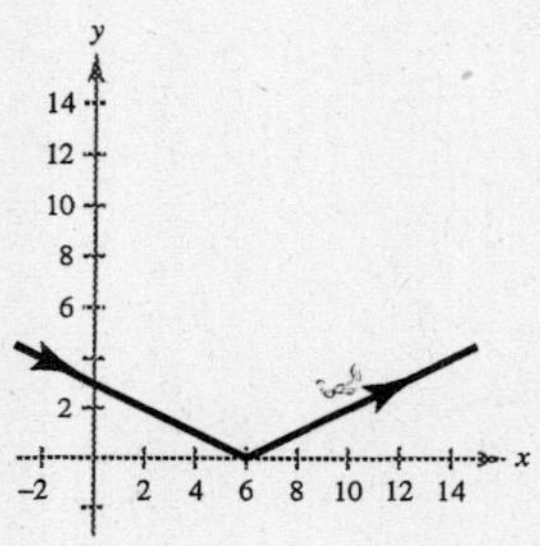

13. $x = 3 \cos \theta \implies \left(\dfrac{x}{3}\right)^2 = \cos^2 \theta$

$y = 3 \sin \theta \implies \left(\dfrac{y}{3}\right)^2 = \sin^2 \theta$

$\left(\dfrac{x}{3}\right)^2 + \left(\dfrac{y}{3}\right)^2 = 1$

$x^2 + y^2 = 9$

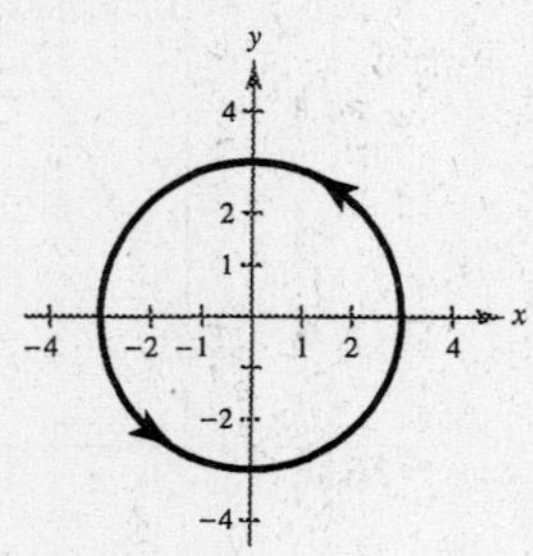

15. $x = 4 \sin 2\theta \implies \left(\dfrac{x}{4}\right)^2 = \sin^2 2\theta$

$y = 2 \cos 2\theta \implies \left(\dfrac{y}{2}\right)^2 = \cos^2 2\theta$

$\left(\dfrac{x}{4}\right)^2 + \left(\dfrac{y}{2}\right)^2 = 1$

$\dfrac{x^2}{16} + \dfrac{y^2}{4} = 1$

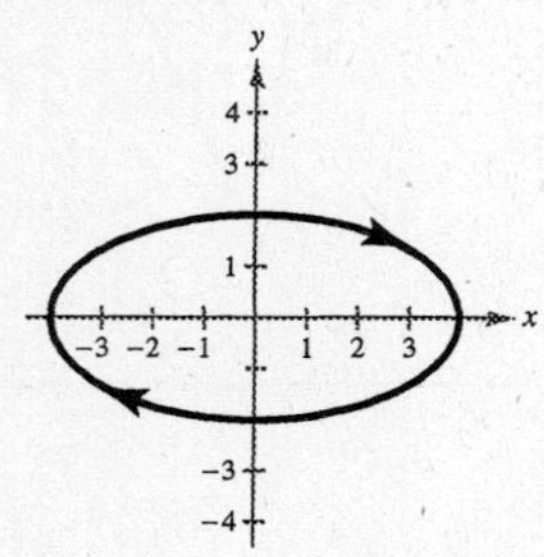

17. $x = 4 + 2 \cos \theta \implies \left(\dfrac{x - 4}{2}\right)^2 = \cos^2 \theta$

$y = -1 + \sin \theta \implies (y + 1)^2 = \sin^2 \theta$

$\dfrac{(x - 4)^2}{4} + \dfrac{(y + 1)^2}{1} = 1$

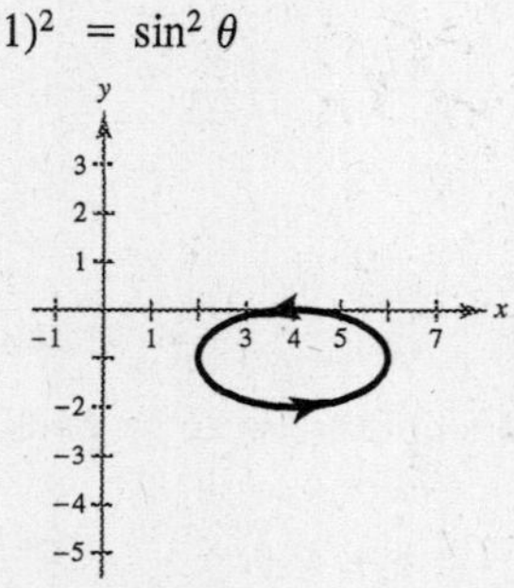

19. $x = e^{-t} \implies \dfrac{1}{x} = e^t$

$y = e^{3t} \implies y = (e^t)^3$

$y = \left(\dfrac{1}{x}\right)^3$

$y = \dfrac{1}{x^3}, \ x > 0, y > 0$

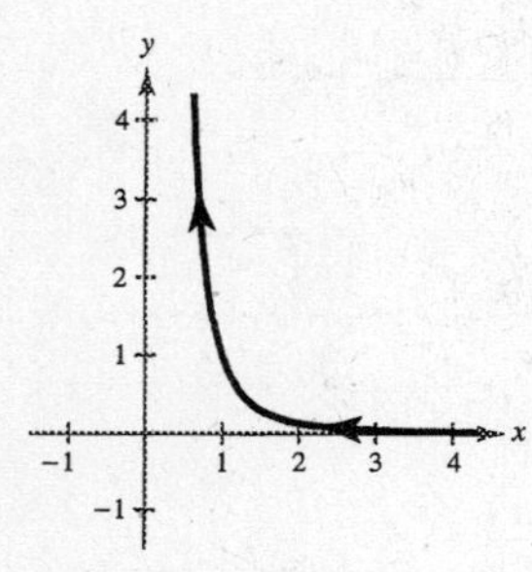

21. $x = t^3 \quad \Rightarrow x^{1/3} = t$

$y = 3 \ln t \Rightarrow y = \ln t^3$

$y = \ln(x^{1/3})^3$

$y = \ln x$

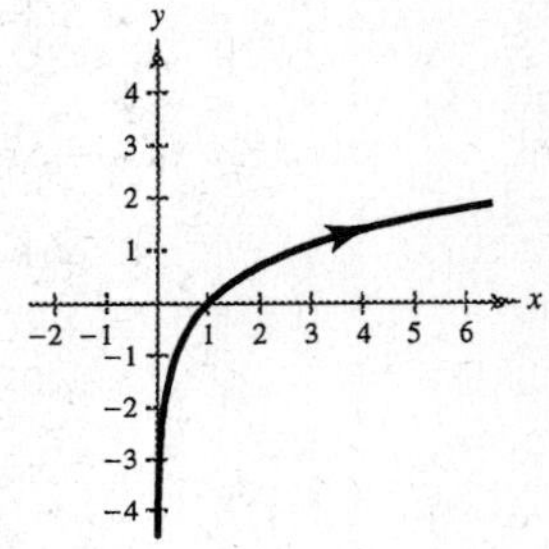

23. By eliminating the parameter, each curve becomes $y = 2x + 1$.

(a) $x = t$

$y = 2t + 1$

There are no restrictions on x and y.

Domain: $(-\infty, \infty)$

Orientation: Left to right

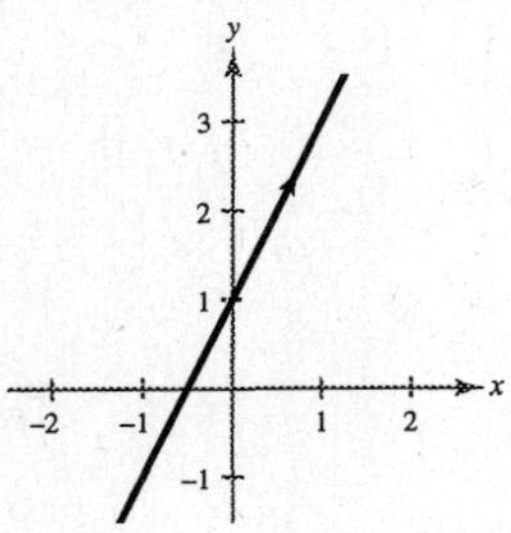

(b) $x = \cos\theta \quad \Rightarrow -1 \le x \le 1$

$y = 2\cos\theta + 1 \Rightarrow -1 \le y \le 3$

The graph oscillates.

Domain: $[-1, 1]$

Orientation: Depends on θ

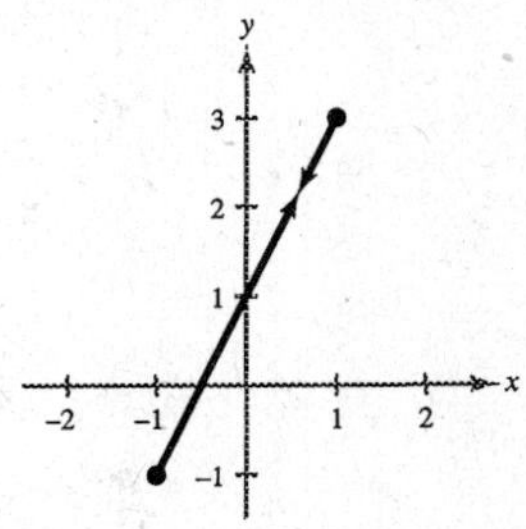

(c) $x = e^{-t} \quad \Rightarrow x > 0$

$y = 2e^{-t} + 1 \Rightarrow y > 1$

Domain: $(0, \infty)$

Orientation: Downward or right to left

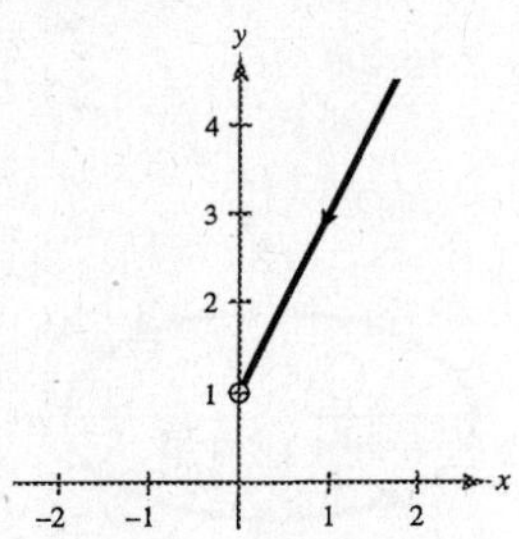

(d) $x = e^{t} \quad \Rightarrow x > 0$

$y = 2e^{t} + 1 \Rightarrow y > 1$

Domain: $(0, \infty)$

Orientation: Upward or left to right

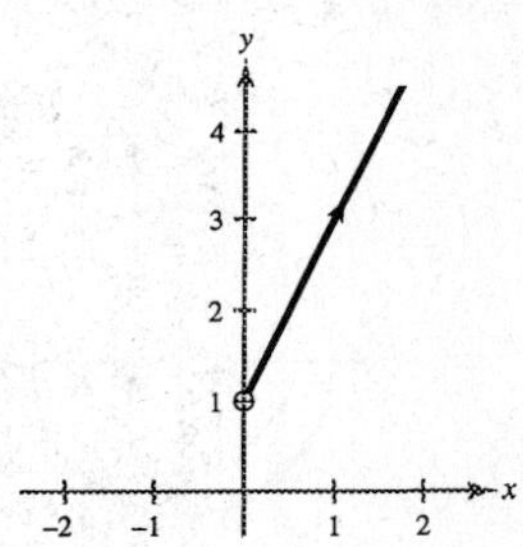

25. $x = x_1 + t(x_2 - x_1), y = y_1 + t(y_2 - y_1)$

$$\frac{x - x_1}{x_2 - x_1} = t$$

$$y = y_1 + \left(\frac{x - x_1}{x_2 - x_1}\right)(y_2 - y_1)$$

$$y - y_1 = \frac{y_2 - y_1}{x_2 - x_1}(x - x_1) = m(x - x_1)$$

27. $x = h + a\cos\theta, y = k + b\sin\theta$

$$\frac{x - h}{a} = \cos\theta, \quad \frac{y - k}{b} = \sin\theta$$

$$\frac{(x - h)^2}{a^2} + \frac{(y - k)^2}{b^2} = 1$$

29. From Exercise 25 we have:

$x = 0 + t(6 - 0) = 6t$

$y = 0 + t(-3 - 0) = -3t$

31. From Exercise 26 we have:

$x = 3 + 4\cos\theta$

$y = 2 + 4\sin\theta$

33. Vertices: $(\pm 4, 0) \Rightarrow (h, k) = (0, 0)$ and $a = 4$

Foci: $(\pm 3, 0) \Rightarrow c = 3$

$c^2 = a^2 - b^2 \Rightarrow 9 = 16 - b^2 \Rightarrow b = \sqrt{7}$

From Exercise 27 we have:

$x = 4 \cos \theta$

$y = \sqrt{7} \sin \theta$

35. Vertices: $(\pm 4, 0) \Rightarrow (h, k) = (0, 0)$ and $a = 4$

Foci: $(\pm 5, 0) \Rightarrow c = 5$

$c^2 = a^2 + b^2 \Rightarrow 25 = 16 + b^2 \Rightarrow b = 3$

From Exercise 28 we have:

$x = 4 \sec \theta$

$y = 3 \tan \theta$

37. $y = 3x - 2$

(a) $t = x \Rightarrow x = t$ and $y = 3t - 2$

(b) $t = 2 - x \Rightarrow x = -t + 2$ and $y = 3(-t + 2) - 2 = -3t + 4$

39. $y = x^2$

(a) $t = x \Rightarrow x = t$ and $y = t^2$

(b) $t = 2 - x \Rightarrow x = -t + 2$ and $y = (-t + 2)^2 = t^2 - 4t + 4$

41. $y = x^2 + 1$

(a) $t = x \Rightarrow x = t$ and $y = t^2 + 1$

(b) $t = 2 - x \Rightarrow x = -t + 2$ and $y = (-t + 2)^2 + 1 = t^2 - 4t + 5$

43. $y = \dfrac{1}{x}$

(a) $t = x \Rightarrow x = t$ and $y = \dfrac{1}{t}$

(b) $t = 2 - x \Rightarrow x = -t + 2$ and $y = \dfrac{1}{-t + 2} = \dfrac{-1}{t - 2}$

45. $x = 4(\theta - \sin \theta)$

$y = 4(1 - \cos \theta)$

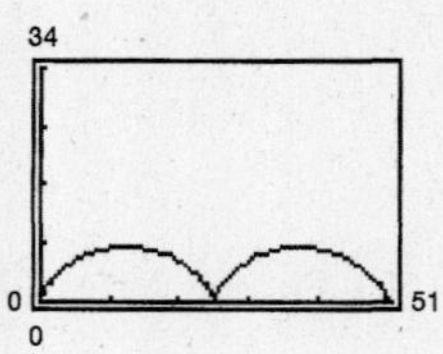

47. $x = \theta - \dfrac{3}{2} \sin \theta$

$y = 1 - \dfrac{3}{2} \cos \theta$

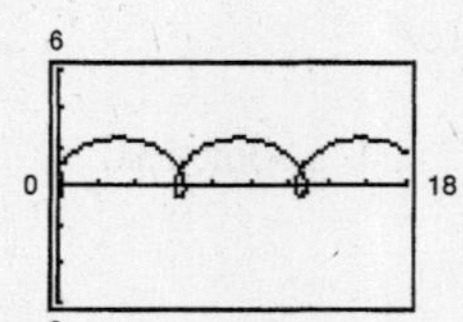

49. $x = 3 \cos^3 \theta$

$y = 3 \sin^3 \theta$

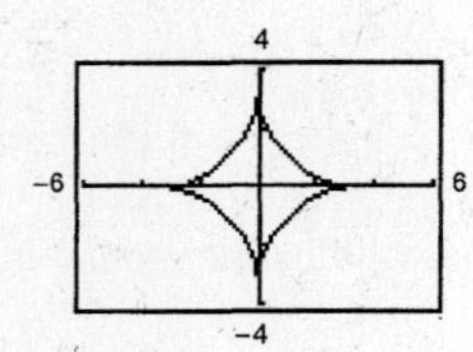

51. $x = 2 \cot \theta$

$y = 2 \sin^2 \theta$

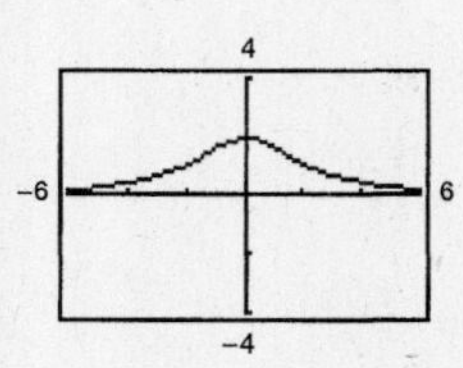

53. $x = 2 \cos\theta \Rightarrow -2 \le x \le 2$

$y = \sin 2\theta \Rightarrow -1 \le y \le 1$

Matches graph (b).

Domain: $[-2, 2]$

Range: $[-1, 1]$

55. $x = \dfrac{1}{2}(\cos \theta + \theta \sin \theta)$

$y = \dfrac{1}{2}(\sin \theta - \theta \cos \theta)$

Matches graph (d).

Domain: $(-\infty, \infty)$

Range: $(-\infty, \infty)$

57. $x = (v_0 \cos \theta)t$ and $y = h + (v_0 \sin \theta)t - 16t^2$

(a) $\theta = 60°,\ v_0 = 88$ ft/sec

$x = (88 \cos 60°)t$ and $y = (88 \sin 60°)t - 16t^2$

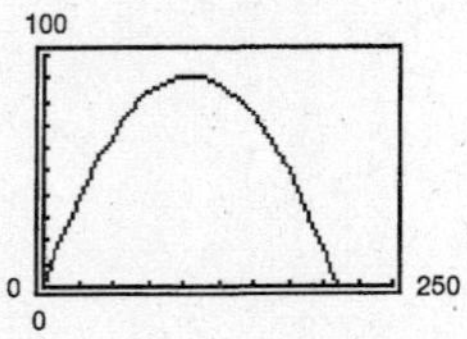

Maximum height: 90.7 feet

Range: 209.6 feet

(b) $\theta = 60°,\ v_0 = 132$ ft/sec

$x = (132 \cos 60°)t$ and $y = (132 \sin 60°)t - 16t^2$

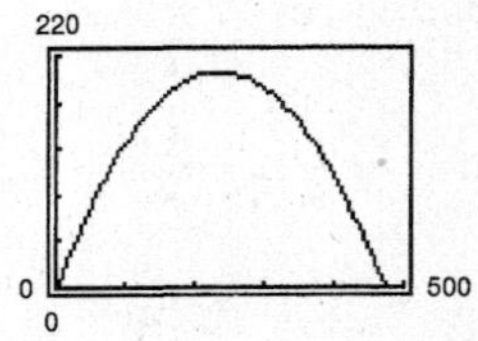

Maximum height: 204.2 feet

Range: 471.6 feet

(c) $\theta = 45°,\ v_0 = 88$ ft/sec

$x = (88 \cos 45°)t$ and $y = (88 \sin 45°)t - 16t^2$

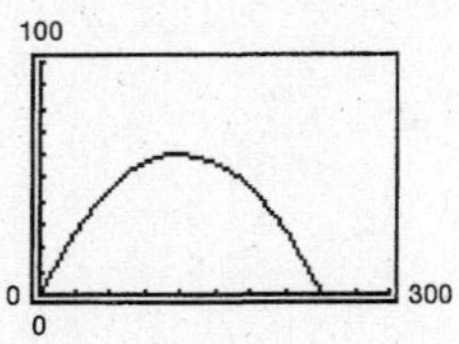

Maximum height: 60.5 ft

Range: 242.0 ft

(d) $\theta = 45°,\ v_0 = 132$ ft/sec

$x = (132 \cos 45°)t$ and $y = (132 \sin 45°)t - 16t^2$

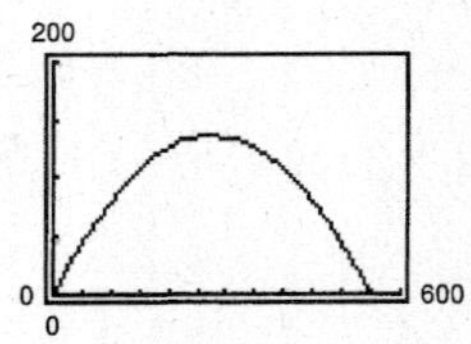

Maximum height: 136.1 ft

Range: 544.5 ft

59. (a) 100 miles per hour $= 100\left(\frac{5280}{3600}\right)$ ft/sec $= \frac{440}{3}$ ft/sec

$$x = \left(\frac{440}{3} \cos \theta\right)t \approx (146.67 \cos \theta)t$$

$$y = 3 + \left(\frac{440}{3} \sin \theta\right)t - 16t^2 \approx 3 + (146.67 \sin \theta)t - 16t^2$$

(b) For $\theta = 15°$, we have:

$$x = \left(\frac{440}{3} \cos 15°\right)t \approx 141.7t$$

$$y = 3 + \left(\frac{440}{3} \sin 15°\right)t - 16t^2 \approx 3 + 38.0t - 16t^2$$

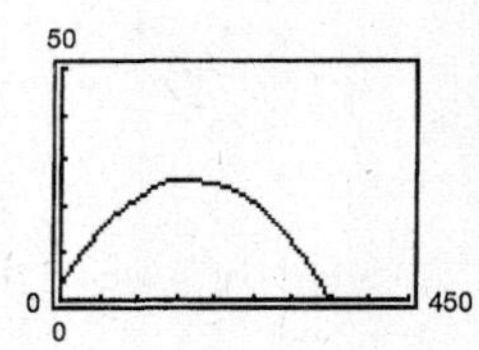

The ball hits the ground inside the ballpark, so it is not a home run.

(c) For $\theta = 23°$, we have:

$$x = \left(\frac{440}{3} \cos 23°\right)t \approx 135.0t$$

$$y = 3 + \left(\frac{440}{3} \sin 23°\right)t - 16t^2 \approx 3 + 57.3t - 16t^2$$

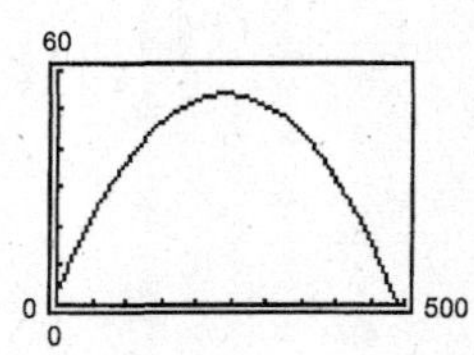

The ball easily clears the 7-foot fence at 408 feet so it is a home run.

(d) Find θ so that $y = 7$ when $x = 408$ by graphing the parametric equations for θ values between 15° and 23°. This occurs when $\theta \approx 19.3°$.

61. $x = (v_0 \cos \theta)t \Rightarrow t = \dfrac{x}{v_0 \cos \theta}$

$$\begin{aligned} y &= h + (v_0 \sin \theta)t - 16t^2 \\ &= h + (v_0 \sin \theta)\left(\frac{x}{v_0 \cos \theta}\right) - 16\left(\frac{x}{v_0 \cos \theta}\right)^2 \\ &= h + (\tan \theta)x - \frac{16x^2}{v_0^2 \cos^2 \theta} \\ &= -\frac{16 \sec^2 \theta}{v_0^2}x^2 + (\tan \theta)x + h \end{aligned}$$

63. When the circle has rolled θ radians, the center is at $(a\theta, a)$.

$$\begin{aligned} \sin \theta &= \sin(180° - \theta) \\ &= \frac{|AC|}{b} = \frac{|BD|}{b} \Rightarrow |BD| = b \sin \theta \end{aligned}$$

$$\begin{aligned} \cos \theta &= -\cos(180° - \theta) \\ &= \frac{|AP|}{-b} \Rightarrow |AP| = -b \cos \theta \end{aligned}$$

Therefore, $x = a\theta - b \sin \theta$ and $y = a - b \cos \theta$.

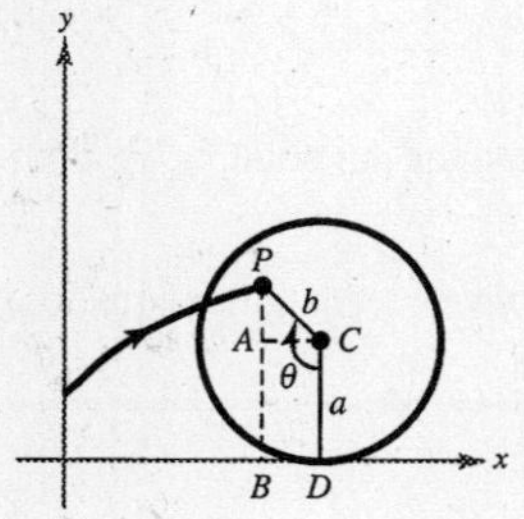

65. True

$x = t$

$y = t^2 + 1 \Rightarrow y = x^2 + 1$

$x = 3t$

$y = 9t^2 + 1 \Rightarrow y = x^2 + 1$

67. $\theta = 105°$

$\theta' = 180° - 105° = 75°$

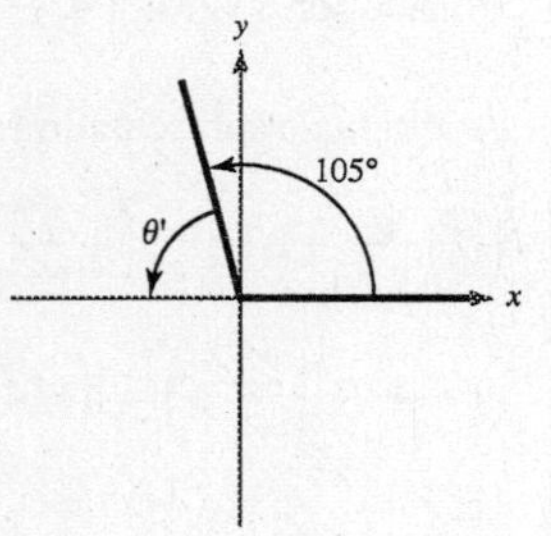

69. $\theta = -\dfrac{2\pi}{3}$

$\theta' = -\dfrac{2\pi}{3} + \pi = \dfrac{\pi}{3}$

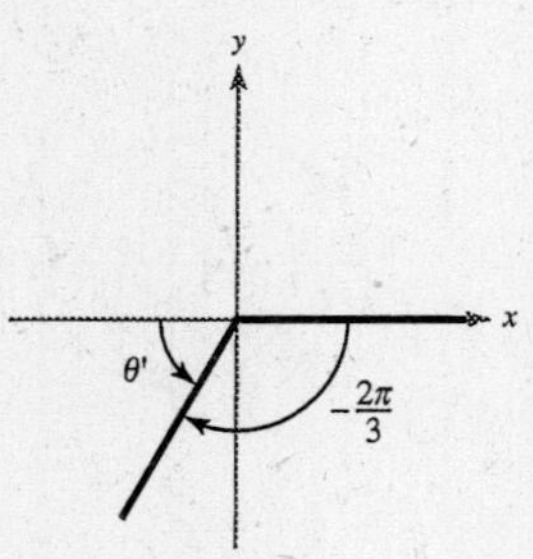

71. $y = \arcsin(x + 1)$

This is the graph of $y = \arcsin x$ shifted to the left one unit

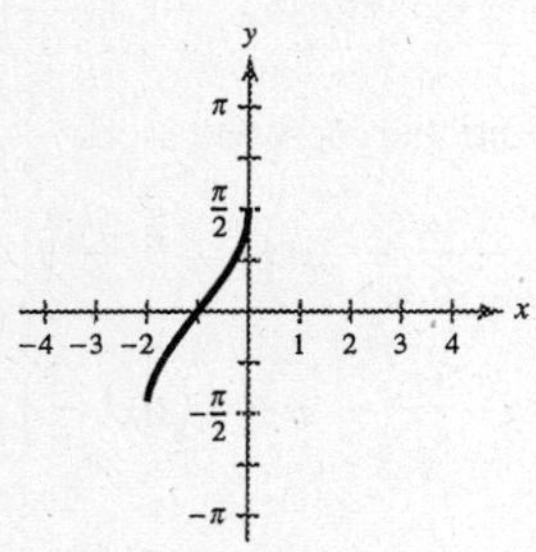

73. $f(x) = 2 \arctan x$

Horizontal asymptotes at $y = \pm\pi$

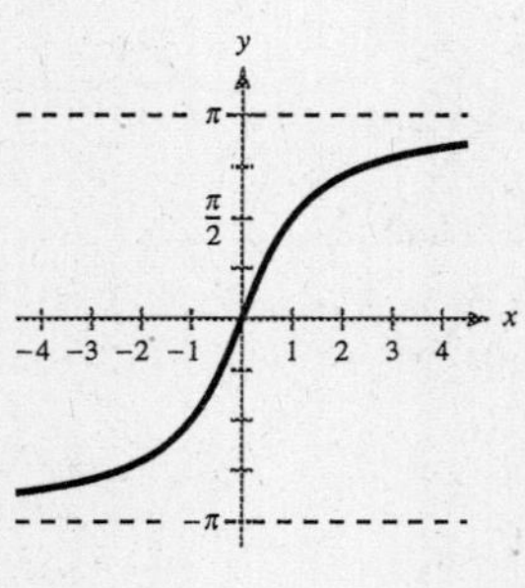

Section 6.7 Polar Coordinates

- In polar coordinates you do not have unique representation of points. The point (r, θ) can be represented by $(r, \theta \pm 2n\pi)$ or by $(-r, \theta \pm (2n + 1)\pi)$ where n is any integer. The pole is represented by $(0, \theta)$ where θ is any angle.
- To convert from polar coordinates to rectangular coordinates, use the following relationships.
  $x = r \cos \theta$
  $y = r \sin \theta$
- To convert from rectangular coordinates to polar coordinates, use the following relationships.
  $r = \pm\sqrt{x^2 + y^2}$
  $\tan \theta = y/x$
  If θ is in the same quadrant as the point (x, y), then r is positive. If θ is in the opposite quadrant as the point (x, y), then r is negative.
- You should be able to convert rectangular equations to polar form and vice versa.

Solutions to Odd-Numbered Exercises

1. Polar Coordinates: $\left(4, -\frac{\pi}{3}\right)$

Additional representations

$$\left(4, -\frac{\pi}{3} + 2\pi\right) = \left(4, \frac{5\pi}{3}\right)$$

$$\left(-4, -\frac{\pi}{3} - \pi\right) = \left(-4, -\frac{4\pi}{3}\right)$$

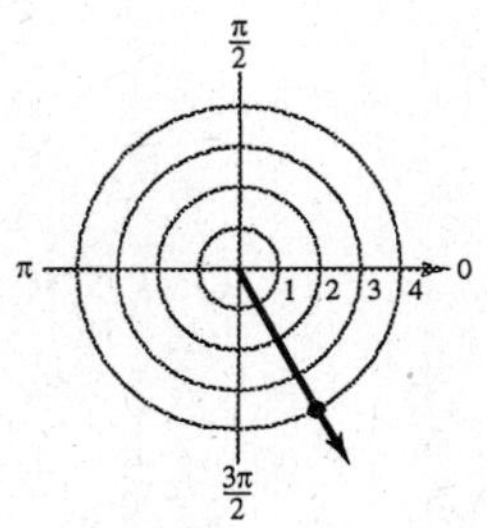

3. Polar Coordinates: $\left(0, -\frac{7\pi}{6}\right)$

Additional representations

$$\left(0, -\frac{7\pi}{6} + 2\pi\right) = \left(0, \frac{5\pi}{6}\right)$$

$$\left(0, -\frac{7\pi}{6} + \pi\right) = \left(0, -\frac{\pi}{6}\right) \text{ or } \left(0, -\frac{13\pi}{6}\right) \text{ or } (0, \theta) \text{ for any } \theta$$

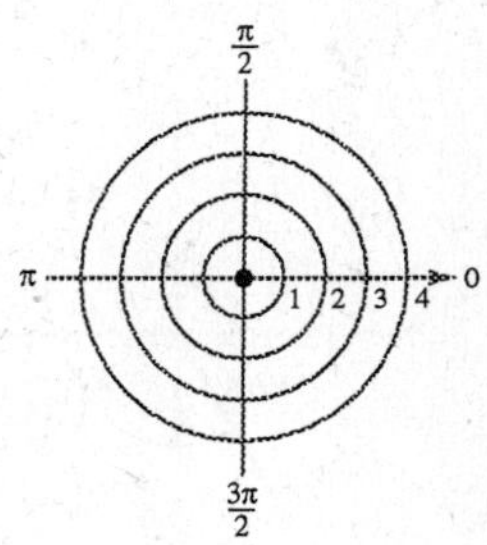

5. Polar Coordinates: $\left(\sqrt{2}, 2.36\right)$

Additional representations

$$\left(\sqrt{2}, 2.36 + 2\pi\right) \approx \left(\sqrt{2}, 8.64\right)$$

$$\left(-\sqrt{2}, 2.36 - \pi\right) \approx \left(-\sqrt{2}, -0.78\right)$$

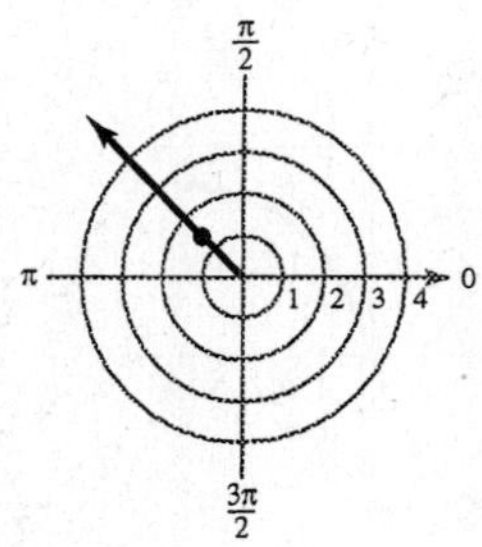

7. Polar Coordinates: $(2\sqrt{2}, 4.71)$

Additional representations

$(2\sqrt{2}, 4.71 + 2\pi) \approx (2\sqrt{2}, 10.99)$

$(-2\sqrt{2}, 4.71 + \pi) \approx (-2\sqrt{2}, 7.85)$

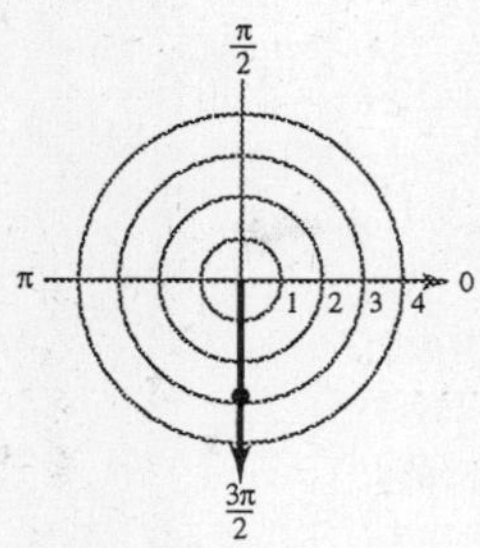

9. Polar Coordinates: $\left(3, \frac{\pi}{2}\right)$

$x = 3\cos\frac{\pi}{2} = 0$

$y = 3\sin\frac{\pi}{2} = 3$

Rectangular Coordinates: $(0, 3)$

11. Polar Coordinates: $\left(-1, \frac{5\pi}{4}\right)$

$x = -1\cos\left(\frac{5\pi}{4}\right) = \frac{\sqrt{2}}{2}, y = -1\sin\left(\frac{5\pi}{4}\right) = \frac{\sqrt{2}}{2}$

Rectangular Coordinates: $\left(\frac{\sqrt{2}}{2}, \frac{\sqrt{2}}{2}\right)$

13. Polar Coordinates: $\left(2, \frac{3\pi}{4}\right)$

$x = 2\cos\frac{3\pi}{4} = -\sqrt{2}$

$y = 2\sin\frac{3\pi}{4} = \sqrt{2}$

Rectangular Coordinates: $(-\sqrt{2}, \sqrt{2})$

15. Polar Coordinates: $(-2.5, 1.1)$

$x = -2.5\cos 1.1 \approx -1.1340$

$y = -2.5\sin 1.1 \approx -2.2280$

Rectangular Coordinates: $(-1.1340, -2.2280)$

17. Rectangular Coordinates: $(1, 1)$

$r = \pm\sqrt{2}, \tan\theta = 1, \theta = \frac{\pi}{4} \text{ or } \frac{5\pi}{4}$

Polar Coordinates: $\left(\sqrt{2}, \frac{\pi}{4}\right), \left(-\sqrt{2}, \frac{5\pi}{4}\right)$

19. Rectangular Coordinates: $(-6, 0)$

$r = \pm 6, \tan\theta = 0, \theta = 0 \text{ or } \pi$

Polar Coordinates: $(6, \pi), (-6, 0)$

21. Rectangular Coordinates: $(-3, 4)$

$r = \pm\sqrt{9 + 16} = \pm 5, \tan\theta = -\frac{4}{3}, \theta \approx 2.2143, 5.3559$

Polar Coordinates: $(5, 2.2143), (-5, 5.3559)$

23. Rectangular Coordinates: $(-\sqrt{3}, -\sqrt{3})$

$r = \pm\sqrt{3 + 3} = \pm\sqrt{6}, \tan\theta = 1, \theta = \frac{\pi}{4} \text{ or } \frac{5\pi}{4}$

Polar Coordinates: $\left(\sqrt{6}, \frac{5\pi}{4}\right), \left(-\sqrt{6}, \frac{\pi}{4}\right)$

25. Rectangular Coordinates: $(6, 9)$

$r = \pm\sqrt{6^2 + 9^2} = \pm\sqrt{117} = \pm 3\sqrt{13}$

$\tan\theta = \frac{9}{6}, \theta \approx 0.9828, 4.1244$

Polar Coordinates: $(3\sqrt{13}, 0.9828), (-3\sqrt{13}, 4.1244)$

27. Rectangular: $(3, -2)$

$(3, -2)$ ▶ Pol

$\approx (3.606, -0.5880)$

or $(\sqrt{13}, -0.5880)$

or $(\sqrt{13}, 5.6952)$

29. Rectangular: $(\sqrt{3}, 2)$

$(\sqrt{3}, 2)$ ▶ Pol

$\approx (2.646, 0.8571)$

or $(\sqrt{7}, 0.8571)$

31. Rectangular: $\left(\frac{5}{2}, \frac{4}{3}\right)$

$\left(\frac{5}{2}, \frac{4}{3}\right)$ ▶ Pol

$\approx (2.833, 0.4900)$

or $\left(\frac{17}{6}, 0.4900\right)$

33. $x^2 + y^2 = 9$

$r = 3$

35. $y = 4$

$r \sin \theta = 4$

$r = 4 \csc \theta$

37. $x = 10$

$r \cos \theta = 10$

$r = 10 \sec \theta$

39. $3x - y + 2 = 0$

$3r \cos \theta - r \sin \theta + 2 = 0$

$r(3 \cos \theta - \sin \theta) = -2$

$$r = \frac{-2}{3 \cos \theta - \sin \theta}$$

41. $xy = 16$

$(r \cos \theta)(r \sin \theta) = 16$

$r^2 = 16 \sec \theta \csc \theta = 32 \csc 2\theta$

43. $y^2 - 8x - 16 = 0$

$r^2 \sin^2 \theta - 8r \cos \theta - 16 = 0$

By the Quadratic Formula, we have:

$$r = \frac{-(-8 \cos \theta) \pm \sqrt{(-8 \cos \theta)^2 - 4(\sin^2 \theta)(-16)}}{2 \sin^2 \theta}$$

$$= \frac{8 \cos \theta \pm \sqrt{64 \cos^2 \theta + 64 \sin^2 \theta}}{2 \sin^2 \theta}$$

$$= \frac{8 \cos \theta \pm \sqrt{64 (\cos^2 \theta + \sin^2 \theta)}}{2 \sin^2 \theta}$$

$$= \frac{8 \cos \theta \pm 8}{2 \sin^2 \theta}$$

$$= \frac{4(\cos \theta \pm 1)}{1 - \cos^2 \theta}$$

$$r = \frac{4(\cos \theta + 1)}{(1 + \cos \theta)(1 - \cos \theta)} = \frac{4}{1 - \cos \theta}$$

or

$$r = \frac{4(\cos \theta - 1)}{(1 + \cos \theta)(1 - \cos \theta)} = \frac{-4}{1 + \cos \theta}$$

45. $x^2 + y^2 = a^2$

$r^2 = a^2$

$r = a$

47. $x^2 + y^2 - 2ax = 0$

$r^2 - 2a\, r \cos \theta = 0$

$r(r - 2a \cos \theta) = 0$

$r - 2a \cos \theta = 0$

$r = 2a \cos \theta$

49. $r = 4 \sin \theta$

$r^2 = 4r \sin \theta$

$x^2 + y^2 = 4y$

$x^2 + y^2 - 4y = 0$

51. $\theta = \frac{2\pi}{3}$

$\tan \theta = \tan \frac{2\pi}{3}$

$\frac{y}{x} = -\sqrt{3}$

$y = -\sqrt{3}x$

$\sqrt{3}x + y = 0$

53.
$$\begin{aligned} r &= 4 \\ r^2 &= 16 \\ x^2 + y^2 &= 16 \end{aligned}$$

55.
$$\begin{aligned} r &= 4 \csc \theta \\ r \sin \theta &= 4 \\ y &= 4 \end{aligned}$$

57.
$$\begin{aligned} r^2 &= \cos \theta \\ r^3 &= r \cos \theta \\ \left(\pm\sqrt{x^2 + y^2}\right)^3 &= x \\ \pm\left(x^2 + y^2\right)^{3/2} &= x \\ (x^2 + y^2)^3 &= x^2 \\ x^2 + y^2 &= x^{2/3} \\ x^2 + y^2 - x^{2/3} &= 0 \end{aligned}$$

59.
$$\begin{aligned} r &= 2 \sin 3\theta \\ r &= 2 \sin(\theta + 2\theta) \\ r &= 2[\sin \theta \cos 2\theta + \cos \theta \sin 2\theta] \\ r &= 2[\sin \theta(1 - 2 \sin^2 \theta) + \cos \theta(2 \sin \theta \cos \theta)] \\ r &= 2[\sin \theta - 2 \sin^3 \theta + 2 \sin \theta \cos^2 \theta] \\ r &= 2[\sin \theta - 2 \sin^3 \theta + 2 \sin \theta(1 - \sin^2 \theta)] \\ r &= 2(3 \sin \theta - 4 \sin^3 \theta) \\ r^4 &= 6r^3 \sin \theta - 8r^3 \sin^3 \theta \\ \left(x^2 + y^2\right)^2 &= 6\left(x^2 + y^2\right)y - 8y^3 \\ \left(x^2 + y^2\right)^2 &= 6x^2y - 2y^3 \end{aligned}$$

61.
$$\begin{aligned} r &= \frac{2}{1 + \sin \theta} \\ r(1 + \sin \theta) &= 2 \\ r + r \sin \theta &= 2 \\ r &= 2 - r \sin \theta \\ \pm\sqrt{x^2 + y^2} &= 2 - y \\ x^2 + y^2 &= (2 - y)^2 \\ x^2 + y^2 &= 4 - 4y + y^2 \\ x^2 + 4y - 4 &= 0 \end{aligned}$$

63.
$$\begin{aligned} r &= \frac{6}{2 - 3 \sin \theta} \\ r(2 - 3 \sin \theta) &= 6 \\ 2r &= 6 + 3r \sin \theta \\ 2\left(\pm\sqrt{x^2 + y^2}\right) &= 6 + 3y \\ 4\left(x^2 + y^2\right) &= (6 + 3y)^2 \\ 4x^2 + 4y^2 &= 36 + 36y + 9y^2 \\ 4x^2 - 5y^2 - 36y - 36 &= 0 \end{aligned}$$

65.
$$\begin{aligned} r &= 6 \\ r^2 &= 36 \\ x^2 + y^2 &= 36 \end{aligned}$$

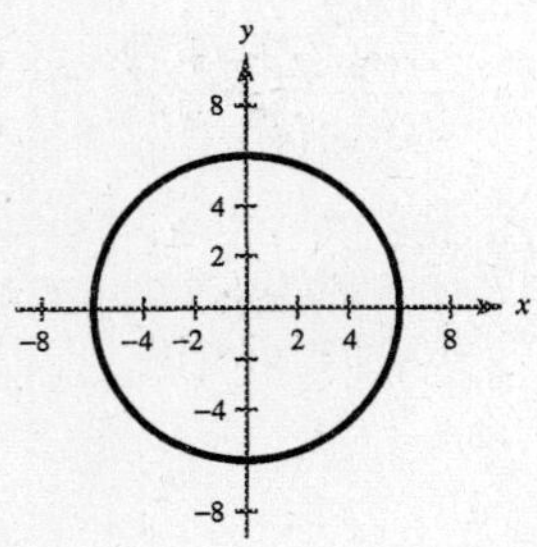

67.
$$\begin{aligned} \theta &= \frac{\pi}{6} \\ \tan \theta &= \tan \frac{\pi}{6} \\ \frac{y}{x} &= \frac{\sqrt{3}}{3} \\ y &= \frac{\sqrt{3}}{3}x \\ 3y &= \sqrt{3}x \\ -\sqrt{3}x + 3y &= 0 \end{aligned}$$

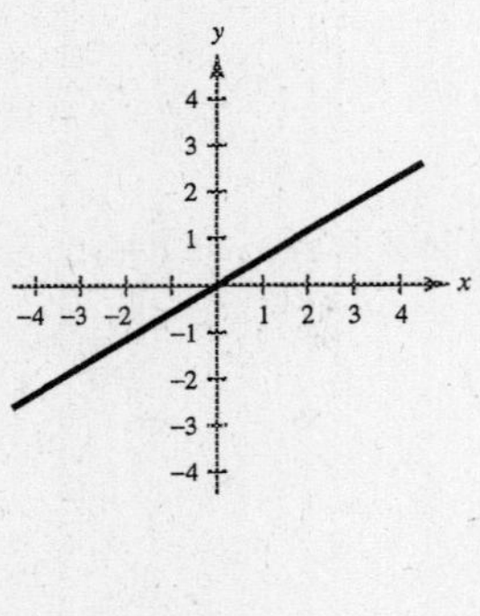

69.
$$\begin{aligned} r &= 3 \sec \theta \\ r \cos\theta &= 3 \\ x &= 3 \\ x - 3 &= 0 \end{aligned}$$

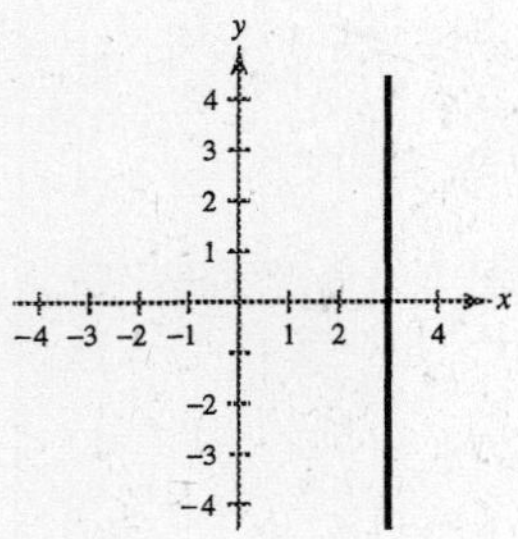

71. True. Because r is a directed distance, then the point (r, θ) can be represented as $(r, \theta \pm 2n\pi)$.

73.

$$r = 2(h\cos\theta + k\sin\theta)$$
$$r = 2\left(h\left(\frac{x}{r}\right) + k\left(\frac{y}{r}\right)\right)$$
$$r = \frac{2hx + 2ky}{r}$$
$$r^2 = 2hx + 2ky$$
$$x^2 + y^2 = 2hx + 2ky$$
$$x^2 - 2hx + y^2 - 2ky = 0$$
$$\left(x^2 - 2hx + h^2\right) + \left(y^2 - 2ky + k^2\right) = h^2 + k^2$$
$$(x - h)^2 + (y - k)^2 = h^2 + k^2$$

Center: (h, k)

Radius: $\sqrt{h^2 + k^2}$

75. (a) $(r_1, \theta_1) = (x_1, y_1)$ where $x_1 = r_1\cos\theta_1$ and $y_1 = r_1\sin\theta_1$.

$(r_2, \theta_2) = (x_2, y_2)$ where $x_2 = r_2\cos\theta_2$ and $y_2 = r_2\sin\theta_2$.

$$\begin{aligned} d &= \sqrt{(x_1 - x_2)^2 + (y_1 - y_2)^2} \\ &= \sqrt{x_1^2 - 2x_1x_2 + x_2^2 + y_1^2 - 2y_1y_2 + y_2^2} \\ &= \sqrt{(x_1^2 + y_1^2) + (x_2^2 + y_2^2) - 2(x_1x_2 + y_1y_2)} \\ &= \sqrt{r_1^2 + r_2^2 - 2(r_1r_2\cos\theta_1\cos\theta_2 + r_1r_2\sin\theta_1\sin\theta_2)} \\ &= \sqrt{r_1^2 + r_2^2 - 2r_1r_2\cos(\theta_1 - \theta_2)} \end{aligned}$$

(b) If $\theta_1 = \theta_2$, then

$$\begin{aligned} d &= \sqrt{r_1^2 + r_2^2 - 2r_1r_2} \\ &= \sqrt{(r_1 - r_2)^2} \\ &= |r_1 - r_2|. \end{aligned}$$

This represents the distance between two points on the line $\theta = \theta_1 = \theta_2$.

(c) If $\theta_1 - \theta_2 = 90°$, then $d = \sqrt{r_1^2 + r_2^2}$. This is the result of the Pythagorean Theorem.

(d) The results should be the same. For example, use the points

$$\left(3, \frac{\pi}{6}\right) \text{ and } \left(4, \frac{\pi}{3}\right).$$

The distance is $d \approx 2.053$. Now use the representations

$$\left(-3, \frac{7\pi}{6}\right) \text{ and } \left(-4, \frac{4\pi}{3}\right).$$

The distance is still $d \approx 2.053$.

77. $$\begin{aligned} \log_6 \frac{x^2z}{3y} &= \log_6 x^2z - \log_6 3y \\ &= \log_6 x^2 + \log_6 z - (\log_6 3 + \log_6 y) \\ &= 2\log_6 x + \log_6 z - \log_6 3 - \log_6 y \end{aligned}$$

79. $$\begin{aligned} \ln x(x + 4)^2 &= \ln x + \ln(x + 4)^2 \\ &= \ln x + 2\ln(x + 4) \end{aligned}$$

81. $\log_7 x - \log_7 3y = \log_7 \dfrac{x}{3y}$

83. $$\begin{aligned} \frac{1}{2}\ln x + \ln(x - 2) &= \ln\sqrt{x} + \ln(x - 2) \\ &= \ln\sqrt{x}\,(x - 2) \end{aligned}$$

Section 6.8 Graphs of Polar Equations

- When graphing polar equations:
  1. Test for symmetry.
     (a) $\theta = \pi/2$: Replace (r, θ) by $(r, \pi - \theta)$ or $(-r, -\theta)$.
     (b) Polar axis: Replace (r, θ) by $(r, -\theta)$ or $(-r, \pi - \theta)$.
     (c) Pole: Replace (r, θ) by $(r, \pi + \theta)$ or $(-r, \theta)$.
     (d) $r = f(\sin\theta)$ is symmetric with respect to the line $\theta = \pi/2$.
     (e) $r = f(\cos\theta)$ is symmetric with respect to the polar axis.
  2. Find the θ values for which $|r|$ is maximum.
  3. Find the θ values for which $r = 0$.
  4. Know the different types of polar graphs.
     (a) Limaçons $(0 < a, 0 < b)$
     $r = a \pm b\cos\theta$
     $r = a \pm b\sin\theta$
     (b) Rose Curves, $n \geq 2$
     $r = a\cos n\theta$
     $r = a\sin n\theta$
     (c) Circles
     $r = a\cos\theta$
     $r = a\sin\theta$
     $r = a$
     (d) Lemniscates
     $r^2 = a^2\cos 2\theta$
     $r^2 = a^2\sin 2\theta$
  5. Plot additional points.

Solutions to Odd-Numbered Exercises

1. $r = 3\cos 2\theta$

Rose curve with 4 petals

3. $r = 3(1 - 2\cos\theta)$

Limaçon with inner loop

5. $r = 6\sin 2\theta$

Rose curve with 4 petals

7. $r = 5 + 4\cos\theta$

$\theta = \dfrac{\pi}{2}$: $-r = 5 + 4\cos(-\theta)$

$-r = 5 + 4\cos\theta$

Not an equivalent equation

Polar axis: $r = 5 + 4\cos(-\theta)$

$r = 5 + 4\cos\theta$

Equivalent equation

Pole: $-r = 5 + 4\cos\theta$

Not an equivalent equation

Answer: Symmetric with respect to polar axis

9. $r = \dfrac{2}{1 + \sin\theta}$

$\theta = \dfrac{\pi}{2}$: $r = \dfrac{2}{1 + \sin(\pi - \theta)}$

$r = \dfrac{2}{1 + \sin\pi\cos\theta - \cos\pi\sin\theta}$

$r = \dfrac{2}{1 + \sin\theta}$

Equivalent equation

Polar axis: $r = \dfrac{2}{1 + \sin(-\theta)}$

$r = \dfrac{2}{1 - \sin\theta}$

Not an equivalent equation

Pole: $-r = \dfrac{2}{1 + \sin\theta}$

Answer: Symmetric with respect to $\theta = \pi/2$

11. $r^2 = 16 \cos 2\theta$

$\theta = \frac{\pi}{2}$: $(-r)^2 = 16 \cos 2(-\theta)$

$r^2 = 16 \cos 2\theta$

Equivalent equation

Polar axis: $r^2 = 16 \cos 2(-\theta)$

$r^2 = 16 \cos 2\theta$

Equivalent equation

Pole: $(-r)^2 = 16 \cos 2\theta$

$r^2 = 16 \cos 2\theta$

Equivalent equation

Answer: Symmetric with respect to $\theta = \frac{\pi}{2}$, the polar axis, and the pole

13. $|r| = |10(1 - \sin \theta)| = 10|1 - \sin \theta| \le 10(2) = 20$

$|1 - \sin \theta| = 2$

$1 - \sin \theta = 2$ or $1 - \sin \theta = -2$

$\sin \theta = -1$ $\qquad$ $\sin \theta = 3$

$\theta = \frac{3\pi}{2}$ $\qquad$ Not possible

Maximum: $|r| = 20$ when $\theta = \frac{3\pi}{2}$.

$0 = 10(1 - \sin \theta)$

$\sin \theta = 1$

$\theta = \frac{\pi}{2}$

Zero: $r = 0$ when $\theta = \frac{\pi}{2}$.

15. $|r| = |4 \cos 3\theta| = 4|\cos 3\theta| \le 4$

$|\cos 3\theta| = 1$

$\cos 3\theta = \pm 1$

$\theta = 0, \frac{\pi}{3}, \frac{2\pi}{3}$

Maximum: $|r| = 4$ when $\theta = 0, \frac{\pi}{3}, \frac{2\pi}{3}$.

$0 = 4 \cos 3\theta$

$\cos 3\theta = 0$

$\theta = \frac{\pi}{6}, \frac{\pi}{2}, \frac{5\pi}{6}$

Zero: $r = 0$ when $\theta = \frac{\pi}{6}, \frac{\pi}{2}, \frac{5\pi}{6}$.

17. Circle: $r = 5$

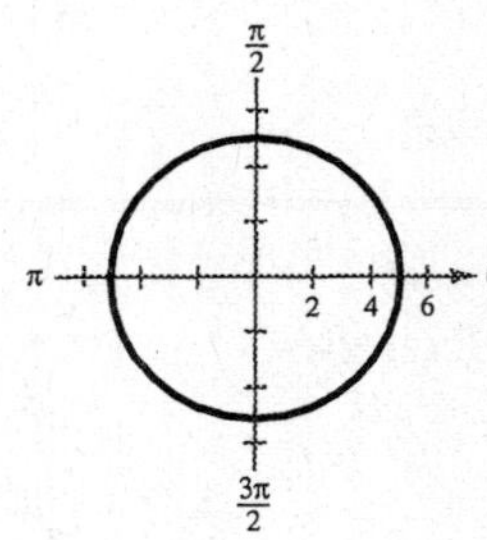

19. Circle: $r = \frac{\pi}{6}$

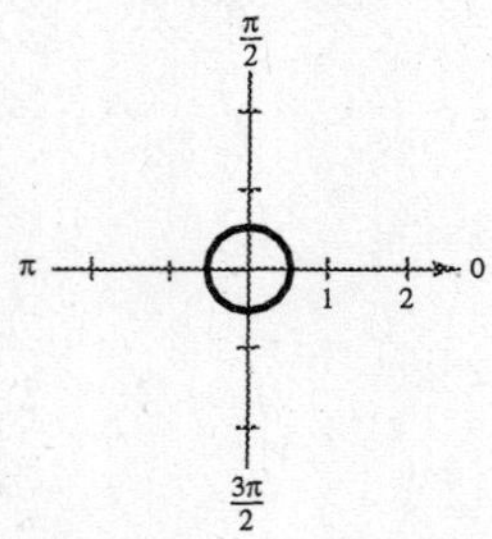

21. $r = 3 \sin \theta$

Symmetric with respect to $\theta = \frac{\pi}{2}$

Circle with a radius of $\frac{3}{2}$

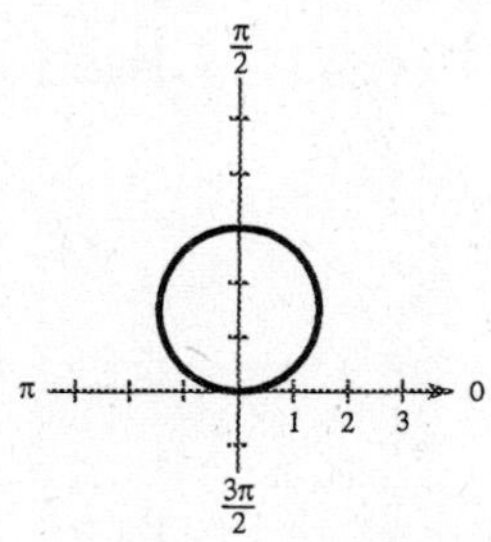

23. $r = 3(1 - \cos \theta)$

Symmetric with respect to the polar axis

$\frac{a}{b} = \frac{3}{3} = 1 \implies$ Cardioid

$|r| = 6$ when $\theta = \pi$.

$r = 0$ when $\pi = 0$.

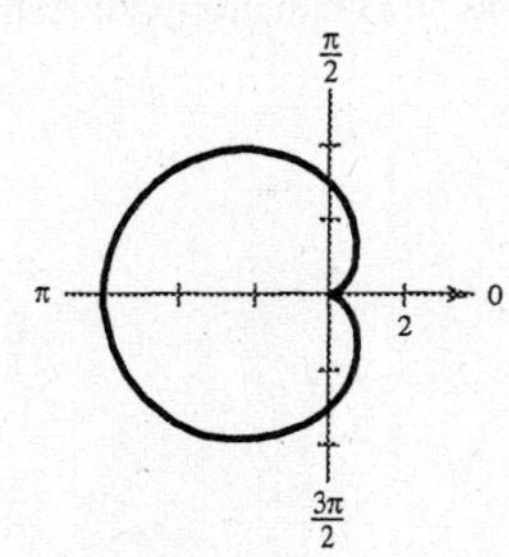

25. $r = 4(1 + \sin\theta)$

Symmetric with respect to $\theta = \dfrac{\pi}{2}$

$\dfrac{a}{b} = \dfrac{4}{4} = 1 \Rightarrow$ Cardioid

$|r| = 8$ when $\theta = \dfrac{\pi}{2}$.

$r = 0$ when $\theta = \dfrac{3\pi}{2}$.

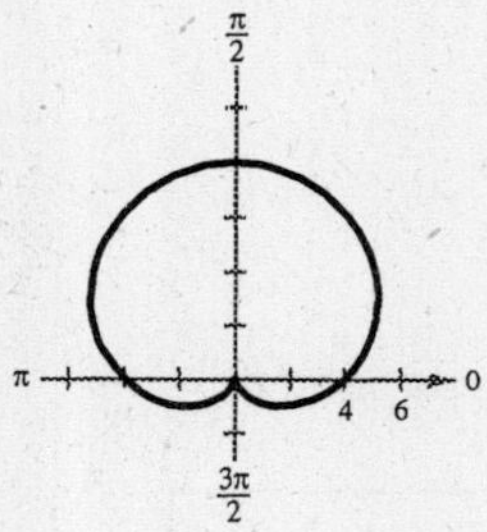

27. $r = 3 + 6\sin\theta$

Symmetric with respect to $\theta = \dfrac{\pi}{2}$

$\dfrac{a}{b} = \dfrac{3}{6} < 1 \Rightarrow$ Limaçon with inner loop

$|r| = 9$ when $\theta = \dfrac{\pi}{2}$

$r = 0$ when $\theta = \dfrac{7\pi}{6}, \dfrac{11\pi}{6}$

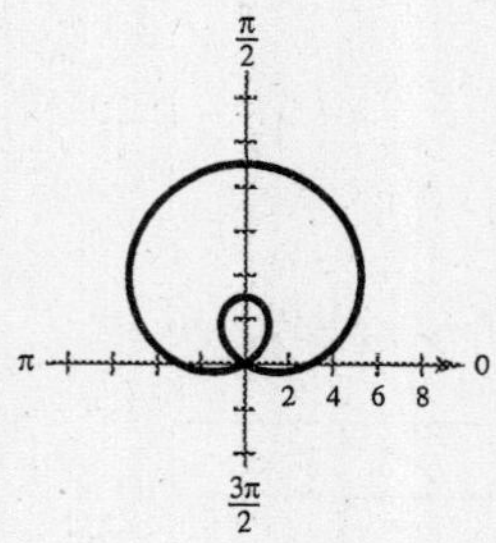

29. $r = 1 - 2\sin\theta$

Symmetric with respect to $\theta = \dfrac{\pi}{2}$

$\dfrac{a}{b} = \dfrac{1}{2} < 1 \Rightarrow$ Limaçon with inner loop

$|r| = 3$ when $\theta = \dfrac{3\pi}{2}$

$r = 0$ when $\theta = \dfrac{\pi}{6}, \dfrac{5\pi}{6}$

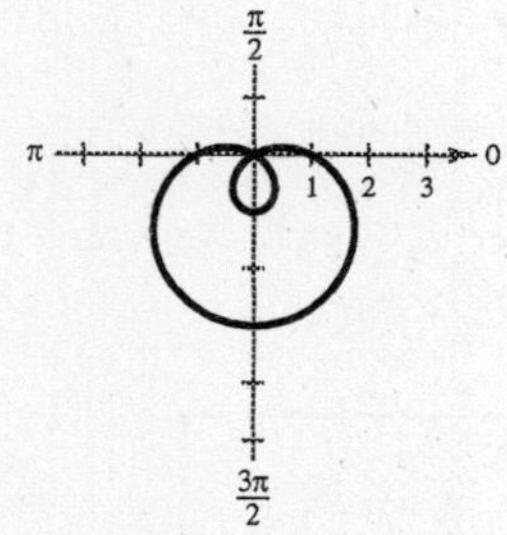

31. $r = 3 - 4\cos\theta$

Symmetric with respect to the polar axis

$\dfrac{a}{b} = \dfrac{3}{4} < 1 \Rightarrow$ Limaçon with inner loop

$|r| = 7$ when $\theta = \pi$.

$r = 0$ when $\cos\theta = \dfrac{3}{4}$ or

$\theta \approx 0.723, 5.560$

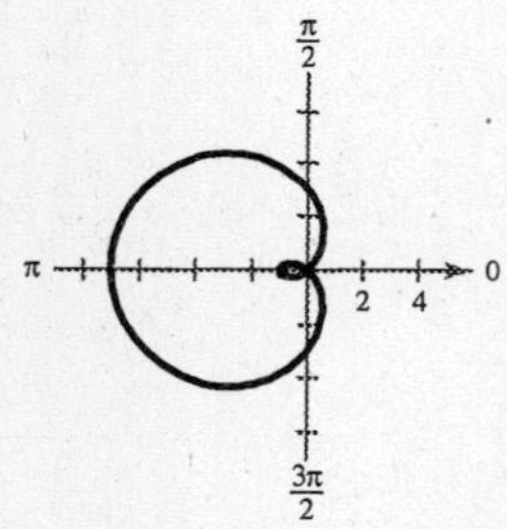

33. $r = 5\sin 2\theta$

Symmetric with respect to $\theta = \dfrac{\pi}{2}$, the polar axis, and the pole.

Rose curve $(n = 2)$ with 4 petals

$|r| = 5$ when $\theta = \dfrac{\pi}{4}, \dfrac{3\pi}{4}, \dfrac{5\pi}{4}, \dfrac{7\pi}{4}$.

$r = 0$ when $\theta = 0, \dfrac{\pi}{2}, \pi$.

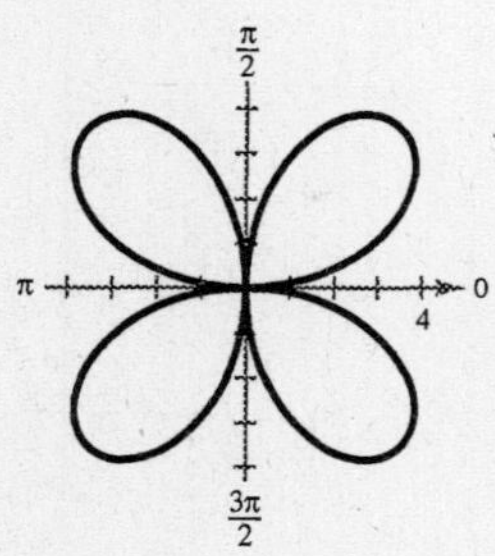

35.

$r = 2\sec\theta$

$r = \dfrac{2}{\cos\theta}$

$r\cos\theta = 2$

$x = 2 \Rightarrow$ Line

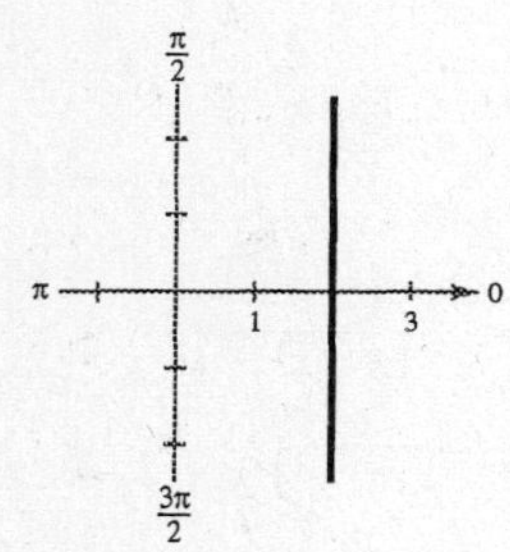

37. $$r = \frac{3}{\sin\theta - 2\cos\theta}$$

$$r(\sin\theta - 2\cos\theta) = 3$$

$$y - 2x = 3$$

$$y = 2x + 3 \implies \text{Line}$$

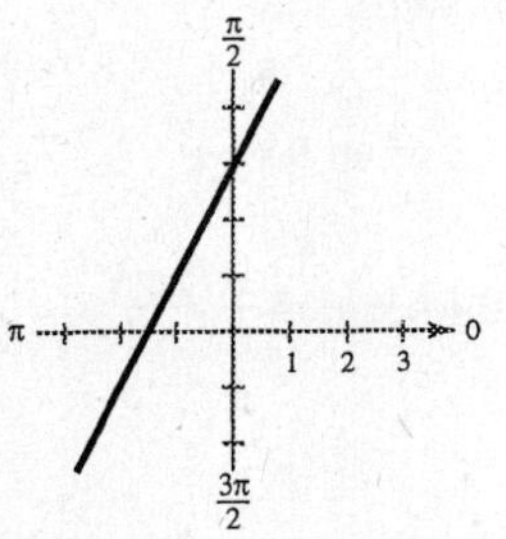

39. $r^2 = 9\cos 2\theta$

Symmetric with respect to the polar axis, $\theta = \frac{\pi}{2}$, and the pole

Lemniscate

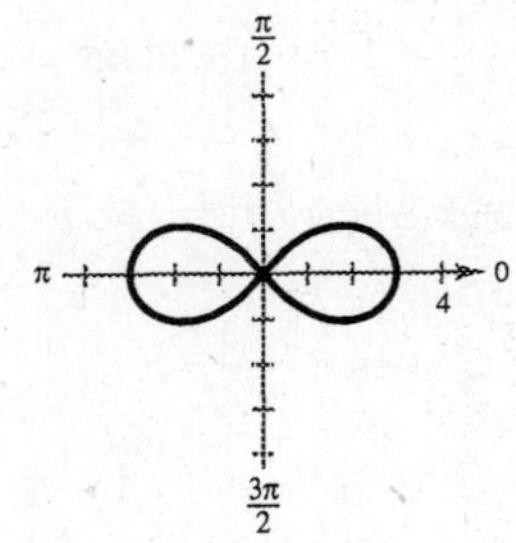

41. $r = 8\cos\theta$

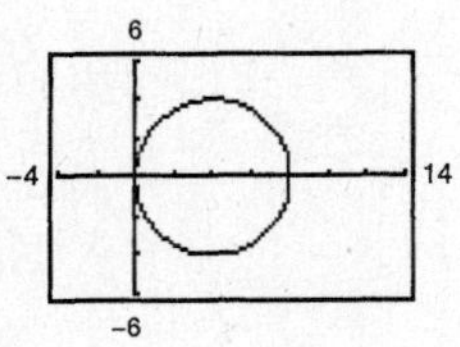

43. $r = 3(2 - \sin\theta)$

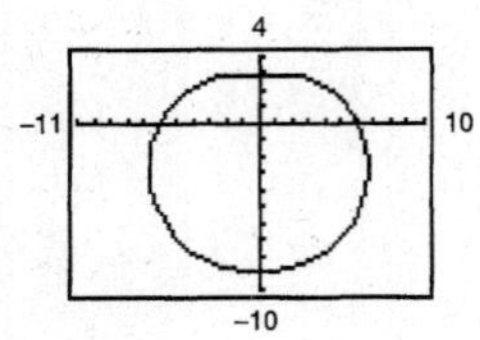

45. $r = 8\sin\theta\cos^2\theta$

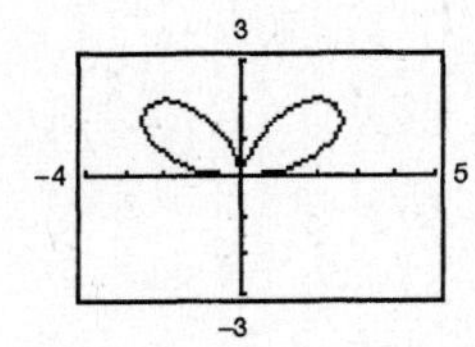

47. $r = 3 - 4\cos\theta$

$0 \le \theta < 2\pi$

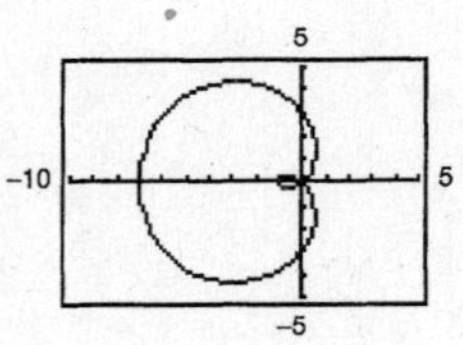

49. $r = 2\cos\left(\frac{3\theta}{2}\right)$

$0 \le \theta < 4\pi$

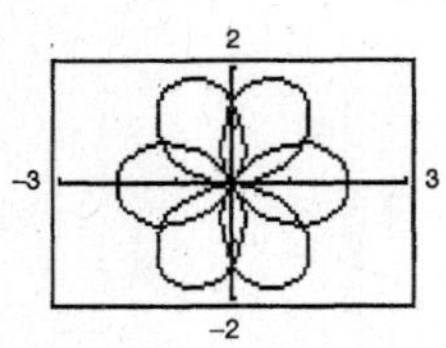

51. $r^2 = 9\sin 2\theta$

$0 \le \theta < \pi$

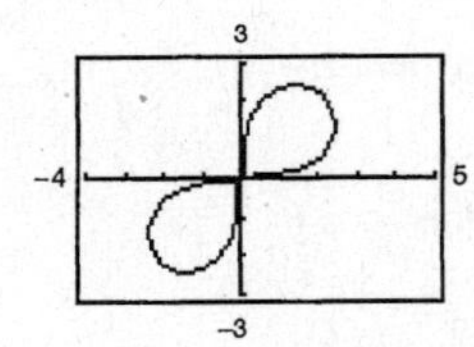

53. $$r = 2 - \sec\theta = 2 - \frac{1}{\cos\theta}$$

$$r\cos\theta = 2\cos\theta - 1$$

$$r(r\cos\theta) = 2r\cos\theta - r$$

$$\left(\pm\sqrt{x^2 + y^2}\right)x = 2x - \left(\pm\sqrt{x^2 + y^2}\right)$$

$$\left(\pm\sqrt{x^2 + y^2}\right)(x + 1) = 2x$$

$$\left(\pm\sqrt{x^2 + y^2}\right) = \frac{2x}{x + 1}$$

$$x^2 + y^2 = \frac{4x^2}{(x + 1)^2}$$

$$y^2 = \frac{4x^2}{(x + 1)^2} - x^2$$

$$= \frac{4x^2 - x^2(x + 1)^2}{(x + 1)^2} = \frac{4x^2 - x^2\left(x^2 + 2x + 1\right)}{(x + 1)^2}$$

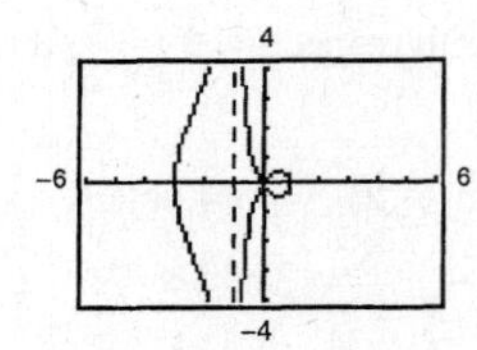

—CONTINUED—

53. —CONTINUED—

$$= \frac{-x^4 - 2x^3 + 3x^2}{(x + 1)^2} = \frac{-x^2(x^2 + 2x - 3)}{(x + 1)^2}$$

$$y = \pm\sqrt{\frac{x^2(3 - 2x - x^2)}{(x + 1)^2}} = \pm\left|\frac{x}{x + 1}\right|\sqrt{3 - 2x - x^2}$$

The graph has an asymptote at $x = -1$.

55. $r = \dfrac{3}{\theta}$

$$\theta = \frac{3}{r} = \frac{3 \sin \theta}{r \sin \theta} = \frac{3 \sin \theta}{y}$$

$$y = \frac{3 \sin \theta}{\theta}$$

As $\theta \to 0, y \to 3$

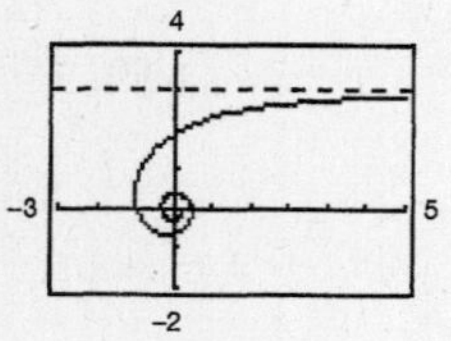

57. True. For a graph to have polar axis symmetry, replace (r, θ) by $(r, -\theta)$ or $(-r, \pi - \theta)$.

59. $r = 6 \cos \theta$

(a) $0 \le \theta \le \dfrac{\pi}{2}$

Upper half of circle

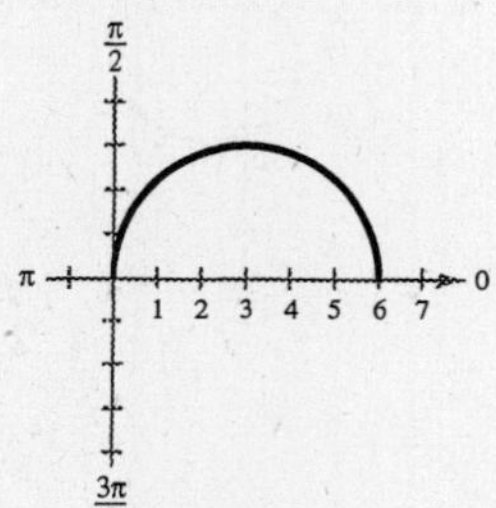

(b) $\dfrac{\pi}{2} \le \theta \le \pi$

Lower half of circle

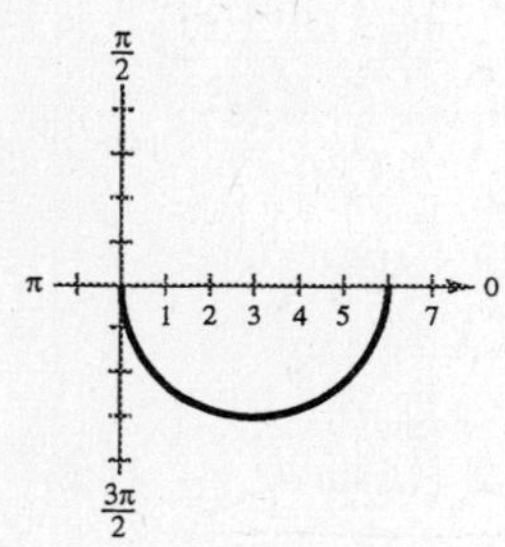

(c) $-\dfrac{\pi}{2} \le \theta \le \dfrac{\pi}{2}$

Entire circle

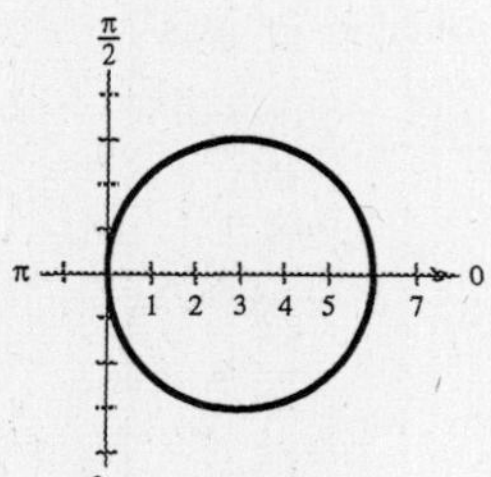

(d) $\dfrac{\pi}{4} \le \theta \le \dfrac{3\pi}{4}$

Left half of circle

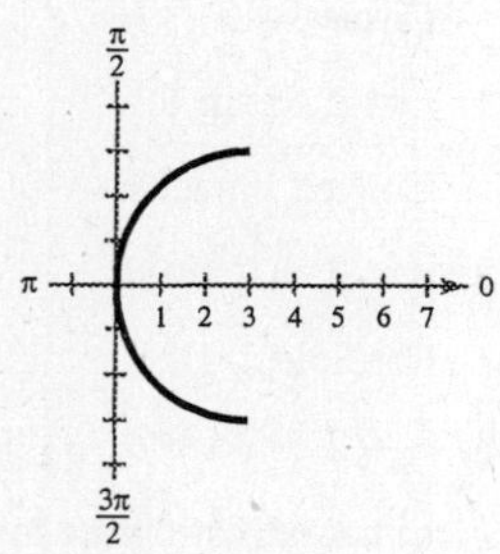

61. Let the curve $r = f(\theta)$ be rotated by ϕ to form the curve $r = g(\theta)$. If (r_1, θ_1) is a point on $r = f(\theta)$, then $(r_1, \theta_1 + \phi)$ is on $r = g(\theta)$. That is, $g(\theta_1 + \phi) = r_1 = f(\theta_1)$. Letting $\theta = \theta_1 + \phi$, or $\theta_1 = \theta - \phi$, we see that $g(\theta) = g(\theta_1 + \phi) = f(\theta_1) = f(\theta - \phi)$.

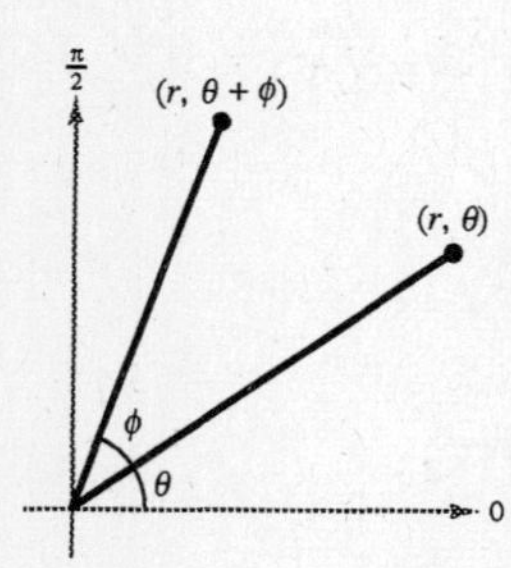

63. (a) $r = 2 - \sin\left(\theta - \dfrac{\pi}{4}\right)$

$= 2 - \left[\sin\theta\cos\dfrac{\pi}{4} - \cos\theta\sin\dfrac{\pi}{4}\right]$

$= 2 - \dfrac{\sqrt{2}}{2}(\sin\theta - \cos\theta)$

(b) $r = 2 - \sin\left(\theta - \dfrac{\pi}{2}\right)$

$= 2 - \left[\sin\theta\cos\dfrac{\pi}{2} - \cos\theta\sin\dfrac{\pi}{2}\right]$

$= 2 + \cos\theta$

(c) $r = 2 - \sin(\theta - \pi)$

$= 2 - [\sin\theta\cos\pi - \cos\theta\sin\pi]$

$= 2 + \sin\theta$

(d) $r = 2 - \sin\left(\theta - \dfrac{3\pi}{2}\right)$

$= 2 - \left[\sin\theta\cos\dfrac{3\pi}{2} - \cos\theta\sin\dfrac{3\pi}{2}\right]$

$= 2 - \cos\theta$

65. (a) $r = 1 - \sin\theta$

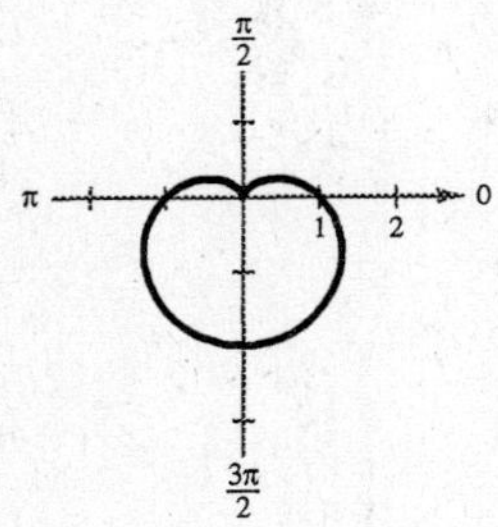

(b) $r = 1 - \sin\left(\theta - \dfrac{\pi}{4}\right)$

Rotate the graph in part (a) through the angle $\dfrac{\pi}{4}$.

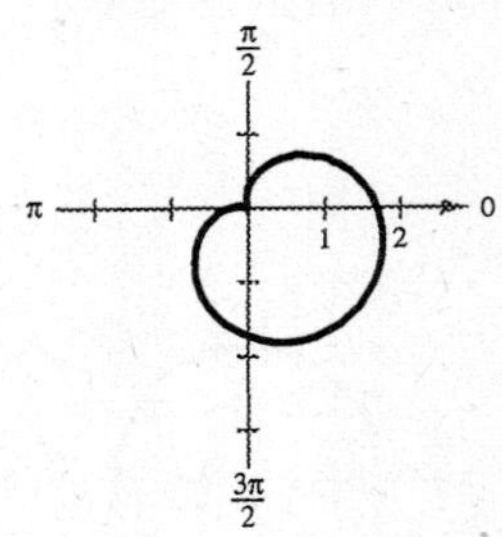

67. $r = 2 + k\sin\theta$

$k = 0$: $r = 2$

Circle

$k = 1$: $r = 2 + \sin\theta$

Convex limaçon

$k = 2$: $r = 2 + 2\sin\theta$

Cardioid

$k = 3$: $r = 2 + 3\sin\theta$

Limaçon with inner loop

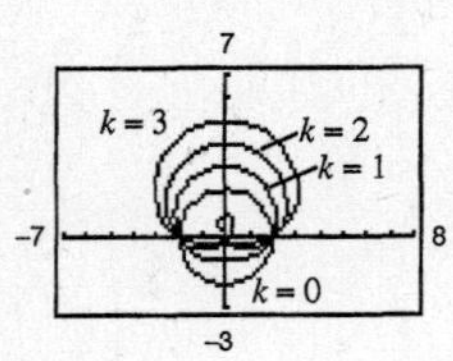

69. Vertices: $(-4, 2), (2, 2) \Rightarrow$ Center at $(-1, 2)$ and $a = 3$

Minor axis of length 4: $2b = 4 \Rightarrow b = 2$

Horizontal major axis

$$\frac{(x-h)^2}{a^2} + \frac{(y-k)^2}{b^2} = 1$$

$$\frac{(x+1)^2}{9} + \frac{(y-2)^2}{4} = 1$$

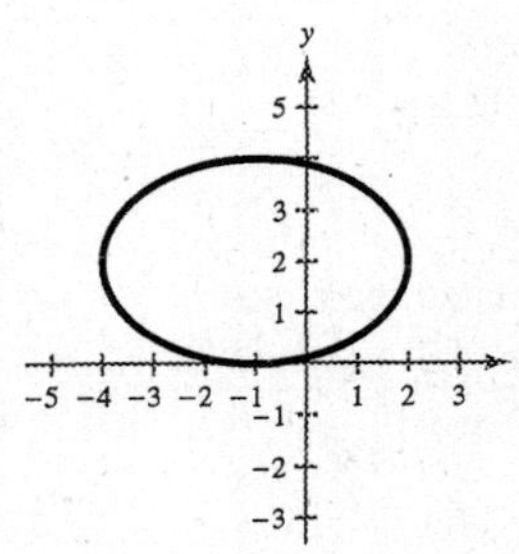

Section 6.9 Polar Equations of Conics

- The graph of a polar equation of the form

$$r = \frac{ep}{1 \pm e \cos \theta} \text{ or } r = \frac{ep}{1 \pm e \sin \theta}$$

is a conic, where $e > 0$ is the eccentricity and $|p|$ is the distance between the focus (pole) and the directrix.

(a) If $e < 1$, the graph is an ellipse.
(b) If $e = 1$, the graph is a parabola.
(c) If $e > 1$, the graph is a hyperbola.

- Guidelines for finding polar equations of conics:

(a) Horizontal directrix above the pole: $r = \dfrac{ep}{1 + e \sin \theta}$

(b) Horizontal directrix below the pole: $r = \dfrac{ep}{1 - e \sin \theta}$

(c) Vertical directrix to the right of the pole: $r = \dfrac{ep}{1 + e \cos \theta}$

(d) Vertical directrix to the left of the pole: $r = \dfrac{ep}{1 - e \cos \theta}$

Solutions to Odd-Numbered Exercises

1. $r = \dfrac{4e}{1 + e \cos \theta}$

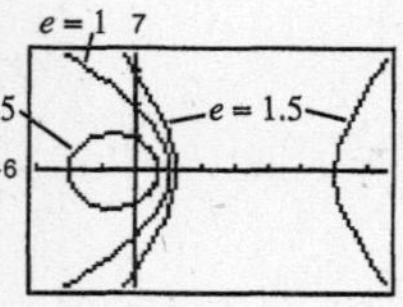

(a) $e = 1, r = \dfrac{4}{1 + \cos \theta}$, parabola

(b) $e = 0.5, r = \dfrac{2}{1 + 0.5 \cos \theta} = \dfrac{4}{2 + \cos \theta}$, ellipse

(c) $e = 1.5, r = \dfrac{6}{1 + 1.5 \cos \theta} = \dfrac{12}{2 + 3 \cos \theta}$, hyperbola

3. $r = \dfrac{4e}{1 - e \sin \theta}$

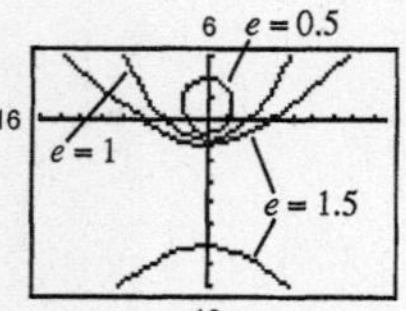

(a) $e = 1, r = \dfrac{4}{1 - \sin \theta}$, parabola

(b) $e = 0.5, r = \dfrac{2}{1 - 0.5 \sin \theta} = \dfrac{4}{2 - \sin \theta}$, ellipse

(c) $e = 1.5, r = \dfrac{6}{1 - 1.5 \sin \theta} = \dfrac{12}{2 - 3 \sin \theta}$, hyperbola

5. $r = \dfrac{2}{1 + \cos \theta}$

$e = 1 \Rightarrow$ Parabola

Vertical directrix to the right of the pole

Matches graph (f).

7. $r = \dfrac{3}{1 + 2 \sin \theta}$

$e = 2 \Rightarrow$ Hyperbola

Matches graph (d).

9. $r = \dfrac{4}{2 + \cos \theta}$

$= \dfrac{2}{1 + 0.5 \cos \theta}$

$e = 0.5 \Rightarrow$ Ellipse

Matches graph (a).

11. $r = \dfrac{2}{1 - \cos\theta}$

$e = 1$, the graph is a parabola.

Vertex: $(1, \pi)$

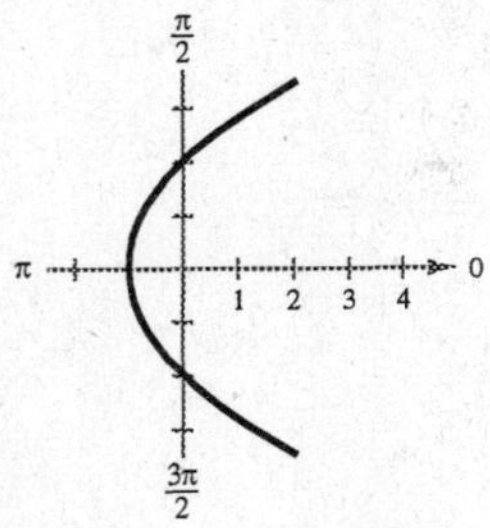

13. $r = \dfrac{5}{1 + \sin\theta}$

$e = 1$, the graph is a parabola.

Vertex: $\left(\dfrac{5}{2}, \dfrac{\pi}{2}\right)$

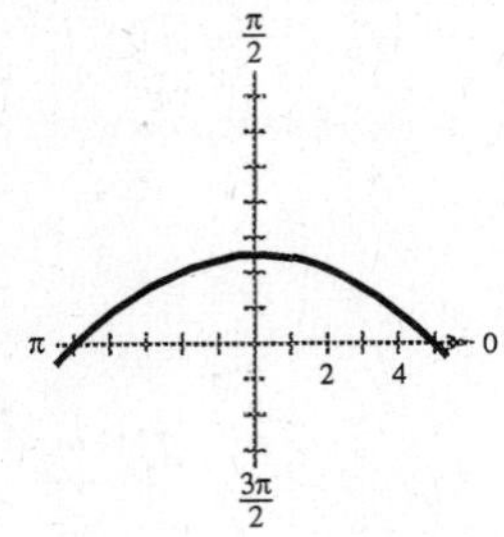

15. $r = \dfrac{2}{2 - \cos\theta} = \dfrac{1}{1 - \frac{1}{2}\cos\theta}$

$e = \dfrac{1}{2} < 1$, the graph is an ellipse.

Vertices: $(2, 0), \left(\dfrac{2}{3}, \pi\right)$

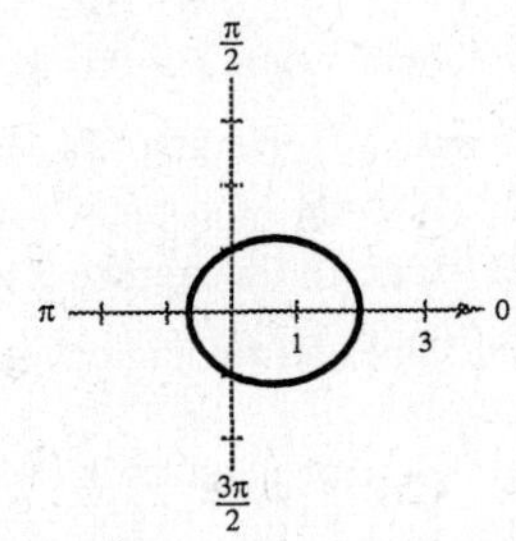

17. $r = \dfrac{6}{2 + \sin\theta} = \dfrac{3}{1 + \frac{1}{2}\sin\theta}$

$e = \dfrac{1}{2} < 1$, the graph is an ellipse.

Vertices: $\left(2, \dfrac{\pi}{2}\right), \left(6, \dfrac{3\pi}{2}\right)$

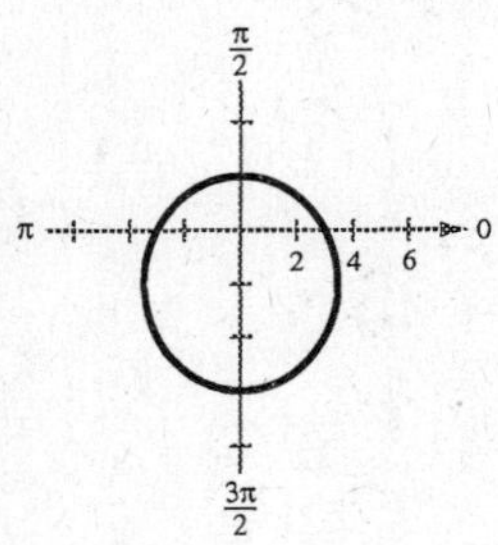

19. $r = \dfrac{3}{2 + 4\sin\theta} = \dfrac{\frac{3}{2}}{1 + 2\sin\theta}$

$e = 2 > 1$, the graph is a hyperbola.

Vertices: $\left(\dfrac{1}{2}, \dfrac{\pi}{2}\right), \left(-\dfrac{3}{2}, \dfrac{3\pi}{2}\right)$

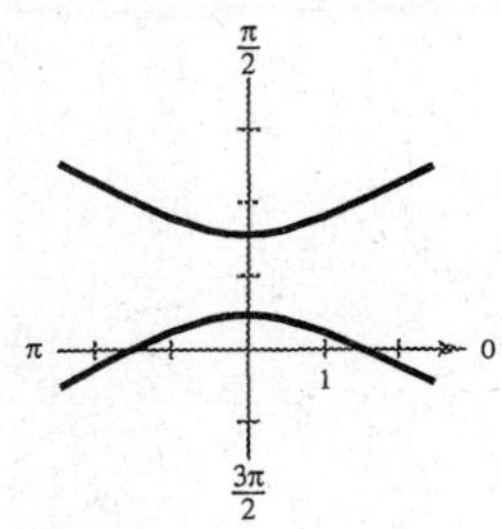

21. $r = \dfrac{3}{2 - 6\cos\theta} = \dfrac{\frac{3}{2}}{1 - 3\cos\theta}$

$e = 3 > 1$, the graph is a hyperbola.

Vertices: $\left(-\dfrac{3}{4}, 0\right), \left(\dfrac{3}{8}, \pi\right)$

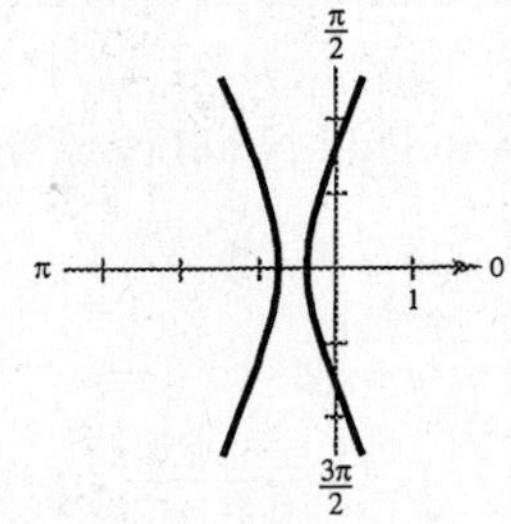

23. $r = \dfrac{4}{2 - \cos\theta} = \dfrac{2}{1 - \frac{1}{2}\cos\theta}$

$e = \dfrac{1}{2} < 1$, the graph is an ellipse.

Vertices: $(4, 0), \left(\dfrac{4}{3}, \pi\right)$

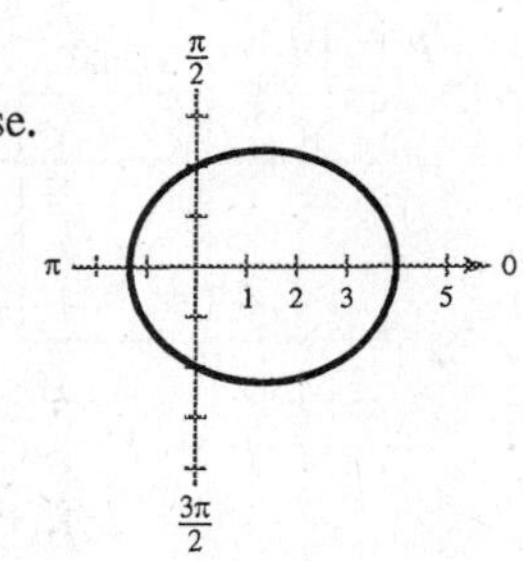

25. $r = \dfrac{-1}{1 - \sin\theta}$

$e = 1 \Rightarrow$ Parabola

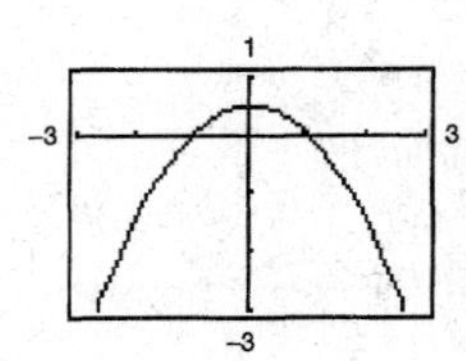

27. $r = \dfrac{3}{-4 + 2\cos\theta}$

$e = \dfrac{1}{2} \Rightarrow$ Ellipse

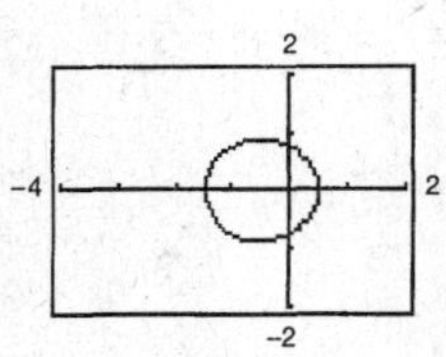

29. $r = \dfrac{2}{1 - \cos\left(\theta - \dfrac{\pi}{4}\right)}$

Rotate the graph in Exercise 11 through the angle $\dfrac{\pi}{4}$.

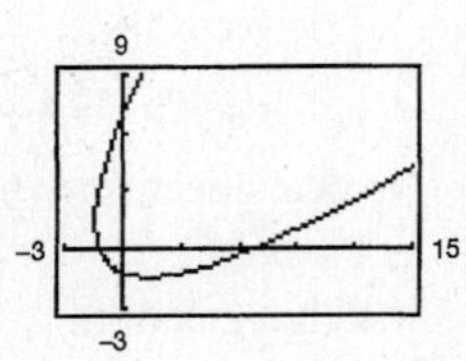

31. $r = \dfrac{6}{2 + \sin\left(\theta + \dfrac{\pi}{6}\right)}$

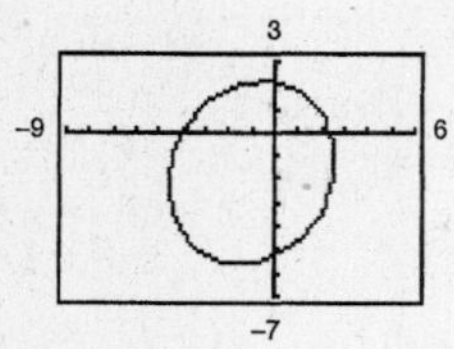

Rotate the graph in Exercise 17 through the angle $-\dfrac{\pi}{6}$.

33. Parabola: $e = 1$

Directrix: $x = -1$

Vertical directrix to the left of the pole

$$r = \frac{1(1)}{1 - 1\cos\theta} = \frac{1}{1 - \cos\theta}$$

35. Ellipse: $e = \dfrac{1}{2}$

Directrix: $y = 1$

$p = 1$

Horizontal directrix above the pole

$$r = \frac{\frac{1}{2}(1)}{1 + \frac{1}{2}\sin\theta} = \frac{1}{2 + \sin\theta}$$

37. Hyperbola: $e = 2$

Directrix: $x = 1$

$p = 1$

Vertical directrix to the right of the pole

$$r = \frac{2(1)}{1 + 2\cos\theta} = \frac{2}{1 + 2\cos\theta}$$

39. Parabola

Vertex: $\left(1, -\dfrac{\pi}{2}\right) \Rightarrow e = 1, p = 2$

Horizontal directrix below the pole

$$r = \frac{1(2)}{1 - 1\sin\theta} = \frac{2}{1 - \sin\theta}$$

41. Parabola

Vertex: $(5, \pi) \Rightarrow e = 1, p = 10$

Vertical directrix to the left of the pole

$$r = \frac{1(10)}{1 - 1\cos\theta} = \frac{10}{1 - \cos\theta}$$

43. Ellipse: Vertices $(2, 0), (10, \pi)$

Center: $(4, \pi)$; $c = 4, a = 6, e = \dfrac{2}{3}$

Vertical directrix to the right of the pole

$$r = \frac{\frac{2}{3}p}{1 + \frac{2}{3}\cos\theta} = \frac{2p}{3 + 2\cos\theta}$$

$$2 = \frac{2p}{3 + 2\cos 0}$$

$p = 5$

$$r = \frac{2(5)}{3 + 2\cos\theta} = \frac{10}{3 + 2\cos\theta}$$

45. Ellipse: Vertices $(20, 0), (4, \pi)$

Center: $(8, 0)$; $c = 8, a = 12, e = \dfrac{2}{3}$

Vertical directrix to the left of the pole

$$r = \frac{\frac{2}{3}p}{1 - \frac{2}{3}\cos\theta} = \frac{2p}{3 - 2\cos\theta}$$

$$20 = \frac{2p}{3 - 2\cos 0}$$

$p = 10$

$$r = \frac{2(10)}{3 - 2\cos\theta} = \frac{20}{3 - 2\cos\theta}$$

47. Hyperbola: Vertices $\left(1, \frac{3\pi}{2}\right), \left(9, \frac{3\pi}{2}\right)$

Center: $\left(5, \frac{3\pi}{2}\right)$; $c = 5, a = 4, e = \frac{5}{4}$

Horizontal directrix below the pole

$$r = \frac{\frac{5}{4}p}{1 - \frac{5}{4}\sin\theta} = \frac{5p}{4 - 5\sin\theta}$$

$$1 = \frac{5p}{4 - 5\sin\frac{3\pi}{2}}$$

$$p = \frac{9}{5}$$

$$r = \frac{5\left(\frac{9}{5}\right)}{4 - 5\sin\theta} = \frac{9}{4 - 5\sin\theta}$$

49. When $\theta = 0, r = c + a = ea + a = a(1 + e)$.

Therefore,

$$a(1 + e) = \frac{ep}{1 - e\cos 0}$$

$$a(1 + e)(1 - e) = ep$$

$$a(1 - e^2) = ep.$$

Thus, $r = \dfrac{ep}{1 - e\cos\theta} = \dfrac{(1 - e^2)a}{1 - e\cos\theta}$.

51. $r = \dfrac{[1 - (0.0167)^2](92.960 \times 10^6)}{1 - 0.0167\cos\theta}$

$\approx \dfrac{9.2934 \times 10^7}{1 - 0.0167\cos\theta}$

Perihelion distance:
$r = 92.960 \times 10^6(1 - 0.0167) \approx 9.1408 \times 10^7$

Aphelion distance:
$r = 92.960 \times 10^6(1 + 0.0167) \approx 9.4512 \times 10^7$

53. $r = \dfrac{[1 - (0.2481)^2](5.9 \times 10^9)}{1 - 0.2481\cos\theta}$

$\approx \dfrac{5.5368 \times 10^9}{1 - 0.2481\cos\theta}$

Perihelion distance:
$r = 5.9 \times 10^9(1 - 0.2481) \approx 4.4362 \times 10^9$

Aphelion distance:
$r = 5.9 \times 10^9(1 + 0.2481) \approx 7.3638 \times 10^9$

55. $r = \dfrac{[1 - (0.0934)^2](141 \times 10^6)}{1 - 0.0934\cos\theta}$

$\approx \dfrac{1.3977 \times 10^8}{1 - 0.0934\cos\theta}$

Perihelion distance:

$r = 141 \times 10^6(1 - 0.0934) \approx 1.2783 \times 10^8$

Aphelion distance:

$r = 141 \times 10^6(1 + 0.0934) \approx 1.5417 \times 10^8$

57. (a) $r = \dfrac{ep}{1 + e\sin\theta}$

Since the graph is a parabola, $e = 1$. The distance between the vertex and the focus (pole) is 4100, so the distance between the focus (pole) and the directrix is $p = 8200$.

$$r = \frac{8200}{1 + \sin\theta}$$

(b)

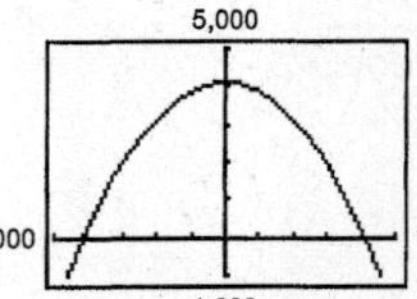

(c) When $\theta = 30°$, $r = \dfrac{8200}{1 + \sin 30°} \approx 5466.7$

Distance between surface of Earth and satellite:

$5466.7 - 4000 \approx 1467$ miles

(d) When $\theta = 60°$, $r = \dfrac{8200}{1 + \sin 60°} \approx 4394.4$

Distance between surface of Earth and satellite:

$4394.4 - 4000 \approx 394$ miles

59. True. The graphs represent the same hyperbola, although the graphs are not traced out in the same order as θ goes from 0 to 2π.

61.
$$\frac{x^2}{a^2} + \frac{y^2}{b^2} = 1$$
$$\frac{r^2 \cos^2 \theta}{a^2} + \frac{r^2 \sin^2 \theta}{b^2} = 1$$
$$\frac{r^2 \cos^2 \theta}{a^2} + \frac{r^2(1 - \cos^2 \theta)}{b^2} = 1$$
$$r^2b^2 \cos^2 \theta + r^2a^2 - r^2a^2 \cos^2 \theta = a^2b^2$$
$$r^2(b^2 - a^2)\cos^2 \theta + r^2a^2 = a^2b^2$$
Since $b^2 - a^2 = -c^2$, we have
$$-r^2c^2 \cos^2 \theta + r^2a^2 = a^2b^2$$
$$-r^2\left(\frac{c}{a}\right)^2 \cos^2 \theta + r^2 = b^2, e = \frac{c}{a}$$
$$-r^2e^2 \cos^2 \theta + r^2 = b^2$$
$$r^2(1 - e^2 \cos^2 \theta) = b^2$$
$$r^2 = \frac{b^2}{1 - e^2 \cos^2 \theta}$$

63. $\dfrac{x^2}{169} + \dfrac{y^2}{144} = 1$

$a = 13, b = 12, c = 5, e = \dfrac{5}{13}$

$$r^2 = \frac{144}{1 - \left(\frac{25}{169}\right) \cos^2 \theta} = \frac{24{,}336}{169 - 25 \cos^2 \theta}$$

65. $\dfrac{x^2}{9} - \dfrac{y^2}{16} = 1$

$a = 3, b = 4, c = 5, e = \dfrac{5}{3}$

$$r^2 = \frac{-16}{1 - \left(\frac{25}{9}\right) \cos^2 \theta} = \frac{144}{25 \cos^2 \theta - 9}$$

67. One focus: $\left(5, \dfrac{\pi}{2}\right)$

Vertices: $\left(4, \dfrac{\pi}{2}\right), \left(4, -\dfrac{\pi}{2}\right)$

$a = 4, c = 5 \implies b = 3$ and $e = \dfrac{5}{4}$

$$\frac{y^2}{16} - \frac{x^2}{9} = 1$$
$$\frac{r^2 \sin^2 \theta}{16} - \frac{r^2 \cos^2 \theta}{9} = 1$$

$$9r^2 (1 - \cos^2 \theta) - 16r^2 \cos^2 \theta = 144$$
$$9r^2 - 25r^2 \cos^2 \theta = 144$$
$$r^2 = \frac{144}{9 - 25 \cos^2 \theta}$$

or

$$9r^2 \sin^2 \theta - 16r^2(1 - \sin^2 \theta) = 144$$
$$25r^2 \sin^2 \theta - 16r^2 = 144$$
$$r^2 = \frac{144}{25 \sin^2 \theta - 16}$$

69. $4\sqrt{3} \tan \theta - 3 = 1$
$$4\sqrt{3} \tan \theta = 4$$
$$\tan \theta = \frac{1}{\sqrt{3}}$$
$$\theta = \frac{\pi}{6} + n\pi$$

71. $12 \sin^2 \theta = 9$
$$\sin^2 \theta = \frac{3}{4}$$
$$\sin \theta = \pm\frac{\sqrt{3}}{2}$$
$$\theta = \frac{\pi}{3} + n\pi, \frac{2\pi}{3} + n\pi$$

73. $2 \cot x = 5 \cos \dfrac{\pi}{2}$
$$2 \cot x = 0$$
$$\cot x = 0$$
$$x = \frac{\pi}{2} + n\pi$$

For 75 and 77 use the following:

u and v are in Quadrant IV

$\sin u = -\frac{3}{5} \Rightarrow \cos u = \frac{4}{5}$

$\cos v = \frac{1}{\sqrt{2}} \Rightarrow \sin v = -\frac{1}{\sqrt{2}}$

75. $\cos(u + v) = \cos u \cos v - \sin u \sin v$

$$= \left(\frac{4}{5}\right)\left(\frac{1}{\sqrt{2}}\right) - \left(-\frac{3}{5}\right)\left(-\frac{1}{\sqrt{2}}\right)$$

$$= \frac{4}{5\sqrt{2}} - \frac{3}{5\sqrt{2}}$$

$$= \frac{1}{5\sqrt{2}}$$

$$= \frac{\sqrt{2}}{10}$$

77. $\cos(u - v) = \cos u \cos v + \sin u \sin v$

$$= \left(\frac{4}{5}\right)\left(\frac{1}{\sqrt{2}}\right) + \left(-\frac{3}{5}\right)\left(-\frac{1}{\sqrt{2}}\right)$$

$$= \frac{4}{5\sqrt{2}} + \frac{3}{5\sqrt{2}}$$

$$= \frac{7}{5\sqrt{2}}$$

$$= \frac{7\sqrt{2}}{10}$$

79. $\sin u = \frac{4}{5}, \frac{\pi}{2} < u < \pi \Rightarrow \cos u = -\frac{3}{5}$

$\sin 2u = 2 \sin u \cos u$

$$= 2\left(\frac{4}{5}\right)\left(-\frac{3}{5}\right)$$

$$= -\frac{24}{25}$$

$\cos 2u = \cos^2 u - \sin^2 u$

$$= \left(-\frac{3}{5}\right)^2 - \left(\frac{4}{5}\right)^2$$

$$= \frac{9}{25} - \frac{16}{25}$$

$$= -\frac{7}{25}$$

$\tan 2u = \frac{\sin 2u}{\cos 2u}$

$$= \frac{-\frac{24}{25}}{-\frac{7}{25}}$$

$$= \frac{24}{7}$$

Review Exercises for Chapter 6

Solutions to Odd-Numbered Exercises

1. Points: $(-1, 2)$ and $(2, 5)$

$$m = \frac{5 - 2}{2 - (-1)} = \frac{3}{3} = 1$$

$\tan \theta = 1 \Rightarrow \theta = \frac{\pi}{4}$ radian $= 45°$

3. $y = 2x + 4 \Rightarrow m = 2$

$\tan \theta = 2 \Rightarrow \theta = \arctan 2 \approx 1.1071$ radians $\approx 63.43°$

5. $4x + y = 2 \Rightarrow y = -4x + 2 \Rightarrow m_1 = -4$

$-5x + y = -1 \Rightarrow y = 5x - 1 \Rightarrow m_2 = 5$

$$\tan \theta = \left|\frac{5 - (-4)}{1 + (-4)(5)}\right| = \frac{9}{19}$$

$\theta = \arctan \frac{9}{19} \approx 0.4424$ radian $\approx 25.35°$

7. $2x - 7y = 8 \Rightarrow y = \frac{2}{7}x - \frac{8}{7} \Rightarrow m_1 = \frac{2}{7}$

$0.4x + y = 0 \Rightarrow y = -0.4x \Rightarrow m_2 = -0.4$

$$\tan \theta = \left|\frac{-0.4 - \frac{2}{7}}{1 + \left(\frac{2}{7}\right)(-0.4)}\right| = \frac{24}{31}$$

$\theta = \arctan\left(\frac{24}{31}\right) \approx 0.6588$ radian $\approx 37.75°$

9. $(1, 2) \Rightarrow x_1 = 1, y_1 = 2$

$x - y - 3 = 0 \Rightarrow A = 1, B = -1, C = -3$

$$d = \frac{|1(1) + (-1)(2) + (-3)|}{\sqrt{1^2 + (-1)^2}} = \frac{4}{\sqrt{2}} = 2\sqrt{2}$$

11. Hyperbola

13. Vertex: $(4, 2) = (h, k)$

Focus: $(4, 0) \Rightarrow p = -2$

$(x - h)^2 = 4p(y - k)$

$(x - 4)^2 = -8(y - 2)$

15. Vertex: $(0, 2) = (h, k)$

Directrix: $x = -3 \Rightarrow p = 3$

$(y - k)^2 = 4p(x - h)$

$(y - 2)^2 = 12x$

17. $x^2 = -2y \Rightarrow p = -\frac{1}{2}$

Focus: $\left(0, -\frac{1}{2}\right)$

$d_1 = b + \frac{1}{2}$

$$d_2 = \sqrt{(2 - 0)^2 + \left(-2 + \frac{1}{2}\right)^2}$$

$$= \sqrt{4 + \frac{9}{4}} = \frac{5}{2}$$

$d_1 = d_2$

$b + \frac{1}{2} = \frac{5}{2}$

$b = 2$

The slope of the line is

$$m = \frac{-2 - 2}{2 - 0} = -2$$

Tangent line: $y = -2x + 2$

x-intercept: $(1, 0)$

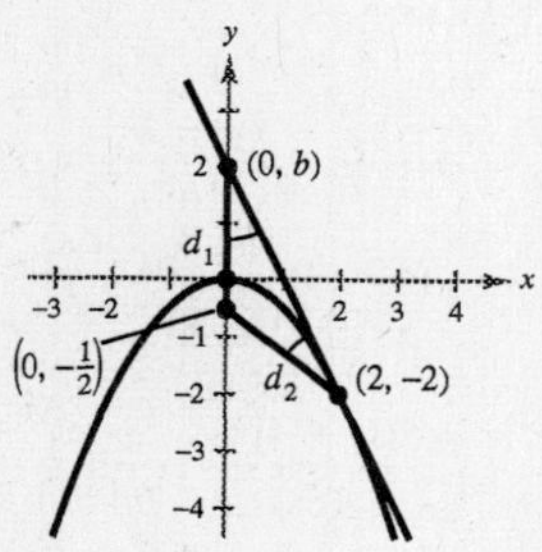

19. Parabola

Opens downward

Vertex: $(0, 12)$

$(x - h)^2 = 4p(y - k)$

$x^2 = 4p(y - 12)$

Solution points: $(\pm 4, 10)$

$16 = 4p(10 - 12)$

$16 = -8p$

$-2 = p$

$x^2 = -8(y - 12)$

To find the x-intercepts, let $y = 0$.

$x^2 = 96$

$x = \pm\sqrt{96} = \pm 4\sqrt{6}$

At the base, the archway is $2(4\sqrt{6}) = 8\sqrt{6}$ meters wide.

21. Vertices: $(-3, 0), (7, 0) \Rightarrow a = 5$
$(h, k) = (2, 0)$

Foci: $(0, 0), (4, 0) \Rightarrow c = 2$

$b^2 = a^2 - c^2 = 25 - 4 = 21$

$$\frac{(x - h)^2}{a^2} + \frac{(y - k)^2}{b^2} = 1$$

$$\frac{(x - 2)^2}{25} + \frac{y^2}{21} = 1$$

23. Vertices: $(0, 1), (4, 1) \Rightarrow a = 2, (h, k) = (2, 1)$
Endpoints of minor axis: $(2, 0), (2, 2) \Rightarrow b = 1$

$$\frac{(x - h)^2}{a^2} + \frac{(y - k)^2}{b^2} = 1$$

$$\frac{(x - 2)^2}{4} + (y - 1)^2 = 1$$

25. $2a = 10 \Rightarrow a = 5$

$b = 4$

$c^2 = a^2 - b^2 = 25 - 16 = 9 \Rightarrow c = 3$

The foci occur 3 feet from the center of the arch on a line connecting the tops of the pillars.

27. $\dfrac{(x+2)^2}{81}+\dfrac{(y-1)^2}{100}=1$

$a=10,\ b=9,\ c=\sqrt{19}$

Center: $(-2, 1)$

Vertices: $(-2, 11)$ and $(-2, -9)$

Foci: $\left(-2, 1 \pm \sqrt{19}\right)$

Eccentricity: $e=\dfrac{\sqrt{19}}{10}$

29. $16x^2+9y^2-32x+72y+16=0$

$16(x^2-2x+1)+9(y^2+8y+16)=-16+16+144$

$16(x-1)^2+9(y+4)^2=144$

$\dfrac{(x-1)^2}{9}+\dfrac{(y+4)^2}{16}=1$

$a=4,\ b=3,\ c=\sqrt{7}$

Center: $(1, -4)$

Vertices: $(1, 0)$ and $(1, -8)$

Foci: $\left(1, -4 \pm \sqrt{7}\right)$

Eccentricity: $e=\dfrac{\sqrt{7}}{4}$

31. Vertices: $(0, \pm 1) \Rightarrow a=1, (h, k)=(0, 0)$

Foci: $(0, \pm 3) \Rightarrow c=3$

$b^2=c^2-a^2=9-1=8$

$\dfrac{(y-k)^2}{a^2}-\dfrac{(x-h)^2}{b^2}=1$

$y^2-\dfrac{x^2}{8}=1$

33. Foci: $(0, 0), (8, 0) \Rightarrow c=4, (h, k)=(4, 0)$

Asymptotes: $y=\pm 2(x-4) \Rightarrow \dfrac{b}{a}=2, b=2a$

$b^2=c^2-a^2 \Rightarrow 4a^2=16-a^2 \Rightarrow a^2=\dfrac{16}{5},$

$b^2=\dfrac{64}{5}$

$\dfrac{(x-h)^2}{a^2}-\dfrac{(y-k)^2}{b^2}=1$

$\dfrac{(x-4)^2}{\frac{16}{5}}-\dfrac{y^2}{\frac{64}{5}}=1$

$\dfrac{5(x-4)^2}{16}-\dfrac{5y^2}{64}=1$

35. $\dfrac{(x-3)^2}{16}-\dfrac{(y+5)^2}{4}=1$

$a=4,\ b=2,\ c=\sqrt{20}=2\sqrt{5}$

Center: $(3, -5)$

Vertices: $(7, -5)$ and $(-1, -5)$

Foci: $\left(3 \pm 2\sqrt{5}, -5\right)$

Asymptotes: $y=-5 \pm \dfrac{1}{2}(x-3)$

$y=\dfrac{1}{2}x-\dfrac{13}{2}$ or $y=-\dfrac{1}{2}x-\dfrac{7}{2}$

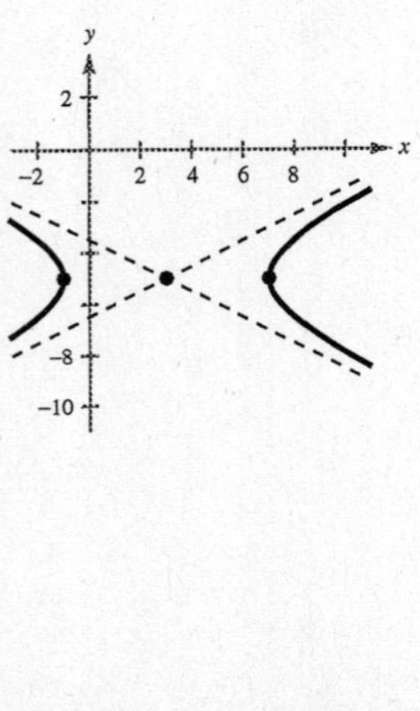

37. $9x^2-16y^2-18x-32y-151=0$

$9(x^2-2x+1)-16(y^2+2y+1)=151+9-16$

$9(x-1)^2-16(y+1)^2=144$

$\dfrac{(x-1)^2}{16}-\dfrac{(y+1)^2}{9}=1$

$a=4,\ b=3,\ c=5$

Center: $(1, -1)$

Vertices: $(5, -1)$ and $(-3, -1)$

Foci: $(6, -1)$ and $(-4, -1)$

Asymptotes: $y=-1 \pm \dfrac{3}{4}(x-1)$

$y=\dfrac{3}{4}x-\dfrac{7}{4}$ or $y=-\dfrac{3}{4}x-\dfrac{1}{4}$

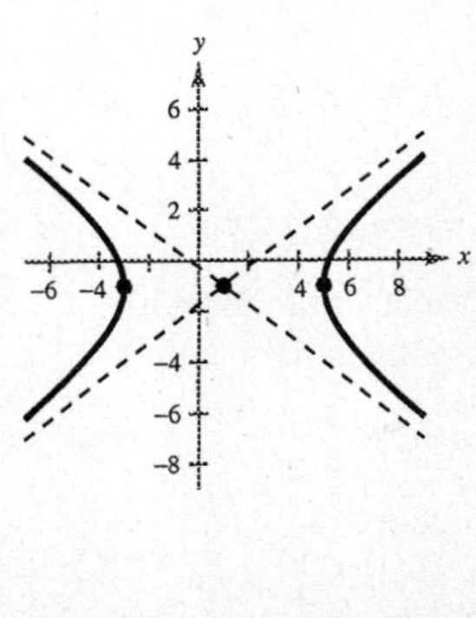

39. Foci: $(\pm 100, 0) \implies c = 100$

Center: $(0, 0)$

$$\frac{d_2}{186{,}000} - \frac{d_1}{186{,}000} = 0.0005 \implies d_2 - d_1 = 93 = 2a \implies a = 46.5$$

$$b^2 = c^2 - a^2 = 100^2 - 46.5^2 = 7837.75$$

$$\frac{x^2}{2162.25} - \frac{y^2}{7837.75} = 1$$

$$y^2 = 7837.75\left(\frac{60^2}{2162.25} - 1\right) \approx 5211.5736$$

$y \approx 72$ miles

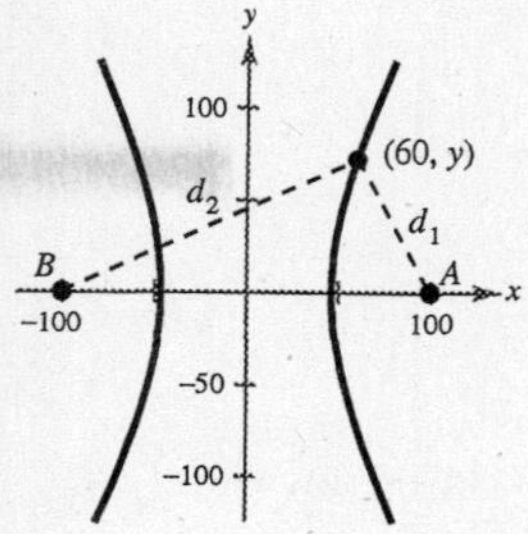

41. $5x^2 - 2y^2 + 10x - 4y + 17 = 0$

$AC = 5(-2) = -10 < 0$

The graph is a hyperbola.

43. $xy - 4 = 0$

$A = C = 0, B = 1$

$B^2 - 4AC = 1^2 - 4(0)(0) = 1 > 0$

The graph is a hyperbola.

$$\cot 2\theta = 0 \implies 2\theta = \frac{\pi}{2} \implies \theta = \frac{\pi}{4}$$

$$x = x'\cos\frac{\pi}{4} - y'\sin\frac{\pi}{4} = \frac{x' - y'}{\sqrt{2}}$$

$$y = x'\sin\frac{\pi}{4} + y'\cos\frac{\pi}{4} = \frac{x' + y'}{\sqrt{2}}$$

$$\left(\frac{x' - y'}{\sqrt{2}}\right)\left(\frac{x' + y'}{\sqrt{2}}\right) - 4 = 0$$

$$\frac{(x')^2 - (y')^2}{2} = 4$$

$$\frac{(x')^2}{8} - \frac{(y')^2}{8} = 1$$

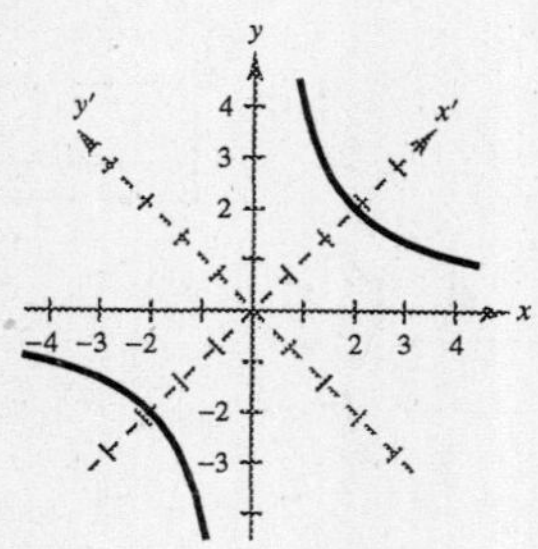

45. $5x^2 - 2xy + 5y^2 - 12 = 0$

$A = C = 5, B = -2$

$B^2 - 4AC = (-2)^2 - 4(5)(5) = -96 < 0$

The graph is an ellipse.

$$\cot 2\theta = 0 \implies 2\theta = \frac{\pi}{2} \implies \theta = \frac{\pi}{4}$$

$$x = x'\cos\frac{\pi}{4} - y'\sin\frac{\pi}{4} = \frac{x' - y'}{\sqrt{2}}$$

$$y = x'\sin\frac{\pi}{4} + y'\cos\frac{\pi}{4} = \frac{x' + y'}{\sqrt{2}}$$

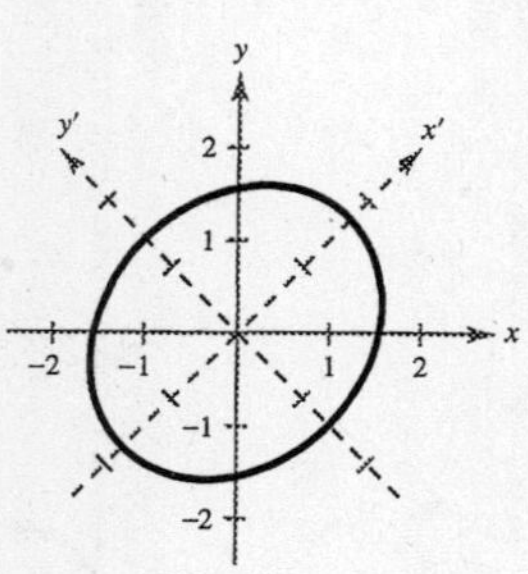

—CONTINUED—

45. —CONTINUED—

$$5\left(\frac{x'-y'}{\sqrt{2}}\right)^2 - 2\left(\frac{x'-y'}{\sqrt{2}}\right)\left(\frac{x'+y'}{\sqrt{2}}\right) + 5\left(\frac{x'+y'}{\sqrt{2}}\right)^2 - 12 = 0$$

$$\frac{5}{2}[(x')^2 - 2(x'y') + (y')^2] - [(x')^2 - (y')^2] + \frac{5}{2}[(x')^2 + 2(x'y') + (y')^2] = 12$$

$$4(x')^2 + 6(y')^2 = 12$$

$$\frac{(x')^2}{3} + \frac{(y')^2}{2} = 1$$

47. $16x^2 - 24xy + 9y^2 - 30x - 40y = 0$

$B^2 - 4AC = (-24)^2 - 4(16)(9) = 0$

The graph is a parabola.

To use a graphing utility, we need to solve for y in terms of x.

$9y^2 + (-24x - 40)y + (16x^2 - 30x) = 0$

$$y = \frac{-(-24x - 40) \pm \sqrt{(-24x - 40)^2 - 4(9)(16x^2 - 30x)}}{2(9)}$$

$$= \frac{(24x + 40) \pm \sqrt{(24x + 40)^2 - 36(16x^2 - 30x)}}{18}$$

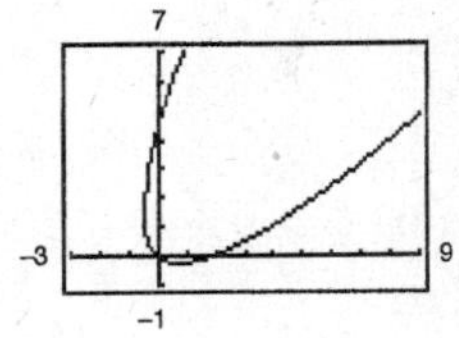

49. $x^2 + y^2 + 2xy + 2\sqrt{2}x - 2\sqrt{2}y + 2 = 0$

$B^2 - 4AC = 2^2 - 4(1)(1) = 0$

The graph is a parabola.

To use a graphing utility, we need to solve for y in terms of x.

$y^2 + (2x - 2\sqrt{2})y + (x^2 + 2\sqrt{2}x + 2)$

$$y = \frac{-(2x - 2\sqrt{2}) \pm \sqrt{(2x - 2\sqrt{2})^2 - 4(x^2 + 2\sqrt{2}x + 2)}}{2}$$

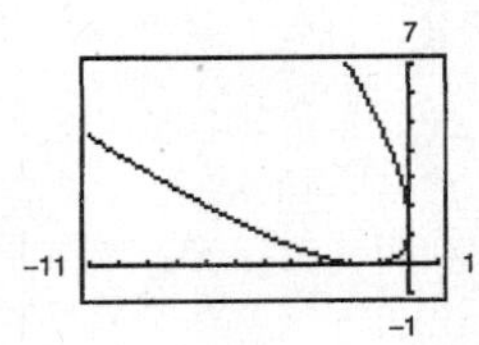

51. $x = 3\cos 0 = 3$

$y = 2\sin^2 0 = 0$

53. $x = 3\cos\frac{\pi}{6} = \frac{3\sqrt{3}}{2}$

$y = 2\sin^2\frac{\pi}{6} = \frac{1}{2}$

55. $x = 2t \implies \frac{x}{2} = t$

$y = 4t \implies y = 4\left(\frac{x}{2}\right) = 2x$

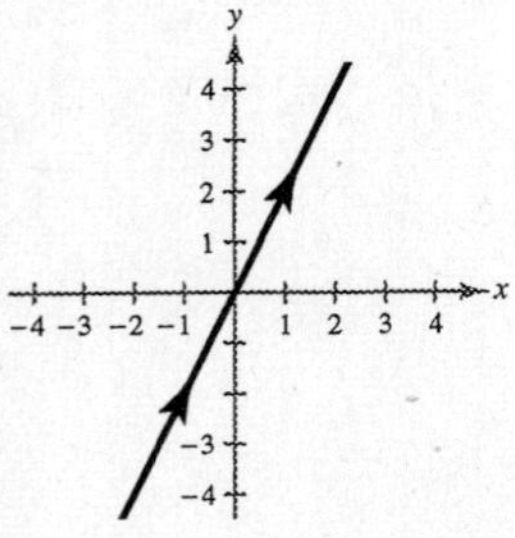

57. $x = t^2,\ x \ge 0$

$y = \sqrt{t} \implies y^2 = t$

$x = (y^2)^2 \implies x = y^4 \implies y = \sqrt[4]{x}$

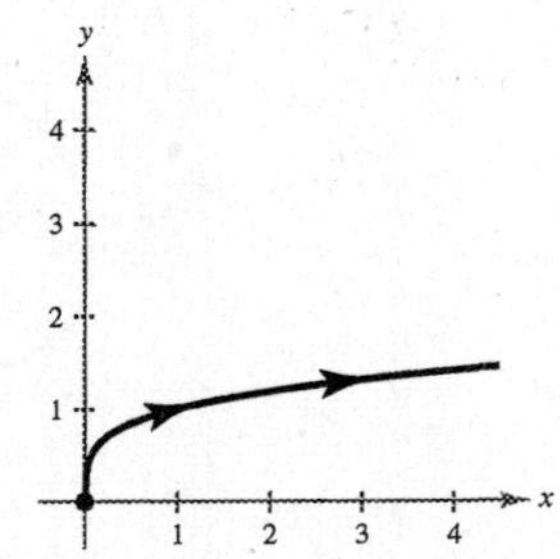

59. $x = 6\cos\theta,\ y = 6\sin\theta$

$\cos\theta = \frac{x}{6},\ \sin\theta = \frac{y}{6}$

$$\frac{x^2}{36} + \frac{y^2}{36} = 1$$

$$x^2 + y^2 = 36$$

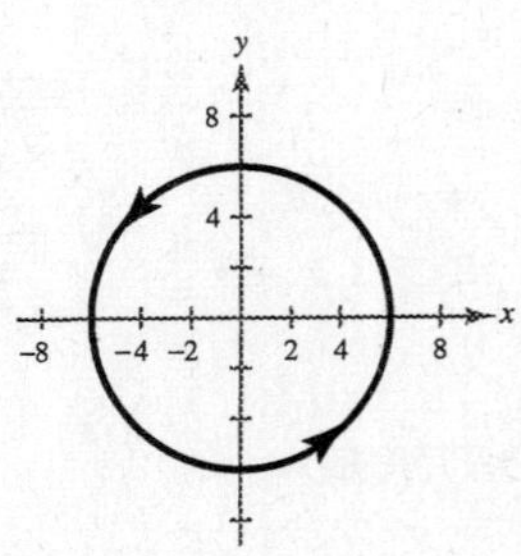

61. Center: $(5, 4)$

Radius: 6

$x = h + r\cos\theta = 5 + 6\cos\theta$

$y = k + r\sin\theta = 4 + 6\sin\theta$

63. Hyperbola

Vertices: $(0, \pm 4)$

Foci: $(0, \pm 5)$

Center: $(0, 0)$

$a = 4, c = 5, b = \sqrt{c^2 - a^2} = 3$

$x = 3\tan\theta, y = 4\sec\theta$

65. Polar coordinates: $\left(2, \frac{\pi}{4}\right)$

Additional polar representations: $\left(2, \frac{9\pi}{4}\right)\left(-2, \frac{5\pi}{4}\right)$

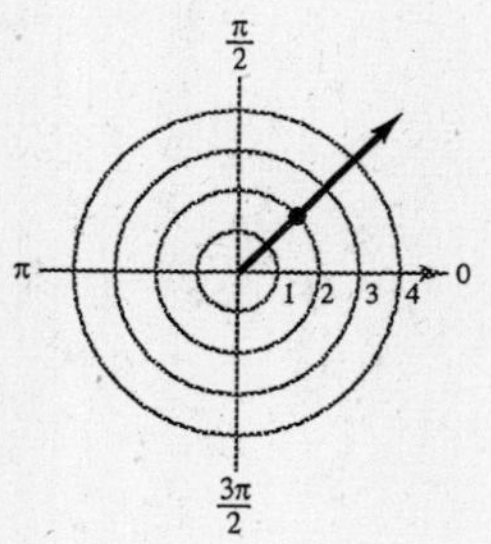

67. Polar coordinates: $(-7, 4.19)$

Additional polar representations: $(7, 1.05), (-7, 10.47)$

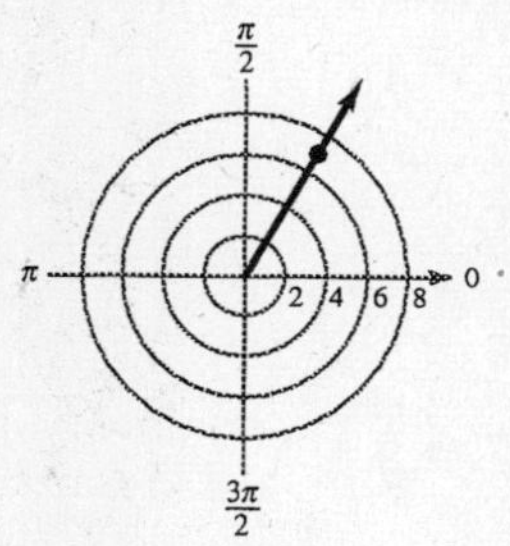

69. Polar coordinates: $\left(-1, \frac{\pi}{3}\right)$

$x = -1\cos\frac{\pi}{3} = -\frac{1}{2}$

$y = -1\sin\frac{\pi}{3} = -\frac{\sqrt{3}}{2}$

Rectangular coordinates: $\left(-\frac{1}{2}, -\frac{\sqrt{3}}{2}\right)$

71. Polar coordinates: $\left(3, \frac{3\pi}{4}\right)$

$x = 3\cos\frac{3\pi}{4} = -\frac{3\sqrt{2}}{2}$

$y = 3\sin\frac{3\pi}{4} = \frac{3\sqrt{2}}{2}$

Rectangular coordinates: $\left(-\frac{3\sqrt{2}}{2}, \frac{3\sqrt{2}}{2}\right)$

73. Rectangular coordinates: $(0, 2)$

$r = \pm\sqrt{0^2 + 2^2} = \pm 2$

$\tan\theta$ is undefined $\Rightarrow \theta = \frac{\pi}{2}, \frac{3\pi}{2}$

Polar coordinates: $\left(2, \frac{\pi}{2}\right)$ or $\left(-2, \frac{3\pi}{2}\right)$

75. Rectangular coordinates: $(4, 6)$

$r = \pm\sqrt{4^2 + 6^2} = \pm\sqrt{52} = \pm 2\sqrt{13}$

$\tan\theta = \frac{6}{4} \Rightarrow \theta \approx 0.9828, 4.1244$

Polar coordinates: $(2\sqrt{13}, 0.9828)$ or $(-2\sqrt{13}, 4.1244)$

77. $x^2 + y^2 - 6y = 0$

$r^2 - 6r\sin\theta = 0$

$r(r - 6\sin\theta) = 0$

$r = 0$ or $r = 6\sin\theta$

Since $r = 6\sin\theta$ contains $r = 0$, we just have $r = 6\sin\theta$

79. $xy = 5$

$(r\cos\theta)(r\sin\theta) = 5$

$r^2 = \frac{5}{\sin\theta\cos\theta} = \frac{10}{\sin 2\theta} = 10\csc 2\theta$

81.

$$r = 3\cos\theta$$
$$r^2 = 3r\cos\theta$$
$$x^2 + y^2 = 3x$$

83.

$$r^2 = \sin\theta$$
$$r^3 = r\sin\theta$$
$$\left(\pm\sqrt{x^2 + y^2}\right)^3 = y$$
$$(x^2 + y^2)^3 = y^2$$
$$x^2 + y^2 = y^{2/3}$$

85. $r = 4$

Circle of radius 4 centered at the pole

Symmetric with respect to $\theta = \pi/2$, the polar axis and the pole

Maximum value of $|r| = 4$ when $\theta = 0, \frac{\pi}{2}, \pi, \frac{3\pi}{2}$

Zeros: None

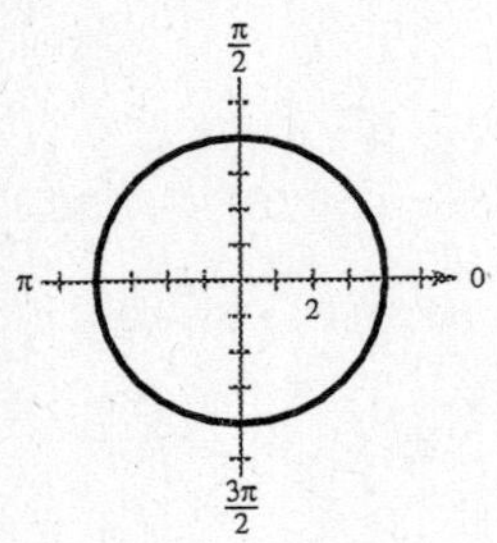

87. $r = 4 \sin 2\theta$

Rose curve ($n = 2$) with 4 petals

Symmetric with respect to $\theta = \pi/2$, the polar axis, and the pole.

Maximum value of $|r| = 4$ when $\theta = \frac{\pi}{4}, \frac{3\pi}{4}, \frac{5\pi}{4}, \frac{7\pi}{4}$

Zeros: $r = 0$ when $\theta = 0, \frac{\pi}{2}, \pi, \frac{3\pi}{2}$

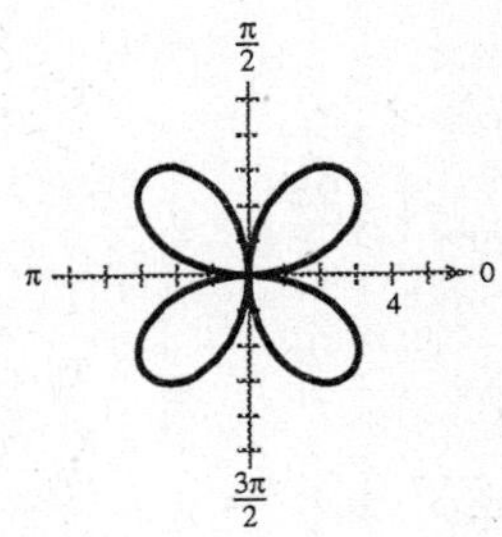

89. $r = -2(1 + \cos \theta)$

Symmetric with respect to the polar axis

Maximum value of $|r| = 4$ when $\theta = 0$

Zeros: $r = 0$ when $\theta = \pi$

$\frac{a}{b} = \frac{2}{2} = 1 \Rightarrow$ Cardioid

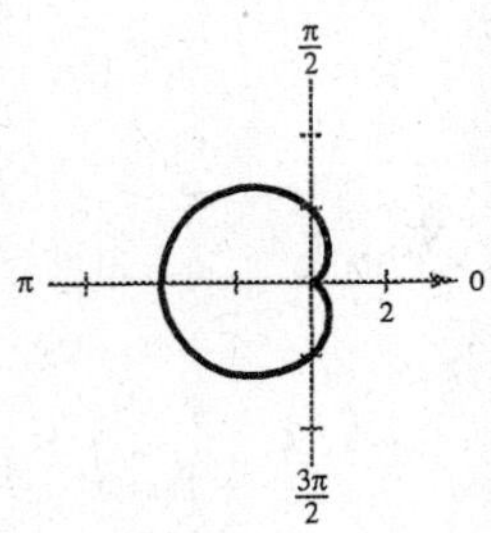

91. $r = 2 + 6 \sin \theta$

Limaçon with inner loop

$r = f(\sin \theta) \Rightarrow \theta = \frac{\pi}{2}$ symmetry

Maximum value: $|r| = 8$ when $\theta = \frac{\pi}{2}$

Zeros: $2 + 6 \sin \theta = 0 \Rightarrow \sin \theta = -\frac{1}{3} \Rightarrow \theta \approx 3.4814, 5.9433$

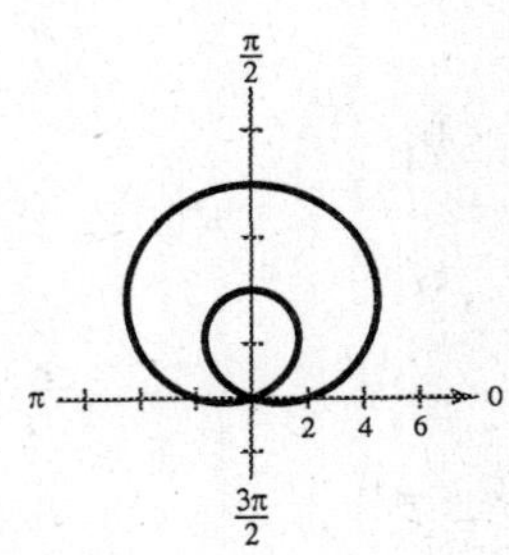

93. $r = -3 \cos 2\theta$

Rose curve with 4 petals

$r = f(\cos \theta) \Rightarrow$ polar axis symmetry

$\theta = \frac{\pi}{2}$: $r = -3 \cos 2(\pi - \theta) = -3 \cos(2\pi - 2\theta) = -3 \cos 2\theta$

Equivalent equation $\Rightarrow \theta = \frac{\pi}{2}$ symmetry

Pole: $r = -3 \cos 2(\pi + \theta) = -3 \cos(2\pi + 2\theta) = -3 \cos 2\theta$

Equivalent equation $\Rightarrow$ pole symmetry

Maximum value: $|r| = 3$ when $\theta = 0, \frac{\pi}{2}, \pi, \frac{3\pi}{2}$

Zeros: $-3 \cos 2\theta = 0$ when $\cos 2\theta = 0 \Rightarrow \theta = \frac{\pi}{4}, \frac{3\pi}{4}, \frac{5\pi}{4}, \frac{7\pi}{4}$

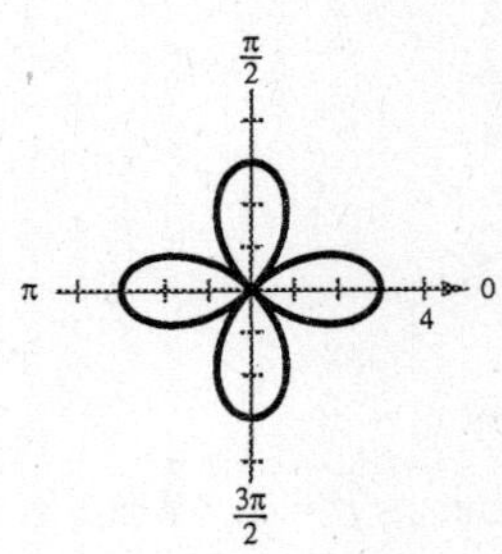

95. $r = 3(2 - \cos\theta)$

$= 6 - 3\cos\theta$

$\frac{a}{b} = \frac{6}{3} = 2$

The graph is a convex limaçon.

97. $r = 4\cos 3\theta$

The graph is a rose curve with 3 petals.

99. $r = \frac{1}{1 + 2\sin\theta}, e = 2$

Hyperbola symmetric with respect to $\theta = \frac{\pi}{2}$ and having vertices at $\left(\frac{1}{3}, \frac{\pi}{2}\right)$ and $\left(-1, \frac{3\pi}{2}\right)$.

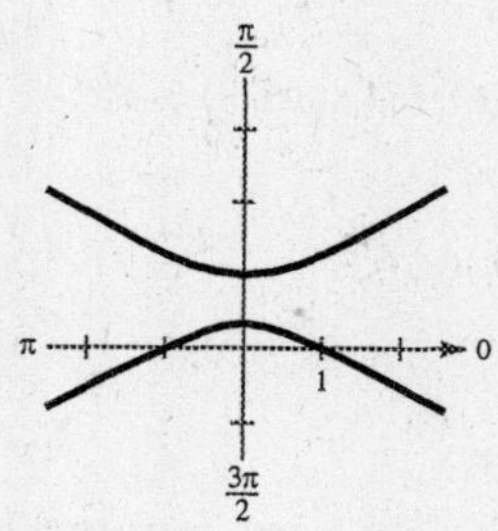

101. $r = \frac{4}{5 - 3\cos\theta}$

$r = \frac{\frac{4}{5}}{1 - \left(\frac{3}{5}\right)\cos\theta}, e = \frac{3}{5}$

Ellipse symmetric with respect to the polar axis and having vertices at (2, 0) and $\left(\frac{1}{2}, \pi\right)$.

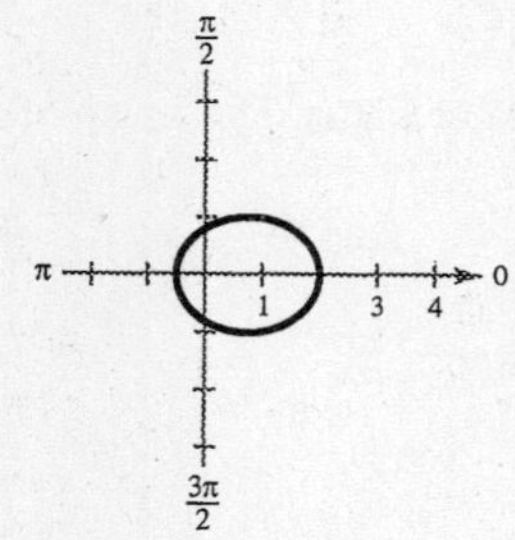

103. Parabola: $r = \frac{ep}{1 - e\cos\theta}, e = 1$

Vertex: $(2, \pi)$

Focus: $(0, 0) \implies p = 4$

$r = \frac{4}{1 - \cos\theta}$

105. Ellipse: $r = \frac{ep}{1 - e\cos\theta}$

Vertices: $(5, 0), (1, \pi) \implies a = 3$

One focus: $(0, 0) \implies c = 2$

$e = \frac{c}{a} = \frac{2}{3}, p = \frac{5}{2}$

$r = \frac{\left(\frac{2}{3}\right)\left(\frac{5}{2}\right)}{1 - \left(\frac{2}{3}\right)\cos\theta} = \frac{\frac{5}{3}}{1 - \left(\frac{2}{3}\right)\cos\theta} = \frac{5}{3 - 2\cos\theta}$

107. $a + c = 122{,}000 + 4000 \implies a + c = 126{,}000$

$a - c = 119 + 4000 \implies a - c = 4{,}119$

$2a = 130{,}119$

$a = 65{,}059.5$

$c = 60{,}940.5$

$e = \frac{c}{a} = \frac{60{,}940.5}{65{,}059.5} \approx 0.937$

$r = \frac{ep}{1 - e\cos\theta} \approx \frac{0.937p}{1 - 0.937\cos\theta}$

$r = 126{,}000$ when $\theta = 0$

$126{,}000 = \frac{ep}{1 - e\cos 0}$

$ep = 126{,}000\left(1 - \frac{60{,}940.5}{65{,}059.5}\right) \approx 7977.2$

Thus, $r \approx \frac{7977.2}{1 - 0.937\cos\theta}$

When $\theta = \frac{\pi}{3}, r \approx \frac{7977.2}{1 - 0.937\cos\frac{\pi}{3}} \approx 15{,}008.8$ miles

The distance from the surface of Earth and the satellite is $15{,}008.8 - 4000 \approx 11{,}008.8$ miles.

109. False. When classifying equations of the form $Ax^2 + Bxy + Cy^2 + Dx + Ey + F = 0$, its graph can be determined by its discriminant. For a graph to be a parabola, its discriminant, $B^2 - 4AC$, must equal zero. So, if $B = 0$, then A **or** C equals 0.

111. False. The following are **two** sets of parametric equations for the line.

$x = t, y = 3 - 2t$

$x = 3t, y = 3 - 6t$

113. $2a = 10 \Rightarrow a = 5$

b must be less than 5; $0 < b < 5$

As b approaches 5, the ellipse becomes more circular and approaches a circle of radius 5.

115. $x = 4 \cos t$ and $y = 3 \sin t$

(a) $x = 4 \cos 2t$ and $y = 3 \sin 2t$

The speed would double.

(b) $x = 5 \cos t$ and $y = 3 \sin t$

The elliptical orbit would be flatter. The length of the major axis is greater.

117. (a) $x^2 + y^2 = 25$

$r = 5$

The graphs are the same.

They are both circles centered at $(0, 0)$ with a radius of 5.

(b) $x - y = 0 \Rightarrow y = x$

$\theta = \dfrac{\pi}{4}$

The graphs are the same.

They are both lines with slope 1 and intercept $(0, 0)$.

Problem Solving for Chapter 6

Solutions to Odd-Numbered Exercises

1. (a) $\theta = \pi - 1.10 - 0.84 \approx 1.2016$ radians

(b) $\sin 0.84 = \dfrac{x}{3250} \Rightarrow x = 3250 \sin 0.84 \approx 2420$ feet

$\sin 1.10 = \dfrac{y}{6700} \Rightarrow y = 6700 \sin 1.10 \approx 5971$ feet

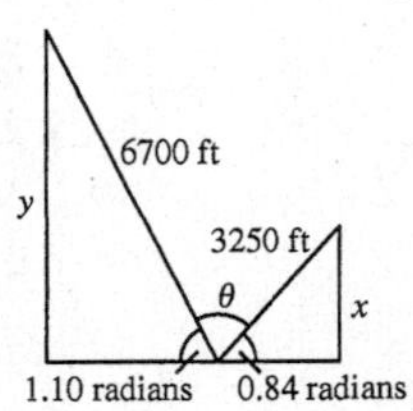

3. Since the axis of symmetry is the x-axis, the vertex is $(h, 0)$ and $y^2 = 4p(x - h)$. Also, since the focus is $(0, 0)$, $0 - h = p \Rightarrow h = -p$ and $y^2 = 4p(x + p)$.

5. (a)

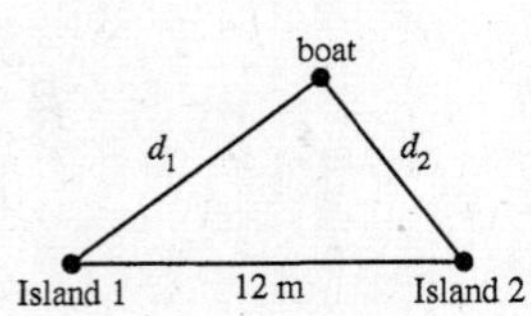

Since $d_1 + d_2 \le 20$, by definition, the outer bound that the boat can travel is an ellipse. The islands are the foci.

(b)

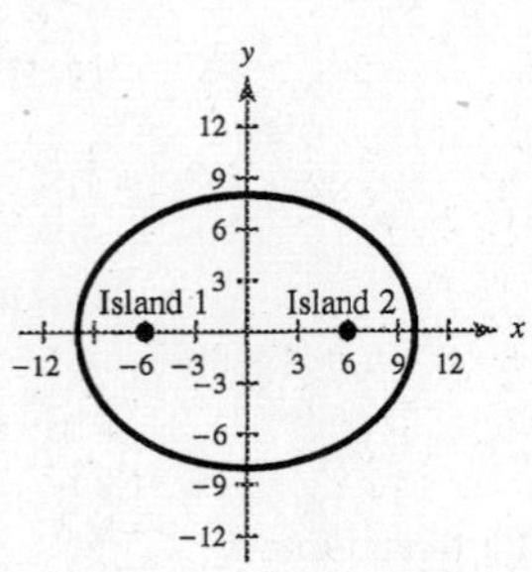

Island 1 is located at $(-6, 0)$ and Island 2 is located at $(6, 0)$.

(c) $d_1 + d_2 = 2a = 20 \Rightarrow a = 10$

The boat traveled 20 miles. The vertex is $(10, 0)$.

(d) $c = 6, a = 10 \Rightarrow b^2 = a^2 - c^2 = 64$

$$\frac{x^2}{100} + \frac{y^2}{64} = 1$$

7. $Ax^2 + Cy^2 + Dx + Ey + F = 0$

Assume that the conic is *not* degenerate.

(a) $A = C, A \neq 0$

$$Ax^2 + Ay^2 + Dx + Ey + F = 0$$

$$x^2 + y^2 + \frac{D}{A}x + \frac{E}{A}y + \frac{F}{A} = 0$$

$$\left(x^2 + \frac{D}{A}x + \frac{D^2}{4A^2}\right) + \left(y^2 + \frac{E}{A}y + \frac{E^2}{4A^2}\right) = -\frac{F}{A} + \frac{D^2}{4A^2} + \frac{E^2}{4A^2}$$

$$\left(x + \frac{D}{2A}\right)^2 + \left(y + \frac{E}{2A}\right)^2 = \frac{D^2 + E^2 - 4AF}{4A^2}$$

This is a circle with center $\left(-\frac{D}{2A}, -\frac{E}{2A}\right)$ and

radius $\frac{\sqrt{D^2 + E^2 - 4AF}}{2|A|}$

(b) $A = 0$ or $C = 0$ (but not both).

Let $C = 0$.

$$Ax^2 + Dx + Ey + F = 0$$

$$x^2 + \frac{D}{A}x = -\frac{E}{A}y - \frac{F}{A}$$

$$x^2 + \frac{D}{A}x + \frac{D^2}{4A^2} = -\frac{E}{A}y - \frac{F}{A} + \frac{D^2}{4A^2}$$

$$\left(x + \frac{D}{2A}\right)^2 = -\frac{E}{A}\left(y + \frac{F}{E} - \frac{D^2}{4AE}\right)$$

This is a parabola with vertex $\left(-\frac{D}{2A}, \frac{D^2 - 4AF}{4AE}\right)$

$A = 0$ yields a similar result.

(c) $AC > 0 \Rightarrow A$ and C are either both positive or are both negative (if that is the case, move the terms to the other side of the equation so that they are both positive).

$$Ax^2 + Cy^2 + Dx + Ey + F = 0$$

$$A\left(x^2 + \frac{D}{A}x + \frac{D^2}{4A^2}\right) + C\left(y^2 + \frac{E}{C}y + \frac{E^2}{4C^2}\right) = -F + \frac{D^2}{4A} + \frac{E^2}{4C}$$

$$A\left(x + \frac{D}{2A}\right)^2 + C\left(y + \frac{E}{2C}\right)^2 = \frac{CD^2 + AE^2 - 4ACF}{4AC}$$

$$\frac{\left(x + \frac{D}{2A}\right)^2}{\frac{CD^2 + AE^2 - 4ACF}{4A^2C}} + \frac{\left(y + \frac{E}{2C}\right)^2}{\frac{CD^2 + AE^2 - 4ACF}{4AC^2}} = 1$$

Since A and C are both positive, $4A^2C$ and $4AC^2$ are both positive. $CD^2 + AE^2 - 4ACF$ must be positive or the conic is degenerate. Thus, we have an ellipse with center $\left(-\frac{D}{2A}, -\frac{E}{2C}\right)$

(d) $AC < 0 \Rightarrow A$ and C have opposite signs. Let's assume that A is positive and C is negative. (If A is negative and C is positive, move the terms to the other side of the equation.)

From part (c) we have

$$\frac{\left(x + \frac{D}{2A}\right)^2}{\frac{CD^2 + AE^2 - 4ACF}{4A^2C}} + \frac{\left(y + \frac{E}{2C}\right)^2}{\frac{CD^2 + AE^2 - 4ACF}{4AC^2}} = 1$$

Since $A > 0$ and $C < 0$, the first denominator is positive if $CD^2 + AE^2 - 4ACF < 0$ and is negative if $CD^2 + AE^2 - 4ACF > 0$, since $4A^2C$ is negative. The second denominator would have the *opposite* sign since $4AC^2 > 0$. Thus, we have a hyperbola with center $\left(-\frac{D}{2A}, -\frac{E}{2C}\right)$

9. To change the orientation, we can just replace t with $-t$.

$x = \cos(-t) = \cos t$

$y = 2\sin(-t) = -2\sin t$

11. (a) $y^2 = \dfrac{t^2(1-t^2)^2}{(1+t^2)^2}, x^2 = \dfrac{(1-t^2)^2}{(1+t^2)^2}$

$$\frac{1-x}{1+x} = \frac{1-\left(\dfrac{1-t^2}{1+t^2}\right)}{1+\left(\dfrac{1-t^2}{1+t^2}\right)} = \frac{2t^2}{2} = t^2$$

Thus, $y^2 = x^2\left(\dfrac{1-x}{1+x}\right)$.

(b)
$$r^2 \sin^2\theta = r^2\cos^2\theta\left(\frac{1-r\cos\theta}{1+r\cos\theta}\right)$$
$$\sin^2\theta(1+r\cos\theta) = \cos^2\theta(1-r\cos\theta)$$
$$r\cos\theta\sin^2\theta + \sin^2\theta = \cos^2\theta - r\cos^3\theta$$
$$r\cos\theta(\sin^2\theta + \cos^2\theta) = \cos^2\theta - \sin^2\theta$$
$$r\cos\theta = \cos 2\theta$$
$$r = \cos 2\theta \cdot \sec\theta$$

(c)

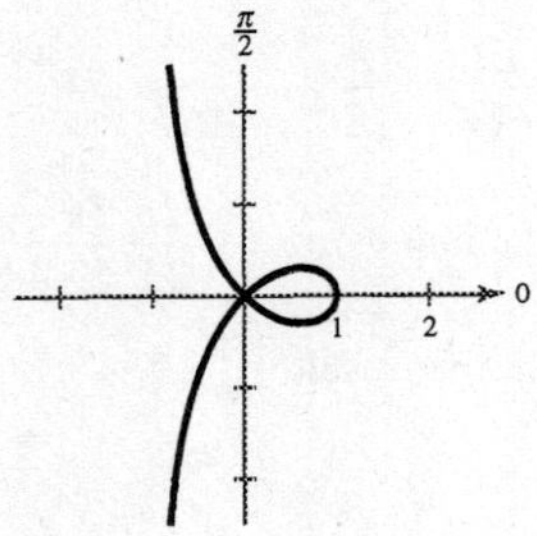

13. $r = a\sin\theta + b\cos\theta$

$r^2 = r(a\sin\theta + b\cos\theta)$

$r^2 = ar\sin\theta + br\cos\theta$

$x^2 + y^2 = ay + bx$

$x^2 + y^2 - bx - ay = 0$

$$\left(x^2 - bx + \frac{b^2}{4}\right) + \left(y^2 - ay + \frac{a^2}{4}\right) = \frac{a^2}{4} + \frac{b^2}{4}$$

$$\left(x - \frac{b}{2}\right)^2 + \left(y - \frac{a}{2}\right)^2 = \frac{a^2+b^2}{4}$$

This represents a circle with center $\left(\dfrac{b}{2}, \dfrac{a}{2}\right)$ and radius $r = \dfrac{1}{2}\sqrt{a^2+b^2}$

15.

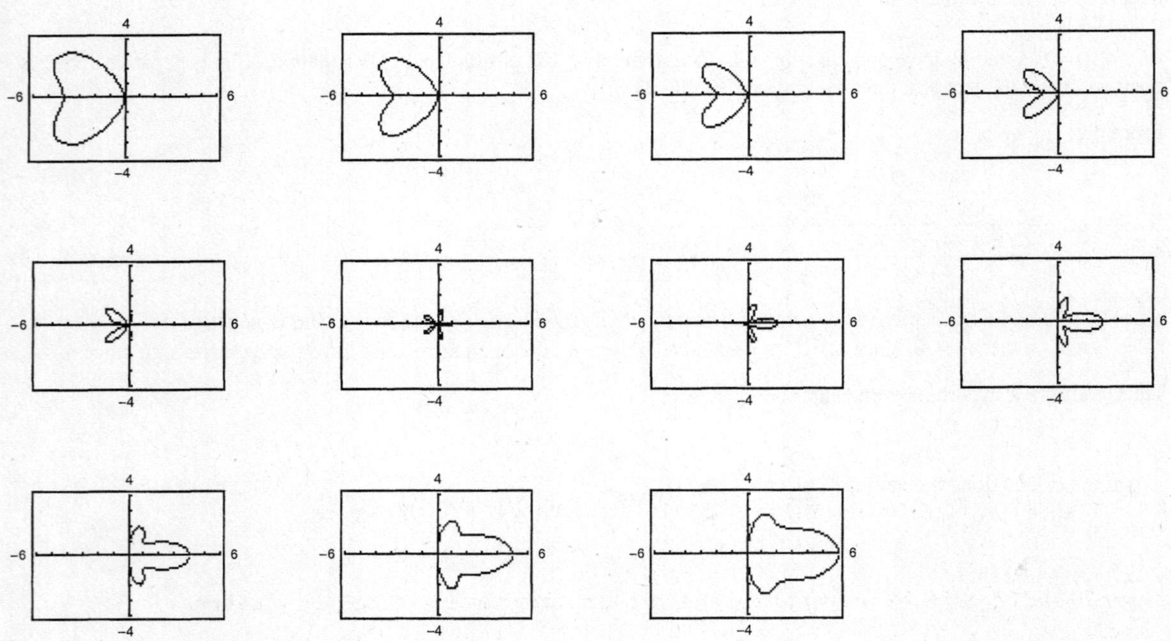

$n = 1, 2, 3, 4, 5$ produce "bells"; $n = -1, -2, -3, -4, -5$ produce "hearts".

Chapter 6 Practice Test

1. Find the angle, θ, between the lines $3x + 4y = 12$ and $4x - 3y = 12$.

2. Find the distance between the point $(5, -9)$ and the line $3x - 7y = 21$.

3. Find the vertex, focus and directrix of the parabola $x^2 - 6x - 4y + 1 = 0$.

4. Find an equation of the parabola with its vertex at $(2, -5)$ and focus at $(2, -6)$.

5. Find the center, foci, vertices, and eccentricity of the ellipse $x^2 + 4y^2 - 2x + 32y + 61 = 0$.

6. Find an equation of the ellipse with vertices $(0, \pm 6)$ and eccentricity $e = \frac{1}{2}$.

7. Find the center, vertices, foci, and asymptotes of the hyperbola $16y^2 - x^2 - 6x - 128y + 231 = 0$.

8. Find an equation of the hyperbola with vertices at $(\pm 3, 2)$ and foci at $(\pm 5, 2)$.

9. Rotate the axes to eliminate the xy-term. Sketch the graph of the resulting equation, showing both sets of axes.

   $5x^2 + 2xy + 5y^2 - 10 = 0$

10. Use the discriminant to determine whether the graph of the equation is a parabola, ellipse, or hyperbola.

    (a) $6x^2 - 2xy + y^2 = 0$ (b) $x^2 + 4xy + 4y^2 - x - y + 17 = 0$

11. Convert the polar point $\left(\sqrt{2}, \frac{3\pi}{4}\right)$ to rectangular coordinates.

12. Convert the rectangular point $\left(\sqrt{3}, -1\right)$ to polar coordinates.

13. Convert the rectangular equation $4x - 3y = 12$ to polar form.

14. Convert the polar equation $r = 5 \cos \theta$ to rectangular form.

15. Sketch the graph of $r = 1 - \cos \theta$.

16. Sketch the graph of $r = 5 \sin 2\theta$.

17. Sketch the graph of $r = \dfrac{3}{6 - \cos \theta}$.

18. Find a polar equation of the parabola with its vertex at $\left(6, \frac{\pi}{2}\right)$ and focus at $(0, 0)$.

For Exercises 19 and 20, eliminate the parameter and write the corresponding rectangular equation.

19. $x = 3 - 2 \sin \theta, y = 1 + 5 \cos \theta$

20. $x = e^{2t}, y = e^{4t}$

Chapter P Practice Test Solutions

1. $5(x-2)-4=3x+8$

$5x-10-4=3x+8$

$5x-14=3x+8$

$2x=22$

$x=11$

2. $y=\sqrt{7-x}$

x	7	6	3	-2
y	0	1	2	3

3. $y=\sqrt{25-x^2}$

Domain: $25-x^2 \geq 0$

$(5-x)(5+x) \geq 0$

Critical numbers: $x=\pm 5$

Test intervals: $(-\infty,-5), (-5,5), (5,\infty)$

Solution set: $[-5,5]$

4. $[x-(-3)]^2+(y-5)^2=6^2$

$(x+3)^2+(y-5)^2=36$

5. $m=\dfrac{-1-4}{3-2}=-5$

$y-4=-5(x-2)$

$y-4=-5x+10$

$y=-5x+14$

6. $y=\dfrac{4}{3}x-3$

7. $2x+3y=0$

$y=-\dfrac{2}{3}x$

$m_1=-\dfrac{2}{3}$

$\perp m_2=\dfrac{3}{2}$ through $(4, 1)$

$y-1=\dfrac{3}{2}(x-4)$

$y-1=\dfrac{3}{2}x-6$

$y=\dfrac{3}{2}x-5$

8. $(5, 32)$ and $(9, 44)$

$m=\dfrac{44-32}{9-5}=\dfrac{12}{4}=3$

$y-32=3(x-5)$

$y-32=3x-15$

$y=3x+17$

When $x=20$, $y=3(20)+17$

$y=\$77.$

9. $f(x - 3) = (x - 3)^2 - 2(x - 3) + 1$

$= x^2 - 6x + 9 - 2x + 6 + 1$

$= x^2 - 8x + 16$

10. $f(3) = 12 - 11 = 1$

$$\frac{f(x) - f(3)}{x - 3} = \frac{(4x - 11) - 1}{x - 3}$$

$$= \frac{4x - 12}{x - 3}$$

$$= \frac{4(x - 3)}{x - 3} = 4, x \neq 3$$

11. $f(x) = \sqrt{36 - x^2} = \sqrt{(6 + x)(6 - x)}$

Domain: $[-6, 6]$

Range: $[0, 6]$, because

$0 \leq (6 + x)(6 - x) \leq 36$ on this interval

12. (a) $6x - 5y + 4 = 0$

$y = \dfrac{6x + 4}{5}$ is a function of x.

(b) $x^2 + y^2 = 9$

$y = \pm\sqrt{9 - x^2}$ is not a function of x.

(c) $y^3 = x^2 + 6$

$y = \sqrt[3]{x^2 + 6}$ is a function of x.

13. Parabola

Vertex: $(0, -5)$

Intercepts: $(0, -5)$, $(\pm\sqrt{5}, 0)$

y-axis symmetry

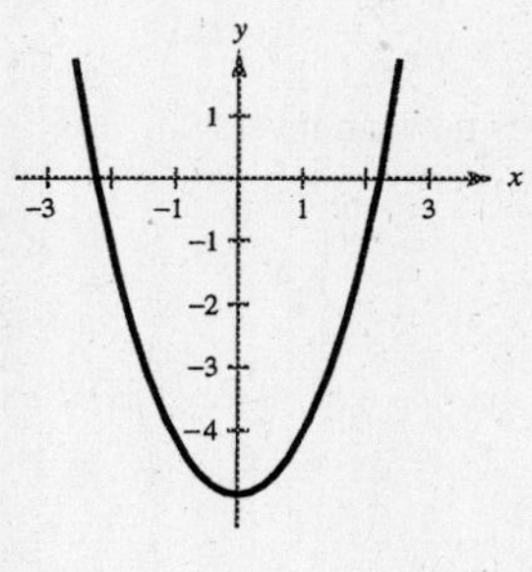

14. Intercepts: $(0, 3)$, $(-3, 0)$

x	0	1	-1	2	-2	-3	-4
y	3	4	2	5	1	0	1

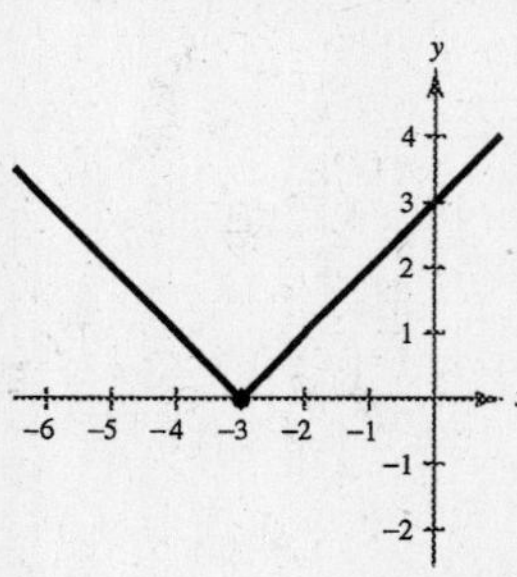

15.

x	0	1	2	3	-1	-2	-3
y	1	3	5	7	2	6	12

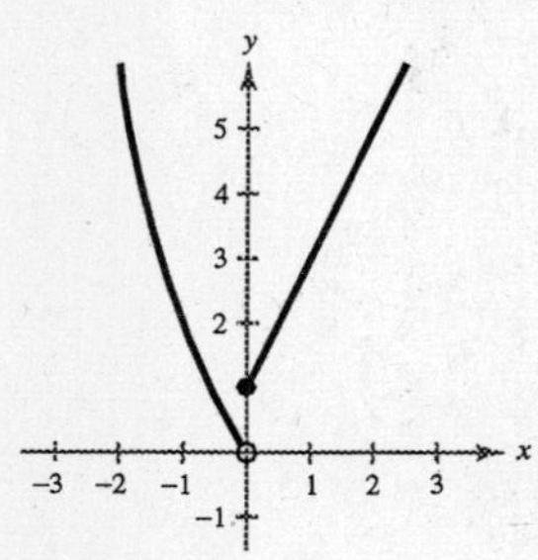

16. (a) $f(x + 2)$

Horizontal shift two units to the left

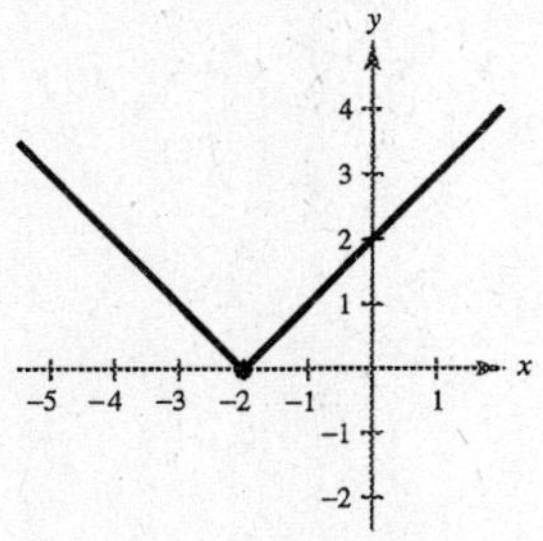

(b) $-f(x) + 2$

Reflection in the x-axis and a vertical shift two units upward

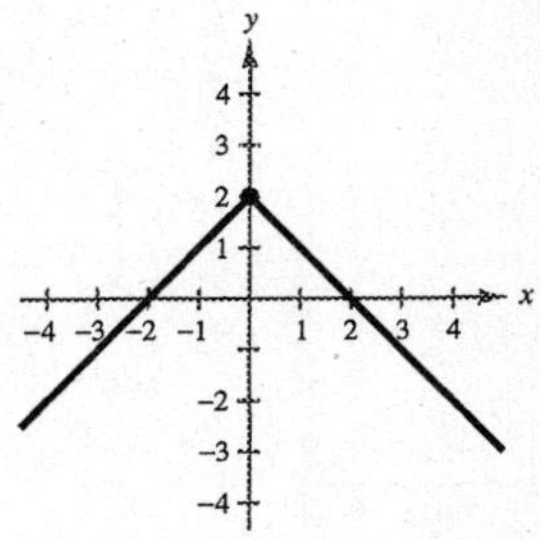

17. (a) $(g - f)(x) = g(x) - f(x)$

$= (2x^2 - 5) - (3x + 7)$

$= 2x^2 - 3x - 12$

(b) $(fg)(x) = f(x)g(x)$

$= (3x + 7)(2x^2 - 5)$

$= 6x^3 + 14x^2 - 15x - 35$

18. $f(g(x)) = f(2x + 3)$

$= (2x + 3)^2 - 2(2x + 3) + 16$

$= 4x^2 + 12x + 9 - 4x - 6 + 16$

$= 4x^2 + 8x + 19$

19. $f(x) = x^3 + 7$

$y = x^3 + 7$

$x = y^3 + 7$

$x - 7 = y^3$

$\sqrt[3]{x - 7} = y$

$f^{-1}(x) = \sqrt[3]{x - 7}$

20. (a) $f(x) = |x - 6|$ does not have an inverse.

Its graph does not pass the horizontal line test.

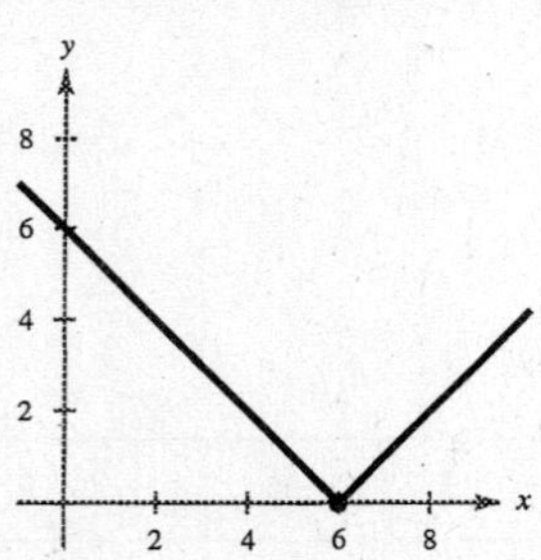

(b) $f(x) = ax + b, a \neq 0$ does have an inverse.

$y = ax + b$

$x = ay + b$

$\frac{x - b}{a} = y$

$f^{-1}(x) = \frac{x - b}{a}$

(c) $f(x) = x^3 - 19$ does have an inverse.

$y = x^3 - 19$

$x = y^3 - 19$

$x + 19 = y^3$

$\sqrt[3]{x + 19} = y$

$f^{-1}(x) = \sqrt[3]{x + 19}$

21.
$$f(x) = \sqrt{\frac{3-x}{x}}, \quad 0 < x \le 3, y \ge 0$$
$$y = \sqrt{\frac{3-x}{x}}$$
$$x = \sqrt{\frac{3-y}{y}}$$
$$x^2 = \frac{3-y}{y}$$
$$x^2y = 3 - y$$
$$x^2y + y = 3$$
$$y(x^2 + 1) = 3$$
$$y = \frac{3}{x^2+1}$$
$$f^{-1}(x) = \frac{3}{x^2+1}, \quad x \ge 0$$

22. False. The slopes of 3 and $\frac{1}{3}$ are not **negative** reciprocals.

23. True. Let $y = (fg)(x)$. Then $x = (f \circ g)^{-1}(y)$.

Also,
$$(f \circ g)(x) = y$$
$$f(g(x)) = y$$
$$g(x) = f^{-1}(y)$$
$$x = g^{-1}(f^{-1}(y))$$
$$x = (g^{-1} \circ f^{-1})(y)$$

Since $x = x$, we have $(f \circ g)^{-1}(y) = (g^{-1} \circ f^{-1})(y)$.

24. True. It must pass the vertical line test to be a function and it must pass the horizontal line test to have an inverse.

25. $y \approx 0.669x + 2.669$

Chapter 1 Practice Test Solutions

1. $350° = 350\left(\frac{\pi}{180}\right) = \frac{35\pi}{18}$

2. $\frac{5\pi}{9} = \frac{5\pi}{9} \cdot \frac{180}{\pi} = 100°$

3.
$$135°14'12'' = \left(135 + \tfrac{14}{60} + \tfrac{12}{3600}\right)°$$
$$\approx 135.2367°$$

4.
$$-22.569° = -(22° + 0.569(60)')$$
$$= -22°34.14'$$
$$= -(22°34' + 0.14(60)'')$$
$$\approx -22°34'8''$$

5. $\cos\theta = \frac{2}{3}$

$x = 2, r = 3, y = \pm\sqrt{9-4} = \pm\sqrt{5}$

$\tan\theta = \frac{y}{x} = \pm\frac{\sqrt{5}}{2}$

6. $\sin\theta = 0.9063$

$\theta = \arcsin(0.9063)$

$\theta = 65° = \frac{13\pi}{36}$ or $\theta = 180° - 65° = 115° = \frac{23\pi}{36}$

7. $\tan 20° = \frac{35}{x}$

$x = \frac{35}{\tan 20°} \approx 96.1617$

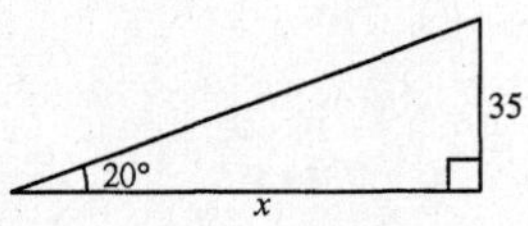

8. $\theta = \frac{6\pi}{5}$, θ is in Quadrant III.

Reference angle: $\frac{6\pi}{5} - \pi = \frac{\pi}{5}$ or $36°$

9. $\csc 3.92 = \frac{1}{\sin 3.92} \approx -1.4242$

10. $\tan\theta = 6 = \frac{6}{1}$, θ lies in Quandrant III.

$y = -6, x = -1, r = \sqrt{36+1} = \sqrt{37}$,

so $\sec\theta = \frac{\sqrt{37}}{-1} \approx -6.0828$.

11. Period: 4π

Amplitude: 3

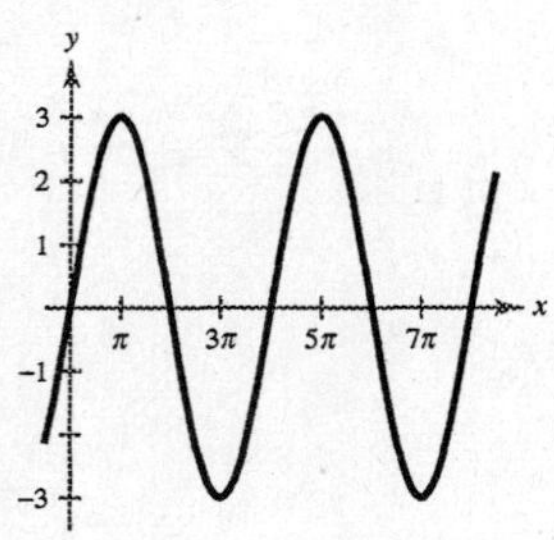

12. Period: 2π

Amplitude: 2

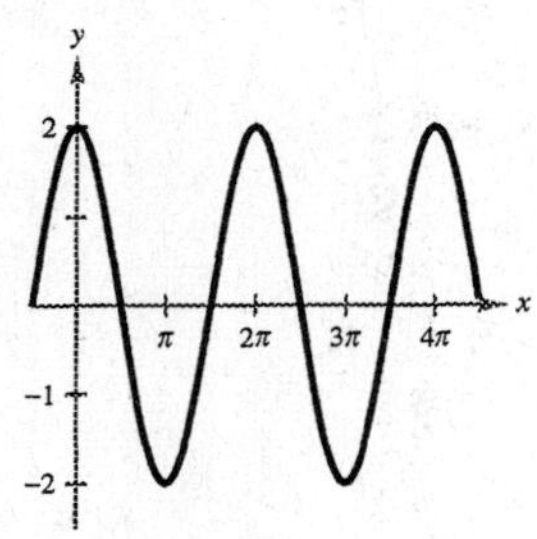

13. Period: $\frac{\pi}{2}$

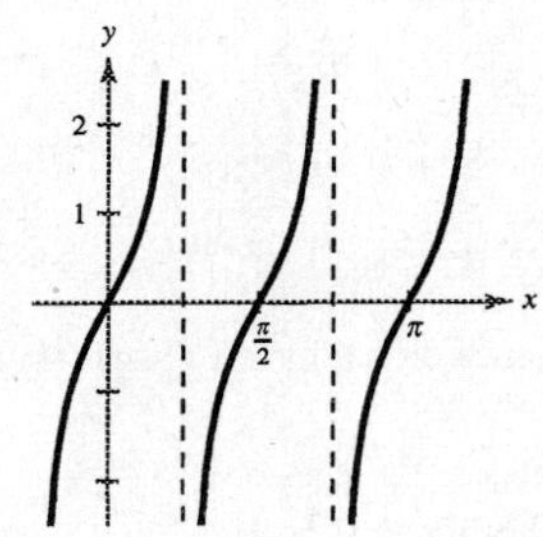

14. Period: 2π

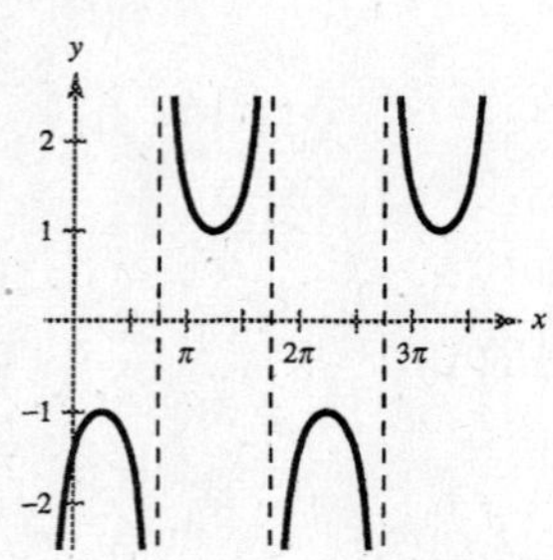

15.

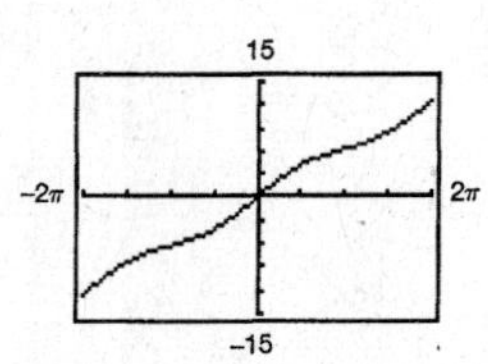

16.

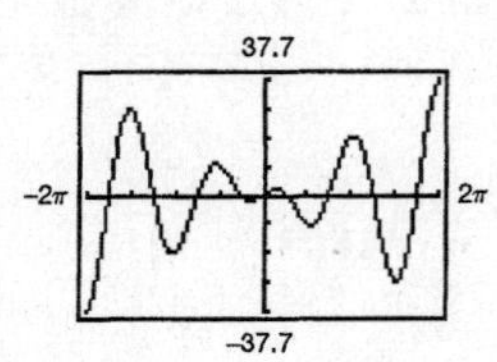

17. $\theta = \arcsin 1$

$\sin \theta = 1$

$\theta = \dfrac{\pi}{2} = 90°$

18. $\theta = \arctan(-3)$

$\tan \theta = -3$

$\theta \approx -1.249 \approx -71.565°$

19. $\sin\left(\arccos \dfrac{4}{\sqrt{35}}\right)$

$\sin \theta = \dfrac{\sqrt{19}}{\sqrt{35}} \approx 0.7368$

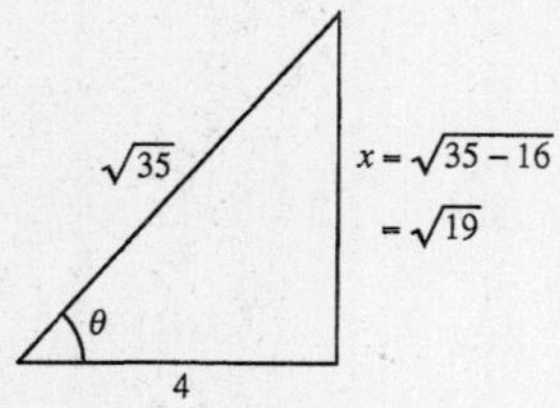

20. $\cos\left(\arcsin \dfrac{x}{4}\right)$

$\cos \theta = \dfrac{\sqrt{16 - x^2}}{4}$

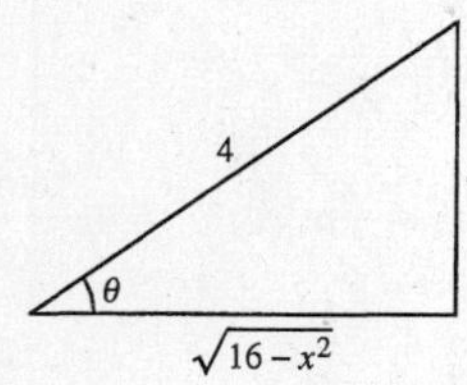

21. Given $A = 40°, c = 12$

$B = 90° - 40° = 50°$

$\sin 40° = \dfrac{a}{12}$

$a = 12 \sin 40° \approx 7.713$

$\cos 40° = \dfrac{b}{12}$

$b = 12 \cos 40° \approx 9.193$

22. Given $B = 6.84°, a = 21.3$

$A = 90° - 6.84° = 83.16°$

$\sin 83.16° = \dfrac{21.3}{c}$

$c = \dfrac{21.3}{\sin 83.16°} \approx 21.453$

$\tan 83.16° = \dfrac{21.3}{b}$

$b = \dfrac{21.3}{\tan 83.16°} \approx 2.555$

23. Given $a = 5, b = 9$

$c = \sqrt{25 + 81} = \sqrt{106} \approx 10.296$

$\tan A = \dfrac{5}{9}$

$A = \arctan \dfrac{5}{9} \approx 29.055°$

$B \approx 90° - 29.055° = 60.945°$

24. $\sin 67° = \dfrac{x}{20}$

$x = 20 \sin 67° \approx 18.41$ feet

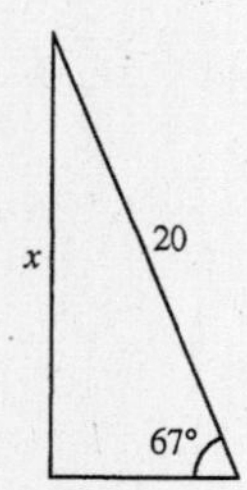

25. $\tan 5° = \dfrac{250}{x}$

$x = \dfrac{250}{\tan 5°}$

≈ 2857.513 feet

≈ 0.541 mi

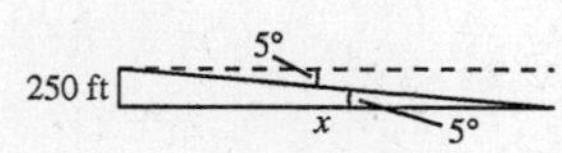

Chapter 2 Practice Test Solutions

1. $\tan x = \dfrac{4}{11}$, $\sec x < 0 \implies x$ is in Quadrant III.

$y = -4,\ x = -11,\ r = \sqrt{16 + 121} = \sqrt{137}$

$\sin x = -\dfrac{4}{\sqrt{137}} = -\dfrac{4\sqrt{137}}{137}$ $\qquad \csc x = -\dfrac{\sqrt{137}}{4}$

$\cos x = -\dfrac{11}{\sqrt{137}} = -\dfrac{11\sqrt{137}}{137}$ $\qquad \sec x = -\dfrac{\sqrt{137}}{11}$

$\tan x = \dfrac{4}{11}$ $\qquad \cot x = \dfrac{11}{4}$

2. $\dfrac{\sec^2 x + \csc^2 x}{\csc^2 x(1 + \tan^2 x)} = \dfrac{\sec^2 x + \csc^2 x}{\csc^2 x + (\csc^2 x)\tan^2 x} = \dfrac{\sec^2 x + \csc^2 x}{\csc^2 x + \dfrac{1}{\sin^2 x} \cdot \dfrac{\sin^2 x}{\cos^2 x}}$

$$= \frac{\sec^2 x + \csc^2 x}{\csc^2 x + \dfrac{1}{\cos^2 x}} = \frac{\sec^2 x + \csc^2 x}{\csc^2 x + \sec^2 x} = 1$$

3. $\ln|\tan\theta| - \ln|\cot\theta| = \ln\dfrac{|\tan\theta|}{|\cot\theta|} = \ln\left|\dfrac{\sin\theta/\cos\theta}{\cos\theta/\sin\theta}\right| = \ln\left|\dfrac{\sin^2\theta}{\cos^2\theta}\right| = \ln|\tan^2\theta| = 2\ln|\tan\theta|$

4. $\cos\left(\dfrac{\pi}{2} - x\right) = \dfrac{1}{\csc x}$ is true since $\cos\left(\dfrac{\pi}{2} - x\right) = \sin x = \dfrac{1}{\csc x}$.

5. $\sin^4 x + (\sin^2 x)\cos^2 x = \sin^2 x(\sin^2 x + \cos^2 x) = \sin^2 x(1) = \sin^2 x$

6. $(\csc x + 1)(\csc x - 1) = \csc^2 x - 1 = \cot^2 x$

7. $\dfrac{\cos^2 x}{1 - \sin x} \cdot \dfrac{1 + \sin x}{1 + \sin x} = \dfrac{\cos^2 x(1 + \sin x)}{1 - \sin^2 x} = \dfrac{\cos^2 x(1 + \sin x)}{\cos^2 x} = 1 + \sin x$

8. $\dfrac{1 + \cos\theta}{\sin\theta} + \dfrac{\sin\theta}{1 + \cos\theta} = \dfrac{(1 + \cos\theta)^2 + \sin^2\theta}{\sin\theta(1 + \cos\theta)}$

$$= \frac{1 + 2\cos\theta + \cos^2\theta + \sin^2\theta}{\sin\theta(1 + \cos\theta)} = \frac{2 + 2\cos\theta}{\sin\theta(1 + \cos\theta)} = \frac{2}{\sin\theta} = 2\csc\theta$$

9. $\tan^4 x + 2\tan^2 x + 1 = (\tan^2 x + 1)^2 = (\sec^2 x)^2 = \sec^4 x$

10. (a) $\sin 105° = \sin(60° + 45°) = \sin 60° \cos 45° + \cos 60° \sin 45°$

$$= \frac{\sqrt{3}}{2} \cdot \frac{\sqrt{2}}{2} + \frac{1}{2} \cdot \frac{\sqrt{2}}{2} = \frac{\sqrt{2}}{4}(\sqrt{3} + 1)$$

(b) $\tan 15° = \tan(60° - 45°) = \dfrac{\tan 60° - \tan 45°}{1 + \tan 60° \tan 45°}$

$$= \frac{\sqrt{3} - 1}{1 + \sqrt{3}} \cdot \frac{1 - \sqrt{3}}{1 - \sqrt{3}} = \frac{2\sqrt{3} - 1 - 3}{1 - 3} = \frac{2\sqrt{3} - 4}{-2} = 2 - \sqrt{3}$$

11. $(\sin 42°)\cos 38° - (\cos 42°)\sin 38° = \sin(42° - 38°) = \sin 4°$

12. $\tan\left(\theta + \dfrac{\pi}{4}\right) = \dfrac{\tan\theta + \tan\left(\dfrac{\pi}{4}\right)}{1 - (\tan\theta)\tan\left(\dfrac{\pi}{4}\right)} = \dfrac{\tan\theta + 1}{1 - \tan\theta(1)} = \dfrac{1 + \tan\theta}{1 - \tan\theta}$

13. $\sin(\arcsin x - \arccos x) = \sin(\arcsin x)\cos(\arccos x) - \cos(\arcsin x)\sin(\arccos x)$

$$= (x)(x) - \left(\sqrt{1 - x^2}\right)\left(\sqrt{1 - x^2}\right) = x^2 - (1 - x^2) = 2x^2 - 1$$

14. (a) $\cos(120°) = \cos[2(60°)] = 2\cos^2 60° - 1 = 2\left(\frac{1}{2}\right)^2 - 1 = -\frac{1}{2}$

(b) $\tan(300°) = \tan[2(150°)] = \dfrac{2\tan 150°}{1 - \tan^2 150°} = \dfrac{-\frac{2\sqrt{3}}{3}}{1 - \left(\frac{1}{3}\right)} = -\sqrt{3}$

15. (a) $\sin 22.5° = \sin\dfrac{45°}{2} = \sqrt{\dfrac{1 - \cos 45°}{2}} = \sqrt{\dfrac{1 - \frac{\sqrt{2}}{2}}{2}} = \dfrac{\sqrt{2 - \sqrt{2}}}{2}$

(b) $\tan\dfrac{\pi}{12} = \tan\dfrac{\frac{\pi}{6}}{2} = \dfrac{\sin\frac{\pi}{6}}{1 + \cos\left(\frac{\pi}{6}\right)} = \dfrac{\frac{1}{2}}{1 + \frac{\sqrt{3}}{2}} = \dfrac{1}{2 + \sqrt{3}} = 2 - \sqrt{3}$

16. $\sin\theta = \frac{4}{5}$, θ lies in Quadrant II $\Rightarrow$ $\cos\theta = -\frac{3}{5}$.

$$\cos\frac{\theta}{2} = \sqrt{\frac{1 + \cos\theta}{2}} = \sqrt{\frac{1 - \frac{3}{5}}{2}} = \sqrt{\frac{2}{10}} = \frac{1}{\sqrt{5}} = \frac{\sqrt{5}}{5}$$

17. $(\sin^2 x)\cos^2 x = \dfrac{1 - \cos 2x}{2} \cdot \dfrac{1 + \cos 2x}{2} = \dfrac{1}{4}[1 - \cos^2 2x] = \dfrac{1}{4}\left[1 - \dfrac{1 + \cos 4x}{2}\right]$

$= \dfrac{1}{8}[2 - (1 + \cos 4x)] = \dfrac{1}{8}[1 - \cos 4x]$

18. $6(\sin 5\theta)\cos 2\theta = 6\left\{\frac{1}{2}[\sin(5\theta + 2\theta) + \sin(5\theta - 2\theta)]\right\} = 3[\sin 7\theta + \sin 3\theta]$

19. $\sin(x + \pi) + \sin(x - \pi) = 2\left(\sin\dfrac{[(x + \pi) + (x - \pi)]}{2}\right)\cos\dfrac{[(x + \pi) - (x - \pi)]}{2}$

$= 2\sin x\cos\pi = -2\sin x$

20. $\dfrac{\sin 9x + \sin 5x}{\cos 9x - \cos 5x} = \dfrac{2\sin 7x\cos 2x}{-2\sin 7x\sin 2x} = -\dfrac{\cos 2x}{\sin 2x} = -\cot 2x$

21. $\frac{1}{2}[\sin(u + v) - \sin(u - v)] = \frac{1}{2}\{(\sin u)\cos v + (\cos u)\sin v - [(\sin u)\cos v - (\cos u)\sin v]\}$

$= \frac{1}{2}[2(\cos u)\sin v] = (\cos u)\sin v$

22. $4\sin^2 x = 1$

$$\sin^2 x = \frac{1}{4}$$

$$\sin x = \pm\frac{1}{2}$$

$$\sin x = \frac{1}{2} \quad \text{or} \quad \sin x = -\frac{1}{2}$$

$$x = \frac{\pi}{6} \text{ or } \frac{5\pi}{6} \qquad x = \frac{7\pi}{6} \text{ or } \frac{11\pi}{6}$$

23. $\tan^2\theta + (\sqrt{3} - 1)\tan\theta - \sqrt{3} = 0$

$$(\tan\theta - 1)(\tan\theta + \sqrt{3}) = 0$$

$$\tan\theta = 1 \quad \text{or} \quad \tan\theta = -\sqrt{3}$$

$$\theta = \frac{\pi}{4} \text{ or } \frac{5\pi}{4} \qquad \theta = \frac{2\pi}{3} \text{ or } \frac{5\pi}{3}$$

24. $\sin 2x = \cos x$

$$2(\sin x)\cos x - \cos x = 0$$

$$\cos x(2\sin x - 1) = 0$$

$$\cos x = 0 \quad \text{or} \quad \sin x = \frac{1}{2}$$

$$x = \frac{\pi}{2} \text{ or } \frac{3\pi}{2} \qquad x = \frac{\pi}{6} \text{ or } \frac{5\pi}{6}$$

25. $\tan^2 x - 6\tan x + 4 = 0$

$$\tan x = \frac{-(-6) \pm \sqrt{(-6)^2 - 4(1)(4)}}{2(1)}$$

$$\tan x = \frac{6 \pm \sqrt{20}}{2} = 3 \pm \sqrt{5}$$

$$\tan x = 3 + \sqrt{5} \quad \text{or} \quad \tan x = 3 - \sqrt{5}$$

$$x \approx 1.3821 \text{ or } 4.5237 \qquad x = 0.6524 \text{ or } 3.7940$$

Chapter 3 Practice Test Solutions

1. $C = 180° - (40° + 12°) = 128°$

$$a = \sin 40°\left(\frac{100}{\sin 12°}\right) \approx 309.164$$

$$c = \sin 128°\left(\frac{100}{\sin 12°}\right) \approx 379.012$$

2. $\sin A = 5\left(\frac{\sin 150°}{20}\right) = 0.125$

$$A \approx 7.181°$$

$$B \approx 180° - (150° + 7.181°) = 22.819°$$

$$b = \sin 22.819°\left(\frac{20}{\sin 150°}\right) \approx 15.513$$

3. Area $= \frac{1}{2}ab\sin C$

$$= \tfrac{1}{2}(3)(5)\sin 130°$$

≈ 5.745 square units

4. $h = b\sin A$

$$= 35\sin 22.5°$$

$$\approx 13.394$$

$$a = 10$$

Since $a < h$ and A is acute, the triangle has no solution.

5. $\cos A = \frac{(53)^2 + (38)^2 - (49)^2}{2(53)(38)} \approx 0.4598$

$$A \approx 62.627°$$

$$\cos B = \frac{(49)^2 + (38)^2 - (53)^2}{2(49)(38)} \approx 0.2782$$

$$B \approx 73.847°$$

$$C \approx 180° - (62.627° + 73.847°)$$

$$= 43.526°$$

6. $c^2 = (100)^2 + (300)^2 - 2(100)(300)\cos 29°$

$$\approx 47522.8176$$

$$c \approx 218$$

$$\cos A = \frac{(300)^2 + (218)^2 - (100)^2}{2(300)(218)} \approx 0.97495$$

$$A \approx 12.85°$$

$$B \approx 180° - (12.85° + 29°) = 138.15°$$

7. $s = \dfrac{a+b+c}{2} = \dfrac{4.1+6.8+5.5}{2} = 8.2$

$\text{Area} = \sqrt{s(s-a)(s-b)(s-c)}$

$= \sqrt{8.2(8.2-4.1)(8.2-6.8)(8.2-5.5)}$

≈ 11.273 square units

8. $x^2 = (40)^2 + (70)^2 - 2(40)(70)\cos 168°$

≈ 11977.6266

$x \approx 190.442$ miles

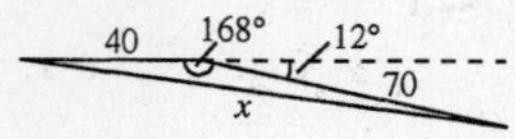

9. $\mathbf{w} = 4(3\mathbf{i} + \mathbf{j}) - 7(-\mathbf{i} + 2\mathbf{j})$

$= 19\mathbf{i} - 10\mathbf{j}$

10. $\dfrac{\mathbf{v}}{\|\mathbf{v}\|} = \dfrac{5\mathbf{i} + 3\mathbf{j}}{\sqrt{25+9}} = \dfrac{5}{\sqrt{34}}\mathbf{i} - \dfrac{3}{\sqrt{34}}\mathbf{j}$

$= \dfrac{5\sqrt{34}}{34}\mathbf{i} - \dfrac{3\sqrt{34}}{34}\mathbf{j}$

11. $\mathbf{u} = 6\mathbf{i} + 5\mathbf{j} \qquad \mathbf{v} = 2\mathbf{i} - 3\mathbf{j}$

$\mathbf{u} \cdot \mathbf{v} = 6(2) + 5(-3) = -3$

$\|\mathbf{u}\| = \sqrt{61} \qquad \|\mathbf{v}\| = \sqrt{13}$

$\cos\theta = \dfrac{-3}{\sqrt{61}\sqrt{13}}$

$\theta \approx 96.116°$

12. $4(\mathbf{i}\cos 30° + \mathbf{j}\sin 30°)$

$= 4\left(\dfrac{\sqrt{3}}{2}\mathbf{i} + \dfrac{1}{2}\mathbf{j}\right)$

$= \langle 2\sqrt{3}, 2\rangle$

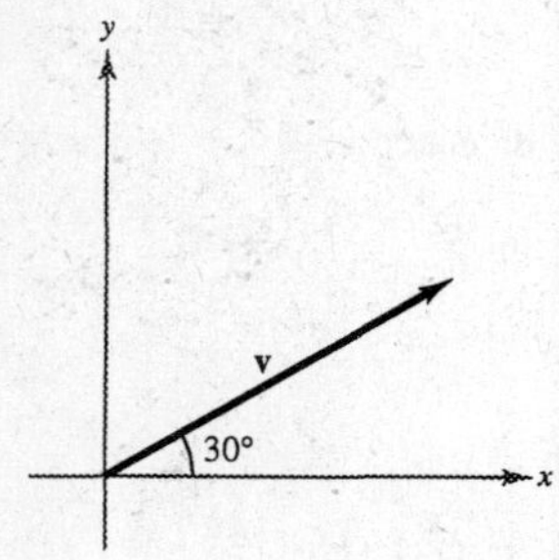

13. $\text{proj}_{\mathbf{v}}\mathbf{u} = \left(\dfrac{\mathbf{u}\cdot\mathbf{v}}{\|\mathbf{v}\|^2}\right)\mathbf{v} = \dfrac{-10}{20}\langle -2, 4\rangle = \langle 1, -2\rangle$

14. $\mathbf{u} = 7\cos 35°\mathbf{i} + 7\sin 35°\mathbf{j}$

$\mathbf{v} = 4\cos 123°\mathbf{i} + 4\sin 123°\mathbf{j}$

$\mathbf{u} + \mathbf{v} = (7\cos 35° + 4\cos 123°)\mathbf{i} + (7\sin 35° + 4\sin 123°)\mathbf{j}$

$\approx \langle 3.56, 7.37\rangle$

15. Answer is not unique. Two possibilities are: $\langle 10, 3\rangle$ and $\langle -10, -3\rangle$

Chapter 4 Practice Test Solutions

1. $4 + \sqrt{-81} - 3i^2 = 4 + 9i + 3 = 7 + 9i$

2. $\dfrac{3+i}{5-4i} \cdot \dfrac{5+4i}{5+4i} = \dfrac{15 + 12i + 5i + 4i^2}{25+16} = \dfrac{11+17i}{41} = \dfrac{11}{41} + \dfrac{17}{41}i$

3. $x = \dfrac{-(-4) \pm \sqrt{(-4)^2 - 4(1)(7)}}{2(1)} = \dfrac{4 \pm \sqrt{-12}}{2} = \dfrac{4 \pm 2\sqrt{3}i}{2} = 2 \pm \sqrt{3}i$

4. False: $\sqrt{-6}\sqrt{-6} = \left(\sqrt{6}i\right)\left(\sqrt{6}i\right) = \sqrt{36}i^2 = 6(-1) = -6$

5. $b^2 - 4ac = (-8)^2 - 4(3)(7) = -20 < 0$

Two complex solutions.

6. $x^4 + 13x^2 + 36 = 0$

$(x^2 + 4)(x^2 + 9) = 0$

$x^2 + 4 = 0 \Rightarrow x^2 = -4 \Rightarrow x = \pm 2i$

$x^2 + 9 = 0 \Rightarrow x^2 = -9 \Rightarrow x = \pm 3i$

7.
$$\begin{aligned} f(x) &= (x-3)[x-(-1+4i)][x-(-1-4i)] \\ &= (x-3)[(x+1)-4i][(x+1)+4i] \\ &= (x-3)[(x+1)^2 - 16i^2] \\ &= (x-3)(x^2+2x+17) \\ &= x^3 - x^2 + 11x - 51 \end{aligned}$$

8. Since $x = 4 + i$ is a zero, so is its conjugate $4 - i$.

$$\begin{aligned} [x-(4+i)][x-(4-i)] &= [(x-4)-i][(x-4)+i] \\ &= (x-4)^2 - i^2 \\ &= x^2 - 8x + 17 \end{aligned}$$

$$\begin{array}{r} x - 2 \\ x^2 - 8x + 17 \overline{\smash{\big)}\, x^3 - 10x^2 + 33x - 34} \\ \underline{x^3 - 8x^2 + 17x } \\ -2x^2 + 16x - 34 \\ \underline{-2x^2 + 16x - 34} \\ 0 \end{array}$$

Thus, $f(x) = (x^2 - 8x + 17)(x - 2)$ and the zeros of $f(x)$ are: $4 \pm i, 2$

9. $r = \sqrt{25 + 25} = \sqrt{50} = 5\sqrt{2}$

$\tan\theta = \dfrac{-5}{5} = -1$

Since z is in Quadrant IV,

$\theta = 315°$

$z = 5\sqrt{2}(\cos 315° + i \sin 315°)$.

10. $\cos 225° = -\dfrac{\sqrt{2}}{2}, \sin 225° = -\dfrac{\sqrt{2}}{2}$

$$\begin{aligned} z &= 6\left(-\frac{\sqrt{2}}{2} - i\frac{\sqrt{2}}{2}\right) \\ &= -3\sqrt{2} - 3\sqrt{2}i \end{aligned}$$

11.
$$\begin{aligned} [7(\cos 23° + i \sin 23°)][4(\cos 7° + i \sin 7°)] &= 7(4)[\cos(23° + 7°) + i \sin(23° + 7°)] \\ &= 28(\cos 30° + i \sin 30°) \\ &= 14\sqrt{3} + 14i \end{aligned}$$

12.
$$\frac{9\left(\cos\frac{5\pi}{4} + i \sin\frac{5\pi}{4}\right)}{3(\cos\pi + i\sin\pi)} = \frac{9}{3}\left[\cos\left(\frac{5\pi}{4} - \pi\right) + i\sin\left(\frac{5\pi}{4} - \pi\right)\right] = 3\left(\cos\frac{\pi}{4} + i\sin\frac{\pi}{4}\right) = \frac{3\sqrt{2}}{2} + \frac{3\sqrt{2}}{2}i$$

13.
$$\begin{aligned} (2+2i)^8 &= \left[2\sqrt{2}(\cos 45° + i \sin 45°)\right]^8 = (2\sqrt{2})^8[\cos(8)(45°) + i\sin(8)(45°)] \\ &= 4096[\cos 360° + i \sin 360°] = 4096 \end{aligned}$$

14. $z = 8\left(\cos\frac{\pi}{3} + i\sin\frac{\pi}{3}\right), n = 3$

The cube roots of z are: $\sqrt[3]{8}\left[\cos\frac{(\pi/3) + 2\pi k}{3} + i\sin\frac{(\pi/3) + 2\pi k}{3}\right], k = 0, 1, 2.$

For $k = 0$, $\sqrt[3]{8}\left[\cos\frac{\pi/3}{3} + i\sin\frac{\pi/3}{3}\right] = 2\left(\cos\frac{\pi}{9} + i\sin\frac{\pi}{9}\right).$

For $k = 1$, $\sqrt[3]{8}\left[\cos\frac{\pi/3 + 2\pi}{3} + i\sin\frac{\pi/3 + 2\pi}{3}\right] = 2\left(\cos\frac{7\pi}{9} + i\sin\frac{7\pi}{9}\right).$

For $k = 2$, $\sqrt[3]{8}\left[\cos\frac{\pi/3 + 4\pi}{3} + i\sin\frac{\pi/3 + 4\pi}{3}\right] = 2\left(\cos\frac{13\pi}{9} + i\sin\frac{13\pi}{9}\right).$

15. $x^4 = -i = 1\left(\cos\frac{3\pi}{2} + i\sin\frac{3\pi}{2}\right)$

The fourth roots are: $\sqrt[4]{1}\left[\cos\frac{(3\pi/2) + 2\pi k}{4} + i\sin\frac{(3\pi/2) + 2\pi k}{4}\right], k = 0, 1, 2, 3$

For $k = 0$, $\cos\left(\frac{3\pi/2}{4}\right) + i\sin\left(\frac{3\pi/2}{4}\right) = \cos\frac{3\pi}{8} + i\sin\frac{3\pi}{8}.$

For $k = 1$, $\cos\left(\frac{3\pi/2 + 2\pi}{4}\right) + i\sin\left(\frac{3\pi/2 + 2\pi}{4}\right) = \cos\frac{7\pi}{8} + i\sin\frac{7\pi}{8}.$

For $k = 2$, $\cos\left(\frac{3\pi/2 + 4\pi}{4}\right) + i\sin\left(\frac{3\pi/2 + 4\pi}{4}\right) = \cos\frac{11\pi}{8} + i\sin\frac{11\pi}{8}.$

For $k = 3$, $\cos\left(\frac{3\pi/2 + 6\pi}{4}\right) + i\sin\left(\frac{3\pi/2 + 6\pi}{4}\right) = \cos\frac{15\pi}{8} + i\sin\frac{15\pi}{8}.$

Chapter 5 Practice Test Solutions

1. $x^{3/5} = 8$

$x = 8^{5/3} = (\sqrt[3]{8})^5 = 2^5 = 32$

2. $3^{x-1} = \frac{1}{81}$

$3^{x-1} = 3^{-4}$

$x - 1 = -4$

$x = -3$

3. $f(x) = 2^{-x} = \left(\frac{1}{2}\right)^x$

x	-2	-1	0	1	2
$f(x)$	4	2	1	$\frac{1}{2}$	$\frac{1}{4}$

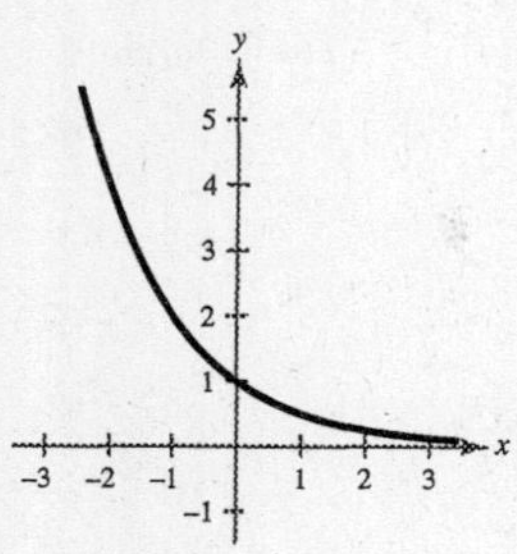

4. $g(x) = e^x + 1$

x	-2	-1	0	1	2
$g(x)$	1.14	1.37	2	3.72	8.39

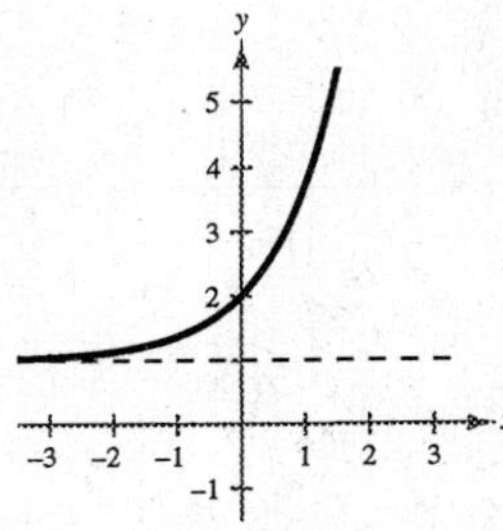

5. (a) $A = P\left(1 + \frac{r}{n}\right)^{nt}$

$A = 5000\left(1 + \frac{0.09}{12}\right)^{12(3)} \approx \$6543.23$

(b) $A = P\left(1 + \frac{r}{n}\right)^{nt}$

$A = 5000\left(1 + \frac{0.09}{4}\right)^{4(3)} \approx \$6530.25$

(c) $A = Pe^{rt}$

$A = 5000e^{(0.09)(3)} \approx \$6549.82$

6. $7^{-2} = \frac{1}{49}$

$\log_7 \frac{1}{49} = -2$

7. $x - 4 = \log_2 \frac{1}{64}$

$2^{x-4} = \frac{1}{64}$

$2^{x-4} = 2^{-6}$

$x - 4 = -6$

$x = -2$

8. $\log_b \sqrt[4]{\frac{8}{25}} = \frac{1}{4} \log_b \frac{8}{25}$

$= \frac{1}{4}[\log_b 8 - \log_b 25]$

$= \frac{1}{4}[\log_b 2^3 - \log_b 5^2]$

$= \frac{1}{4}[3 \log_b 2 - 2 \log_b 5]$

$= \frac{1}{4}[3(0.3562) - 2(0.8271)]$

$= -0.1464$

9. $5 \ln x - \frac{1}{2} \ln y + 6 \ln z = \ln x^5 - \ln \sqrt{y} + \ln z^6 = \ln\left(\frac{x^5 z^6}{\sqrt{y}}\right), z > 0$

10. $\log_9 28 = \frac{\log 28}{\log 9} \approx 1.5166$

11. $\log N = 0.6646$

$N = 10^{0.6646} \approx 4.62$

12.

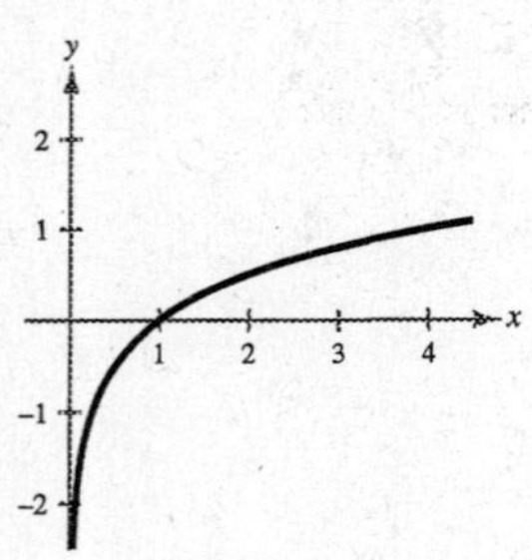

13. Domain:

$x^2 - 9 > 0$

$(x + 3)(x - 3) > 0$

$x < -3$ or $x > 3$

14.

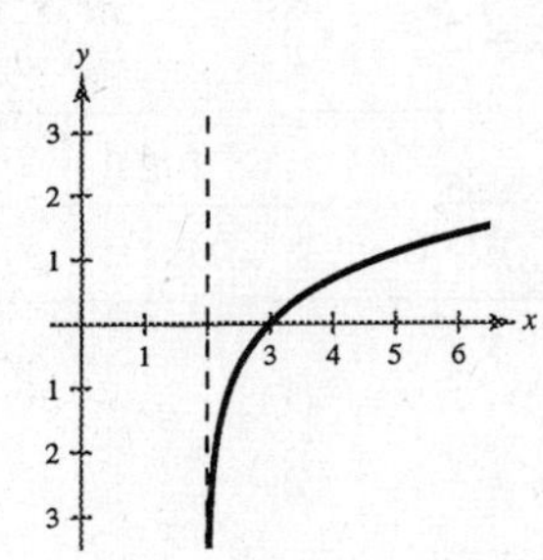

15. False. $\frac{\ln x}{\ln y} \neq \ln(x - y)$ since $\frac{\ln x}{\ln y} = \log_y x$.

16. $5^3 = 41$

$x = \log_5 41 = \frac{\ln 41}{\ln 5} \approx 2.3074$

17. $x - x^2 = \log_5 \frac{1}{25}$

$$5^{x-x^2} = \frac{1}{25}$$
$$5^{x-x^2} = 5^{-2}$$
$$x - x^2 = -2$$
$$0 = x^2 - x - 2$$
$$0 = (x + 1)(x - 2)$$
$$x = -1 \text{ or } x = 2$$

18. $\log_2 x + \log_2(x - 3) = 2$

$$\log_2[x(x - 3)] = 2$$
$$x(x - 3) = 2^2$$
$$x^2 - 3x = 4$$
$$x^2 - 3x - 4 = 0$$
$$(x + 1)(x - 4) = 0$$
$$x = 4$$
$$x = -1 \text{ (extraneous)}$$

$x = 4$ is the only solution.

19. $\dfrac{e^x + e^{-x}}{3} = 4$

$$e^x(e^x + e^{-x}) = 12e^x$$
$$e^{2x} + 1 = 12e^x$$
$$e^{2x} - 12e^x + 1 = 0$$
$$e^x = \frac{12 \pm \sqrt{144 - 4}}{2}$$

$e^x \approx 11.9161$ or $e^x \approx 0.0839$

$x = \ln 11.9161$ $\quad$ $x = \ln 0.0839$

$x \approx 2.478$ $\quad$ $x \approx -2.478$

20. $A = Pe^{et}$

$$12{,}000 = 6000e^{0.13t}$$
$$2 = e^{0.13t}$$
$$0.13t = \ln 2$$
$$t = \frac{\ln 2}{0.13}$$

$t \approx 5.3319$ years or 5 years 4 months

Chapter 6 Practice Test Solutions

1. $3x + 4y = 12 \Rightarrow y = -\frac{3}{4}x + 3 \Rightarrow m_1 = -\frac{3}{4}$

$4x - 3y = 12 \Rightarrow y = \frac{4}{3}x - 4 \Rightarrow m_2 = \frac{4}{3}$

$$\tan\theta = \left|\frac{\frac{4}{3} - \left(-\frac{3}{4}\right)}{1 + \left(\frac{4}{3}\right)\left(-\frac{3}{4}\right)}\right| = \left|\frac{\frac{25}{12}}{0}\right|$$

Since $\tan\theta$ is undefined, the lines are perpendicular (note that $m_2 = -1/m_1$) and $\theta = 90°$.

2. $x_1 = 5, x_2 = -9, A = 3, B = -7, C = -21$

$$d = \frac{|3(5) + (-7)(-9) + (-21)|}{\sqrt{3^2 + (-7)^2}} = \frac{57}{\sqrt{58}} \approx 7.484$$

3. $x^2 - 6x - 4y + 1 = 0$

$$x^2 - 6x + 9 = 4y - 1 + 9$$
$$(x - 3)^2 = 4y + 8$$
$$(x - 3)^2 = 4(1)(y + 2) \Rightarrow p = 1$$

Vertex: $(3, -2)$

Focus: $(3, -1)$

Directrix: $y = -3$

4. Vertex: $(2, -5)$

Focus: $(2, -6)$

Vertical axis; opens downward with $p = -1$

$$(x - h)^2 = 4p(y - k)$$
$$(x - 2)^2 = 4(-1)(y + 5)$$
$$x^2 - 4x + 4 = -4y - 20$$
$$x^2 - 4x + 4y + 24 = 0$$

5.
$$x^2 + 4y^2 - 2x + 32y + 61 = 0$$
$$(x^2 - 2x + 1) + 4(y^2 + 8y + 16) = -61 + 1 + 64$$
$$(x - 1)^2 + 4(y + 4)^2 = 4$$
$$\frac{(x-1)^2}{4} + \frac{(y+4)^2}{1} = 1$$

$a = 2, b = 1, c = \sqrt{3}$

Horizontal major axis

Center: $(1, -4)$

Foci: $\left(1 \pm \sqrt{3}, -4\right)$

Vertices: $(3, -4), (-1, -4)$

Eccentricity: $e = \frac{\sqrt{3}}{2}$

6. Vertices: $(0, \pm 6)$

Eccentricity: $e = \frac{1}{2}$

Center: $(0, 0)$

Vertical major axis

$a = 6, e = \frac{c}{a} = \frac{c}{6} = \frac{1}{2} \Rightarrow c = 3$

$b^2 = (6)^2 - (3)^2 = 27$

$\frac{x^2}{27} + \frac{y^2}{36} = 1$

7.
$$16y^2 - x^2 - 6x - 128y + 231 = 0$$
$$16(y^2 - 8y + 16) - (x^2 + 6x + 9) = -231 + 256 - 9$$
$$16(y - 4)^2 - (x + 3)^2 = 16$$
$$\frac{(y-4)^2}{1} - \frac{(x+3)^2}{16} = 1$$

$a = 1, b = 4, c = \sqrt{17}$

Center: $(-3, 4)$

Vertical transverse axis

Vertices: $(-3, 5), (-3, 3)$

Foci: $\left(-3, 4 \pm \sqrt{17}\right)$

Asymptotes: $y = 4 \pm \frac{1}{4}(x + 3)$

8. Vertices: $(\pm 3, 2)$

Foci: $(\pm 5, 2)$

Center: $(0, 2)$

Horizontal transverse axis

$a = 3, c = 5, b = 4$

$$\frac{(x-0)^2}{9} - \frac{(y-2)^2}{16} = 1$$
$$\frac{x^2}{9} - \frac{(y-2)^2}{16} = 1$$

9. $5x^2 + 2xy + 5y^2 - 10 = 0$

$A = 5, B = 2, C = 5$

$\cot 2\theta = \frac{5-5}{2} = 0$

$2\theta = \frac{\pi}{2} \Rightarrow \theta = \frac{\pi}{4}$

$x = x' \cos\frac{\pi}{4} - y' \sin\frac{\pi}{4}$

$= \frac{x' - y'}{\sqrt{2}}$

$x = x' \cos\frac{\pi}{4} + y' \sin\frac{\pi}{4}$

$= \frac{x' + y'}{\sqrt{2}}$

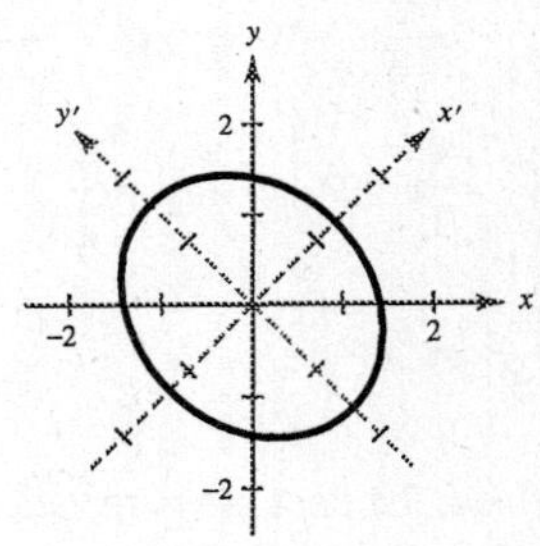

$$5\left(\frac{x'-y'}{\sqrt{2}}\right)^2 + 2\left(\frac{x'-y'}{\sqrt{2}}\right)\left(\frac{x'+y'}{\sqrt{2}}\right) + 5\left(\frac{x'+y'}{\sqrt{2}}\right)^2 - 10 = 0$$
$$\frac{5(x')^2}{2} - \frac{10x'y'}{2} + \frac{5(y')^2}{2} + (x')^2 - (y')^2 + \frac{5(x')^2}{2} + \frac{10x'y'}{2} + \frac{5(y')^2}{2} - 10 = 0$$
$$6(x')^2 + 4(y')^2 - 10 = 0$$
$$\frac{3(x')^2}{5} + \frac{2(y')^2}{5} = 1$$
$$\frac{(x')^2}{5/3} + \frac{(y')^2}{5/2} = 1$$

Ellipse centered at the origin

10. (a) $6x^2 - 2xy + y^2 = 0$

$A = 6, B = -2, C = 1$

$B^2 - 4AC = (-2)^2 - 4(6)(1) = -20 < 0$

Ellipse

(b) $x^2 + 4xy + 4y^2 - x - y + 17 = 0$

$A = 1, B = 4, C = 4$

$B^2 - 4AC = (4)^2 - 4(1)(4) = 0$

Parabola

11. Polar: $\left(\sqrt{2}, \frac{3\pi}{4}\right)$

$$x = \sqrt{2}\cos\frac{3\pi}{4} = \sqrt{2}\left(-\frac{1}{\sqrt{2}}\right) = -1$$

$$y = \sqrt{2}\sin\frac{3\pi}{4} = \sqrt{2}\left(\frac{1}{\sqrt{2}}\right) = 1$$

Rectangular: $(-1, 1)$

12. Rectangular: $\left(\sqrt{3}, -1\right)$

$$r = \pm\sqrt{\left(\sqrt{3}\right)^2 + (-1)^2} = \pm 2$$

$$\tan\theta = \frac{\sqrt{3}}{-1} = -\sqrt{3}$$

$$\theta = \frac{2\pi}{3} \text{ or } \theta = \frac{5\pi}{3}$$

Polar: $\left(-2, \frac{2\pi}{3}\right)$ or $\left(2, \frac{5\pi}{3}\right)$

13. Rectangular: $4x - 3y = 12$

Polar: $4r\cos\theta - 3r\sin\theta = 12$

$$r(4\cos\theta - 3\sin\theta) = 12$$

$$r = \frac{12}{4\cos\theta - 3\sin\theta}$$

14. Polar: $r = 5\cos\theta$

$$r^2 = 5r\cos\theta$$

Rectangular: $x^2 + y^2 = 5x$

$$x^2 + y^2 - 5x = 0$$

15. $r = 1 - \cos\theta$

Cardioid

Symmetry: Polar axis

Maximum value of $|r|$: $r = 2$ when $\theta = \pi$.

Zero of r: $r = 0$ when $\theta = 0$

θ	0	$\frac{\pi}{2}$	π	$\frac{3\pi}{2}$
r	0	1	2	1

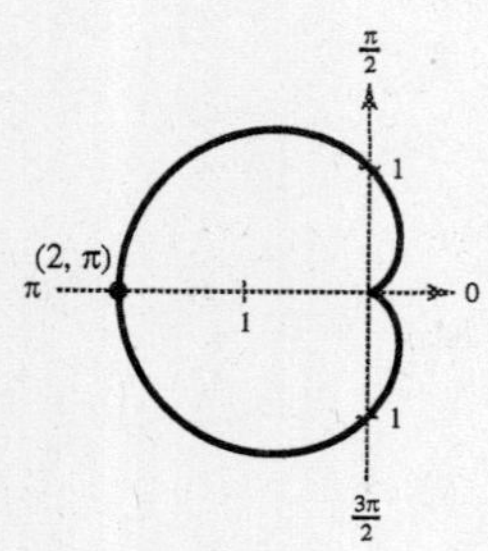

16. $r = 5\sin 2\theta$

Rose curve with four petals

Symmetry: Polar axis, $\theta = \frac{\pi}{2}$, and pole

Maximum value of $|r|$: $|r| = 5$ when $\theta = \frac{\pi}{4}, \frac{3\pi}{4}, \frac{5\pi}{4}, \frac{7\pi}{4}$

Zeros of r: $r = 0$ when $\theta = 0, \frac{\pi}{2}, \pi, \frac{3\pi}{2}$

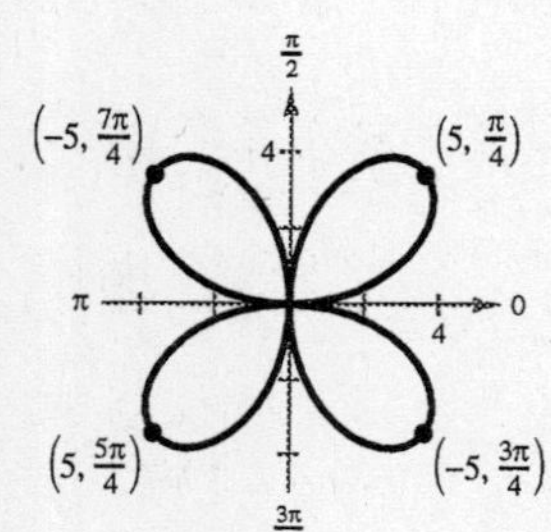

17. $r = \dfrac{3}{6 - \cos\theta}$

$r = \dfrac{\frac{1}{2}}{1 - \frac{1}{6}\cos\theta}$

$e = \dfrac{1}{6} < 1$, so the graph is an ellipse.

θ	0	$\frac{\pi}{2}$	π	$\frac{3\pi}{2}$
r	$\frac{3}{5}$	$\frac{1}{2}$	$\frac{3}{7}$	$\frac{1}{2}$

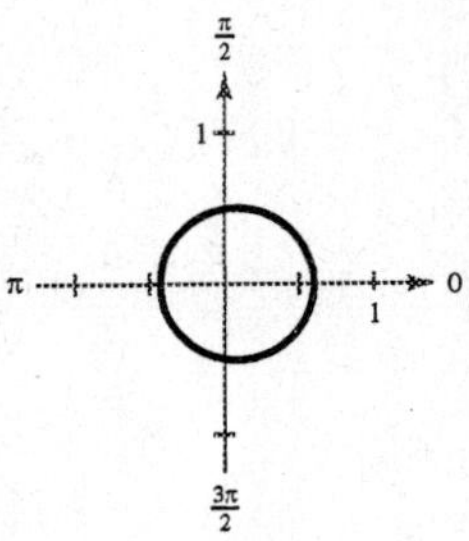

18. Parabola

Vertex: $\left(6, \dfrac{\pi}{2}\right)$

Focus: (0, 0)

$e = 1$

$r = \dfrac{ep}{1 + e\sin\theta}$

$r = \dfrac{p}{1 + \sin\theta}$

$6 = \dfrac{p}{1 + \sin(\pi/2)}$

$6 = \dfrac{p}{2}$

$12 = p$

$r = \dfrac{12}{1 + \sin\theta}$

19. $x = 3 - 2\sin\theta,\ y = 1 + 5\cos\theta$

$\dfrac{x-3}{-2} = \sin\theta, \dfrac{y-1}{5} = \cos\theta$

$\left(\dfrac{x-3}{-2}\right)^2 + \left(\dfrac{y-1}{5}\right)^2 = 1$

$\dfrac{(x-3)^2}{4} + \dfrac{(y-1)^2}{25} = 1$

20. $x = e^{2t}, y = e^{4t}$

$x > 0, y > 0$

$y = (e^{2t})^2 = (x)^2 = x^2, x > 0, y > 0$

PART II

Chapter P Chapter Test

1. $-\frac{10}{3} = -3\frac{1}{3}$

$-|-4| = -4$

$-\frac{10}{3} > -|-4|$

2. $d\left(3\frac{3}{4}, -5.4\right) = |3.75 - (-5.4)|$

$= 9.15$

3. $(5 - x) + 0 = 5 - x$

This equation illustrates the Additive Identity Property.

4. $y = 4 - |x|$

Symmetry

$y = 4 - |-x| \Rightarrow y = 4 - |x| \Rightarrow$ y-axis symmetry

$-y = 4 - |x| \Rightarrow y = -4 + |x| \Rightarrow$ No x-axis symmetry

$-y = 4 - |-x| \Rightarrow y = -4 + |x| \Rightarrow$ No origin symmetry

x-intercepts: $0 = 4 - |x|$

$|x| = 4$

$x = \pm 4$

$(4, 0), (-4, 0)$

y-intercept: $y = 4 - |0|$

$y = 4$

$(0, 4)$

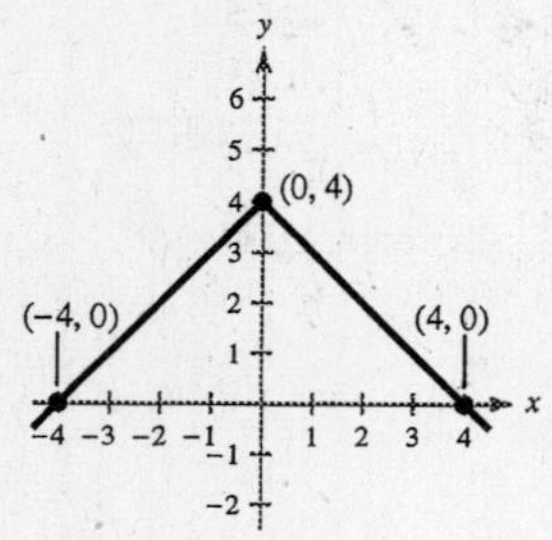

5. $y = x^2 - 1$

Symmetry

$y = (-x)^2 - 1 \Rightarrow y = x^2 - 1 \Rightarrow$ y-axis symmetry

$-y = x^2 - 1 \Rightarrow y = -x^2 + 1 \Rightarrow$ No x-axis symmetry

$-y = (-x)^2 - 1 \Rightarrow y = -x^2 + 1 \Rightarrow$ No origin symmetry

x-intercepts: $0 = x^2 - 1$

$x^2 = 1$

$x = \pm 1$

$(1, 0), (-1, 0)$

y-intercept: $y = 0^2 - 1$

$y = -1$

$(0, -1)$

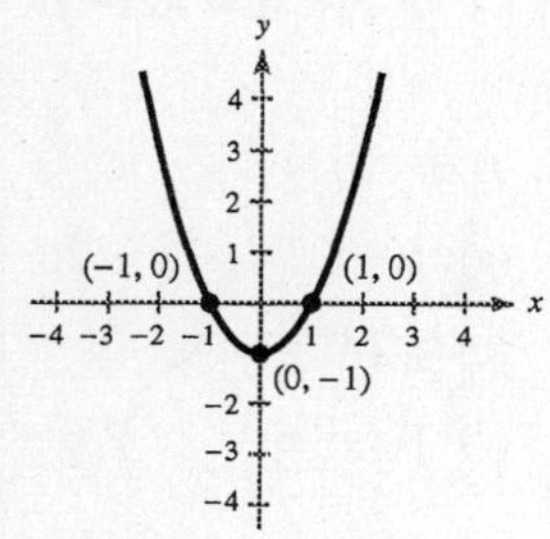

6. $(x - 3)^2 + y^2 = 9$

Circle

Center: $(3, 0)$

Radius: $r = 3$

x-axis symmetry

Intercepts: $(0, 0), (6, 0)$

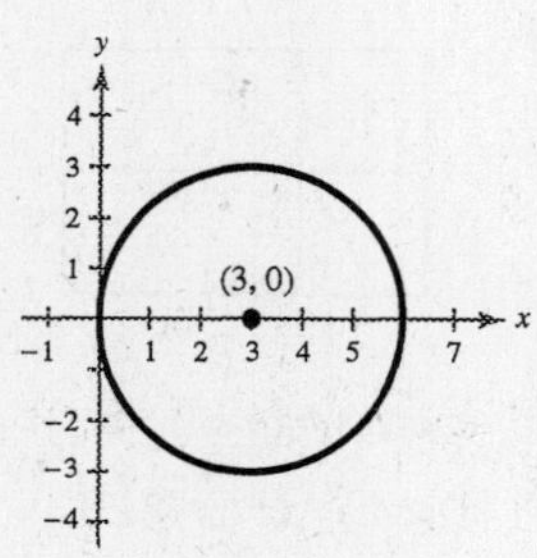

7. $\frac{2}{3}(x - 1) + \frac{1}{4}x = 10$

$12\left[\frac{2}{3}(x - 1) + \frac{1}{4}x\right] = 12(10)$

$8(x - 1) + 3x = 120$

$8x - 8 + 3x = 120$

$11x = 128$

$x = \frac{128}{11}$

8. $(x - 3)(x + 2) = 14$

$x^2 - x - 6 = 14$

$x^2 - x - 20 = 0$

$(x + 4)(x - 5) = 0$

$x + 4 = 0$ or $x - 5 = 0$

$x = -4$ $\quad x = 5$

9. $2\sqrt{x} - \sqrt{2x + 1} = 1$

$-\sqrt{2x + 1} = 1 - 2\sqrt{x}$

$\left(-\sqrt{2x + 1}\right)^2 = \left(1 - 2\sqrt{x}\right)^2$

$2x + 1 = 1 - 4\sqrt{x} + 4x$

$-2x = -4\sqrt{x}$

$x = 2\sqrt{x}$

$x^2 = 4x$

$x^2 - 4x = 0$

$x(x - 4) = 0$

$x = 0$ or $x = 4$

Only $x = 4$ is a solution to the original equation. $x = 0$ is extraneous.

10. $|3x - 1| = 7$

$$3x - 1 = 7 \quad \text{or} \quad 3x - 1 = -7$$
$$3x = 8 \qquad\qquad 3x = -6$$
$$x = \tfrac{8}{3} \qquad\qquad x = -2$$

11. $-4x + 7y = -5$

$$7y = 4x - 5$$
$$y = \tfrac{4}{7}x - \tfrac{5}{7} \Rightarrow m_1 = \tfrac{4}{7}$$

(a) Parallel line: $m_2 = \frac{4}{7}$

$$y - 8 = \tfrac{4}{7}(x - 3)$$
$$7(y - 8) = 4(x - 3)$$
$$7y - 56 = 4x - 12$$
$$0 = 4x - 7y + 44$$

(b) Perpendicular line: $m_2 = -\frac{7}{4}$

$$y - 8 = -\tfrac{7}{4}(x - 3)$$
$$4(y - 8) = -7(x - 3)$$
$$4y - 32 = -7x + 21$$
$$7x + 4y - 53 = 0$$

12. $f(x) = |x + 2| - 15$

(a) $f(-8) = |-8 + 2| - 15 = 6 - 15 = -9$

(b) $f(14) = |14 + 2| - 15 = 16 - 15 = 1$

(c) $f(x - 6) = |x - 6 + 2| - 15 = |x - 4| - 15$

13. $f(x) = \sqrt{100 - x^2}$

Domain: $100 - x^2 \geq 0 \Rightarrow -10 \leq x \leq 10$ or $[-10, 10]$.

14. $f(x) = 2x^6 + 5x^4 - x^2$

(a)

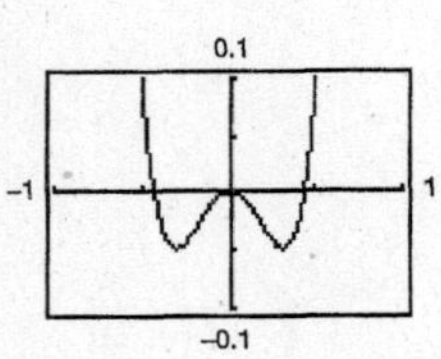

(b) Increasing on $(-0.31, 0)$ and $(0.31, \infty)$

Decreasing on $(-\infty, -0.31)$ and $(0, 0.31)$

(c) y-axis symmetry $\Rightarrow$ The function is even.

15. $f(x) = 4x\sqrt{3 - x}$

(a)

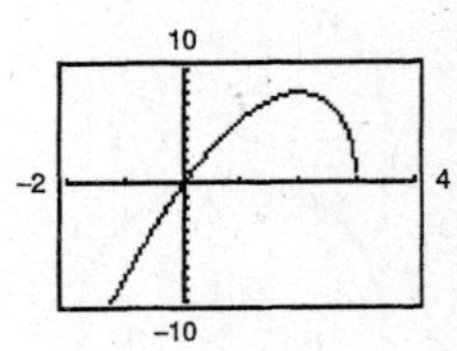

(b) Increasing on $(-\infty, 2)$

Decreasing on $(2, 3)$

(c) The function is neither odd nor even.

16. $f(x) = |x + 5|$

(a)

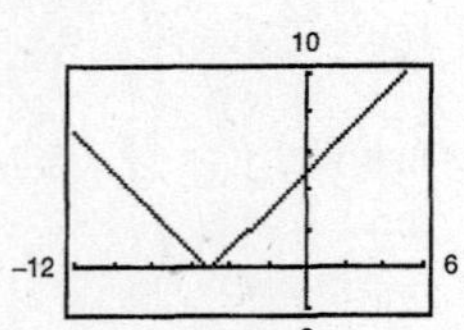

(b) increasing on $(-5, \infty)$

decreasing on $(-\infty, -5)$

(c) The function is neither odd nor even.

17. $f(x) = \begin{cases} 3x + 7, x \leq -3 \\ 4x^2 - 1, x > -3 \end{cases}$

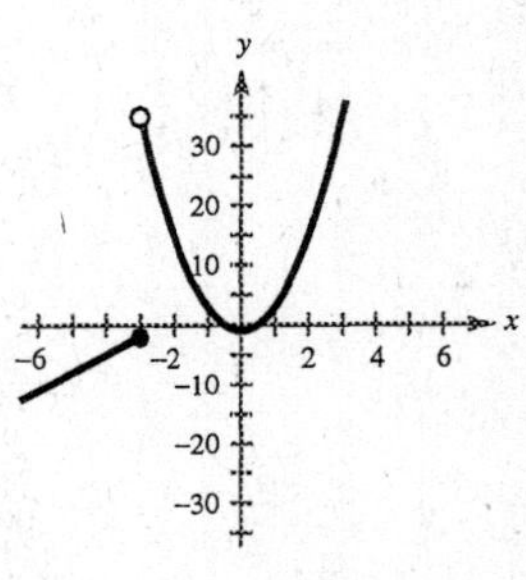

18. $h(x) = -x^3 - 7$

Common function: $f(x) = x^3$

Transformation:
Reflection in the x-axis and a vertical shift 7 units downward.

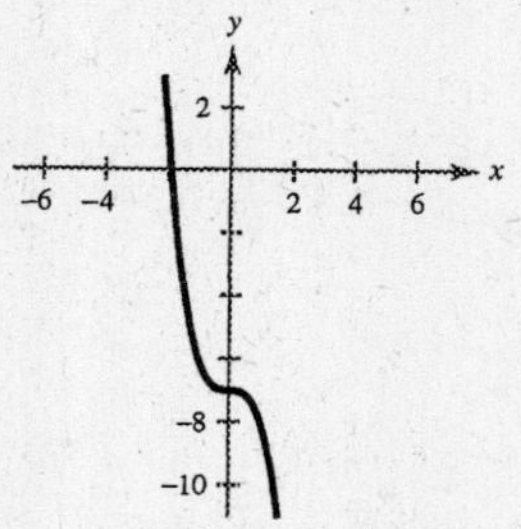

19. $h(x) = -\sqrt{x+5} + 8$

Common function: $f(x) = \sqrt{x}$

Transformation:
Reflection in the x-axis, a horizontal shift 5 units to the left, and a vertical shift 8 units upward.

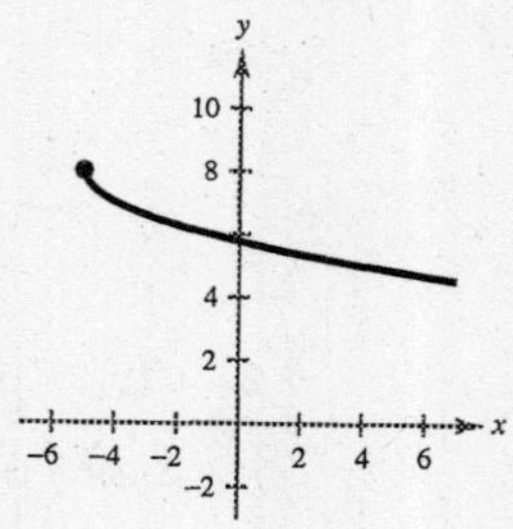

20. $h(x) = \frac{1}{4}|x + 1| - 3$

Common function: $f(x) = |x|$

Transformation:
Vertical shrink, horizontal shift 1 unit to the left, vertical shift 3 units down.

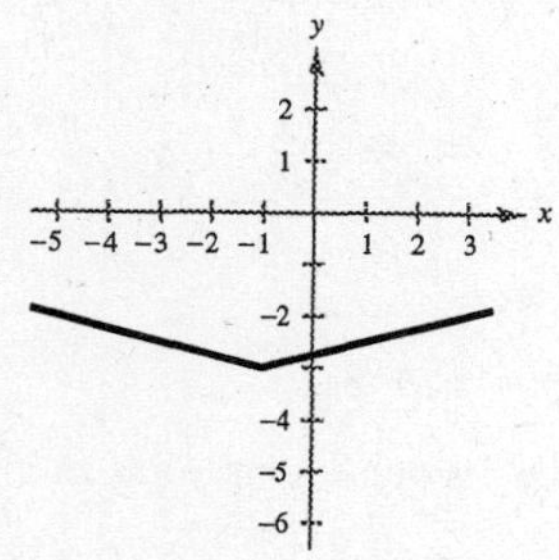

21. $f(x) = 3x^2 - 7,\ g(x) = -x^2 - 4x + 5$

(a) $(f + g)(x) = f(x) + g(x)$

$= (3x^2 - 7) + (-x^2 - 4x + 5)$

$= 2x^2 - 4x - 2$

$= 2(x^2 - 2x - 1)$

(b) $(f - g)(x) = f(x) - g(x)$

$= (3x^2 - 7) - (-x^2 - 4x + 5)$

$= 4x^2 + 4x - 12$

$= 4(x^2 + x - 3)$

(c) $(fg)(x) = f(x)g(x)$

$= (3x^2 - 7)(-x^2 - 4x + 5)$

$= -3x^4 - 12x^3 + 15x^2 + 7x^2 + 28x - 35$

$= -3x^4 - 12x^3 + 22x^2 + 28x - 35$

(d) $\left(\frac{f}{g}\right)(x) = \frac{f(x)}{g(x)}$

$= \frac{3x^2 - 7}{-x^2 - 4x + 5}$

$= -\frac{3x^2 - 7}{(x + 5)(x - 1)},\ x \neq -5, 1$

(e) $(f \cdot g)(x) = f(g(x))$

$= f(-x^2 - 4x + 5)$

$= 3(-x^2 - 4x + 5)^2 - 7$

$= 3(x^4 + 8x^3 + 6x^2 - 40x + 25) - 7$

$= 3x^4 + 24x^3 + 18x^2 - 120x + 68$

(f) $(g \cdot f)(x) = g(f(x))$

$= g(3x^2 - 7)$

$= -(3x^2 - 7)^2 - 4(3x^2 - 7) + 5$

$= -(9x^4 - 42x^2 + 49) - 12x^2 + 28 + 5$

$= -9x^4 + 30x^2 - 16$

22. $f(x) = \frac{1}{x},\ g(x) = 2\sqrt{x}$

(a) $(f + g)(x) = f(x) + g(x)$

$= \frac{1}{x} + 2\sqrt{x}$

$= \frac{1 + 2x\sqrt{x}}{x}$

(b) $(f - g)(x) = f(x) - g(x)$

$= \frac{1}{x} - 2\sqrt{x}$

$= \frac{1 - 2x\sqrt{x}}{x}$

(c) $(fg)(x) = f(x)g(x)$

$= \left(\frac{1}{x}\right)(2\sqrt{x})$

$= \frac{2\sqrt{x}}{x}$

—CONTINUED—

22. —CONTINUED—

(d) $\left(\frac{f}{g}\right)(x) = \frac{f(x)}{g(x)}$

$= \frac{\frac{1}{x}}{2\sqrt{x}}$

$= \frac{1}{2x\sqrt{x}}$

$= \frac{\sqrt{x}}{2x^2}$

(e) $(f \cdot g)(x) = f(g(x))$

$= f(2\sqrt{x})$

$= \frac{1}{2\sqrt{x}}$

$= \frac{\sqrt{x}}{2x}$

(f) $(g \cdot f)(x) = g(f(x))$

$= g\left(\frac{1}{x}\right)$

$= 2\sqrt{\frac{1}{x}}$

$= \frac{2}{\sqrt{x}}$

$= \frac{2\sqrt{x}}{x}$

23. $f(x) = x^3 + 8$

Since f is one-to-one, f has an inverse.

$y = x^3 + 8$

$x = y^3 + 8$

$x - 8 = y^3$

$\sqrt[3]{x - 8} = y$

$f^{-1}(x) = \sqrt[3]{x - 8}$

24. $f(x) = |x^2 - 3| + 6$

Since f is not on-to-one, f does not have an inverse.

25. $f(x) = \frac{3x\sqrt{x}}{8} = \frac{3}{8}x^{3/2}$

Since f is one-to-one, f has an inverse.

$y = \frac{3}{8}x^{3/2}$

$x = \frac{3}{8}y^{3/2}$

$\frac{8}{3}x = y^{3/2}$

$\left(\frac{8}{3}x\right)^{2/3} = y, x \geq 0$

$f^{-1}(x) = \left(\frac{8}{3}x\right)^{2/3}, x \geq 0$

26. (6, 58), (10, 78)

$m = \frac{78 - 58}{10 - 6} = \frac{20}{4} = 5$

$C - 58 = 5(x - 6)$

$C - 58 = 5x - 30$

$C = 5x + 28$

When $x = 25$; $C = 5(25) + 28 = \$153$

Chapter 1 Chapter Test

1. $\theta = \frac{5\pi}{4}$

(a)

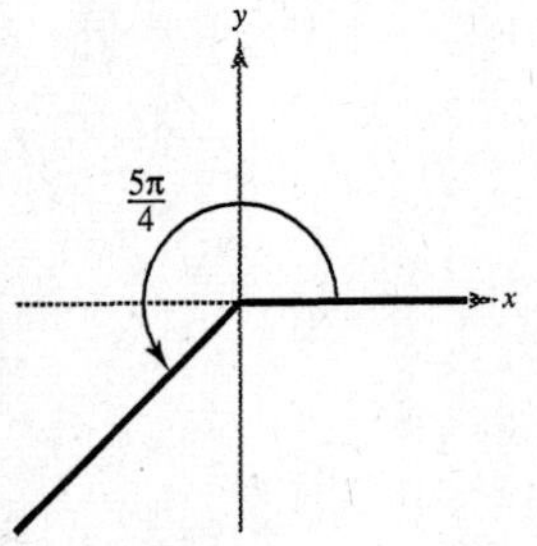

(b) $\frac{5\pi}{4} + 2\pi = \frac{13\pi}{4}$

$\frac{5\pi}{4} - 2\pi = -\frac{3\pi}{4}$

(c) $\frac{5\pi}{4}\left(\frac{180°}{\pi}\right) = 225°$

2. $90\frac{\text{km}}{\text{hr}} \times \frac{1\text{ hr}}{60\text{ min}} \times \frac{1000\text{ m}}{1\text{ km}} = 1500$ meters per minute

$\frac{\text{Revolutions}}{\text{minute}} = \frac{1500}{\pi}$

Circumference $= 2\pi\left(\frac{1}{2}\right) = \pi = \pi$ meters

Angular speed $= \left(\frac{1500\text{ revolutions}}{\pi\text{ minute}}\right)\left(\frac{2\pi\text{ radians}}{\text{revolution}}\right)$

$= 3000$ radians per minute

3. $x = -2, y = 6$

$r = \sqrt{(-2)^2 + (6)^2} = 2\sqrt{10}$

$\sin\theta = \frac{y}{r} = \frac{6}{2\sqrt{10}} = \frac{3}{\sqrt{10}} = \frac{3\sqrt{10}}{10}$

$\cos\theta = \frac{x}{r} = \frac{-2}{2\sqrt{10}} = -\frac{1}{\sqrt{10}} = -\frac{\sqrt{10}}{10}$

$\tan\theta = \frac{y}{x} = \frac{6}{-2} = -3$

$\csc\theta = \frac{r}{y} = \frac{2\sqrt{10}}{6} = \frac{\sqrt{10}}{3}$

$\sec\theta = \frac{r}{x} = \frac{2\sqrt{10}}{-2} = -\sqrt{10}$

$\cot\theta = \frac{x}{y} = \frac{-2}{6} = -\frac{1}{3}$

4.

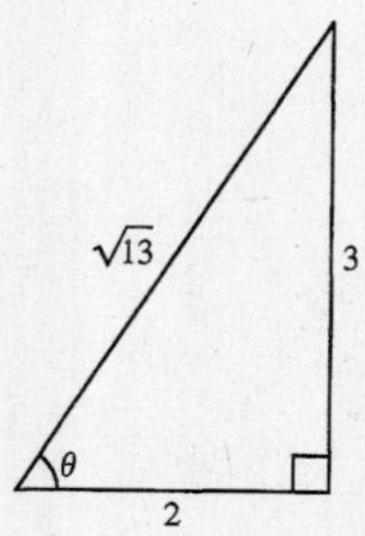

For $0 \le \theta < \frac{\pi}{2}$, we have

$\sin\theta = \frac{\text{opp}}{\text{hyp}} = \frac{3}{\sqrt{13}} = \frac{3\sqrt{13}}{13}$

$\cos\theta = \frac{\text{adj}}{\text{hyp}} = \frac{2}{\sqrt{13}} = \frac{2\sqrt{13}}{13}$

$\csc\theta = \frac{\text{hyp}}{\text{opp}} = \frac{\sqrt{13}}{3}$

$\sec\theta = \frac{\text{hyp}}{\text{adj}} = \frac{\sqrt{13}}{2}$

$\cot\theta = \frac{\text{adj}}{\text{opp}} = \frac{2}{3}$

For $\pi \le \theta < \frac{3\pi}{2}$, we have

$\sin\theta = -\frac{3\sqrt{13}}{13}$

$\cos\theta = -\frac{2\sqrt{13}}{13}$

$\csc\theta = -\frac{\sqrt{13}}{3}$

$\sec\theta = -\frac{\sqrt{13}}{2}$

$\cot\theta = \frac{2}{3}$

5. $\theta = 290°$

$\theta' = 360° - 290° = 70°$

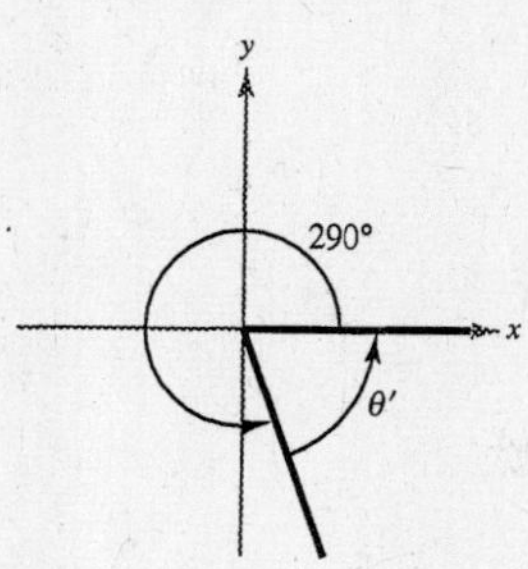

6. $\sec\theta < 0$ and $\tan\theta > 0$

$\frac{r}{x} < 0$ and $\frac{y}{x} > 0$

Quandrant III

7. $\cos\theta = -\frac{\sqrt{3}}{2}$

Reference angle is 30° and θ is in Quandrant II or III.

$\theta = 150°$ or $210°$

8. $\csc\theta = 1.030$

$\frac{1}{\sin\theta} = 1.030$

$\sin\theta = \frac{1}{1.030}$

$\theta = \arcsin\frac{1}{1.030}$

$\theta \approx 1.33$ and $\pi - 1.33 \approx 1.81$

9. $\cos\theta = \frac{3}{5}$, $\tan\theta < 0 \Longrightarrow \theta$ lies in Quadrant IV

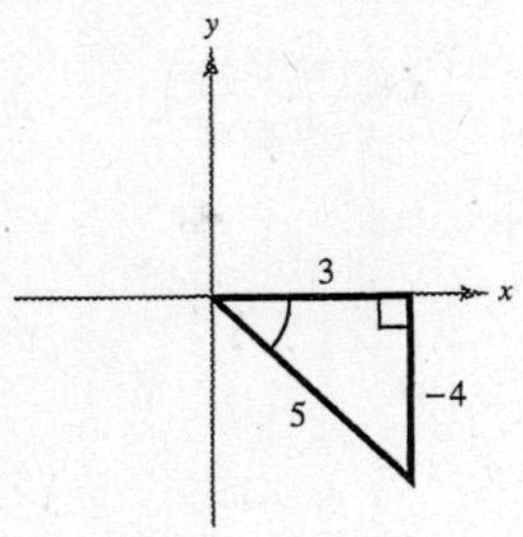

Let $x = 3$, $r = 5 \Longrightarrow y = -4$

$\sin\theta = -\frac{4}{5}$ $\qquad \csc\theta = -\frac{5}{4}$

$\cos\theta = \frac{3}{5}$ $\qquad \sec\theta = \frac{5}{3}$

$\tan\theta = -\frac{4}{3}$ $\qquad \cot\theta = -\frac{3}{4}$

10. $\sec\theta = -\frac{17}{8}$, $\sin\theta > 0 \Longrightarrow \theta$ lies in Quadrant II

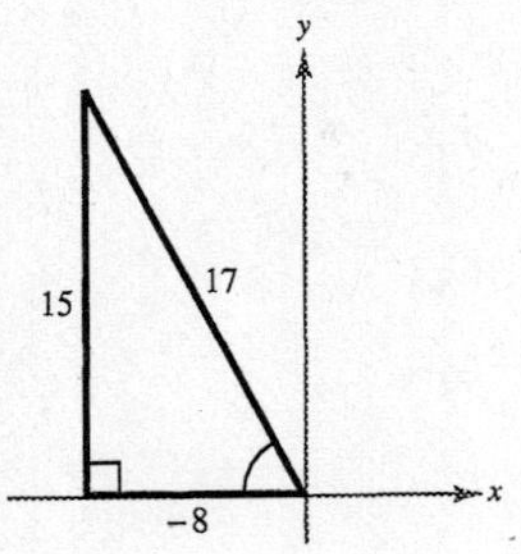

Let $r = 17$, $x = -8 \Longrightarrow y = 15$

$\sin\theta = \frac{15}{17}$ $\qquad \csc\theta = \frac{17}{15}$

$\cos\theta = -\frac{8}{17}$ $\qquad \sec\theta = -\frac{17}{8}$

$\tan\theta = -\frac{15}{8}$ $\qquad \cot\theta = -\frac{8}{15}$

11. $g(x) = -2\sin\left(x - \frac{\pi}{4}\right)$

Period: 2π

Amplitude: $|-2| = 2$

Shifted to the right by $\frac{\pi}{4}$ units and reflected in the x-axis.

x	0	$\frac{\pi}{4}$	$\frac{\pi}{2}$	$\frac{3\pi}{4}$	π
y	$\sqrt{2}$	0	$-\sqrt{2}$	-2	$-\sqrt{2}$

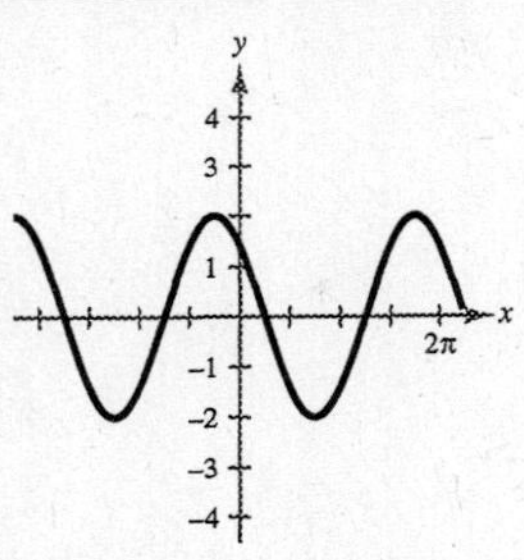

12. $f(\alpha) = \frac{1}{2}\tan 2\alpha$

Period: $\frac{\pi}{2}$

Asymptotes: $x = -\frac{\pi}{4}$, $x = \frac{\pi}{4}$

α	$-\frac{\pi}{8}$	0	$\frac{\pi}{8}$
$f(\alpha)$	$-\frac{1}{2}$	0	$\frac{1}{2}$

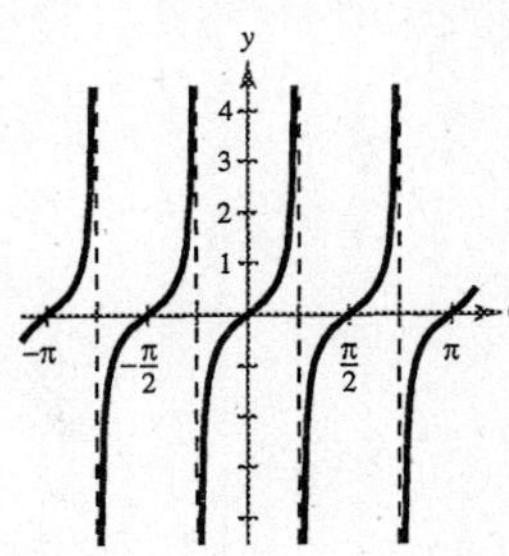

13. $y = \sin 2\pi x + 2\cos\pi x$

Periodic: period $= 2$

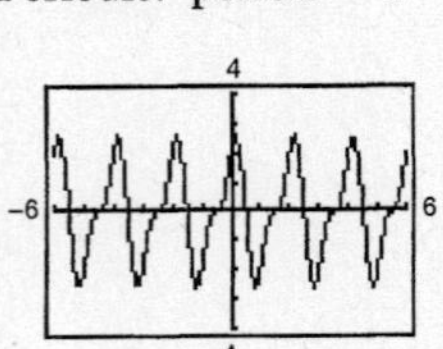

14.

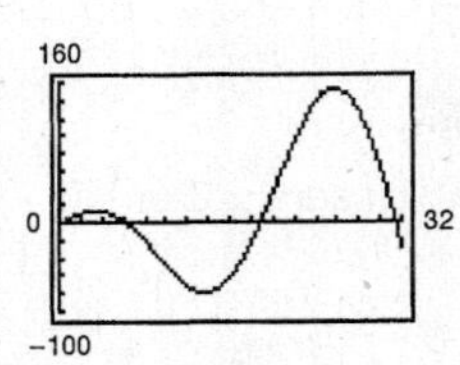

Not periodic

15. $f(x) = a \sin(bx + c)$

Amplitude: $2 \Longrightarrow |a| = 2$

Reflected in the x-axis: $a = -2$

Period: $4\pi = \dfrac{2\pi}{b} \Longrightarrow b = \dfrac{1}{2}$

Phase shift: $\dfrac{c}{b} = -\dfrac{\pi}{2} \Longrightarrow c = -\dfrac{\pi}{4}$

$f(x) = -2 \sin\left(\dfrac{x}{2} - \dfrac{\pi}{4}\right)$

16. Let $u = \arccos \dfrac{2}{3}$,

$\cos u = \dfrac{2}{3}$.

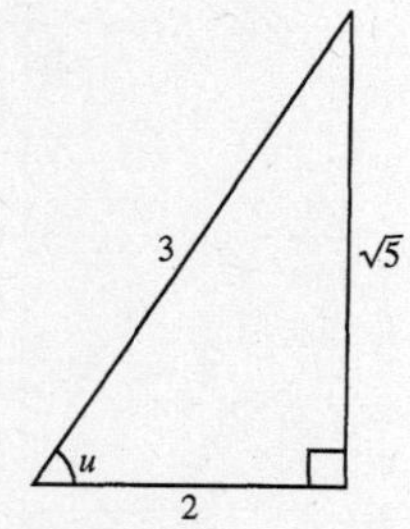

$\tan\left(\arccos \dfrac{2}{3}\right) = \tan u = \dfrac{\sqrt{5}}{2}$

17. $f(x) = 2 \arcsin\left(\dfrac{1}{2}x\right)$

Domain: $[-2, 2]$

Range: $[-\pi, \pi]$

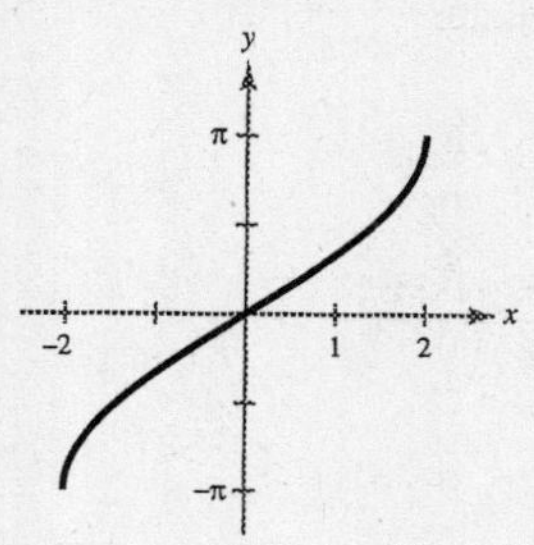

18.

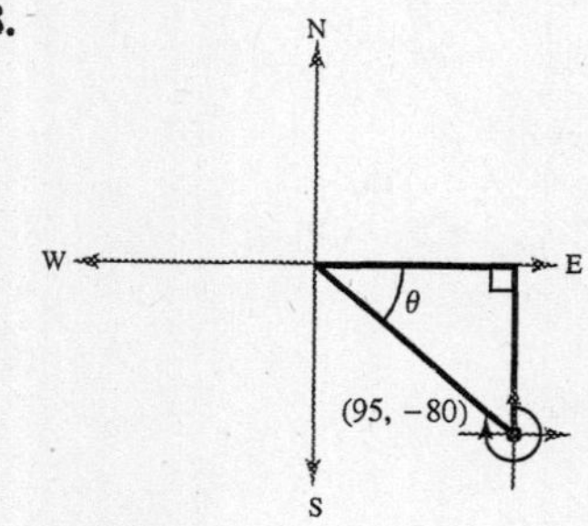

$\tan \theta = -\frac{80}{95} \Longrightarrow \theta \approx -40.1°$

Bearing: $90° - 40.1° = 49.9°$

19. $d = a \cos bt$

$a = -6$

$\dfrac{2\pi}{b} = 2 \Longrightarrow b = \pi$

$d = -6 \cos \pi t$

Chapter 2 Chapter Test

1. $\tan \theta = \dfrac{3}{2}$ and $\cos \theta < 0$

θ is in Quadrant III.

$\sec \theta = -\sqrt{1 + \tan^2 \theta} = -\sqrt{1 + \left(\dfrac{3}{2}\right)^2} = -\dfrac{\sqrt{13}}{2}$

$\cos \theta = \dfrac{1}{\sec \theta} = -\dfrac{2}{\sqrt{13}} = -\dfrac{2\sqrt{13}}{13}$

$\sin \theta = \tan \theta \cos \theta = \left(\dfrac{3}{2}\right)\left(-\dfrac{2}{\sqrt{13}}\right) = -\dfrac{3}{\sqrt{13}} = -\dfrac{3\sqrt{13}}{13}$

$\csc \theta = \dfrac{1}{\sin \theta} = -\dfrac{\sqrt{13}}{3}$

$\cot \theta = \dfrac{1}{\tan \theta} = \dfrac{2}{3}$

2. $\csc^2 \beta (1 - \cos^2 \beta) = \dfrac{1}{\sin^2 \beta}(\sin^2 \beta) = 1$

3. $\dfrac{\sec^4 x - \tan^4 x}{\sec^2 x + \tan^2 x} = \dfrac{(\sec^2 x + \tan^2 x)(\sec^2 x - \tan^2 x)}{\sec^2 x + \tan^2 x}$

$= \sec^2 x - \tan^2 x = 1$

4. $\dfrac{\cos \theta}{\sin \theta} + \dfrac{\sin \theta}{\cos \theta} = \dfrac{\cos^2 \theta + \sin^2 \theta}{\sin \theta \cos \theta} = \dfrac{1}{\sin \theta \cos \theta}$

$= \csc \theta \sec \theta$

5. $y = \tan \theta, y = -\sqrt{\sec^2 \theta - 1}$

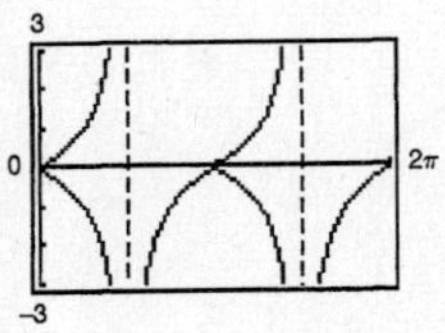

$\tan \theta = -\sqrt{\sec^2 \theta - 1}$ on

$\theta = 0, \dfrac{\pi}{2} < \theta \leq \pi, \dfrac{3\pi}{2} < \theta < 2\pi.$

6. $y_1 = \cos x + \sin x \tan x, y_2 = \sec x$

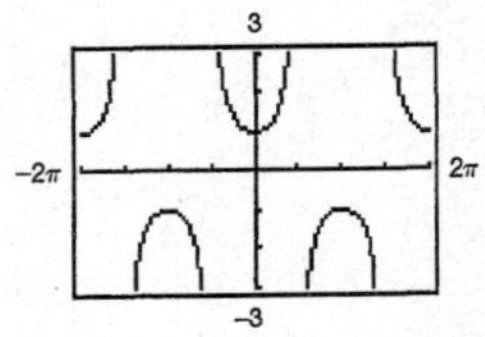

It appears that $y_1 = y_2$.

$\cos x + \sin x \tan x = \cos + \sin x \dfrac{\sin x}{\cos x}$

$= \cos + \dfrac{\sin^2 x}{\cos x}$

$= \dfrac{\cos^2 x + \sin^2 x}{\cos x}$

$= \dfrac{1}{\cos x} = \sec x$

7. $\sin \theta \sec \theta = \sin \theta \dfrac{1}{\cos \theta} = \dfrac{\sin \theta}{\cos \theta} = \tan \theta$

8. $\sec^2 x \tan^2 x + \sec^2 x = \sec^2 x (\sec^2 x - 1) + \sec^2 x$

$= \sec^4 x - \sec^2 x + \sec^2 x$

$= \sec^4 x$

9. $\dfrac{\csc \alpha + \sec \alpha}{\sin \alpha + \cos \alpha} = \dfrac{\dfrac{1}{\sin \alpha} + \dfrac{1}{\cos \alpha}}{\sin \alpha + \cos \alpha} = \dfrac{\dfrac{\cos \alpha + \sin \alpha}{\sin \alpha \cos \alpha}}{\sin \alpha + \cos \alpha} = \dfrac{1}{\sin \alpha \cos \alpha}$

$= \dfrac{\cos^2 \alpha + \sin^2 \alpha}{\sin \alpha \cos \alpha} = \dfrac{\cos^2 \alpha}{\sin \alpha \cos \alpha} + \dfrac{\sin^2 \alpha}{\sin \alpha \cos \alpha}$

$= \dfrac{\cos \alpha}{\sin \alpha} + \dfrac{\sin \alpha}{\cos \alpha} = \cot \alpha + \tan \alpha$

10. $\cos\left(x + \dfrac{\pi}{2}\right) = \cos\left(\dfrac{\pi}{2} - (-x)\right) = \sin(-x) = -\sin x$

11. $\sin(n\pi + \theta) = (-1)^n \sin \theta$, n is an integer.

For n odd: $\sin(n\pi + \theta) = \sin n\pi \cos \theta + \cos n\pi \sin \theta$

$= (0) \cos \theta + (-1) \sin \theta = -\sin \theta$

For n even: $\sin(n\pi + \theta) = \sin n\pi \cos \theta + \cos n\pi \sin \theta$

$= (0) \cos \theta + (1) \sin \theta = \sin \theta$

When n is odd, $(-1)^n = -1$. When n is even $(-1)^n = 1$.

Thus, $\sin(n\pi + \theta) = (-1)^n \sin \theta$ for any integer n.

12. $(\sin x + \cos x)^2 = \sin^2 x + 2 \sin x \cos x + \cos^2 x$

$= 1 + 2 \sin x \cos x$

$= 1 + \sin 2x$

13. $\sin^4 x \tan^2 x = \sin^4 x\left(\dfrac{\sin^2 x}{\cos^2 x}\right) = \dfrac{\sin^6 x}{\cos^2 x} = \dfrac{(\sin^2 x)^3}{\cos^2 x}$

$$= \frac{\left(\dfrac{1 - \cos 2x}{2}\right)^3}{\dfrac{1 + \cos 2x}{2}}$$

$$= \frac{\dfrac{1 - 3\cos 2x + 3\cos^2 2x - \cos^3 2x}{8}}{\dfrac{1 + \cos 2x}{2}}$$

$$= \frac{\frac{1}{4}\left[1 - 3\cos 2x + 3\left(\dfrac{1 + \cos 4x}{2}\right) - \cos 2x\left(\dfrac{1 + \cos 4x}{2}\right)\right]}{1 + \cos 2x}$$

$$= \frac{\frac{1}{8}[2 - 6\cos 2x + 3 + 3\cos 4x - \cos 2x - \cos 2x \cos 4x]}{1 + \cos 2x}$$

$$= \frac{1}{8}\left[\frac{5 - 7\cos 2x + 3\cos 4x - \frac{1}{2}(\cos(-2x) + \cos(6x))}{1 + \cos 2x}\right]$$

$$= \frac{1}{16}\left[\frac{10 - 14\cos 2x + 6\cos 4x - \cos 2x - \cos 6x}{1 + \cos 2x}\right]$$

$$= \frac{1}{16}\left[\frac{10 - 15\cos 2x + 6\cos 4x - \cos 6x}{1 + \cos 2x}\right]$$

14. $\dfrac{\sin 4\theta}{1 + \cos 4\theta} = \tan\dfrac{4\theta}{2} = \tan 2\theta$

15. $4\cos 2\theta \sin 4\theta = 4\left(\dfrac{1}{2}\right)[\sin(2\theta + 4\theta) - \sin(2\theta - 4\theta)]$

$$= 2[\sin 6\theta - \sin(-2\theta)]$$

$$= 2(\sin 6\theta + \sin 2\theta)$$

16. $\sin 3\theta - \sin 4\theta = 2\cos\left(\dfrac{3\theta + 4\theta}{2}\right)\sin\left(\dfrac{3\theta - 4\theta}{2}\right)$

$$= 2\cos\frac{7\theta}{2}\sin\left(\frac{-\theta}{2}\right)$$

$$= -2\cos\frac{7\theta}{2}\sin\frac{\theta}{2}$$

17. $\tan^2 x + \tan x = 0$

$\tan x(\tan x + 1) = 0$

$\tan x = 0$ or $\tan x + 1 = 0$

$\tan x = -1$

$x = 0, \pi$ $\quad x = \dfrac{3\pi}{4}, \dfrac{7\pi}{4}$

18. $\sin 2\alpha - \cos\alpha = 0$

$2\sin\alpha\cos\alpha - \cos\alpha = 0$

$\cos\alpha(2\sin\alpha - 1) = 0$

$\cos\alpha = 0$ or $2\sin\alpha - 1 = 0$

$\alpha = \dfrac{\pi}{2}, \dfrac{3\pi}{2}$ $\quad \sin\alpha = \dfrac{1}{2}$

$\alpha = \dfrac{\pi}{6}, \dfrac{5\pi}{6}$

19. $4\cos^2 x - 3 = 0$

$\cos^2 x = \dfrac{3}{4}$

$\cos x = \pm\sqrt{\dfrac{3}{4}} = \pm\dfrac{\sqrt{3}}{2}$

$x = \dfrac{\pi}{6}, \dfrac{5\pi}{6}, \dfrac{7\pi}{6}, \dfrac{11\pi}{6}$

20. $\csc^2 x - \csc x - 2 = 0$

$(\csc x - 2)(\csc x + 1) = 0$

$\csc x - 2 = 0$ or $\csc x + 1 = 0$

$\csc x = 2 \qquad \csc = -1$

$\dfrac{1}{\sin x} = 2 \qquad \dfrac{1}{\sin x} = -1$

$\sin x = \dfrac{1}{2} \qquad \sin x = -1$

$x = \dfrac{\pi}{6}, \dfrac{5\pi}{6} \qquad x = \dfrac{3\pi}{2}$

21. $3 \cos x - x = 0$

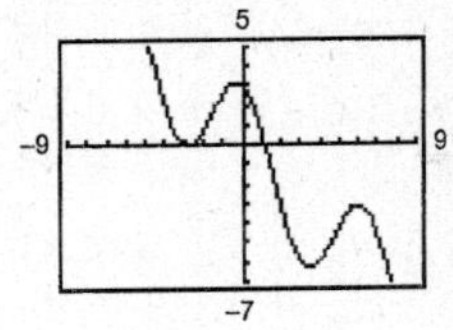

$x \approx -2.938, -2.663, 1.170$

22. $105° = 135° - 30°$

$$\begin{aligned}\cos 105° &= \cos(135° - 30°)\\ &= \cos 135° \cos 30° + \sin 135° \sin 30°\\ &= -\cos 45° \cos 30° + \sin 45° \sin 30°\\ &= \left(-\frac{\sqrt{2}}{2}\right)\left(\frac{\sqrt{3}}{2}\right) + \left(\frac{\sqrt{2}}{2}\right)\left(\frac{1}{2}\right)\\ &= \frac{-\sqrt{6} + \sqrt{2}}{4} = \frac{\sqrt{2} - \sqrt{6}}{4}\end{aligned}$$

23. $x = 1, y = 2, r = \sqrt{5}$

$$\begin{aligned}\sin 2u &= 2 \sin u \cos u\\ &= 2\left(\frac{2}{\sqrt{5}}\right)\left(\frac{1}{\sqrt{5}}\right) = \frac{4}{5}\end{aligned}$$

$$\tan 2u = \frac{2 \tan u}{1 - \tan^2 u} = \frac{2(2)}{1 - (2)^2} = \frac{4}{-3} = -\frac{4}{3}$$

24. Let $y_1 = 31 \sin\left(\dfrac{2\pi t}{365} - 1.4\right)$ and $y_2 = 20$.

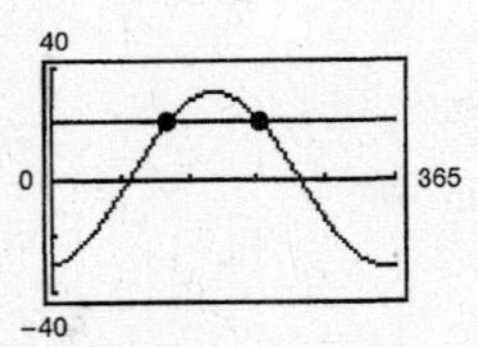

The points of intersection occur when $t \approx 123$ and $t \approx 223$. The number of days that $D > 20°$ is 100, from day 123 to day 223.

Chapter 3 Chapter Test

1. $A = 24°, B = 68°, a = 12.2$

$C = 180° - 24° - 68° = 88°$

$b = \dfrac{a \sin B}{\sin A} = \dfrac{12.2 \sin 68°}{\sin 24°} \approx 27.81$

$c = \dfrac{a \sin C}{\sin A} = \dfrac{12.2 \sin 88°}{\sin 24°} \approx 29.98$

2. $B = 104°, C = 33°, a = 18.1$

$A = 180° - 104° - 33° = 43°$

$b = \dfrac{a \sin B}{\sin A} = \dfrac{18.1 \sin 104°}{\sin 43°} \approx 25.75$

$c = \dfrac{a \sin C}{\sin A} = \dfrac{18.1 \sin 33°}{\sin 43°} \approx 14.45$

3. $A = 24°, a = 11.2, b = 13.4$

$$\sin B = \frac{b \sin A}{a} = \frac{13.4 \sin 24°}{11.2} \approx 0.4866$$

Two Solutions

$B \approx 29.12°$ or $B \approx 150.88°$

$C \approx 126.88°$ $\quad C \approx 5.12°$

$c = \frac{a \sin C}{\sin A} = \frac{11.2 \sin 126.88°}{\sin 24°}$ $\quad c = \frac{11.2 \sin 5.12°}{\sin 24°}$

$c \approx 22.03$ $\quad c \approx 2.46$

4. $a = 4.0, b = 7.3, c = 12.4$

$$\cos C = \frac{a^2 + b^2 - c^2}{2ab} = \frac{4^2 + 7.3^2 - 12.4^2}{2(4)(7.3)} \approx -1.4464 < -1$$

No solution

5. $B = 100°, a = 15, b = 23$

$$\sin A = \frac{a \sin B}{b} = \frac{15 \sin 100°}{23} \Rightarrow A \approx 39.96°$$

$$C \approx 180° - 100° - 39.96° = 40.04°$$

$$c \approx \frac{b \sin C}{\sin B} = \frac{23 \sin 40.04°}{\sin 100°} \approx 15.02$$

6. $C = 123°, a = 41, b = 57$

$$c^2 = 41^2 + 57^2 - 2(41)(57)\cos 123° \Rightarrow c \approx 86.46$$

$$\sin A = \frac{a \sin C}{c} = \frac{41 \sin 123°}{86.46} \Rightarrow A \approx 23.43°$$

$$B \approx 180° - 23.43° - 123° = 33.57°$$

7. $a = 60, b = 70, c = 82$

$$s = \frac{60 + 70 + 82}{2} = 106$$

$$\text{Area} = \sqrt{106(46)(36)(24)} \approx 2052.5 \text{ square meters}$$

8.

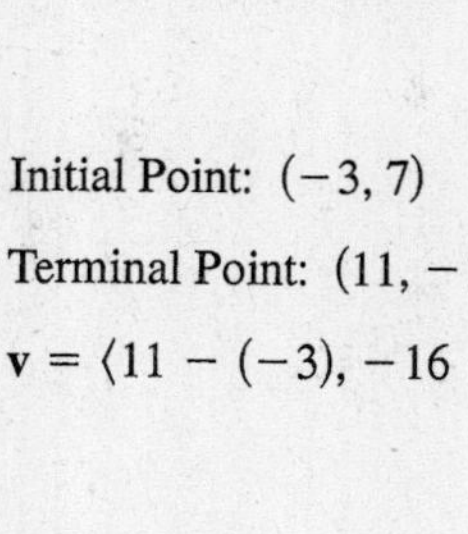

$$b^2 = 370^2 + 240^2 - 2(370)(240)\cos 167°$$

$$b \approx 606.3 \text{ miles}$$

$$\sin A = \frac{a \sin B}{b} = \frac{240 \sin 167°}{606.3}$$

$$A \approx 5°$$

Bearing: N 24° + 5° E = N 29° E

9. Initial Point: $(-3, 7)$

Terminal Point: $(11, -16)$

$$\mathbf{v} = \langle 11 - (-3), -16 - 7 \rangle = \langle 14, -23 \rangle$$

10. $$\mathbf{v} = 12\left(\frac{\mathbf{u}}{\|\mathbf{u}\|}\right) = 12\left(\frac{\langle 3, -5 \rangle}{\sqrt{3^2 + (-5)^2}}\right) = \frac{12}{\sqrt{34}}\langle 3, -5 \rangle$$

$$= \frac{6\sqrt{34}}{17}\langle 3, -5 \rangle = \left\langle \frac{18\sqrt{34}}{17}, -\frac{30\sqrt{34}}{17} \right\rangle$$

11. $\mathbf{u} + \mathbf{v} = \langle 3, 5\rangle + \langle -7, 1\rangle = \langle -4, 6\rangle$

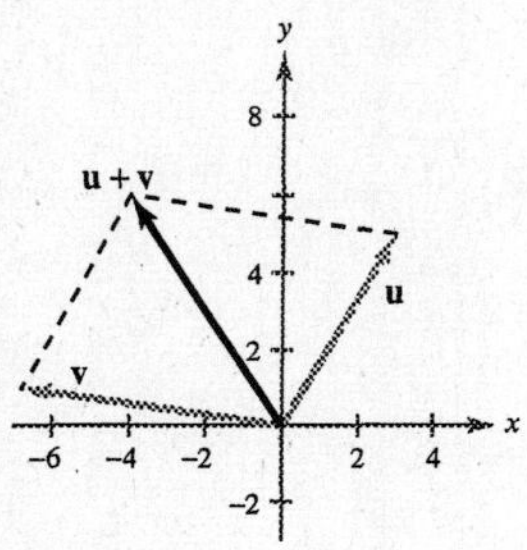

12. $\mathbf{u} - \mathbf{v} = \langle 3, 5\rangle - \langle -7, 1\rangle = \langle 10, 4\rangle$

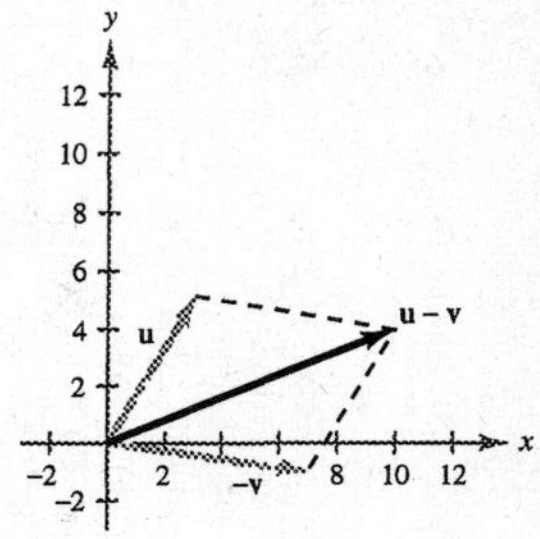

13. $5\mathbf{u} - 3\mathbf{v} = 5\langle 3, 5\rangle - 3\langle -7, 1\rangle = \langle 15, 25\rangle + \langle 21, -3\rangle$

$= \langle 36, 22\rangle$

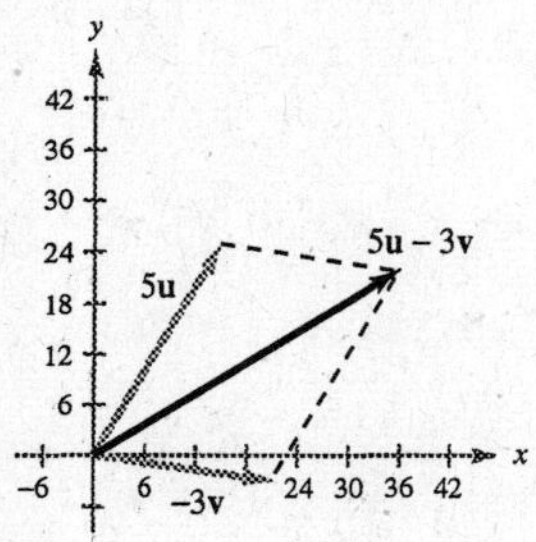

14. $\dfrac{\mathbf{u}}{\|\mathbf{u}\|} = \dfrac{\langle 4, -3\rangle}{\sqrt{4^2 + (-3)^2}} = \dfrac{1}{5}\langle 4, -3\rangle = \left\langle \dfrac{4}{5}, -\dfrac{3}{5}\right\rangle$

15.

$$\mathbf{u} = 250(\cos 45^\circ\,\mathbf{i} + \sin 45^\circ\,\mathbf{j})$$

$$\mathbf{v} = 130(\cos(-60^\circ)\mathbf{i} + \sin(-60^\circ)\mathbf{j})$$

$$\mathbf{R} = \mathbf{u} + \mathbf{v} \approx 241.7767\,\mathbf{i} + 64.1934\,\mathbf{j}$$

$$\|\mathbf{R}\| \approx \sqrt{241.7767^2 + 64.1934^2} \approx 250.15 \text{ pounds}$$

$$\tan\theta \approx \frac{64.1934}{241.7767} \implies \theta \approx 14.9^\circ$$

16. $\mathbf{u} = \langle -1, 5\rangle, \mathbf{v} = \langle 3, -2\rangle$

$$\cos\theta = \frac{\mathbf{u}\cdot\mathbf{v}}{\|\mathbf{u}\|\,\|\mathbf{v}\|} = \frac{-13}{\sqrt{26}\sqrt{13}} \implies \theta = 135^\circ$$

17. $\mathbf{u} = \langle 6, 10\rangle, \mathbf{v} = \langle 2, 3\rangle$

$\mathbf{u}\cdot\mathbf{v} = 42 \neq 0 \implies$ **u** and **v** are not orthogonal.

18. $\mathbf{u} = \langle 6, 7\rangle, \mathbf{v} = \langle -5, -1\rangle$

$$\mathbf{w}_1 = \text{proj}_{\mathbf{v}}\,\mathbf{u} = \left(\frac{\mathbf{u}\cdot\mathbf{v}}{\|\mathbf{v}\|^2}\right)\mathbf{v} = -\frac{37}{26}\langle -5, -1\rangle = \frac{37}{26}\langle 5, 1\rangle$$

$$\mathbf{w}_2 = \mathbf{u} - \mathbf{w}_1 = \langle 6, 7\rangle - \frac{37}{26}\langle 5, 1\rangle$$

$$= \left\langle -\frac{29}{26}, \frac{145}{26}\right\rangle$$

$$= \frac{29}{26}\langle -1, 5\rangle$$

Chapters 1–3 Cumulative Test

1. (a)

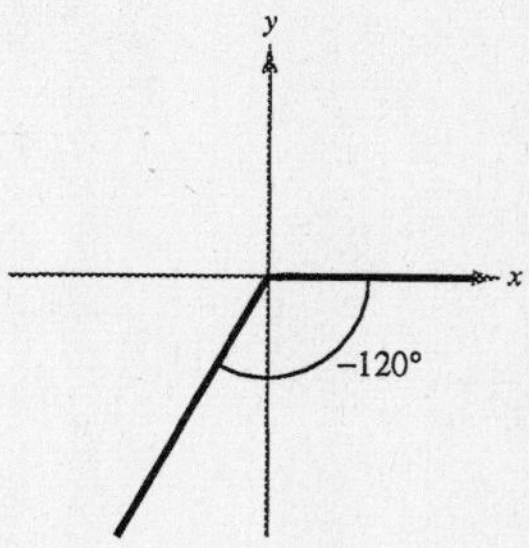

(b) $-120° + 360° = 240°$

(c) $-120\left(\dfrac{\pi}{180°}\right) = -\dfrac{2\pi}{3}$

(d) $-120°$ is located in Quadrant III.

$240° - 180° = 60°$

(e) $\sin(-120°) = -\sin 60° = -\dfrac{\sqrt{3}}{2}$

$\cos(-120°) = -\cos 60° = -\dfrac{1}{2}$

$\tan(-120°) = \tan 60° = \sqrt{3}$

$\csc(-120°) = \dfrac{1}{-\sin 60°} = -\dfrac{2\sqrt{3}}{3}$

$\sec(-120°) = \dfrac{1}{-\cos 60°} = -2$

$\cot(-120°) = \dfrac{1}{\tan 60°} = \dfrac{\sqrt{3}}{3}$

2. $2.35\left(\dfrac{180°}{\pi}\right) \approx 134.6°$

3. $\tan\theta = \dfrac{y}{x} = -\dfrac{4}{3} \Rightarrow r = 5$

Since $\sin\theta < 0$ θ is in Quadrant IV, $\Rightarrow x = 3$.

$\cos\theta = \dfrac{x}{r} = \dfrac{3}{5}$

4. $f(x) = 3 - 2\sin\pi x$

Period: $\dfrac{2\pi}{\pi} = 2$

Amplitude: $|a| = |-2| = 2$

Upward shift of 3 units (reflected in x-axis prior to shift)

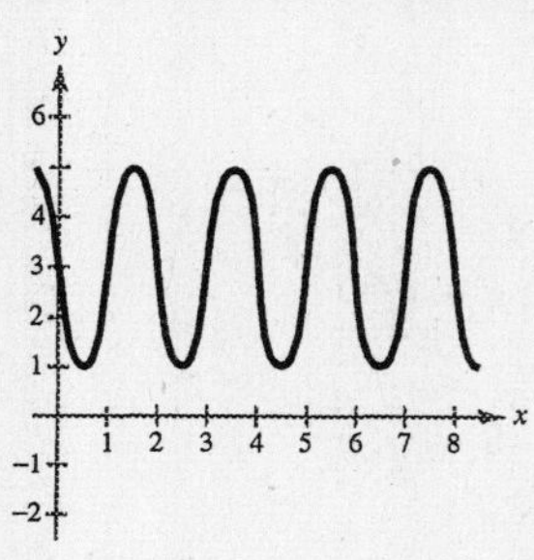

5. $g(x) = \dfrac{1}{2}\tan\left(x - \dfrac{\pi}{2}\right)$

Period: π

Asymptotes: $x = 0, x = \pi$

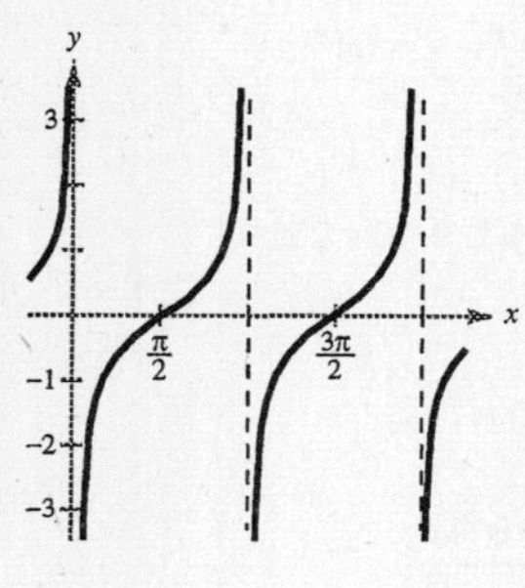

6. $h(x) = a\cos(bx + c)$

Graph is reflected in x-axis.

Amplitude: $a = -3$

Period: $2 = \dfrac{2\pi}{\pi} \Rightarrow b = \pi$

No phase shift: $c = 0$

$h(x) = -3\cos(\pi x)$

7. $f(x) = \dfrac{x}{2}\sin x,\ -3\pi \le x \le 3\pi$

$-\dfrac{x}{2} \le f(x) \le \dfrac{x}{2}$

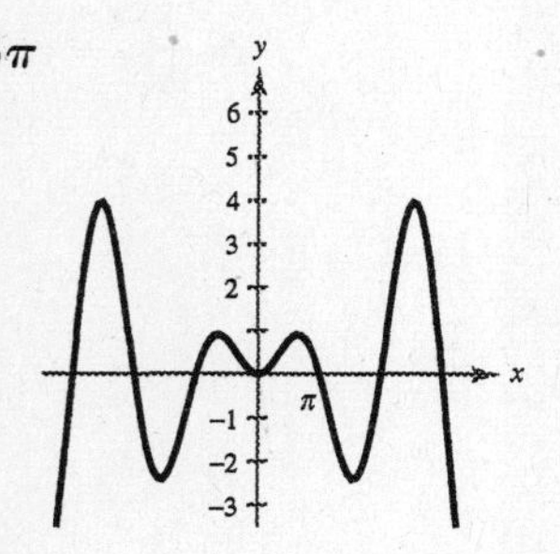

8. $\tan(\arctan 6.7) = 6.7$

9. $\tan\left(\arcsin \frac{3}{5}\right) = \frac{3}{4}$

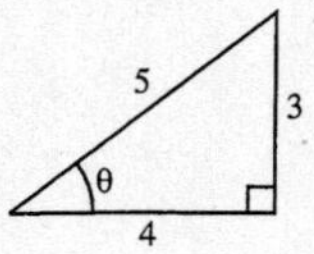

10. $y = \arccos(2x)$

$\sin y = \sin(\arccos(2x)) = \sqrt{1 - 4x^2}$

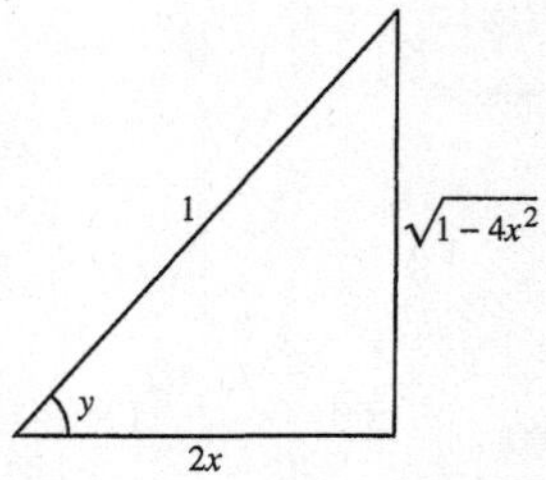

11. $\cos\left(\frac{\pi}{2} - x\right)\csc x = \sin x\left(\frac{1}{\sin x}\right) = 1$

12. $\frac{\sin\theta - 1}{\cos\theta} - \frac{\cos\theta}{\sin\theta - 1} = \frac{\sin\theta - 1}{\cos\theta} - \frac{\cos\theta(\sin\theta + 1)}{\sin^2\theta - 1}$

$$= \frac{\sin\theta - 1}{\cos\theta} + \frac{\cos\theta(\sin\theta + 1)}{\cos^2\theta} = \frac{\sin\theta - 1}{\cos\theta} + \frac{\sin\theta + 1}{\cos\theta} = \frac{2\sin\theta}{\cos\theta} = 2\tan\theta$$

13. $\cot^2\alpha(\sec^2\alpha - 1) = \cot^2\alpha\tan^2\alpha = 1$

14. $\sin(x + y)\sin(x - y) = \frac{1}{2}[\cos(x + y - (x - y)) - \cos(x + y + x - y)]$

$$= \frac{1}{2}[\cos 2y - \cos 2x] = \frac{1}{2}[1 - 2\sin^2 y - (1 - 2\sin^2 x)] = \sin^2 x - \sin^2 y$$

15. $\sin^2 x\cos^2 x = \left(\frac{1 - \cos 2x}{2}\right)\left(\frac{1 + \cos 2x}{2}\right)$

$$= \frac{1}{4}(1 - \cos 2x)(1 + \cos 2x)$$

$$= \frac{1}{4}(1 - \cos^2 2x)$$

$$= \frac{1}{4}\left(1 - \frac{1 + \cos 4x}{2}\right)$$

$$= \frac{1}{8}(2 - (1 + \cos 4x))$$

$$= \frac{1}{8}(1 - \cos 4x)$$

16. $2\cos^2\beta - \cos\beta = 0$

$\cos\beta(2\cos\beta - 1) = 0$

$\cos\beta = 0 \qquad 2\cos\beta - 1 = 0$

$\beta = \frac{\pi}{2}, \frac{3\pi}{2} \qquad \cos\beta = \frac{1}{2}$

$\beta = \frac{\pi}{3}, \frac{5\pi}{3}$

Answer: $\frac{\pi}{3}, \frac{\pi}{2}, \frac{3\pi}{2}, \frac{5\pi}{3}$

17. $3\tan\theta - \cot\theta = 0$

$$3\tan\theta - \frac{1}{\tan\theta} = 0$$

$$\frac{3\tan^2\theta - 1}{\tan\theta} = 0$$

$$3\tan^2\theta - 1 = 0$$

$$\tan^2\theta = \frac{1}{3}$$

$$\tan\theta = \pm\frac{\sqrt{3}}{3}$$

$$\theta = \frac{\pi}{6}, \frac{5\pi}{6}, \frac{7\pi}{6}, \frac{11\pi}{6}$$

18. $\sin^2 x + 2\sin x + 1 = 0$

$$(\sin x + 1)(\sin x + 1) = 0$$

$$\sin x + 1 = 0$$

$$\sin x = -1$$

$$x = \frac{3\pi}{2}$$

19. $\sin u = \frac{12}{13} \Rightarrow \cos u = \frac{5}{13}$ and $\tan u = \frac{12}{5}$ since u is in Quadrant I.

$\cos v = \frac{3}{5} \Rightarrow \sin v = \frac{4}{5}$ and $\tan v = \frac{4}{3}$ since v is in Quadrant I.

$$\tan(u - v) = \frac{\tan u - \tan v}{1 + \tan u \tan v} = \frac{\frac{12}{5} - \frac{4}{3}}{1 + \left(\frac{12}{5}\right)\left(\frac{4}{3}\right)} = \frac{16}{63}$$

20. $\tan\theta = \frac{1}{2}$

$$\tan 2\theta = \frac{2\tan\theta}{1 - \tan^2\theta} = \frac{2\left(\frac{1}{2}\right)}{1 - \left(\frac{1}{2}\right)^2} = \frac{4}{3}$$

21. $\tan\theta = \frac{4}{3} \Rightarrow \cos\theta = \pm\frac{3}{5}$

$$\sin\frac{\theta}{2} = \sqrt{\frac{1 - \cos\theta}{2}} = \sqrt{\frac{1 - \frac{3}{5}}{2}} = \frac{\sqrt{5}}{5}$$

$$\text{or } = \sqrt{\frac{1 + \frac{3}{5}}{2}} = \frac{2\sqrt{5}}{5}$$

22. $$5\sin\frac{3\pi}{4}\cos\frac{7\pi}{4} = \frac{5}{2}\left[\sin\left(\frac{3\pi}{4} + \frac{7\pi}{4}\right) + \sin\left(\frac{3\pi}{4} - \frac{7\pi}{4}\right)\right]$$

$$= \frac{5}{2}\left[\sin\frac{5\pi}{2} + \sin(-\pi)\right]$$

$$= \frac{5}{2}\left(\sin\frac{5\pi}{2} - \sin\pi\right)$$

23. Given: $A = 30°, a = 9, b = 8$

$$\frac{\sin B}{8} = \frac{\sin 30°}{9}$$

$$\sin B = \frac{8}{9}\left(\frac{1}{2}\right)$$

$$B = \arcsin\left(\frac{4}{9}\right)$$

$$B \approx 26.4°$$

$$C = 180° - A - B \approx 123.6°$$

$$\frac{c}{\sin 123.6°} = \frac{9}{\sin 30°}$$

$$c \approx 15.0$$

24. Given: $A = 30°, b = 8, c = 10$

$$a^2 = 8^2 + 10^2 - 2(8)(10)\cos 30°$$

$$a^2 \approx 25.4$$

$$a \approx 5.0$$

$$\cos B = \frac{5.0^2 + 10^2 - 8^2}{2(5.0)(10)}$$

$$\cos B = 0.61$$

$$B \approx 52.4°$$

$$C = 180° - A - B \approx 97.6°$$

25. Given: $A = 30°, C = 90°, b = 10$

$$B = 180° - 30° - 90° = 60°$$

$$\tan 30° = \frac{a}{10} \Rightarrow a = 10 \tan 30° \approx 5.8$$

$$\cos 30° = \frac{10}{c} \Rightarrow c = \frac{10}{\cos 30°} \approx 11.5$$

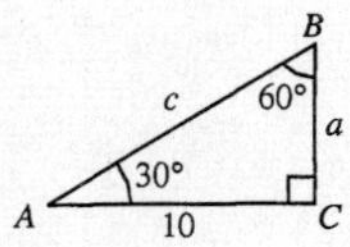

26. $a = 4, b = 8, c = 9$

$$\cos C = \frac{4^2 + 8^2 - 9^2}{2(4)(8)} = \frac{-1}{64} \Rightarrow C \approx 90.9°$$

$$\sin A \approx \frac{4 \sin 90.9°}{9} \Rightarrow A \approx 26.4°$$

$$B \approx 180° - 26.4° - 90.9° = 62.7°$$

27. Area $= \frac{1}{2}(7)(12) \sin 60° \approx 36.4$ square inches

28. $s = \dfrac{11 + 16 + 17}{2} = 22$

Area $= \sqrt{22(11)(6)(5)} \approx 85.2$ square inches

29. $\mathbf{u} = \langle 3, 5 \rangle = 3\mathbf{i} + 5\mathbf{j}$

30. $\mathbf{u} = 3\mathbf{i} + 4\mathbf{j}, \mathbf{v} = \mathbf{i} - 2\mathbf{j}$

$$\mathbf{u} \cdot \mathbf{v} = 3(1) + 4(-2) = -5$$

31. $\mathbf{u} = \langle 8, -2 \rangle, \mathbf{v} = \langle 1, 5 \rangle$

$$\mathbf{w}_1 = \text{proj}_{\mathbf{v}}\, \mathbf{u} = \left(\frac{\mathbf{u} \cdot \mathbf{v}}{\|\mathbf{v}\|^2}\right)\mathbf{v} = \frac{-2}{26}\langle 1, 5 \rangle = -\frac{1}{13}\langle 1, 5 \rangle$$

$$\mathbf{w}_2 = \mathbf{u} - \mathbf{w}_1 = \langle 8, -2 \rangle - \left\langle -\frac{1}{13}, -\frac{5}{13} \right\rangle = \left\langle \frac{105}{13}, -\frac{21}{13} \right\rangle$$

$$= \frac{21}{13}\langle 5, -1 \rangle$$

32. Height of smaller triangle:

$$\tan 16° 45' = \frac{h_1}{200}$$

$$h_1 = 200 \tan 16.75° \approx 60.2 \text{ feet}$$

Height of larger triangle:

$$\tan 18° = \frac{h_2}{200}$$

$$h_2 = 200 \tan 18° \approx 65.0 \text{ feet}$$

Height of flag:

$h_2 - h_1 = 65.0 - 60.2 \approx 5$ feet

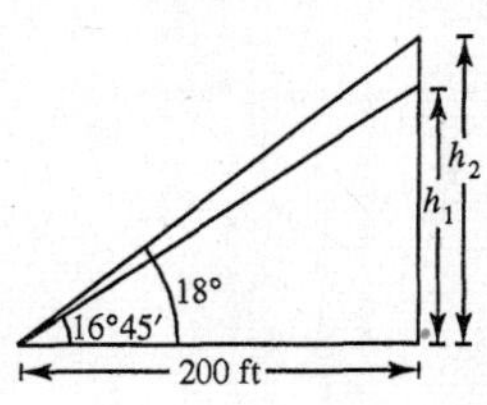

Not drawn to scale

33. Number of revolutions per minute: $\dfrac{3142}{2\pi} \approx 500$ rev/min

Time to make 10,000 revolutions: $\dfrac{10{,}000}{500} = 20$ minutes

34. $\tan \theta = \dfrac{5}{12} \Rightarrow \theta \approx 22.6°$

35. $d = a \cos bt$

$$|a| = 4 \Rightarrow a = 4$$

$$\frac{2\pi}{b} = 8 \Rightarrow b = \frac{\pi}{4}$$

$$d = 4 \cos \frac{\pi}{4}t$$

36. $\mathbf{v}_1 = 500\langle \cos 60°, \sin 60° \rangle = \langle 250, 250\sqrt{3} \rangle$

$\mathbf{v}_2 = 50\langle \cos 30°, \sin 30° \rangle = \langle 25\sqrt{3}, 25 \rangle$

$\mathbf{v} = \mathbf{v}_1 + \mathbf{v}_2 = \langle 250 + 25\sqrt{3}, 250\sqrt{3} + 25 \rangle$

$\approx \langle 293.3, 458.0 \rangle$

$\|\mathbf{v}\| = \sqrt{(293.3)^2 + (458.0)^2} \approx 543.9$

$\tan \theta = \dfrac{458.0}{293.3} \approx 1.56 \implies \theta \approx 57.4°$

The plane is traveling N 32.6° E at 543.9 kilometers per hour.

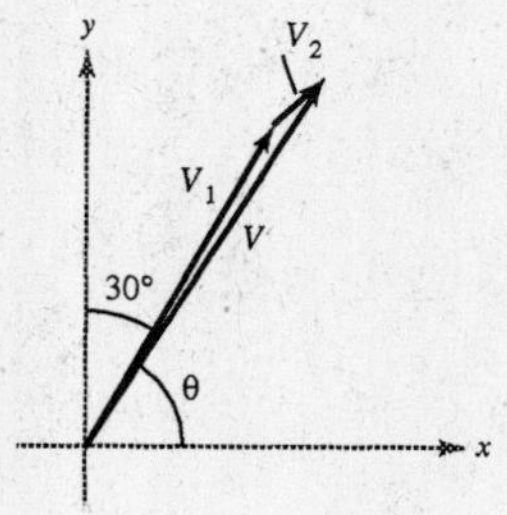

Chapter 4 Chapter Test

1. $-3 + \sqrt{-81} = -3 + 9i$

2. $10i - \left(3 + \sqrt{-25}\right) = 10i - (3 + 5i) = -3 + 5i$

3. $(2 + 6i)^2 = 4 + 24i + 36i^2 = -32 + 24i$

4. $\left(2 + \sqrt{3}i\right)\left(2 - \sqrt{3}i\right) = 4 - 3i^2 = 4 + 3 = 7.$

5. $\dfrac{5}{2 + i} = \dfrac{5}{2 + i} \cdot \dfrac{2 - i}{2 - i} = \dfrac{5(2 - i)}{4 + 1} = 2 - i$

6. $2x^2 - 2x + 3 = 0$

$$x = \frac{-(-2) \pm \sqrt{(-2)^2 - 4(2)(3)}}{2(2)} = \frac{2 \pm \sqrt{-20}}{4}$$

$$= \frac{2 \pm 2\sqrt{5}i}{4} = \frac{1}{2} \pm \frac{\sqrt{5}}{2}i$$

7. Since $x^5 + x^3 - x + 1 = 0$ is a fifth degree polynomial equation, it has 5 solutions in the complex number system.

8. Since $x^4 - 3x^3 + 2x^2 - 4x - 5 = 0$ is a fourth degree polynomial equation, it has 4 solutions in the complex number system.

9. $x^3 - 6x^2 + 5x - 30 = 0$

$x^2(x - 6) + 5(x - 6) = 0$

$(x - 6)(x^2 + 5) = 0$

$x - 6 = 0$ or $x^2 + 5 = 0$

$x = 6$ or $x = \pm\sqrt{5}i$

10. $x^4 - 2x^2 - 24 = 0$

$(x^2 - 6)(x^2 + 4) = 0$

$x^2 - 6 = 0$ or $x^2 + 4 = 0$

$x = \pm\sqrt{6}$ or $x = \pm 2i$

11. $h(x) = x^4 - 2x^2 - 8$

Zeros: $x = \pm 2 \Rightarrow (x - 2)(x + 2) = x^2 - 4$ is a factor of $h(x)$.

$$\begin{array}{r} x^2 + 2 \\ x^2 + 0x - 4 \overline{) x^4 + 0x^3 - 2x^2 + 0x - 8} \\ \underline{x^4 + 0x^3 - 4x^2} \\ 2x^2 + 0x - 8 \\ \underline{2x^2 + 0x - 8} \\ 0 \end{array}$$

Thus, $h(x) = (x^2 - 4)(x^2 + 2) = (x + 2)(x - 2)(x + \sqrt{2}i)(x - \sqrt{2}i)$.

The zeros of $h(x)$ are: $x = \pm 2, \pm\sqrt{2}i$.

12. $g(v) = 2v^3 - 11v^2 + 22v - 15$

Zero: $\frac{3}{2} \Rightarrow 2v - 3$ is a factor of $g(v)$

$$\begin{array}{r} v^2 - 4v + 5 \\ 2v - 3 \overline{) 2v^3 - 11v^2 + 22v - 15} \\ \underline{2v^3 - 3v^2} \\ -8v^2 + 22v \\ \underline{-8v^2 + 12v} \\ 10v - 15 \\ \underline{10v - 15} \\ 0 \end{array}$$

Thus, $g(v) = (2v - 3)(v^2 - 4v + 5)$. By the Quadratic Formula, the zeros of $v^2 - 4v + 5$ are $2 \pm i$.

The zeros of $g(v)$ are: $v = \frac{3}{2}, 2 \pm i$.

$g(v) = (2v - 3)(v - 2 - i)(v - 2 + i)$

13. $f(x) = x(x - 3)[x - (3 + i)][x - (3 - i)]$

$= (x^2 - 3x)[(x - 3) - i][(x - 3) + i]$

$= (x^2 - 3x)[(x - 3)^2 - i^2]$

$= (x^2 - 3x)(x^2 - 6x + 10)$

$= x^4 - 9x^3 + 28x^2 - 30x$

14. $f(x) = [x - (1 + \sqrt{6}i)][x - (1 - \sqrt{6}i)](x - 3)(x - 3)$

$= [(x - 1) - \sqrt{6}i][(x - 1) + \sqrt{6}i](x^2 - 6x + 9)$

$= [(x - 1)^2 - 6i^2](x^2 - 6x + 9)$

$= (x^2 - 2x + 7)(x^2 - 6x + 9)$

$= x^4 - 8x^3 + 28x^2 - 60x + 63$

15. No, complex zeros occur in conjugate pairs for polynomial functions with *real* coefficients. If $a + bi$ is a zero, so is $a - bi$.

16. $z = 5 - 5i$

$|z| = \sqrt{5^2 + (-5)^2} = \sqrt{50} = 5\sqrt{2}$

$\tan\theta = \dfrac{-5}{5} = -1$ and θ is in Quadrant IV $\Rightarrow \theta = \dfrac{7\pi}{4}$

$z = 5\sqrt{2}\left(\cos\dfrac{7\pi}{4} + i\sin\dfrac{7\pi}{4}\right)$

17. $z = 6(\cos 120° + i\sin 120°) = 6\left(-\dfrac{1}{2} + \dfrac{\sqrt{3}}{2}i\right) = -3 + 3\sqrt{3}i$

18. $\left[3\left(\cos\frac{7\pi}{6} + i\sin\frac{7\pi}{6}\right)\right]^8 = 3^8\left(\cos\frac{28\pi}{3} + i\sin\frac{28\pi}{3}\right)$

$$= 6561\left(-\frac{1}{2} - \frac{\sqrt{3}}{2}i\right) = -\frac{6561}{2} - \frac{6561\sqrt{3}}{2}i$$

19. $(3 - 3i)^6 = \left[3\sqrt{2}\left(\cos\frac{7\pi}{4} + i\sin\frac{7\pi}{4}\right)\right]^6$

$$= \left(3\sqrt{2}\right)^6\left(\cos\frac{21\pi}{2} + i\sin\frac{21\pi}{2}\right)$$

$$= 5832(0 + i)$$

$$= 5832i$$

20. $z = 256\left(1 + \sqrt{3}i\right)$

$|z| = 256\sqrt{1^2 + \left(\sqrt{3}\right)^2} = 256\sqrt{4} = 512$

$\tan\theta = \frac{\sqrt{3}}{1} \Rightarrow \theta = \frac{\pi}{3}$

$z = 512\left(\cos\frac{\pi}{3} + i\sin\frac{\pi}{3}\right)$

Fourth roots of z: $\sqrt[4]{512}\left[\cos\left(\frac{\frac{\pi}{3} + 2\pi k}{4}\right) + i\sin\left(\frac{\frac{\pi}{3} + 2\pi k}{4}\right)\right], k = 0, 1, 2, 3$

$k = 0$: $4\sqrt[4]{2}\left(\cos\frac{\pi}{12} + i\sin\frac{\pi}{12}\right)$

$k = 1$: $4\sqrt[4]{2}\left(\cos\frac{7\pi}{12} + i\sin\frac{7\pi}{12}\right)$

$k = 2$: $4\sqrt[4]{2}\left(\cos\frac{13\pi}{12} + i\sin\frac{13\pi}{12}\right)$

$k = 3$: $4\sqrt[4]{2}\left(\cos\frac{19\pi}{12} + i\sin\frac{19\pi}{12}\right)$

21. $x^3 - 27i = 0 \Rightarrow x^3 = 27i$

The solutions to the equation are the cube roots of $z = 27i = 27\left(\cos\frac{\pi}{2} + i\sin\frac{\pi}{2}\right)$.

Cube roots of z: $\sqrt[3]{27}\left[\cos\left(\frac{\frac{\pi}{2} + 2\pi k}{3}\right) + i\sin\left(\frac{\frac{\pi}{2} + 2\pi k}{3}\right)\right], k = 0, 1, 2$

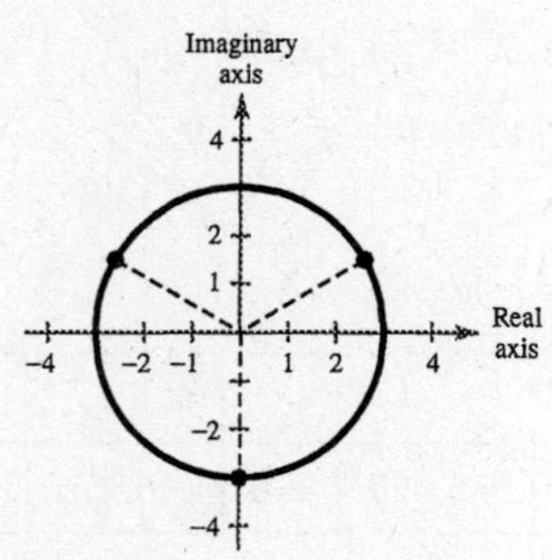

$k = 0$: $3\left(\cos\frac{\pi}{6} + i\sin\frac{\pi}{6}\right)$

$k = 1$: $3\left(\cos\frac{5\pi}{6} + i\sin\frac{5\pi}{6}\right)$

$k = 2$: $3\left(\cos\frac{3\pi}{2} + i\sin\frac{3\pi}{2}\right)$

Chapter 5 Chapter Test

1. $12.4^{2.79} \approx 1123.690$

2. $4^{3\pi/2} \approx 687.291$

3. $e^{-7/10} \approx 0.497$

4. $e^{3.1} \approx 22.198$

5. $f(x) = 10^{-x}$

x	-1	$-\frac{1}{2}$	0	$\frac{1}{2}$	1
$f(x)$	10	3.162	1	0.316	0.1

Asymptote: $y = 0$

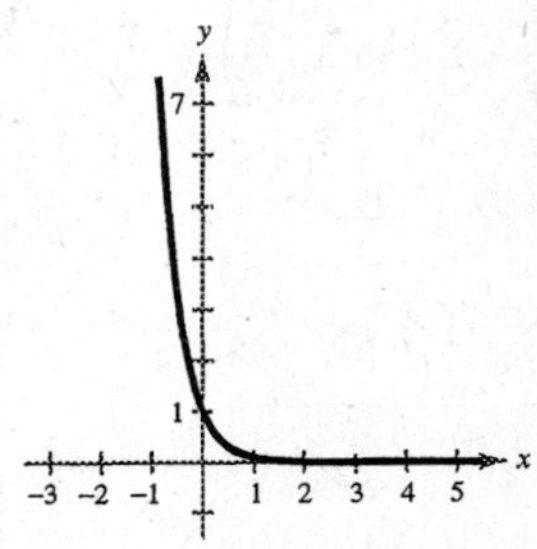

6. $f(x) = -6^{x-2}$

x	-1	0	1	2	3
$f(x)$	-0.005	-0.028	-0.167	-1	-6

Asymptote: $y = 0$

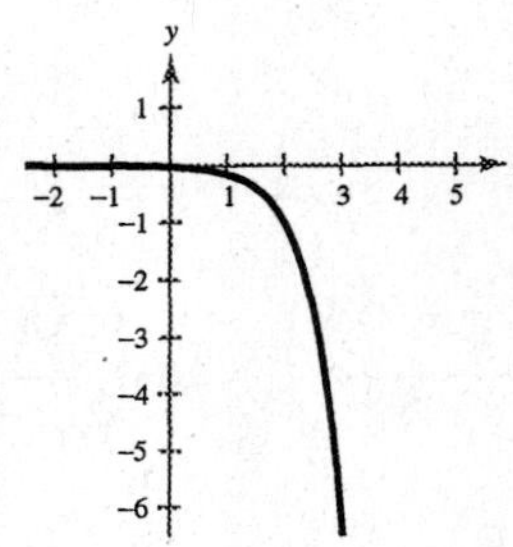

7. $f(x) = 1 - e^{2x}$

x	-1	$-\frac{1}{2}$	0	$\frac{1}{2}$	1
$f(x)$	0.865	0.632	0	-1.718	-6.389

Asymptote: $y = 1$

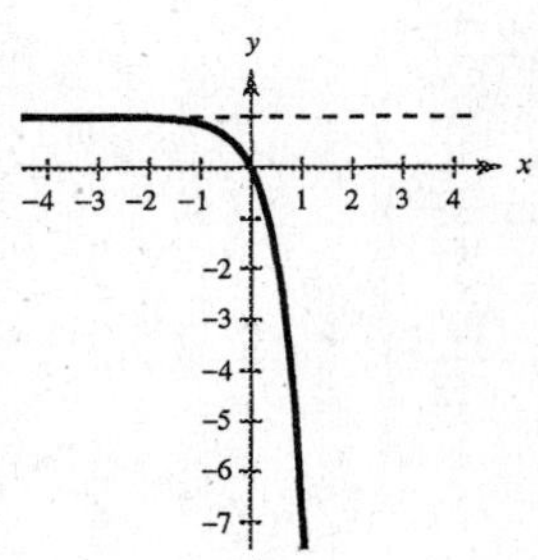

8. (a) $\log_7 7^{-0.89} = -0.89$

(b) $4.6 \ln e^2 = 4.6(2) = 9.2$

9. $f(x) = -\log_{10} x - 6$

x	$\frac{1}{2}$	1	$\frac{3}{2}$	2	4
$f(x)$	-5.699	-6	-6.176	-6.301	-6.602

Asymptote: $x = 0$

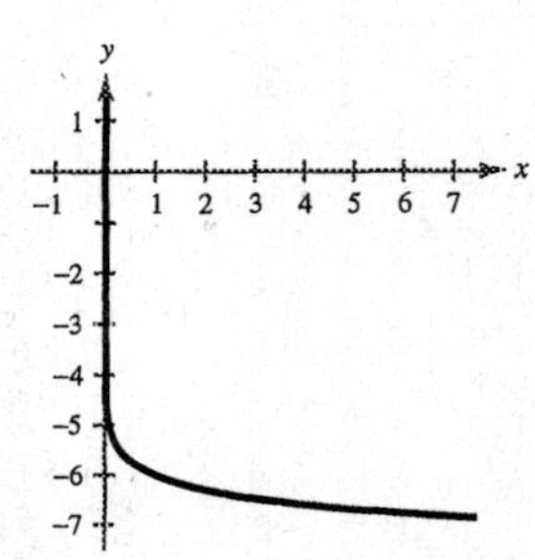

10. $f(x) = \ln(x - 4)$

x	5	7	9	11	13
$f(x)$	0	1.099	1.609	1.946	2.197

Asymptote: $x = 4$

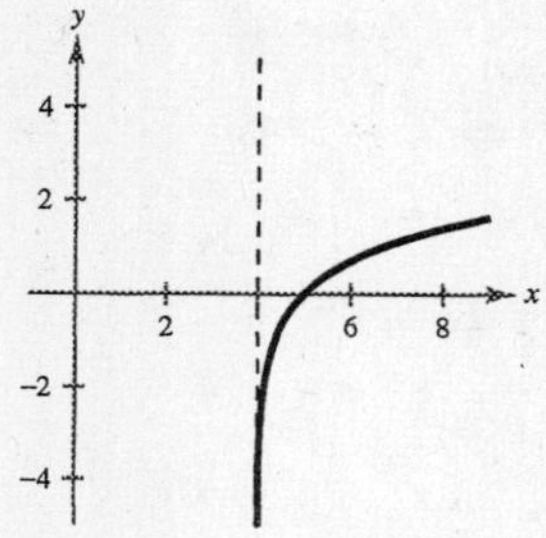

11. $f(x) = 1 + \ln(x + 6)$

x	−5	−3	−1	0	1
$f(x)$	1	2.099	2.609	2.792	2.946

Asymptote: $x = -6$

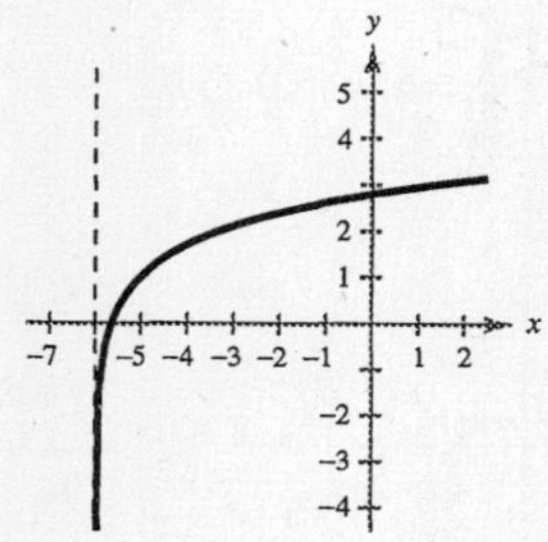

12. $\log_7 44 = \dfrac{\ln 44}{\ln 7} = \dfrac{\log_{10} 44}{\log_{10} 7} \approx 1.945$

13. $\log_{2/5} 0.9 = \dfrac{\ln 0.9}{\ln (2/5)} = \dfrac{\log_{10} 0.9}{\log_{10}(2/5)} \approx 0.115$

14. $\log_{24} 68 = \dfrac{\ln 68}{\ln 24} = \dfrac{\log_{10} 68}{\log_{10} 24} \approx 1.328$

15. $\log_2 3a^4 = \log_2 3 + \log_2 a^4 = \log_2 3 + 4 \log_2|a|$

16. $\ln \dfrac{5\sqrt{x}}{6} = \ln(5\sqrt{x}) - \ln 6 = \ln 5 + \ln\sqrt{x} - \ln 6$

$= \ln 5 + \frac{1}{2}\ln x - \ln 6$

17. $\log_3 13 + \log_3 y = \log_3 13y$

18. $4 \ln x - 4 \ln y = \ln x^4 - \ln y^4 = \ln\left(\dfrac{x^4}{y^4}\right), x > 0, y > 0$

19. $\dfrac{1025}{8 + e^{4x}} = 5$

$1025 = 5(8 + e^{4x})$

$205 = 8 + e^{4x}$

$197 = e^{4x}$

$\ln 197 = 4x$

$\dfrac{\ln 197}{4} = x$

$x \approx 1.321$

20. $\log_{10} x - \log_{10}(8 - 5x) = 2$

$\log_{10} \dfrac{x}{8 - 5x} = 2$

$\dfrac{x}{8 - 5x} = 10^2$

$x = 100(8 - 5x)$

$x = 800 - 500x$

$510x = 800$

$x = \dfrac{800}{501} \approx 1.597$

21. $y = Ce^{kt}$

$(0, 2745)$: $2745 = Ce^{k(0)} \Rightarrow C = 2745$

$y = 2745e^{kt}$

$(9, 11{,}277)$: $11{,}277 = 2745e^{k(9)}$

$$\frac{11{,}277}{2745} = e^{9k}$$

$$\ln\left(\frac{11277}{2745}\right) = 9k$$

$$\frac{1}{9}\ln\left(\frac{11277}{2745}\right) = k \Rightarrow k \approx 0.1570$$

Thus, $y = 2745e^{0.1570t}$

22.

$$y = Ce^{kt}$$

$$\frac{1}{2}C = Ce^{k(22)}$$

$$\frac{1}{2} = e^{22k}$$

$$\ln\left(\frac{1}{2}\right) = 22k$$

$$\frac{\ln(1/2)}{22} = k \Rightarrow k \approx -0.0315$$

$$y = Ce^{-0.0315t}$$

When $t = 19$: $y = Ce^{-0.0315(19)} \approx 0.55C$

Thus, 55% will remain after 19 years.

23. $H = 70.228 + 5.104x + 9.222 \ln x, \frac{1}{4} \le x \le 6$

(a)

x	H(cm)
$\frac{1}{4}$	58.720
$\frac{1}{2}$	66.388
1	75.332
2	86.828
3	95.671
4	103.43
5	110.59
6	117.38

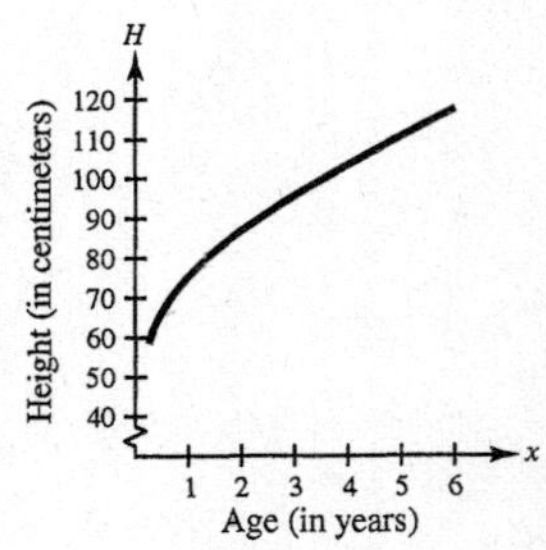

(b) When $x = 4$, $H \approx 103.43$ cm.

Chapter 6 Chapter Test

1. $2x - 7y + 3 = 0$

$$y = \frac{2}{7}x + \frac{3}{7}$$

$$\tan\theta = \frac{2}{7}$$

$$\theta \approx 15.9°$$

2. $3x + 2y - 4 = 0 \Rightarrow y = -\frac{3}{2}x + 2 \Rightarrow m_1 = -\frac{3}{2}$

$4x - y + 6 = 0 \Rightarrow y = 4x + 6 \quad \Rightarrow m_2 = 4$

$$\tan\theta = \left|\frac{4 - \left(-\frac{3}{2}\right)}{1 + 4\left(-\frac{3}{2}\right)}\right| = \frac{11}{10}$$

$$\theta \approx 47.7°$$

3. $y = 5 - x \Rightarrow x + y - 5 = 0 \Rightarrow A = 1, B = 1, C = -5$

$(x_1, y_1) = (7, 5)$

$$d = \frac{|(1)(7) + (1)(5) + (-5)|}{\sqrt{1^2 + 1^2}} = \frac{7}{\sqrt{2}} = \frac{7\sqrt{2}}{2}$$

4. $y^2 - 4x + 4 = 0$

$$y^2 = 4(x - 1)$$

Parabola

Vertex: (1, 0)

Focus: (2, 0)

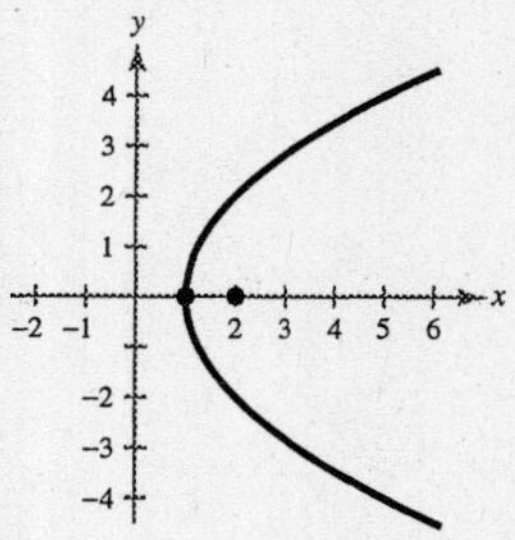

5. $x^2 - 4y^2 - 4x = 0$

$$(x - 2)^2 - 4y^2 = 4$$

$$\frac{(x - 2)^2}{4} - \frac{y^2}{1} = 1$$

Hyperbola

Center: (2, 0)

Horizontal transverse axis

$a = 2, b = 1, c^2 = 1 + 4 = 5 \Rightarrow c = \sqrt{5}$

Vertices: (0, 0), (4, 0)

Foci: $\left(2 \pm \sqrt{5}, 0\right)$

Asymptotes: $y = \pm\frac{1}{2}(x - 2)$

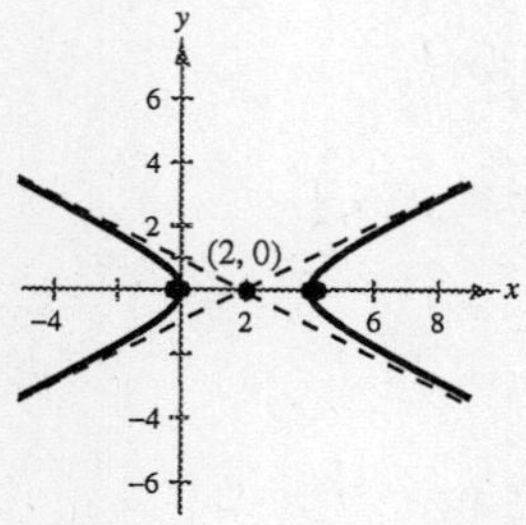

6.

$$9x^2 + 16y^2 + 54x - 32y - 47 = 0$$

$$9(x^2 + 6x + 9) + 16(y^2 - 2y + 1) = 47 + 81 + 16$$

$$9(x + 3)^2 + 16(y - 1)^2 = 144$$

$$\frac{(x + 3)^2}{16} + \frac{(y - 1)^2}{9} = 1$$

Ellipse

Center: $(-3, 1)$

$a = 4, b = 3, c = \sqrt{7}$

Foci: $\left(-3 \pm \sqrt{7}, 1\right)$

Vertices: $(1, 1), (-7, 1)$

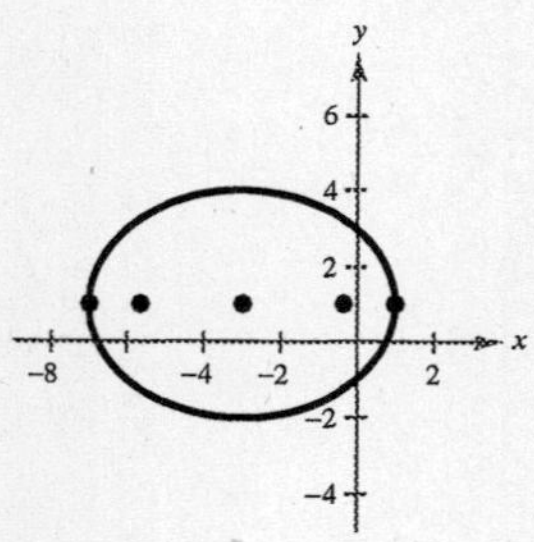

7.

$$2x^2 + 2y^2 - 8x - 4y + 9 = 0$$

$$2(x^2 - 4x + 4) + 2(y^2 - 2y + 1) = -9 + 8 + 2$$

$$2(x - 2)^2 + 2(y - 1)^2 = 1$$

$$(x - 2)^2 + (y - 1)^2 = \frac{1}{2}$$

Circle

Center: (2, 1)

Radius: $\sqrt{\frac{1}{2}} = \frac{\sqrt{2}}{2} \approx 0.707$

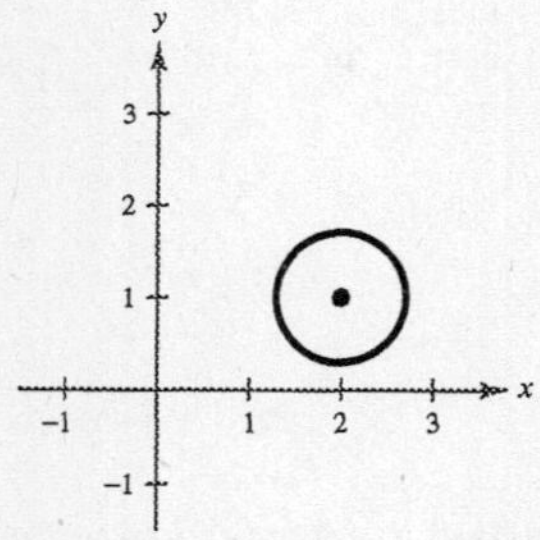

8. Parabola

Vertex: $(3, -2)$

Vertical axis

Point: $(0, 4)$

$$(x-h)^2 = 4p(y-k)$$
$$(x-3)^2 = 4p(y+2)$$
$$(0-3)^2 = 4p(4+2)$$
$$9 = 24p$$
$$p = \frac{9}{24} = \frac{3}{8}$$

Equation: $(x-3)^2 = 4\left(\frac{3}{8}\right)(y+2)$

$$(x-3)^2 = \frac{3}{2}(y+2)$$

9. Hyperbola

Foci: $(0, 0)$ and $(0, 4) \implies c = 2$

Asymptotes: $y = \pm\frac{1}{2}x + 2$

Vertical transverse axis

Center: $(0, 2) = (h, k)$

$$\frac{a}{b} = \frac{1}{2} \implies 2a = b$$
$$c^2 = a^2 + b^2$$
$$4 = a^2 + (2a)^2$$
$$4 = 5a^2$$
$$\frac{4}{5} = a^2$$
$$b^2 = (2a)^2 = 4a^2 = \frac{16}{5}$$
$$\frac{(y-k)^2}{a^2} - \frac{(x-h)^2}{b^2} = 1$$
$$\frac{(y-2)^2}{\frac{4}{5}} - \frac{x^2}{\frac{16}{5}} = 1$$
$$\frac{5(y-2)^2}{4} - \frac{5x^2}{16} = 1$$

10. (a) $x^2 + 6xy + y^2 - 6 = 0$

$A = 1, B = 6, C = 1$

$$\cot 2\theta = \frac{1-1}{6} = 0$$
$$2\theta = 90^\circ$$
$$\theta = 45^\circ$$

(b) $x = x'\cos 45^\circ - y'\sin 45^\circ$

$$= \frac{x'-y'}{\sqrt{2}}$$

$y = x'\sin 45^\circ + y'\cos 45^\circ$

$$= \frac{x'+y'}{\sqrt{2}}$$

$$\left(\frac{x'-y'}{\sqrt{2}}\right)^2 + 6\left(\frac{x'-y'}{\sqrt{2}}\right)\left(\frac{x'+y'}{\sqrt{2}}\right) + \left(\frac{x'+y'}{\sqrt{2}}\right)^2 - 6 = 0$$

$$\frac{1}{2}((x')^2 - 2(x')(y') + (y')^2) + 3((x')^2 - (y')^2) + \frac{1}{2}((x')^2 + 2(x')(y') + (y')^2) - 6 = 0$$

$$4(x')^2 - 2(y')^2 = 6$$
$$\frac{2(x')^2}{3} - \frac{(y')^2}{3} = 1$$

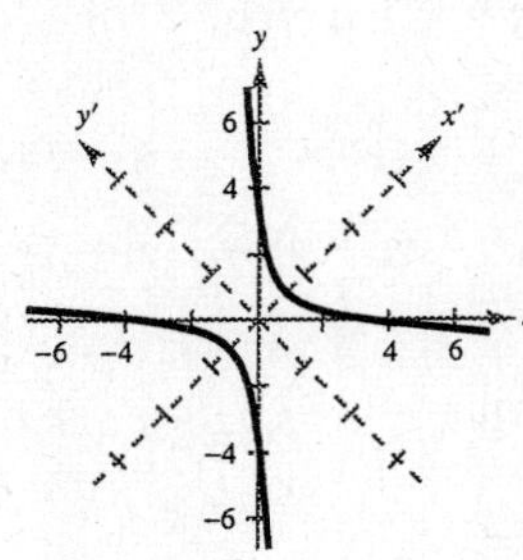

For the graphing utility, we need to solve for y in terms of x.

$$y^2 + 6xy + 9x^2 = 6 - x^2 + 9x^2$$
$$(y+3x)^2 = 6 + 8x^2$$
$$y + 3x = \pm\sqrt{6+8x^2}$$
$$y = -3x \pm \sqrt{6+8x}$$

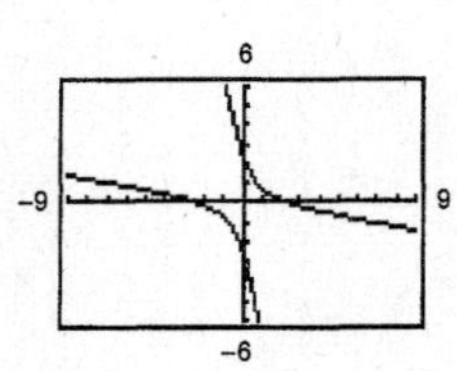

11. $x = 2 + 3\cos\theta$

$y = 2\sin\theta$

θ	0	$\pi/2$	π	$3\pi/2$
x	5	2	-1	2
y	0	2	0	-2

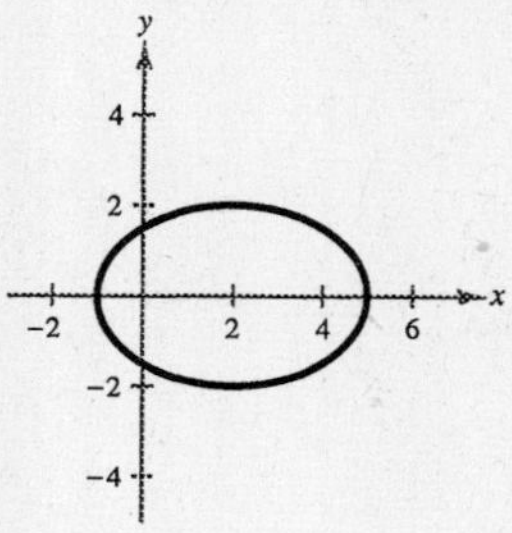

$$x = 2 + 3\cos\theta \Longrightarrow \frac{x-2}{3} = \cos\theta$$

$$y = 2\sin\theta \Longrightarrow \frac{y}{2} = \sin\theta$$

$$\cos^2\theta + \sin^2\theta = 1$$

$$\frac{(x-2)^2}{9} + \frac{y^2}{4} = 1$$

12. $(6, 4), (2, -3)$

$x = x_1 + t(x_2 - x_1) = 6 + t(2 - 6) = 6 - 4t$

$y = y_1 + t(y_2 - y_1) = 4 + t(-3 - 4) = 4 - 7t$

Answers are not unique. Another possible set:

$x = 6 + 4t$

$y = 4 + 7t$

13. Polar Coordinates: $\left(-2, \frac{5\pi}{6}\right)$

$x = -2\cos\frac{5\pi}{6} = -2\left(-\frac{\sqrt{3}}{2}\right) = \sqrt{3}$

$y = -2\sin\frac{5\pi}{6} = -2\left(\frac{1}{2}\right) = -1$

Rectangular Coordinates: $\left(\sqrt{3}, -1\right)$

14. Rectangular Coordinates: $(2, -2)$

$r = \pm\sqrt{2^2 + (-2)^2} = \pm\sqrt{8} = \pm 2\sqrt{2}$

$\tan\theta = -1 \Rightarrow \theta = \frac{3\pi}{4}, \frac{7\pi}{4}$

Polar Coordinates:

$\left(2\sqrt{2}, \frac{7\pi}{4}\right), \left(-2\sqrt{2}, \frac{3\pi}{4}\right), \left(2\sqrt{2}, -\frac{\pi}{4}\right)$

15. $x^2 + y^2 - 4y = 0$

$r^2 - 4r\sin\theta = 0$

$r^2 = 4r\sin\theta$

$r = 4\sin\theta$

16. $r = \dfrac{4}{1 + \cos\theta}$

$e = 1 \Rightarrow$ Parabola

Vertex: $(2, 0)$

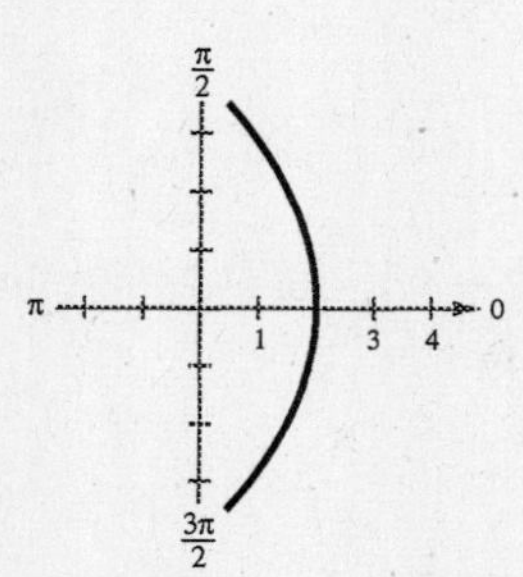

17. $r = \dfrac{4}{2 + \cos\theta} = \dfrac{2}{1 + \frac{1}{2}\cos\theta}$

$e = \frac{1}{2} \Rightarrow$ Ellipse

Vertex: $\left(\frac{4}{3}, 0\right), (4, \pi)$

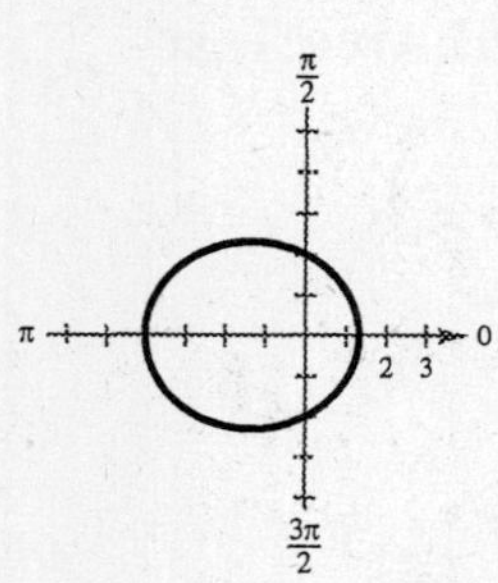

18. $r = 2 + 3 \sin \theta$

$\frac{a}{b} = \frac{2}{3} < 1$ Limaçon with inner loop

θ	0	$\frac{\pi}{2}$	π	$\frac{3\pi}{2}$
r	2	5	2	-1

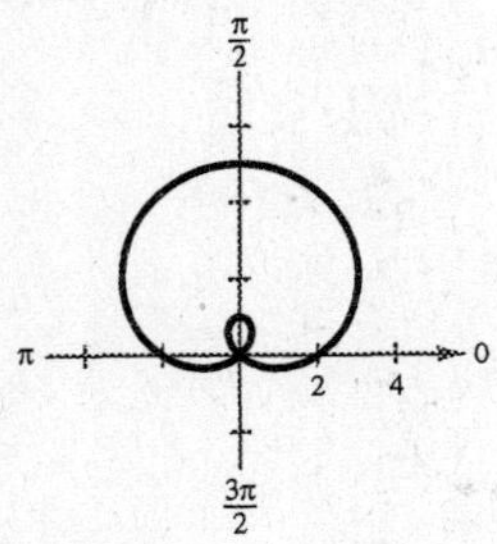

19. $r = 3 \sin 2\theta$

Rose curve ($n = 2$) with four petals

$|r| = 3$ when $\theta = \frac{\pi}{4}, \frac{3\pi}{4}, \frac{5\pi}{4}, \frac{7\pi}{4}$

$r = 0$ when $\theta = 0, \frac{\pi}{2}, \pi, \frac{3\pi}{2}$

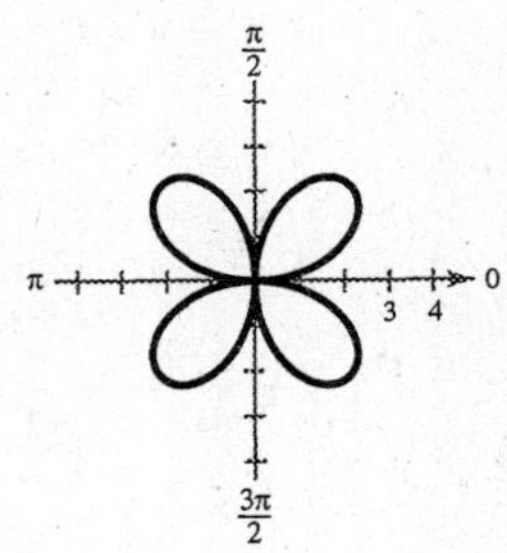

20.

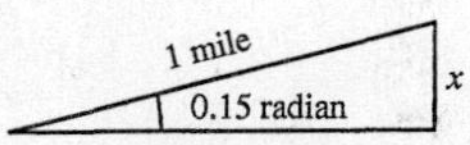

Not drawn to scale

Slope: $m = \tan 0.15 \approx 0.1511$

$$\sin 0.15 = \frac{x}{5280 \text{ feet}}$$

$$x = 5280 \sin 0.15 \approx 789 \text{ feet}$$

21. $x = (115 \cos \theta)t$ and $y = 3 + (115 \sin \theta)t - 16t^2$

When $\theta = 30°$: $x = (115 \cos 30°)t$

$y = 3 + (115 \sin 30°)t - 16t^2$

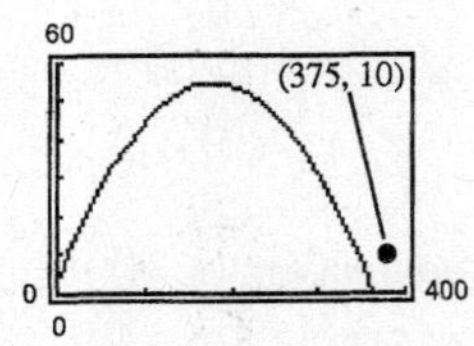

The ball hits the ground inside the ballpark, so it is not a home run.

When $\theta = 35°$: $x = (115 \cos 35°)t$

$y = 3 + (115 \sin 35°)t - 16t^2$

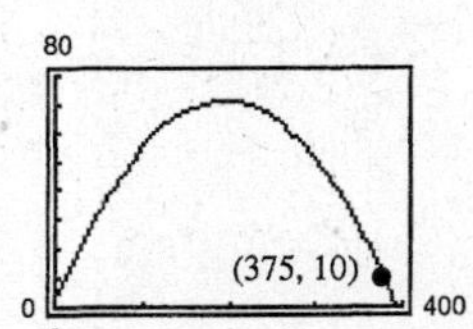

The ball clears the 10 foot fence at 375 feet, so it is a home run.

Chapters 4–6 Cumulative Test

1. $3 - \sqrt{-25} = 3 - 5i$

2. $6i - \left(2 + \sqrt{-81}\right) = 6i - (2 + 9i) = 6i - 2 - 9i$
$= -2 - 3i$

3. $(2i - 3)^2 = (2i)^2 - 2(2i)(3) + 3^2 = 4i^2 - 12i + 9$
$= -4 - 12i + 9 = 5 - 12i$

4. $\left(\sqrt{3} + i\right)\left(\sqrt{3} - i\right) = \left(\sqrt{3}\right)^2 - i^2 = 3 + 1 = 4$

5. $\dfrac{4i}{1+2i} = \dfrac{4i}{1+2i}\cdot\dfrac{1-2i}{1-2i} = \dfrac{4i-8i^2}{1-4i^2} = \dfrac{8+4i}{1+4}$

$= \dfrac{8}{5} + \dfrac{4}{5}i$

6. $f(x) = x^3 + 2x^2 + 4x + 8$

$x^3 + 2x^2 + 4x + 8 = 0$

$x^2(x+2) + 4(x+2) = 0$

$(x+2)(x^2+4) = 0$

$(x+2)(x+2i)(x-2i) = 0$

$x = -2$ or $x = \pm 2i$

7. $f(x) = x^4 + 4x^3 - 21x^2$

$x^4 + 4x^3 - 21x^2 = 0$

$x^2(x^2 + 4x - 21) = 0$

$x^2(x+7)(x-3) = 0$

$x = 0, x = -7$, or $x = 3$

8. Zeros: $-5, -2$ and $2 + \sqrt{3}i$

Since $2 + \sqrt{3}i$ is a zero, so is $2 - \sqrt{3}i$.

$f(x) = (x+5)(x+2)\left[x - \left(2 + \sqrt{3}i\right)\right]\left[x - \left(2 - \sqrt{3}i\right)\right]$

$= (x^2 + 7x + 10)\left[(x-2) - \sqrt{3}i\right]\left[(x-2) + \sqrt{3}i\right]$

$= (x^2 + 7x + 10)(x^2 - 4x + 7)$

$= x^4 + 3x^3 - 11x^2 + 9x + 70$

9. $r = |-2 + 2i| = \sqrt{(-2)^2 + (2)^2} = 2\sqrt{2}$

$\tan\theta = \dfrac{2}{-2} = -1$ and θ is in Quadrant II $\Longrightarrow \theta = \dfrac{3\pi}{4}$

Thus, $-2 + 2i = 2\sqrt{2}\left(\cos\dfrac{3\pi}{4} + i\sin\dfrac{3\pi}{4}\right)$.

10. $[4(\cos 30^\circ + i\sin 30^\circ)][6(\cos 120^\circ + i\sin 120^\circ)] = (4)(6)[\cos(30^\circ + 120^\circ) + i\sin(30^\circ + 120^\circ)]$

$= 24(\cos 150^\circ + i\sin 150^\circ)$

$= 24\left(-\dfrac{\sqrt{3}}{2} + \dfrac{1}{2}i\right)$

$= -12\sqrt{3} + 12i$

11. $\left[2\left(\cos\dfrac{2\pi}{3} + i\sin\dfrac{2\pi}{3}\right)\right]^4 = 2^4\left(\cos\dfrac{8\pi}{3} + i\sin\dfrac{8\pi}{3}\right)$

$= 16\left(\cos\dfrac{2\pi}{3} + i\sin\dfrac{2\pi}{3}\right)$

$= 16\left(-\dfrac{1}{2} + \dfrac{\sqrt{3}}{2}i\right)$

$= -8 + 8\sqrt{3}i$

12. $(-\sqrt{3} - i)^6 = \left[2\left(\cos\dfrac{7\pi}{6} + i\sin\dfrac{7\pi}{6}\right)\right]^6$

$= 2^6(\cos 7\pi + i\sin 7\pi)$

$= 64(-1 + 0i)$

$= -64$

13. $1 = 1(\cos 0 + i \sin 0)$

Cube roots of 1: $\sqrt[3]{1}\left[\cos\left(\frac{0+2\pi k}{3}\right) + i\sin\left(\frac{0+2\pi k}{3}\right)\right], k = 0, 1, 2$

$k = 0$: $\sqrt[3]{1}\left[\cos\left(\frac{0+2\pi(0)}{3}\right) + i\sin\left(\frac{0+2\pi(0)}{3}\right)\right] = \cos 0 + i \sin 0 = 1$

$k = 1$: $\sqrt[3]{1}\left[\cos\left(\frac{0+2\pi(1)}{3}\right) + i\sin\left(\frac{0+2\pi(1)}{3}\right)\right] = \cos\frac{2\pi}{3} + i\sin\frac{2\pi}{3} = -\frac{1}{2} + \frac{\sqrt{3}}{2}i$

$k = 2$: $\sqrt[3]{1}\left[\cos\left(\frac{0+2\pi(2)}{3}\right) + i\sin\left(\frac{0+2\pi(2)}{3}\right)\right] = \cos\frac{4\pi}{3} + i\sin\frac{4\pi}{3} = -\frac{1}{2} - \frac{\sqrt{3}}{2}i$

14. $x^4 - 81i = 0 \Rightarrow x^4 = 81i$

The solutions to the equation are the fourth roots of

$z = 81i = 81\left(\cos\frac{\pi}{2} + i\sin\frac{\pi}{2}\right)$ which are:

$$\sqrt[4]{81}\left[\cos\left(\frac{\frac{\pi}{2}+2\pi k}{4}\right) + i\sin\left(\frac{\frac{\pi}{2}+2\pi k}{4}\right)\right], k = 0, 1, 2, 3$$

$k = 0$: $3\left(\cos\frac{\pi}{8} + i\sin\frac{\pi}{8}\right)$

$k = 1$: $3\left(\cos\frac{5\pi}{8} + i\sin\frac{5\pi}{8}\right)$

$k = 2$: $3\left(\cos\frac{9\pi}{8} + i\sin\frac{9\pi}{8}\right)$

$k = 3$: $3\left(\cos\frac{13\pi}{8} + i\sin\frac{13\pi}{8}\right)$

15. $f(x) = \left(\frac{2}{5}\right)^x$

$g(x) = -\left(\frac{2}{5}\right)^{-x+3}$

g is a reflection in the x-axis, a reflection in the y-axis, and a horizontal shift 3 units to the right of the graph of f.

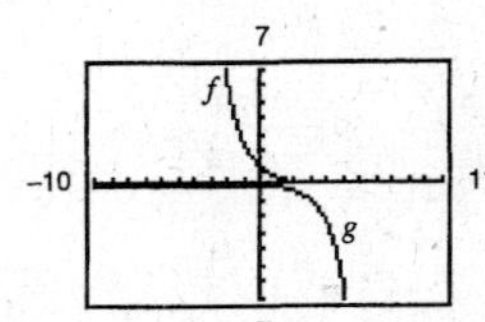

16. $f(x) = 2.2^x$

$g(x) = -2.2^x + 4$

g is a reflection in the x-axis, and a vertical shift 4 units upward of the graph of f.

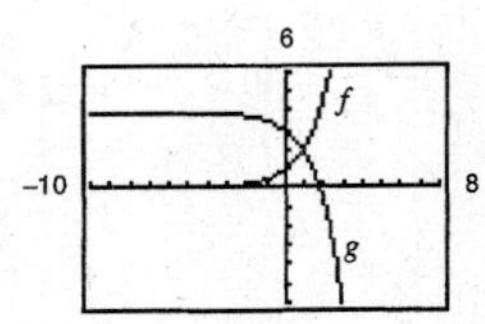

17. $\log_{10} 98 \approx 1.991$

18. $\log_{10}\left(\frac{6}{7}\right) \approx -0.067$

19. $\ln\sqrt{31} \approx 1.717$

20. $\ln\left(\sqrt{40} - 5\right) \approx 0.281$

21. $\log_7 1.8 = \frac{\log_{10} 1.8}{\log_{10} 7} = \frac{\ln 1.8}{\ln 7} \approx 0.302$

22. $\log_3 0.149 = \frac{\log_{10} 0.149}{\log_{10} 3} = \frac{\ln 0.149}{\ln 3} \approx -1.733$

23. $\log_{1/2} 17 = \frac{\log_{10} 17}{\log_{10}\left(\frac{1}{2}\right)} = \frac{\ln 17}{\ln\left(\frac{1}{2}\right)} \approx -4.087$

24. $\ln\left(\frac{x^2-16}{x^4}\right) = \ln(x^2 - 16) - \ln x^4$

$= \ln(x + 4)(x - 4) - 4\ln x$

$= \ln(x + 4) + \ln(x - 4) - 4\ln x, x > 4$

25. $2\ln x - \frac{1}{2}\ln(x + 5) = \ln x^2 - \ln\sqrt{x+5}$

$= \ln\frac{x^2}{\sqrt{x+5}}, x > 0$

26. $6e^{2x} = 72$

$$e^{2x} = 12$$
$$2x = \ln 12$$
$$x = \frac{\ln 12}{2} \approx 1.242$$

27. $4^{x-5} + 21 = 30$

$$4^{x-5} = 9$$
$$\ln 4^{x-5} = \ln 9$$
$$(x-5)\ln 4 = \ln 9$$
$$x - 5 = \frac{\ln 9}{\ln 4}$$
$$x = 5 + \frac{\ln 9}{\ln 4}$$
$$x \approx 6.5850$$

28. $\log_2 x + \log_2 5 = 6$

$$\log_2 5x = 6$$
$$5x = 2^6$$
$$x = \frac{64}{5}$$

29. $\ln(4x) - \ln 2 = 8$

$$\ln\left(\frac{4x}{2}\right) = 8$$
$$\ln 2x = 8$$
$$2x = e^8$$
$$x = \frac{e^8}{2}$$
$$x \approx 1490.4790$$

30. $f(x) = \dfrac{1000}{1 + 4e^{-0.2x}}$

Horizontal asymptotes:
$y = 0$ and $y = 1000$

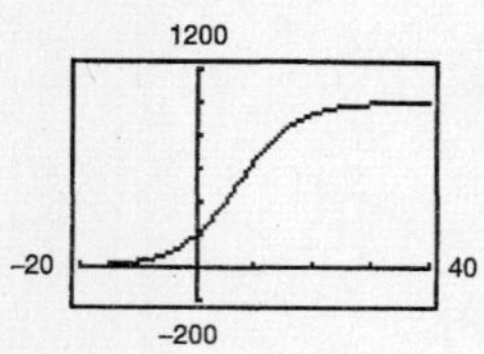

31.

$$N = 175e^{kt}$$
$$420 = 175e^{k(8)}$$
$$2.4 = e^{8k}$$
$$\ln 2.4 = 8k$$
$$\frac{\ln 2.4}{8} = k$$
$$k \approx 0.1094$$
$$N = 175e^{0.1094t}$$
$$350 = 175e^{0.1094t}$$
$$2 = e^{0.1094t}$$
$$\ln 2 = 0.1094t$$
$$t = \frac{\ln 2}{0.1094} \approx 6.3 \text{ hours to double}$$

32. $2x + y - 3 = 0 \Rightarrow y = -2x + 3 \Rightarrow m_1 = -2$

$$x - 3y + 6 = 0 \Rightarrow y = \tfrac{1}{3}x + 2 \Rightarrow m_2 = \tfrac{1}{3}$$
$$\tan\theta = \left|\frac{m_2 - m_1}{1 + m_1 m_2}\right| = \left|\frac{\frac{1}{3} - (-2)}{1 + (-2)\left(\frac{1}{3}\right)}\right| = \left|\frac{\frac{7}{3}}{\frac{1}{3}}\right| = 7$$
$$\theta = \arctan 7 \approx 81.87^\circ$$

33. $y = 2x - 4 \Rightarrow 2x - y - 4 = 0 \Rightarrow$

$A = 2, B = -1, C = -4$

$(6, -3) \Rightarrow x_1 = 6$ and $y_1 = -3$

$$d = \frac{|Ax_1 + By_1 + C|}{\sqrt{A^2 + B^2}} = \frac{|2(6) + (-1)(-3) + (-4)|}{\sqrt{(2)^2 + (-1)^2}}$$
$$= \frac{11}{\sqrt{5}} = \frac{11\sqrt{5}}{5}$$

34. $9x^2 + 4y^2 - 36x + 8y + 4 = 0$

$AC > 0 \Longrightarrow$ The conic is an ellipse.

$$9x^2 - 36x + 4y^2 + 8y = -4$$

$$9(x^2 - 4x + 4) + 4(y^2 + 2y + 1) = -4 + 36 + 4$$

$$9(x - 2)^2 + 4(y + 1)^2 = 36$$

$$\frac{(x - 2)^2}{4} + \frac{(y + 1)^2}{9} = 1$$

Center: $(2, -1)$

$a = 3, b = 2, c^2 = 9 - 4 = 5 \Longrightarrow c = \sqrt{5}$

Vertical Major Axis

Vertices: $(2, -1 \pm 3) \Longrightarrow (2, 2)$ and $(2, -4)$

Foci: $\left(2, -1 \pm \sqrt{5}\right)$

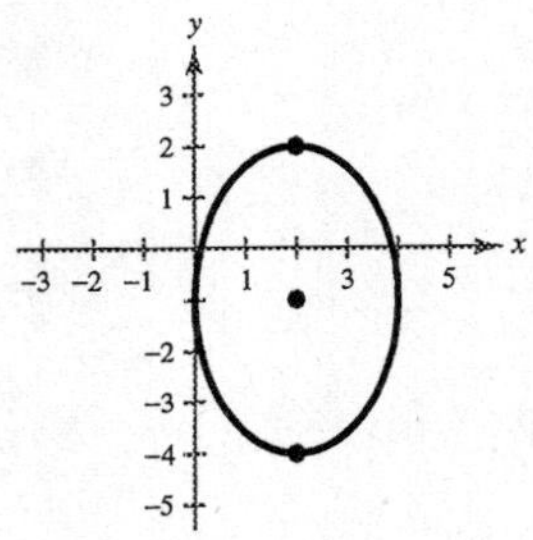

35. $4x^2 - y^2 - 4 = 0$

$AC < 0 \Longrightarrow$ The conic is a hyperbola.

$4x^2 - y^2 = 4$

$\dfrac{x^2}{1} - \dfrac{y^2}{4} = 1$

Center: $(0, 0)$

$a = 1, b = 2, c^2 = 1 + 4 = 5 \Longrightarrow c = \sqrt{5}$

Horizontal Transverse Axis

Vertices: $(\pm 1, 0)$

Foci: $\left(\pm\sqrt{5}, 0\right)$

Asymptotes: $y = \pm 2x$

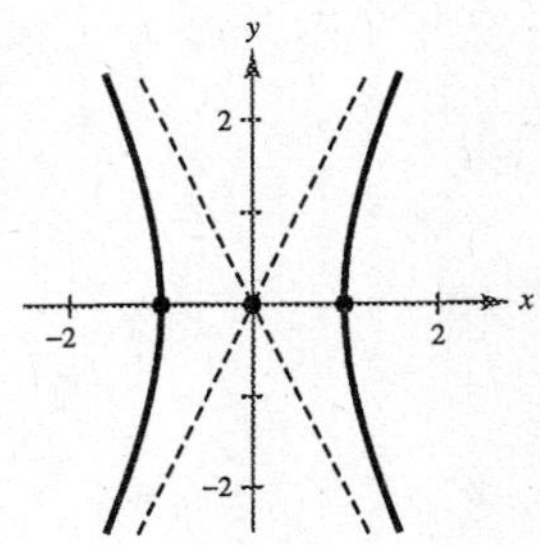

36. $x^2 + y^2 + 2x - 6y - 12 = 0$

$A = C \Longrightarrow$ The conic is a circle.

$$x^2 + 2x + y^2 - 6y = 12$$

$$(x^2 + 2x + 1) + (y^2 - 6y + 9) = 12 + 1 + 9$$

$$(x + 1)^2 + (y - 3)^2 = 22$$

Center: $(-1, 3)$

Radius: $\sqrt{22}$

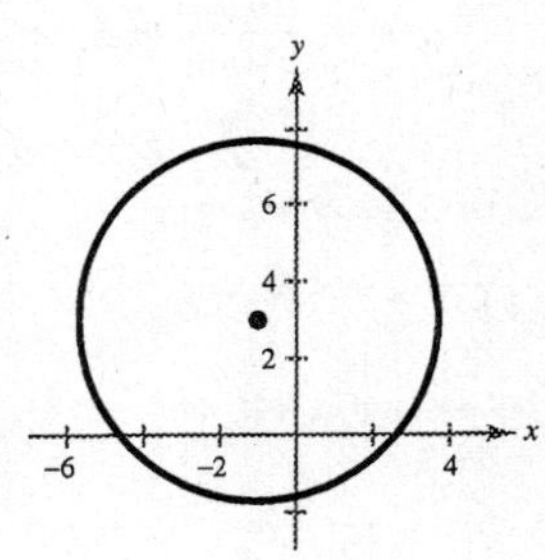

37. $y^2 + 2x + 2 = 0$

$AC = 0 \Longrightarrow$ The conic is a parabola.

$y^2 = -2x - 2$

$y^2 = -2(x + 1)$

$y^2 = 4\left(-\tfrac{1}{2}\right)(x + 1)$

Vertex: $(-1, 0)$

Opens to the left since $p < 0$.

Focus: $\left(-1 - \tfrac{1}{2}, 0\right) = \left(-\tfrac{3}{2}, 0\right)$

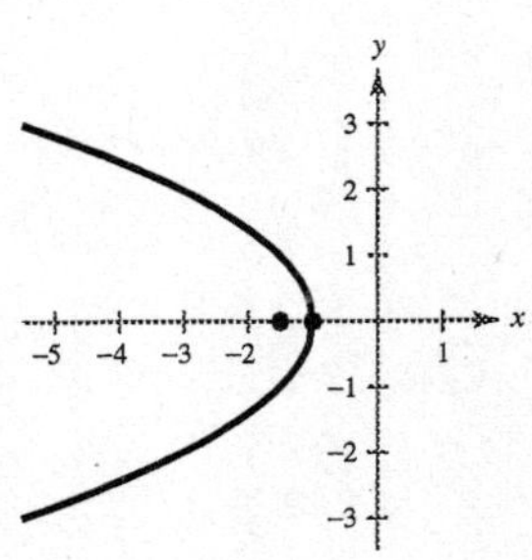

38. Circle

Center: $(2, -4)$

Point on circle: $(0, 4)$

$$(x-2)^2 + (y+4)^2 = r^2$$

$$(0-2)^2 + (4+4)^2 = r^2 \Rightarrow r^2 = 68$$

$$(x-2)^2 + (y+4)^2 = 68$$

$$x^2 - 4x + 4 + y^2 + 8x + 16 - 68 = 0$$

$$x^2 + y^2 - 4x + 8y - 48 = 0$$

39. Hyperbola

Foci: $(0, 0)$ and $(0, 6) \Rightarrow$ Vertical transverse axis

Center: $(0, 3)$ and $c = 3$

Asymptotes: $y = \pm\frac{2\sqrt{5}}{5}x + 3 \Rightarrow \pm\frac{a}{b} = \pm\frac{2\sqrt{5}}{5} \Rightarrow \frac{a}{b} = \frac{2}{\sqrt{5}} \Rightarrow b = \frac{\sqrt{5}a}{2}$

Since $c^2 = a^2 + b^2 \Rightarrow 9 = a^2 + \left(\frac{\sqrt{5}a}{2}\right)^2 \Rightarrow 9 = \frac{9}{4}a^2$

$$a^2 = 4 \Rightarrow b^2 = \frac{5a^2}{4} = \frac{5(4)}{4} = 5$$

Equation: $\frac{(y-k)^2}{a^2} - \frac{(x-h)^2}{b^2} = 1$

$$\frac{(y-3)^2}{4} - \frac{x^2}{5} = 1$$

$$5(y-3)^2 - 4x^2 = 20$$

$$5(y^2 - 6y + 9) - 4x^2 - 20 = 0$$

$$5y^2 - 30y + 45 - 4x^2 - 20 = 0$$

$$5y^2 - 4x^2 - 30y + 25 = 0$$

40. $x^2 + xy + y^2 + 2x - 3y - 30 = 0$

(a) $\cot 2\theta = \frac{1-1}{1} = 0 \Rightarrow 2\theta = \frac{\pi}{2} \Rightarrow \theta = \frac{\pi}{4}$ or $45°$

(b) Since $b^2 - 4ac < 0$, the graph is an ellipse.

$$x = x'\cos 45° - y'\sin 45° = \frac{x' - y'}{\sqrt{2}}$$

$$y = x'\sin 45° + y'\cos 45° = \frac{x' - y'}{\sqrt{2}}$$

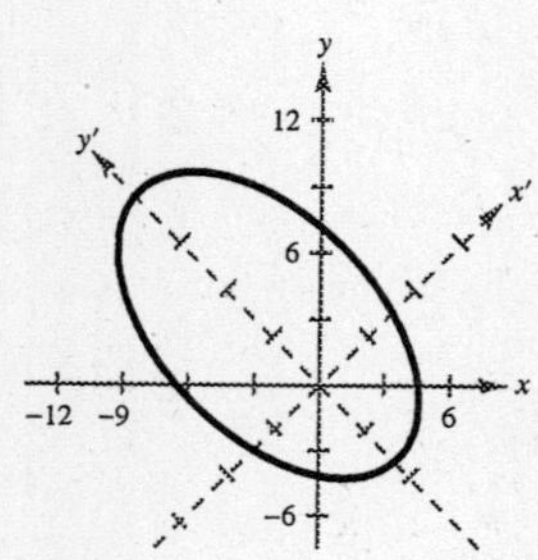

—CONTINUED—

40. —CONTINUED—

$$x^2 + xy + y^2 + 2x - 3y - 30 = 0$$

$$\left(\frac{x' - y'}{\sqrt{2}}\right)^2 + \left(\frac{x' - y'}{\sqrt{2}}\right)\left(\frac{x' + y'}{\sqrt{2}}\right) + \left(\frac{x' + y'}{\sqrt{2}}\right)^2 + 2\left(\frac{x' - y'}{\sqrt{2}}\right) - 3\left(\frac{x' + y'}{\sqrt{2}}\right) - 30 = 0$$

$$\frac{1}{2}[(x')^2 - 2x'y' + (y')^2] + \frac{1}{2}[(x')^2 - (y')^2] + \frac{1}{2}[(x')^2 + 2x'y' + (y')^2] + \sqrt{2}(x' - y') - \frac{3\sqrt{2}}{2}(x' + y') - 30 = 0$$

$$\frac{3}{2}(x')^2 + \frac{1}{2}(y')^2 - \frac{\sqrt{2}}{2}x' - \frac{5\sqrt{2}}{2}y' - 30 = 0$$

$$3(x')^2 + (y')^2 - \sqrt{2}x' - 5\sqrt{2}y' - 60 = 0$$

$$3\left[(x')^2 - \frac{\sqrt{2}}{3}x' + \frac{2}{36}\right] + \left[(y')^2 - 5\sqrt{2}y' + \frac{50}{4}\right] = 60 + \frac{1}{6} + \frac{25}{2}$$

$$3\left(x' - \frac{\sqrt{2}}{6}\right)^2 + \left(y' - \frac{5\sqrt{2}}{2}\right)^2 = \frac{218}{3}$$

$$\frac{\left(x' - \frac{\sqrt{2}}{6}\right)^2}{\frac{218}{9}} + \frac{\left(y' - \frac{5\sqrt{2}}{2}\right)^2}{\frac{218}{3}} = 1$$

To use a graphing utility, solve for y in terms of x.

$$y^2 + y(x - 3) + (x^2 + 2x - 30) = 0$$

$$y_1 = \frac{-(x - 3) + \sqrt{(x - 3)^2 - 4(x^2 + 2x - 30)}}{2}$$

$$y_2 = \frac{-(x - 3) - \sqrt{(x - 3)^2 - 4(x^2 + 2x - 30)}}{2}$$

41. $x = 3 + 4\cos\theta \Longrightarrow \cos\theta = \dfrac{x - 3}{4}$

$y = \sin\theta$

$$\cos^2\theta + \sin^2\theta = 1$$

$$\left(\frac{x - 3}{4}\right)^2 + (y)^2 = 1$$

$$\frac{(x - 3)^2}{16} + y^2 = 1$$

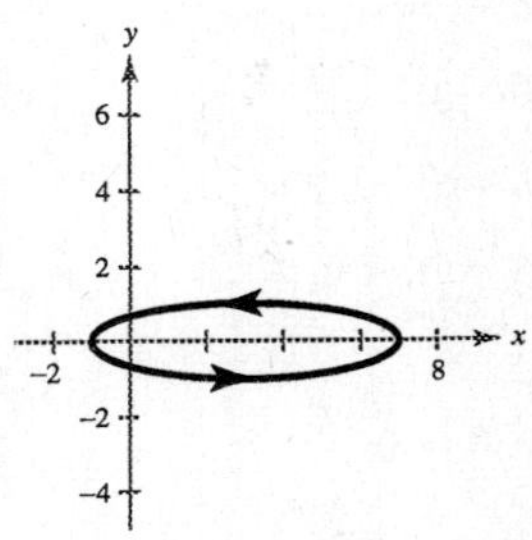

The graph is an ellipse with a horizontal major axis.

Center: $(3, 0)$

$a = 4$ and $b = 1$

Vertices: $(3 \pm 4, 0) \Longrightarrow (7, 0)$ and $(-1, 0)$

42. Line through $(3, -2)$ and $(-3, 4)$

$x = x_1 + t(x_2 - x_1) = 3 - 6t$

$y = y_1 + t(y_2 - y_1) = -2 + 6t$

43. $x^2 + y^2 - 6y = 0$

$r^2 - 6r\sin\theta = 0$

$r(r - 6\sin\theta) = 0$

$r = 6\sin\theta$

44. $r = \dfrac{3}{2 + \cos\theta} = \dfrac{\frac{3}{2}}{1 + \frac{1}{2}\cos\theta}$

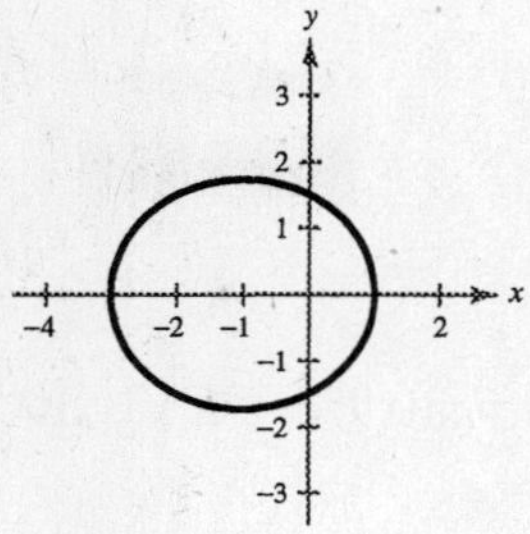

$e = \dfrac{1}{2}$

Ellipse with a vertical directrix to the right of the pole.

θ	0	$\pi/2$	π	$3\pi/2$
r	1	3/2	3	3/2

45. $r = \dfrac{4}{1 + \sin\theta}$

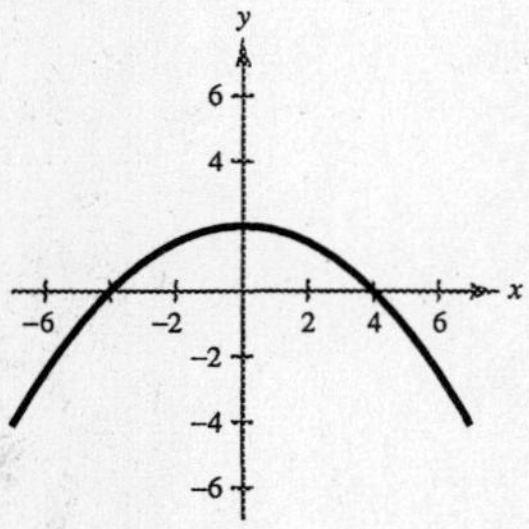

$e = 1$

Parabola with a horizontal directrix above the pole.

θ	0	$\pi/2$	π
r	4	2	4

46. (a) $r = 2 + 3 \sin\theta$ is a limaçon. Matches (iii).

(b) $r = 3 \sin\theta$ is a circle. Matches (i).

(c) $r = 3 \sin 2\theta$ is a rose curve. Matches (ii).

PART III

CHAPTER P
Prerequisites

CHAPTER P
Prerequisites

Section P.1 Review of Real Numbers and Their Properties

Solutions to Even-Numbered Exercises

2. $\sqrt{5}, -7, -\frac{7}{3}, 0, 3.12, \frac{5}{4}, -3, 12, 5$

(a) Natural numbers: 12, 5

(b) Integers: $-7, 0, -3, 12, 5$

(c) Rational numbers: $-7, -\frac{7}{3}, 0, 3.12, \frac{5}{4}, -3, 12, 5$

(d) Irrational numbers: $\sqrt{5}$

4. $2.3030030003\ldots, 0.7575, -4.63, \sqrt{10}, -75, 4$

(a) Natural numbers: 4

(b) Integers: $-75, 4$

(c) Rational numbers: $0.7575, -4.63, -75, 4$

(d) Irrational numbers: $2.3030030003\ldots, \sqrt{10}$

6. $25, -17, -\frac{12}{5}, \sqrt{9}, 3.12, \frac{1}{2}\pi, 7, -11.1, 13$

(a) Natural numbers: $25, \sqrt{9}, 7, 13$

(b) Integers: $25, -17, \sqrt{9}, 7, 13$

(c) Rational numbers: $25, -17, -\frac{12}{5}, \sqrt{9}, 3.12, 7, -11.1, 13$

(d) Irrational numbers: $\frac{1}{2}\pi$

8. $\frac{1}{3} = 0.\overline{3}$

10. $\frac{6}{11} = 0.\overline{54}$

12. $-6 < -2.5$

14. $-3.5 < 1$

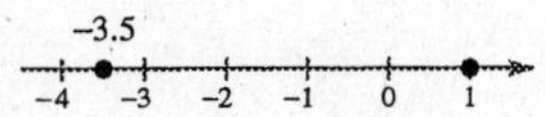

16. $1 < \frac{16}{3}$

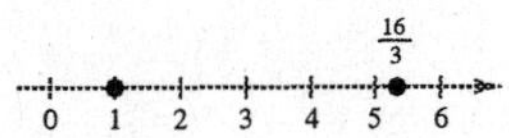

18. $-\frac{8}{7} < -\frac{3}{7}$

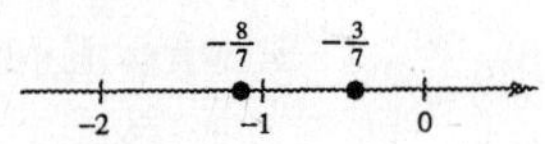

20. The inequality $x \geq -2$ denotes the set of all real numbers greater than or equal to -2. The interval is unbounded.

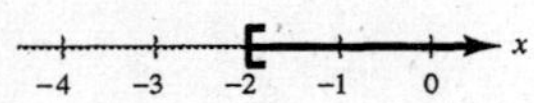

22. The inequality $x > 3$ denotes the set of all real numbers greater than 3. The interval is unbounded.

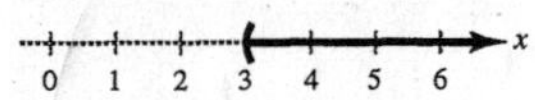

24. The inequality $x < 2$ denotes the set of all real numbers less than 2. The interval is unbounded.

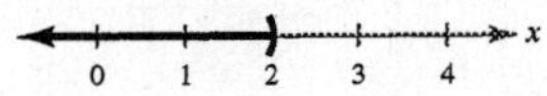

26. The inequality $0 \leq x \leq 5$ denotes the set of all real numbers greater than or equal to zero and less than or equal to 5. The interval is bounded.

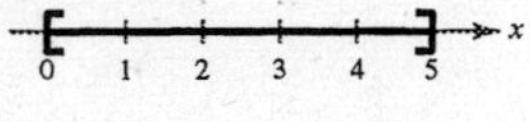

28. The inequality $0 < x \leq 6$ denotes the set of all real numbers greater than zero and less than or equal to 6. The interval is bounded.

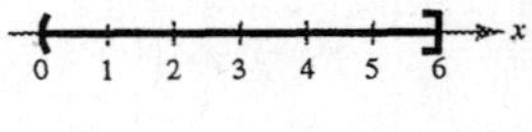

30. $-6 \leq y < 0$

32. $y \leq 25$

34. $-3 \leq k < 5$

36. $2.5\% \leq r \leq 5\%$

38. This interval consists of all real numbers greater than or equal to -5 and less than or equal to 7.

40. This interval consists of all real numbers less than or equal to 4.

42. $|0| = 0$

44. $|4 - 1| = |3| = 3$

46. $-3 - |-3| = -3 - (3) = -6$

48. $-3|-3| = -3(3) = -9$

50. If $x > 1$, then $x - 1$ is positive.

Thus, $\dfrac{|x-1|}{x-1} = \dfrac{x-1}{x-1} = 1.$

52. $|-4| = |4|$ since $|-4| = 4$ and $|4| = 4$.

54. $-|-6| < |-6|$ since $|-6| = 6$ and $-|-6| = -(6) = -6$.

56. $-(-2) > -2$ since $-(-2) = 2$.

58. $d\left(-4, -\frac{3}{2}\right) = \left|-4 - \left(-\frac{3}{2}\right)\right| = \left|-4 + \frac{3}{2}\right| = \frac{5}{2}$

60. $d(-126, -75) = |-126 - (-75)| = 51$

62. $d\left(\frac{1}{4}, \frac{11}{4}\right) = \left|\frac{1}{4} - \frac{11}{4}\right| = \frac{5}{2}$

64. $d(9.34, -5.65) = |9.34 - (-5.65)| = 14.99$

66.

Budgeted Expense, b	*Actual Expense, a*	$\|a - b\|$	*0.05b*
\$9400	\$9772	\$372	$0.05(9400) = \$470$

The actual expense difference is less than \$500 (but is less than 5% of the budget) so the actual expense does pass the test.

68.

Budgeted Expense, b	*Actual Expense, a*	$\|a - b\|$	*0.05b*
\$2575	\$2613	\$38	$0.05(2575) = \$128.75$

Since \$38 < \$500, and \$38 < \$128.75, the actual expense passes the "budget variance test."

70. Total: $1 + 386 + 181 + 25{,}573 + 272{,}793 + 63{,}579 + 114{,}514 + 6261 = 483{,}288$

Civil War: $\dfrac{1}{483{,}288} \approx 0.00000207 = 0.000207\% \times 360° \approx 0°$

Indian War: $\dfrac{0}{483{,}288} = 0 = 0\% \times 360° = 0°$

Spanish-American War: $\dfrac{386}{483{,}288} \approx 0.000799 = 0.0799\% \times 360° \approx 0.3°$

Mexican Border War: $\dfrac{181}{483{,}288} \approx 0.000375 = 0.0375\% \times 360° \approx 0.1°$

World War I: $\dfrac{25{,}573}{483{,}288} \approx 0.0529 = 5.29\% \times 360° \approx 19.0°$

World War II: $\dfrac{272{,}793}{483{,}288} \approx 0.564 = 56.4\% \times 360° \approx 203.0°$

Korean War: $\dfrac{63{,}579}{483{,}288} \approx 0.132 = 13.2\% \times 360° \approx 47.5°$

Vietnam War: $\dfrac{114{,}514}{483{,}288} \approx 0.237 = 23.7\% \times 360° \approx 85.3°$

Gulf War: $\dfrac{6261}{483{,}288} \approx 0.0130 = 1.30\% \times 360° \approx 4.7°$

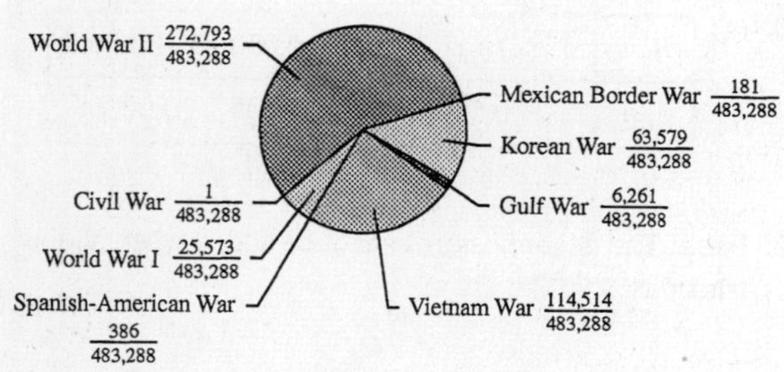

72. $|326 - 351| = |-25| = 25$

You travel 25 miles.

74. $d(48°, 82°) = |82° - 48°| = 34°$

76. $d(x, -10) = |x + 10|$, and $d(x, -10) \geq 6$, thus, $|x + 10| \geq 6$.

78. $d(y, a) = |y - a|$ and $d(y, a) \leq 2$, thus $|y - a| \leq 2$.

80. $6x^3 - 5x$

Terms: $6x^3, -5x$

Coefficients: $6, -5$

82. $3\sqrt{3}x^2 + 1$

Terms: $3\sqrt{3}x^2, 1$

Coefficients: $3\sqrt{3}$

84. $3x^4 - \dfrac{x^2}{4}$

Terms: $3x^4, -\dfrac{x^2}{4}$

Coefficients: $3, -\dfrac{1}{4}$

86. $9 - 7x$

(a) $9 - 7(-3) = 9 + 21 = 30$

(b) $9 - 7(3) = 9 - 21 = -12$

88. $-x^2 + 5x = 4$

(a) $-(-1)^2 + 5(-1) - 4 = -1 - 5 - 4 = -10$

(b) $-(1)^2 + 5(1) - 4 = -1 + 5 - 4 = 0$

90. $\dfrac{x}{x+2}$

(a) $\dfrac{2}{2+2} = \dfrac{2}{4} = \dfrac{1}{2}$

(b) $\dfrac{-2}{-2+2} = \dfrac{2}{0}$

Division by 0 is undefined.

92. $2\left(\dfrac{1}{2}\right) = 1$

Multiplicative Inverse Property

94. $(x + 3) - (x + 3) = 0$

Additive Inverse Property

96. $(z - 2) + 0 = z - 2$

Additive Identity Property

98. $x + (y + 10) = (x + y) + 10$

Associative Property of Addition

100. $\frac{1}{7}(7 \cdot 12) = \left(\frac{1}{7} \cdot 7\right)12$ Associative Property of Multiplication

$= 1 \cdot 12$ Multiplicative Inverse Property

$= 12$ Multiplicative Identity Property

102. $\dfrac{6}{7} - \dfrac{4}{7} = \dfrac{6-4}{7} = \dfrac{2}{7}$

104. $\dfrac{10}{11} + \dfrac{6}{33} - \dfrac{13}{66} = \dfrac{60}{66} + \dfrac{12}{66} - \dfrac{13}{66} = \dfrac{59}{66}$

106. $-\left(6 \cdot \dfrac{4}{8}\right) = -6 \cdot \dfrac{1}{2} = -3$

108. $\dfrac{5x}{6} \cdot \dfrac{2}{9} = \dfrac{5x}{3} \cdot \dfrac{1}{9} = \dfrac{5x}{27}$

110. (a)

n	1	10	100	10,000	100,000
$5/n$	5	0.5	0.05	0.0005	0.00005

(b) The value of $5/n$ approaches 0 as n increases without bound.

112. False. The denominators cannot be added when adding fractions.

114. Yes. y is nonnegative if $y \ge 0$. y is positive if $y > 0$.

116.

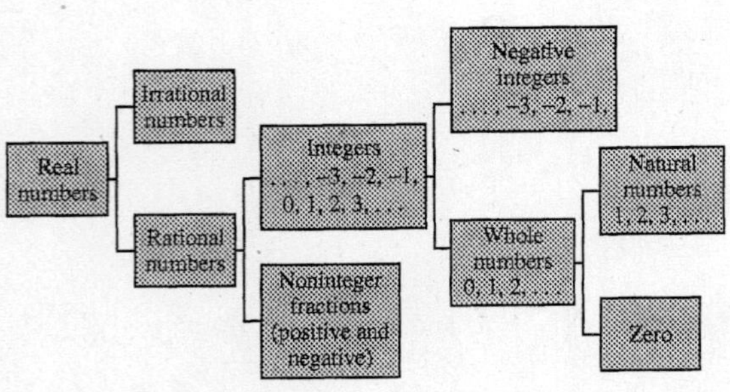

118. (a) Since $C < 0$, $-C > 0$. The expression is positive.

(b) Since $A > C$, $A - C > 0$. The expression is positive.

Section P.2 Solving Equations

Solutions to Even-Numbered Exercises

2. $3(x + 2) = 5x + 4$ is *conditional*. There are real values of x for which the equation is not true (for example, $x = 0$).

4. $3(x + 2) - 5 = 3x + 1$ is an *identity* by simplification. It is true for all real values of x.
$3(x + 2) - 5 = 3x + 6 - 5 = 3x + 1$

6. $-7(x - 3) + 4x = 3(7 - x)$ is an *identity* by simplification. It is true for all real values of x.
$-7(x - 3) + 4x = -7x + 21 + 4x = 21 - 3x = 3(7 - x)$

8. $x^2 + 2(3x - 2) = x^2 + 6x - 4$ is an *identity* by simplification. It is true for all real values of x.

10. $\frac{5}{x} + \frac{3}{x} = 24$ is *conditional*. There are real values of x for which the equation is not true (for example, $x = 1$).

12.
$$\begin{aligned} 7 - x &= 19 \\ 7 - x + x &= 19 + x \\ 7 &= 19 + x \\ 7 - 19 &= 19 + x - 19 \\ -12 &= x \end{aligned}$$

14.
$$\begin{aligned} 7x + 2 &= 23 \\ 7x + 2 - 2 &= 23 - 2 \\ 7x &= 21 \\ \frac{7x}{7} &= \frac{21}{7} \\ x &= 3 \end{aligned}$$

16.
$$\begin{aligned} 7x + 3 &= 3x - 17 \\ 7x + 3 - 3 - 3x &= 3x - 17 - 3 - 3x \\ 4x &= -20 \\ x &= -5 \end{aligned}$$

18.
$$\begin{aligned} 3(x + 3) &= 5(1 - x) - 1 \\ 3x + 9 &= 5 - 5x - 1 \\ 3x + 9 &= 4 - 5x \\ 8x &= -5 \\ x &= -\tfrac{5}{8} \end{aligned}$$

20.
$$\begin{aligned} 9x - 10 &= 5x + 2(2x - 5) \\ 9x - 10 &= 5x + 4x - 10 \\ 9x - 10 &= 9x - 10 \end{aligned}$$

The solution is the set of all real numbers.

22.
$$\begin{aligned} \frac{x}{5} - \frac{x}{2} &= 3 + \frac{3x}{10} \\ 10\left(\frac{x}{5} - \frac{x}{2}\right) &= 10\left(3 + \frac{3x}{10}\right) \\ 2x - 5x &= 30 + 3x \\ -6x &= 30 \\ x &= -5 \end{aligned}$$

24.
$$\begin{aligned} \frac{3x}{2} + \frac{1}{4}(x - 2) &= 10 \\ (4)\left(\frac{3x}{2}\right) + (4)\frac{1}{4}(x - 2) &= (4)10 \\ 6x + (x - 2) &= 40 \\ 7x - 2 &= 40 \\ 7x &= 42 \\ x &= 6 \end{aligned}$$

26.
$$\begin{aligned} 0.60x + 0.40(100 - x) &= 50 \\ 0.60x + 40 - 0.40x &= 50 \\ 0.20x &= 10 \\ x &= 50 \end{aligned}$$

28. $8(x + 2) - 3(2x + 1) = 2(x + 5)$

$$8x + 16 - 6x - 3 = 2x + 10$$
$$2x + 13 = 2x + 10$$
$$13 = 10$$

False equation $\Rightarrow$ no solution

30.
$$\frac{17 + y}{y} + \frac{32 + y}{y} = 100$$
$$(y)\frac{17 + y}{y} + (y)\frac{32 + y}{y} = 100(y)$$
$$17 + y + 32 + y = 100y$$
$$49 + 2y = 100y$$
$$49 = 98y$$
$$\frac{1}{2} = y$$

32.
$$\frac{10x + 3}{5x + 6} = \frac{1}{2}$$
$$2(10x + 3) = 1(5x + 6)$$
$$20x + 6 = 5x + 6$$
$$15x = 0$$
$$x = 0$$

34.
$$\frac{15}{x} - 4 = \frac{6}{x} + 3$$
$$\frac{15}{x} - \frac{6}{x} = 7$$
$$\frac{9}{x} = 7$$
$$9 = 7x$$
$$\frac{9}{7} = x$$

36.
$$\frac{1}{x} + \frac{2}{x - 5} = 0$$

Multiply both sides by $x(x - 5)$

$$1(x - 5) + 2x = 0$$
$$3x - 5 = 0$$
$$3x = 5$$
$$x = \frac{5}{3}$$

38.
$$\frac{7}{2x + 1} - \frac{8x}{2x - 1} = -4$$
$$(2x + 1)(2x - 1)\frac{7}{2x + 1} - (2x + 1)(2x - 1)\frac{8x}{2x - 1} = -4(2x + 1)(2x - 1)$$
$$7(2x - 1) - 8x(2x + 1) = -4(2x + 1)(2x - 1)$$
$$14x - 7 - 16x^2 - 8x = -16x^2 + 4$$
$$6x = 11$$
$$x = \frac{11}{6}$$

40.
$$\frac{4}{x - 1} + \frac{6}{3x + 1} = \frac{15}{3x + 1}$$
$$(x - 1)(3x + 1)\frac{4}{x - 1} + (x - 1)(3x + 1)\frac{6}{3x + 1} = (x - 1)(3x + 1)\frac{15}{3x + 1}$$
$$4(3x + 1) + 6(x - 1) = 15(x - 1)$$
$$12x + 4 + 6x - 6 = 15x - 15$$
$$18x - 2 = 15x - 15$$
$$3x = -13$$
$$x = -\frac{13}{3}$$

42. $$\frac{1}{x-2}+\frac{3}{x+3}=\frac{4}{x^2+x-6}$$

$$(x^2+x-6)\frac{1}{x-2}+(x^2+x-6)\frac{3}{x+3}=(x^2+x-6)\frac{4}{x^2+x-6}$$

$$(x+3)+3(x-2)=4$$

$$x+3+3x-6=4$$

$$4x-3=4$$

$$4x=7$$

$$x=\frac{7}{4}$$

44. $$\frac{6}{x}-\frac{2}{x+3}=\frac{3(x+5)}{x^2+3x}$$

$$(x^2+3x)\frac{6}{x}-(x^2+3x)\frac{2}{x+3}=(x^2+3x)\frac{3(x+5)}{x^2+3x}$$

$$6(x+3)-2x=3(x+5)$$

$$6x+18-2x=3x+15$$

$$4x+18=3x+15$$

$$x=-3$$

Check: $$\frac{6}{-3}-\frac{2}{-3+3}=\frac{3(-3+5)}{[(-3)^2+3(-3)]}$$

$$-2-\frac{2}{0}=\frac{-6}{0}$$

Division by zero is undefined. Thus, $x=-3$ is not a solution, and the original equation has no solution.

46. $$(x+1)^2+2(x-2)=(x+1)(x-2)$$

$$x^2+2x+1+2x-4=x^2-x-2$$

$$5x=1$$

$$x=\tfrac{1}{5}$$

48. $$(2x+1)^2=4(x^2+x+1)$$

$$4x^2+4x+1=4x^2+4x+4$$

$$1=4$$

This is a false equation. Thus, the equation has no solution.

50. $x^2=16x$

General form: $x^2-16x=0$

52. $$13-3(x+7)^2=0$$

$$13-3(x^2+14x+49)=0$$

$$13-3x^2-42x-147=0$$

General form: $-3x^2-42x-134=0$

$$3x^2+42x+134=0$$

54. $$x(x+2)=5x^2+1$$

$$x^2+2x=5x^2+1$$

$$-4x^2+2x-1=0$$

$$(-1)(-4x^2+2x-1)=-1(0)$$

General form: $4x^2-2x+1=0$

56. $$9x^2-1=0$$

$$(3x+1)(3x-1)=0$$

$$3x+1=0 \Rightarrow x=-\tfrac{1}{3}$$

$$3x-1=0 \Rightarrow x=\tfrac{1}{3}$$

58. $x^2-10x+9=0$

$$(x-9)(x-1)=0$$

$$x-9=0 \Rightarrow x=9$$

$$x-1=0 \Rightarrow x=1$$

60. $4x^2+12x+9=0$

$$(2x+3)(2x+3)=0$$

$$2x+3=0 \Rightarrow x=-\tfrac{3}{2}$$

62. $$2x^2=19x+33$$

$$2x^2-19x-33=0$$

$$(2x+3)(x-11)=0$$

$$2x+3=0 \Rightarrow x=-\tfrac{3}{2}$$

$$x-11=0 \Rightarrow x=11$$

64.
$$-x^2 + 8x = 12$$
$$-x^2 + 8x - 12 = 0$$
$$(-1)(-x^2 + 8x - 12) = (-1)(0)$$
$$x^2 - 8x + 12 = 0$$
$$(x - 6)(x - 2) = 0$$
$$x - 6 = 0 \Rightarrow x = 6$$
$$x - 2 = 0 \Rightarrow x = 2$$

66.
$$\tfrac{1}{8}x^2 - x - 16 = 0$$
$$x^2 - 8x - 128 = 0$$
$$(x - 16)(x + 8) = 0$$
$$x - 16 = 0 \Rightarrow x = 16$$
$$x + 8 = 0 \Rightarrow x = -8$$

68.
$$(x + a)^2 - b^2 = 0$$
$$x^2 + 2ax + a^2 - b^2 = 0$$
$$[x + (a + b)][x + (a - b)] = 0$$
$$x + (a + b) = 0 \Rightarrow x = -a - b$$
$$x + (a - b) = 0 \Rightarrow x = -a + b$$

70. $x^2 = 169$
$$x = \pm\sqrt{169} = \pm 13 = \pm 13.00$$

72. $x^2 = 32$
$$x = \pm\sqrt{32} = \pm 4\sqrt{2} = \pm 5.66$$

74. $9x^2 = 36$
$$x^2 = 4$$
$$x = \pm\sqrt{4} = \pm 2 = \pm 2.00$$

76. $(x + 13)^2 = 25$
$$x + 13 = \pm\sqrt{25}$$
$$x + 13 = \pm 5$$
$$x = -13 \pm 5 = -18, -8 = -18.00, -8.00$$

78. $(x - 5)^2 = 30$
$$x - 5 = \pm\sqrt{30}$$
$$x = 5 \pm \sqrt{30} = -0.48, 10.48$$

80. $(4x + 7)^2 = 44$
$$4x + 7 = \pm\sqrt{44}$$
$$4x = -7 \pm 2\sqrt{11}$$
$$x = \frac{-7 \pm 2\sqrt{11}}{4} = -\frac{7}{4} \pm \frac{\sqrt{11}}{2} = -3.41, -0.09$$

82. $(x + 5)^2 = (x + 4)^2$
$$x + 5 = \pm(x + 4)$$
$$x + 5 = +(x + 4)$$
$$5 = 4$$
No solution
$$x + 5 = -(x + 4)$$
$$x + 5 = -x - 4$$
$$2x = -9$$
$$x = -\tfrac{9}{2} = -4.50$$

84.
$$x^2 + 4x = 0$$
$$x^2 + 4x + 2^2 = 2^2$$
$$(x + 2)^2 = 4$$
$$x + 2 = \pm\sqrt{4}$$
$$x = -2 \pm 2 = 0, -4$$

86.
$$x^2 - 2x - 3 = 0$$
$$x^2 - 2x = 3$$
$$x^2 - 2x + (-1)^2 = 3 + 1$$
$$(x - 1)^2 = 4$$
$$x - 1 = \pm\sqrt{4}$$
$$x = 1 \pm 2 = 3, -1$$

88. $x^2 + 8x + 14 = 0$
$$x^2 + 8x = -14$$
$$x^2 + 8x + 4^2 = -14 + 16$$
$$(x + 4)^2 = 2$$
$$x + 4 = \pm\sqrt{2}$$
$$x = -4 \pm \sqrt{2}$$

90.
$$\begin{aligned} 9x^2 - 12x &= 14 \\ x^2 - \tfrac{4}{3}x &= \tfrac{14}{9} \\ x^2 - \tfrac{4}{3}x + \left(\tfrac{2}{3}\right)^2 &= \tfrac{14}{9} + \tfrac{4}{9} \\ \left(x - \tfrac{2}{3}\right)^2 &= \tfrac{18}{9} \\ \left(x - \tfrac{2}{3}\right)^2 &= 2 \\ x - \tfrac{2}{3} &= \pm\sqrt{2} \\ x &= \tfrac{2}{3} \pm \sqrt{2} \end{aligned}$$

92.
$$\begin{aligned} 4x^2 - 4x - 99 &= 0 \\ x^2 - x &= \tfrac{99}{4} \\ x^2 - x + \left(\tfrac{1}{2}\right)^2 &= \tfrac{99}{4} + \tfrac{1}{4} \\ \left(x - \tfrac{1}{2}\right)^2 &= \tfrac{100}{4} \\ \left(x - \tfrac{1}{2}\right)^2 &= 25 \\ x - \tfrac{1}{2} &= \pm\sqrt{25} \\ x &= \tfrac{1}{2} \pm 5 = \tfrac{11}{2}, -\tfrac{9}{2} \end{aligned}$$

94. $2x^2 - x - 1 = 0$
$$\begin{aligned} x &= \frac{-b \pm \sqrt{b^2 - 4ac}}{2a} \\ &= \frac{-(-1) \pm \sqrt{(-1)^2 - 4(2)(-1)}}{2(2)} \\ &= \frac{1 \pm \sqrt{1 + 8}}{4} \\ &= \frac{1 \pm 3}{4} = 1, -\frac{1}{2} \end{aligned}$$

96. $25x^2 - 20x + 3 = 0$
$$\begin{aligned} x &= \frac{-b \pm \sqrt{b^2 - 4ac}}{2a} \\ &= \frac{-(-20) \pm \sqrt{(-20)^2 - 4(25)(3)}}{2(25)} \\ &= \frac{20 \pm \sqrt{400 - 300}}{50} \\ &= \frac{20 \pm 10}{50} = \frac{3}{5}, \frac{1}{5} \end{aligned}$$

98. $x^2 - 10x + 22 = 0$
$$\begin{aligned} x &= \frac{-b \pm \sqrt{b^2 - 4ac}}{2a} \\ &= \frac{-(-10) \pm \sqrt{(-10)^2 - 4(1)(22)}}{2(1)} \\ &= \frac{10 \pm \sqrt{100 - 88}}{2} \\ &= \frac{10 \pm 2\sqrt{3}}{2} = 5 \pm \sqrt{3} \end{aligned}$$

100.
$$\begin{aligned} 6x &= 4 - x^2 \\ x^2 + 6x - 4 &= 0 \\ x &= \frac{-b \pm \sqrt{b^2 - 4ac}}{2a} \\ &= \frac{-6 \pm \sqrt{6^2 - 4(1)(-4)}}{2(1)} \\ &= \frac{-6 \pm \sqrt{36 + 16}}{2} \\ &= \frac{-6 \pm 2\sqrt{13}}{2} \\ &= -3 \pm \sqrt{13} \end{aligned}$$

102. $4x^2 - 4x - 4 = 0$
$$\begin{aligned} x^2 - x - 1 &= 0 \\ x &= \frac{-b \pm \sqrt{b^2 - 4ac}}{2a} \\ &= \frac{-(-1) \pm \sqrt{(-1)^2 - 4(1)(-1)}}{2(1)} \\ &= \frac{1 \pm \sqrt{1 + 4}}{2} \\ &= \frac{1}{2} \pm \frac{\sqrt{5}}{2} \end{aligned}$$

104.
$$\begin{aligned} 16x^2 + 22 &= 40x \\ 8x^2 - 20x + 11 &= 0 \\ x &= \frac{-b \pm \sqrt{b^2 - 4ac}}{2a} \\ &= \frac{-(-20) \pm \sqrt{(-20)^2 - 4(8)(11)}}{2(8)} \\ &= \frac{20 \pm \sqrt{400 - 352}}{16} = \frac{20 \pm 4\sqrt{3}}{16} \\ &= \frac{5}{4} \pm \frac{\sqrt{3}}{4} \end{aligned}$$

106. $36x^2 + 24x - 7 = 0$

$$x = \frac{-b \pm \sqrt{b^2 - 4ac}}{2a}$$

$$= \frac{-24 \pm \sqrt{24^2 - 4(36)(-7)}}{2(36)}$$

$$= \frac{-24 \pm \sqrt{576 + 1008}}{72}$$

$$= \frac{-24 \pm 12\sqrt{11}}{72}$$

$$= -\frac{1}{3} \pm \frac{\sqrt{11}}{6}$$

108. $16x^2 - 40x + 5 = 0$

$$x = \frac{-b \pm \sqrt{b^2 - 4ac}}{2a}$$

$$= \frac{-(-40) \pm \sqrt{(-40)^2 - 4(16)(5)}}{2(16)}$$

$$= \frac{40 \pm \sqrt{1600 - 320}}{32}$$

$$= \frac{40 \pm 16\sqrt{5}}{32}$$

$$= \frac{5}{4} \pm \frac{\sqrt{5}}{2}$$

110. $3x + x^2 - 1 = 0$

$$x^2 + 3x - 1 = 0$$

$$x = \frac{-b \pm \sqrt{b^2 - 4ac}}{2a}$$

$$= \frac{-3 \pm \sqrt{3^2 - 4(1)(-1)}}{2(1)}$$

$$= \frac{-3 \pm \sqrt{13}}{2}$$

$$= -\frac{3}{2} \pm \frac{\sqrt{13}}{2}$$

112. $25h^2 + 80h + 61 = 0$

$$h = \frac{-b \pm \sqrt{b^2 - 4ac}}{2a}$$

$$= \frac{-80 \pm \sqrt{80^2 - 4(25)(61)}}{2(25)}$$

$$= \frac{-80 \pm \sqrt{6400 - 6100}}{50}$$

$$= -\frac{8}{5} \pm \frac{10\sqrt{3}}{50}$$

$$= -\frac{8}{5} \pm \frac{\sqrt{3}}{5}$$

114. $(z + 6)^2 = -2z$

$$z^2 + 12z + 36 = -2z$$

$$z^2 + 14z + 36 = 0$$

$$z = \frac{-b \pm \sqrt{b^2 - 4ac}}{2a}$$

$$= \frac{-14 \pm \sqrt{14^2 - 4(1)(36)}}{2(1)}$$

$$= \frac{-14 \pm \sqrt{52}}{2}$$

$$= -7 \pm \sqrt{13}$$

116. $\left(\frac{5}{7}x - 14\right)^2 = 8x$

$$\frac{25}{49}x^2 - 20x + 196 = 8x$$

$$\frac{25}{49}x^2 - 28x + 196 = 0$$

$$25x^2 - 1372x + 9604 = 0$$

$$x = \frac{-b \pm \sqrt{b^2 - 4ac}}{2a} = \frac{-(-1372) \pm \sqrt{(-1372)^2 - 4(25)(9604)}}{2(25)}$$

$$= \frac{1372 \pm \sqrt{921{,}984}}{50} = \frac{686 \pm 196\sqrt{6}}{25} = \frac{686}{25} \pm \frac{196\sqrt{6}}{25}$$

118. $2x^2 - 2.50x - 0.42 = 0$

$$x = \frac{-b \pm \sqrt{b^2 - 4ac}}{2a}$$

$$= \frac{-(-2.50) \pm \sqrt{(-2.50)^2 - 4(2)(-0.42)}}{2(2)}$$

$$= 1.400, -0.150$$

120. $-0.005x^2 + 0.101x - 0.193 = 0$

$$x = \frac{-b \pm \sqrt{b^2 - 4ac}}{2a}$$

$$= \frac{-0.101 \pm \sqrt{(0.101)^2 - 4(-0.005)(-0.193)}}{2(-0.005)}$$

$$\approx 2.137, 18.063$$

122. $1100x^2 + 326x - 715 = 0$

$$x = \frac{-b \pm \sqrt{b^2 - 4ac}}{2a} = \frac{-326 \pm \sqrt{(326)^2 - 4(1100)(-715)}}{2(1100)}$$

$$\approx 0.672, -0.968$$

124. $-3.22x^2 - 0.08x + 28.651 = 0$

$$x = \frac{-b \pm \sqrt{b^2 - 4ac}}{2a} = \frac{-(-0.08) \pm \sqrt{(-0.08)^2 - 4(-3.22)(28.651)}}{2(-3.22)}$$

$$\approx -2.995, 2.971$$

126. $11x^2 + 33x = 0$

$11x(x + 3) = 0$

$11x = 0 \Rightarrow x = 0$

$x + 3 = 0 \Rightarrow x = -3$

128. $x^2 - 14x + 49 = 0$

$(x - 7)^2 = 0$

$x - 7 = 0 \Rightarrow x = 7$

130. $x^2 + 3x - \frac{3}{4} = 0$

$x^2 + 3x + \left(\frac{3}{2}\right)^2 = \frac{3}{4} + \frac{9}{4}$

$\left(x + \frac{3}{2}\right)^2 = 3$

$x + \frac{3}{2} = \pm\sqrt{3}$

$x = -\frac{3}{2} \pm \sqrt{3}$

132. $a^2x^2 - b^2 = 0$

$(ax + b)(ax - b) = 0$

$ax + b = 0 \Rightarrow x = -\dfrac{b}{a}$

$ax - b = 0 \Rightarrow x = \dfrac{b}{a}$

134. $4x^2 + 2x + 4 = 2x + 8$

$4x^2 - 4 = 0$

$4(x^2 - 1) = 0$

$4(x + 1)(x - 1) = 0$

$x + 1 = 0 \Rightarrow x = -1$

$x - 1 = 0 \Rightarrow x = 1$

136. $20x^3 - 125x = 0$

$5x(4x^2 - 25) = 0$

$5x(2x + 5)(2x - 5) = 0$

$5x = 0 \Rightarrow x = 0$

$2x + 5 = 0 \Rightarrow x = -\frac{5}{2}$

$2x - 5 = 0 \Rightarrow x = \frac{5}{2}$

138. $x^6 - 64 = 0$

$(x^3 - 8)(x^3 + 8) = 0$

$(x - 2)(x^2 + 2x + 4)(x + 2)(x^2 - 2x + 4) = 0$

$x - 2 = 0 \Rightarrow x = 2$

$x^2 + 2x + 4 = 0 \Rightarrow$ No real solutions

$x + 2 = 0 \Rightarrow x = -2$

$x^2 - 2x + 4 = 0 \Rightarrow$ No real solutions

Real solutions: $x = 2, -2$

140. $27x^3 - 512 = 0$

$(3x - 8)(9x^2 + 24x + 64) = 0$

$3x - 8 = 0 \Rightarrow x = \frac{8}{3}$

$9x^2 + 24x + 64 = 0 \Rightarrow$ No real solutions

Real solution: $x = \frac{8}{3}$

142. $9x^4 - 24x^3 + 16x^2 = 0$

$x^2(9x^2 - 24x + 16) = 0$

$x^2(3x - 4)^2 = 0$

$x^2 = 0 \Rightarrow x = 0$

$3x - 4 = 0 \Rightarrow x = \frac{4}{3}$

144. $x^3 + 2x^2 + 3x + 6 = 0$

$x^2(x + 2) + 3(x + 2) = 0$

$(x + 2)(x^2 + 3) = 0$

$x + 2 = 0 \Rightarrow x = -2$

$x^2 + 3 = 0 \Rightarrow$ No real solutions

Real solution: $x = -2$

146. $x^4 + 2x^3 - 8x - 16 = 0$

$x^3(x + 2) - 8(x + 2) = 0$

$(x^3 - 8)(x + 2) = 0$

$(x - 2)(x^2 + 2x + 4)(x + 2) = 0$

$x - 2 = 0 \Rightarrow x = 2$

$x^2 + 2x + 4 = 0 \Rightarrow$ No real solutions

$x + 2 = 0 \Rightarrow x = -2$

Real solutions: $x = 2, -2$

148. $x^4 + 5x^2 - 36 = 0$

$(x^2 + 9)(x^2 - 4) = 0$

$(x^2 + 9)(x + 2)(x - 2) = 0$

$x^2 + 9 = 0 \Rightarrow$ No real solutions

$x + 2 = 0 \Rightarrow x = -2$

$x - 2 = 0 \Rightarrow x = 2$

Real solutions: $x = 2, -2$

150. $36t^4 + 29t^2 - 7 = 0$

$(36t^2 - 7)(t^2 + 1) = 0$

$(6t + \sqrt{7})(6t - \sqrt{7})(t^2 + 1) = 0$

$6t + \sqrt{7} = 0 \Rightarrow t = -\dfrac{\sqrt{7}}{6}$

$6t - \sqrt{7} = 0 \Rightarrow t = \dfrac{\sqrt{7}}{6}$

$t^2 + 1 = 0 \Rightarrow$ No real solutions

Real solutions: $t = \pm\dfrac{\sqrt{7}}{6}$

152. $x^6 + 3x^3 + 2 = 0$

$(x^3 + 2)(x^3 + 1) = 0$

$(x + \sqrt[3]{2})[x^2 - \sqrt[3]{2}x + (\sqrt[3]{2})^2](x + 1)(x^2 - x + 1) = 0$

$x + \sqrt[3]{2} = 0 \Rightarrow x = -\sqrt[3]{2}$

$x^2 - \sqrt[3]{2}x + (\sqrt[3]{2})^2 = 0 \Rightarrow$ No real solutions

$x + 1 = 0 \Rightarrow x = -1$

$x^2 - x + 1 = 0 \Rightarrow$ No real solutions

Real solutions: $x = -\sqrt[3]{2}, -1$

154. $4\sqrt{x} - 3 = 0$

$4\sqrt{x} = 3$

$16x = 9$

$x = \frac{9}{16}$

156. $\sqrt{5 - x} - 3 = 0$

$\sqrt{5 - x} = 3$

$5 - x = 9$

$x = -4$

158. $\sqrt[3]{3x + 1} - 5 = 0$

$\sqrt[3]{3x + 1} = 5$

$3x + 1 = 125$

$3x = 124$

$x = \frac{124}{3}$

160. $x + \sqrt{31 - 9x} = 5$

$\sqrt{31 - 9x} = 5 - x$

$31 - 9x = 25 - 10x + x^2$

$0 = x^2 - x - 6$

$0 = (x - 3)(x + 2)$

$0 = x - 3 \Rightarrow x = 3$

$0 = x + 2 \Rightarrow x = -2$

162. $\sqrt{x + 5} = \sqrt{x - 5}$

$x + 5 = x - 5$

$5 = -5$

No solution

164. $(x+3)^{3/2} = 8$

$$(x+3)^3 = 8^2$$
$$(x+3)^3 = 64$$
$$x+3 = \sqrt[3]{64}$$
$$x = -3 + 4 = 1$$

166. $(x+2)^{2/3} = 9$

$$(x+2)^2 = 9^3$$
$$x + 2 = \pm\sqrt{729}$$
$$x = -2 \pm 27 = -29, 25$$

168. $(x^2 - x - 22)^{3/2} = 27$

$$x^2 - x - 22 = 27^{2/3}$$
$$x^2 - x - 22 = 9$$
$$x^2 - x - 31 = 0$$
$$x = \frac{-(-1) \pm \sqrt{(-1)^2 - 4(1)(-31)}}{2(1)}$$
$$= \frac{1 \pm 5\sqrt{5}}{2} = \frac{1}{2} \pm \frac{5\sqrt{5}}{2}$$

170. $4x^2(x-1)^{1/3} + 6x(x-1)^{4/3} = 0$

$$2x[2x(x-1)^{1/3} + 3(x-1)^{4/3}] = 0$$
$$2x(x-1)^{1/3}[2x + 3(x-1)] = 0$$
$$2x(x-1)^{1/3}(5x-3) = 0$$
$$2x = 0 \Rightarrow x = 0$$
$$x - 1 = 0 \Rightarrow x = 1$$
$$5x - 3 = 0 \Rightarrow x = \frac{3}{5}$$

172. $\frac{4}{x} - \frac{5}{3} = \frac{x}{6}$

$$(6x)\frac{4}{x} - (6x)\frac{5}{3} = (6x)\frac{x}{6}$$
$$24 - 10x = x^2$$
$$x^2 + 10x - 24 = 0$$
$$(x+12)(x-2) = 0$$
$$x + 12 = 0 \Rightarrow x = -12$$
$$x - 2 = 0 \Rightarrow x = 2$$

174. $\frac{4}{x+1} - \frac{3}{x+2} = 1$

$$4(x+2) - 3(x+1) = (x+1)(x+2), x \neq -2, -1$$
$$4x + 8 - 3x - 3 = x^2 + 3x + 2$$
$$x^2 + 2x - 3 = 0$$
$$(x-1)(x+3) = 0$$
$$x - 1 = 0 \Rightarrow x = 1$$
$$x + 3 = 0 \Rightarrow x = -3$$

176. $4x + 1 = \frac{3}{x}$

$$(x)4x + (x)1 = (x)\frac{3}{x}$$
$$4x^2 + x = 3$$
$$4x^2 + x - 3 = 0$$
$$(4x-3)(x+1) = 0$$
$$4x - 3 = 0 \Rightarrow x = \frac{3}{4}$$
$$x + 1 = 0 \Rightarrow x = -1$$

178. $\frac{x+1}{3} - \frac{x+1}{x+2} = 0$

$$3(x+2)\frac{x+1}{3} - 3(x+2)\frac{x+1}{x+2} = 0$$
$$(x+2)(x+1) - 3(x+1) = 0$$
$$x^2 + 3x + 2 - 3x - 3 = 0$$
$$x^2 - 1 = 0$$
$$(x+1)(x-1) = 0$$
$$x + 1 = 0 \Rightarrow x = -1$$
$$x - 1 = 0 \Rightarrow x = 1$$

180. $|3x + 2| = 7$

$$3x + 2 = 7 \Rightarrow x = \tfrac{5}{3}$$
$$-(3x+2) = 7$$
$$-3x - 2 = 7 \Rightarrow x = -3$$

182. $|x^2 + 6x| = 3x + 18$

First equation:

$$x^2 + 6x = 3x + 18$$
$$x^2 + 3x - 18 = 0$$
$$(x - 3)(x + 6) = 0$$
$$x - 3 = 0 \Rightarrow x = 3$$
$$x + 6 = 0 \Rightarrow x = -6$$

Second equation:

$$-(x^2 + 6x) = 3x + 18$$
$$0 = x^2 + 9x + 18$$
$$0 = (x + 3)(x + 6)$$
$$0 = x + 3 \Rightarrow x = -3$$
$$0 = x + 6 \Rightarrow x = -6$$

184. $|x - 10| = x^2 - 10x$

First equation:

$$x - 10 = x^2 - 10x$$
$$0 = x^2 - 11x + 10$$
$$0 = (x - 1)(x - 10)$$
$0 = x - 1 \Rightarrow x = 1$, not a solution
$$0 = x - 10 \Rightarrow x = 10$$

Second equation:

$$-(x - 10) = x^2 - 10x$$
$$0 = x^2 - 9x - 10$$
$$0 = (x - 10)(x + 1)$$
$$0 = x - 10 \Rightarrow x = 10$$
$$0 = x + 1 \Rightarrow x = -1$$

186. $10{,}000 = 0.32m + 2500$

$$7{,}500 = 0.32m$$
$$\frac{7{,}500}{0.32} = m$$
$m = 23{,}437.5$ miles

188. (a)

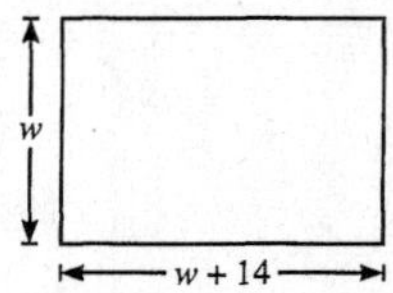

(b) $$w(w + 14) = 1632$$
$$w^2 + 14w - 1632 = 0$$

(c) $(w - 34)(w + 48) = 0$

$w = 34$ or $w = -48$ Extraneous

$l = 34 + 14 = 48$

width: 34 feet

length: 48 feet

190. $x^2 = 10^2 + \left(\frac{1}{2}x\right)^2$

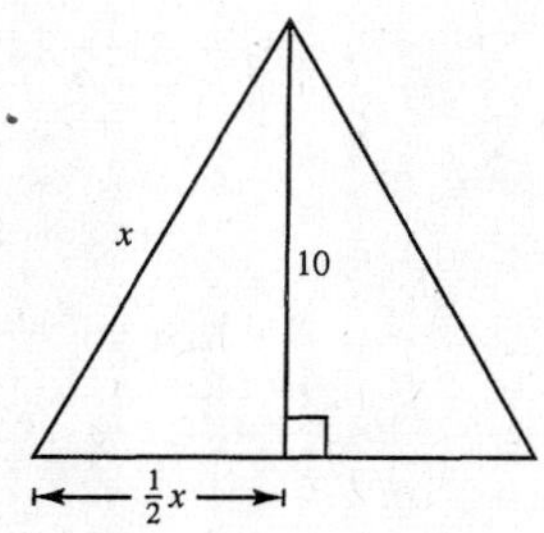

$$x^2 = 100 + \frac{1}{4}x^2$$
$$\frac{3}{4}x^2 = 100$$
$$x^2 = \frac{400}{3}$$
$$x = \pm\sqrt{\frac{400}{3}} = \pm\frac{20}{\sqrt{3}} = \pm\frac{20\sqrt{3}}{3}$$

Choose the positive value for x. So, the length of each side is $\frac{20\sqrt{3}}{3}$ inches.

192. $C = \sqrt{0.2x + 1} = 2.5$

$$0.2x + 1 = (2.5)^2$$
$$0.2x = 6.25 - 1$$
$$x = 26.25$$

26,250 passengers

194.
$$p = 40 - \sqrt{0.0001x + 1}$$
$$34.70 = 40 - \sqrt{0.0001x + 1}$$
$$40 - 34.70 = \sqrt{0.0001x + 1}$$
$$(5.30)^2 = 0.0001x + 1$$
$$x = 270{,}900$$

The demand is approximately 270,900 units per day.

196. False. If both sides of the equation are graphed, you can see that they intersect, which means that the equation has a real solution.

198. The student should have subtracted $15x$ from both sides so that the equation is equal to zero. By factoring out an x, there are **two** solutions.

$x = 0$ or $x = 6$

200. Remove symbols of grouping, combine like terms, reduce fractions.

Add (or subtract) the same quantity to (from) both sides of the equation.

Multiply (or divide) both sides of the equation by the same nonzero quantity.

Interchange the two sides of the equation.

202. (a) $ax^2 + bx = 0,\ a \neq 0, b \neq 0$

$x(ax + b) = 0$

$x = 0$ or $ax + b = 0$

$ax = -b$

$x = -\dfrac{b}{a}$

(b) $ax^2 - ax = 0,\ a \neq 0$

$ax(x - 1) = 0$

$ax = 0$ or $x - 1 = 0$

$x = 0$ $\quad$ $x = 1$

204. Isolate the absolute value by subtracting x from both sides of the equation. The expression inside the absolute value signs can be positive or negative, so two separate equations must be solved.

206. Isolate the radical by subtracting x from each side of the equation. Square each side and solve the resulting equation.

Section P.3 The Cartesian Plane and Graphs of Equations

Solutions to Even-Numbered Exercises

2. $A: \left(\frac{3}{2}, -4\right);\ B: (0, -2);$
$C: \left(-3, \frac{5}{2}\right),\ D: (-6, 0)$

4. $(4, -8)$

6. $x < 0$ and $y < 0$ in Quadrant III.

8. $x > 2$ and $y = 3$ in Quadrant I.

10. $x > 4$ in Quadrants I and IV.

12. If $(-x, y)$ is in Quadrant IV, then (x, y) must be in Quadrant III.

14. If $xy < 0$, then x and y have opposite signs. This happens in Quadrants II and IV.

16.

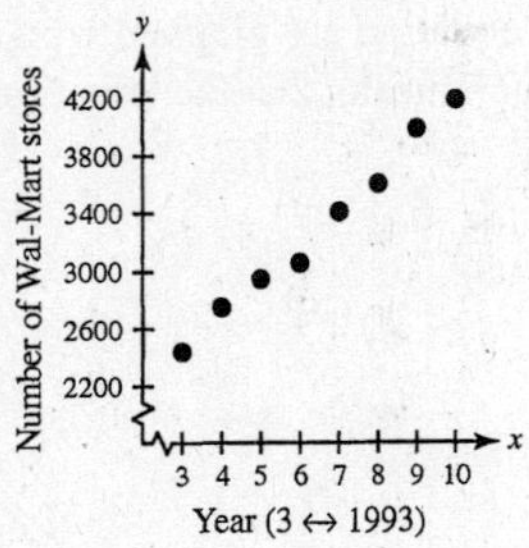

18. (a) $(1, 0), (13, 5)$

$$\text{Distance} = \sqrt{(13-1)^2 + (5-0)^2}$$
$$= \sqrt{12^2 + 5^2} = \sqrt{169} = 13$$

$(13, 5), (13, 0)$

$$\text{Distance} = |5 - 0| = |5| = 5$$

$(1, 0), (13, 0)$

$$\text{Distance} = |1 - 13| = |-12| = 12$$

(b) $5^2 + 12^2 = 25 + 144 = 169 = 13^2$

20. (a) $(1, 5), (5, -2)$

$$\text{Distance} = \sqrt{(1-5)^2 + (5-(-2))^2}$$
$$= \sqrt{(-4)^2 + (7)^2} = \sqrt{16 + 49} = \sqrt{65}$$

$(1, 5), (1, -2)$

$$\text{Distance} = |5 - (-2)| = |5 + 2| = |7| = 7$$

$(1, -2), (5, -2)$

$$\text{Distance} = |1 - 5| = |-4| = 4$$

(b) $4^2 + 7^2 = 16 + 49 = 65 = \left(\sqrt{65}\right)^2$

22. (a)

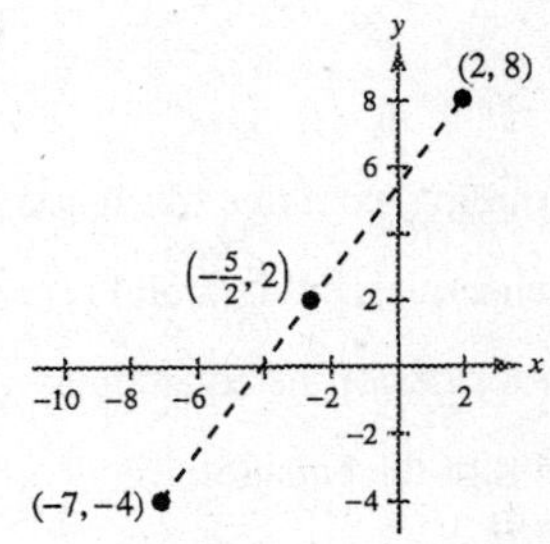

(b) $d = \sqrt{(-7-2)^2 + (-4-8)^2}$

$$= \sqrt{81 + 144} = 15$$

(c) $\left(\dfrac{-7+2}{2}, \dfrac{-4+8}{2}\right) = \left(-\dfrac{5}{2}, 2\right)$

24. (a)

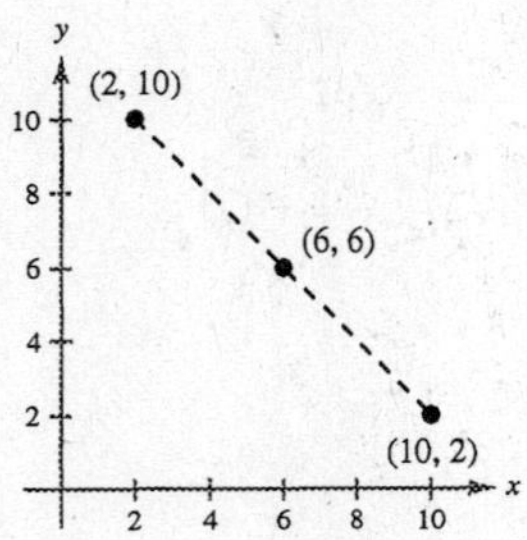

(b) $d = \sqrt{(2-10)^2 + (10-2)^2}$

$$= \sqrt{64 + 64} = 8\sqrt{2}$$

(c) $\left(\dfrac{2+10}{2}, \dfrac{10+2}{2}\right) = (6, 6)$

26. (a)

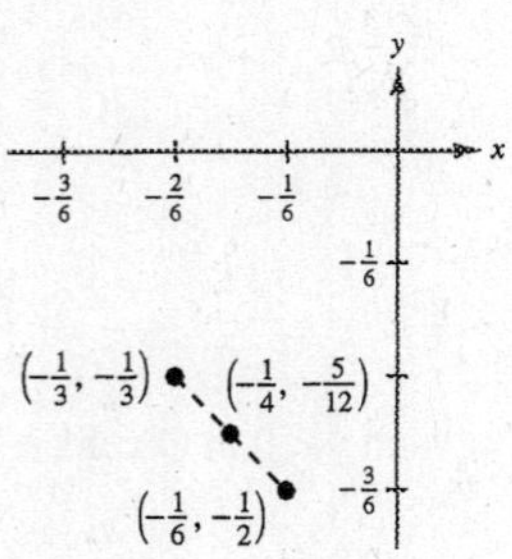

(b) $d = \sqrt{\left(-\dfrac{1}{3} + \dfrac{1}{6}\right)^2 + \left(-\dfrac{1}{3} + \dfrac{1}{2}\right)^2}$

$$= \sqrt{\frac{1}{36} + \frac{1}{36}} = \frac{\sqrt{2}}{6}$$

(c) $\left(\dfrac{-\frac{1}{3} + \left(-\frac{1}{6}\right)}{2}, \dfrac{-\frac{1}{3} + \left(-\frac{1}{2}\right)}{2}\right) = \left(-\dfrac{1}{4}, -\dfrac{5}{12}\right)$

28. $(-6.8, 1.3), (5.6, 4.9)$

(a)

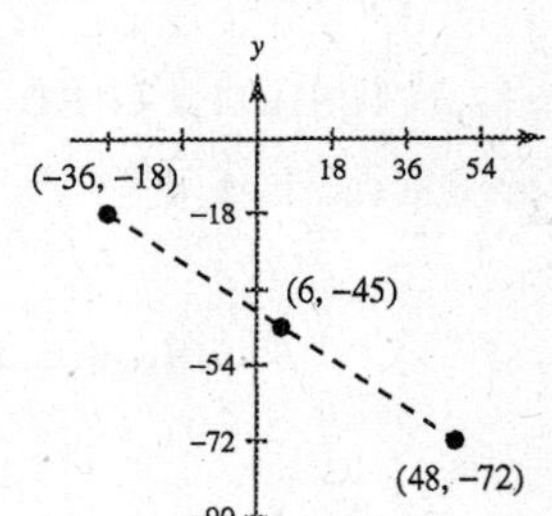

(b) $d = \sqrt{(-6.8 - 5.6)^2 + (1.3 - 4.9)^2}$

$$= \sqrt{153.76 + 12.96} = \sqrt{166.72}$$

(c) $\left(\dfrac{-6.8 + 5.6}{2}, \dfrac{1.3 + 4.9}{2}\right) = (-0.6, 3.1)$

30.

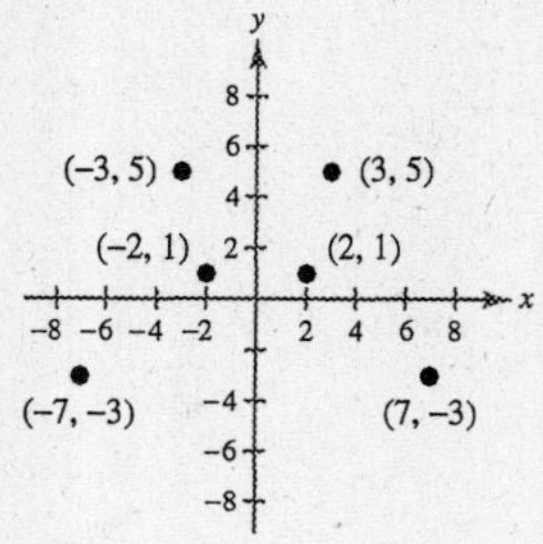

(a) The point is reflected through the y-axis.

(b) The point is reflected through the x-axis.

(c) The point is reflected through the origin.

32. Let (0, 0) represent the coordinates of Naples and let (−120, 150) represent the coordinates of Rome.

$$\text{Distance} = \sqrt{(0-(-120))^2 + (0-150)^2}$$
$$= \sqrt{14{,}400 + 22{,}500}$$
$$= \sqrt{36{,}900}$$
$$= 30\sqrt{41} \approx 192 \text{ kilometers}$$

34. (1999, 1428.9) and (2001, 2068.2)

$$\left(\frac{1999 + 1428.9}{2}, \frac{2001 + 2068.2}{2}\right) = (2000, 1748.55)$$

2000 sales: \$1748.55 million

36. $y = x^2 - 3x + 2$

(a) $(2, 0)$: $(2)^2 - 3(2) + 2 \stackrel{?}{=} 0$

$$4 - 6 + 2 \stackrel{?}{=} 0$$
$$0 = 0$$

Yes, the point *is* on the graph.

(b) $(-2, 8)$: $(-2)^2 - 3(-2) + 2 \stackrel{?}{=} 8$

$$4 + 6 + 2 \stackrel{?}{=} 8$$
$$12 \neq 8$$

No, the point *is not* on the graph.

38. $y = \frac{1}{3}x^3 - 2x^2$

(a) $\left(2, -\frac{16}{3}\right)$: $\frac{1}{3}(2)^3 - 2(2)^2 \stackrel{?}{=} -\frac{16}{3}$

$$\frac{1}{3} \cdot 8 - 2 \cdot 4 \stackrel{?}{=} -\frac{16}{3}$$
$$\frac{8}{3} - 8 \stackrel{?}{=} -\frac{16}{3}$$
$$\frac{8}{3} - \frac{24}{3} \stackrel{?}{=} -\frac{16}{3}$$
$$-\frac{16}{3} = -\frac{16}{3}$$

Yes, the point *is* on the graph.

(b) $(-3, 9)$: $\frac{1}{3}(-3)^3 - 2(-3)^2 \stackrel{?}{=} 9$

$$\frac{1}{3}(-27) - 2(9) \stackrel{?}{=} 9$$
$$-9 - 18 \stackrel{?}{=} 9$$
$$-27 \neq 9$$

No, the point *is not* on the graph.

40. $y = x^2 - 3x$

x	-1	0	1	2	3
y	4	0	-2	-2	0
(x, y)	$(-1, 4)$	$(0, 0)$	$(1, -2)$	$(2, -2)$	$(3, 0)$

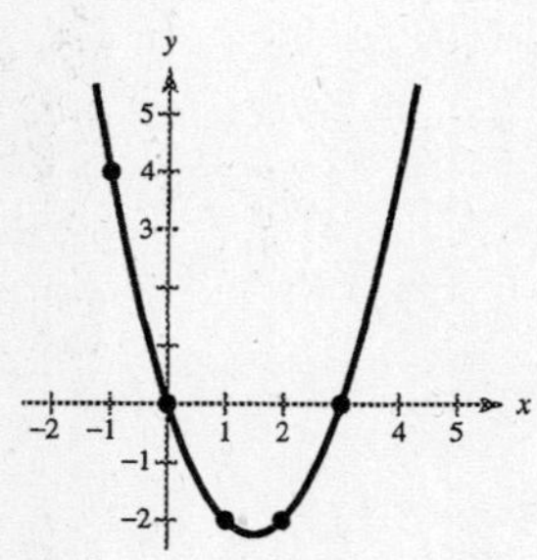

42. $y = (x + 3)^2$

x-intercept: $0 = (x + 3)^2$

$0 = x + 3$

$x = -3$

$(-3, 0)$

y-intercept: $y = (0 + 3)^2$

$y = 9$

$(0, 9)$

44. $y = 8 - 3x$

x-intercept: $0 = 8 - 3x$

$-8 = -3x$

$x = \frac{8}{3}$

$\left(\frac{8}{3}, 0\right)$

y-intercept: $y = 8 - 3(0)$

$y = 8$

$(0, 8)$

46. $y = \sqrt{2x - 1}$

x-intercept: $0 = \sqrt{2x - 1}$

$0 = 2x - 1$

$x = \frac{1}{2}$

$\left(\frac{1}{2}, 0\right)$

y-intercept: $y = \sqrt{2(0) - 1}$

$y = \sqrt{-1}$

No y-intercept

48. $y = -|x + 10|$

x-intercept: $0 = -|x + 10|$

$0 = x + 10$

$x = -10$

$(-10, 0)$

y-intercept: $y = -|0 + 10|$

$y = -10$

$(0, -10)$

50. $y = x^4 - 25$

x-intercept: $0 = x^4 - 25$

$0 = (x^2 + 5)(x^2 - 5)$

$0 = (x^2 + 5)\left(x + \sqrt{5}\right)\left(x - \sqrt{5}\right)$

$x = -\sqrt{5}, \sqrt{5}$

$\left(-\sqrt{5}, 0\right), \left(\sqrt{5}, 0\right)$

y-intercept: $y = (0^4 - 25)$

$y = -25$

$(0, -25)$

52. $y^2 = x + 1$

x-intercept: $0 = x + 1$

$x = -1$

$(-1, 0)$

y-intercepts: $y^2 = 0 + 1$

$y = \pm 1$

$(0, 1), (0, -1)$

54. $x - (-y)^2 = 0 \Rightarrow x - y^2 = 0$

x-axis symmetry

56. $y = x^4 - x^2 + 3$

$y = (-x)^4 - (-x)^2 + 3$

$y = x^4 - x^2 + 3$

y-axis symmetry

58. $y = \dfrac{1}{x^2 + 1}$

$y = \dfrac{1}{(-x)^2 + 1}$

$y = \dfrac{1}{x^2 + 1}$

y-axis symmetry

60. $xy = 4$

$(-x)(-y) = 4$

$xy = 4$

Origin symmetry

62.

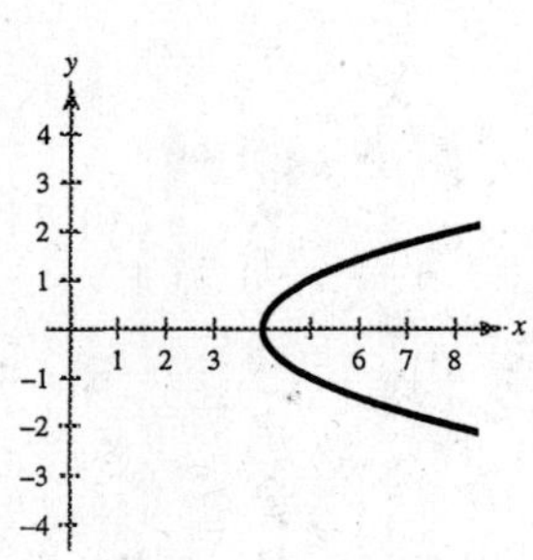

64.

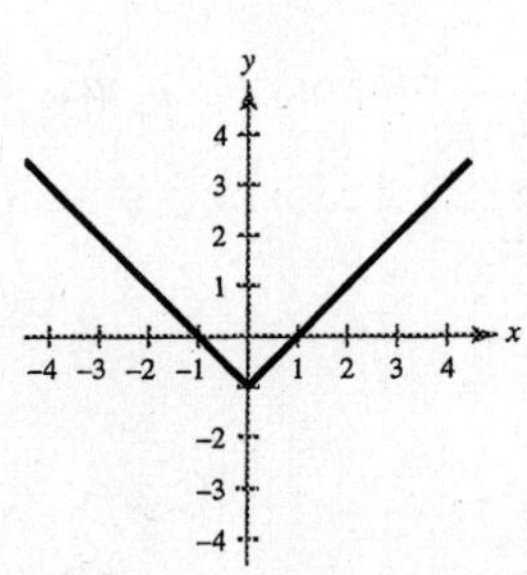

66. $y = 2x - 3$

x-intercept: $\left(\frac{3}{2}, 0\right)$

y-intercept: $(0, -3)$

No symmetry

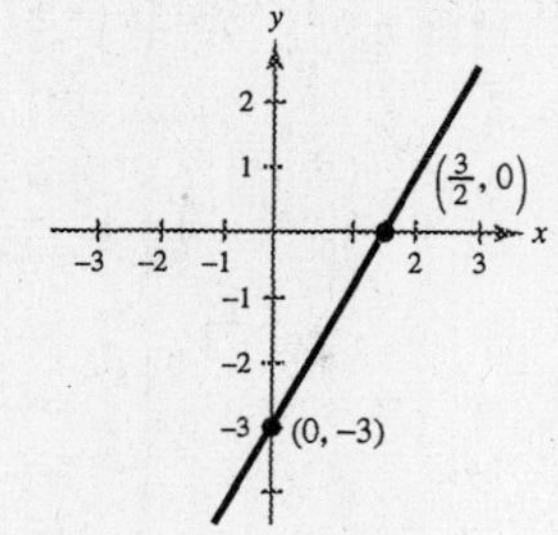

68. $y = -x^2 - 2x$

x-intercept: $(-2, 0), (0, 0)$

y-intercept: $(0, 0)$

No symmetry

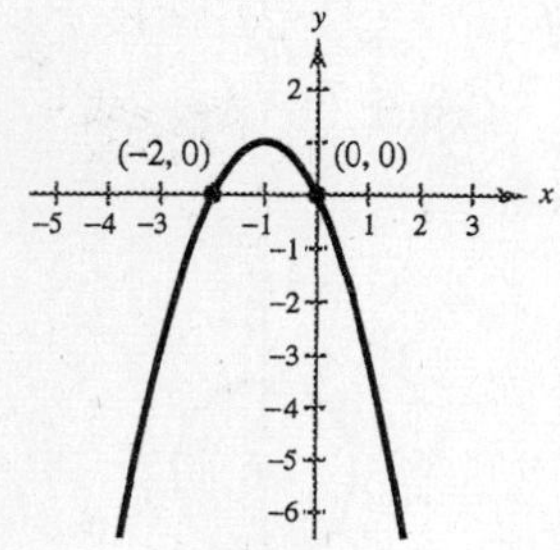

70. $y = x^3 - 1$

x-intercept: $(1, 0)$

y-intercept: $(0, -1)$

No symmetry

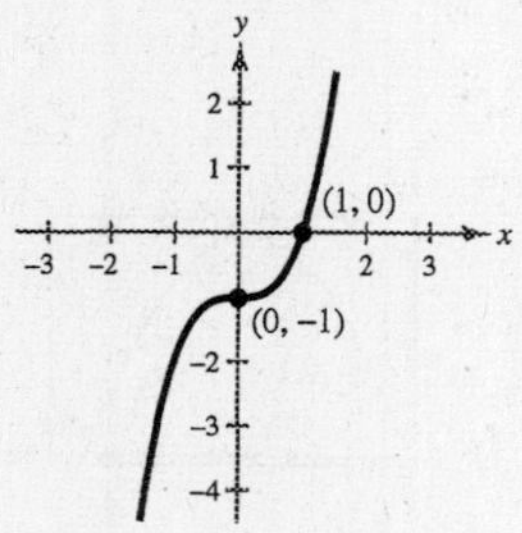

72. $y = \sqrt{1 - x}$

Domain: $(-\infty, 1]$

x-intercept: $(1, 0)$

y-intercept: $(0, 1)$

No symmetry

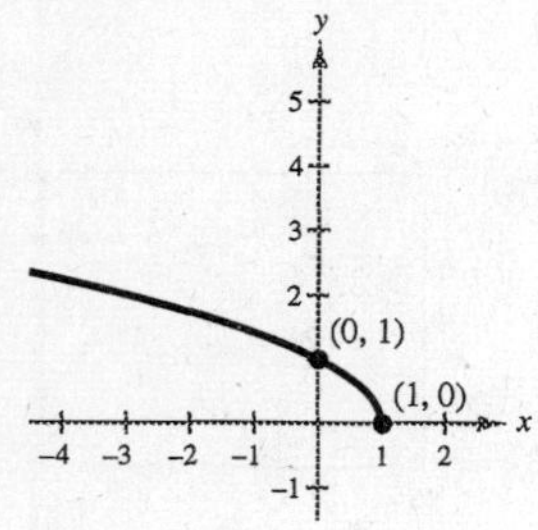

74. $y = 1 - |x|$

x-intercepts: $(\pm 1, 0)$

y-intercept: $(0, 1)$

y-axis symmetry

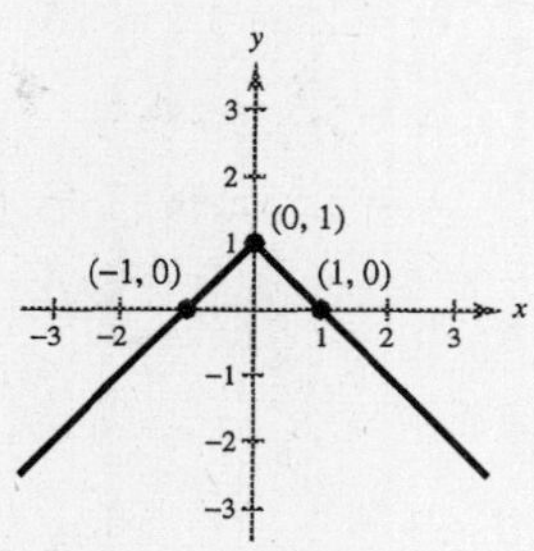

76. $x = y^2 - 5$

x-intercept: $(-5, 0)$

y-intercept: $\left(0, \pm\sqrt{5}\right)$

x-axis symmetry

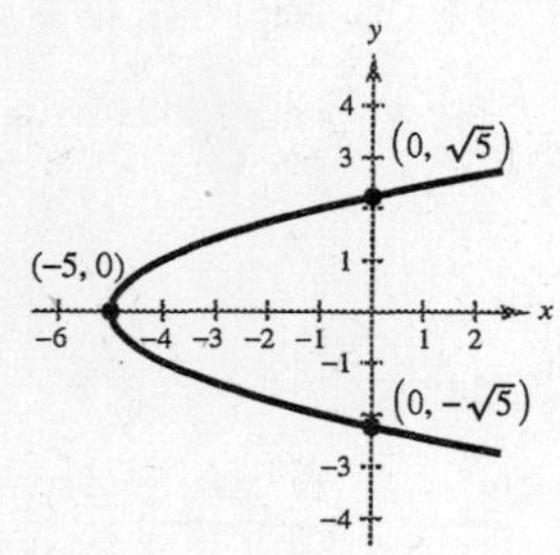

78. $(x - (-7))^2 + (y - (-4))^2 = 7^2$

$$(x + 7)^2 + (y + 4)^2 = 49$$

80. $r = \sqrt{(3 - (-1))^2 + (-2 - 1)^2}$

$$= \sqrt{4^2 + (-3)^2} = \sqrt{25} = 5$$

$$(x - 3)^2 + (y - (-2))^2 = 5^2$$

$$(x - 3)^2 + (y + 2)^2 = 25$$

82. $r = \frac{1}{2}\sqrt{(-4 - 4)^2 + (-1 - 1)^2}$

$$= \frac{1}{2}\sqrt{(-8)^2 + (-2)^2}$$

$$= \frac{1}{2}\sqrt{64 + 4}$$

$$= \frac{1}{2}\sqrt{68} = \left(\frac{1}{2}\right)(2)\sqrt{17} = \sqrt{17}$$

Midpoint of diameter (center of circle):

$$\left(\frac{-4 + 4}{2}, \frac{-1 + 1}{2}\right) = (0, 0)$$

$$(x - 0)^2 + (y - 0)^2 = \left(\sqrt{17}\right)^2$$

$$x^2 + y^2 = 17$$

84. $x^2 + y^2 = 16$

Center: $(0, 0)$, radius: 4

86. $x^2 + (y - 1)^2 = 1$

Center: $(0, 1)$, radius: 1

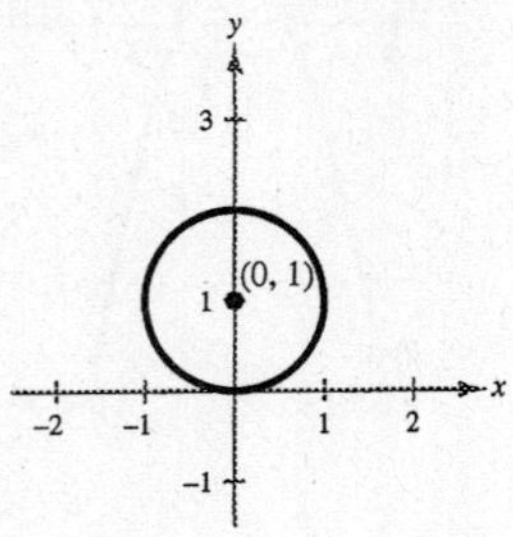

88. $(x - 2)^2 + (y + 1)^2 = 3$

Center: $(2, -1)$; radius: $\sqrt{3}$

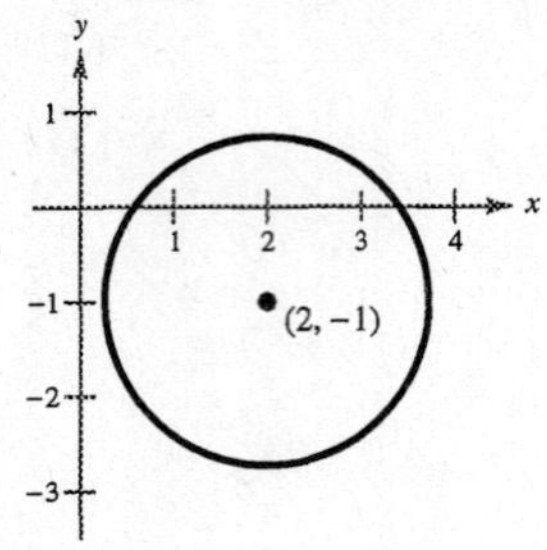

90. $y = 8100 - 929t, 0 \le t \le 6$

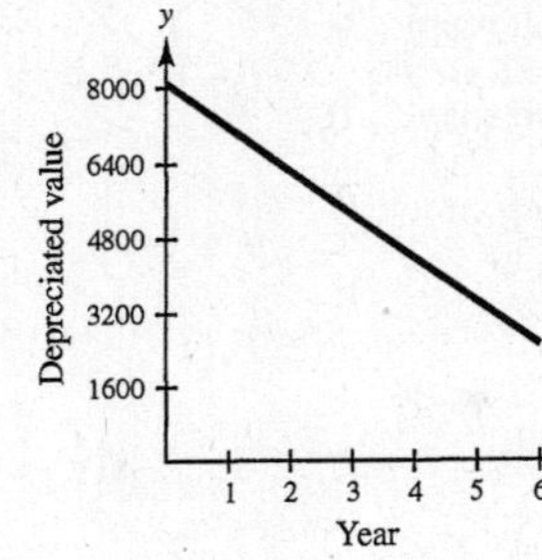

92. (a)

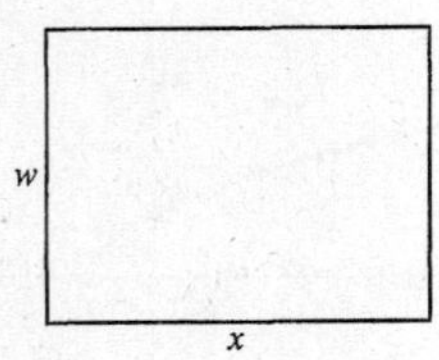

(b) $2x + 2w = 22$

$x + w = 11 \Rightarrow w = 11 - x$

$A = x \cdot w = x(11 - x)$

(c)

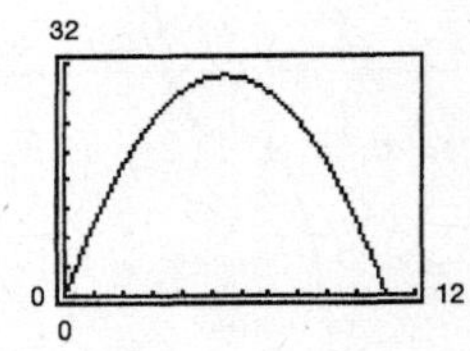

(d) The area is maximized when $x = 5.5$ and

$W = 11 - 5.5 = 5.5$

$x = 5.5$ yards

$w = 5.5$ yards

94. $y = \dfrac{10{,}770}{x^2} - 0.37, 5 \le x \le 100$

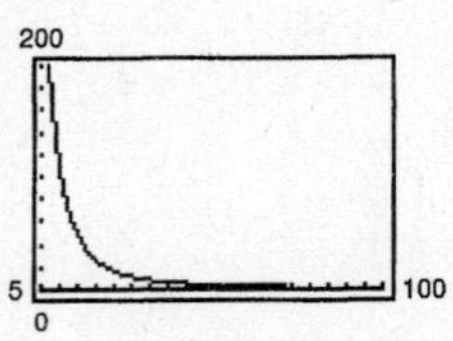

When $x = 50$, $y = 3.938$. The resistance is about 3.94 ohms.

96. True

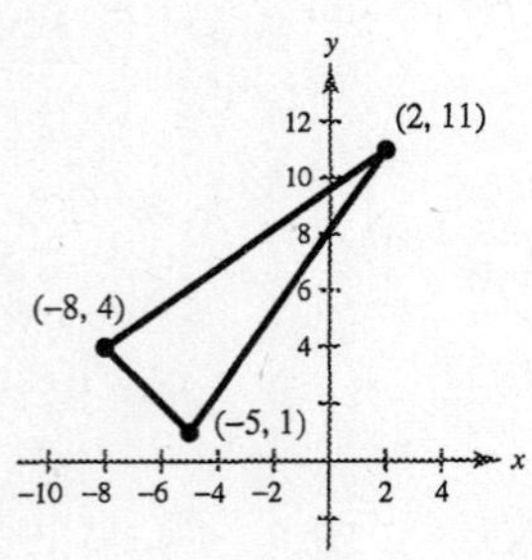

The distance between $(-8, 4)$ and $(2, 11)$ is:

$$d_1 = \sqrt{(2 - (-8))^2 + (11 - 4)^2} = \sqrt{149}$$

The distance between $(-5, 1)$ and $(2, 11)$ is:

$$d_2 = \sqrt{(2 - (-5))^2 + (11 - 1)^2} = \sqrt{149}$$

Since $d_1 = d_2$, the triangle is isosceles.

(Note: The length of the third side is $d_3 = \sqrt{18}$.)

98. True. Horizontal lines have no x-intercepts and all other lines cross the y-axis one time.

100. No. It depends on the magnitude of the quantities measured.

102. The viewing window is incorrect. Change the viewing window. For example, the graph of $y = x^2 + 25$ will not appear on a standard window with $-10 \le x \le 10$ and $-10 \le y \le 10$. By noting a few points on the graph, say $(-5, 50)$, $(0,0)$, and $(5, 50)$, you can pick a more appropriate window, such as $-10 \le x \le 10, 0 \le y \le 100$.

Section P.4 Linear Equations in Two Variables

Solutions to Even-Numbered Exercises

2. (a) $m = 0$. The line is horizontal. Matches L_2.

(b) $m = -\frac{3}{4}$. Because the slope is negative, the line falls. Matches L_1.

(c) $m = 1$. Because the slope is positive, the line rises. Matches L_3.

4.

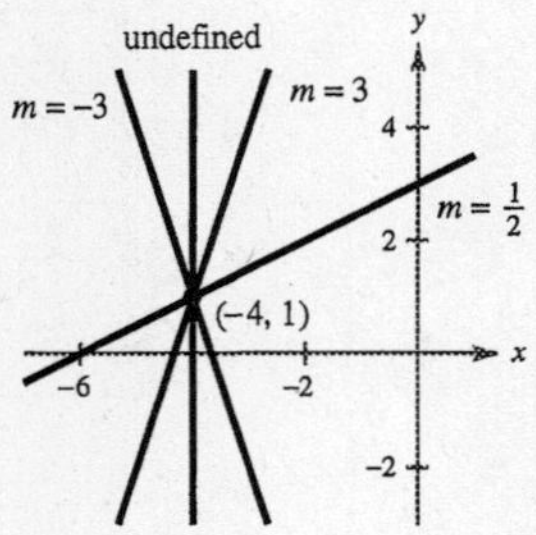

6. The line appears to go through $(1, 0)$ and $(4, 8)$.

$$\text{Slope} = \frac{y_2 - y_1}{x_2 - x_1} = \frac{8 - 0}{4 - 1} = \frac{8}{3}$$

8. The line appears to go through $(0, 7)$ and $(7, 0)$.

$$\text{Slope} = \frac{y_2 - y_1}{x_2 - x_1} = \frac{0 - 7}{7 - 0} = -1$$

10. $y = x - 10$

Slope: $m = 1$

y-intercept: $(0, -10)$

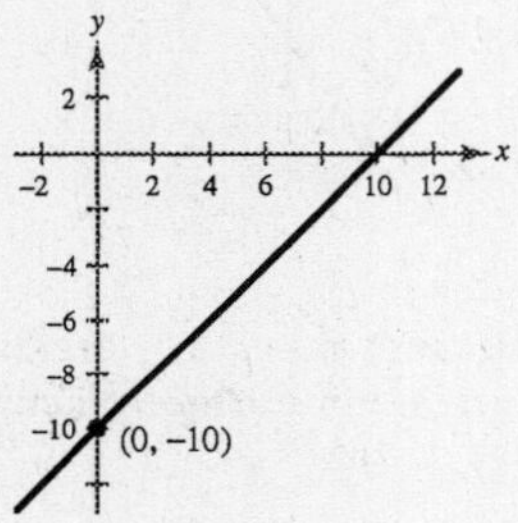

12. $y = -\frac{3}{2}x + 6$

Slope: $m = -\frac{3}{2}$

y-intercept: $(0, 6)$

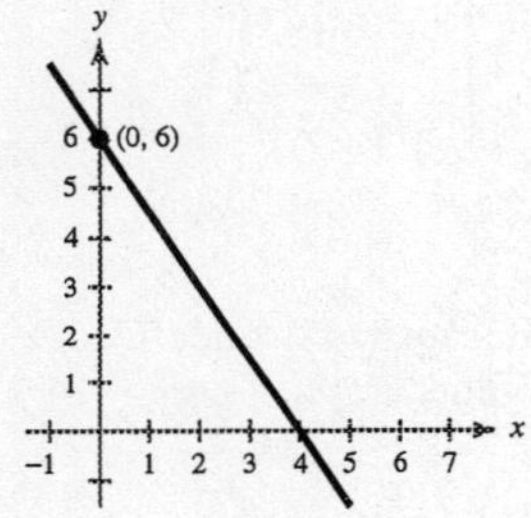

14. $3y + 5 = 0$

$$3y = -5$$

$$y = -\frac{5}{3}$$

Slope: $m = 0$

y-intercept: $\left(0, -\frac{5}{3}\right)$

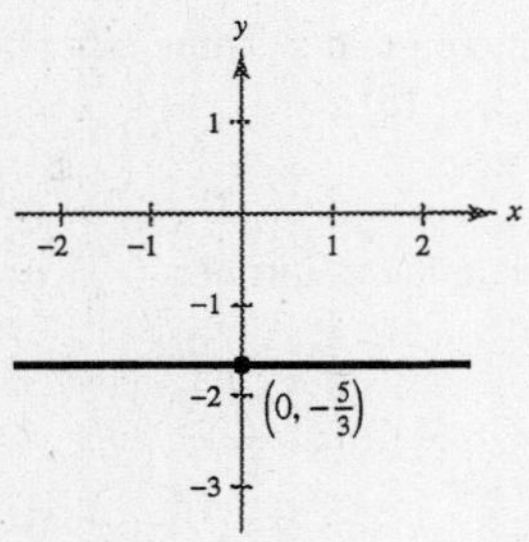

16. $2x + 3y = 9$

$$3y = -2x + 9$$

$$y = -\frac{2}{3}x + 3$$

Slope: $m = -\frac{2}{3}$

y-intercept: $(0, 3)$

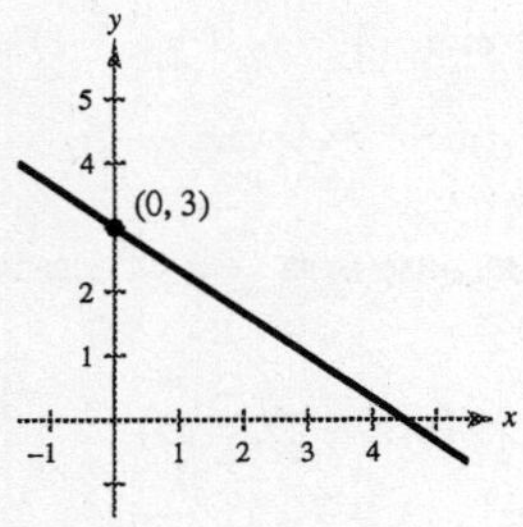

18. $y + 4 = 0$

$$y = -4$$

Slope: $m = 0$

y-intercept: $(0, -4)$

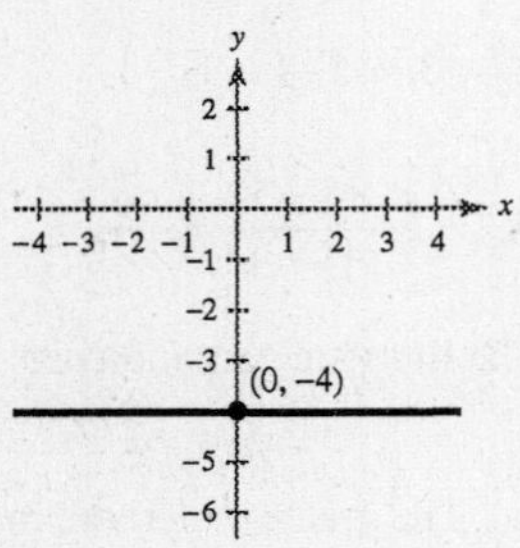

20. $x - 2 = 0$

$$x = 2$$

Slope: undefined (vertical line)

y-intercept: none

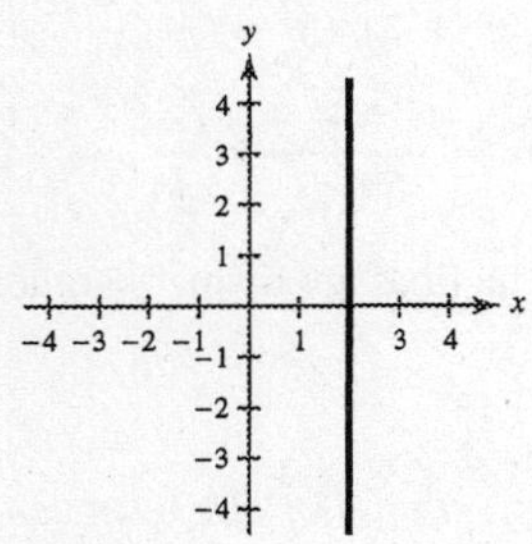

22. Slope $= \dfrac{-4-4}{4-2} = -4$

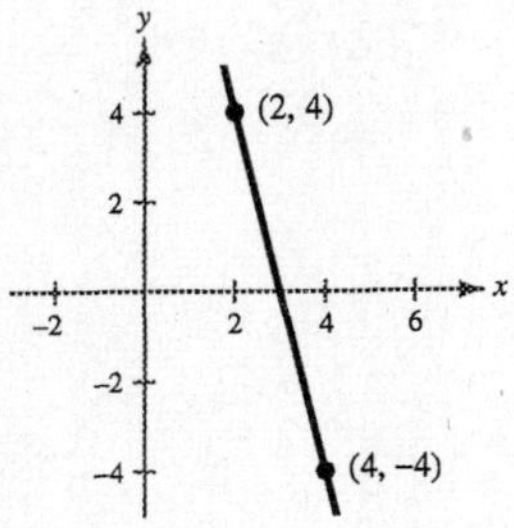

24. Slope $= \dfrac{0-(-10)}{-4-0} = -\dfrac{5}{2}$

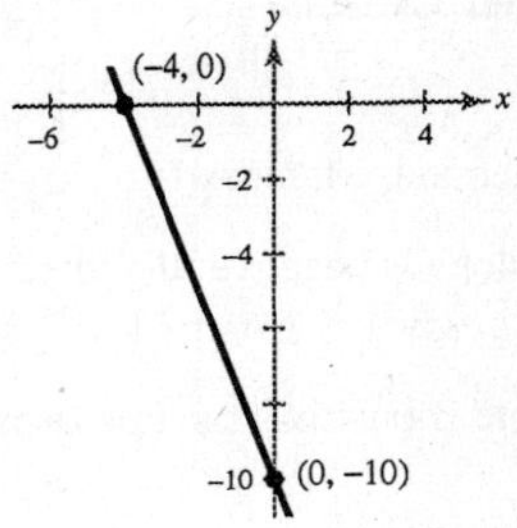

26. Slope $= \dfrac{-\frac{1}{4}-\frac{3}{4}}{\frac{5}{4}-\frac{7}{8}} = \dfrac{-1}{\frac{3}{8}} = -\dfrac{8}{3}$

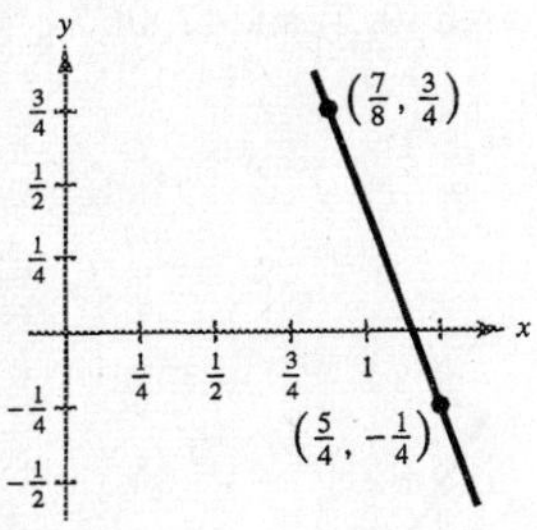

28. Slope $= \dfrac{-2.6-(-8.3)}{2.25-(-1.75)} = 1.425$

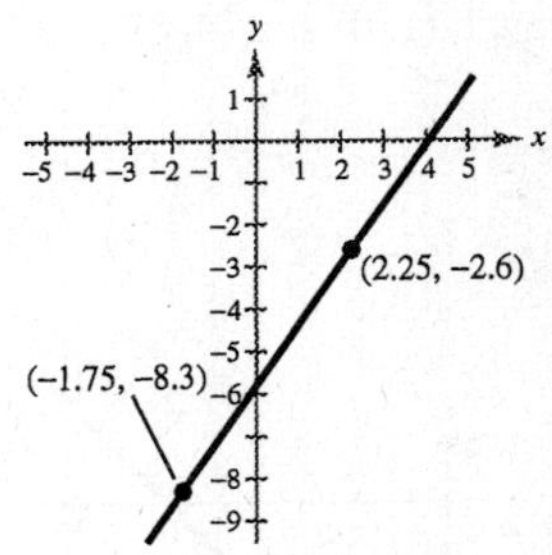

30. Because m is undefined, x does not change. Three other points are: $(-4, 0), (-4, 3), (-4, 5)$.

32. Because $m = -1$, y decreases by 1 for every one unit increase in x. Three other points are: $(0, 4), (9, -5), (11, -7)$.

34. Because $m = 0$, y does not change. Three other points are: $(-4, -1), (-2, -1), (0, -1)$.

36. Because $m = -2$, y decreases by 2 for every one unit increase in x. Three other points are: $(-2, -5), (1, -11), (3, -15)$.

38. Because $m = -\frac{1}{2}$, y decreases by 1 for every 2 unit increase in x. Three other points are: $(-3, -5), (1, -7), (5, -9)$.

40. L_1: $(-2, -1), (1, 5)$

$$m_1 = \frac{5-(-1)}{1-(-2)} = \frac{6}{3} = 2$$

L_2: $(1, 3), (5, -5)$

$$m_2 = \frac{-5-3}{5-1} = \frac{-8}{4} = -2$$

The lines are neither parallel nor perpendicular.

42. L_1: $(4, 8), (-4, 2)$

$$m_1 = \frac{2-8}{-4-4} = \frac{-6}{-8} = \frac{3}{4}$$

L_2: $(3, -5), \left(-1, \frac{1}{3}\right)$

$$m_2 = \frac{\frac{1}{3}-(-5)}{-1-3} = \frac{\frac{16}{3}}{-4} = -\frac{4}{3}$$

The lines are perpendicular.

44. (a) $m = 400$. The revenues are increasing \$400 per day.

(b) $m = 100$. The revenues are increasing \$100 per day.

(c) $m = 0$. There is no change in revenue. (Revenue remains constant.)

46. (a) The greatest increase in the graph is between 1998 and 1999. The least increase in the graph is between 1996 and 1997.

(b) Slope $= \dfrac{135.0-5.8}{11-1} = 12.92$

(c) Each year the net profit increases by \$12.92 million.

48. $y = \frac{6}{100}x$

$y = \frac{6}{100}(200) = 12$ feet

50. $(3, 156), m = 4.50$

$V - 156 = 4.50(t - 3)$

$V - 156 = 4.50t - 13.50$

$V = 4.5t + 142.5$

52. The y-intercept is 8.5 and the slope is 2, which represents the increase in hourly wage per unit produced. Matches graph (c).

54. The y-intercept is 750 and the slope is -100, which represents the decrease in the value of the word processor each year. Matches graph (d).

56. $m = -1, (0, 10)$

$y - 10 = -1(x - 0)$

$y - 10 = -x$

$y = -x + 10$

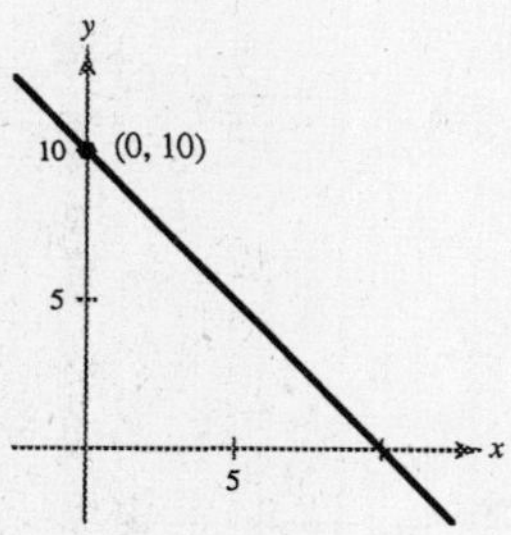

58. $m = 4, (0, 0)$

$y - 0 = 4(x - 0)$

$y = 4x$

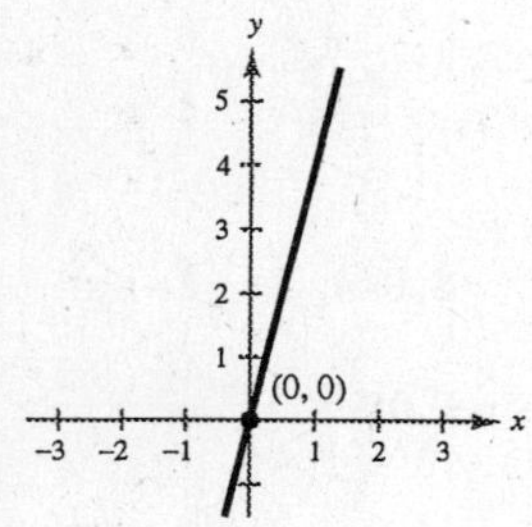

60. $m = \frac{3}{4}, (-2, -5)$

$y + 5 = \frac{3}{4}(x + 2)$

$4y + 20 = 3x + 6$

$4y = 3x - 14$

$y = \frac{3}{4}x - \frac{7}{2}$

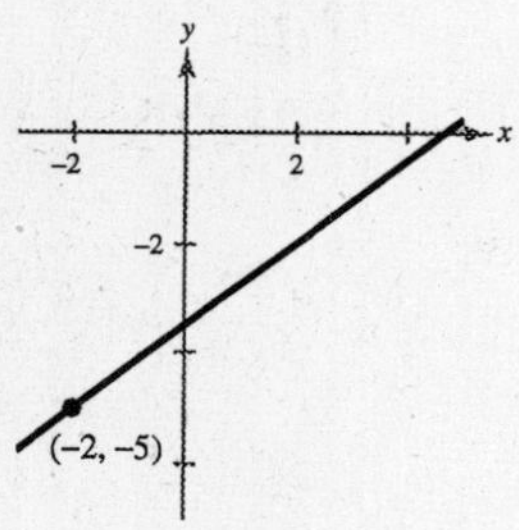

62. m is undefined, $(-10, 4)$

Because the slope is undefined, the line is a vertical line passing through $x = -10$, which is the equation.

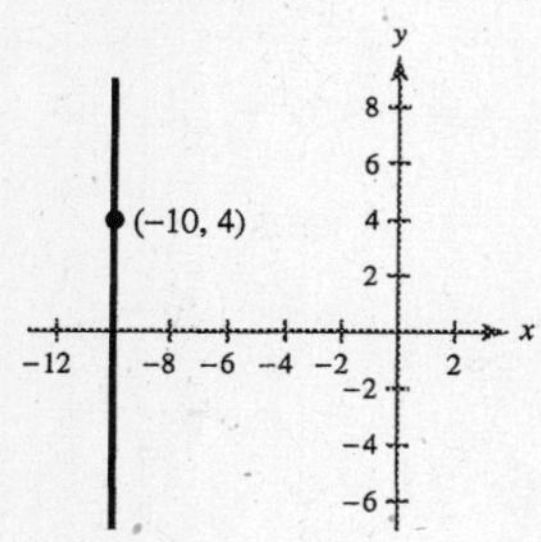

64. $m = 0, \left(-\frac{1}{2}, \frac{3}{2}\right)$

$y - \frac{3}{2} = 0\left(x + \frac{1}{2}\right)$

$y - \frac{3}{2} = 0$

$y = \frac{3}{2}$

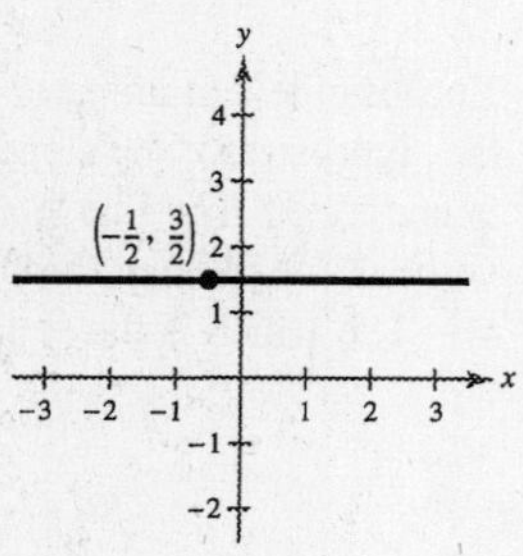

66. $m = -\frac{5}{2}, (2.3, -8.5)$

$y - (-8.5) = -\frac{5}{2}(x - 2.3)$

$y + 8.5 = -2.5x + 5.75$

$y = -2.5x - 2.75$

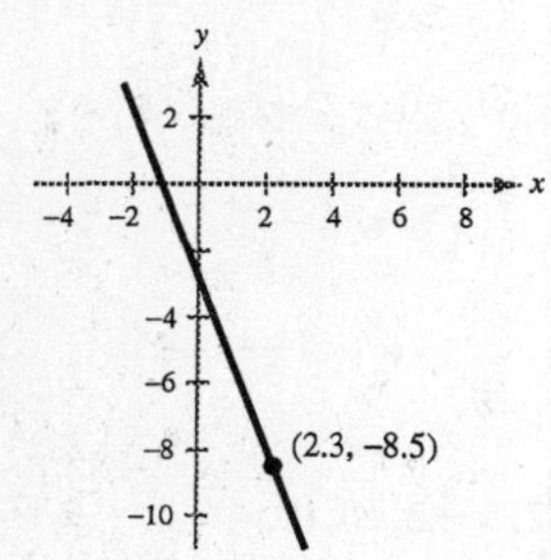

68. $(4, 3), (-4, -4)$

$y - 3 = \frac{-4 - 3}{-4 - 4}(x - 4)$

$y - 3 = \frac{7}{8}(x - 4)$

$y - 3 = \frac{7}{8}x - \frac{7}{2}$

$y = \frac{7}{8}x - \frac{1}{2}$

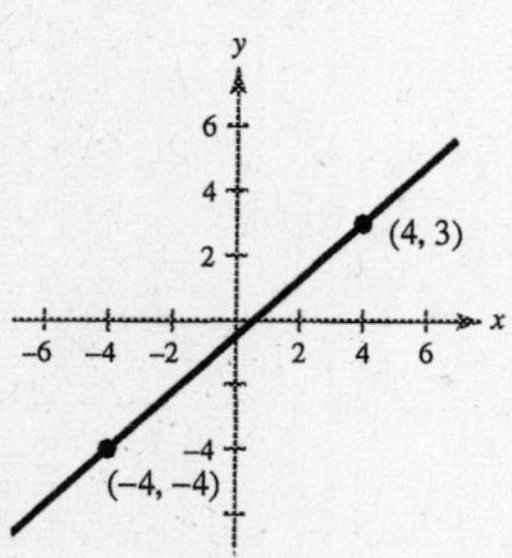

70. $(-1, 4), (6, 4)$

$y - 4 = \frac{4 - 4}{6 - (-1)}(x + 1)$

$y - 4 = 0(x + 1)$

$y - 4 = 0$

$y = 4$

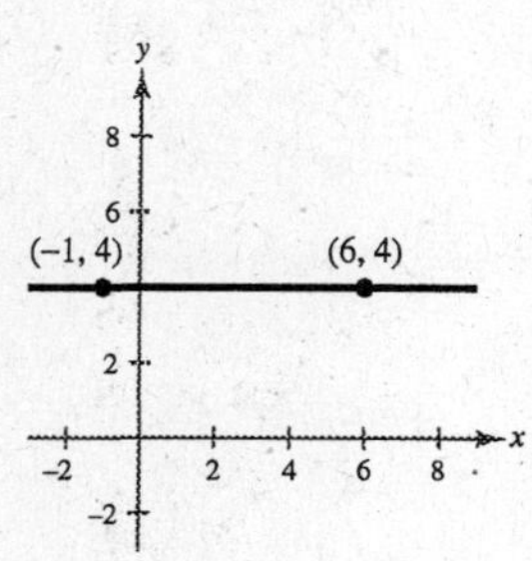

72. $(1, 1), \left(6, -\frac{2}{3}\right)$

$$y - 1 = \frac{-\frac{2}{3} - 1}{6 - 1}(x - 1)$$

$$y - 1 = -\frac{1}{3}(x - 1)$$

$$y - 1 = -\frac{1}{3}x + \frac{1}{3}$$

$$y = -\frac{1}{3}x + \frac{4}{3}$$

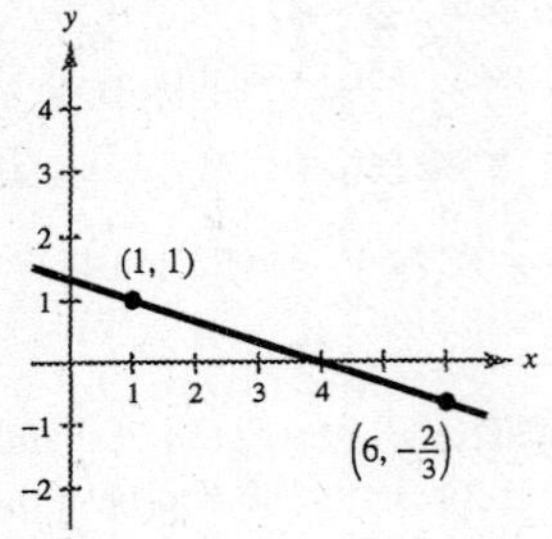

74. $\left(\frac{3}{4}, \frac{3}{2}\right), \left(-\frac{4}{3}, \frac{7}{4}\right)$

$$y - \frac{3}{2} = \frac{\frac{7}{4} - \frac{3}{2}}{-\frac{4}{3} - \frac{3}{4}}\left(x - \frac{3}{4}\right)$$

$$y - \frac{3}{2} = \frac{\frac{1}{4}}{-\frac{25}{12}}\left(x - \frac{3}{4}\right)$$

$$y - \frac{3}{2} = -\frac{3}{25}\left(x - \frac{3}{4}\right)$$

$$y - \frac{3}{2} = -\frac{3}{25}x + \frac{9}{100}$$

$$y = -\frac{3}{25}x + \frac{159}{100}$$

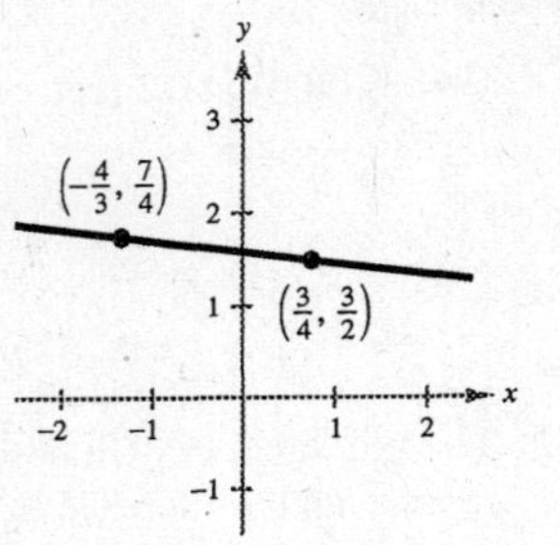

76. $(-8, 0.6), (2, -2.4)$

$$y - 0.6 = \frac{-2.4 - 0.6}{2 - (-8)}(x + 8)$$

$$y - 0.6 = -\frac{3}{10}(x + 8)$$

$$10y - 6 = -3(x + 8)$$

$$10y - 6 = -3x - 24$$

$$10y = -3x - 18$$

$$y = -\frac{3}{10}x - \frac{9}{5} \quad \text{or} \quad y = -0.3x - 1.8$$

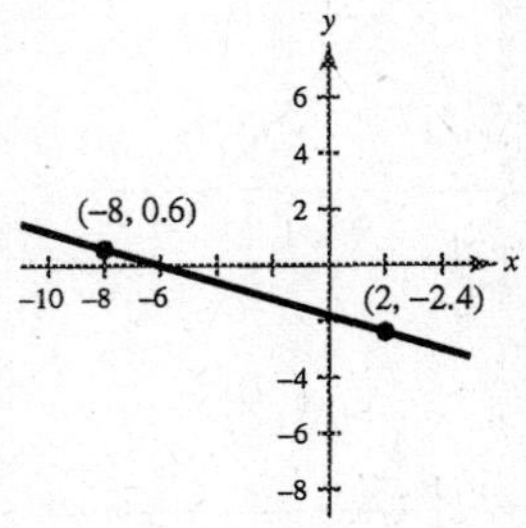

78. $\left(\frac{1}{5}, -2\right), (-6, -2)$

$$y + 2 = \frac{-2 - (-2)}{-6 - \frac{1}{5}}(x + 6)$$

$$y + 2 = \frac{0}{-6 - \frac{1}{5}}(x + 6)$$

$$y + 2 = 0$$

$$y = -2$$

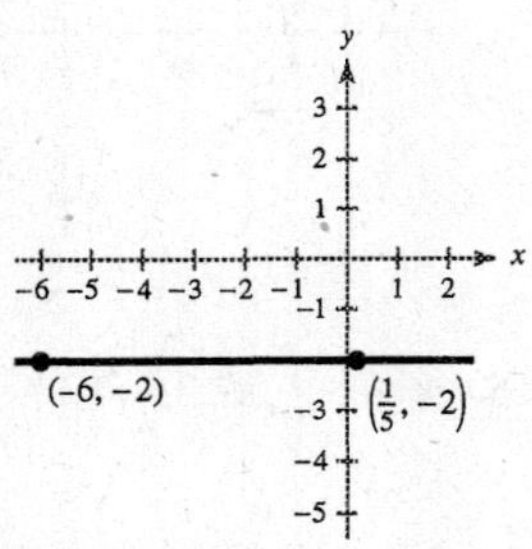

80. $(1.5, -2), (1.5, 0.2)$

$$y + 2 = \frac{-2 - 0.2}{1.5 - 1.5}(x - 1.5)$$

$$y + 2 = \frac{-2 - 0.2}{0}(x - 1.5)$$

Because the denominator on the right is zero, the slope is is undefined. The line is a vertical line passing through $x = 1.5$, which is the equation.

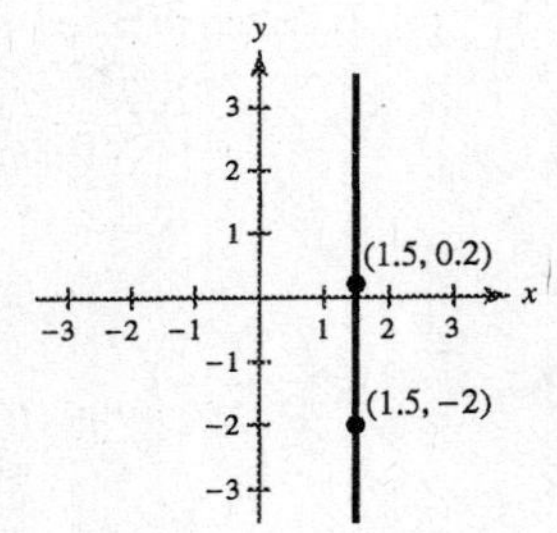

82. $(-3, 0), (0, 4)$

$$\frac{x}{-3} + \frac{y}{4} = 1$$

$$(-12)\frac{x}{-3} + (-12)\frac{y}{4} = (-12) \cdot 1$$

$$4x - 3y + 12 = 0$$

84. $\left(\frac{2}{3}, 0\right), (0, -2)$

$$\frac{x}{\frac{2}{3}} + \frac{y}{-2} = 1$$

$$\frac{3x}{2} - \frac{y}{2} = 1$$

$$3x - y - 2 = 0$$

86. $(d, 0), (0, d), (-3, 4)$

$$\frac{x}{d} + \frac{y}{d} = 1$$

$$x + y = d$$

$$-3 + 4 = d$$

$$1 = d$$

$$x + y = 1$$

$$x + y - 1 = 0$$

88. $x + y = 7$

$y = -x + 7$

Slope: $m = -1$

(a) $m = -1, (-3, 2)$

$y - 2 = -1(x + 3)$

$y - 2 = -x - 3$

$y = -x - 1$

(b) $m = 1, (-3, 2)$

$y - 2 = 1(x + 3)$

$y = x + 5$

90. $5x + 3y = 0$

$3y = -5x$

$y = -\frac{5}{3}x$

Slope: $m = -\frac{5}{3}$

(a) $m = -\frac{5}{3}, \left(\frac{7}{8}, \frac{3}{4}\right)$

$y - \frac{3}{4} = -\frac{5}{3}\left(x - \frac{7}{8}\right)$

$24y - 18 = -40\left(x - \frac{7}{8}\right)$

$24y - 18 = -40x + 35$

$24y = -40x + 53$

$y = -\frac{5}{3}x + \frac{53}{24}$

(b) $m = \frac{3}{5}, \left(\frac{7}{8}, \frac{3}{4}\right)$

$y - \frac{3}{4} = \frac{3}{5}\left(x - \frac{7}{8}\right)$

$40y - 30 = 24\left(x - \frac{7}{8}\right)$

$40y - 30 = 24x - 21$

$40y = 24x + 9$

$y = \frac{3}{5}x + \frac{9}{40}$

92. $y = 1$

Slope: $m = 0$

(a) $m = 0, (4, -2)$

$y + 2 = 0(x - 4)$

$y + 2 = 0$

$y = -2$

(b) The reciprocal of 0 is undefined. The line is vertical, passing through $(4, -2)$. The equation is $x = 4$.

94. $x = -2$

Slope: undefined

(a) The original line is the vertical line through $x = -2$. The line parallel to this line containing $(-5, 1)$ is the vertical line $x = -5$.

(b) A perpendicular to a vertical line is a horizontal line, whose slope is 0. The horizontal line containing $(-5, 1)$ is the line $y = 1$.

96. $6x + 2y = 9$

$2y = -6x + 9$

$y = -3x + \frac{9}{2}$

Slope: $m = -3$

(a) $(-3.9, -1.4), m = -3$

$y - (-1.4) = -3(x - (-3.9))$

$y + 1.4 = -3x - 11.7$

$y = -3x - 13.1$

(b) $(-3.9, -1.4), m = \frac{1}{3}$

$y - (-1.4) = \frac{1}{3}(x - (-3.9))$

$y + 1.4 = \frac{1}{3}x + 1.3$

$y = \frac{1}{3}x - 0.1$

98. (a) $y = \frac{2}{3}x$ (b) $y = -\frac{3}{2}x$ (c) $y = \frac{2}{3}x + 2$

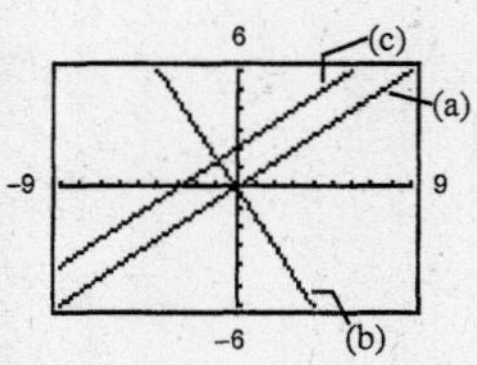

(a) is parallel to (c). (b) is perpendicular to (a) and (c).

100. (a) $y = x - 8$ (b) $y = x + 1$ (c) $y = -x + 3$

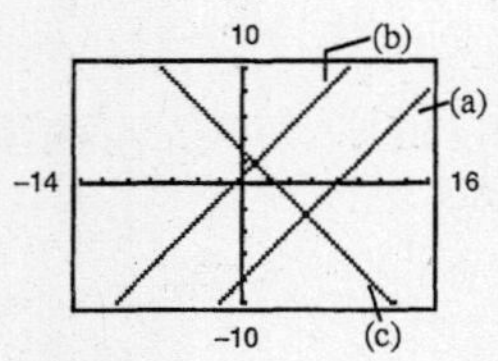

(a) is parallel to (b). (c) is perpendicular to (a) and (b).

102. Set the distance between (6, 5) and (x, y) equal to the distance between $(1, -8)$ and (x, y).

$$\sqrt{(x-6)^2+(y-5)^2} = \sqrt{(x-1)^2+(y-(-8))^2}$$
$$(x-6)^2+(y-5)^2 = (x-1)^2+(y+8)^2$$
$$x^2-12x+36+y^2-10y+25 = x^2-2x+1+y^2+16y+64$$
$$x^2+y^2-12x-10y+61 = x^2+y^2-2x+16y+65$$
$$-12x-10y+61 = -2x+16y+65$$
$$-10x-26y-4=0$$
$$-2(5x+13y+2)=0$$
$$5x+13y+2=0$$

104. Set the distance between $\left(-\frac{1}{2}, -4\right)$ and (x, y) equal to the distance between $\left(\frac{7}{2}, \frac{5}{4}\right)$ and (x, y).

$$\sqrt{\left(x-\left(-\tfrac{1}{2}\right)\right)^2+(y-(-4))^2} = \sqrt{\left(x-\tfrac{7}{2}\right)^2+\left(y-\tfrac{5}{4}\right)^2}$$
$$\left(x+\tfrac{1}{2}\right)^2+(y+4)^2 = \left(x-\tfrac{7}{2}\right)^2+\left(y-\tfrac{5}{4}\right)^2$$
$$x^2+x+\tfrac{1}{4}+y^2+8y+16 = x^2-7x+\tfrac{49}{4}+y^2-\tfrac{5}{2}y+\tfrac{25}{16}$$
$$x^2+y^2+x+8y+\tfrac{65}{4} = x^2+y^2-7x-\tfrac{5}{2}y+\tfrac{221}{16}$$
$$x+8y+\tfrac{65}{4} = -7x-\tfrac{5}{2}y+\tfrac{221}{16}$$
$$8x+\tfrac{21}{2}y+\tfrac{39}{16}=0$$
$$128x+168y+39=0$$

106. $t = 0$ represents 1996

(0, 3927) and (4, 3800)

$$m = \frac{3800-3927}{4-0} = \frac{-127}{4} = -31.75$$

$N = -31.75t + 3927$

$t = 9$ represents 2005: $N = -31.75(9) + 3927 \approx 3641$ stores.

$t = 14$ represents 2010: $N = -31.75(14) + 3927 \approx 3483$ stores.

108. Let $t = 0$ represent 1999.

(0, 27,913) and (2, 28,197)

$$m = \frac{28{,}197 - 27{,}913}{2-0} = 142$$

$N = 142t + 27{,}913$

$t = 6$ represents 2005: $N = 142(6) + 27{,}913 = 28{,}765$ students.

110. (0, 25,000) and (10, 2000)

$$m = \frac{2000-25000}{10-0} = -2300$$

$V = -2300t + 25{,}000, \quad 0 \le t \le 10$

112. $W = 0.75x + 11.50$

114. (580, 50) and (625, 47)

(a) $m = \dfrac{47 - 50}{625 - 580} = \dfrac{-3}{45} = -\dfrac{1}{15}$

$$x - 50 = -\frac{1}{15}(p - 580)$$

$$x - 50 = -\frac{1}{15}p + \frac{116}{3}$$

$$x = -\frac{1}{15}p + \frac{266}{3}$$

(b) $x = -\dfrac{1}{15}(655) + \dfrac{266}{3} = 45$ units

(c) $x = -\dfrac{1}{15}(595) + \dfrac{266}{3} = 49$ units

116. $W = 0.07S + 2500$

118. Two approximate points on the line are (6, 1175) and (12, 2380).

$$m = \frac{2380 - 1175}{12 - 6} \approx 201$$

$$y - 2380 = 201(t - 12)$$

$$y - 2380 = 201t - 2412$$

$$y = 201t - 32$$

This answer may vary depending on the points and estimates used.

120. (a) and (b)

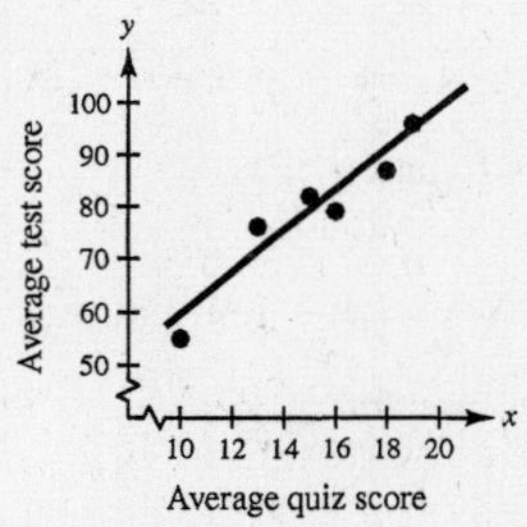

(c) Two approximate points on the line are (10, 59) and (19, 95).

$$m = \frac{95 - 59}{19 - 10} = 4$$

$$y - 59 = 4(x - 10)$$

$$y - 59 = 4x - 40$$

$$y = 4x + 19$$

(d) $y = 4(17) + 19 = 87$

(e) Each point will shift four units upward, so the best-fitting line will move four units upward. The slope remains the same, as the new line is parallel to the old, but the y-intercept becomes $(0, 19 + 4) = (0, 23)$, so the new equation is $y = 4x + 23$.

122. $(-8, 2)$ and $(-1, 4) : m_1 = \dfrac{4 - 2}{-1 - (-8)} = \dfrac{2}{7}$

$(0, -4)$ and $(-7, 7) : m_2 = \dfrac{7 - (-4)}{-7 - 0} = \dfrac{11}{-7}$

False, the lines are not parallel.

124. On a vertical line, all the points have the same x-value, so when you evaluate $m = \dfrac{y_2 - y_1}{x_2 - x_1}$, you would have a zero in the denominator, and division by zero is undefined.

126. Since $|-4| > \left|\frac{5}{2}\right|$, the steeper line is the one with a slope of -4. The slope with the greatest magnitude corresponds to the steepest line.

128. No, the slopes of two perpendicular lines have opposite signs. (Assume that neither line is vertical or horizontal.)

Section P.5 Functions

Solutions to Even-Numbered Exercises

2. No, it is not a function. The domain value of -1 is matched with two output values.

4. Yes, it is a function. Each domain value is matched with only one range value.

6. No, the table does not represent a function. The input values of 0 and 1 are each matched with two different output values.

8. Yes, the table does represent a function. Each input value is matched with only one output value.

10. (a) The element c in A is matched with two elements, 2 and 3 of B, so it is not a function.

(b) Each element of A is matched with exactly one element of B, so it does represent a function.

(c) This is not a function from A to B (it represents a function from B to A instead).

(d) Each element of A is matched with exactly one element of B, so it does represent a function.

12. Reading from the graph, $f(1998)$ is approximately 11 million.

14. $x = y^2 \Rightarrow y = \pm\sqrt{x}$

Thus, y is *not* a function of x.

16. $x + y^2 = 4 \Rightarrow y = \pm\sqrt{4 - x}$

Thus, y is *not* a function of x.

18. $(x - 2)^2 + y^2 = 4 \Rightarrow y = \pm\sqrt{4 - (x - 2)^2}$

Thus, y is *not* a function of x.

20. $y = \sqrt{x + 5}$

This is a function of x.

22. $|y| = 4 - x \Rightarrow y = 4 - x$ or $y = -(4 - x)$

Thus, y is *not* a function of x.

24. $g(y) = 7 - 3y$

(a) $g(0) = 7 - 3(0) = 7$

(b) $g\left(\frac{7}{3}\right) = 7 - 3\left(\frac{7}{3}\right) = 0$

(c) $g(s + 2) = 7 - 3(s + 2)$

$= 7 - 3s - 6 = 1 - 3s$

26. $h(t) = t^2 - 2t$

(a) $h(2) = 2^2 - 2(2) = 0$

(b) $h(1.5) = (1.5)^2 - 2(1.5) = -0.75$

(c) $h(x + 2) = (x + 2)^2 - 2(x + 2) = x^2 + 2x$

28. $f(x) = \sqrt{x + 8} + 2$

(a) $f(-8) = \sqrt{(-8) + 8} + 2 = 2$

(b) $f(1) = \sqrt{(1) + 8} + 2 = 5$

(c) $f(x - 8) = \sqrt{(x - 8) + 8} + 2 = \sqrt{x} + 2$

30. $q(t) = \dfrac{2t^2 + 3}{t^2}$

(a) $q(2) = \dfrac{2(2)^2 + 3}{(2)^2} = \dfrac{8 + 3}{4} = \dfrac{11}{4}$

(b) $q(0) = \dfrac{2(0)^2 + 3}{(0)^2}$

Division by zero is undefined.

(c) $q(-x) = \dfrac{2(-x)^2 + 3}{(-x)^2} = \dfrac{2x^2 + 3}{x^2}$

32. $f(x) = |x| + 4$

(a) $f(2) = |2| + 4 = 6$

(b) $f(-2) = |-2| + 4 = 6$

(c) $f(x^2) = |x^2| + 4 = x^2 + 4$

34. $f(x) = \begin{cases} x^2 + 2, & x \le 1 \\ 2x^2 + 2, & x > 1 \end{cases}$

(a) $f(-2) = (-2)^2 + 2 = 6$

(b) $f(1) = (1)^2 + 2 = 3$

(c) $f(2) = 2(2)^2 + 2 = 10$

36. $f(x) = \begin{cases} 4 - 5x, & x \le -2 \\ 0, & -2 < x \le 2 \\ x^2 + 1 & x > 2 \end{cases}$

(a) $f(-3) = 4 - 5(-3) = 19$

(b) $f(4) = (4)^2 + 1 = 17$

(c) $f(-1) = 0$

38. $g(x) = \sqrt{x - 3}$

$g(3) = \sqrt{3 - 3} = 0$

$g(4) = \sqrt{4 - 3} = 1$

$g(5) = \sqrt{5 - 3} = \sqrt{2}$

$g(6) = \sqrt{6 - 3} = \sqrt{3}$

$g(7) = \sqrt{7 - 3} = 2$

x	3	4	5	6	7
$g(x)$	0	1	$\sqrt{2}$	$\sqrt{3}$	2

40. $f(s) = \dfrac{|s - 2|}{s - 2}$

$$f(0) = \frac{|0 - 2|}{0 - 2} = \frac{2}{-2} = -1$$

$$f(1) = \frac{|1 - 2|}{1 - 2} = \frac{1}{-1} = -1$$

$$f\left(\frac{3}{2}\right) = \frac{\left|\frac{3}{2} - 2\right|}{\frac{3}{2} - 2} = \frac{\frac{1}{2}}{-\frac{1}{2}} = -1$$

$$f\left(\frac{5}{2}\right) = \frac{\left|\frac{5}{2} - 2\right|}{\frac{5}{2} - 2} = \frac{\frac{1}{2}}{\frac{1}{2}} = 1$$

$$f(4) = \frac{|4 - 2|}{4 - 2} = \frac{2}{2} = 1$$

s	0	1	$\frac{3}{2}$	$\frac{5}{2}$	4
$f(s)$	-1	-1	-1	1	1

42. $h(x) = \begin{cases} 9 - x^2, & x < 3 \\ x - 3, & x \ge 3 \end{cases}$

$h(1) = 9 - (1)^2 = 8$

$h(2) = 9 - (2)^2 = 5$

$h(3) = (3) - 3 = 0$

$h(4) = (4) - 3 = 1$

$h(5) = (5) - 3 = 2$

s	1	2	3	4	5
$h(x)$	8	5	0	1	2

44.
$$\begin{aligned} f(x) &= 5x + 1 \\ 5x + 1 &= 0 \\ x &= -\frac{1}{5} \end{aligned}$$

46.
$$\begin{aligned} f(x) &= \frac{12 - x^2}{5} \\ \frac{12 - x^2}{5} &= 0 \\ x^2 &= 12 \\ x &= \pm\sqrt{12} = \pm 2\sqrt{3} \end{aligned}$$

48.
$$\begin{aligned} f(x) &= x^2 - 8x + 15 \\ x^2 - 8x + 15 &= 0 \\ (x - 5)(x - 3) &= 0 \\ x - 5 = 0 &\Rightarrow x = 5 \\ x - 3 = 0 &\Rightarrow x = 3 \end{aligned}$$

50.
$$\begin{aligned} f(x) &= x^3 - x^2 - 4x + 4 \\ x^3 - x^2 - 4x + 4 &= 0 \\ x^2(x - 1) - 4(x - 1) &= 0 \\ (x - 1)(x^2 - 4) &= 0 \\ x - 1 = 0 &\Rightarrow x = 1 \\ x^2 - 4 = 0 &\Rightarrow x = \pm 2 \end{aligned}$$

52.
$$\begin{aligned} f(x) &= g(x) \\ x^4 - 2x^2 &= 2x^2 \\ x^4 - 4x^2 &= 0 \\ x^2(x^2 - 4) &= 0 \\ x^2(x + 2)(x - 2) &= 0 \\ x^2 = 0 &\Rightarrow x = 0 \\ x + 2 = 0 &\Rightarrow x = -2 \\ x - 2 = 0 &\Rightarrow x = 2 \end{aligned}$$

54.
$$f(x) = g(x)$$
$$\sqrt{x} - 4 = 2 - x$$
$$x + \sqrt{x} - 6 = 0$$
$$(\sqrt{x} + 3)(\sqrt{x} - 2) = 0$$
$\sqrt{x} + 3 = 0 \Rightarrow \sqrt{x} = -3$, which is a contradiction, since $\sqrt{x}$ represents the principal square root.

$\sqrt{x} - 2 = 0 \Rightarrow \sqrt{x} = 2 \Rightarrow x = 4$

56. $f(x) = 1 - 2x^2$

Because $f(x)$ is a polynomial, the domain is all real numbers x.

58. $s(y) = \dfrac{3y}{y + 5}$

$y + 5 \neq 0$

$y \neq -5$

The domain is all real numbers $y \neq -5$.

60. $f(t) = \sqrt[3]{t + 4}$

Because $f(t)$ is a cube root, the domain is all real numbers t.

62. $f(x) = \sqrt[4]{x^2 + 3x}$

$x^2 + 3x \geq 0$

$x(x + 3) \geq 0$

Test intervals:
$(-\infty, -3), (-3, 0), (0, \infty)$

The domain is all real numbers $x \leq -3, x \geq 0$.

64. $h(x) = \dfrac{10}{x^2 - 2x}$

$x^2 - 2x \neq 0$

$x(x - 2) \neq 0$

The domain is all real numbers $x \neq 0$ and $x \neq 2$.

66. $f(x) = \dfrac{\sqrt{x + 6}}{6 + x}$

$6 + x \neq 0$

$x \neq -6$

The domain is all real numbers $x \neq -6$.

68. $f(x) = \dfrac{x - 5}{\sqrt{x^2 - 9}}$

$x^2 - 9 > 0$

$(x + 3)(x - 3) > 0$

Test intervals: $(-\infty, -3), (-3, 3), (3, \infty)$

The domain is all real numbers $x < -3$ or $x > 3$.

70. $f(x) = x^2 - 3$

$f(-2) = (-2)^2 - 3 = 1$

$f(-1) = (-1)^2 - 3 = -2$

$f(0) = (0)^2 - 3 = -3$

$f(1) = (1)^2 - 3 = -2$

$f(2) = (2)^2 - 3 = 1$

$\{(-2, 1), (-1, -2), (0, -3), (1, -2), (2, 1)\}$

72. $f(x) = |x + 1|$

$\{(-2, 1), (-1, 0), (0, 1), (1, 2), (2, 3)\}$

74. By plotting the data, you can see that they represent a line, or $f(x) = cx$. Because $(0, 0)$ and $\left(1, \frac{1}{4}\right)$ are on the line, the slope is $\frac{1}{4}$. Thus, $f(x) = \frac{1}{4}x$.

76. By plotting the data, you can see that they represent $h(x) = c\sqrt{|x|}$. Because $\sqrt{|-4|} = 2$ and $\sqrt{|-1|} = 1$, and the corresponding y values are 6 and 3, $c = 3$ and $h(x) = 3\sqrt{|x|}$.

78.
$$\begin{aligned}
f(x) &= 5x - x^2\\
f(5+h) &= 5(5+h) - (5+h)^2\\
&= 25 + 5h - (25 + 10h + h^2)\\
&= 25 + 5h - 25 - 10h - h^2\\
&= -h^2 - 5h\\
f(5) &= 5(5) - (5)^2\\
&= 25 - 25 = 0\\
\frac{f(5+h) - f(5)}{h} &= \frac{-h^2 - 5h}{h}\\
&= \frac{-h(h+5)}{h} = -(h+5)
\end{aligned}$$

80.
$$\begin{aligned}
f(x) &= 4x^2 - 2x\\
f(x+h) &= 4(x+h)^2 - 2(x+h)\\
&= 4(x^2 + 2xh + h^2) - 2x - 2h\\
&= 4x^2 + 8xh + 4h^2 - 2x - 2h\\
\frac{f(x+h) - f(x)}{h} &= \frac{4x^2 + 8xh + 4h^2 - 2x - 2h - 4x^2 + 2x}{h}\\
&= \frac{8xh + 4h^2 - 2h}{h}\\
&= \frac{h(8x + 4h - 2)}{h}\\
&= 8x + 4h - 2, \quad h \neq 0
\end{aligned}$$

82.
$$\begin{aligned}
f(t) &= \frac{1}{t-2}\\
f(1) &= \frac{1}{1-2}\\
&= -1\\
\frac{f(t) - f(1)}{t-1} &= \frac{\frac{1}{t-2} - (-1)}{t-1}\\
&= \frac{1 + (t-2)}{(t-2)(t-1)}\\
&= \frac{(t-1)}{(t-2)(t-1)}\\
&= \frac{1}{t-2}, \quad t \neq 1
\end{aligned}$$

84.
$$\begin{aligned}
f(x) &= x^{2/3} + 1\\
f(8) &= 8^{2/3} + 1 = 5\\
\frac{f(x) - f(8)}{x-8} &= \frac{x^{2/3} + 1 - 5}{x-8} = \frac{x^{2/3} - 4}{x-8}
\end{aligned}$$

86. $A = \pi r^2, \quad C = 2\pi r$

$$r = \frac{C}{2\pi}$$

$$A = \pi\left(\frac{C}{2\pi}\right)^2 = \frac{C^2}{4\pi}$$

88. (a) The maximum profit is \$3375.

(b)

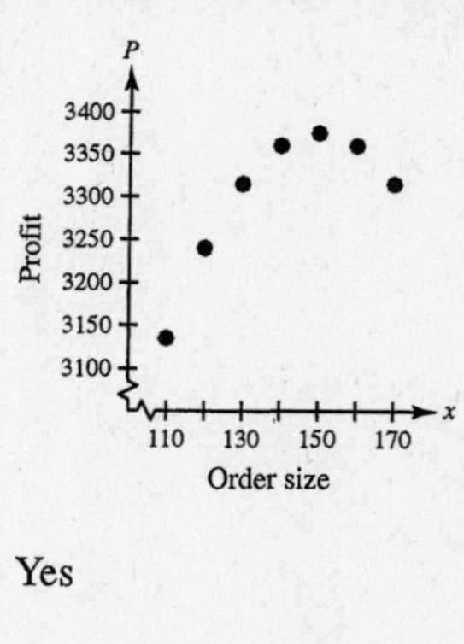

Yes

(c)
$$\begin{aligned}
\text{Profit} &= \text{Revenue} - \text{Cost}\\
&= (\text{price per unit})(\text{number of units}) - (\text{cost})(\text{number of units})\\
&= [90 - (x - 100)(0.15)]x - 60x, \quad x > 100\\
&= (90 - 0.15x + 15)x - 60x\\
&= (105 - 0.15x)x - 60x\\
&= 105x - 0.15x^2 - 60x\\
&= 45x - 0.15x^2, \quad x > 100
\end{aligned}$$

90. $A = l \cdot w = (2x)y = 2xy$

But $y = \sqrt{36 - x^2}$, so $A = 2x\sqrt{36 - x^2}$, $0 < x < 6$.

92. (a) $V = l \cdot w \cdot h = x \cdot y \cdot x = x^2y$ where $4x + y = 108$.

Thus, $y = 108 - 4x$ and $V = x^2(108 - 4x) = 108x^2 - 4x^3$

Domain: $0 < x < 27$

(b)

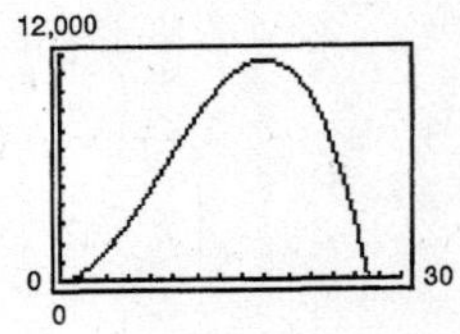

(c) The dimensions that will maximize the volume of the package are $18 \times 18 \times 36$. From the graph, the maximum volume occurs when $x = 18$. To find the dimension for y, use the equation $y = 108 - 4x$.

$y = 108 - 4x = 108 - 4(18) = 108 - 72 = 36.$

94. (a) *Model:* (Total cost) = (Fixed costs) + (Variable costs)

Labels: Total cost = C

Fixed cost = 6000

Variable costs = $0.95x$

Equation: $C = 6000 + 0.95x$

(b) $\overline{C} = \frac{C}{x} = \frac{6000 + 0.95x}{x}$

96. $F(y) = 149.76\sqrt{10}y^{5/2}$

(a)

y	5	10	20	30	40
$F(y)$	26,474.08	149,760.00	847,170.49	2,334,527.36	4,792,320

The force, in tons, of the water against the dam increases with the depth of the water.

(b) It appears that approximately 21 feet of water would produce 1,000,000 tons of force.

(c) $1{,}000{,}000 = 149.76\sqrt{10}y^{5/2}$

$$\frac{1{,}000{,}000}{149.76\sqrt{10}} = y^{5/2}$$

$$2111.56 \approx y^{5/2}$$

$$21.37 \approx y$$

98. $y = -\frac{1}{10}x^2 + 3x + 6$

$y(30) = -\frac{1}{10}(30)^2 + 3(30) + 6 = 6$ feet

If the child holds a glove at a height of 5 feet, then the ball *will* be over the child's head since it will be at a height of 6 feet.

100. False. The range is $(-1, \infty)$.

102. The domain is the set of inputs of the function, and the range is the set of outputs.

Section P.6 Analyzing Graphs of Functions

Solutions to Even-Numbered Exercises

2. Domain: $(-\infty, \infty)$

Range: $[0, \infty)$

4. Domain: $(-\infty, 1), (1, \infty)$

Range: $-1, 1$

6. (a) $f(-1) = 4$

(b) $f(2) = 4$

(c) $f(0) = 2$

(d) $f(1) = 0$

8. (a) $f(2) = 0$

(b) $f(1) = 1$

(c) $f(3) = 2$

(d) $f(-1) = 3$

10. $y = \frac{1}{4}x^3$

A vertical line intersects the graph no more than once, so y is a function of x.

12. $x^2 + y^2 = 25$

A vertical line intersects the graph more than once, so y is not a function of x.

14. $x = |y + 2|$

A vertical line intersects the graph more than once, so y is not a function of x.

16.

$$f(x) = 3x^2 + 22x - 16$$
$$0 = (3x - 2)(x + 8)$$
$$3x - 2 = 0 \Rightarrow x = \tfrac{2}{3}$$
$$x + 8 = 0 \Rightarrow x = -8$$

18.

$$f(x) = \frac{x^2 - 9x + 14}{4x}$$
$$0 = \frac{x^2 - 9x + 14}{4x}$$
$$0 = (x - 7)(x - 2)$$
$$x - 7 = 0 \Rightarrow x = 7$$
$$x - 2 = 0 \Rightarrow x = 2$$

20.

$$f(x) = x^3 - 4x^2 - 9x + 36$$
$$0 = x^3 - 4x^2 - 9x + 36$$
$$0 = x^2(x - 4) - 9(x - 4)$$
$$0 = (x - 4)(x^2 - 9)$$
$$x - 4 = 0 \Rightarrow x = 4$$
$$x^2 - 9 = 0 \Rightarrow x = \pm 3$$

22.

$$f(x) = 9x^4 - 25x^2$$
$$0 = 9x^4 - 25x^2$$
$$0 = x^2(9x^2 - 25)$$
$$x^2 = 0 \Rightarrow x = 0$$
$$9x^2 - 25 = 0 \Rightarrow x = \pm\tfrac{5}{3}$$

24.

$$f(x) = \sqrt{3x + 2}$$
$$0 = \sqrt{3x + 2}$$
$$0 = 3x + 2$$
$$-\tfrac{2}{3} = x$$

26.

$$f(x) = x(x - 7)$$
$$0 = x(x - 7)$$
$$x = 0$$
$$x - 7 = 0 \Rightarrow x = 7$$

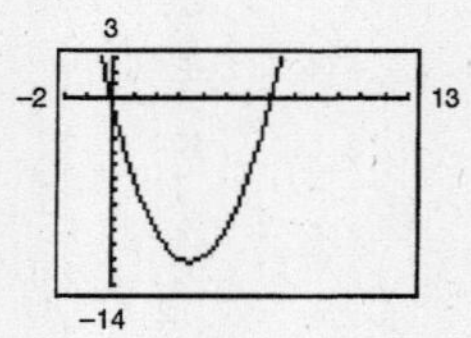

28.

$$f(x) = \sqrt{3x - 14} - 8$$
$$0 = \sqrt{3x - 14} - 8$$
$$8 = \sqrt{3x - 14}$$
$$64 = 3x - 14$$
$$x = 26$$

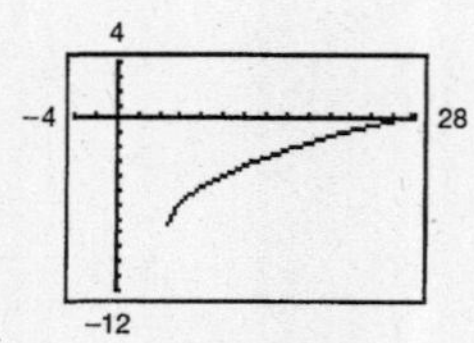

30.

$$f(x) = \frac{2x^2 - 9}{3 - x}$$
$$0 = \frac{2x^2 - 9}{3 - x}$$
$$2x^2 - 9 = 0 \Rightarrow x = \pm\frac{3\sqrt{2}}{2}$$

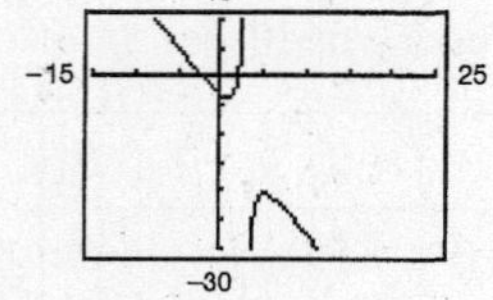

32. $f(x) = x^2 - 4x$

The graph is decreasing on $(-\infty, 2)$ and increasing on $(2, \infty)$.

34. $f(x) = \sqrt{x^2 - 1}$

The graph is decreasing on $(-\infty, -1)$ and increasing on $(1, \infty)$.

36. $f(x) = \begin{cases} 2x + 1, & x \le -1 \\ x^2 - 2, & x > -1 \end{cases}$

The graph is decreasing on $(-1, 0)$ and increasing on $(-\infty, -1)$ and $(0, \infty)$.

38. The graph is decreasing on $(-2, -1)$ and $(-1, 0)$ and increasing on $(-\infty, -2)$ and $(0, \infty)$.

40. $g(x) = x$

(a)

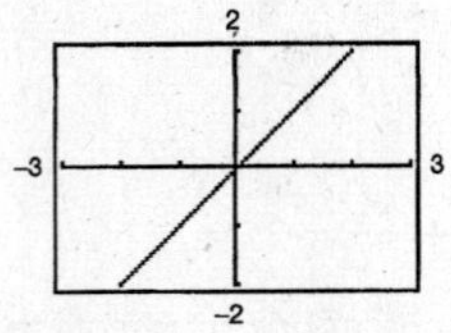

Increasing on $(-\infty, \infty)$

(b)

x	-2	-1	0	1	2
$g(x)$	-2	-1	0	1	2

42. $h(x) = x^2 - 4$

(a)

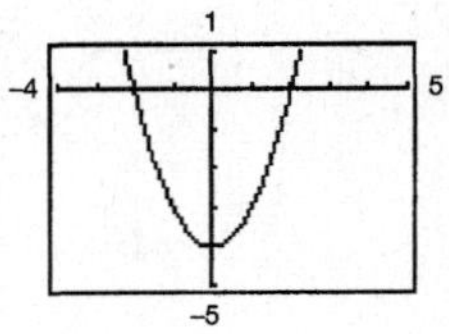

Decreasing on $(-\infty, 0)$; Increasing on $(0, \infty)$

(b)

x	-2	-1	0	1	2
$h(x)$	0	-3	-4	-3	0

44. $f(x) = 3x^4 - 6x^2$

(a)

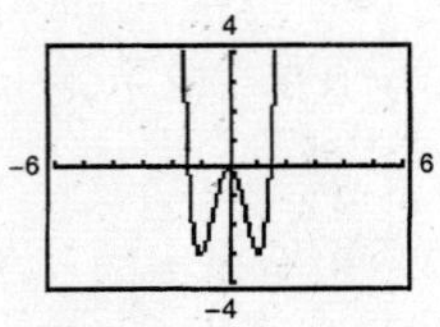

Increasing on $(-1, 0)$, $(1, \infty)$

Decreasing on $(-\infty, -1)$, $(0, 1)$

(b)

x	-2	-1	0	1	2
$f(x)$	24	-3	0	-3	24

46. $f(x) = x\sqrt{x + 3}$

(a)

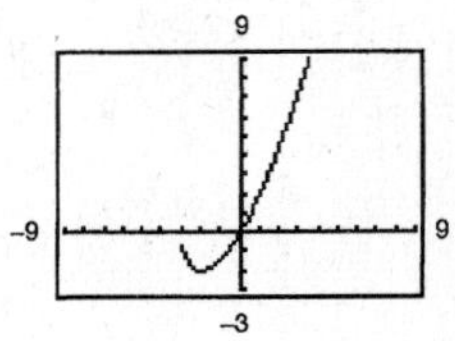

Increasing on $(-2, \infty)$

Decreasing on $(-3, -2)$

(b)

x	-3	-2	-1	0	1
$f(x)$	0	-2	-1.414	0	2

48. $f(x) = x^{2/3}$

(a)

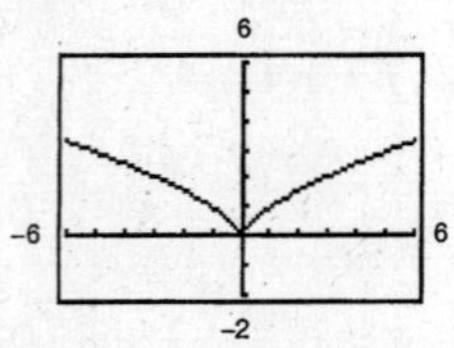

Decreasing on $(-\infty, 0)$

Increasing on $(0, \infty)$

(b)

x	-2	-1	0	1	2
$f(x)$	1.59	1	0	1	1.59

50. $f(x) = 3x^2 - 2x - 5$

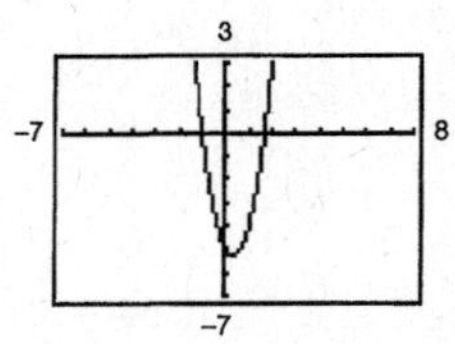

Relative minimum: $\left(\frac{1}{3}, -\frac{16}{3}\right)$

52. $f(x) = x^3 - 3x^2 - x + 1$

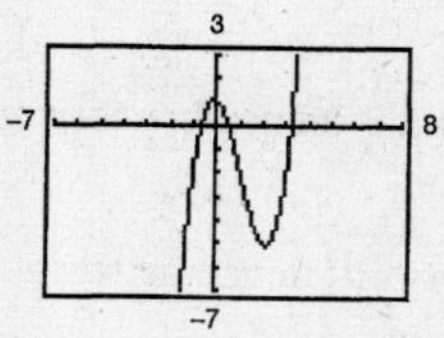

Relative maximum: $(-0.15, 1.08)$

Relative minimum: $(2.15, -5.08)$

54. $f(x) = 4x + 2$

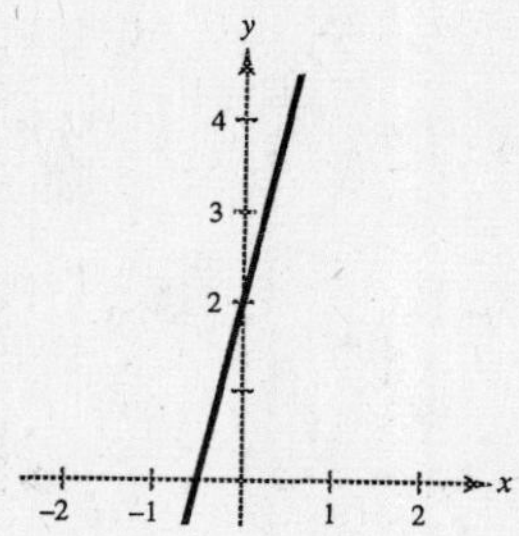

$$f(x) \geq 0$$
$$4x + 2 \geq 0$$
$$4x \geq -2$$
$$x \geq -\tfrac{1}{2}$$
$$\left[-\tfrac{1}{2}, \infty\right)$$

56. $f(x) = x^2 - 4x$

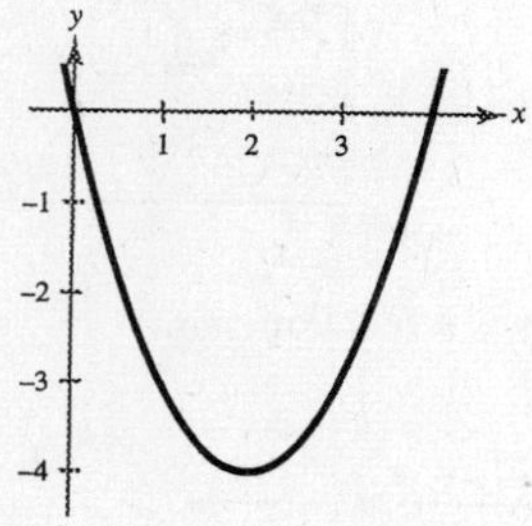

$$f(x) \geq 0$$
$$x^2 - 4x \geq 0$$
$$x(x - 4) \geq 0$$
$$(-\infty, 0], [4, \infty)$$

58. $f(x) = \sqrt{x + 2}$

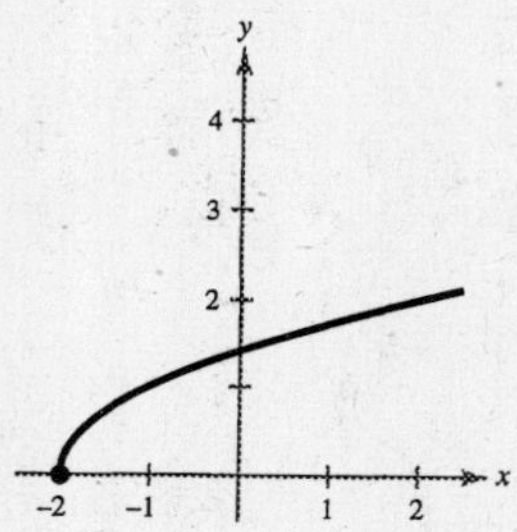

$$f(x) \geq 0$$
$$\sqrt{x + 2} \geq 0$$
$$x + 2 \geq 0$$
$$x \geq -2$$
$$[-2, \infty)$$

60. $f(x) = \frac{1}{2}(2 + |x|)$

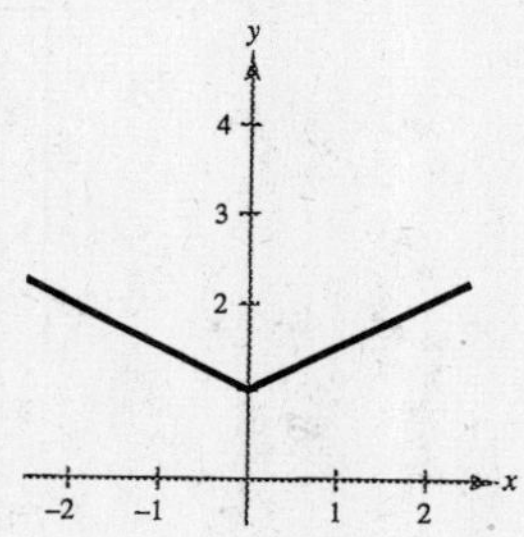

$f(x)$ is always greater than 0.
$(-\infty, \infty)$

62. $h(x) = x^3 - 5$

$$h(-x) = (-x)^3 - 5$$
$$= -x^3 - 5$$
$$\neq h(x)$$
$$\neq -h(x)$$

The function is neither odd nor even.

64. $f(x) = x\sqrt{1 - x^2}$

$$f(-x) = -x\sqrt{1 - (-x)^2}$$
$$= -x\sqrt{1 - x^2}$$
$$= -f(x)$$

The function is odd.

66. $g(s) = 4s^{2/3}$

$$g(-s) = 4(-s)^{2/3}$$
$$= 4s^{2/3}$$
$$= g(s)$$

The function is even.

68. $h = \text{top} - \text{bottom}$

$$= 3 - (4x - x^2)$$
$$= 3 - 4x + x^2$$

70. $h = \text{top} - \text{bottom}$

$$= 2 - \sqrt[3]{x}$$

72. $L = \text{right} - \text{left}$

$$= 2 - \sqrt[3]{2y}$$

74. $L = \text{right} - \text{left}$

$$= \frac{2}{y} - 0$$
$$= \frac{2}{y}$$

76. (a) and (e)

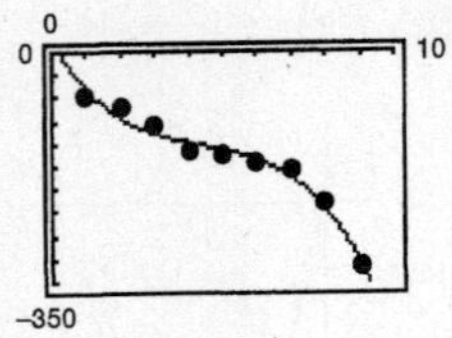

(b) y is a function of x.

(c) $y = -1.3767x^3 + 18.156x^2 - 89.40x + 16.5$

(d) $1 \le x \le 9$

(f) Most accurate in 1996 (least difference between actual value and model value)

Least accurate in 1992 (most difference between actual value and model value)

78.

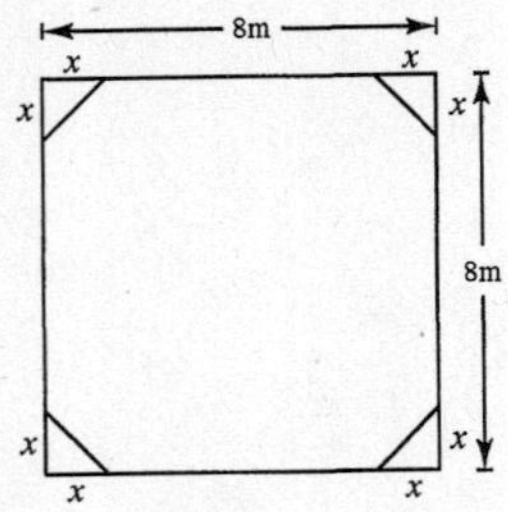

(a) $A = (8)(8) - 4\left(\frac{1}{2}\right)(x)(x)$

$= 64 - 2x^2$

Domain: $0 \le x \le 4$

(b)

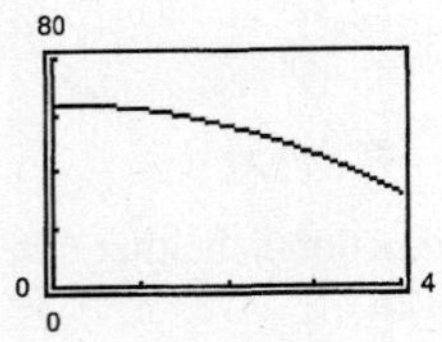

Range: $32 \le A \le 64$

(c) When $x = 4$, the resulting figure is a square.

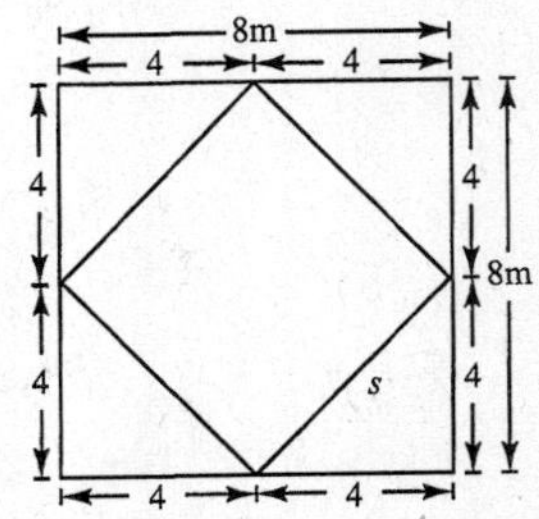

By the Pythagorean Theorem,
$4^2 + 4^2 = s^2 \implies s = \sqrt{32} = 4\sqrt{2}$ meters.

80. False. An odd function is symmetric with respect to the origin, so its domain must include negative values.

82. Yes, the graph in Exercise 11 does represent x as a function of y. Each y-value corresponds to only one x-value.

84. $\left(-\frac{5}{3}, -7\right)$

(a) If f is even, another point is $\left(\frac{5}{3}, -7\right)$.

(b) If f is odd, another point is $\left(\frac{5}{3}, 7\right)$.

86. $(5, -1)$

(a) If f is even, another point is $(-5, -1)$.

(b) If f is odd, another point is $(-5, 1)$.

88. The graph of $y = x^7$ will pass through the origin and will be symmetric with the origin.
The graph of $y = x^8$ will pass through the origin and will be symmetric with respect to the y-axis.

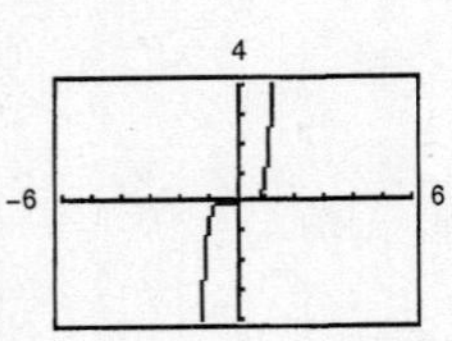

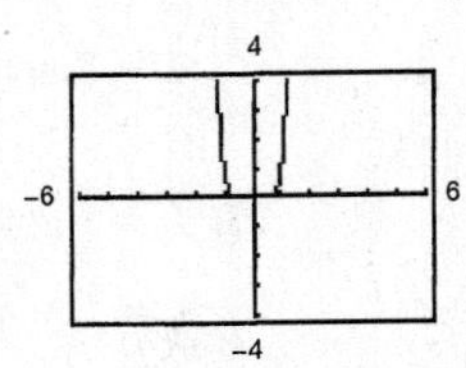

Section P.7 A Library of Functions

Solutions to Even-Numbered Exercises

2. $f(-3) = -8, f(1) = 2$

$(-3, -8), (1, 2)$

$$m = \frac{2 - (-8)}{1 - (-3)} = \frac{10}{4} = \frac{5}{2}$$

$$f(x) - 2 = \frac{5}{2}(x - 1)$$

$$f(x) = \frac{5}{2}x - \frac{1}{2}$$

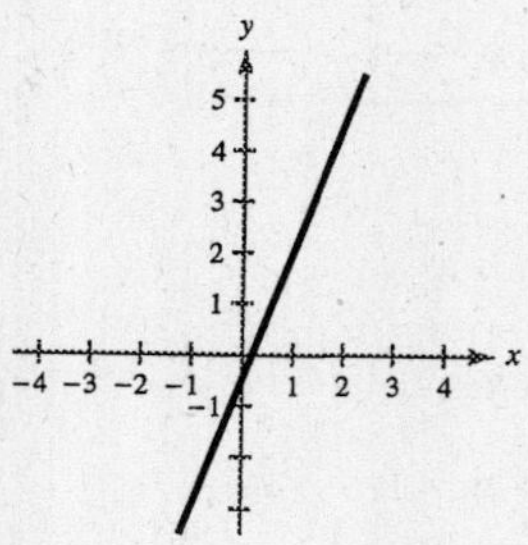

4. $f(3) = 9, f(-1) = -11$

$(3, 9), (-1, -11)$

$$m = \frac{-11 - 9}{-1 - 3} = \frac{-20}{-4} = 5$$

$$f(x) - 9 = 5(x - 3)$$

$$f(x) = 5x - 6$$

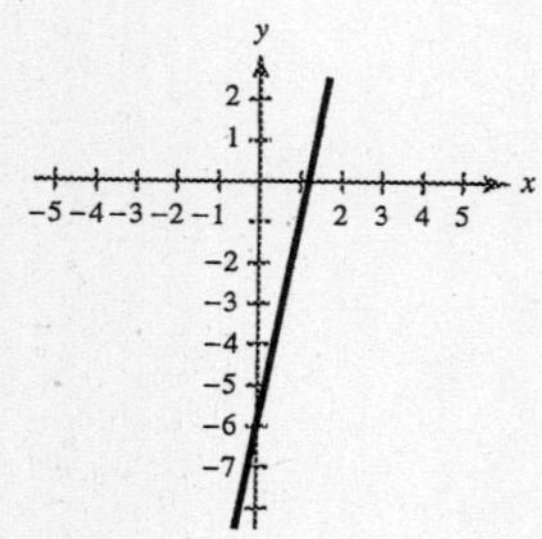

6. $f(-10) = 12, f(16) = -1$

$(-10, 12), (16, -1)$

$$m = \frac{-1 - 12}{16 - (-10)} = \frac{-13}{26} = -\frac{1}{2}$$

$$f(x) - (-1) = -\frac{1}{2}(x - 16)$$

$$f(x) = -\frac{1}{2}x + 7$$

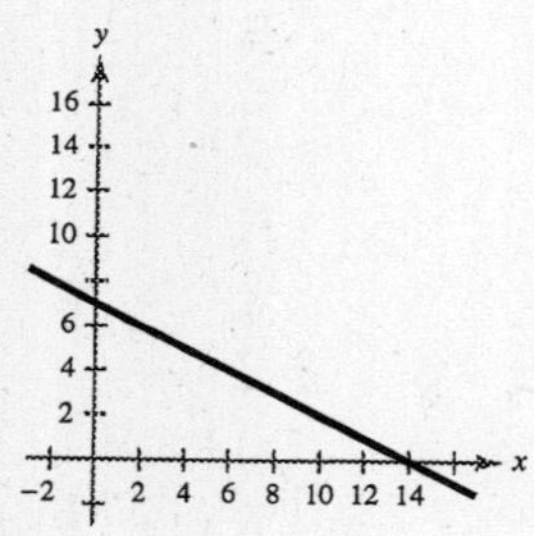

8. $f\left(\frac{2}{3}\right) = -\frac{15}{2}, f(-4) = -11$

$\left(\frac{2}{3}, -\frac{15}{2}\right), (-4, -11)$

$$m = \frac{-11 - \left(-\frac{15}{2}\right)}{-4 - \frac{2}{3}} = \frac{-\frac{7}{2}}{-\frac{14}{3}} = \left(-\frac{7}{2}\right) \cdot \left(-\frac{3}{14}\right) = \frac{3}{4}$$

$$f(x) - (-11) = \frac{3}{4}(x - (-4))$$

$$f(x) = \frac{3}{4}x - 8$$

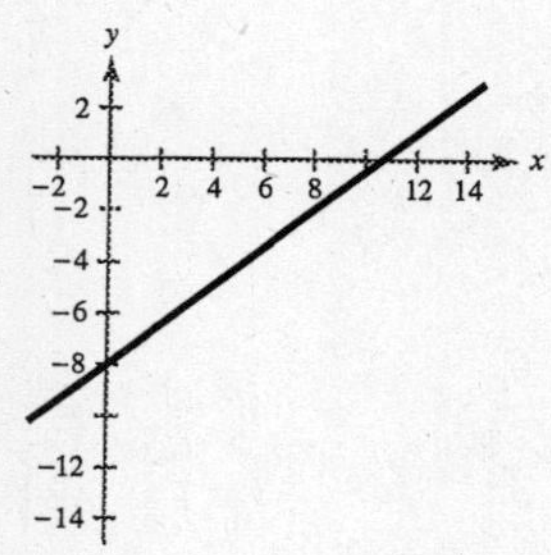

10. $f(x) = 3x - \frac{5}{2}$

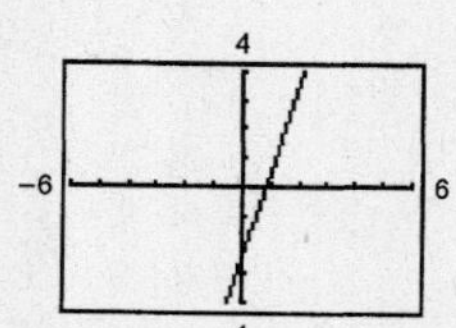

12. $f(x) = \frac{5}{6} - \frac{2}{3}x$

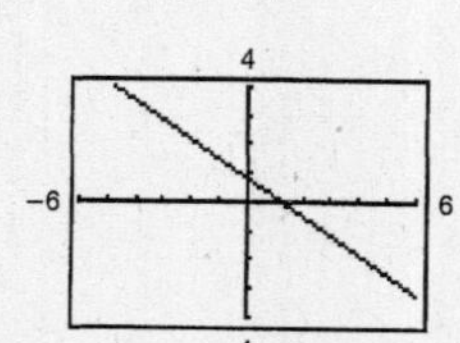

14. $f(x) = -x^2 + 8x$

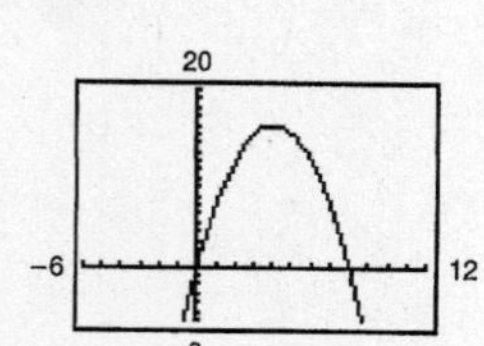

16. $g(x) = x^2 - 6x - 16$

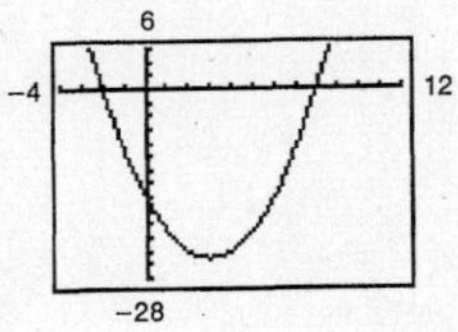

18. $f(x) = 8 - x^3$

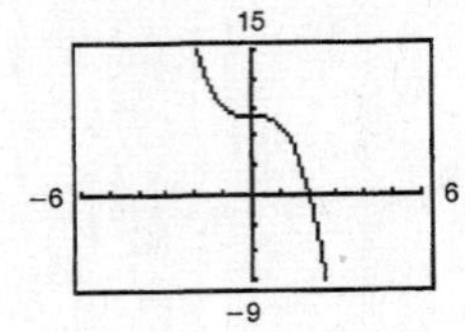

20. $g(x) = 2(x + 3)^3 + 1$

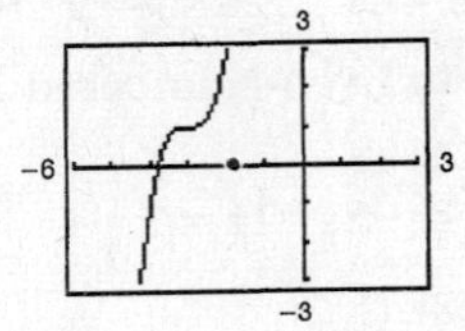

22. $f(x) = 4 - 2\sqrt{x}$

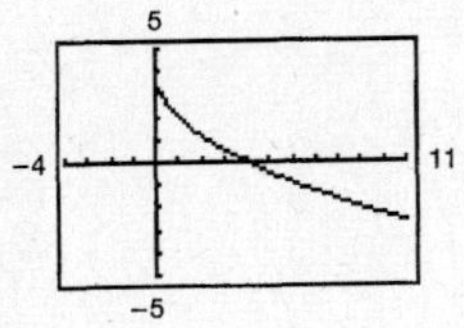

24. $h(x) = \sqrt{x + 2} + 3$

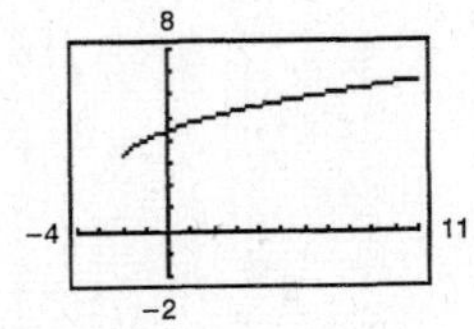

26. $f(x) = 4 + \dfrac{1}{x}$

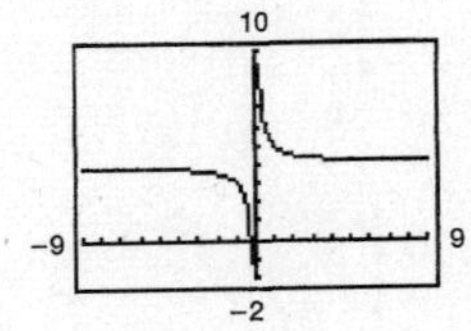

28. $k(x) = \dfrac{1}{x - 3}$

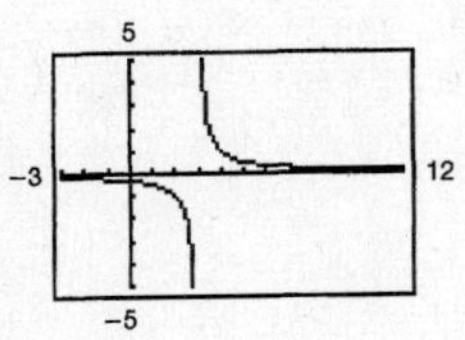

30. $g(x) = 2\llbracket x \rrbracket$

(a) $g(-3) = 2\llbracket -3 \rrbracket = 2(-3) = -6$

(b) $g(0.25) = 2\llbracket 0.25 \rrbracket = 2(0) = 0$

(c) $g(9.5) = 2\llbracket 9.5 \rrbracket = 2(9) = 18$

(d) $g\left(\dfrac{11}{3}\right) = 2\left\llbracket \dfrac{11}{3} \right\rrbracket = 2(3) = 6$

32. $f(x) = 4\llbracket x \rrbracket + 7$

(a) $f(0) = 4\llbracket 0 \rrbracket + 7 = 4(0) + 7 = 7$

(b) $f(-1.5) = 4\llbracket -1.5 \rrbracket + 7 = 4(-2) + 7 = -1$

(c) $f(6) = 4\llbracket 6 \rrbracket + 7 = 4(6) + 7 = 31$

(d) $f\left(\dfrac{5}{3}\right) = 4\left\llbracket \dfrac{5}{3} \right\rrbracket + 7 = 4(1) + 7 = 11$

34. $k(x) = \left\llbracket \dfrac{1}{2}x + 6 \right\rrbracket$

(a) $k(5) = \left\llbracket \dfrac{1}{2}(5) + 6 \right\rrbracket = \llbracket 8.5 \rrbracket = 8$

(b) $k(-6.1) = \left\llbracket \dfrac{1}{2}(-6.1) + 6 \right\rrbracket = \llbracket 2.95 \rrbracket = 2$

(c) $k(0.1) = \left\llbracket \dfrac{1}{2}(0.1) + 6 \right\rrbracket = \llbracket 6.05 \rrbracket = 6$

(d) $k(15) = \left\llbracket \dfrac{1}{2}(15) + 6 \right\rrbracket = \llbracket 13.5 \rrbracket = 13$

36. $g(x) = -7\llbracket x + 4 \rrbracket + 6$

(a) $g\left(\dfrac{1}{8}\right) = -7\left\llbracket \dfrac{1}{8} + 4 \right\rrbracket + 6 = -7\left\llbracket 4\dfrac{1}{8} \right\rrbracket + 6 = -7(4) + 6 = -22$

(b) $g(9) = -7\llbracket 9 + 4 \rrbracket + 6 = -7\llbracket 13 \rrbracket + 6 = -7(13) + 6 = -85$

(c) $g(-4) = -7\llbracket -4 + 4 \rrbracket + 6 = -7\llbracket 0 \rrbracket + 6 = -7(0) + 6 = 6$

(d) $g\left(\dfrac{3}{2}\right) = -7\left\llbracket \dfrac{3}{2} + 4 \right\rrbracket + 6 = -7\left\llbracket 5\dfrac{1}{2} \right\rrbracket + 6 = -7(5) + 6 = -29$

38.

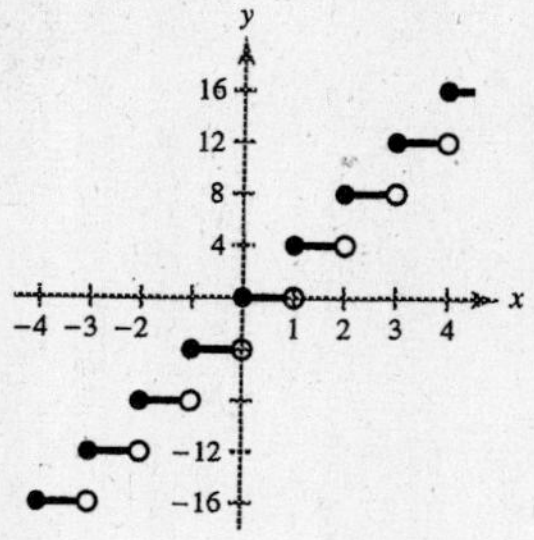

40.

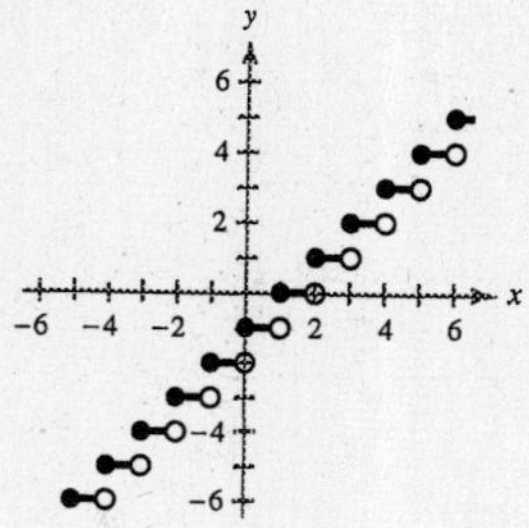

42.

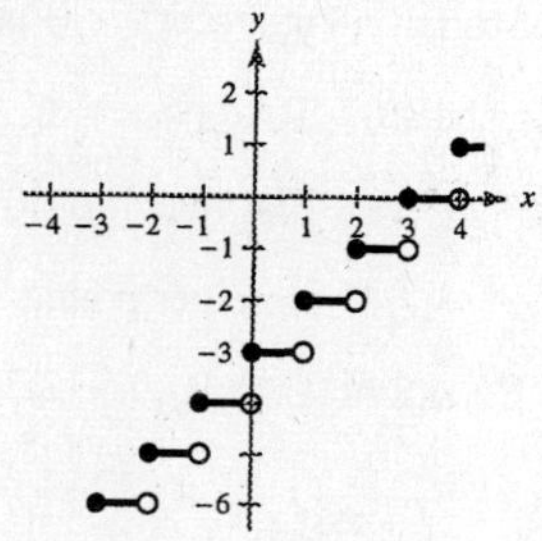

44.

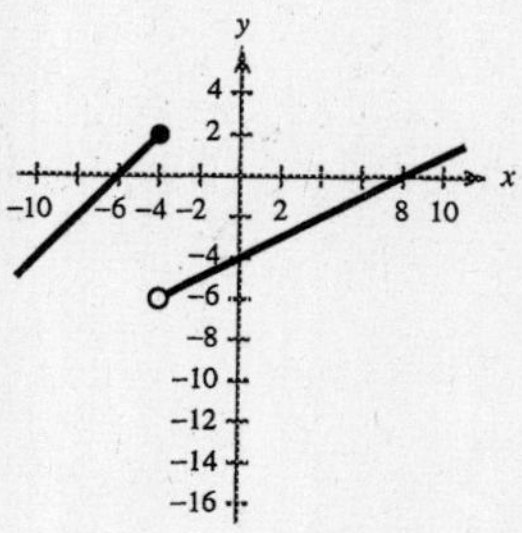

46. $f(x) = \begin{cases} 1 - (x-1)^2, & x \le 2 \\ \sqrt{x-2}, & x > 2 \end{cases}$

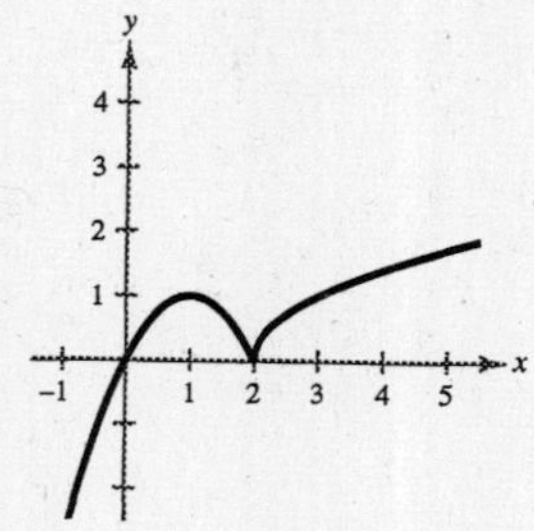

48.

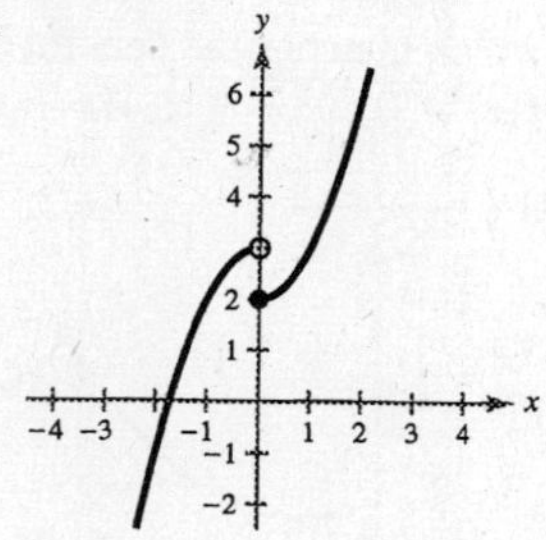

50.

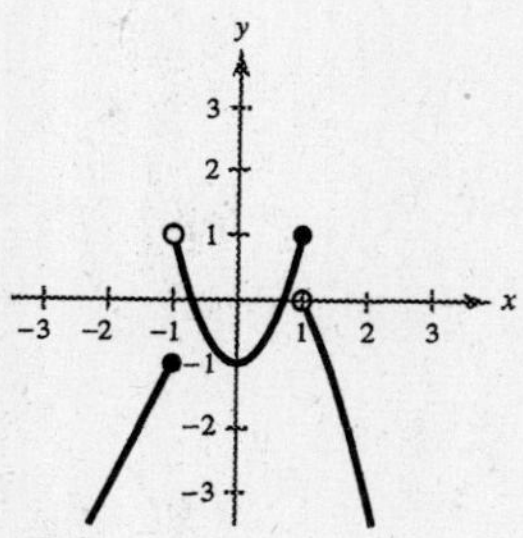

52. $g(x) = 2\left(\frac{1}{4}x - \left[\!\left[\frac{1}{4}x\right]\!\right]\right)^2$

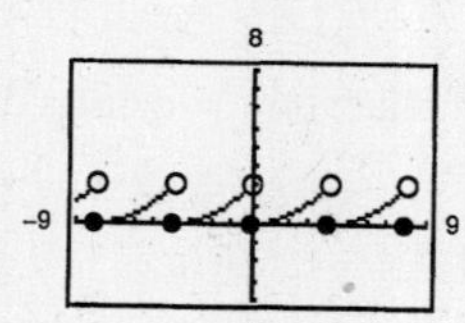

Domain: $(-\infty, \infty)$

Range: $(0, 2)$

Pattern: sawtooth

54. $y = \sqrt{x}$

56. $y = \dfrac{1}{x}$

58. $y = x^2$

60. $y = [\![x]\!]$

62. $y = x^3$

64. (a) $C_2(t) = 1.05 - 0.38[\![-(t-1)]\!]$ is the appropriate model since the cost does not increase until after the next minute of conversation has started.

(b)

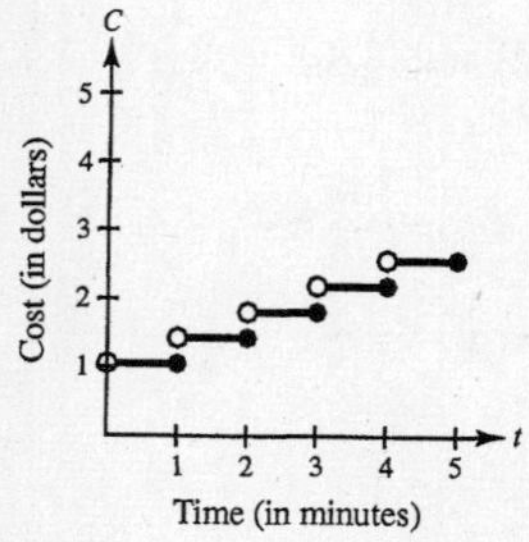

$C = 1.05 - 0.38[\![-17.75]\!] = \$7.89$

66. (a) Model: (Total cost) = (Flat rate) + (Rate per pound)

Labels: Total cost $= C$

Flat rate $= 9.80$

Rate per pound $= 2.50[\![x]\!],\ x > 0$

Equation: $C = 9.80 + 2.50[\![x]\!],\ x > 0$

(b)

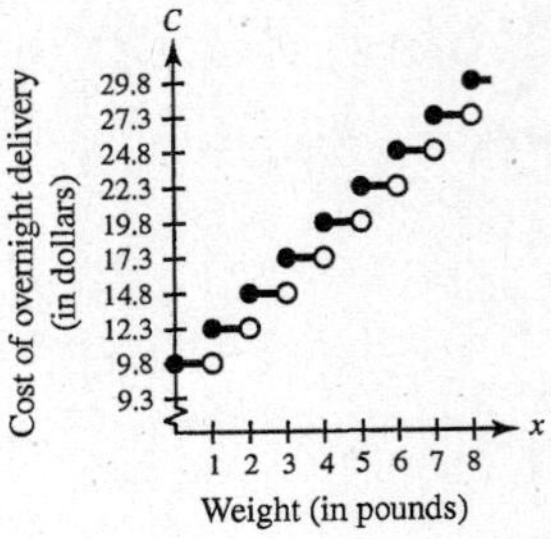

68. (a) The domain of $f(x) = -1.97x + 26.3$ is $6 < x \le 12$. One way to see this is to notice that this is the equation of a line with negative slope, so the function values are decreasing as x increases, which matches the data for the corresponding part of the table. The domain of $f(x) = 0.505x^2 - 1.47x + 6.3$ is then $1 \le x \le 6$.

(b)

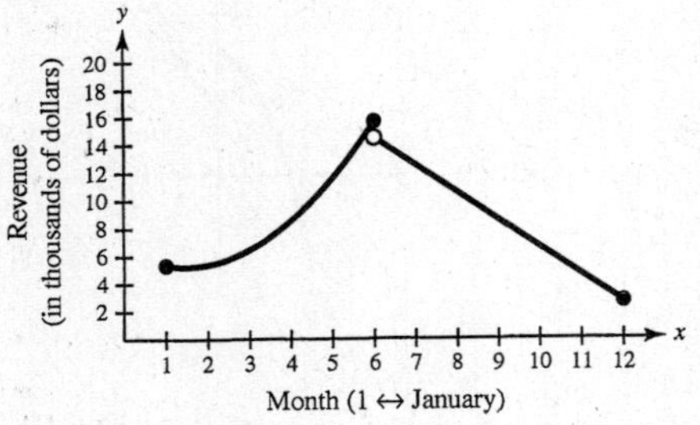

(c) $f(5) = 0.505(5)^2 - 1.47(5) + 6.3 = 0.505(25) - 7.35 + 6.3 = 11.575$

$f(11) = -1.97(11) + 26.3 = 4.63$

These values represent the income in thousands of dollars for the months of May and November, respectively.

(d) The model values are very close to the actual values.

70. False. A piecewise-defined function is a function that is defined by two or more equations over a specified domain. That domain may or may not include x- and y-intercepts.

72. For the line through (0, 6) and (3, 2):

$$m = \frac{6-2}{0-3} = -\frac{4}{3}$$

$$y - 6 = -\frac{4}{3}(x - 0) \Rightarrow y = -\frac{4}{3}x + 6$$

For the line through (3, 2) and (8, 0):

$$m = \frac{2-0}{3-8} = -\frac{2}{5}$$

$$y - 0 = -\frac{2}{5}(x - 8) \Rightarrow y = -\frac{2}{5}x + \frac{16}{5}$$

$$f(x) = \begin{cases} -\frac{4}{3}x + 6, & 0 \le x \le 3 \\ -\frac{2}{5}x + \frac{16}{5}, & 3 < x \le 8 \end{cases}$$

Note that the respective domains can also be $0 \le x < 3$ and $3 \le x \le 8$.

Section P.8 Shifting, Reflecting, and Stretching Graphs

Solutions to Even-Numbered Exercises

2. (a) $f(x) = \sqrt{x} + c$

$c = -3 : f(x) = \sqrt{x} - 3$

$c = -1 : f(x) = \sqrt{x} - 1$

$c = 1 : f(x) = \sqrt{x} + 1$

$c = 3 : f(x) = \sqrt{x} + 3$

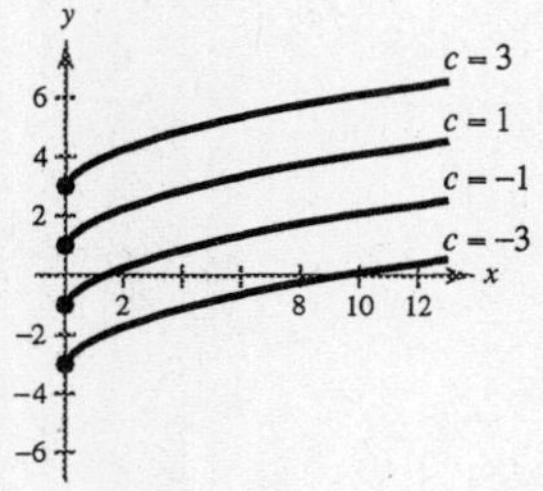

(b) $f(x) = \sqrt{x - c}$

$c = -3 : f(x) = \sqrt{x + 3}$

$c = -1 : f(x) = \sqrt{x + 1}$

$c = 1 : f(x) = \sqrt{x - 1}$

$c = 3 : f(x) = \sqrt{x - 3}$

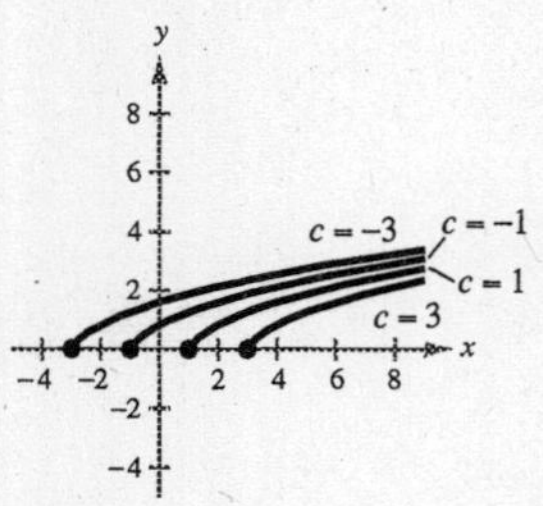

(c) $f(x) = \sqrt{x - 3} + c$

$c = -3 : f(x) = \sqrt{x - 3} - 3$

$c = -1 : f(x) = \sqrt{x - 3} - 1$

$c = 1 : f(x) = \sqrt{x - 3} + 1$

$c = 3 : f(x) = \sqrt{x - 3} + 3$

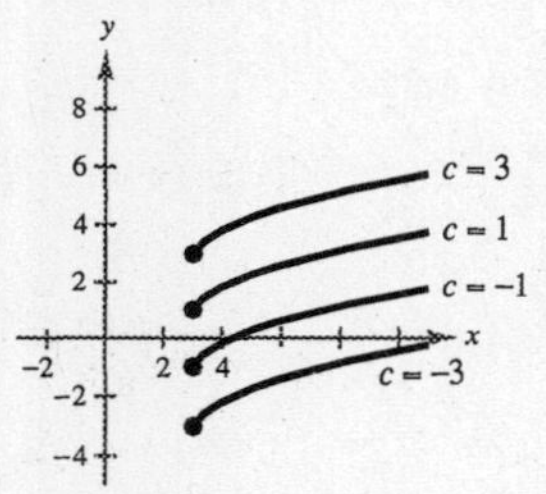

4. (a) $f(x) = \begin{cases} x^2 + c, & x < 0 \\ -x^2 + c, & x \geq 0 \end{cases}$

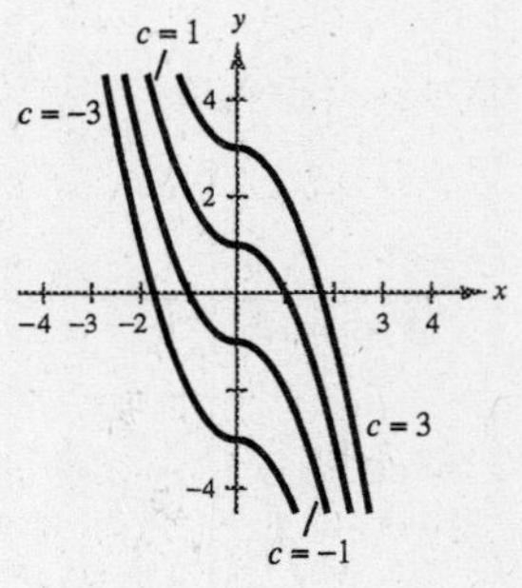

(b) $f(x) = \begin{cases} (x + c)^2, & x < 0 \\ -(x + c)^2, & x \geq 0 \end{cases}$

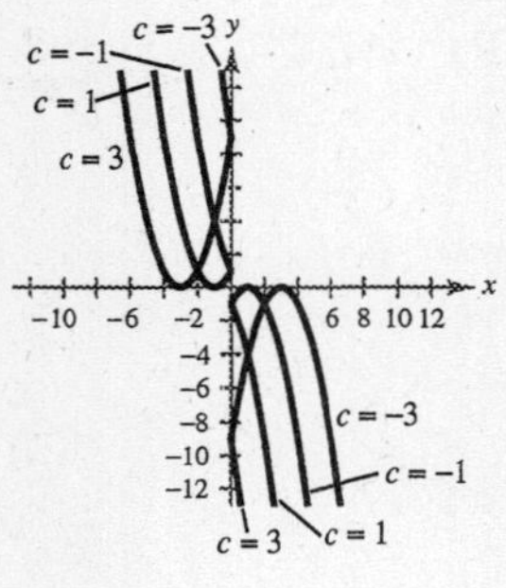

6. (a) $y = f(-x)$

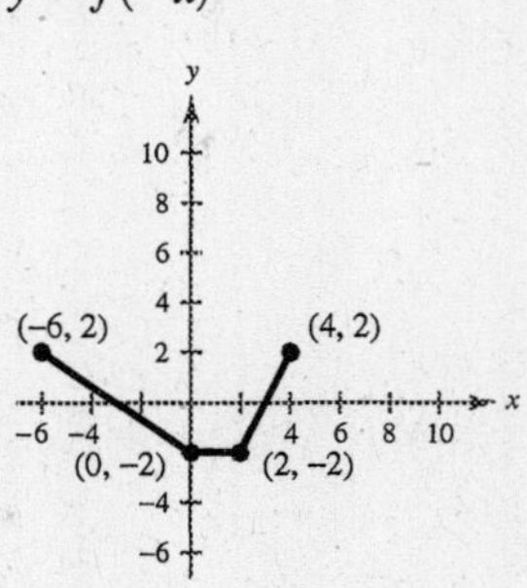

(b) $y = f(x) + 4$

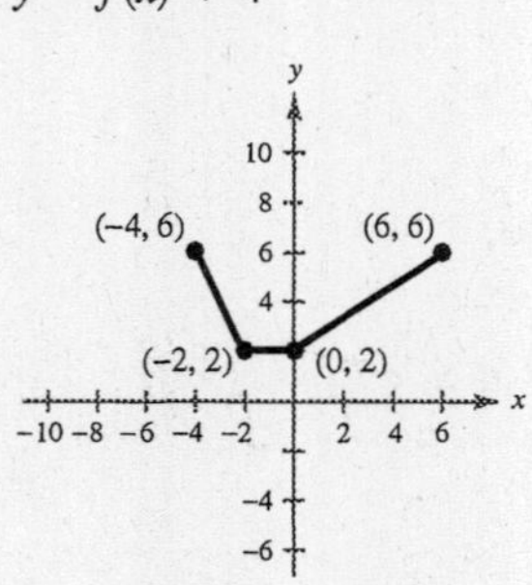

(c) $y = 2f(x)$

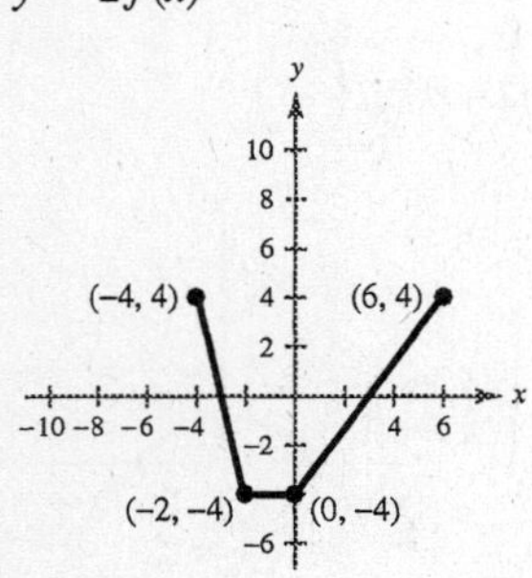

—CONTINUED—

6. —CONTINUED—

(d) $y = -f(x - 4)$

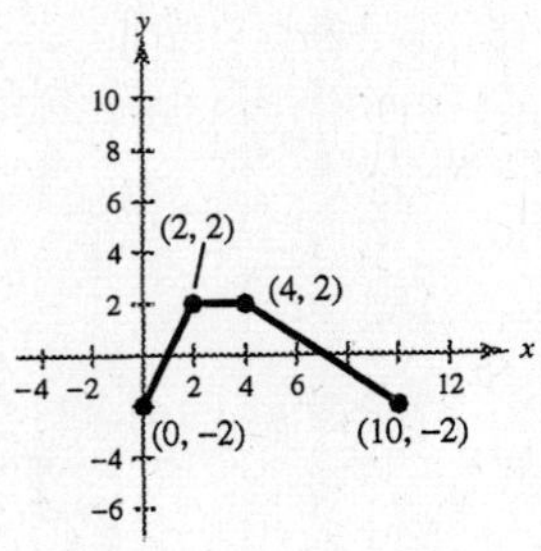

(e) $y = f(x) - 3$

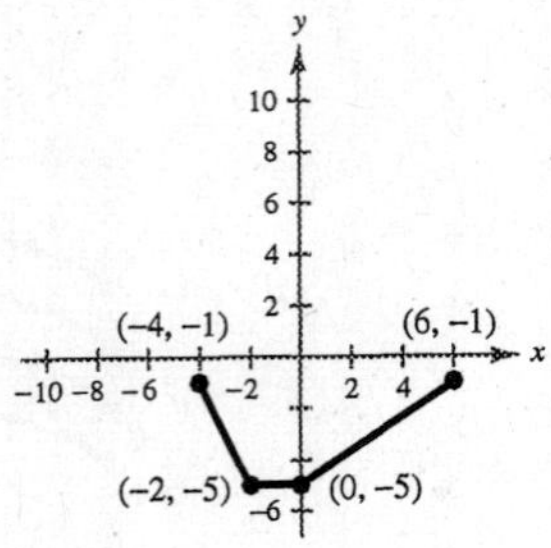

(f) $y = -f(x) - 1$

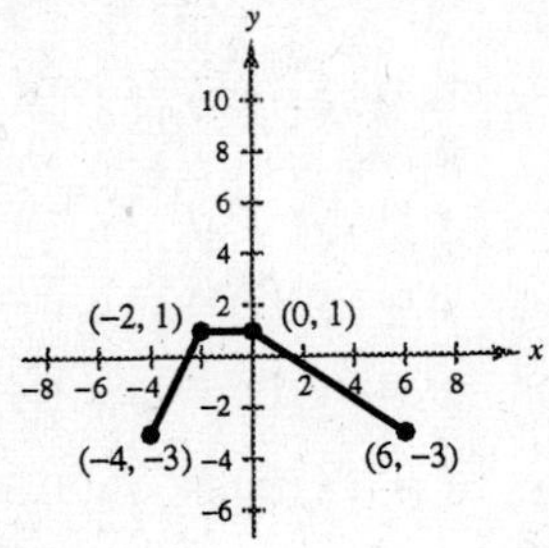

(g) $y = f(2x)$

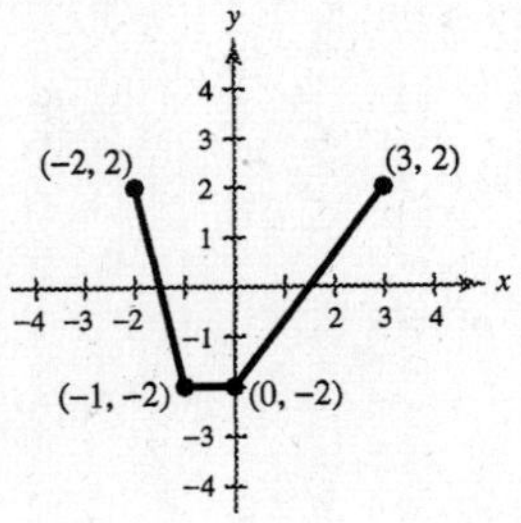

8. (a) $y = f(x - 5)$

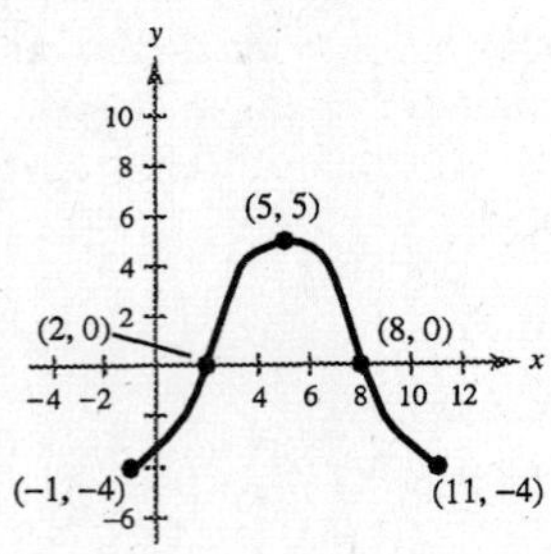

(b) $y = -f(x) + 3$

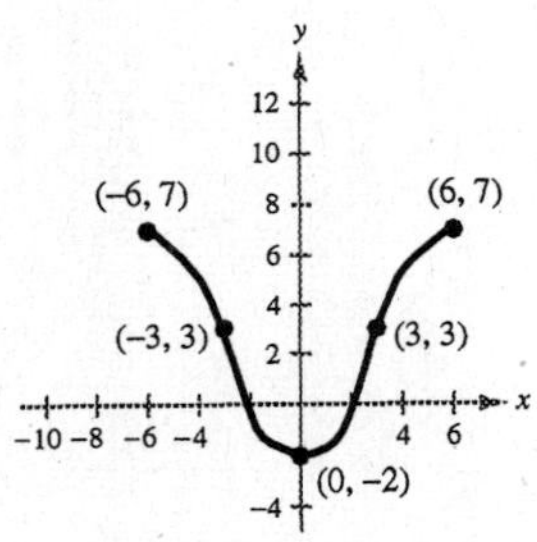

(c) $y = \frac{1}{3}f(x)$

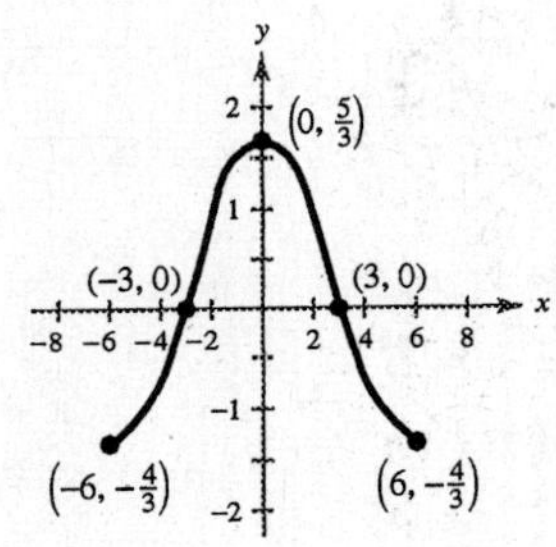

(d) $y = -f(x + 1)$

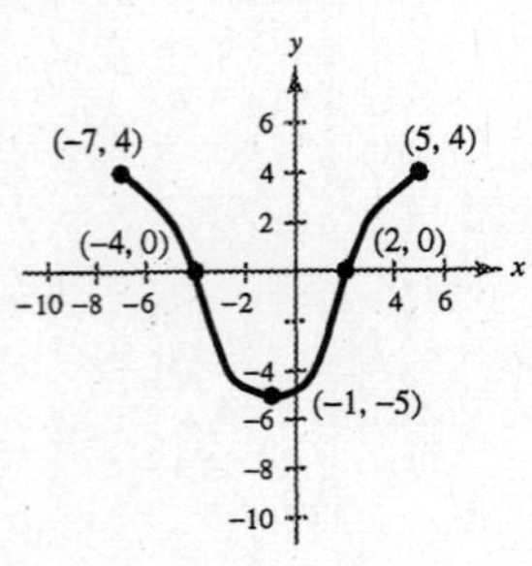

(e) $y = f(-x)$

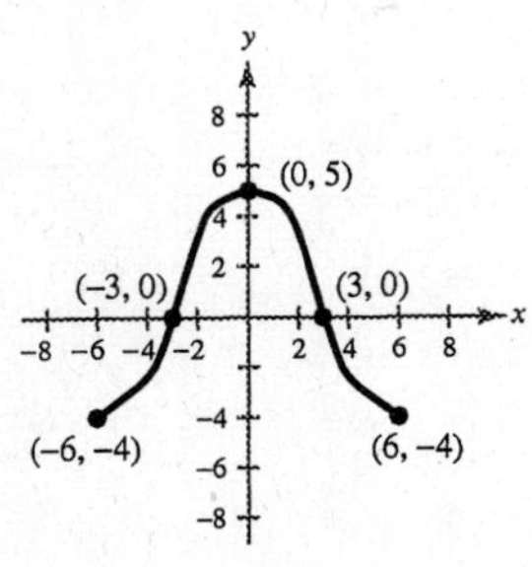

(f) $y = f(x) - 10$

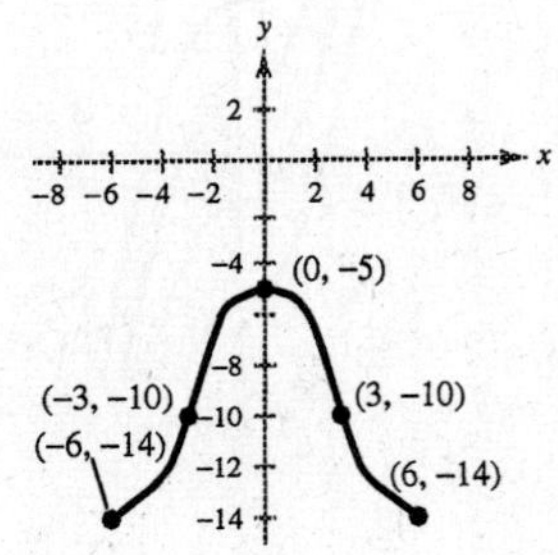

(g) $y = f\left(\frac{1}{3}x\right)$

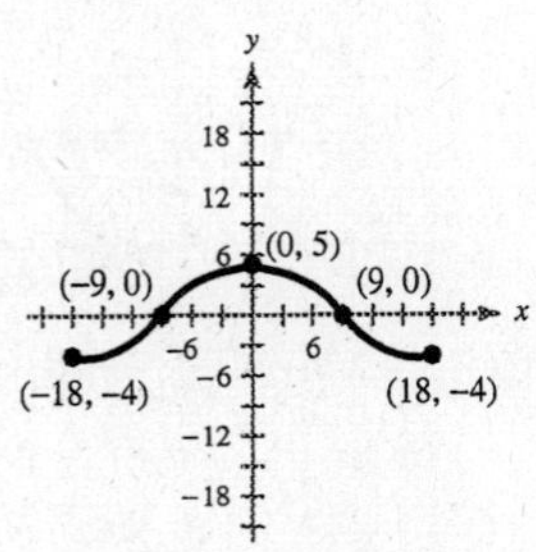

10. (a) The graph of $f(x) = x^3$ was reflected in the x-axis and shifted upward one unit.

$y = -x^3 + 1 = 1 - x^3$

(b) The graph of $f(x) = x^3$ was shifted to the right one unit and upward one unit.

$y = (x - 1)^3 + 1$

(c) The graph of $f(x) = x^3$ was reflected in the x-axis and shifted to the left three units and downward one unit.

$y = -(x + 3)^3 - 1$

(d) The graph of $f(x) = x^3$ was shifted to the right ten units and downward four units.

$y = (x - 10)^3 - 4$

12. (a) The graph of $f(x) = \sqrt{x}$ was shifted down three units.

$y = \sqrt{x} - 3$

(b) The graph of $f(x) = \sqrt{x}$ was shifted downward seven units and to the left one unit.

$y = \sqrt{x + 1} - 7$

(c) The graph of $f(x) = \sqrt{x}$ was reflected in the x-axis and shifted to the right five units and upward five units.

$y = -\sqrt{x - 5} + 5$

(d) The graph of $f(x) = \sqrt{x}$ was reflected about the x- and y-axis and shifted to the right three units and downward four units.

$y = -\sqrt{-x + 3} - 4 = -\sqrt{-(x - 3)} - 4$

14. Common function: $y = x$

Transformation: vertical shrink

Formula: $y = \frac{1}{2}x$

16. Common function: $y = [\![x]\!]$

Transformation: vertical shift

Formula: $y = [\![x]\!] + 4$

18. Common function: $y = |x|$

Transformation: horizontal shift

Formula: $y = |x + 2|$

20. $f(x) = (x - 8)^2$

Horizontal shift of eight units to the right.

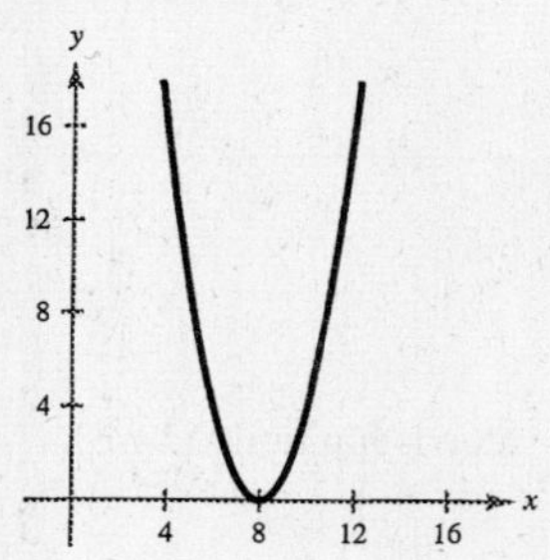

22. $f(x) = -x^3 - 1$

Reflection in the x-axis; vertical shift of one unit downward.

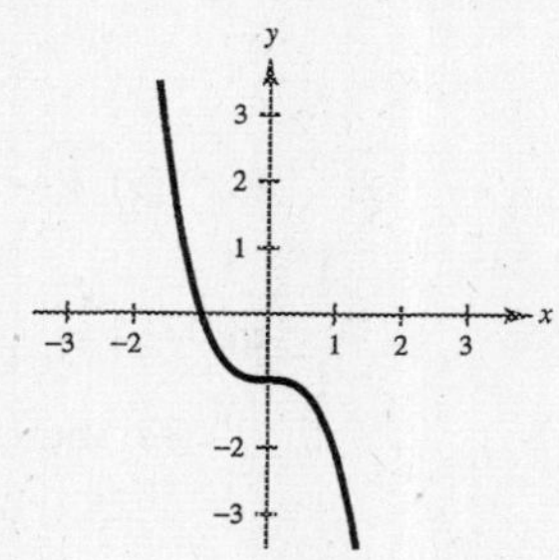

24. $f(x) = -(x + 10)^2 + 5$

Reflection in the x-axis; horizontal shift of 10 units to the left; vertical shift of 5 units upward.

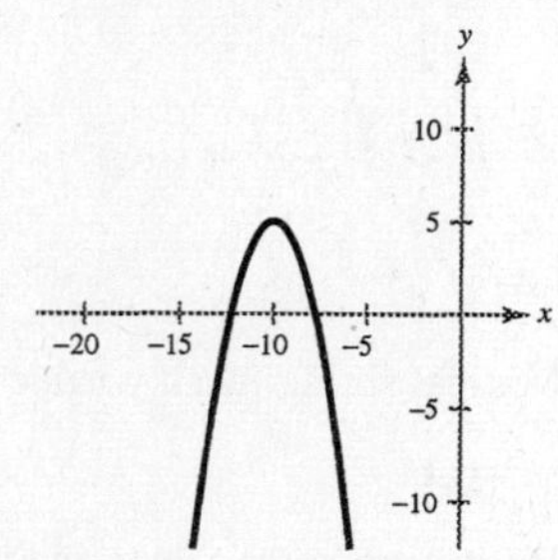

26. $f(x) = (x + 3)^3 - 10$

Horizontal shift of 3 units to the left; vertical shift of 10 units downward.

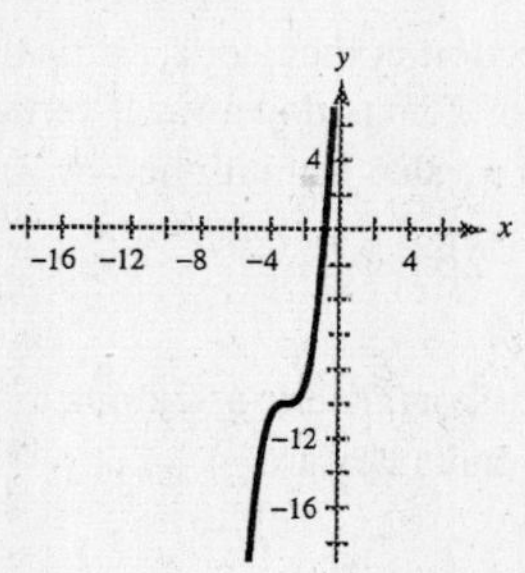

28. $f(x) = 6 - |x + 5|$

Reflection in the x-axis; horizontal shift of 5 units to the left; vertical shift of 6 units upward.

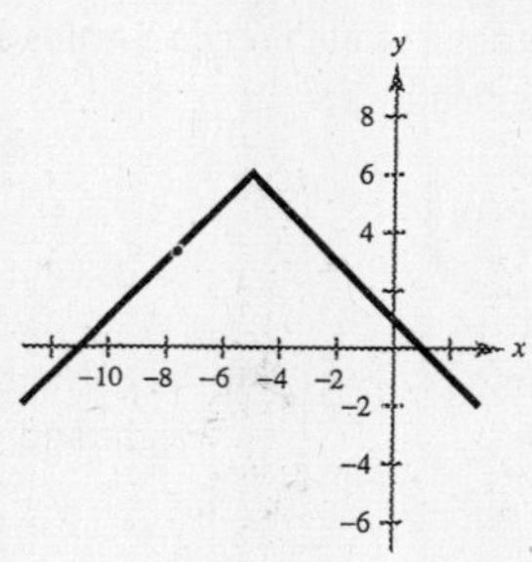

30. $f(x) = |-x + 3| + 9$

Reflection in the y-axis; horizontal shift of 3 units to the left; vertical shift of 9 units upward.

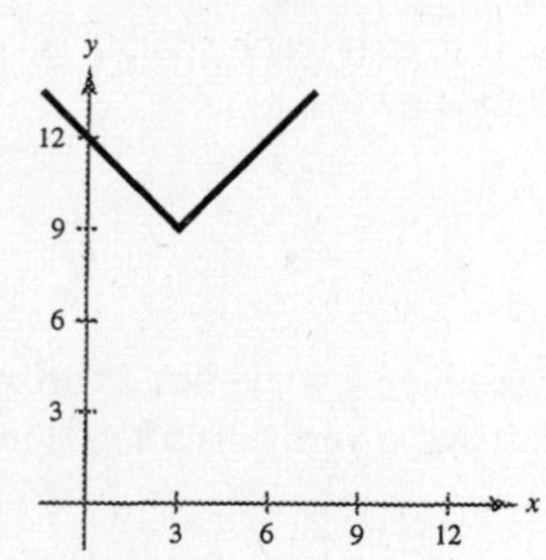

32. $f(x) = 2[\![x + 5]\!]$

Horizontal shift of 5 units to the left; vertical stretch (each y-value is multiplied by 2)

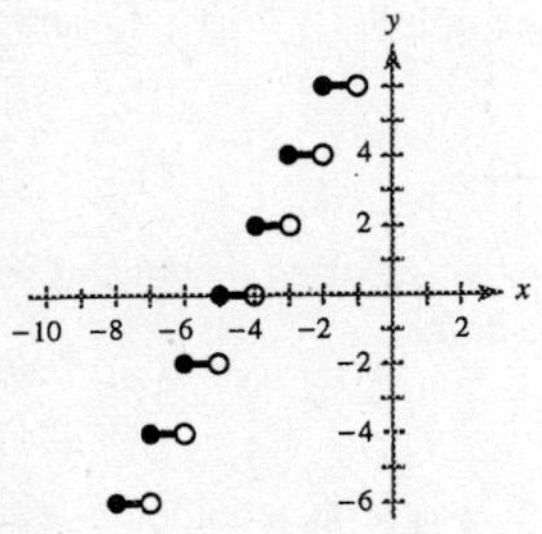

34. $f(x) = \sqrt{x + 4} + 8$

Horizontal shift of 4 units to the left; vertical shift of 8 units upward.

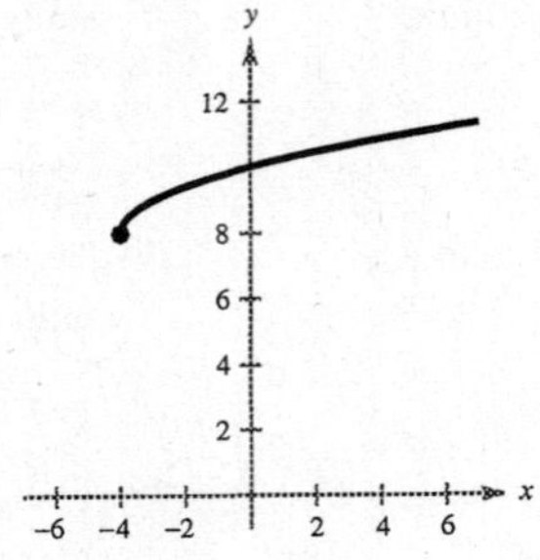

36. $f(x) = -\sqrt{x + 1} - 6$

Reflection in the x-axis; horizontal shift of 1 unit to the left; vertical shift of 6 units downward.

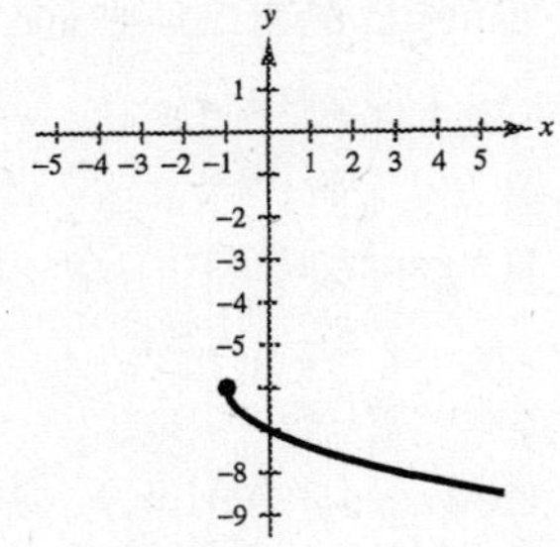

38. $f(x) = \sqrt{3x} + 1$

Horizontal shrink (each x-value is multiplied by $\frac{1}{3}$); vertical shift of 1 unit upward

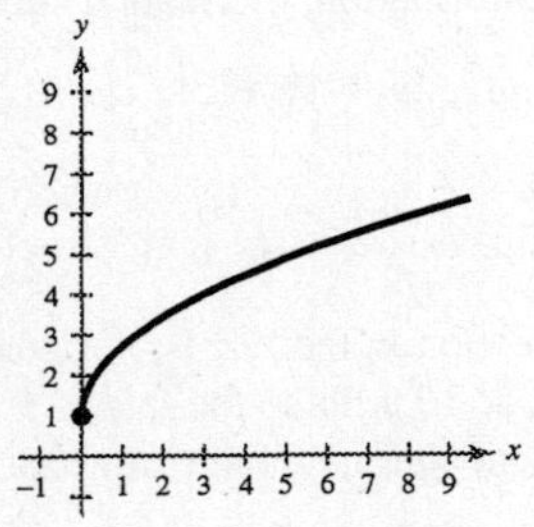

40. $f(x) = -(x + 3)^2 + 7$

42. $f(x) = (-x - 6)^3 - 6$

44. $f(x) = |x + 1| - 7$

46. $f(x) = -\sqrt{-x} - 9$

48. $f(x) = x^3$

(a) Vertical shrink (each y-value is multiplied by $\frac{1}{4}$).

$y = \frac{1}{4}x^3$

(b) Reflection in the x-axis and a vertical stretch (each y-value is multiplied by 2).

$y = -2x^3$

50. $f(x) = \sqrt{x}$

(a) Vertical stretch (each y-value is multiplied by 8).

$y = 8\sqrt{x}$

(b) Reflection in the x-axis and a vertical shrink (each y-value is multiplied by $\frac{1}{4}$).

$y = -\frac{1}{4}\sqrt{x}$

52. Common function: $f(x) = |x|$

Vertical stretch (each y-value is multiplied by 6).

$y = 6|x|$

54. Common function: $y = [\![x]\!]$

Horizontal stretch (each x-value is multiplied by 2).

$y = \left[\!\left[\frac{1}{2}x\right]\!\right]$

56. Common function: $f(x) = |x|$

Reflection in the x-axis; vertical shift of 2 units downward; vertical (each y-value is multiplied by 2).

$y = -2|x| - 2$

58. The graph of $y = |x|$ has a horizontal shift of 4 units to the left and a vertical shift of 2 units downward.

$y = |x + 4| - 2$

60. The graph of $y = x^2$ has a horizontal shift of 2 units to the right and a vertical shift of 4 units upward.

$y = (x - 2)^2 + 4$

62. (a) $g(x) = f(x) - 5$

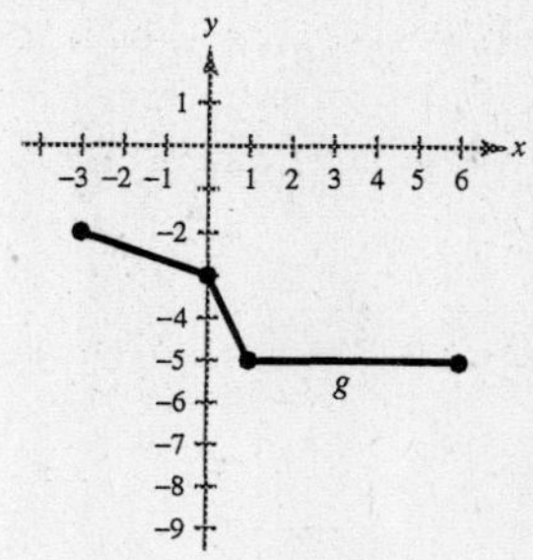

(b) $g(x) = f(x) + \frac{1}{2}$

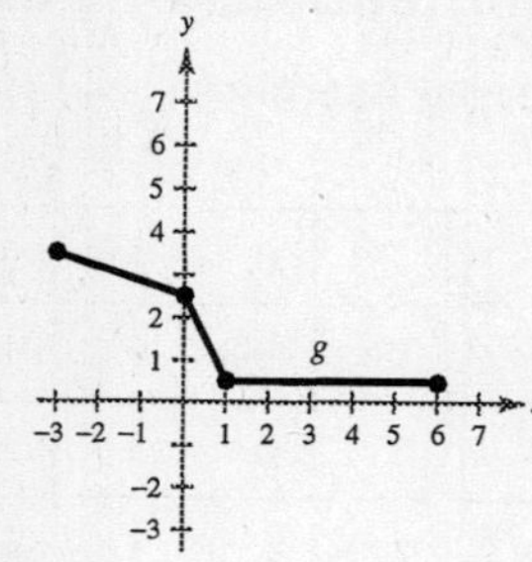

(c) $g(x) = f(-x)$

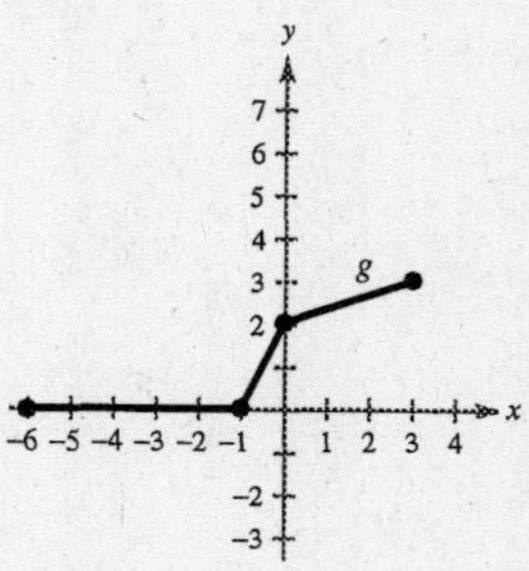

(d) $g(x) = -4f(x)$

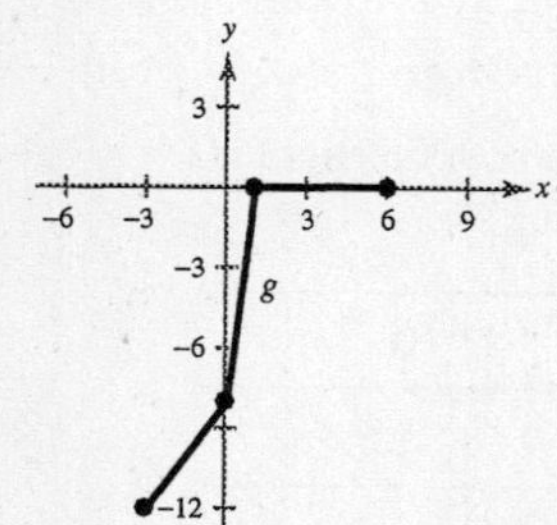

(e) $g(x) = f(2x) + 1$

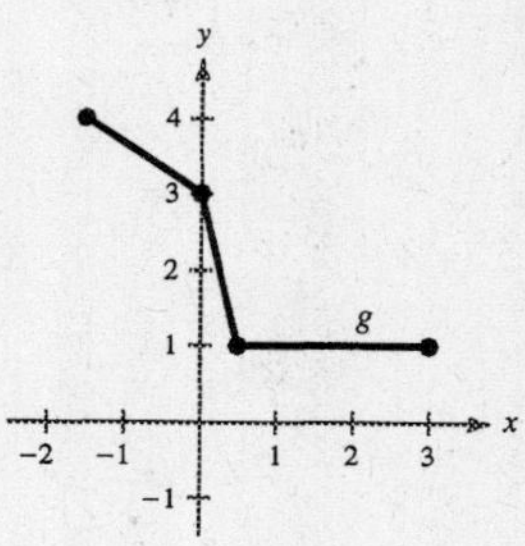

(f) $g(x) = f\left(\frac{1}{4}x\right) - 2$

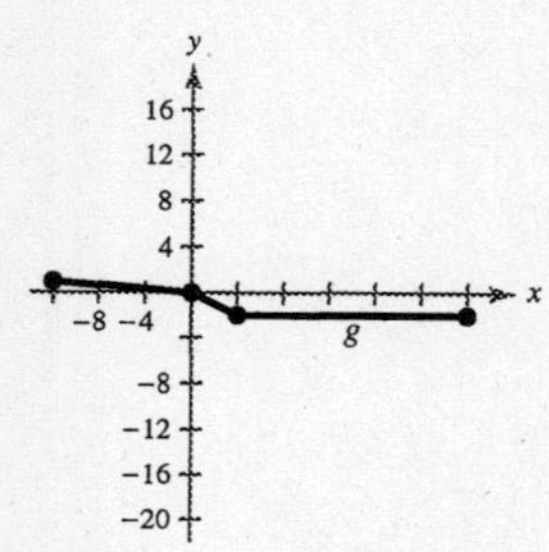

64. (a) The graph is a horizontal shift 14.979 units to the left of the graph of the common function $f(x) = x^2$ and a vertical shrink by a factor of 0.0037.

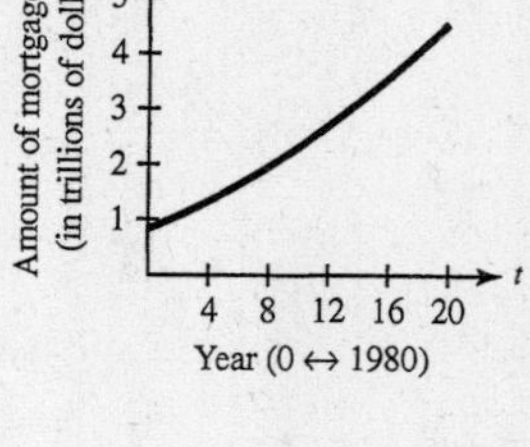

(b) $f(t) = 0.0037(t + 24.979)^2$. By shifting the graph 10 units to the left, you obtain $t = 0$ represents 1990.

66. False. The point $(-2, -61)$ lies on the transformation.

68. If you consider the x-axis to be a mirror, the graph of $y = -f(x)$ is the mirror image of the graph of $y = f(x)$.

70. Answers will vary.

Section P.9 Combinations of Functions

Solutions to Even-Numbered Exercises

2.

x	-2	-1	0	1	2
$f(x)$	-2	0	-1	-1	1
$g(x)$	1	1	0	2	2
$h(x) = (f + g)(x)$	-1	1	-1	1	3

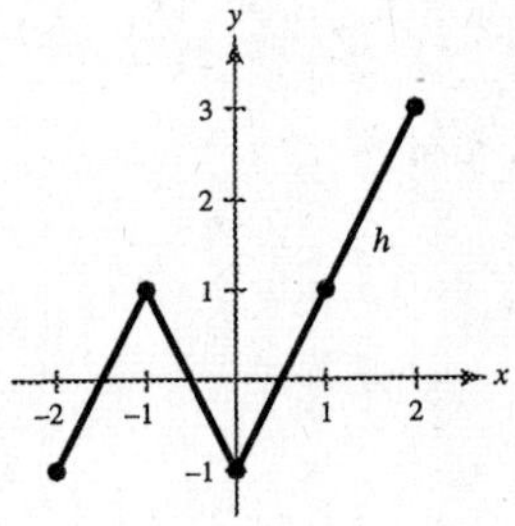

4. The domain common to both functions is $[-1, 1]$, which is the domain of the sum.

x	-1	0	1
$f(x)$	0	1.5	3
$g(x)$	-1	-2	1
$h(x) = f(x) + g(x)$	-1	-0.5	4

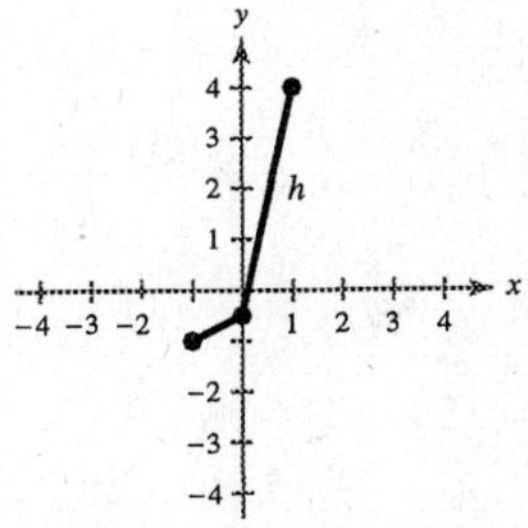

6. $f(x) = 2x - 5,\ g(x) = 2 - x$

(a) $(f + g)(x) = 2x - 5 + 2 - x$

$= x - 3$

(b) $(f - g)(x) = 2x - 5 - (2 - x)$

$= 2x - 5 - 2 + x$

$= 3x - 7$

(c) $(fg)(x) = (2x - 5)(2 - x)$

$= 4x - 2x^2 - 10 + 5x$

$= -2x^2 + 9x - 10$

(d) $\left(\dfrac{f}{g}\right)(x) = \dfrac{2x - 5}{2 - x}$

Domain: $2 - x \neq 0$

$x \neq 2$

8. $f(x) = 2x - 5,\ g(x) = 4$

(a) $(f + g)(x) = 2x - 5 + 4 = 2x - 1$

(b) $(f - g)(x) = 2x - 5 - 4 = 2x - 9$

(c) $(fg)(x) = (2x - 5)(4) = 8x - 20$

(d) $\left(\dfrac{f}{g}\right)(x) = \dfrac{2x - 5}{4} = \dfrac{1}{2}x - \dfrac{5}{4}$

Domain: $-\infty < x < \infty$

10. $f(x) = \sqrt{x^2 - 4},\ g(x) = \dfrac{x^2}{x^2 + 1}$

(a) $(f + g)(x) = \sqrt{x^2 - 4} + \dfrac{x^2}{x^2 + 1}$

(b) $(f - g)(x) = \sqrt{x^2 - 4} - \dfrac{x^2}{x^2 + 1}$

(c) $(fg)(x) = \sqrt{x^2 - 4}\left(\dfrac{x^2}{x^2 + 1}\right)$

$= \dfrac{x^2\sqrt{x^2 - 4}}{x^2 + 1}$

(d) $\left(\dfrac{f}{g}\right)(x) = \sqrt{x^2 - 4} \div \dfrac{x^2}{x^2 + 1}$

$= \dfrac{(x^2 + 1)\sqrt{x^2 - 4}}{x^2}$

Domain: $x^2 - 4 \geq 0$

$x^2 \geq 4 \implies x \geq 2 \text{ or } x \leq -2$

Domain: $|x| \geq 2$

12. $f(x) = \dfrac{x}{x+1}$, $g(x) = x^3$

(a) $(f+g)(x) = \dfrac{x}{x+1} + x^3 = \dfrac{x + x^4 + x^3}{x+1}$

(b) $(f-g)(x) = \dfrac{x}{x+1} - x^3 = \dfrac{x - x^4 - x^3}{x+1}$

(c) $(fg)(x) = \dfrac{x}{x+1} \cdot x^3 = \dfrac{x^4}{x+1}$

(d) $\left(\dfrac{f}{g}\right)(x) = \dfrac{x}{x+1} \div x^3$

$= \dfrac{x}{x+1} \cdot \dfrac{1}{x^3} = \dfrac{1}{x^2(x+1)}$

Domain: $x \neq 0, -1$

14. $(f-g)(-1) = f(-1) - g(-1)$

$= (-1)^2 + 1 - (-1 - 4)$

$= 1 + 1 - (-5)$

$= 7$

16. $(f+g)(1) = f(1) + g(1)$

$= (1)^2 + 1 + (1) - 4$

$= -1$

18. $(f+g)(t-2) = f(t-2) + g(t-2)$

$= (t-2)^2 + 1 + (t-2) - 4$

$= t^2 - 4t + 4 + 1 + t - 2 - 4$

$= t^2 - 3t - 1$

20. $(fg)(-6) = f(-6) \cdot g(-6)$

$= [(-6)^2 + 1][(-6) - 4]$

$= (37)(-10)$

$= -370$

22. $\left(\dfrac{f}{g}\right)(0) = \dfrac{f(0)}{g(0)} = \dfrac{0^2 + 1}{0 - 4} = -\dfrac{1}{4}$

24. $(fg)(5) + f(4) = f(5)g(5) + f(4) = (5^2 + 1)(5 - 4) + (4^2 + 1) = 26 \cdot 1 + 17 = 43$

26. $f(x) = \frac{1}{3}x$, $g(x) = -x + 4$

$(f+g)(x) = \frac{1}{3}x - x + 4 = -\frac{2}{3}x + 4$

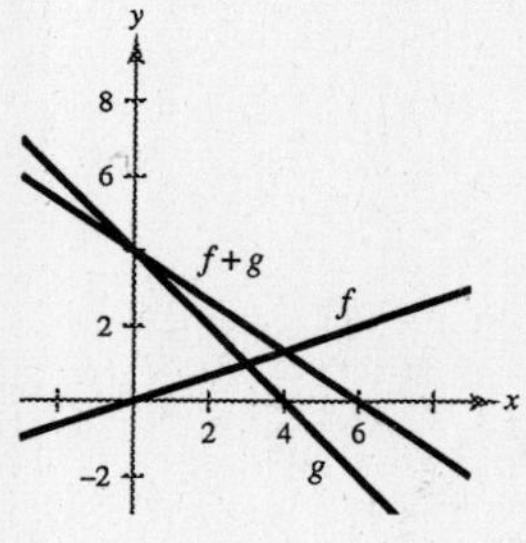

28. $f(x) = 4 - x^2$, $g(x) = x$

$(f+g)(x) = 4 - x^2 + x = 4 + x - x^2$

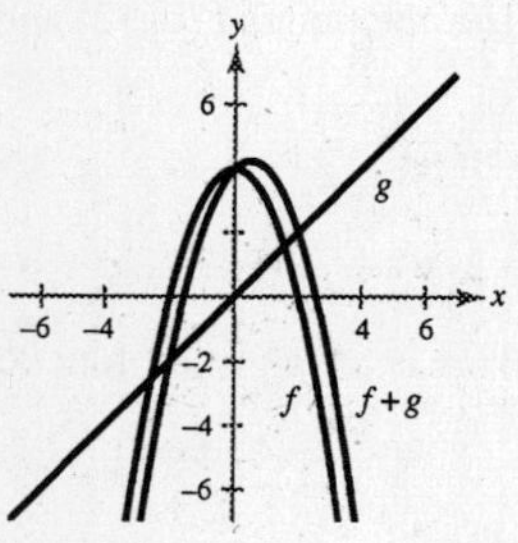

30. $f(x) = \frac{x}{2}$, $g(x) = \sqrt{x}$

$(f + g)(x) = \frac{x}{2} + \sqrt{x}$

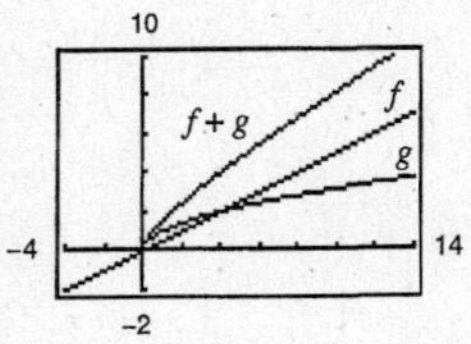

$g(x)$ contributes most to the magnitude of the sum for $0 \le x \le 2$. $f(x)$ contributes most to the magnitude of the sum for $x > 6$.

32. Total sales $= R_1 + R_2$

$= 480 - 8t - 0.8t^2 + 254 + 0.78t$

$= 734 - 7.22t - 0.8t^2$

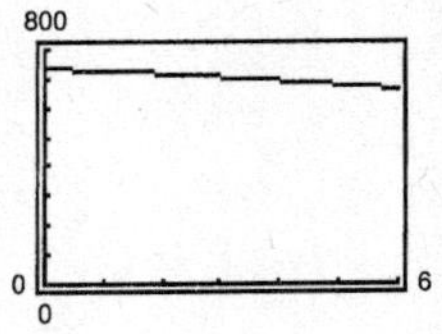

34. (a) T is a function of t since for each time t there corresponds one and only one temperature T.

(b) $T(4) \approx 60°$

$T(15) \approx 72°$

(c) $H(t) = T(t - 1)$; All the temperature changes would be one hour later.

(d) $H(t) = T(t) - 1$; The temperature would be decreased by one degree.

(e) The points at the endpoints of the individual functions that form each "piece" appear to be (0, 60), (6, 60), (7, 72), (20, 72), (21, 60), and (24, 60). Note that the value $t = 24$ is chosen for the last ordered pair because that is when the day ends and the cycle starts over.

From $t = 0$ to $t = 6$: This is the constant function $T(t) = 60$.

From $t = 6$ to $t = 7$: Use the points (6, 60) and (7, 72).

$$m = \frac{72 - 60}{7 - 6} = 12$$

$$y - 60 = 12(x - 6) \implies y = 12x - 12, \text{ or } T(t) = 12t - 12$$

From $t = 7$ to $t = 20$: This is the constant function $T(t) = 72$.

From $t = 20$ to $t = 21$: Use the points (20, 72) and (21, 60).

$$m = \frac{72 - 60}{20 - 21} = -12$$

$$y - 60 = -12(x - 21) \implies y = -12x + 312, \text{ or } T(t) = -12t + 312$$

From $t = 21$ to $t = 24$: This is the constant function $T(t) = 60$.

A piecewise-defined function is $T(t) = \begin{cases} 60, & 0 \le t \le 6 \\ 12t - 12, & 6 < t < 7 \\ 72, & 7 \le t \le 20 \\ -12t + 312, & 20 < t < 21 \\ 60, & 21 \le t \le 24 \end{cases}$.

Note that the endpoints of each domain interval can be ascribed to the function on either side of it.

36. $f(x) = 3x + 5$, $g(x) = 5 - x$

(a) $(f \circ g)(x) = f(g(x)) = f(5 - x) = 3(5 - x) + 5 = 20 - 3x$

(b) $(g \circ f)(x) = g(f(x)) = g(3x + 5) = 5 - (3x + 5) = -3x$

(c) $(f \circ f)(x) = f(f(x)) = f(3x + 5) = 3(3x + 5) + 5 = 9x + 20$

38. $f(x) = x^3,\ g(x) = \dfrac{1}{x}$

(a) $(f \circ g)(x) = f(g(x))$
$= f\left(\dfrac{1}{x}\right)$
$= \left(\dfrac{1}{x}\right)^3 = \dfrac{1}{x^3}$

(b) $(g \circ f)(x) = g(f(x))$
$= g(x^3)$
$= \dfrac{1}{x^3}$

(c) $(f \circ f)(x) = f(f(x))$
$= f(x^3)$
$= (x^3)^3 = x^9$

40. $f(x) = \sqrt[3]{x - 5},\ g(x) = x^3 + 1$

(a) $(f \circ g)(x) = f(g(x))$
$= f(x^3 + 1)$
$= \sqrt[3]{x^3 + 1 - 5}$
$= \sqrt[3]{x^3 - 4}$

Domain of $f, g, f \circ g, g \circ f$: all real numbers

(b) $(g \circ f)(x) = g(f(x))$
$= g\left(\sqrt[3]{x - 5}\right)$
$= \left(\sqrt[3]{x - 5}\right)^3 + 1$
$= x - 5 + 1 = x - 4$

42. $f(x) = x^{2/3},\ g(x) = x^6$

(a) $(f \circ g)(x) = f(g(x))$
$= f(x^6)$
$= (x^6)^{2/3} = x^4$

Domain of $f, g, f \circ g, g \circ f$: all real numbers

(b) $(g \circ f)(x) = g(f(x))$
$= g(x^{2/3})$
$= (x^{2/3})^6 = x^4$

44. $f(x) = |x - 4|,\ g(x) = 3 - x$

(a) $(f \circ g)(x) = f(g(x))$
$= f(3 - x)$
$= |(3 - x) - 4|$
$= |-x - 1|$

Domain of $f, g, f \circ g, g \circ f$: all real numbers

(b) $(g \circ f)(x) = g(f(x))$
$= g(|x - 4|)$
$= 3 - (|x - 4|)$
$= 3 - |x - 4|$

46. $f(x) = \dfrac{3}{x^2 - 1},\ g(x) = x + 1$

(a) $(f \circ g)(x) = f(g(x))$
$= f(x + 1)$
$= \dfrac{3}{(x + 1)^2 - 1}$
$= \dfrac{3}{x^2 + 2x + 1 - 1}$
$= \dfrac{3}{x^2 + 2x}$

Domain of $f, g \circ f$: all real numbers $x \neq \pm 1$

Domain of g: all real numbers

Domain of $f \circ g$: all real numbers $x \neq 0, -2$

(b) $(g \circ f)(x) = g(f(x))$
$= g\left(\dfrac{3}{x^2 - 1}\right)$
$= \dfrac{3}{x^2 - 1} + 1$
$= \dfrac{3 + x^2 - 1}{x^2 - 1}$
$= \dfrac{x^2 + 2}{x^2 - 1}$

48. (a) $(f - g)(1) = f(1) - g(1)$
$= 2 - 3 = -1$

(b) $(fg)(4) = f(4) \cdot g(4) = 0$
$= 4 \cdot 0 = 0$

50. (a) $(f \circ g)(1) = f(g(1))$
$= f(3) = 2$

(b) $(g \circ f)(3) = g(f(3))$
$= g(2) = 2$

52. $h(x) = (1 - x)^3$

One possibility: Let $g(x) = 1 - x$ and $f(x) = x^3$.

$(f \circ g)(x) = f(1 - x) = (1 - x)^3 = h(x)$

54. $h(x) = \sqrt{9 - x}$

One possibility: Let $g(x) = 9 - x$ and $f(x) = \sqrt{x}$.

$(f \circ g)(x) = f(9 - x) = \sqrt{9 - x} = h(x)$

56. $h(x) = \dfrac{4}{(5x + 2)^2}$

One possibility: Let $g(x) = 5x + 2$ and $f(x) = \dfrac{4}{x^2}$.

$(f \circ g)(x) = f(5x + 2) = \dfrac{4}{(5x + 2)^2} = h(x)$

58. $h(x) = \dfrac{27x^3 + 6x}{10 - 27x^3}$

One possibility: Let $g(x) = x^3$ and $f(x) = \dfrac{27x + 6\sqrt[3]{x}}{10 - 27x}$.

$(f \circ g)(x) = f(x^3) = \dfrac{27x^3 + 6\sqrt[3]{x^3}}{10 - 27x^3} = h(x)$

60. $(A \circ r)(t) = A(r(t)) = A(0.6t) = \pi(0.6t)^2 = 0.36\pi t^2$

$A \circ r$ represents the area of the circle at time t.

62. True. The range of g must be a subset of the domain of f for $(f \circ g)(x)$ to be defined.

64. Let $f(x)$ and $g(x)$ be two odd functions and define $h(x) = f(x)g(x)$. Then

$$\begin{aligned} h(-x) &= f(-x)g(-x) \\ &= [-f(x)][-g(x)] \quad \text{Since } f \text{ and } g \text{ are odd} \\ &= f(x)g(x) \\ &= h(x) \end{aligned}$$

Thus, $h(x)$ is even.

Let $f(x)$ and $g(x)$ be two even functions and define $h(x) = f(x)g(x)$. Then

$$\begin{aligned} h(-x) &= f(-x)g(-x) \\ &= f(x)g(x) \quad \text{Since } f \text{ and } g \text{ are even} \\ &= h(x) \end{aligned}$$

Thus, $h(x)$ is even.

Section P.10 Inverse Functions

Solutions to Even-Numbered Exercises

2. The inverse is a line through (0, 6) and (6, 0). Matches graph (b).

4. The inverse is a third-degree equation through (0, 0). Matches graph (d).

6.
$$\begin{aligned} f(x) &= \tfrac{1}{3}x \\ f^{-1}(x) &= 3x \\ f(f^{-1}(x)) &= f(3x) = \tfrac{1}{3}(3x) = x \\ f^{-1}(f(x)) &= f^{-1}(\tfrac{1}{3}x) = 3(\tfrac{1}{3}x) = x \end{aligned}$$

8.
$$\begin{aligned} f(x) &= x - 4 \\ f^{-1}(x) &= x + 4 \\ f(f^{-1}(x)) &= f(x + 4) = x + 4 - 4 = x \\ f^{-1}(f(x)) &= f^{-1}(x - 4) = x - 4 + 4 = x \end{aligned}$$

10. $f(x) = \dfrac{x-1}{5}$

$f^{-1}(x) = 5x + 1$

$f(f^{-1}(x)) = f(5x + 1) = \dfrac{5x + 1 - 1}{5} = \dfrac{5x}{5} = x$

$f^{-1}(f(x)) = f^{-1}\left(\dfrac{x-1}{5}\right) = 5\left(\dfrac{x-1}{5}\right) + 1 = x - 1 + 1 = x$

12. $f(x) = x^5$

$f^{-1}(x) = \sqrt[5]{x}$

$f(f^{-1}(x)) = f(\sqrt[5]{x}) = (\sqrt[5]{x})^5 = x$

$f^{-1}(f(x)) = f^{-1}(x^5) = \sqrt[5]{x^5} = x$

14. (a) $f(x) = x - 5,\ g(x) = x + 5$

$f(g(x)) = f(x + 5) = (x + 5) - 5 = x$

$g(f(x)) = g(x - 5) = (x - 5) + 5 = x$

(b)

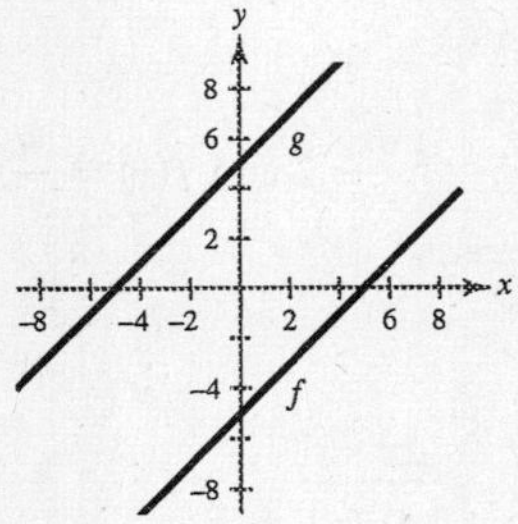

16. (a) $f(x) = 3 - 4x,\ g(x) = \dfrac{3-x}{4}$

$f(g(x)) = f\left(\dfrac{3-x}{4}\right) = 3 - 4\left(\dfrac{3-x}{4}\right) = 3 - (3 - x) = x$

$g(f(x)) = g(3 - 4x) = \dfrac{3 - (3 - 4x)}{4} = \dfrac{4x}{4} = x$

(b)

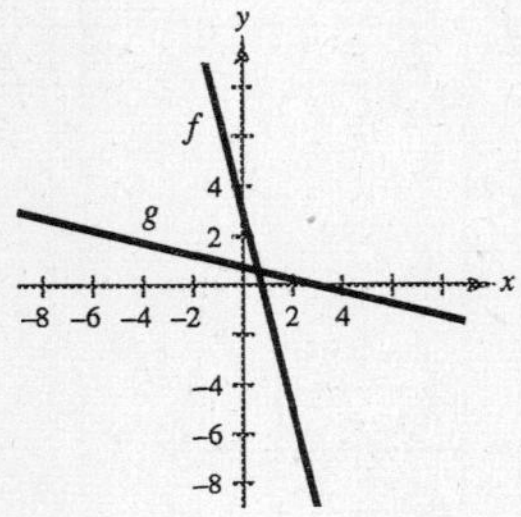

18. (a) $f(x) = \dfrac{1}{x},\ g(x) = \dfrac{1}{x}$

$f(g(x)) = f\left(\dfrac{1}{x}\right) = \dfrac{1}{\frac{1}{x}} = 1 \div \dfrac{1}{x} = 1 \cdot \dfrac{x}{1} = x$

$g(f(x)) = g\left(\dfrac{1}{x}\right) = \dfrac{1}{\frac{1}{x}} = 1 \div \dfrac{1}{x} = 1 \cdot \dfrac{x}{1} = x$

(b)

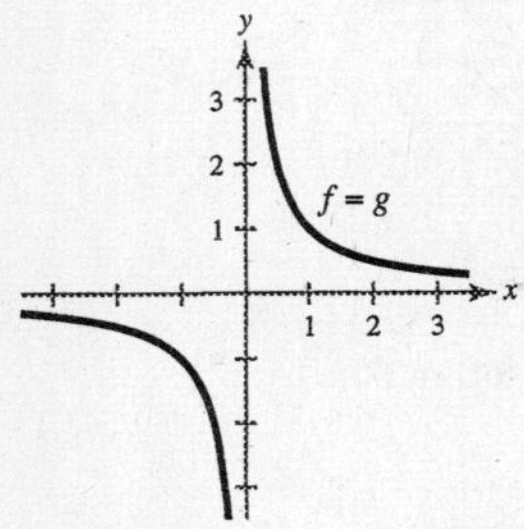

20. (a) $f(x) = 1 - x^3,\ g(x) = \sqrt[3]{1-x}$

$f(g(x)) = f(\sqrt[3]{1-x}) = 1 - (\sqrt[3]{1-x})^3 = 1 - (1 - x) = x$

$g(f(x)) = g(1 - x^3) = \sqrt[3]{1 - (1 - x^3)} = \sqrt[3]{x^3} = x$

(b)

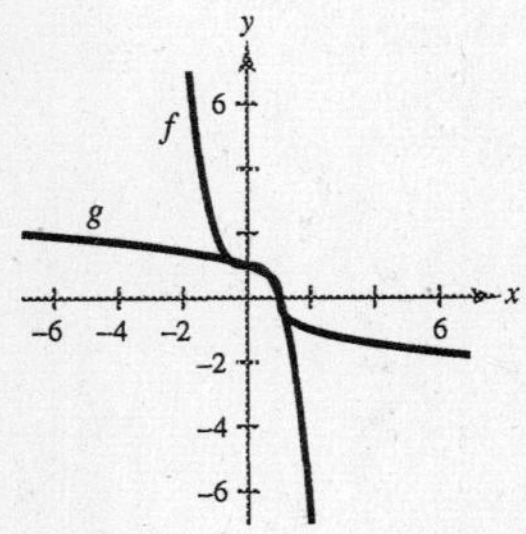

22. (a) $f(x) = \dfrac{1}{1+x},\ x \ge 0;\ g(x) = \dfrac{1-x}{x},\ 0 < x \le 1$

$$f(g(x)) = f\left(\frac{1-x}{x}\right) = \frac{1}{1+\left(\frac{1-x}{x}\right)} = \frac{1}{\frac{x}{x}+\frac{1-x}{x}} = \frac{1}{\frac{1}{x}} = x$$

$$g(f(x)) = g\left(\frac{1}{1+x}\right) = \frac{1-\left(\frac{1}{1+x}\right)}{\left(\frac{1}{1+x}\right)} = \frac{\frac{1+x}{1+x}-\frac{1}{1+x}}{\frac{1}{1+x}} = \frac{\frac{x}{1+x}}{\frac{1}{1+x}} = \frac{x}{1+x}\cdot\frac{x+1}{1} = x$$

(b)

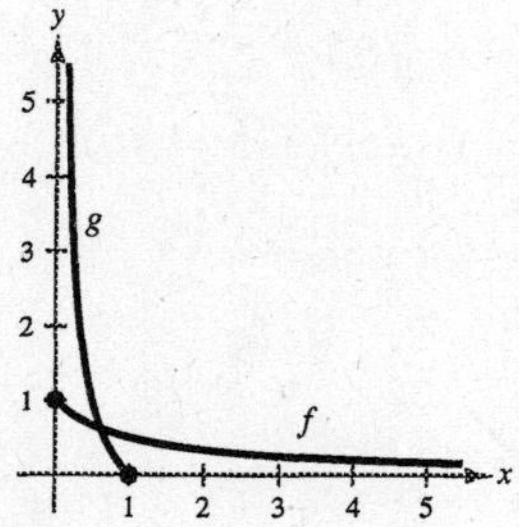

24. (a) $f(x) = \dfrac{x+3}{x-2},\ g(x) = \dfrac{2x+3}{x-1}$

$$f(g(x)) = f\left(\frac{2x+3}{x-1}\right) = \frac{\frac{2x+3}{x-1}+3}{\frac{2x+3}{x-1}-2} = \frac{\frac{2x+3+3x-3}{x-1}}{\frac{2x+3-2x+2}{x-1}}$$

$$= \frac{5x}{5} = x$$

$$g(f(x)) = g\left(\frac{x+3}{x-2}\right) = \frac{2\left(\frac{x+3}{x-2}\right)+3}{\frac{x+3}{x-2}-1}$$

$$= \frac{\frac{2x+6+3x-6}{x-2}}{\frac{x+3-x+2}{x-2}} = \frac{5x}{5} = x$$

(b)

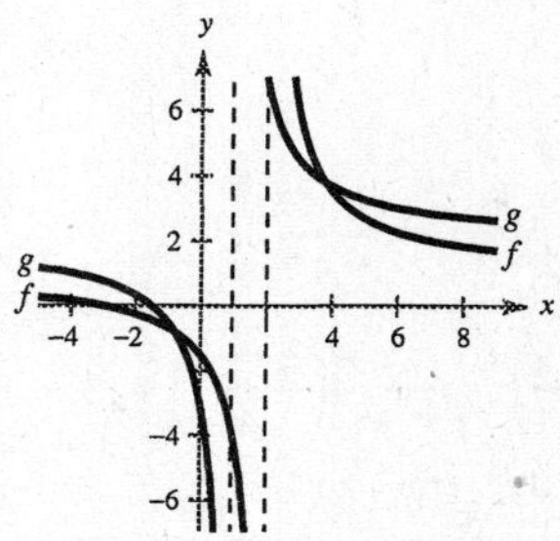

26. Yes, $\{(10, -3), (6, -2), (4, -1), (1, 0), (-3, 2), (-10, 2)\}$ does represent a function.

28.

x	-10	-7	-4	-1	2	5
$f^{-1}(x)$	-3	-2	-1	0	1	2

30. No, because some horizontal lines intersect the graph twice, f does not have an inverse.

32. Yes, because no horizontal lines intersect the graph at more than one point, f has an inverse.

34. $f(x) = 10$

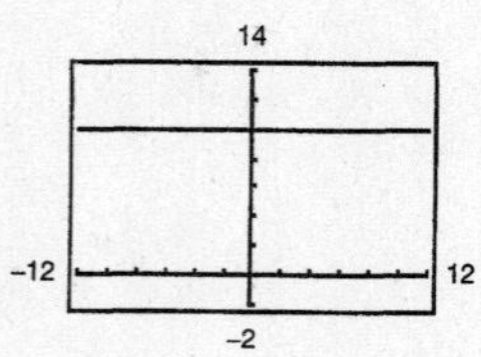

f does not pass the horizontal line test, so f has no inverse.

36. $g(x) = (x + 5)^3$

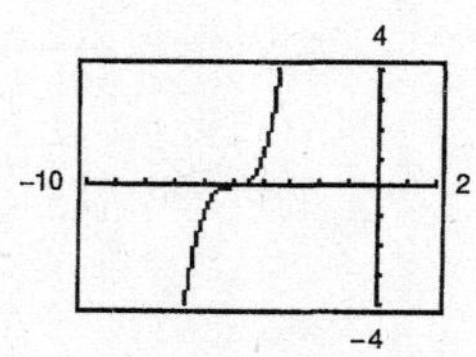

g passes the horizontal line test, so g has an inverse.

38. $f(x) = \frac{1}{8}(x + 2)^2 - 1$

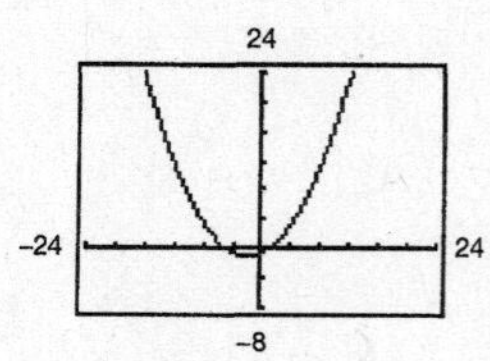

f does not pass the horizontal line test, so f has no inverse.

40. $f(x) = 3x + 1$

$$y = 3x + 1$$
$$x = 3y + 1$$
$$\frac{x-1}{3} = y$$
$$f^{-1}(x) = \frac{x-1}{3}$$

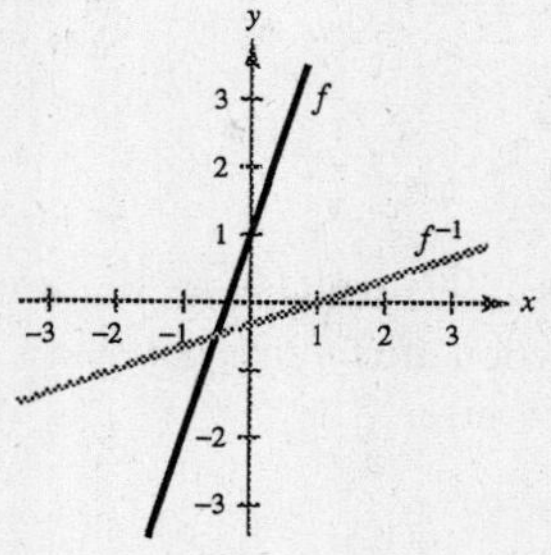

42. $f(x) = x^3 + 1$

$$y = x^3 + 1$$
$$x = y^3 + 1$$
$$x - 1 = y^3$$
$$\sqrt[3]{x-1} = y$$
$$f^{-1}(x) = \sqrt[3]{x-1}$$

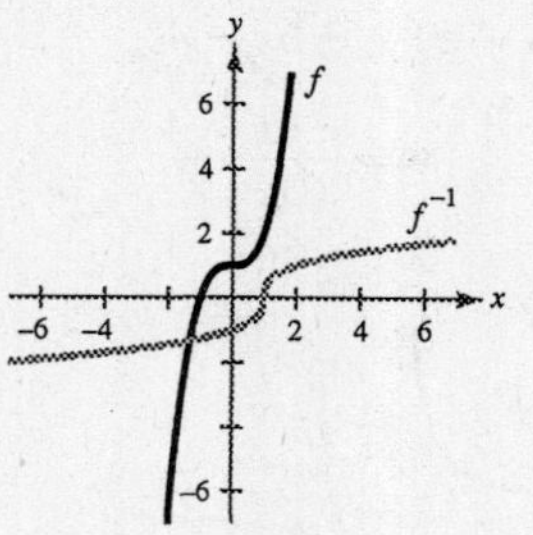

44. $f(x) = x^2,\ x \ge 0$

$$y = x^2$$
$$x = y^2$$
$$\sqrt{x} = y$$
$$f^{-1}(x) = \sqrt{x}$$

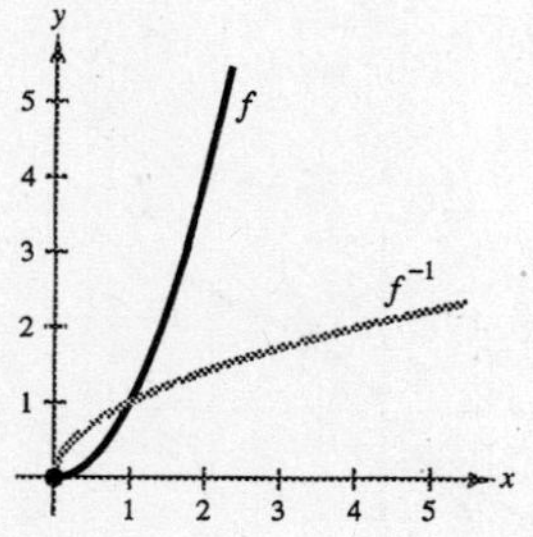

46. $f(x) = x^2 - 2, x \le 0$

$$y = x^2 - 2$$
$$x = y^2 - 2$$
$$\pm\sqrt{x+2} = y$$
$$f^{-1}(x) = -\sqrt{x+2}$$

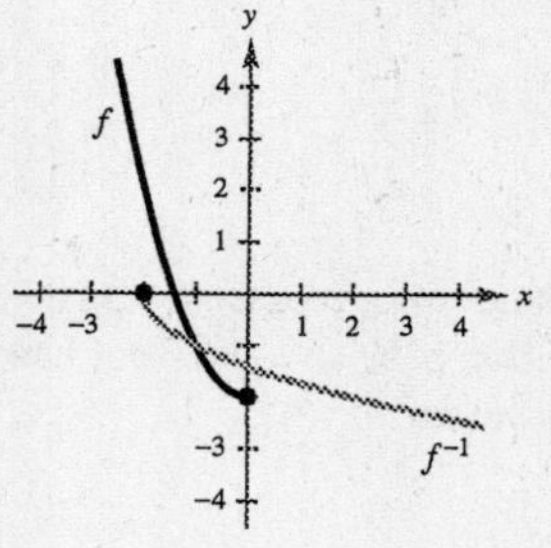

48. $f(x) = -\frac{2}{x}$

$$y = -\frac{2}{x}$$
$$x = -\frac{2}{y}$$
$$y = -\frac{2}{x}$$
$$f^{-1}(x) = -\frac{2}{x}$$

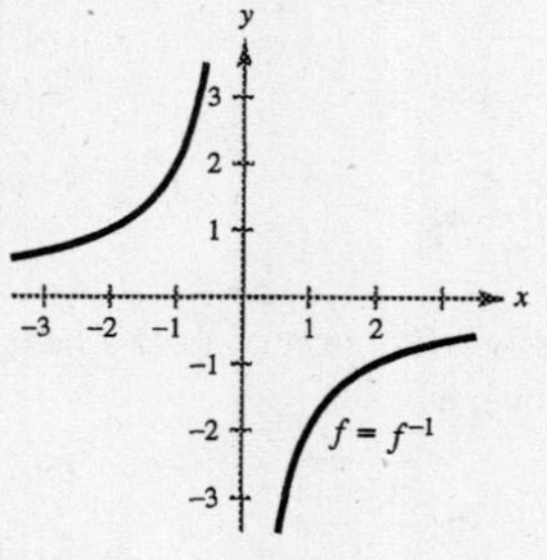

50. $f(x) = \frac{x-3}{x+2}$

$$y = \frac{x-3}{x+2}$$
$$x = \frac{y-3}{y+2}$$
$$xy + 2x - y + 3 = 0$$
$$y(x-1) = -2x - 3$$
$$y = \frac{-2x-3}{x-1}$$
$$f^{-1}(x) = \frac{-2x-3}{x-1}$$

52. $f(x) = x^{3/5}$

$y = x^{3/5}$

$x = y^{3/5}$

$x^{5/3} = (y^{3/5})^{5/3}$

$x^{5/3} = y$

$f^{-1}(x) = x^{5/3}$

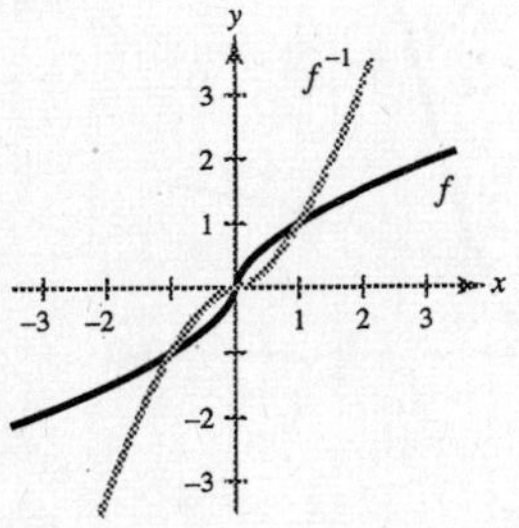

54. $f(x) = \dfrac{8x - 4}{2x + 6}$

$y = \dfrac{8x - 4}{2x + 6}$

$x = \dfrac{8y - 4}{2y + 6}$

$2xy + 6x = 8y - 4$

$y(2x - 8) = -6x - 4$

$y = \dfrac{-6x - 4}{2x - 8}$

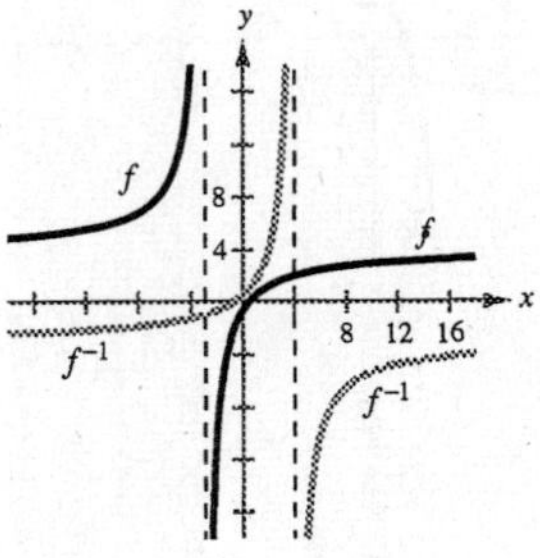

56. $f(x) = \dfrac{1}{x^2}$

$y = \dfrac{1}{x^2}$

$x = \dfrac{1}{y^2}$

$y^2 = \dfrac{1}{x}$

$y = \pm\sqrt{\dfrac{1}{x}}$

This does not represent y as a function of x. f does not have an inverse.

58. $f(x) = 3x + 5$

$y = 3x + 5$

$x = 3y + 5$

$x - 5 = 3y$

$\dfrac{x - 5}{3} = y$

This is a function of x, so f has an inverse.

$f^{-1}(x) = \dfrac{x - 5}{3}$

60. $f(x) = \dfrac{3x + 4}{5}$

$y = \dfrac{3x + 4}{5}$

$x = \dfrac{3y + 4}{5}$

$5x = 3y + 4$

$5x - 4 = 3y$

$\dfrac{5x - 4}{3} = y$

This is a function of x, so f has an inverse.

$f^{-1}(x) = \dfrac{5x - 4}{3}$

62. $q(x) = (x - 5)^2$

$y = (x - 5)^2$

$x = (y - 5)^2$

$\pm\sqrt{x} = y - 5$

$5 \pm \sqrt{x} = y$

This does not represent y as a function of x, so q does not have an inverse.

64. $f(x) = \begin{cases} -x, & x \le 0 \\ x^2 - 3x, & x > 0 \end{cases}$

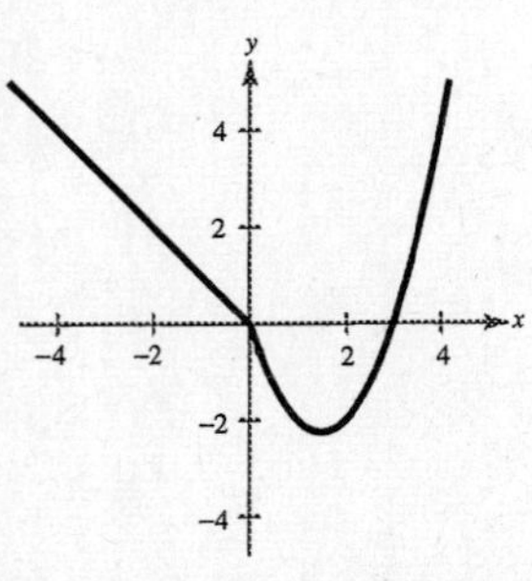

The graph fails the horizontal line test, so f does not have an inverse.

66. $f(x) = |x - 2|,\ x \le 2 \Rightarrow y \ge 0$

$y = |x - 2|,\ x \le 2,\ y \ge 0$

$x = |y - 2|,\ y \le 2,\ x \ge 0$

$x = y - 2$ or $-x = y - 2$

$2 + x = y$ or $2 - x = y$

The portion that satisfies the conditions $y \le 2$ and $x \ge 0$ is $2 - x = y$. This is a function of x, so f has an inverse.

$f^{-1}(x) = 2 - x,\ x \ge 0$

68. $f(x) = \sqrt{x-2} \Rightarrow x \geq 2,\ y \geq 0$

$y = \sqrt{x-2},\ x \geq 2,\ y \geq 0$

$x = \sqrt{y-2},\ y \geq 2,\ x \geq 0$

$x^2 = y - 2,\ x \geq 0,\ y \geq 2$

$x^2 + 2 = y,\ x \geq 0,\ y \geq 2$

This is a function of x, so f has an inverse.

$f^{-1}(x) = x^2 + 2,\ x \geq 0$

70. $(g^{-1} \circ f^{-1})(-3) = g^{-1}(f^{-1}(-3))$

$= g^{-1}(8(-3+3))$

$= g^{-1}(0) = \sqrt[3]{0} = 0$

72. $(g^{-1} \circ g^{-1})(-4) = g^{-1}(g^{-1}(-4))$

$= g^{-1}(\sqrt[3]{-4}) = \sqrt[3]{\sqrt[3]{-4}} = \sqrt[9]{-4}$

or $= \sqrt[9]{4}$

74. $g^{-1} \circ f^{-1} = g^{-1}(f^{-1}(x))$

$= g^{-1}(8(x+3))$

$= \sqrt[3]{8(x+3)}$

$= 2\sqrt[3]{x+3}$

76. $f^{-1} \circ g^{-1}(x) = f^{-1}(g^{-1}(x))$

$= f^{-1}\left(\dfrac{x+5}{2}\right)$

$= \dfrac{x+5}{2} - 4$

$= \dfrac{x+5-8}{2}$

$= \dfrac{x-3}{2}$

78. $(g \circ f)(x) = g(f(x))$

$= g(x+4)$

$= 2(x+4) - 5$

$= 2x + 8 - 5$

$= 2x + 3$

$y = 2x + 3$

$x = 2y + 3$

$x - 3 = 2y$

$\dfrac{x-3}{2} = y$

$(g \circ f)^{-1}(x) = \dfrac{x-3}{2}$

80. (a) Yes, f^{-1} exists.

(b) It represents the year for a given per capita consumption of bottled water.

(c) $f^{-1}(16.0) = 8$.

82. (a) $y = 8 + 0.75x$

$x = 8 + 0.75y$

$x - 8 = 0.75y$

$\dfrac{x-8}{0.75} = y$

$f^{-1}(x) = \dfrac{x-8}{0.75}$

(b) x = hourly wage

y = number of units produced

(c) $y = \dfrac{22.25 - 8}{0.75} = 19$ units

84. (a)
$$x = 1.25y + 1.60(50 - y)$$
$$x = 1.25y + 80 - 1.60y$$
$$x - 80 = -0.35y$$
$$\frac{x - 80}{-0.35} = y$$
$$y = \frac{80 - x}{0.35}$$

x = total cost

y = number of pounds of less expensive ground beef

(b)
$$0 \le y \le 50$$
$$0 \le \frac{80 - x}{0.35} \le 50$$
$$0 \le 80 - x \le 17.5$$
$$-80 \le -x \le -62.5$$
$$62.5 \le x \le 80$$

(c) $\dfrac{80 - 73}{0.35} = y = 20$ pounds

86. True. If $f(x) = x - 6$ and $f^{-1}(x) = x + 6$, then the y-intercept of f is $(0, -6)$ and the x-intercept of f^{-1} is $(-6, 0)$.

88.

x	$f(x)$
-2	-5
-1	-2
1	2
3	3

x	$f^{-1}(x)$
-5	-2
-2	-1
2	1
3	3

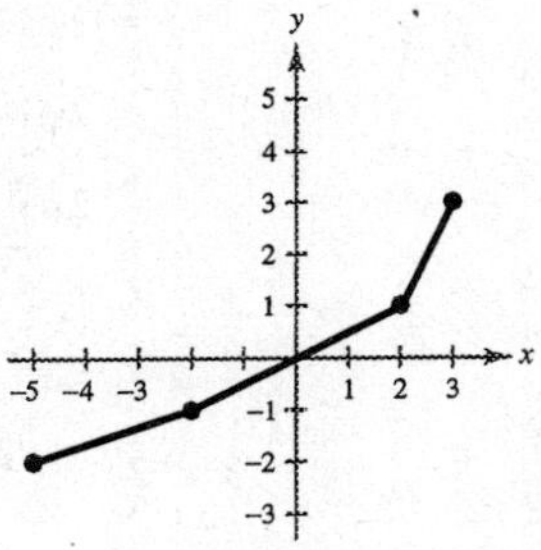

90.

x	$f(x)$
-4	3
-2	4
0	0
3	-1

The graph does not pass the Horizontal Line Test, so $f^{-1}(x)$ does not exist.

92.
$$f(x) = k(x^3 + 3x - 4)$$
$$y = k(x^3 + 3x - 4)$$
$$x = k(y^3 + 3y - 4)$$
$$-5 = k[(2)^3 + 3(2) - 4]$$
$$-5 = 10k$$
$$-\tfrac{1}{2} = k$$

Review Exercises for Chapter P

Solutions to Even-Numbered Exercises

2. $\left\{\sqrt{15}, -22, -\frac{10}{3}, 0, 5.2, \frac{3}{7}\right\}$

(a) Natural numbers: none

(b) Integers: $-22, 0$

(c) Rational numbers: $-22, -\frac{10}{3}, 0, 5.2, \frac{3}{7}$

(d) Irrational numbers: $\sqrt{15}$

4. (a) $\frac{9}{25} = 0.36$

(b) $\frac{5}{7} = 0.\overline{714285}$

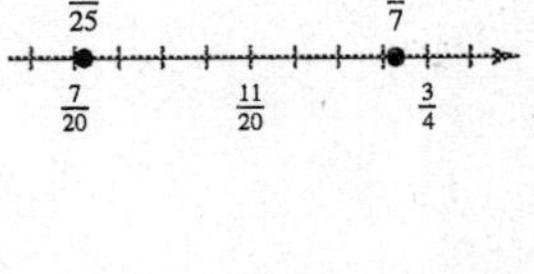

6. $x > 1$

The set consists of all real numbers greater than 1.

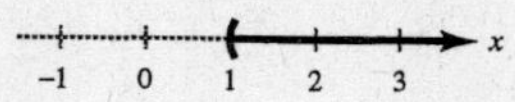

8. $d(-112, -6) = |-6 - (-112)| = 106$

10. $d(z, -16) = |z - (-16)| = |z + 16|$ and $d(z, -16) > 8$.
Thus, $|z + 16| > 8$.

12. $\dfrac{x}{x - 3}$

(a) $\dfrac{-3}{-3 - 3} = \dfrac{1}{2}$

(b) $\dfrac{3}{3 - 3}$ is undefined.

14. $(t + 4)(2t) = (2t)(t + 4)$ illustrates the Commutative Property of Multiplication.

16. $\dfrac{2}{y + 4} \cdot \dfrac{y + 4}{2} = 1, \quad y \neq -4$
Illustrates the Multiplicative Inverse Property

18. $\dfrac{|-10|}{-10} = \dfrac{10}{-10} = -1$

20. $(16 - 8) \div 4 = 8 \div 4 = 2$

22. $-4[16 - 3(7 - 10)] = -4[16 - 3(-3)] = -4[16 + 9] = -4(25) = -100$

24. $3(x - 2) + 2x = 2(x + 3)$

$3x - 6 + 2x = 2x + 6$

$3x = 12$

Conditional equation

26. $4x + 2(7 - x) = 5$

$4x + 14 - 2x = 5$

$2x = -9$

$x = -\dfrac{9}{2}$

28. $\dfrac{1}{2}(x - 3) - 2(x + 1) = 5$

$x - 3 - 4x - 4 = 10$

$-3x = 17$

$x = -\dfrac{17}{3}$

30. $\dfrac{4x - 3}{6} + \dfrac{x}{4} = x - 2$

$(12)\dfrac{4x - 3}{6} + (12)\dfrac{x}{4} = 12(x - 2)$

$8x - 6 + 3x = 12x - 24$

$18 = x$

32. $\dfrac{5}{x - 2} = \dfrac{13}{2x - 3}$

$5(2x - 3) = 13(x - 2)$

$10x - 15 = 13x - 26$

$11 = 3x$

$\dfrac{11}{3} = x$

34. $2x^2 - x - 28 = 0$

$(2x + 7)(x - 4) = 0$

$2x + 7 = 0$ or $x - 4 = 0$

$2x = -7 \qquad x = 4$

$x = -\dfrac{7}{2}$

36. $16x^2 = 25$

$x^2 = \dfrac{25}{16}$

$x = \pm\sqrt{\dfrac{25}{16}} = \pm\dfrac{5}{4}$

38. $(x - 8)^2 = 15$

$x - 8 = \pm\sqrt{15}$

$x = 8 \pm \sqrt{15}$

40. $x^2 + 6x - 3 = 0$

$a = 1, b = 6, c = -3$

$x = \dfrac{-6 \pm \sqrt{(6)^2 - 4(1)(-3)}}{2(1)}$

$x = \dfrac{-6 \pm \sqrt{48}}{2}$

$x = \dfrac{-6 \pm 4\sqrt{3}}{2}$

$x = -3 \pm 2\sqrt{3}$

42. $-20 - 3x + 3x^2 = 0$

$a = 3, b = -3, c = -20$

$x = \dfrac{-(-3) \pm \sqrt{(-3)^2 - 4(3)(-20)}}{2(3)}$

$x = \dfrac{3 \pm \sqrt{249}}{6}$

$x = \dfrac{1}{2} \pm \dfrac{\sqrt{249}}{6}$

44. $4x^3 - 6x^2 = 0$

$2x^2(2x - 3) = 0$

$2x^2 = 0$ or $2x - 3 = 0$

$x = 0$ or $x = \dfrac{3}{2}$

46. $9x^4 + 27x^3 - 4x^2 - 12x = 0$

$x(9x^3 + 27x^2 - 4x - 12) = 0$

$x = 0$ is a solution.

-3	9	27	-4	-12
		-27	0	12
	9	0	-4	0

$x = -3$ is a solution.

$9x^2 - 4 = 0$

$9x^2 = 4$

$x^2 = \dfrac{4}{9}$

$x = \pm\sqrt{\dfrac{4}{9}} = \pm\dfrac{2}{3}$

$x = -\dfrac{2}{3}$ and $x = \dfrac{2}{3}$ are solutions.

48. $\sqrt{x - 2} - 8 = 0$

$\sqrt{x - 2} = 8$

$(\sqrt{x - 2})^2 = 8^2$

$x - 2 = 64$

$x = 66$

50. $5\sqrt{x} - \sqrt{x - 1} = 6$

$5\sqrt{x} = 6 + \sqrt{x - 1}$

$(5\sqrt{x})^2 = (6 + \sqrt{x - 1})^2$

$25x = 36 + x - 1 + 12\sqrt{x - 1}$

$24x - 35 = 12\sqrt{x - 1}$

$(24x - 35)^2 = (12\sqrt{x - 1})^2$

$576x^2 - 1680x + 1225 = 144(x - 1)$

$576x^2 - 1824x + 1369 = 0$

—CONTINUED—

50. —CONTINUED—

$$x = \frac{-(-1824) \pm \sqrt{(-1824)^2 - 4(576)(1369)}}{2(576)}$$

$$x = \frac{1824 \pm \sqrt{172,800}}{1152}$$

$$x = \frac{1824 \pm 240\sqrt{3}}{1152}$$

$$x = \frac{38 \pm 5\sqrt{3}}{24}$$

$x = \dfrac{38 - 5\sqrt{3}}{24}$ does not check as a solution. The solution is $x = \dfrac{38 + 5\sqrt{3}}{24}$.

52. $(x + 2)^{3/4} = 27$

$x + 2 = 27^{4/3}$

$x + 2 = 81$

$x = 79$

54. $|2x + 3| = 7$

$2x + 3 = 7$ or $2x + 3 = -7$

$2x = 4$ $\qquad$ $2x = -10$

$x = 2$ $\qquad$ $x = -5$

56. $y > 0$

Quadrants I and II

58. $xy = 4$

Quadrants I and III

60. (a)

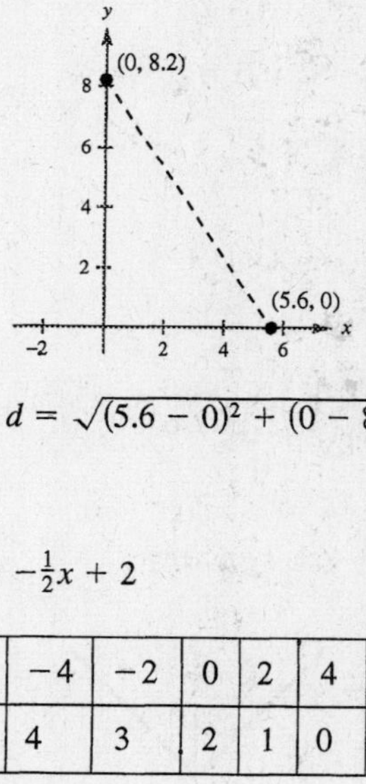

(b) $d = \sqrt{(5.6 - 0)^2 + (0 - 8.2)^2} \approx 9.9$

62. (a)

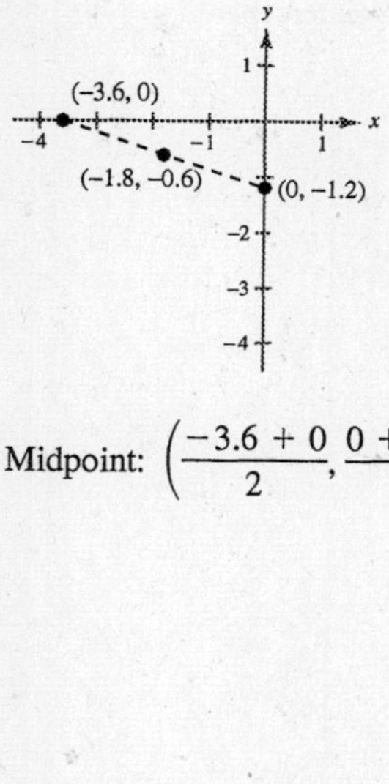

(b) Midpoint: $\left(\dfrac{-3.6 + 0}{2}, \dfrac{0 + (-1.2)}{2}\right) = (-1.8, -0.6)$

64. $y = -\frac{1}{2}x + 2$

x	-4	-2	0	2	4
y	4	3	2	1	0

66. $y = 2x^2 - x - 9$

x	-2	-1	0	1	2	3
y	1	-6	-9	-8	-3	6

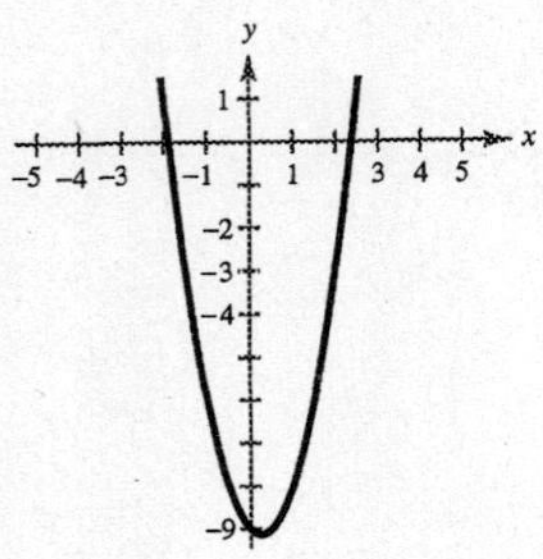

68. $y = \sqrt{x + 2}$

x	-2	-1	0	2	4
y	0	1	1.41	2	2.45

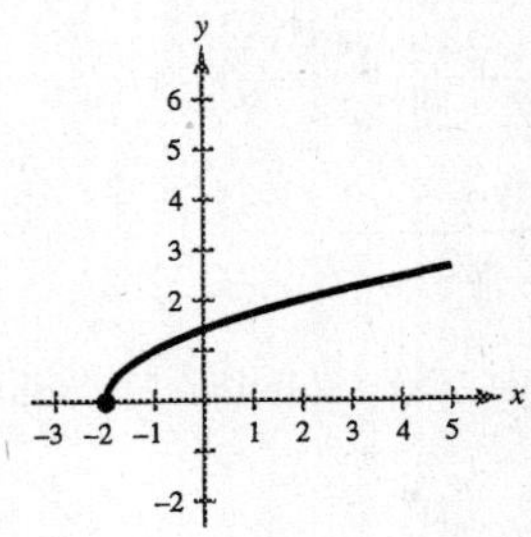

70. $y = 4 - (x - 4)^2$

x-intercepts:

$0 = 4 - (x - 4)^2$

$(x - 4)^2 = 4$

$x - 4 = \pm 2$

$x = 4 \pm 2 = 6, 2$

$(2, 0)$ and $(6, 0)$ are the x-intercepts

y-intercept:

$y = 4 - (0 - 4)^2$

$y = 4 - 16 = -12$

$(0, -12)$ is the y-intercept

72. $y = |x + 2| + |3 - x|$

x-intercepts:

$0 = |x + 2| + |3 - x|$

$0 = x + 2 + 3 - x$

$0 \neq 5$

No x-intercepts

y-intercept:

$y = |0 + 2| + |3 - 0| = 2 + 3 = 5$

$(0, 5)$ is the y-intercept

74. $y^2 = 2 - x^2$

$(-y)^2 = 2 - x^2$

$y^2 = 2 - x^2$

x-axis symmetry

$y^2 = 2 - (-x)^2$

$y^2 = 2 - x^2$

y-axis symmetry

$(-y)^2 = 2 - (-x)^2$

$y^2 = 2 - x^2$

origin symmetry

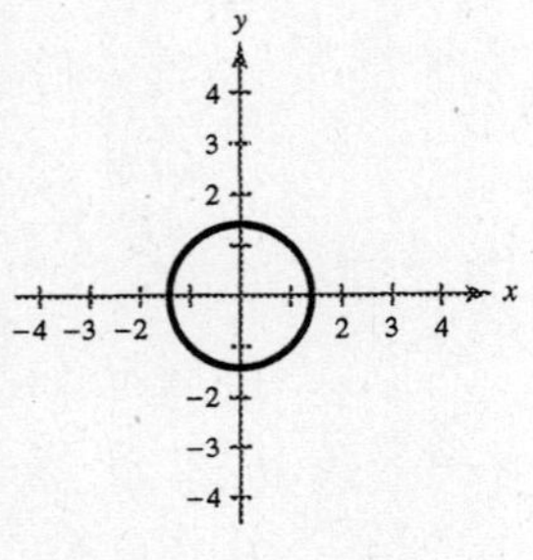

76. $y = -6 - x^3$

No symmetry

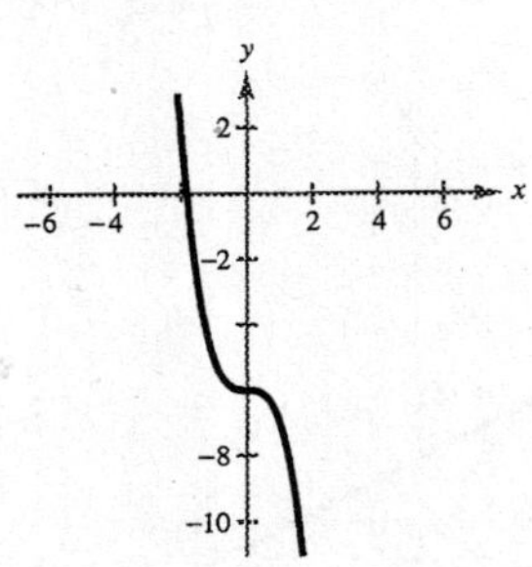

78. $y = |x| + 9$

$y = |-x| + 9$

$y = |x| + 9$

y-axis symmetry

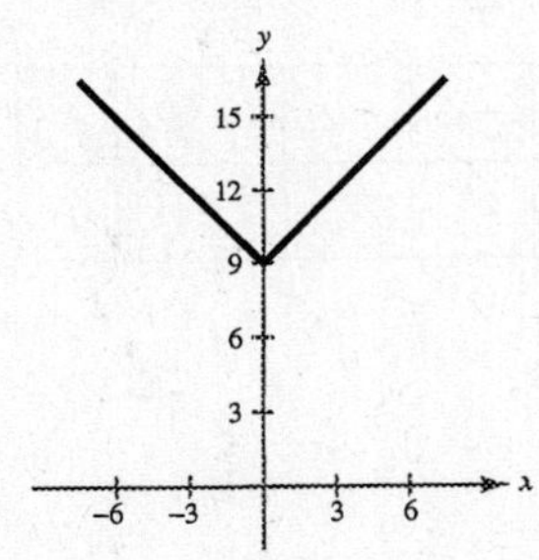

80. $x^2 + (y - 8)^2 = 81$

Center: $(0, 8)$

Radius: 9

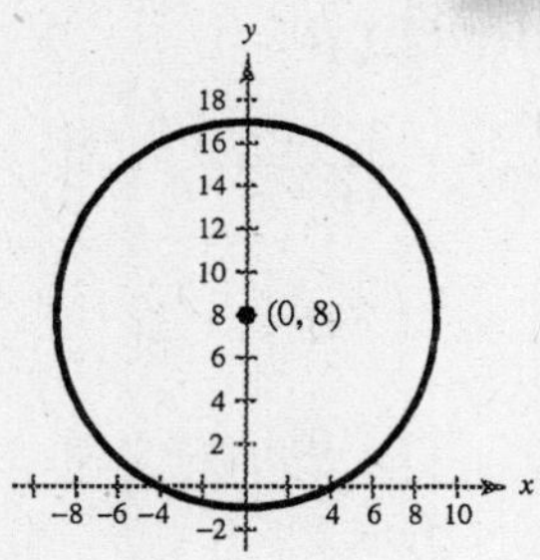

82. $(x + 4)^2 + \left(y - \frac{3}{2}\right)^2 = 100$

$(x - (-4))^2 + \left(y - \frac{3}{2}\right)^2 = 100$

Center: $\left(-4, \frac{3}{2}\right)$

Radius: 10

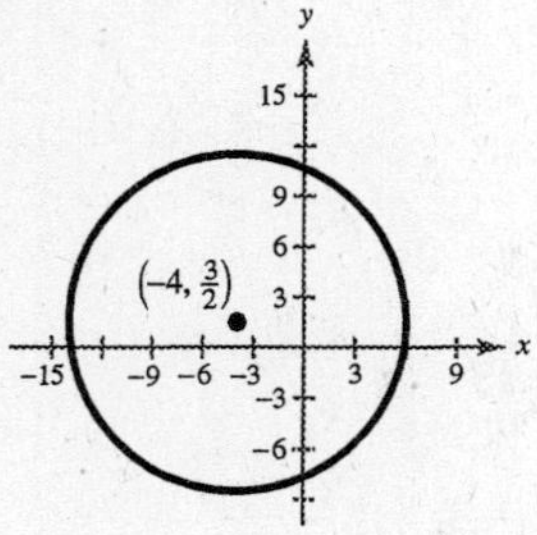

84. Endpoints of diameter: $(-2, -3), (4, -10)$

Center: $\left(\dfrac{-2 + 4}{2}, \dfrac{-3 + (-10)}{2}\right) = \left(1, -\dfrac{13}{2}\right)$

Radius: $\dfrac{1}{2}\sqrt{(-2 - 4)^2 + (-3 - (-10))^2} = \dfrac{1}{2}\sqrt{85}$

$$(x - 1)^2 + \left(y + \frac{13}{2}\right)^2 = \frac{85}{4}$$

86. $y = 4x - 3$

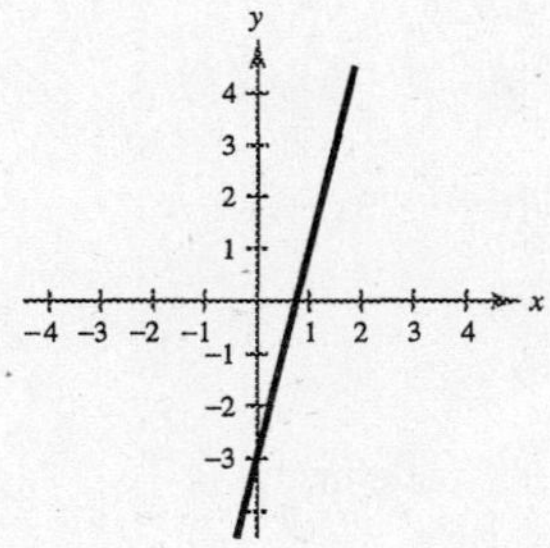

88. $x = -3$

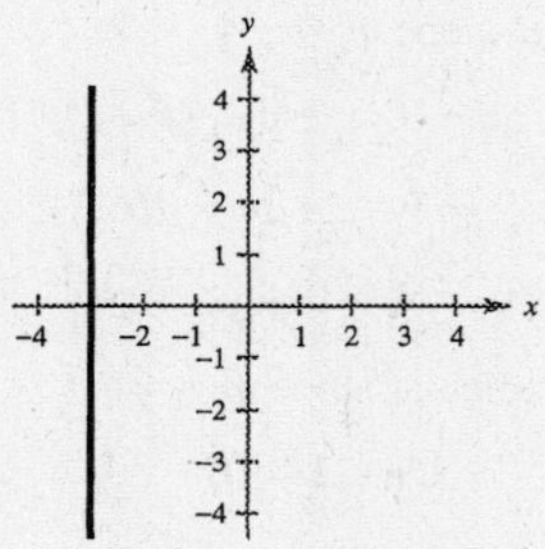

90. $y = -10x + 9$

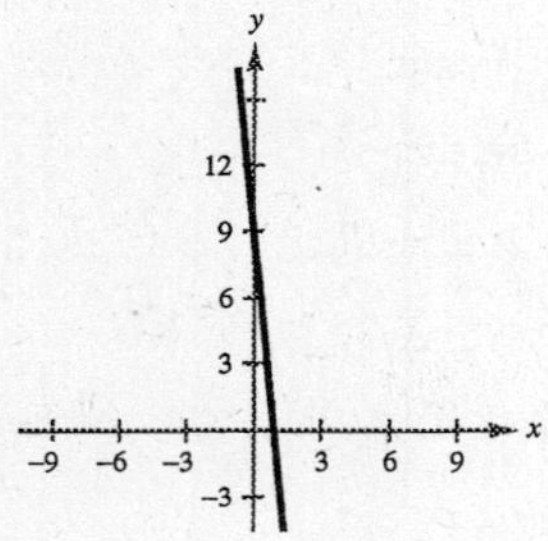

92. $y = \frac{5}{6}x + 5$

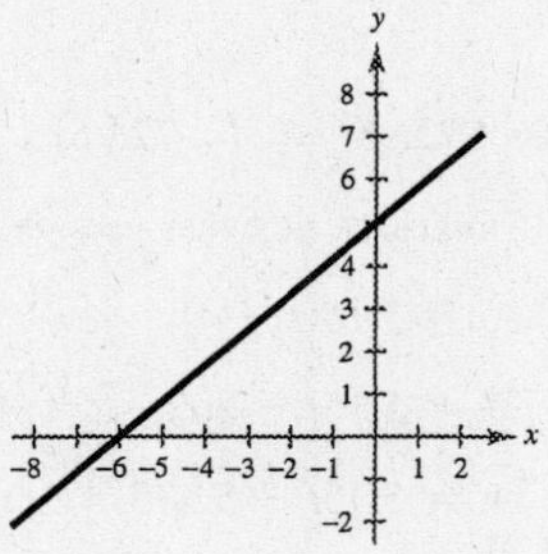

94. $(-1, 8), (6, 5)$

Slope:

$$m = \frac{8 - 5}{-1 - 6} = -\frac{3}{7}$$

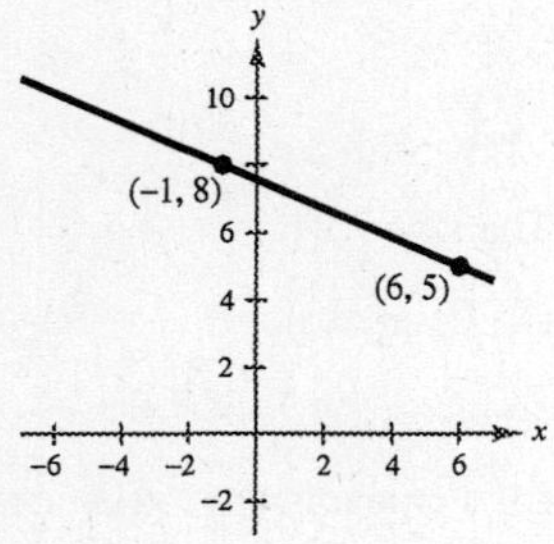

96. $(-3, 2), (8, 2)$

Slope:

$$m = \frac{2 - 2}{-3 - 8} = \frac{0}{-11} = 0$$

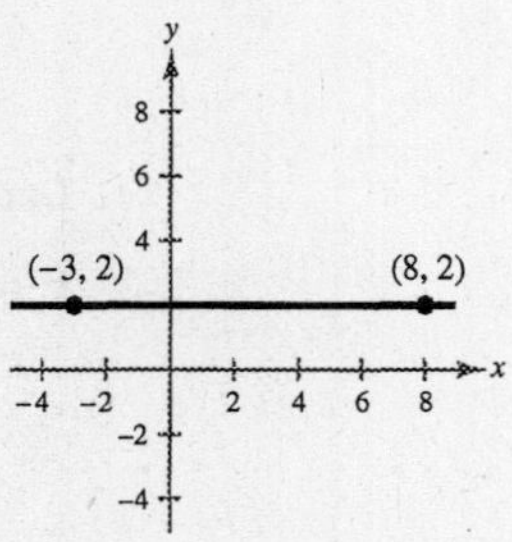

98. $(7, 3.8), (5.2, 0)$

$$m = \frac{3.8 - 0}{7 - 5.2} = \frac{3.8}{1.8} = \frac{19}{9}$$

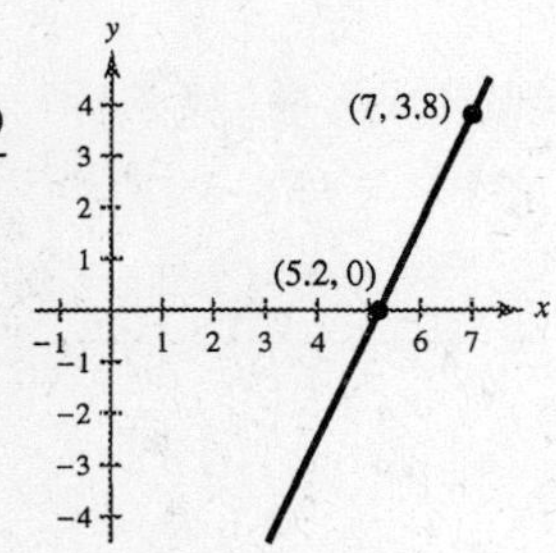

100. $(2, 5), (-2, -1)$

$$m = \frac{5 - (-1)}{2 - (-2)} = \frac{6}{4} = \frac{3}{2}$$

$$y - 5 = \frac{3}{2}(x - 2)$$

$$2y - 10 = 3x - 6$$

$$3x - 2y + 4 = 0$$

102. $(11, -2), (6, -1)$

$$m = \frac{-2 - (-1)}{11 - 6} = -\frac{1}{5}$$

$$y - (-2) = -\frac{1}{5}(x - 11)$$

$$5y + 10 = -x + 11$$

$$x + 5y - 1 = 0$$

104. $(-3, 4), (-7, 9)$

$$m = \frac{9 - 4}{-7 - (-3)} = -\frac{5}{4}$$

$$y - 4 = -\frac{5}{4}(x - (-3))$$

$$4y - 16 = -5x - 15$$

$$5x + 4y - 1 = 0$$

106. $(-2, 6),\ m = 0$

$$y - 6 = 0(x - (-2))$$

$$y - 6 = 0$$

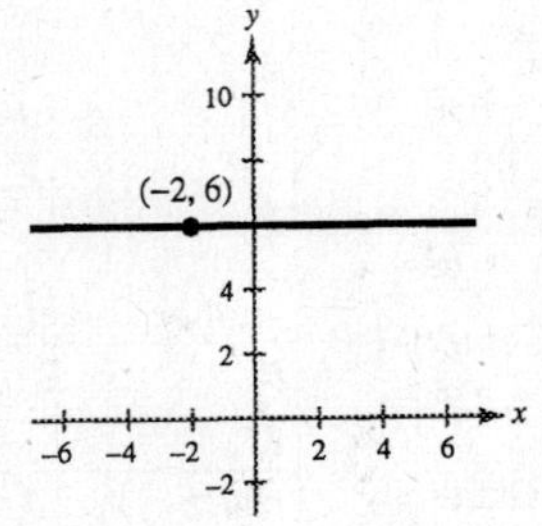

108. $(-8, 5),\ m$ is undefined

$$x + 8 = 0$$

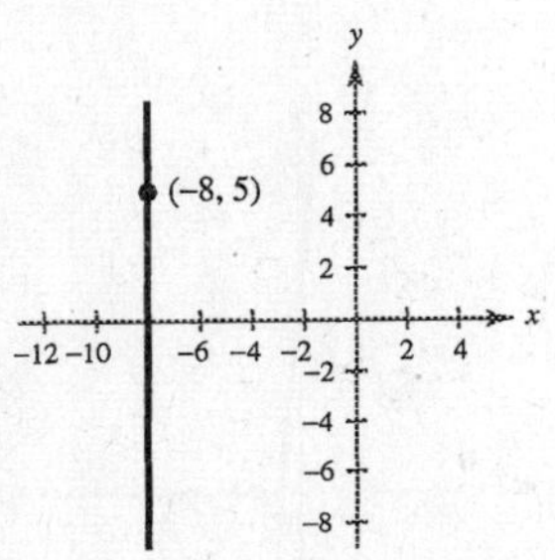

110. $(-8, 3),\ 2x + 3y = 5$

$$3y = 5 - 2x$$

$$y = \tfrac{5}{3} - \tfrac{2}{3}x$$

(a) Parallel slope $m = -\frac{2}{3}$

$$y - 3 = -\tfrac{2}{3}(x - (-8))$$

$$3y - 9 = -2x - 16$$

$$2x + 3y + 7 = 0$$

(b) Perpendicular slope $m = \frac{3}{2}$

$$y - 3 = \tfrac{3}{2}(x - (-8))$$

$$2y - 6 = 3x + 24$$

$$0 = 3x - 2y + 30$$

112. $(-2, 5),\ y = -4$

The slope of $y = -4$ is 0.

(a) Parallel slope: $y = 0$

$$y - 5 = 0(x - (-2)) \Longrightarrow y = 5$$

(b) Perpendicular slope: undefined, since a line perpendicular to a horizontal line is vertical

The vertical line through $(-2, 5)$ is $x = -2$.

114. 2004 value: \$72.95 $\Longrightarrow (4, 72.95)$

Rate: \$5.15 increase per year $\Longrightarrow m = 5.15$

$$V - 72.95 = 5.15(t - 4)$$

$$V - 72.95 = 5.15t - 20.6$$

$$V = 5.15t + 52.35,\ 4 \le t \le 9$$

116. Point: $(5, 85),\ m = 3.75$

(a) $V - 85 = 3.75(t - 5)$

$$V - 85 = 3.75t - 18.75$$

$$V = 3.75t + 66.25$$

(b)

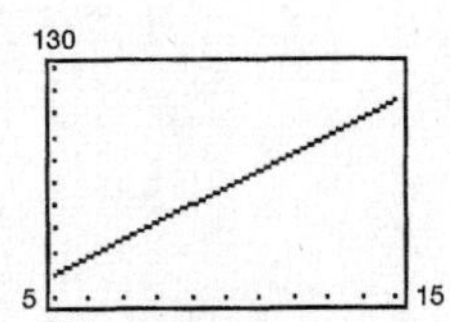

(c) Dollar value of product in 2010:

$$V = \$103.75$$

118. (a) $\{(v, -1), (u, 2), (w, 0), (u, -2)\}$ is not a function from A to B because u is matched with two elements in the range.

(b) $\{(u, -2), (v, 2), (w, 1)\}$ is a function from A to B.

(c) $\{(u, 2), (v, 2), (w, 1), (w, 1)\}$ is a function from A to B.

(d) $\{(w, -2), (v, 0), (w, 2)\}$ is not a function from A to B because w is matched with two elements in the range.

120. $2x - y - 3 = 0$

$$y = 2x - 3$$

y is a function of x.

122. $|y| = x + 2$

$y = x + 2$ or $y = -x - 2$

y is not a function of x. Each x-value corresponds to two y-values.

124. $f(x) = \dfrac{4}{x^2 + 1}$

(a) $f(1) = \dfrac{4}{1^2 + 1} = \dfrac{4}{2} = 2$

(b) $f(-5) = \dfrac{4}{(-5)^2 + 1} = \dfrac{4}{26} = \dfrac{2}{13}$

(c) $f(-t) = \dfrac{4}{(-t)^2 + 1} = \dfrac{4}{t^2 + 1}$

(d) $f(0) = \dfrac{4}{0^2 + 1} = \dfrac{4}{1} = 4$

126. $f(x) = 3x + 4$

Domain: All real numbers

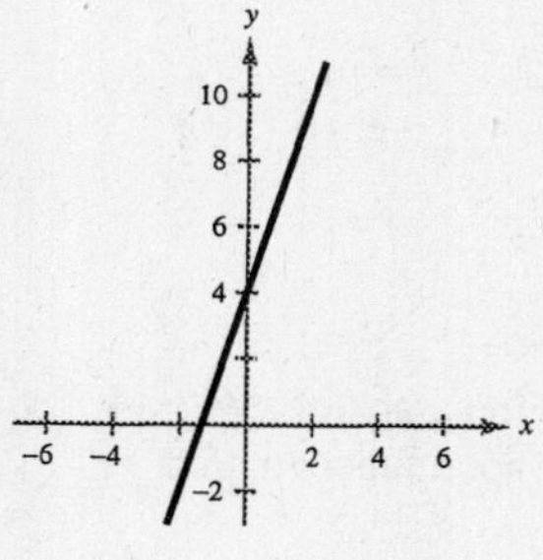

128. $f(x) = \sqrt{x^2 + 8x}$

Domain: $x^2 + 8x \geq 0$

$x(x + 8) \geq 0$

$x \leq -8$ or $x \geq 0$

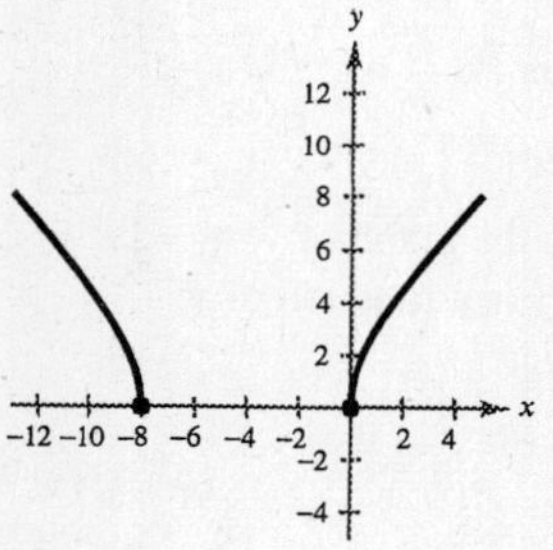

130. $h(t) = |t + 1|$

Domain: All real numbers

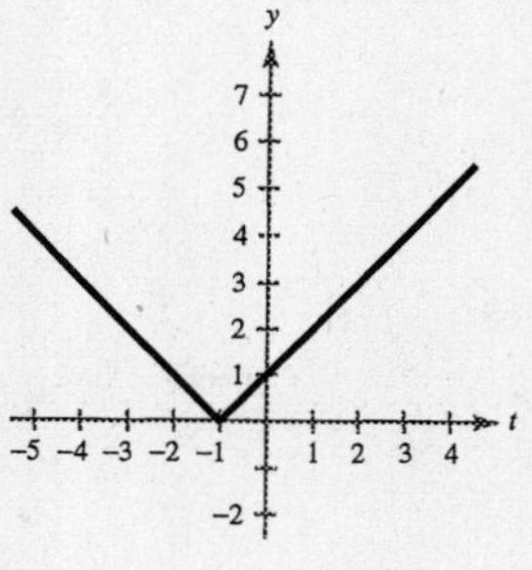

132. $f(x) = x^3 - 5x^2 + x$

$$\frac{f(x+h) - f(x)}{h} = \frac{(x+h)^3 - 5(x+h)^2 + (x+h) - (x^3 - 5x^2 + x)}{h}$$

$$= \frac{x^3 + 3x^2h + 3xh^2 + h^3 - 5x^2 - 10xh - 5h^2 + x + h - x^3 + 5x^2 - x}{h}$$

$$= \frac{3x^2h + 3xh^2 + h^3 - 10xh - 5h^2 + h}{h}$$

$$= 3x^2 + 3xh + h^2 - 10x - 5h + 1, h \neq 0$$

134. (a) $C(x) = 5.35x + 16{,}000$

(b) $P(x) = \text{Revenue} - \text{Cost}$

$= 8.20x - (5.35x + 16{,}000)$

$= 8.20x - 5.35x - 16{,}000$

$= 2.85x - 16{,}000$

136. (a) (40% of $(50 - x)$) + (100% of x) = (amount of acid in final mixture)

Amount of acid in final mixture $= f(x)$

$f(x) = 0.4(50 - x) + 1.0x = 20 + 0.6x$

(b) Domain: $0 \leq x \leq 50$

Range: $20 \leq f(x) \leq 50$

(c) $20 + 0.6x = 50\%(50)$

$20 + 0.6x = 25$

$0.6x = 5$

$x = 8\frac{1}{3}$ liters

138. A vertical line intersects the graph of $y = -\frac{3}{5}x^3 - 2x + 1$ no more than once $\Rightarrow$ y is a function of x.

140. A vertical line intersects the graph of $x = -|4 - y|$ more than once $\Rightarrow$ y is not a function of x.

142. $\dfrac{8x + 3}{11 - x} = 0$

$8x + 3 = 0$

$x = -\dfrac{3}{8}$

144. $f(x) = x^3 - x^2 - 25x + 25$

$x^3 - x^2 - 25x + 25 = 0$

$x^2(x - 1) - 25(x - 1) = 0$

$(x^2 - 25)(x - 1) = 0$

$(x + 5)(x - 5)(x - 1) = 0$

Zeros: $x = -5, 1, 5$

146. $f(x) = (x^2 - 4)^2$

From the graph, $f(x)$ is increasing on $(-2, 0)$ and $(2, \infty)$, and decreasing on $(-\infty, -2)$ and $(0, 2)$.

148. $f(x) = x^4 - 20x^2$

$f(-x) = (-x)^4 - 20(-x)^2 = x^4 - 20x^2 = f(x)$

$f(x)$ is an even function.

150. $f(x) = \sqrt[5]{6x^2}$

$f(-x) = \sqrt[5]{6(-x)^2} = \sqrt[5]{6x^2} = f(x)$

$f(x)$ is an even function.

152. $f(0) = -5,\ f(4) = -8$

$(0, -5), (4, -8)$

$$m = \frac{-5 - (-8)}{0 - 4} = -\frac{3}{4}$$

$$f(x) - (-5) = -\frac{3}{4}(x - 0)$$

$$f(x) = -\frac{3}{4}x - 5$$

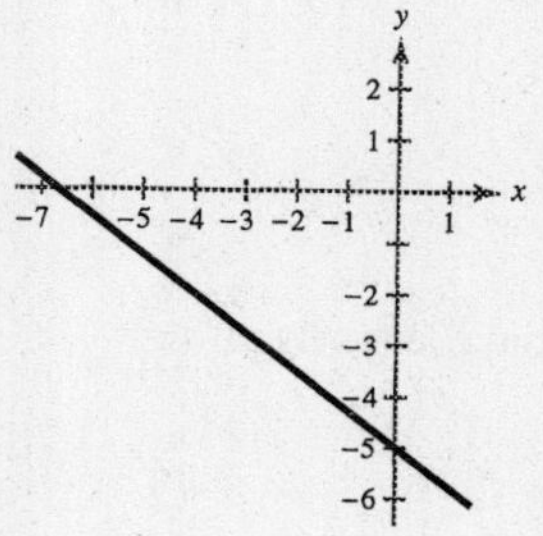

154. $f(3.3) = 5.6,\ f(-4.7) = -1.4$

$(3.3, 5.6), (-4.7, -1.4)$

$$m = \frac{5.6 - (-1.4)}{3.3 - (-4.7)} = \frac{7}{8}$$

$$f(x) - 5.6 = \frac{7}{8}(x - 3.3)$$

$$f(x) = \frac{7}{8}x + \frac{21.7}{8}$$

$$f(x) = \frac{7}{8}x + \frac{217}{80}$$

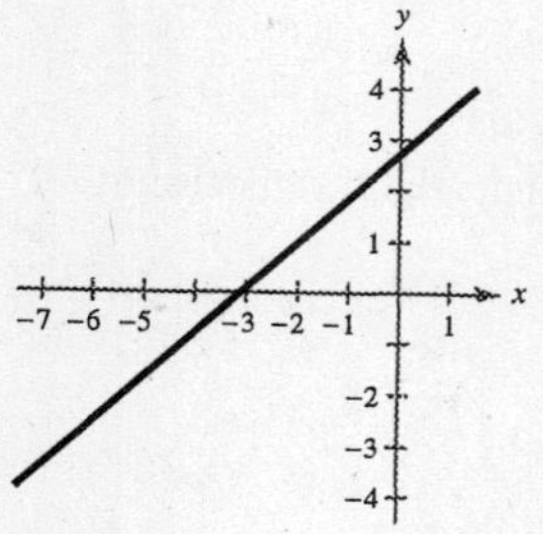

156. $h(x) = x^3 - 2$

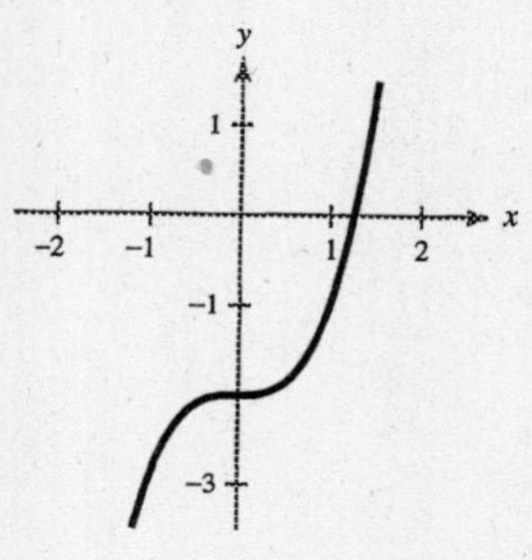

158. $f(x) = \sqrt{x + 1}$

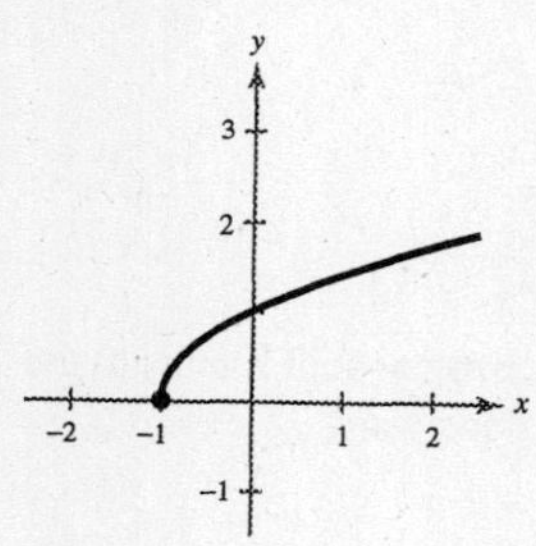

160. $g(x) = \dfrac{1}{x + 5}$

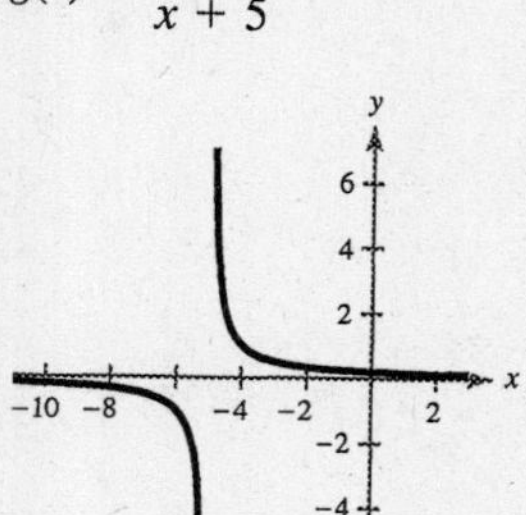

162. $g(x) = [\![x + 4]\!]$

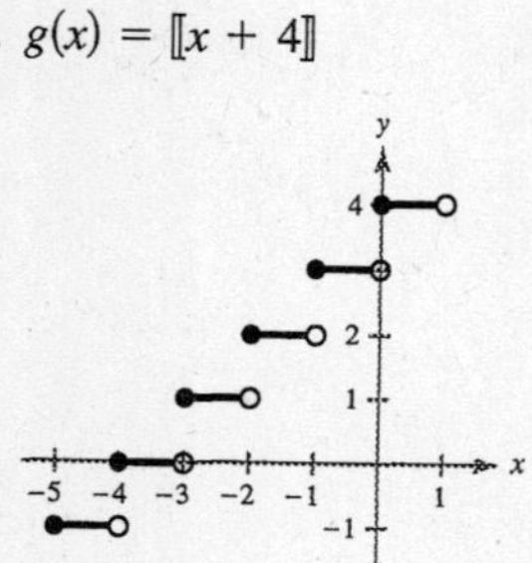

164. $f(x) = \begin{cases} x^2 - 2, & x < -2 \\ 5, & -2 \le x \le 0 \\ 8x - 5, & x > 0 \end{cases}$

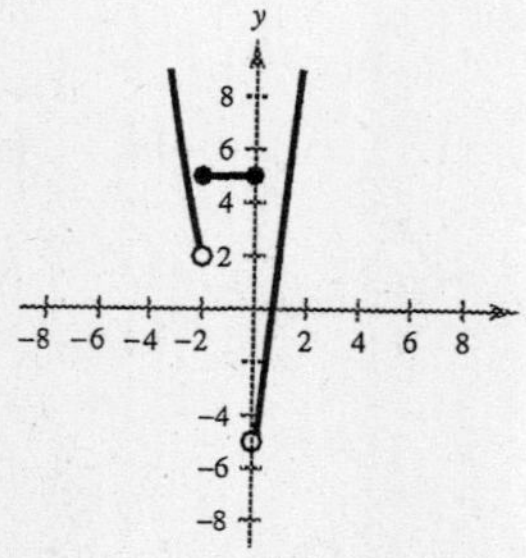

166. Common function: $f(x) = \sqrt{x}$

The graph of $f(x) = \sqrt{x}$ was shifted upward 4 units.

168. $f(x) = x^3,\ h(x) = (x - 2)^3 + 2$

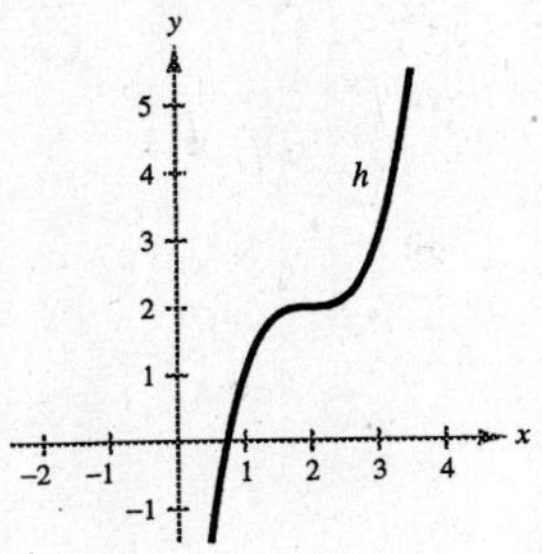

Horizontal shift 2 units to the right; vertical shift 2 units upward

170. $f(x) = |x|,\ h(x) = |x + 3| - 5$

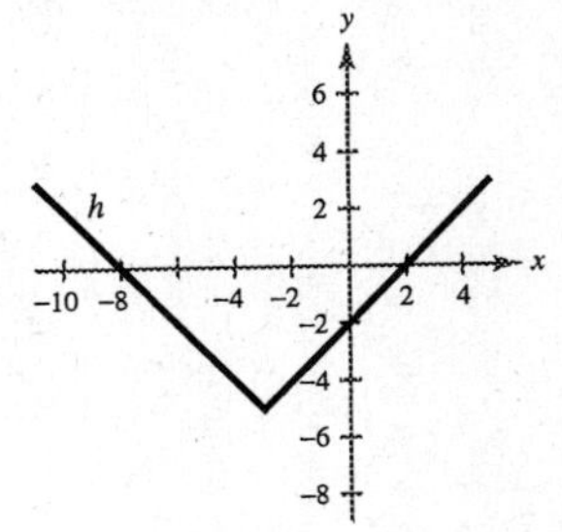

Horizontal shift 3 units to the left; vertical shift 5 units downward

172. $f(x) = x^3,\ h(x) = -(x - 5)^3 - 5$

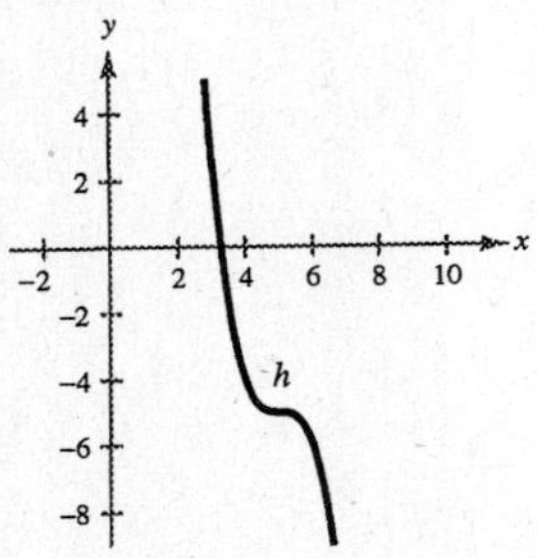

Reflection in the x-axis; horizontal shift 5 units to the right; vertical shift 5 units downward

174. $f(x) = \sqrt{x}$

$h(x) = -\sqrt{x + 1} + 9$

Reflection in the x-axis, a horizontal shift 1 unit to the left, and a vertical shift 9 units upward.

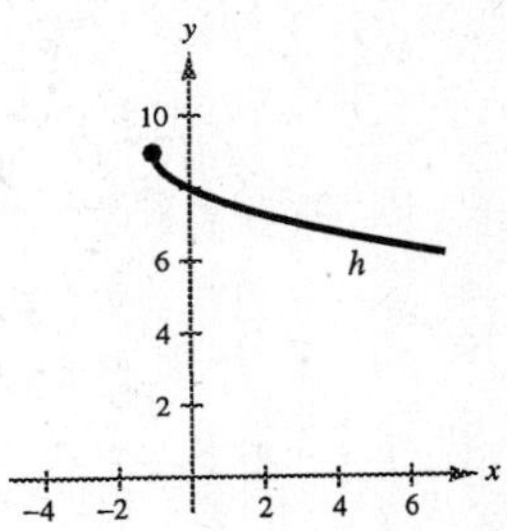

176. $f(x) = x^2,\ h(x) = -(x + 1)^2 - 3$

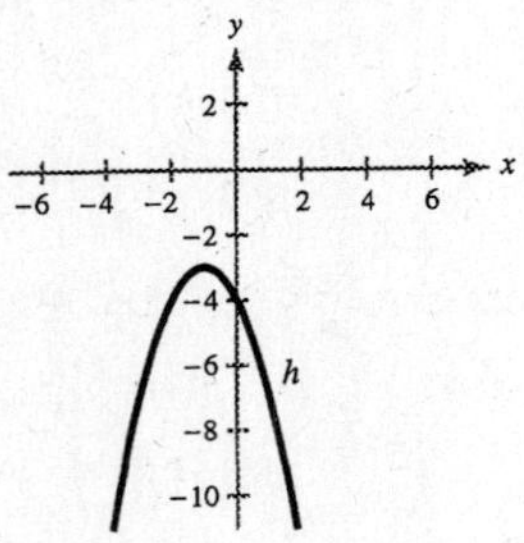

Reflection in the x-axis; horizontal shift 1 unit to the left; vertical shift 3 units downward

178. $f(x) = x^3,\ h(x) = -\frac{1}{3}x^3$

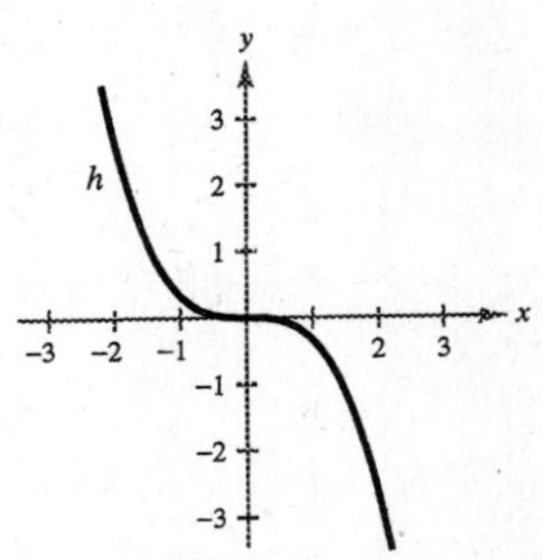

Reflection in the x-axis; vertical shrink by a factor of 1/3

180. $f(x) = |x|,\ h(x) = \frac{1}{2}|x| - 1$

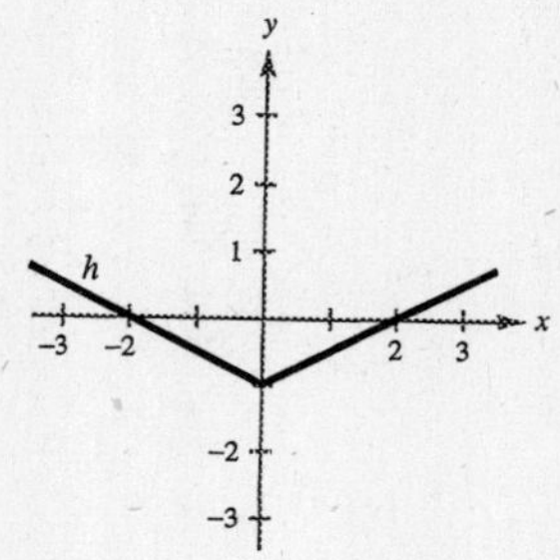

Vertical shrink by a factor of 1/2; vertical shift 1 unit downward

182. $f(x) = x^2 - 4, g(x) = \sqrt{3 - x}$

(a) $(f + g)(x) = x^2 - 4 + \sqrt{3 - x}$

(b) $(f - g)(x) = x^2 - 4 - \sqrt{3 - x}$

(c) $(fg)(x) = (x^2 - 4)\sqrt{3 - x} = x^2\sqrt{3 - x} - 4\sqrt{3 - x}$

(d) $(f/g)(x) = \frac{x^2 - 4}{\sqrt{3 - x}}$, Domain: $x < 3$

184. $f(x) = x^3 - 4, g(x) = \sqrt[3]{x + 7}$

The domains of $f(x)$ and $g(x)$ are all real numbers.

(a) $(f \circ g)(x) = \left(\sqrt[3]{x + 7}\right)^3 - 4 = x + 7 - 4 = x + 3$, Domain: all real numbers

(b) $(g \circ f)(x) = \sqrt[3]{(x^3 - 4) + 7} = \sqrt[3]{x^3 + 3}$, Domain: all real numbers

186. $h(x) = \sqrt[3]{x + 2}$

One possibility: Let $g(x) = x + 2$ and $f(x) = \sqrt[3]{x}$.

$(f \circ g)(x) = f(x + 2) = \sqrt[3]{x + 2} = h(x)$

188.

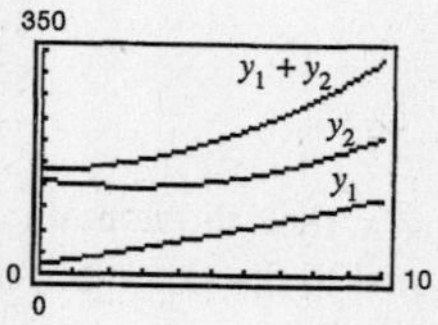

$$(y_1 + y_2)(x) = 0.2071x^2 + 8.65x + 14.21 + 1.414x^2 - 7.28x + 146.9$$
$$= 1.6211x^2 + 1.37x + 161.11$$

$(y_1 + y_2)(15) = 1.6211(15)^2 + 1.37(15) + 161.11 \approx \$546.4$

An estimate of the total value of the imports in 2005 is \$546.4 billion.

190.
$$f(x) = x + 5$$
$$y = x + 5$$
$$x = y + 5$$
$$x - 5 = y$$
$$f^{-1}(x) = x - 5$$
$$f(f^{-1}(x)) = f(x - 5) = (x - 5) + 5 = x$$
$$f^{-1}(f(x)) = f^{-1}(x + 5) = (x + 5) - 5 = x$$

192. No, because some horizontal lines intersect the graph twice

194. No, because some horizontal lines intersect the graph twice

196. Yes, because no horizontal lines intersect the graph at more than one point

198. $f(x) = 5x - 7$

(a) $y = 5x - 7$

$x = 5y - 7$

$x + 7 = 5y$

$\frac{x+7}{5} = y$

$f^{-1}(x) = \frac{x+7}{5}$

(b)

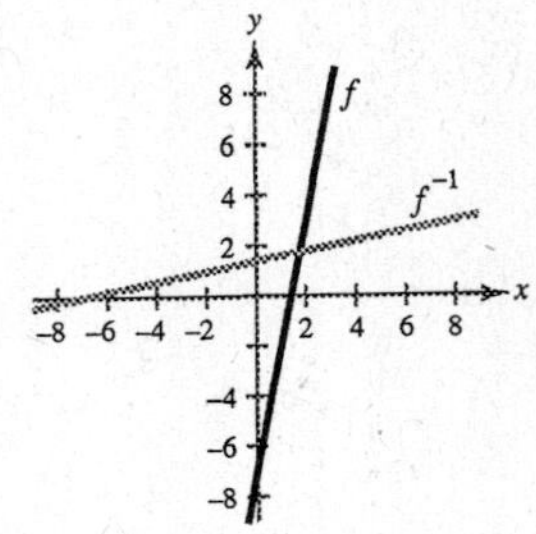

(c) $f^{-1}(f(x)) = f^{-1}(5x - 7) = \frac{(5x-7)+7}{5} = \frac{5x}{5} = x$

$f(f^{-1}(x)) = f\left(\frac{x+7}{5}\right) = 5\left(\frac{x+7}{5}\right) - 7 = x + 7 - 7 = x$

200. $f(x) = x^3 + 2$

(a) $y = x^3 + 2$

$x = y^3 + 2$

$x - 2 = y^3$

$\sqrt[3]{x-2} = y$

$f^{-1}(x) = \sqrt[3]{x-2}$

(b)

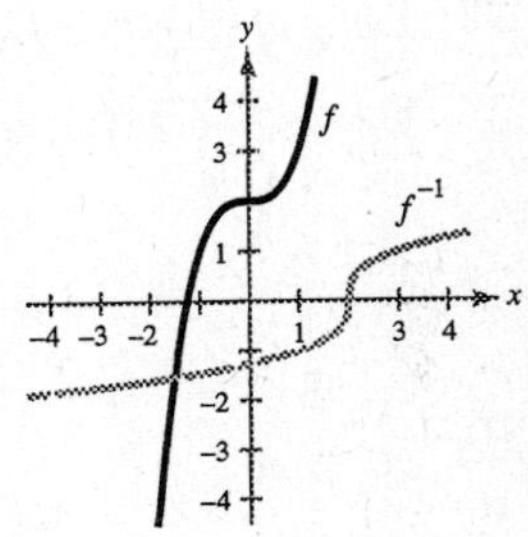

(c) $f^{-1}(f(x)) = f^{-1}(x^3 + 2) = \sqrt[3]{(x^3+2) - 2} = \sqrt[3]{x^3} = x$

$f(f^{-1}(x)) = f\left(\sqrt[3]{x-2}\right) = \left(\sqrt[3]{x-2}\right)^3 + 2 = x - 2 + 2 = x$

202. $f(x) = |x - 2|$

Increasing on $[2, \infty)$

Let $f(x) = x - 2, x \geq 2, y \geq 0$.

$y = x - 2$

$x = y - 2, x \geq 0, y \geq 2$

$x + 2 = y, x \geq 0, y \geq 2$

$f^{-1}(x) = x + 2, x \geq 0$

204. True. If $f(x) = x^3$ and $g(x) = \sqrt[3]{x}$, then the domain of g is all real numbers, which is equal to the range of f and vice versa.

Problem Solving for Chapter P

Solutions to Even-Numbered Exercises

2. Mapping numbers onto letters is *not* a function. Each number between 2 and 9 is mapped to more than one letter.

$\{(2, A), (2, B), (2, C), (3, D), (3, E), (3, F), (4, G), (4, H), (4, I), (5, J), (5, K), (5, L),$
$(6, M), (6, N), (6, O), (7, P), (7, Q), (7, R), (7, S), (8, T), (8, U), (8, V), (9, W), (9, X), (9, Y), (9, Z)\}$

Mapping letters onto numbers *is* a function. Each letter is only mapped to one number.

$\{(A, 2), (B, 2), (C, 2), (D, 3), (E, 3), (F, 3), (G, 4), (H, 4), (I, 4), (J, 5), (K, 5), (L, 5),$
$(M, 6), (N, 6), (O, 6), (P, 7), (Q, 7), (R, 7), (S, 7), (T, 8), (U, 8), (V, 8), (W, 9), (X, 9), (Y, 9), (Z, 9)\}$

4. $f(x) = x$ $\qquad$ $g(x) = x$

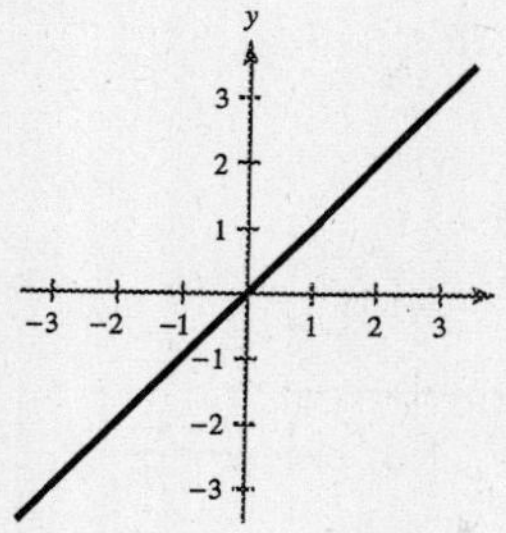

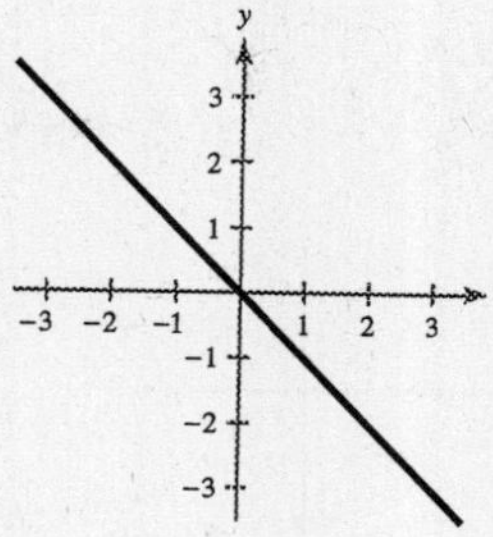

$(f \circ f)(x) = x$ and $(g \circ g)(x) = x$

These are the only two linear functions that are their own inverse functions since m has to equal $\dfrac{1}{m}$ for this to be true.

6. It appears, from the drawing in the text, that $(x, y) = (6, 8)$

The line between $(2.5, 2)$ and $(6, 8)$ is $y = \frac{12}{7}x - \frac{16}{7}$

The line between $(9.5, 2)$ and $(6, 8)$ is $y = -\frac{12}{7}x + \frac{128}{7}$

The path of the ball is:

$$f(x) = \begin{cases} \frac{12}{7}x - \frac{16}{7}, & 2.5 \le x \le 6 \\ -\frac{12}{7}x + \frac{128}{7}, & 6 < x \le 9.5 \end{cases}$$

8. (a)–(d) Use $f(x) = 4x$ and $g(x) = x + 6$

(a) $(fg)(x) = f(x + 6) = 4(x + 6) = 4x + 24$

(b) $(f \circ g)^{-1}(x) = \dfrac{x - 24}{4} = \dfrac{1}{4}x - 6$

(c) $f^{-1}(x) = \dfrac{1}{4}x$

$g^{-1}(x) = x - 6$

(d) $(g^{-1} \circ f^{-1})(x) = g^{-1}\left(\dfrac{1}{4}x\right) = \dfrac{1}{4}x - 6$

(e) $f(x) = x^3 + 1$ and $g(x) = 2x$

$(f \circ g)(x) = f(2x) = (2x)^3 + 1 = 8x^3 + 1$

$(f \circ g)^{-1}(x) = \sqrt[3]{\dfrac{x - 1}{8}} = \dfrac{1}{2}\sqrt[3]{x - 1}$

$f^{-1}(x) = \sqrt[3]{x - 1}$

$g^{-1}(x) = \dfrac{1}{2}x$

$(g^{-1} \circ f^{-1})(x) = g^{-1}\left(\sqrt[3]{x - 1}\right) = \dfrac{1}{2}\sqrt[3]{x - 1}$

(f) Answers will vary

(g) Conjecture: $(f \circ g)^{-1}(x) = (g^{-1} \circ f^{-1})(x)$

10. $H(x) = \begin{cases} 1, & x \ge 0 \\ 0, & x < 0 \end{cases}$

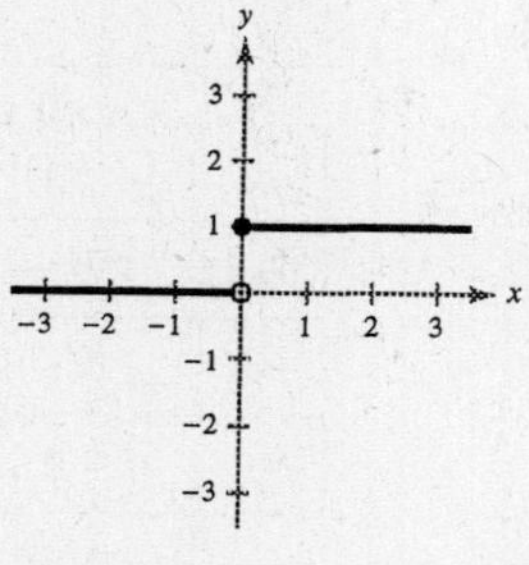

—CONTINUED—

10. —CONTINUED—

(a) $H(x) - 2$

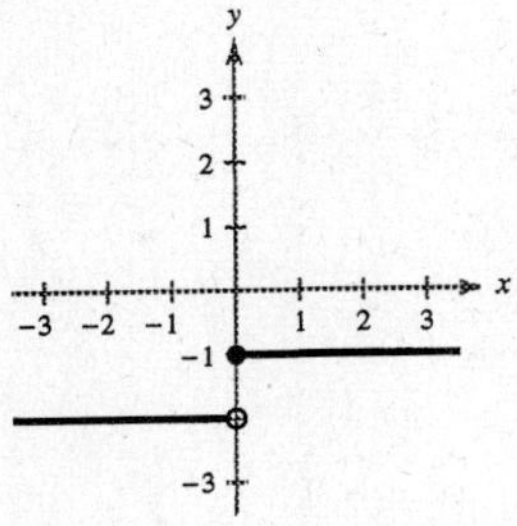

(b) $H(x - 2)$

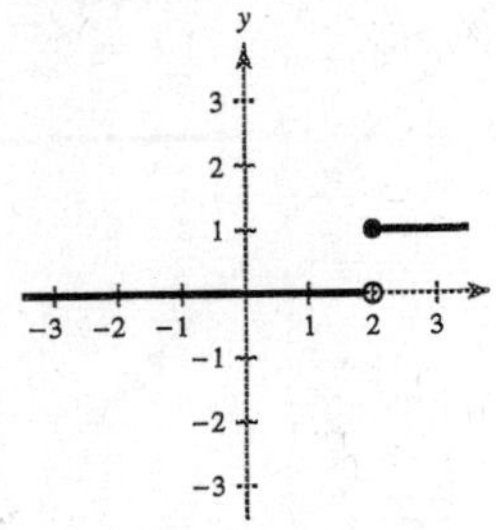

(c) $-H(x)$

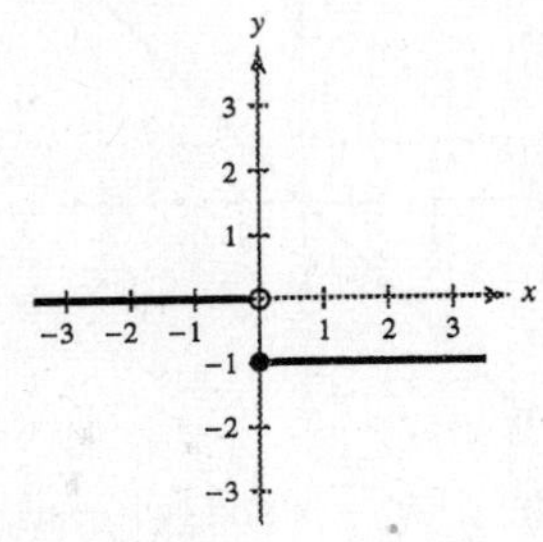

(d) $H(-x)$

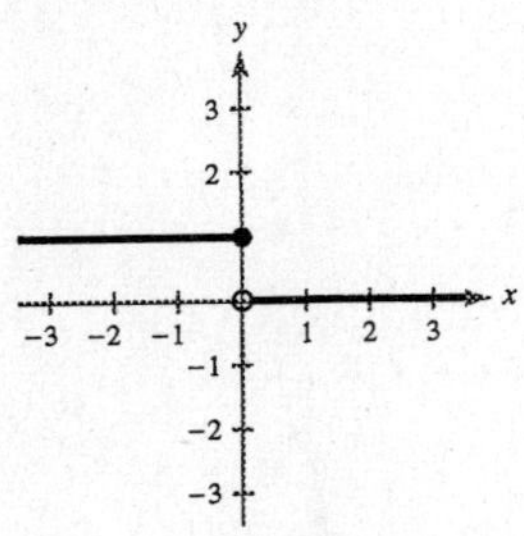

(e) $\frac{1}{2}H(x)$

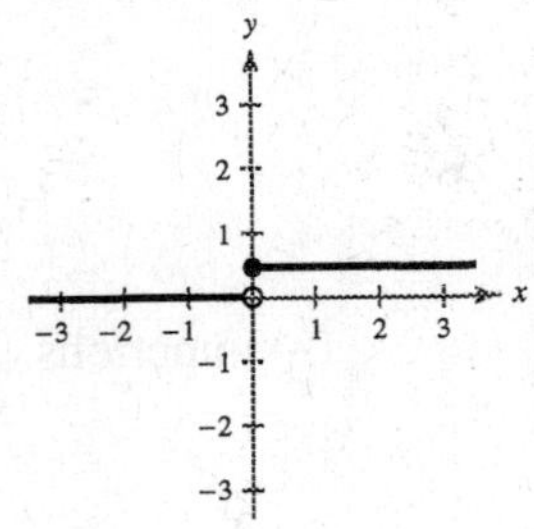

(f) $-H(x - 2) + 2$

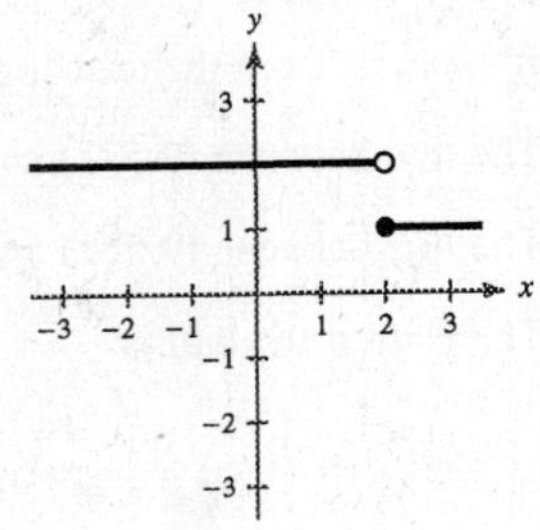

12.
$$\begin{aligned}(f \circ (g \circ h))(x) &= f((g \circ h)(x))\\ &= f(g(h(x)))\\ &= (f \circ g \circ h)(x)\\ ((f \circ g) \circ h)(x) &= (f \circ g)(h(x))\\ &= f(g(h(x)))\\ &= (f \circ g \circ h)(x)\end{aligned}$$

14.

x	$f(x)$	$f^{-1}(x)$
-4	—	2
-3	4	1
-2	1	0
-1	0	—
0	-2	-1
1	-3	-2
2	-4	—
3	—	—
4	—	-3

(a)

x	$f(f^{-1}(x))$
-4	$f(f^{-1}(-4)) = f(2) = -4$
-2	$f(f^{-1}(-2)) = f(0) = -2$
0	$f(f^{-1}(0)) = f(-1) = 0$
4	$f(f^{-1}(4)) = f(-3) = 4$

(b)

x	$(f + f^{-1})(x)$
-3	$f(-3) + f^{-1}(-3) = 4 + 1 = 5$
-2	$f(-2) + f^{-1}(-2) = 1 + 0 = 1$
0	$f(0) + f^{-1}(0) = -2 + (-1) = -3$
1	$f(1) + f^{-1}(1) = -3 + (-2) = -5$

(c)

x	$(f \cdot f^{-1})(x)$
-3	$f(-3)f^{-1}(-3) = (4)(1) = 4$
-2	$f(-2)f^{-1}(-2) = (1)(0) = 0$
0	$f(0)f^{-1}(0) = (-2)(-1) = 2$
1	$f(1)f^{-1}(1) = (-3)(-2) = 6$

(d)

x	$\lvert f^{-1}(x)\rvert$
-4	$\lvert f^{-1}(-4)\rvert = \lvert 2\rvert = 2$
-3	$\lvert f^{-1}(-3)\rvert = \lvert 1\rvert = 1$
0	$\lvert f^{-1}(0)\rvert = \lvert -1\rvert = 1$
4	$\lvert f^{-1}(4)\rvert = \lvert -3\rvert = 3$

CHAPTER 1
Trigonometry

CHAPTER 1
Trigonometry

Section 1.1 Radian and Degree Measure

Solutions to Even-Numbered Exercises

2. 5.5 radians

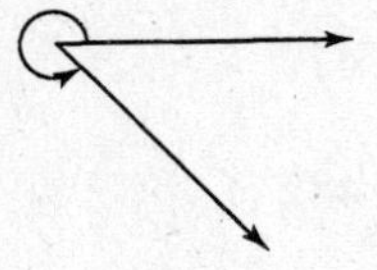

4. -4 radians

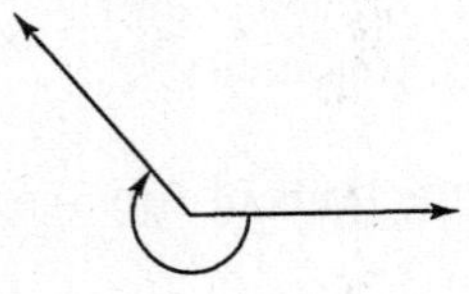

6. 6 radians

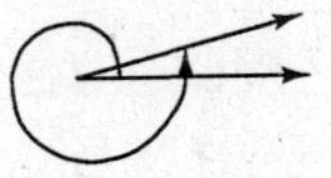

8. (a) Since $\pi < \frac{11\pi}{8} < \frac{3\pi}{2}$; $\frac{11\pi}{8}$ lies in Quadrant III.

(b) Since $\pi < \frac{9\pi}{8} < \frac{3\pi}{2}$; $\frac{9\pi}{8}$ lies in Quadrant III.

10. (a) Since $-\frac{\pi}{2} < -1 < 0$; -1 lies in Quadrant IV.

(b) Since $-\frac{3\pi}{2} < -\frac{11\pi}{9} < -\pi$; $-\frac{11\pi}{9}$ lies in Quadrant II.

12. (a) Since $\frac{3\pi}{2} < 6.02 < 2\pi$; 6.02 lies in Quadrant IV.

(b) Since $-\frac{3\pi}{2} < -4.25 < -\pi$; 4.25 lies in Quadrant II.

14. (a) $-\frac{7\pi}{4}$

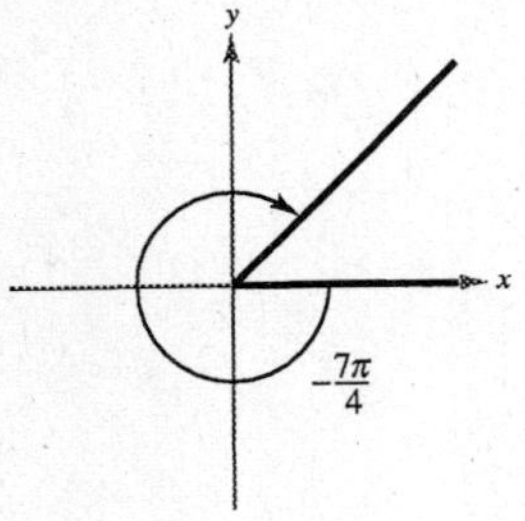

(b) $\frac{5\pi}{2}$

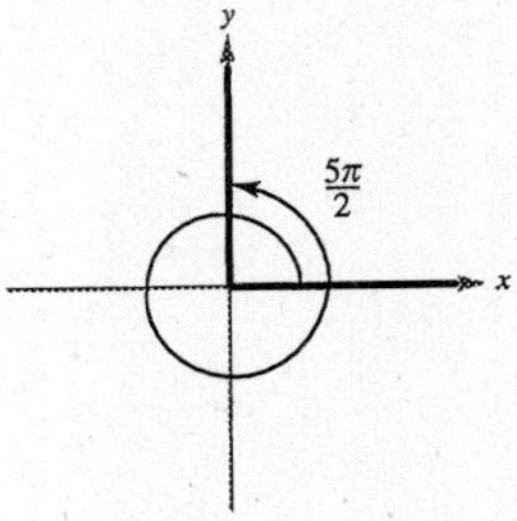

16. (a) 4

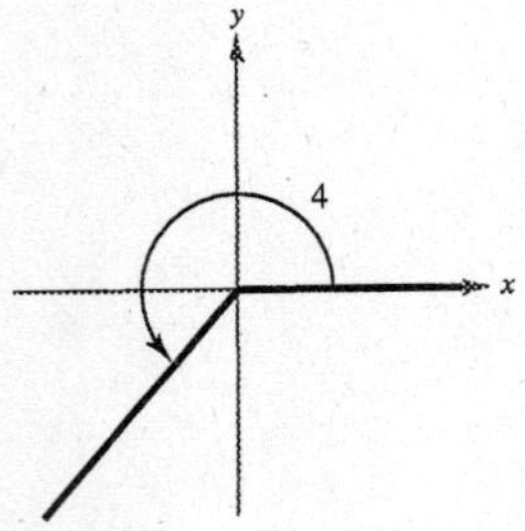

(b)

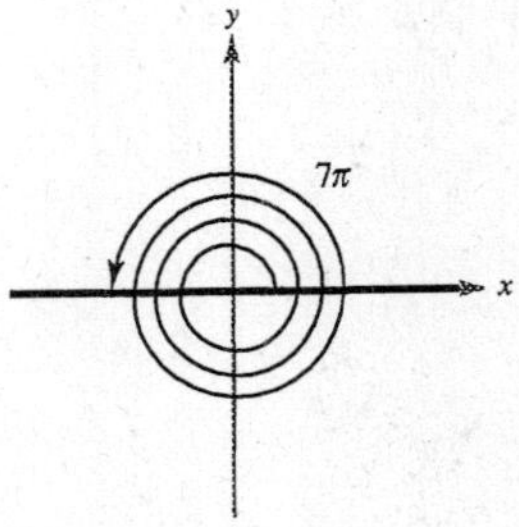

18. (a) $\frac{7\pi}{6} + 2\pi = \frac{19\pi}{6}$

$\frac{7\pi}{6} - 2\pi = -\frac{5\pi}{6}$

(b) $-\frac{11\pi}{6} + 2\pi = \frac{\pi}{6}$

$-\frac{11\pi}{6} - 2\pi = -\frac{23\pi}{6}$

20. (a) $-\frac{9\pi}{4} + 2\pi = -\frac{\pi}{4}$

$-\frac{9\pi}{4} + 4\pi = \frac{7\pi}{4}$

(b) $-\frac{2\pi}{15} + 2\pi = \frac{28\pi}{15}$

$-\frac{2\pi}{15} - 2\pi = -\frac{32\pi}{15}$

22. (a) Complement: $\frac{\pi}{2} - \frac{\pi}{12} = \frac{5\pi}{12}$

Supplement: $\pi - \frac{\pi}{12} = \frac{11\pi}{12}$

(b) Complement: none $\left(\frac{11\pi}{12} > \frac{\pi}{2}\right)$

Supplement: $\pi - \frac{11\pi}{12} = \frac{\pi}{12}$

24. (a) Complement: none $\left(3 > \dfrac{\pi}{2}\right)$

Supplement: $\pi - 3 \approx 0.14$

(b) Complement: $\dfrac{\pi}{2} - 1.5 \approx 0.07$

Supplement: $\pi - 1.5 \approx 1.64$

26. (a) $315° = 315°\left(\dfrac{\pi}{180°}\right) = \dfrac{7\pi}{4}$

(b) $120° = 120°\left(\dfrac{\pi}{180°}\right) = \dfrac{2\pi}{3}$

28. (a) $-270° = -270°\left(\dfrac{\pi}{180°}\right) = -\dfrac{3\pi}{2}$

(b) $144° = 144°\left(\dfrac{\pi}{180°}\right) = \dfrac{4\pi}{5}$

30. $87.4° = 87.4°\left(\dfrac{\pi}{180°}\right) \approx 1.525$ radians

32. $-48.27° = -48.27°\left(\dfrac{\pi}{180°}\right) \approx -0.842$ radians

34. $345° = 345°\left(\dfrac{\pi}{180°}\right) \approx 6.021$ radians

36. $0.54° = 0.54°\left(\dfrac{\pi}{180°}\right) \approx 0.009$ radians

38. (a) $-\dfrac{7\pi}{12} = -\dfrac{7\pi}{12}\left(\dfrac{180°}{\pi}\right) = -105°$

(b) $\dfrac{\pi}{9} = \dfrac{\pi}{9}\left(\dfrac{180°}{\pi}\right) = 20°$

40. (a) $\dfrac{11\pi}{6} = \dfrac{11\pi}{6}\left(\dfrac{180°}{\pi}\right) = 330°$

(b) $\dfrac{34\pi}{15} = \dfrac{34\pi}{15}\left(\dfrac{180°}{\pi}\right) = 408°$

42. $\dfrac{5\pi}{11} = \dfrac{5\pi}{11}\left(\dfrac{180°}{\pi}\right) \approx 81.818°$

44. $\dfrac{13\pi}{2} = \dfrac{13\pi}{2}\left(\dfrac{180°}{\pi}\right) = 1170°$

46. $4.8\pi = 4.8\pi\left(\dfrac{180°}{\pi}\right) = 864°$

48. $-0.57 = -0.57\left(\dfrac{180°}{\pi}\right) \approx -32.659°$

50. 120°

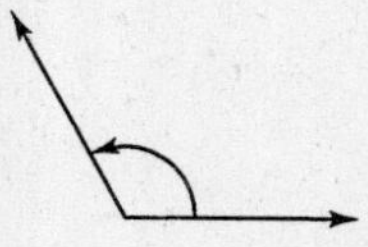

52. −330°

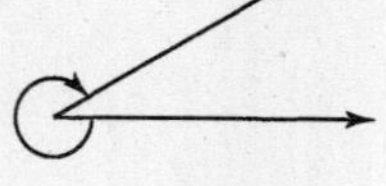

54. 10°

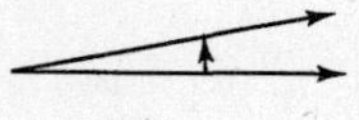

56. (a) Since $0° < 8.3° < 90°$; 8.3° lies in Quadrant I.

(b) Since $180° < 257° 30' < 270°$; $257° 30'$ lies in Quadrant III.

58. (a) Since $-270° < -260° < -180°$; $-260°$ lies in Quadrant II.

(b) Since $-90° < -3.4° < 0°$; $-3.4°$ lies in Quadrant IV.

60. (a) −270°

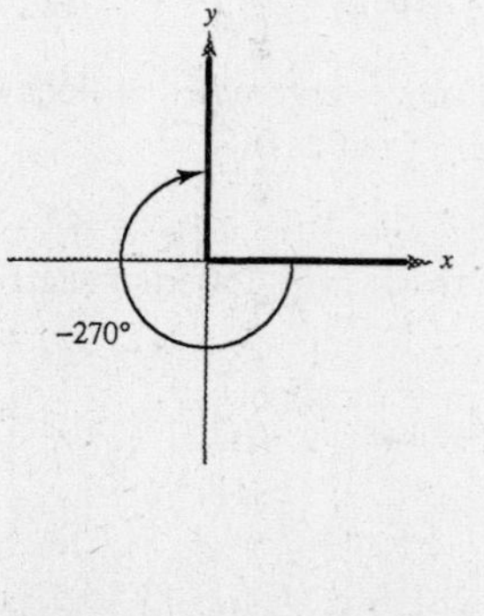

(b) −120°

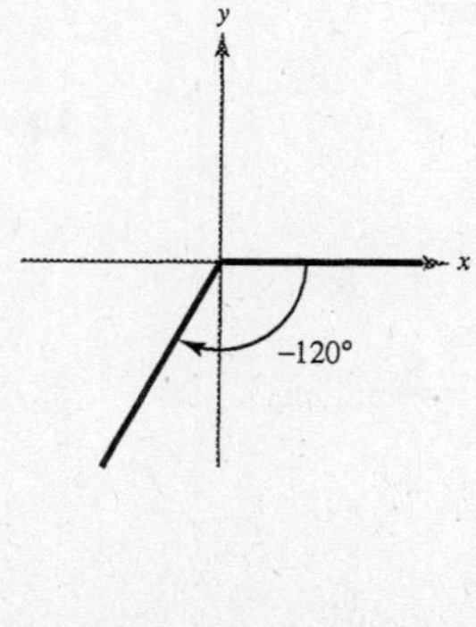

62. (a) $-750°$

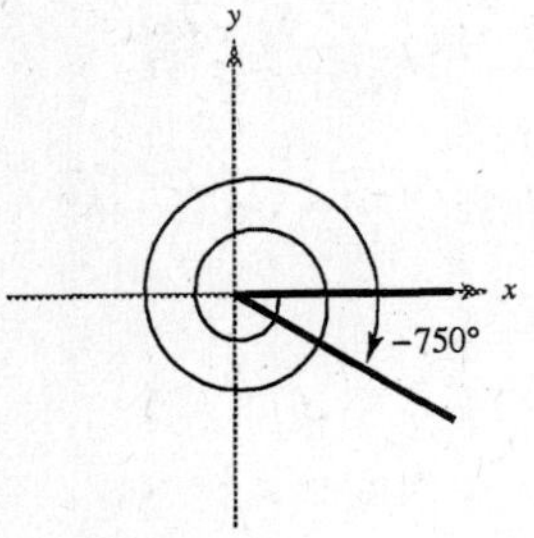

(b) $-600°$

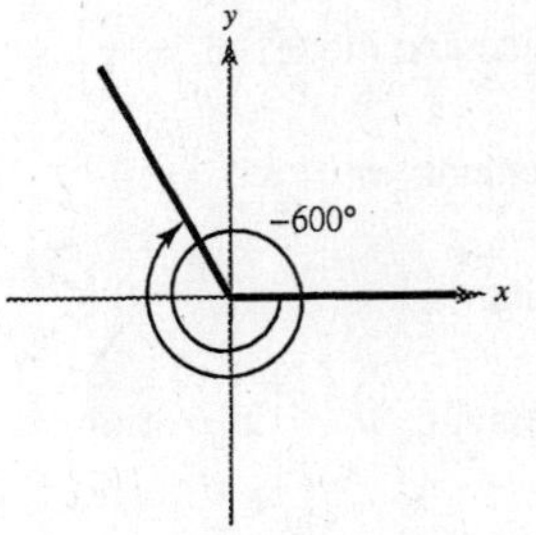

64. (a) $120° + 360° = 480°$

$120° - 360° = -240°$

(b) $-420° + 720° = 300°$

$-420° + 360° = -60°$

66. (a) $-420° + 720° = 300°$

$-420° + 360° = -60°$

(b) $230° + 360° = 590°$

$230° - 360° = -130°$

68. (a) Complement: $90° - 3° = 87°$

Supplement: $180° - 3° = 177°$

(b) Complement: $90° - 64° = 26°$

Supplement: $180° - 64° = 116°$

70. (a) Complement: none $(130° > 90°)$

Supplement: $180° - 130° = 50°$

(b) Complement: none $(170° > 90°)$

Supplement: $180° - 170° = 10°$

72. (a) $245°10' = 245° + \left(\frac{10}{60}\right)°$

$\approx 245° + 0.167°$

$= 245.167°$

(b) $2°12' = 2° + \left(\frac{12}{60}\right)°$

$= 2° + 0.2° = 2.2°$

74. (a) $-135°36'' = -135° - \left(\frac{36}{3600}\right)°$

$= -135° - 0.01° = -135.01°$

(b) $-408°16'20'' = -\left(408° + \left(\frac{16}{60}\right)° + \left(\frac{20}{3600}\right)°\right)$

$\approx -(408° + 0.2667° + 0.0056°)$

$\approx -408.272°$

76. (a) $-345.12° = -(345° + (0.12)(60'))$

$= -(345° + 7' + 0.2(60''))$

$= -345°\,7'\,12''$

(b) $0.45° = 0° + (0.45)(60')$

$= 0° + 27'$

$= 0°27'$

78. (a) $-0.355° = -(0° + (0.355)(60'))$

$= -(0° + 21' + (0.3)(60''))$

$= -(0° + 21' + 18'')$

$= -0°21'18''$

(b) $0.7865 = 0° + (0.7865)(60')$

$= 0° + 47' + (0.19)(60'')$

$= 0° + 47' + 11.4''$

$= 0°47'11.4''$

80. $s = r\theta$

$29 = 10\,\theta$

$\theta = \frac{29}{10}$ radians

82. $s = r\theta$

$60 = 75\theta$

$\theta = \frac{60}{75} = \frac{4}{5}$ radians

Because the angle represented is clockwise, this angle is $-\frac{4}{5}$ radians.

84. $r = 14$ feet, $s = 8$ feet

$$\theta = \frac{s}{r} = \frac{8}{14} = \frac{4}{7} \text{ radians}$$

86. $r = 80$ kilometers, $s = 160$ kilometers

$$\theta = \frac{s}{r} = \frac{160}{80} = 2 \text{ radians}$$

88. $r = 9$ feet, $\theta = 60° = \dfrac{\pi}{3}$

$$s = r\theta = 9\left(\frac{\pi}{3}\right) = 3\pi \text{ feet}$$

90. $r = 20$ centimeters, $\theta = \dfrac{\pi}{4}$

$$s = r\theta = 20\left(\frac{\pi}{4}\right) = 5\pi \text{ centimeters}$$

92. $r = 4000$ miles

$$\theta = 47°\,36'32'' - 37°\,46'39'' = 9°\,49'53''$$
$$\approx 0.1716 \text{ radian}$$
$$s = r\theta \approx (4000)(0.1716) = 686.4 \text{ miles}$$

94. $r = 4000$ miles

$$\theta = 31°\,47' + 26°\,10' = 57°\,57'$$
$$\approx 1.01142 \text{ radians}$$
$$s = r\theta \approx (4000)(1.01142) \approx 4045.7 \text{ miles}$$

96. $r = 3189$ kilometers,

$$s = r\,\theta$$
$$400 = 6378\,\theta$$
$$\frac{400}{6378} = \theta$$
$$0.062716 \approx \theta$$

The difference in latitude is about 0.062716 radians $\approx 3.59°$.

98. $\theta = \dfrac{s}{r} = \dfrac{24}{5} = 4.8 \text{ radians} = 4.8\left(\dfrac{180°}{\pi}\right) \approx 275°$

100. Linear velocity for either pulley: $1700(2\pi) = 3400\pi$ in/min

(a) Angular speed of motor pulley: $\omega = \dfrac{v}{r} = \dfrac{3400\pi}{1} = 3400\pi$ rad/min

Angular speed of the saw arbor: $\omega = \dfrac{v}{r} = \dfrac{3400\pi}{2} = 1700\pi$ rad/min

(b) Revolutions per minute of the saw arbor: $\dfrac{1700\pi}{2\pi} = 850$ rev/min

102. Circumference: $C = 2\pi(1.68) = 3.36\pi$ inches

360 rev/min = 6 rev/sec

Linear speed: $(3.36\pi)(6) = 20.16\pi$ inches/sec

104. True. If α and β are coterminal angles, then $\alpha = \beta + n(360°)$ where n is an integer.

The difference between α and β is $\alpha - \beta = n(360°)$
$$= 2\pi n$$

106. (a) An angle is in standard position if its vertex is at the origin and its initial side is on the positive x-axis.

(b) A negative angle is generated by a clockwise rotation of the terminal side.

(c) Two angles in standard position with the same terminal sides are coterminal.

(d) An obtuse angle measures between 90° and 180°.

108. 1 Radian $= \left(\dfrac{180}{\pi}\right)^{\circ} \approx 57.3°$, so one radian is much larger than one degree.

110. The area of a circle is $A = \pi r^2 \Rightarrow \pi = \dfrac{A}{r^2}$.

The circumference of a circle is $C = 2\pi r$.

$$C = 2\left(\frac{A}{r^2}\right)r$$

$$C = \frac{2A}{r}$$

$$\frac{Cr}{2} = A$$

For a sector, $C = s = r\theta$

Thus, $A = \dfrac{(r\theta)r}{2} = \dfrac{1}{2}\theta r^2$ for a sector.

112. $f(x) = x^5 - 4$

Vertical shift 4 units downward

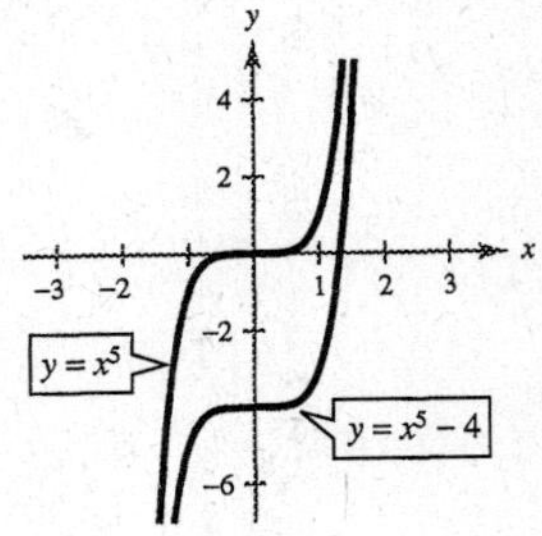

114. $f(x) = -(x + 3)^5$

Reflection in the x-axis and a horizontal shift 3 units to the left.

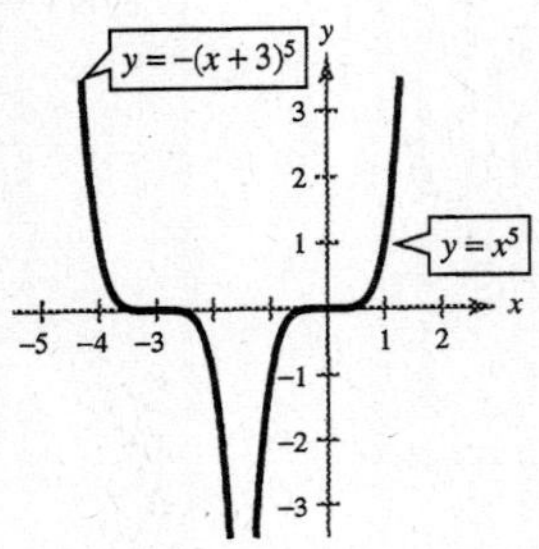

116. $\dfrac{2}{\sqrt{3}} = \dfrac{2}{\sqrt{3}} \cdot \dfrac{\sqrt{3}}{\sqrt{3}} = \dfrac{2\sqrt{3}}{3}$

118. $\dfrac{5\sqrt{5}}{2\sqrt{10}} = \dfrac{5}{2}\sqrt{\dfrac{5}{10}} = \dfrac{5}{2}\sqrt{\dfrac{1}{2}} = \dfrac{5}{2\sqrt{2}} \cdot \dfrac{\sqrt{2}}{\sqrt{2}} = \dfrac{5\sqrt{2}}{4}$

120. $\sqrt{18^2 + 12^2} = \sqrt{324 + 144} = \sqrt{468}$

$= \sqrt{36 \cdot 13} = 6\sqrt{13}$

122. $\sqrt{17^2 - 9^2} = \sqrt{289 - 81} = \sqrt{208}$

$= \sqrt{16 \cdot 13} = 4\sqrt{13}$

Section 1.2 Trigonometric Functions: The Unit Circle

Solutions to Even-Numbered Exercises

2. $x = 12, y = 5$

$r = \sqrt{12^2 + 5^2} = 13$

$\sin\theta = \dfrac{y}{r} = \dfrac{5}{13}$ $\csc\theta = \dfrac{r}{y} = \dfrac{13}{5}$

$\cos\theta = \dfrac{x}{r} = \dfrac{12}{13}$ $\sec\theta = \dfrac{r}{x} = \dfrac{13}{12}$

$\tan\theta = \dfrac{y}{x} = \dfrac{5}{12}$ $\cot\theta = \dfrac{x}{y} = \dfrac{12}{5}$

4. $x = -4, y = -3$

$r = \sqrt{(-4)^2 + (-3)^2} = 5$

$\sin\theta = \dfrac{y}{r} = -\dfrac{3}{5}$ $\csc\theta = \dfrac{r}{y} = -\dfrac{5}{3}$

$\cos\theta = \dfrac{x}{r} = -\dfrac{4}{5}$ $\sec\theta = \dfrac{r}{x} = -\dfrac{5}{4}$

$\tan\theta = \dfrac{y}{x} = \dfrac{3}{4}$ $\cot\theta = \dfrac{x}{y} = \dfrac{4}{3}$

6. $t = \dfrac{\pi}{3}, (x\ y) = \left(\dfrac{1}{2}, \dfrac{\sqrt{3}}{2}\right)$

8. $t = \dfrac{5\pi}{4}, (x, y) = \left(-\dfrac{\sqrt{2}}{2}, -\dfrac{\sqrt{2}}{2}\right)$

10. $t = \dfrac{5\pi}{3}, (x, y) = \left(\dfrac{1}{2}, -\dfrac{\sqrt{3}}{2}\right)$

12. $t = \pi, (x, y) = (-1, 0)$

14. $t = \frac{\pi}{3}, (x, y) = \left(\frac{1}{2}, \frac{\sqrt{3}}{2}\right)$

$\sin \frac{\pi}{3} = \frac{\sqrt{3}}{2}$

$\cos \frac{\pi}{3} = \frac{1}{2}$

$\tan \frac{\pi}{3} = \frac{\frac{\sqrt{3}}{2}}{\frac{1}{2}} = \sqrt{3}$

16. $t = -\frac{\pi}{4}, (x, y) = \left(\frac{\sqrt{2}}{2}, -\frac{\sqrt{2}}{2}\right)$

$\sin\left(-\frac{\pi}{4}\right) = -\frac{\sqrt{2}}{2}$

$\cos\left(-\frac{\pi}{4}\right) = \frac{\sqrt{2}}{2}$

$\tan\left(-\frac{\pi}{4}\right) = \frac{-\frac{\sqrt{2}}{2}}{\frac{\sqrt{2}}{2}} = -1$

18. $t = -\frac{4\pi}{3}, (x, y) = \left(-\frac{1}{2}, \frac{\sqrt{3}}{2}\right)$

$\sin\left(-\frac{4\pi}{3}\right) = \frac{\sqrt{3}}{2}$

$\cos\left(-\frac{4\pi}{3}\right) = -\frac{1}{2}$

$\tan\left(-\frac{4\pi}{3}\right) = \frac{\frac{\sqrt{3}}{2}}{-\frac{1}{2}} = -\sqrt{3}$

20. $t = \frac{5\pi}{3}, (x, y) = \left(\frac{1}{2}, -\frac{\sqrt{3}}{2}\right)$

$\sin \frac{5\pi}{3} = -\frac{\sqrt{3}}{2}$

$\cos \frac{5\pi}{3} = \frac{1}{2}$

$\tan \frac{5\pi}{3} = \frac{-\frac{\sqrt{3}}{2}}{\frac{1}{2}} = -\sqrt{3}$

22. $t = -2\pi, (x, y) = (1, 0)$

$\sin(-2\pi) = 0$

$\cos(-2\pi) = 1$

$\tan(-2\pi) = \frac{0}{1} = 0$

24. $t = \frac{5\pi}{6}, (x, y) = \left(-\frac{\sqrt{3}}{2}, \frac{1}{2}\right)$

$\sin \frac{5\pi}{6} = \frac{1}{2}$

$\cos \frac{5\pi}{6} = -\frac{\sqrt{3}}{2}$

$\tan \frac{5\pi}{6} = \frac{\frac{1}{2}}{-\frac{\sqrt{3}}{2}} = -\frac{\sqrt{3}}{3}$

$\csc \frac{5\pi}{6} = \frac{1}{\sin t} = 2$

$\sec \frac{5\pi}{6} = \frac{1}{\cos t} = -\frac{2\sqrt{3}}{3}$

$\cot \frac{5\pi}{6} = \frac{1}{\tan t} = -\sqrt{3}$

26. $t = \frac{3\pi}{2}, (x, y) = (0, -1)$

$\sin \frac{3\pi}{2} = -1$

$\cos \frac{3\pi}{2} = 0$

$\tan \frac{3\pi}{2}$ is undefined.

$\csc \frac{3\pi}{2} = \frac{1}{\sin t} = -1$

$\sec \frac{3\pi}{2}$ is undefined.

$\cot \frac{3\pi}{2} = \frac{0}{-1} = 0$

28. $t = \frac{7\pi}{4}, (x, y) = \left(\frac{\sqrt{2}}{2}, -\frac{\sqrt{2}}{2}\right)$

$\sin \frac{7\pi}{4} = -\frac{\sqrt{2}}{2}$

$\cos \frac{7\pi}{4} = \frac{\sqrt{2}}{2}$

$\tan \frac{7\pi}{4} = \frac{-\frac{\sqrt{2}}{2}}{\frac{\sqrt{2}}{2}} = -1$

$\csc \frac{7\pi}{4} = \frac{1}{\sin t} = -\sqrt{2}$

$\sec \frac{7\pi}{4} = \frac{1}{\cos t} = \sqrt{2}$

$\cot \frac{7\pi}{4} = \frac{1}{\tan t} = -1$

30. $\cos 5\pi = \cos \pi = -1$

32. $\sin \frac{9\pi}{4} = \sin \frac{\pi}{4} = \frac{\sqrt{2}}{2}$

34. $\sin \frac{19\pi}{6} = \sin \frac{7\pi}{6} = -\frac{1}{2}$

36. $\cos\left(-\frac{8\pi}{3}\right) = \cos \frac{4\pi}{3} = -\frac{1}{2}$

38. $\sin(-t) = \frac{3}{8}$

(a) $\sin t = -\sin(-t) = -\frac{3}{8}$

(b) $\csc t = \frac{1}{\sin t} = -\frac{8}{3}$

40. $\cos t = -\frac{3}{4}$

(a) $\cos(-t) = \cos t = -\frac{3}{4}$

(b) $\sec(-t) = \sec t = \frac{1}{\cos t} = -\frac{4}{3}$

42. $\cos t = \frac{4}{5}$

(a) $\cos(\pi - t) = -\cos t = -\frac{4}{5}$

(b) $\cos(t + \pi) = -\cos t = -\frac{4}{5}$

44. $\tan \frac{\pi}{3} \approx 1.7321$

46. $\cot 1 \approx 0.6421$

48. $\cos(-2.5) \approx -0.8011$

50. $\sec 1.8 \approx -4.4014$

52. $\sin(-0.9) \approx -0.7833$

54. (a) $\sin 0.75 = y \approx 0.7$

(b) $\cos 2.5 = x \approx -0.8$

56. (a) $\sin t = -0.75$

$t \approx 4.0$ or $t \approx 5.4$

(b) $\cos t = 0.75$

$t \approx 0.7$ or $t \approx 5.6$

58. $y(t) = \frac{1}{4}\cos 6t$

(a) $y(0) = \frac{1}{4}\cos 0 = 0.25$ feet

(b) $y\left(\frac{1}{4}\right) = \frac{1}{4}\cos \frac{3}{2} \approx 0.02$ feet

(c) $y\left(\frac{1}{2}\right) = \frac{1}{4}\cos 3 \approx -0.25$ feet

60. True. $\tan a = \tan(a - 6\pi)$ since the period of tan is π.

62. $\cos \theta = x = \cos(-\theta)$

$\sec \theta = \frac{1}{x} = \sec(-\theta)$

$\sin \theta = y$

$\sin(-\theta) = -y = -\sin \theta$

$\csc \theta = \frac{1}{y}$

$\csc(-\theta) = -\frac{1}{y} = -\csc \theta$

$\tan \theta = \frac{y}{x}$

$\tan(-\theta) = \frac{-y}{x} = -\tan \theta$

$\cot \theta = \frac{x}{y}$

$\cot(-\theta) = \frac{x}{-y} = -\cot \theta$

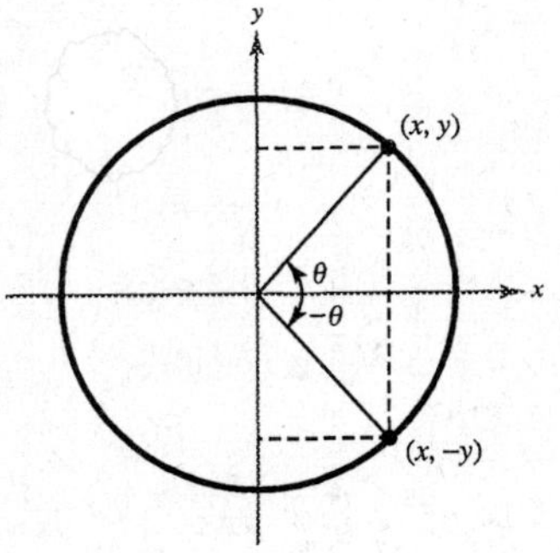

64. $h(t) = f(t)g(t)$ is an odd function, since

$$\begin{aligned} h(-t) &= f(-t)g(-t) \\ &= -f(t)g(t) \\ &= -h(t) \end{aligned}$$

66. $f(x) = \frac{1}{4}x^3 + 1$

$$\begin{aligned} y &= \tfrac{1}{4}x^3 + 1 \\ x &= \tfrac{1}{4}y^3 + 1 \\ x - 1 &= \tfrac{1}{4}y^3 \\ 4(x - 1) &= y^3 \\ y &= \sqrt[3]{4(x - 1)} \\ f^{-1}(x) &= \sqrt[3]{4(x - 1)} \end{aligned}$$

68. $f(x) = \dfrac{x + 2}{x - 4}$

$$\begin{aligned} y &= \frac{x + 2}{x - 4} \\ x &= \frac{y + 2}{y - 4} \\ x(y - 4) &= y + 2 \\ xy - 4x &= y + 2 \\ xy - y &= 4x + 2 \\ y(x - 1) &= 4x + 2 \\ y &= \frac{2(2x + 1)}{x - 1} \\ f^{-1}(x) &= \frac{2(2x + 1)}{x - 1} \end{aligned}$$

Section 1.3 Right Triangle Trigonometry

Solutions to Even-Numbered Exercises

2.

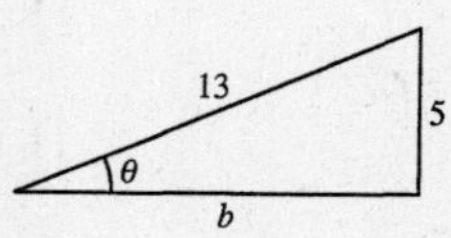

$\text{adj} = \sqrt{13^2 - 5^2} = \sqrt{169 - 25} = 12$

$\sin\theta = \dfrac{\text{opp}}{\text{hyp}} = \dfrac{5}{13}$ $\csc\theta = \dfrac{\text{hyp}}{\text{opp}} = \dfrac{13}{5}$

$\cos\theta = \dfrac{\text{adj}}{\text{hyp}} = \dfrac{12}{13}$ $\sec\theta = \dfrac{\text{hyp}}{\text{adj}} = \dfrac{13}{12}$

$\tan\theta = \dfrac{\text{opp}}{\text{adj}} = \dfrac{5}{12}$ $\cot\theta = \dfrac{\text{adj}}{\text{opp}} = \dfrac{12}{5}$

4.

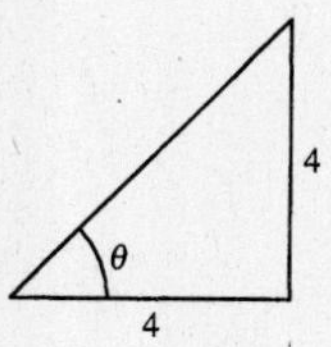

$\text{hyp} = \sqrt{4^2 + 4^2} = \sqrt{32} = 4\sqrt{2}$

$\sin\theta = \dfrac{\text{opp}}{\text{hyp}} = \dfrac{4}{4\sqrt{2}} = \dfrac{1}{\sqrt{2}} = \dfrac{\sqrt{2}}{2}$ $\csc\theta = \dfrac{\text{hyp}}{\text{opp}} = \dfrac{4\sqrt{2}}{4} = \sqrt{2}$

$\cos\theta = \dfrac{\text{adj}}{\text{hyp}} = \dfrac{4}{4\sqrt{2}} = \dfrac{1}{\sqrt{2}} = \dfrac{\sqrt{2}}{2}$ $\sec\theta = \dfrac{\text{hyp}}{\text{adj}} = \dfrac{4\sqrt{2}}{4} = \sqrt{2}$

$\tan\theta = \dfrac{\text{opp}}{\text{adj}} = \dfrac{4}{4} = 1$ $\cot\theta = \dfrac{\text{adj}}{\text{opp}} = \dfrac{4}{4} = 1$

6.

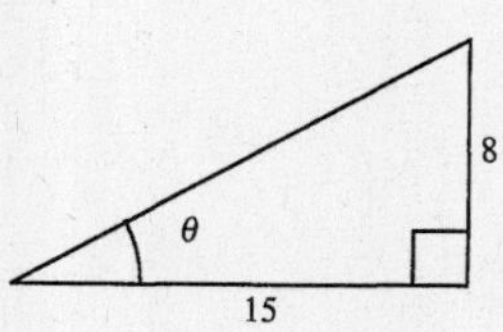

$\text{hyp} = \sqrt{15^2 + 8^2} = \sqrt{289} = 17$

$\sin\theta = \dfrac{\text{opp}}{\text{hyp}} = \dfrac{8}{17}$ $\csc\theta = \dfrac{\text{hyp}}{\text{opp}} = \dfrac{17}{8}$

$\cos\theta = \dfrac{\text{adj}}{\text{hyp}} = \dfrac{15}{17}$ $\sec\theta = \dfrac{\text{hyp}}{\text{adj}} = \dfrac{17}{15}$

$\tan\theta = \dfrac{\text{opp}}{\text{adj}} = \dfrac{8}{15}$ $\cot\theta = \dfrac{\text{adj}}{\text{opp}} = \dfrac{15}{8}$

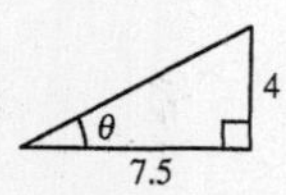

$\text{hyp} = \sqrt{7.5^2 + 4^2} = \dfrac{17}{2}$

$\sin\theta = \dfrac{\text{opp}}{\text{hyp}} = \dfrac{4}{\frac{17}{2}} = \dfrac{8}{17}$ $\csc\theta = \dfrac{\text{hyp}}{\text{opp}} = \dfrac{\frac{17}{2}}{4} = \dfrac{17}{8}$

$\cos\theta = \dfrac{\text{adj}}{\text{hyp}} = \dfrac{7.5}{\frac{17}{2}} = \dfrac{15}{17}$ $\sec\theta = \dfrac{\text{hyp}}{\text{adj}} = \dfrac{\frac{17}{2}}{7.5} = \dfrac{17}{15}$

$\tan\theta = \dfrac{\text{opp}}{\text{adj}} = \dfrac{4}{7.5} = \dfrac{8}{15}$ $\cot\theta = \dfrac{\text{adj}}{\text{opp}} = \dfrac{7.5}{4} = \dfrac{15}{8}$

The function values are the same because the triangles are similar, and corresponding sides are proportional.

8. 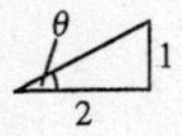

$\text{hyp} = \sqrt{1^2 + 2^2} = \sqrt{5}$

$\sin\theta = \dfrac{\text{opp}}{\text{hyp}} = \dfrac{1}{\sqrt{5}} = \dfrac{\sqrt{5}}{5}$ $\qquad \csc\theta = \dfrac{\text{hyp}}{\text{opp}} = \dfrac{\sqrt{5}}{1} = \sqrt{5}$

$\cos\theta = \dfrac{\text{adj}}{\text{hyp}} = \dfrac{2}{\sqrt{5}} = \dfrac{2\sqrt{5}}{5}$ $\qquad \sec\theta = \dfrac{\text{hyp}}{\text{adj}} = \dfrac{\sqrt{5}}{2}$

$\tan\theta = \dfrac{\text{opp}}{\text{adj}} = \dfrac{1}{2}$ $\qquad \cot\theta = \dfrac{\text{adj}}{\text{opp}} = \dfrac{2}{1} = 2$

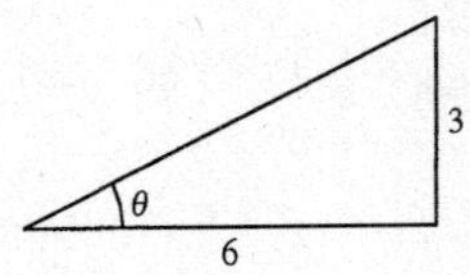

$\text{hyp} = \sqrt{3^2 + 6^2} = 3\sqrt{5}$

$\sin\theta = \dfrac{3}{3\sqrt{5}} = \dfrac{1}{\sqrt{5}} = \dfrac{\sqrt{5}}{5}$ $\qquad \csc\theta = \dfrac{3\sqrt{5}}{3} = \sqrt{5}$

$\cos\theta = \dfrac{6}{3\sqrt{5}} = \dfrac{2}{\sqrt{5}} = \dfrac{2\sqrt{5}}{5}$ $\qquad \sec\theta = \dfrac{3\sqrt{5}}{6} = \dfrac{\sqrt{5}}{2}$

$\tan\theta = \dfrac{3}{6} = \dfrac{1}{2}$ $\qquad \cot\theta = \dfrac{6}{3} = 2$

The function values are the same because the triangles are similar, and corresponding sides are proportional.

10. 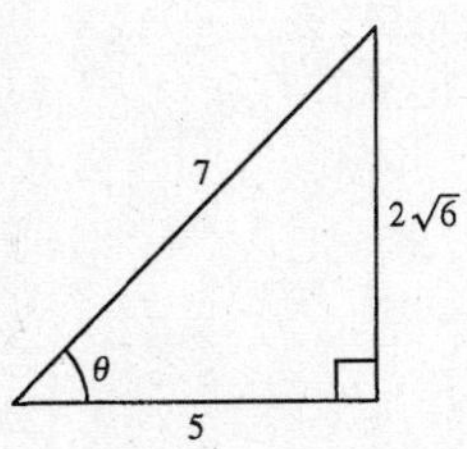

$\text{hyp} = \sqrt{7^2 - 5^2} = \sqrt{24} = 2\sqrt{6}$

$\sin\theta = \dfrac{\text{opp}}{\text{hyp}} = \dfrac{2\sqrt{6}}{7}$

$\tan\theta = \dfrac{\text{opp}}{\text{adj}} = \dfrac{2\sqrt{6}}{5}$

$\csc\theta = \dfrac{1}{\sin\theta} = \dfrac{7}{2\sqrt{6}} = \dfrac{7\sqrt{6}}{12}$

$\sec\theta = \dfrac{1}{\cos\theta} = \dfrac{7}{5}$

$\cot\theta = \dfrac{1}{\tan\theta} = \dfrac{5}{2\sqrt{6}} = \dfrac{5\sqrt{6}}{12}$

12. 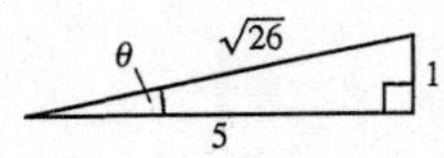

$\text{hyp} = \sqrt{5^2 + 1^2} = \sqrt{26}$

$\sin\theta = \dfrac{\text{opp}}{\text{hyp}} = \dfrac{1}{\sqrt{26}} = \dfrac{\sqrt{26}}{26}$

$\cos\theta = \dfrac{\text{adj}}{\text{hyp}} = \dfrac{5}{\sqrt{26}} = \dfrac{5\sqrt{26}}{26}$

$\tan\theta = \dfrac{\text{opp}}{\text{adj}} = \dfrac{1}{5}$

$\csc\theta = \dfrac{\text{hyp}}{\text{opp}} = \dfrac{\sqrt{26}}{1} = \sqrt{26}$

$\sec\theta = \dfrac{\text{hyp}}{\text{adj}} = \dfrac{\sqrt{26}}{5}$

14. 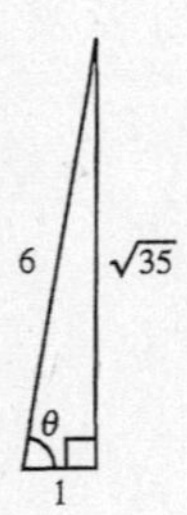

$\text{opp} = \sqrt{6^2 - 1^2} = \sqrt{35}$

$\sin\theta = \dfrac{\text{opp}}{\text{hyp}} = \dfrac{\sqrt{35}}{6}$

$\cos\theta = \dfrac{\text{adj}}{\text{hyp}} = \dfrac{1}{6}$

$\tan\theta = \dfrac{\text{opp}}{\text{adj}} = \dfrac{\sqrt{35}}{1} = \sqrt{35}$

$\csc\theta = \dfrac{\text{hyp}}{\text{opp}} = \dfrac{6}{\sqrt{35}} = \dfrac{6\sqrt{35}}{35}$

$\cot\theta = \dfrac{\text{adj}}{\text{opp}} = \dfrac{1}{\sqrt{35}} = \dfrac{\sqrt{35}}{35}$

16. 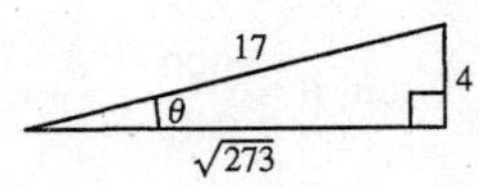

$\text{adj} = \sqrt{17^2 - 4^2} = \sqrt{273}$

$\sin\theta = \dfrac{\text{opp}}{\text{hyp}} = \dfrac{4}{17}$

$\cos\theta = \dfrac{\text{adj}}{\text{hyp}} = \dfrac{\sqrt{273}}{17}$

$\tan\theta = \dfrac{\text{opp}}{\text{adj}} = \dfrac{4}{\sqrt{273}} = \dfrac{4\sqrt{273}}{273}$

$\sec\theta = \dfrac{1}{\cos\theta} = \dfrac{17}{\sqrt{273}} = \dfrac{17\sqrt{273}}{273}$

$\cot\theta = \dfrac{1}{\tan\theta} = \dfrac{\sqrt{273}}{4}$

18. $\sin 30^\circ = \frac{1}{2}, \tan 30^\circ = \frac{\sqrt{3}}{3}$

(a) $\csc 30^\circ = \frac{1}{\sin 30^\circ} = 2$

(b) $\cot 60^\circ = \tan(90^\circ - 60^\circ) = \tan 30^\circ = \frac{\sqrt{3}}{3}$

(c) $\cos 30^\circ = \frac{\sin 30^\circ}{\tan 30^\circ} = \frac{\frac{1}{2}}{\frac{\sqrt{3}}{3}} = \frac{3}{2\sqrt{3}} = \frac{\sqrt{3}}{2}$

(d) $\cot 30^\circ = \frac{1}{\tan 30^\circ} = \frac{3}{\sqrt{3}} = \frac{3\sqrt{3}}{3} = \sqrt{3}$

20. $\sec\theta = 5, \tan\theta = 2\sqrt{6}$

(a) $\cos\theta = \frac{1}{\sec\theta} = \frac{1}{5}$

(b) $\cot\theta = \frac{1}{\tan\theta} = \frac{1}{2\sqrt{6}} = \frac{\sqrt{6}}{12}$

(c) $\cot(90^\circ - \theta) = \tan\theta = 2\sqrt{6}$

(d) $\sin\theta = \tan\theta\cos\theta = (2\sqrt{6})\left(\frac{1}{5}\right) = \frac{2\sqrt{6}}{5}$

22. $\tan\beta = 5$

(a) $\cot\beta = \frac{1}{\tan\beta} = \frac{1}{5}$

(b) $\cos\beta = \frac{1}{\sqrt{1 + \tan^2\beta}} = \frac{1}{\sqrt{1 + 5^2}}$

$= \frac{1}{\sqrt{26}} = \frac{\sqrt{26}}{26}$

(c) $\tan(90^\circ - \beta) = \cot\beta = \frac{1}{\tan\beta} = \frac{1}{5}$

(d) $\csc\beta = \sqrt{1 + \cot^2\beta}$

$= \sqrt{1 + \left(\frac{1}{5}\right)^2}$

$= \sqrt{1 + \frac{1}{25}} = \sqrt{\frac{26}{25}} = \frac{\sqrt{26}}{5}$

24. (a) $\cot 45^\circ = \frac{1}{1} = 1$

(b) $\cos 45^\circ = \frac{1}{\sqrt{2}} = \frac{\sqrt{2}}{2}$

(c) $\csc 45^\circ = \frac{\sqrt{2}}{1} = \sqrt{2}$

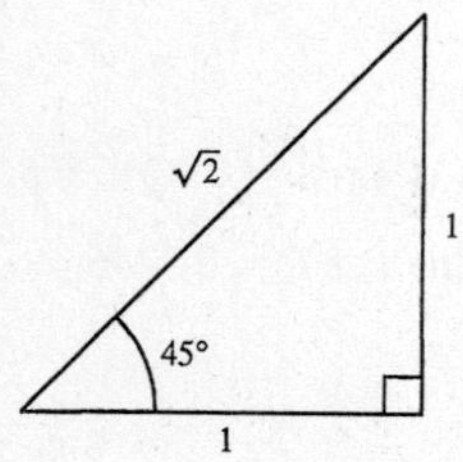

26. (a) $\sin 60^\circ = \frac{\sqrt{3}}{2}$

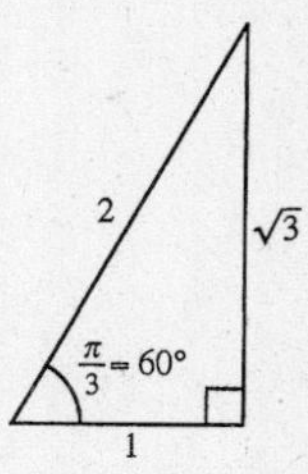

(b) $\tan 45^\circ = \frac{1}{1} = 1$

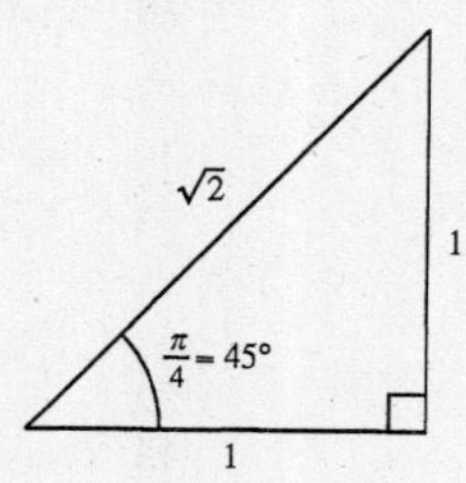

(c) $\sec 30^\circ = \frac{2}{\sqrt{3}} = \frac{2\sqrt{3}}{3}$

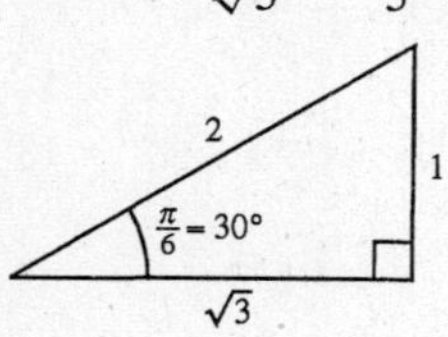

28. (a) $\tan 23.5^\circ \approx 0.4348$

(b) $\cot 66.5^\circ = \frac{1}{\tan 66.5^\circ} \approx 0.4348$

30. (a) $\cos 16^\circ 18' = \cos\left(16 + \frac{18}{60}\right)^\circ \approx 0.9598$

(b) $\sin 73^\circ 56' = \sin\left(73 + \frac{56}{60}\right)^\circ \approx 0.9609$

32. (a) $\cos 4^\circ 50' 15'' = \cos\left(4 + \frac{50}{60} + \frac{15}{3600}\right)^\circ$

≈ 0.9964

(b) $\sec 4^\circ 50' 15'' = \frac{1}{\cos 4^\circ 50' 15''}$

≈ 1.0036

34. (a) $\sec 56^\circ 8' 10'' = \sec\left(56 + \frac{8}{60} = \frac{10}{3600}\right)^\circ \approx 1.7946$

(b) $\cos 56^\circ 8' 10'' = \cos\left(56 + \frac{8}{60} + \frac{10}{3600}\right)^\circ \approx 0.5572$

36. (a) $\sec\left(\frac{9}{5} \cdot 20 + 32\right)^{\circ} \approx 2.6695$

(b) $\cot\left(\frac{9}{5} \cdot 30 + 32\right)^{\circ} \approx 0.0699$

38. (a) $\cos\theta = \frac{\sqrt{2}}{2} \Rightarrow \theta = 45^{\circ} = \frac{\pi}{4}$

(b) $\tan\theta = 1 \Rightarrow \theta = 45^{\circ} = \frac{\pi}{4}$

40. (a) $\tan\theta = \sqrt{3} \Rightarrow \theta = 60^{\circ} = \frac{\pi}{3}$

(b) $\cos\theta = \frac{1}{2} \Rightarrow \theta = 60^{\circ} = \frac{\pi}{3}$

42. (a) $\cot\theta = \frac{\sqrt{3}}{3}$

$\tan\theta = \frac{3}{\sqrt{3}} = \sqrt{3} \Rightarrow \theta = 60^{\circ} = \frac{\pi}{3}$

(b) $\sec\theta = \sqrt{2}$

$\cos\theta = \frac{1}{\sqrt{2}} = \frac{\sqrt{2}}{2} \Rightarrow \theta = 45^{\circ} = \frac{\pi}{4}$

44. (a) $\cos\theta = 0.9848 \Rightarrow \theta \approx 10^{\circ} \approx 0.175$

(b) $\cos\theta = 0.8746 \Rightarrow \theta \approx 29^{\circ} \approx 0.506$

46. (a) $\sin\theta = 0.3746 \Longrightarrow \theta \approx 22^{\circ} \approx 0.384$

(b) $\cos\theta = 0.3746 \Longrightarrow \theta \approx 68^{\circ} \approx 1.187$

48. $\cos\theta \sec\theta = \cos\theta\left(\frac{1}{\cos\theta}\right) = 1$

50. $\cot\alpha \sin\alpha = \frac{\cos\alpha}{\sin\alpha} \sin\alpha = \cos\alpha$

52. $(1 + \sin\theta)(1 - \sin\theta) = 1 - \sin^2\theta = \cos^2\theta$

54. $\sin^2\theta - \cos^2\theta = \sin^2\theta - (1 - \sin^2\theta)$

$= \sin^2\theta - 1 + \sin^2\theta$

$= 2\sin^2\theta - 1$

56. $\frac{\tan\beta + \cot\beta}{\tan\beta} = \frac{\tan\beta}{\tan\beta} + \frac{\cot\beta}{\tan\beta}$

$= 1 + \frac{\cot\beta}{\left(\frac{1}{\cot\beta}\right)}$

$= 1 + \cot^2\beta = \csc^2\beta$

58. $\sin 60^{\circ} = \frac{y}{18}$

$y = 18\sin 60^{\circ} = 18\left(\frac{\sqrt{3}}{2}\right)$

$= 9\sqrt{3}$

60. $\sin 45^{\circ} = \frac{20}{r}$

$r = \frac{20}{\sin 45^{\circ}} = \frac{20}{\frac{\sqrt{2}}{2}} = 20\sqrt{2}$

62. $\frac{h}{30} = \frac{6}{10}$

$h = 30\left(\frac{6}{10}\right) = 18$ feet

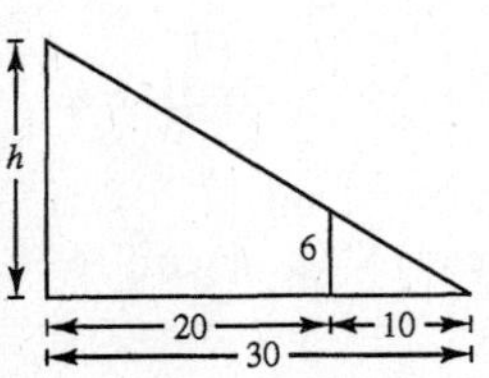

64. (a)

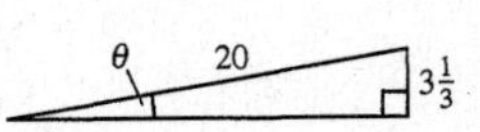

(b) $\sin\theta = \frac{\text{opp}}{\text{hyp}}$

$\sin\theta = \frac{3\frac{1}{3}}{20}$

(c) $\sin\theta = \frac{1}{6} \Rightarrow \theta = 9.59^{\circ}$

66. Let $h =$ the height of the mountain.

Let $x =$ the horizontal distance from where the 9° angle of elevation is sighted to the point at that level directly below the mountain peak.

Then $\tan 3.5° = \dfrac{h}{x + 13}$ and $\tan 9° = \dfrac{h}{x}$.

$\tan 9° = \dfrac{h}{x} \Rightarrow x = \dfrac{h}{\tan 9°}$

Substitute $x = \dfrac{h}{\tan 9°}$ into the expression for tan 3.5°.

$$\tan 3.5° = \frac{h}{\dfrac{h}{\tan 9°} + 13}$$

$$\tan 3.5° = \frac{h \tan 9°}{h + 13 \tan 9°}$$

$$h \tan 3.5° + 13 \tan 9° \tan 3.5° = h \tan 9°$$

$$13 \tan 9° \tan 3.5° = h(\tan 9° - \tan 3.5°)$$

$$\frac{13 \tan 9° \tan 3.5°}{\tan 9° - \tan 3.5°} = h$$

$$1.2953 \approx h$$

The mountain is about 1.3 miles high.

68. $\tan 3° = \dfrac{x}{15}$

$x = 15 \tan 3°$

$d = 5 + 2x = 5 + 2(15 \tan 3°) \approx 6.57$ centimeters

70. $x \approx 2.588,\ y \approx 9.659$

$\sin 75° = \dfrac{y}{10} \approx 0.97$ $\qquad \csc 75° = \dfrac{10}{y} \approx 1.04$

$\cos 75° = \dfrac{x}{10} \approx 0.26$ $\qquad \sec 75° = \dfrac{10}{x} \approx 3.86$

$\tan 75° = \dfrac{y}{x} \approx 3.73$ $\qquad \cot 75° = \dfrac{x}{y} \approx 0.27$

72. $\sec 30° = \csc 60°$

True, because $\sec(90° - \theta) = \csc \theta$.

74. $\cot^2 10° - \csc^2 10° = -1$

True, because

$$1 + \cot^2 \theta = \csc^2 \theta$$

$$\cot^2 \theta = \csc^2 \theta - 1$$

$$\cot^2 \theta - \csc^2 \theta = -1.$$

76. $\tan[(5°)^2] = \tan^2(5°)$

False.

$\tan[(5°)^2] = \tan 25° \approx 0.4663$

$\tan^2(5°) = (\tan 5°)(\tan 5°)$

≈ 0.0077

78. Yes. Given $\tan \theta$, $\sec \theta$ can be found from the identity $1 + \tan^2\theta = \sec^2\theta$.

80. (a)

θ	0°	18°	36°	54°	72°	90°
$\sin \theta$	0	0.3090	0.5878	0.8090	0.9511	1
$\cos \theta$	1	0.9511	0.8090	0.5878	0.3090	0

(b) $\sin \theta$ increases from 0 to 1.

(c) $\cos \theta$ decreases from 1 to 0.

(d) As the angle increases the length of the side opposite the angle increases relative to the length of the hypotenuse and the length of the side adjacent to the angle decreases relative to the length of the hypotenuse. Thus the sine increases and the cosine decreases.

82. $\dfrac{2t^2 + 5t - 12}{9 - 4t^2} \div \dfrac{t^2 - 16}{4t^2 + 12t + 9} = \dfrac{2t^2 + 5t - 12}{9 - 4t^2} \cdot \dfrac{4t^2 + 12t + 9}{t^2 - 16}$

$$= \frac{(2t - 3)(t + 4)}{(3 + 2t)(3 - 2t)} \cdot \frac{(2t + 3)(2t + 3)}{(t + 4)(t - 4)} = -\frac{(2t + 3)}{(t - 4)} = \frac{2t + 3}{4 - t}, t \neq \pm\frac{3}{2}, -4$$

84. $\dfrac{\left(\dfrac{3}{x} - \dfrac{1}{4}\right)}{\left(\dfrac{12}{x} - 1\right)} = \dfrac{\dfrac{12 - x}{4x}}{\dfrac{12 - x}{x}} = \dfrac{12 - x}{4x} \cdot \dfrac{x}{12 - x} = \dfrac{1}{4}, x \neq 0, 12$

86. $\dfrac{3x + 2}{x^2 + x - 2} = \dfrac{4}{x + 2} - \dfrac{2}{1 - x}$

$$\frac{3x + 2}{(x + 2)(x - 1)} = \frac{4}{x + 2} + \frac{2}{x - 1}$$

$$\frac{3x + 2}{(x + 2)(x - 1)} = \frac{4(x - 1) + 2(x + 2)}{(x + 2)(x - 1)}$$

$$3x + 2 = 4x - 4 + 2x + 4$$

$$-3x + 2 = 0$$

$$x = \frac{2}{3}$$

Section 1.4 Trigonometric Functions of Any Angle

Solutions to Even-Numbered Exercises

2. (a) $x = -12, y = -5$

$r = \sqrt{(-12)^2 + (-5)^2} = 13$

$\sin\theta = \dfrac{y}{r} = -\dfrac{5}{13}$

$\cos\theta = \dfrac{x}{r} = -\dfrac{12}{13}$

$\tan\theta = \dfrac{y}{x} = \dfrac{-5}{-12} = \dfrac{5}{12}$

$\csc\theta = \dfrac{r}{y} = \dfrac{13}{-5} = \dfrac{13}{-5}$

$\sec\theta = \dfrac{r}{x} = \dfrac{13}{-12} = -\dfrac{13}{12}$

$\cot\theta = \dfrac{x}{y} = \dfrac{-12}{-5} = \dfrac{12}{5}$

(b) $x = -1, y = 1$

$r = \sqrt{(-1)^2 + 1^2} = \sqrt{2}$

$\sin\theta = \dfrac{y}{r} = \dfrac{1}{\sqrt{2}} = \dfrac{\sqrt{2}}{2}$

$\cos\theta = \dfrac{x}{r} = \dfrac{-1}{\sqrt{2}} = -\dfrac{\sqrt{2}}{2}$

$\tan\theta = \dfrac{y}{x} = \dfrac{1}{-1} = -1$

$\csc\theta = \dfrac{r}{y} = \dfrac{\sqrt{2}}{1} = \sqrt{2}$

$\sec\theta = \dfrac{r}{x} = \dfrac{\sqrt{2}}{-1} = -\sqrt{2}$

$\cot\theta = \dfrac{x}{y} = \dfrac{-1}{1} = -1$

4. (a) $x = 3, y = 1$

$r = \sqrt{3^2 + 1^2} = \sqrt{10}$

$\sin\theta = \dfrac{y}{r} = \dfrac{1}{\sqrt{10}} = \dfrac{\sqrt{10}}{10}$

$\cos\theta = \dfrac{x}{r} = \dfrac{3}{\sqrt{10}} = \dfrac{3\sqrt{10}}{10}$

$\tan\theta = \dfrac{y}{x} = \dfrac{1}{3}$

$\csc\theta = \dfrac{r}{y} = \dfrac{\sqrt{10}}{1} = \sqrt{10}$

$\sec\theta = \dfrac{r}{x} = \dfrac{\sqrt{10}}{3}$

$\cot\theta = \dfrac{x}{y} = \dfrac{3}{1} = 3$

(b) $x = 4, y = -4$

$r = \sqrt{4^2 + (-4)^2} = 4\sqrt{2}$

$\sin\theta = \dfrac{y}{r} = \dfrac{-4}{4\sqrt{2}} = -\dfrac{\sqrt{2}}{2}$

$\cos\theta = \dfrac{x}{r} = \dfrac{4}{4\sqrt{2}} = \dfrac{\sqrt{2}}{2}$

$\tan\theta = \dfrac{y}{x} = \dfrac{-4}{4} = -1$

$\csc\theta = \dfrac{r}{y} = \dfrac{4\sqrt{2}}{-4} = -\sqrt{2}$

$\sec\theta = \dfrac{r}{x} = \dfrac{4\sqrt{2}}{4} = \sqrt{2}$

$\cot\theta = \dfrac{x}{y} = \dfrac{4}{-4} = -1$

6. $x = 8, y = 15$

$r = \sqrt{8^2 + 15^2} = 17$

$\sin\theta = \frac{y}{r} = \frac{15}{17}$ $\qquad$ $\csc\theta = \frac{r}{y} = \frac{17}{15}$

$\cos\theta = \frac{x}{r} = \frac{8}{17}$ $\qquad$ $\sec\theta = \frac{r}{x} = \frac{17}{8}$

$\tan\theta = \frac{y}{x} = \frac{15}{8}$ $\qquad$ $\cot\theta = \frac{x}{y} = \frac{8}{15}$

8. $x = -5, y = -2$

$r = \sqrt{(-5)^2 + (-2)^2} = \sqrt{29}$

$\sin\theta = \frac{y}{r} = \frac{-2}{\sqrt{29}} = -\frac{2\sqrt{29}}{29}$ $\qquad$ $\csc\theta = \frac{r}{y} = \frac{\sqrt{29}}{-2} = -\frac{\sqrt{29}}{2}$

$\cos\theta = \frac{x}{r} = \frac{-5}{\sqrt{29}} = -\frac{5\sqrt{29}}{29}$ $\qquad$ $\sec\theta = \frac{r}{x} = \frac{\sqrt{29}}{-5} = -\frac{\sqrt{29}}{5}$

$\tan\theta = \frac{y}{x} = \frac{-2}{-5} = \frac{2}{5}$ $\qquad$ $\cot\theta = \frac{x}{y} = \frac{-5}{-2} = \frac{5}{2}$

10. $x = 3\frac{1}{2} = \frac{7}{2}, y = -7\frac{3}{4} = -\frac{31}{4}$

$r = \sqrt{\left(\frac{7}{2}\right)^2 + \left(-\frac{31}{4}\right)^2} = \frac{\sqrt{1157}}{4}$

$\sin\theta = \frac{x}{r} = \frac{-\frac{31}{4}}{\frac{\sqrt{1157}}{4}} = -\frac{31\sqrt{1157}}{1157}$ $\qquad$ $\csc\theta = \frac{r}{y} = \frac{\frac{\sqrt{1157}}{4}}{-\frac{31}{4}} = -\frac{\sqrt{1157}}{31}$

$\cos\theta = \frac{x}{r} = \frac{\frac{7}{2}}{\frac{\sqrt{1157}}{4}} = \frac{14\sqrt{1157}}{1157}$ $\qquad$ $\sec\theta = \frac{r}{x} = \frac{\frac{\sqrt{1157}}{4}}{\frac{7}{2}} = \frac{\sqrt{1157}}{14}$

$\tan\theta = \frac{y}{x} = \frac{-\frac{31}{4}}{\frac{7}{2}} = -\frac{31}{14}$ $\qquad$ $\cot\theta = \frac{x}{y} = \frac{\frac{7}{2}}{-\frac{31}{4}} = -\frac{14}{31}$

12. $\sin\theta > 0$ and $\cos\theta > 0$

$\frac{y}{r} > 0$ and $\frac{x}{r} > 0$

Quadrant I

14. $\sec\theta > 0$ and $\cot\theta < 0$

$\frac{r}{x} > 0$ and $\frac{x}{y} < 0$

Quadrant IV

16. $\cos\theta = \frac{x}{r} = \frac{-4}{5} \Rightarrow y = |3|$

θ in Quadrant III $\Rightarrow y = -3$

$\sin\theta = \frac{y}{r} = -\frac{3}{5}$ $\qquad$ $\csc\theta = -\frac{5}{3}$

$\cos\theta = \frac{x}{r} = -\frac{4}{5}$ $\qquad$ $\sec\theta = -\frac{5}{4}$

$\tan\theta = \frac{y}{x} = \frac{3}{4}$ $\qquad$ $\cot\theta = \frac{4}{3}$

18. $\cos\theta = \frac{x}{r} = \frac{8}{17} \Rightarrow y = |15|$

$\tan\theta < 0 \Rightarrow y = -15$

$\sin\theta = \frac{y}{r} = \frac{-15}{17} = -\frac{15}{17}$ $\qquad$ $\csc\theta = -\frac{17}{15}$

$\cos\theta = \frac{x}{r} = \frac{8}{17}$ $\qquad$ $\sec\theta = \frac{17}{8}$

$\tan\theta = \frac{y}{x} = \frac{-15}{8} = -\frac{15}{8}$ $\qquad$ $\cot\theta = -\frac{8}{15}$

20. $\csc\theta = \frac{r}{y} = \frac{4}{1} \Rightarrow x = \left|\sqrt{15}\right|$

$\cot\theta < 0 \Rightarrow x = -\sqrt{15}$

$\sin\theta = \frac{y}{r} = \frac{1}{4}$ $\qquad \csc\theta = 4$

$\cos\theta = \frac{x}{r} = -\frac{\sqrt{15}}{4}$ $\qquad \sec\theta = -\frac{4\sqrt{15}}{15}$

$\tan\theta = \frac{y}{x} = -\frac{\sqrt{15}}{15}$ $\qquad \cot\theta = -\sqrt{15}$

22. $\sin\theta = 0 \Rightarrow \theta = 0 + \pi n$

$\sec\theta = -1 \Rightarrow \theta = \pi + 2\pi n$

$y = 0, x = -r$

$\sin\theta = 0$ $\qquad \csc\theta = \frac{r}{y}$ is undefined

$\cos\theta = \frac{x}{r} = \frac{-r}{r} = -1$ $\qquad \sec\theta = \frac{r}{x} = -1$

$\tan\theta = \frac{y}{x} = 0$ $\qquad \cot\theta = \frac{x}{y}$ is undefined

24. $\tan\theta$ is undefined $\Rightarrow \theta = n\pi + \frac{\pi}{2}$

$\pi \le \theta \le 2\pi \Rightarrow \theta = \frac{3\pi}{2}, x = 0, y = -r$

$\sin\theta = \frac{y}{r} = \frac{-r}{r} = -1$ $\qquad \csc\theta = \frac{r}{y} = -1$

$\cos\theta = \frac{x}{r} = \frac{0}{r} = 0$ $\qquad \sec\theta = \frac{r}{x}$ is undefined.

$\tan\theta = \frac{y}{x}$ is undefined. $\qquad \cot\theta = \frac{x}{y} = \frac{0}{y} = 0$

26. Let $x > 0$.

$\left(-x, -\frac{1}{3}x\right)$ Quadrant III

$$r = \sqrt{x^2 + \frac{1}{9}x^2} = \frac{\sqrt{10}x}{3}$$

$$\sin\theta = \frac{y}{r} = \frac{\left(-\frac{1}{3}\right)x}{\frac{\sqrt{10}x}{3}} = -\frac{\sqrt{10}}{10}$$

$$\cos\theta = \frac{x}{r} = \frac{-x}{\frac{\sqrt{10}x}{3}} = -\frac{3\sqrt{10}}{10}$$

$$\tan\theta = \frac{y}{x} = \frac{\left(-\frac{1}{3}\right)x}{-x} = \frac{1}{3}$$

$$\csc\theta = \frac{r}{y} = \frac{\frac{\sqrt{10}x}{3}}{\left(-\frac{1}{3}\right)x} = -\sqrt{10}$$

$$\sec\theta = \frac{r}{x} = \frac{\frac{\sqrt{10}x}{3}}{-x} = -\frac{\sqrt{10}}{3}$$

$$\cot\theta = \frac{x}{y} = \frac{-x}{\left(-\frac{1}{3}\right)x} = 3$$

28. Let $x > 0$.

$4x + 3y = 0 \Rightarrow y = -\frac{4}{3}x$

$\left(x, -\frac{4}{3}x\right)$ Quadrant IV

$$r = \sqrt{x^2 + \frac{16}{9}x^2} = \frac{5}{3}x$$

$\sin\theta = \frac{y}{r} = \frac{\left(-\frac{4}{3}\right)x}{\frac{5}{3}x} = -\frac{4}{5}$ $\qquad \csc\theta = -\frac{5}{4}$

$\cos\theta = \frac{x}{r} = \frac{x}{\frac{5}{3}x} = \frac{3}{5}$ $\qquad \sec\theta = \frac{5}{3}$

$\tan\theta = \frac{y}{x} = \frac{\left(-\frac{4}{3}\right)x}{x} = -\frac{4}{3}$ $\qquad \tan\theta = -\frac{3}{4}$

30. $\cos\frac{3\pi}{2} = \frac{x}{r} = \frac{0}{1} = 0$

since $\frac{3\pi}{2}$ corresponds to $(0, -1)$.

32. $(x, y) = (-1, 0), r = 1$

$\sec\pi = \frac{r}{x} = \frac{1}{-1} = -1$

34. $\tan \pi = \frac{y}{x} = \frac{0}{-1} = 0$

since π corresponds to $(-1, 0)$.

36. $(x, y) = (0, 1)$

$\cot \frac{\pi}{2} = \frac{x}{y} = \frac{0}{1} = 0$

38. $\theta = 309°$

$\theta' = 360° - 309° = 51°$

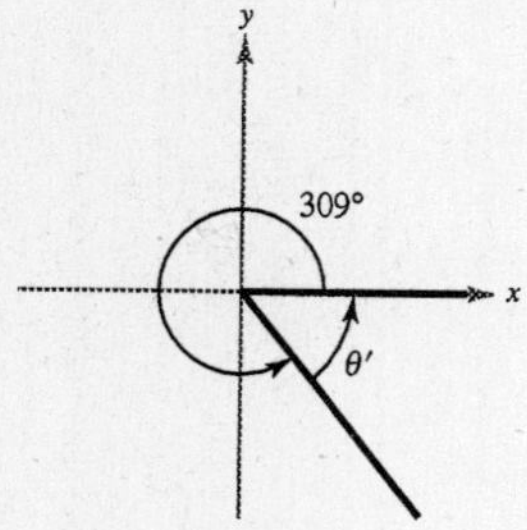

40. $\theta = -145°$ is coterminal with $215°$.

$\theta' = 215° - 180° = 35°$

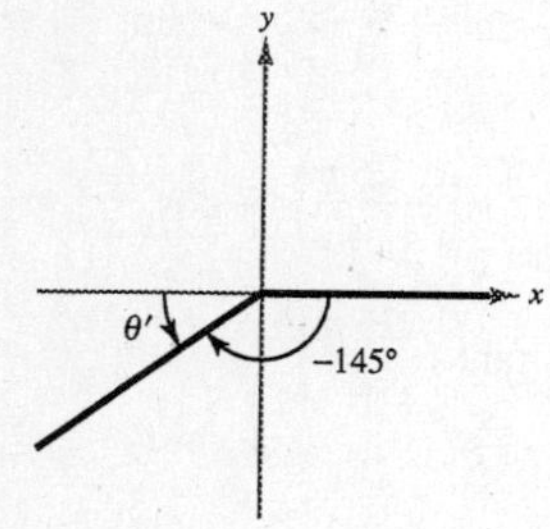

42. $\theta = \frac{7\pi}{4}$

$\theta' = 2\pi - \frac{7\pi}{4} = \frac{\pi}{4}$

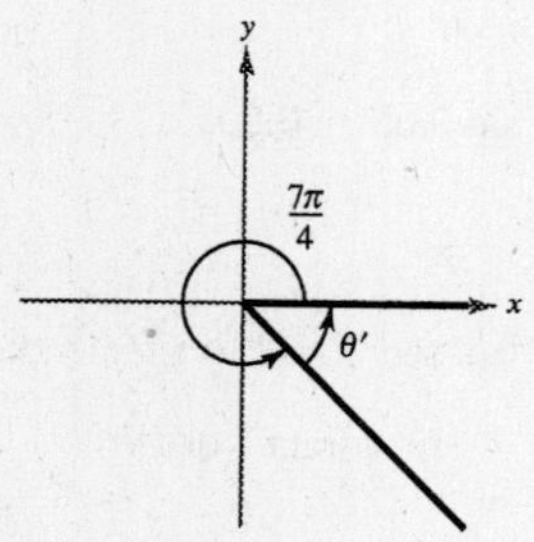

44. $\theta = \frac{11\pi}{3}$ is coterminal with $\frac{5\pi}{3}$.

$\theta' = 2\pi - \frac{5\pi}{3} = \frac{\pi}{3}$

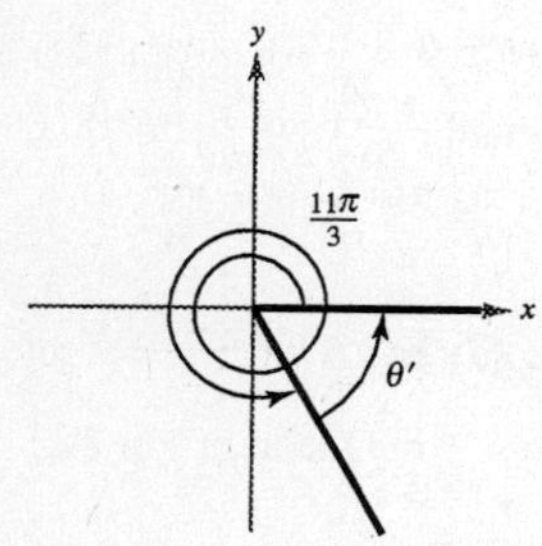

46. $\theta = 300°$, $\theta' = 360° - 300° = 60°$, Quadrant IV

$\sin 300° = -\sin 60° = -\frac{\sqrt{3}}{2}$

$\cos 300° = \cos 60° = \frac{1}{2}$

$\tan 300° = -\tan 60° = -\sqrt{3}$

48. $\theta = -405°$ is coterminal with $315°$.

$\theta' = 360° - 315° = 45°$ in Quadrant IV

$\sin(-405°) = -\sin 45° = -\frac{\sqrt{2}}{2}$

$\cos(-405°) = \cos 45° = \frac{\sqrt{2}}{2}$

$\tan(-405°) = -\tan 45° = -1$

50. $\theta = -840°$ is coterminal with $240°$.

$\theta' = 240° - 180° = 60°$ in Quadrant III.

$\sin(-840°) = -\sin 60° = -\frac{\sqrt{3}}{2}$

$\cos(-840°) = -\cos 60° = -\frac{1}{2}$

$\tan(-840°) = \tan 60° = \sqrt{3}$

52. $\theta = \frac{\pi}{4}$, $\theta' = \frac{\pi}{4}$ in Quadrant I

$\sin \frac{\pi}{4} = \frac{\sqrt{2}}{2}$

$\cos \frac{\pi}{4} = \frac{\sqrt{2}}{2}$

$\tan \frac{\pi}{4} = 1$

54. $\theta = -\frac{\pi}{2}$ is coterminal with $\frac{3\pi}{2}$.

$\sin\left(-\frac{\pi}{2}\right) = \sin \frac{3\pi}{2} = -1$

$\cos\left(-\frac{\pi}{2}\right) = \cos \frac{3\pi}{2} = 0$

$\tan\left(-\frac{\pi}{2}\right) = \tan \frac{3\pi}{2}$ is undefined.

56. $\theta = \frac{10\pi}{3}$ is coterminal with $\frac{4\pi}{3}$.

$\theta' = \frac{4\pi}{3} - \pi = \frac{\pi}{3}$ in Quadrant III.

$\sin \frac{10\pi}{3} = -\sin \frac{\pi}{3} = -\frac{\sqrt{3}}{2}$

$\cos \frac{10\pi}{3} = -\cos \frac{\pi}{3} = -\frac{1}{2}$

$\tan \frac{10\pi}{3} = \tan \frac{\pi}{3} = \sqrt{3}$

58. $\theta = -\frac{25\pi}{4}$ is coterminal with $\frac{7\pi}{4}$.

$\theta' = 2\pi - \frac{7\pi}{4} = \frac{\pi}{4}$ in Quadrant IV.

$$\sin\left(-\frac{25\pi}{4}\right) = -\sin\left(\frac{\pi}{4}\right) = -\frac{\sqrt{2}}{2}$$

$$\cos\left(-\frac{25\pi}{4}\right) = \cos\left(\frac{\pi}{4}\right) = \frac{\sqrt{2}}{2}$$

$$\tan\left(-\frac{25\pi}{4}\right) = -\tan\left(\frac{\pi}{4}\right) = -1$$

60. $\sec 225° = \frac{1}{\cos 225°} \approx -1.4142$

62. $\csc(-330°) = \frac{1}{\sin(-330°)} = 2.0000$

64. $\cot 1.35 = \frac{1}{\tan 1.35} \approx 0.2245$

66. $\tan\left(-\frac{\pi}{9}\right) \approx -0.3640$

68. $\sin 0.65 \approx 0.6052$

70. (a) $\cos\theta = \frac{\sqrt{2}}{2} \Rightarrow$ reference angle is 45° or $\frac{\pi}{4}$ and θ is in Quadrant I or IV.

Values in degrees: 45°, 315°

Values in radians: $\frac{\pi}{4}, \frac{7\pi}{4}$

(b) $\cos\theta = -\frac{\sqrt{2}}{2} \Rightarrow$ reference angle is 45° or $\frac{\pi}{4}$ and θ is in Quadrant II or III.

Values in degrees: 135°, 225°

Values in radians: $\frac{3\pi}{4}, \frac{5\pi}{4}$

72. (a) $\sec\theta = 2 \Rightarrow$ reference angle is 60° or $\frac{\pi}{3}$ and θ is in Quadrant I or IV.

Values in degrees: 60°, 300°

Values in radians: $\frac{\pi}{3}, \frac{5\pi}{3}$

(b) $\sec\theta = -2 \Rightarrow$ reference angle is 60° or $\frac{\pi}{3}$ and θ is in Quadrant II or III.

Values in degrees: 120°, 240°

Values in radians: $\frac{2\pi}{3}, \frac{4\pi}{3}$

74. (a) $\sin\theta = \frac{\sqrt{3}}{2} \Rightarrow$ reference angle is 60° or $\frac{\pi}{3}$ and θ is in Quadrant I or II.

Values in degrees: 60°, 120°

Values in radians: $\frac{\pi}{3}, \frac{2\pi}{3}$

(b) $\sin\theta = -\frac{\sqrt{3}}{2} \Rightarrow$ reference angle is 60° or $\frac{\pi}{3}$ and θ is in Quadrant III or IV.

Values in degrees: 240°, 300°

Values in radians: $\frac{4\pi}{3}, \frac{5\pi}{3}$

76. $\cos\theta = 0.8746$

Quadrant I: $\theta = \cos^{-1} 0.8746 \approx 29.00°$

Quadrant IV: $\theta = 360° - 29.00° = 331.00°$

78. $\sin\theta = -0.6514$

$\sin^{-1}(-0.6514) \approx -40.65°$

Quadrant III: $\theta = 180° + 40.65° = 220.65°$

Quadrant IV: $\theta = 360° - 40.65° = 319.35°$

80. $\sin\theta = 0.0175$

Quadrant I: $\theta = \sin^{-1} 0.0175 \approx 0.018$

Quadrant II: $\theta = \pi - 0.018 = 3.124$

82. $\cot\theta = 5.671 \Rightarrow \dfrac{1}{\tan\theta} = 5.671$

Quadrant I: $\theta = \tan^{-1}\left(\dfrac{1}{5.671}\right) \approx 0.175$

Quadrant III: $\theta = \pi + 0.1745 \approx 3.316$

84. $\cos\theta = -0.3214$

Quadrant II: $\theta = \cos^{-1}(-0.3214) \approx 1.898$

Quadrant III: $\theta = 2\pi - 1.898 \approx 4.385$

86. $\cot\theta = -3$

$1 + \cot^2\theta = \csc^2\theta$

$1 + (-3)^2 = \csc^2\theta$

$10 = \csc^2\theta$

$\csc\theta > 0$ in Quadrant II.

$\sqrt{10} = \csc\theta$

$\csc\theta = \dfrac{1}{\sin\theta}$

$\sin\theta = \dfrac{1}{\csc\theta} = \dfrac{1}{\sqrt{10}} = \dfrac{\sqrt{10}}{10}$

88. $\csc\theta = -2$

$1 + \cot^2\theta = \csc^2\theta$

$\cot^2\theta = \csc^2\theta - 1$

$\cot^2\theta = (-2)^2 - 1$

$\cot^2\theta = 3$

$\cot\theta < 0$ in Quadrant IV.

$\cot\theta = -\sqrt{3}$

90. $\sec\theta = -\dfrac{9}{4}$

$1 + \tan^2\theta = \sec^2\theta$

$\tan^2\theta = \sec^2\theta - 1$

$\tan^2\theta = \left(-\dfrac{9}{4}\right)^2 - 1$

$\tan^2\theta = \dfrac{65}{16}$

$\tan\theta > 0$ in Quadrant III.

$\tan\theta = \dfrac{\sqrt{65}}{4}$

92. $S = 23.1 + 0.442t + 4.3\cos\dfrac{\pi t}{6}$

(a) For February 2003, $t = 2$.

$$S = 23.1 + 0.442(2) + 4.3\cos\frac{\pi(2)}{6} \approx 26{,}134 \text{ units.}$$

(b) For February 2004, $t = 14$.

$$S = 23.1 + 0.442(14) + 4.3\cos\frac{\pi(14)}{6} \approx 31{,}438 \text{ units}$$

(c) For June 2003, $t = 6$.

$$S = 23.1 + 0.442(6) + 4.3\cos\frac{\pi(6)}{6} \approx 21{,}452 \text{ units}$$

(d) For June 2004, $t = 18$.

$$S = 23.1 + 0.442(18) + 4.3\cos\frac{\pi(18)}{6} \approx 26{,}756 \text{ units}$$

94. $\sin\theta = \dfrac{6}{d} \Rightarrow d = \dfrac{6}{\sin\theta}$

(a) $\theta = 30°$

$$d = \frac{6}{\sin 30°} = \frac{6}{\frac{1}{2}} = 12 \text{ miles}$$

(b) $\theta = 90°$

$$d = \frac{6}{\sin 90°} = \frac{6}{1} = 6 \text{ miles}$$

(c) $\theta = 120°$

$$d = \frac{6}{\sin 120°} \approx 6.9 \text{ miles}$$

96. False. Let $n = 1$ and $\theta = 225°$. $0 \le 135 \le 360$, but $360n - \theta = 135$ is not the reference angle. The reference angle would be 45°.

98. Determine the trigonometric function of the reference angle and prefix the appropriate sign.

100. $y = 6 - 7x$

x-intercept: $\left(\frac{6}{7}, 0\right)$

y-intercept: $(0, 6)$

No asymptotes

Domain: All real numbers

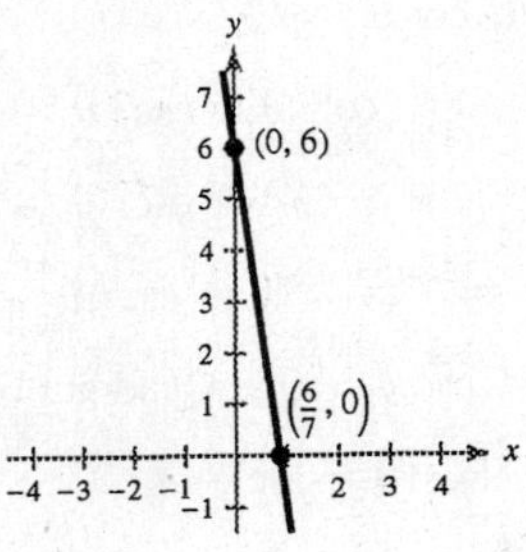

102. $y = 2x^2 - 5x = x(2x - 5)$

x-intercepts: $(0, 0), \left(\frac{5}{2}, 0\right)$

y-intercept: $(0, 0)$

No asymptotes

Domain: All real numbers

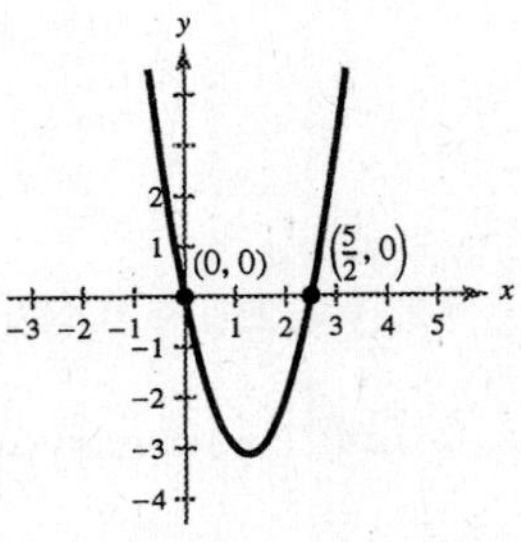

104. $g(x) = x^4 + 2x^2 - 3 = (x^2 + 3)(x^2 - 1) = (x^2 + 3)(x + 1)(x - 1)$

x-intercepts: $(-1, 0), (1, 0)$

y-intercept: $(0, -3)$

No asymptotes

Domain: All real numbers

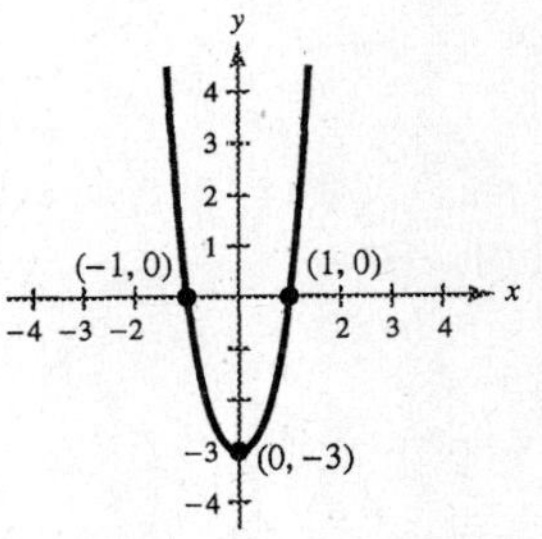

106. $f(x) = \sqrt{4x - 1}$

x-intercept: $\left(\frac{1}{4}, 0\right)$

No y-intercept

No asymptotes

Domain: $x \ge \frac{1}{4}$

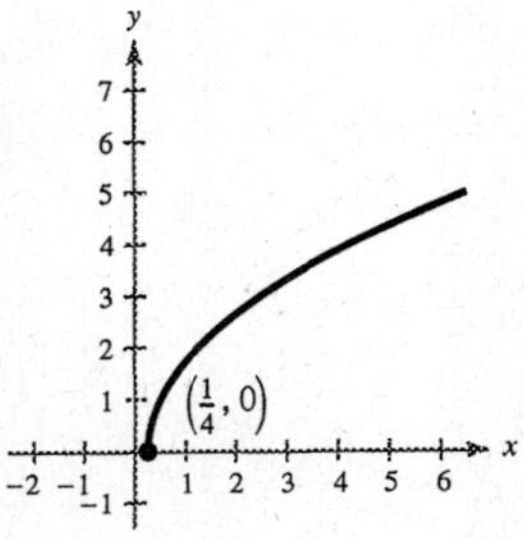

Section 1.5 Graphs of Sine and Cosine Functions

Solutions to Even-Numbered Exercises

2. $y = 2\cos 3x$

Period $= \dfrac{2\pi}{b} = \dfrac{2\pi}{3}$

Amplitude $= |a| = 2$

4. $y = -3\sin\dfrac{x}{3}$

Period $= \dfrac{2\pi}{b} = \dfrac{2\pi}{\frac{1}{3}} = 6\pi$

Amplitude $= |a| = |-3| = 3$

6. $y = \dfrac{3}{2}\cos\dfrac{\pi x}{2}$

Period $= \dfrac{2\pi}{b} = \dfrac{2\pi}{\frac{\pi}{2}} = 4$

Amplitude $= |a| = \dfrac{3}{2}$

8. $y = -\cos\dfrac{2x}{3}$

Period $= \dfrac{2\pi}{b} = \dfrac{2\pi}{\frac{2}{3}} = 3\pi$

Amplitude $= |a| = |-1| = 1$

10. $y = \dfrac{1}{3}\sin 8x$

Period $= \dfrac{2\pi}{b} = \dfrac{2\pi}{8} = \dfrac{\pi}{4}$

Amplitude $= |a| = \dfrac{1}{3}$

12. $y = \dfrac{5}{2}\cos\dfrac{x}{4}$

Period $= \dfrac{2\pi}{b} = \dfrac{2\pi}{\frac{1}{4}} = 8\pi$

Amplitude $= |a| = \dfrac{5}{2}$

14. $y = \dfrac{2}{3}\cos\dfrac{\pi x}{10}$

Period $= \dfrac{2\pi}{b} = \dfrac{2\pi}{\frac{\pi}{10}} = 20$

Amplitude $= |a| = \dfrac{2}{3}$

16. $f(x) = \cos x$, $g(x) = \cos(x + \pi)$
g is a horizontal shift of f π units to the left.

18. $f(x) = \sin 3x$, $g(x) = \sin(-3x)$
g is a reflection of f about the y-axis.

20. $f(x) = \sin x$, $g(x) = \sin 3x$
The period of g is one-third the period of f.

22. $f(x) = \cos 4x$, $g(x) = -2 + \cos 4x$
g is a vertical shift of f two units downward.

24. The period of g is one-third the period of f.

26. Shift the graph of f two units upward to obtain the graph of g.

28. $f(x) = \sin x$, $g(x) = \sin\dfrac{x}{3}$

x	0	$\frac{\pi}{2}$	π	$\frac{3\pi}{2}$	2π
$\sin x$	0	1	0	-1	0
$\sin\frac{x}{3}$	0	$\frac{1}{2}$	$\frac{\sqrt{3}}{2}$	1	$\frac{\sqrt{3}}{2}$

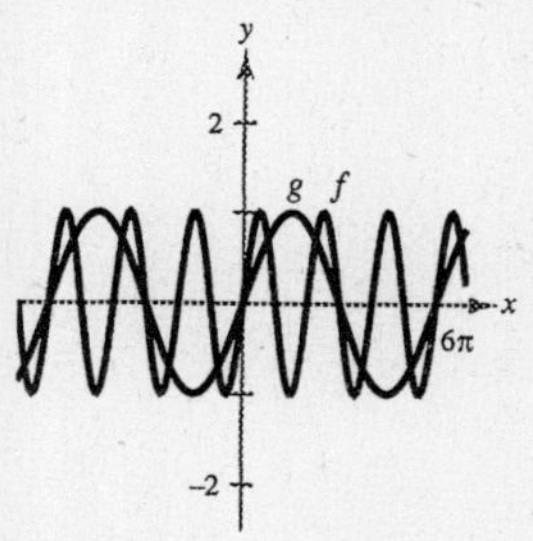

30. $f(x) = 2\cos 2x$, $g(x) = -\cos 4x$

x	0	$\frac{\pi}{4}$	$\frac{\pi}{2}$	$\frac{3\pi}{4}$	π
$2\cos 2x$	2	0	-2	0	2
$-\cos 4x$	-1	1	-1	1	-1

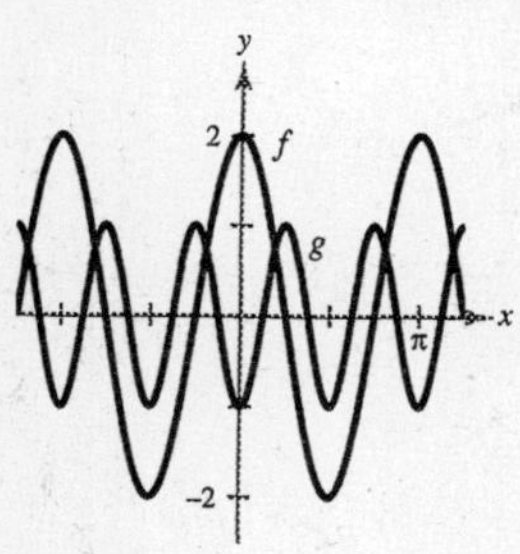

32. $f(x) = 4 \sin \pi x,\ g(x) = 4 \sin \pi x - 3$

x	0	$\frac{1}{2}$	1	$\frac{3}{2}$	2
$f(x)$	0	4	0	-4	0
$g(x)$	-3	1	-3	-7	-3

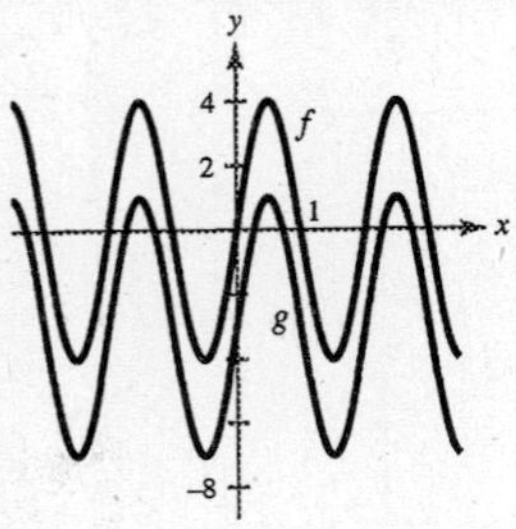

34. $f(x) = -\cos x$

$g(x) = -\cos(x - \pi)$

x	0	$\frac{\pi}{2}$	π	$\frac{3\pi}{2}$	2π
$-\cos x$	-1	0	1	0	-1
$-\cos(x - \pi)$	1	0	-1	0	1

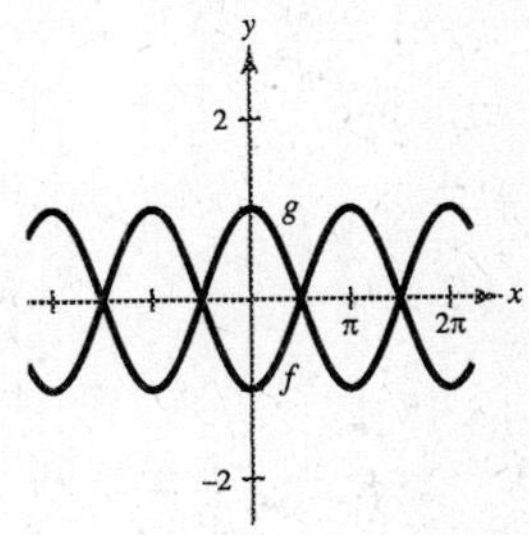

36. $y = -3 \cos 4x$

$$\text{Period} = \frac{2\pi}{4} = \frac{\pi}{2}$$

Amplitude $= 3$

x	0	$\frac{\pi}{8}$	$\frac{\pi}{4}$	$\frac{3\pi}{8}$	$\frac{\pi}{2}$
y	-3	0	3	0	-3

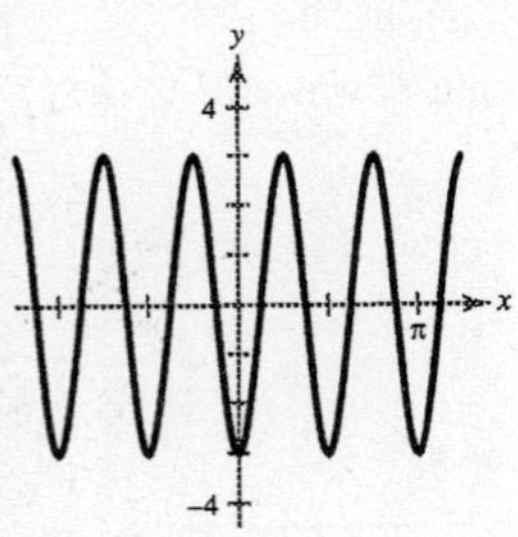

38. $y = \sin \dfrac{\pi x}{4}$

$$\text{Period} = \frac{2\pi}{\frac{\pi}{4}} = 8$$

Amplitude $= 1$

x	0	2	4	6	8
y	0	1	0	-1	0

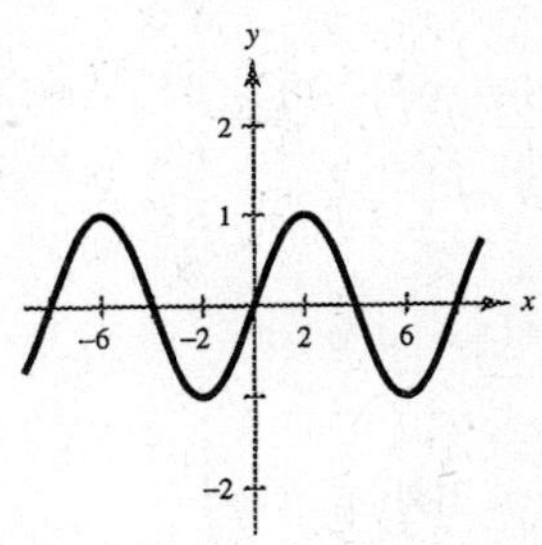

40. $y = -10 \cos \dfrac{\pi x}{6}$

$$\text{Period} = \frac{2\pi}{\frac{\pi}{6}} = 12$$

Amplitude $= 10$

x	0	3	6	9	12
y	-10	0	10	0	-10

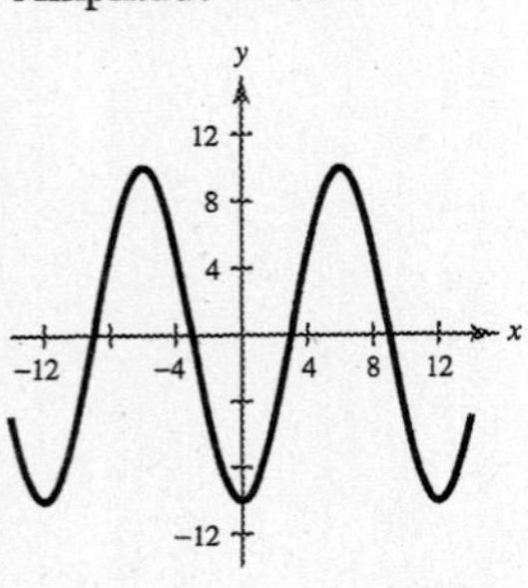

42. $y = \sin(x - \pi)$

Period $= 2\pi$

Amplitude $= 1$

x	0	$\frac{\pi}{2}$	π	$\frac{3\pi}{2}$	2π
y	0	-1	0	1	0

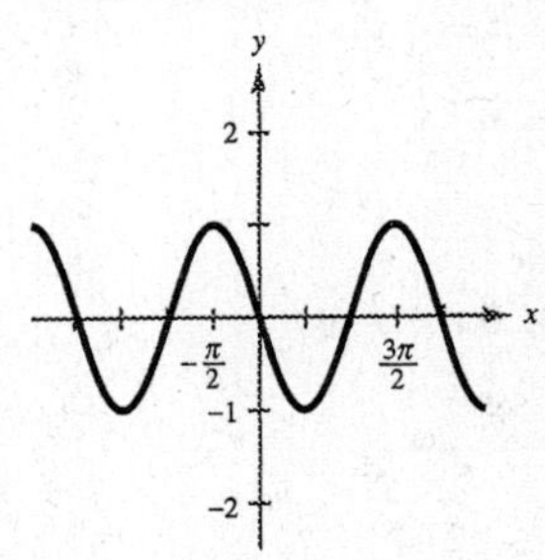

44. $y = 4\cos\left(x + \dfrac{\pi}{4}\right)$

Period $= 2\pi$

Amplitude $= 4$

x	0	$\dfrac{\pi}{2}$	π	$\dfrac{3\pi}{2}$	2π
y	$2\sqrt{2}$	$-2\sqrt{2}$	$-2\sqrt{2}$	$2\sqrt{2}$	$2\sqrt{2}$

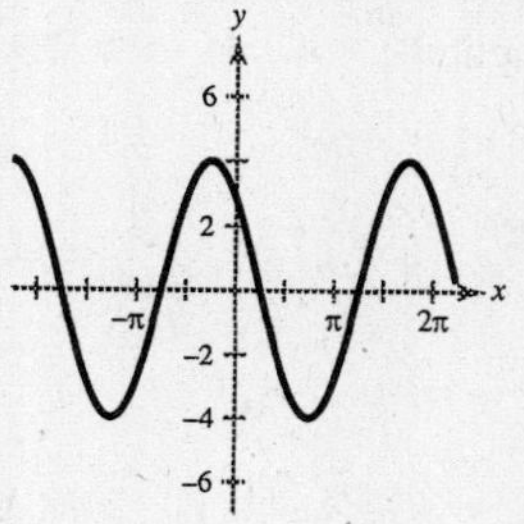

46. $y = -3 + 5\cos\dfrac{\pi t}{12}$

Period $= \dfrac{2\pi}{\dfrac{\pi}{12}} = 24$

Amplitude $= 5$

x	0	6	12	18	24
y	2	-3	-8	-3	2

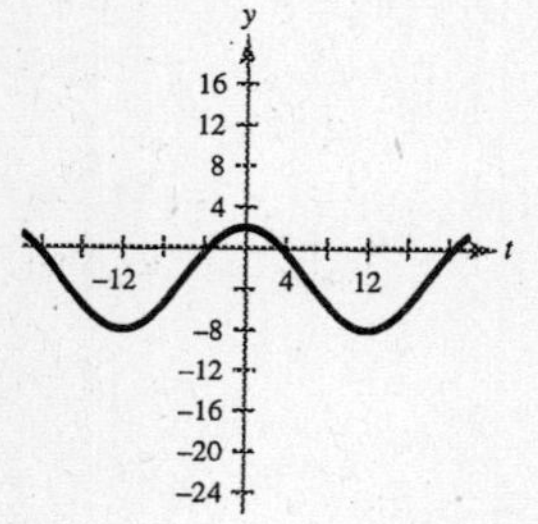

48. $y = 2\cos x - 3$

Period $= 2\pi$

Amplitude $= 2$

x	0	$\dfrac{\pi}{2}$	π	$\dfrac{3\pi}{2}$	2π
y	-1	-3	-5	-3	-1

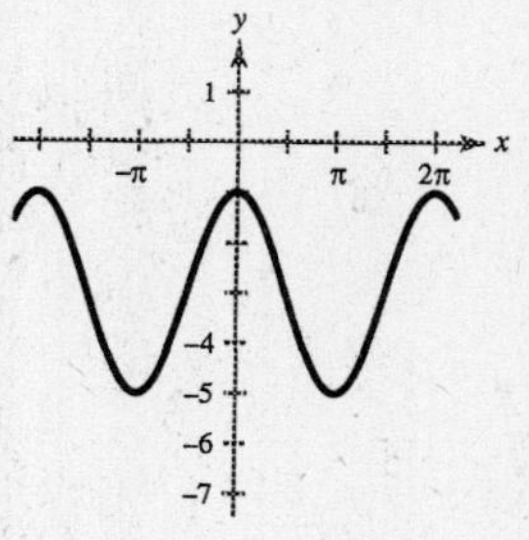

50. $y = 4\cos\left(x + \dfrac{\pi}{4}\right) + 4$

Period $= 2\pi$

Amplitude $= 4$

x	$-\dfrac{\pi}{4}$	$\dfrac{\pi}{4}$	$\dfrac{3\pi}{4}$	$\dfrac{5\pi}{4}$	$\dfrac{7\pi}{4}$
y	8	4	0	4	8

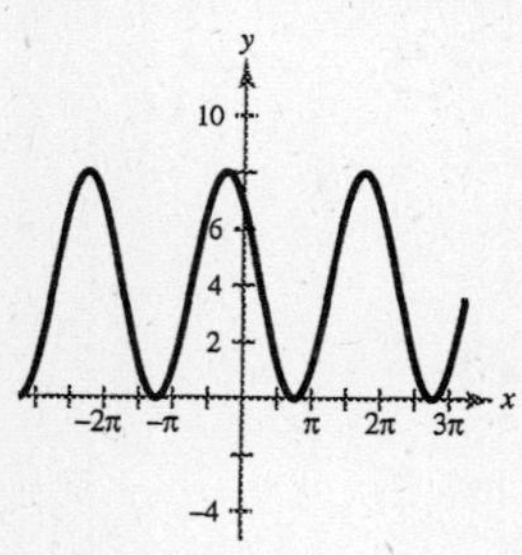

52. $y = -3\cos(6x + \pi)$

Period $= \dfrac{2\pi}{6} = \dfrac{\pi}{3}$

Amplitude $= 3$

x	0	$\dfrac{\pi}{12}$	$\dfrac{\pi}{6}$	$\dfrac{3\pi}{12}$	$\dfrac{\pi}{3}$
y	3	0	-3	0	3

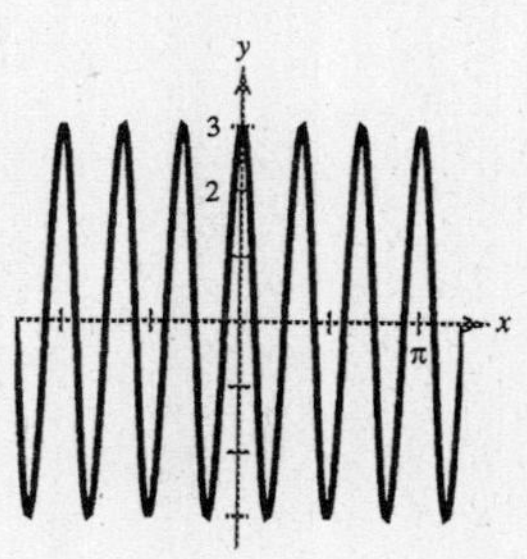

54. $y = -4\sin\left(\frac{2}{3}x - \frac{\pi}{3}\right)$

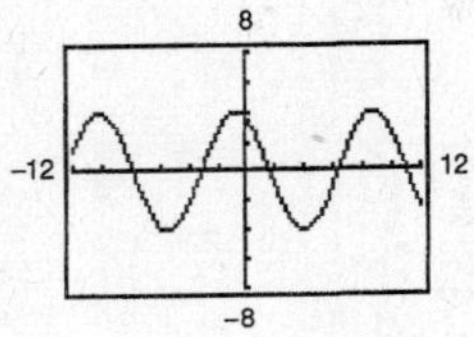

56. $y = 3\cos\left(\frac{\pi x}{2} + \frac{\pi}{2}\right) - 2$

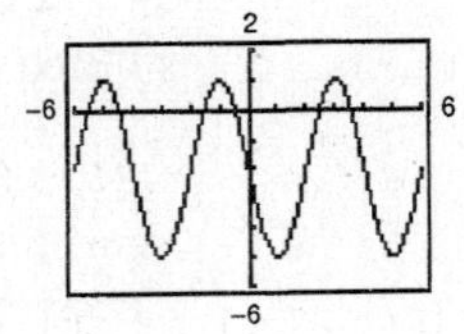

58. $y = \frac{1}{100}\sin 120\pi t$

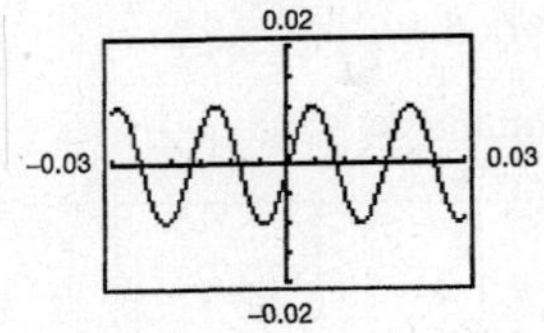

60. $f(x) = a\cos x + d$

Amplitude $= \frac{1-(-3)}{2} = 2$

$1 = 2\cos 0 + d$

$d = 1 - 2 = -1$

$a = 2, d = -1$

62. $f(x) = a\cos x + d$

Amplitude $= \frac{-2-(-4)}{2} = 1$

Reflected in the x-axis: $a = -1$

$-4 = -1\cos 0 + d$

$d = -3$

$a = -1, d = -3$

64. $y = a\sin(bx - c)$

Amplitude $= 2 \implies a = 2$

Period $= 4\pi$

$\frac{2\pi}{b} = 4\pi \implies b = \frac{1}{2}$

Phase shift: $c = 0$

$a = 2, b = \frac{1}{2}, c = 0$

66. $y = a\sin(bx - c)$

Amplitude $= 2 \implies a = 2$

Period $= 4$

$\frac{2\pi}{b} = 4 \implies b = \frac{\pi}{2}$

Phase shift: $\frac{c}{b} = -1 \implies c = -\frac{\pi}{2}$

$a = 2, b = \frac{\pi}{2}, c = -\frac{\pi}{2}$

68. $y_1 = \cos x$

$y_2 = -1$

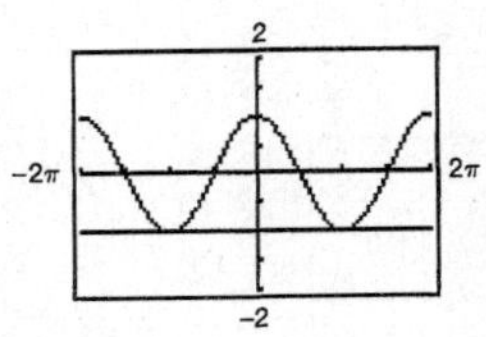

$y_1 = y_2$ when $x = \pi, -\pi$

70. $v = 1.75\sin\frac{\pi t}{2}$

(a) Period $= \frac{2\pi}{\frac{\pi}{2}} = 4$ seconds

(b) $\frac{1 \text{ cycle}}{4 \text{ seconds}} \cdot \frac{60 \text{ seconds}}{1 \text{ minute}} = 15$ cycles per minute

(c)

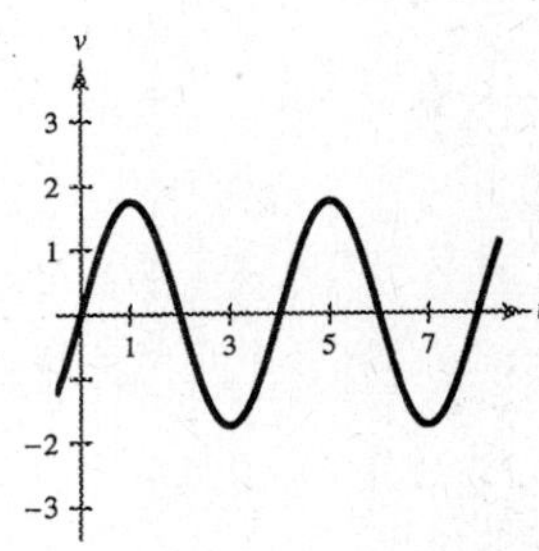

72. $P = 100 - 20\cos\frac{5\pi t}{3}$

(a) Period $= \frac{2\pi}{\frac{5\pi}{3}} = \frac{6}{5}$ seconds

(b) $\frac{1 \text{ heartbeat}}{\frac{6}{5} \text{ seconds}} \cdot \frac{60 \text{ seconds}}{1 \text{ minute}} = 50$ heartbeats per minute

74. (a) and (c)

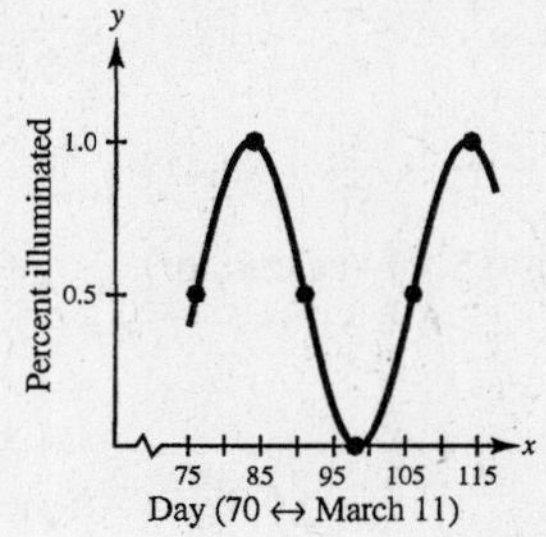

The model is a good fit for the data.

(b) Amplitude: 0.5

Period: 30 $\Rightarrow \dfrac{2\pi}{30} = \dfrac{\pi}{15}$

Horizontal shift of sine function: 76 units (days) to the right

Vertical shift: up $\dfrac{1}{2}$

An equation is $y = \dfrac{1}{2} + \dfrac{1}{2}\sin\left[\dfrac{\pi}{15}(x - 76)\right]$.

(d) 30 days

(e) For May 8, 2005, $x = 128$: $y = \dfrac{1}{2} + \dfrac{1}{2}\sin\left[\dfrac{\pi}{15}(128 - 76)\right] = \dfrac{1}{2} + \dfrac{1}{2}\sin\dfrac{52\pi}{15} \approx 0.003$, or about 0.3%. This would very nearly correspond to the point (128, 0.0).

76. False. The graph of $\sin(x + 2\pi)$ is the graph of $\sin(x)$ translated to the *left* by one period, and the graphs are indeed identical.

78. True. Because $\cos x = \sin\left(x + \dfrac{\pi}{2}\right)$, $y = -\cos x$ is a reflection in the x-axis of $y = \sin\left(x + \dfrac{\pi}{2}\right)$.

80. $f(x) = \sin x$, $g(x) = -\cos\left(x + \dfrac{\pi}{2}\right)$

x	0	$\frac{\pi}{2}$	π	$\frac{3\pi}{2}$	2π
$\sin x$	0	1	0	-1	0
$-\cos\left(x + \frac{\pi}{2}\right)$	0	1	0	-1	0

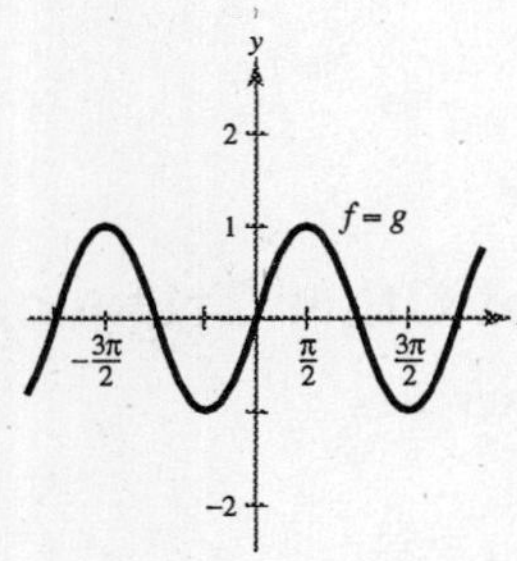

Conjecture: $\sin x = -\cos\left(x + \dfrac{\pi}{2}\right)$

82. $y = 2 + \sin x$

$y = 3.5 + \sin x$

$y = -2 + \sin x$

Each value of d produces a vertical shift of $y = \sin x$ upward (or downward) by d units.

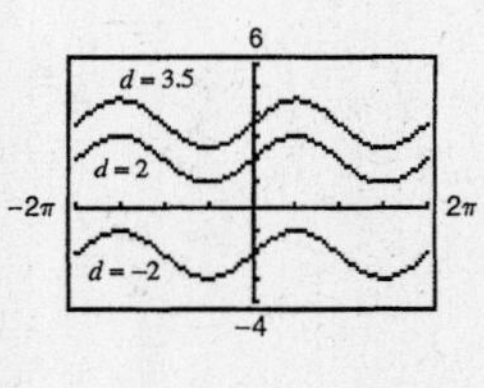

84. $y = \sin(x - 1)$

$y = \sin(x - 3)$

$y = \sin(x - (-2)) = \sin(x + 2)$

Each value of c produces a horizontal shift of $y = \sin x$ to the left (or right) by c units.

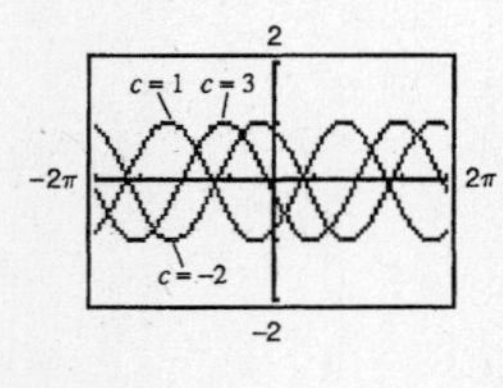

86. (a) $\sin\frac{1}{2} \approx \frac{1}{2} - \frac{\left(\frac{1}{2}\right)^3}{3!} + \frac{\left(\frac{1}{2}\right)^5}{5!} \approx 0.4794$

$\sin\frac{1}{2} \approx 0.4794$ (by calculator)

(b) $\sin 1 \approx 1 - \frac{1}{3!} + \frac{1}{5!} \approx 0.8417$

$\sin 1 \approx 0.8415$ (by calculator)

(c) $\sin\frac{\pi}{6} \approx 1 - \frac{\left(\frac{\pi}{6}\right)^3}{3!} + \frac{\left(\frac{\pi}{6}\right)^5}{5!} \approx 0.5000$

$\sin\frac{\pi}{6} = 0.5$ (by calculator)

(d) $\cos(-0.5) \approx 1 - \frac{(-0.5)^2}{2!} + \frac{(-0.5)^4}{4!} \approx 0.8776$

$\cos(-0.5) \approx 0.8776$ (by calculator)

(e) $\cos 1 \approx 1 - \frac{1}{2!} + \frac{1}{4!} \approx 0.5417$

$\cos 1 \approx 0.5403$ (by calculator)

(f) $\cos\frac{\pi}{4} \approx 1 - \frac{\left(\frac{\pi}{4}\right)^2}{2!} + \frac{\left(\frac{\pi}{4}\right)^4}{4!} = 0.7074$

$\cos\frac{\pi}{4} \approx 0.7071$ (by calculator)

The error in the approximation is not the same in each case. The error appears to increase as x moves farther away from 0.

88. $\frac{2}{x+5} - \frac{2}{x-5}$

$$\frac{2}{x+5} - \frac{2}{x-5} = \frac{2(x-5) - 2(x+5)}{(x+5)(x-5)} = \frac{2x - 10 - 2x - 10}{(x+5)(x-5)} = -\frac{20}{(x+5)(x-5)}$$

90. $\frac{x}{x-5} + \frac{1}{2}$

$$\frac{x}{x-5} + \frac{1}{2} = \frac{2x + x - 5}{2(x-5)} = \frac{3x-5}{2(x-5)}$$

92. $f(x) = \frac{\sqrt{x-3}}{x-8}$

Domain: $x \geq 3, x \neq 8$

94. $f(x) = \sqrt[3]{4 - x^2}$

Domain: All real numbers

Section 1.6 Graphs of Other Trigonometric Functions

Solutions to Even-Numbered Exercises

2. $y = \tan\frac{x}{2}$

$$\text{Period} = \frac{\pi}{b} = \frac{\pi}{\frac{1}{2}} = 2\pi$$

Asymptotes: $x = -\pi, x = \pi$

Matches graph (c).

4. $y = -\csc x$

Period: 2π

Matches graph (d).

6. $y = -2\sec\frac{\pi x}{2}$

$$\text{Period} = \frac{2\pi}{b} = \frac{2\pi}{\frac{\pi}{2}} = 4$$

Asymptotes: $x = -1, x = 1$

Reflected in x-axis

Matches graph (b).

8. $y = \frac{1}{4}\tan x$

Period $= \pi$

Asymptotes: $x = -\frac{\pi}{2}, x = \frac{\pi}{2}$

x	$-\frac{\pi}{4}$	0	$\frac{\pi}{4}$
y	$-\frac{1}{4}$	0	$\frac{1}{4}$

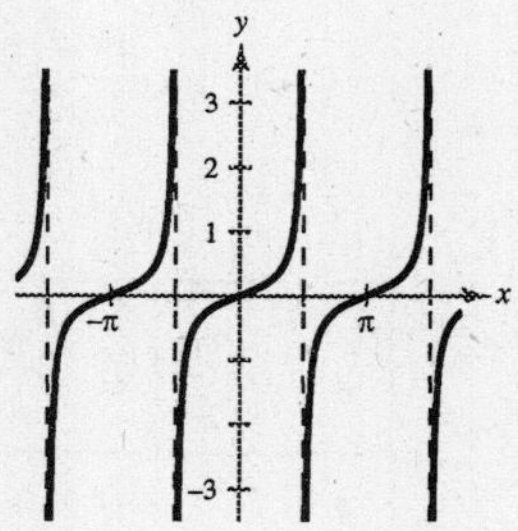

10. $y = -3\tan \pi x$

Period $= \frac{\pi}{\pi} = 1$

Asymptotes: $x = -\frac{1}{2}, x = \frac{1}{2}$

x	$-\frac{1}{4}$	0	$\frac{1}{4}$
y	3	0	-3

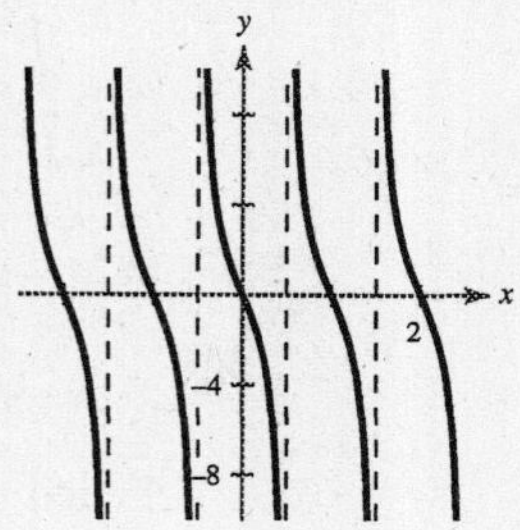

12. $y = \frac{1}{4}\sec x$

Period $= 2\pi$

Asymptotes: $x = -\frac{\pi}{2}, x = \frac{\pi}{2}$

x	$-\frac{\pi}{4}$	0	$\frac{\pi}{4}$
y	0.354	$\frac{1}{4}$	0.354

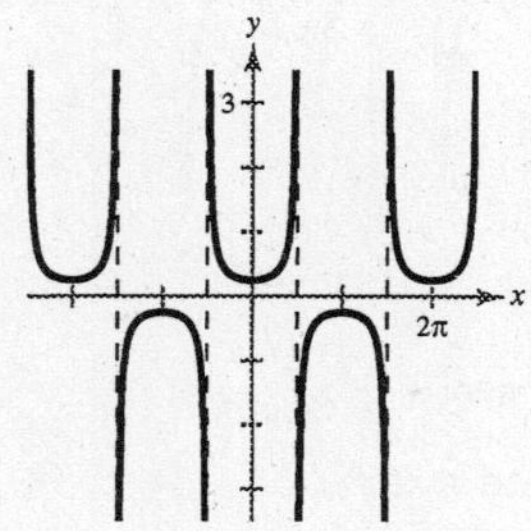

14. $y = 3\csc 4x$

Period $= \frac{2\pi}{4} = \frac{\pi}{2}$

Asymptotes: $x = 0, x = \frac{\pi}{4}$

x	$\frac{\pi}{16}$	$\frac{\pi}{8}$	$\frac{3\pi}{16}$
y	4.243	3	4.243

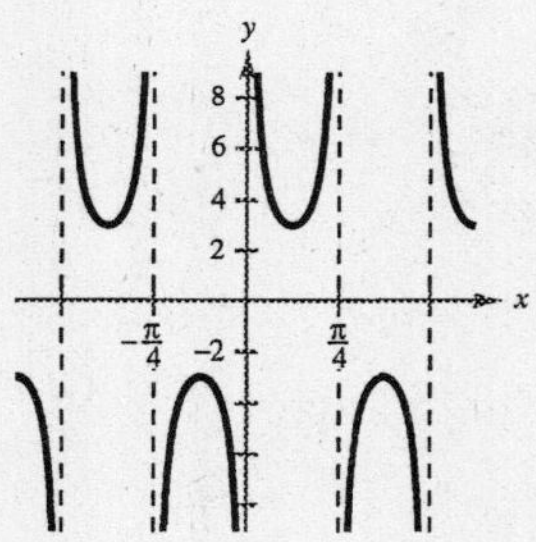

16. $y = -2\sec 4x + 2$

Period $= \frac{2\pi}{4} = \frac{\pi}{2}$

Asymptotes: $x = -\frac{\pi}{8}, x = \frac{\pi}{8}$

x	$-\frac{\pi}{16}$	0	$\frac{\pi}{16}$
y	-0.828	0	-0.828

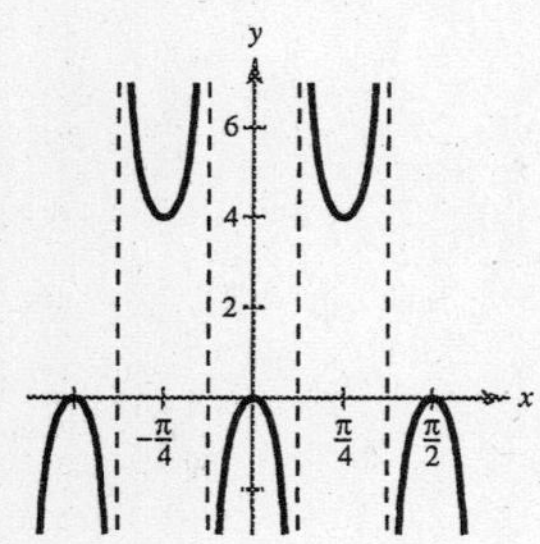

18. $y = \csc \frac{x}{3}$

Period $= \frac{2\pi}{\frac{1}{3}} = 6\pi$

Asymptotes: $x = 0, x = 3\pi$

x	π	2π	4π
y	1.155	1.155	-1.155

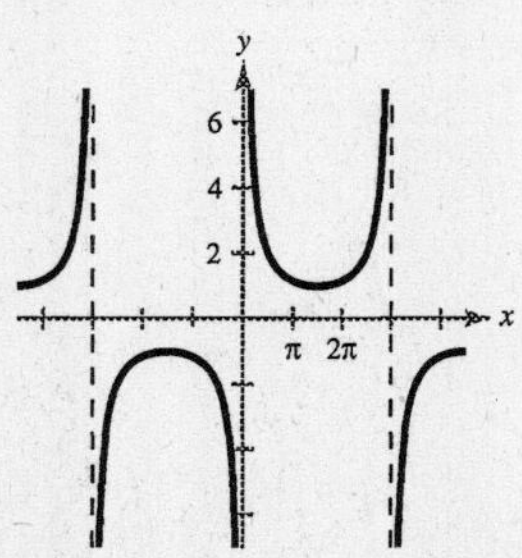

20. $y = 3\cot \frac{\pi x}{2}$

Period $= \frac{\pi}{\frac{\pi}{2}} = 2$

Asymptotes: $x = 0, x = 2$

x	$\frac{1}{4}$	1	$\frac{3}{2}$
y	7.243	0	-3

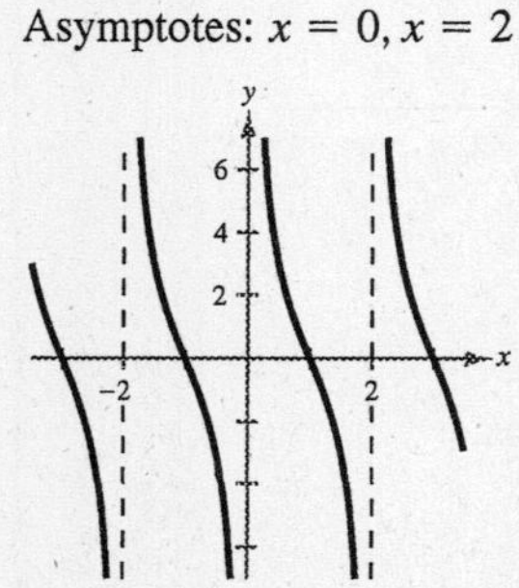

22. $y = -\frac{1}{2}\tan x$

Period $= \pi$

Asymptotes: $x = -\frac{\pi}{2}, x = \frac{\pi}{2}$

x	$-\frac{\pi}{4}$	0	$\frac{\pi}{4}$
y	$\frac{1}{2}$	0	$-\frac{1}{2}$

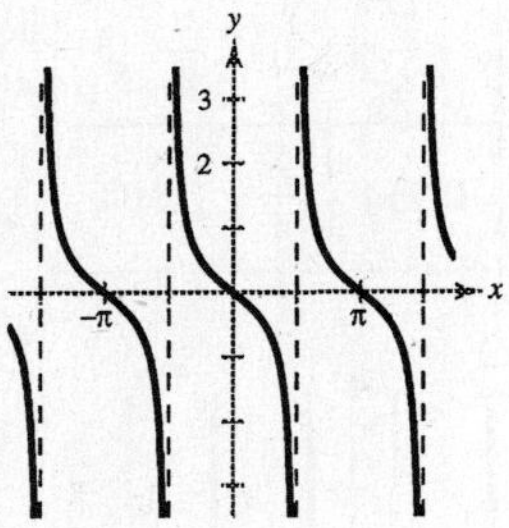

24. $y = \tan(x + \pi)$

Period $= \pi$

Asymptotes: $x = -\frac{\pi}{2}, x = \frac{\pi}{2}$

x	$-\frac{\pi}{4}$	0	$\frac{\pi}{4}$
y	-1	0	1

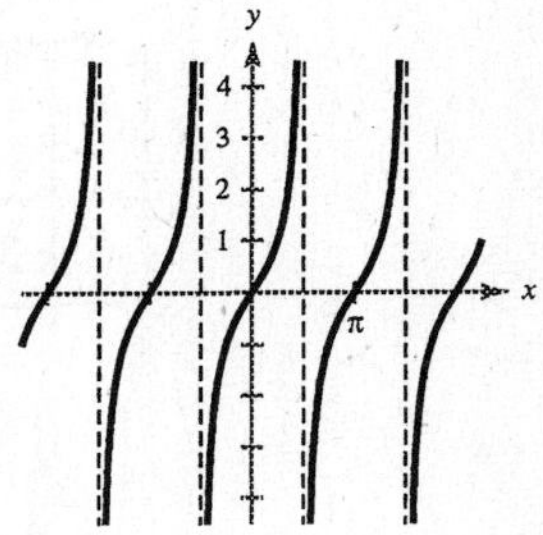

26. $y = \sec(\pi - x)$

Period $= 2\pi$

Asymptotes: $x = -\frac{\pi}{2}, x = \frac{\pi}{2}$

x	$-\frac{\pi}{4}$	0	$\frac{\pi}{4}$
y	-1.414	-1	-1.414

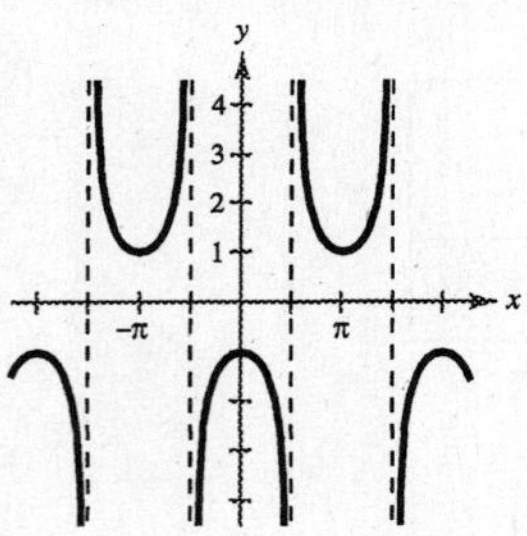

28. $y = 2\cot\left(x + \frac{\pi}{2}\right)$

Period $= \pi$

Asymptotes: $x = -\frac{\pi}{2}, x = \frac{\pi}{2}$

x	$-\frac{\pi}{4}$	0	$\frac{\pi}{4}$
y	2	0	-2

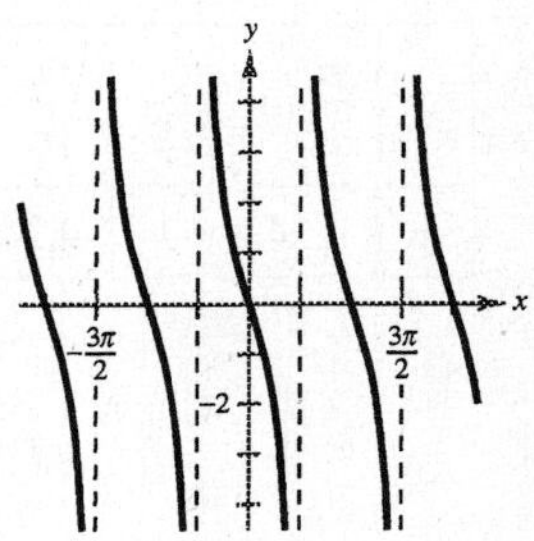

30. $y = -\tan 2x$

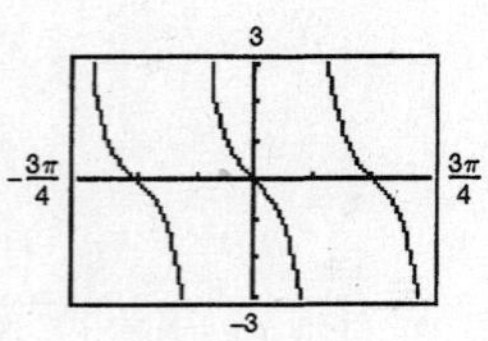

32. $y = \sec \pi x \implies y = \frac{1}{\cos(\pi x)}$

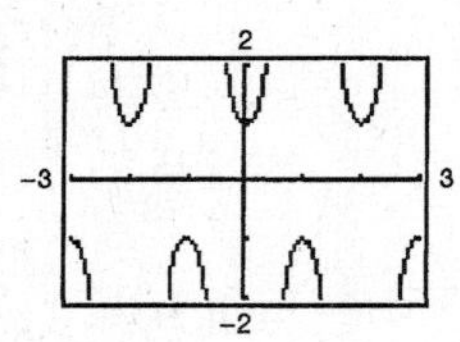

34. $y = \frac{1}{4}\cot\left(x - \frac{\pi}{2}\right)$

$= \frac{1}{4\tan\left(x - \frac{\pi}{2}\right)}$

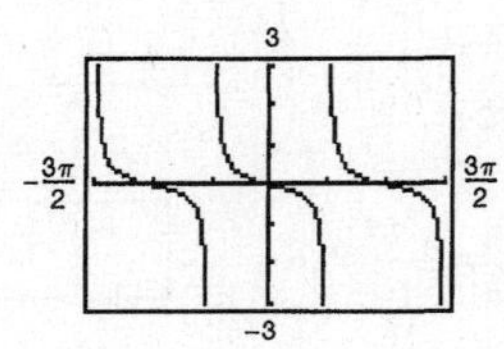

36. $y = 2\sec(2x - \pi) \implies y = \frac{2}{\cos(2x - \pi)}$

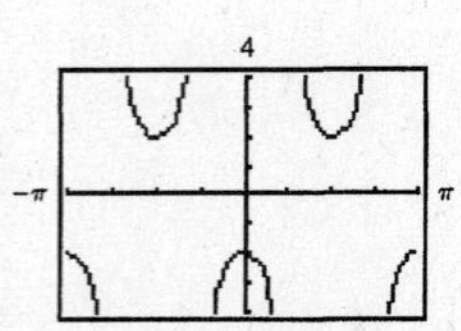

38. $y = \frac{1}{3}\sec\left(\frac{\pi x}{2} + \frac{\pi}{2}\right) \implies y = \frac{1}{3\cos\left(\frac{\pi x}{2} + \frac{\pi}{2}\right)}$

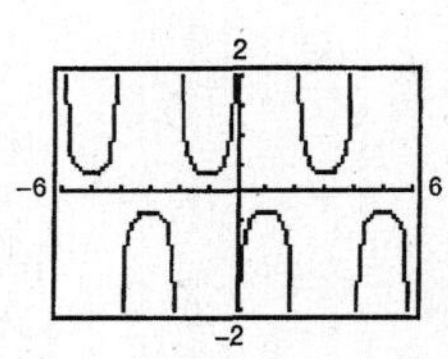

40. $\tan x = \sqrt{3}$

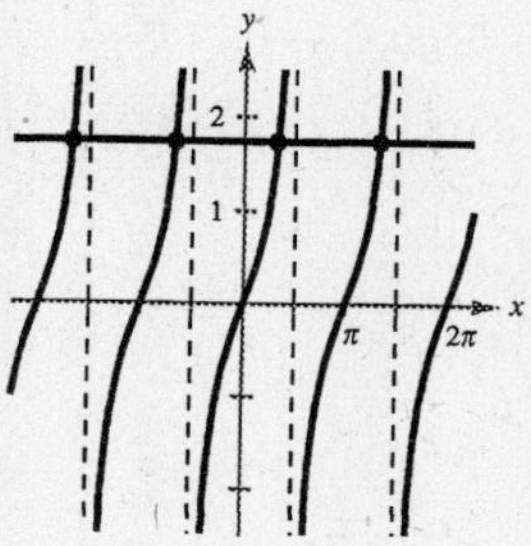

$$x = -\frac{5\pi}{3}, -\frac{2\pi}{3}, \frac{\pi}{3}, \frac{4\pi}{3}$$

42. $\cot x = 1$

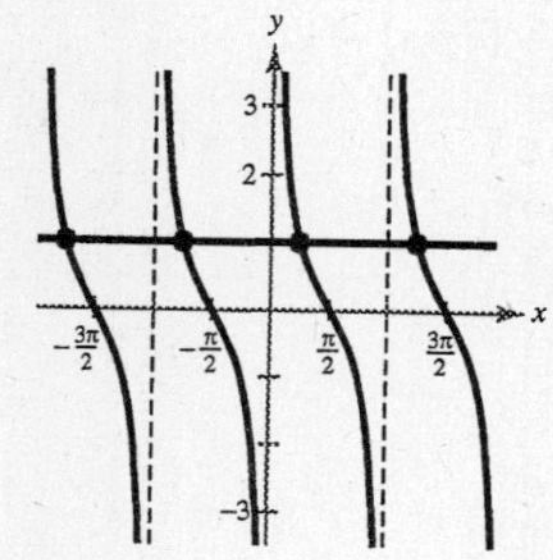

$$x = -\frac{7\pi}{4}, -\frac{3\pi}{4}, \frac{\pi}{4}, \frac{5\pi}{4}$$

44. $\sec x = 2$

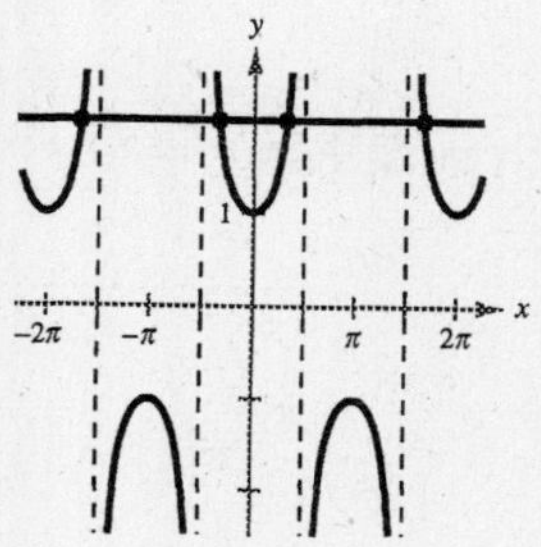

$$x = -\frac{5\pi}{3}, -\frac{\pi}{3}, \frac{\pi}{3}, \frac{5\pi}{3}$$

46. $\csc x = -\dfrac{2\sqrt{3}}{3}$

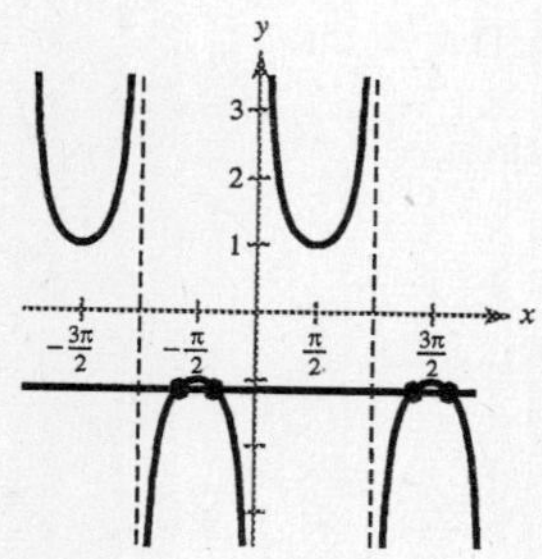

$$x = -\frac{2\pi}{3}, -\frac{\pi}{3}, \frac{4\pi}{3}, \frac{5\pi}{3}$$

48. $f(x) = \tan x$

$\tan(-x) = -\tan x$

Thus, the function is odd and the graph of $y = \tan x$ is symmetric with the origin.

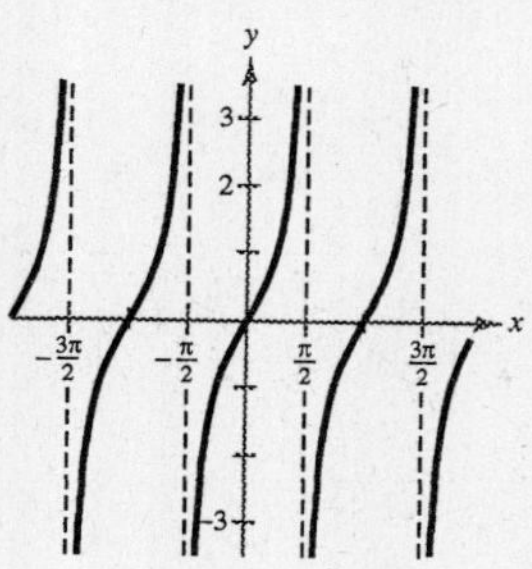

50. $f(x) = \tan \dfrac{\pi x}{2}$, $g(x) = \dfrac{1}{2} \sec \dfrac{\pi x}{2}$

(a)

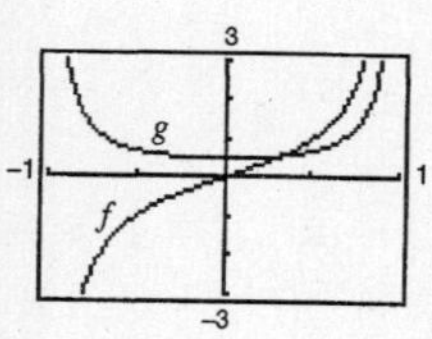

(b) The interval in which $f < g$ is $\left(-1, \frac{1}{3}\right)$.

(c) The interval in which $2f < 2g$ is $\left(-1, \frac{1}{3}\right)$, which is the same interval as part (b).

52. $y_1 = \sin x \sec x$, $y_2 = \tan x$

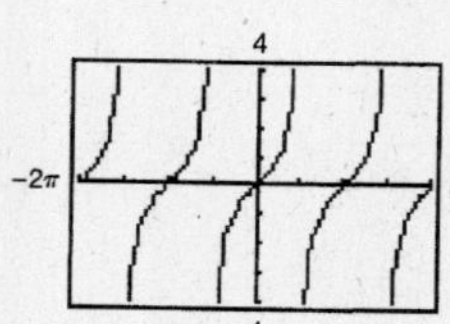

It appears that $y_1 = y_2$.

$$\sin x \sec x = \sin x \frac{1}{\cos x} = \frac{\sin x}{\cos x} = \tan x$$

54. $y_1 = \sec^2 x - 1$, $y_2 = \tan^2 x$

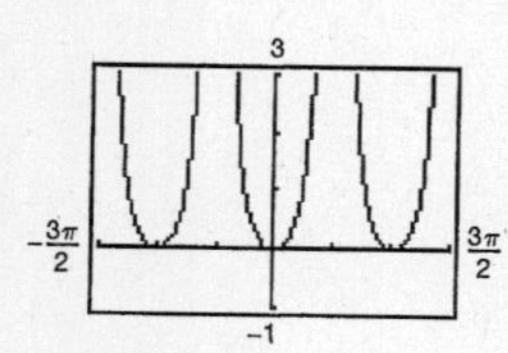

It appears that $y_1 = y_2$.

$$1 + \tan^2 x = \sec^2 x$$

$$\tan^2 x = \sec^2 x - 1$$

56. $f(x) = x \sin x$

Matches graph (a) as $x \longrightarrow 0, f(x) \longrightarrow 0$.

58. $g(x) = |x| \cos x$

Matches graph (c) as $x \longrightarrow 0$, $g(x) \longrightarrow 0$.

60. $f(x) = \sin x - \cos\left(x + \frac{\pi}{2}\right)$

$g(x) = 2 \sin x$

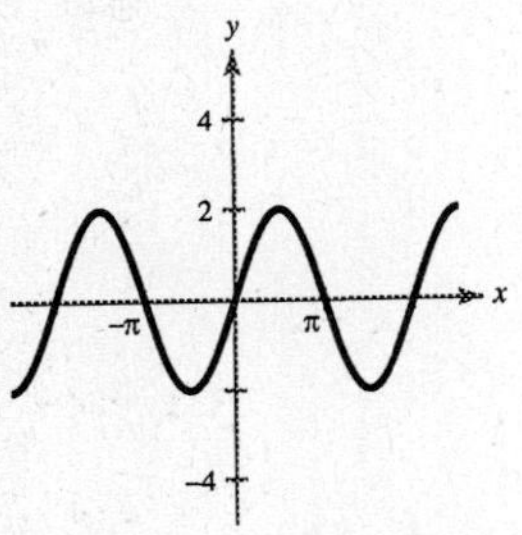

It appears that $f(x) = g(x)$. That is, that

$\sin x - \cos\left(x + \frac{\pi}{2}\right) = 2 \sin x$.

62. $f(x) = \cos^2 \frac{\pi x}{2}$

$g(x) = \frac{1}{2}(1 + \cos \pi x)$

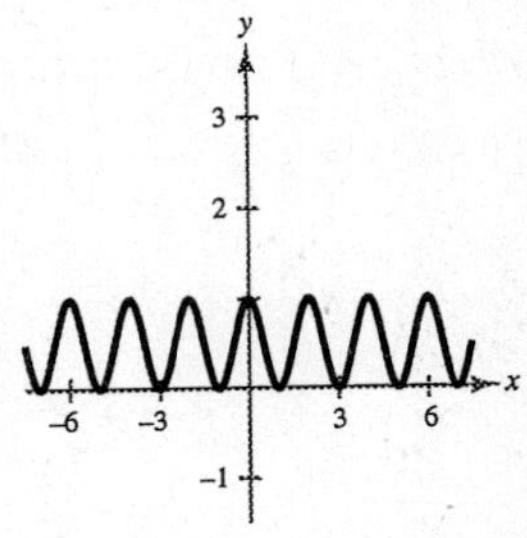

It appears that $f(x) = g(x)$. That is, that

$\cos^2 \frac{\pi x}{2} = \frac{1}{2}(1 + \cos \pi x)$.

64. $f(x) = x^2 \cos x$

Damping factor: x^2

As $x \to \infty$, $f(x)$ oscillates and approaches $-\infty$ and ∞.

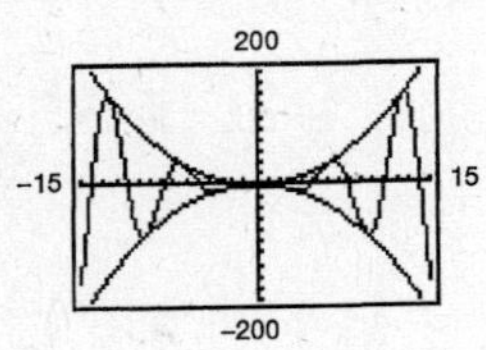

66. $h(x) = x^3 \cos x$

Damping factor: x^3

As $x \to \infty$, $h(x)$ oscillates and approaches $-\infty$ and ∞.

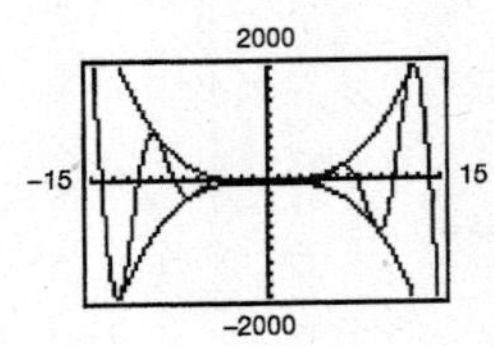

68. $y = \frac{4}{x} + \sin 2x, x > 0$

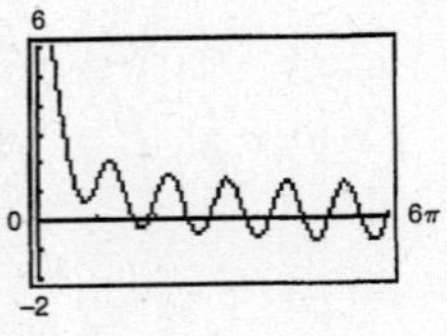

As $x \longrightarrow 0, y \longrightarrow \infty$

70. $f(x) = \frac{1 - \cos x}{x}$

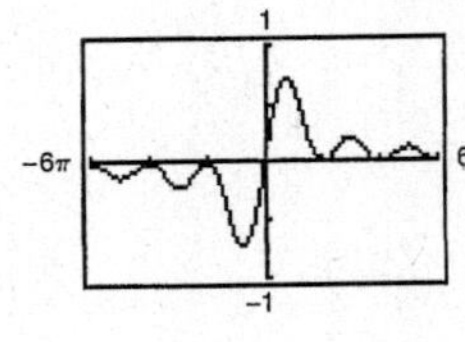

As $x \longrightarrow 0, f(x) \longrightarrow 0$.

72. $h(x) = x \sin \frac{1}{x}$

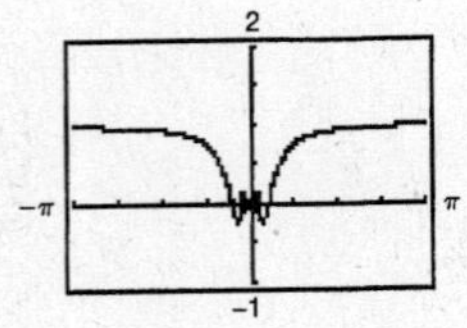

As $x \longrightarrow 0$, $h(x)$ oscillates.

74. $\cos x = \dfrac{27}{d}$

$$d = \frac{27}{\cos x} = 27 \sec x, \; -\frac{\pi}{2} < x < \frac{\pi}{2}$$

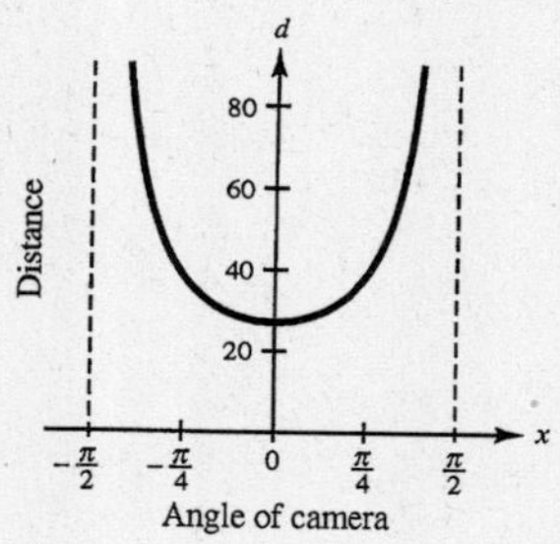

76. $S = 74 + 3t - 40 \cos \dfrac{\pi t}{6}$

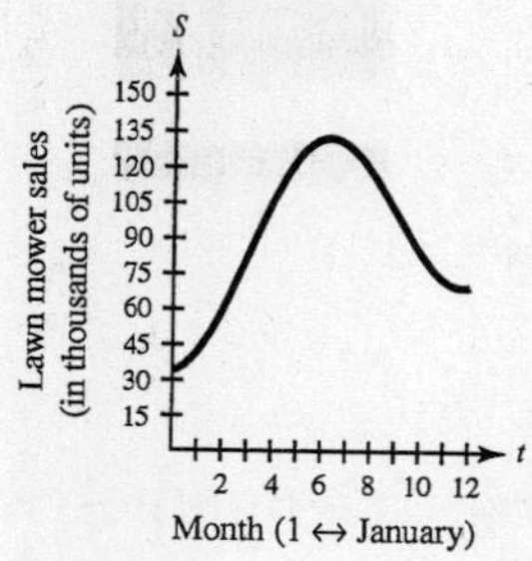

78. $y = \frac{1}{2}t \cos 4t, \; t > 0$

(a)

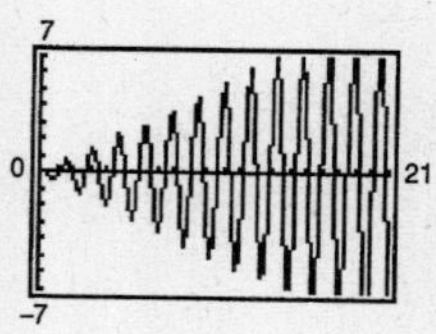

(b) Damped cosine wave; oscillates and goes to $-\infty$ and ∞ as t increases.

80. True. The graph of $y = \sec x$ can be obtained by translating the graph of $y = \csc x$ to the left by $\dfrac{\pi}{2}$ units.

Thus, $y = \sec x = \csc\left(x + \dfrac{\pi}{2}\right) = \dfrac{1}{\sin\left(x + \dfrac{\pi}{2}\right)}$

82. As $x \longrightarrow \pi$ from the left, $f(x) = \csc x \longrightarrow \infty$.

As $x \longrightarrow \pi$ from the right, $f(x) = \csc x \longrightarrow -\infty$.

84.

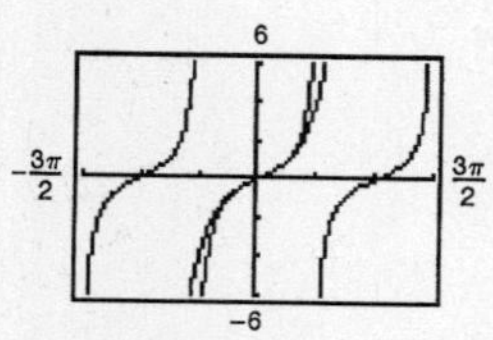

$y = \tan x$

$y = x + \dfrac{2x^3}{3!} + \dfrac{16x^5}{5!}$

The graphs are nearly the same for $-1.1 < x < 1.1$

86. (a) $y_1 = \dfrac{4}{\pi}\left(\sin \pi x + \dfrac{1}{3}\sin 3\pi x\right)$

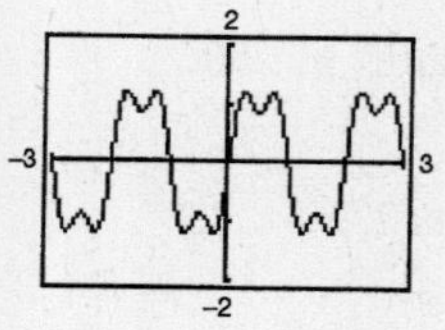

$y_2 = \dfrac{4}{\pi}\left(\sin \pi x + \dfrac{1}{3}\sin 3\pi x + \dfrac{1}{5}\sin 5\pi x\right)$

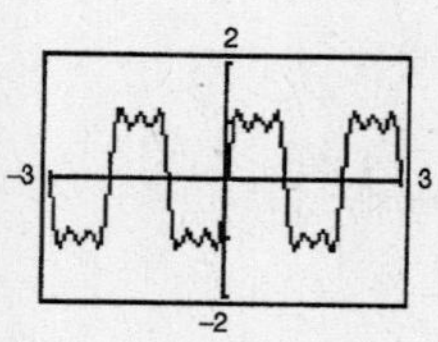

(b) $y_3 = \dfrac{4}{\pi}\left(\sin \pi x + \dfrac{1}{3}\sin 3\pi x + \dfrac{1}{5}\sin 5\pi x + \dfrac{1}{7}\sin 7\pi x\right)$

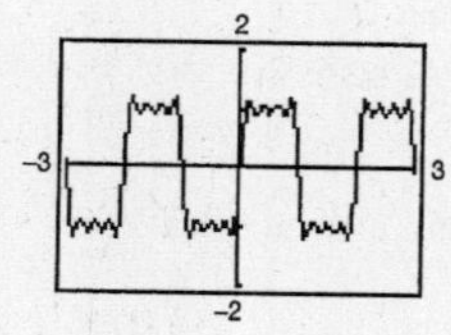

(c) $y_4 = \dfrac{4}{\pi}\left(\sin \pi x + \dfrac{1}{3}\sin 3\pi x + \dfrac{1}{5}\sin 5\pi x + \dfrac{1}{7}\sin 7\pi x + \dfrac{1}{9}\sin 9\pi x\right)$

88. $(x - 5)^2 = 8$

$x - 5 = \pm\sqrt{8}$

$x = 5 \pm 2\sqrt{2}$

90. $9x^2 + 12x + 3 = 0$

$(9x + 3)(x + 1) = 0$

$9x + 3 = 0 \Rightarrow x = -\frac{1}{3}$

$x + 1 = 0 \Rightarrow x = -1$

92. $2x^2 - 4x - 6 = 0$

$2(x^2 - 2x - 3) = 0$

$2(x + 1)(x - 3) = 0$

$x + 1 = 0 \Rightarrow x = -1$

$x - 3 = 0 \Rightarrow x = 3$

94. $2x^2 + 4x - 9 = 2(x - 1)^2$

$2x^2 + 4x - 9 = 2(x^2 - 2x + 1)$

$2x^2 + 4x - 9 = 2x^2 - 4x + 2$

$8x - 11 = 0$

$8x = 11$

$x = \frac{11}{8}$

Section 1.7 Inverse Trigonometric Functions

Solutions to Even-Numbered Exercises

2. $y = \arcsin 0 \Rightarrow \sin y = 0$ for $-\dfrac{\pi}{2} \le y \le \dfrac{\pi}{2} \Rightarrow y = 0$

4. $y = \arccos 0 \Rightarrow \cos y = 0$ for $0 \le y \le \pi \Rightarrow y = \dfrac{\pi}{2}$

6. $y = \arctan(-1) \Rightarrow \tan y = -1$ for $-\dfrac{\pi}{2} < y < \dfrac{\pi}{2} \Rightarrow y = -\dfrac{\pi}{4}$

8. $y = \arcsin\left(-\dfrac{\sqrt{2}}{2}\right) \Rightarrow \sin y = -\dfrac{\sqrt{2}}{2}$ for $-\dfrac{\pi}{2} \le y \le \dfrac{\pi}{2} \Rightarrow y = -\dfrac{\pi}{4}$

10. $y = \arctan(\sqrt{3}) \Rightarrow \tan y = \sqrt{3}$ for $-\dfrac{\pi}{2} < y < \dfrac{\pi}{2} \Rightarrow y = \dfrac{\pi}{3}$

12. $y = \arcsin \dfrac{\sqrt{2}}{2} \Rightarrow \sin y = \dfrac{\sqrt{2}}{2}$ for $-\dfrac{\pi}{2} \le y \le \dfrac{\pi}{2} \Rightarrow y = \dfrac{\pi}{4}$

14. $y = \arctan\left(-\dfrac{\sqrt{3}}{3}\right) \Rightarrow \tan y = -\dfrac{\sqrt{3}}{3}$ for $-\dfrac{\pi}{2} < y < \dfrac{\pi}{2} \Rightarrow y = -\dfrac{\pi}{6}$

16. $y = \arccos 1 \Rightarrow \cos y = 1$ for $0 \le y \le \pi \Rightarrow y = 0$

18. $\arcsin 0.45 \approx 0.47$

20. $\arccos(-0.7) \approx 2.35$

22. $\arctan 15 \approx 1.50$

24. $\arccos 0.26 \approx 1.31$

26. $\arcsin(-0.125) \approx -0.13$

28. $\arctan 2.8 \approx 1.23$

30. $\arccos\left(-\dfrac{1}{3}\right) \approx 1.91$

32. $\arctan\left(-\dfrac{95}{7}\right) \approx -1.50$

34. $\arccos(-1) = \pi$

$\arccos\left(-\dfrac{1}{2}\right) = \dfrac{2\pi}{3}$

$\cos\left(\dfrac{\pi}{6}\right) = \dfrac{\sqrt{3}}{2}$

36. $f(x) = \sin x$

$g(x) = \arcsin x$

$y = x$

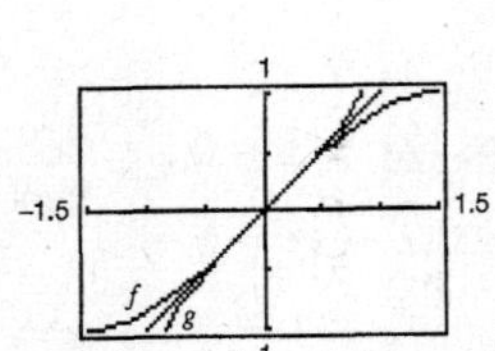

38. $\cos\theta = \dfrac{4}{x}$

$\theta = \arccos \dfrac{4}{x}$

40. $\tan\theta = \dfrac{x+1}{10}$

$$\theta = \arctan\left(\frac{x+1}{10}\right)$$

42. $\tan\theta = \dfrac{x-1}{x^2-1} = \dfrac{1}{x+1}$

$$\theta = \arctan\frac{1}{x+1}$$

$$x \neq 1$$

44. $\tan(\arctan 25) = 25$

46. $\sin[\arcsin(-0.2)] = -0.2$

48. $\arccos\left(\cos\dfrac{7\pi}{2}\right) = \arccos 0 = \dfrac{\pi}{2}$

Note: $\dfrac{7\pi}{2}$ is not in the range of the arccosine function.

50. Let $u = \arcsin\dfrac{4}{5}$,

$\sin u = \dfrac{4}{5},\ 0 < u < \dfrac{\pi}{2}.$

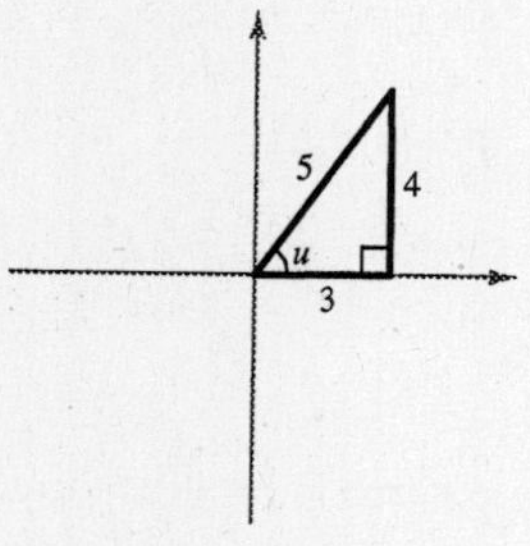

$$\sec\left(\arcsin\frac{4}{5}\right) = \sec u = \frac{\text{hyp}}{\text{adj}} = \frac{5}{3}$$

52. Let $u = \arccos\dfrac{\sqrt{5}}{5}$,

$\cos u = \dfrac{\sqrt{5}}{5},\ 0 < u < \dfrac{\pi}{2}.$

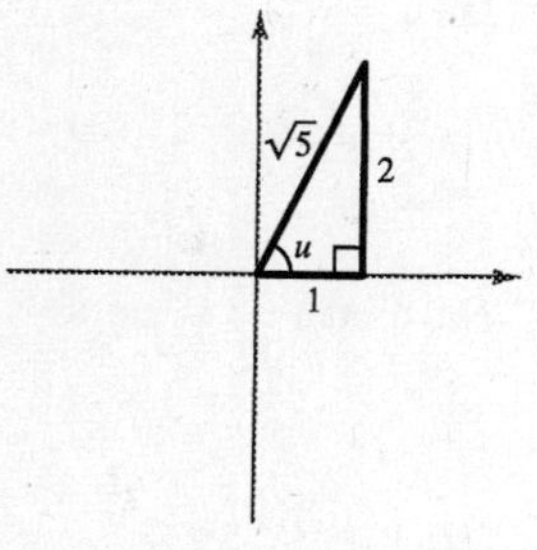

$$\sin\left(\arccos\frac{\sqrt{5}}{5}\right) = \sin u = \frac{2}{\sqrt{5}} = \frac{2\sqrt{5}}{5}$$

54. Let $u = \arctan\left(-\dfrac{5}{12}\right)$,

$\tan u = -\dfrac{5}{12},\ -\dfrac{\pi}{2} < u < 0.$

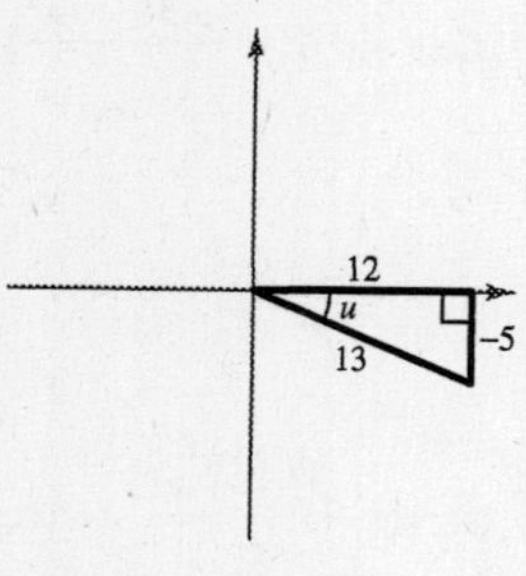

$$\csc\left[\arctan\left(-\frac{5}{12}\right)\right] = \csc u = \frac{\text{hyp}}{\text{opp}} = -\frac{13}{5}$$

56. Let $u = \arcsin\left(-\dfrac{3}{4}\right)$,

$\sin u = -\dfrac{3}{4},\ -\dfrac{\pi}{2} < u < 0.$

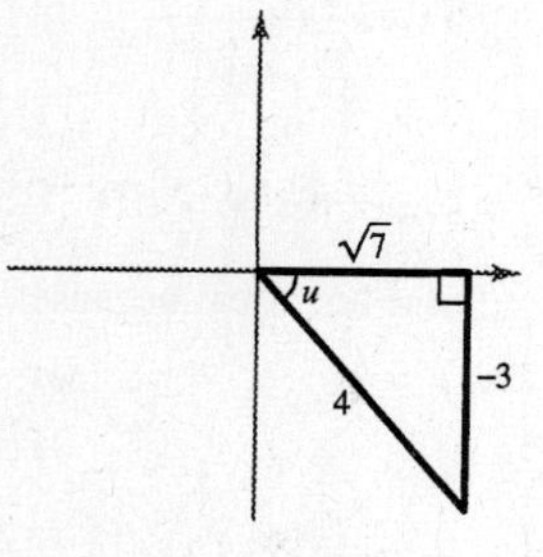

$$\tan\left[\arcsin\left(-\frac{3}{4}\right)\right] = \tan u = -\frac{3}{\sqrt{7}} = -\frac{3\sqrt{7}}{7}$$

58. Let $u = \arctan\dfrac{5}{8}$,

$\tan u = \dfrac{5}{8},\ 0 < u < \dfrac{\pi}{2}.$

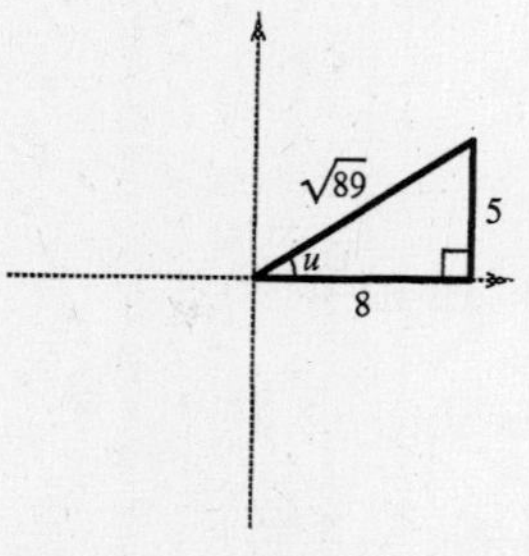

$$\cot\left(\arctan\frac{5}{8}\right) = \cot u = \frac{\text{adj}}{\text{opp}} = \frac{8}{5}$$

60. Let $u = \arctan x$,

$\tan u = x = \dfrac{x}{1}.$

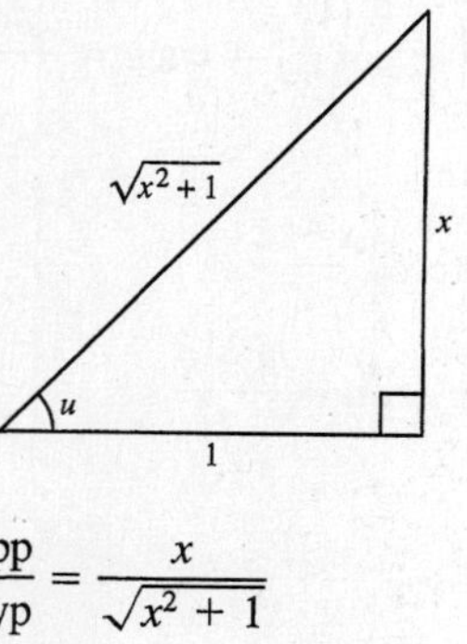

$$\sin(\arctan x) = \sin u = \frac{\text{opp}}{\text{hyp}} = \frac{x}{\sqrt{x^2+1}}$$

62. Let $u = \arctan 3x$,

$\tan u = 3x = \dfrac{3x}{1}$.

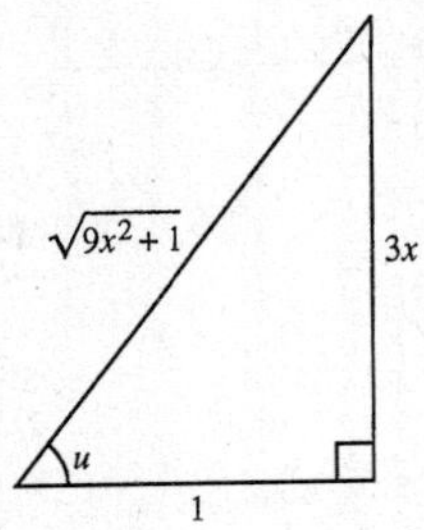

$\sec(\arctan 3x) = \sec u = \dfrac{\text{hyp}}{\text{adj}} = \sqrt{9x^2 + 1}$

64. Let $u = \arcsin(x - 1)$,

$\sin u = x - 1 = \dfrac{x-1}{1}$.

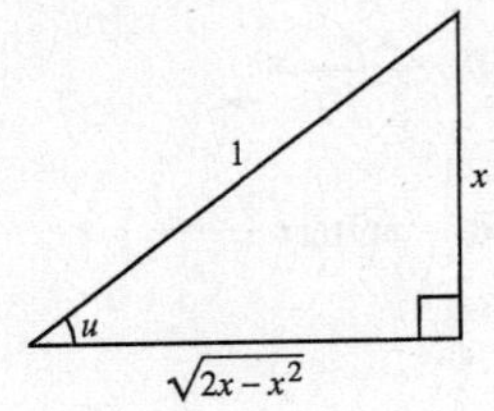

$\sec[\arcsin(x - 1)] = \sec u = \dfrac{\text{hyp}}{\text{adj}} = \dfrac{1}{\sqrt{2x - x^2}}$

66. Let $u = \arctan \dfrac{1}{x}$,

$\tan u = \dfrac{1}{x}$.

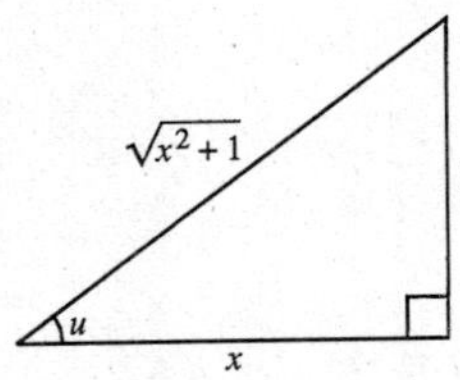

$\cot\left(\arctan \dfrac{1}{x}\right) = \cot u = \dfrac{\text{adj}}{\text{opp}} = x$

68. Let $u = \arcsin \dfrac{x-h}{r}$,

$\sin u = \dfrac{x-h}{r}$.

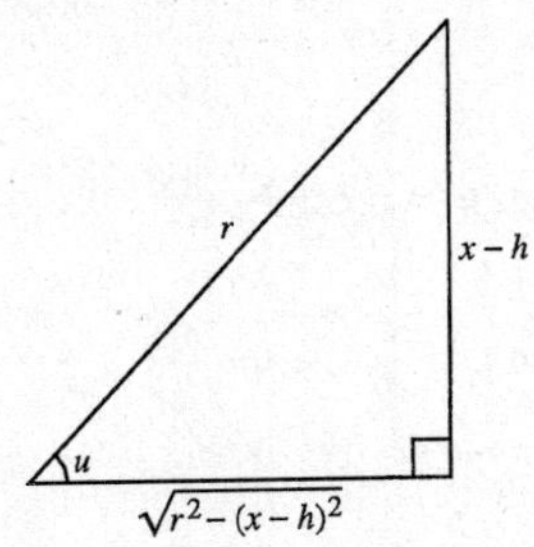

$\cos\left(\arcsin \dfrac{x-h}{r}\right) = \cos u = \dfrac{\sqrt{r^2 - (x-h)^2}}{r}$

70. $f(x) = \tan\left(\arccos \dfrac{x}{2}\right)$

$g(x) = \dfrac{\sqrt{4 - x^2}}{x}$

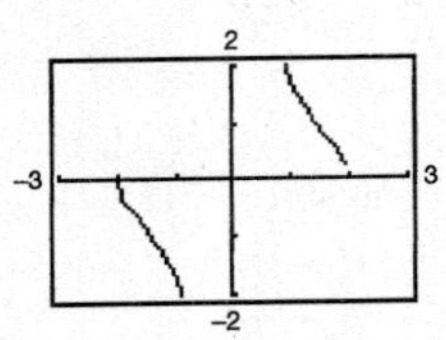

Asymptote: $x = 0$

These are equal because:

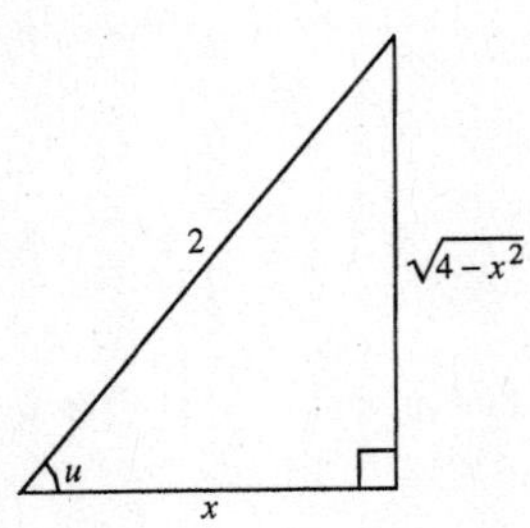

Let $u = \arccos \dfrac{x}{2}$.

$\tan\left(\arccos \dfrac{x}{2}\right) = \tan u = \dfrac{\sqrt{4 - x^2}}{x}$

72. If $\arcsin \dfrac{\sqrt{36 - x^2}}{6} = u$,

then $\sin u = \dfrac{\sqrt{36 - x^2}}{6}$.

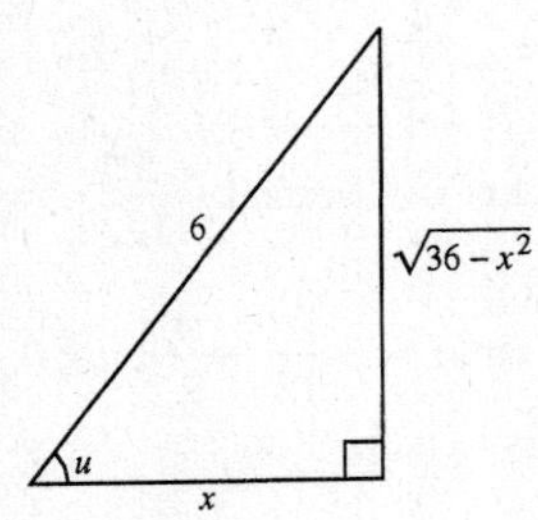

$\arcsin \dfrac{\sqrt{36 - x^2}}{6} = \arccos \dfrac{x}{6}$

74. If $\arccos \dfrac{x-2}{2} = u$,

then $\cos u = \dfrac{x-2}{2}$.

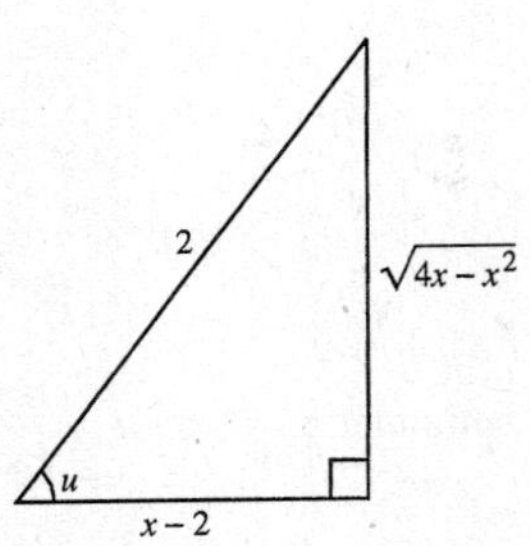

$\arccos \dfrac{x-2}{2} = \arctan \dfrac{\sqrt{4x - x^2}}{x - 2}$

76. $y = \arcsin \dfrac{x}{2}$

Domain: $[-2, 2]$

Range: $\left[-\dfrac{\pi}{2}, \dfrac{\pi}{2}\right]$

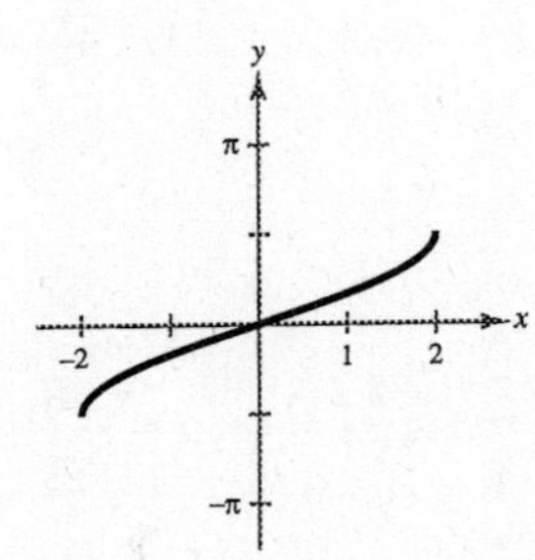

78. $g(t) = \arccos(t + 2)$

Domain: $[-3, -1]$

Range: $[0, \pi]$

This is the graph of $y = \arccos t$ shifted two units to the left.

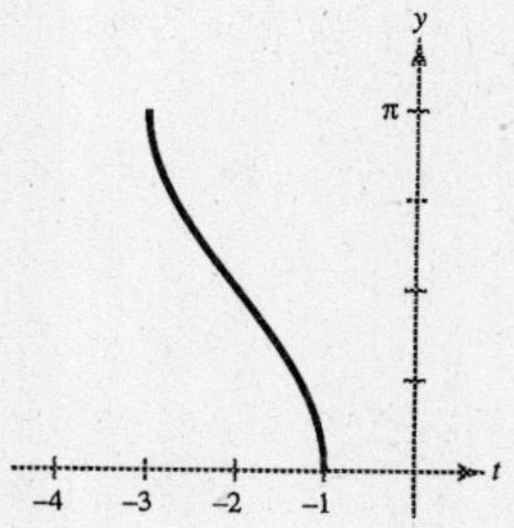

80. $f(x) = \dfrac{\pi}{2} + \arctan x$

Domain: $(-\infty, \infty)$

Range: $(0, \pi)$

This is the graph of $y = \arctan x$ shifted upward $\dfrac{\pi}{2}$ units.

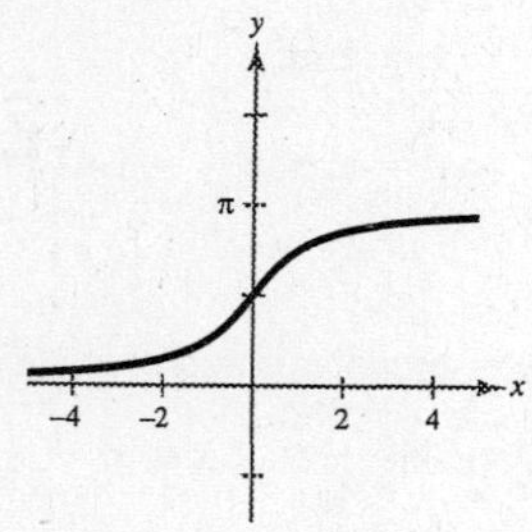

82. $f(x) = \arccos \dfrac{x}{4}$

Domain: $[-4, 4]$

Range: $[0, \pi]$

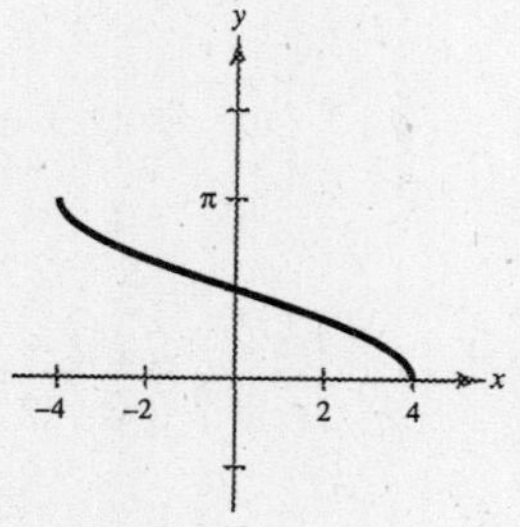

84. $f(x) = \pi \arcsin (4x)$

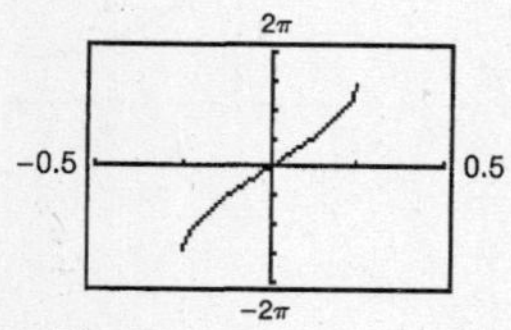

86. $f(x) = -3 + \arctan (\pi x)$

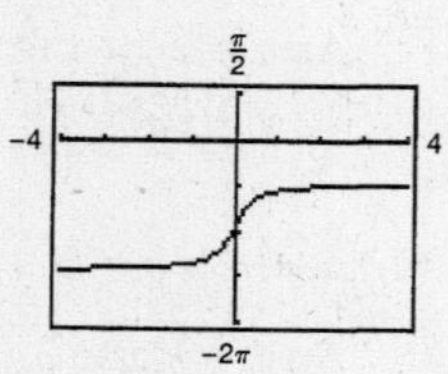

88. $f(x) = \dfrac{\pi}{2} + \arccos\left(\dfrac{1}{\pi}\right) \approx 2.82$

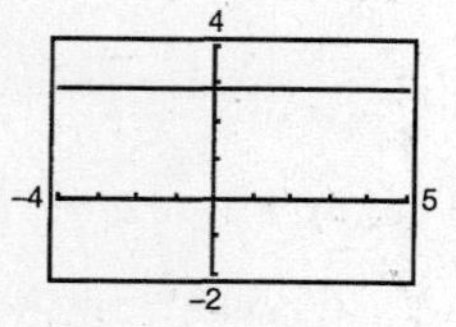

90. $f(t) = 4 \cos \pi t + 3 \sin \pi t$

$$= \sqrt{4^2 + 3^2} \sin\left(\pi t + \arctan \frac{4}{3}\right)$$

$$= 5 \sin\left(\pi t + \arctan \frac{4}{3}\right)$$

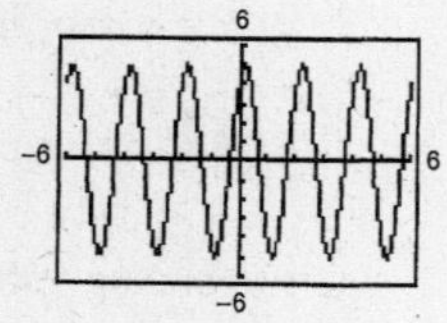

The graph implies that

$$A \cos \omega t + B \sin \omega t = \sqrt{A^2 + B^2} \sin\left(\omega t + \arctan \frac{A}{B}\right)$$

is true.

92. (a) $\tan \theta = \dfrac{s}{750}$

$$\theta = \arctan \frac{s}{750}$$

(b) When $s = 300$,

$$\theta = \arctan \frac{300}{750} \approx 0.38 \approx 21.8°.$$

When $s = 1200$,

$$\theta = \arctan \frac{1200}{750} \approx 1.01 \approx 58.0°.$$

94. (a) $\tan \theta = \dfrac{11}{17}$

$$\theta = \arctan \frac{11}{17} \approx 0.5743 \approx 32.9°$$

(b) $r = \dfrac{1}{2}(40) = 20$

$$\tan \theta = \frac{h}{r} = \frac{h}{20}$$

$$h = 20 \tan \theta = 20 \cdot \frac{11}{17} \approx 12.94 \text{ feet}$$

96. (a) $\tan\theta = \frac{6}{x}$

$\theta = \arctan\frac{6}{x}$

(b) $x = 7$ miles

$\theta = \arctan\frac{6}{7} \approx 0.71 \approx 40.6^\circ$

$x = 1$ mile

$\theta = \arctan\frac{6}{1} \approx 1.41 \approx 80.5^\circ$

98. False.

$\frac{5\pi}{6}$ is not in the range of arcsin (x).

$\arcsin\frac{1}{2} = \frac{\pi}{6}$

100. $y = \text{arccot}\, x$ if and only if $\cot y = x$.

Domain: $-\infty < x < \infty$

Range: $0 < x < \pi$

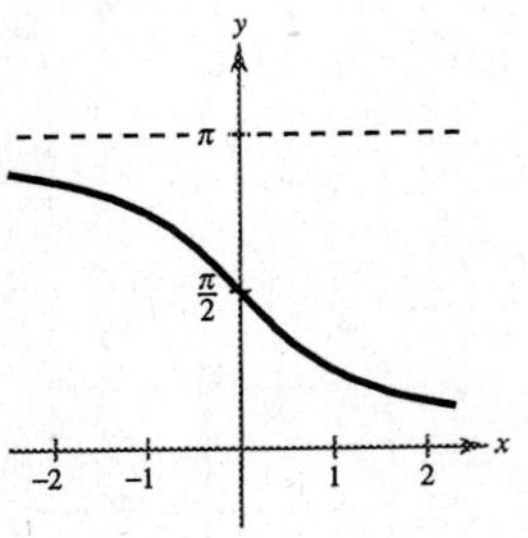

102. $y = \text{arccsc}\, x$ if and only if $\csc y = x$.

Domain: $(-\infty, -1] \cup [1, \infty)$

Range: $\left[-\frac{\pi}{2}, 0\right) \cup \left(0, \frac{\pi}{2}\right]$

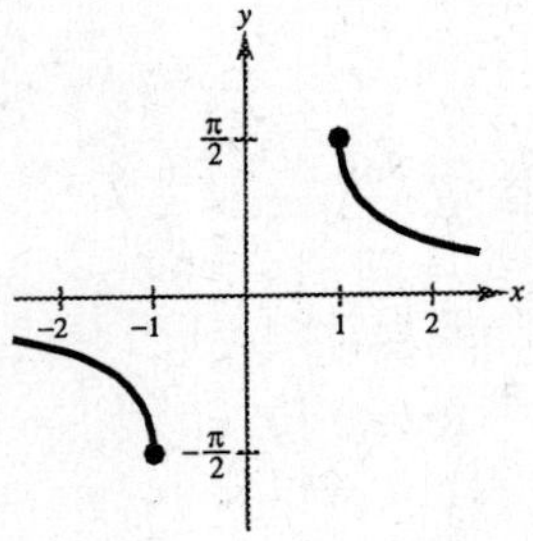

104. Area = arctan b − arctan a

(a) $a = 0, b = 1$

Area = $\arctan 1 - \arctan 0 = \frac{\pi}{4} - 0 = \frac{\pi}{4}$

(b) $a = -1, b = 1$

Area = $\arctan 1 - \arctan(-1)$

$= \frac{\pi}{4} - \left(-\frac{\pi}{4}\right) = \frac{\pi}{2}$

(c) $a = 0, b = 3$

Area = $\arctan 3 - \arctan 0$

$\approx 1.25 - 0 = 1.25$

$= 1.25$

(d) $a = -1, b = 3$

Area = $\arctan 3 - \arctan(-1)$

$\approx 1.25 - \left(-\frac{\pi}{4}\right) \approx 2.03$

106. (a) $f \cdot f^{-1} = \sin(\arcsin x)$

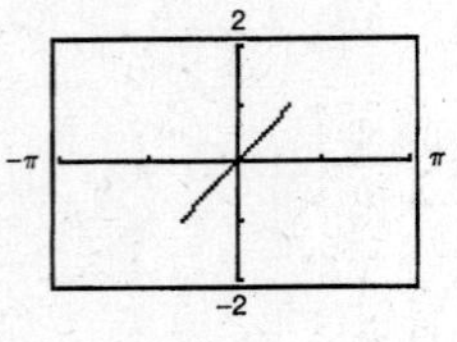

$f^{-1} \cdot f = \arcsin(\sin x)$

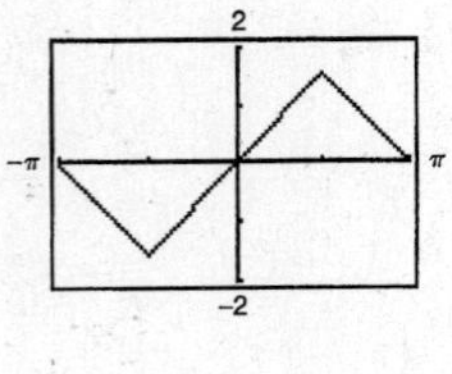

(b) The graphs coincide with the graph of $y = x$ only for certain values of x.

$f \cdot f^{-1} = x$ over its entire domain, $-1 \le x \le 1$.

$f^{-1} \cdot f = x$ over the region $-\frac{\pi}{2} \le x \le \frac{\pi}{2}$,

corresponding to the region where $\sin x$ is one-to-one and thus has an inverse.

108.

$$y = \arctan(-x)$$

$$\tan y = -x, \ -\frac{\pi}{2} < y < \frac{\pi}{2}$$

$$-\tan y = x$$

$$\tan(-y) = x, \ -\frac{\pi}{2} < -y < \frac{\pi}{2}$$

$$\arctan(\tan(-y)) = \arctan x$$

$$-y = \arctan x$$

$$y = -\arctan x$$

110. $y_2 = \dfrac{\pi}{2} - y_1$

$$\arctan x + \arctan \frac{1}{x} = y_1 + y_2$$

$$= y_1 + \left(\frac{\pi}{2} - y_1\right) = \frac{\pi}{2}$$

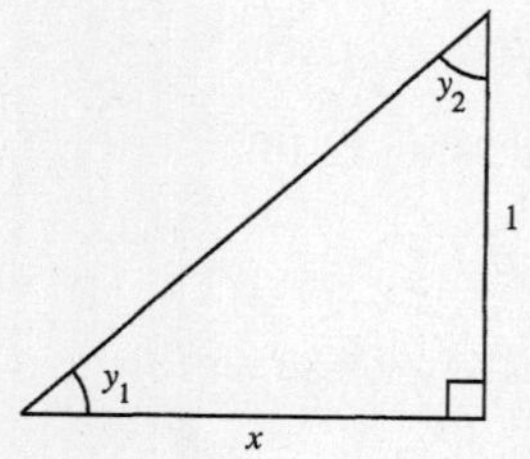

112. $\arcsin x = \arcsin \dfrac{x}{1} = \arctan \dfrac{x}{\sqrt{1 - x^2}}$

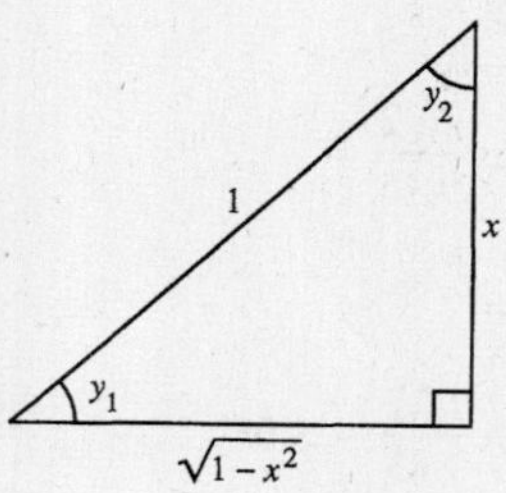

114. $\tan \theta = 2$

$\text{hyp} = \sqrt{1^2 + 2^2} = \sqrt{5}$

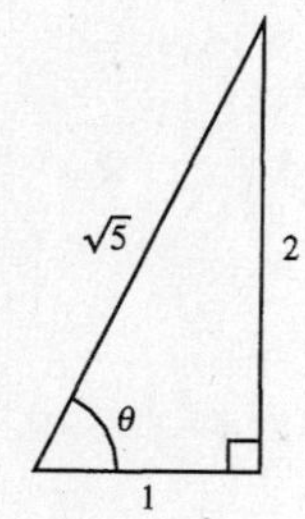

116. $\sec \theta = 3$

$\text{opp} = \sqrt{3^2 - 1^2} = 2\sqrt{2}$

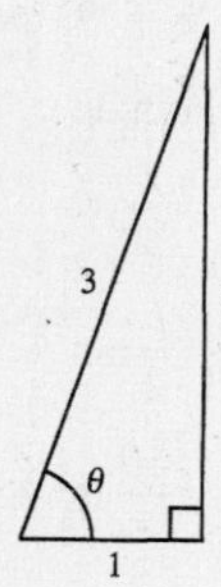

118. Rate downstream: $18 + x$

Rate upstream: $18 - x$

rate $\times$ time = distance $\Rightarrow t = \dfrac{d}{r}$

(Time to go upstream) + (Time to go downstream) = 4

$$\frac{35}{18 - x} + \frac{35}{18 + x} = 4$$

$$35(18 + x) + 35(18 - x) = 4(18 - x)(18 + x)$$

$$630 + 35x + 630 - 35x = 4(324 - x^2)$$

$$1260 = 4(324 - x^2)$$

$$315 = 324 - x^2$$

$$x^2 = 9$$

$$x = \pm 3$$

The speed of the current is 3 miles per hour.

Section 1.8 Applications and Models

Solutions to Even-Numbered Exercises

2. $B = 54°, c = 15$

$A = 90° - B = 90° - 54° = 36°$

$\sin B = \frac{b}{c} \Rightarrow b = c \sin B$

$= 15 \sin 54° \approx 12.14$

$\cos B = \frac{a}{c} \Rightarrow a = c \cos B$

$= 15 \cos 54° \approx 8.82$

4. $A = 8.4°, a = 40.5$

$B = 90° - A = 90° - 8.4° = 81.6°$

$\tan A = \frac{a}{b} \Rightarrow b = \frac{a}{\tan A}$

$= \frac{40.5}{\tan 8.4°} \approx 274.27$

$\sin A = \frac{a}{c} \Rightarrow c = \frac{a}{\sin A}$

$= \frac{40.5}{\sin 8.4°} \approx 277.24$

6. $a = 25, c = 35$

$b = \sqrt{c^2 - a^2} = \sqrt{35^2 - 25^2} = \sqrt{600} \approx 24.49$

$\sin A = \frac{a}{c} \Rightarrow A = \arcsin \frac{a}{c}$

$= \arcsin \frac{25}{35} \approx 45.58°$

$\cos B = \frac{a}{c} \Rightarrow B = \arccos \frac{a}{c}$

$= \arccos \frac{25}{35} \approx 44.42°$

8. $b = 1.32, c = 9.45$

$a = \sqrt{c^2 - b^2} = \sqrt{87.5601} \approx 9.36$

$\cos A = \frac{b}{c} \Rightarrow A = \arccos \frac{b}{c}$

$= \arccos \frac{1.32}{9.45} \approx 81.97°$

$\sin B = \frac{b}{c} \Rightarrow B = \arcsin \frac{b}{c}$

$= \arcsin \frac{1.32}{9.45} \approx 8.03°$

10. $B = 65° 12', a = 14.2$

$A = 90° - B = 90° - 65° 12' = 24° 48'$

$\cos B = \frac{a}{c} \Rightarrow c = \frac{a}{\cos B}$

$= \frac{14.2}{\cos 65° 12'} \approx 33.85$

$\tan B = \frac{b}{a} \Rightarrow b = a \tan B$

$= 14.2 \tan 65° 12'$

≈ 30.73

12. $\theta = 18°, b = 10$ meters

$\tan \theta = \frac{\text{altitude}}{\frac{b}{2}}$

$\text{altitude} = \frac{b}{2} \tan \theta$

$= \frac{10}{2} \tan 18° \approx 1.62$ meters

14. $\theta = 27°, b = 11$ feet

$\tan \theta = \frac{\text{altitude}}{\frac{b}{2}}$

$\text{altitude} = \frac{b}{2} \tan \theta$

$= \frac{11}{2} \tan 27° \approx 2.80$ feet

16. $\tan 20° = \frac{600}{x}$

$x = \frac{600}{\tan 20°} \approx 1648.5$ feet

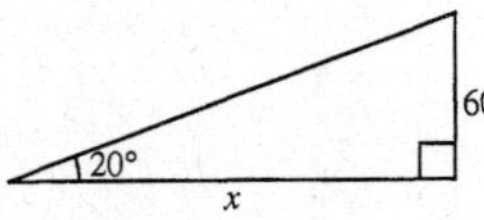

18. $\tan 33° = \frac{h}{125}$

$h = 125 \tan 33° \approx 81.2$ feet

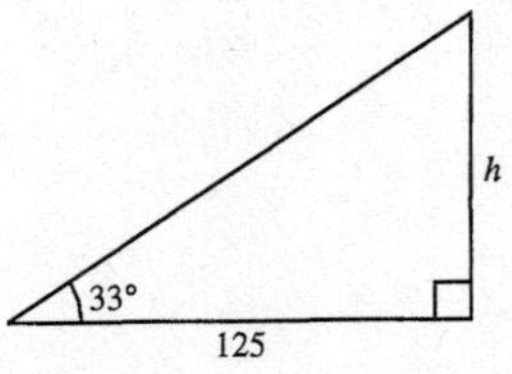

20. $\tan 51° = \dfrac{h}{100}$

$h = 100 \tan 51°$

≈ 123.5 feet

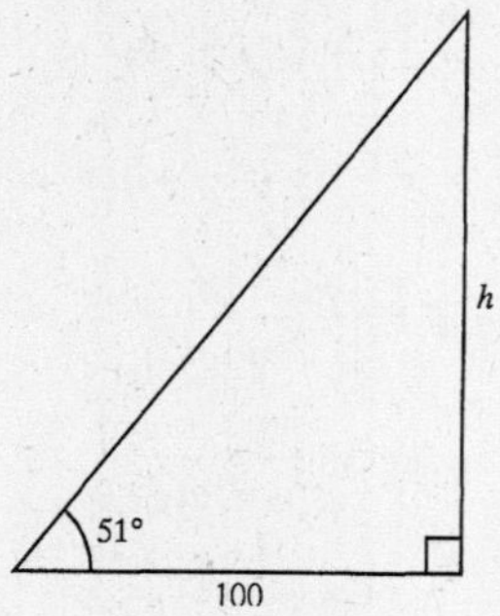

22. $\tan \theta = \dfrac{75}{50}$

$\theta = \arctan \dfrac{3}{2} \approx 56.3°$

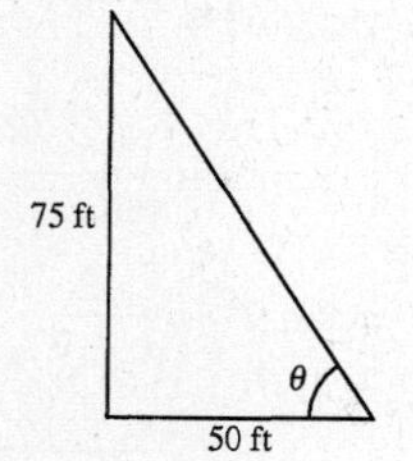

24. $12{,}500 + 4000 = 16{,}500$

$\sin \theta = \dfrac{4000}{16{,}500}$

$\theta = \arcsin\left(\dfrac{4000}{16{,}500}\right)$

$\theta \approx 14.03°$

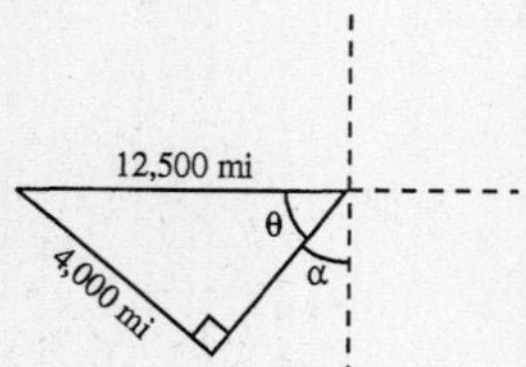

Not drawn to scale

Angle of depression

$= \alpha \approx 90° - 14.03° = 75.97°$

26. (a) Since the airplane speed is

$$\left(275\frac{\text{ft}}{\text{sec}}\right)\left(60\frac{\text{sec}}{\text{min}}\right) = 16{,}500\frac{\text{ft}}{\text{min}},$$

after one minute its distance travelled is 16,500 feet.

$\sin 18° = \dfrac{a}{16{,}500}$

$a = 16{,}500 \sin 18°$

≈ 5099 ft

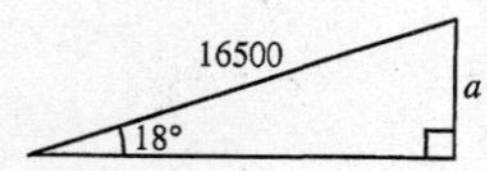

(b) $\sin 18° = \dfrac{10{,}000}{275\,s}$

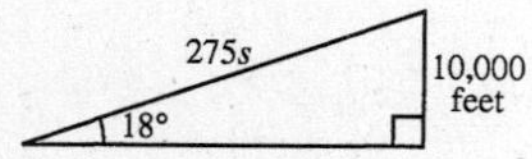

$s = \dfrac{10{,}000}{275(\sin 18°)} \approx 117.7$ seconds

28.

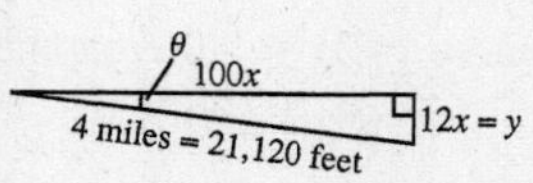

Angle of grade: $\tan \theta = \dfrac{12x}{100x}$

$\theta = \arctan 0.12 \approx 6.8°$

Change in elevation:

$\sin \theta = \dfrac{y}{21{,}120}$

$y = 21{,}120 \sin \theta$

$= 21{,}120 \sin(\arctan 0.12)$

≈ 2516.3 feet

30. $\theta = 32°,\ \phi = 68°$

(a) $\alpha = 90° - 32° = 58°$

(b) Bearing from A to C: N 58° E

$\beta = \theta = 32°$

$\gamma = 90° - \phi = 22°$

$C = \beta + \gamma = 54°$

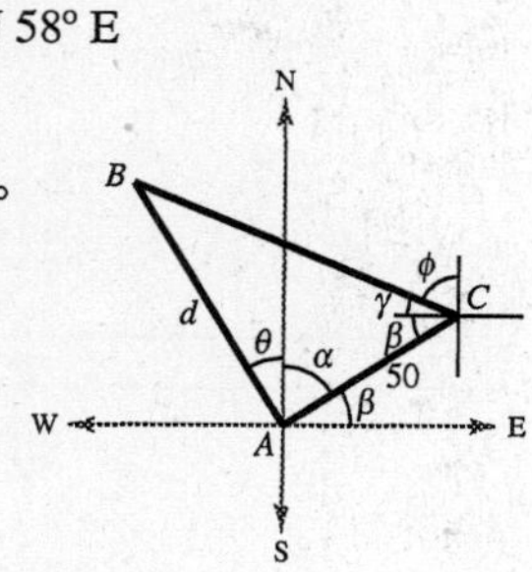

$\tan C = \dfrac{d}{50} \Rightarrow \tan 54° = \dfrac{d}{50} \Rightarrow d \approx 68.82$ meters

32. $\tan \theta = \dfrac{45}{30} \Rightarrow \theta \approx 56.3°$

Bearing: N 56.3° W

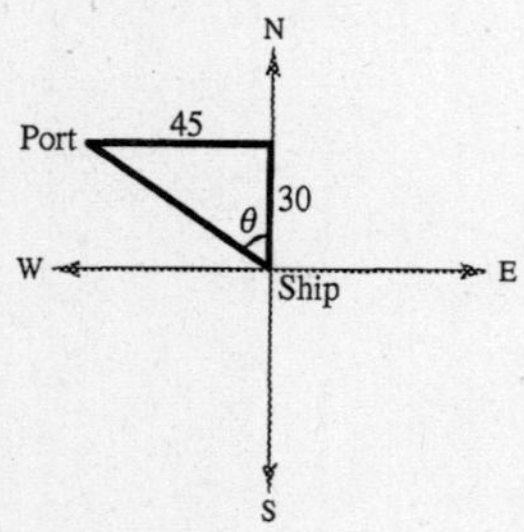

34.

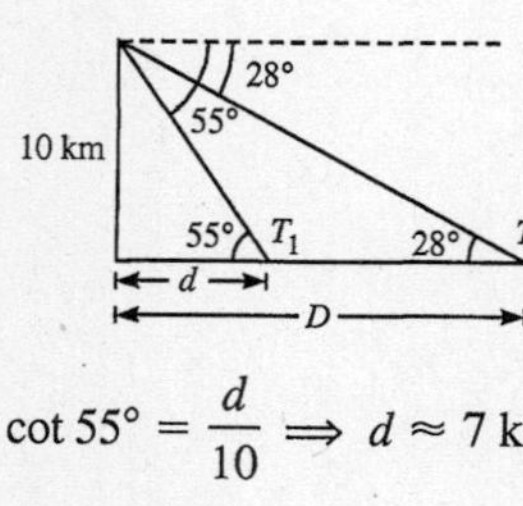

$\cot 55° = \dfrac{d}{10} \Rightarrow d \approx 7$ kilometers

$\cot 28° = \dfrac{D}{10} \Rightarrow D \approx 18.8$ kilometers

Distance between towns:

$D - d = 18.8 - 7 = 11.8$ kilometers

36. $\tan 2.5^\circ = \dfrac{h}{x}$

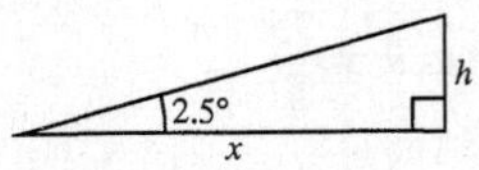

Not drawn to scale

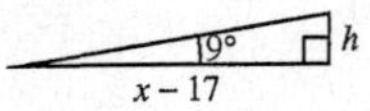

$$x = \frac{h}{\tan 2.5^\circ}$$

$$\tan 9^\circ = \frac{h}{x - 17}$$

$$x = \frac{h}{\tan 9^\circ} + 17$$

$$\frac{h}{\tan 2.5^\circ} = \frac{h}{\tan 9^\circ} + 17$$

$$h = \frac{17}{\left(\dfrac{1}{\tan 2.5^\circ} - \dfrac{1}{\tan 9^\circ}\right)} \approx 1.025 \text{ miles}$$

$$\approx 5410 \text{ feet}$$

38. $L_1 = 2x - y = 8 \Rightarrow m_1 = 2$

$L_2 = x - 5y = -4 \Rightarrow m_2 = \dfrac{1}{5}$

$$\tan \alpha = \left|\frac{m_2 - m_1}{1 + m_2 m_1}\right|$$

$$\alpha = \arctan\left|\frac{m_2 - m_1}{1 + m_2 m_1}\right|$$

$$= \arctan\left|\frac{\frac{1}{5} - 2}{1 + \frac{1}{5}(2)}\right|$$

$$= \arctan\left(\frac{9}{7}\right) \approx 52.1^\circ$$

40.

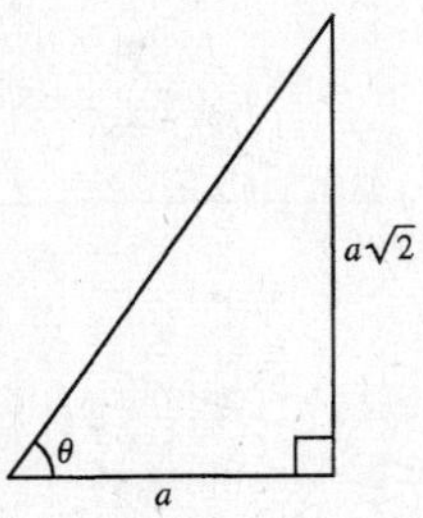

$$\tan \theta = \frac{a\sqrt{2}}{a} = \sqrt{2}$$

$$\theta = \arctan \sqrt{2} \approx 54.7^\circ$$

42.

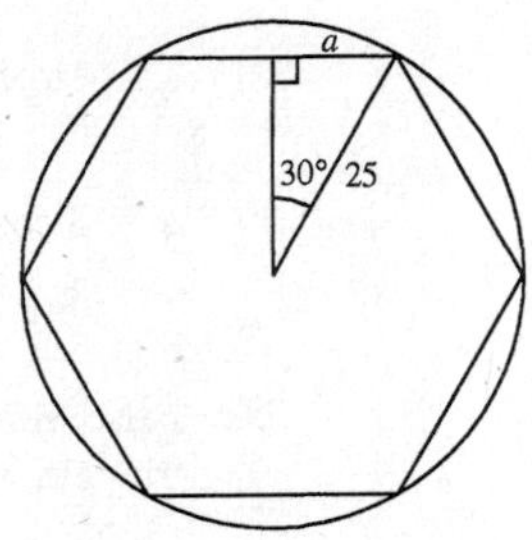

$$\sin 30^\circ = \frac{a}{25}$$

$$a = 25 \sin 30^\circ = 12.5$$

Length of side $= 2a = 2(12.5)$

$= 25$ inches

44.

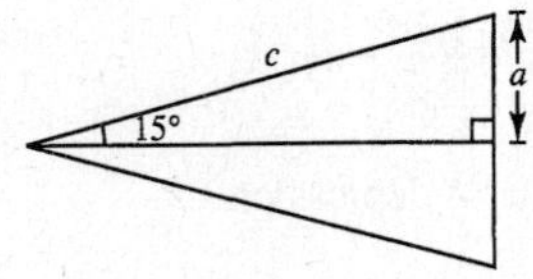

$$c = \frac{35}{2} = 17.5$$

$$\sin 15^\circ = \frac{a}{c}$$

$$a = c \sin 15^\circ = 17.5 \sin 15^\circ$$

$$\approx 4.53$$

Distance $= 2a \approx 9.06$ centimeters

46.

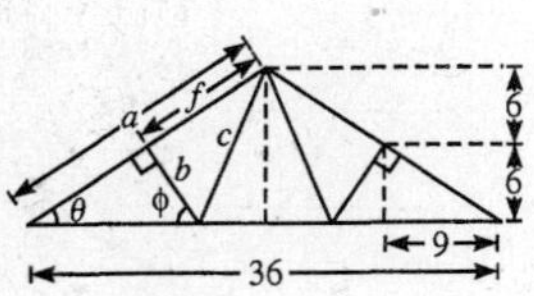

$$\tan \theta = \frac{12}{18}$$

$$\theta = \arctan \frac{2}{3} = 0.588 \text{ rad} \approx 33.7^\circ$$

$$\cos \theta = \frac{18}{a}$$

$$a = \frac{18}{\cos \theta} \approx 21.6 \text{ feet}$$

$$f \approx \frac{21.6}{2} = 10.8 \text{ feet}$$

$$\phi \approx 90 - 33.7 = 56.3^\circ$$

$$\sin \phi = \frac{6}{b}$$

$$b = \frac{6}{\sin \phi} \approx 7.2 \text{ feet}$$

$$c = \sqrt{10.8^2 + 7.2^2} \approx 13 \text{ feet}$$

48. $d = \dfrac{1}{2} \cos 20\,\pi t$

(a) Maximum displacement: $|a| = \left|\dfrac{1}{2}\right| = \dfrac{1}{2}$

(b) Frequency

$$\frac{\omega}{2\pi} = \frac{20\pi}{2\pi} = 10$$

(c) Least positive value for t for which $d = 0$

$$\frac{1}{2} \cos 20\,\pi t = 0$$

$$\cos 20\pi t = 0$$

$$20\pi t = \arccos 0$$

$$20\pi t = \frac{\pi}{2}$$

$$t = \frac{\pi}{2} \cdot \frac{1}{20\pi} = \frac{1}{40}$$

50. $d = \frac{1}{64}\sin 792\pi t$

(a) Maximum displacement:

$$|a| = \left|\frac{1}{64}\right| = \frac{1}{64}$$

(b) Frequency:

$$\frac{\omega}{2\pi} = \frac{792\pi}{2\pi} = 396$$

(c) Least positive value for t for which $d = 0$

$$\frac{1}{64}\sin 792\pi t = 0$$

$$\sin 792\pi t = 0$$

$$792\pi t = \arcsin 0$$

$$792\pi t = \pi$$

$$t = \frac{\pi}{792\pi} = \frac{1}{792}$$

52. Displacement at $t = 0$ is $0 \Rightarrow d = a\sin\omega t$

Amplitude: $|a| = 3$

Period: $\frac{2\pi}{\omega} = 6 \Rightarrow \omega = \frac{\pi}{3}$

$$d = 3\sin\left(\frac{\pi t}{3}\right)$$

54. Displacement at $t = 0$ is $2 \Rightarrow d = a\cos\omega t$

Amplitude: $|a| = 2$

Period: $\frac{2\pi}{\omega} = 10 \Rightarrow \omega = \frac{\pi}{5}$

$$d = 2\cos\left(\frac{\pi t}{5}\right)$$

56. At $t = 0$, buoy is at its high point $\Rightarrow d = a\cos\omega t$.

Distance from high to low $= 2|a| = 3.5$

$$|a| = \frac{7}{4}$$

Returns to high point every 10 seconds:

Period $= \frac{2\pi}{\omega} = 10 \Rightarrow \omega = \frac{\pi}{5}$

$$d = \frac{7}{4}\cos\frac{\pi t}{5}$$

58. False. Since the tower is not exactly vertical, a right triangle with sides 191 ft and d is not formed.

60. (a)

θ	L_1	L_2	$L_1 + L_2$
0.1	$\frac{2}{\sin 0.1}$	$\frac{3}{\cos 0.1}$	23.0
0.2	$\frac{2}{\sin 0.2}$	$\frac{3}{\cos 0.2}$	13.1
0.3	$\frac{2}{\sin 0.3}$	$\frac{3}{\cos 0.3}$	9.9
0.4	$\frac{2}{\sin 0.4}$	$\frac{3}{\cos 0.4}$	8.4

(b)

θ	L_1	L_2	$L_1 + L_2$
0.5	$\frac{2}{\sin 0.5}$	$\frac{3}{\cos 0.5}$	7.6
0.6	$\frac{2}{\sin 0.6}$	$\frac{3}{\cos 0.6}$	7.2
0.7	$\frac{2}{\sin 0.7}$	$\frac{3}{\cos 0.7}$	7.0
0.8	$\frac{2}{\sin 0.8}$	$\frac{3}{\cos 0.8}$	7.1

The minimum length of the elevator is 7.0 meters.

(c) $L = L_1 + L_2 = \frac{2}{\sin\theta} + \frac{3}{\cos\theta}$

(d)

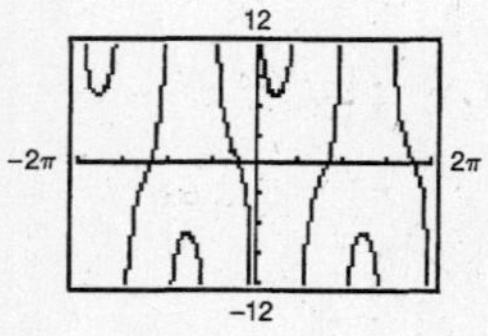

From the graph, it appears that the minimum length is 7.0 meters, which agrees with the estimate of part (b).

62. (a)

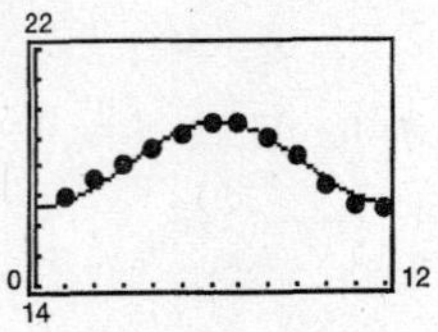

(b) Period $= \dfrac{2\pi}{b} = \dfrac{2\pi}{\frac{\pi}{6}} = 12$

Yes, this is what is expected there are 12 months in one year.

(c) Amplitude: $|a| = |1.41| = 1.41$, which represents the maximum change in time of sunset from the average time ($d = 18.09$) of sunset.

64. No. N 24° E means 24° east of north.

66. $5y - 3x = 12$

$$5y = 3x + 12$$

$$y = \frac{3}{5}x + \frac{12}{5}$$

The graph is a line with

$m = \dfrac{3}{5}$ and y-intercept $\left(0, \dfrac{12}{5}\right)$.

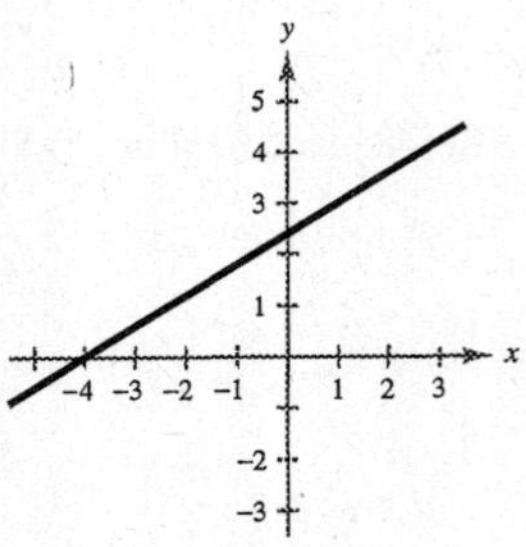

68. $(x + 3)^2 = 5y - 8$

$$(x + 3)^2 = 5\left(y - \frac{8}{5}\right)$$

Parabola with vertex

$\left(-3, \dfrac{8}{5}\right)$

x	0	-6
y	$\frac{17}{5}$	$\frac{17}{5}$

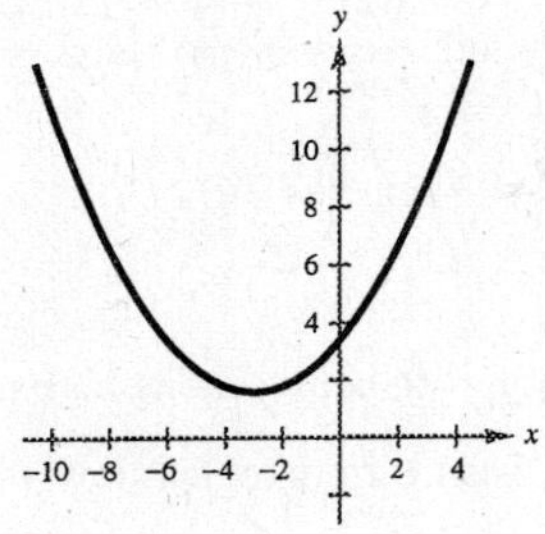

70. $2x^2 + y^2 - 4 = 0$

$$2x^2 + y^2 = 4$$

$$\frac{x^2}{2} + \frac{y^2}{4} = 1$$

Ellipse with center $(0, 0)$

Vertical major axis

$a = 2, b = \sqrt{2}$

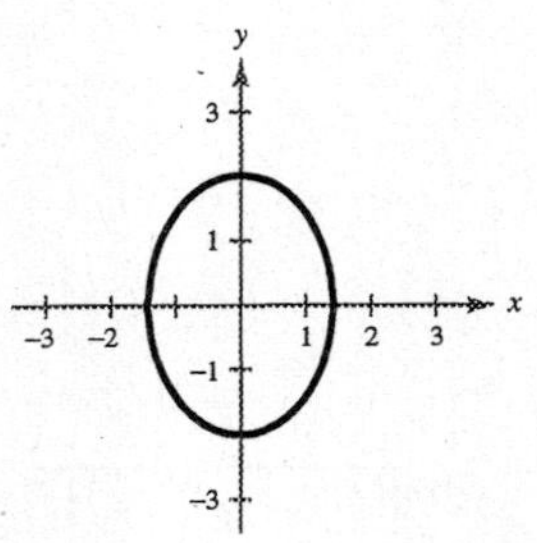

72. $(x - 2)^2 + y^2 = 25$

Circle with center $(2, 0)$

and radius $r = 5$

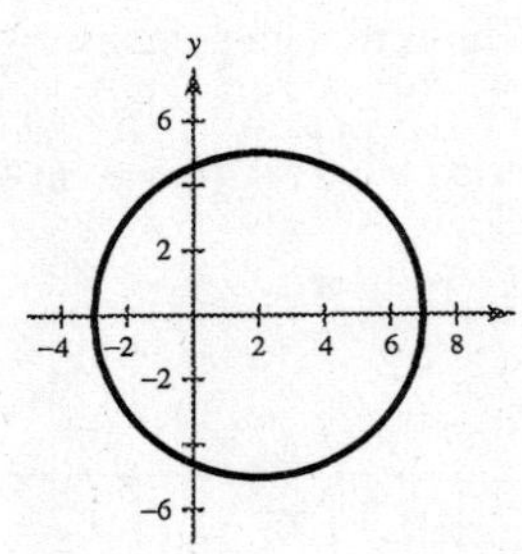

Review Exercises for Chapter 1

Solutions to Even-Numbered Exercises

2.

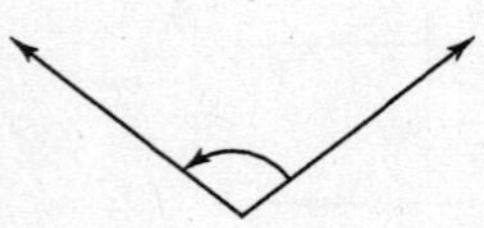

2 radians

4.

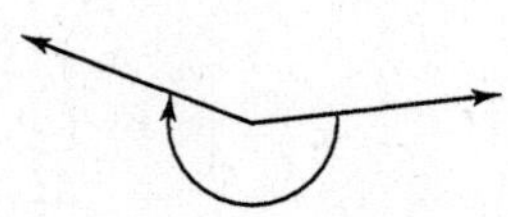

-3.5 radians

6.

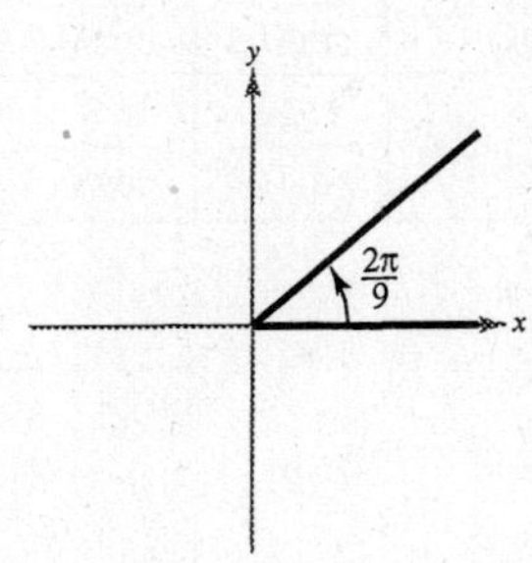

$$\frac{2\pi}{9} + 2\pi = \frac{20\pi}{9}$$

$$\frac{2\pi}{9} - 2\pi = -\frac{16\pi}{9}$$

8.

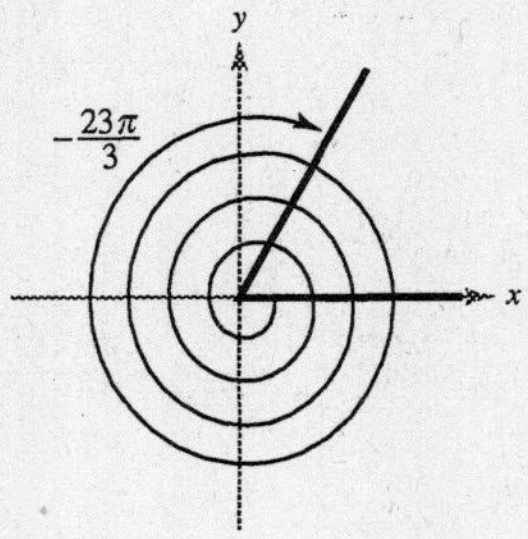

$$-\frac{23\pi}{3} + 8\pi = \frac{\pi}{3}$$

$$-\frac{23\pi}{3} + 2\pi = -\frac{17\pi}{3}$$

10.

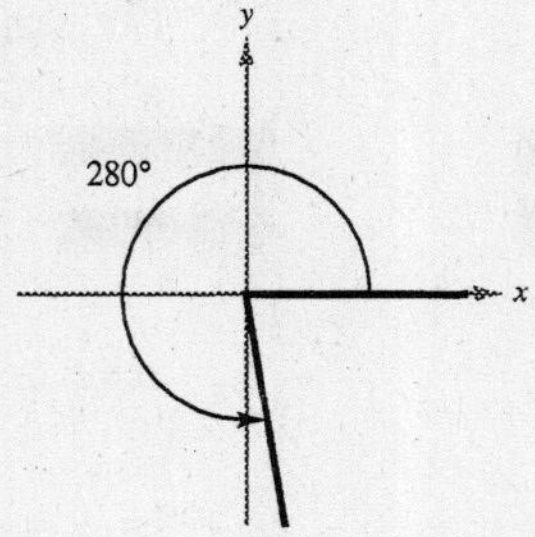

$$280° + 360° = 640°$$

$$280° - 360° = -80°$$

12.

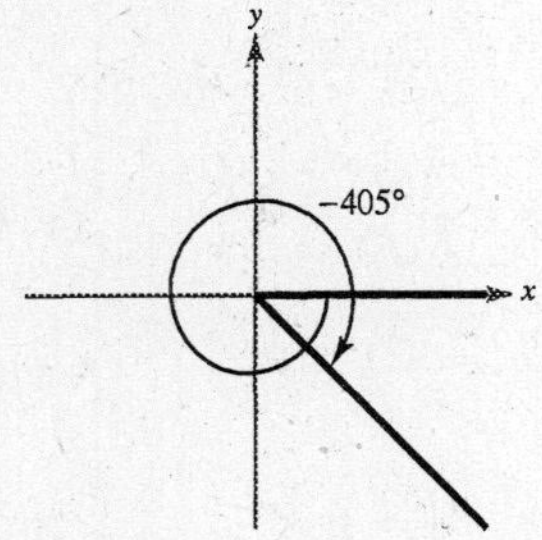

$$-405° + 720° = 315°$$

$$-405° + 360° = -45°$$

14. $-\frac{11\pi}{6} \cdot \frac{180°}{\pi} = -330.00°$

16. $5.7 \cdot \frac{180°}{\pi} \approx 326.59°$

18. $-127.5° \cdot \frac{\pi}{180°} \approx -2.2253$

20. $196° \, 77' = \left(196 + \frac{77}{60}\right)^{\circ} \cdot \frac{\pi}{180°} \approx 3.4432$

22. (linear speed) = (angular speed) · (radius)

= (5π rad/s) · (13.5 inches)

= 67.5π inches per second

≈ 212.1 inches per second

≈ 12.05 miles per hour

24. $t = \frac{3\pi}{4}, (x, y) = \left(-\frac{\sqrt{2}}{2}, \frac{\sqrt{2}}{2}\right)$

26. $t = -\frac{4\pi}{3}, (x, y) = \left(-\frac{1}{2}, \frac{\sqrt{3}}{2}\right)$

28. $t = \frac{\pi}{4}$

$\sin\frac{\pi}{4} = \frac{\sqrt{2}}{2}$ $\qquad \csc\frac{\pi}{4} = \sqrt{2}$

$\cos\frac{\pi}{4} = \frac{\sqrt{2}}{2}$ $\qquad \sec\frac{\pi}{4} = \sqrt{2}$

$\tan\frac{\pi}{4} = 1$ $\qquad \cot\frac{\pi}{4} = 1$

30. $t = 2\pi$

$\sin 2\pi = 0$ $\qquad$ $\csc 2\pi$ is undefined.

$\cos 2\pi = 1$ $\qquad$ $\sec 2\pi = 1$

$\tan 2\pi = 0$ $\qquad$ $\cot 2\pi$ is undefined.

32. $\cos 4\pi = \cos 0 = 1$

34. $\cos\left(-\frac{13\pi}{3}\right) = \cos\left(\frac{5\pi}{3}\right) = \frac{1}{2}$

36. $\csc 10.5 \approx -1.14$

38. $\sin\left(-\frac{\pi}{9}\right) \approx -0.34$

40. adj = 6, opp = 6

$\text{hyp} = \sqrt{6^2 + 6^2} = 6\sqrt{2}$

$\sin\theta = \dfrac{6}{6\sqrt{2}} = \dfrac{\sqrt{2}}{2}$ $\qquad \csc\theta = \dfrac{6\sqrt{2}}{6} = \sqrt{2}$

$\cos\theta = \dfrac{6}{6\sqrt{2}} = \dfrac{\sqrt{2}}{2}$ $\qquad \sec\theta = \dfrac{6\sqrt{2}}{6} = \sqrt{2}$

$\tan\theta = \dfrac{6}{6} = 1$ $\qquad \cot\theta = \dfrac{6}{6} = 1$

42. opp = 5, hyp = 9

$\text{adj} = \sqrt{9^2 - 5^2} = 2\sqrt{14}$

$\sin\theta = \dfrac{5}{9}$ $\qquad \csc\theta = \dfrac{9}{5}$

$\cos\theta = \dfrac{2\sqrt{14}}{9}$ $\qquad \sec\theta = \dfrac{9}{2\sqrt{14}} = \dfrac{9\sqrt{14}}{28}$

$\tan\theta = \dfrac{5}{2\sqrt{14}} = \dfrac{5\sqrt{14}}{28}$ $\qquad \cot = \dfrac{2\sqrt{14}}{5}$

44. $\tan\theta = 4$

(a) $\cot\theta = \dfrac{1}{\tan\theta} = \dfrac{1}{4}$

(b) $\sec\theta = \sqrt{1 + \tan^2\theta} = \sqrt{1 + 16} = \sqrt{17}$

(c) $\cos\theta = \dfrac{1}{\sec\theta} = \dfrac{1}{\sqrt{17}} = \dfrac{\sqrt{17}}{17}$

(d) $\csc\theta = \sqrt{1 + \cot^2\theta} = \sqrt{1 + \dfrac{1}{16}} = \dfrac{\sqrt{17}}{4}$

46. $\csc\theta = 5$

(a) $\sin\theta = \dfrac{1}{\csc\theta} = \dfrac{1}{5}$

(b) $\cot\theta = \sqrt{\csc^2\theta - 1} = \sqrt{25 - 1} = 2\sqrt{6}$

(c) $\tan\theta = \dfrac{1}{\cot\theta} = \dfrac{1}{2\sqrt{6}} = \dfrac{\sqrt{6}}{12}$

(d) $\sec(90° - \theta) = \csc\theta = 5$

48. $\csc 11° \approx 5.24$

50. $\sec 79.3° \approx 5.39$

52. $\cos 78°\,11'\,58'' = \cos\left(78 + \dfrac{11}{60} + \dfrac{58}{3600}\right)° \approx 0.20$

54.

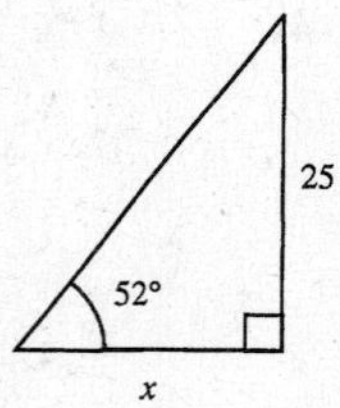

$\tan 52° = \dfrac{25}{x}$

$x = \dfrac{25}{\tan 52°} \approx 19.5$ feet

56. $(x, y) = (3, -4)$

$r = \sqrt{3^2 + (-4)^2} = 5$

$\sin\theta = \dfrac{y}{r} = -\dfrac{4}{5}$ $\qquad \csc\theta = \dfrac{r}{y} = -\dfrac{5}{4}$

$\cos\theta = \dfrac{x}{r} = \dfrac{3}{5}$ $\qquad \sec\theta = \dfrac{r}{x} = \dfrac{5}{3}$

$\tan\theta = \dfrac{y}{x} = -\dfrac{4}{3}$ $\qquad \cot\theta = \dfrac{x}{y} = -\dfrac{3}{4}$

58. $(x, y) = \left(-\dfrac{10}{3}, -\dfrac{2}{3}\right)$

$r = \sqrt{\left(-\dfrac{10}{3}\right)^2 + \left(-\dfrac{2}{3}\right)^2} = \dfrac{2\sqrt{26}}{3}$

$\sin\theta = \dfrac{y}{r} = \dfrac{-\frac{2}{3}}{\frac{2\sqrt{26}}{3}} = -\dfrac{\sqrt{26}}{26}$ $\qquad \csc\theta = \dfrac{r}{y} = \dfrac{\frac{2\sqrt{26}}{3}}{-\frac{2}{3}} = -\sqrt{26}$

$\cos\theta = \dfrac{x}{r} = \dfrac{-\frac{10}{3}}{\frac{2\sqrt{26}}{3}} = -\dfrac{5\sqrt{26}}{26}$ $\qquad \sec\theta = \dfrac{r}{x} = \dfrac{\frac{2\sqrt{26}}{3}}{-\frac{10}{3}} = -\dfrac{\sqrt{26}}{5}$

$\tan\theta = \dfrac{y}{x} = \dfrac{-\frac{2}{3}}{-\frac{10}{3}} = \dfrac{1}{5}$ $\qquad \cot\theta = \dfrac{x}{y} = \dfrac{-\frac{10}{3}}{-\frac{2}{3}} = 5$

60. $(x, y) = (0.3, 0.4)$

$r = \sqrt{(0.3)^2 + (0.4)^2} = 0.5$

$\sin\theta = \frac{y}{r} = \frac{0.4}{0.5} = \frac{4}{5} = 0.8$

$\cos\theta = \frac{x}{r} = \frac{0.3}{0.5} = \frac{3}{5} = 0.6$

$\tan\theta = \frac{y}{x} = \frac{0.4}{0.3} = \frac{4}{3} \approx 1.33$

$\csc\theta = \frac{r}{y} = \frac{0.5}{0.4} = \frac{5}{4} = 1.25$

$\sec\theta = \frac{r}{x} = \frac{0.5}{0.3} = \frac{5}{3} \approx 1.67$

$\cot\theta = \frac{x}{y} = \frac{0.3}{0.4} = \frac{3}{4} = 0.75$

62. $(x', y') = (-2x, -3x), x > 0$

$r = \sqrt{(-2x)^2 + (-3x)^2} = \sqrt{13}x$

$\sin\theta = \frac{y'}{r} = \frac{-3x}{\sqrt{13}x} = -\frac{3\sqrt{13}}{13}$

$\cos\theta = \frac{x'}{r} = \frac{-2x}{\sqrt{13}x} = -\frac{2\sqrt{13}}{13}$

$\tan\theta = \frac{y'}{x'} = \frac{-3x}{-2x} = \frac{3}{2}$

$\csc\theta = \frac{r}{y'} = \frac{\sqrt{13}x}{-3x} = -\frac{\sqrt{13}}{3}$

$\sec\theta = \frac{r}{x'} = \frac{\sqrt{13}x}{-2x} = -\frac{\sqrt{13}}{2}$

$\cot\theta = \frac{x'}{y'} = \frac{-2x}{-3x} = \frac{2}{3}$

64. $\csc\theta = \frac{3}{2}, \cos\theta < 0$

θ is in Quadrant II.

$\sin\theta = \frac{1}{\csc\theta} = \frac{2}{3}$

$\cos\theta = -\sqrt{1 - \sin^2\theta} = -\frac{\sqrt{5}}{3}$

$\tan\theta = \frac{\sin\theta}{\cos\theta} = -\frac{2\sqrt{5}}{5}$

$\sec\theta = \frac{1}{\cos\theta} = -\frac{3\sqrt{5}}{5}$

$\cot\theta = \frac{1}{\tan\theta} = -\frac{\sqrt{5}}{2}$

66. $\tan\theta = \frac{5}{4}, \cos\theta < 0$

θ is in Quadrant III.

$\sec\theta = -\sqrt{1 + \tan^2\theta} = -\sqrt{1 + \left(\frac{25}{16}\right)} = -\frac{\sqrt{41}}{4}$

$\cos\theta = \frac{1}{\sec\theta} = -\frac{4\sqrt{41}}{41}$

$\sin\theta = -\sqrt{1 - \cos^2\theta} = -\sqrt{1 - \frac{16}{41}} = -\frac{5\sqrt{41}}{41}$

$\csc\theta = \frac{1}{\sin\theta} = -\frac{\sqrt{41}}{5}$

$\cot\theta = \frac{1}{\tan\theta} = \frac{4}{5}$

68. $\sin\theta = -\frac{2}{4} = -\frac{1}{2}, \cos\theta > 0$

θ is in Quadrant IV.

$\csc\theta = \frac{1}{\sin\theta} = -2$

$\cos\theta = \sqrt{1 - \sin^2\theta} = \sqrt{1 - \left(\frac{1}{4}\right)} = \frac{\sqrt{3}}{2}$

$\sec\theta = \frac{1}{\cos\theta} = \frac{2\sqrt{3}}{3}$

$\tan\theta = \frac{\sin\theta}{\cos\theta} = -\frac{\sqrt{3}}{3}$

$\cot\theta = \frac{1}{\tan\theta} = -\sqrt{3}$

70. $\sin\frac{\pi}{4} = \frac{\sqrt{2}}{2}$

$\cos\frac{\pi}{4} = \frac{\sqrt{2}}{2}$

$\tan\frac{\pi}{4} = \frac{\frac{\sqrt{2}}{2}}{\frac{\sqrt{2}}{2}} = 1$

72. $\sin\left(-\frac{5\pi}{4}\right) = \frac{\sqrt{2}}{2}$

$\cos\left(-\frac{5\pi}{4}\right) = -\frac{\sqrt{2}}{2}$

$\tan\left(-\frac{5\pi}{4}\right) = \frac{\frac{\sqrt{2}}{2}}{-\frac{\sqrt{2}}{2}} = -1$

74. $\sin(-150°) = -\frac{1}{2}$

$\cos(-150°) = -\frac{\sqrt{3}}{2}$

$\tan(-150°) = \frac{-\frac{1}{2}}{-\frac{\sqrt{3}}{2}} = \frac{\sqrt{3}}{3}$

76. $\sin(315°) = -\dfrac{\sqrt{2}}{2}$

$\cos(315°) = \dfrac{\sqrt{2}}{2}$

$\tan(315°) = \dfrac{-\dfrac{\sqrt{2}}{2}}{\dfrac{\sqrt{2}}{2}} = -1$

78. $\tan 3 \approx -0.14$

80. $\cot(-4.8) \approx 0.09$

82. $\tan\left(\dfrac{-25\pi}{7}\right) \approx 4.38$

84. $y = \cos x$

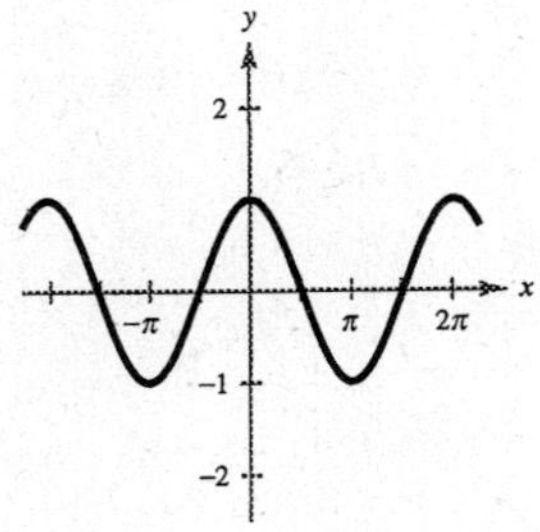

86. $f(x) = 8\cos\left(-\dfrac{x}{4}\right)$

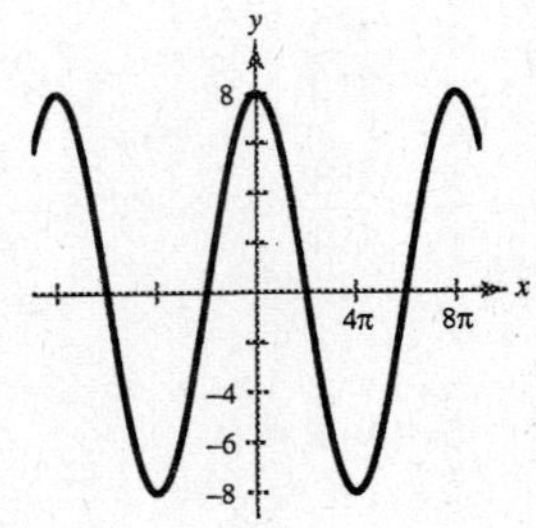

88. $y = -4 - \cos \pi x$

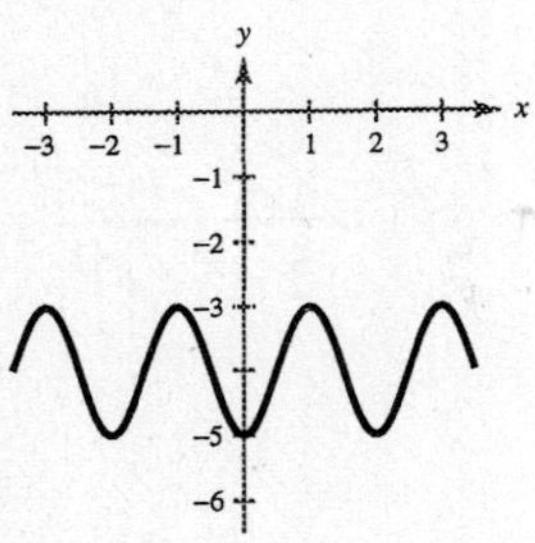

90. $g(t) = 3\cos(t + \pi)$

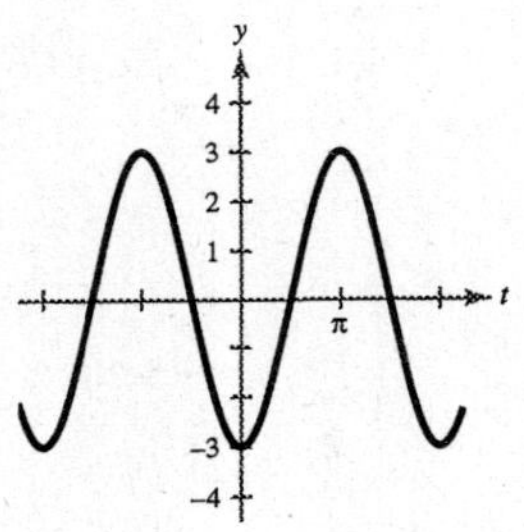

92. $y = a \cos bx$

Amplitude: $2 \Rightarrow a = 2$

Period:

$\dfrac{1}{264}$ second $\Rightarrow b = \dfrac{2\pi}{\frac{1}{264}} = 528\pi$

$y = 2\cos(528\pi x)$

94. $f(t) = \tan\left(t - \dfrac{\pi}{4}\right)$

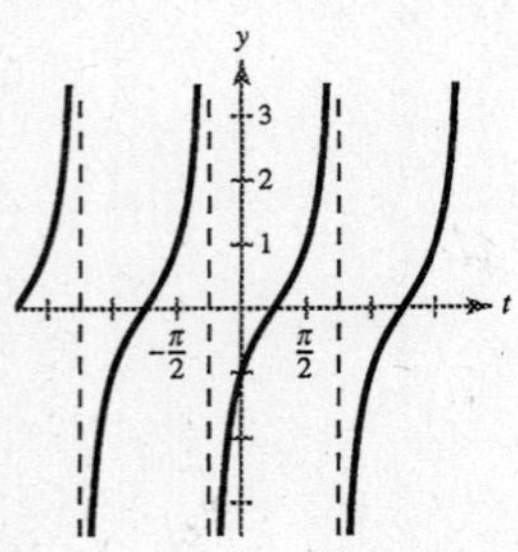

96. $g(t) = 2 \cot 2t$

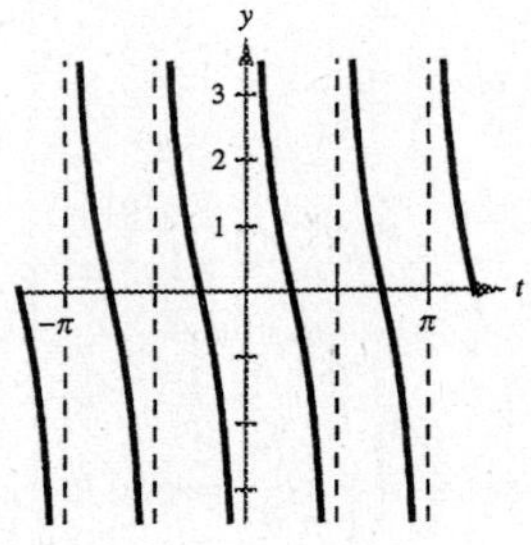

98. $h(t) = \sec\left(t - \dfrac{\pi}{4}\right)$

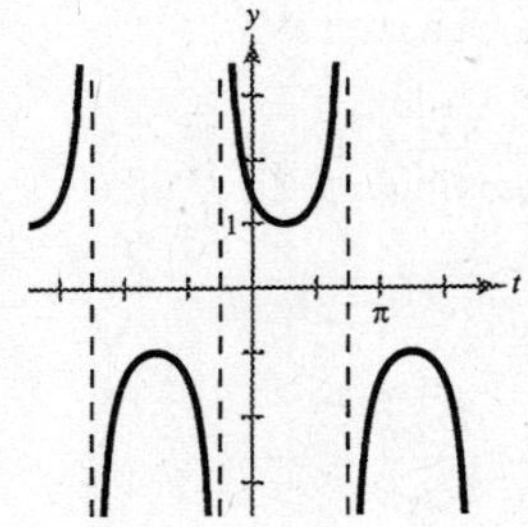

100. $f(t) = 3\csc\left(2t + \frac{\pi}{4}\right)$

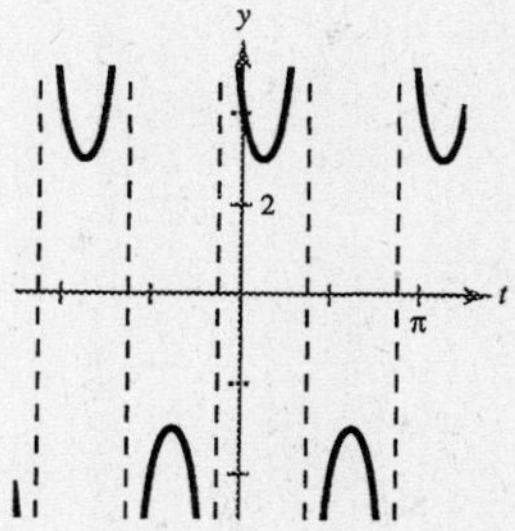

102. $g(x) = x^4 \cos x$

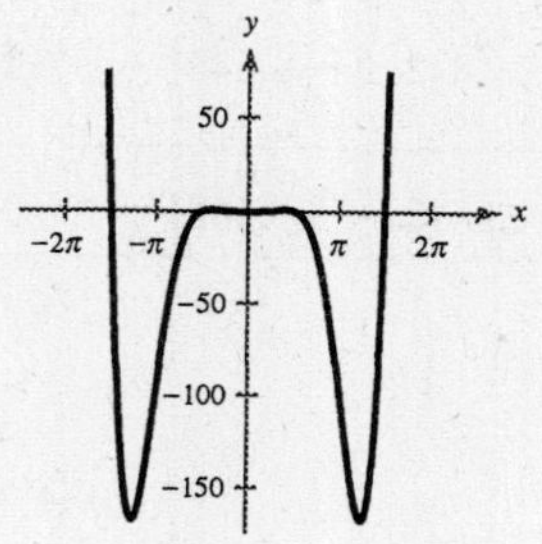

104. $\arcsin(-1) = -\frac{\pi}{2}$

106. $\arcsin(0.213) \approx 0.21$

108. $\sin^{-1}(0.89) \approx 1.10$

110. $\arccos\left(\frac{\sqrt{2}}{2}\right) = \frac{\pi}{4}$

112. $\cos^{-1}\left(\frac{\sqrt{3}}{2}\right) = \frac{\pi}{6}$

114. $\arccos(-0.888) \approx 2.66$

116. $\arctan(2.34) \approx 1.17$

118. $\arctan(99.1) \approx 1.56$

120. $\tan^{-1}(8.2) \approx 1.45$

122. $\cos(\arccos 0.25) = 0.25$

124. $\arccos[\cos(-5\pi)] = \arccos(-1) = \pi$

126. Let $u = \arccos \frac{3}{5}$

$\tan\left(\arccos \frac{3}{5}\right) = \tan u = \frac{4}{3}$

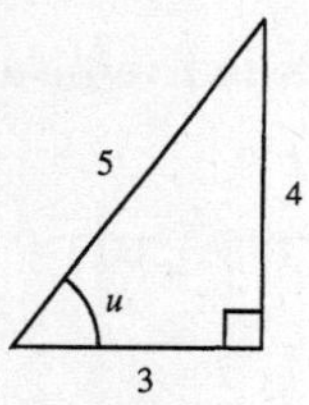

128. Let $u = \arcsin\left(-\frac{12}{13}\right)$

$\cot\left[\arcsin\left(-\frac{12}{13}\right)\right] = \cot u$

$= -\frac{5}{12}$

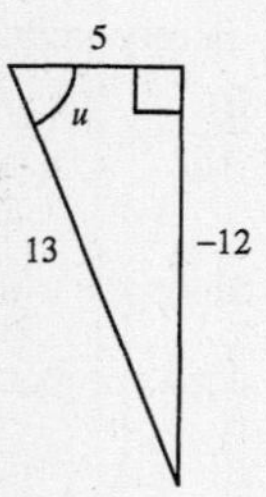

130. $\tan 21° = \frac{h}{25}$

$h = 25 \tan 21° \approx 9.6$ feet

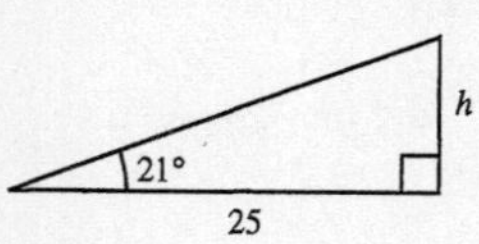

132. Amplitude: $\frac{1.5}{2} = 0.75$ inches

Period: 3 seconds

$d = a \cos bt$

$a = 0.75$

$b = \frac{2\pi}{3}$

$d = 0.75 \cos\left(\frac{2\pi t}{3}\right)$

134. False. The inverse sine, $y = \arcsin x$, cannot be defined as a function over any interval that is greater than the interval defined as $-\frac{\pi}{2} \le x \le \frac{\pi}{2}$.

136. False. The range of arctan is $\left(-\frac{\pi}{2}, \frac{\pi}{2}\right)$, so $\arctan(-1) = -\frac{\pi}{4}$.

138. $y = -3 \sin x$ matches graph (a).

Period: 2π

Amplitude: 3

140. $y = 2 \sin \frac{x}{2}$ matches graph (c).

Period: 4π

Amplitude: 2

142. (a)

θ	0.1	0.4	0.7	1.0	1.3
$\tan\left(\theta - \frac{\pi}{2}\right)$	-9.9666	-2.3652	-1.1872	-0.6421	-0.2776
$-\cot\theta$	-9.9666	-2.3652	-1.1872	-0.6421	-0.2776

(b) $\tan\left(\theta - \frac{\pi}{2}\right) = -\cot\theta$

144. $A = \frac{1}{2}r^2\theta,\ s = r\theta$

(a) $A = \frac{1}{2}r^2(0.8) = 0.4r^2, r > 0$

$s = r(0.8) = 0.8r, r > 0$

As r increases, the area function increases more rapidly.

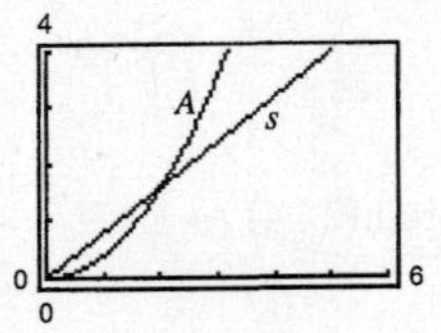

(b) $A = \frac{1}{2}(10)^2\theta = 50\theta, \theta > 0$

$s = 10\theta, \theta > 0$

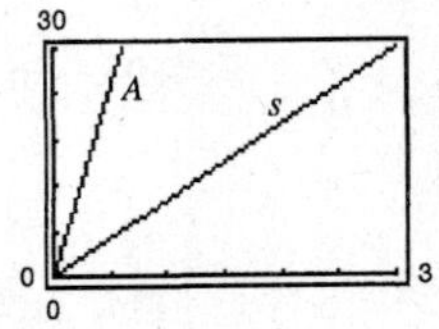

Problem Solving for Chapter 1

Solutions to Even-Numbered Exercises

2. Gear 1: $\frac{24}{32}(360°) = 270° = \frac{\pi}{2}$ radians

Gear 2: $\frac{24}{26}(360°) \approx 332.308° \approx 5.80$ radians

Gear 3: $\frac{24}{22}(360°) \approx 392.727° \approx 6.85$ radians

Gear 4: $\frac{40}{32}(360°) = 450° = \frac{5\pi}{2}$ radians

Gear 5: $\frac{24}{19}(360°) \approx 454.737° \approx 7.94$ radians

4. (a) $\triangle ABC$, $\triangle ADE$, and $\triangle AFG$ are all similar triangles since they all have the same angles. $\angle A$ is part of all three triangles and $\angle C = \angle E = \angle G = 90°$. Thus, $\angle B = \angle D = \angle F$.

(b) Since the triangles are similar, the ratios of corresponding sides are equal.

$$\frac{BC}{AB} = \frac{DE}{AD} = \frac{FG}{AF}$$

(c) Since the ratios: $\frac{\text{opp}}{\text{hyp}} = \frac{BC}{AB} = \frac{DE}{AD} = \frac{FG}{AF} = \sin A$ it does not matter which triangle is used to calculate $\sin A$.Any triangle similar to these three triangles could be used to find $\sin A$. The value of $\sin A$ would not change.

(d) Since the values of all six trigonometric functions can be found by taking the ratios of the sides of a right triangle, similar triangles would yield the same values.

6. Given: f is an even function and g is an odd function.

(a) $h(x) = [f(x)]^2$

$h(-x) = [f(-x)]^2$

$= [f(x)]^2$ since f is even

$= h(x)$

Thus, h is an even function.

(b) $h(x) = [g(x)]^2$

$h(-x) = [g(-x)]^2$

$= [-g(x)]^2$ since g is odd

$= [g(x)]^2$

$= h(x)$

Thus, h is an even function.

Conjecture: The square of either an even function or an odd function is an even function.

8. Physical (23 days): $P = \sin \frac{2\pi t}{23}, t \geq 0$

Emotional (28 days): $E = \sin \frac{2\pi t}{28}, t \geq 0$

Intellectual (33 days): $I = \sin \frac{2\pi t}{33}, t \geq 0$

(a)

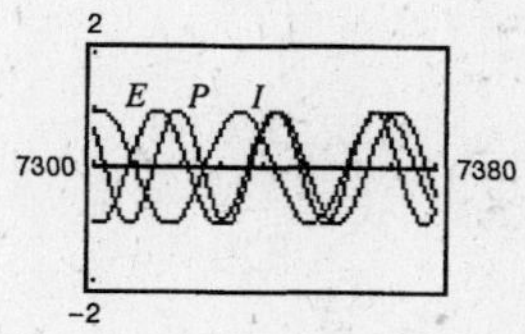

(b) Number of days since birth until September 1, 2004:

$t = 365 \times 20$	+	5	+	11	+	31	+	1
20 years	+	leap years	+	remaining July days	+	August days		day in September

$t = 7348$

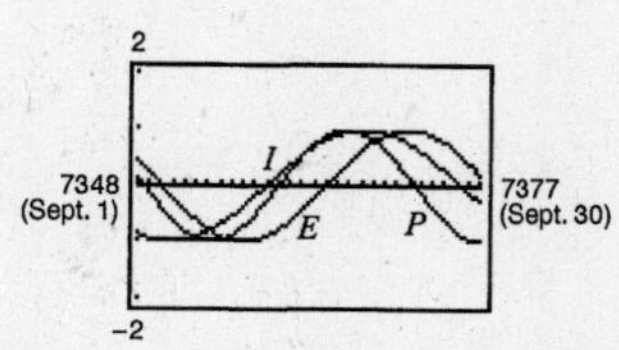

All three drop early in the month, then peak toward the middle of the month, and drop again toward the latter part of the month.

(c) For September 22, 2004, use $t = 7369$.

$P \approx 0.631$

$E \approx 0.901$

$I \approx 0.945$

10. (a) Both graphs have a period of 2 and intersect when $x = 5.35$. They should also intersect when $x = 5.35 - 2 = 3.35$ and $x = 5.35 + 2 = 7.35$.

(b) The graphs intersect when $x = 5.35 - 3(2) = -0.65$.

(c) Since $13.35 = 5.35 + 4(2)$ and $-4.65 = 5.35 - 5(2)$ the graphs will intersect again at these values. Therefore $f(13.35) = g(-4.65)$.

12.

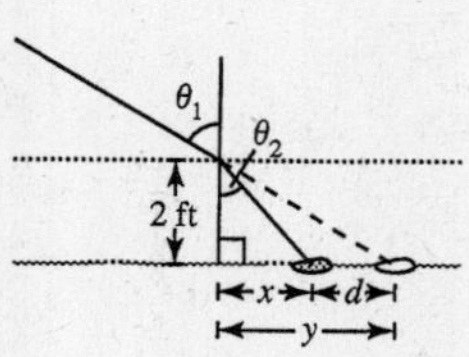

(a) $\dfrac{\sin \theta_1}{\sin \theta_2} = 1.333$

$\sin \theta_2 = \dfrac{\sin \theta_1}{1.333} = \dfrac{\sin 60°}{1.333} \approx 0.6497$

$\theta_2 \approx 40.52°$

(b) $\tan \theta_2 = \dfrac{x}{2} \Rightarrow x = 2 \tan 40.52° \approx 1.71$ feet

$\tan \theta_1 = \dfrac{y}{2} \Rightarrow y = 2 \tan 60° \approx 3.46$ feet

(c) $d = y - x = 3.46 - 1.71 = 1.75$ feet

(d) As you more closer to the rock, θ_1 decreases, which causes y to decrease, which in turn causes d to decrease.

CHAPTER 2
Analytic Trigonometry

CHAPTER 2
Analytic Trigonometry

Section 2.1 Using Fundamental Identities

Solutions to Even-Numbered Exercises

2. $\tan x = \dfrac{\sqrt{3}}{3}, \cos x = -\dfrac{\sqrt{3}}{2}$

x is in Quadrant III.

$$\sin x = -\sqrt{1 - \left(-\frac{\sqrt{3}}{2}\right)^2} = -\sqrt{\frac{1}{4}} = -\frac{1}{2}$$

$$\csc x = \frac{1}{\sin x} = -2$$

$$\sec x = \frac{1}{\cos x} = -\frac{2}{\sqrt{3}} = -\frac{2\sqrt{3}}{3}$$

$$\cot x = \frac{1}{\tan x} = \frac{3}{\sqrt{3}} = \sqrt{3}$$

4. $\csc \theta = \dfrac{5}{3}, \tan \theta = \dfrac{3}{4}$

θ is in Quadrant I.

$$\sin \theta = \frac{1}{\csc \theta} = \frac{3}{5}$$

$$\cos \theta = \frac{\sin \theta}{\tan \theta} = \frac{3}{5} \cdot \frac{4}{3} = \frac{4}{5}$$

$$\sec \theta = \frac{1}{\cos \theta} = \frac{5}{4}$$

$$\cot \theta = \frac{1}{\tan \theta} = \frac{4}{3}$$

6. $\cot \phi = -3, \sin \phi = \dfrac{\sqrt{10}}{10}$

ϕ is in Quadrant II.

$$\cos \phi = \cot \phi \sin \phi = -\frac{3\sqrt{10}}{10}$$

$$\tan \phi = \frac{1}{\cot \phi} = -\frac{1}{3}$$

$$\csc \phi = \frac{1}{\sin \phi} = \sqrt{10}$$

$$\sec \phi = \frac{1}{\cos \phi} = -\frac{10}{3\sqrt{10}} = -\frac{\sqrt{10}}{3}$$

8. $\cos\left(\dfrac{\pi}{2} - x\right) = \dfrac{3}{5}, \cos x = \dfrac{4}{5}$

x is in Quadrant I.

$$\sin x = \sqrt{1 - \left(\frac{4}{5}\right)^2} = \frac{3}{5}$$

$$\tan x = \frac{\sin x}{\cos x} = \frac{3}{5} \cdot \frac{5}{4} = \frac{3}{4}$$

$$\csc x = \frac{1}{\sin x} = \frac{5}{3}$$

$$\sec x = \frac{1}{\cos x} = \frac{5}{4}$$

$$\cot x = \frac{1}{\tan x} = \frac{4}{3}$$

10. $\sec x = 4, \sin x > 0$

x is in Quadrant I.

$$\cos x = \frac{1}{\sec x} = \frac{1}{4}$$

$$\sin x = \sqrt{1 - \left(\frac{1}{4}\right)^2} = \frac{\sqrt{15}}{4}$$

$$\tan x = \frac{\sin x}{\cos x} = \frac{\sqrt{15}}{4} \cdot \frac{4}{1} = \sqrt{15}$$

$$\csc x = \frac{1}{\sin x} = \frac{4}{\sqrt{15}} = \frac{4\sqrt{15}}{15}$$

$$\cot x = \frac{1}{\tan x} = \frac{1}{\sqrt{15}} = \frac{\sqrt{15}}{15}$$

14. $\tan \theta$ is undefined, $\sin \theta > 0$.

$$\theta = \frac{\pi}{2}$$

$$\tan \theta = \frac{\sin \theta}{\cos \theta} \text{ is undefined} \Rightarrow \cos \theta = 0$$

$$\sin \theta = \sqrt{1 - 0^2} = 1$$

$$\csc \theta = \frac{1}{\sin \theta} = 1$$

$$\sec \theta = \frac{1}{\cos \theta} \text{ is undefined.}$$

$$\cot \theta = \frac{\cos \theta}{\sin \theta} = \frac{0}{1} = 0$$

18. $(1 - \cos^2 x)(\csc x) = (\sin^2 x)\dfrac{1}{\sin x} = \sin x$

Matches (f).

22. $\cos^2 x(\sec^2 x - 1) = \cos^2 x(\tan^2 x)$

$$= \cos^2 x\left(\frac{\sin^2 x}{\cos^2 x}\right)$$

$$= \sin^2 x$$

Matches (c).

26. $\dfrac{\cos^2\left[\left(\frac{\pi}{2}\right) - x\right]}{\cos x} = \dfrac{\sin^2 x}{\cos x} = \dfrac{\sin x}{\cos x}\sin x$

$$= \tan x \sin x$$

Matches (d).

12. $\csc \theta = -5, \cos \theta < 0$

θ is in Quadrant III.

$$\sin \theta = \frac{1}{\csc \theta} = -\frac{1}{5}$$

$$\cos \theta = -\sqrt{1 - \left(-\frac{1}{5}\right)^2} = \frac{-2\sqrt{6}}{5}$$

$$\tan \theta = \frac{\sin \theta}{\cos \theta} = -\frac{1}{5} \cdot -\frac{5}{2\sqrt{6}} = \frac{\sqrt{6}}{12}$$

$$\sec \theta = \frac{1}{\cos \theta} = -\frac{5}{2\sqrt{6}} = -\frac{5\sqrt{6}}{12}$$

$$\cot \theta = \frac{1}{\tan \theta} = \frac{12}{\sqrt{6}} = 2\sqrt{6}$$

16. $\tan x \csc x = \dfrac{\sin x}{\cos x} \cdot \dfrac{1}{\sin x} = \dfrac{1}{\cos x} = \sec x$

Matches (a).

20. $\dfrac{\sin\left[\left(\frac{\pi}{2}\right) - x\right]}{\cos\left[\left(\frac{\pi}{2}\right) - x\right]} = \dfrac{\cos x}{\sin x} = \cot x$

Matches (c).

24. $\cot x \sec x = \dfrac{\cos x}{\sin x} \cdot \dfrac{1}{\cos x} = \dfrac{1}{\sin x} = \csc x$

Matches (a).

28. $\cos \beta \tan \beta = \cos \beta \dfrac{\sin \beta}{\cos \beta} = \sin \beta$

30. $\sec^2 x(1 - \sin^2 x) = \sec^2 x - \sec^2 x \sin^2 x$

$$= \sec^2 x - \frac{1}{\cos^2 x} \cdot \sin^2 x$$

$$= \sec^2 x - \frac{\sin^2 x}{\cos^2 x}$$

$$= \sec^2 x - \tan^2 x$$

$$= 1$$

32. $\dfrac{\csc \theta}{\sec \theta} = \dfrac{\dfrac{1}{\sin \theta}}{\dfrac{1}{\cos \theta}} = \dfrac{\cos \theta}{\sin \theta} = \cot \theta$

34. $\dfrac{1}{\tan^2 x + 1} = \dfrac{1}{\sec^2 x} = \dfrac{1}{\dfrac{1}{\cos^2 x}} = \cos^2 x$

36. $\dfrac{\tan^2 \theta}{\sec^2 \theta} = \dfrac{\sin^2 \theta}{\cos^2 \theta} \cdot \dfrac{1}{\sec^2 \theta} = \dfrac{\sin^2 \theta}{\cos^2 \theta} \cdot \dfrac{1}{\dfrac{1}{\cos^2 \theta}} = \dfrac{\sin^2 \theta \cos^2 \theta}{\cos^2 \theta} = \sin^2 \theta$

38. $\cot\left(\dfrac{\pi}{2} - x\right) \cos x = \tan x \cos x = \dfrac{\sin x}{\cos x} \cdot \cos x = \sin x$

40. $(\cos t)(1 + \tan^2 t) = (\cos t)(\sec^2 t) = \dfrac{\cos t}{\cos^2 t} = \dfrac{1}{\cos t} = \sec t$

42. $\csc \phi \tan \phi + \sec \phi = \dfrac{1}{\sin \phi} \cdot \dfrac{\sin \phi}{\cos \phi} + \sec \phi$

$$= \frac{1}{\cos \phi} + \sec \phi = 2 \sec \phi$$

44. $\sin \theta \sec \theta + \cos \theta \csc \theta = \dfrac{\sin \theta}{\cos \theta} + \dfrac{\cos \theta}{\sin \theta}$

$$= \frac{\sin^2 \theta + \cos^2 \theta}{\cos \theta \sin \theta}$$

$$= \frac{1}{\cos \theta \sin \theta}$$

$$= \sec \theta \csc \theta$$

46. $\sin^2 x \csc^2 x - \sin^2 x = \sin^2 x(\csc^2 x - 1)$

$$= \sin^2 x \cot^2 x$$

$$= \sin^2 x \cdot \frac{\cos^2 x}{\sin^2 x}$$

$$= \cos^2 x$$

48. $\cos^2 x + \cos^2 x \tan^2 x = \cos^2 x(1 + \tan^2 x)$

$$= \cos^2 x + (\sec^2 x)$$

$$= \cos^2 x\left(\frac{1}{\cos^2 x}\right)$$

$$= 1$$

50. $\dfrac{\cos^2 x - 4}{\cos x - 2} = \dfrac{(\cos x + 2)(\cos x - 2)}{\cos x - 2}$

$$= \cos x + 2$$

52. $1 - 2\cos^2 x + \cos^4 x = (1 - \cos^2 x)^2$

$$= (\sin^2 x)^2$$

$$= \sin^4 x$$

54. $\sec^4 x - \tan^4 x = (\sec^2 x + \tan^2 x)(\sec^2 x - \tan^2 x)$

$$= (\sec^2 x + \tan^2 x)(1)$$

$$= \sec^2 x + \tan^2 x$$

56. $\sec^3 x - \sec^2 x - \sec x + 1 = \sec^2 x(\sec x - 1) - (\sec x - 1)$
$= (\sec^2 x - 1)(\sec x - 1)$
$= \tan^2 x(\sec x - 1)$

58. $(\cot x + \csc x)(\cot x - \csc x) = \cot^2 x - \csc^2 x$
$= -1$

60. $(3 - 3\sin x)(3 + 3\sin x) = 9 - 9\sin^2 x$
$= 9(1 - \sin^2 x)$
$= 9\cos^2 x$

62. $\dfrac{1}{\sec x + 1} - \dfrac{1}{\sec x - 1} = \dfrac{\sec x - 1 - (\sec x + 1)}{(\sec x + 1)(\sec x - 1)}$
$= \dfrac{\sec x - 1 - \sec x - 1}{\sec^2 x - 1}$
$= \dfrac{-2}{\tan^2 x}$
$= -2\left(\dfrac{1}{\tan^2 x}\right) = -2\cot^2 x$

64. $\tan x - \dfrac{\sec^2 x}{\tan x} = \dfrac{\tan^2 x - \sec^2 x}{\tan x}$
$= \dfrac{-1}{\tan x} = -\cot x$

66. $\dfrac{5}{\tan x + \sec x} \cdot \dfrac{\tan x - \sec x}{\tan x - \sec x} = \dfrac{5(\tan x - \sec x)}{\tan^2 x - \sec^2 x}$
$= \dfrac{5(\tan x - \sec x)}{-1}$
$= 5(\sec x - \tan x)$

68. $\dfrac{\tan^2 x}{\csc x + 1} \cdot \dfrac{\csc x - 1}{\csc x - 1} = \dfrac{\tan^2 x(\csc x - 1)}{\csc^2 x - 1}$
$= \dfrac{\tan^2 x(\csc x - 1)}{\cot^2 x}$
$= \tan^2 x(\csc x - 1)\tan^2 x$
$= \tan^4 x(\csc x - 1)$

70. $y_1 = \sec x - \cos x$, $y_2 = \sin x \tan x$

x	0.2	0.4	0.6	0.8	1.0	1.2	1.4
y_1	0.0403	0.1646	0.3863	0.7386	1.3105	2.3973	5.7135
y_2	0.0403	0.1646	0.3863	0.7386	1.3105	2.3973	5.7135

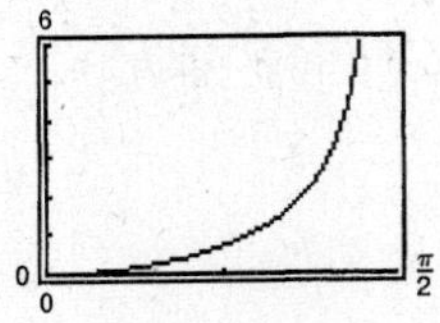

It appears that $y_1 = y_2$.

72. $y_1 = \sec^4 x - \sec^2 x$, $y_2 = \tan^2 x + \tan^4 x$

x	0.2	0.4	0.6	0.8	1.0	1.2	1.4
y_1	0.0428	0.2107	0.6871	2.1841	8.3087	50.3869	1163.6143
y_2	0.0428	0.2107	0.6871	2.1841	8.3087	50.3869	1163.6143

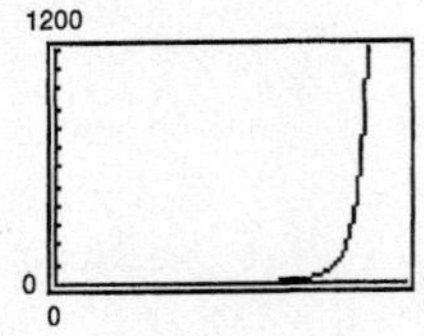

It appears that $y_1 = y_2$.

74. $y_1 = \sec x \csc x - \tan x$

y_1 and $y_2 = \sin x$

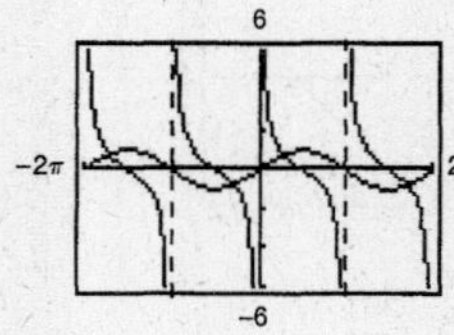

y_1 and $y_2 = \cos x$

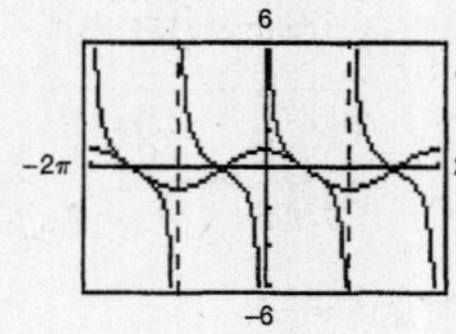

y_1 and $y_2 = \tan x$

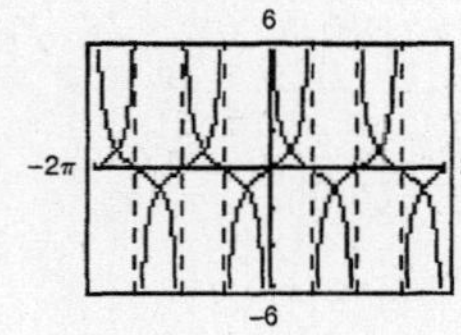

y_1 and $y_2 = \csc x$

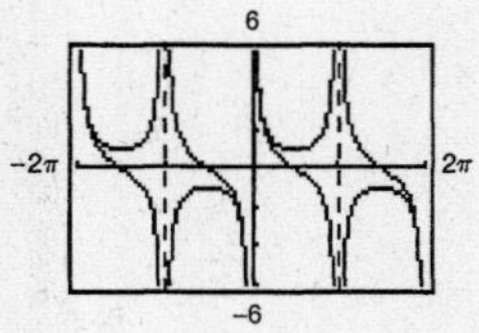

y_1 and $y_2 = \sec x$

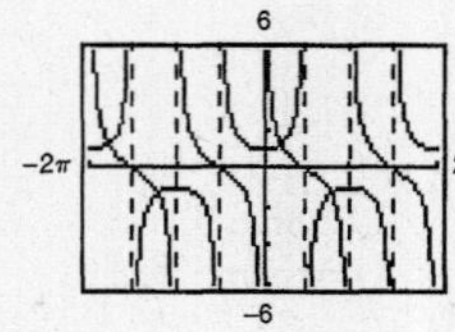

y_1 and $y_2 = \cot x$

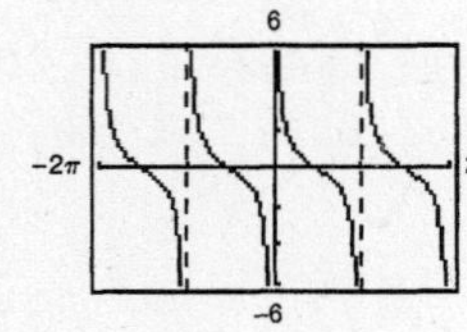

It appears that $\sec x \csc x - \tan x = \cot x$.

76. $y_1 = \dfrac{1}{2}\left(\dfrac{1 + \sin\theta}{\cos\theta} + \dfrac{\cos\theta}{1 + \sin\theta}\right)$

y_1 and $y_2 = \sin\theta$

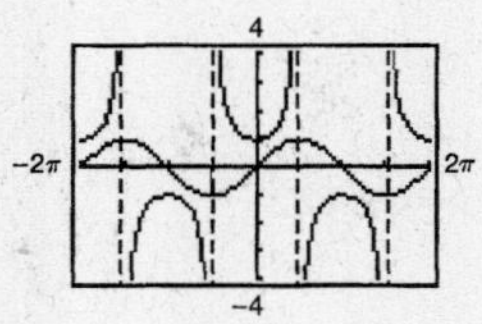

y_1 and $y_2 = \cos\theta$

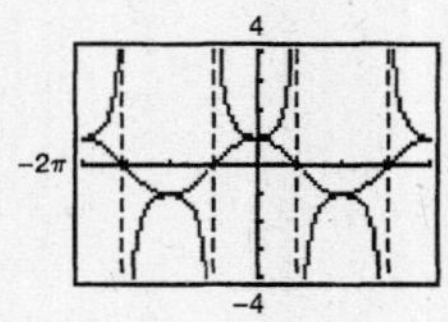

y_1 and $y_2 = \tan\theta$

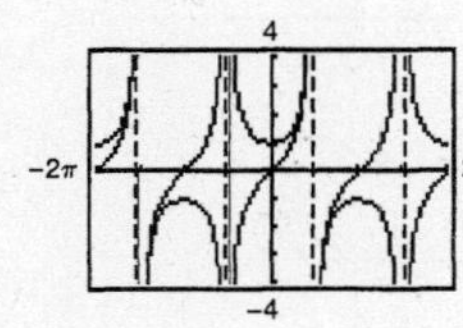

y_1 and $y_2 = \dfrac{1}{\sin\theta} = \csc\theta$

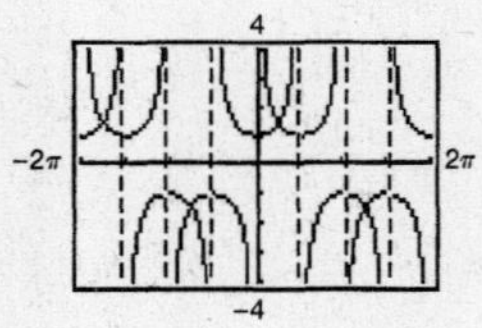

y_1 and $y_2 = \dfrac{1}{\cos\theta} = \sec\theta$

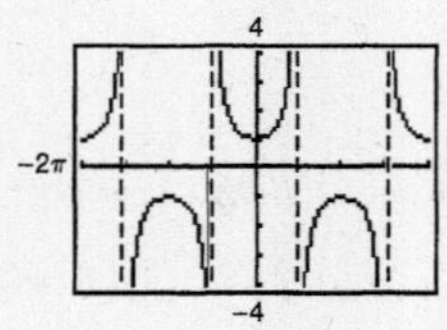

y_1 and $y_2 = \dfrac{1}{\tan\theta} = \cot\theta$

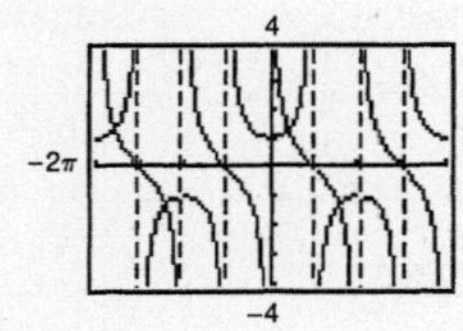

It appears that $\dfrac{1}{2}\left(\dfrac{1 + \sin\theta}{\cos\theta} + \dfrac{\cos\theta}{1 + \sin\theta}\right) = \sec\theta.$

78. Let $x = 2\cos\theta$

$$\begin{aligned}\sqrt{64 - 16x^2} &= \sqrt{64 - 16(2\cos\theta)^2}\\ &= \sqrt{64(1 - \cos^2\theta)}\\ &= \sqrt{64\sin^2\theta}\\ &= 8\sin\theta\end{aligned}$$

80. Let $x = 2\sec\theta$.

$$\begin{aligned}\sqrt{x^2 - 4} &= \sqrt{(2\sec\theta)^2 - 4}\\ &= \sqrt{4(\sec^2\theta - 1)}\\ &= \sqrt{4\tan^2\theta}\\ &= 2\tan\theta\end{aligned}$$

82. Let $x = 10\tan\theta$.

$$\begin{aligned}\sqrt{x^2 + 100} &= \sqrt{(10\tan\theta)^2 + 100}\\ &= \sqrt{100(\tan^2\theta + 1)}\\ &= \sqrt{100\sec^2\theta}\\ &= 10\sec\theta\end{aligned}$$

84. $x = 6\sin\theta$

$$3 = \sqrt{36 - x^2}$$
$$= \sqrt{36 - (6\sin\theta)^2}$$
$$= \sqrt{36(1 - \sin^2\theta)}$$
$$= \sqrt{36\cos^2\theta}$$
$$= 6\cos\theta$$

$$\cos\theta = \frac{3}{6} = \frac{1}{2}$$

$$\sin\theta = \pm\sqrt{1 - \cos^2\theta}$$
$$= \pm\sqrt{1 - \left(\frac{1}{2}\right)^2}$$
$$= \pm\sqrt{\frac{3}{4}}$$
$$= \pm\frac{\sqrt{3}}{2}$$

86. $x = 10\cos\theta$

$$-5\sqrt{3} = \sqrt{100 - x^2}$$
$$= \sqrt{100 - (10\cos\theta)^2}$$
$$= \sqrt{100(1 - \cos^2\theta)}$$
$$= \sqrt{100\sin^2\theta}$$
$$= 10\sin\theta$$

$$\sin\theta = -\frac{5\sqrt{3}}{10}$$
$$= -\frac{\sqrt{3}}{2}$$

$$\cos\theta = \sqrt{1 - \sin^2\theta}$$
$$= \sqrt{1 - \left(-\frac{\sqrt{3}}{2}\right)^2}$$
$$= \frac{1}{2}$$

88. $\cos\theta = -\sqrt{1 - \sin^2\theta}$

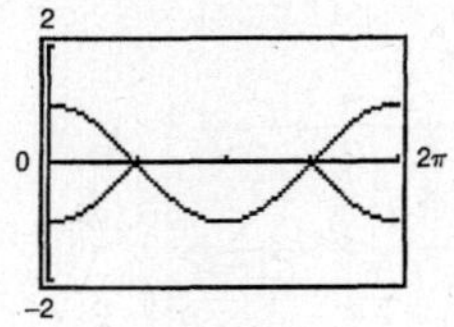

$$\frac{\pi}{2} \le \theta \le \frac{3\pi}{2}$$

90. $\csc\theta = \sqrt{1 + \cot^2\theta}$

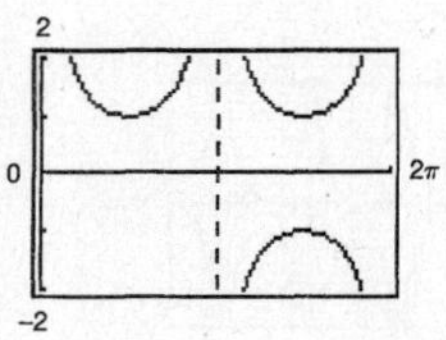

$$0 < \theta < \pi$$

92. $\tan^2\theta + 1 = \sec^2\theta$

(a) $\theta = 346°$

$$(\tan 346°)^2 + 1 \approx 1.0622$$

$$(\sec 346°)^2 = \left(\frac{1}{\cos 346°}\right)^2 \approx 1.0622$$

(b) $\theta = 3.1$

$$(\tan 3.1)^2 + 1 \approx 1.00173$$

$$(\sec 3.1)^2 = \left(\frac{1}{\cos 3.1}\right)^2 \approx 1.00173$$

94. $\sin(-\theta) = -\sin\theta$

(a) $\theta = 250°$

$$\sin(-250°) \approx 0.9397$$
$$-(\sin 250°) \approx 0.9397$$

(b) $\theta = \frac{1}{2}$

$$\sin\left(-\tfrac{1}{2}\right) \approx -0.4794$$
$$-\left(\sin\tfrac{1}{2}\right) \approx -0.4794$$

96. $1 + \cot^2\theta = \csc^2\theta$

(a) $\theta = 240°$

$$1 + \cot^2 240° = 1 + \frac{\cos^2 240°}{\sin^2 240°} \approx 1.333$$

$$\csc^2 240° = \frac{1}{\sin^2 240°} \approx 1.333$$

(b) $\theta = 2.2$

$$1 + \cot^2 2.2 = 1 + \frac{\cos^2 2.2}{\sin^2 2.2} \approx 1.5298$$

$$\csc^2 2.2 = \frac{1}{\sin^2 2.2} \approx 1.5298$$

98. $\cos(-\theta) = \cos\theta$

(a) $\theta = 125°$

$$\cos(-125°) \approx -0.5736$$
$$\cos 125° \approx -0.5736$$

(b) $\theta = \frac{5\pi}{6}$

$$\cos\left(-\frac{5\pi}{6}\right) \approx 0.8660$$

$$\cos\frac{5\pi}{6} \approx -0.8660$$

100. $\csc x \cot x - \cos x = \frac{1}{\sin x}\left(\frac{\cos x}{\sin x}\right) - \cos x$

$= \frac{\cos x}{\sin^2 x} - \cos x$

$= \frac{\cos x - \sin^2 x \cos x}{\sin^2 x}$

$= \frac{\cos x(1 - \sin^2 x)}{\sin^2 x}$

$= \frac{\cos x(1 - \sin^2 x)}{\sin^2 x}$

$= \frac{\cos x \cos^2 x}{\sin^2 x}$

$= \cos x \cot^2 x$

102. False. A cofunction identity can be used to transform a tangent function so that it can be represented by a cotangent function.

104. As $x \longrightarrow 0^+$,

$\cos x \longrightarrow 1$ and $\sec x = \frac{1}{\cos x} \longrightarrow 1.$

106. As $x \longrightarrow \pi^+$,

$\sin x \longrightarrow 0$ and $\csc x = \frac{1}{\sin x} \longrightarrow -\infty.$

108. The equation is **not** an identity.

$\cot \theta = \pm\sqrt{\csc^2 \theta - 1}$

110. The equation is **not** an identity.

$\frac{1}{5 \cos \theta} = \frac{1}{5}\left(\frac{1}{\cos \theta}\right) = \frac{1}{5} \sec \theta \neq 5 \sec \theta$

112. The equation is **not** an identity because the angles may not be the same.

$\sin \theta \csc \phi = \sin \theta\left(\frac{1}{\sin \phi}\right) \neq 1$ unless $\theta = \phi$

114. Divide both sides of $\sin^2 \theta + \cos^2 \theta = 1$ by $\cos^2 \theta$:

$\frac{\sin^2 \theta}{\cos^2 \theta} + \frac{\cos^2 \theta}{\cos^2 \theta} = \frac{1}{\cos^2 \theta}$

$\tan^2 \theta + 1 = \sec^2 \theta$

Divide both sides of $\sin^2 \theta + \cos^2 \theta = 1$ by $\sin^2 \theta$:

$\frac{\sin^2 \theta}{\sin^2 \theta} + \frac{\cos^2 \theta}{\sin^2 \theta} = \frac{1}{\sin^2 \theta}$

$1 + \cot^2 \theta = \csc^2 \theta$

Discussion for remembering identities will vary, but one key is first to learn the identities that concern the sine and cosine functions thoroughly, and then to use these as a basis to establish the other identities when necessary.

116. $(2\sqrt{z} + 3)^2 = (2\sqrt{z})^2 + 2(2\sqrt{z})(3) + (3)^2$

$= 4z + 12\sqrt{z} + 9$

118. $\frac{6x}{x - 4} - \frac{3}{4 - x} = \frac{6x}{x - 4} + \frac{3}{x - 4}$

$= \frac{6x + 3}{x - 4}$

$= \frac{3(2x + 1)}{x - 4}$

120. $\frac{x}{x^2 - 25} + \frac{x^2}{x - 5} = \frac{x}{(x - 5)(x + 5)} + \frac{x^2(x + 5)}{(x - 5)(x + 5)}$

$= \frac{x + x^3 + 5x^2}{(x - 5)(x + 5)}$

$= \frac{x(1 + x^2 + 5x)}{(x - 5)(x + 5)}$

$= \frac{x(x^2 + 5x + 1)}{x^2 - 25}$

Section 2.2 Verifying Trigonometric Identities

Solutions to Even-Numbered Exercises

2. $\sec y \cos y = \left(\dfrac{1}{\cos y}\right)\cos y = 1$

4. $\cot^2 y(\sec^2 y - 1) = \cot^2 y \tan^2 y = 1$

6. $\cos^2 \beta - \sin^2 \beta = \cos^2 \beta - (1 - \cos^2 \beta)$

$= 2\cos^2 \beta - 1$

8. $2 - \sec^2 z = 2 - (1 + \tan^2 z)$

$= 1 - \tan^2 z$

10. $\cos x + \sin x \tan x = \cos x + \sin x\left(\dfrac{\sin x}{\cos x}\right)$

$= \dfrac{\cos^2 x + \sin^2 x}{\cos x}$

$= \dfrac{1}{\cos x}$

$= \sec x$

12. $\cos t(\csc^2 t - 1) = \cos t \cot^2 t$

$= \sin t\left(\dfrac{\cos t}{\sin t}\right)\cot^2 t$

$= \dfrac{1}{\csc t}\cot^3 t$

$= \dfrac{\cot^3 t}{\csc t}$

14. $\dfrac{1}{\tan \beta} + \tan \beta = \dfrac{1 + \tan^2 \beta}{\tan \beta}$

$= \dfrac{\sec^2 \beta}{\tan \beta}$

16. $\sec^6 x(\sec x \tan x) - \sec^4 x(\sec x \tan x) = \sec^4 x(\sec x \tan x)(\sec^2 x - 1)$

$= \sec^4 x(\sec x \tan x)\tan^2 x$

$= \sec^5 x \tan^3 x$

18. $\dfrac{\sec \theta - 1}{1 - \cos \theta} = \dfrac{\sec \theta - 1}{1 - (1/\sec \theta)} \cdot \dfrac{\sec \theta}{\sec \theta}$

$= \dfrac{\sec \theta(\sec \theta - 1)}{\sec \theta - 1}$

$= \sec \theta$

20. $\sec x - \cos x = \dfrac{1}{\cos x} - \cos x$

$= \dfrac{1 - \cos^2 x}{\cos x}$

$= \dfrac{\sin^2 x}{\cos x}$

$= \sin x \cdot \dfrac{\sin x}{\cos x}$

$= \sin x \tan x$

22. $\dfrac{1}{\sin x} - \dfrac{1}{\csc x} = \dfrac{\csc x - \sin x}{\sin x \csc x} = \dfrac{\csc x - \sin x}{1} = \csc x - \sin x$

24. $\dfrac{1 + \sin\theta}{\cos\theta} + \dfrac{\cos\theta}{1 + \sin\theta} = \dfrac{(1 + \sin\theta)^2 + \cos^2\theta}{\cos\theta(1 + \sin\theta)}$

$$= \frac{1 + 2\sin\theta + \sin^2\theta + \cos^2\theta}{\cos\theta(1 + \sin\theta)} = \frac{2 + 2\sin\theta}{\cos\theta(1 + \sin\theta)}$$

$$= \frac{2(1 + \sin\theta)}{\cos\theta(1 + \sin\theta)} = \frac{2}{\cos\theta}$$

$$= 2\sec\theta$$

26. $\cos x - \dfrac{\cos x}{1 - \tan x} = \dfrac{\cos x(1 - \tan x) - \cos x}{1 - \tan x}$

$$= \frac{-\cos x\tan x}{1 - \tan x}$$

$$= \frac{-\cos x(\sin x/\cos x)}{1 - (\sin x/\cos x)} \cdot \frac{\cos x}{\cos x}$$

$$= \frac{-\sin x\cos x}{\cos x - \sin x}$$

$$= \frac{\sin x\cos x}{\sin x - \cos x}$$

28. $\dfrac{\cos[(\pi/2) - x]}{\sin[(\pi/2) - x]} = \dfrac{\sin x}{\cos x} = \tan x$

30. $(1 + \sin y)[1 + \sin(-y)] = (1 + \sin y)(1 - \sin y)$

$$= 1 - \sin^2 y$$

$$= \cos^2 y$$

32. $\dfrac{\tan x + \tan y}{1 - \tan x\tan y} = \dfrac{\dfrac{1}{\cot x} + \dfrac{1}{\cot y}}{1 - \dfrac{1}{\cot x} \cdot \dfrac{1}{\cot y}} \cdot \dfrac{\cot x\cot y}{\cot x\cot y}$

$$= \frac{\cot y + \cot x}{\cot x\cot y - 1}$$

34. $\dfrac{\cos x - \cos y}{\sin x + \sin y} + \dfrac{\sin x - \sin y}{\cos x + \cos y} = \dfrac{(\cos x - \cos y)(\cos x + \cos y) + (\sin x - \sin y)(\sin x + \sin y)}{(\sin x + \sin y)(\cos x + \cos y)}$

$$= \frac{\cos^2 x - \cos^2 y + \sin^2 x - \sin^2 y}{(\sin x + \sin y)(\cos x + \cos y)}$$

$$= \frac{(\cos^2 x + \sin^2 x) - (\cos^2 y + \sin^2 y)}{(\sin x + \sin y)(\cos x + \cos y)}$$

$$= 0$$

36. $\sqrt{\dfrac{1 - \cos\theta}{1 + \cos\theta}} = \sqrt{\dfrac{1 - \cos\theta}{1 + \cos\theta} \cdot \dfrac{1 - \cos\theta}{1 - \cos\theta}}$

$$= \sqrt{\frac{(1 - \cos\theta)^2}{1 - \cos^2\theta}}$$

$$= \sqrt{\frac{(1 - \cos\theta)^2}{\sin^2\theta}}$$

$$= \frac{1 - \cos\theta}{|\sin\theta|}$$

38. $\sec^2 y - \cot^2\left(\dfrac{\pi}{2} - y\right) = \sec^2 y - \tan^2 y = 1$

40. $\sec^2\left(\dfrac{\pi}{2} - x\right) - 1 = \csc^2 x - 1 = \cot^2 x$

42.

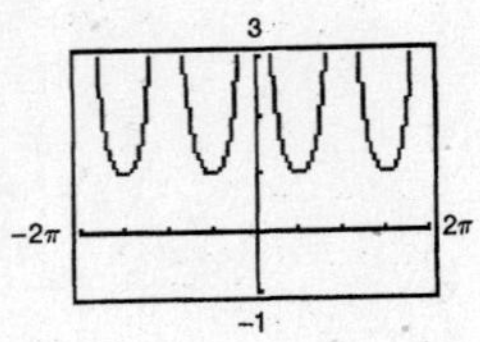

$$\csc x(\csc x - \sin x) + \frac{\sin x - \cos x}{\sin x} + \cot x = \csc^2 x - \csc x \sin x + 1 - \frac{\cos x}{\sin x} + \cot x$$
$$= \csc^2 x - 1 + 1 - \cot x + \cot x$$
$$= \csc^2 x$$

44.

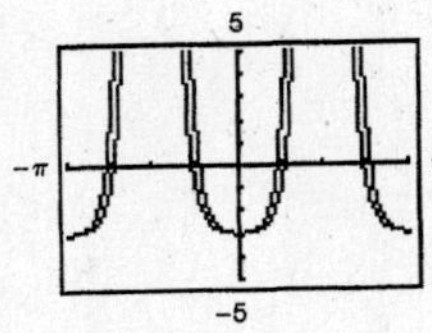

Not an identity

46.

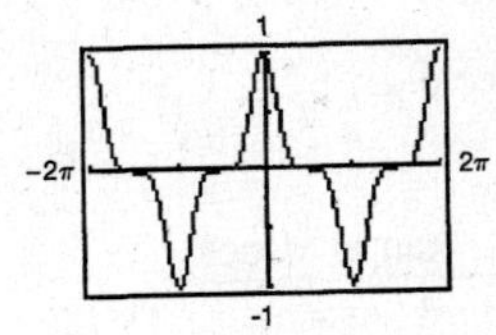

$$(\sin^4 \beta - 2\sin^2 \beta + 1)\cos \beta = (\sin^2 \beta - 1)^2 \cos \beta$$
$$= (-\cos^2 \beta)^2 \cos \beta$$
$$= \cos^5 \beta$$

48.

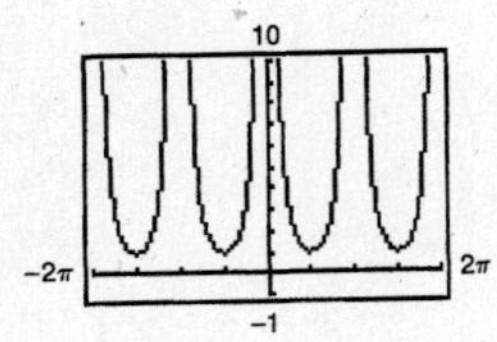

$$\csc^4 \theta - \cot^4 \theta = (\csc^2 \theta - \cot^2 \theta)(\csc^2 \theta + \cot^2 \theta)$$
$$= \csc^2 \theta + \cot^2 \theta$$
$$= \csc^2 \theta + (\csc^2 \theta - 1)$$
$$= 2\csc^2 \theta - 1$$

50.

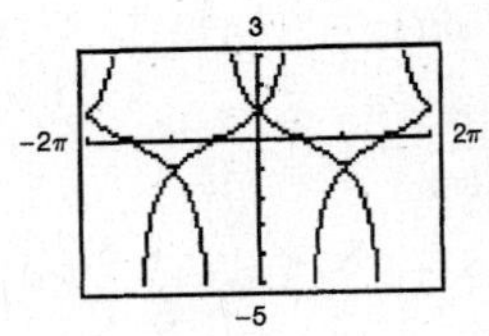

Not an identity

52.

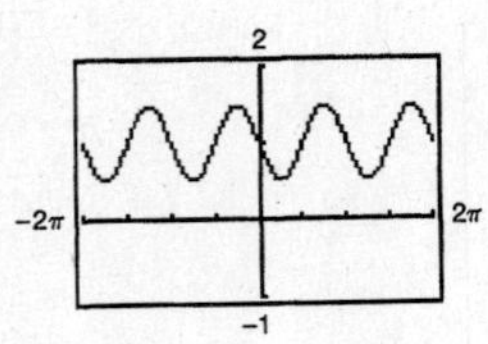

$$\frac{\sin^3 \beta + \cos^3 \beta}{\sin \beta + \cos \beta} = \frac{(\sin \beta + \cos \beta)(\sin^2 \beta - \sin \beta \cos \beta + \cos^2 \beta)}{\sin \beta + \cos \beta}$$
$$= \sin^2 \beta + \cos^2 \beta - \sin \beta \cos \beta$$
$$= 1 - \sin \beta \cos \beta$$

54. $\cos^2 55° + \cos^2 35° = \cos^2 55° + \sin^2(90° - 35°)$
$$= \cos^2 55° + \sin^2 55°$$
$$= 1$$

56. $\sin^2 12° + \sin^2 40° + \sin^2 50° + \sin^2 78° = \sin^2 12° + \sin^2 78° + \sin^2 40° + \sin^2 50°$
$$= \cos^2(90° - 12°) + \sin^2 78° + \cos^2(90° - 40°) + \sin^2 50°$$
$$= \cos^2 78° + \sin^2 78° + \cos^2 50° + \sin^2 50°$$
$$= 1 + 1 = 2$$

58. (a) $\dfrac{h\sin(90° - \theta)}{\sin\theta} = \dfrac{h\cos\theta}{\sin\theta} = h\cot\theta$

(b)

θ	10°	20°	30°	40°	50°	60°	70°	80°	90°
s	28.36	13.74	8.66	5.96	4.20	2.89	1.82	0.88	0

(c) Greatest: 10°, Least: 90°

(d) Noon

60. True. An identity is an equation that is true for all real values in the domain of the variable.

62. $\tan\theta = \sqrt{\sec^2\theta - 1}$

True identity: $\tan\theta = \pm\sqrt{\sec^2\theta - 1}$

$\tan\theta = \sqrt{\sec^2\theta - 1}$ is not true for $\pi/2 < \theta < \pi$ or $3\pi/2 < \theta < 2\pi$. Thus, the equation is not true for $\theta = 3\pi/4$.

64. $x^2 + 5x - 7 = 0$

$a = 1, b = 5, c = -7$

$$x = \frac{-5 \pm \sqrt{(5)^2 - 4(1)(-7)}}{2(1)}$$

$$x = \frac{-5 \pm \sqrt{53}}{2}$$

66. $8x^2 - 4x - 3 = 0$

$a = 8, b = -4, c = -3$

$$x = \frac{-(-4) \pm \sqrt{(-4)^2 - 4(8)(-3)}}{2(8)}$$

$$x = \frac{4 \pm \sqrt{112}}{16}$$

$$x = \frac{4 \pm 4\sqrt{7}}{16}$$

$$x = \frac{1}{4} \pm \frac{\sqrt{7}}{4}$$

Section 2.3 Solving Trigonometric Equations

Solutions to Even-Numbered Exercises

2. $\sec x - 2 = 0$

(a) $x = \dfrac{\pi}{3}$

$$\sec\frac{\pi}{3} - 2 = \frac{1}{\cos\frac{\pi}{3}} - 2 = \frac{1}{\frac{1}{2}} - 2 = 2 - 2 = 0$$

(b) $x = \dfrac{5\pi}{3}$

$$\sec\frac{5\pi}{3} - 2 = \frac{1}{\cos\frac{5\pi}{3}} - 2 = \frac{1}{\frac{1}{2}} - 2 = 2 - 2 = 0$$

4. $2\cos^2 4x - 1 = 0$

(a) $x = \dfrac{\pi}{16}$

$$2\cos^2\left[4\left(\frac{\pi}{16}\right)\right] - 1 = 2\cos^2\frac{\pi}{4} - 1$$
$$= 2\left(\frac{\sqrt{2}}{2}\right)^2 - 1$$
$$= 2\left(\frac{1}{2}\right) - 1 = 1 - 1 = 0$$

(b) $x = \dfrac{3\pi}{16}$

$$2\cos^2\left[4\left(\frac{3\pi}{16}\right)\right] - 1 = 2\cos^2\frac{3\pi}{4} - 1$$
$$= 2\left(-\frac{\sqrt{2}}{2}\right)^2 - 1$$
$$= 2\left(\frac{1}{2}\right) - 1 = 0$$

6. $\csc^4 x - 4\csc^2 x = 0$

(a) $x = \dfrac{\pi}{6}$

$$\csc^4\frac{\pi}{6} - 4\csc^2\frac{\pi}{6} = \frac{1}{\sin^4\frac{\pi}{6}} - \frac{4}{\sin^2\frac{\pi}{6}}$$
$$= \frac{1}{\left(\frac{1}{2}\right)^4} - \frac{4}{\left(\frac{1}{2}\right)^2}$$
$$= 16 - 16 = 0$$

(b) $x = \dfrac{5\pi}{6}$

$$\csc^4\frac{5\pi}{6} - 4\csc\frac{5\pi}{6} = \frac{1}{\sin^4\frac{5\pi}{6}} - \frac{4}{\sin^2\frac{5\pi}{6}}$$
$$= \frac{1}{\left(\frac{1}{2}\right)^4} - \frac{4}{\left(\frac{1}{2}\right)^2}$$
$$= 16 - 16 = 0$$

8. $2\sin x + 1 = 0$

$$\sin x = -\frac{1}{2}$$
$$x = \frac{7\pi}{6} + 2n\pi$$
or
$$x = \frac{11\pi}{6} + 2n\pi$$

10. $\tan x + \sqrt{3} = 0$

$$\tan x = -\sqrt{3}$$
$$x = \frac{2\pi}{3} + n\pi$$

12. $3\cot^2 x - 1 = 0$

$$\cot^2 x = \frac{1}{3}$$
$$\cot x = \pm\frac{1}{\sqrt{3}}$$
$$x = \frac{\pi}{3} + n\pi$$
or
$$x = \frac{2\pi}{3} + n\pi$$

14. $(3\tan^2 x - 1)(\tan^2 x - 3) = 0$

$3\tan^2 x - 1 = 0$ or $\tan^2 x - 3 = 0$

$$\tan x = \pm\frac{1}{\sqrt{3}}$$
$$x = \frac{\pi}{6} + n\pi$$
or
$$x = \frac{5\pi}{6} + n\pi$$

$$\tan x = \pm\sqrt{3}$$
$$x = \frac{\pi}{3} + n\pi$$
or
$$x = \frac{2\pi}{3} + n\pi$$

16.
$$\sin^2 x = 3\cos^2 x$$
$$\sin^2 x - 3(1 - \sin^2 x) = 0$$
$$4\sin^2 x = 3$$
$$\sin x = \pm\frac{\sqrt{3}}{2}$$
$$x = \frac{\pi}{3} + n\pi$$
or
$$x = \frac{2\pi}{3} + n\pi$$

18. $\tan^2 3x = 3$

$$\tan 3x = \pm\sqrt{3}$$

$$3x = \frac{\pi}{3} + n\pi \implies x = \frac{\pi}{9} + \frac{n\pi}{3} \text{ or } 3x = \frac{2\pi}{3} + n\pi \implies x = \frac{2\pi}{9} + \frac{n\pi}{3}$$

20. $\cos 2x(2\cos x + 1) = 0$

$$\cos 2x = 0 \quad \text{or} \quad 2\cos x + 1 = 0$$

$$2x = \frac{\pi}{2} + n\pi \qquad \cos x = -\frac{1}{2}$$

$$x = \frac{\pi}{4} + \frac{n\pi}{2} \qquad x = \frac{2\pi}{3} + 2n\pi$$

or

$$x = \frac{4\pi}{3} + 2n\pi$$

22. $\sec^2 x - 1 = 0$

$$\sec^2 x = 1$$

$$\sec x = \pm 1$$

$$x = 0 \text{ or } x = \pi$$

24.

$$2\sin^2 x = 2 + \cos x$$

$$2 - 2\cos^2 x = 2 + \cos x$$

$$2\cos^2 x + \cos x = 0$$

$$\cos x(2\cos x + 1) = 0$$

$$\cos x = 0 \quad \text{or} \quad 2\cos x + 1 = 0$$

$$x = \frac{\pi}{2}, \frac{3\pi}{2} \qquad 2\cos x = -1$$

$$\cos x = -\frac{1}{2}$$

$$x = \frac{2\pi}{3}, \frac{4\pi}{3}$$

26.

$$\sec x \csc x = 2\csc x$$

$$\sec x \csc x - 2\csc x = 0$$

$$\csc x(\sec x - 2) = 0$$

$$\csc x = 0 \quad \text{or} \quad \sec x - 2 = 0$$

No solution

$$\sec x = 2$$

$$x = \frac{\pi}{3}, \frac{5\pi}{3}$$

28.

$$\sec x + \tan x = 1$$

$$\frac{1}{\cos x} + \frac{\sin x}{\cos x} = 1$$

$$1 + \sin x = \cos x$$

$$(1 + \sin x)^2 = \cos^2 x$$

$$1 + 2\sin x + \sin^2 x = \cos^2 x$$

$$1 + 2\sin x + \sin^2 x = 1 - \sin^2 x$$

$$2\sin^2 x + 2\sin x = 0$$

$$2\sin x(\sin x + 1) = 0$$

$$\sin x = 0 \quad \text{or} \quad \sin x + 1 = 0$$

$$x = 0, \pi \qquad \sin x = -1$$

$$x = \frac{3\pi}{2}$$

(π is extraneous) $\left(\frac{3\pi}{2} \text{ is extraneous}\right)$

$x = 0$ is the only solution.

30. $2\sin^2 x + 3\sin x + 1 = 0$

$(2\sin x + 1)(\sin x + 1) = 0$

$2\sin x + 1 = 0$ or $\sin x + 1 = 0$

$\sin x = -\dfrac{1}{2}$ $\qquad$ $\sin x = -1$

$x = \dfrac{7\pi}{6}, \dfrac{11\pi}{6}$ $\qquad$ $x = \dfrac{3\pi}{2}$

32. $\cos x + \sin x \tan x = 2$

$$\cos x + \sin x\left(\frac{\sin x}{\cos x}\right) = 2$$

$$\frac{\cos^2 x + \sin^2 x}{\cos x} = 2$$

$$\frac{1}{\cos x} = 2$$

$$\cos x = \frac{1}{2}$$

$$x = \frac{\pi}{3}, \frac{5\pi}{3}$$

34. $\sin 2x = -\dfrac{\sqrt{3}}{2}$

$2x = \dfrac{4\pi}{3} + 2n\pi$ or $2x = \dfrac{5\pi}{3} + 2n\pi$

$x = \dfrac{2\pi}{3} + n\pi$ $\qquad$ $x = \dfrac{5\pi}{6} + n\pi$

36. $\sec 4x = 2$

$4x = \dfrac{\pi}{3} + 2n\pi$ or $4x = \dfrac{5\pi}{3} + 2n\pi$

$x = \dfrac{\pi}{12} + \dfrac{n\pi}{2}$ $\qquad$ $x = \dfrac{5\pi}{12} + \dfrac{n\pi}{2}$

38. $\sin\dfrac{x}{2} = -\dfrac{\sqrt{3}}{2}$

$\dfrac{x}{2} = \dfrac{4\pi}{3} + 2n\pi$ or $\dfrac{x}{2} = \dfrac{5\pi}{3} + 2n\pi$

$x = \dfrac{8\pi}{3} + 4n\pi$ $\qquad$ $x = \dfrac{10\pi}{3} + 4n\pi$

40. $y = \sin \pi x + \cos \pi x$

$$\sin \pi x + \cos \pi x = 0$$

$$\sin \pi x = -\cos \pi x$$

$$\pi x = -\frac{\pi}{4} + n\pi$$

$$x = -\frac{1}{4} + n$$

For $-1 < x < 3$ the intercepts are $-\frac{1}{4}, \frac{3}{4}, \frac{7}{4}, \frac{11}{4}$

42. $y = \sec^4\left(\dfrac{\pi x}{8}\right) - 4$

$$\sec^4\left(\frac{\pi x}{8}\right) - 4 = 0$$

$$\sec^4\left(\frac{\pi x}{8}\right) = 4$$

$$\sec\left(\frac{\pi x}{8}\right) = \pm\sqrt{2}$$

$$\frac{\pi x}{8} = \frac{\pi}{4} + \frac{\pi}{2}n$$

$$x = 2 + 4n$$

For $-3 < x < 3$ the intercepts are -2 and 2.

44. $y^2 + y - 20 = 0$

$(y + 5)(y - 4) = 0$

$y + 5 = 0 \quad \text{or} \quad y - 4 = 0$

$y = -5 \qquad\qquad y = 4$

$\sin^2 x + \sin x - 20 = 0$

$(\sin x + 5)(\sin x - 4) = 0$

$\sin x = -5 \quad \text{or} \quad \sin x = 4$

No solution No solution

46. $4 \sin^3 x + 2 \sin^2 x - 2 \sin x - 1 = 0$

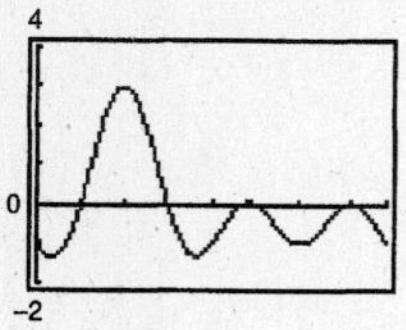

$x \approx 0.7854, 2.3562, 3.6652, 3.9270, 5.4978, 5.7596$

48. $\dfrac{\cos x \cot x}{1 - \sin x} = 3$

$y_1 = \left(\dfrac{\dfrac{\cos x}{\tan x}}{1 - \sin x}\right) - 3$

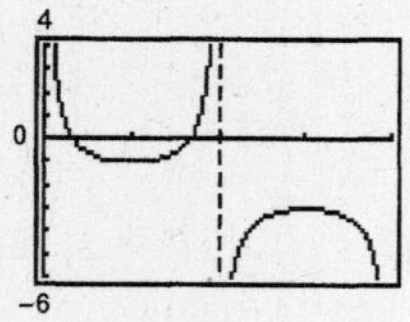

$x \approx 0.5236, 2.6180$

50. $x \cos x - 1 = 0$

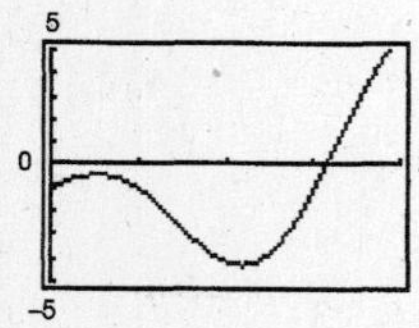

$x \approx 4.9172$

52. $\csc^2 x + 0.5 \cot x - 5 = 0$

$y_1 = \left(\dfrac{1}{\sin x}\right)^2 + \dfrac{1}{2 \tan x} - 5$

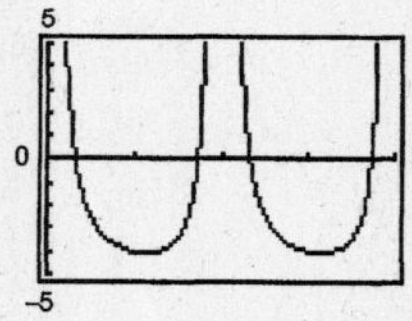

$x \approx 0.5153, 2.7259, 3.6569, 5.8675$

54. $6 \sin^2 x - 7 \sin x + 2 = 0$

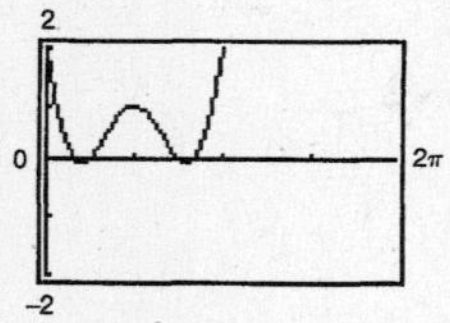

$x \approx 0.5236, 0.7297, 2.4119, 2.6180$

56. $3 \tan^2 x + 4 \tan x - 4 = 0$

$\tan x = \dfrac{-4 \pm \sqrt{4^2 - 4(3)(-4)}}{2(3)} = \dfrac{-4 \pm \sqrt{64}}{6} = -2, \dfrac{2}{3}$

$\tan x = -2 \qquad\qquad \tan x = \frac{2}{3}$

$x = \arctan(-2) + n\pi \qquad x = \arctan\left(\dfrac{2}{3}\right) + n\pi$

$\approx -1.1071 + n\pi \qquad \approx 0.5880 + n\pi$

The values of x in $[0, 2\pi)$ are 0.5880, 3.7296, 2.0344, 5.1760.

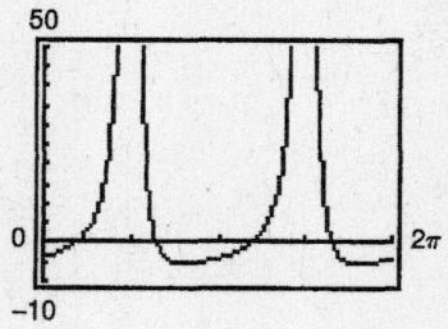

58. $4 \cos^2 x - 4 \cos x - 1 = 0$

$\cos x = \dfrac{4 \pm \sqrt{(-4)^2 - 4(4)(-1)}}{2(4)}$

$= \dfrac{4 \pm \sqrt{32}}{8} = \dfrac{1 \pm \sqrt{2}}{2}$

$\cos x = \dfrac{1 - \sqrt{2}}{2} \qquad \cos x = \dfrac{1 + \sqrt{2}}{2}$

$x = \arccos\left(\dfrac{1 - \sqrt{2}}{2}\right) \qquad$ No solution

$\approx 1.7794 \qquad \left(\dfrac{1 + \sqrt{2}}{2} > 1\right)$

Solutions in $[0, 2\pi)$ are $\arccos\left(\dfrac{1 - \sqrt{2}}{2}\right)$ and $2\pi - \arccos\left(\dfrac{1 - \sqrt{2}}{2}\right)$: 1.7794, 4.5038.

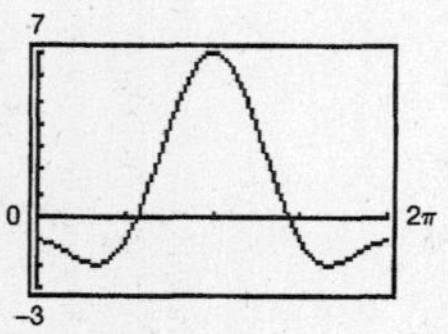

60. $\sec^2 x + \tan x - 3 = 0$

$1 + \tan^2 x + \tan x - 3 = 0$

$\tan^2 x + \tan x - 2 = 0$

$(\tan x + 2)(\tan x - 1) = 0$

$\tan x + 2 = 0$ $\qquad$ $\tan x - 1 = 0$

$\tan x = -2$ $\qquad$ $\tan x = 1$

$x = \arctan(-2) + n\pi$ $\qquad$ $x = \arctan(1) + n\pi$

$\approx -1.1071 + n\pi$ $\qquad$ $= \dfrac{\pi}{4} + n\pi$

Solutions in $[0, 2\pi)$ are $\arctan(-2) + \pi$, $\arctan(-2) + 2\pi$, $\dfrac{\pi}{4}, \dfrac{5\pi}{4}$.

62. $2\sin^2 x - 7\sin x + 3 = 0$

$(\sin x - 3)(2\sin x - 1) = 0$

$\sin x - 3 = 0$ $\qquad$ $2\sin x - 1 = 0$

No solution $\qquad$ $\sin x = \dfrac{1}{2}$

$x = \dfrac{\pi}{6}, \dfrac{5\pi}{6}$

Solutions in $[0, 2\pi)$ are $\dfrac{\pi}{6}, \dfrac{5\pi}{6}$.

64. (a) $f(x) = 2\sin x + \cos 2x$

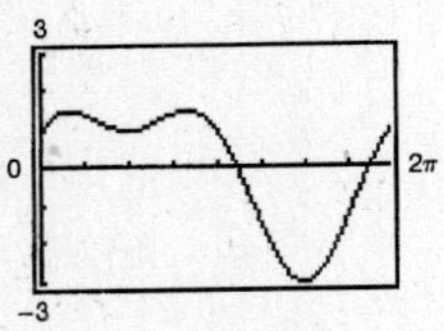

max: $(0.5236, 1.5)$ $\qquad$ min: $(1.5708, 1.0)$

max: $(2.6180, 1.5)$ $\qquad$ min: $(4.7124, -3.0)$

(b) $2\cos x - 4\sin x\cos x = 0$

$2\cos x(1 - 2\sin x) = 0$

$2\cos x = 0$ $\qquad$ $1 - 2\sin x = 0$

$x = \dfrac{\pi}{2}, \dfrac{3\pi}{2}$ $\qquad$ $x = \dfrac{\pi}{2}, \dfrac{3\pi}{2}$

$\approx 1.5708, 4.7124$

$\sin x = \dfrac{1}{2}$

$x = \dfrac{\pi}{6}, \dfrac{5\pi}{6}$

$\approx 0.5236, 2.6180$

66. Graph $y = \cos x$ and $y = x$ on the same set of axes. Their point of intersection gives the value of c such that $f(c) = c \Rightarrow \cos c = c$.

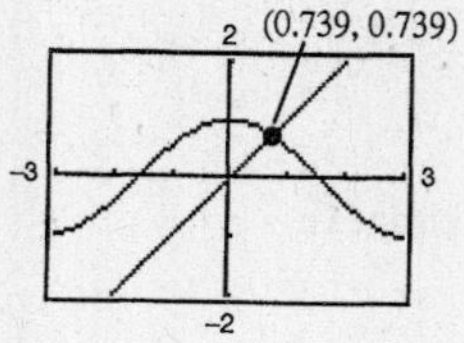

$c \approx 0.739$

68. $f(x) = \dfrac{\sin x}{x}$

(a) Domain: all real numbers except $x = 0$.

(b) The graph has y-axis symmetry.

(c) As $x \longrightarrow 0, f(x) \longrightarrow 1$.

(d) $\dfrac{\sin x}{x} = 0$ has four solutions in the interval $[-8, 8]$.

$$(\sin x)\left(\frac{1}{x}\right) = 0$$

$$\sin x = 0$$

$$x = -2\pi, -\pi, \pi, 2\pi$$

70. $y = 1.56t^{-1/2}(\cos 1.9t)$

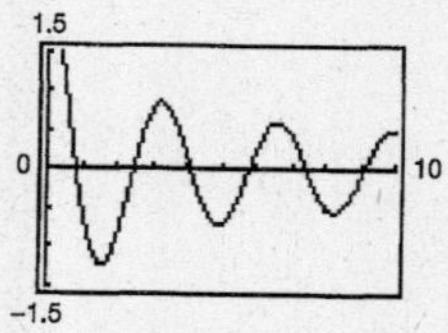

The time beyond which the displacement does not exceed 1 foot from equilibrium is $t \approx 1.91$ seconds.

72.

$$\text{range} = 300 \text{ feet}$$

$$v_0 = 100 \text{ feet per second}$$

$$r = \tfrac{1}{32}v_0^2 \sin 2\theta$$

$$\tfrac{1}{32}(100)^2 \sin 2\theta = 300$$

$$\sin 2\theta = 0.96$$

$$2\theta = \arcsin(0.96) \approx 73.74^\circ$$

$$\theta \approx 36.9^\circ$$

or

$$2\theta = 180^\circ - \arcsin(0.96) \approx 106.26^\circ$$

$$\theta \approx 53.1^\circ$$

74. $A = 2x \cos x,\ 0 < x < \dfrac{\pi}{2}$

(a)

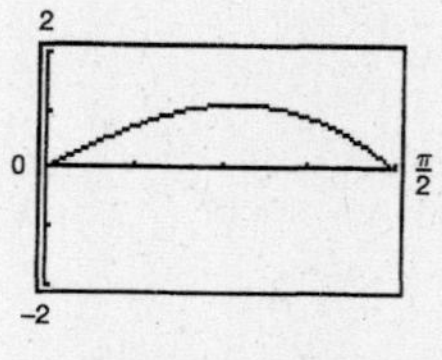

The maximum area of $A \approx 1.12$ occurs when $x \approx 0.86$.

(b) $A \geq 1$ for $0.6 < x < 1.1$

76. $f(x) = 3 \sin(0.6x - 2)$

(a) Zero: $\sin(0.6x - 2) = 0$

$$0.6x - 2 = 0$$

$$0.6x = 2$$

$$x = \frac{2}{0.6} = \frac{10}{3}$$

(b) $g(x) = -0.45x^2 + 5.52x - 13.70$

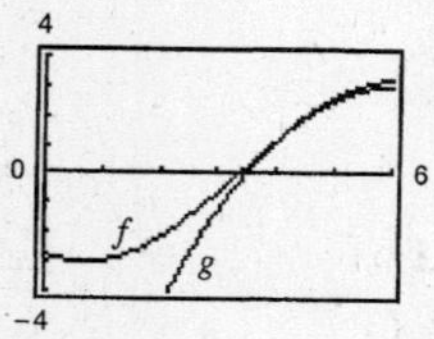

For $3.5 \leq x \leq 6$ the approximation appears to be good.

(c) $-0.45x^2 + 5.52x - 13.70 = 0$

$$x = \frac{-5.52 \pm \sqrt{(5.52)^2 - 4(-0.45)(-13.70)}}{2(-0.45)}$$

$x \approx 3.46, 8.81$

The zero of g on $[0, 6]$ is 3.46. The zero is close to the zero $\frac{10}{3} \approx 3.33$ of f.

78. False. $\sin x = 3.4$ has no solution since 3.4 is outside the range of sin. Also, 3.4 is outside the domain of arcsin, so $x = \arcsin(3.4)$ is an invalid equation.

80. By inspecting the graphs of y_1 and y_2, it appears they intersect at three points.

82.

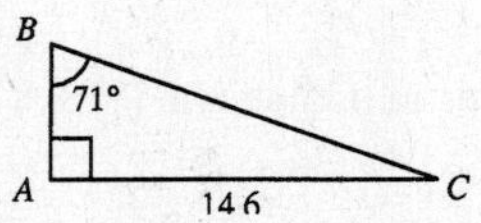

Given: $A = 90°, B = 71°, b = 14.6$

$$C = 90° - 71° = 19°$$

$$\sin 71° = \frac{14.6}{a}$$

$$a = \frac{14.6}{\sin 71} \approx 15.4$$

$$\tan 71° = \frac{14.6}{c}$$

$$c = \frac{14.6}{\tan 71°} \approx 5.0$$

84. 570°

$570° - 360° = 210°$ Quadrant III

Reference angle: 30°

$$\sin 570° = -\sin 30° = -\frac{1}{2}$$

$$\cos 570° = -\cos 30° = -\frac{\sqrt{3}}{2}$$

$$\tan 570° = \tan 30° = \frac{\sqrt{3}}{3}$$

86. $-1410°$

$-1410° + 4(360°) = 30°$ Quadrant I

$$\sin(-1410°) = \sin 30° = \frac{1}{2}$$

$$\cos(-1410°) = \cos 30° = \frac{\sqrt{3}}{2}$$

$$\tan(-1410°) = \tan 30° = \frac{\sqrt{3}}{3}$$

88.

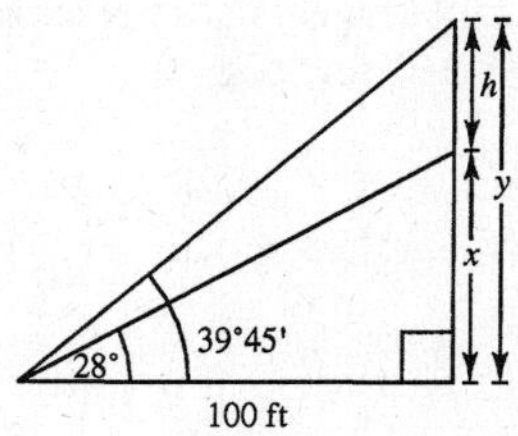

$$h = y - x$$

$$\tan 39.75° = \frac{y}{100}$$

$$100 \tan 39.75° = y$$

$$\tan 28° = \frac{x}{100}$$

$$100 \tan 28° = x$$

$$h = 100 \tan 39.75° - 100 \tan 28°$$

$$h \approx 30 \text{ feet}$$

Section 2.4 Sum and Difference Formulas

Solutions to Even-Numbered Exercises

2. (a) $\sin\left(\frac{3\pi}{4} + \frac{5\pi}{6}\right) = \sin\frac{3\pi}{4}\cos\frac{5\pi}{6} + \cos\frac{3\pi}{4}\sin\frac{5\pi}{6}$

$$= \left(\frac{\sqrt{2}}{2}\right)\left(-\frac{\sqrt{3}}{2}\right) + \left(-\frac{\sqrt{2}}{2}\right)\left(\frac{1}{2}\right)$$

$$= -\frac{\sqrt{6} + \sqrt{2}}{4}$$

(b) $\sin\frac{3\pi}{4} + \sin\frac{5\pi}{6} = \frac{\sqrt{2}}{2} + \frac{1}{2} = \frac{\sqrt{2} + 1}{2}$

4. (a) $\cos\left(\frac{2\pi}{3} - \frac{\pi}{6}\right) = \cos\frac{2\pi}{3}\cos\frac{\pi}{6} + \sin\frac{2\pi}{3}\sin\frac{\pi}{6}$

$= \left(-\frac{1}{2}\right)\left(\frac{\sqrt{3}}{2}\right) + \left(\frac{\sqrt{3}}{2}\right)\left(\frac{1}{2}\right) = 0$

(b) $\cos\frac{2\pi}{3} + \cos\frac{\pi}{6} = -\frac{1}{2} + \frac{\sqrt{3}}{2} = \frac{\sqrt{3}-1}{2}$

6. (a) $\sin(135° - 30°) = \sin 135° \cos 30° - \cos 135° \sin 30°$

$= \left(\frac{\sqrt{2}}{2}\right)\left(\frac{\sqrt{3}}{2}\right) - \left(-\frac{\sqrt{2}}{2}\right)\left(\frac{1}{2}\right)$

$= \frac{\sqrt{6}+\sqrt{2}}{4}$

(b) $\sin 135° - \cos 30° = \frac{\sqrt{2}}{2} - \frac{\sqrt{3}}{2}$

$= \frac{\sqrt{2}-\sqrt{3}}{2}$

8. $165° = 135° + 30°$

$\sin 165° = \sin(135° + 30°)$

$= \sin 135° \cos 30° + \sin 30° \cos 135°$

$= \sin 45° \cos 30° - \sin 30° \cos 45°$

$= \frac{\sqrt{2}}{2} \cdot \frac{\sqrt{3}}{2} - \frac{1}{2} \cdot \frac{\sqrt{2}}{2}$

$= \frac{\sqrt{2}}{4}(\sqrt{3} - 1)$

$\cos 165° = \cos(135° + 30°)$

$= \cos 135° \cos 30° - \sin 135° \sin 30°$

$= -\cos 45° \cos 30° - \sin 45° \sin 30°$

$= -\frac{\sqrt{2}}{2} \cdot \frac{\sqrt{3}}{2} - \frac{\sqrt{2}}{2} \cdot \frac{1}{2}$

$= -\frac{\sqrt{2}}{4}(\sqrt{3} + 1)$

$\tan 165° = \tan(135° + 30°)$

$= \frac{\tan 135° + \tan 30°}{1 - \tan 135° \tan 30°}$

$= \frac{-\tan 45° + \tan 30°}{1 + \tan 45° \tan 30°}$

$= \frac{-1 + \frac{\sqrt{3}}{3}}{1 + \frac{\sqrt{3}}{3}}$

$= -2 + \sqrt{3}$

10. $255° = 300° - 45°$

$\sin 255° = \sin(300° - 45°)$

$= \sin 300° \cos 45° - \sin 45° \cos 300°$

$= -\sin 60° \cos 45° - \sin 45° \cos 60°$

$= -\frac{\sqrt{3}}{2} \cdot \frac{\sqrt{2}}{2} - \frac{\sqrt{2}}{2} \cdot \frac{1}{2}$

$= -\frac{\sqrt{2}}{4}(\sqrt{3} + 1)$

$\cos 255° = \cos(300° - 45°)$

$= \cos 300° \cos 45° + \sin 300° \sin 45°$

$= \cos 60° \cos 45° - \sin 60° \sin 45°$

$= \frac{1}{2} \cdot \frac{\sqrt{2}}{2} - \frac{\sqrt{3}}{2} \cdot \frac{\sqrt{2}}{2}$

$= \frac{\sqrt{2}}{4}(1 - \sqrt{3})$

$\tan 255° = \tan(300° - 45°)$

$= \frac{\tan 300° - \tan 45°}{1 + \tan 300° \tan 45°}$

$= \frac{-\tan 60° - \tan 45°}{1 - \tan 60° \tan 45°}$

$= \frac{-\sqrt{3} - 1}{1 - \sqrt{3}} = 2 + \sqrt{3}$

12. $\frac{7\pi}{12} = \frac{\pi}{3} + \frac{\pi}{4}$

$$\sin\frac{7\pi}{12} = \sin\left(\frac{\pi}{3} + \frac{\pi}{4}\right)$$
$$= \sin\frac{\pi}{3}\cos\frac{\pi}{4} + \sin\frac{\pi}{4}\cos\frac{\pi}{3}$$
$$= \frac{\sqrt{3}}{2}\cdot\frac{\sqrt{2}}{2} + \frac{\sqrt{2}}{2}\cdot\frac{1}{2}$$
$$= \frac{\sqrt{2}}{4}(\sqrt{3} + 1)$$

$$\cos\frac{7\pi}{12} = \cos\left(\frac{\pi}{3} + \frac{\pi}{4}\right)$$
$$= \cos\frac{\pi}{3}\cos\frac{\pi}{4} - \sin\frac{\pi}{3}\sin\frac{\pi}{4}$$
$$= \frac{1}{2}\cdot\frac{\sqrt{2}}{2} - \frac{\sqrt{3}}{2}\cdot\frac{\sqrt{2}}{2}$$
$$= \frac{\sqrt{2}}{4}(1 - \sqrt{3})$$

$$\tan\frac{7\pi}{12} = \tan\left(\frac{\pi}{3} + \frac{\pi}{4}\right)$$
$$= \frac{\tan\frac{\pi}{3} + \tan\frac{\pi}{4}}{1 - \tan\frac{\pi}{3}\tan\frac{\pi}{4}}$$
$$= \frac{\sqrt{3} + 1}{1 - \sqrt{3}}$$
$$= -2 - \sqrt{3}$$

14. $-\frac{\pi}{12} = \frac{\pi}{6} - \frac{\pi}{4}$

$$\sin\left(-\frac{\pi}{12}\right) = \sin\left(\frac{\pi}{6} - \frac{\pi}{4}\right)$$
$$= \sin\frac{\pi}{6}\cos\frac{\pi}{4} - \sin\frac{\pi}{4}\cos\frac{\pi}{6}$$
$$= \frac{1}{2}\cdot\frac{\sqrt{2}}{2} - \frac{\sqrt{2}}{2}\cdot\frac{\sqrt{3}}{2}$$
$$= \frac{\sqrt{2}}{4}(1 - \sqrt{3})$$

$$\cos\left(-\frac{\pi}{12}\right) = \cos\left(\frac{\pi}{6} - \frac{\pi}{4}\right)$$
$$= \cos\frac{\pi}{6}\cos\frac{\pi}{4} + \sin\frac{\pi}{6}\sin\frac{\pi}{4}$$
$$= \frac{\sqrt{3}}{2}\cdot\frac{\sqrt{2}}{2} + \frac{1}{2}\cdot\frac{\sqrt{2}}{2}$$
$$= \frac{\sqrt{2}}{4}(\sqrt{3} + 1)$$

$$\tan\left(-\frac{\pi}{12}\right) = \tan\left(\frac{\pi}{6} - \frac{\pi}{4}\right)$$
$$= \frac{\tan\frac{\pi}{6} - \tan\frac{\pi}{4}}{1 + \tan\frac{\pi}{6}\tan\frac{\pi}{4}}$$
$$= \frac{\frac{\sqrt{3}}{3} - 1}{1 + \frac{\sqrt{3}}{3}}$$
$$= -2 + \sqrt{3}$$

16. $-105 = 30° - 135°$

$$\sin(30° - 135°) = \sin 30°\cos 135° - \cos 30°\sin 135°$$
$$= \sin 30°(-\cos 45°) - \cos 30°\sin 45°$$
$$= \left(\frac{1}{2}\right)\left(-\frac{\sqrt{2}}{2}\right) - \left(\frac{\sqrt{3}}{2}\right)\left(\frac{\sqrt{2}}{2}\right)$$
$$= -\frac{\sqrt{2}}{4}(1 + \sqrt{3})$$

$$\cos(30° - 135°) = \cos 30°\cos 135° + \sin 30°\sin 135°$$
$$= \cos 30°(-\cos 45°) + \sin 30°\sin 45°$$
$$= \left(\frac{\sqrt{3}}{2}\right)\left(-\frac{\sqrt{2}}{2}\right) + \left(\frac{1}{2}\right)\left(\frac{\sqrt{2}}{2}\right)$$
$$= \frac{\sqrt{2}}{4}(1 - \sqrt{3})$$

$$\tan(30° - 135°) = \frac{\tan 30° - \tan 135°}{1 + \tan 30°\tan 135°}$$
$$= \frac{\tan 30° - (-\tan 45°)}{1 + \tan 30°(-\tan 45°)}$$
$$= \frac{\frac{\sqrt{3}}{3} - (-1)}{1 + \left(\frac{\sqrt{3}}{3}\right)(-1)} = 2 + \sqrt{3}$$

18. $15° = 45° - 30°$

$$\sin 15° = \sin(45° - 30°) = \sin 45° \cos 30° - \cos 45° \sin 30°$$

$$= \left(\frac{\sqrt{2}}{2}\right)\left(\frac{\sqrt{3}}{2}\right) - \left(\frac{\sqrt{2}}{2}\right)\left(\frac{1}{2}\right) = \frac{\sqrt{2}(\sqrt{3}-1)}{4} = \frac{\sqrt{2}}{4}(\sqrt{3}-1)$$

$$\cos 15° = \cos(45° - 30°) = \cos 45° \cos 30° + \sin 45° \sin 30°$$

$$= \left(\frac{\sqrt{2}}{2}\right)\left(\frac{\sqrt{3}}{2}\right) + \left(\frac{\sqrt{2}}{2}\right)\left(\frac{1}{2}\right) = \frac{\sqrt{2}(\sqrt{3}+1)}{4} = \frac{\sqrt{2}}{4}(\sqrt{3}+1)$$

$$\tan 15° = \tan(45° - 30°) = \frac{\tan 45° - \tan 30°}{1 + \tan 45° \tan 30°}$$

$$= \frac{1 - \frac{\sqrt{3}}{3}}{1 + (1)\left(\frac{\sqrt{3}}{3}\right)} = \frac{\frac{3-\sqrt{3}}{3}}{\frac{3+\sqrt{3}}{3}} = \frac{3-\sqrt{3}}{3+\sqrt{3}} \cdot \frac{3-\sqrt{3}}{3-\sqrt{3}} = \frac{12-6\sqrt{3}}{6} = 2 - \sqrt{3}$$

20. $-\frac{7\pi}{12} = -\frac{\pi}{3} - \frac{\pi}{4}$

$$\sin\left(-\frac{7\pi}{12}\right) = \sin\left(-\frac{\pi}{3} - \frac{\pi}{4}\right) = \sin\left(-\frac{\pi}{3}\right)\cos\left(\frac{\pi}{4}\right) - \cos\left(-\frac{\pi}{3}\right)\sin\left(\frac{\pi}{4}\right)$$

$$= \left(-\frac{\sqrt{3}}{2}\right)\left(\frac{\sqrt{2}}{2}\right) - \left(\frac{1}{2}\right)\left(\frac{\sqrt{2}}{2}\right) = -\frac{\sqrt{2}}{4}(\sqrt{3}+1)$$

$$\cos\left(-\frac{7\pi}{12}\right) = \cos\left(-\frac{\pi}{3} - \frac{\pi}{4}\right) = \cos\left(-\frac{\pi}{3}\right)\cos\left(\frac{\pi}{4}\right) + \sin\left(-\frac{\pi}{3}\right)\sin\left(\frac{\pi}{4}\right)$$

$$= \left(\frac{1}{2}\right)\left(\frac{\sqrt{2}}{2}\right) + \left(-\frac{\sqrt{3}}{2}\right)\left(\frac{\sqrt{2}}{2}\right) = \frac{\sqrt{2}}{4}(1-\sqrt{3})$$

$$\tan\left(-\frac{7\pi}{12}\right) = \tan\left(-\frac{\pi}{3} - \frac{\pi}{4}\right) = \frac{\tan\left(-\frac{\pi}{3}\right) - \tan\left(\frac{\pi}{4}\right)}{1 + \tan\left(-\frac{\pi}{3}\right)\tan\left(\frac{\pi}{4}\right)}$$

$$= \frac{-\sqrt{3}-1}{1 + (-\sqrt{3})(1)} = 2 + \sqrt{3}$$

22. $\frac{5\pi}{12} = \frac{\pi}{4} + \frac{\pi}{6}$

$$\sin\left(\frac{\pi}{4} + \frac{\pi}{6}\right) = \sin\frac{\pi}{4}\cos\frac{\pi}{6} + \cos\frac{\pi}{4}\sin\frac{\pi}{6}$$

$$= \left(\frac{\sqrt{2}}{2}\right)\left(\frac{\sqrt{3}}{2}\right) + \left(\frac{\sqrt{2}}{2}\right)\left(\frac{1}{2}\right) = \frac{\sqrt{2}}{4}(\sqrt{3}+1)$$

$$\cos\left(\frac{\pi}{4} + \frac{\pi}{6}\right) = \cos\frac{\pi}{4}\cos\frac{\pi}{6} - \sin\frac{\pi}{4}\sin\frac{\pi}{6}$$

$$= \left(\frac{\sqrt{2}}{2}\right)\left(\frac{\sqrt{3}}{2}\right) - \left(\frac{\sqrt{2}}{2}\right)\left(\frac{1}{2}\right) = \frac{\sqrt{2}}{4}(\sqrt{3}-1)$$

$$\tan\left(\frac{\pi}{4} + \frac{\pi}{6}\right) = \frac{\tan\frac{\pi}{4} + \tan\frac{\pi}{6}}{1 - \tan\frac{\pi}{4}\tan\frac{\pi}{6}} = \frac{1 + \frac{\sqrt{3}}{3}}{1 - (1)\left(\frac{\sqrt{3}}{3}\right)} = \sqrt{3} + 2$$

24. $\sin 140° \cos 50° + \cos 140° \sin 50° = \sin(140° + 50°) = \sin 190°$

26. $\dfrac{\tan 140° - \tan 60°}{1 + \tan 140° \tan 60°} = \tan(140° - 60°) = \tan 80°$

28. $\cos\dfrac{\pi}{7}\cos\dfrac{\pi}{5} - \sin\dfrac{\pi}{7}\sin\dfrac{\pi}{5} = \cos\left(\dfrac{\pi}{7} + \dfrac{\pi}{5}\right)$

$$= \cos\frac{12\pi}{35}$$

30. $\cos 3x \cos 2y + \sin 3x \sin 2y = \cos(3x - 2y)$

32. $\cos 15° \cos 60° + \sin 15° \sin 60° = \cos(15° - 60°)$

$$= \cos(-45°) = \frac{\sqrt{2}}{2}$$

34. $\cos\dfrac{\pi}{16}\cos\dfrac{3\pi}{16} - \sin\dfrac{\pi}{16}\sin\dfrac{3\pi}{16} = \cos\left(\dfrac{\pi}{16} + \dfrac{3\pi}{16}\right)$

$$= \cos\frac{\pi}{4} = \frac{\sqrt{2}}{2}$$

36. $\dfrac{\tan\left(\frac{5\pi}{4}\right) - \tan\left(\frac{\pi}{12}\right)}{1 + \tan\left(\frac{5\pi}{4}\right)\tan\left(\frac{\pi}{12}\right)} = \tan\left(\dfrac{5\pi}{4} - \dfrac{\pi}{12}\right)$

$$= \tan\left(\frac{7\pi}{6}\right)$$

$$= \tan\left(\frac{\pi}{6}\right) = \frac{\sqrt{3}}{3}$$

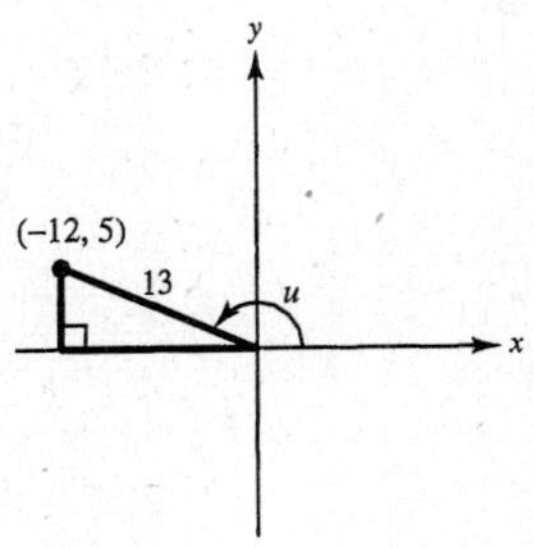

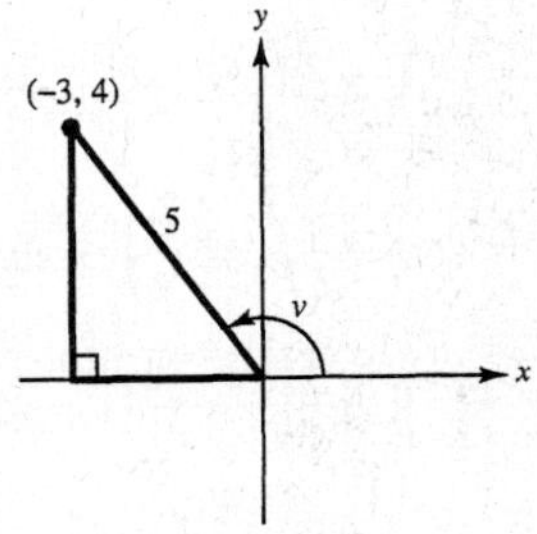

Figures for Exercises 38–44

38. $\cos(u - v) = \cos u \cos v + \sin u \sin v$

$$= \left(-\frac{12}{13}\right)\left(-\frac{3}{5}\right) + \left(\frac{5}{13}\right)\left(\frac{4}{5}\right)$$

$$= \frac{36}{65} + \frac{20}{65} = \frac{56}{65}$$

40. $\sin(v - u) = \sin v \cos u - \cos v \sin u$

$$= \left(\frac{4}{5}\right)\left(-\frac{12}{13}\right) - \left(-\frac{3}{5}\right)\left(\frac{5}{13}\right)$$

$$= -\frac{48}{65} + \frac{15}{65} = -\frac{33}{65}$$

42. $\csc(u - v) = \dfrac{1}{\sin(u - v)} = \dfrac{1}{-\sin(v - u)}$

$$= \frac{1}{-\left(-\frac{33}{65}\right)} = \frac{65}{33}$$

44. $\tan(u + v) = \dfrac{\tan u + \tan v}{1 - \tan u \tan v} = \dfrac{\left(-\frac{5}{12}\right) + \left(-\frac{4}{3}\right)}{1 - \left(-\frac{5}{12}\right)\left(-\frac{4}{3}\right)}$

$$= \frac{-\frac{7}{4}}{\frac{4}{9}} = -\frac{63}{16}$$

$$\cot(u + v) = \frac{1}{\tan(u + v)} = \frac{1}{-\frac{63}{16}} = -\frac{16}{63}$$

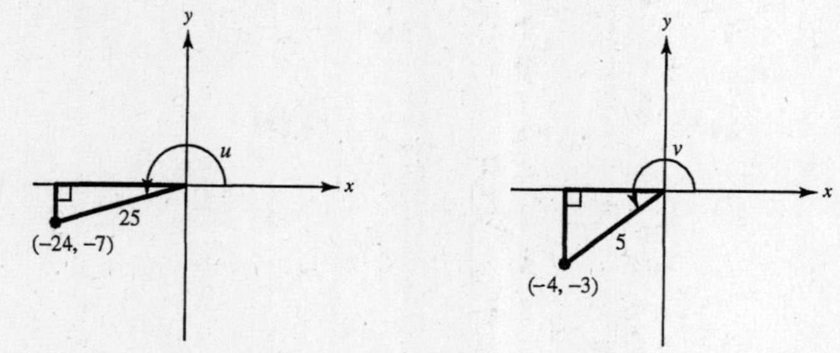

Figures for Exercises 46–50

46. $\sin(u + v) = \sin u \cos v + \cos u \sin v$

$$= \left(-\frac{7}{25}\right)\left(-\frac{4}{5}\right) + \left(-\frac{24}{25}\right)\left(-\frac{3}{5}\right)$$

$$= \frac{28}{125} + \frac{72}{125} = \frac{100}{125} = \frac{4}{5}$$

48. $\tan(v - u) = \dfrac{\tan v - \tan u}{1 + \tan v \tan u} = \dfrac{\left(\frac{3}{4}\right) - \left(\frac{7}{24}\right)}{1 + \left(\frac{3}{4}\right)\left(\frac{7}{24}\right)}$

$$= \frac{\frac{11}{24}}{\frac{39}{32}} = \frac{44}{117}$$

$$\cot(v - u) = \frac{1}{\tan(v - u)} = \frac{1}{\frac{44}{117}} = \frac{117}{44}$$

50. $\sin(u - v) = \sin u \cos v - \cos u \sin v = \left(-\frac{7}{25}\right)\left(-\frac{4}{5}\right) - \left(-\frac{24}{25}\right)\left(-\frac{3}{5}\right)$

$$= \frac{28}{125} - \frac{72}{125} = -\frac{44}{125}$$

$$\csc(u - v) = \frac{1}{\sin(u - v)} = \frac{1}{-\frac{44}{125}} = -\frac{125}{44}$$

52. Let:

$u = \arctan 2x$ and $v = \arccos x$

$\tan u = 2x$ $\qquad \cos v = x$

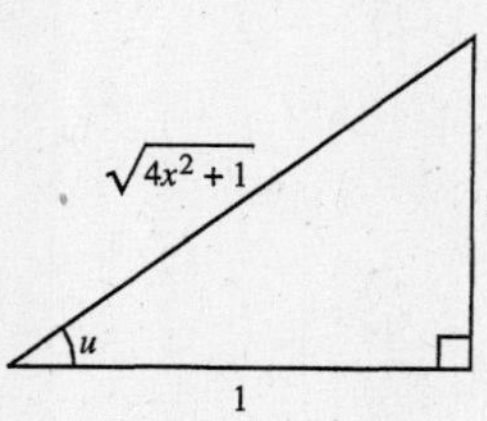

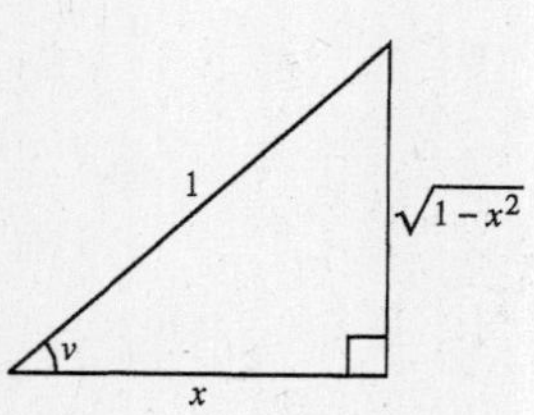

$$\sin(\arctan 2x - \arccos x) = \sin(u - v)$$

$$= \sin u \cos v - \cos u \sin v$$

$$= \frac{2x}{\sqrt{4x^2 + 1}}(x) - \frac{1}{\sqrt{4x^2 + 1}}\left(\sqrt{1 - x^2}\right)$$

$$= \frac{2x^2 - \sqrt{1 - x^2}}{\sqrt{4x^2 + 1}}$$

54. Let:

$u = \arccos x$ and $v = \arctan x$

$\cos u = x$ $\tan v = x$

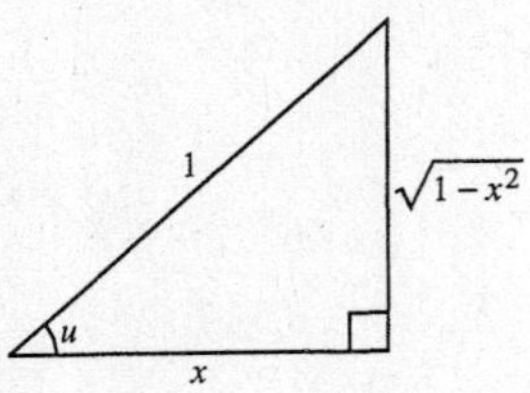

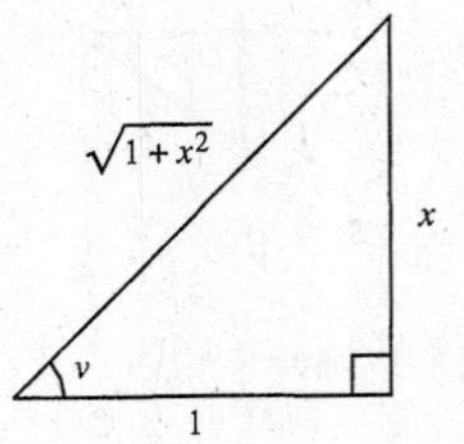

$$\begin{aligned}\cos(\arccos x - \arctan x) &= \cos(u - v)\\ &= \cos u \cos v + \sin u \sin v\\ &= (x)\left(\frac{1}{\sqrt{1+x^2}}\right) + \left(\sqrt{1-x^2}\right)\left(\frac{x}{\sqrt{1+x^2}}\right)\\ &= \frac{x + x\sqrt{1-x^2}}{\sqrt{1+x^2}}\end{aligned}$$

56. $$\begin{aligned}\sin\left(\frac{\pi}{2} + x\right) &= \sin\frac{\pi}{2}\cos x + \sin x \cos\frac{\pi}{2}\\ &= (1)(\cos x) + (\sin x)(0)\\ &= \cos x\end{aligned}$$

58. $$\begin{aligned}\cos\left(\frac{5\pi}{4} - x\right) &= \cos\frac{5\pi}{4}\cos x + \sin\frac{5\pi}{4}\sin x\\ &= -\frac{\sqrt{2}}{2}(\cos x + \sin x)\end{aligned}$$

60. $$\tan\left(\frac{\pi}{4} - \theta\right) = \frac{\tan\frac{\pi}{4} - \tan\theta}{1 + \tan\frac{\pi}{4}\tan\theta} = \frac{1 - \tan\theta}{1 + \tan\theta}$$

62. $$\begin{aligned}\sin(x + y)\sin(x - y) &= (\sin x \cos y + \sin y \cos x)(\sin x \cos y - \sin y \cos x)\\ &= \sin^2 x \cos^2 y - \sin^2 y \cos^2 x\\ &= \sin^2 x(1 - \sin^2 y) - \sin^2 y \cos^2 x\\ &= \sin^2 x - \sin^2 x \sin^2 y - \sin^2 y \cos^2 x\\ &= \sin^2 x - \sin^2 y(\sin^2 x + \cos^2 x)\\ &= \sin^2 x - \sin^2 y\end{aligned}$$

64. $$\begin{aligned}\cos(x + y) + \cos(x - y) &= \cos x \cos y - \sin x \sin y + \cos x \cos y + \sin x \sin y\\ &= 2\cos x \cos y\end{aligned}$$

66. $$\begin{aligned}\cos(\pi + x) &= \cos\pi\cos x - \sin\pi\sin x\\ &= (-1)\cos x - (0)\sin x\\ &= -\cos x\end{aligned}$$

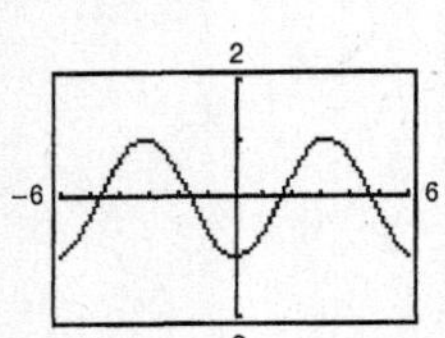

68. $$\begin{aligned}\tan(\pi + \theta) &= \frac{\tan\pi + \tan\theta}{1 - \tan\pi\tan\theta}\\ &= \frac{0 + \tan\theta}{1 - (0)\tan\theta}\\ &= \tan\theta\end{aligned}$$

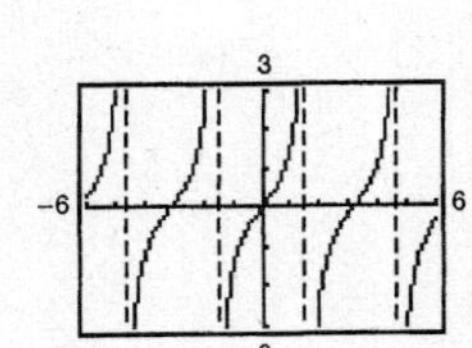

70.

$$\sin\left(x+\frac{\pi}{6}\right)-\sin\left(x-\frac{\pi}{6}\right)=\frac{1}{2}$$

$$\sin x\cos\frac{\pi}{6}+\cos x\sin\frac{\pi}{6}-\left(\sin x\cos\frac{\pi}{6}-\cos x\sin\frac{\pi}{6}\right)=\frac{1}{2}$$

$$2\cos x(0.5)=\frac{1}{2}$$

$$\cos x=\frac{1}{2}$$

$$x=\frac{\pi}{3},\frac{5\pi}{3}$$

72.

$$\tan(x+\pi)+2\sin(x+\pi)=0$$

$$\frac{\tan x+\tan\pi}{1-\tan x\tan\pi}+2(\sin x\cos\pi+\cos x\sin\pi)=0$$

$$\frac{\tan x+0}{1-\tan x(0)}+2[\sin x(-1)+\cos x(0)]=0$$

$$\frac{\tan x}{1}-2\sin x=0$$

$$\frac{\sin x}{\cos x}=2\sin x$$

$$\sin x=2\sin x\cos x$$

$$\sin x(1-2\cos x)=0$$

$$\sin x=0 \quad \text{or} \quad \cos x=\frac{1}{2}$$

$$x=0,\pi \qquad x=\frac{\pi}{3},\frac{5\pi}{3}$$

74. $\tan(x+\pi)-\cos\left(x+\frac{\pi}{2}\right)=0$

Answers: $(0, 0), (3.14, 0) \Rightarrow x=0, \pi$

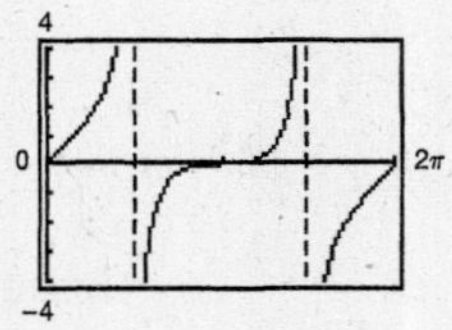

76.

$$y_1=A\cos 2\pi\left(\frac{t}{T}-\frac{x}{\lambda}\right)$$

$$y_2=A\cos 2\pi\left(\frac{t}{T}+\frac{x}{\lambda}\right)$$

$$y_1+y_2=A\cos 2\pi\left(\frac{t}{T}-\frac{x}{\lambda}\right)+A\cos 2\pi\left(\frac{t}{T}+\frac{x}{\lambda}\right)$$

$$y_1+y_2=A\left[\cos 2\pi\frac{t}{T}\cos 2\pi\frac{x}{\lambda}+\sin 2\pi\frac{t}{T}\sin 2\pi\frac{x}{\lambda}\right]+A\left[\cos 2\pi\frac{t}{T}\cos 2\pi\frac{x}{\lambda}-\sin 2\pi\frac{t}{T}\sin 2\pi\frac{x}{\lambda}\right]$$

$$=2A\cos 2\pi\frac{t}{T}\cos 2\pi\frac{x}{\lambda}$$

78. False.

$\cos(u \pm v) = \cos u \cos v \mp \sin u \sin v$

80. True.

$\sin\left(x - \frac{\pi}{2}\right) = -\sin\left(\frac{\pi}{2} - x\right) = -\cos x$

82. $\sin(n\pi + \theta) = \sin n\pi \cos\theta + \sin\theta \cos n\pi$

$= (0)(\cos\theta) + (\sin\theta)(-1)^n$

$= (-1)^n(\sin\theta)$, where n is an integer.

84. $C = \arctan\frac{a}{b} \Rightarrow \sin C = \frac{a}{\sqrt{a^2+b^2}}, \cos C = \frac{b}{\sqrt{a^2+b^2}}$

$$\sqrt{a^2+b^2}\cos(B\theta - C) = \sqrt{a^2+b^2}\left(\cos B\theta \cdot \frac{b}{\sqrt{a^2+b^2}} + \sin B\theta \cdot \frac{a}{\sqrt{a^2+b^2}}\right)$$

$= b\cos B\theta + a\sin B\theta$

$= a\sin B\theta + b\cos B\theta$

86. $3\sin 2\theta + 4\cos 2\theta$

$a = 3, b = 4, B = 2$

(a) $C = \arctan\frac{b}{a} = \arctan\frac{4}{3} \approx 0.9273$

$3\sin 2\theta + 4\cos 2\theta = \sqrt{a^2+b^2}\sin(B\theta + C)$

$\approx 5\sin(2\theta + 0.9273)$

(b) $C = \arctan\frac{a}{b} = \arctan\frac{3}{4} \approx 0.6435$

$3\sin 2\theta + 4\cos 2\theta = \sqrt{a^2+b^2}\cos(B\theta - C)$

$\approx 5\cos(2\theta - 0.6435)$

88. $\sin 2\theta - \cos 2\theta$

$a = 1, b = -1, B = 2$

(a) $C = \arctan\frac{b}{a} = \arctan(-1) = -\frac{\pi}{4}$

$\sin 2\theta - \cos 2\theta = \sqrt{a^2+b^2}\sin(B\theta + C)$

$= \sqrt{2}\sin\left(2\theta - \frac{\pi}{4}\right)$

(b) $C = \arctan\frac{a}{b} = \arctan(-1) = -\frac{\pi}{4}$

$\sin 2\theta - \cos 2\theta = \sqrt{a^2+b^2}\cos(B\theta - C)$

$= \sqrt{2}\cos\left(2\theta + \frac{\pi}{4}\right)$

90. $C = \arctan\frac{a}{b} = -\frac{3\pi}{4} \Rightarrow a = b, a < 0, b < 0$

$\sqrt{a^2+b^2} = 5 \Rightarrow a = b = \frac{-5\sqrt{2}}{2}$

$B = 1$

$5\cos\left(\theta + \frac{3\pi}{4}\right) = -\frac{5\sqrt{2}}{2}\sin\theta - \frac{5\sqrt{2}}{2}\cos\theta$

92. For $m_2 > m_1 > 0$, the angle θ between the lines is

$\theta = \arctan\left(\frac{m_2 - m_1}{1 + m_1 m_2}\right).$

$m_2 = 1$

$m_1 = \frac{1}{\sqrt{3}}$

$$\theta = \arctan\left(\frac{1 - \frac{1}{\sqrt{3}}}{1 + \frac{1}{\sqrt{3}}}\right) = \arctan(2 - \sqrt{3}) = 15°$$

94. From the figure, it appears that $u + v = w$. Assume that u, v, and w are all in Quadrant I. From the figure:

$$\tan u = \frac{s}{3s} = \frac{1}{3}$$

$$\tan v = \frac{s}{2s} = \frac{1}{2}$$

$$\tan w = \frac{s}{s} = 1$$

$$\tan(u + v) = \frac{\tan u + \tan v}{1 - \tan u \tan v} = \frac{1/3 + 1/2}{1 - (1/3)(1/2)} = \frac{5/6}{1 - (1/6)} = 1 = \tan w.$$

Thus, $\tan(u + v) = \tan w$. Because u, v, and w are all in Quadrant I, we have

$$\arctan[\tan(u + v)] = \arctan[\tan w]$$

$$u + v = w.$$

96. First, prove $\cos(u - v) = \cos u \cos v + \sin u \sin v$ using the figure containing points

$A(1, 0)$

$B(\cos(u - v), \sin(u - v))$

$C(\cos v, \sin v)$

$D(\cos u, \sin u)$

on the unit circle.

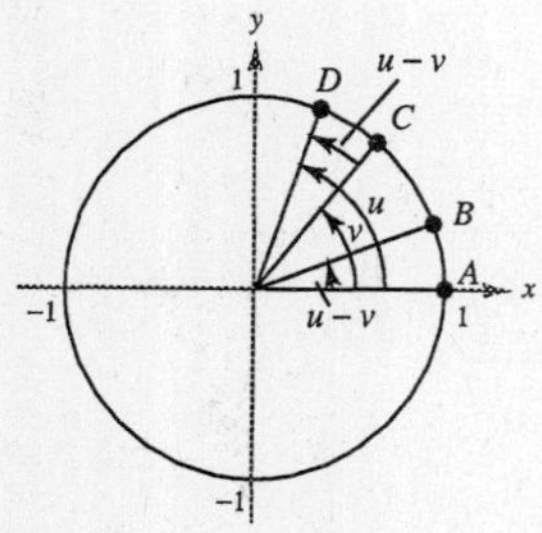

Since chords AB and CD are each subtended by angle $u - v$, their lengths are equal. Equating $[d(A, B)]^2 = [d(C, D)]^2$ we have $(\cos(u - v) - 1)^2 + \sin^2(u - v) = (\cos u - \cos v)^2 + (\sin u - \sin v)^2$. Simplifying and solving for $\cos(u - v)$, we have $\cos(u - v) = \cos u \cos v + \sin u \sin v$.

Using $\sin \theta = \cos\left(\frac{\pi}{2} - \theta\right)$ we have

$$\sin(u - v) = \cos\left[\frac{\pi}{2} - (u - v)\right] = \cos\left[\left(\frac{\pi}{2} - u\right) - (-v)\right]$$

$$= \cos\left(\frac{\pi}{2} - u\right)\cos(-v) + \sin\left(\frac{\pi}{2} - u\right)\sin(-v)$$

$$= \sin u \cos v - \cos u \sin v.$$

98.

$$f(x) = \frac{7 - x}{8}$$

$$y = \frac{7 - x}{8}$$

$$8y = 7 - x$$

$$x = 7 - 8y \Longrightarrow f^{-1}(x) = -8x + 7$$

$$f(f^{-1}(x)) = \frac{7 - f^{-1}(x)}{8}$$

$$= \frac{7 - (-8x + 7)}{8}$$

$$= x$$

$$f^{-1}(f(x)) = -8\left(\frac{7 - x}{8}\right) + 7$$

$$= x$$

100.

$$f(x) = \sqrt{x - 16}$$

$$y = \sqrt{x - 16}$$

$$y^2 = x - 16$$

$$x = y^2 + 16 \Longrightarrow f^{-1}(x) = x^2 + 16$$

$$f(f^{-1}(x)) = \sqrt{(x^2 + 16) - 16} = x$$

$$f^{-1}(f(x)) = \left(\sqrt{x - 16}\right)^2 + 16 = x$$

Section 2.5 Multiple-Angle and Product-to-Sum Formulas

Solutions to Even-Numbered Exercises

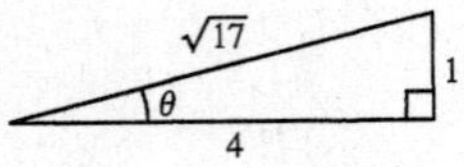

Figure for Exercises 2–8

2. $\tan\theta = \dfrac{1}{4}$

4. $\sin 2\theta = 2\sin\theta\cos\theta = 2\left(\dfrac{1}{\sqrt{17}}\right)\left(\dfrac{4}{\sqrt{17}}\right) = \dfrac{8}{17}$

6.
$$\sec 2\theta = \frac{1}{\cos 2\theta}$$
$$= \frac{1}{\cos^2\theta - \sin^2\theta}$$
$$= \frac{1}{\left(\frac{4}{\sqrt{17}}\right)^2 - \left(\frac{1}{\sqrt{17}}\right)^2} = \frac{1}{\frac{16}{17} - \frac{1}{17}} = \frac{17}{15}$$

8.
$$\cot 2\theta = \frac{1}{\tan 2\theta} = \frac{1 - \tan^2\theta}{2\tan\theta} = \frac{1 - \left(\frac{1}{4}\right)^2}{2\left(\frac{1}{4}\right)}$$
$$= \frac{15}{8}$$

10.
$$\sin 2x + \cos x = 0$$
$$2\sin x\cos x + \cos x = 0$$
$$\cos x(2\sin x + 1) = 0$$
$$\cos x = 0 \quad \text{or} \quad 2\sin x + 1 = 0$$
$$x = \frac{\pi}{2}, \frac{3\pi}{2} \qquad \sin x = -\frac{1}{2}$$
$$x = \frac{7\pi}{6}, \frac{11\pi}{6}$$

12.
$$\sin 2x\sin x = \cos x$$
$$2\sin x\cos x\sin x - \cos x = 0$$
$$\cos x(2\sin^2 x - 1) = 0$$
$$\cos x = 0 \quad \text{or} \quad 2\sin^2 x - 1 = 0$$
$$x = \frac{\pi}{2}, \frac{3\pi}{2} \qquad \sin^2 x = \frac{1}{2}$$
$$\sin x = \pm\frac{\sqrt{2}}{2}$$
$$x = \frac{\pi}{4}, \frac{3\pi}{4}, \frac{5\pi}{4}, \frac{7\pi}{4}$$

14.
$$\cos 2x + \sin x = 0$$
$$1 - 2\sin^2 x + \sin x = 0$$
$$2\sin^2 x - \sin x - 1 = 0$$
$$(2\sin x + 1)(\sin x - 1) = 0$$
$$2\sin x + 1 = 0 \quad \text{or} \quad \sin x - 1 = 0$$
$$\sin x = -\frac{1}{2} \qquad \sin x = 1$$
$$x = \frac{7\pi}{6}, \frac{11\pi}{6} \qquad x = \frac{\pi}{2}$$

16.

$$\tan 2x - 2\cos x = 0$$

$$\frac{2\tan x}{1-\tan^2 x} = 2\cos x$$

$$2\tan x = 2\cos x(1-\tan^2 x)$$

$$2\tan x = 2\cos x - 2\cos x\tan^2 x$$

$$2\tan x = 2\cos x - 2\cos x\frac{\sin^2 x}{\cos^2 x}$$

$$2\tan x = 2\cos x - 2\frac{\sin^2 x}{\cos x}$$

$$\tan x = \cos x - \frac{\sin^2 x}{\cos x}$$

$$\frac{\sin x}{\cos x} = \cos x - \frac{\sin^2 x}{\cos x}$$

$$\frac{\sin x}{\cos x} + \frac{\sin^2 x}{\cos x} - \cos x = 0$$

$$\frac{\sin x + \sin^2 x - \cos^2 x}{\cos x} = 0$$

$$\frac{1}{\cos x}[\sin x + \sin^2 x - (1-\sin^2 x)] = 0$$

$$\sec x[2\sin^2 x + \sin x - 1] = 0$$

$$\sec x(2\sin x - 1)(\sin x + 1) = 0$$

$\sec x = 0$ or $2\sin x - 1 = 0$ or $\sin x + 1 = 0$

No solution; $\sin x = \frac{1}{2}$, $x = \frac{\pi}{6}, \frac{5\pi}{6}$; $\sin x = -1$, $x = \frac{3\pi}{2}$

Also, values for which $\cos x = 0$ need to be checked.

$\frac{\pi}{2}, \frac{3\pi}{2}$ are solutions.

$$x = \frac{\pi}{6}, \frac{\pi}{2}, \frac{5\pi}{6}, \frac{3\pi}{2}$$

18.

$$(\sin 2x + \cos 2x)^2 = 1$$

$$\sin^2 2x + 2\sin 2x\cos 2x + \cos^2 2x = 1$$

$$2\sin 2x\cos 2x = 0$$

$$\sin 4x = 0$$

$$4x = n\pi$$

$$x = \frac{n\pi}{4}$$

$$x = 0, \frac{\pi}{4}, \frac{\pi}{2}, \frac{3\pi}{4}, \pi, \frac{5\pi}{4}, \frac{3\pi}{2}, \frac{7\pi}{4}$$

20. $6\cos^2 x - 3 = 3(2\cos^2 x - 1)$

$= 3\cos 2x$

22. $(\cos x + \sin x)(\cos x - \sin x) = \cos^2 x - \sin^2 x$

$= \cos 2x$

24. $\cos u = -\frac{2}{3}, \frac{\pi}{2} < u < \pi$

$\sin 2u = 2\sin u\cos u = 2 \cdot \frac{\sqrt{5}}{3}\left(-\frac{2}{3}\right) = -\frac{4\sqrt{5}}{9}$

$\cos 2u = \cos^2 u - \sin^2 u = \frac{4}{9} - \frac{5}{9} = -\frac{1}{9}$

$\tan 2u = \frac{2\tan u}{1 - \tan^2 u} = \frac{2\left(-\frac{\sqrt{5}}{2}\right)}{1 - \frac{5}{4}} = 4\sqrt{5}$

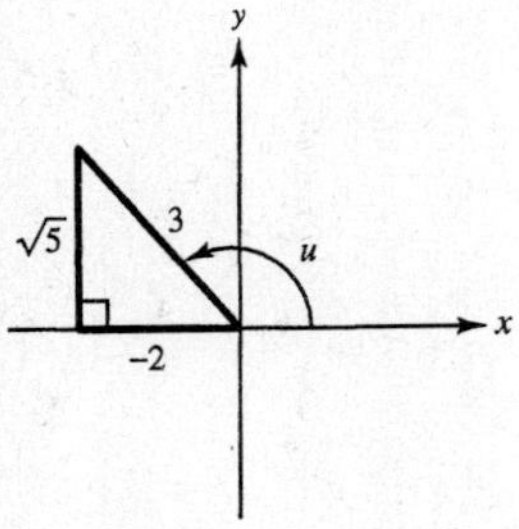

26. $\cot u = -4, \frac{3\pi}{2} < u < 2\pi$

$\sin 2u = 2\sin u\cos u = 2\left(-\frac{1}{\sqrt{17}}\right)\left(\frac{4}{\sqrt{17}}\right) = -\frac{8}{17}$

$\cos 2u = \cos^2 u - \sin^2 u$

$= \left(\frac{4}{\sqrt{17}}\right)^2 - \left(-\frac{1}{\sqrt{17}}\right)^2 = \frac{15}{17}$

$\tan 2u = \frac{2\tan u}{1 - \tan^2 u} = \frac{2\left(-\frac{1}{4}\right)}{1 - \left(-\frac{1}{4}\right)^2} = -\frac{8}{15}$

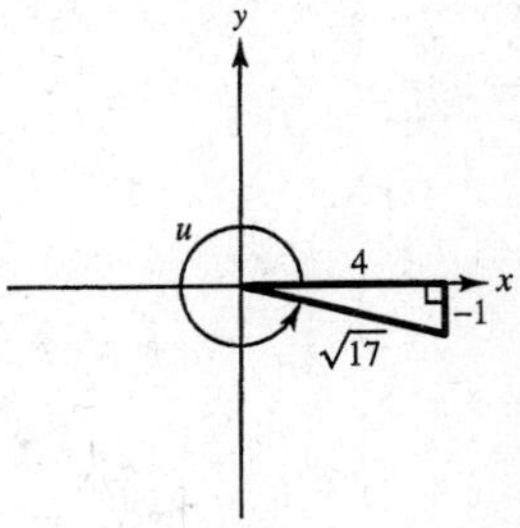

$\csc u = 3, \frac{\pi}{2} < u < \pi$

28. $\sin 2u = 2\sin u\cos u = 2 \cdot \frac{1}{3}\left(-\frac{2\sqrt{2}}{3}\right) = -\frac{4\sqrt{2}}{9}$

$\cos 2u = \cos^2 u - \sin^2 u = \left(-\frac{2\sqrt{2}}{3}\right)^2 - \left(\frac{1}{3}\right)^2 = \frac{7}{9}$

$\tan 2u = \frac{2\tan u}{1 - \tan^2 u} = \frac{2\left(-\frac{\sqrt{2}}{4}\right)}{1 - \left(-\frac{\sqrt{2}}{4}\right)^2} = -\frac{4\sqrt{2}}{7}$

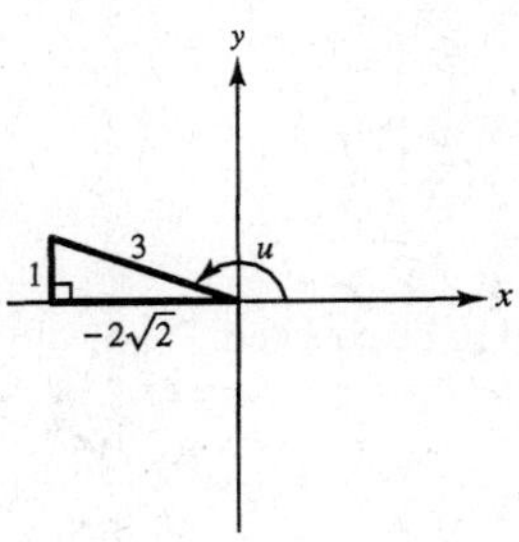

30. $\sin^8 x = \sin^4 x \sin^4 x = (\sin^2 x \sin^2 x)(\sin^2 x \sin^2 x)$

$\sin^2 x = \frac{1 - \cos 2x}{2}$

$\sin^4 x = \left(\frac{1 - \cos 2x}{2}\right)\left(\frac{1 - \cos 2x}{2}\right)$

$= \frac{1}{4}(1 - 2\cos 2x + \cos^2 2x)$

$= \frac{1}{4}\left(1 - 2\cos 2x + \left(\frac{1 + \cos 4x}{2}\right)\right)$

$= \frac{1}{8}(3 - 4\cos 2x + \cos 4x)$

—CONTINUED—

30. —CONTINUED—

$$\sin^8 x = \sin^4 x \sin^4 x$$

$$= \frac{1}{64}(3 - 4\cos 2x + \cos 4x)(3 - 4\cos 2x + \cos 4x)$$

$$= \frac{1}{64}(9 - 24\cos 2x + 16\cos^2 2x + 6\cos 4x - 8\cos 2x\cos 4x + \cos^2 4x)$$

$$= \frac{1}{64}\left[9 - 24\cos 2x + 16\left(\frac{1+\cos 4x}{2}\right) + 6\cos 4x - 8\frac{1}{2}(\cos 6x + \cos 2x) + \left(\frac{1+\cos 8x}{2}\right)\right]$$

$$= \frac{1}{64}\left[\frac{35}{2} - 28\cos 2x + 14\cos 4x - 4\cos 6x + \frac{1}{2}\cos 8x\right]$$

$$= \frac{1}{128}[35 - 56\cos 2x + 28\cos 4x - 8\cos 6x + \cos 8x]$$

In the above, we used $\cos 2x \cos 4x = \frac{1}{2}(\cos 6x + \cos 2x)$ which follows from $\cos x(4x \pm 2x) = \cos 4x \cos 2x \pm \sin 4x \sin 2x$.

32. $\sin^4 x \cos^4 x = \sin^2 x \sin^2 x \cos^2 x \cos^2 x$

$$= (\sin^2 x \cos^2 x)(\sin^2 x \cos^2 x)$$

$$= \left(\frac{1}{4}\sin^2 2x\right)\left(\frac{1}{4}\sin^2 2x\right)$$

$$= \left[\frac{1}{4}\left(\frac{1-\cos 4x}{2}\right)\right]\left[\frac{1}{4}\left(\frac{1-\cos 4x}{2}\right)\right]$$

$$= \frac{1}{64}[1 - 2\cos 4x + \cos^2 4x]$$

$$= \frac{1}{64}\left[1 - 2\cos 4x + \left(\frac{1+\cos 8x}{2}\right)\right]$$

$$= \frac{1}{64}\left[\frac{3}{2} - 2\cos 4x + \frac{1}{2}\cos 8x\right]$$

$$= \frac{1}{128}(3 - 4\cos 4x + \cos 8x)$$

34. $\sin^4 x \cos^2 x = \sin^2 x \sin^2 x \cos^2 x$

$$= \left(\frac{1-\cos 2x}{2}\right)\left(\frac{1-\cos 2x}{2}\right)\left(\frac{1+\cos 2x}{2}\right)$$

$$= \frac{1}{8}(1 - \cos 2x)(1 - \cos^2 2x)$$

$$= \frac{1}{8}(1 - \cos 2x - \cos^2 2x + \cos^3 2x)$$

$$= \frac{1}{8}\left[1 - \cos 2x - \left(\frac{1+\cos 4x}{2}\right) + \cos 2x\left(\frac{1+\cos 4x}{2}\right)\right]$$

$$= \frac{1}{16}[2 - 2\cos 2x - 1 - \cos 4x + \cos 2x + \cos 2x\cos 4x]$$

$$= \frac{1}{16}\left[1 - \cos 2x - \cos 4x + \frac{1}{2}\cos 2x + \frac{1}{2}\cos 6x\right]$$

$$= \frac{1}{32}[2 - 2\cos 2x - 2\cos 4x + \cos 2x + \cos 6x]$$

$$= \frac{1}{32}[2 - \cos 2x - 2\cos 4x + \cos 6x]$$

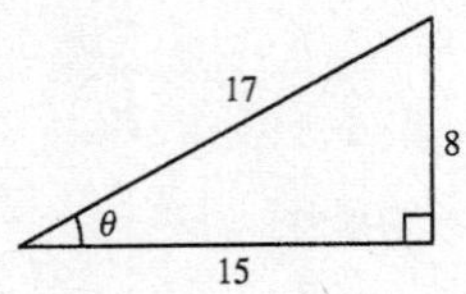

Figure for Exercises 36–40

36. $\sin\frac{\theta}{2} = \sqrt{\frac{1-\cos\theta}{2}} = \sqrt{\frac{1-\frac{15}{17}}{2}}$

$= \sqrt{\frac{\frac{2}{17}}{2}} = \frac{1}{\sqrt{17}} = \frac{\sqrt{17}}{17}$

38. $\sec\frac{\theta}{2} = \frac{1}{\cos\frac{\theta}{2}} = \frac{1}{\sqrt{\frac{1+\cos\theta}{2}}}$

$= \frac{1}{\sqrt{\frac{1+\frac{15}{17}}{2}}} = \frac{1}{\sqrt{\frac{16}{17}}}$

$= \frac{\sqrt{17}}{4}$

40. $\cot\frac{\theta}{2} = \frac{1}{\tan\frac{\theta}{2}} = \frac{\sin\theta}{1-\cos\theta} = \frac{\frac{8}{17}}{1-\frac{15}{17}}$

$= \frac{\frac{8}{17}}{\frac{2}{17}} = 4$

42. $\sin 165° = \sin\left(\frac{1}{2}\cdot 330°\right) = \sqrt{\frac{1-\cos 330°}{2}} = \sqrt{\frac{1-\frac{\sqrt{3}}{2}}{2}} = \frac{1}{2}\sqrt{2-\sqrt{3}}$

$\cos 165° = \cos\left(\frac{1}{2}\cdot 330°\right) = -\sqrt{\frac{1+\cos 330°}{2}} = -\sqrt{\frac{1+\frac{\sqrt{3}}{2}}{2}} = -\frac{1}{2}\sqrt{2+\sqrt{3}}$

$\tan 165° = \tan\left(\frac{1}{2}\cdot 330°\right) = \frac{\sin 330°}{1+\cos 330°} = \frac{-\frac{1}{2}}{1+\frac{\sqrt{3}}{2}} = \frac{-1}{2+\sqrt{3}} = \sqrt{3}-2$

44. $\sin 67° 30' = \sin\left(\frac{1}{2}\cdot 135°\right) = \sqrt{\frac{1-\cos 135°}{2}} = \sqrt{\frac{1+\frac{\sqrt{2}}{2}}{2}} = \frac{1}{2}\sqrt{2+\sqrt{2}}$

$\cos 67° 30' = \cos\left(\frac{1}{2}\cdot 135°\right) = \sqrt{\frac{1+\cos 135°}{2}} = \sqrt{\frac{1-\frac{\sqrt{2}}{2}}{2}} = \frac{1}{2}\sqrt{2-\sqrt{2}}$

$\tan 67° 30' = \tan\left(\frac{1}{2}\cdot 135°\right) = \frac{\sin 135°}{1+\cos 135°} = \frac{\frac{\sqrt{2}}{2}}{1-\frac{\sqrt{2}}{2}} = 1+\sqrt{2}$

46. $\sin \dfrac{\pi}{12} = \sin\left[\dfrac{1}{2}\left(\dfrac{\pi}{6}\right)\right] = \sqrt{\dfrac{1 - \cos \dfrac{\pi}{6}}{2}} = \sqrt{\dfrac{1 - \dfrac{\sqrt{3}}{2}}{2}} = \dfrac{1}{2}\sqrt{2 - \sqrt{3}}$

$\cos \dfrac{\pi}{12} = \cos\left[\dfrac{1}{2}\left(\dfrac{\pi}{6}\right)\right] = \sqrt{\dfrac{1 + \cos \dfrac{\pi}{6}}{2}} = \dfrac{1}{2}\sqrt{2 + \sqrt{3}}$

$\tan \dfrac{\pi}{12} = \tan\left[\dfrac{1}{2}\left(\dfrac{\pi}{6}\right)\right]\dfrac{\sin\dfrac{\pi}{6}}{1 + \cos\dfrac{\pi}{6}} = \dfrac{\dfrac{1}{2}}{1 + \dfrac{\sqrt{3}}{2}} = 2 - \sqrt{3}$

48. $\sin \dfrac{7\pi}{12} = \sin\left[\dfrac{1}{2}\left(\dfrac{7\pi}{6}\right)\right] = \sqrt{\dfrac{1 - \cos(7\pi(6))}{2}} = \sqrt{\dfrac{1 + \dfrac{\sqrt{3}}{2}}{2}} = \dfrac{1}{2}\sqrt{2 + \sqrt{3}}$

$\cos \dfrac{7\pi}{12} = \cos\left[\dfrac{1}{2}\left(\dfrac{7\pi}{6}\right)\right] = -\sqrt{\dfrac{1 + \cos \dfrac{7\pi}{6}}{2}} = -\sqrt{\dfrac{1 - \dfrac{\sqrt{3}}{2}}{2}} = -\dfrac{1}{2}\sqrt{2 - \sqrt{3}}$

$\tan \dfrac{7\pi}{12} = \tan\left[\dfrac{1}{2}\left(\dfrac{7\pi}{6}\right)\right] = \dfrac{\sin\dfrac{7\pi}{6}}{1 + \cos\dfrac{7\pi}{6}} = \dfrac{-\dfrac{1}{2}}{1 - \dfrac{\sqrt{3}}{2}} = -2 - \sqrt{3}$

50. $\cos u = \dfrac{3}{5}, 0 < u < \dfrac{\pi}{2}$

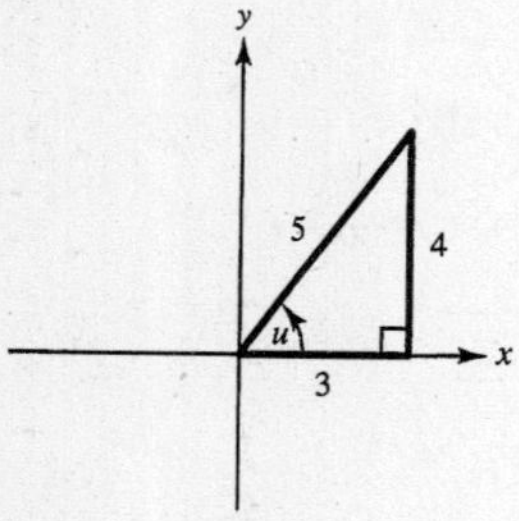

$\sin\left(\dfrac{u}{2}\right) = \sqrt{\dfrac{1 - \cos u}{2}} = \sqrt{\dfrac{1 - \dfrac{3}{5}}{2}} = \dfrac{\sqrt{5}}{5}$

$\cos\left(\dfrac{u}{2}\right) = \sqrt{\dfrac{1 + \cos u}{2}} = \sqrt{\dfrac{1 + \dfrac{3}{5}}{2}} = \dfrac{2\sqrt{5}}{5}$

$\tan\left(\dfrac{u}{2}\right) = \dfrac{\sin u}{1 + \cos u} = \dfrac{\dfrac{4}{5}}{1 + \dfrac{3}{5}} = \dfrac{1}{2}$

52. $\cot u = 3, \pi < u < \dfrac{3\pi}{2}$

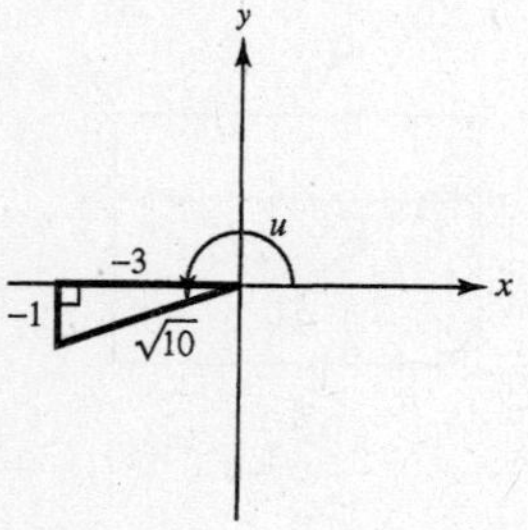

$\sin\left(\dfrac{u}{2}\right) = \sqrt{\dfrac{1 - \cos u}{2}} = \sqrt{\dfrac{1 + \dfrac{3}{\sqrt{10}}}{2}} = \sqrt{\dfrac{10 + 3\sqrt{10}}{20}} = \dfrac{1}{2}\sqrt{\dfrac{10 + 3\sqrt{10}}{5}}$

$\cos\left(\dfrac{u}{2}\right) = -\sqrt{\dfrac{1 + \cos u}{2}} = -\sqrt{\dfrac{1 - \dfrac{3}{\sqrt{10}}}{2}} = -\sqrt{\dfrac{10 - 3\sqrt{10}}{20}} = -\dfrac{1}{2}\sqrt{\dfrac{10 - 3\sqrt{10}}{5}}$

$\tan\left(\dfrac{u}{2}\right) = \dfrac{1 - \cos u}{\sin u} = \dfrac{1 + \dfrac{3}{\sqrt{10}} - \dfrac{1}{\sqrt{10}}}{-\dfrac{1}{\sqrt{10}}} = -3 - \sqrt{10}$

54. $\sec u = -\frac{7}{2}, \frac{\pi}{2} < u < \pi$

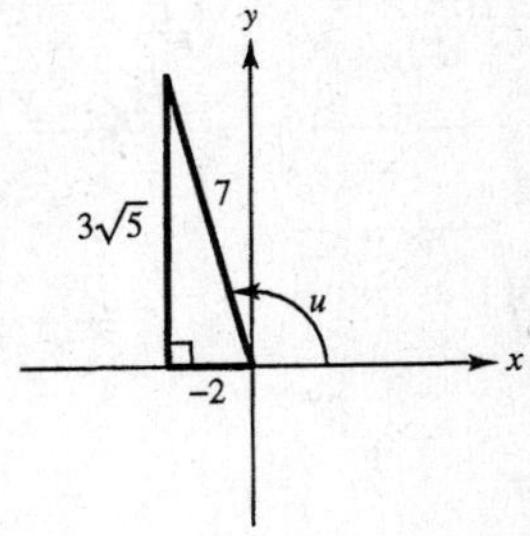

$$\sin\left(\frac{u}{2}\right) = \sqrt{\frac{1 - \cos u}{2}} = \sqrt{\frac{1 + \frac{2}{7}}{2}} = \frac{3\sqrt{14}}{14}$$

$$\cos\left(\frac{u}{2}\right) = \sqrt{\frac{1 + \cos u}{2}} = \sqrt{\frac{1 - \frac{2}{7}}{2}} = \frac{\sqrt{70}}{14}$$

$$\tan\left(\frac{u}{2}\right) = \frac{1 - \cos u}{\sin u} = \frac{1 + \frac{2}{7}}{\frac{3\sqrt{5}}{7}} = \frac{3\sqrt{5}}{5}$$

56. $\sqrt{\frac{1 + \cos 4x}{2}} = \left|\cos \frac{4x}{2}\right| = |\cos 2x|$

58. $-\sqrt{\frac{1 - \cos(x - 1)}{2}} = -\left|\sin\left(\frac{x - 1}{2}\right)\right|$

60. $h(x) = \sin \frac{x}{2} + \cos x - 1$

$$\sin \frac{x}{2} + \cos x - 1 = 0$$

$$\pm\sqrt{\frac{1 - \cos x}{2}} = 1 - \cos x$$

$$\frac{1 - \cos x}{2} = 1 - 2\cos x + \cos^2 x$$

$$1 - \cos x = 2 - 4\cos x + 2\cos^2 x$$

$$2\cos^2 x - 3\cos x + 1 = 0$$

$$(2\cos x - 1)(\cos x - 1) = 0$$

$2\cos x - 1 = 0$ or $\cos x - 1 = 0$

$\cos x = \frac{1}{2}$ $\qquad \cos x = 1$

$x = \frac{\pi}{3}, \frac{5\pi}{3}$ $\qquad x = 0$

$0, \frac{\pi}{3}$, and $\frac{5\pi}{3}$ are all solutions to the equation.

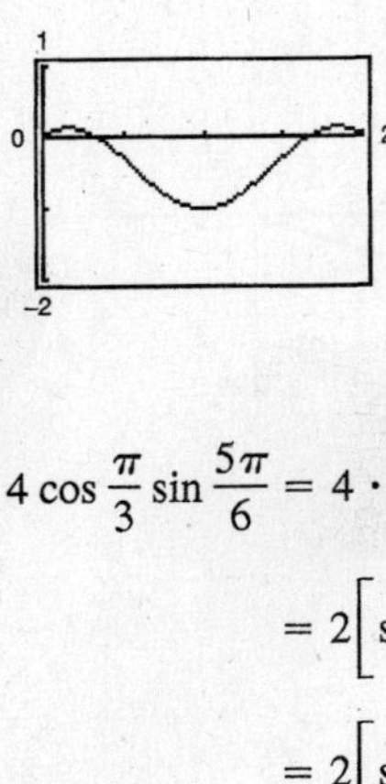

62. $g(x) = \tan \frac{x}{2} - \sin x$

$$\tan \frac{x}{2} - \sin x = 0$$

$$\frac{1 - \cos x}{\sin x} = \sin x$$

$$1 - \cos x = \sin^2 x$$

$$1 - \cos x = 1 - \cos^2 x$$

$$\cos^2 x - \cos x = 0$$

$$\cos x(\cos x - 1) = 0$$

$\cos x = 0$ or $\cos x - 1 = 0$

$x = \frac{\pi}{2}, \frac{3\pi}{2}$ $\qquad \cos x = 1$

$x = 0$

$0, \frac{\pi}{2}$, and $\frac{3\pi}{2}$ are all solutions to the equation.

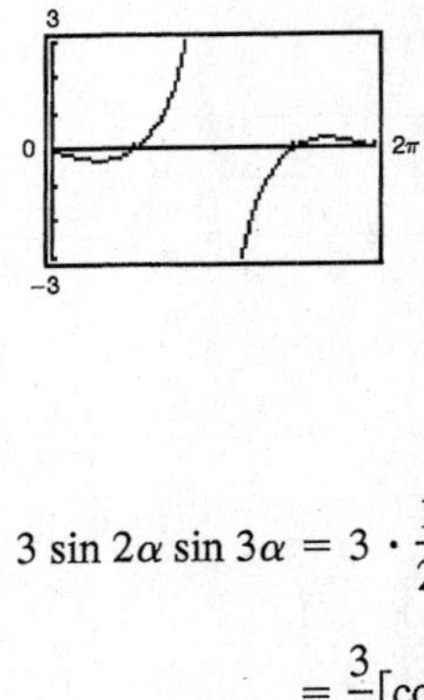

64. $4\cos \frac{\pi}{3} \sin \frac{5\pi}{6} = 4 \cdot \frac{1}{2}\left[\sin\left(\frac{\pi}{3} + \frac{5\pi}{6}\right) - \sin\left(\frac{\pi}{3} - \frac{5\pi}{6}\right)\right]$

$$= 2\left[\sin\left(\frac{7\pi}{6}\right) - \sin\left(-\frac{\pi}{2}\right)\right]$$

$$= 2\left[\sin\left(\frac{7\pi}{6}\right) + \sin\left(\frac{\pi}{2}\right)\right]$$

66. $3\sin 2\alpha \sin 3\alpha = 3 \cdot \frac{1}{2}[\cos(2\alpha - 3\alpha) - \cos(2\alpha + 3\alpha)$

$$= \frac{3}{2}[\cos(-\alpha) - \cos 5\alpha]$$

$$= \frac{3}{2}(\cos \alpha - \cos 5\alpha)$$

68. $\cos 2\theta \cos 4\theta = \frac{1}{2}[\cos(2\theta - 4\theta) + \cos(2\theta + 4\theta)]$
$= \frac{1}{2}[\cos(-2\theta) + \cos 6\theta]$
$= \frac{1}{2}(\cos 2\theta + \cos 6\theta)$

70. $\sin(x + y)\cos(x - y) = \frac{1}{2}(\sin 2x + \sin 2y)$

72. $\sin(\theta + \pi)\sin(\theta - \pi) = \frac{1}{2}(\cos 2\pi - \cos 2\theta)$

74. $6 \sin 45° \cos 15° = 6\left(\frac{1}{2}\right)(\sin 60° + \sin 30°)$
$= 3(\sin 60° + \sin 30°)$

76. $\cos 120° + \cos 30° = 2\cos\left(\frac{120° + 30°}{2}\right)\cos\left(\frac{120° - 30°}{2}\right)$
$= 2\cos 75° \cos 45°$

78. $\sin\frac{5\pi}{4} - \sin\frac{3\pi}{4} = 2\cos\left(\frac{\frac{5\pi}{4} + \frac{3\pi}{4}}{2}\right)\sin\left(\frac{\frac{5\pi}{4} - \frac{3\pi}{4}}{2}\right)$
$= 2\cos\pi \sin\frac{\pi}{4}$

80. $\sin 3\theta + \sin\theta = 2\sin\left(\frac{3\theta + \theta}{2}\right)\cos\left(\frac{3\theta - \theta}{2}\right)$
$= 2\sin 2\theta \cos\theta$

82. $\sin x + \sin 5x = 2\sin\left(\frac{x + 5x}{2}\right)\cos\left(\frac{x - 5x}{2}\right)$
$= 2\sin 3x \cos(-2x)$
$= 2\sin 3x \cos 2x$

84. $\cos(\phi + 2\pi) + \cos\phi = 2\cos\left(\frac{\phi + 2\pi + \phi}{2}\right)\cos\left(\frac{\phi + 2\pi - \phi}{2}\right)$
$= 2\cos(\phi + \pi)\cos(\pi)$

86. $\sin\left(x + \frac{\pi}{2}\right) + \sin\left(x - \frac{\pi}{2}\right) = 2\sin\left(\frac{x + \frac{\pi}{2} + x - \frac{\pi}{2}}{2}\right)\cos\left(\frac{x + \frac{\pi}{2} - \left(x - \frac{\pi}{2}\right)}{2}\right)$
$= 2\sin x \cos\frac{\pi}{2}$

88. $h(x) = \cos 2x - \cos 6x$
$\cos 2x - \cos 6x = 0$
$-2\sin 4x \sin(-2x) = 0$
$2\sin 4x \sin 2x = 0$

$\sin 4x = 0$ or $\sin 2x = 0$

$4x = n\pi$ $\qquad$ $2x = n\pi$

$x = \frac{n\pi}{4}$ $\qquad$ $x = \frac{n\pi}{2}$

$x = 0, \frac{\pi}{4}, \frac{\pi}{2}, \frac{3\pi}{4}, \pi, \frac{5\pi}{4}, \frac{3\pi}{2}, \frac{7\pi}{4}$ $\qquad$ $x = 0, \frac{\pi}{2}, \pi, \frac{3\pi}{2}$

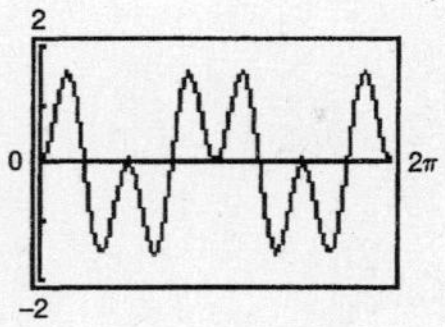

90. $f(x) = \sin^2 3x - \sin^2 x$

$$\sin^2 3x - \sin^2 x = 0$$
$$(\sin 3x + \sin x)(\sin 3x - \sin x) = 0$$
$$(2 \sin 2x \cos x)(2 \cos 2x \sin x) = 0$$

$\sin 2x = 0 \Rightarrow x = 0, \frac{\pi}{2}, \pi, \frac{3\pi}{2}$ or

$\cos x = 0 \Rightarrow x = \frac{\pi}{2}, \frac{3\pi}{2}$ or

$\cos 2x = 0 \Rightarrow x = \frac{\pi}{4}, \frac{3\pi}{4}, \frac{5\pi}{4}, \frac{7\pi}{4}$ or

$\sin x = 0 \Rightarrow x = 0, \pi$

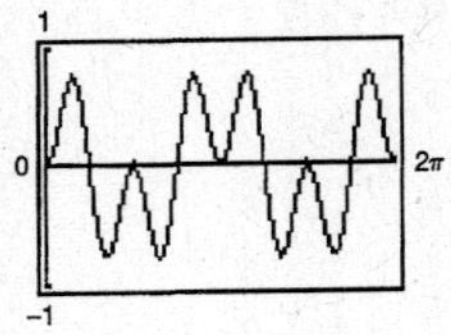

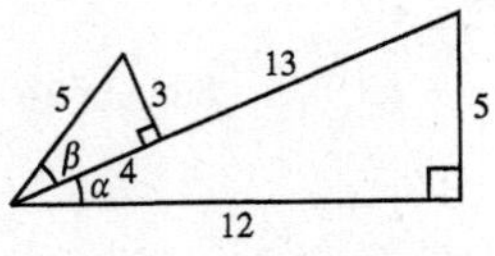

Figure for Exercises 92 and 94

92. $\cos^2 \alpha = (\cos \alpha)^2 = \left(\frac{12}{13}\right)^2 = \frac{144}{169}$

$$\cos^2 \alpha = 1 - \sin^2 \alpha$$
$$= 1 - \left(\frac{5}{13}\right)^2$$
$$= 1 - \frac{25}{169} = \frac{144}{169}$$

94. $\cos \alpha \sin \beta = \left(\frac{12}{13}\right)\left(\frac{3}{5}\right) = \frac{36}{65}$

$$\cos \alpha \sin \beta = \sin\left(\frac{\pi}{2} - \alpha\right)\cos\left(\frac{\pi}{2} - \beta\right)$$
$$= \left(\frac{12}{13}\right)\left(\frac{3}{5}\right) = \frac{36}{65}$$

96. $\sec 2\theta = \frac{1}{\cos 2\theta} = \frac{1}{\cos^2 \theta - \sin^2 \theta}$

$$= \frac{\frac{1}{\cos^2 \theta}}{1 - \frac{\sin^2 \theta}{\cos^2 \theta}}$$
$$= \frac{\sec^2 \theta}{1 - \tan^2 \theta}$$
$$= \frac{\sec^2 \theta}{1 - (\sec^2 \theta - 1)}$$
$$= \frac{\sec^2 \theta}{2 - \sec^2 \theta}$$

98. $\cos^4 x - \sin^4 x = (\cos^2 x - \sin^2 x)(\cos^2 x + \sin^2 x)$

$$= (\cos 2x)(1)$$
$$= \cos 2x$$

100. $\sin\left(\frac{\alpha}{3}\right)\cos\left(\frac{\alpha}{3}\right) = \frac{1}{2}\left[2\left(\sin\left(\frac{\alpha}{3}\right)\cos\left(\frac{\alpha}{3}\right)\right)\right]$

$$= \frac{1}{2}\sin\frac{2\alpha}{3}$$

102. $\frac{\cos 3\beta}{\cos \beta} = \frac{\cos^3 \beta - 3\sin^2 \beta \cos \beta}{\cos \beta}$

$$= \cos^2 \beta - 3\sin^2 \beta$$
$$= 1 - \sin^2 \beta - 3\sin^2 \beta$$
$$= 1 - 4\sin^2 \beta$$

104. $\tan\dfrac{u}{2} = \dfrac{1-\cos u}{\sin u}$

$= \dfrac{1}{\sin u} - \dfrac{\cos u}{\sin u}$

$= \csc u - \cot u$

106. $\dfrac{\sin x \pm \sin y}{\cos x + \cos y} = \dfrac{2\sin\left(\dfrac{x \pm y}{2}\right)\cos\left(\dfrac{x \mp y}{2}\right)}{2\cos\left(\dfrac{x+y}{2}\right)\cos\left(\dfrac{x-y}{2}\right)}$

$= \tan\left(\dfrac{x \pm y}{2}\right)$

108. $\dfrac{\cos t + \cos 3t}{\sin 3t - \sin t} = \dfrac{2\cos\left(\dfrac{4t}{2}\right)\cos\left(-\dfrac{2t}{2}\right)}{2\cos\left(\dfrac{4t}{2}\right)\sin\left(\dfrac{2t}{2}\right)}$

$= \dfrac{\cos(-t)}{\sin(t)}$

$= \dfrac{\cos(t)}{\sin(t)}$

$= \cot t$

110. $\cos\left(\dfrac{\pi}{3} + x\right) + \cos\left(\dfrac{\pi}{3} - x\right) = 2\cos\left(\dfrac{\dfrac{\pi}{3} + x + \dfrac{\pi}{3} - x}{2}\right)\cos\left(\dfrac{\dfrac{\pi}{3} + x - \left(\dfrac{\pi}{3} - x\right)}{2}\right)$

$= 2\cos\left(\dfrac{\pi}{3}\right)\cos(x)$

$= 2\left(\dfrac{1}{2}\right)\cos x = \cos x$

112. $\sin 4\beta = 2\sin 2\beta\cos 2\beta$

$= 2(2\sin\beta\cos\beta)(1 - 2\sin^2\beta)$

$= 4\sin\beta\cos\beta(1 - 2\sin^2\beta)$

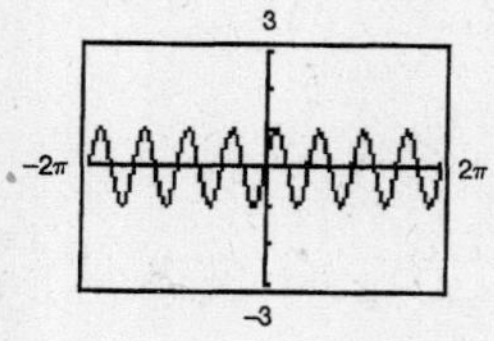

114. $\dfrac{\cos 3x - \cos x}{\sin 3x - \sin x} = \dfrac{-2\sin\left(\dfrac{3x+x}{2}\right)\sin\left(\dfrac{3x-x}{2}\right)}{2\cos\left(\dfrac{3x+x}{2}\right)\sin\left(\dfrac{3x-x}{2}\right)}$

$= \dfrac{-2\sin 2x\sin x}{2\cos 2x\sin x}$

$= -\tan 2x$

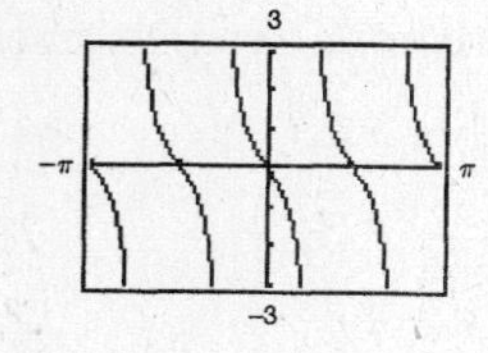

116. $f(x) = \cos^2 x = \dfrac{1 + \cos 2x}{2} = \dfrac{1}{2} + \dfrac{\cos 2x}{2}$

Shifted upward by $\dfrac{1}{2}$ unit.

Amplitude: $|a| = \dfrac{1}{2}$

Period: $\dfrac{2\pi}{2} = \pi$

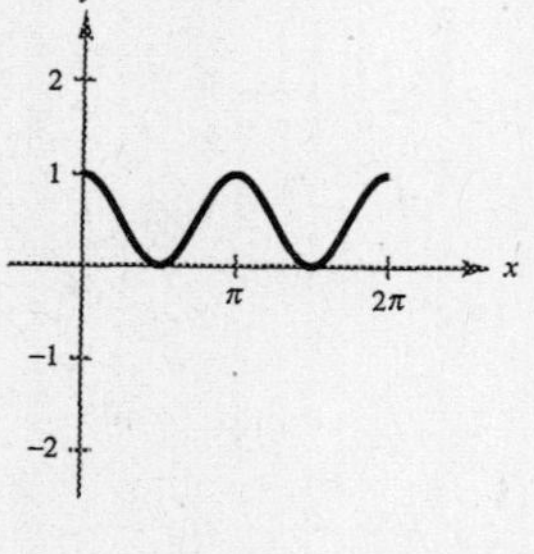

118. $\cos(2\arccos x) = \cos^2(\arccos x) - \sin^2(\arccos x)$

$= x^2 - (1 - x^2) = 2x^2 - 1$

120. $r = \frac{1}{32}v_0^2 \sin 2\theta$

$= \frac{1}{32}v_0^2\, 2(\sin\theta\cos\theta)$

$= \frac{1}{16}v_0^2 \sin\theta\cos\theta$

122. $\frac{x}{2} = 2r\sin^2\frac{\theta}{2} = 2r\left(\frac{1-\cos\theta}{2}\right)$

$= r(1-\cos\theta)$

So, $x = 2r(1-\cos\theta)$

124. False.

$\sin\frac{u}{2} = \sqrt{\frac{1-\cos u}{2}}$

when u is in the second quadrant.

126. $f(x) = \cos 2x - 2\sin x$

(a)

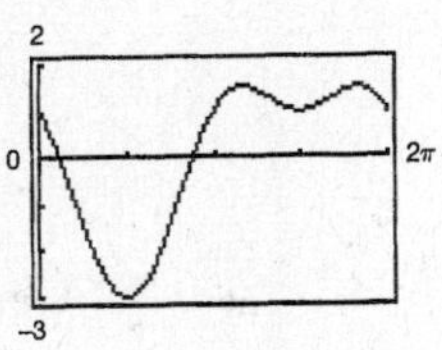

Maximum points: (3.6652, 1.5), (5.7596, 1.5)

Minimum points: (1.5708, −3), (4.7124, 1)

(b) $-2\cos x(2\sin x + 1) = 0$

$-2\cos x = 0$ or $2\sin x + 1 = 0$

$\cos x = 0$ $\qquad \sin x = -\frac{1}{2}$

$x = \frac{\pi}{2}, \frac{3\pi}{2}$ $\qquad x = \frac{7\pi}{6}, \frac{11\pi}{6}$

$\frac{\pi}{2} \approx 1.5708$ $\qquad \frac{7\pi}{6} \approx 3.6652$

$\frac{3\pi}{2} \approx 4.7124$ $\qquad \frac{11\pi}{2} \approx 5.7596$

128. (a)

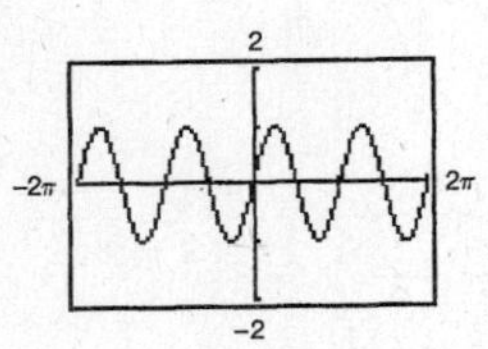

(b) The graph appears to be that of $\sin 2x$.

(c) $2\sin x\left[2\cos^2\left(\frac{x}{2}\right) - 1\right] = 2\sin x\cos x$

$= \sin 2x$

130. Let x = number of gallons of 100% concentrate.

$0.30(55 - x) + 1.00x = 0.50(55)$

$16.50 - 0.30x + x = 27.50$

$0.70x = 11$

$x \approx 15.7$ gallons

Review Exercises for Chapter 2

Solutions to Even-Numbered Exercises

2. $\frac{1}{\sin x} = \csc x$

4. $\frac{1}{\tan x} = \cot x$

6. $\sqrt{1+\tan^2 x} = \sqrt{\sec^2 x} = |\sec x|$

8. $\tan\theta = \frac{2}{3}, \sec\theta = \frac{\sqrt{13}}{3}$

θ is in Quadrant I.

$\cos\theta = \frac{1}{\sec\theta} = \frac{3}{\sqrt{13}} = \frac{3\sqrt{13}}{13}$

$\sin\theta = \sqrt{1-\cos^2\theta} = \sqrt{1-\frac{9}{13}} = \sqrt{\frac{4}{13}} = \frac{2\sqrt{13}}{13}$

$\csc\theta = \frac{1}{\sin\theta} = \frac{\sqrt{13}}{2}$

$\cot\theta = \frac{1}{\tan\theta} = \frac{3}{2}$

10. $\csc\left(\frac{\pi}{2} - \theta\right) = \sec\theta = 9, \sin\theta = \frac{4\sqrt{5}}{9}$

θ is in Quadrant I.

$\cos\theta = \frac{1}{\sec\theta} = \frac{1}{9}$

$\tan\theta = \frac{\sin\theta}{\cos\theta} = \frac{\frac{4\sqrt{5}}{9}}{\frac{1}{9}} = 4\sqrt{5}$

$\csc\theta = \frac{1}{\sin\theta} = \frac{9}{4\sqrt{5}} = \frac{9\sqrt{5}}{20}$

$\cot\theta = \frac{1}{\tan\theta} = \frac{1}{4\sqrt{5}} = \frac{\sqrt{5}}{20}$

12. $\dfrac{\tan\theta}{1-\cos^2\theta} = \dfrac{\dfrac{\sin\theta}{\cos\theta}}{\sin^2\theta} = \dfrac{1}{\sin\theta\cos\theta}$

$= \csc\theta\sec\theta$

14. $\cot^2 x(\sin^2 x) = \dfrac{\cos^2 x}{\sin^2 x}\sin^2 x = \cos^2 x$

16. $\dfrac{\cot\left(\dfrac{\pi}{2}-u\right)}{\cos u} = \dfrac{\tan u}{\cos u} = \tan u\sec u$

18. $\tan^2\theta\csc^2\theta - \tan^2\theta = \tan^2\theta(\csc^2\theta - 1)$

$= \tan^2\theta\cot^2\theta = 1$

20. $(\sec x - \tan x)^2 = \sec^2 x - 2\sec x\tan x + \tan^2 x$

$= 1 + \tan^2 x - 2\sec x\tan x + \tan^2 x$

$= 1 - 2\sec x\tan x + 2\tan^2 x$

22. $\dfrac{\cos^2 x}{1-\sin x} = \dfrac{\cos^2 x}{(1-\sin x)}\cdot\dfrac{(1+\sin x)}{(1+\sin x)}$

$= \dfrac{\cos^2 x(1+\sin x)}{1-\sin^2 x}$

$= 1 + \sin x$

24. $\sin^{-1/2} x\cos x = \dfrac{1}{\sqrt{\sin x}}(\cos x) = \dfrac{\sqrt{\sin x}}{\sin x}(\cos x)$

$= \sqrt{\sin x}\left(\dfrac{\cos x}{\sin x}\right) = \sqrt{\sin x}\cot x$

26. $\sec^2 x\cot x - \cot x = \cot x(\sec^2 x - 1) = \cot x\tan^2 x$

$= \left(\dfrac{1}{\tan x}\right)\tan^2 x = \tan x$

28. $\cot\left(\dfrac{\pi}{2}-x\right) = \tan x$ by the Cofunction Identity

30. $\dfrac{1}{\tan x\csc x\sin x} = \dfrac{1}{(\tan x)\left(\dfrac{1}{\sin x}\right)(\sin x)} = \dfrac{1}{\tan x}$

$= \cot x$

32. $\cos^3 x\sin^2 x = \cos x\cos^2 x\sin^2 x$

$= \cos x(1-\sin^2 x)\sin^2 x$

$= \cos x(\sin^2 x - \sin^4 x)$

$= (\sin^2 x - \sin^4 x)\cos x$

34. $4\cos\theta = 1 + 2\cos\theta$

$2\cos\theta = 1$

$\cos\theta = \dfrac{1}{2}$

$\theta = \dfrac{\pi}{3} + 2n\pi$ or $\dfrac{5\pi}{3} + 2n\pi$

36. $\dfrac{1}{2}\sec x - 1 = 0$

$\dfrac{1}{2}\sec x = 1$

$\sec x = 2$

$\cos x = \dfrac{1}{2}$

$x = \dfrac{\pi}{3} + 2n\pi$ or $\dfrac{5\pi}{3} + 2n\pi$

38. $4\tan^2 u - 1 = \tan^2 u$

$3\tan^2 u - 1 = 0$

$\tan^2 u = \dfrac{1}{3}$

$\tan u = \pm\dfrac{1}{\sqrt{3}} = \pm\dfrac{\sqrt{3}}{3}$

$u = \dfrac{\pi}{6} + n\pi$ or $\dfrac{5\pi}{6} + n\pi$

40. $2\sin^2 x - 3\sin x = -1$

$2\sin^2 x - 3\sin x + 1 = 0$

$(2\sin x - 1)(\sin x - 1) = 0$

$2\sin x - 1 = 0$ or $\sin x - 1 = 0$

$\sin x = \frac{1}{2}$ $\quad$ $\sin x = 1$

$x = \frac{\pi}{6}, \frac{5\pi}{6}$ $\quad$ $x = \frac{\pi}{2}$

42. $\sin^2 x + 2\cos x = 2$

$1 - \cos^2 x + 2\cos x = 2$

$0 = \cos^2 x - 2\cos x + 1$

$0 = (\cos x - 1)^2$

$\cos x - 1 = 0$

$\cos x = 1$

$x = 0$

44. $\sqrt{3}\tan 3x = 0$

$\tan 3x = 0$

$3x = 0, \pi, 2\pi, 3\pi, 4\pi, 5\pi$

$x = 0, \frac{\pi}{3}, \frac{2\pi}{3}, \pi, \frac{4\pi}{3}, \frac{5\pi}{3}$

46. $3\csc^2 5x = -4$

$\csc^2 5x = -\frac{4}{3}$

$\csc 5x = \pm\sqrt{-\frac{4}{3}}$

No real solution

48. $2\cos^2 x + 3\cos x = 0$

$\cos x(2\cos x + 3) = 0$

$\cos x = 0$ or $2\cos x + 3 = 0$

$x = \frac{\pi}{2}, \frac{3\pi}{2}$ $\quad$ $2\cos x = -3$

$\cos x = -\frac{3}{2}$

No solution

50. $\sec^2 x - 6\tan x + 4 = 0$

$1 + \tan^2 x + 6\tan x + 4 = 0$

$\tan^2 x + 6\tan x + 5 = 0$

$(\tan x + 5)(\tan x + 1) = 0$

$\tan x + 5 = 0$ or $\tan x + 1 = 0$

$\tan x = -5$ $\quad$ $\tan x = -1$

$x = \arctan(-5) + \pi$ $\quad$ $x = \frac{3\pi}{4}, \frac{7\pi}{4}$

$x = \arctan(-5) + 2\pi$

52. $\sin(345°) = \sin(300° + 45°)$

$= \sin 300° \cos 45° + \cos 300° \sin 45°$

$= -\frac{\sqrt{3}}{2}\cdot\frac{\sqrt{2}}{2} + \frac{1}{2}\cdot\frac{\sqrt{2}}{2}$

$= \frac{\sqrt{2}}{4}(-\sqrt{3} + 1) = \frac{\sqrt{2}}{4}(1 - \sqrt{3})$

$\cos(345°) = \cos(300° + 45°)$

$= \cos 300° \cos 45° - \sin 300° \sin 45°$

$= \frac{1}{2}\cdot\frac{\sqrt{2}}{2} - \left(-\frac{\sqrt{3}}{2}\right)\frac{\sqrt{2}}{2}$

$= \frac{\sqrt{2}}{4}(1 + \sqrt{3})$

$\tan(345°) = \tan(300° + 45°)$

$= \frac{\tan 300° + \tan 45°}{1 - \tan 300° \tan 45°} = \frac{-\sqrt{3} + 1}{1 + \sqrt{3}(1)}\cdot\frac{1 - \sqrt{3}}{1 - \sqrt{3}}$

$= \frac{4 - 2\sqrt{3}}{-2} = -2 + \sqrt{3}$

54. $\sin\left(\frac{19\pi}{12}\right) = \sin\left(\frac{11\pi}{6} - \frac{\pi}{4}\right)$

$= \sin\frac{11\pi}{6}\cos\frac{\pi}{4} - \cos\frac{11\pi}{6}\sin\frac{\pi}{4}$

$= -\frac{1}{2}\cdot\frac{\sqrt{2}}{2} - \frac{\sqrt{3}}{2}\cdot\frac{\sqrt{2}}{2}$

$= -\frac{\sqrt{2}}{4}(1 + \sqrt{3}) = -\frac{\sqrt{2}}{4}(\sqrt{3} + 1)$

$\cos\left(\frac{19\pi}{12}\right) = \cos\left(\frac{11\pi}{6} - \frac{\pi}{4}\right)$

$= \cos\frac{11\pi}{6}\cos\frac{\pi}{4} + \sin\frac{11\pi}{6}\sin\frac{\pi}{4}$

$= \frac{\sqrt{3}}{2}\cdot\frac{\sqrt{2}}{2} + \left(-\frac{1}{2}\right)\frac{\sqrt{2}}{2}$

$= \frac{\sqrt{2}}{4}(\sqrt{3} - 1)$

—CONTINUED—

54. —CONTINUED—

$$\tan\left(\frac{19\pi}{12}\right) = \tan\left(\frac{11\pi}{6} - \frac{\pi}{4}\right)$$

$$= \frac{\tan\frac{11\pi}{6} - \tan\frac{\pi}{4}}{1 + \tan\frac{11\pi}{6}\tan\frac{\pi}{4}}$$

$$= \frac{-\frac{\sqrt{3}}{3} - 1}{1 + \left(-\frac{\sqrt{3}}{3}\right)(1)} = \frac{-\sqrt{3} - 3}{3 - \sqrt{3}} \cdot \frac{3 + \sqrt{3}}{3 + \sqrt{3}}$$

$$= \frac{-(12 + 6\sqrt{3})}{6} = -2 - \sqrt{3}$$

56. $\cos 45° \cos 120° - \sin 45° \sin 120° = \cos(45° + 120°) = \cos 165°$

58. $\dfrac{\tan 68° - \tan 115°}{1 + \tan 68° \tan 115°} = \tan(68° - 115°) = \tan(-47°)$

Figures for Exercises 60–64

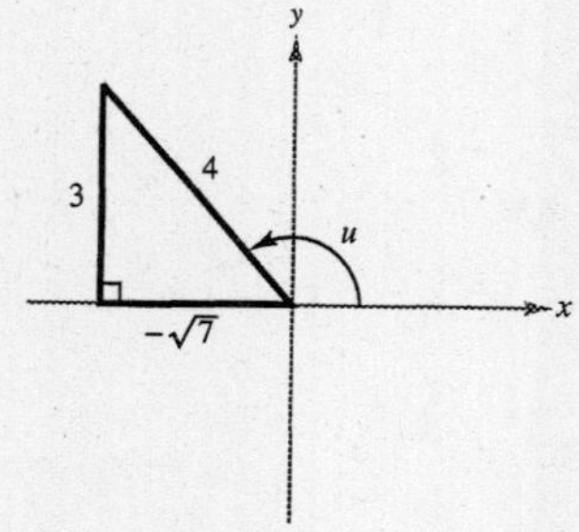

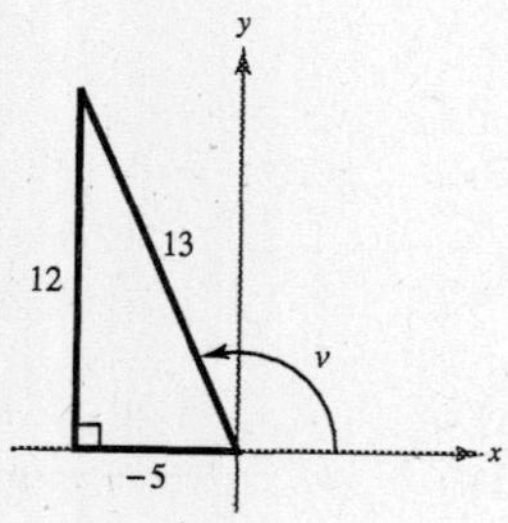

60. $\tan(u + v) = \dfrac{\tan u + \tan v}{1 - \tan u \tan v} = \dfrac{\left(-\frac{3}{\sqrt{7}}\right) + \left(-\frac{12}{5}\right)}{1 - \left(-\frac{3}{\sqrt{7}}\right)\left(-\frac{12}{5}\right)}$

$$= \frac{15 + 12\sqrt{7}}{36 - 5\sqrt{7}} \cdot \frac{36 + 5\sqrt{7}}{36 + 5\sqrt{7}} = \frac{960 + 507\sqrt{7}}{1121}$$

62. $\sin(u - v) = \sin u \cos v - \cos u \sin v$

$$= \left(\frac{3}{4}\right)\left(-\frac{5}{13}\right) - \left(-\frac{\sqrt{7}}{4}\right)\left(\frac{12}{13}\right)$$

$$= \frac{-15 + 12\sqrt{7}}{52} = \frac{12\sqrt{7} - 15}{52}$$

64. $\tan(u - v) = \dfrac{\tan u - \tan v}{1 + \tan u \tan v} = \dfrac{\left(-\frac{3}{\sqrt{7}}\right) - \left(-\frac{12}{5}\right)}{1 + \left(-\frac{3}{\sqrt{7}}\right)\left(-\frac{12}{5}\right)}$

$$= \frac{-15 + 12\sqrt{7}}{36 + 5\sqrt{7}} \cdot \frac{36 - 5\sqrt{7}}{36 - 5\sqrt{7}} = \frac{-960 + 507\sqrt{7}}{1121}$$

66.

$$\cos\left(x + \frac{\pi}{6}\right) - \cos\left(x - \frac{\pi}{6}\right) = 1$$

$$\left(\cos x \cos \frac{\pi}{6} - \sin x \sin \frac{\pi}{6}\right) - \left(\cos x \cos \frac{\pi}{6} + \sin x \sin \frac{\pi}{6}\right) = 1$$

$$-2 \sin x \sin \frac{\pi}{6} = 1$$

$$-2 \sin x\left(\frac{1}{2}\right) = 1$$

$$\sin x = -1$$

$$x = \frac{3\pi}{2}$$

68.

$$\cos\left(x + \frac{3\pi}{4}\right) - \cos\left(x - \frac{3\pi}{4}\right) = 0$$

$$\left(\cos x \cos \frac{3\pi}{4} - \sin x \sin \frac{3\pi}{4}\right) - \left(\cos x \cos \frac{3\pi}{4} + \sin x \sin \frac{3\pi}{4}\right) = 0$$

$$-2 \sin x \sin \frac{3\pi}{4} = 0$$

$$-2 \sin x\left(\frac{\sqrt{2}}{2}\right) = 0$$

$$-\sqrt{2} \sin x = 0$$

$$\sin x = 0$$

$$x = 0, \pi$$

70. $\dfrac{1 - \cos 2x}{1 + \cos 2x} = \dfrac{1 - (1 - 2\sin^2 x)}{1 + (2\cos x^2 - 1)}$

$= \dfrac{2\sin^2 x}{2\cos^2 x}$

$= \tan^2 x$

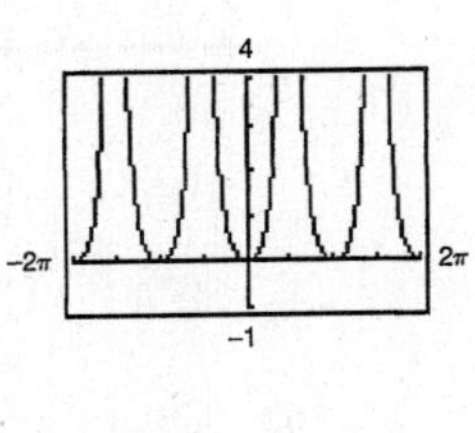

72. $\cos u = -\dfrac{2}{\sqrt{5}}, \dfrac{\pi}{2} < u < \pi \Rightarrow \sin u = \dfrac{1}{\sqrt{5}}$ and $\tan u = -\dfrac{1}{2}$

$$\sin 2u = 2 \sin u \cos u = 2\left(\frac{1}{\sqrt{5}}\right)\left(-\frac{2}{\sqrt{5}}\right) = -\frac{4}{5}$$

$$\cos 2u = \cos^2 u - \sin^2 u = \left(-\frac{2}{\sqrt{5}}\right)^2 - \left(\frac{1}{\sqrt{5}}\right)^2 = \frac{3}{5}$$

$$\tan 2u = \frac{2 \tan u}{1 - \tan^2 u} = \frac{2\left(-\frac{1}{2}\right)}{1 - \left(-\frac{1}{2}\right)^2} = \frac{-1}{\frac{3}{4}} = -\frac{4}{3}$$

74.
$$r = \frac{1}{32}v_0^2 \sin 2\theta$$
$$77 = \frac{1}{32}(50^2)\sin 2\theta$$
$$77 = \frac{625}{8}\sin 2\theta$$
$$0.9856 = \sin 2$$
$$80.26° \approx 2\theta$$
$$40° \approx \theta$$

76. $\cos^2 3x = \dfrac{1 + \cos 6x}{2}$

78. $\cos^2 x \tan^2 x = \sin^2 x = \dfrac{1 - \cos 2x}{2}$

80. $\sin 15° = \sin\left(\dfrac{30°}{2}\right) = \sqrt{\dfrac{1 - \cos 30°}{2}} = \sqrt{\dfrac{1 - \frac{\sqrt{3}}{2}}{2}}$
$$= \frac{1}{2}\sqrt{2 - \sqrt{3}}$$

$\cos 15° = \cos\left(\dfrac{30°}{2}\right) = \sqrt{\dfrac{1 + \cos 30°}{2}} = \sqrt{\dfrac{1 + \frac{\sqrt{3}}{2}}{2}}$
$$= \frac{1}{2}\sqrt{2 + \sqrt{3}}$$

$\tan 15° = \tan\left(\dfrac{30°}{2}\right) = \dfrac{1 - \cos 30°}{\sin 30°} = \dfrac{1 - \frac{\sqrt{3}}{2}}{\frac{1}{2}}$
$$= 2 - \sqrt{3}$$

82. $\sin\left(-\dfrac{17\pi}{12}\right) = \sin\left(-\dfrac{\frac{17\pi}{6}}{2}\right) = \sqrt{\dfrac{1 + \cos\left(-\frac{17\pi}{6}\right)}{2}} = \sqrt{\dfrac{1 + \left(-\frac{\sqrt{3}}{2}\right)}{2}}$
$$= \frac{1}{2}\sqrt{2 + \sqrt{3}}$$

$\cos\left(-\dfrac{17\pi}{12}\right) = \cos\left(\dfrac{-\frac{17\pi}{6}}{2}\right) = -\sqrt{\dfrac{1 - \cos\left(-\frac{17\pi}{6}\right)}{2}} = -\sqrt{\dfrac{1 - \left(-\frac{\sqrt{3}}{2}\right)}{2}}$
$$= -\frac{1}{2}\sqrt{2 - \sqrt{3}}$$

$\tan\left(-\dfrac{17\pi}{12}\right) = \tan\left(\dfrac{-\left(\frac{17\pi}{6}\right)}{2}\right) = \dfrac{1 - \cos\left(-\frac{17\pi}{6}\right)}{\sin\left(-\frac{17\pi}{6}\right)} = \dfrac{1 - \left(-\frac{\sqrt{3}}{2}\right)}{-\frac{1}{2}} = -2 - \sqrt{3}$

84. $\dfrac{\sin 6x}{1 + \cos 6x} = \tan 3x$

86. Volume V of the trough will be the area A of the isosceles triangle times the length l of the trough.

$V = A \cdot l$

(a) $A = \frac{1}{2}bh$

$\cos\frac{\theta}{2} = \frac{h}{0.5} \Rightarrow h = 0.5\cos\frac{\theta}{2}$

$\sin\frac{\theta}{2} = \frac{\frac{b}{2}}{0.5} \Rightarrow \frac{b}{2} = 0.5\sin\frac{\theta}{2}$

$A = 0.5\sin\frac{\theta}{2}\,0.5\cos\frac{\theta}{2}$

$= (0.5)^2\sin\frac{\theta}{2}\cos\frac{\theta}{2}$

$= 0.25\sin\frac{\theta}{2}\cos\frac{\theta}{2}$ square meters

$V = (0.25)(4)\sin\frac{\theta}{2}\cos\frac{\theta}{2}$ cubic meters

$= \sin\frac{\theta}{2}\cos\frac{\theta}{2}$ cubic meters

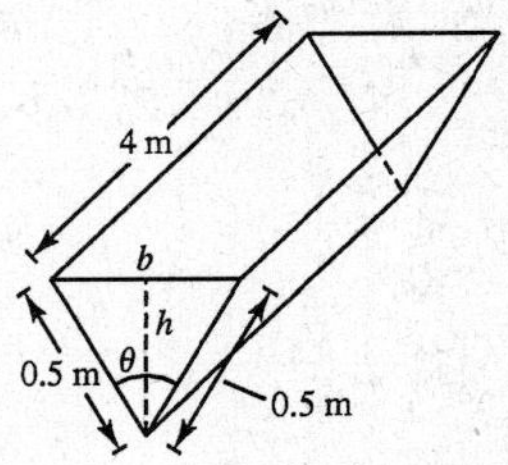

Not to scale

(b) $V = \sin\frac{\theta}{2}\cos\frac{\theta}{2}$

$= \frac{1}{2}\left(2\sin\frac{\theta}{2}\cos\frac{\theta}{2}\right)$

$= \frac{1}{2}\sin\theta$ cubic meters

Volume is maximum when $\theta = \frac{\pi}{2}$.

88. $6\sin 15^\circ \sin 45^\circ = 6\left(\tfrac{1}{2}\right)[\cos(15^\circ - 45^\circ) - \cos(15^\circ + 45^\circ)]$

$= 3[\cos(-30^\circ) - \cos 60^\circ]$

$= 3(\cos 30^\circ - \cos 60^\circ)$

90. $4\sin 3\alpha\cos 2\alpha = 4\left(\tfrac{1}{2}\right)[\sin(3\alpha + 2\alpha) + \sin(3\alpha - 2\alpha)]$

$= 2(\sin 5\alpha + \sin\alpha)$

92. $\cos 3\theta + \cos 2\theta = 2\cos\left(\frac{3\theta + 2\theta}{2}\right)\cos\left(\frac{3\theta - 2\theta}{2}\right)$

$= 2\cos\frac{5\theta}{2}\cos\frac{\theta}{2}$

94. $\sin\left(x + \frac{\pi}{4}\right) - \sin\left(x - \frac{\pi}{4}\right) = 2\cos\left[\frac{\left(x + \frac{\pi}{4}\right) + \left(x - \frac{\pi}{4}\right)}{2}\right]\sin\left[\frac{\left(x + \frac{\pi}{4}\right) - \left(x - \frac{\pi}{4}\right)}{2}\right]$

$= 2\cos x\sin\frac{\pi}{4}$

96. False. If $\frac{\pi}{2} < \theta < \pi$, then $\frac{\pi}{4} < \frac{\theta}{2} < \frac{\pi}{2}$ and $\frac{\theta}{2}$ is in Quadrant I. $\cos\frac{\theta}{2} > 0$

98. True. $4\sin(-x)\cos(-x) = 4(-\sin x)\cos x$

$$= -4\sin x\cos x = -2(2\sin x\cos x)$$

$$= -2\sin 2x$$

100. Reciprocal Identities: $\sin\theta = \dfrac{1}{\csc\theta}$ $\quad\csc\theta = \dfrac{1}{\sin\theta}$

$$\cos\theta = \frac{1}{\sec\theta} \qquad \sec\theta = \frac{1}{\cos\theta}$$

$$\tan\theta = \frac{1}{\cot\theta} \qquad \cot\theta = \frac{1}{\tan\theta}$$

Quotient Identities: $\tan\theta = \dfrac{\sin\theta}{\cos\theta}$ $\quad\cot\theta = \dfrac{\cos\theta}{\sin\theta}$

Pythagorean Identities: $\sin^2\theta + \cos^2\theta = 1$

$$1 + \tan^2\theta = \sec^2\theta$$

$$1 + \cot^2\theta = \csc^2\theta$$

102. $a\sin x - b = 0$

$$\sin x = \frac{b}{a}$$

If $|b| > |a|$, then $\left|\dfrac{b}{a}\right| > 1$ and there is no solution

since $|\sin x| \le 1$ for all x.

104. $y_1 = \dfrac{\cos 3x}{\cos x}$ $\qquad y_2 = (2\sin x)^2$

If the graph of y_2 is reflected in the x-axis and then shifted upward by one unit, it coincides with the graph of y_1. Therefore,

$$\frac{\cos 3x}{\cos x} = -(2\sin x)^2 + 1.$$

So, $y_1 = 1 - y_2$.

106. $y = 2 = \dfrac{1}{2}x^2 + 3\sin\dfrac{\pi x}{2}$

Approximate roots:
$-3.1395, -2.0000, -0.4378, 2.0000$

$$y = 2 = \frac{1}{2}x^2 + 3\sin\frac{\pi x}{2}$$

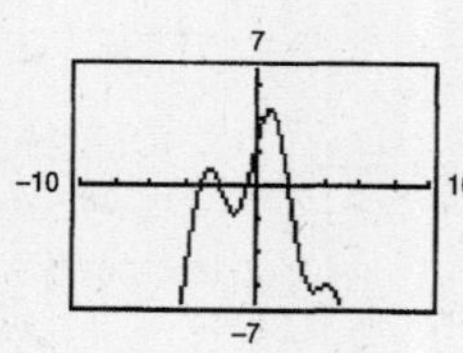

Problem Solving for Chapter 2

Solutions to Even-Numbered Exercises

2. $\cos\left[\dfrac{(2n+1)\pi}{2}\right] = \cos\left(\dfrac{2n\pi + \pi}{2}\right)$

$$= \cos\left(n\pi + \frac{\pi}{2}\right)$$

$$= \cos n\pi\cos\frac{\pi}{2} - \sin n\pi\sin\frac{\pi}{2}$$

$$= (\pm 1)(0) - (0)(1)$$

$$= 0$$

Thus, $\cos\left[\dfrac{(2n+1)\pi}{2}\right] = 0$ for all integers n.

4. $p(t) = \dfrac{1}{4\pi}[p_1(t) + 30p_2(t) + p_3(t) + p_5(t) + 30p_6(t)]$

(a) $p_1(t) = \sin(524\pi t)$

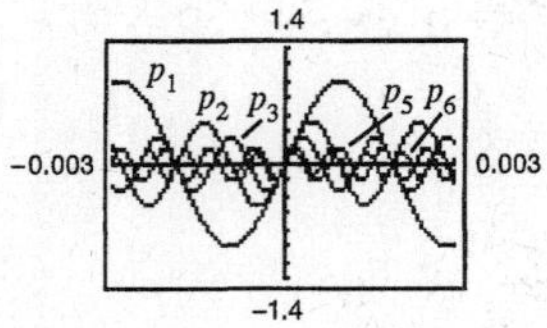

$p_2(t) = \dfrac{1}{2}\sin(1048\pi t)$

$p_3(t) = \dfrac{1}{3}\sin(1572\pi t)$

$p_5(t) = \dfrac{1}{5}\sin(2620\pi t)$

$p_6(t) = \dfrac{1}{6}\sin(3144\pi t)$

The graph of $p(t) = \dfrac{1}{4\pi}\left[\sin(524\pi t) + 15\sin(1048\pi t) + \dfrac{1}{3}\sin(1572\pi t) + \dfrac{1}{5}\sin(2620\pi t) + 5\sin(3144\pi t)\right]$ yields the graph shown in the text and below.

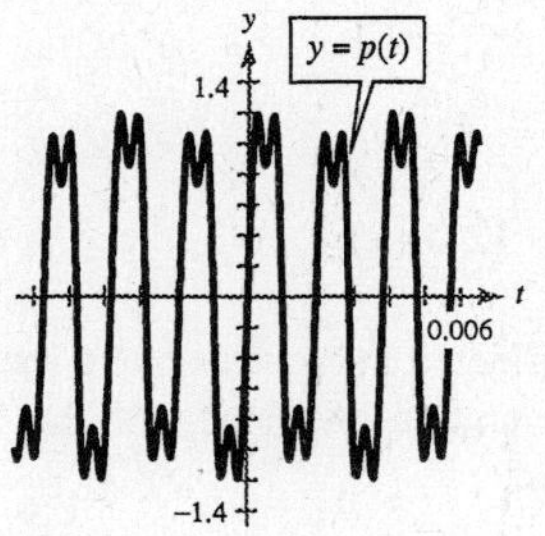

(b)

Function	Period
$p_1(t)$	$\dfrac{2\pi}{524\pi} = \dfrac{1}{262} \approx 0.0038$
$p_2(t)$	$\dfrac{2\pi}{1048\pi} = \dfrac{1}{524} \approx 0.0019$
$p_3(t)$	$\dfrac{2\pi}{1572\pi} = \dfrac{1}{786} \approx 0.0013$
$p_5(t)$	$\dfrac{2\pi}{2620\pi} = \dfrac{1}{1310} \approx 0.0008$
$p_6(t)$	$\dfrac{2\pi}{3144\pi} = \dfrac{1}{1572} \approx 0.0006$

The graph of p appears to be periodic with a period of $\dfrac{1}{262} \approx 0.0038$

(c)

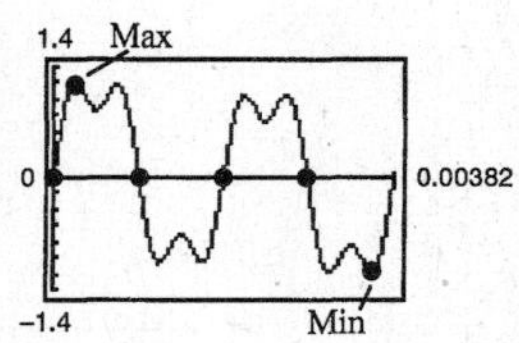

Over one cycle, $0 \le t < \dfrac{1}{262}$, we have four t-intercepts:

$t = 0$, $t \approx 0.00096$, $t \approx 0.00191$, and $t \approx 0.00285$

(d) The absolute maximum value of p over one cycle is $p \approx 1.1952$, and the absolute minimum value of p over one cycle is $p \approx -1.1952$.

6.
$$\frac{\cos(x+h) - \cos x}{h} = \frac{\cos x \cos h - \sin x \sin h - \cos x}{h}$$
$$= \frac{\cos x \cos h - \cos x - \sin x \sin h}{h}$$
$$= \frac{\cos x\,(\cos h - 1)}{h} - \frac{\sin x \sin h}{h}$$
$$= \cos x \frac{(\cos h - 1)}{h} - \sin x\left(\frac{\sin h}{h}\right)$$

8. $F = \dfrac{0.6\,W \sin(\theta + 90°)}{\sin 12°}$

(a) $F = \dfrac{0.6\,W(\sin\theta\cos 90° + \cos\theta\sin 90°)}{\sin 12°}$

$= \dfrac{0.6\,W[(\sin\theta)(0) + (\cos\theta)(1)]}{\sin 12°}$

$= \dfrac{0.6\,W\cos\theta}{\sin 12°}$

(b) Let $y_1 = \dfrac{0.6(185)\cos x}{\sin 12°}$

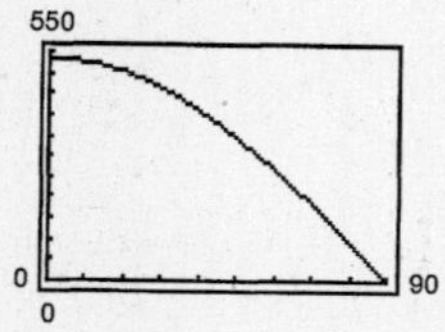

(c) The force is maximum (533.88 pounds) when $\theta = 0°$. The force is minimum (0 pounds) when $\theta = 90°$.

10. $d = 35 - 28\cos\dfrac{\pi}{6.2}t$ when $t = 0$ corresponds to 12:00 A.M.

(a) The high tides occur when $\cos\dfrac{\pi}{6.2}t = -1$. Solving yields $t = 6.2$ or $t = 18.6$. These t-values correspond to 6:12 P.M. and 6:36 A.M.

The low tide occurs when $\cos\dfrac{\pi}{6.2}t = 1$. Solving yields $t = 0$ and $t = 12.4$ which corresponds to 12:00 A.M. and 12:24 P.M.

(b) The water depth is never 3.5 feet. At low tide the depth is $d = 35 - 28 = 7$ feet.

(c)

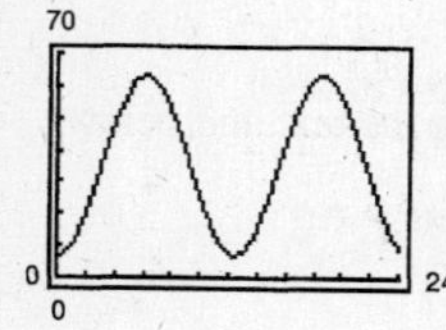

12. (a) $n = \dfrac{\sin\left(\frac{\theta}{2} + \frac{\alpha}{2}\right)}{\sin\frac{\theta}{2}}$

$= \dfrac{\sin\left(\frac{\theta}{2}\right)\cos\left(\frac{\alpha}{2}\right) + \cos\left(\frac{\theta}{2}\right)\sin\left(\frac{\alpha}{2}\right)}{\sin\left(\frac{\theta}{2}\right)}$

$= \cos\left(\frac{\alpha}{2}\right) + \cot\left(\frac{\theta}{2}\right)\sin\left(\frac{\alpha}{2}\right)$

For $\alpha = 60°$, $n = \cos 30° + \cot\left(\frac{\theta}{2}\right)\sin 30°$

$n = \dfrac{\sqrt{3}}{2} + \dfrac{1}{2}\cot\left(\dfrac{\theta}{2}\right)$

(b) For glass, $n = 1.50$

$1.50 = \dfrac{\sqrt{3}}{2} + \dfrac{1}{2}\cot\left(\dfrac{\theta}{2}\right)$

$2\left(1.50 - \dfrac{\sqrt{3}}{2}\right) = \cot\left(\dfrac{\theta}{2}\right)$

$\dfrac{1}{3 - \sqrt{3}} = \tan\left(\dfrac{\theta}{2}\right)$

$\theta = 2\tan^{-1}\left(\dfrac{1}{3 - \sqrt{3}}\right)$

$\theta \approx 76.52°$

14. (a) $\cos(3\theta) = \cos(2\theta + \theta)$

$= \cos 2\theta\cos\theta - \sin 2\theta\sin\theta$

$= (1 - 2\sin^2\theta)\cos\theta - (2\sin\theta\cos\theta)\sin\theta$

$= \cos\theta - 4\sin^2\theta\cos\theta$

$= \cos\theta\,(1 - 4\sin^2\theta)$

(b) $\cos(4\theta) = \cos(2\theta + 2\theta)$

$= \cos 2\theta\cos 2\theta - \sin 2\theta\sin 2\theta$

$= \cos^2 2\theta - \sin^2 2\theta$

$= (1 - \sin^2 2\theta) - \sin^2 2\theta$

$= 1 - 2\sin^2 2\theta$

$= 1 - 2(2\sin\theta\cos\theta)^2$

$= 1 - 8\sin^2\theta\cos^2\theta$

CHAPTER 3
Additional Topics in Trigonometry

CHAPTER 3
Additional Topics in Trigonometry

Section 3.1 Law of Sines

Solutions to Even-Numbered Exercises

2.

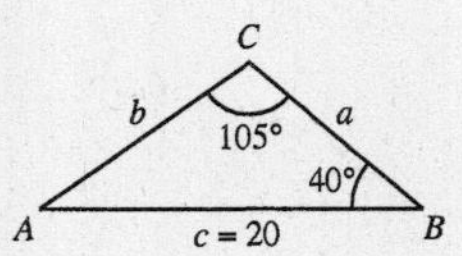

Given: $B = 40°, C = 105°, c = 20$

$A = 180° - B - C = 35°$

$$a = \frac{c}{\sin C}(\sin A) = \frac{20 \sin 35°}{\sin 105°} \approx 11.88$$

$$b = \frac{c}{\sin C}(\sin B) = \frac{20 \sin 40°}{\sin 105°} \approx 13.31$$

4.

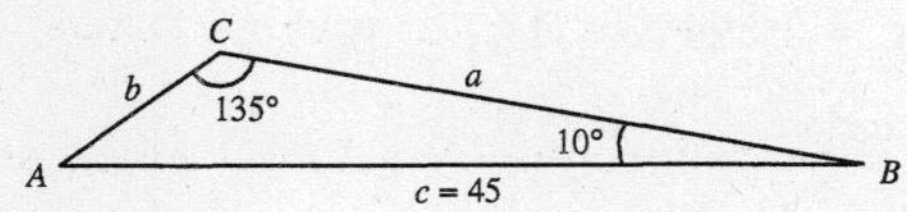

Given: $B = 10°, C = 135°, c = 45$

$A = 180° - B - C = 35°$

$$a = \frac{c}{\sin C}(\sin A) = \frac{45 \sin 35°}{\sin 135°} \approx 36.50$$

$$b = \frac{c}{\sin C}(\sin B) = \frac{45 \sin 10°}{\sin 135°} \approx 11.05$$

6. Given: $A = 60°, a = 9, c = 10$

$$\sin C = \frac{c \sin A}{a} = \frac{10 \sin 60°}{9} \approx 0.9623 \implies C \approx 74.21° \text{ or } C \approx 105.79°$$

<u>Case 1</u>

$C \approx 74.21°$

$B = 180° - A - C \approx 45.79°$

$$b = \frac{a}{\sin A}(\sin B) \approx \frac{9 \sin 45.79°}{\sin 60°} \approx 7.45$$

<u>Case 2</u>

$C \approx 105.79°$

$B = 180° - A - C \approx 14.21°$

$$b = \frac{a}{\sin A}(\sin B) \approx \frac{9 \sin 14.21°}{\sin 60°} \approx 2.55$$

8. Given: $A = 24.3°, C = 54.6°, c = 2.68$

$B = 180° - A - C = 101.1°$

$$a = \frac{c}{\sin C}(\sin A) = \frac{2.68 \sin 24.3°}{\sin 54.6°} \approx 1.35$$

$$b = \frac{c}{\sin C}(\sin B) = \frac{2.68 \sin 101.1°}{\sin 54.6°} \approx 3.23$$

10. Given: $A = 5° 40', B = 8° 15', b = 4.8$

$C = 180° - A - B = 166° 5'$

$$a = \frac{b}{\sin B}(\sin A) = \frac{4.8 \sin 5° 40'}{\sin 8° 15'} \approx 3.30$$

$$c = \frac{b}{\sin B}(\sin C) = \frac{4.8 \sin 166° 5'}{\sin 8° 15'} \approx 8.05$$

12. Given: $B = 2° 45', b = 6.2, c = 5.8$

$$\sin C = \frac{c \sin B}{b} = \frac{5.8 \sin 2° 45'}{6.2} \approx 0.04488 \implies C \approx 2.57° \text{ or } 2° 34'$$

$A = 180 - B - C \approx 174.68°$, or $174° 41'$

$$a = \frac{b}{\sin B}(\sin A) \approx \frac{6.2 \sin 174.68°}{\sin 2° 45'} \approx 11.99$$

14. Given: $A = 100°, a = 125, c = 10$

$$\sin C = \frac{c \sin A}{a} = \frac{10 \sin 100°}{125} \approx 0.07878 \Rightarrow C \approx 4.52°$$

$$B = 180° - A - C \approx 75.48°$$

$$b = \frac{a}{\sin A}(\sin B) \approx \frac{125 \sin 75.48°}{\sin 100°} \approx 122.87$$

16. Given: $C = 85° 20', a = 35, c = 50$

$$\sin A = \frac{a \sin C}{c} = \frac{35 \sin 85° 20'}{50} \approx 0.6977 \Rightarrow A \approx 44.24°, \text{ or } 44° 14'$$

$$B = 180° - A - C \approx 50.43°, \text{ or } 50° 26'$$

$$b = \frac{C \sin B}{\sin C} \approx \frac{50 \sin 50.43°}{\sin 85° 20'} \approx 38.67$$

18. Given: $B = 28°, C = 104°, a = 3\frac{5}{8}$

$$A = 180° - B - C = 48°$$

$$b = \frac{a \sin B}{\sin A} = \frac{3\frac{5}{8} \sin 28°}{\sin 48°} \approx 2.29$$

$$c = \frac{a \sin C}{\sin A} = \frac{3\frac{5}{8} \sin 104°}{\sin 48°} \approx 4.73$$

20. Given: $A = 58°, a = 11.4, c = 12.8$

$$\sin B = \frac{b \sin A}{a} = \frac{12.8 \sin 58°}{11.4} \approx 0.9522 \Rightarrow B \approx 72.21° \text{ or } B \approx 107.79°$$

Case 1

$B \approx 72.21°$

$C = 180° - A - B \approx 49.79°$

$$c = \frac{a}{\sin A}(\sin C) \approx \frac{11.4 \sin 49.79°}{\sin 58°} \approx 10.27$$

Case 2

$B \approx 107.79°$

$C = 180° - A - B \approx 14.21°$

$$c = \frac{a}{\sin A}(\sin C) \approx \frac{11.4 \sin 14.21°}{\sin 58°} \approx 3.30$$

22. Given: $A = 76°, a = 34, b = 21$

$$\sin B = \frac{b \sin A}{a} = \frac{21 \sin 76°}{34} \approx 0.5993 \Rightarrow B \approx 36.82°$$

$$C = 180° - A - B \approx 67.18°$$

$$c = \frac{a \sin C}{\sin A} \approx \frac{34 \sin 67.18°}{\sin 76°} \approx 32.30$$

24. Given: $A = 110°, a = 125, b = 100$

$$\sin B = \frac{b \sin A}{a} = \frac{100 \sin 110°}{125} \approx 0.75175 \Rightarrow B \approx 48.74°$$

$$C = 180° - A - B \approx 21.26°$$

$$c = \frac{a \sin C}{\sin A} \approx \frac{125 \sin 21.26°}{\sin 110°} \approx 48.23$$

26. Given: $A = 60°, a = 10$

(a) One solution if $b \le 10$ or $b = \dfrac{10}{\sin 60°}$.

(b) Two solutions if $10 < b < \dfrac{10}{\sin 60°}$.

(c) No solutions if $b > \dfrac{10}{\sin 60°}$.

28. Given: $A = 88°, a = 315.6$

(a) One solution if $b \le 315.6$ or $b = \dfrac{315.6}{\sin 88°}$

(b) Two solutions if $315.6 < b < \dfrac{315.6}{\sin 88°}$

(c) No solutions if $b > \dfrac{315.6}{\sin 88°}$

30. Area $= \frac{1}{2}ac \sin B = \frac{1}{2}(62)(20) \sin 130° \approx 474.9$

32. $A = 5° 15', b = 4.5, c = 22$

Area $= \frac{1}{2}bc \sin A$

$= \left(\frac{1}{2}\right)(4.5)(22) \sin 5.25° \approx 4.529$

34. $C = 84° 30', a = 16, b = 20$

Area $= \frac{1}{2}ab \sin C$

$= \left(\frac{1}{2}\right)(16)(20) \sin 84.5° \approx 159.3$

36. (a)

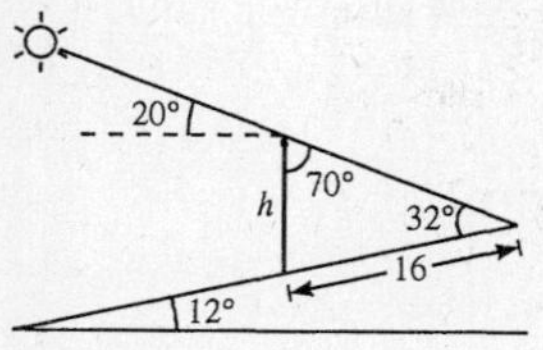

(b) $\frac{h}{\sin 32°} = \frac{16}{\sin 70°}$

(c) $h = \frac{16 \sin 32°}{\sin 70°} \approx 9.0$ meters

38.

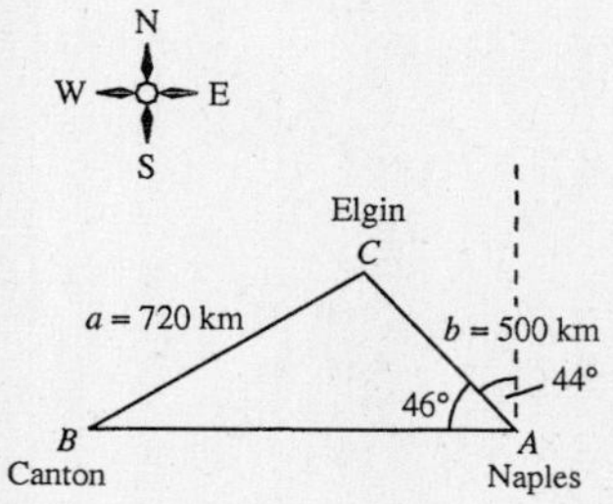

Given: $A = 46°, a = 720, b = 500$

$\sin B = \frac{b \sin A}{a} = \frac{500 \sin 46°}{720} \approx 0.50 \implies B \approx 30°$

The bearing from C to B is $S\,60°\,W$, or 240°.

40.

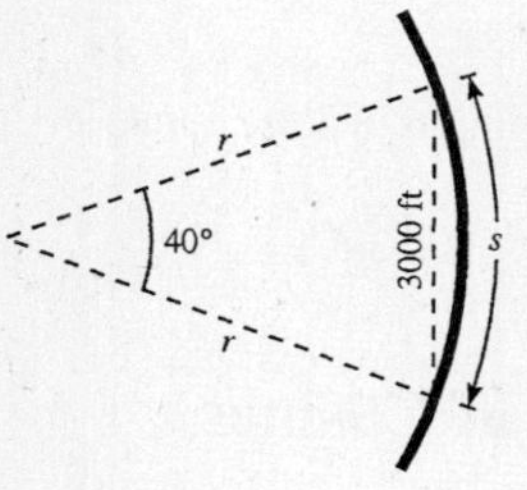

(b) $r = \frac{3000 \sin\left[\frac{1}{2}(180° - 40°)\right]}{\sin 40°} \approx 4385.71$ feet

(c) $s \approx 40\left(\frac{\pi}{180}\right)4385.71 \approx 3061.80$ feet

42. Given: $A = 15°, B = 135°, c = 30$

$C = 180° - A - B = 30°$

From Pine Knob:

$b = \frac{c \sin B}{\sin C} = \frac{30 \sin 135°}{\sin 30°} \approx 42.4$ kilometers

From Colt Station:

$a = \frac{c \sin A}{\sin C} = \frac{30 \sin 15°}{\sin 30°} \approx 15.5$ kilometers

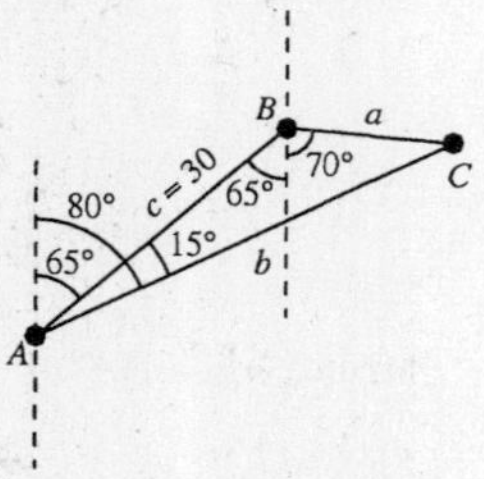

44. (a) $\sin \alpha = \frac{5.45}{58.36} \approx 0.09339 \implies \alpha \approx 5.36°$

(b) $90° + \alpha + \beta + \theta = 180° \implies \beta \approx 90° - 5.36° - \theta \implies \beta \approx 84.64° - \theta$

(c) $\frac{d}{\sin(84.64° - \theta)} = \frac{58.36}{\sin \theta}$, or $d = \sin(84.64° - \theta)\left(\frac{58.36}{\sin \theta}\right)$

(d)

θ	10°	20°	30°	40°	50°	60°
d	324.1	154.2	95.2	63.8	43.3	28.1

46. False. Two sides and one opposite angle do not necessarily determine a unique triangle.

48. (a) $A = \frac{1}{2}(30)(20)\sin\left(\theta + \frac{\theta}{2}\right) - \frac{1}{2}(8)(20)\sin\frac{\theta}{2} - \frac{1}{2}(8)(30)\sin\theta$

$= 300\sin\frac{3\theta}{2} - 80\sin\frac{\theta}{2} - 120\sin\theta$

$= 20\left[15\sin\frac{3\theta}{2} - 4\sin\frac{\theta}{2} - 6\sin\theta\right]$

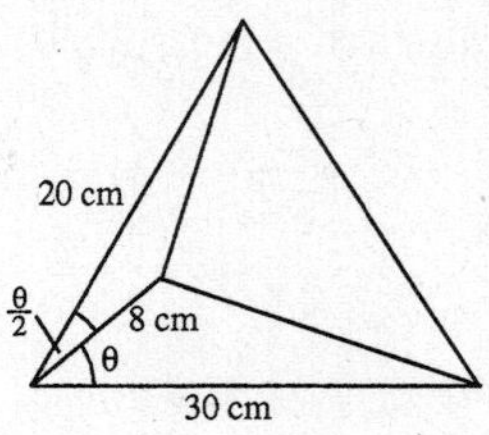

(b)

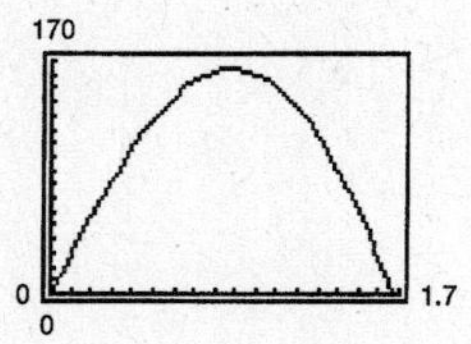

(c) Domain: $0 \le \theta \le 1.6690$

The domain would increase in length and the area would increase if the 8-centimeter line segment were decreased.

50. $\tan x \cos x \sec x = \tan x \cos x \frac{1}{\cos x} = \tan x$

52. $1 + \cot^2\left(\frac{\pi}{2} - x\right) = 1 + \tan^2 x = \sec^2 x$

54. $2\cos 5\theta \sin 2\theta = 2 \cdot \frac{1}{2}[\sin(5\theta + 2\theta) - \sin(5\theta - 2\theta)]$

$= \sin 7\theta - \sin 3\theta$

Section 3.2 Law of Cosines

Solutions to Even-Numbered Exercises

2. Given: $a = 8, b = 3, c = 9$

$\cos A = \frac{b^2 + c^2 - a^2}{2bc} = \frac{3^2 + 9^2 - 8^2}{2(3)(9)} \approx 0.48148 \Rightarrow A \approx 61.22°$

$\cos c = \frac{a^2 + b^2 - c^2}{2ab} = \frac{8^2 + 3^2 - 9^2}{2(8)(3)} \approx -0.16667 \Rightarrow C \approx 99.59°$

$B \approx 180° - 61.22° - 99.59° \approx 19.19°$

4. Given: $C = 105°, a = 10, b = 4.5$

$c^2 = a^2 + b^2 - 2ab\cos C = 10^2 + 4.5^2 - 2(10)(4.5)\cos 105° \approx 143.5437 \Rightarrow c \approx 11.98$

$\cos B = \frac{a^2 + c^2 - b^2}{2ac} \approx \frac{10^2 + (12.0)^2 - (4.5)^2}{2(10)(12.0)} \approx 0.93187 \Rightarrow B \approx 21.27°$

$A = 180° - 105° - 21.27° \approx 53.73°$

6. Given: $a = 55, b = 25, c = 72$

$\cos C = \frac{a^2 + b^2 - c^2}{2ab} = \frac{55^2 + 25^2 - 72^2}{2(55)(25)} \approx -0.5578 \Rightarrow C \approx 123.91°$

$\cos A = \frac{b^2 + c^2 - a^2}{2bc} = \frac{25^2 + 72^2 - 55^2}{2(25)(72)} \approx 0.7733 \Rightarrow A \approx 39.35°$

$B = 180° - 123.91° - 39.35° \approx 16.75°$

8. Given: $a = 1.42, b = 0.75, c = 1.25$

$$\cos A = \frac{b^2 + c^2 - a^2}{2bc} = \frac{(0.75)^2 + (1.25)^2 - (1.42)^2}{2(0.75)(1.25)} = 0.05792 \Rightarrow A \approx 86.68°$$

$$\cos B = \frac{a^2 + c^2 - b^2}{2ac} = \frac{(1.42)^2 + (1.25)^2 - (0.75)^2}{2(1.42)(1.25)} \approx 0.84969 \Rightarrow B \approx 31.82°$$

$$C = 180° - 86.68° - 31.82° \approx 61.50°$$

10. Given: $A = 55°, b = 3, c = 10$

$$a^2 = b^2 + c^2 - 2bc \cos A = 3^2 + 10^2 - 2(3)(10) \cos 55° \approx 74.585 \Rightarrow a \approx 8.64$$

$$\sin B = \frac{b \sin A}{a} \approx \frac{3 \sin 55°}{8.636} \approx 0.2846 \Rightarrow A \approx 16.53°$$

$$C \approx 180° - 16.53° - 55° \approx 108.47°$$

12. Given: $B = 75° 20', a = 6.2, c = 9.5$

$$b^2 = a^2 + c^2 - 2ac \cos B = (6.2)^2 + (9.5)^2 - 2(6.2)(9.5) \cos 75° 20' \approx 98.8636 \Rightarrow b \approx 9.94$$

$$\sin A = \frac{a \sin B}{b} \approx \frac{6.2 \sin 75° 20'}{9.94} \approx 0.6032 \Rightarrow A \approx 37.1°, \text{ or } 37° 6'$$

$$C \approx 180° - 75° 20' - 37° 6' \approx 67° 34'$$

14. Given: $C = 15° 15', a = 6.25, b = 2.15$

$$c^2 = a^2 + b^2 - 2ab \cos C = (6.25)^2 + (2.15)^2 - 2(6.25)(2.15) \cos 15° 15' \approx 17.7563 \Rightarrow c \approx 4.21$$

$$\cos A = \frac{b^2 + c^2 - a^2}{2bc} \approx \frac{(2.15)^2 + (4.21)^2 - (6.25)^2}{2(2.15)(4.21)} \approx -0.92076 \Rightarrow A \approx 157.0°$$

$$B \approx 180° - 15° 15' - 157.4° \approx 7.7°$$

16. Given: $C = 103°, a = \frac{3}{8}, b = \frac{3}{4}$

$$c^2 = a^2 + b^2 - 2ab \cos C = \left(\frac{3}{8}\right)^2 + \left(\frac{3}{4}\right)^2 - 2\left(\frac{3}{8}\right)\left(\frac{3}{4}\right) \cos 103° \approx 0.8297 \Rightarrow c \approx 0.91$$

$$\cos A = \frac{b^2 + c^2 - a^2}{2bc} \approx \frac{\left(\frac{3}{4}\right)^2 + (0.91)^2 - \left(\frac{3}{8}\right)^2}{2\left(\frac{3}{4}\right)(0.91)} \approx 0.9160 \Rightarrow A \approx 23.6°$$

$$B \approx 180° - 23.6° - 103° \approx 53.4°$$

18.

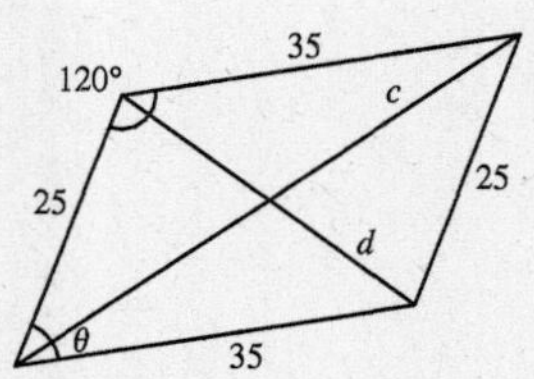

$$c^2 = 25^2 + 35^2 - 2(25)(35) \cos 120°$$
$$= 2725 \Rightarrow c \approx 52.20$$

$$\theta = \frac{1}{2}[360° - 2(120°)] = 60°$$

$$d^2 = 25^2 + 35^2 - 2(25)(35) \cos 60°$$
$$= 975 \Rightarrow d \approx 31.22$$

20.

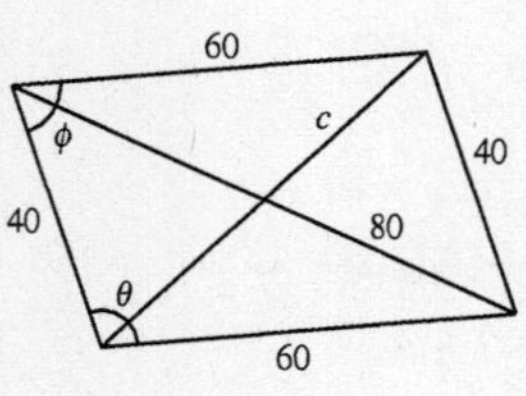

$$\cos \theta = \frac{40^2 + 60^2 - 80^2}{2(40)(60)} = -\frac{1}{4} \Rightarrow \theta \approx 104.5°$$

$$\phi \approx \frac{1}{2}[360° - 2(104.5°)] \approx 75.5°$$

$$c^2 \approx 40^2 + 60^2 - 2(40)(60) \cos 75.5° = 4000$$

$$c \approx 63.25$$

22. $\cos \alpha = \dfrac{25^2 + 17.5^2 - 25^2}{2(25)(17.5)}$

$\alpha \approx 69.512°$

$\beta \approx 180 - \alpha \approx 110.488°$

$a^2 = 17.5^2 + 25^2 - 2(17.5)(25)\cos 110.488°$

$a \approx 35.18$

$z = 180 - 2\alpha \approx 40.975$

$\cos \mu = \dfrac{25^2 + 35.18^2 - 17.5^2}{2(25)(35.18)}$

$\mu \approx 27.775°$

$\theta = \mu + z \approx 68.7°$

$\omega = 180° - \mu - \beta \approx 41.738°$

$\phi = \omega + \alpha \approx 111.3°$

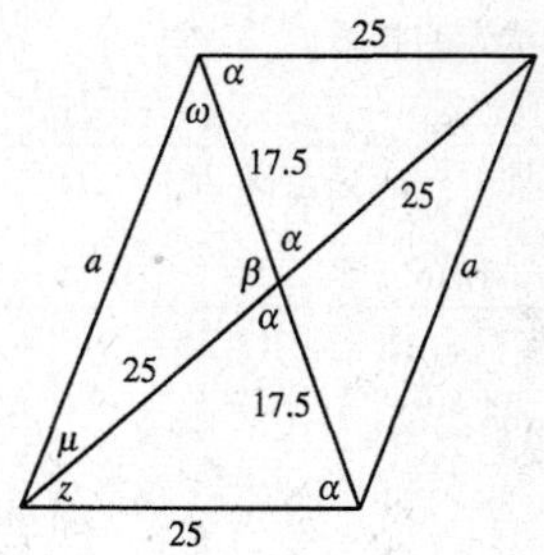

24. $a = 12,\ b = 15,\ c = 9 \implies s = \dfrac{12 + 15 + 9}{2} = 18$

Area $= \sqrt{18(6)(3)(9)} = 54$

26. Given: $a = 75.4, b = 52, c = 52$

$s = \dfrac{75.4 + 52 + 52}{2} = 89.7$

Area $= \sqrt{s(s-a)(s-b)(s-c)}$

$= \sqrt{89.7(14.3)(37.7)(37.7)} \approx 1350$

28. Given: $a = 3.05, b = 0.75, c = 2.45$

$s = \dfrac{3.05 + 0.75 + 2.45}{2} = 3.125$

Area $= \sqrt{s(s-a)(s-b)(s-c)}$

$= \sqrt{3.125(0.075)(2.375)(0.675)}$

≈ 0.61

30. Distance from Niagara to Rosemount:

$d = \sqrt{810^2 + 648^2 - 2(810)(648)\cos(137°)}$

≈ 1357.8 miles

Bearing from Niagara to Rosemount:

N $(75° - \theta)$ E

$\cos \theta \approx \dfrac{(1357.8)^2 + 810^2 - 648^2}{2(1357.8)(810)}$

≈ 0.9456

$\theta \approx 19.0°$

N 56.0° E

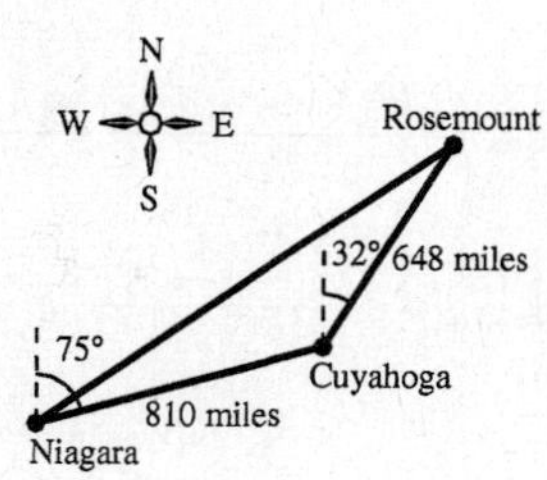

32. $\cos A = \dfrac{115^2 + 76^2 - 92^2}{2(115)(76)} \approx 0.6028 \implies A \approx 52.9°$

$\cos C = \dfrac{115^2 + 92^2 - 76^2}{2(115)(92)} \approx 0.75203 \implies c \approx 41.2°$

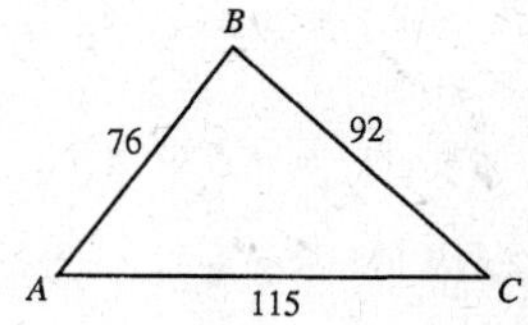

34. $\cos \theta = \dfrac{2^2 + 3^2 - (4.5)^2}{2(2)(3)} \approx -0.60417$

$\theta \approx 127.2°$

36. The angles at the base of the tower are 96° and 84°. The longer guy wire g_1 is given by:

$$g_1^2 = 75^2 + 100^2 - 2(75)(100)\cos 96° \approx 17{,}192.9 \Rightarrow g_1 \approx 131.1 \text{ feet}$$

The shorter guy wire g_2 is given by:

$$g_2^2 = 75^2 + 100^2 - 2(75)(100)\cos 84° \approx 14{,}057.1 \Rightarrow g_2 \approx 118.6 \text{ feet}$$

38.

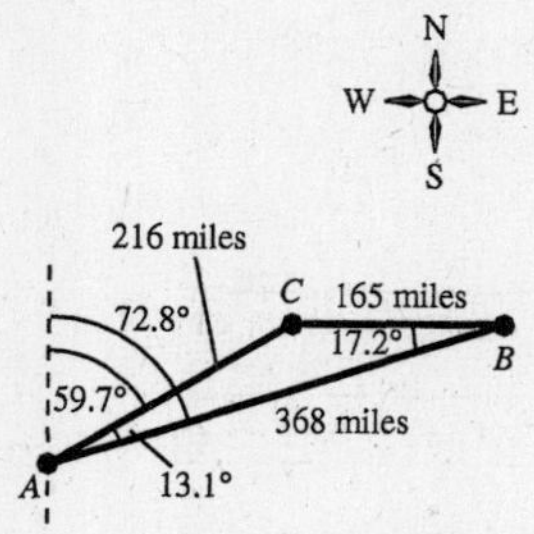

$a = 165, b = 216, c = 368$

$$\cos B = \frac{165^2 + 368^2 - 216^2}{2(165)(368)} \approx 0.9551$$

$B \approx 17.2°$

$$\cos A = \frac{216^2 + 368^2 - 165^2}{2(216)(368)} \approx 0.9741$$

$A \approx 13.1°$

(a) Bearing of Minneapolis (C) from Phoenix (A)

N (90° − 17.2° − 13.1°) E

N 59.7 E

(b) Bearing of Albany (B) from Phoenix (A)

N (90° − 17.2°) E

N 72.8° E

40. $d = \sqrt{330^2 + 420^2 - 2(330)(420)\cos 8°}$

≈ 103.9 feet

42. $a = \sqrt{20^2 + 20^2 - 2(20)(20)\cos 11°}$

≈ 3.8 miles

44. (a)

$$7^2 = 1.5^2 + x^2 - 2(1.5)x\cos\theta$$

$$49 = 2.25 + x^2 - 3x\cos\theta$$

$$x^2 - 3x\cos\theta - 46.75 = 0$$

(b) $$x = \frac{3\cos\theta \pm \sqrt{(-3\cos\theta)^2 - 4(1)(-46.75)}}{2(1)}$$

$$x = \frac{1}{2}\left(3\cos\theta + \sqrt{9\cos^2\theta + 187}\right)$$

(c)

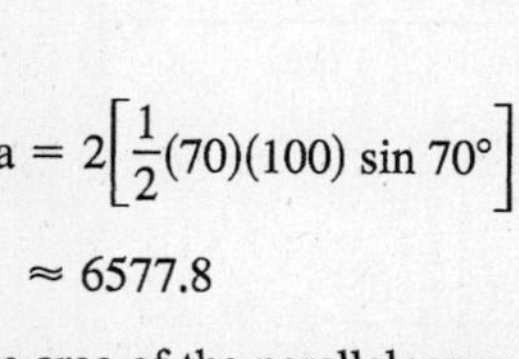

(d) Maximum: 8.5 inches

46.

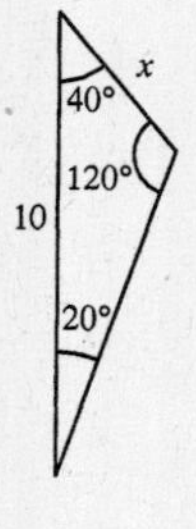

$$\frac{x}{\sin 20°} = \frac{10}{\sin 120°}$$

$$x = \frac{10\sin 20°}{\sin 120°} \approx 3.95 \text{ feet}$$

48. area $= 2\left[\frac{1}{2}(70)(100)\sin 70°\right]$

≈ 6577.8

(The area of the parallelogram is the sum of the areas of two triangles.)

50. False. To solve an SSA triangle, the Law of Sines is needed.

52. (a) Working with $\triangle OBC$, we have $\cos\alpha = \dfrac{\frac{a}{2}}{R}$.

This implies that $2R = \dfrac{a}{\cos\alpha}$.

Since we know that

$$\frac{a}{\sin A} = \frac{b}{\sin B} = \frac{c}{\sin C},$$

we can complete the proof by showing that $\cos\alpha = \sin A$. The solution of the system

$$A + B + C = 180°$$
$$\alpha - C + A = \beta$$
$$\alpha + \beta = B$$

is $\alpha = 90° - A$. Therefore:

$$2R = \frac{a}{\cos\alpha} = \frac{a}{\cos(90° - A)} = \frac{a}{\sin A}.$$

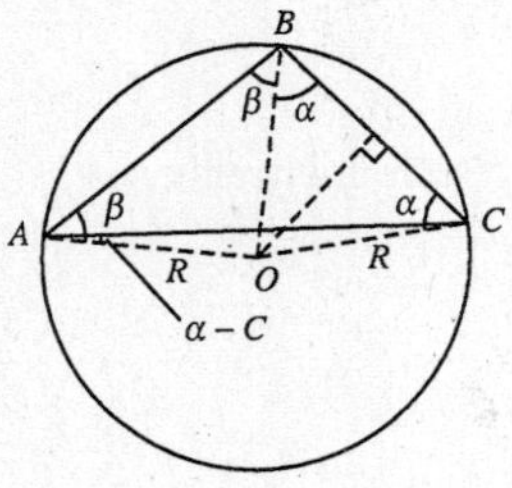

(b) By Heron's Formula, the area of the triangle is

$$\text{Area} = \sqrt{s(s-a)(s-b)(s-c)}.$$

We can also find the area by dividing the area into six triangles and using the fact that the area is $\frac{1}{2}$ the base times the height. Using the figure as given, we have

$$\begin{aligned}\text{Area} &= \frac{1}{2}xr + \frac{1}{2}xr + \frac{1}{2}yr + \frac{1}{2}yr + \frac{1}{2}zr + \frac{1}{2}zr\\ &= r(x + y + z)\\ &= rs.\end{aligned}$$

Therefore: $rs = \sqrt{s(s-a)(s-b)(s-c)} \Rightarrow$

$$r = \sqrt{\frac{(s-a)(s-b)(s-c)}{s}}.$$

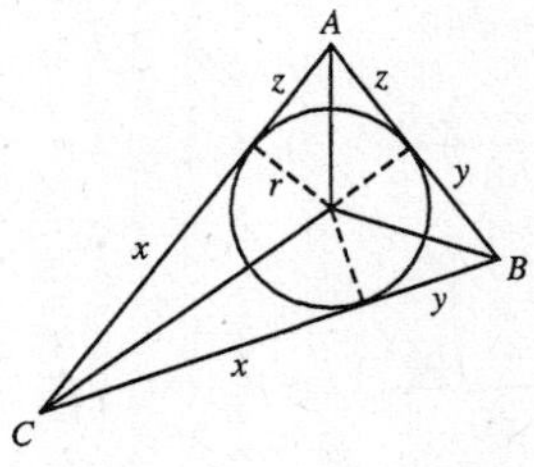

54. Given: $a = 200$ ft, $b = 250$ ft, $c = 325$ ft

$$s = \frac{200 + 250 + 325}{2} \approx 387.5$$

Radius of the inscribed circle: $r = \sqrt{\dfrac{(s-a)(s-b)(s-c)}{s}} = \sqrt{\dfrac{(187.5)(137.5)(62.5)}{387.5}} \approx 64.5$ ft

Circumference of an inscribed circle: $C = 2\pi r \approx 2\pi(64.5) \approx 405.3$ ft

56.
$$\begin{aligned}\frac{1}{2}bc(1 - \cos A) &= \frac{1}{2}bc\left[1 + \frac{a^2 - (b^2 + c^2)}{2bc}\right]\\ &= \frac{1}{2}bc\left[\frac{2bc + a^2 - b^2 - c^2}{2bc}\right]\\ &= \frac{a^2 - (b^2 - 2bc + c^2)}{4}\\ &= \frac{a^2 - (b-c)^2}{4}\\ &= \left(\frac{a - (b-c)}{2}\right)\left(\frac{a + (b-c)}{2}\right)\\ &= \frac{a - b + c}{2}\cdot\frac{a + b - c}{2}\end{aligned}$$

58. $\arccos 0 = \dfrac{\pi}{2}$

60. $\arctan(-\sqrt{3}) = -\arctan\sqrt{3} = -\dfrac{\pi}{3}$

62. $\arccos\left(-\dfrac{\sqrt{3}}{2}\right) = \pi - \arccos\dfrac{\sqrt{3}}{2} = \pi - \dfrac{\pi}{6} = \dfrac{5\pi}{6}$

64. Let $u = \arccos 3x$

$\cos u = 3x = \frac{3x}{1}.$

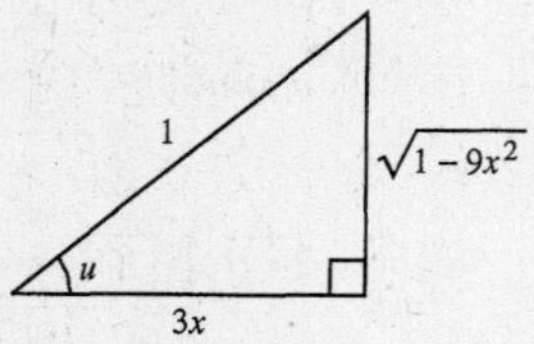

$\tan(\arccos 3x) = \tan u = \frac{\sqrt{1-9x^2}}{3x}$

66. Let $u = \arcsin \frac{x-1}{2}$

$\sin u = \frac{x-1}{2}.$

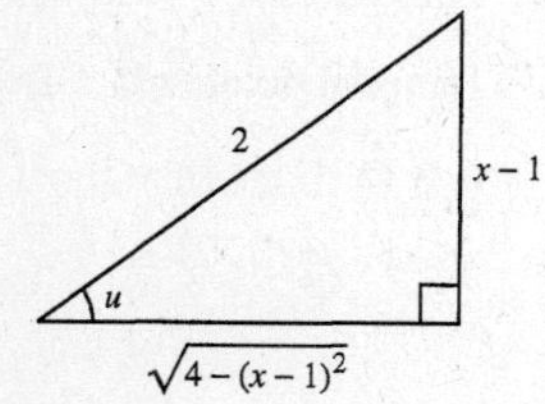

$$\cos\left(\arcsin \frac{x-1}{2}\right) = \cos u = \frac{\sqrt{4-(x-1)^2}}{2}$$

68. $x = 2\cos\theta, -\frac{\pi}{2} < \theta < \frac{\pi}{2}$

$$-\sqrt{2} = \sqrt{4-x^2}$$
$$-\sqrt{2} = \sqrt{4-(2\cos\theta)^2}$$
$$-\sqrt{2} = \sqrt{4-4\cos^2\theta}$$
$$-\sqrt{2} = \sqrt{4(1-\cos^2\theta)}$$
$$-\sqrt{2} = \sqrt{4\sin^2\theta}$$
$$-\sqrt{2} = 2\sin\theta$$
$$-\frac{\sqrt{2}}{2} = \sin\theta \Rightarrow \cos\theta = \frac{\sqrt{2}}{2} \Rightarrow x = 2\left(\frac{\sqrt{2}}{2}\right) = \sqrt{2}$$
$$\sec\theta = \frac{1}{\cos\theta} = \frac{1}{\frac{\sqrt{2}}{2}} = \sqrt{2}$$
$$\csc\theta = \frac{1}{\sin\theta} = \frac{1}{-\frac{\sqrt{2}}{2}} = -\sqrt{2}$$

70. $x = 6\tan\theta, -\frac{\pi}{2} < \theta < \frac{\pi}{2}$

$$12 = \sqrt{36+x^2}$$
$$12 = \sqrt{36+(6\tan\theta)^2}$$
$$12 = \sqrt{36+36\tan^2\theta}$$
$$12 = \sqrt{36(1+\tan^2\theta)}$$
$$12 = \sqrt{36\sec^2\theta}$$
$$12 = 6\sec\theta$$
$$2 = \sec\theta$$
$$\cos\theta = \frac{1}{2}$$
$$\sin^2\theta + \left(\frac{1}{2}\right)^2 = 1$$
$$\sin^2\theta = 1 - \frac{1}{4} = \frac{3}{4}$$
$$\sin\theta = \pm\sqrt{\frac{3}{4}} = \pm\frac{\sqrt{3}}{2}$$
$$\csc\theta = \frac{1}{\sin\theta} = \frac{1}{\pm\frac{\sqrt{3}}{2}} = \pm\frac{2}{\sqrt{3}} = \pm\frac{2\sqrt{3}}{3}$$

72. $\sin\left(x - \frac{\pi}{2}\right) - \sin\left(x + \frac{\pi}{2}\right) = 2\cos\left(\frac{x - \frac{\pi}{2} + x + \frac{\pi}{2}}{2}\right)\sin\left(\frac{x - \frac{\pi}{2} - \left(x + \frac{\pi}{2}\right)}{2}\right)$

$$= 2\cos\left(\frac{2x}{2}\right)\sin\left(\frac{-\pi}{2}\right)$$
$$= 2\cos x \sin\left(-\frac{\pi}{2}\right)$$

Section 3.3 Vectors in the Plane

Solutions to Even-Numbered Exercises

2. Initial point: $(0, 0)$

Terminal point: $(-4, -2)$

$\mathbf{v} = \langle -4-0, -2-0 \rangle = \langle -4, -2 \rangle$

$\|\mathbf{v}\| = \sqrt{(-4)^2 + (-2)^2} = \sqrt{20} = 2\sqrt{5}$

4. Initial point: $(-1, -1)$

Terminal point: $(3, 5)$

$\mathbf{v} = \langle 3-(-1), 5-(-1) \rangle = \langle 4, 6 \rangle$

$\|\mathbf{v}\| = \sqrt{4^2 + 6^2} = \sqrt{52} = 2\sqrt{13}$

6. Initial point: $(-4, -1)$

Terminal point: $(3, -1)$

$\mathbf{v} = \langle 3 - (-4), -1 - (-1) \rangle = \langle 7, 0 \rangle$

$\|\mathbf{v}\| = \sqrt{7^2 + 0^2} = 7$

8. Initial point: $(1, 11)$

Terminal point: $(9, 3)$

$\mathbf{v} = \langle 9 - 1, 3 - 11 \rangle = \langle 8, -8 \rangle$

$\|\mathbf{v}\| = \sqrt{8^2 + (-8)^2} = 8\sqrt{2}$

10. Initial point: $(-3, 11)$

Terminal point: $(9, 40)$

$\mathbf{v} = \langle 9 - (-3), 40 - 11 \rangle = \langle 12, 29 \rangle$

$\|\mathbf{v}\| = \sqrt{12^2 + 29^2} = \sqrt{985}$

12. Initial point: $(-2, 7)$

Terminal point: $(5, -17)$

$\mathbf{v} = \langle 5 - (-2), -17 - 7 \rangle = \langle 7, -24 \rangle$

$\|\mathbf{v}\| = \sqrt{7^2 + (-24)^2} = 25$

14. $5\mathbf{v}$

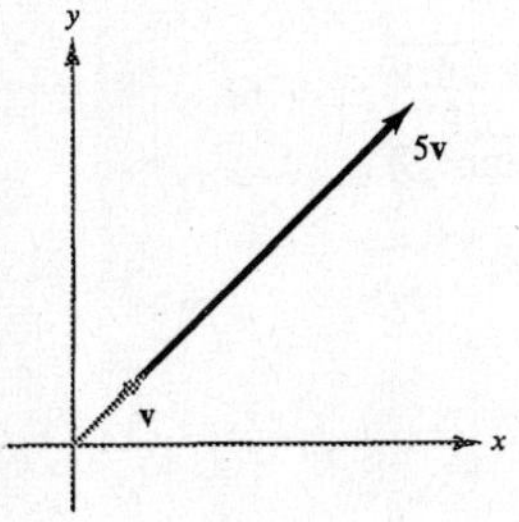

16. $\mathbf{u} - \mathbf{v}$

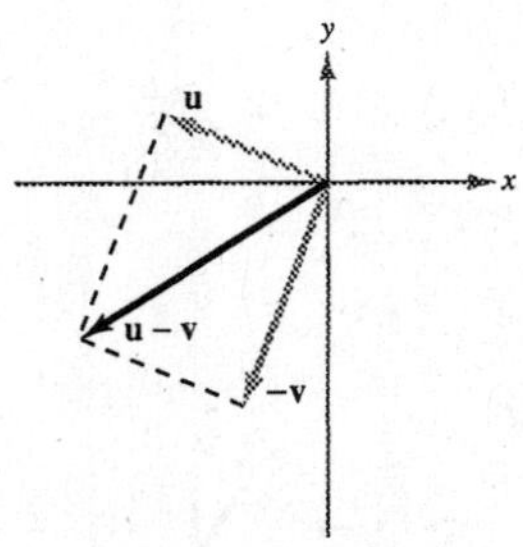

18. $\mathbf{v} - \frac{1}{2}\mathbf{u}$

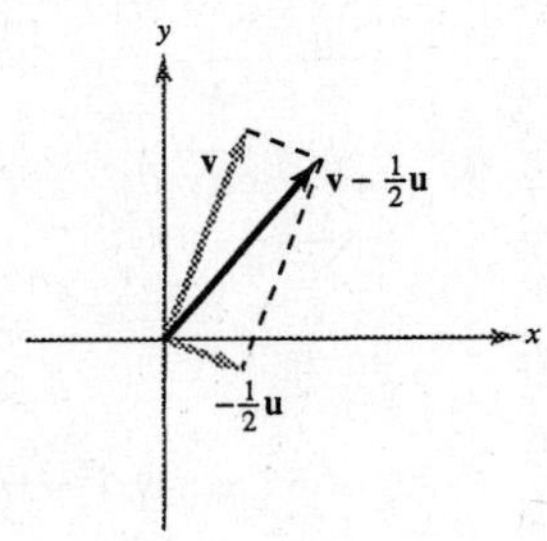

20. $\mathbf{u} = \langle 2, 3 \rangle, \mathbf{v} = \langle 4, 0 \rangle$

(a) $\mathbf{u} + \mathbf{v} = \langle 6, 3 \rangle$

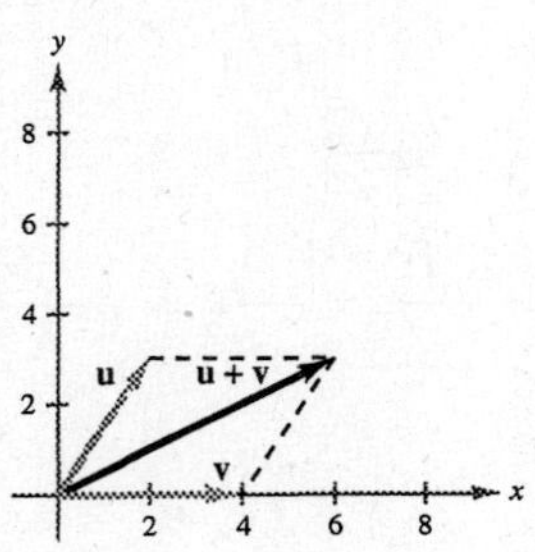

(b) $\mathbf{u} - \mathbf{v} = \langle -2, 3 \rangle$

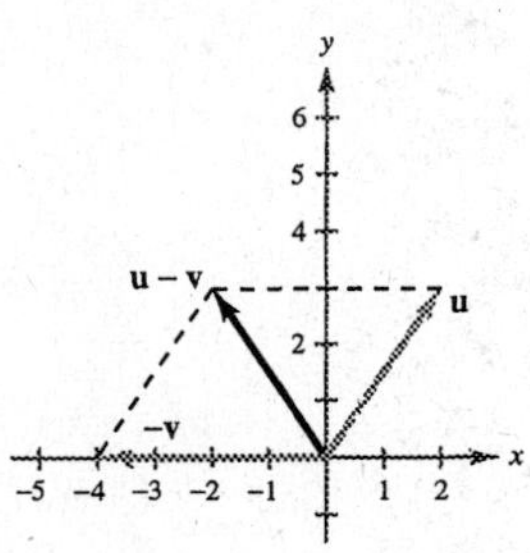

(c) $2\mathbf{u} - 3\mathbf{v} = \langle 4, 6 \rangle - \langle 12, 0 \rangle$

$= \langle -8, 6 \rangle$

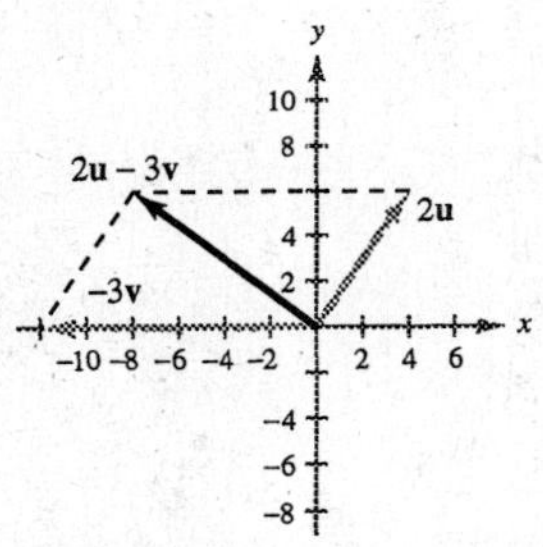

22. $\mathbf{u} = \langle 0, 0 \rangle, \mathbf{v} = \langle 2, 1 \rangle$

(a) $\mathbf{u} + \mathbf{v} = \langle 2, 1 \rangle$

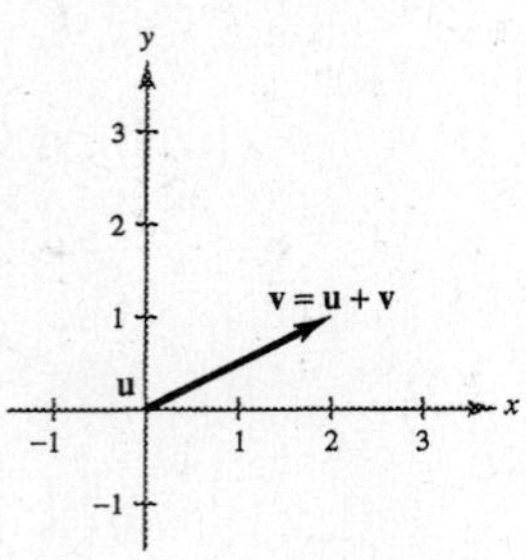

(b) $\mathbf{u} - \mathbf{v} = \langle -2, -1 \rangle$

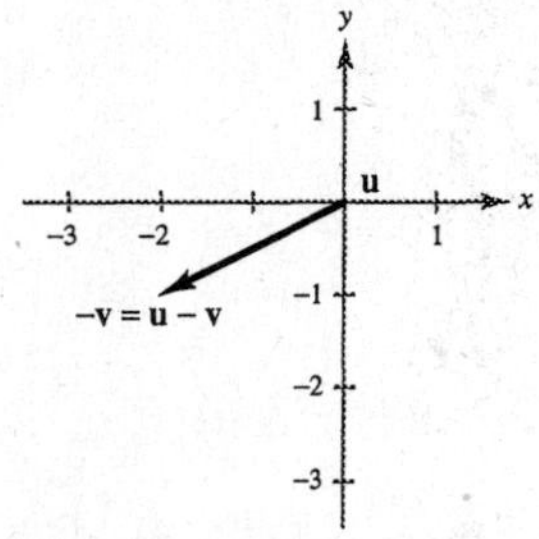

(c) $2\mathbf{u} - 3\mathbf{v} = \langle 0, 0 \rangle - \langle 6, 3 \rangle$

$= \langle -6, -3 \rangle$

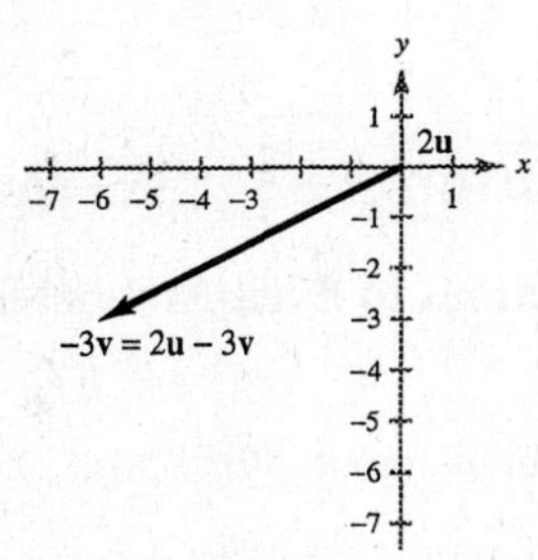

24. $\mathbf{u} = -2\mathbf{i} + \mathbf{j}, \mathbf{v} = -\mathbf{i} + 2\mathbf{j}$

(a) $\mathbf{u} + \mathbf{v} = -3\mathbf{i} + 3\mathbf{j}$

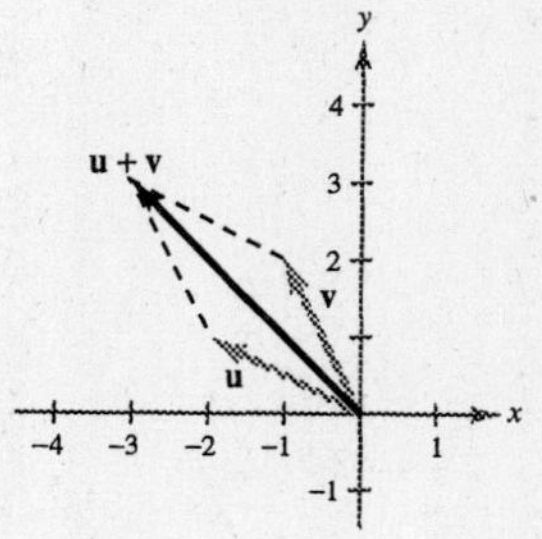

(b) $\mathbf{u} - \mathbf{v} = -\mathbf{i} - \mathbf{j}$

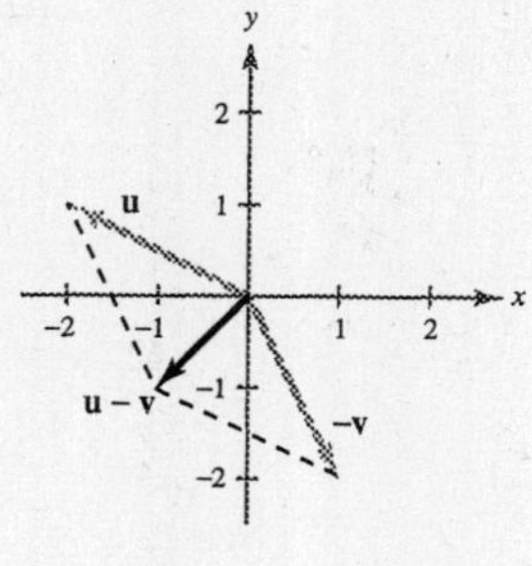

(c) $2\mathbf{u} - 3\mathbf{v} = (-4\mathbf{i} + 2\mathbf{j}) - (-3\mathbf{i} + 6\mathbf{j})$

$= -\mathbf{i} - 4\mathbf{j}$

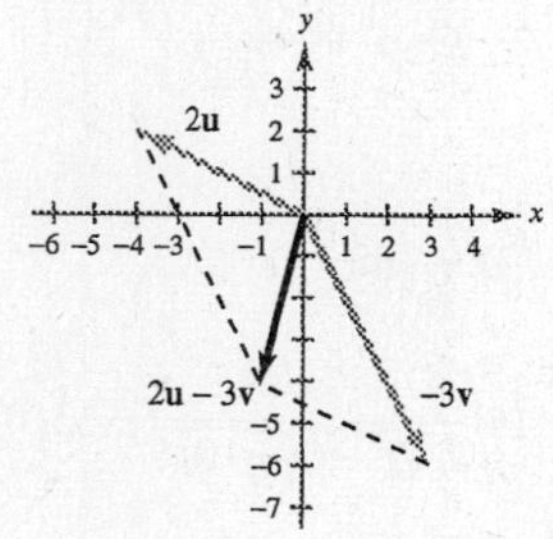

26. $\mathbf{u} = 3\mathbf{j}, \mathbf{v} = 2\mathbf{i}$

(a) $\mathbf{u} + \mathbf{v} = 2\mathbf{i} + 3\mathbf{j}$

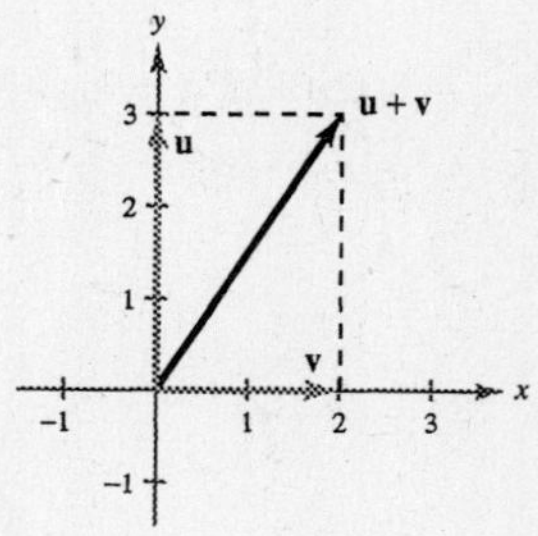

(b) $\mathbf{u} - \mathbf{v} = -2\mathbf{i} + 3\mathbf{j}$

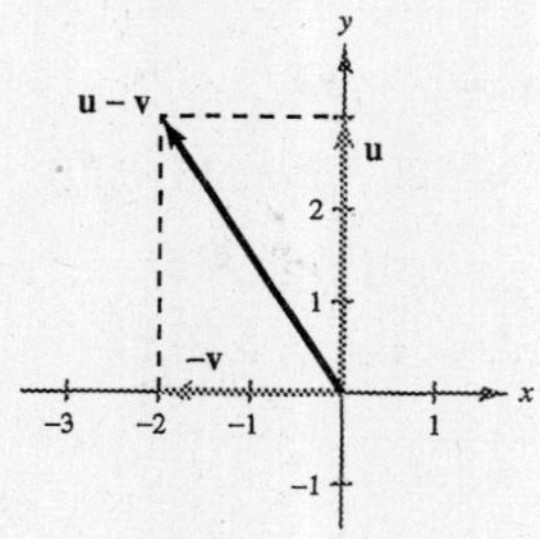

(c) $2\mathbf{u} - 3\mathbf{v} = 6\mathbf{j} - 6\mathbf{i}$

$= -6\mathbf{i} + 6\mathbf{j}$

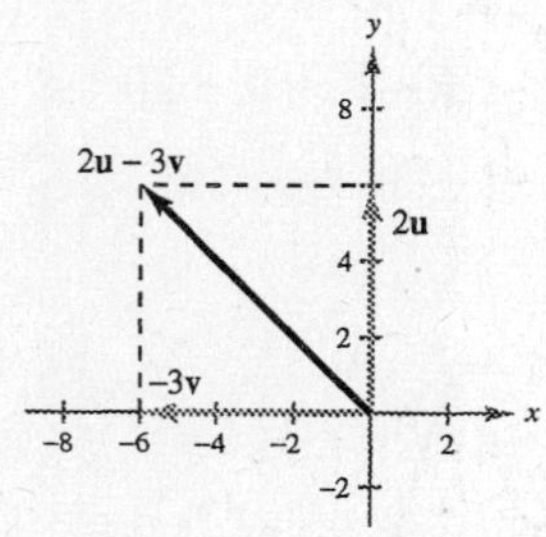

28. $\mathbf{u} = \langle 0, -2 \rangle$

$$\mathbf{v} = \frac{1}{\|\mathbf{u}\|}\mathbf{u} = \frac{1}{\sqrt{0^2 + (-2)^2}}\langle 0, -2 \rangle$$

$$= \frac{1}{2}\langle 0, -2 \rangle = \langle 0, -1 \rangle$$

30. $\mathbf{v} = \langle 5, -12 \rangle$

$$\mathbf{u} = \frac{1}{\|\mathbf{v}\|}\mathbf{v} = \frac{1}{\sqrt{5^2 + (-12)^2}}\langle 5, -12 \rangle$$

$$= \frac{1}{13}\langle 5, -12 \rangle$$

$$= \left\langle \frac{5}{13}, -\frac{12}{13} \right\rangle$$

32. $\mathbf{v} = \mathbf{i} + \mathbf{j}$

$$\mathbf{u} = \frac{1}{\|\mathbf{v}\|}\mathbf{v}$$

$$= \frac{1}{\sqrt{1^2 + 1^2}}(\mathbf{i} + \mathbf{j})$$

$$= \frac{1}{\sqrt{2}}(\mathbf{i} + \mathbf{j})$$

$$= \frac{\sqrt{2}}{2}\mathbf{i} + \frac{\sqrt{2}}{2}\mathbf{j}$$

34. $\mathbf{w} = -6\mathbf{i}$

$$\mathbf{v} = \frac{1}{\|\mathbf{w}\|}\mathbf{w} = \frac{1}{\sqrt{(-6)^2 + 0^2}}(-6\mathbf{i})$$

$$= \frac{1}{6}(-6\mathbf{i}) = -\mathbf{i}$$

36. $\mathbf{w} = 7\mathbf{j} - 3\mathbf{i}$

$$\mathbf{v} = \frac{1}{\|\mathbf{w}\|}\mathbf{w} = \frac{1}{\sqrt{(-3)^2 + 7^2}}(-3\mathbf{i} + 7\mathbf{j})$$

$$= -\frac{3}{\sqrt{58}}\mathbf{i} + \frac{7}{\sqrt{58}}\mathbf{j}$$

38. $\mathbf{v} = 6\left(\frac{1}{\|\mathbf{u}\|}\mathbf{u}\right) = 6\left(\frac{1}{\sqrt{(-3)^2 + 3^2}}\langle -3, 3\rangle\right)$

$$= 6\left(\frac{1}{3\sqrt{2}}\langle -3, 3\rangle\right) = \left\langle -\frac{6}{\sqrt{2}}, \frac{6}{\sqrt{2}}\right\rangle$$

40. $\mathbf{v} = 10\left(\frac{1}{\|\mathbf{u}\|}\mathbf{u}\right)$

$$= 10\left(\frac{1}{\sqrt{0^2 + (-10)^2}}\langle -10, 0\rangle\right)$$

$$= 10\left(\frac{1}{10}\langle -10, 0\rangle\right)$$

$$= \langle -10, 0\rangle$$

42. $\mathbf{v} = \frac{3}{4}\mathbf{w} = \frac{3}{4}(\mathbf{i} + 2\mathbf{j})$

$$= \frac{3}{4}\mathbf{i} + \frac{3}{2}\mathbf{j} = \left\langle \frac{3}{4}, \frac{3}{2}\right\rangle$$

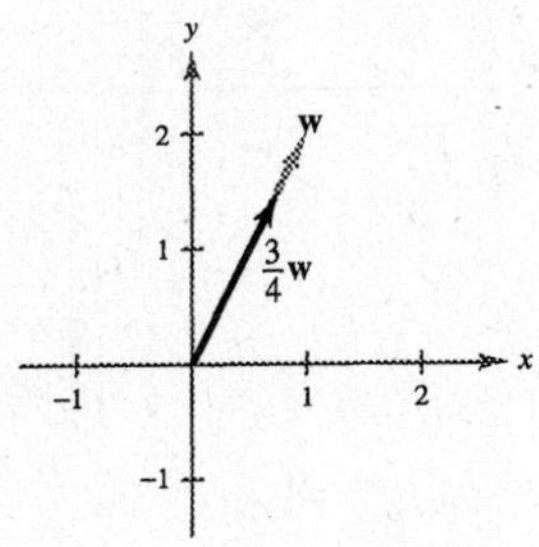

44. $\mathbf{v} = -\mathbf{u} + \mathbf{w}$

$$= -(2\mathbf{i} - \mathbf{j}) + (\mathbf{i} + 2\mathbf{j})$$

$$= -\mathbf{i} + 3\mathbf{j} = \langle -1, 3\rangle$$

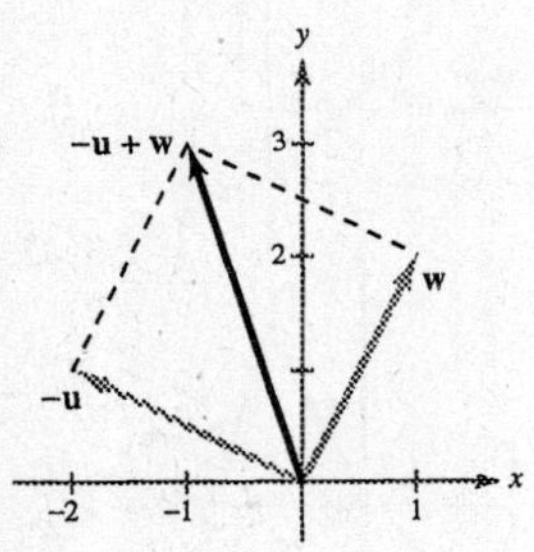

46. $\mathbf{v} = \mathbf{u} - 2\mathbf{w}$

$$= (2\mathbf{i} - \mathbf{j}) - 2(\mathbf{i} + 2\mathbf{j})$$

$$= -5\mathbf{j} = \langle 0, -5\rangle$$

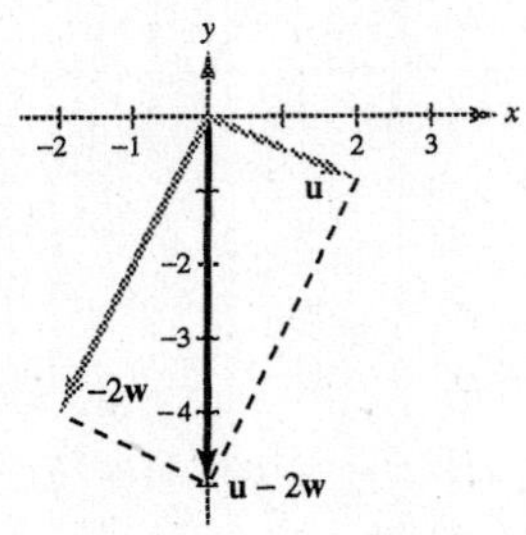

48. $\mathbf{v} = 8(\cos 135°\,\mathbf{i} + \sin 135°\,\mathbf{j})$

$\|\mathbf{v}\| = 8,\ \theta = 135°$

50. $\mathbf{v} = -5\mathbf{i} + 4\mathbf{j}$

$$\|\mathbf{v}\| = \sqrt{(-5)^2 + 4^2} = \sqrt{41}$$

$$\tan\theta = -\frac{4}{5}$$

Since $\mathbf{v}$ lies in Quadrant II, $\theta = 141.3°$.

52. $\mathbf{v} = \langle \cos 45°, \sin 45°\rangle$

$$= \left\langle \frac{\sqrt{2}}{2}, \frac{\sqrt{2}}{2}\right\rangle$$

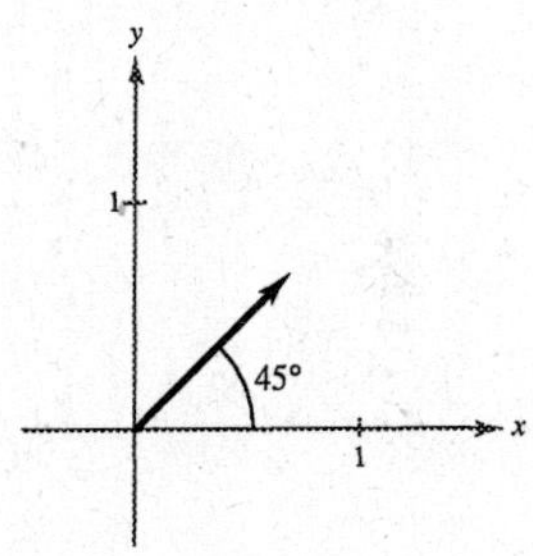

54. $\mathbf{v} = \left\langle \frac{5}{2}\cos 45°, \frac{5}{2}\sin 45°\right\rangle$

$$= \left\langle \frac{5\sqrt{2}}{4}, \frac{5\sqrt{2}}{4}\right\rangle$$

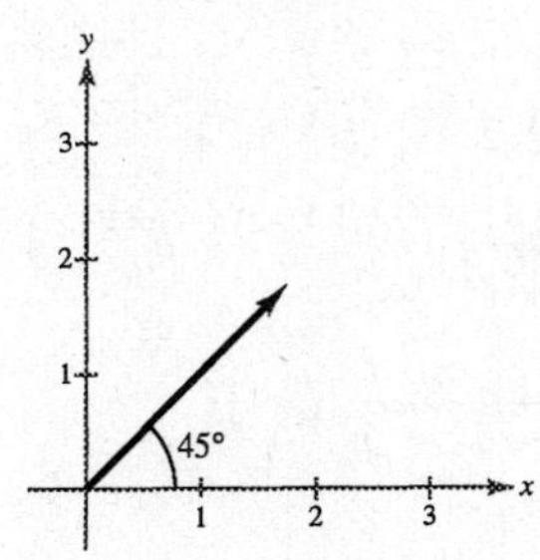

56. $\mathbf{v} = \left\langle 4\sqrt{3}\cos 90°, 4\sqrt{3}\sin 90°\right\rangle$

$= \left\langle 0, 4\sqrt{3}\right\rangle$

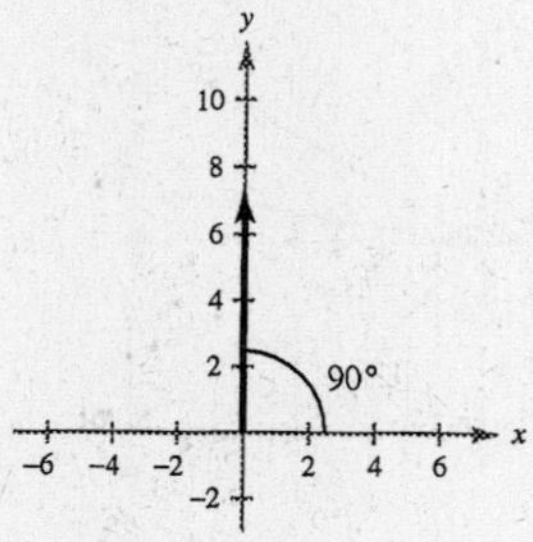

58. $\mathbf{v} = 3\left(\dfrac{1}{\sqrt{3^2 + 4^2}}\right)(3\mathbf{i} + 4\mathbf{j})$

$= \dfrac{3}{5}(3\mathbf{i} + 4\mathbf{j})$

$= \dfrac{9}{5}\mathbf{i} + \dfrac{12}{5}\mathbf{j} = \left\langle \dfrac{9}{5}, \dfrac{12}{5}\right\rangle$

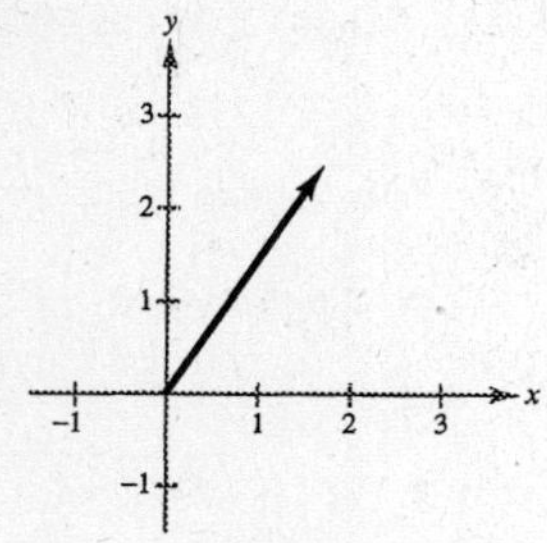

60. $\mathbf{u} = \langle 4\cos 60°, 4\sin 60°\rangle = \left\langle 2, 2\sqrt{3}\right\rangle$

$\mathbf{v} = \langle 4\cos 90°, 4\sin 90°\rangle = \langle 0, 4\rangle$

$\mathbf{u} + \mathbf{v} = \left\langle 2, 4 + 2\sqrt{3}\right\rangle$

62. $\mathbf{u} = \langle 50\cos 30°, 50\sin 30°\rangle = \left\langle 25\sqrt{3}, 25\right\rangle$

$\approx \langle 43.301, 25\rangle$

$\mathbf{v} = \langle 30\cos 110°, 30\sin 110°\rangle \approx \langle -10.261, 28.191\rangle$

$\mathbf{u} + \mathbf{v} \approx \langle 33.04, 53.19\rangle$

64. $\mathbf{v} = 3\mathbf{i} - 2\mathbf{j}$

$\mathbf{w} = 2\mathbf{i} + 2\mathbf{j}$

$\mathbf{u} = \mathbf{v} - \mathbf{w} = \mathbf{i} - 4\mathbf{j}$

$\cos\theta = \dfrac{\|\mathbf{v}\|^2 + \|\mathbf{w}\|^2 - \|\mathbf{v} - \mathbf{w}\|^2}{2\|\mathbf{v}\|\,\|\mathbf{w}\|} = \dfrac{13 + 8 - 17}{2\sqrt{13}\sqrt{8}}$

$= \dfrac{1}{\sqrt{13}\sqrt{2}}$

$\theta \approx 78.7°$

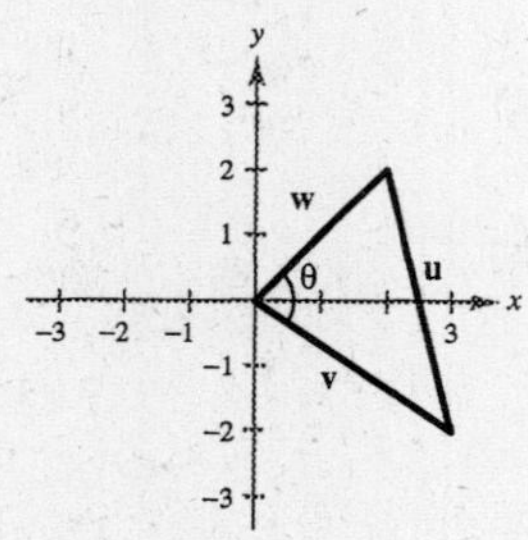

66. $\mathbf{v} = \mathbf{i} + 2\mathbf{j}$

$\mathbf{w} = 2\mathbf{i} - \mathbf{j}$

$\mathbf{u} = \mathbf{v} - \mathbf{w} = -\mathbf{i} + 3\mathbf{j}$

$\cos\theta = \dfrac{\|\mathbf{v}\|^2 + \|\mathbf{w}\|^2 - \|\mathbf{v} - \mathbf{w}\|^2}{2\|\mathbf{v}\|\,\|\mathbf{w}\|} = \dfrac{5 + 5 - 10}{2\sqrt{5}\sqrt{5}} = 0$

$\theta = 90°$

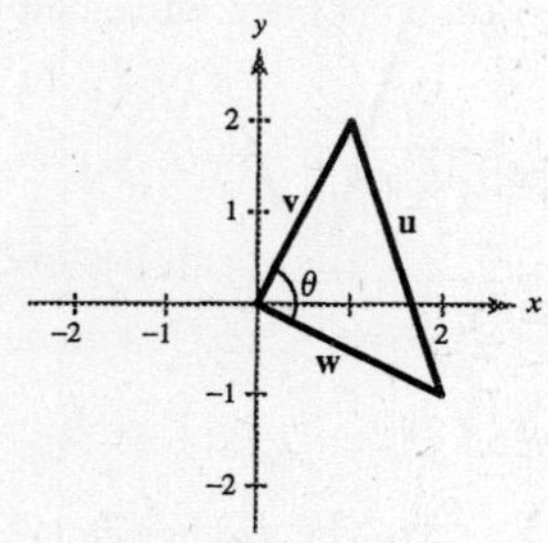

68. Force One: $\mathbf{u} = 3000\mathbf{i}$

Force Two: $\mathbf{v} = 1000\cos\theta\mathbf{i} + 1000\sin\theta\mathbf{j}$

Resultant Force: $\mathbf{u} + \mathbf{v} = (3000 + 1000\cos\theta)\mathbf{i} + 1000\sin\theta\mathbf{j}$

$\|\mathbf{u} + \mathbf{v}\| = \sqrt{(3000 + 1000\cos\theta)^2 + (1000\sin\theta)^2} = 3750$

$$9{,}000{,}000 + 6{,}000{,}000\cos\theta + 1{,}000{,}000 = 14{,}062{,}500$$

$$6{,}000{,}000\cos\theta = 4{,}062{,}500$$

$$\cos\theta = \frac{4{,}062{,}500}{6{,}000{,}000} \approx 0.6771$$

$$\theta \approx 47.4°$$

70.

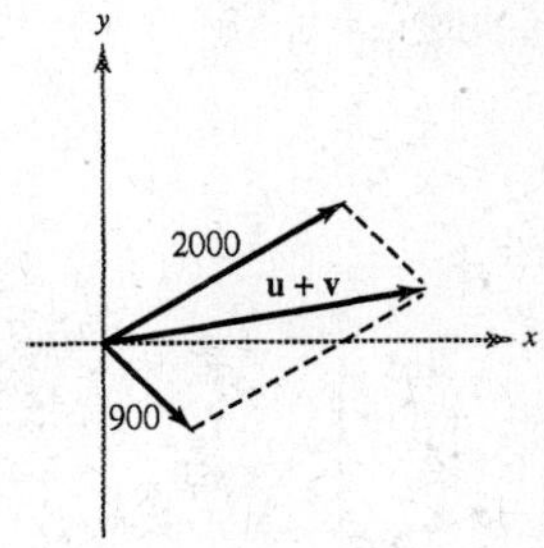

$$\mathbf{u} = (2000 \cos 30°)\mathbf{i} + (2000 \sin 30°\mathbf{j})$$
$$\approx 1732.05\mathbf{i} + 1000\mathbf{j}$$
$$\mathbf{v} = (900 \cos(-45°))\mathbf{i} + (900 \sin(-45°)\mathbf{j})$$
$$\approx 636.4\mathbf{i} + -636.4\mathbf{j}$$
$$\mathbf{u} + \mathbf{v} \approx 2368.4\mathbf{i} + 363.6\mathbf{j}$$
$$\|\mathbf{u} + \mathbf{v}\| \approx \sqrt{(2368.4)^2 + (363.6)^2} \approx 2396.19$$
$$\tan\theta = \frac{363.6}{2368.4} \approx 0.1535 \implies \theta \approx 8.7°$$

72.

$$\mathbf{u} = (70 \cos 30°)\mathbf{i} - (70 \sin 30°)\mathbf{j} \approx 60.62\mathbf{i} - 35\mathbf{j}$$
$$\mathbf{v} = (40 \cos 45°)\mathbf{i} + (40 \sin 45°)\mathbf{j} \approx 28.28\mathbf{i} + 28.28\mathbf{j}$$
$$\mathbf{w} = (60 \cos 135°)\mathbf{i} + (60 \sin 135°)\mathbf{j} \approx -42.43\mathbf{i} + 42.43\mathbf{j}$$
$$\mathbf{u} + \mathbf{v} + \mathbf{w} = 46.48\mathbf{i} + 35.71\mathbf{j}$$
$$\|\mathbf{u} + \mathbf{v} + \mathbf{w}\| \approx 58.61 \text{ pounds}$$
$$\tan\theta \approx \frac{35.71}{46.47} \approx 0.7683$$
$$\theta \approx 37.5°$$

74. Horizontal component of velocity: $1200 \cos 6° \approx 1193.4$ ft/sec

Vertical component of velocity: $1200 \sin 6° \approx 125.4$ ft/sec

76. Rope $\overrightarrow{AC}$: $\mathbf{u} = 10\mathbf{i} - 24\mathbf{j}$

The vector lies in Quadrant IV and its reference angle is $\arctan\left(\frac{12}{5}\right)$.

$$\mathbf{u} = \|\mathbf{u}\|\left[\cos\left(\arctan \tfrac{12}{5}\right)\mathbf{i} - \sin\left(\arctan \tfrac{12}{5}\right)\mathbf{j}\right]$$

Rope $\overrightarrow{BC}$: $\mathbf{v} = -20\mathbf{i} - 24\mathbf{j}$

The vector lies in Quadrant III and its reference angle is $\arctan\left(\frac{6}{5}\right)$.

$$\mathbf{v} = \|\mathbf{v}\|\left[-\cos\left(\arctan \tfrac{6}{5}\right)\mathbf{i} - \sin\left(\arctan \tfrac{6}{5}\right)\mathbf{j}\right]$$

Resultant: $\mathbf{u} + \mathbf{v} = -5000\mathbf{j}$

$$\|\mathbf{u}\| \cos\left(\arctan \tfrac{12}{5}\right) - \|\mathbf{v}\| \cos\left(\arctan \tfrac{6}{5}\right) = 0$$
$$-\|\mathbf{u}\| \sin\left(\arctan \tfrac{12}{5}\right) - \|\mathbf{v}\| \sin\left(\arctan \tfrac{6}{5}\right) = -5000$$

Solving this system of equations yields: $T_{AC} = \|\mathbf{u}\| \approx 3611.1$ pounds

$T_{BC} = \|\mathbf{v}\| \approx 2169.5$ pounds

78. Rope 1: $\mathbf{u} = \|\mathbf{u}\|(\cos 70°\mathbf{i} - \sin 70°\mathbf{j})$

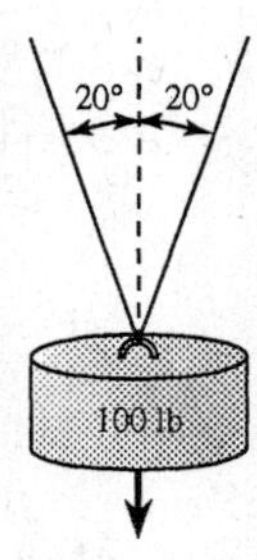

Rope 2: $\mathbf{v} = \|\mathbf{u}\|(-\cos 70°\mathbf{i} - \sin 70°\mathbf{j})$

Resultant: $\mathbf{u} + \mathbf{v} = -100\mathbf{j}$

$$-\|\mathbf{u}\| \sin 70° - \|\mathbf{u}\| \sin 70° = -100$$
$$\|\mathbf{u}\| \approx 53.2$$

Therefore, the tension of each rope is $\|\mathbf{u}\| \approx 53.2$ pounds.

80. (a)

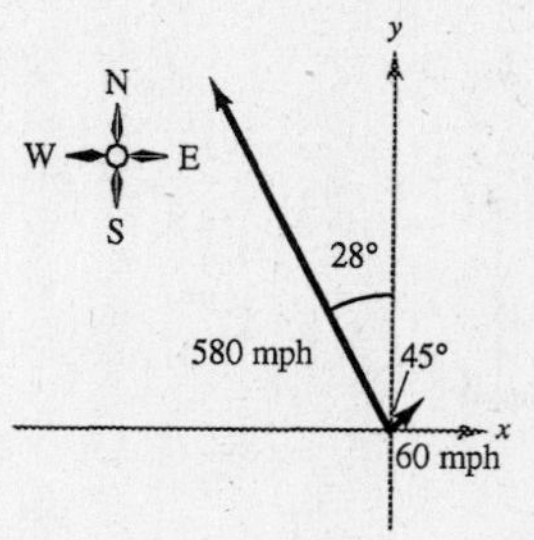

(b) The velocity vector $\mathbf{v}_w$ of the wind has a magnitude of 60 and a direction angle of 45°.

$$\begin{aligned}\mathbf{v}_w &= \|\mathbf{v}_w\|(\cos\theta)\mathbf{i} + \|\mathbf{v}_w\|(\sin\theta)\mathbf{j}\\ &= 60(\cos 45°)\mathbf{i} + 60(\sin 45°)\mathbf{j}\\ &= 60[(\cos 45°)\mathbf{i} + (\sin 45°)\mathbf{j}]\\ &= 60\langle\cos 45°, \sin 45°\rangle, \text{ or } \langle 30\sqrt{2}, 30\sqrt{2}\rangle\end{aligned}$$

(c) The velocity vector $\mathbf{v}_j$ of the jet has a magnitude of 580 and a direction angle of 118°.

$$\begin{aligned}\mathbf{v}_j &= \|\mathbf{v}_j\|(\cos\theta)\mathbf{i} + \|\mathbf{v}_j\|(\sin\theta)\mathbf{j}\\ &= 580(\cos 118°)\mathbf{i} + 580(\sin 118°)\mathbf{j}\\ &= 580[(\cos 118°)\mathbf{i} + (\sin 118°)\mathbf{j}]\\ &= 580\langle\cos 118°, \sin 118°\rangle\end{aligned}$$

(d) The velocity of the jet (in the wind) is

$$\begin{aligned}\mathbf{v} &= \mathbf{v}_w + \mathbf{v}_j\\ &= 60\langle\cos 45°, \sin 45°\rangle + 580\langle\cos 118°, \sin 118°\rangle\\ &= \langle 60\cos 45° + 580\cos 118°, 60\sin 45° + 580\sin 118°\rangle\\ &\approx \langle -229.87, 554.54\rangle\end{aligned}$$

The resultant speed of the jet is

$$\begin{aligned}\|\mathbf{v}\| &= \sqrt{(-229.87)^2 + (554.54)^2}\\ &\approx 600.3 \text{ miles per hour}\end{aligned}$$

(e) If θ is the direction of the flight path, then

$$\tan\theta = \frac{554.54}{-229.87} \approx -2.4124$$

Because θ lies in the Quadrant II, $\theta = 180° + \arctan(-2.4124) \approx 180° - 67.5° = 112.5°$. The true bearing of the jet is $112.5° - 90° = 22.5°$ west of north, or $360° - 22.5° = 337.5°$.

82. Horizontal force: $\mathbf{u} = \|\mathbf{u}\|\mathbf{i}$

Weight: $\mathbf{w} = -\mathbf{j}$

Rope: $\mathbf{t} = \|\mathbf{t}\|(\cos 135°\mathbf{i} + \sin 135°\mathbf{j})$

$$\mathbf{u} + \mathbf{w} + \mathbf{t} = \mathbf{0} \Longrightarrow \begin{aligned}\|\mathbf{u}\| + \|\mathbf{t}\|\cos 135° &= 0\\ -1 + \|\mathbf{t}\|\sin 135° &= 0\end{aligned}$$

$\|\mathbf{t}\| \approx \sqrt{2}$ pounds

$\|\mathbf{u}\| \approx 1$ pound

84. True.

$\|\mathbf{u}\| = \sqrt{a^2 + b^2} = 1 \Longrightarrow a^2 + b^2 = 1$

86. $\mathbf{F}_1 = \langle 10, 0\rangle, \mathbf{F}_2 = 5\langle \cos\theta, \sin\theta\rangle$

(a) $\mathbf{F}_1 + \mathbf{F}_2 = \langle 10 + 5\cos\theta, 5\sin\theta\rangle$

$$\|\mathbf{F}_1 + \mathbf{F}_2\| = \sqrt{(10 + 5\cos\theta)^2 + (5\sin\theta)^2}$$
$$= \sqrt{100 + 100\cos\theta + 25\cos^2\theta + 25\sin^2\theta}$$
$$= 5\sqrt{4 + 4\cos\theta + \cos^2\theta + \sin^2\theta}$$
$$= 5\sqrt{4 + 4\cos\theta + 1}$$
$$= 5\sqrt{5 + 4\cos\theta}$$

(b)

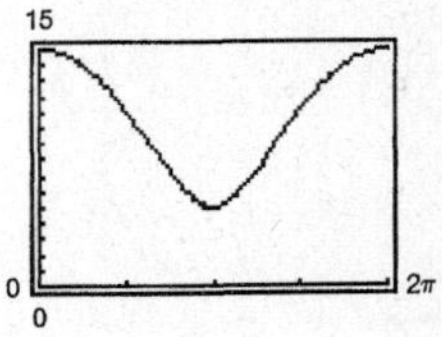

(c) Range: $[5, 15]$
Maximum is 15 when $\theta = 0$.
Minimum is 5 when $\theta = \pi$.

(d) The magnitude of the resultant is never 0 because the magnitudes of $\mathbf{F}_1$ and $\mathbf{F}_2$ are not the same.

88. The following program is written for a *TI-82* or *TI-83* or *TI-83 Plus* graphing calculator. The program sketches two vectors $\mathbf{u} = a\mathbf{i} + b\mathbf{j}$ and $\mathbf{v} = c\mathbf{i} + d\mathbf{j}$ in standard position, and then sketches the vector difference $\mathbf{u} - \mathbf{v}$ using the parallelogram law.

```
PROGRAM: SUBVECT
:Input "ENTER A", A
:Input "ENTER B", B
:Input "ENTER C", C
:Input "ENTER D", D
:Line (0, 0, A, B)
:Line (0, 0, C, D)
:Pause
:A-C→E
:B-D→F
:Line (A, B, C, D)
:Line (A, B, E, F)
:Line (0, 0, E, F)
:Pause
:ClrDraw
:Stop
```

90.
$$\mathbf{u} = \langle 80 - 10, 80 - 60\rangle = \langle 70, 20\rangle$$
$$\mathbf{v} = \langle -20 - (-100), 70 - 0\rangle = \langle 80, 70\rangle$$
$$\mathbf{u} - \mathbf{v} = \langle 70 - 80, 20 - 70\rangle = \langle -10, -50\rangle$$
$$\mathbf{v} - \mathbf{u} = \langle 80 - 70, 70 - 20\rangle = \langle 10, 50\rangle$$

92. $x = 8\sin\theta$

$$\sqrt{64 - x^2} = \sqrt{64 - (8\sin^2\theta)}$$
$$= \sqrt{64 - 64\sin^2\theta}$$
$$= 8\sqrt{1 - \sin^2\theta}$$
$$= 8\sqrt{\cos^2\theta}$$
$$= 8\cos\theta$$

94. $x = 5\sec\theta$

$$\sqrt{(x^2 - 25)^3} = \sqrt{[(5\sec\theta)^2 - 25]^3}$$
$$= \sqrt{(25\sec^2\theta - 25)^3}$$
$$= \sqrt{[25(\sec^2\theta - 1)]^3}$$
$$= \sqrt{(25\tan^2\theta)^3}$$
$$= \sqrt{15{,}625\tan^6\theta}$$
$$= 125\tan^3\theta$$

96. $\sin x\left(2\sin x + \sqrt{2}\right) = 0$

$\sin x = 0$ $\qquad$ $2\sin x + \sqrt{2} = 0$

$x = 0 + n\pi$ $\qquad$ $\sin x = -\dfrac{\sqrt{2}}{2}$

$$x = \frac{5\pi}{4} + 2\pi n, \frac{7\pi}{4} + 2\pi n$$

$$x = n\pi, \frac{5\pi}{4} + 2\pi n, \frac{7\pi}{4} + 2\pi n$$

98. $\cos x \csc x + \cos x\sqrt{2} = 0$

$\cos x\left(\csc x + \sqrt{2}\right) = 0$

$\cos x = 0 \qquad \csc x + \sqrt{2} = 0$

$x = \dfrac{\pi}{2} + n\pi \qquad \csc x = -\sqrt{2}$

$x = \dfrac{5\pi}{4} + 2\pi n, \dfrac{7\pi}{4} + 2\pi n$

$x = \dfrac{\pi}{2} + n\pi, \dfrac{5\pi}{4} + 2\pi n, \dfrac{7\pi}{4} + 2\pi n$

Section 3.4 Vectors and Dot Products

Solutions to Even-Numbered Exercises

2. $\mathbf{u} = \langle 5, 12\rangle, \mathbf{v} = \langle -3, 2\rangle$

$\mathbf{u} \cdot \mathbf{v} = 5(-3) + 12(2) = 9$

4. $\mathbf{u} = 3\mathbf{i} + 4\mathbf{j}, \mathbf{v} = 7\mathbf{i} - 2\mathbf{j}$

$\mathbf{u} \cdot \mathbf{v} = 3(7) + 4(-2) = 13$

6. $\mathbf{v} = \langle -3, 4\rangle$

$\|\mathbf{v}\| + 3 = \sqrt{\mathbf{v} \cdot \mathbf{v}} + 3$

$= \sqrt{(-3)(-3) + 4(4)} + 3 = 8$

The result is a scalar.

8. $\mathbf{u} = \langle 2, 2\rangle, \mathbf{v} = \langle -3, 4\rangle$

$3\mathbf{u} \cdot \mathbf{v} = 3[(2)(-3) + (2)(4)] = 3(2) = 6$

The result is a scalar.

10. $\mathbf{u} = \langle 2, -4\rangle$

$\|\mathbf{u}\| = \sqrt{\mathbf{u} \cdot \mathbf{u}}$

$= \sqrt{2(2) + (-4)(-4)}$

$= \sqrt{20} = 2\sqrt{5}$

12. $\mathbf{u} = 12\mathbf{i} - 16\mathbf{j}$

$\|\mathbf{u}\| = \sqrt{\mathbf{u} \cdot \mathbf{u}} = \sqrt{12(12) + (-16)(-16)}$

$= \sqrt{400} = 20$

14. $\mathbf{u} = -21\mathbf{i}$

$\|\mathbf{u}\| = \sqrt{\mathbf{u} \cdot \mathbf{u}} = \sqrt{(-21)(-21) + 0(0)}$

$= \sqrt{21^2} = 21$

16. $\mathbf{u} = \langle 3, 2\rangle, \mathbf{v} = \langle 4, 0\rangle$

$\cos\theta = \dfrac{\mathbf{u} \cdot \mathbf{v}}{\|\mathbf{u}\|\,\|\mathbf{v}\|} = \dfrac{3(4) + 2(0)}{\sqrt{13}\,(4)}$

$= \dfrac{3}{\sqrt{13}} \approx 0.83205$

$\theta \approx 33.69°$

18. $\mathbf{u} = 2\mathbf{i} - 3\mathbf{j}, \mathbf{v} = \mathbf{i} - 2\mathbf{j}$

$\cos\theta = \dfrac{\mathbf{u} \cdot \mathbf{v}}{\|\mathbf{u}\|\,\|\mathbf{v}\|}$

$= \dfrac{2(1) + (-3)(-2)}{\sqrt{2^2 + 3^2}\sqrt{1^2 + 2^2}}$

$= \dfrac{8}{\sqrt{65}} \approx 0.992278$

$\theta \approx 7.13°$

20. $\mathbf{u} = -6\mathbf{i} - 3\mathbf{j}, \mathbf{v} = -8\mathbf{i} + 4\mathbf{j}$

$\cos\mathbf{u} = \dfrac{\mathbf{u} \cdot \mathbf{v}}{\|\mathbf{u}\|\,\|\mathbf{v}\|} = \dfrac{-6(-8) + (-3)(4)}{\sqrt{45}\sqrt{80}}$

$= \dfrac{36}{60} = 0.6$

$\theta \approx 53.13°$

22. $\mathbf{u} = 2\mathbf{i} - 3\mathbf{j}, \mathbf{v} = 4\mathbf{i} + 3\mathbf{j}$

$$\cos\theta = \frac{\mathbf{u}\cdot\mathbf{v}}{\|\mathbf{u}\|\,\|\mathbf{v}\|} = \frac{2(4) + (-3)(3)}{\sqrt{13}\sqrt{25}} \approx -0.0555$$

$$\theta \approx 93.18°$$

24. $\mathbf{u} = \cos\left(\frac{\pi}{4}\right)\mathbf{i} + \sin\left(\frac{\pi}{4}\right)\mathbf{j} = \frac{\sqrt{2}}{2}\mathbf{i} + \frac{\sqrt{2}}{2}\mathbf{j}$

$$\mathbf{v} = \cos\left(\frac{\pi}{2}\right)\mathbf{i} + \sin\left(\frac{\pi}{2}\right)\mathbf{j} = \mathbf{j}$$

$$\cos\theta = \frac{\mathbf{u}\cdot\mathbf{v}}{\|\mathbf{u}\|\,\|\mathbf{v}\|} = \frac{\frac{\sqrt{2}}{2}(0) + \frac{\sqrt{2}}{2}(1)}{1\cdot 1} = \frac{\sqrt{2}}{2}$$

$$\theta = \frac{\pi}{4}$$

26. $P = (-3, -4), Q = (1, 7), R = (8, 2)$

$\overrightarrow{PQ} = \langle 4, 11\rangle, \overrightarrow{QR} = \langle 7, -5\rangle,$

$\overrightarrow{PR} = \langle 11, 6\rangle, \overrightarrow{QP} = \langle -4, -11\rangle$

$$\cos\alpha = \frac{\overrightarrow{PQ}\cdot\overrightarrow{PR}}{\|\overrightarrow{PQ}\|\,\|\overrightarrow{PR}\|} = \frac{110}{(\sqrt{137})(\sqrt{157})}$$

$\Rightarrow \alpha \approx 41.4°$

$$\cos\beta = \frac{\overrightarrow{QR}\cdot\overrightarrow{QP}}{\|\overrightarrow{QR}\|\,\|\overrightarrow{QP}\|} = \frac{27}{(\sqrt{74})(\sqrt{137})}$$

$\Rightarrow \beta \approx 74.4°$

$\alpha \approx 180° - 41.4° - 74.4° = 64.2°$

28. $P = (-3, 5), Q = (-1, 9), R = (7, 9)$

$\overrightarrow{PQ} = \langle 2, 4\rangle, \overrightarrow{QR} = \langle 8, 0\rangle,$

$\overrightarrow{PR} = \langle 10, 4\rangle, \overrightarrow{QP} = \langle -2, -4\rangle$

$$\cos\alpha = \frac{\overrightarrow{PQ}\cdot\overrightarrow{PR}}{\|\overrightarrow{PQ}\|\,\|\overrightarrow{PR}\|} = \frac{36}{(\sqrt{20})(\sqrt{116})}$$

$\Rightarrow \alpha \approx 41.6°$

$$\cos\beta = \frac{\overrightarrow{QR}\cdot\overrightarrow{QP}}{\|\overrightarrow{QR}\|\,\|\overrightarrow{QP}\|} = \frac{-16}{8(\sqrt{20})}$$

$\Rightarrow \beta \approx 116.6°$

$y \approx 180° - 41.6° - 116.6° = 21.8°$

30. $\|\mathbf{u}\| = 100, \|\mathbf{v}\| = 250, \theta = \frac{\pi}{6}$

$$\mathbf{u}\cdot\mathbf{v} = \|\mathbf{u}\|\,\|\mathbf{v}\|\cos\theta$$

$$= (100)(250)\cos\frac{\pi}{6}$$

$$= 25{,}000\cdot\frac{\sqrt{3}}{2}$$

$$= 12{,}500\sqrt{3}$$

32. $\mathbf{u} = \langle 3, 15\rangle, \mathbf{v} = \langle -1, 5\rangle$

$\mathbf{u} \neq k\mathbf{v} \Rightarrow$ Not parallel

$\mathbf{u}\cdot\mathbf{v} \neq 0 \Rightarrow$ Not orthogonal

Neither

34. $\mathbf{u} = 1, \mathbf{v} = -2\mathbf{i} + 2\mathbf{j}$

$\mathbf{u} \neq k\mathbf{v} \Rightarrow$ Not parallel

$\mathbf{u}\cdot\mathbf{v} \neq 0 \Rightarrow$ Not orthogonal

Neither

36. $\mathbf{u} = \langle\cos\theta, \sin\theta\rangle$

$\mathbf{v} = \langle\sin\theta, -\cos\theta\rangle$

$\mathbf{u}\cdot\mathbf{v} = 0 \Rightarrow \mathbf{u}$ and $\mathbf{v}$ are orthogonal.

38. $\mathbf{u} = \langle 4, 2\rangle, \mathbf{v} = \langle 1, -2\rangle$

$$\mathbf{w}_1 = \text{proj}_{\mathbf{v}}\mathbf{u} = \left(\frac{\mathbf{u}\cdot\mathbf{v}}{\|\mathbf{v}\|^2}\right)\mathbf{v} = 0\langle 1, -2\rangle = \langle 0, 0\rangle$$

$$\mathbf{w}_2 = \mathbf{u} - \mathbf{w}_1 = \langle 4, 2\rangle - \langle 0, 0\rangle = \langle 4, 2\rangle$$

40. $\mathbf{u} = \langle -3, -2\rangle, \mathbf{v} = \langle -4, -1\rangle$

$$\mathbf{w}_1 = \text{proj}_{\mathbf{v}}\mathbf{u} = \left(\frac{\mathbf{u}\cdot\mathbf{v}}{\|\mathbf{v}\|^2}\right)\mathbf{v} = \left(\frac{14}{17}\right)\langle -4, -1\rangle$$

$$\mathbf{w}_2 = \mathbf{u} - \mathbf{w}_1 = \langle -3, -2\rangle - \frac{14}{17}\langle -4, -1\rangle$$

$$= \frac{5}{17}\langle 1, -4\rangle$$

42. $\mathbf{u} = \langle -8, 3 \rangle$

For $\mathbf{v}$ to be orthogonal to $\mathbf{u}$, $\mathbf{u} \cdot \mathbf{v}$ must be equal to 0.

Two possibilities: $\langle 3, 8 \rangle, \langle -3, -8 \rangle$

44. $\mathbf{u} = -\frac{5}{2}\mathbf{i} - 3\mathbf{j}$

For $\mathbf{v}$ to be orthogonal to $\mathbf{u}$, $\mathbf{u} \cdot \mathbf{v}$ must be equal to 0.

Two possibilities: $\mathbf{v} = 3\mathbf{i} - \frac{5}{2}\mathbf{j}$

$\mathbf{v} = -3\mathbf{i} + \frac{5}{2}\mathbf{j}$

46. $P = (1, 3), Q = (-3, 5), \mathbf{v} = -2\mathbf{i} + 3\mathbf{j}$

$$\begin{aligned}\text{work} &= \mathbf{v} \cdot \overrightarrow{PQ}\\ &= (-2\mathbf{i} + 3\mathbf{j}) \cdot (-4\mathbf{i} + 2\mathbf{j})\\ &= (-2)(-4) + 3(2) = 14\end{aligned}$$

48. (a) Force due to gravity: $\mathbf{F} = -30{,}000\mathbf{j}$

Unit vector along hill: $\mathbf{v} = (\cos d)\mathbf{i} + (\sin d)\mathbf{j}$

projection of $\mathbf{F}$ unto v:

$$\begin{aligned}\mathbf{w}_1 &= \text{proj}_{\mathbf{v}}\mathbf{F}\\ &= \left(\frac{\mathbf{F} \cdot \mathbf{v}}{\|\mathbf{v}\|^2}\right)\mathbf{v}\\ &= (\mathbf{F} \cdot \mathbf{v})\mathbf{v}\\ &= -30{,}000 \sin d\, \mathbf{v}\end{aligned}$$

The magnitude of the force is $30{,}000 \sin d$.

(b)

d	0°	1°	2°	3°	4°	5°	6°	7°	8°	9°	10°
Force	0	523.6	1047.0	1570.1	2092.7	2614.7	3135.9	3656.1	4175.2	4693.0	5209.4

(c) Force perpendicular to the hill when $d = 5°$:

Force $= \sqrt{(30{,}000)^2 - (2614.7)^2} \approx 29{,}885.8$ pounds

50. work $= (2400)(5) = 12{,}000$ foot-pounds

52. work $= (\cos 35°)(15{,}691)(800)$

$\approx 10{,}282{,}652$ newton-meters

54. True.

$W = \mathbf{F} \cdot \overrightarrow{PQ} = 0$ if $\mathbf{F}$ and $\overrightarrow{PQ}$ are orthogonal.

56. (a) $\text{proj}_{\mathbf{v}}\mathbf{u} = \mathbf{u} \Rightarrow \mathbf{u}$ and $\mathbf{v}$ are parallel.

(b) $\text{proj}_{\mathbf{v}}\mathbf{u} = 0 \Rightarrow \mathbf{u}$ and $\mathbf{v}$ are orthogonal.

58. Let $\mathbf{u} = \langle u_1, u_2 \rangle$ and $\mathbf{v} = \langle v_1, v_2 \rangle$.

$\mathbf{u} - \mathbf{v} = \langle u_1 - v_1, u_2 - v_2 \rangle$

$$\begin{aligned}\|\mathbf{u} - \mathbf{v}\|^2 &= (u_1 - v_1)^2 + (u_2 - v_2)^2\\ &= u_1^2 - 2u_1v_1 + v_1^2 + u_2^2 - 2u_2v_2 + v_2^2\\ &= u_1^2 + u_2^2 + v_1^2 + v_2^2 - 2u_1v_1 - 2u_2v_2\\ &= \|\mathbf{u}\|^2 + \|\mathbf{v}\|^2 - 2(u_1v_1 + u_2v_2)\\ &= \|\mathbf{u}\|^2 + \|\mathbf{v}\|^2 - 2\mathbf{u} \cdot \mathbf{v}\end{aligned}$$

60.
$$\begin{aligned}\sqrt{18} \cdot \sqrt{112} &= \sqrt{18 \cdot 112}\\ &= \sqrt{2 \cdot 3^2 \cdot 2^4 \cdot 7}\\ &= (3 \cdot 2^2)\sqrt{2 \cdot 7}\\ &= 12\sqrt{14}\end{aligned}$$

62. $\sin 2x + \sqrt{2}\cos x = 0$

$2\sin x\cos x + \sqrt{2}\cos x = 0$

$\cos x(2\sin x + \sqrt{2}) = 0$

$\cos x = 0$ $\qquad$ $2\sin x + \sqrt{2} = 0$

$x = \frac{\pi}{2}, \frac{3\pi}{2}$ $\qquad$ $\sin x = -\frac{\sqrt{2}}{2}$

$x = \frac{5\pi}{4}, \frac{7\pi}{4}$

$x = \frac{\pi}{2}, \frac{5\pi}{4}, \frac{3\pi}{2}, \frac{7\pi}{4}$

64. $\cos 2x - 3\sin x = 2$

$1 - 2\sin^2 x - 3\sin x - 2 = 0$

$2\sin^2 x + 3\sin x + 1 = 0$

$(2\sin x + 1)(\sin x + 1) = 0$

$2\sin x + 1 = 0$ $\qquad$ $\sin x + 1 = 0$

$\sin x = -\frac{1}{2}$ $\qquad$ $\sin x = -1$

$x = \frac{7\pi}{6}, \frac{11\pi}{6}$ $\qquad$ $x = \frac{3\pi}{2}$

$x = \frac{7\pi}{6}, \frac{3\pi}{2}, \frac{11\pi}{6}$

66. $\sin u = -\frac{12}{13}, \cos u = \sqrt{1 - \left(-\frac{12}{13}\right)^2} = \frac{5}{13}$

$\cos v = \frac{24}{25}, \sin v = -\sqrt{1 - \left(\frac{24}{25}\right)^2} = -\frac{7}{25}$

$\sin(u + v) = \sin u\cos v + \cos u\sin v$

$= \left(-\frac{12}{13}\right)\left(\frac{24}{25}\right) + \left(\frac{5}{13}\right)\left(-\frac{7}{25}\right)$

$= -\frac{323}{325}$

68. $\sin u = -\frac{12}{13}, \cos u = \frac{5}{13}, \tan u = -\frac{12}{5}$

$\cos v = \frac{24}{25}, \sin v = -\frac{7}{25}, \tan v = -\frac{7}{24}$

$\tan(u - v) = \frac{\tan u - \tan v}{1 + \tan u\tan v}$

$= \frac{\left(-\frac{12}{5}\right) - \left(-\frac{7}{24}\right)}{1 + \left(-\frac{12}{5}\right)\left(-\frac{7}{24}\right)} = \frac{-\frac{253}{120}}{\frac{17}{10}}$

$= -\frac{253}{204}$

Review Exercises for Chapter 3

Solutions to Even-Numbered Exercises

2. Given: $A = 22°, B = 121°, a = 17$

$C = 180 - A - B = 37°$

$b = \frac{a\sin B}{\sin A} = \frac{17\sin 121°}{\sin 22°} \approx 38.90$

$c = \frac{a\sin C}{\sin A} = \frac{17\sin 37°}{\sin 22°} \approx 27.31$

4. Given: $B = 10°, C - \ = 20°, c = 33$

$A = 180° - B - C = 150°$

$a = \frac{c\sin A}{\sin C} = \frac{33\sin 150°}{\sin 20°} \approx 48.24$

$b = \frac{c\sin B}{\sin C} = \frac{33\sin 10°}{\sin 20°} \approx 16.75$

6. Given: $A = 95°, B = 45°, c = 104.8$

$C = 180° - A - B = 40°$

$a = \frac{c\sin A}{\sin C} = \frac{104.8\sin 95°}{\sin 40°} \approx 162.42$

$b = \frac{c\sin B}{\sin C} = \frac{104.8\sin 45°}{\sin 40°} \approx 115.29$

8. Given: $B = 64°, C = 36°, a = 367$

$A = 180° - B - C = 80°$

$b = \frac{a\sin B}{\sin A} = \frac{367\sin 64°}{\sin 80°} \approx 334.95$

$c = \frac{a\sin C}{\sin A} = \frac{367\sin 36°}{\sin 80°} \approx 219.04$

10. Given: $B = 150°, a = 10, b = 3$

$$\sin A = \frac{a \sin B}{b} = \frac{10 \sin 150°}{3} \approx 1.67 > 1$$

No solution

12. Given: $B = 25°,\ a = 6.2,\ b = 4$

$$\sin A = \frac{a \sin B}{b} \approx 0.65506 \implies A \approx 40.92° \text{ or } 139.08°$$

Case 1: $A \approx 40.92°$

$C \approx 180° - 25° - 40.92° = 114.08°$

$c \approx 8.64$

Case 2: $A \approx 139.08°$

$C \approx 180° - 25° - 139.08° = 15.92°$

$c \approx 2.60$

14. $B = 80°, a = 4, c = 8$

Area $= \frac{1}{2}ac \sin B = \frac{1}{2}(4)(8)(0.9848) = 15.76$

16. $A = 11°, b = 22, c = 21$

Area $= \frac{1}{2}bc \sin A \approx \frac{1}{2}(22)(21)(0.1908)$

$= 44.08$

18. $16^2 = w^2 + 12^2 - 2w(12) \cos 140°$

$w^2 - (24 \cos 140°)w - 112 = 0 \implies w \approx 4.83$

20. The triangle of base 400 feet formed by the two angles of sight to the tree has base angles of $90° - 22° 30' = 67° 30'$, or $67.5°$, and $90° - 15° = 75°$. The angle at the tree measures $180° - 67.5° - 75° = 37.5°$.

$$b = \frac{400 \sin 75°}{\sin 37.5°} \approx 634.683$$

$$c = \frac{400 \sin 67.5°}{\sin 37.5°} \approx 607.055$$

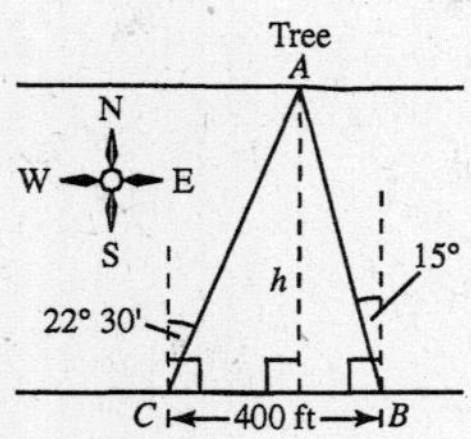

$$\text{Area of } \triangle ABC = \frac{1}{2}ab \sin A$$

$$\text{Area of } \triangle ABC = \frac{1}{2}(400)h$$

$$\frac{1}{2}ab \sin A = \frac{1}{2}(400)h$$

$$(634.683)(607.055) \sin 37.5° = 400h$$

$$\frac{(634.683)(607.055) \sin 37.5°}{400} = h$$

$$586.4 = h$$

The width of the river is about 586.4 feet.

22. Given: $a = 80, b = 60, c = 100$

$$\cos C = \frac{a^2 + b^2 - c^2}{2ab} = \frac{6400 + 3600 - 10{,}000}{2(80)(60)} = 0 \implies C = 90°$$

$$\sin A = \frac{80}{100} = 0.8 \implies A \approx 53.13°$$

$$\sin B = \frac{60}{100} = 0.6 \implies B \approx 36.87°$$

24. Given: $a = 16.4, b = 8.8, c = 12.2$

$$\cos A = \frac{b^2 + c^2 - a^2}{2bc} = \frac{8.8^2 + 12.2^2 - 16.4^2}{2(8.8)(12.2)} \approx -0.1988 \implies A \approx 101.47°$$

$$\sin B = \frac{b \sin A}{a} \approx \frac{8.8 \sin 101.47°}{16.4} \approx 0.5259 \implies B \approx 31.73°$$

$$C \approx 180° - 101.47° - 31.73° = 46.80°$$

26. Given: $B = 150°, a = 10, c = 20$

$$b^2 = 10^2 + 20^2 - 2(10)(20)\cos 150° \implies b \approx 29.09$$

$$\sin A = \frac{a \sin B}{b} \approx \frac{10 \sin 150°}{29.09} \implies A \approx 9.90°$$

$$C \approx 180° - 150° - 9.90° = 20.10°$$

28. Given: $A = 62°, b = 11.34, c = 19.52$

$$a^2 = 11.34^2 + 19.52^2 - 2(11.34)(19.52)\cos 62° \implies a \approx 17.37$$

$$\sin B = \frac{b \sin A}{a} \approx \frac{11.34 \sin 62°}{17.37} \implies B \approx 35.20°$$

$$C \approx 180° - 62° - 35.20° = 82.80°$$

30. $d^2 = 850^2 + 1060^2 - 2(850)(1060) \cos 72°$

$\approx 1{,}289{,}251$

$d \approx 1135$ miles

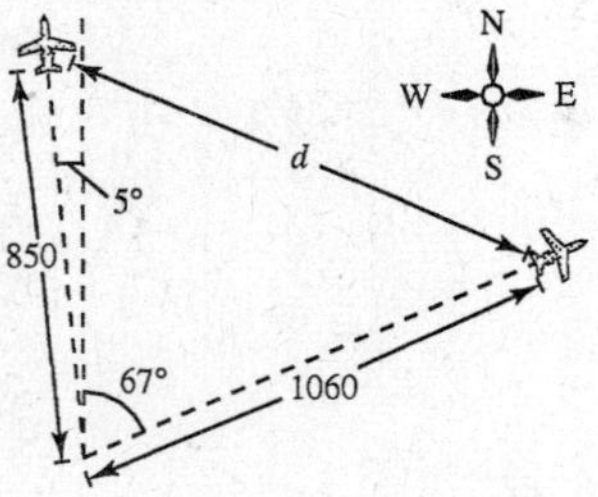

32. $a = 15, b = 8, c = 10$

$$s = \frac{15 + 8 + 10}{2} = 16.5$$

$$\text{Area} = \sqrt{16.5(1.5)(8.5)(6.5)} \approx 36.979$$

34. $a = 38.1, b = 26.7, c = 19.4$

$$s = \frac{38.1 + 26.7 + 19.4}{2} = 42.1$$

$$\text{Area} = \sqrt{42.1(4)(15.4)(22.7)} \approx 242.630$$

36. Initial point: $(3, 4)$

Terminal point: $(-5, -7)$

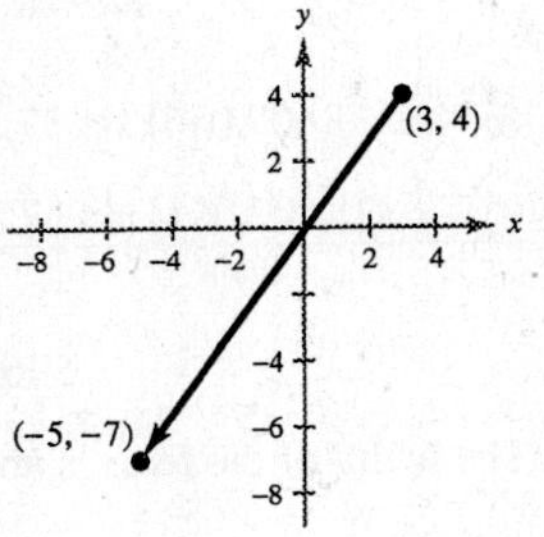

38. Initial point: $(-6, -8)$

Terminal point: $(8, 3)$

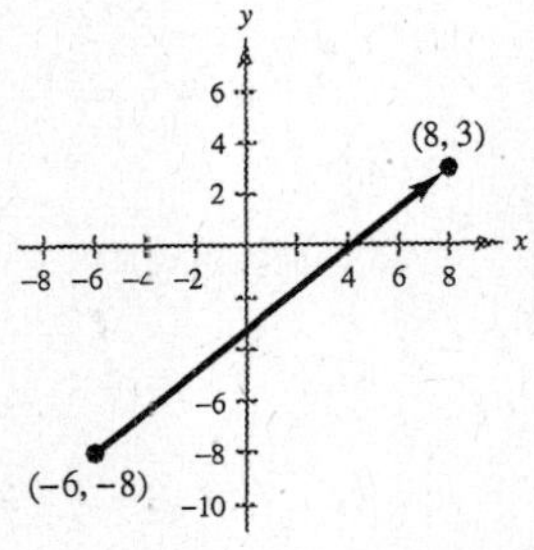

40. Initial point: $(0, 1)$

Terminal point: $\left(6, \frac{7}{2}\right)$

$$\mathbf{v} = \left\langle 6 - 0, \frac{7}{2} - 1 \right\rangle = \left\langle 6, \frac{5}{2} \right\rangle$$

42. Initial point: $(1, 5)$

Terminal point: $(15, 9)$

$\mathbf{v} = \langle 15 - 1, 9 - 5 \rangle = \langle 14, 4 \rangle$

44. $\|\mathbf{v}\| = \frac{1}{2}, \theta = 225°$

$$\left\langle \frac{1}{2}\cos 225°, \frac{1}{2}\sin 225° \right\rangle = \left\langle -\frac{\sqrt{2}}{4}, -\frac{\sqrt{2}}{4} \right\rangle$$

46. $\mathbf{u} = 6\mathbf{i} - 5\mathbf{j},\ \mathbf{v} = 10\mathbf{i} + 3\mathbf{j}$

$$4\mathbf{u} - 5\mathbf{v} = (24\mathbf{i} - 20\mathbf{j}) - (50\mathbf{i} + 15\mathbf{j}) = -26\mathbf{i} - 35\mathbf{j} = \langle -26, -35 \rangle$$

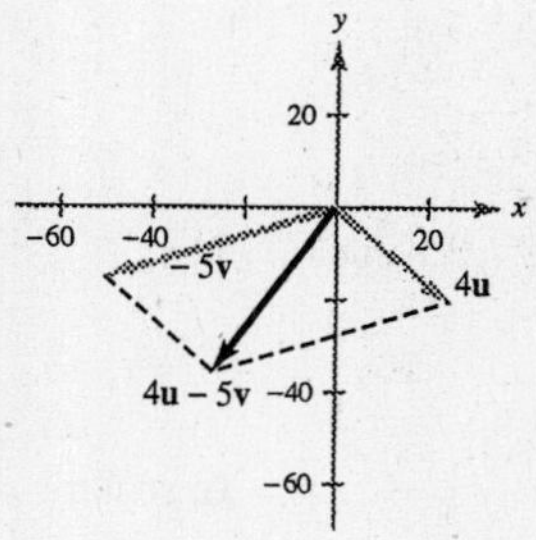

48. $\mathbf{v} = 10\mathbf{i} + 3\mathbf{j}$

$\frac{1}{2}\mathbf{v} = 5\mathbf{i} + \frac{3}{2}\mathbf{j} = \left\langle 5, \frac{3}{2} \right\rangle$

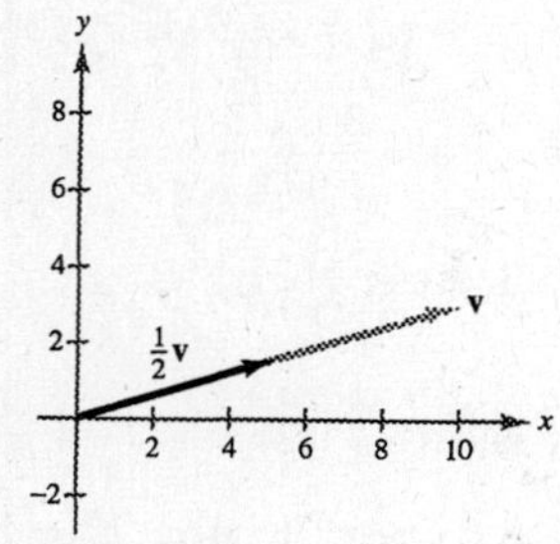

50. $\mathbf{u} = \langle -6, -8 \rangle = -6\mathbf{i} - 8\mathbf{j}$

52. Initial point: $(-2, 7)$

Terminal point: $(5, -9)$

$\mathbf{u} = \langle 5 - (-2), -9 - 7 \rangle = \langle 7, -16 \rangle = 7\mathbf{i} - 16\mathbf{j}$

54. $\mathbf{v} = 4\mathbf{i} - \mathbf{j}$

$\|\mathbf{v}\| = \sqrt{4^2 + (-1)^2} = \sqrt{17}$

$\tan\theta = \frac{-1}{4}$, θ in Quadrant IV $\Rightarrow \theta \approx 346°$

$\mathbf{v} \approx \sqrt{17}(\cos 346°\,\mathbf{i} + \sin 346°\,\mathbf{j})$

56. $\mathbf{v} = 3(\cos 150°\mathbf{i} + \sin 150°\,\mathbf{j})$

$\|\mathbf{v}\| = 3, \theta = 150°$

58. $\mathbf{v} = -4\mathbf{i} + 7\mathbf{j}$

$\|\mathbf{v}\| = \sqrt{(-4)^2 + 7^2} = \sqrt{65}$

$\tan\theta = \frac{7}{-4}$, θ in Quadrant II $\Rightarrow \theta \approx 119.7°$

60. $\mathbf{v} = 8\mathbf{i} - \mathbf{j}$

$\|\mathbf{v}\| = \sqrt{8^2 + (-1)^2} = \sqrt{65}$

$\tan\theta = \frac{-1}{8}$, θ in Quadrant IV $\Rightarrow \theta \approx 352.9°$

62. Rope One: $\mathbf{u} = \|\mathbf{u}\|(\cos 30°\mathbf{i} - \sin 30°\mathbf{j}) = \|\mathbf{u}\|\left(\frac{\sqrt{3}}{2}\mathbf{i} - \frac{1}{2}\mathbf{j}\right)$

Rope Two: $\mathbf{v} = \|\mathbf{u}\|(-\cos 30°\mathbf{i} - \sin 30°\mathbf{j}) = \|\mathbf{u}\|\left(-\frac{\sqrt{3}}{2}\mathbf{i} - \frac{1}{2}\mathbf{j}\right)$

Resultant: $\mathbf{u} + \mathbf{v} = -\|\mathbf{u}\|\mathbf{j} = -180\mathbf{j}$

$\|\mathbf{u}\| = 180$

Therefore, the tension on each rope is $\|\mathbf{u}\| = 180$ lb.

64. Airspeed: $\mathbf{u} = 724(\cos 60°\mathbf{i} + \sin 60°\mathbf{j})$

$= 362(\mathbf{i} + \sqrt{3}\mathbf{j})$

Wind: $\mathbf{w} = 32\mathbf{i}$

Groundspeed $= \mathbf{u} + \mathbf{w} = (394\mathbf{i} + 362\sqrt{3}\mathbf{j})$

$\|\mathbf{u} + \mathbf{w}\| = \sqrt{(394)^2 + (362\sqrt{3})^2} \approx 740.5$ km/hr

$\tan\theta = \dfrac{362\sqrt{3}}{394} \implies \theta \approx 57.9°$

Bearing: N 32.1° E

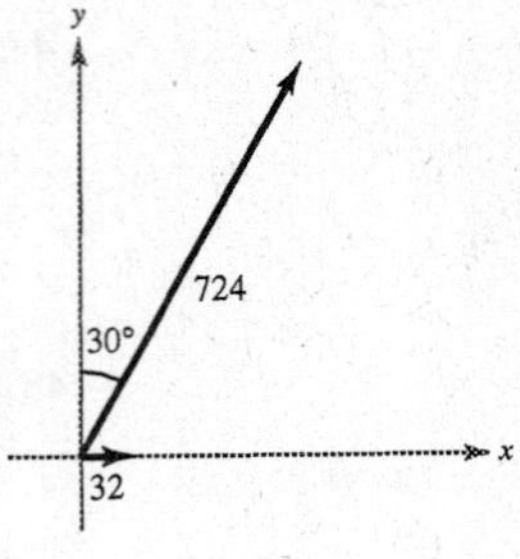

66. $\mathbf{u} = \langle -7, 12\rangle, \mathbf{v} = \langle -4, -14\rangle$

$\mathbf{u}\cdot\mathbf{v} = -7(-4) + 12(-14) = -140$

68. $\mathbf{u} = -7\mathbf{i} + 2\mathbf{j}, \mathbf{v} = 16\mathbf{i} - 12\mathbf{j}$

$\mathbf{u}\cdot\mathbf{v} = -7(16) + 2(-12) = -136$

70. $\mathbf{v} = \langle 2, 1\rangle$

$\|\mathbf{v}\|^2 = \mathbf{v}\cdot\mathbf{v} = 2^2 + 1^2 = 5$; scalar

72. $\mathbf{u} = \langle -3, 4\rangle, \mathbf{v} = \langle 2, 1\rangle$

$3\mathbf{u}\cdot\mathbf{v} = 3(-3(2) + 4(1)) = 3(-2) = -6$; scalar

74. $\mathbf{u} = \cos 45°\,\mathbf{i} + \sin 45°\,\mathbf{j}$

$\mathbf{v} = \cos 300°\,\mathbf{i} + \sin 300°\,\mathbf{j}$

Angle between $\mathbf{u}$ and $\mathbf{v}$: $60° + 45° = 105°$

76. $\mathbf{u} = \langle 3, \sqrt{3}\rangle, \mathbf{v} = \langle 4, 3\sqrt{3}\rangle$

$\cos\theta = \dfrac{\mathbf{u}\cdot\mathbf{v}}{\|\mathbf{u}\|\,\|\mathbf{v}\|} = \dfrac{21}{\sqrt{12}\sqrt{43}}$

$\implies \theta \approx 22.4°$

78. $\mathbf{u} = \langle \frac{1}{4}, -\frac{1}{2}\rangle, \mathbf{v} = \langle -2, 4\rangle$

$\mathbf{v} = -8\mathbf{u} \implies$ Parallel

80. $\mathbf{u} = -2\mathbf{i} + \mathbf{j}, \mathbf{v} = 3\mathbf{i} + 6\mathbf{j}$

$\mathbf{u}\cdot\mathbf{v} = 0 \implies$ Orthogonal

82. $\mathbf{u} = \langle 5, 6\rangle, \mathbf{v} = \langle 10, 0\rangle$

$\text{proj}_{\mathbf{v}}\mathbf{u} = \left(\dfrac{\mathbf{u}\cdot\mathbf{v}}{\|\mathbf{v}\|^2}\right)\mathbf{v} = \dfrac{50}{100}\langle 10, 0\rangle = \langle 5, 0\rangle$

$\mathbf{u} - \text{proj}_{\mathbf{v}}\mathbf{u} = \langle 0, 6\rangle$

84. $\mathbf{u} = \langle -3, 5\rangle, \mathbf{v} = \langle -5, 2\rangle$

$\text{proj}_{\mathbf{v}}\mathbf{u} = \left(\dfrac{\mathbf{u}\cdot\mathbf{v}}{\|\mathbf{v}\|^2}\right)\mathbf{v} = \dfrac{25}{29}\langle -5, 2\rangle$

$\mathbf{u} - \text{proj}_{\mathbf{v}}\mathbf{u} = \langle -3, 5\rangle - \dfrac{25}{29}\langle -5, 2\rangle$

$= \dfrac{19}{29}\langle 2, 5\rangle$

86. work $= \mathbf{v}\cdot\overrightarrow{PQ}$

$= (3\mathbf{i} - 6\mathbf{j})\cdot(-10\mathbf{i} + 17\mathbf{j})$

$= -30 - 102$

$= -132$

88. False. There may be no solution, one solution, or two solutions.

90. False, $a = b = 0$.

92. $a^2 = b^2 + c^2 - 2bc\cos A$

$b^2 = a^2 + c^2 - 2ac\cos B$

$c^2 = a^2 + b^2 - 2ab\cos C$

94. A and C appear to have the same magnitude and direction.

96. If $k > 0$, the direction of $k\mathbf{u}$ is the same, and the magnitude is $k\|\mathbf{u}\|$.

If $k < 0$, the direction of $k\mathbf{u}$ is the opposite direction of $\mathbf{u}$ and the magnitude is $|k|\,\|\mathbf{u}\|$.

Problem Solving for Chapter 3

Solutions to Even-Numbered Exercises

2. 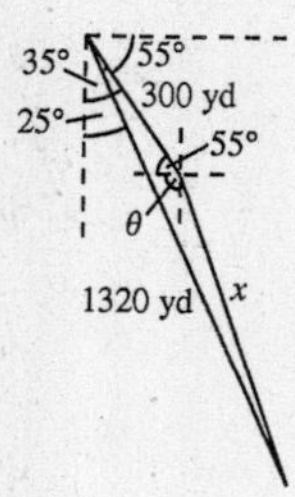

$\frac{3}{4}$ mile $= 1320$ yards

$x^2 = 1320^2 + 300^2 - 2(1320)(300)\cos 10°$

$x \approx 1025.881$ yards ≈ 0.58 mile

$\frac{\sin\theta}{1320} = \frac{\sin 10°}{1025.881}$

$\sin\theta \approx 0.2234$

$\theta = 180° - \sin^{-1}(0.2234)$

$\theta \approx 167.09°$

Bearing: $\theta - 55° - 90° \approx 22.09°$

S 22.09° E

4. (a) 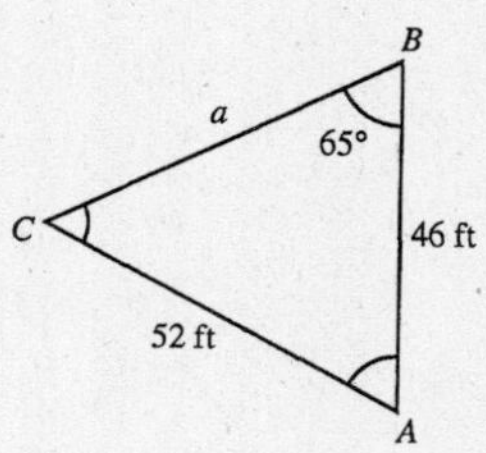

(b) $\frac{\sin C}{46} = \frac{\sin 65°}{52}$

$\sin C = \frac{46 \sin 65°}{52} \approx 0.801734$

$C \approx 53.296°$

$A = 180° - B - C = 61.704°$

$\frac{a}{\sin 61.704°} = \frac{52}{\sin 65°}$

$a = \frac{52 \sin 61.704°}{\sin 65°}$

$a \approx 50.52$ feet

(c) Area $= \frac{1}{2}(46)(52)\sin 61.704°$

≈ 1053.09 square feet

Number of bags: $\frac{1053.09}{50} \approx 21.06$

To entirely cover the courtyard, you would need to buy 22 bags.

6. (a) $\mathbf{u} = -120\mathbf{j}$

$\mathbf{v} = 40\mathbf{i}$

(b) $\mathbf{s} = \mathbf{u} + \mathbf{v} = 40\mathbf{i} - 120\mathbf{j}$

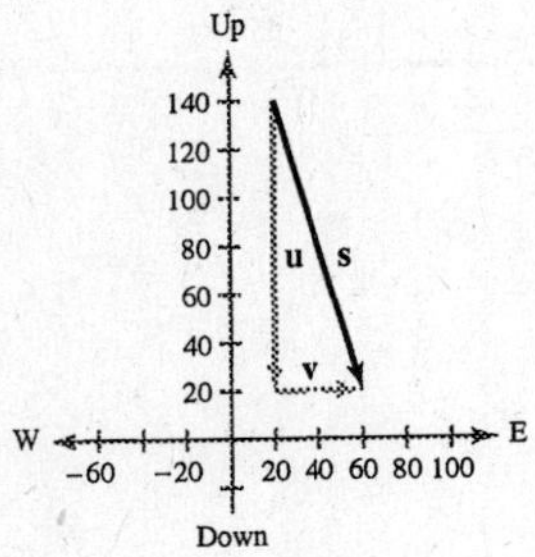

(c) $\|\mathbf{s}\| = \sqrt{40^2 + (-120)^2} = \sqrt{16000} = 40\sqrt{10}$

≈ 126.49 miles per hour

This represents the actual rate of the skydiver's fall.

(d) $\tan\theta = \dfrac{120}{40} \Rightarrow \theta = \tan^{-1} 3 \Rightarrow \theta \approx 71.565°$

(e)

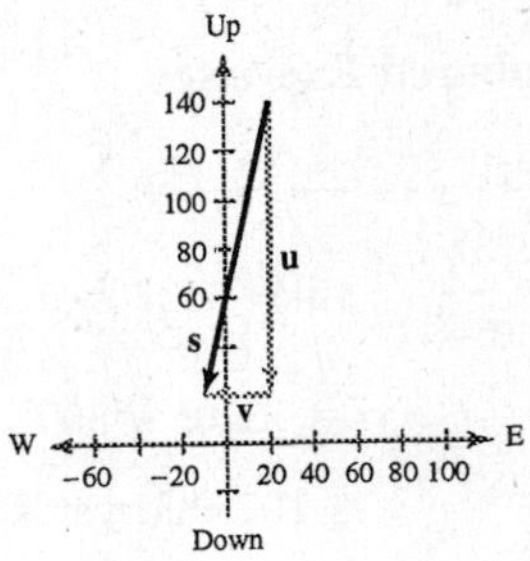

$\mathbf{s} = 30\mathbf{i} - 120\mathbf{j}$

$\|\mathbf{s}\| = \sqrt{30^2 + (-120)^2}$

$= \sqrt{15300}$

≈ 123.69 miles per hour

8. Let $\mathbf{u} \cdot \mathbf{v} = 0$ and $\mathbf{u} \cdot \mathbf{w} = 0$.

Then, $\mathbf{u} \cdot (c\mathbf{v} + d\mathbf{w}) = \mathbf{u} \cdot c\mathbf{v} + \mathbf{u} \cdot d\mathbf{w}$

$= c\mathbf{u} \cdot \mathbf{v} + d\mathbf{u} \cdot \mathbf{w}$

$= c(0) + d(0)$

$= 0.$

Thus for all scalars c and d, $\mathbf{u}$ is orthogonal to $c\mathbf{v} + d\mathbf{w}$.

10. (a)

θ	$100 \sin \theta$	$100 \cos \theta$
0.5°	0.8727	99.9962
1.0°	1.7452	99.9848
1.5°	2.6177	99.9657
2.0°	3.4899	99.9391
2.5°	4.3619	99.9048
3.0°	5.2336	99.8630

(b) *No*, the airplane's speed does *not* equal the sum of the vertical and horizontal components of its velocity. To find speed:

speed $= \sqrt{(\|\mathbf{v}\| \sin\theta)^2 + (\|\mathbf{v}\| \cos\theta)^2}$

(c) (i) speed $= \sqrt{5.235^2 + 149.909^2} \approx 150$ miles per hour

(ii) speed $= \sqrt{10.463^2 + 149.634^2} \approx 150$ miles per hour

CHAPTER 4
Complex Numbers

CHAPTER 4
Complex Numbers

Section 4.1 Complex Numbers

Solutions to Even-Numbered Exercises

2. $a + bi = 13 + 4i$

$a = 13$

$b = 4$

4. $(a + 6) + 2bi = 6 - 5i$

$2b = -5$

$b = -\frac{5}{2}$

$a + 6 = 6$

$a = 0$

6. $3 + \sqrt{-16} = 3 + 4i$

8. $1 + \sqrt{-8} = 1 + 2\sqrt{2}i$

10. $\sqrt{-4} = 2i$

12. 45

14. $-4i^2 + 2i = -4(-1) + 2i$

$= 4 + 2i$

16. $\sqrt{-0.0004} = 0.02i$

18. $(13 - 2i) + (-5 + 6i) = 8 + 4i$

20. $(3 + 2i) - (6 + 13i) = 3 + 2i - 6 - 13i$

$= -3 - 11i$

22. $\left(8 + \sqrt{-18}\right) - \left(4 + 3\sqrt{2}i\right) = 8 + 3\sqrt{2}i - 4 - 3\sqrt{2}i$

$= 4$

24. $22 + (-5 + 8i) + 10i = 17 + 18i$

26. $(1.6 + 3.2i) + (-5.8 + 4.3i) = -4.2 + 7.5i$

28. $\sqrt{-5} \cdot \sqrt{-10} = \left(\sqrt{5}i\right)\left(\sqrt{10}i\right)$

$= \sqrt{50}i^2 = 5\sqrt{2}(-1) = -5\sqrt{2}$

30. $\left(\sqrt{-75}\right)^2 = \left(\sqrt{75}i\right)^2 = 75i^2 = -75$

32. $(6 - 2i)(2 - 3i) = 12 - 18i - 4i + 6i^2$

$= 12 - 22i - 6 = 6 - 22i$

34. $-8i(9 + 4i) = -72i - 32i^2$

$= 32 - 72i$

36. $\left(3 + \sqrt{-5}\right)\left(7 - \sqrt{-10}\right) = \left(3 + \sqrt{5}i\right)\left(7 - \sqrt{10}i\right)$

$= 21 - 3\sqrt{10}i + 7\sqrt{5}i - \sqrt{50}i^2$

$= 21 + \sqrt{50} + 7\sqrt{5}i - 3\sqrt{10}i$

$= \left(21 + 5\sqrt{2}\right) + \left(7\sqrt{5} - 3\sqrt{10}\right)i$

38. $(2 - 3i)^2 = 4 - 12i + 9i^2$

$= 4 - 9 - 12i$

$= -5 - 12i$

40. $(1 - 2i)^2 - (1 + 2i)^2 = 1 - 4i + 4i^2 - (1 + 4i + 4i^2)$

$= 1 - 4i + 4i^2 - 1 - 4i - 4i^2$

$= -8i$

42. The complex conjugate of $7 - 12i$ is $7 + 12i$.

$$\begin{aligned}(7 - 12i)(7 + 12i) &= 49 - 144i^2 \\ &= 49 - (-144) \\ &= 193\end{aligned}$$

44. The complex conjugate of $-3 + \sqrt{2}i$ is $-3 - \sqrt{2}i$.

$$\begin{aligned}\left(-3 + \sqrt{2}i\right)\left(-3 - \sqrt{2}i\right) &= 9 - 2i^2 \\ &= 9 - (-2) \\ &= 11\end{aligned}$$

46. The complex conjugate of $\sqrt{-15} = \sqrt{15}i$ is $-\sqrt{15}i$.

$$\left(\sqrt{15}i\right)\left(-\sqrt{15}i\right) = -15i^2 = -(-15) = 15$$

48. The complex conjugate of $1 + \sqrt{8}$ is $1 + \sqrt{8}$.

$$\begin{aligned}\left(1 + \sqrt{8}\right)\left(1 + \sqrt{8}\right) &= 1 + 2\sqrt{8} + 8 \\ &= 9 + 4\sqrt{2}\end{aligned}$$

50. $-\dfrac{14}{2i} \cdot \dfrac{-2i}{-2i} = \dfrac{28i}{-4i^2} = \dfrac{28i}{4} = 7i$

52. $\dfrac{5}{1-i} \cdot \dfrac{1+i}{1+i} = \dfrac{5+5i}{1-i^2} = \dfrac{5+5i}{2} = \dfrac{5}{2} + \dfrac{5}{2}i$

54.
$$\begin{aligned}\frac{6-7i}{1-2i} \cdot \frac{1+2i}{1+2i} &= \frac{6 + 12i - 7i - 14i^2}{1 - 4i^2} \\ &= \frac{20 + 5i}{5} = \frac{20}{5} + \frac{5}{5}i = 4 + i\end{aligned}$$

56. $\dfrac{8+16i}{2i} \cdot \dfrac{-2i}{-2i} = \dfrac{-16i - 32i^2}{-4i^2} = 8 - 4i$

58.
$$\begin{aligned}\frac{5i}{(2+3i)^2} &= \frac{5i}{4 + 12i + 9i^2} = \frac{5i}{-5 + 12i} \cdot \frac{-5 - 12i}{-5 - 12i} \\ &= \frac{-25i - 60i^2}{25 - 144i^2} = \frac{60 - 25i}{169} = \frac{60}{169} - \frac{25}{169}i\end{aligned}$$

60.
$$\begin{aligned}\frac{2i}{2+i} + \frac{5}{2-i} &= \frac{2i(2-i)}{(2+i)(2-i)} + \frac{5(2+i)}{(2+i)(2-i)} \\ &= \frac{4i - 2i^2 + 10 + 5i}{4 - i^2} \\ &= \frac{12 + 9i}{5} \\ &= \frac{12}{5} + \frac{9}{5}i\end{aligned}$$

62.
$$\begin{aligned}\frac{1+i}{i} - \frac{3}{4-i} &= \frac{(1+i)(4-i) - 3i}{i(4-i)} \\ &= \frac{4 - i + 4i - i^2 - 3i}{4i - i^2} \\ &= \frac{5}{1+4i} \cdot \frac{1-4i}{1-4i} \\ &= \frac{5 - 20i}{1 - 16i^2} \\ &= \frac{5}{17} - \frac{20}{17}i\end{aligned}$$

64. $x^2 + 6x + 10 = 0;\quad a = 1, b = 6, c = 10$

$$\begin{aligned}x &= \frac{-6 \pm \sqrt{6^2 - 4(1)(10)}}{2(1)} \\ &= \frac{-6 \pm \sqrt{-4}}{2} \\ &= -3 \pm i\end{aligned}$$

66. $9x^2 - 6x + 37 = 0; a = 9, b = -6, c = 37$

$$\begin{aligned}x &= \frac{-(-6) \pm \sqrt{(-6)^2 - 4(9)(37)}}{2(9)} \\ &= \frac{6 \pm \sqrt{-1296}}{18} \\ &= \frac{1}{3} \pm \frac{36i}{18} \\ &= \frac{1}{3} \pm 2i\end{aligned}$$

68. $16t^2 - 4t + 3 = 0;\ a = 16,\ b = -4,\ c = 3$

$$\begin{aligned}t &= \frac{-(-4) \pm \sqrt{(-4)^2 - 4(16)(3)}}{2(16)} \\ &= \frac{4 \pm \sqrt{-176}}{32} \\ &= \frac{4 \pm 4\sqrt{11}i}{32} \\ &= \frac{1}{8} \pm \frac{\sqrt{11}}{8}i\end{aligned}$$

70. $\frac{7}{8}x^2 - \frac{3}{4}x + \frac{5}{16} = 0$

$14x^2 - 12x + 5 = 0;\ a = 14,\ b = -12,\ c = 5$

$$x = \frac{-(-12) \pm \sqrt{(-12)^2 - 4(14)(5)}}{2(14)}$$

$$= \frac{12 \pm \sqrt{-136}}{28}$$

$$= \frac{12 \pm 2i\sqrt{34}}{28}$$

$$= \frac{3}{7} \pm \frac{\sqrt{34}}{14}i$$

72. $4.5x^2 - 3x + 12 = 0;\ a = 4.5,\ b = -3,\ c = 12$

$$x = \frac{-(-3) \pm \sqrt{(-3)^2 - 4(4.5)(12)}}{2(4.5)}$$

$$= \frac{3 \pm \sqrt{-207}}{9} = \frac{3 \pm 3i\sqrt{23}}{9} = \frac{1}{3} \pm \frac{\sqrt{23}}{3}i$$

74. $4i^2 - 2i^3 = -4 + 2i$

76. $(-i)^3 = (-1)(i^3) = (-1)(-i) = i$

78. $\left(\sqrt{-2}\right)^6 = \left(\sqrt{2}i\right)^6 = 8i^6 = 8i^4i^2 = -8$

80. $\frac{1}{(2i)^3} = \frac{1}{8i^3} = \frac{1}{-8i} \cdot \frac{8i}{8i} = \frac{8i}{-64i^2} = \frac{1}{8}i$

82. (a) $2^4 = 16$

(b) $(-2)^4 = 16$

(c) $(2i)^4 = 2^4i^4 = 16(1) = 16$

(d) $(-2i)^4 = (-2)^4i^4 = 16(1) = 16$

84. (a) $z_1 = 9 + 16i,\ z_2 = 20 - 10i$

(b) $\frac{1}{z} = \frac{1}{z_1} + \frac{1}{z_2}$

$$= \frac{1}{9 + 16i} + \frac{1}{20 - 10i}$$

$$= \frac{20 - 10i + 9 + 16i}{(9 + 16i)(20 - 10i)}$$

$$= \frac{29 + 6i}{340 + 230i}$$

$$z = \left(\frac{340 + 230i}{29 + 6i}\right)\left(\frac{29 - 6i}{29 - 6i}\right)$$

$$= \frac{11{,}240 + 4630i}{877} = \frac{11{,}240}{877} + \frac{4630}{877}i$$

86. True.

$$x^4 - x^2 + 14 = 56$$

$$\left(-i\sqrt{6}\right)^4 - \left(-i\sqrt{6}\right)^2 + 14 \stackrel{?}{=} 56$$

$$36 + 6 + 14 \stackrel{?}{=} 56$$

$$56 = 56$$

88. $\sqrt{-6}\sqrt{-6} = \sqrt{6}i\sqrt{6}i = 6i^2 = -6$

90. Answers will vary.

92. $(x^3 - 3x^2) - (6 - 2x - 4x^2) = x^3 - 3x^2 - 6 + 2x + 4x^2$

$$= x^3 + x^2 + 2x - 6$$

94. $(2x - 5)^2 = (2x)^2 - 2(2x)(5) + (5)^2$

$$= 4x^2 - 20x + 25$$

96. $8 - 3x = -34$

$$-3x = -42$$

$$x = 14$$

98. $5[x - (3x + 11)] = 20x - 15$

$5x - 15x - 55 = 20x - 15$

$-30x = 40$

$x = \frac{40}{-30} = -\frac{4}{3}$

100. $F = \alpha\frac{m_1m_2}{r^2}$

$r^2 = \alpha\frac{m_1m_2}{F}$

$r = \sqrt{\frac{\alpha m_1m_2}{F}} = \frac{\sqrt{\alpha m_1m_2}}{\sqrt{F}} \cdot \frac{\sqrt{F}}{\sqrt{F}} = \frac{\sqrt{\alpha m_1m_2F}}{F}$

Section 4.2 Complex Solutions of Equations

Solutions to Even-Numbered Exercises

2. $x^6 + 4x^2 + 12 = 0$

The degree of the polynomial is 6, so the equation has 6 solutions in the complex number system.

4. $14 - x + 4x^2 - 7x^5 = 0$

$-7x^5 + 4x^2 - x + 14 = 0$

The degree of the polynomial is 5, so the equation has 5 solutions in the complex number system.

6. $2x^2 - x - 1 = 0$

$a = 2, b = -1, c = -1$

$b^2 - 4ac = (-1)^2 - 4(2)(-1) = 9 > 0$

2 real solutions

8. $\frac{1}{3}x^2 - 5x + 25 = 0$

$a = \frac{1}{3}, b = -5, c = 25$

$b^2 - 4ac = (-5)^2 - 4\left(\frac{1}{3}\right)(25) = -\frac{25}{3} < 0$

No real solutions

10. $-2x^2 + 11x - 2 = 0$

$a = -2, b = 11, c = -2$

$b^2 - 4ac = 11^2 - 4(-2)(-2) = 105 > 0$

2 real solutions

12. $x^2 - 4x + 53 = 0$

$a = 1, b = -4, c = 53$

$b^2 - 4ac = (-4)^2 - 4(1)(53) = -196 < 0$

No real solutions

14. $3x^2 - 1 = 0$

$x^2 = \frac{1}{3}$

$x = \pm\frac{\sqrt{3}}{3}$

16. $16 - (x - 1)^2 = 0$

$x - 1 = \pm\sqrt{16}$

$x = 1 \pm 4$

$x = 5, -3$

18. $4x^2 + 4x + 1 = 0$

$(2x + 1)^2 = 0$

$x = -\frac{1}{2}$

20. $54 + 16x - x^2 = 0$

$-x^2 + 16x + 54 = 0$

$x = \frac{-16 \pm \sqrt{16^2 - 4(-1)(54)}}{2(-1)} = \frac{-16 \pm \sqrt{472}}{-2}$

$= 8 \pm \sqrt{118}$

22. $4x^2 - 4x + 21 = 0$

$x = \frac{-(-4) \pm \sqrt{(-4)^2 - 4(4)(21)}}{2(4)} = \frac{4 \pm \sqrt{-320}}{8}$

$= \frac{1}{2} \pm \frac{i\,8\sqrt{5}}{8}$

$= \frac{1}{2} \pm \sqrt{5}i$

24. $125 - 30x + 0.4x^2 = 0$

$0.4x^2 - 30x + 125 = 0$

$$x = \frac{-(-30) \pm \sqrt{(-30)^2 - 4(0.4)(125)}}{2(0.4)}$$

$$= \frac{30 \pm \sqrt{700}}{0.8}$$

$$= \frac{300 \pm 100\sqrt{7}}{8}$$

$$= \frac{75 \pm 25\sqrt{7}}{2}$$

26. $6 - (x - 1)^2 = 0$

$x - 1 = \pm\sqrt{6}$

$x = 1 \pm \sqrt{6}$

28. $f(x) = x^3 - 4x^2 - 4x + 16$

$= x^2(x - 4) - 4(x - 4)$

$= (x^2 - 4)(x - 4)$

$= (x + 2)(x - 2)(x - 4)$

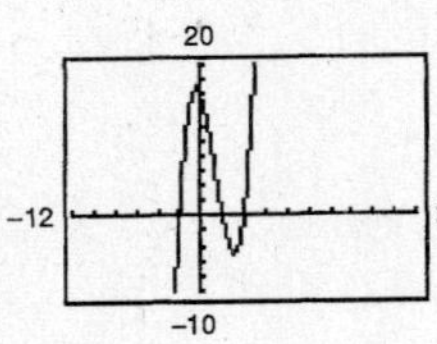

The zeros are $x = -2$, $x = 2$, and $x = 4$. Yes, the number of real zeros (3) is the same as the number of x-intercepts.

30. $f(x) = x^4 - 3x^2 - 4$

$= (x^2 - 4)(x^2 + 1)$

$= (x + 2)(x - 2)(x + i)(x - i)$

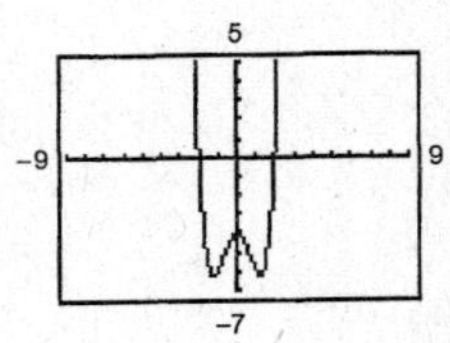

The zeros are $x = -2$, $x = 2$, $x = -i$, and $x = i$. Yes, the number of real zeros (2) is the same as the number of x-intercepts.

32. $f(x) = x^2 - x + 56$

By the Quadratic Formula, the zeros of $f(x)$ are

$$x = \frac{-(-1) \pm \sqrt{(-1)^2 - 4(1)(56)}}{2(1)} = \frac{1 \pm \sqrt{-223}}{2}$$

$$= \frac{1 \pm \sqrt{223}i}{2}$$

$$f(x) = \left(x - \frac{1 - \sqrt{223}i}{2}\right)\left(x - \frac{1 + \sqrt{223}i}{2}\right)$$

34. $g(x) = x^2 + 10x + 23$

By the Quadratic Formula, the zeros of $g(x)$ are

$$x = \frac{-10 \pm \sqrt{10^2 - 4(1)(23)}}{2(1)} = \frac{-10 \pm \sqrt{8}}{2}$$

$$= -5 \pm \sqrt{2}$$

$$g(x) = \left(x + 5 + \sqrt{2}\right)\left(x + 5 - \sqrt{2}\right)$$

36. $f(y) = y^4 - 625$

$y^4 - 625 = 0$

$(y^2 + 25)(y^2 - 25) = 0$

$y^2 + 25 = 0 \Rightarrow y = \pm\sqrt{-25} = \pm 5i$

$y^2 - 25 = 0 \Rightarrow y = \pm\sqrt{25} = \pm 5$

Zeros: $y = -5i, 5i, -5, 5$

$f(y) = (y + 5)(y - 5)(y + 5i)(y - 5i)$

38. $h(x) = x^2 - 6x - 10$

$$x = \frac{-(-6) \pm \sqrt{(-6)^2 - 4(1)(-10)}}{2(1)}$$

$$x = \frac{6 \pm \sqrt{76}}{2}$$

$$x = \frac{6 \pm 2\sqrt{19}}{2}$$

$x = 3 \pm \sqrt{19}$

The zeros of $h(x)$ are $x = 3 \pm \sqrt{19}$.

$h(x) = \left(x - 3 - \sqrt{19}\right)\left(x - 3 + \sqrt{19}\right)$

40. $f(x) = x^3 - 8x^2 - 12x + 96$

$= x^2(x - 8) - 12(x - 8)$

$= (x^2 - 12)(x - 8)$

$x^2 - 12 \Rightarrow x = \pm 2\sqrt{3}$

$x - 8 \Rightarrow x = 8$

The zeros of $f(x)$ are $x = \pm 2\sqrt{3}$ and $x = 8$.

$f(x) = \left(x + 2\sqrt{3}\right)\left(x - 2\sqrt{3}\right)(x - 8)$

42. $h(x) = x^3 + 5x^2 + 2x + 10$

$= x^2(x + 5) + 2(x + 5)$

$= (x^2 + 2)(x + 5)$

$x^2 + 2 \Rightarrow x = \pm\sqrt{2}i$

$x + 5 \Rightarrow x = -5$

The zeros of $h(x)$ are $x = \pm\sqrt{2}i$ and $x = -5$.

$h(x) = \left(x + \sqrt{2}i\right)\left(x - \sqrt{2}i\right)(x + 5)$

44. $g(x) = 4x^3 + 3x^2 + 96x + 72$

$= x^2(4x + 3) + 24(4x + 3)$

$= (x^2 + 24)(4x + 3)$

$x^2 + 24 \Rightarrow x = \pm 2\sqrt{6}i$

$4x + 3 \Rightarrow x = -\frac{3}{4}$

The zeros of $g(x)$ are $x = \pm 2\sqrt{6}i$ and $x = -\frac{3}{4}$.

$g(x) = \left(x + 2\sqrt{6}i\right)\left(x - 2\sqrt{6}i\right)(4x + 3)$

46. $h(x) = x^4 + x^3 + 100x^2 + 100x$

$= x(x^3 + x^2 + 100x + 100)$

$= x[x^2(x + 1) + 100(x + 1)]$

$= x(x^2 + 100)(x + 1)$

$x^2 + 100 \Rightarrow x = \pm 100$

$x + 1 \Rightarrow x = -1$

$x \Rightarrow x = 0$

The zeros of $h(x)$ are $x = 0, -1$, and $\pm 10i$.

$h(x) = x(x + 10i)(x - 10i)(x + 1)$

48. $f(x) = x^4 + 29x^2 + 100$

$x^4 + 29x^2 + 100 = 0$

$(x^2 + 4)(x^2 + 25) = 0$

$x^2 + 4 = 0 \Rightarrow x = \pm\sqrt{-4} = \pm 2i$

$x^2 + 25 = 0 \Rightarrow x = \pm\sqrt{-25} = \pm 5i$

$x = \pm 2i, \pm 5i$

$f(x) = (x + 2i)(x - 2i)(x + 5i)(x - 5i)$

50. $f(x) = x^3 + x^2 + 9x + 9$

Since $3i$ is a zero, so is $-3i$.

$$\begin{array}{r|rrrr} 3i & 1 & 1 & 9 & 9 \\ & & 3i & -9+3i & -9 \\ \hline & 1 & 1+3i & 3i & 0 \end{array}$$

$$\begin{array}{r|rrr} -3i & 1 & 1+3i & 3i \\ & & -3i & -3i \\ \hline & 1 & 1 & 0 \end{array}$$

The zero of $x + 1$ is $x = -1$.

The zeros of f are $x = -1, \pm 3i$.

52. $g(x) = x^3 - 7x^2 - x + 87$

Since $5 + 2i$ is a zero, so is $5 - 2i$.

$$\begin{array}{r|rrrr} 5+2i & 1 & -7 & -1 & 87 \\ & & 5+2i & -14+6i & -87 \\ \hline & 1 & -2+2i & -15+6i & 0 \end{array}$$

$$\begin{array}{r|rrr} 5-2i & 1 & -2+2i & -15+6i \\ & & 5-2i & 15-6i \\ \hline & 1 & 3 & 0 \end{array}$$

The zero of $x + 3$ is $x = -3$.

The zeros of g are $x = -3, 5 \pm 2i$.

54. $h(x) = 3x^3 - 4x^2 + 8x + 8$

Since $1 - \sqrt{3}i$ is a zero, so is $1 + \sqrt{3}i$.

$$\begin{array}{r|rrrr} 1-\sqrt{3}i & 3 & -4 & 8 & 8 \\ & & 3-3\sqrt{3}i & -10-2\sqrt{3}i & -8 \\ \hline & 3 & -1-3\sqrt{3}i & -2-2\sqrt{3}i & \end{array}$$

$$\begin{array}{r|rrr} 1+\sqrt{3}i & 3 & -1-3\sqrt{3}i & -2-2\sqrt{3}i \\ & & 3+3\sqrt{3}i & 2+2\sqrt{3}i \\ \hline & 3 & 2 & 0 \end{array}$$

The zero of $3x + 2$ is $x = -\frac{2}{3}$.

The zeros of h are $x = -\frac{2}{3}, 1 \pm \sqrt{3}i$.

56. $f(x) = x^3 + 4x^2 + 14x + 20$

Since $-1 - 3i$ is a zero, so is $-1 + 3i$.

$$\begin{array}{r|rrrr} -1-3i & 1 & 4 & 14 & 20 \\ & & -1-3i & -12-6i & -20 \\ \hline & 1 & 3-3i & 2-6i & 0 \end{array}$$

$$\begin{array}{r|rrr} -1+3i & 1 & 3-3i & 2-6i \\ & & -1+3i & -2+6i \\ \hline & 1 & 2 & 0 \end{array}$$

The zero of $x + 2$ is $x = -2$.

The zeros of f are $x = -2, -1 \pm 3i$.

58. $f(x) = (x - 4)(x - 3i)(x + 3i)$

$= (x - 4)(x^2 + 9)$

$= x^3 - 4x^2 + 9x - 36$

Note: $f(x) = a(x^3 - 4x^2 + 9x - 36)$, where a is any real number, has the zeros 4, $3i$ and $-3i$.

60. $f(x) = (x - 2)(x - 4 - i)(x - 4 + i)$

$= (x - 2)(x^2 - 8x + 17)$

$= x^3 - 10x^2 + 33x - 34$

Note: $f(x) = a(x^3 - 10x^2 + 33x - 34)$ where a is any real number, has the zeros 2, $4 \pm i$.

62. $f(x) = (x + 5)^2(x - 1 - \sqrt{3}i)(x - 1 + \sqrt{3}i)$

$= (x^2 + 10x + 25)(x^2 - 2x + 4)$

$= x^4 + 8x^3 + 9x^2 - 10x + 100$

Note: $f(x) = a(x^4 + 8x^3 + 9x^2 - 10x + 100)$, where a is any real number, has the zeros $-5, -5, 1 \pm \sqrt{3}i$.

64. (a) $R = -265.193t^2 + 5621.38t - 19{,}147.5$

(b)

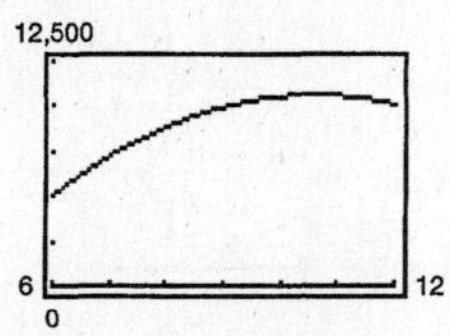

(c) From the graph of the model, the revenue never reached \$11 billion. The maximum for the model is about \$10.642 billion.

(d) The maximum occurs for $t = -\dfrac{b}{2a} = -\dfrac{5621.38}{2(-265.193)} \approx 10.599$.

$R(10.599) = -265.193(10.599)^2 + 5621.38(10.599) - 19{,}147.5 \approx 10{,}642$

The revenue never reached \$11 billion.

66. False. f does not have real coefficients.

68. $g(x) = 3f(x)$. This function has the same zeros as f because it is a vertical stretch of f. The zeros of g are r_1, r_2, and r_3.

70. $g(x) = f(2x)$. Note that x is a zero of g if and only if $2x$ is a zero of f. The zeros of g are $\dfrac{r_1}{2}, \dfrac{r_2}{2}$, and $\dfrac{r_3}{2}$.

72. $g(x) = f(-x)$. Note that x is a zero of g if and only if $-x$ is a zero of f. The zeros of g are $-r_1$, $-r_2$, and $-r_3$.

74. (a) No. The graph is not of $f(x) = x^2(x + 2)(x - 3.5)$ because this function has a $(0, 0)$ as an intercept and the given graph does not go through $(0, 0)$.

(b) No. The graph is not of $g(x) = (x + 2)(x - 3.5)$ because this function's graph is a parabola and the graph given is not. The function shown in the graph must be at least a fourth-degree polynomial.

(c) Yes. The function shown in the graph is $h(x) = (x + 2)(x - 3.5)(x^2 + 1)$.

(d) The graph is not of $k(x) = (x + 1)(x + 2)(x - 3.5)$ because this function has $(-1, 0)$ as an intercept and the given graph does not.

76. $(12 - 5i) + 16i = 12 + 11i$

78. $(9 - 5i)(9 + 5i) = 81 - 25i^2 = 81 + 25 = 106$

80. $g(x) = f(x) - 2$

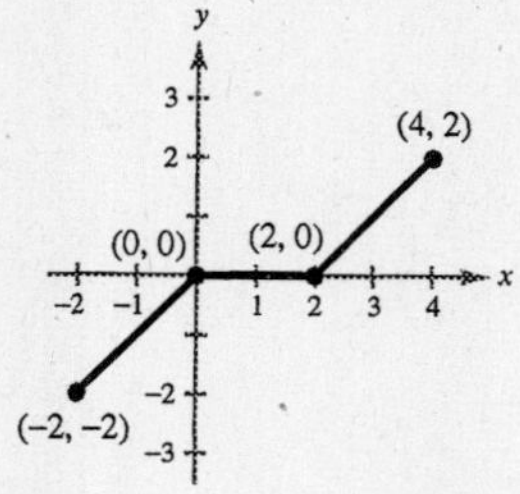

82. $g(x) = f(-x)$

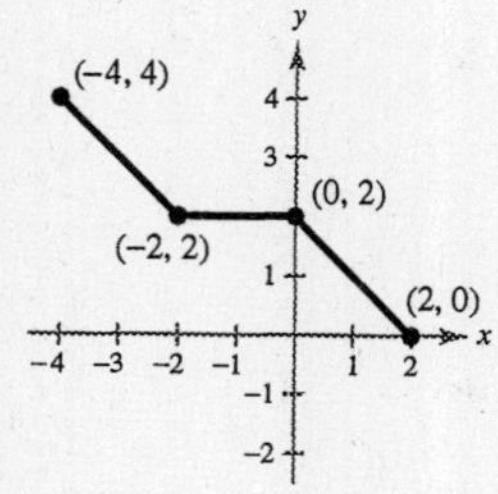

84. $g(x) = f\left(\frac{1}{2}x\right)$

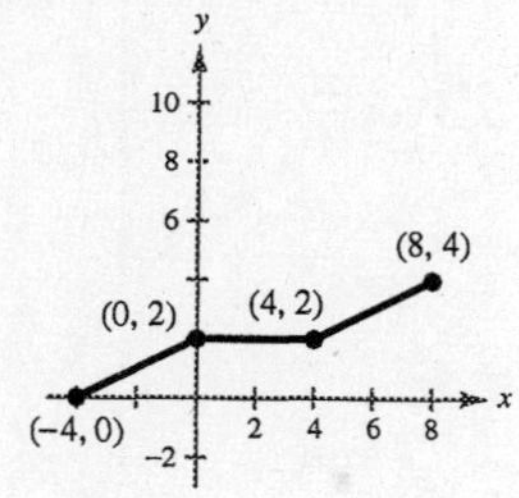

Section 4.3 Trigonometric Form of a Complex Number

Solutions to Even-Numbered Exercises

2. $|-7| = \sqrt{(-7)^2 + 0^2} = \sqrt{49} = 7$

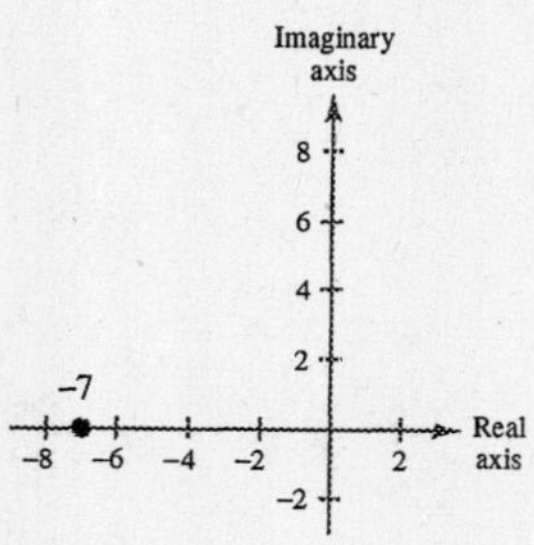

4. $|5 - 12i| = \sqrt{5^2 + (-12)^2}$

$= \sqrt{169} = 13$

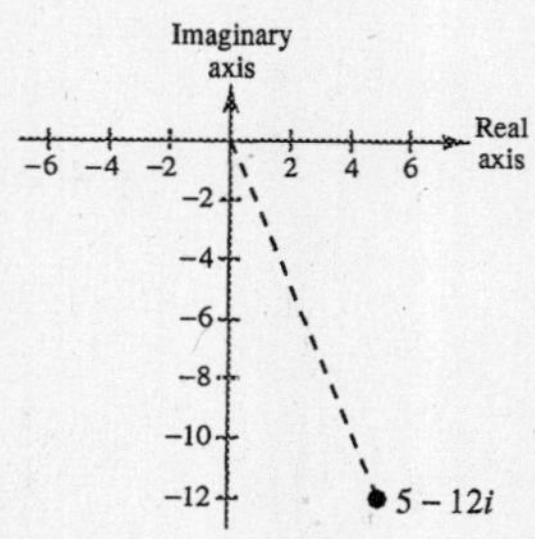

6. $|-8 + 3i| = \sqrt{(-8)^2 + (3)^2}$

$= \sqrt{73}$

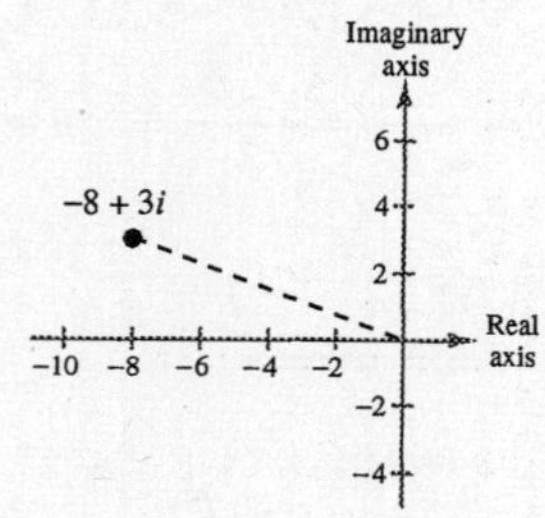

8. $z = -2$

$r = \sqrt{(-2)^2 + 0^2} = \sqrt{4} = 2$

$\tan \theta = \dfrac{0}{-2} \Rightarrow \theta = \pi$

$z = 2(\cos \pi + i \sin \pi)$

10. $z = -1 + \sqrt{3}i$

$r = \sqrt{(-1)^2 + \left(\sqrt{3}\right)^2} = \sqrt{4} = 2$

$\tan \theta = \dfrac{\sqrt{3}}{-1} = -\sqrt{3} \Rightarrow \theta = \dfrac{2\pi}{3}$

$z = 2\left(\cos \dfrac{2\pi}{3} + i \sin \dfrac{2\pi}{3}\right)$

12. $z = 2 + 2i$

$r = \sqrt{2^2 + 2^2} = \sqrt{8} = 2\sqrt{2}$

$\tan\theta = \dfrac{2}{2} = 1 \Rightarrow \theta = \dfrac{\pi}{4}$

$z = 2\sqrt{2}\left(\cos\dfrac{\pi}{4} + i\sin\dfrac{\pi}{4}\right)$

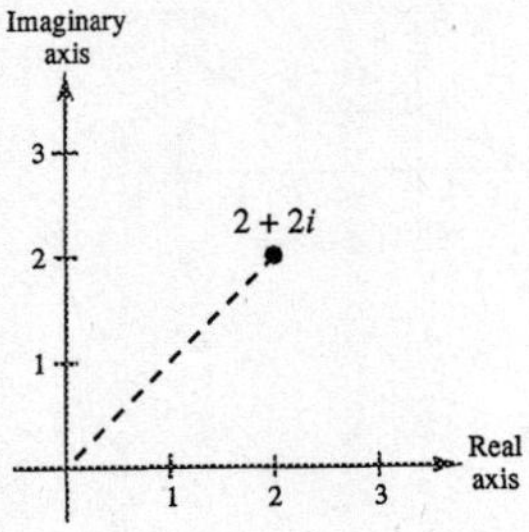

14. $z = 4 - 4\sqrt{3}i$

$r = \sqrt{4^2 + (-4\sqrt{3})^2} = 8$

$\tan\theta = \dfrac{-4\sqrt{3}}{4} = -\sqrt{3} \Rightarrow \theta = \dfrac{5\pi}{3}$

$z = 8\left(\cos\dfrac{5\pi}{3} + i\sin\dfrac{5\pi}{3}\right)$

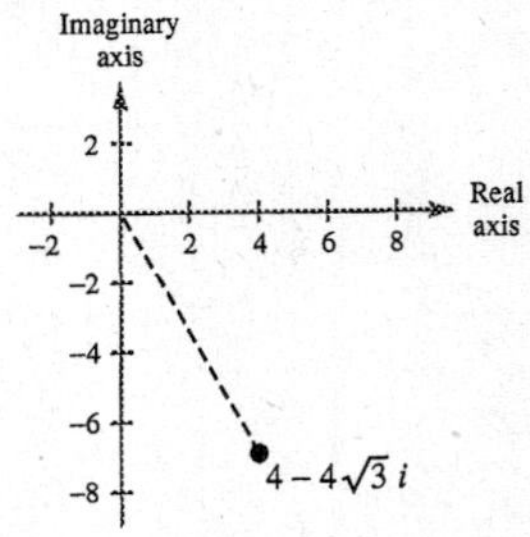

16. $z = \dfrac{5}{2}(\sqrt{3} - i)$

$r = \sqrt{\left(\dfrac{5}{2}\sqrt{3}\right)^2 + \left(\dfrac{5}{2}(-1)\right)^2} = \sqrt{\dfrac{100}{4}} = \sqrt{25} = 5$

$\tan\theta = \dfrac{-1}{\sqrt{3}} = \dfrac{-\sqrt{3}}{3} \Rightarrow \theta = \dfrac{11\pi}{6}$

$z = 5\left(\cos\dfrac{11\pi}{6} + i\sin\dfrac{11\pi}{6}\right)$

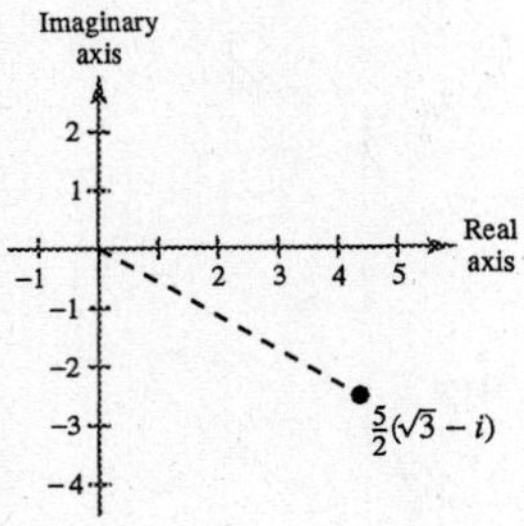

18. $z = 4i$

$r = \sqrt{0^2 + 4^2} = \sqrt{16^2} = 4$

$\tan\theta = \dfrac{4}{0}$, undefined $\Rightarrow \theta = \dfrac{\pi}{2}$

$z = 4\left(\cos\dfrac{\pi}{2} + i\sin\dfrac{\pi}{2}\right)$

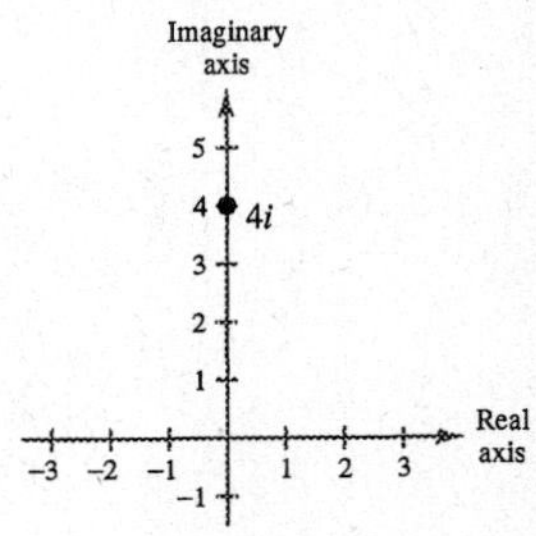

20. $z = 3 - i$

$r = \sqrt{(3)^2 + (-1)^2} = \sqrt{10}$

$\tan\theta = \dfrac{-1}{3} = \theta \approx -0.32 \approx 5.96$

$z = \sqrt{10}(\cos 5.96 + i\sin 5.96)$

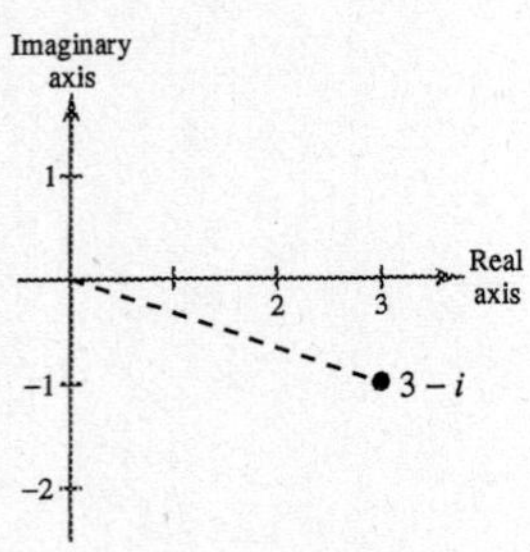

22. $z = 4$

$r = \sqrt{4^2 + 0^2} = \sqrt{16} = 4$

$\tan\theta = \dfrac{0}{4} = 0 \Rightarrow \theta = 0$

$z = 4(\cos 0 + i\sin 0)$

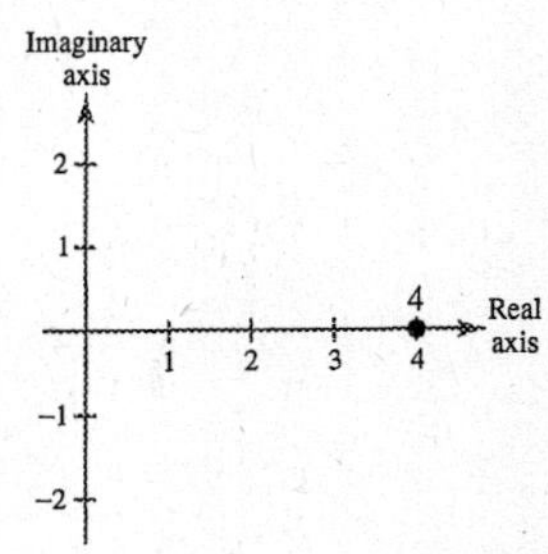

24. $z = 2\sqrt{2} - i$

$r = \sqrt{(2\sqrt{2})^2 + (-1)^2} = \sqrt{9} = 3$

$\tan\theta = \dfrac{-1}{2\sqrt{2}} = -\dfrac{\sqrt{2}}{4} \implies \theta \approx -0.34 \approx 5.94$

$z = 3(\cos 5.94 + i \sin 5.94)$

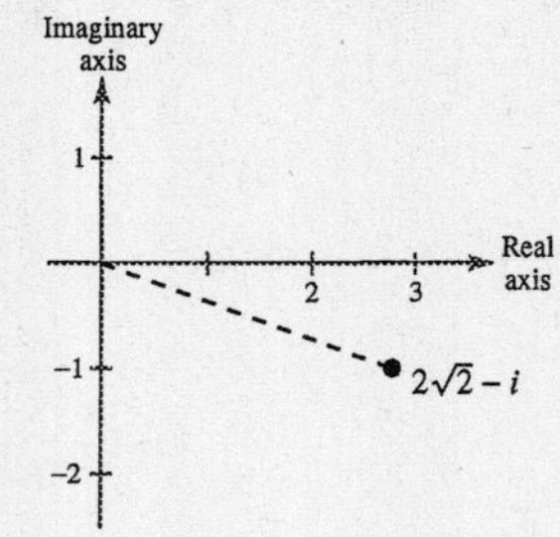

26. $z = 1 + 3i$

$r = \sqrt{1^2 + 3^2} = \sqrt{10}$

$\tan\theta = \frac{3}{1} = 3 \implies \theta \approx 1.25$

$z \approx \sqrt{10}(\cos 1.25 + i \sin 1.25)$

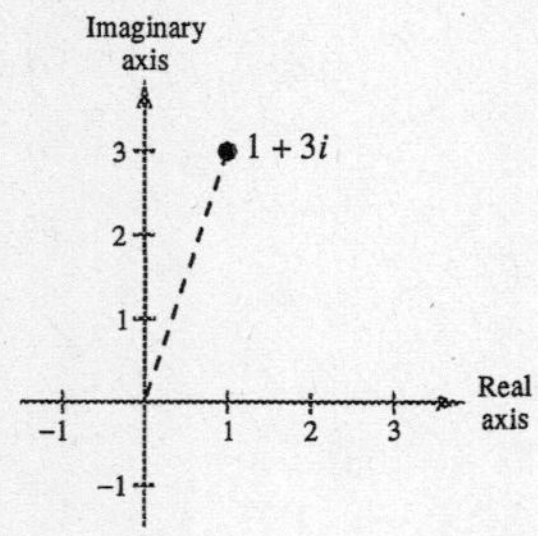

28. $8 + 3i \approx 8.54, \theta \approx 0.36$

$= 8.54(\cos 0.36 + i \sin 0.36)$

30. $-5 - i \approx 5.10, \theta \approx 3.34$

$= 5.10(\cos 3.34 + i \sin 3.34)$

32. $4\sqrt{5} - 4i \approx 9.80, \theta \approx 5.86$

$= 9.80(\cos 5.86 + i \sin 5.86)$

34. $-9 - 2\sqrt{10}i \approx 11, \theta \approx 3.75$

$= 11(\cos 3.75 + i \sin 3.75)$

36. $5(\cos 135° + i \sin 135°) = 5\left[-\dfrac{\sqrt{2}}{2} + i\left(\dfrac{\sqrt{2}}{2}\right)\right]$

$= -\dfrac{5\sqrt{2}}{2} + \dfrac{5\sqrt{2}}{2}i$

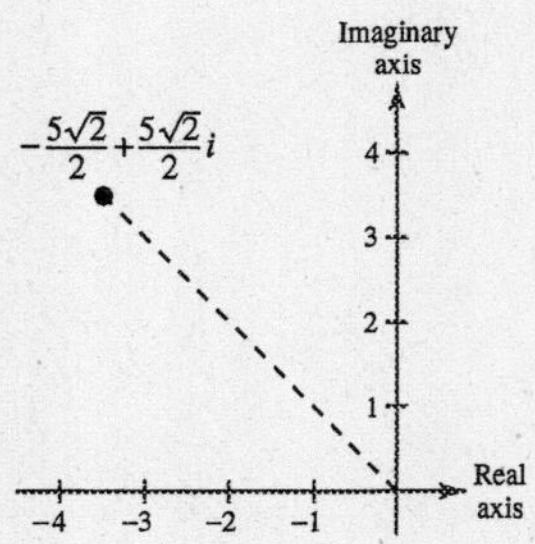

38. $\dfrac{1}{4}(\cos 225° + i \sin 225°) = \dfrac{1}{4}\left(-\dfrac{\sqrt{2}}{2} + i\left(-\dfrac{\sqrt{2}}{2}\right)\right)$

$= -\dfrac{\sqrt{2}}{8} - \dfrac{\sqrt{2}}{8}i$

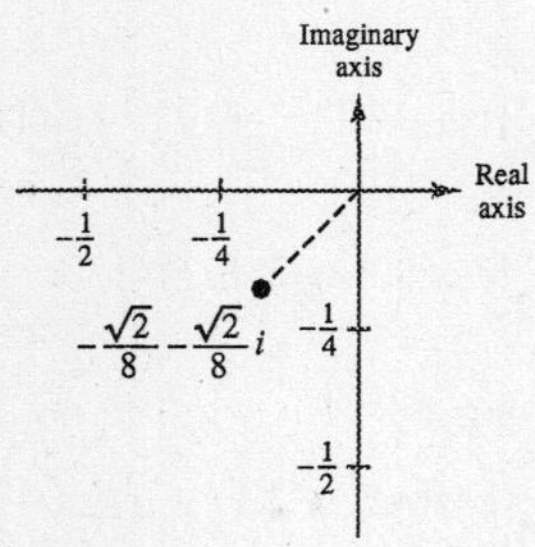

40. $6\left(\cos\dfrac{5\pi}{12} + i \sin\dfrac{5\pi}{12}\right) \approx 1.5529 + i\,5.7956$

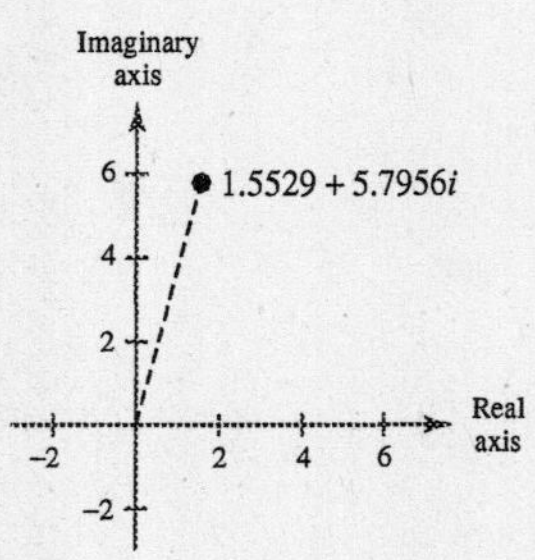

42. $7(\cos 0° + i \sin 0°) = 7$

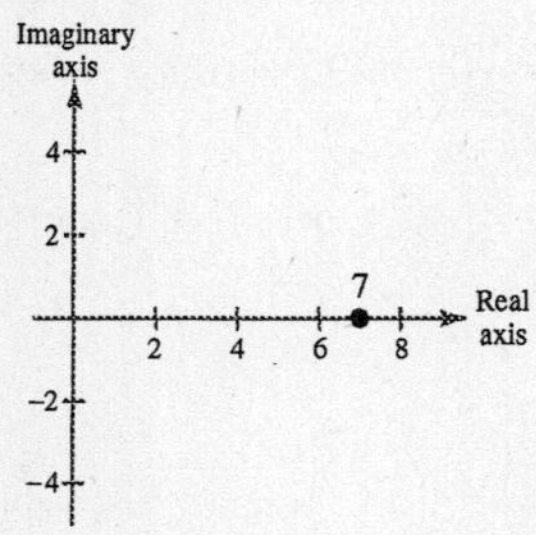

44. $6[\cos(230° 30') + i\sin(230° 30')] \approx -3.8165 - 4.6297i$

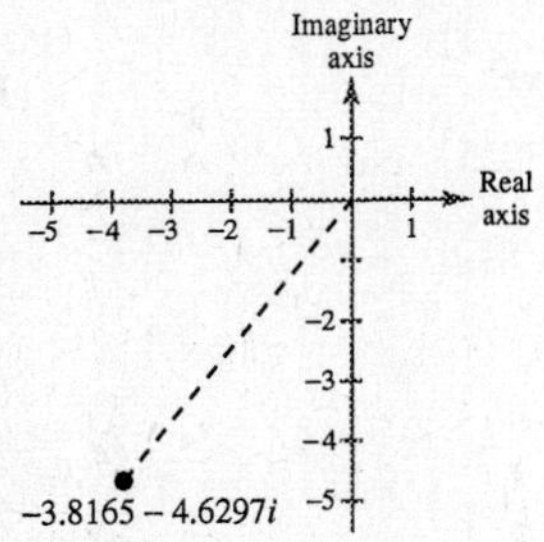

46. $10\left(\cos\frac{2\pi}{5} + i\sin\frac{2\pi}{5}\right) \approx 3.09 + 9.51i$

48. $9(\cos 58° + i\sin 58°) \approx 4.77 + 7.63i$

50. $$\left[\frac{3}{4}\left(\cos\frac{\pi}{3} + i\sin\frac{\pi}{3}\right)\right]\left[4\left(\cos\frac{3\pi}{4} + i\sin\frac{3\pi}{4}\right)\right] = \left(\frac{3}{4}\right)(4)\left[\cos\left(\frac{\pi}{3} + \frac{3\pi}{4}\right) + i\sin\left(\frac{\pi}{3} + \frac{3\pi}{4}\right)\right]$$
$$= 3\left(\cos\frac{13\pi}{12} + i\sin\frac{13\pi}{12}\right)$$

52. $$[0.5(\cos 100° + i\sin 100°)][0.8(\cos 300° + i\sin 300°)] = (0.5)(0.8)[\cos(100° + 300°) + i\sin(100° + 300°)]$$
$$= 0.4(\cos 400° + i\sin 400°)$$
$$= 0.4(\cos 40° + i\sin 40°)$$

54. $$(\cos 5° + i\sin 5°)(\cos 20° + i\sin 20°) = \cos(5° + 20°) + i\sin(5° + 20°)$$
$$= \cos 25° + i\sin 25°$$

56. $$\frac{2(\cos 120° + i\sin 120°)}{4(\cos 40° + i\sin 40°)} = \frac{2}{4}[\cos(120° - 40°) + i\sin(120° - 40°)]$$
$$= \frac{1}{2}(\cos 80° + i\sin 80°)$$

58. $$\frac{5[\cos(4.3) + i\sin(4.3)]}{4[\cos(2.1) + i\sin(2.1)]} = \frac{5}{4}[\cos(4.3 - 2.1) + i\sin(4.3 - 2.1)]$$
$$= \frac{5}{4}[\cos(2.2) + i\sin(2.2)]$$

60. $$\frac{6[\cos 40° + i\sin 40°]}{7[\cos 100° + i\sin 100°]} = \frac{6}{7}[\cos(40° - 100°) + i\sin(40° - 100°)]$$
$$= \frac{6}{7}[\cos 300° + i\sin 300°]$$

62. (a) $\sqrt{3} + i = 2\left(\cos\frac{\pi}{6} + i\sin\frac{\pi}{6}\right)$

$1 + i = \sqrt{2}\left(\cos\frac{\pi}{4} + i\sin\frac{\pi}{4}\right)$

(b) $(\sqrt{3} + i)(1 + i) = \left[2\left(\cos\frac{\pi}{6} + i\sin\frac{\pi}{6}\right)\right]\left[\sqrt{2}\left(\cos\frac{\pi}{4} + i\sin\frac{\pi}{4}\right)\right]$

$= 2\sqrt{2}\left(\cos\frac{5\pi}{12} + i\sin\frac{5\pi}{12}\right)$

$\approx 0.732 + 2.732i$

(c) $(\sqrt{3} + i)(1 + i) = \sqrt{3} + (\sqrt{3} + 1)i + i^2 = (\sqrt{3} - 1) + (\sqrt{3} + 1)i \approx 0.732 + 2.732i$

64. (a) $4 = 4(\cos 0 + i\sin 0)$

$1 - \sqrt{3}i = 2\left(\cos\left(\frac{5\pi}{3}\right) + i\sin\left(\frac{5\pi}{3}\right)\right)$

(b) $4(1 - \sqrt{3}i) = 8\left(\cos\left(\frac{5\pi}{3}\right) + i\sin\left(\frac{5\pi}{3}\right)\right)$

$= 8\left(\frac{1}{2} + i\left(-\frac{\sqrt{13}}{2}\right)\right)$

$= 4 - 4\sqrt{3}i$

(c) $4(1 - \sqrt{3}i) = 4 - 4\sqrt{3}i$

66. (a) $1 + \sqrt{3}i = 2\left(\cos\frac{\pi}{3} + i\sin\frac{\pi}{3}\right)$

$6 - 3i \approx 3\sqrt{5}[\cos(5.82) + i\sin(5.82)]$

(b) $\frac{1 + \sqrt{3}i}{6 - 3i} \approx \frac{2}{3\sqrt{5}}\left[\cos\left(\frac{\pi}{3} - 5.82\right) + i\sin\left(\frac{\pi}{3} - 5.82\right)\right]$

$\approx \frac{2\sqrt{5}}{15}[\cos(-4.77) + i\sin(-4.77)]$

$\approx \frac{2\sqrt{5}}{15}[\cos 1.51 + i\sin 1.51]$

$\approx 0.018 + 0.298i$

(c) $\frac{1 + \sqrt{3}i}{6 - 3i} \cdot \frac{6 + 3i}{6 + 3i} = \frac{(6 - 3\sqrt{3}) + (3 + 6\sqrt{3})i}{45}$

$= \frac{2 - \sqrt{3}}{15} + \frac{1 + 2\sqrt{3}}{15}i$

$\approx 0.018 + 0.298i$

68. (a) $4i = 4\left(\cos\frac{\pi}{2} + i\sin\frac{\pi}{2}\right)$

$-4 + 2i = 2\sqrt{5}(\cos 2.68 + i\sin 2.68)$

(b) $$\frac{4i}{-4+2i} = \frac{4\left(\cos\frac{\pi}{2} + i\sin\frac{\pi}{2}\right)}{2\sqrt{5}(\cos 2.68 + i\sin 2.68)}$$

$$= \frac{2\sqrt{5}}{5}(\cos 5.17 + i\sin 5.17)$$

$$\approx 0.400 - 0.800i$$

(c) $$\frac{4i}{-4+2i} = \frac{4i}{-4+2i}\cdot\frac{-4-i}{-4-2i}$$

$$= \frac{8-16i}{20}$$

$$= \frac{2}{5} - \frac{4}{5}i$$

$$= 0.400 - 0.800i$$

70. $|z| = 3$

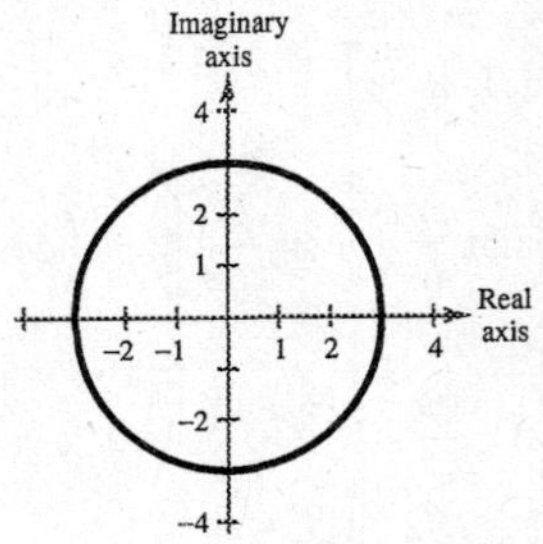

72. $\theta = \frac{5\pi}{4}$

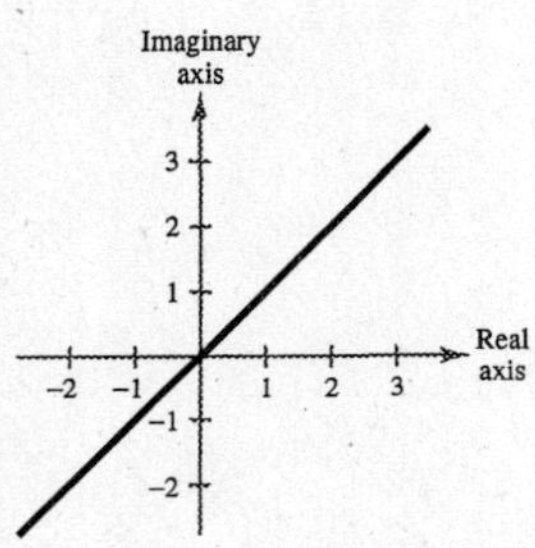

74. True.

$z_1z_2 = r_1r_2(\cos(\theta_1 + \theta_2) + i\sin(\theta_1 + \theta_2))$ and z_1z_2 is zero only when $r_1r_2 = 0$. So, $r_1 = 0$ and/or $r_2 = 0$.

76. $\bar{z} = r[\cos(-\theta) + i\sin(-\theta)]$

$= r[\cos\theta - i\sin\theta]$

$= r\cos\theta - ir\sin\theta$

which is the complex conjugate of $z = r(\cos\theta + i\sin\theta) = r\cos\theta + ir\sin\theta$.

78. $-z = -r(\cos\theta + i\sin\theta)$

$= r(-\cos\theta + i(-\sin\theta))$

$= r(\cos(\theta + \pi) + i\sin(\theta + \pi))$

80. $B = 66°, a = 33.5$

$A = 90° - 66° = 24°$

$$\frac{\sin 66°}{b} = \frac{\sin 24°}{33.5}$$

$$b = \frac{a\sin B}{\sin A} = \frac{(33.5)\sin 66°}{\sin 24°} \approx 75.24$$

$$c = \frac{a\sin C}{\sin A} = \frac{(33.5)\sin 90°}{\sin 24°} \approx 82.36$$

82. $B = 6°, b = 211.2$

$A = 90° - 6° = 84°$

$$a = \frac{b\sin A}{\sin B} = \frac{(211.2)\sin 84°}{\sin 6°} \approx 2009.43$$

$$c = \frac{b\sin C}{\sin B} = \frac{(211.2)\sin 90°}{\sin 6°} \approx 2020.50$$

84. $B = 81°30', c = 6.8$

$B = 81.5°$

$A = 90° - 81.5° = 8.5° = 8°30'$

$a = \dfrac{c \sin A}{\sin C} = \dfrac{(6.8)\sin 8.5°}{1} \approx 1.01$

$b = \dfrac{c \sin B}{\sin C} = \dfrac{(6.8)\sin 81.5°}{1} \approx 6.73$

86. $f = \dfrac{1}{8}\cos 12\pi t$

Maximum displacement: $\dfrac{1}{8}$

$d = 0$ when

$12\pi t = \dfrac{\pi}{2}$

$t = \dfrac{1}{24}$

88. $d = \dfrac{1}{12}\sin 60\pi t$

Maximum displacement: $\dfrac{1}{12}$

$d = 0$ when

$60\pi t = 0$

$t = 0$

Section 4.4 DeMoivre's Theorem

Solutions to Even-Numbered Exercises

2. $$\begin{aligned}(2 + 2i)^6 &= \left[2\sqrt{2}\left(\cos\frac{\pi}{4} + i\sin\frac{\pi}{4}\right)\right]^6\\ &= \left(2\sqrt{2}\right)^6\left(\cos\frac{6\pi}{4} + i\sin\frac{6\pi}{4}\right)\\ &= 512\left(\cos\frac{3\pi}{2} + i\sin\frac{3\pi}{2}\right)\\ &= -512i\end{aligned}$$

4. $$\begin{aligned}(3 - 2i)^8 &= \left[\sqrt{13}\left(\cos\left(-\arctan\left(\tfrac{2}{3}\right)\right) + i\sin\left(-\arctan\left(\tfrac{2}{3}\right)\right)\right)\right]^8\\ &= \left(\sqrt{13}\right)^8\left[\cos\left(-8\arctan\left(\tfrac{2}{3}\right)\right) + i\sin\left(-8\arctan\left(\tfrac{2}{3}\right)\right)\right]\\ &= -239 + 28{,}560i\end{aligned}$$

6. $$\begin{aligned}4\left(1 - \sqrt{3}i\right)^3 &= 4\left[2\left(\cos\frac{5\pi}{3} + i\sin\frac{5\pi}{3}\right)\right]^3\\ &= 4[2^3(\cos 5\pi + i\sin 5\pi)]\\ &= 32(-1)\\ &= -32\end{aligned}$$

8. $$\begin{aligned}[3(\cos 150° + i\sin 150°)]^4 &= 3^4(\cos 600° + i\sin 600°)\\ &= 81(\cos 240° + i\sin 240°)\\ &= 81(-\cos 60° - i\sin 60°)\\ &= -\frac{81}{2} - \frac{81\sqrt{3}}{2}i\end{aligned}$$

10. $$\begin{aligned}\left[2\left(\cos\frac{\pi}{2} + i\sin\frac{\pi}{2}\right)\right]^8 &= 2^8(\cos 4\pi + i\sin 4\pi)\\ &= 256(\cos 0 + i\sin 0)\\ &= 256\end{aligned}$$

12. $$\begin{aligned}(\cos 0 + i\sin 0)^{20} &= \cos 0 + i\sin 0\\ &= 1\end{aligned}$$

14. $(2+5i)^6 = \left[\sqrt{29}\left(\cos\left(\arctan\left(\frac{5}{2}\right)\right) + i\sin\left(\arctan\left(\frac{5}{2}\right)\right)\right)\right]^6$

$= (\sqrt{29})^6\left[\cos\left(6\arctan\left(\frac{5}{2}\right)\right) + i\sin\left(6\arctan\left(\frac{5}{2}\right)\right)\right]$

$= 15{,}939 + 18{,}460\,i$

16. $(\sqrt{3}+2i)^4 = \left[\sqrt{7}\left(\cos\left(\arctan\left(\frac{2}{\sqrt{3}}\right)\right) + i\sin\left(\arctan\left(\frac{2}{\sqrt{3}}\right)\right)\right)\right]^4$

$= (\sqrt{7})^4\left[\cos\left(4\arctan\left(\frac{2}{\sqrt{3}}\right)\right) + i\sin\left(4\arctan\left(\frac{2}{\sqrt{3}}\right)\right)\right]$

$\approx -47 - 13.8564\,i$

18. $[2(\cos 10° + i\sin 10°)]^8 = 256(\cos 80° + i\sin 80°)$

$\approx 44.4539 + 252.1108i$

20. $[4(\cos 110° + i\sin 110°)]^4 = 256(\cos 440° + i\sin 440°)$

$\approx 44.4539 + 252.1108i$

22. $\left[2\left(\cos\frac{\pi}{8} + i\sin\frac{\pi}{8}\right)\right]^6 = 64\left(\cos\frac{3\pi}{4} + i\sin\frac{3\pi}{4}\right)$

$= -32\sqrt{2} + 32\sqrt{2}i$

24. $\left[3\left(\cos\frac{\pi}{12} + i\sin\frac{\pi}{12}\right)\right]^5 = 243\left(\cos\frac{5\pi}{12} + i\sin\frac{5\pi}{12}\right)$

$\approx 62.8930 + 234.7200i$

26. (a) Square roots of $16(\cos 60° + i\sin 60°)$:

$\sqrt{16}\left[\cos\left(\frac{60° + k\,360°}{2}\right) + i\sin\left(\frac{60° + k\,360°}{2}\right)\right]$

$k = 0, 1$

$k = 0$: $4(\cos 30° + i\sin 30°)$

$k = 1$: $4(\cos 210° + i\sin 210°)$

(b)

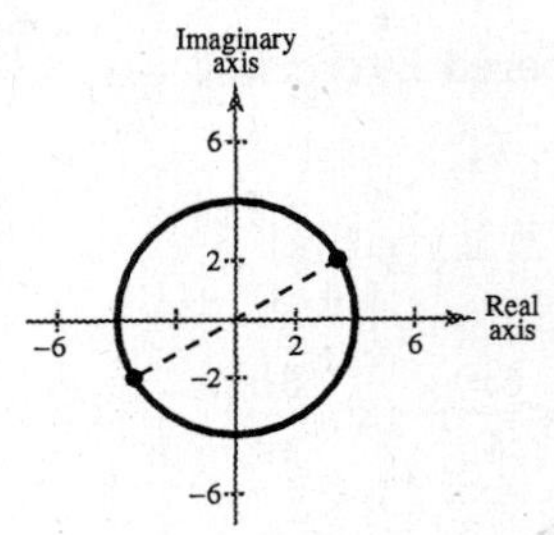

(c) $2\sqrt{3} + 2i, -2\sqrt{3} - 2i$

28. (a) Cube roots of $64\left(\cos\frac{\pi}{3} + i\sin\frac{\pi}{3}\right)$:

$\sqrt[3]{64}\left[\cos\left(\frac{\frac{\pi}{3} + 2k\pi}{3}\right) + i\sin\left(\frac{\frac{\pi}{3} + 2k\pi}{3}\right)\right]$

$k = 0, 1, 2$

$k = 0$: $4\left(\cos\frac{\pi}{9} + i\sin\frac{\pi}{9}\right)$

$k = 1$: $4\left(\cos\frac{7\pi}{9} + i\sin\frac{7\pi}{9}\right)$

$k = 2$: $4\left(\cos\frac{13\pi}{9} + i\sin\frac{13\pi}{9}\right)$

(b)

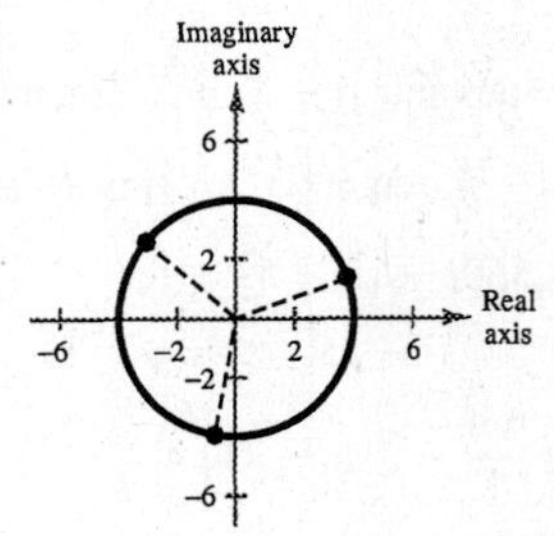

(c) $3.7588 + 1.3681i, -3.0642 + 2.5712i,$
$-0.6946 - 3.9392i$

30. (a) Fifth roots of $32\left(\cos\frac{5\pi}{6} + i\sin\frac{5\pi}{6}\right)$:

$$\sqrt[5]{32}\left[\cos\left(\frac{\frac{5\pi}{6} + 2k\pi}{5}\right) + i\sin\left(\frac{\frac{5\pi}{6} + 2k\pi}{5}\right)\right]$$

$k = 0, 1, 2, 3, 4$

$k = 0$: $2\left(\cos\frac{\pi}{6} + i\sin\frac{\pi}{6}\right)$

$k = 1$: $2\left(\cos\frac{17\pi}{30} + i\sin\frac{17\pi}{30}\right)$

$k = 2$: $2\left(\cos\frac{29\pi}{30} + i\sin\frac{29\pi}{30}\right)$

$k = 3$: $2\left(\cos\frac{41\pi}{30} + i\sin\frac{41\pi}{30}\right)$

$k = 4$: $2\left(\cos\frac{53\pi}{30} + i\sin\frac{53\pi}{30}\right)$

(b)

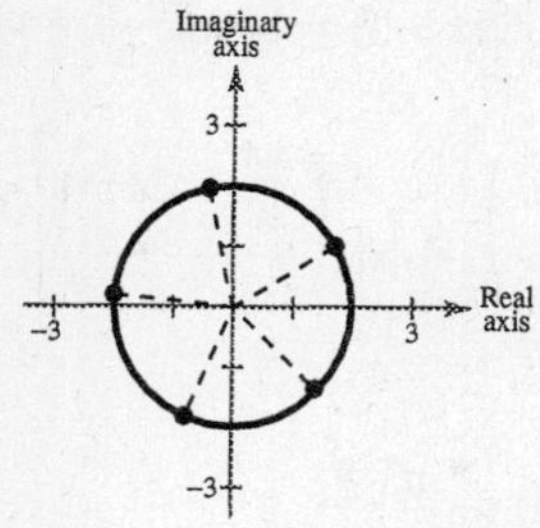

(c) $\sqrt{3} + i, -0.4158 + 1.9563i, -1.9890 + 0.2091i,$
$-0.8135 - 1.8271i, 1.4863 - 1.3383i$

32. (a) Square roots of $-36i = 36\left(\cos\frac{3\pi}{2} + i\sin\frac{3\pi}{2}\right)$:

$$\sqrt{36}\left[\cos\left(\frac{\frac{3\pi}{2} + 2k\pi}{2}\right) + i\sin\left(\frac{\frac{3\pi}{2} + 2k\pi}{2}\right)\right]$$

$k = 0, 1$

$k = 0$: $6\left(\cos\frac{3\pi}{4} + i\sin\frac{3\pi}{4}\right)$

$k = 1$: $6\left(\cos\frac{7\pi}{4} + i\sin\frac{7\pi}{4}\right)$

(b)

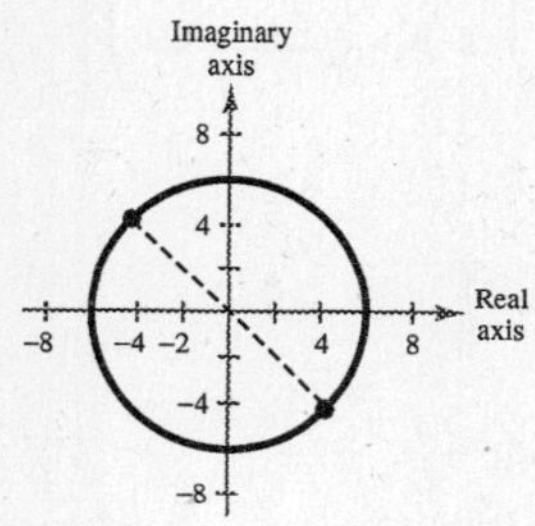

(c) $-3\sqrt{2} + 3\sqrt{2}\,i, 3\sqrt{2} - 3\sqrt{2}\,i$

34. (a) Fourth roots of $625i = 625\left(\cos\frac{\pi}{2} + i\sin\frac{\pi}{2}\right)$:

$$\sqrt[4]{625}\left[\cos\left(\frac{\frac{\pi}{2} + 2k\pi}{4}\right) + i\sin\left(\frac{\frac{\pi}{2} + 2k\pi}{4}\right)\right]$$

$k = 0, 1, 2, 3$

$k = 0$: $5\left(\cos\frac{\pi}{8} + i\sin\frac{\pi}{8}\right)$

$k = 1$: $5\left(\cos\frac{5\pi}{8} + i\sin\frac{5\pi}{8}\right)$

$k = 2$: $5\left(\cos\frac{9\pi}{8} + i\sin\frac{9\pi}{8}\right)$

$k = 3$: $5\left(\cos\frac{13\pi}{8} + i\sin\frac{13\pi}{8}\right)$

(b)

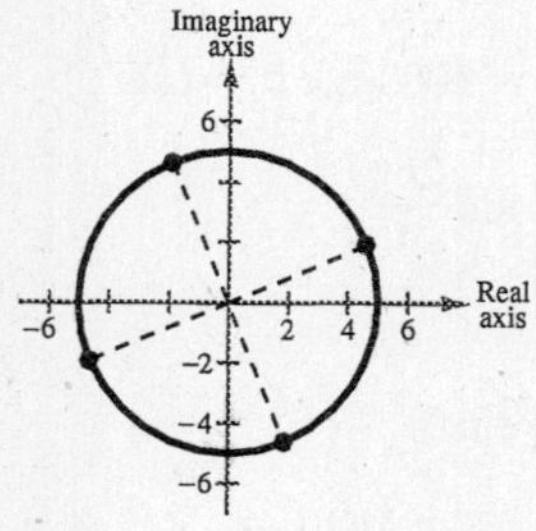

(c) $4.6194 + 1.9134i, -1.9134 + 4.6194i,$
$-4.6194 - 1.9134i, 1.9134 - 4.6194i$

36. (a) Cube roots of $-4\sqrt{2}(1 - i) = 8\left(\cos\frac{3\pi}{4} + i\sin\frac{3\pi}{4}\right)$:

$$\sqrt[3]{8}\left[\cos\left(\frac{\frac{3\pi}{4} + 2k\pi}{3}\right) + i\sin\left(\frac{\frac{3\pi}{4} + 2k\pi}{3}\right)\right]$$

$k = 0, 1, 2$

$k = 0$: $2\left(\cos\frac{\pi}{4} + i\sin\frac{\pi}{4}\right)$

$k = 1$: $2\left(\cos\frac{11\pi}{12} + i\sin\frac{11\pi}{12}\right)$

$k = 2$: $2\left(\cos\frac{19\pi}{12} + i\sin\frac{19\pi}{12}\right)$

(b)

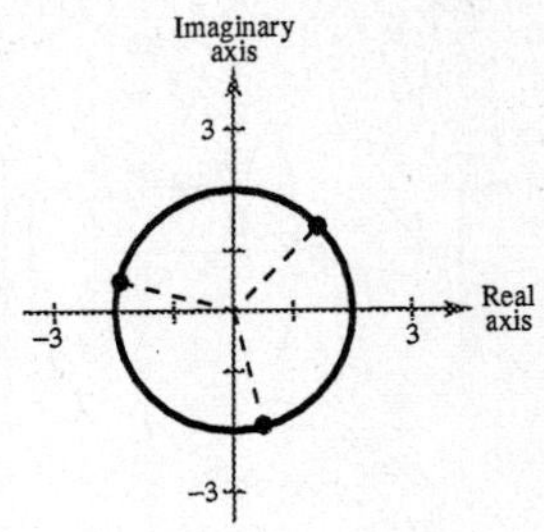

(c) $\sqrt{2} + \sqrt{2}i, -1.9319 + 0.5176i, 0.5176 - 1.9319i$

38. (a) Fourth roots of $i = \cos\frac{\pi}{2} + i\sin\frac{\pi}{2}$:

$$\sqrt[4]{1}\left[\cos\left(\frac{\frac{\pi}{2} + 2k\pi}{4}\right) + i\sin\left(\frac{\frac{\pi}{2} + 2k\pi}{4}\right)\right]$$

$k = 0, 1, 2, 3$

$k = 0$: $\cos\frac{\pi}{8} + i\sin\frac{\pi}{8}$

$k = 1$: $\cos\frac{5\pi}{8} + i\sin\frac{5\pi}{8}$

$k = 2$: $\cos\frac{9\pi}{8} + i\sin\frac{9\pi}{8}$

$k = 3$: $\cos\frac{13\pi}{8} + i\sin\frac{13\pi}{8}$

(b)

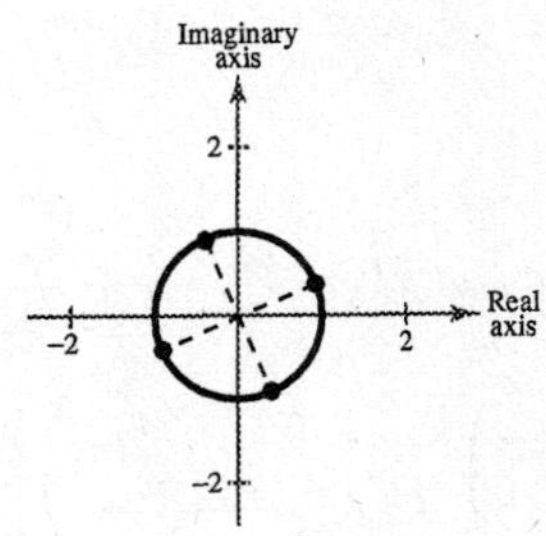

(c) $0.9239 + 0.3827i, -0.3827 + 0.9239i,$
$-0.9239 - 0.3827i, 0.3827 - 0.9239i$

40. (a) Cube roots of $1000 = 1000(\cos 0 + i\sin 0)$:

$$\sqrt[3]{1000}\left(\cos\frac{2k\pi}{3} + i\sin\frac{2k\pi}{3}\right)$$

$k = 0, 1, 2$

$k = 0$: $10(\cos 0 + i\sin 0)$

$k = 1$: $10\left(\cos\frac{2\pi}{3} + i\sin\frac{2\pi}{3}\right)$

$k = 2$: $10\left(\cos\frac{4\pi}{3} + i\sin\frac{4\pi}{3}\right)$

(b)

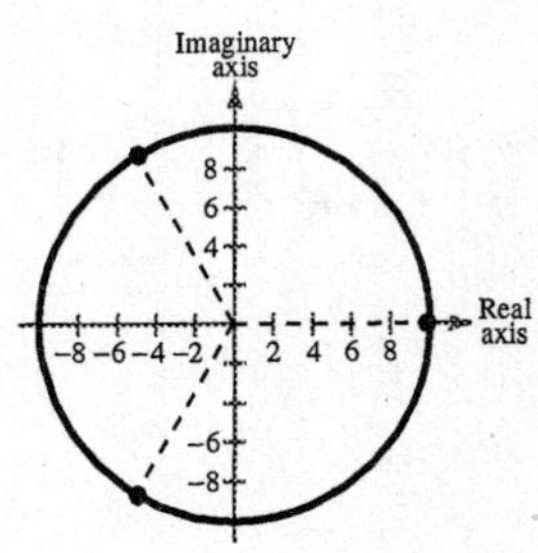

(c) $10, -5 + 5\sqrt{3}i, -5 - 5\sqrt{3}i$

42. (a) Fourth roots of $-4 = 4(\cos \pi + i \sin \pi)$:

$$\sqrt[4]{4}\left[\cos\left(\frac{\pi + 2k\pi}{4}\right) + i \sin\left(\frac{\pi + 2k\pi}{4}\right)\right]$$

$k = 0, 1, 2, 3$

$k = 0$: $\sqrt{2}\left(\cos \frac{\pi}{4} + i \sin \frac{\pi}{4}\right)$

$k = 1$: $\sqrt{2}\left(\cos \frac{3\pi}{4} + i \sin \frac{3\pi}{4}\right)$

$k = 2$: $\sqrt{2}\left(\cos \frac{5\pi}{4} + i \sin \frac{5\pi}{4}\right)$

$k = 3$: $\sqrt{2}\left(\cos \frac{7\pi}{4} + i \sin \frac{7\pi}{4}\right)$

(b)

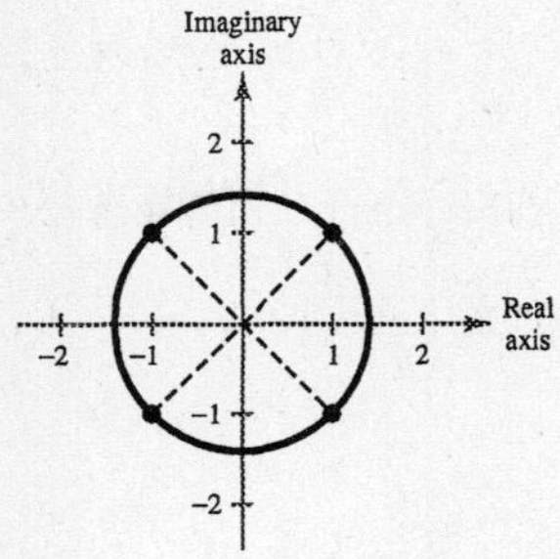

(c) $1 + i, -1 + i, -1 - i, 1 - i$

44. (a) Sixth roots of $64i = 64\left(\cos \frac{\pi}{2} + i \sin \frac{\pi}{2}\right)$:

$$\sqrt[6]{64}\left[\cos\left(\frac{\frac{\pi}{2} + 2k\pi}{6}\right) + i \sin\left(\frac{\frac{\pi}{2} + 2k\pi}{6}\right)\right]$$

$k = 0, 1, 2, 3, 4, 5$

$k = 0$: $2\left(\cos \frac{\pi}{12} + i \sin \frac{\pi}{12}\right)$

$k = 1$: $2\left(\cos \frac{5\pi}{12} + i \sin \frac{5\pi}{12}\right)$

$k = 2$: $2\left(\cos \frac{3\pi}{4} + i \sin \frac{3\pi}{4}\right)$

$k = 3$: $2\left(\cos \frac{13\pi}{12} + i \sin \frac{13\pi}{12}\right)$

$k = 4$: $2\left(\cos \frac{17\pi}{12} + i \sin \frac{17\pi}{12}\right)$

$k = 5$: $2\left(\cos \frac{7\pi}{4} + i \sin \frac{7\pi}{4}\right)$

(b)

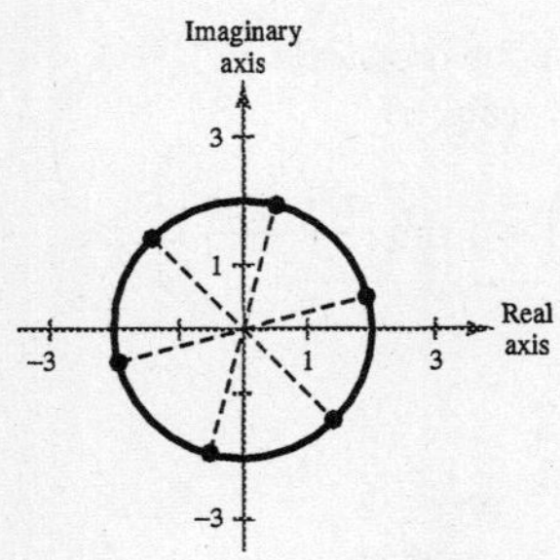

(c) $1.9319 + 0.5176i, 0.5176 + 1.9319i, -\sqrt{2} + \sqrt{2}i,$
$-1.9319 - 0.5176i, -0.5176 - 1.9319i, \sqrt{2} - \sqrt{2}i$

46. $x^3 - i = 0$

$x^3 = i$

The solutions are the cube roots of $i = \cos \frac{\pi}{2} + i \sin \frac{\pi}{2}$:

$$\cos\left(\frac{\frac{\pi}{2} + 2k\pi}{3}\right) + i \sin\left(\frac{\frac{\pi}{2} + 2k\pi}{3}\right)$$

$k = 0, 1, 2$

$k = 0$: $\cos \frac{\pi}{6} + i \sin \frac{\pi}{6} = \frac{\sqrt{3}}{2} + \frac{1}{2}i$

$k = 1$: $\cos \frac{5\pi}{6} + i \sin \frac{5\pi}{6} = -\frac{\sqrt{3}}{2} + \frac{1}{2}i$

$k = 2$: $\cos \frac{3\pi}{2} + i \sin \frac{3\pi}{2} = -i$

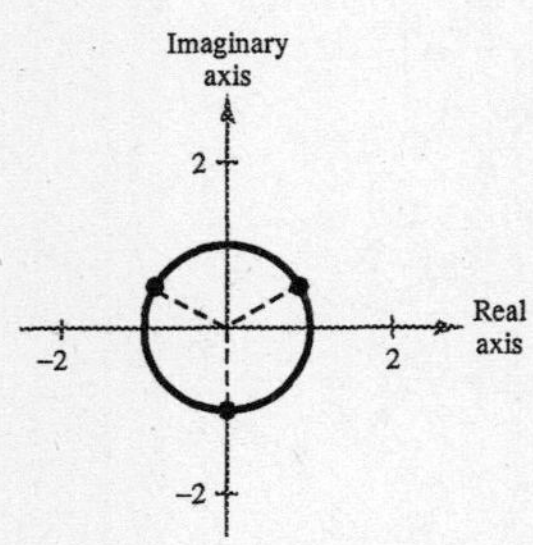

48. $x^3 + 1 = 0$

$x^3 = -1$

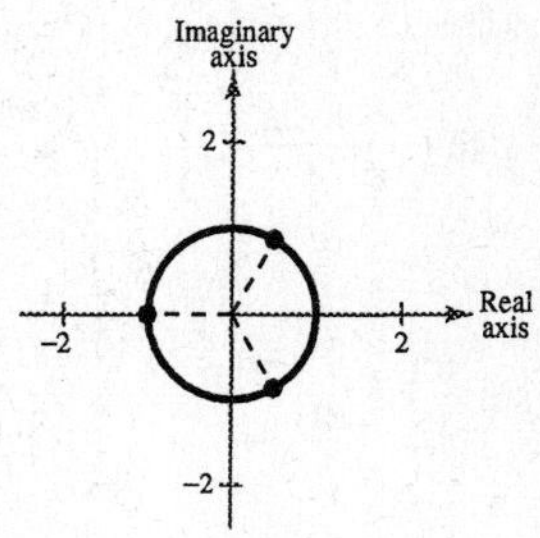

The solutions are the cube roots of $-1 = \cos \pi + i \sin \pi$:

$$\cos\left(\frac{\pi + 2k\pi}{3}\right) + i \sin\left(\frac{\pi + 2k\pi}{3}\right)$$

$k = 0, 1, 2$

$k = 0$: $\cos \frac{\pi}{3} + i \sin \frac{\pi}{3} = \frac{1}{2} + \frac{\sqrt{3}}{2}i$

$k = 1$: $\cos \pi + i \sin \pi = -1$

$k = 2$: $\cos \frac{5\pi}{3} + i \sin \frac{5\pi}{3} = \frac{1}{2} - \frac{\sqrt{3}}{2}i$

50. $x^3 + 125 = 0$

$x^3 = -125$

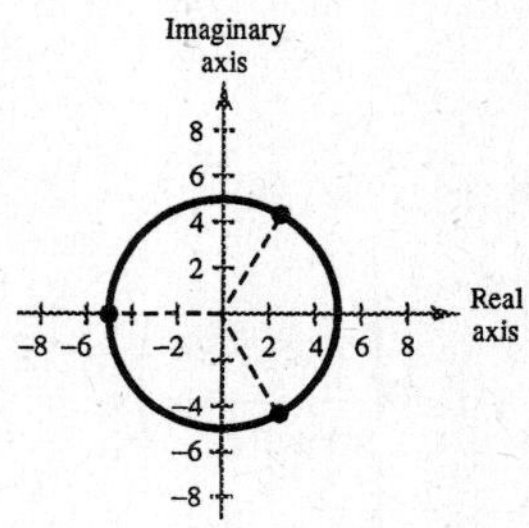

The solutions are the cube roots of $-125 = 125(\cos \pi + i \sin \pi)$:

$$\sqrt[3]{125}\left[\cos\left(\frac{\pi + 2k\pi}{3}\right) + i \sin\left(\frac{\pi + 2k\pi}{3}\right)\right]$$

$k = 0, 1, 2$

$k = 0$: $5\left(\cos \frac{\pi}{3} + i \sin \frac{\pi}{3}\right) = \frac{5}{2} + \frac{5\sqrt{3}}{2}i$

$k = 1$: $5(\cos \pi + i \sin \pi) = -5$

$k = 2$: $5\left(\cos \frac{5\pi}{3} + i \sin \frac{5\pi}{3}\right) = \frac{5}{2} - \frac{5\sqrt{3}}{2}i$

52. $x^3 - 27 = 0$

$x^3 = 27$

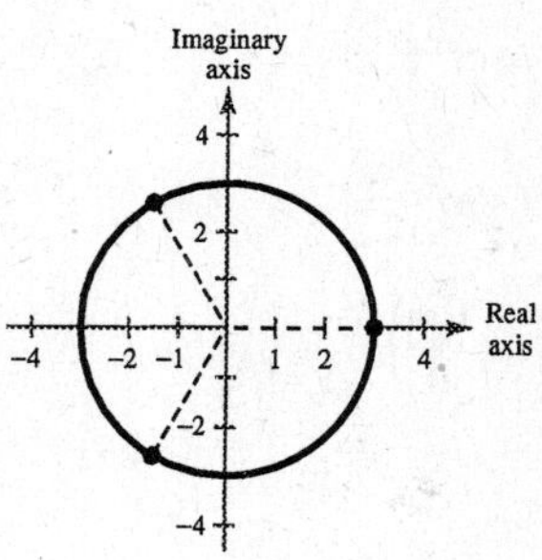

The solutions are the cube roots of $27 = 27(\cos 0 + i \sin 0)$:

$$\sqrt[3]{27}\left[\cos\left(\frac{2k\pi}{3}\right) + i \sin\left(\frac{2k\pi}{3}\right)\right]$$

$k = 0, 1, 2$

$k = 0$: $3(\cos 0 + i \sin 0) = 3$

$k = 1$: $3\left(\cos \frac{2\pi}{3} + i \sin \frac{2\pi}{3}\right) = -\frac{3}{2} + \frac{3\sqrt{3}}{2}$

$k = 2$: $3\left(\cos \frac{4\pi}{3} + i \sin \frac{4\pi}{3}\right) = -\frac{3}{2} - \frac{3\sqrt{3}}{2}$

54. $x^3 + 27i = 0$

$x^3 = -27i$

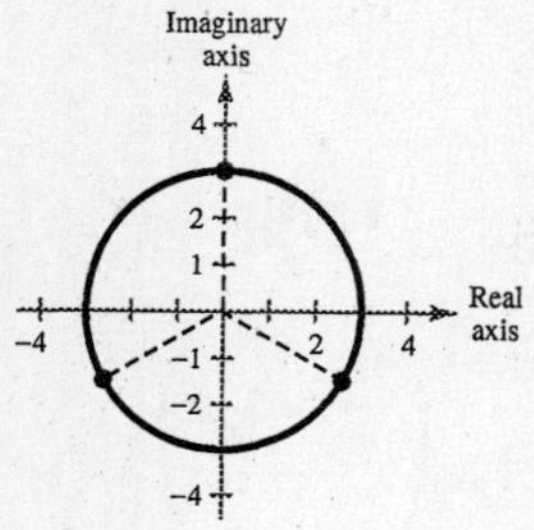

The solutions are the cube roots of $-27i = 27\left(\cos\frac{3\pi}{2} + i\sin\frac{3\pi}{2}\right)$:

$$\sqrt[3]{27}\left[\cos\left(\frac{\frac{3\pi}{2} + 2k\pi}{3}\right) + i\sin\left(\frac{\frac{3\pi}{2} + 2k\pi}{3}\right)\right]$$

$k = 0, 1, 2$

$k = 0$: $3\left(\cos\frac{\pi}{2} + i\sin\frac{\pi}{2}\right) = 3i$

$k = 1$: $3\left(\cos\frac{7\pi}{6} + i\sin\frac{7\pi}{6}\right) = -\frac{3\sqrt{3}}{2} - \frac{3}{2}i$

$k = 2$: $3\left(\cos\frac{11\pi}{6} + i\sin\frac{11\pi}{6}\right) = \frac{3\sqrt{3}}{2} - \frac{3}{2}i$

56. $x^6 + 64i = 0$

$x^6 = -64i$

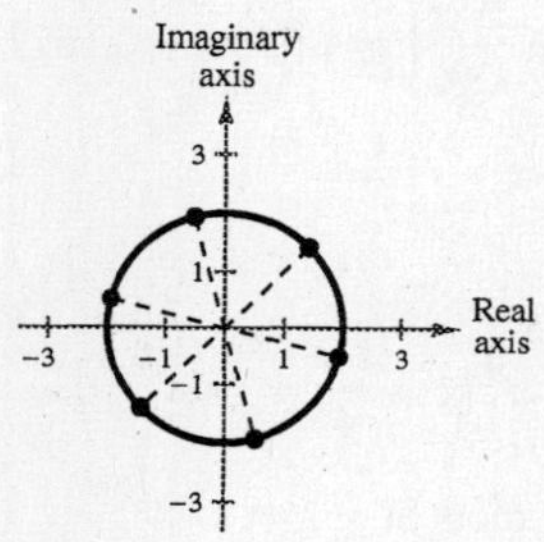

The solutions are the sixth roots of $64i$:

$$\sqrt[6]{64}\left[\cos\left(\frac{\frac{3\pi}{2} + 2k\pi}{6}\right) + i\sin\left(\frac{\frac{3\pi}{2} + 2k\pi}{6}\right)\right]$$

$k = 0, 1, 2, 3, 4, 5$

$k = 0$: $2\left(\cos\frac{\pi}{4} + i\sin\frac{\pi}{4}\right) \approx 1.4142 + 1.4142i$

$k = 1$: $2\left(\cos\frac{7\pi}{12} + i\sin\frac{7\pi}{12}\right) \approx -0.5176 + 1.9319i$

$k = 2$: $2\left(\cos\frac{11\pi}{12} + i\sin\frac{11\pi}{12}\right) \approx -1.9319 + 0.5176i$

$k = 3$: $2\left(\cos\frac{5\pi}{4} + i\sin\frac{5\pi}{4}\right) \approx -1.4142 - 0.4142i$

$k = 4$: $2\left(\cos\frac{19\pi}{12} + i\sin\frac{19\pi}{12}\right) \approx 0.5176 - 1.9319i$

$k = 5$: $2\left(\cos\frac{23\pi}{12} + i\sin\frac{23\pi}{12}\right) \approx 1.9319 - 0.5176i$

58. $x^5 - (1 - i) = 0$

$$x^5 = 1 - i$$

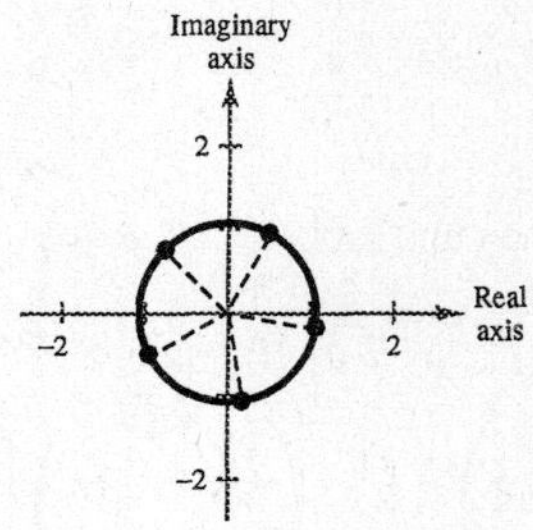

The solutions are the fifth roots of $1 - i = \sqrt{2}\left(\cos\frac{7\pi}{4} + i\sin\frac{7\pi}{4}\right)$:

$$\sqrt[5]{\sqrt{2}}\left[\cos\left(\frac{\frac{7\pi}{4} + 2k\pi}{5}\right) + i\sin\left(\frac{\frac{7\pi}{4} + 2k\pi}{5}\right)\right]$$

$k = 0, 1, 2, 3, 4$

$k = 0$: $\sqrt[10]{2}\left(\cos\frac{7\pi}{20} + i\sin\frac{7\pi}{20}\right) \approx 0.4866 + 0.9550i$

$k = 1$: $\sqrt[10]{2}\left(\cos\frac{3\pi}{4} + i\sin\frac{3\pi}{4}\right) \approx -0.7579 + 0.7579i$

$k = 2$: $\sqrt[10]{2}\left(\cos\frac{23\pi}{20} + i\sin\frac{23\pi}{20}\right) \approx -0.9550 - 0.4866i$

$k = 3$: $\sqrt[10]{2}\left(\cos\frac{31\pi}{20} + i\sin\frac{31\pi}{20}\right) \approx -0.1677 - 1.0586i$

$k = 4$: $\sqrt[10]{2}\left(\cos\frac{39\pi}{20} + i\sin\frac{39\pi}{20}\right) \approx 1.0586 - 0.1677i$

60. $x^4 + (1 + i) = 0$

$$x^4 = -1 - i = \sqrt{2}\left(\cos\frac{5\pi}{4} + i\sin\frac{5\pi}{4}\right)$$

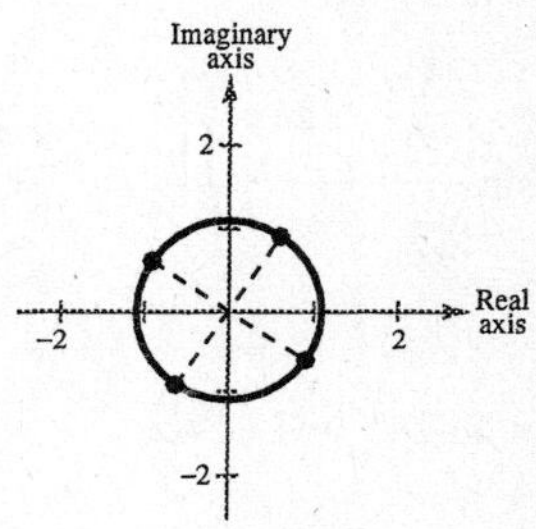

The solutions are the fourth roots of $-1 - i$:

$$\sqrt[4]{\sqrt{2}}\left[\cos\left(\frac{\frac{5\pi}{4} + 2k\pi}{4}\right) + i\sin\left(\frac{\frac{5\pi}{4} + 2k\pi}{4}\right)\right]$$

$k = 0, 1, 2, 3$

$k = 0$: $\sqrt[8]{2}\left(\cos\frac{5\pi}{16} + i\sin\frac{5\pi}{16}\right) \approx 0.6059 + 0.9067i$

$k = 1$: $\sqrt[8]{2}\left(\cos\frac{13\pi}{16} + i\sin\frac{13\pi}{16}\right) \approx -0.9067 + 0.6059i$

$k = 2$: $\sqrt[8]{2}\left(\cos\frac{21\pi}{16} + i\sin\frac{21\pi}{16}\right) \approx -0.6059 - 0.9067i$

$k = 3$: $\sqrt[8]{2}\left(\cos\frac{29\pi}{16} + i\sin\frac{29\pi}{16}\right) \approx 0.9067 - 0.6059i$

62. False. The complex number must be converted to trigonometric form before applying DeMoivre's Theorem.

$(4 + \sqrt{6}i)^8 = [\sqrt{22}(\cos 0.55 + i\sin 0.55)]^8$

64.
$$2^{-1/4}(1-i) = 2^{-1/4}\left[\sqrt{2}\left(\cos\frac{7\pi}{4} + i\sin\frac{7\pi}{4}\right)\right]$$
$$= 2^{1/4}\left(\cos\frac{7\pi}{4} + i\sin\frac{7\pi}{4}\right)$$
$$\left[2^{1/4}\left(\cos\frac{7\pi}{4} + i\sin\frac{7\pi}{4}\right)\right]^4 = (2^{1/4})^4(\cos 7\pi + i\sin 7\pi)$$
$$= 2(\cos\pi + i\sin\pi)$$
$$= -2$$

66. (a) $3(\cos 45° + i\sin 45°)$

$3(\cos 135° + i\sin 135°)$

$3(\cos 225° + i\sin 225°)$

$3(\cos 315° + i\sin 315°)$

(b) These are the fourth roots of -81.

(c) The fourth roots of -81:

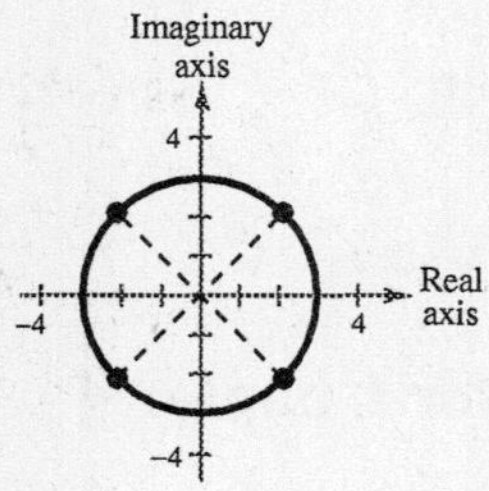

68.
$$\sin(\alpha+\beta) = \sin\alpha\cos\beta + \cos\alpha\sin\beta$$
$$= \left(\frac{3}{\sqrt{34}}\right)\left(\frac{\sqrt{7}}{4}\right) + \left(\frac{5}{\sqrt{34}}\right)\left(\frac{3}{4}\right)$$
$$= \frac{3\sqrt{7}+15}{4\sqrt{34}}$$
$$= \frac{3\sqrt{34}}{136}(\sqrt{7}+5)$$

70.
$$\cos(\beta-\alpha) = \cos\beta\cos\alpha + \sin\beta\sin\alpha$$
$$= \left(\frac{\sqrt{7}}{4}\right)\left(\frac{5}{\sqrt{34}}\right) + \left(\frac{3}{4}\right)\left(\frac{3}{\sqrt{34}}\right)$$
$$= \frac{5\sqrt{7}+9}{4\sqrt{34}}$$
$$= \frac{\sqrt{34}}{136}(5\sqrt{7}+9)$$

72.
$$\tan(\beta-\alpha) = \frac{\tan\beta - \tan\alpha}{1+\tan\beta\tan\alpha}$$
$$= \frac{\left(\frac{3}{\sqrt{7}}\right) - \left(\frac{3}{5}\right)}{1 + \left(\frac{3}{\sqrt{7}}\right)\left(\frac{3}{5}\right)}$$
$$= \frac{15-3\sqrt{7}}{5\sqrt{7}+9} = \frac{-120+51\sqrt{7}}{47}$$

74.
$$\sin 2\beta = 2\sin\beta\cos\beta = 2\left(\frac{3}{4}\right)\left(\frac{\sqrt{7}}{4}\right) = \frac{3\sqrt{7}}{8}$$

76.
$$\cos\frac{\alpha}{2} = \pm\sqrt{\frac{1+\cos\alpha}{2}}$$
$$= \pm\sqrt{\frac{1+(5/\sqrt{34})}{2}}$$
$$= \pm\sqrt{\frac{\sqrt{34}+5}{2\sqrt{34}}}$$
$$= \pm\sqrt{\frac{1}{4}\left(\frac{34-5\sqrt{34}}{17}\right)}$$
$$= \pm\frac{1}{2}\sqrt{\frac{34+5\sqrt{34}}{17}}$$

78.
$$\mathbf{v} = \langle -3, 7\rangle$$
$$\|\mathbf{v}\| = \sqrt{(-3)^2+7^2} = \sqrt{58}$$
$$\mathbf{u} = \left\langle \frac{-3}{\sqrt{58}}, \frac{7}{\sqrt{58}}\right\rangle = \left\langle -\frac{3\sqrt{58}}{58}, \frac{7\sqrt{58}}{58}\right\rangle$$

80. $\mathbf{w} = 8\mathbf{j} = 8\langle 0, 1\rangle = \langle 0, 8\rangle$

$\|\mathbf{w}\| = \sqrt{0^2 + 8^2} = 8$

$\mathbf{u} = \left\langle \frac{0}{8}, \frac{8}{8} \right\rangle = \langle 0, 1\rangle = \mathbf{j}$

82. $\mathbf{u} = \langle -5, 7\rangle$

$\|\mathbf{u}\| = \sqrt{\mathbf{u} \cdot \mathbf{u}} = \sqrt{(-5)(-5) + (7)(7)} = \sqrt{74}$

84. $\mathbf{u} = \langle -5, -12\rangle$

$\|\mathbf{u}\| = \sqrt{\mathbf{u} \cdot \mathbf{u}} = \sqrt{(-5)(-5) + (-12)(-12)} = \sqrt{169}$

$= 13$

86. $\mathbf{u} = 16\mathbf{i} + 4\mathbf{j}$

$\|\mathbf{u}\| = \sqrt{\mathbf{u} \cdot \mathbf{u}} = \sqrt{(16)(16) + (4)(4)} = \sqrt{272} = 4\sqrt{17}$

88. $\mathbf{u} = 24\mathbf{i} + 16\mathbf{j}$

$\|\mathbf{u}\|^2 = \mathbf{u} \cdot \mathbf{u} = (24\mathbf{i} + 16\mathbf{j}) \cdot (24\mathbf{i} + 16\mathbf{j}) = 832$

$\|\mathbf{u}\| = \sqrt{832} = 8\sqrt{13}$

Review Exercises for Chapter 4

Solutions to Even-Numbered Exercises

2. $3 - \sqrt{-25} = 3 - 5i$

4. $-5i + i^2 = -1 - 5i$

6. $\left(\frac{\sqrt{2}}{2} - \frac{\sqrt{2}}{2}i\right) - \left(\frac{\sqrt{2}}{2} + \frac{\sqrt{2}}{2}i\right) = -\sqrt{2}i$

8. $(1 + 6i)(5 - 2i) = 17 + 28i$

10. $i(6 + i)(3 - 2i) = i(20 - 9i) = 9 + 20i$

12. $\frac{3 + 2i}{5 + i} = \frac{3 + 2i}{5 + i} \cdot \frac{5 - i}{5 - i} = \frac{17 + 7i}{26} = \frac{17}{26} + \frac{7}{26}i$

14. $\frac{1}{(2 + i)} - \frac{5}{1 + 4i} = \frac{1}{2 + i} \cdot \frac{2 - i}{2 - i} - \frac{5}{1 + 4i} \cdot \frac{1 - 4i}{1 - 4i}$

$= \frac{2 - i}{5} - \frac{5 - 20i}{17}$

$= \frac{9}{85} + \frac{83}{85}i$

16. $2 + 8x^2 = 0$

$x^2 = -\frac{1}{4}$

$x = \pm\frac{1}{2}i$

18. $6x^2 + 3x + 27 = 0$

$2x^2 + x + 9 = 0$

$x = \frac{-1 \pm \sqrt{1^2 - 4(2)(9)}}{2(2)} = \frac{-1 \pm \sqrt{-71}}{4}$

$= \frac{-1 \pm \sqrt{71}i}{4}$

20. $-2x^6 + 7x^3 + x^2 + 4x - 19 = 0$

6 solutions

22. $\frac{3}{4}x^3 + \frac{1}{2}x^2 + \frac{3}{2}x + 2 = 0$

3 solutions

24. $9x^2 - 12x + 4 = 0$

$a = 9, b = -12, c = 4$

$b^2 - 4ac = (-12)^2 - 4(9)(4) = 0$

1 real repeated solution

26. $4x^2 + \frac{4}{3}x + \frac{1}{9} = 0$

$36x^2 + 12x + 1 = 0$

$a = 36, b = 12, c = 1$

$b^2 - 4ac = 12^2 - 4(36)(1) = 0$

1 real repeated solution

28. $f(x) = 6x - x^2$

$6x - x^2 = 0$

$x(6 - x) = 0$

Zeros: $x = 0, 6$

30. $h(x) = 3 + 4x - x^2$

By the Quadratic Formula, the zeros of $h(x)$ are

$$x = \frac{-4 \pm \sqrt{4^2 - 4(-1)(3)}}{2(-1)} = \frac{-4 \pm \sqrt{28}}{-2}$$

$$= 2 \pm \sqrt{7}$$

32. $s(x) = 2x^2 + 5x + 4$

By the Quadratic Formula, the zeros of $s(x)$ are

$$x = \frac{-5 \pm \sqrt{5^2 - 4(2)(4)}}{2(2)} = \frac{-5 \pm \sqrt{-7}}{4}$$

$$= \frac{-5 \pm \sqrt{7}i}{4}$$

34. $f(x) = 4x^3 - x^2 + 128x - 32$

$= x^2(4x - 1) + 32(4x - 1)$

$= (x^2 + 32)(4x - 1)$

$= \left(x + 4\sqrt{2}i\right)\left(x - 4\sqrt{2}i\right)(4x - 1)$

The zeros of $f(x)$ are $x = \pm 4\sqrt{2}i$ and $x = \frac{1}{4}$.

36. $f(x) = 5x^4 + 126x^2 + 25$

$5x^4 + 126x^2 + 25 = 0$

$(5x^2 + 1)(x^2 + 25) = 0$

$5x^2 + 1 = 0 \Rightarrow x = \pm\frac{\sqrt{5}}{5}i$

$x^2 + 25 = 0 \Rightarrow x = \pm 5i$

Zeros: $x = \pm\frac{\sqrt{5}}{5}i, \pm 5i$

38. $f(x) = 10x^3 + 21x^2 - x - 6$

-2	10	21	-1	-6
		-20	-2	6
	10	1	-3	0

$10x^2 + x - 3 = 0$

$(2x - 1)(5x + 3) = 0$

$x = \frac{1}{2}, -\frac{3}{5}$

Zeros: $x = -2, -\frac{3}{5}, \frac{1}{2}$

$f(x) = (x + 2)(5x + 3)(2x - 1)$

40. $g(x) = x^3 - 8x^2 + 29x - 52$

4	1	-8	29	-52
		4	-16	52
	1	-4	13	0

$x^2 - 4x + 13 = 0$

$$x = \frac{-(-4) \pm \sqrt{(-4)^2 - 4(1)(13)}}{2(1)} = \frac{4 \pm \sqrt{-36}}{2}$$

$$= 2 \pm 3i$$

Zeros: $x = 4, 2 \pm 3i$

$g(x) = (x - 4)(x - 2 - 3i)(x - 2 + 3i)$

42. $f(x) = 5x^3 - 4x^2 + 20x - 16$

Since $2i$ is a zero, so is $-2i$.

$2i$	5	-4	20	-16
		$10i$	$-20 - 8i$	16
	5	$-4 + 10i$	$-8i$	0

$-2i$	5	$-4 + 10i$	$-8i$
		$-10i$	$8i$
	5	-4	0

$5x - 4 = 0$

$x = \frac{4}{5}$

Zeros: $x = \pm 2i, \frac{4}{5}$

$f(x) = (5x - 4)(x - 2i)(x + 2i)$

44. $g(x) = x^4 - 6x^3 + 18x^2 - 26x + 21$

Since $2 + \sqrt{3}i$ is a zero, so is $2 - \sqrt{3}i$.

$2+\sqrt{3}i$	1	-6	18	-26	21
		$2+\sqrt{3}i$	$-11-2\sqrt{3}i$	$20+3\sqrt{3}i$	-21
	1	$-4+\sqrt{3}i$	$7-2\sqrt{3}i$	$-6+3\sqrt{3}i$	0

$2-\sqrt{3}i$	1	$-4+\sqrt{3}i$	$7-2\sqrt{3}i$	$-6+3\sqrt{3}i$
		$2-\sqrt{3}i$	$-4+2\sqrt{3}i$	$6-3\sqrt{3}i$
	1	-2	3	0

By the Quadratic Formula, the zeros of $x^2 - 2x + 3$ are

$$x = \frac{-(-2) \pm \sqrt{(-2)^2 - 4(1)(3)}}{2(1)} = \frac{2 \pm \sqrt{-8}}{2}$$

$$= 1 \pm \sqrt{2}i$$

Zeros: $x = 2 \pm \sqrt{3}i,\ 1 \pm \sqrt{2}i$

$g(x) = \left(x - 2 - \sqrt{3}i\right)\left(x - 2 + \sqrt{3}i\right)\left(x - 1 - \sqrt{2}i\right)\left(x - 1 + \sqrt{2}i\right)$

46. Zeros: $x = -2, 2, 3, 3$

$f(x) = (x + 2)(x - 2)(x - 3)^2$

$= x^4 - 6x^3 + 5x^2 + 24x - 36$

Note: Any multiple of $f(x)$ has the given zeros.

48. Zeros: $x = 5, 1 - \sqrt{2}, 1 + \sqrt{2}$

$f(x) = (x - 5)\left(x - 1 + \sqrt{2}\right)\left(x - 1 - \sqrt{2}\right)$

$= x^3 - 7x^2 + 9x + 5$

Note: Any multiple of $f(x)$ has the given zeros.

50. Zeros: $x = 2, -3, 1 - 2i, 1 + 2i$

$f(x) = (x - 2)(x + 3)(x - 1 + 2i)(x - 1 - 2i)$

$= x^4 - x^3 - 3x^2 + 17x - 30$

Note: Any multiple of $f(x)$ has the given zeros.

52. Zeros: $x = -2i, 2i, -4i, 4i$

$f(x) = (x + 2i)(x - 2i)(x + 4i)(x - 4i)$

$= x^4 + 20x^2 + 64$

Note: Any multiple of $f(x)$ has the given zeros.

54. $P = xp - C$

$9{,}000{,}000 = x(140 - 0.0001x) - (80x + 150{,}000)$

$9{,}000{,}000 = 140x - 0.0001x^2 - 80x - 150{,}000$

$0.0001x^2 - 60x + 9{,}150{,}000 = 0$

$b^2 - 4ac = (-60)^2 - 4(0.0001)(9{,}150{,}000) = -60 < 0$

There are no real solutions. There is no price for which the profit will reach \$9 million.

56. $|-6i| = \sqrt{0^2 + (-6)^2} = \sqrt{36} = 6$

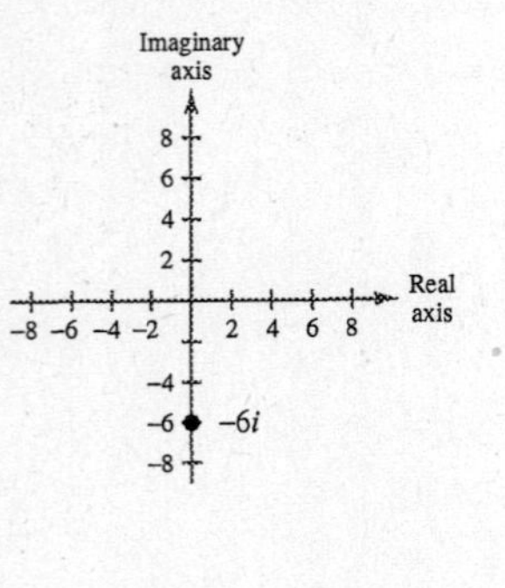

58. $|-10 - 4i| = \sqrt{(-10)^2 + (-4)^2} = \sqrt{116} = 2\sqrt{29}$

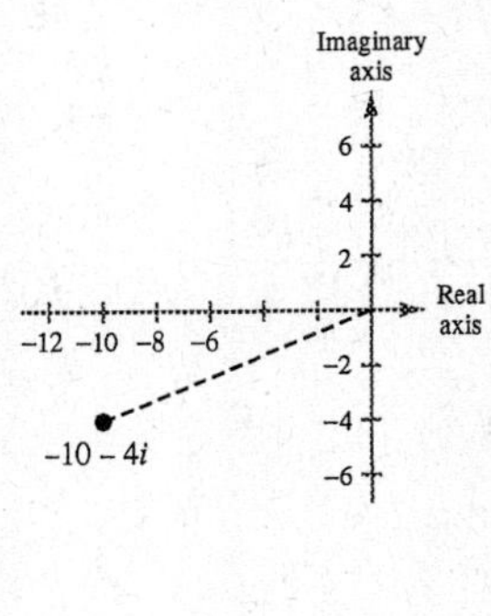

60. $z = 5 + 12i$

$r = \sqrt{5^2 + 12^2} = \sqrt{169} = 13$

$\tan\theta = \frac{12}{5} \Rightarrow \theta \approx 1.176$

$z \approx 13(\cos 1.176 + i \sin 1.176)$

62. $z = -9$

$r = \sqrt{(-9)^2 + 0^2} = 9$

$\tan\theta = \dfrac{0}{-9} = 0 \Rightarrow \theta = \pi$

$z = 9(\cos\pi + i\sin\pi)$

64. (a) $z_1 = -3(1 + i) = -3 - 3i$

$r = \sqrt{(-3)^2 + (-3)^2} = \sqrt{18} = 3\sqrt{2}$

$\tan\theta = \dfrac{-3}{-3} = 1$ and z_1 is in Quadrant III. $\Rightarrow \theta = \dfrac{5\pi}{4}$

$z_1 = 3\sqrt{2}\left(\cos\dfrac{5\pi}{4} + i\sin\dfrac{5\pi}{4}\right)$

$z_2 = 2(\sqrt{3} + i) = 2\sqrt{3} + 2i$

$r = \sqrt{\left(2\sqrt{3}\right)^2 + 2^2} = \sqrt{12 + 4} = \sqrt{16} = 4$

$\tan\theta = \dfrac{2}{2\sqrt{3}} = \dfrac{1}{\sqrt{3}}$ and z_2 is in Quadrant I. $\Rightarrow \theta = \dfrac{\pi}{6}$

$z_2 = 4\left(\cos\dfrac{\pi}{6} + i\sin\dfrac{\pi}{6}\right)$

(b) $$z_1z_2 = \left[3\sqrt{2}\left(\cos\frac{5\pi}{4} + i\sin\frac{5\pi}{4}\right)\right]\left[4\left(\cos\frac{\pi}{6} + i\sin\frac{\pi}{6}\right)\right]$$

$$= (3\sqrt{2})(4)\left[\cos\left(\frac{5\pi}{4} + \frac{\pi}{6}\right) + i\sin\left(\frac{5\pi}{4} + \frac{\pi}{6}\right)\right]$$

$$= 12\sqrt{2}\left(\cos\frac{17\pi}{12} + i\sin\frac{17\pi}{12}\right)$$

$$\frac{z_1}{z_2} = \frac{3\sqrt{2}\left(\cos\frac{5\pi}{4} + i\sin\frac{5\pi}{4}\right)}{4\left(\cos\frac{\pi}{6} + i\sin\frac{\pi}{6}\right)}$$

$$= \frac{3\sqrt{2}}{4}\left[\cos\left(\frac{5\pi}{4} - \frac{\pi}{6}\right) + i\sin\left(\frac{5\pi}{4} - \frac{\pi}{6}\right)\right]$$

$$= \frac{3\sqrt{2}}{4}\left(\cos\frac{13\pi}{12} + i\sin\frac{13\pi}{12}\right)$$

66. $$\left[2\left(\cos\frac{4\pi}{15} + i\sin\frac{4\pi}{15}\right)\right]^5 = 2^5\left[\cos 5\left(\frac{4\pi}{15}\right) + i\sin 5\left(\frac{4\pi}{15}\right)\right]$$

$$= 32\left(\cos\frac{4\pi}{3} + i\sin\frac{4\pi}{3}\right)$$

$$= 32\left(-\frac{1}{2} - \frac{\sqrt{3}}{2}i\right)$$

$$= -16 - 16\sqrt{3}i$$

68. $(1-i)^8 = \left[\sqrt{2}\left(\cos\frac{7\pi}{4} + i\sin\frac{7\pi}{4}\right)\right]^8$

$= (\sqrt{2})^8\left[\cos 8\left(\frac{7\pi}{4}\right) + i\sin 8\left(\frac{7\pi}{4}\right)\right]$

$= 16(\cos 14\pi + i\sin 14\pi)$

$= 16(1 + 0i)$

$= 16$

70. Fourth roots of 256

$256 = 256(\cos 0 + i\sin 0)$

$\sqrt[4]{256} = \sqrt[4]{256}\left(\cos\frac{2\pi k}{4} + i\sin\frac{2\pi k}{4}\right), k = 0, 1, 2, 3$

$k = 0$: $4(\cos 0 + i\sin 0) = 4$

$k = 1$: $4\left(\cos\frac{\pi}{2} + i\sin\frac{\pi}{2}\right) = 4i$

$k = 2$: $4(\cos\pi + i\sin\pi) = -4$

$k = 3$: $4\left(\cos\frac{3\pi}{2} + i\sin\frac{3\pi}{2}\right) = -4i$

72. $x^5 - 32 = 0$

$x^5 = 32$

The solutions of the equation are the fifth roots of 32.

$32 = 32(\cos 0 + i\sin 0)$

$\sqrt[5]{32} = \sqrt[5]{32}\left(\cos\frac{2\pi k}{5} + i\sin\frac{2\pi k}{5}\right), k = 0, 1, 2, 3, 4.$

$k = 0$: $2(\cos 0 + i\sin 0) = 2$

$k = 1$: $2\left(\cos\frac{2\pi}{5} + i\sin\frac{2\pi}{5}\right) \approx 0.6180 + 1.9021i$

$k = 2$: $2\left(\cos\frac{4\pi}{5} + i\sin\frac{4\pi}{5}\right) \approx -1.6180 + 1.1756i$

$k = 3$: $2\left(\cos\frac{6\pi}{5} + i\sin\frac{6\pi}{5}\right) \approx -1.6180 - 1.1756i$

$k = 4$: $2\left(\cos\frac{8\pi}{5} + i\sin\frac{8\pi}{5}\right) \approx 0.6180 - 1.9021i$

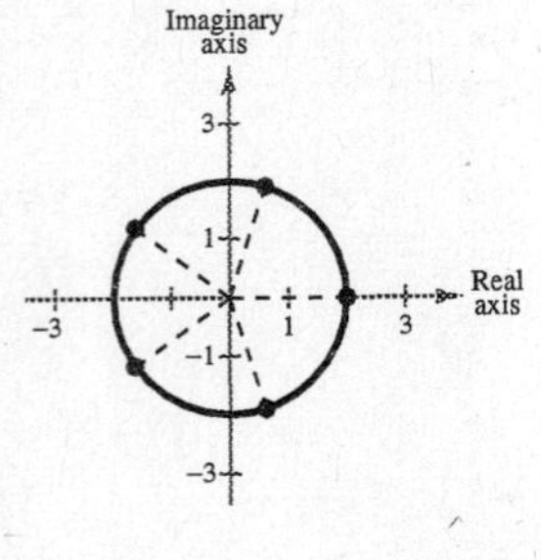

74. $(x^3 - 1)(x^2 + 1) = 0$

$x^3 - 1 = 0$ or $x^2 + 1 = 0$

$x^3 = 1$ or $x^2 = -1$

The solutions of the equation are the cube roots of 1 or the square roots of -1.

$1 = \cos 0 + i\sin 0$

$\sqrt[3]{1} = \cos\left(\frac{2\pi k}{3}\right) + i\sin\left(\frac{2\pi k}{3}\right), k = 0, 1, 2$

$k = 0$: $\cos 0 + i\sin 0 = 1$

$k = 1$: $\cos\frac{2\pi}{3} + i\sin\frac{2\pi}{3} = -\frac{1}{2} + \frac{\sqrt{3}}{2}i$

$k = 2$: $\cos\frac{4\pi}{3} + i\sin\frac{4\pi}{3} = -\frac{1}{2} - \frac{\sqrt{3}}{2}i$

$-1 = \cos\pi + i\sin\pi$

$\sqrt{-1} = \cos\left(\frac{\pi + 2\pi k}{2}\right) + i\sin\left(\frac{\pi + 2\pi k}{2}\right), k = 0, 1$

$k = 0$: $\cos\frac{\pi}{2} + i\sin\frac{\pi}{2} = i$

$k = 1$: $\cos\frac{3\pi}{2} + i\sin\frac{3\pi}{2} = -i$

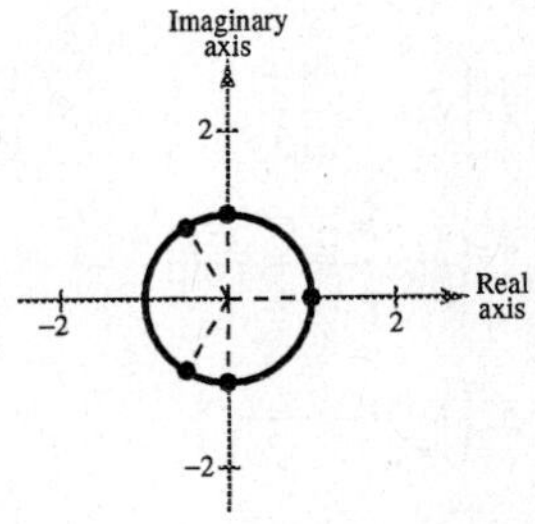

76. False. The equation $325x^2 - 717x + 398 = 0$ has no *real* solution. By the Quadratic Formula, the complex solutions are:

$x = \frac{717}{650} \pm \frac{\sqrt{3311}}{650}i.$

78. Answers will vary.

(a) Two distinct real solutions. $\Rightarrow b^2 - 4ac > 0$

$x^2 - 5x + 6 = 0$

$3x^2 + 8x - 40 = 0$

(b) Two complex solutions. **All** quadratic equations have two complex solutions since the degree of any quadratic equation is 2.

(c) No real solutions. $\Rightarrow b^2 - 4ac < 0$

$x^2 + 5x + 14 = 0$

$6x^2 - 3x + 11 = 0$

80. (a) There are four roots shown and they each have an absolute value of 4.

$4(\cos 60° + i \sin 60°)$

$4(\cos 150° + i \sin 150°)$

$4(\cos 240° + i \sin 240°)$

$4(\cos 330° + i \sin 330°)$

(b) $[4(\cos 60° + i \sin 60°)]^4 = 256(\cos 240° + i \sin 240°)$

$= 256\left(-\frac{1}{2} - \frac{\sqrt{3}}{2}i\right)$

$= -128 - 128\sqrt{3}i$

(c) By using a calculator, show that each of the roots raised to the fourth power yields $-128 - 128\sqrt{3}i$.

82. (a) There are 4 fourth roots of z in the complex number system, so 3 roots are not shown.

(b) The four roots are equally spaced around the circle, so they are 90° apart. The 4 roots are:

$2(\cos 30° + i \sin 30°)$

$2(\cos 120° + i \sin 120°)$

$2(\cos 210° + i \sin 210°)$

$2(\cos 300° + i \sin 300°)$

Problem Solving for Chapter 4

Solutions to Even-Numbered Exercises

2. (a) $z_m = \frac{1}{z} = \frac{1}{1+i} = \frac{1}{1+i} \cdot \frac{1-i}{1-i} = \frac{1-i}{2} = \frac{1}{2} - \frac{1}{2}i$

(b) $z_m = \frac{1}{z} = \frac{1}{3-i} = \frac{1}{3-i} \cdot \frac{3+i}{3+i} = \frac{3+i}{10} = \frac{3}{10} + \frac{1}{10}i$

(c) $z_m = \frac{1}{z} = \frac{1}{-2+8i} = \frac{1}{-2+8i} \cdot \frac{-2-8i}{-2-8i} = \frac{-2-8i}{68} = -\frac{1}{34} - \frac{2}{17}i$

4. $z = a + bi$, $\bar{z} = a - bi$, $w = c + di$, and $\bar{w} = c - di$

(a) $\overline{z + w} = \overline{(a + bi) + (c + di)}$

$= \overline{(a + c) + (b + d)i}$

$= (a + c) - (b + d)i$

$= (a - bi) + (c - di)$

$= \bar{z} + \bar{w}$

(b) $\overline{z - w} = \overline{(a + bi) - (c + di)}$

$= \overline{(a - c) + (b - d)i}$

$= (a - c) - (b - d)i$

$= (a - bi) + (-c + di)$

$= (a - bi) - (c - di)$

$= \bar{z} - \bar{w}$

(c) $\overline{zw} = \overline{(a + bi)(c + di)}$

$= \overline{(ac - bd) + (ad - bc)i}$

$= (ac - bd) - (ad + bc)i$

$= (a - bi)(c - di)$

$= \bar{z}\,\bar{w}$

—CONTINUED—

4. —CONTINUED—

(d) $\overline{\left(\dfrac{z}{w}\right)} = \overline{\left(\dfrac{a+bi}{c+di}\right)}$

$= \overline{\dfrac{(a+bi)(c-di)}{c^2-d^2}}$

$= \overline{\dfrac{(ac+bd)+(bc-ad)i}{c^2-d^2}}$

$= \dfrac{(ac+bd)-(bc-ad)i}{c^2-d^2}$

$= \dfrac{(ac+bd)+(ad-bc)i}{c^2-d^2}$

$= \dfrac{(a-bi)(c+di)}{(c-di)(c+di)}$

$= \dfrac{a-bi}{c-di}$

$= \dfrac{\overline{z}}{\overline{w}}$

(e) $(\overline{z})^2 = (a-bi)^2$

$= a^2 - 2abi - b^2$

$= (a^2-b^2) - 2abi$

$= \overline{(a^2-b^2)+2abi}$

$= \overline{(a+bi)^2}$

$= \overline{z^2}$

(f) $\overline{\overline{z}} = \overline{\overline{(a+bi)}}$

$= \overline{(a-bi)}$

$= a+bi$

$= z$

(g) $\overline{z} = z \Rightarrow a+bi = a-bi$

$\Rightarrow b = -b$

$\Rightarrow b = 0$

$\Rightarrow z = a$

$\Rightarrow z$ is real

6. $f(x) = x^4 - 4x^2 + k$

$x^2 = \dfrac{-(-4) \pm \sqrt{(-4)^2 - 4(1)(k)}}{2(1)} = \dfrac{4 \pm 2\sqrt{4-k}}{2}$

$= 2 \pm \sqrt{4-k}$

$x = \pm\sqrt{2 \pm \sqrt{4-k}}$

(a) For there to be four distinct real roots, both $4-k$ and $2 \pm \sqrt{4-k}$ must be positive. This occurs when $0 < k < 4$. Thus, some possible k-values are $k = 1$, $k = 2$, $k = 3$, $k = \frac{1}{2}$, $k = \sqrt{2}$, etc.

(b) For there to be two real roots, each of multiplicity 2, $4-k$ must equal zero. Thus, $k = 4$.

(c) For there to be four complex zeros, $2 \pm \sqrt{4-k}$ must be nonreal. This occurs when $k > 4$. Some possible k-values are $k = 5$, $k = 6$, $k = 7.4$, etc.

8. Zeros: $-2, \frac{1}{2}, 3$

$f(x) = -(x+2)(2x-1)(x-3)$

$= -2x^3 + 3x^2 + 11x - 6$

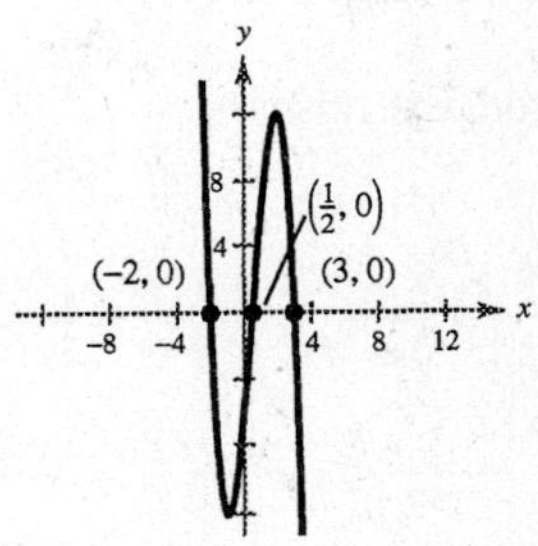

Any nonzero scalar multiple of f would have the same three zeros.

Let $g(x) = af(x)$, $a > 0$.

There are infinitely many possible functions for f.

10. (a) $-2, 0, 2$

(b) The graph touches the x-axis at $x = 0$.

(c) The least possible degree of f is 4 because there are at least 4 real zeros (0 is repeated) and a function can have at most the number of real zeros equal to the degree of the function. The degree cannot be odd by the definition of multiplicity.

(d) The leading coefficient of f is negative. From the information in the table, you can conclude that the graph will eventually fall to the left and fall to the right.

—CONTINUED—

10. —CONTINUED—

(e) $f(x) = -(x - 0)^2[x - (-2)](x - 2)$

$= -x^2(x + 2)(x - 2) = -x^2(x^2 - 4) = -x^4 + 4x^2$

(f)

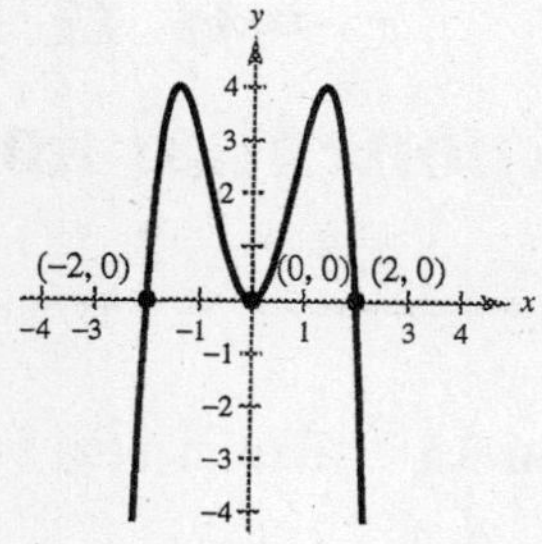

12. (a) $x^2 - (4 + 2i)x + 2 + 4i = 0$

$$x = \frac{4 + 2i \pm \sqrt{(-4 - 2i)^2 - 4(2 + 4i)}}{2}$$

$$= \frac{4 + 2i \pm \sqrt{12 + 16i - 8 - 16i}}{2}$$

$$= \frac{4 + 2i \pm 2}{2}$$

$$= 2 + i \pm 1$$

$x = 3 + i$ or $x = 1 + i$

(b) $x^2 - (3 + 2i)x + 5 + i = 0$

$$x = \frac{3 + 2i \pm \sqrt{(-3 - 2i)^2 - 4(5 + i)}}{2}$$

$$= \frac{3 + 2i \pm \sqrt{5 + 12i - 20 - 4i}}{2}$$

$$= \frac{3 + 2i \pm \sqrt{-15 + 8i}}{2}$$

$$= \frac{3 + 2i \pm (1 + 4i)}{2}$$

$x = 2 + 3i$ or $x = 1 - i$

Note: The square roots of $-15 + 8i$ are $1 + 4i$ and $-1 - 4i$, by DeMoivre's Theorem. Both of these roots yield the same two solutions to the equation.

(c) $2x^2 + (5 - 8i)x - 13 - i = 0$

$$x = \frac{-(5 - 8i) \pm \sqrt{(5 - 8i)^2 + 8(13 + i)}}{4}$$

$$= \frac{-5 + 8i \pm \sqrt{-39 - 80i + 104 + 8i}}{4}$$

$$= \frac{-5 + 8i \pm \sqrt{65 - 72i}}{4}$$

$$= \frac{-5 + 8i \pm (9 - 4i)}{4}$$

$x = 1 + i$ or $x = -\frac{7}{2} + 3i$

Note: The square roots of $65 - 72i$ are $9 - 4i$ and $-9 + 4i$, by DeMoivre's Theorem. Both of these roots yield the same two solutions to the equation.

(d) $3x^2 - (11 + 14i)x + 1 - 9i = 0$

$$x = \frac{11 + 14i \pm \sqrt{(-11 - 14i)^2 - 12(1 - 9i)}}{6}$$

$$= \frac{11 + 14i \pm \sqrt{-75 + 308i - 12 + 108i}}{6}$$

$$= \frac{11 + 14i \pm \sqrt{-87 + 416i}}{6}$$

$$= \frac{11 + 14i \pm (13 + 16i)}{6}$$

$x = 4 + 5i$ or $x = -\frac{1}{3} - \frac{1}{3}i$

Note: The square roots of $-87 + 416i$ are $13 + 16i$ and $-13 - 16i$, by DeMoivre's Theorem. Both of these roots yield the same two solutions to the equation.

14. Let $z = a + bi$ and $\overline{z} = a - bi$

$z^2 - \overline{z}^2 = 0$

$(a + bi)^2 - (a - bi)^2 = 0$

$4abi = 0 \Rightarrow a = 0$ or $b = 0$

$z = a$, a pure real number or

$z = bi$, a pure imaginary number

$z^2 + \overline{z}^2 = 0$

$(a + bi)^2 + (a - bi)^2 = 0$

$2a^2 - 2b^2 = 0 \Rightarrow b = \pm a$

$z = a + ai$ or $z = a - ai$, both are complex solutions

CHAPTER 5
Exponential and Logarithmic Functions

CHAPTER 5
Exponential and Logarithmic Functions

Section 5.1 Exponential Functions and Their Graphs

Solutions to Even-Numbered Exercises

2. $f(x) = 2.3^x = 2.3^{3/2} \approx 3.488$

4. $g(x) = 5000(2^x) = 5000(2^{-1.5}) \approx 1767.767$

6. $f(x) = e^x = e^{3.2} \approx 24.533$

8. $f(x) = 2^x + 1$ rises to the right.

Asymptote: $y = 1$

Intercept: $(0, 2)$

Matches graph (c).

10. $f(x) = 2^{x-2}$ rises to the right.

Asymptote: $y = 0$

Intercept: $\left(0, \frac{1}{4}\right)$

Matches graph (b).

12. $f(x) = 4^x$, $g(x) = 4^x + 1$

Because $g(x) = f(x) + 1$, the graph of g can be obtained by shifting the graph of f one unit upward.

14. $f(x) = 10^x$, $g(x) = 10^{-x+3}$

Because $g(x) = f(-x + 3)$, the graph of g can be obtained by reflecting the graph of f in the y-axis and shifting f three units to the right. (Note: This is equivalent to shifting f three units to the left and then reflecting the graph in the y-axis.)

16. $f(x) = \left(\frac{7}{2}\right)^x$, $g(x) = -\left(\frac{7}{2}\right)^{-x+6}$

Because $g(x) = -f(-x + 6)$, the graph of g can be obtained by reflecting the graph of f in the x-axis and y-axis and shifting f six units to the right. (Note: This is equivalent to shifting f six units to the left and then reflecting the graph in the x-axis and y-axis.)

18. $f(x) = 3.6^x$, $g(x) = -3.6^{-x} + 8$

Because $g(x) = -f(-x) + 8$, the graph of g can be obtained by reflecting the graph of f in the x-axis and y-axis and shifting f eight units upward.

20. $f(x) = \left(\frac{1}{2}\right)^{-x} = 2^x$

x	-2	-1	0	1	2
$f(x)$	0.25	0.5	1	2	4

Asymptote: $y = 0$

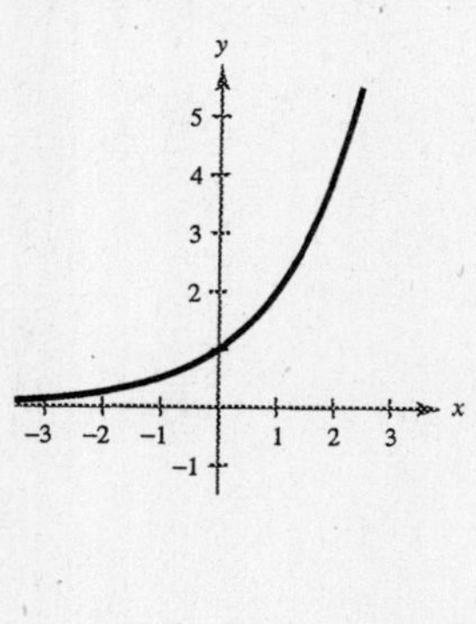

22. $f(x) = 6^x$

x	-2	-1	0	1	2
$f(x)$	0.028	0.167	1	6	36

Asymptote: $y = 0$

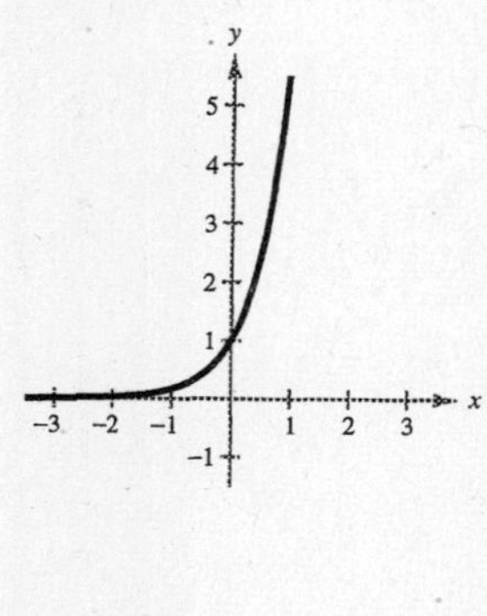

24. $f(x) = 3^{x+2}$

x	-4	-3	-2	-1	0
$f(x)$	0.111	0.333	1	3	9

Asymptote: $y = 0$

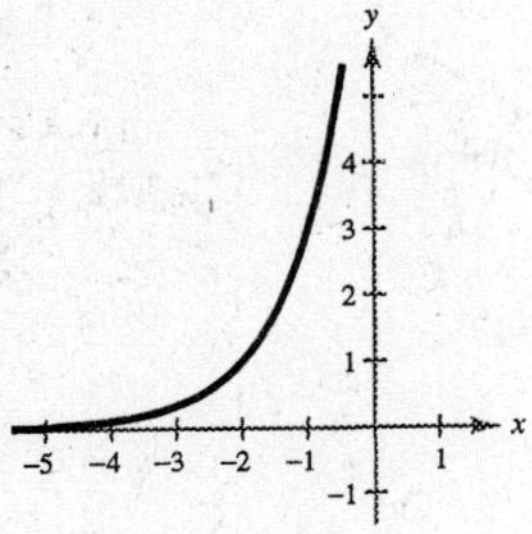

26. $f(x) = e^{-x}$

x	-2	-1	0	1	2
$f(x)$	7.389	2.718	1	0.368	0.135

Asymptote: $y = 0$

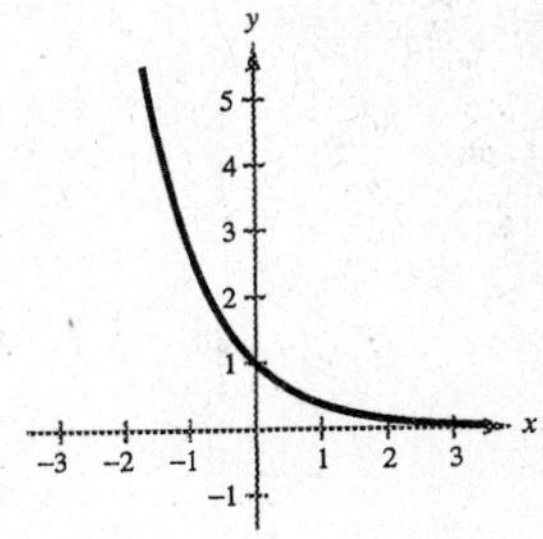

28. $f(x) = 2e^{-0.5x}$

x	-2	-1	0	1	2
$f(x)$	5.437	3.297	2	1.213	0.736

Asymptote: $y = 0$

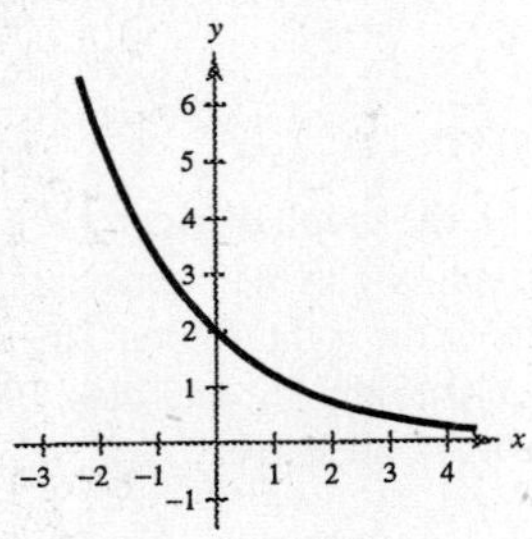

30. $f(x) = 2 + e^{x-5}$

x	0	2	4	5	6
$f(x)$	2.007	2.050	2.368	3	4.718

Asymptote: $y = 2$

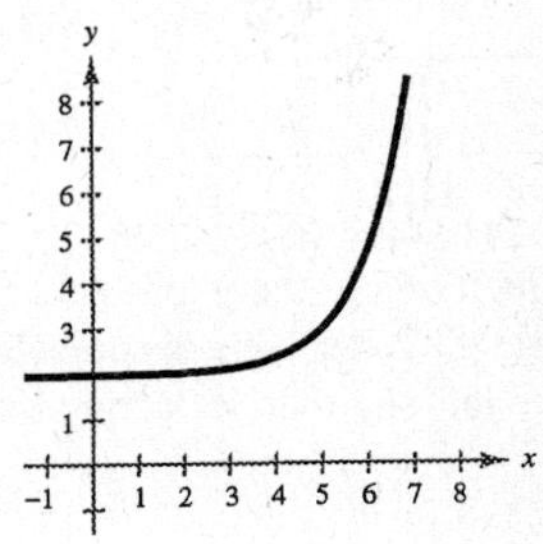

32. $f(x) = -4^{x-3} - 3$

x	-1	0	1	3	4
$f(x)$	-3.004	-3.016	-3.063	-4	-7

Asymptote: $y = -3$

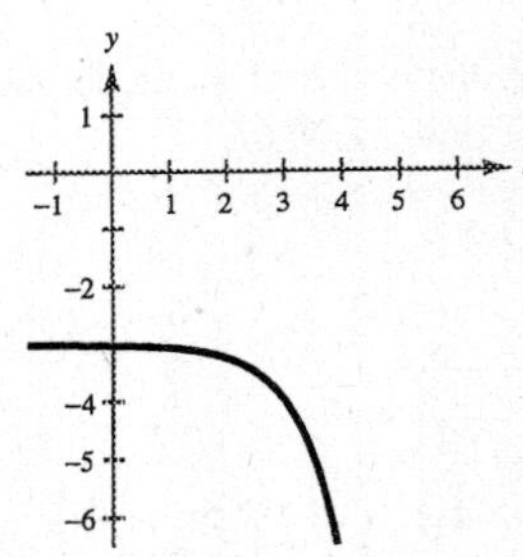

34. $y = 3^{-|x|}$

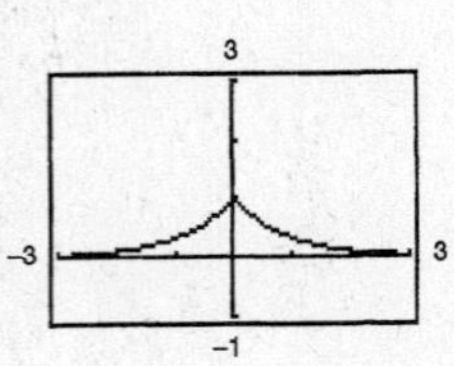

36. $y = 4^{x+1} - 2$

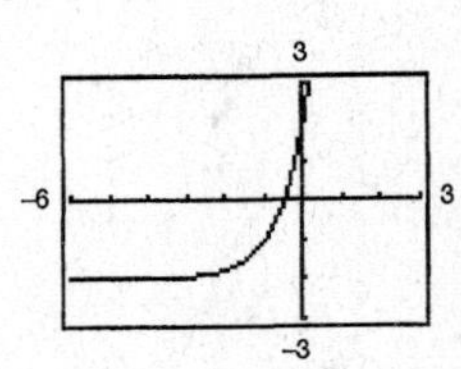

38. $y = 1.08^{5x}$

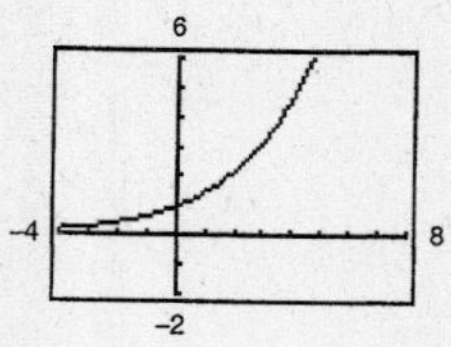

40. $s(t) = 3e^{-0.2t}$

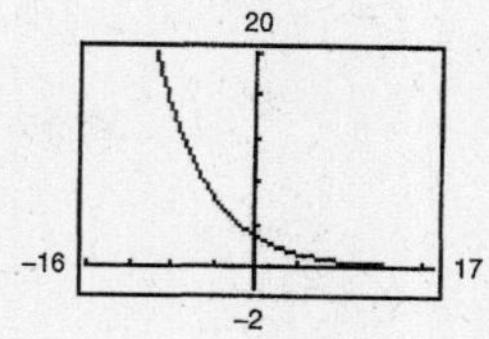

42. $h(x) = e^{x-2}$

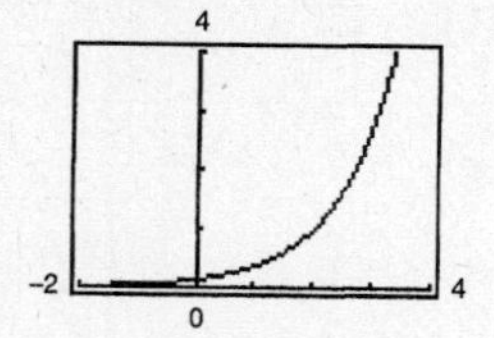

44. $P = \$1000, r = 6\%, t = 10$ years

Compounded n times per year: $A = 1000\left(1 + \frac{0.06}{n}\right)^{10n}$

Compounded continuously: $A = 1000e^{0.06(10)}$

n	1	2	4	12	365	Continuous
A	\$1790.85	\$1806.11	\$1814.02	\$1819.40	\$1822.03	\$1822.12

46. $P = \$1000, r = 6\%, t = 40$ years

Compounded n times per year: $A = 1000\left(1 + \frac{0.06}{n}\right)^{40n}$

Compounded continuously: $A = 1000e^{0.06(40)}$

n	1	2	4	12	365	Continuous
A	\$10,285.72	\$10,640.89	\$10,828.46	\$10,957.45	\$11,021.00	\$11,023.18

48. $A = Pe^{rt}$

$A = 12{,}000e^{0.06t}$

t	1	10	20	30	40	50
A	\$12,742.04	\$21,865.43	\$39,841.40	\$72,595.77	\$132,278.12	\$241,026.44

50. $A = Pe^{rt}$

$A = 12{,}000e^{0.075t}$

t	1	10	20	30	40	50
A	\$12,934.61	\$25,404.00	\$53,780.27	\$113,852.83	\$241,026.44	\$510,252.98

52. $A = 5000e^{(0.075)(50)} \approx \$212{,}605.41$

54. $p = 5000\left(1 - \frac{4}{4 + e^{-0.002x}}\right)$

(a)

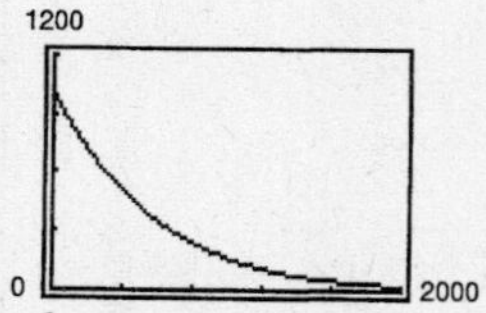

(b) When $x = 500$:

$$p = 5000\left(1 - \frac{4}{4 + e^{-0.002(500)}}\right) \approx \$421.12$$

(c) Since (600, 350.13) is on the graph in part (a), it appears that the greatest price that will still yield a demand of at least 600 units is about \$350.

56. $P(t) = 2500e^{0.0293t}$

(a) $P(10) = 2500e^{0.0293(10)} \approx 3351$

(b) $P(20) = 2500e^{0.0293(20)} \approx 4492$

58. $Q = 10\left(\frac{1}{2}\right)^{t/5730}$

(a) When $t = 0$: $Q = 10\left(\frac{1}{2}\right)^{0/5730}$

$= 10(1) = 10$ grams

(b) When $t = 2000$: $Q = 10\left(\frac{1}{2}\right)^{2000/5730}$

≈ 7.85 grams

(c)

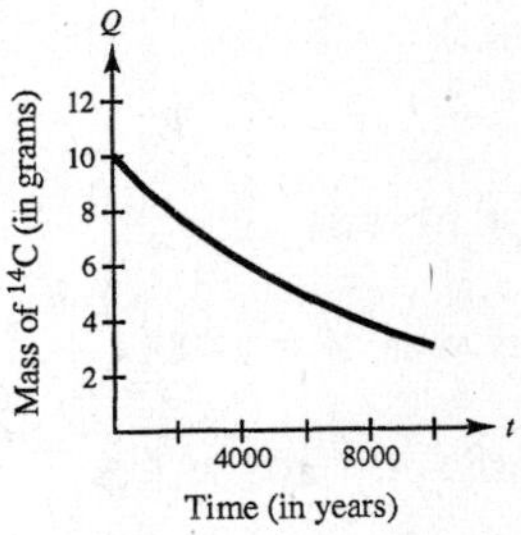

60. (a)

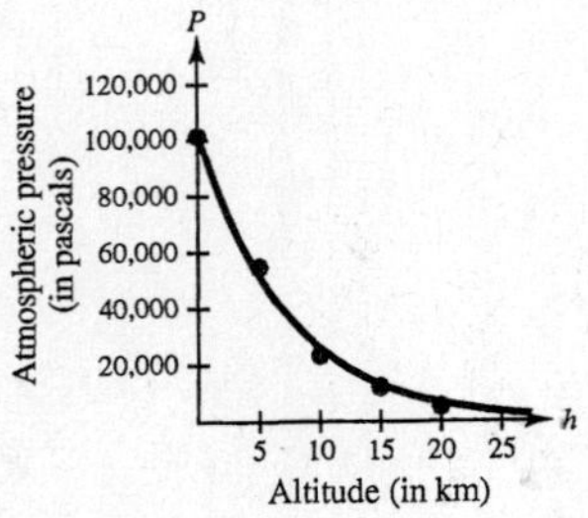

(b) $p = 102{,}303e^{-0.137(8)} \approx 34{,}190$ Pascals

62. False, $e \neq \dfrac{271{,}801}{99{,}990}$.

e is an irrational number.

64. $g(x) = 2^{2x+6}$

$= 2^{2x} \cdot 2^6$

$= 64(2^{2x})$

$= 64(2^2)^x$

$= 64(4^x)$

$= h(x)$

Thus, $g(x) = h(x)$ but $g(x) \neq f(x)$.

66. $f(x) = 5^{-x} + 3$

$g(x) = 5^{3-x} = 5^3 \cdot 5^{-x}$

$h(x) = -5^{x-3} = -(5^x \cdot 5^{-3})$

Thus, none are equal.

68. $y = \left(\frac{1}{2}\right)^x$ and $y = \left(\frac{1}{4}\right)^x$

x	-2	-1	0	1	2
$\left(\frac{1}{2}\right)^x$	4	2	1	$\frac{1}{2}$	$\frac{1}{4}$
$\left(\frac{1}{4}\right)^x$	16	4	1	$\frac{1}{4}$	$\frac{1}{16}$

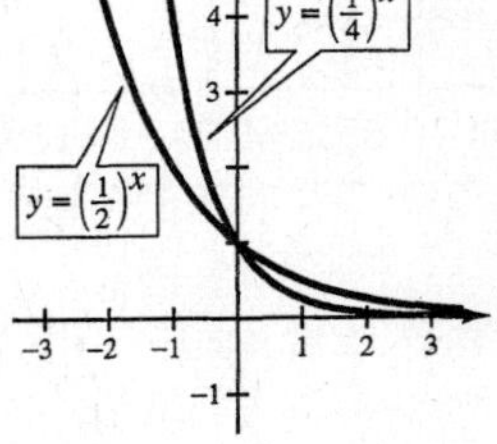

(a) $\left(\frac{1}{4}\right)^x < \left(\frac{1}{2}\right)^x$ when $x > 0$.

(b) $\left(\frac{1}{4}\right)^x > \left(\frac{1}{2}\right)^x$ when $x < 0$.

70. (a) $f(x) = x^2e^{-x}$

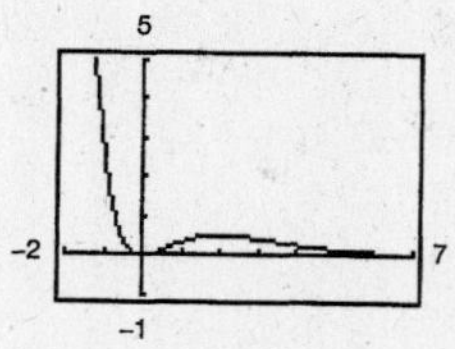

Decreasing: $(-\infty, 0),\ (2, \infty)$

Increasing: $(0, 2)$

Relative maximum: $(2, 4e^{-2})$

Relative minimum: $(0, 0)$

(b) $g(x) = x2^{3-x}$

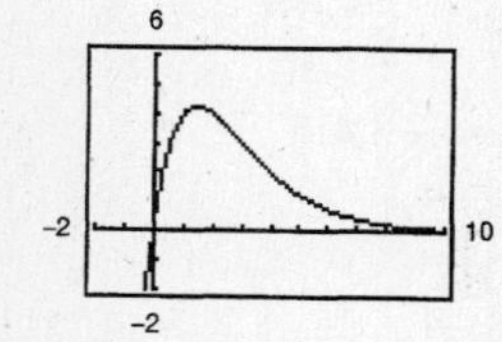

Decreasing: $(1.44, \infty)$

Increasing: $(-\infty, 1.44)$

Relative maximum: $(1.44, 4.25)$

72. In Exercise 71 $f(x) = \left[1 + \dfrac{0.5}{x}\right]^x$ appears to approach $g(x) = e^{0.5}$ as x increases without bound. Therefore, the value of $\left[1 + \dfrac{r}{x}\right]^x$ approaches e^r as x increases without bound.

x	1	10	100	200	500	1100	10,000
$\left[1 + \left(\frac{1}{x}\right)\right]^x$	2	2.5937	2.7048	2.7115	2.7156	2.7170	2.718

$e^1 \approx 2.718281828\ldots$

74. Since $\sqrt{2} \approx 1.414$ we know that $1 < \sqrt{2} < 2$.

Thus, $2^1 < 2^{\sqrt{2}} < 2^2$

$2 < 2^{\sqrt{2}} < 4.$

76. $x^2 + 3y = 4$

$3y = 4 - x^2$

$y = \frac{1}{3}(4 - x^2)$

78. $x - |y| = 2$

$x - 2 = |y|$

$y = x - 2$ and $y = -(x - 2), x \ge 2$

Section 5.2 Logarithmic Functions and Their Graphs

Solutions to Even-Numbered Exercises

2. $\log_3 81 = 4 \Rightarrow 3^4 = 81$

4. $\log_{10} \frac{1}{1000} = -3 \Rightarrow 10^{-3} = \frac{1}{1000}$

6. $\log_{16} 8 = \frac{3}{4} \Rightarrow 16^{3/4} = 8$

8. $\ln 4 = 1.386\ldots \Rightarrow e^{1.386\ldots} = 4$

10. $8^2 = 64 \Rightarrow \log_8 64 = 2$

12. $9^{3/2} = 27 \Rightarrow \log_9 27 = \frac{3}{2}$

14. $10^{-3} = 0.001 \Rightarrow \log_{10} 0.001 = -3$

16. $e^{1/2} = 1.6487\ldots \Rightarrow \ln 1.6487\ldots = \dfrac{1}{2}$

18. $u^v = w \Rightarrow \log_u w = v$

20. $f(x) = \log_{16} x = \log_{16} 4 = \dfrac{1}{2}$

22. $f(x) = \log_{10} x = \log_{10} 10 = 1$

24. $g(x) = \ln x = \ln e^{-2} = -2$

26. $g(x) = \log_b x = \log_b b^{-3} = -3$

28. $f(x) = \log_{10} x = \log_{10} 12.5 \approx 1.097$

30. $f(x) = 3 \ln x = 3 \ln 0.32 \approx -3.418$

32. $g(x) = -\ln x = -\ln \frac{1}{2} \approx 0.693$

34. $f(x) = -\log_3 x$

Asymptote: $x = 0$

Point on graph: $(1, 0)$

Matches graph (f).

36. $f(x) = \log_3(x - 1)$

Asymptote: $x = 1$

Point on graph: $(2, 0)$

Matches graph (e).

38. $f(x) = -\log_3(-x)$

Asymptote: $x = 0$

Point on graph: $(-1, 0)$

Matches graph (a).

40. $g(x) = \log_6 x$

Domain: $(0, \infty)$

x-intercept: $(1, 0)$

Vertical asymptote: $x = 0$

$y = \log_6 x \Rightarrow 6^y = x$

x	$\frac{1}{6}$	1	$\sqrt{6}$	6
y	-1	0	$\frac{1}{2}$	1

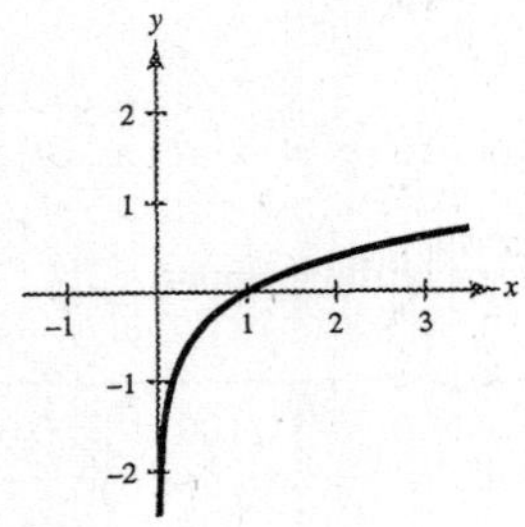

42. $h(x) = \log_4(x - 3)$

Domain: $x - 3 > 0 \Rightarrow x > 3$

The domain is $(3, \infty)$.

x-intercept: $\log_4(x - 3) = 0$

$$4^0 = x - 3$$
$$1 = x - 3$$
$$4 = x$$

The x-intercept is $(4, 0)$.

Vertical asymptote: $x - 3 = 0 \Rightarrow x = 3$

$y = \log_4(x - 3) \Rightarrow 4^y + 3 = x$

x	$3\frac{1}{4}$	4	7	19
y	-1	0	1	2

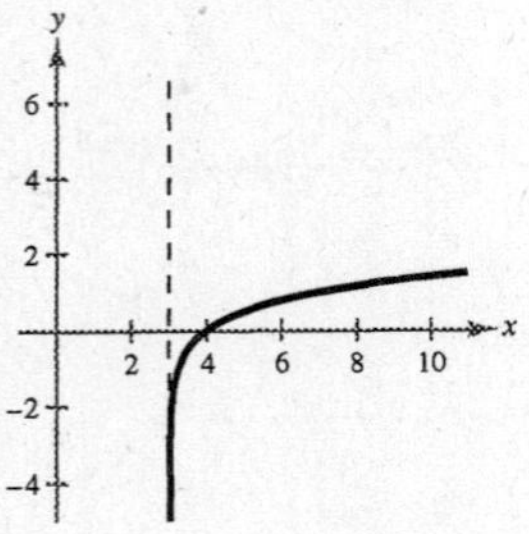

44. $y = \log_5(x - 1) + 4$

Domain: $x - 1 > 0 \Rightarrow x > 1$

The domain is $(1, \infty)$.

x-intercept: $\log_5(x - 1) + 4 = 0$

$$\log_5(x - 1) = -4$$
$$5^{-4} = x - 1$$
$$\frac{1}{625} = x - 1$$
$$\frac{626}{625} = x$$

The x-intercept is $\left(\frac{626}{625}, 0\right)$.

Vertical asymptote: $x - 1 = 0 \Rightarrow x = 1$

$y = \log_5(x - 1) + 4 \Rightarrow 5^{y-4} + 1 = x$

x	1.00032	1.0016	1.008	1.04	1.2
y	-1	0	1	2	3

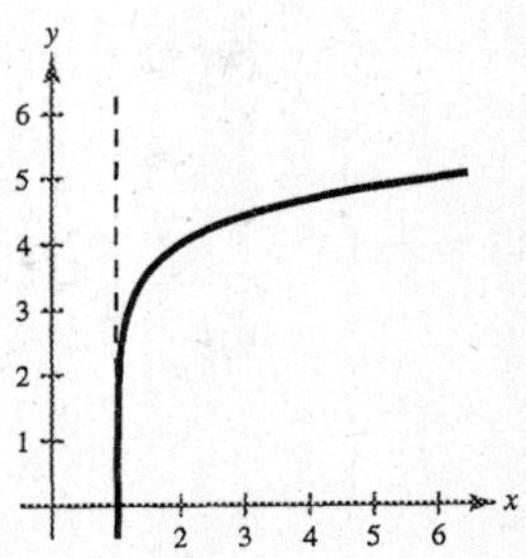

46. $y = \log_{10}(-x)$

Domain: $-x > 0 \implies x < 0$

The domain is $(-\infty, 0)$.

x-intercept: $\log_{10}(-x) = 0$

$$10^0 = -x$$

$$-1 = x$$

The x-intercept is $(-1, 0)$.

Vertical asymptote: $x = 0$

$y = \log_{10}(-x) \implies -10^y = x$

x	$-\frac{1}{100}$	$-\frac{1}{10}$	-1	-10
y	-2	-1	0	1

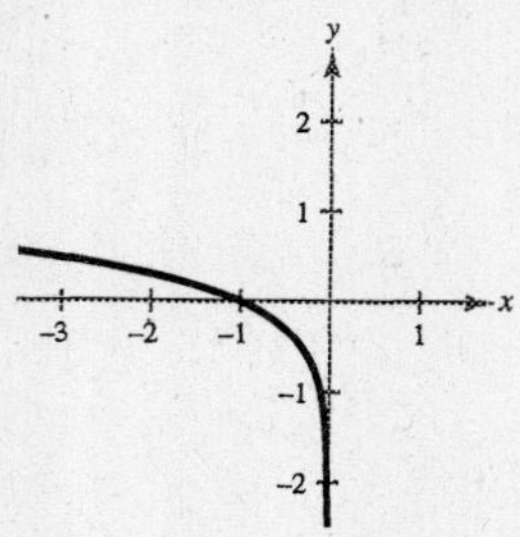

48. $h(x) = \ln(x + 1)$

Domain: $x + 1 > 0 \implies x > -1$

The domain is $(-1, \infty)$.

x-intercept: $\ln(x + 1) = 0$

$$e^0 = x + 1$$

$$1 = x + 1$$

$$0 = x$$

The x-intercept is $(0, 0)$.

Vertical asymptote: $x + 1 = 0 \implies x = -1$

$y = \ln(x + 1) \implies e^y - 1 = x$

x	-0.39	0	1.72	6.39	19.09
y	$-\frac{1}{2}$	0	1	2	3

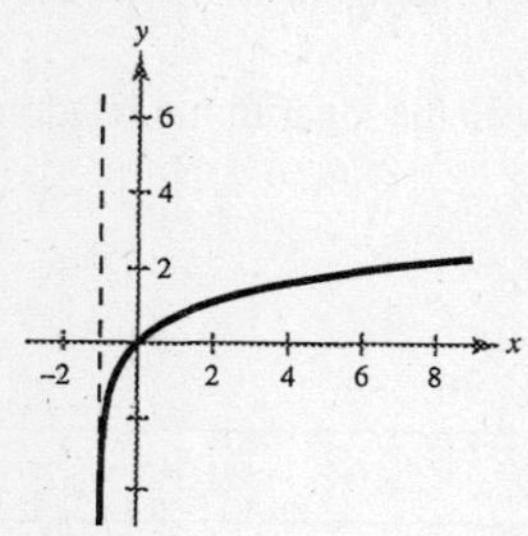

50. $f(x) = \ln(3 - x)$

Domain: $3 - x > 0 \implies x < 3$

The domain is $(-\infty, 3)$.

x-intercept: $\ln(3 - x) = 0$

$$e^0 = 3 - x$$

$$1 = 3 - x$$

$$2 = x$$

The x-intercept is $(2, 0)$.

Vertical asymptote: $3 - x = 0 \implies x = 3$

$y = \ln(3 - x) \implies 3 - e^y = x$

x	2.95	2.86	2.63	2	0.28
y	-3	-2	-1	0	1

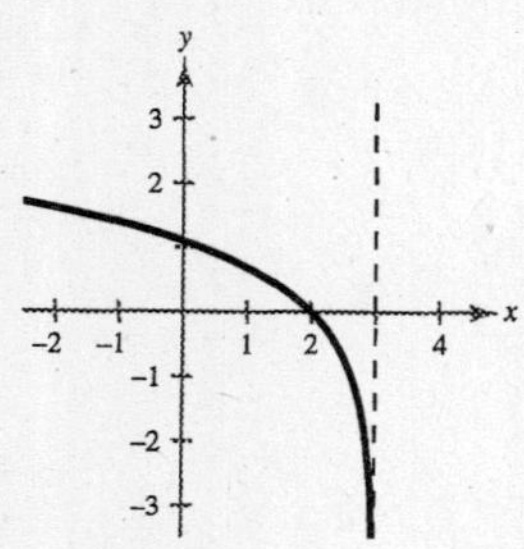

52. $f(x) = \log_{10}(x - 1)$

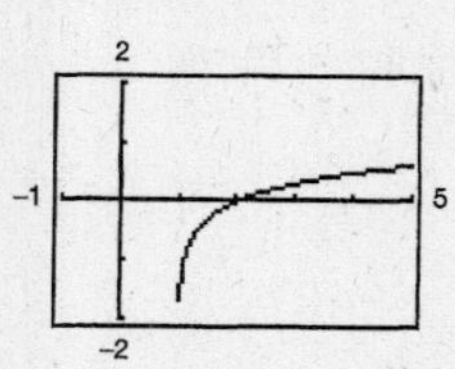

54. $f(x) = \ln(x + 2)$

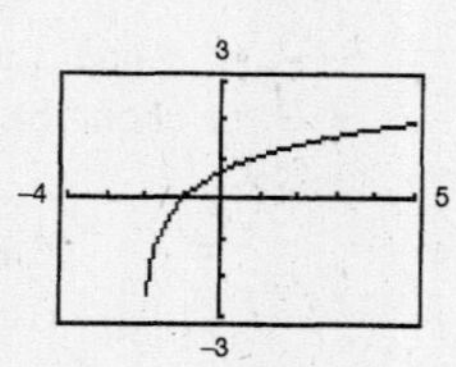

56. $f(x) = 3\ln x - 1$

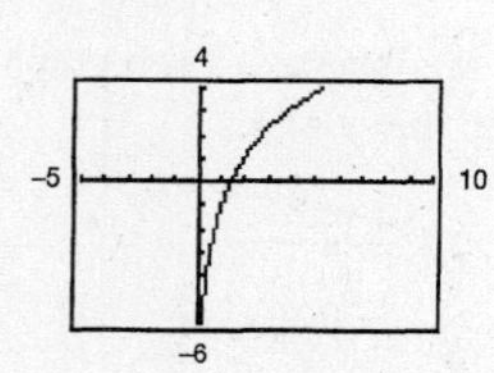

58. $t = \dfrac{\ln K}{0.095}$

(a)

K	1	2	4	6	8	10	12
t	0	7.3	14.6	18.9	21.9	24.2	26.2

The number of years required to multiply the original investment by K increases with K. However, the larger the value of K, the fewer the years required to increase the value of the investment by an additional multiple of the original investment.

(b)

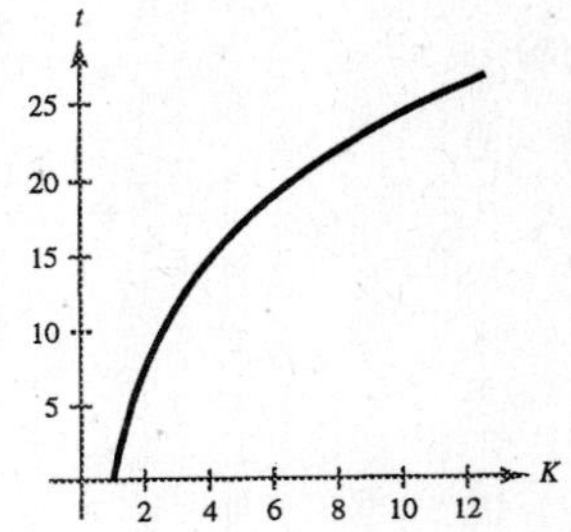

60. $\beta = 10 \log_{10}\left(\dfrac{I}{10^{-12}}\right)$

(a) $\beta = 10 \log_{10}\left(\dfrac{1}{10^{-12}}\right) = 10 \log_{10}(10^{12}) = 120$ decibels

(b) $\beta = 10 \log_{10}\left(\dfrac{10^{-2}}{10^{-12}}\right) = 10 \log_{10}(10^{10}) = 100$ decibels

(c) No, the difference is due to the logarithmic relationship between intensity and number of decibels.

62. $f(x) = \dfrac{\ln x}{x}$

(a)

x	1	5	10	10^2	10^4	10^6
$f(x)$	0	0.322	0.230	0.046	0.00092	0.0000138

(b) As $x \to \infty$, $f(x) \to 0$.

(c)

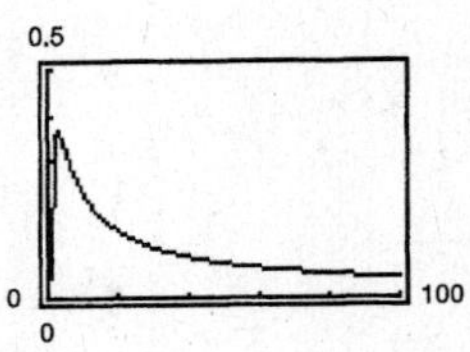

64. True, $\log_3 27 = 3 \Rightarrow 3^3 = 27$.

66. $f(x) = 5^x$, $g(x) = \log_5 x$

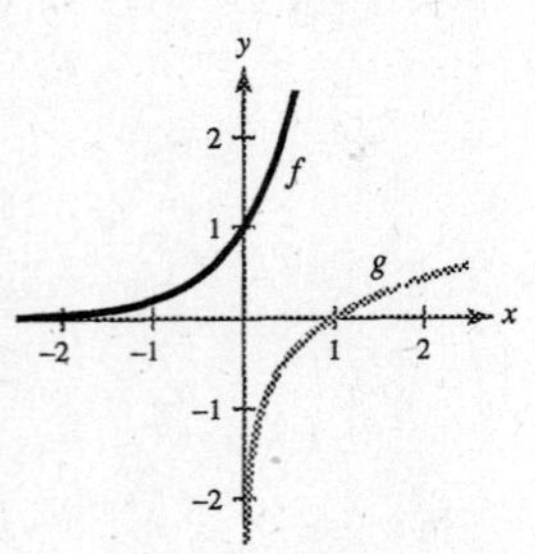

f and g are inverses. Their graphs are reflected about the line $y = x$.

68. $f(x) = 10^x$, $g(x) = \log_{10} x$

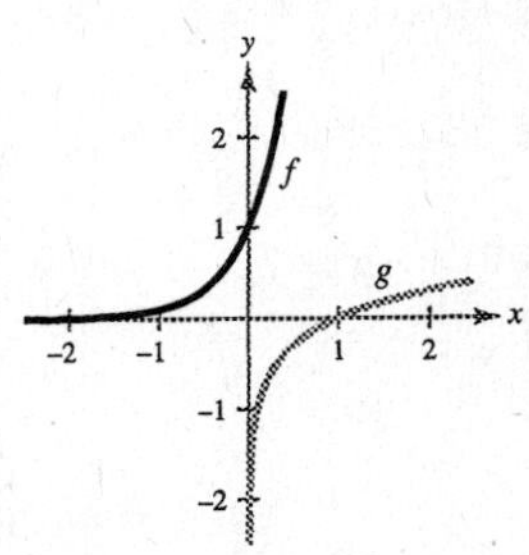

f and g are inverses. Their graphs are reflected about the line $y = x$.

70. (a) False. If y were an exponential function of x, then $y = a^x$, but $a^1 = a$, not 0. Because one point is $(1, 0)$, y is not an exponential function of x.

(b) True. $y = \log_a x$

For $a = 2, y = \log_2 x$.

$x = 1, \log_2 1 = 0$

$x = 2, \log_2 2 = 1$

$x = 8, \log_2 8 = 3$

(c) True. $x = a^y$

For $a = 2, x = 2^y$.

$y = 0, 2^0 = 1$

$y = 1, 2^1 = 2$

$y = 3, 2^3 = 8$

(d) False. If y were a linear function of x, the slope between $(1, 0)$ and $(2, 1)$ and the slope between $(2, 1)$ and $(8, 3)$ would be the same.

However,

$$m_1 = \frac{1-0}{2-1} = 1 \text{ and } m_2 = \frac{3-1}{8-2} = \frac{2}{6} = \frac{1}{3}.$$

Therefore, y is not a linear function of x.

72. $f^{1}(x) = 10^x$

74. $y = \log_a x \implies a^y = x$, so, for example, if $a = -2$, there is no value of y for which $(-2)^y = -4$. If $a = 1$, then every power of a is equal to 1, so x could only be 1. So, $\log_a x$ is defined only for $0 < a < 1$ and $a > 1$.

76. (a) $h(x) = \ln(x^2 + 1)$

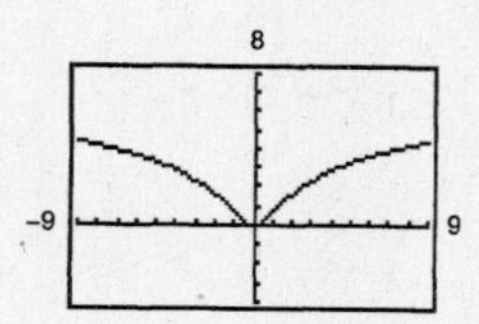

(b) Increasing on $(0, \infty)$.

Decreasing on $(-\infty, 0)$.

(c) Relative minimum: $(0, 0)$

78. $A = l \cdot w$

$= (10 + w)w$

$= 10w + w^2$

Section 5.3 Properties of Logarithms

Solutions to Even-Numbered Exercises

2. (a) $\log_3 x = \dfrac{\log_{10} x}{\log_{10} 3}$

(b) $\log_3 x = \dfrac{\ln x}{\ln 3}$

4. (a) $\log_{1/3} x = \dfrac{\log_{10} x}{\log_{10} \frac{1}{3}}$

(b) $\log_{1/3} x = \dfrac{\ln x}{\ln \frac{1}{3}}$

6. (a) $\log_x \dfrac{3}{4} = \dfrac{\log_{10} \frac{3}{4}}{\log_{10} x}$

(b) $\log_x \dfrac{3}{4} = \dfrac{\ln \frac{3}{4}}{\ln x}$

8. (a) $\log_{7.1} x = \dfrac{\log_{10} x}{\log_{10} 7.1}$

(b) $\log_{7.1} x = \dfrac{\ln x}{\ln 7.1}$

10. $\log_7 4 = \dfrac{\log_{10} 4}{\log_{10} 7} = \dfrac{\ln 4}{\ln 7} \approx 0.712$

12. $\log_{1/4} 5 = \dfrac{\log_{10} 5}{\log_{10} \frac{1}{4}} = \dfrac{\ln 5}{\ln \frac{1}{4}} \approx -1.161$

14. $\log_{20} 0.125 = \dfrac{\log_{10} 0.125}{\log_{10} 20} = \dfrac{\ln 0.125}{\ln 20} \approx -0.694$

16. $\log_3 0.015 = \dfrac{\log_{10} 0.015}{\log_{10} 3} = \dfrac{\ln 0.015}{\ln 3} \approx -3.823$

18. $\log_3 10z = \log_3 10 + \log_3 z$

20. $\log_{10} \frac{y}{2} = \log_{10} y - \log_{10} 2$

22. $\log_6 z^{-3} = -3 \log_6 z$

24. $\ln \sqrt[3]{t} = \ln t^{1/3} = \frac{1}{3} \ln t$

26.
$$\begin{aligned}\log_{10} 4x^2y &= \log_{10} 4 + \log_{10} x^2 + \log_{10} y \\ &= \log_{10} 4 + 2 \log_{10} x + \log_{10} y\end{aligned}$$

28.
$$\begin{aligned}\ln\left(\frac{x^2 - 1}{x^3}\right) &= \ln(x^2 - 1) - \ln x^3 \\ &= \ln[(x + 1)(x - 1)] - \ln x^3 \\ &= \ln(x + 1) + \ln(x - 1) - 3 \ln x\end{aligned}$$

30.
$$\begin{aligned}\ln \frac{6}{\sqrt{x^2 + 1}} &= \ln 6 - \ln\sqrt{x^2 + 1} \\ &= \ln 6 - \ln(x^2 + 1)^{1/2} \\ &= \ln 6 - \frac{1}{2} \ln(x^2 + 1)\end{aligned}$$

32.
$$\begin{aligned}\ln \sqrt{\frac{x^2}{y^3}} = \ln\left(\frac{x^2}{y^3}\right)^{1/2} &= \frac{1}{2} \ln\left(\frac{x^2}{y^3}\right) \\ &= \frac{1}{2}(\ln x^2 - \ln y^3) \\ &= \frac{1}{2}(2 \ln x - 3 \ln y)\end{aligned}$$

34.
$$\begin{aligned}\log_2 \frac{\sqrt{x}\, y^4}{z^4} &= \log_2 \sqrt{x}\, y^4 - \log_2 z^4 \\ &= \log_2 \sqrt{x} + \log_2 y^4 - \log_2 z^4 \\ &= \frac{1}{2} \log_2 x + 4 \log_2 y - 4 \log_2 z\end{aligned}$$

36.
$$\begin{aligned}\log_{10} \frac{xy^4}{z^5} &= \log_{10} xy^4 - \log_{10} z^5 \\ &= \log_{10} x + \log_{10} y^4 - \log_{10} z^5 \\ &= \log_{10} x + 4 \log_{10} y - 5 \log_{10} z\end{aligned}$$

38.
$$\begin{aligned}\ln\sqrt{x^2(x + 2)} &= \ln[x^2(x + 2)]^{1/2} \\ &= \ln[x(x + 2)^{1/2}] \\ &= \ln x + \ln(x + 2)^{1/2} \\ &= \ln x + \frac{1}{2} \ln(x + 2)\end{aligned}$$

40. $\ln y + \ln t = \ln yt = \ln ty$

42. $\log_5 8 - \log_5 t = \log_5 \frac{8}{t}$

44. $\frac{2}{3} \log_7(z - 2) = \log_7(z - 2)^{2/3}$

46. $-4 \log_6 2x = \log_6(2x)^{-4} = \log_6 \frac{1}{16x^4}$

48.
$$\begin{aligned}2 \ln 8 + 5 \ln(z - 4) &= \ln 8^2 + \ln(z - 4)^5 \\ &= \ln 64 + \ln(z - 4)^5 \\ &= \ln 64(z - 4)^5\end{aligned}$$

50.
$$\begin{aligned}3 \log_3 x + 4 \log_3 y - 4 \log_3 z &= \log_3 x^3 + \log_3 y^4 - \log_3 z^4 \\ &= \log_3 x^3y^4 - \log_3 z^4 \\ &= \log_3 \frac{x^3y^4}{z^4}\end{aligned}$$

52.
$$\begin{aligned}4[\ln z + \ln(z + 5)] - 2 \ln(z - 5) &= 4[\ln z(z + 5)] - \ln(z - 5)^2 \\ &= \ln[z(z + 5)]^4 - \ln(z - 5)^2 \\ &= \ln \frac{z^4(z + 5)^4}{(z - 5)^2}\end{aligned}$$

54. $2[3\ln x - \ln(x+1) - \ln(x-1)] = 2[\ln x^3 - \ln(x+1) - \ln(x-1)]$

$= 2[\ln x^3 - [\ln(x+1) + \ln(x-1)]]$

$= 2[\ln x^3 - [\ln(x+1)(x-1)]]$

$= 2\ln\dfrac{x^3}{x^2-1}$

$= \ln\left(\dfrac{x^3}{x^2-1}\right)^2$

56. $\frac{1}{2}[\log_4(x+1) + 2\log_4(x-1)] + 6\log_4 x = \frac{1}{2}[\log_4(x+1) + \log_4(x-1)^2]\log_4 x^6$

$= \frac{1}{2}[\log_4(x+1)(x-1)^2] + \log_4 x^6$

$= \log_4\left[\sqrt{x+1}(x-1)\right] + \log_4 x^6$

$= \log_4\left[x^6(x-1\sqrt{x+1}\right]$

58. $\log_7\sqrt{70} = \frac{1}{2}\log_7 70 = \frac{1}{2}[\log_7 7 + \log_7 10]$

$= \frac{1}{2}[1 + \log_7 10]$

$= \frac{1}{2} + \frac{1}{2}\log_7 10$

$= \frac{1}{2} + \log_7\sqrt{10}$ by Property 1 and Property 3

60. $\log_5 \frac{1}{125} = \log_5 5^{-3} = -3\log_5 5 = -3(1) = -3$

62. $\log_6\sqrt[3]{6} = \log_6 6^{1/3} = \frac{1}{3}\log_6 6 = \frac{1}{3}(1) = \frac{1}{3}$

64. $\log_3 81^{-0.2} = -0.2\log_3 81$

$= -0.2\log_3 3^4 = -0.2(4) = -0.8$

66. $\log_2(-16)$ is undefined because -16 is not in the domain of $\log_2 x$.

68. $3\ln e^4 = (3)(4)\ln e$

$= 12(1)$

$= 12$

70. $\ln\sqrt[4]{e^3} = \ln e^{3/4}$

$= \frac{3}{4}\ln e$

$= \frac{3}{4}(1)$

$= \frac{3}{4}$

72. $2\ln e^6 - \ln e^5 = \ln e^{12} - \ln e^5$

$= \ln\dfrac{e^{12}}{e^5}$

$= \ln e^7$

$= 7$

74. $\log_4 2 + \log_4 32 = \log_4 4^{1/2} + \log_4 4^{5/2}$

$= \frac{1}{2}\log_4 4 + \frac{5}{2}\log_4 4$

$= \frac{1}{2}(1) + \frac{5}{2}(1)$

$= 3$

76. $\log_2(4^2 \cdot 3^4) = \log_2 4^2 + \log_2 3^4$

$= 2\log_2 4 + 4\log_2 3$

$= 2\log_2 2^2 + 4\log_2 3$

$= 4\log_2 2 + 4\log_2 3$

$= 4 + 4\log_2 3$

78. $\log_{10}\dfrac{9}{300} = \log_{10}\dfrac{3}{100}$

$= \log_{10} 3 - \log_{10} 100$

$= \log_{10} 3 - \log_{10} 10^2$

$= \log_{10} 3 - 2\log_{10} 10$

$= \log_{10} 3 - 2$

80. $\ln\dfrac{6}{e^2} = \ln 6 - \ln e^2$

$= \ln 6 - 2\ln e$

$= \ln 6 - 2$

82. $\beta = 10 \log_{10}\left(\dfrac{I}{10^{-12}}\right)$

$= 10[\log_{10} I - \log_{10} 10^{-12}]$

$= 10[\log_{10} I - (-12)\log_{10} 10]$

$= 10[\log_{10} I + 12]$

For $I = 10^{-6}$, $\beta = 10[\log_{10} 10^{-6} + 12] = 60$ decibels

84. $f(ax) = f(a) + f(x)$, $a > 0$, $x > 0$

True, because

$f(ax) = \ln ax = \ln a + \ln x$

$= f(a) + f(x)$.

86. $\sqrt{f(x)} = \frac{1}{2}f(x)$; False.

$\sqrt{f(x)} = \sqrt{\ln x}$ can't be simplified further.

$f(\sqrt{x}) = \ln\sqrt{x} = \ln x^{1/2} = \frac{1}{2}\ln x = \frac{1}{2}f(x)$

88. If $f(x) < 0$, then $0 < x < 1$.

True.

90. Let $x = \log_b u$, then $u = b^x$ and $u^n = b^{nx}$.

$\log_b u^n = \log_b b^{nx} = nx = n \log_b u$

92. $f(x) = \log_4 x = \dfrac{\log_{10} x}{\log_{10} 4} = \dfrac{\ln x}{\ln 4}$

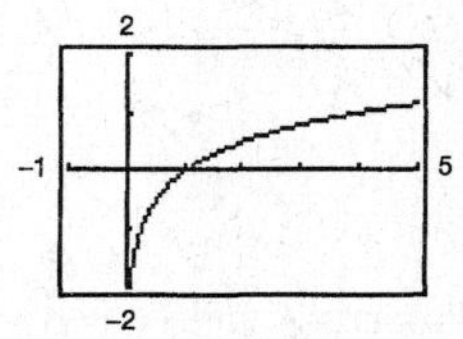

94. $f(x) = \log_{1/4} x = \dfrac{\log_{10} x}{\log_{10} \frac{1}{4}} = \dfrac{\ln x}{\ln \frac{1}{4}}$

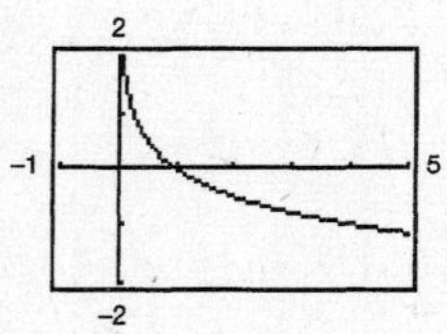

96. $f(x) = \log_{12.4} x = \dfrac{\log_{10} x}{\log_{10} 12.4} = \dfrac{\ln x}{\ln 12.4}$

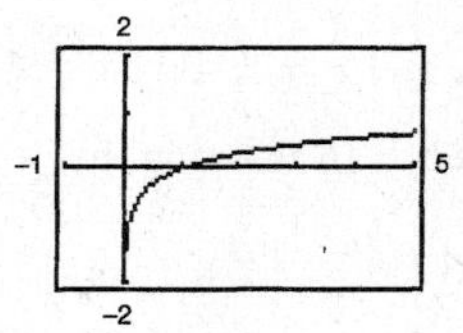

98. $\ln 2 \approx 0.6931$, $\ln 3 \approx 1.0986$, $\ln 5 \approx 1.6094$

$\ln 2 \approx 0.6931$

$\ln 3 \approx 1.0986$

$\ln 4 = \ln(2 \cdot 2) = \ln 2 + \ln 2 \approx 0.6931 + 0.6931 = 1.3862$

$\ln 5 \approx 1.6094$

$\ln 6 = \ln(2 \cdot 3) = \ln 2 + \ln 3 \approx 0.6931 + 1.0986 = 1.7917$

$\ln 8 = \ln 2^3 = 3 \ln 2 \approx 3(0.6931) = 2.0793$

$\ln 9 = \ln 3^2 = 2 \ln 3 \approx 2(1.0986) = 2.1972$

$\ln 10 = \ln(5 \cdot 2) = \ln 5 + \ln 2 \approx 1.6094 + 0.6931 = 2.3025$

$\ln 12 = \ln(2^2 \cdot 3) = \ln 2^2 + \ln 3 = 2 \ln 2 + \ln 3 \approx 2(0.6931) + 1.0986 = 2.4848$

$\ln 15 = \ln(5 \cdot 3) = \ln 5 + \ln 3 \approx 1.6094 + 1.0986 = 2.7080$

$\ln 16 = \ln 2^4 = 4 \ln 2 \approx 4(0.6931) = 2.7724$

$\ln 18 = \ln(3^2 \cdot 2) = \ln 3^2 + \ln 2 = 2 \ln 3 + \ln 2 \approx 2(1.0986) + 0.6931 = 2.8903$

$\ln 20 = \ln(5 \cdot 2^2) = \ln 5 + \ln 2^2 = \ln 5 + 2 \ln 2 \approx 1.6094 + 2(0.6931) = 2.9956$

100. $\left(\frac{2x^2}{3y}\right)^{-3} = \left(\frac{3y}{2x^2}\right)^3$

$= \frac{(3y)^3}{(2x^2)^3}$

$= \frac{27y^3}{8x^6}$

102. $xy(x^{-1} + y^{-1})^{-1} = \frac{xy}{x^{-1} + y^{-1}}$

$= \frac{xy}{\frac{1}{x} + \frac{1}{y}}$

$= \frac{xy}{\frac{y + x}{xy}}$

$= \frac{(xy)^2}{x + y}$

104. $4x^2 - 5x + 1 = 0$

$(4x - 1)(x - 1) = 0$

$4x - 1 = 0 \Rightarrow x = \frac{1}{4}$

$x - 1 = 0 \Rightarrow x = 1$

The zeros are $x = \frac{1}{4}, 1$.

106. $\frac{5}{x - 1} = \frac{2x}{3}$

$5(3) = 2x(x - 1)$

$15 = 2x^2 - 2x$

$0 = 2x^2 - 2x - 15$

$\frac{-(-2) \pm \sqrt{(-2)^2 - 4(2)(-15)}}{2(2)} = x$

$\frac{2 \pm \sqrt{124}}{4} = x$

$\frac{1 \pm \sqrt{31}}{2} = x$

The zeros are $\frac{1 \pm \sqrt{31}}{2}$.

Section 5.4 Exponential and Logarithmic Equations

Solutions to Even-Numbered Exercises

2. $2^{3x+1} = 32$

(a) $x = -1$

$2^{3(-1)+1} = 2^{-2} = \frac{1}{4}$

No, $x = -1$ is not a solution.

(b) $x = 2$

$2^{3(2)+1} = 2^7 = 128$

No, $x = 2$ is not a solution.

4. $5^{2x+3} = 812$

(a) $x = -\frac{3}{2} + \log_5\sqrt{812}$

$5^{2(-3/2 + \log_5\sqrt{812}) + 3} = 5^{-3 + 2\log_5\sqrt{812} + 3}$

$= 5^{2\log_5 812^{1/2}}$

$= 5^{\log_5 (812^{1/2})^2}$

$= 5^{\log_5 812} = 812$

Yes, $x = -1.5 + \log_5\sqrt{812}$ is a solution.

(b) $x \approx 0.581$

$5^{2(0.581)+3} = 5^{4.162} \approx 811.2$

Yes, $x \approx 0.581$ is an approximate solution.

(c) $x = -1.5 + \frac{\ln 812}{\ln 5}$

$5^{2[-1.5 + (\ln 812/\ln 5)] + 3} = 5^{-3 + 2(\ln 812/\ln 5) + 3}$

$= 5^{2(\ln 812/\ln 5)}$

$= 5^{2\log_5 812}$

$= 5^{\log_5 812^2}$

$= 812^2 = 659{,}344$

No, $x = -1.5 + \frac{\ln 812}{\ln 5}$ is not a solution.

6. $\ln(x - 1) = 3.8$

(a) $x = 1 + e^{3.8}$

$\ln(1 + e^{3.8} - 1) = \ln e^{3.8} = 3.8$

Yes, $x = 1 + e^{3.8}$ is a solution.

(b) $x \approx 45.701$

$\ln(45.701 - 1) = \ln(44.701) \approx 3.8$

Yes, $x \approx 45.701$ is an approximate solution.

(c) $x = 1 + \ln 3.8$

$\ln(1 + \ln 3.8 - 1) = \ln(\ln 3.8) \approx 0.289$

No, $x = 1 + \ln 3.8$ is not a solution.

8. $3^x = 243$

$3^x = 3^5$

$x = 5$

10. $3^x = 729$

$3^x = 3^6$

$x = 6$

12. $8^x = 4$

$8^x = 8^{2/3}$

$x = \frac{2}{3}$

14. $\left(\frac{1}{4}\right)^x = 64$

$4^{-x} = 4^3$

$-x = 3$

$x = -3$

16. $\left(\frac{2}{3}\right)^x = \frac{4}{9}$

$\left(\frac{2}{3}\right)^x = \left(\frac{2}{3}\right)^2$

$x = 2$

18. $\ln x - \ln 5 = 0$

$\ln x = \ln 5$

$x = 5$

20. $e^x = 4$

$\ln e^x = \ln 4$

$x = \ln 4$

$x \approx 1.386$

22. $\ln x = -7$

$e^{\ln x} = e^{-7}$

$x = e^{-7}$

$x \approx 0.000912$

24. $\log_5 x = -3$

$x = 5^{-3}$

$x = \frac{1}{125}$, or 0.008

26. $\log_{10} x - 2 = 0$

$\log_{10} x = 2$

$10^{\log_{10} x} = 10^2$

$x = 10^2$

$x = 100$

28. $f(x) = g(x)$

$27^x = 9$

$27^x = 27^{2/3}$

$x = \frac{2}{3}$

Point of intersection: $\left(\frac{2}{3}, 9\right)$

30. $f(x) = g(x)$

$\ln(x - 4) = 0$

$e^{\ln(x-4)} = e^0$

$x - 4 = 1$

$x = 5$

Point of intersection: $(5, 0)$

32. $2(5^x) = 32$

$5^x = 16$

$x = \log_5 16$

$x = \frac{\ln 16}{\ln 5}$

$x \approx 1.723$

34. $4e^x = 91$

$e^x = \frac{91}{4}$

$\ln e^x = \ln \frac{91}{4}$

$x = \ln \frac{91}{4} \approx 3.125$

36. $6^x + 10 = 47$

$6^x = 37$

$x = \log_6 37$

$x = \frac{\ln 37}{\ln 6}$

$x \approx 2.015$

38. $6^{5x} = 3000$

$\ln 6^{5x} = \ln 3000$

$(5x) \ln 6 = \ln 3000$

$5x = \frac{\ln 3000}{\ln 6}$

$x = \frac{\ln 3000}{5 \ln 6}$

≈ 0.894

40. $4^{-3t} = 0.10$

$\ln 4^{-3t} = \ln 0.10$

$(-3t) \ln 4 = \ln 0.10$

$-3t = \frac{\ln 0.10}{\ln 4}$

$t = -\frac{\ln 0.10}{3 \ln 4} \approx 0.554$

42. $2^{x-3} = 32$

$x - 3 = \log_2 32$

$x - 3 = 5$

$x = 8$

44.
$$\begin{aligned}
8^{-2-x} &= 431\\
\ln 8^{-2-x} &= \ln 431\\
(-2-x)\ln 8 &= \ln 431\\
-2\ln 8 - x\ln 8 &= \ln 431\\
-x\ln 8 &= \ln 431 + \ln 8^2\\
x\ln 8 &= -\ln 431 - \ln 64\\
x &= \frac{-\ln 431 - \ln 64}{\ln 8} \approx -4.917
\end{aligned}$$

46.
$$\begin{aligned}
5(10^{x-6}) &= 7\\
10^{x-6} &= \tfrac{7}{5}\\
\log_{10} 10^{x-6} &= \log_{10} \tfrac{7}{5}\\
x-6 &= \log_{10} \tfrac{7}{5}\\
x &= 6 + \log_{10} \tfrac{7}{5} \approx 6.146
\end{aligned}$$

48.
$$\begin{aligned}
8(3^{6-x}) &= 40\\
3^{6-x} &= 5\\
\ln 3^{6-x} &= \ln 5\\
(6-x)\ln 3 &= \ln 5\\
6-x &= \frac{\ln 5}{\ln 3}\\
-x &= \frac{\ln 5}{\ln 3} - 6\\
x &= 6 - \frac{\ln 5}{\ln 3}\\
&\approx 4.535
\end{aligned}$$

50.
$$\begin{aligned}
e^{2x} &= 50\\
\ln e^{2x} &= \ln 50\\
2x &= \ln 50\\
x &= \frac{\ln 50}{2} \approx 1.956
\end{aligned}$$

52.
$$\begin{aligned}
1000e^{-4x} &= 75\\
e^{-4x} &= \frac{3}{40}\\
\ln e^{-4x} &= \ln\frac{3}{40}\\
-4x &= \ln\frac{3}{40}\\
x &= -\frac{1}{4}\ln\frac{3}{40} \approx 0.648
\end{aligned}$$

54.
$$\begin{aligned}
-14 + 3e^x &= 11\\
3e^x &= 25\\
e^x &= \frac{25}{3}\\
\ln e^x &= \ln\frac{25}{3}\\
x &= \ln\frac{25}{3} \approx 2.120
\end{aligned}$$

56.
$$\begin{aligned}
8(4^{6-2x}) + 13 &= 41\\
8(4^{6-2x}) &= 28\\
4^{6-2x} &= 3.5\\
6-2x &= \log_4 3.5\\
6-2x &= \frac{\ln 3.5}{\ln 4}\\
-2x &= 6 + \frac{\ln 3.5}{\ln 4}\\
x &\doteq 3 - \frac{\ln 3.5}{2\ln 4}\\
x &\approx 2.548
\end{aligned}$$

58.
$$\begin{aligned}
e^{2x} - 5e^x + 6 &= 0\\
(e^x - 2)(e^x - 3) &= 0
\end{aligned}$$

$e^x = 2$ or $e^x = 3$

$x = \ln 2 \approx 0.693$ or $x = \ln 3 \approx 1.099$

60.
$$\begin{aligned}
e^{2x} + 9e^x + 36 &= 0\\
(e^x)^2 + 9e^x + 36 &= 0
\end{aligned}$$

Because the discriminant is $9^2 - 4(1)(36) = -63$, there is no solution.

62. $\dfrac{400}{1+e^{-x}} = 350$

$$400 = 350(1+e^{-x})$$
$$\frac{8}{7} = 1 + e^{-x}$$
$$\frac{8}{7} - 1 = e^{-x}$$
$$\frac{1}{7} = e^{-x}$$
$$\ln\frac{1}{7} = \ln e^{-x}$$
$$-x = \ln\frac{1}{7}$$
$$-x = \ln 7^{-1}$$
$$-x = -\ln 7$$
$$x = \ln 7 \approx 1.946$$

64. $\dfrac{119}{e^{6x} - 14} = 7$

$$119 = 7(e^{6x} - 14)$$
$$17 = e^{6x} - 14$$
$$31 = e^{6x}$$
$$\ln 31 = \ln e^{6x}$$
$$\ln 31 = 6x$$
$$x = \frac{\ln 31}{6} \approx 0.572$$

66. $\left(4 - \dfrac{2.471}{40}\right)^{9t} = 21$

$$3.938225^{9t} = 21$$
$$\ln 3.938225^{9t} = \ln 21$$
$$9t \ln 3.938225 = \ln 21$$
$$t = \frac{\ln 21}{9 \ln 3.938225}$$
$$\approx 0.247$$

68. $\left(16 - \dfrac{0.878}{26}\right)^{3t} = 30$

$$\ln\left(16 - \frac{0.878}{26}\right)^{3t} = \ln 30$$
$$3t \ln\left(16 - \frac{0.878}{26}\right) = \ln 30$$
$$t = \frac{\ln 30}{3\ln\left(16 - \frac{0.878}{26}\right)} \approx 0.409$$

70. $f(x) = -4e^{-x-1} + 15$

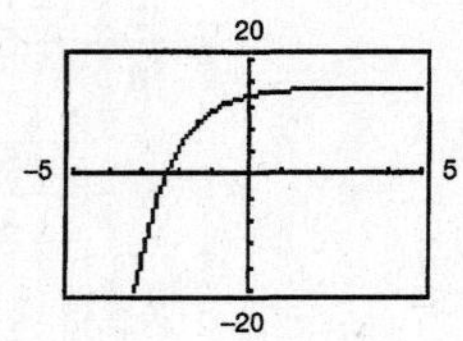

$$0 = -4e^{-x-1} + 15$$
$$-15 = -4e^{-x-1}$$
$$3.75 = e^{-x-1}$$
$$\ln 3.75 = -x - 1$$
$$1 + \ln 3.75 = -x$$
$$-1 - \ln 3.75 = x$$
$$-2.322 \approx x$$

The zero is -2.322.

72. $g(x) = 8e^{-2x/3} - 11$

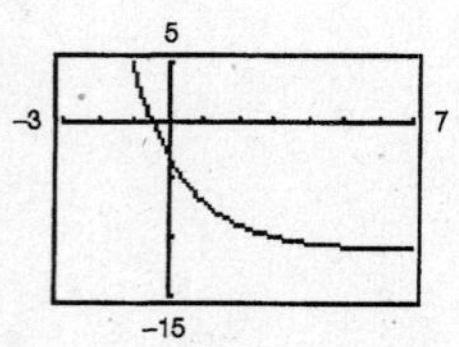

$$8e^{-2x/3} = 11$$
$$e^{-2x/3} = 1.375$$
$$-\frac{2x}{3} = \ln 1.375$$
$$x = -1.5 \ln 1.375$$
$$x \approx -0.478$$

The zero is -0.478.

74. $f(x) = -e^{1.8x} + 7$

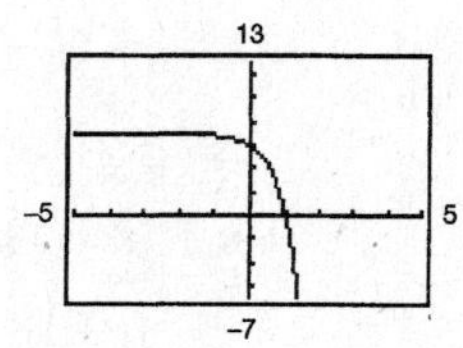

$$-e^{1.8x} + 7 = 0$$
$$-e^{1.8x} = -7$$
$$e^{1.8x} = 7$$
$$1.8x = \ln 7$$
$$x = \frac{\ln 7}{1.8}$$
$$x \approx 1.081$$

The zero is 1.081.

76. $f(x) = e^{2.724x} - 29$

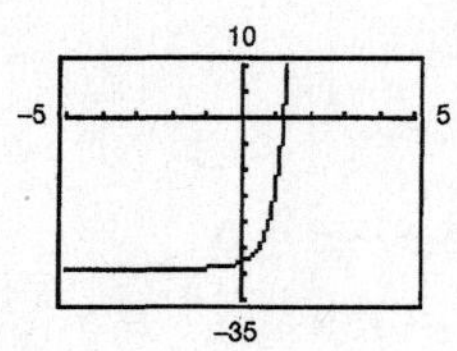

$$e^{2.724x} = 29$$
$$2.724x = \ln 29$$
$$x = \frac{\ln 29}{2.724}$$
$$x \approx 1.236$$

The zero is 1.236.

78. $\ln x = 2$

$$e^{\ln x} = e^2$$

$$x = e^2 \approx 7.389$$

80. $\ln 4x = 1$

$$e^{\ln 4x} = e^1$$

$$4x = e$$

$$x = \frac{e}{4} \approx 0.680$$

82. $\log_{10} 3z = 2$

$$10^{\log_{10} 3z} = 10^2$$

$$3z = 100$$

$$z = \frac{100}{3} \approx 33.333$$

84. $5 \log_{10}(x - 2) = 11$

$$\log_{10}(x - 2) = \frac{11}{5}$$

$$10^{\log_{10}(x-2)} = 10^{11/5}$$

$$x - 2 = 10^{11/5}$$

$$x = 10^{11/5} + 2 \approx 160.489$$

86. $2 \ln x = 7$

$$\ln x = \frac{7}{2}$$

$$e^{\ln x} = e^{7/2}$$

$$x = e^{7/2} \approx 33.115$$

88. $\ln\sqrt{x - 8} = 5$

$$e^{\ln\sqrt{x-8}} = e^5$$

$$\sqrt{x - 8} = e^5$$

$$x - 8 = e^{10}$$

$$x = e^{10} + 8 \approx 22{,}034.466$$

90. $2 - 6 \ln x = 10$

$$-6 \ln x = 8$$

$$\ln x = -\frac{4}{3}$$

$$e^{\ln x} = e^{-4/3}$$

$$x = e^{-4/3}$$

$$x \approx 0.264$$

92. $\ln x + \ln(x + 1) = 1$

$$\ln[x(x + 1)] = 1$$

$$e^{\ln[x(x+1)]} = e^1$$

$$x(x + 1) = e^1$$

$$x^2 + x - e = 0$$

$$x = \frac{-1 \pm \sqrt{1 + 4e}}{2}$$

The only solution is $x = \dfrac{-1 \pm \sqrt{1 + 4e}}{2} \approx 1.223.$

94. $\ln x + \ln(x + 3) = 1$

$$\ln[x(x + 3)] = 1$$

$$e^{\ln[x(x+3)]} = e^1$$

$$x(x + 3) = e^1$$

$$x^2 + 3x - e = 0$$

$$x = \frac{-3 \pm \sqrt{9 + 4e}}{2}$$

The only solution is $x = \dfrac{-3 + \sqrt{9 + 4e}}{2} \approx 0.729.$

96. $\ln(x + 1) - \ln(x - 2) = \ln x$

$$\ln\left(\frac{x + 1}{x - 2}\right) = \ln x$$

$$\frac{x + 1}{x - 2} = x$$

$$x + 1 = x^2 - 2x$$

$$0 = x^2 - 3x - 1$$

$$\frac{-(-3) \pm \sqrt{(-3)^2 - 4(1)(-1)}}{2(1)} = x$$

$$\frac{3 \pm \sqrt{13}}{2} = x$$

$$3.303 \approx x$$

(The negative apparent solution is extraneous.)

98. $\log_{10}(x - 6) = \log_{10}(2x + 1)$

$$x - 6 = 2x + 1$$

$$-7 = x$$

The apparent solution $x = -7$ is extraneous, because the domain of the logarithm function is positive numbers, and $-7 - 6$ and $2(-7) + 1$ are negative. There is no solution.

100. $\log_2 x + \log_2(x + 2) = \log_2(x + 6)$

$$\log_2[x(x + 2)] = \log_2(x + 6)$$
$$x(x + 2) = x + 6$$
$$x^2 + x - 6 = 0$$
$$(x + 3)(x - 2) = 0$$

$x = -3$ or $x = 2$

The value $x = -3$ is extraneous.

The only solution is $x = 2$.

102. $\log_3 x + \log_3(x - 8) = 2$

$$\log_3[x(x - 8)] = 2$$
$$3^{\log_3(x^2 - 8x)} = 3^2$$
$$x^2 - 8x = 9$$
$$x^2 - 8x - 9 = 0$$
$$(x - 9)(x + 1) = 0$$

$x = 9$ or $x = -1$

The value $x = -1$ is extraneous.

The only solution is $x = 9$.

104. $\log_{10} 4x - \log_{10}(12 + \sqrt{x}) = 2$

$$\log_{10}\left(\frac{4x}{12 + \sqrt{x}}\right) = 2$$
$$10^{\log_{10}(4x/(12 + \sqrt{x}))} = 10^2$$
$$\frac{4x}{12 + \sqrt{x}} = 100$$
$$4x = 100(12 + \sqrt{x})$$
$$4x = 1200 + 100\sqrt{x}$$
$$4x - 1200 = 100\sqrt{x}$$
$$x - 300 = 25\sqrt{x}$$
$$(x - 300)^2 = (25\sqrt{x})^2$$
$$x^2 - 600x + 90{,}000 = 625x$$
$$x^2 - 1225x + 90{,}000 = 0$$
$$x = \frac{1225 \pm \sqrt{(-1225)^2 - 4(1)(90{,}000)}}{2}$$
$$x = \frac{1225 \pm \sqrt{1{,}140{,}625}}{2}$$
$$x = \frac{1225 \pm 125\sqrt{73}}{2}$$

$x \approx 78.500$ (extraneous) or $x \approx 1146.500$

The only solution is $x = \dfrac{1225 + 125\sqrt{73}}{2} \approx 1146.500.$

106.

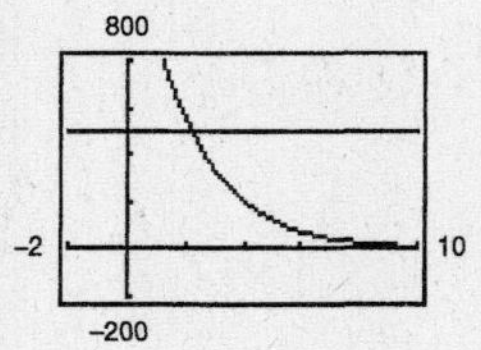

$$500 = 1500e^{-x/2}$$
$$\frac{1}{3} = e^{-x/2}$$
$$\ln\frac{1}{3} = -\frac{x}{2}$$
$$-2\ln\frac{1}{3} = x$$
$$2.197 \approx x$$

The solution is $x \approx 2.197$.

108.

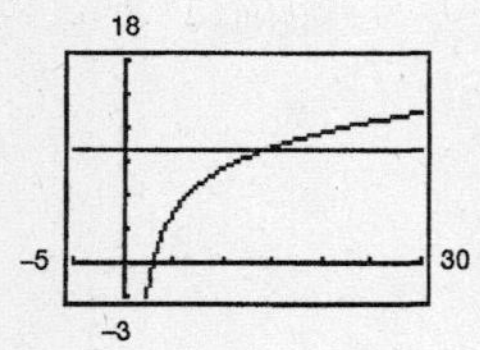

$$10 - 4\ln(x-2) = 0$$
$$-4\ln(x-2) = -10$$
$$\ln(x-2) = 2.5$$
$$e^{\ln(x-2)} = e^{2.5}$$
$$x - 2 = e^{2.5}$$
$$x = e^{2.5} + 2$$
$$x \approx 14.182$$

The solution is $x \approx 14.182$.

110.

$$r = 0.12$$
$$A = Pe^{rt}$$
$$2000 = 1000e^{0.12t}$$
$$2 = e^{0.12t}$$
$$\ln 2 = \ln e^{0.12t}$$
$$\ln 2 = 0.12t$$
$$\frac{\ln 2}{0.12} = t$$
$$t \approx 5.8 \text{ years}$$

112.

$$r = 0.12$$
$$A = Pe^{rt}$$
$$3000 = 1000e^{0.12t}$$
$$3 = e^{0.12t}$$
$$\ln 3 = \ln e^{0.12t}$$
$$\ln 3 = 0.12t$$
$$\frac{\ln 3}{0.12} = t$$
$$t = 9.2 \text{ years}$$

114. $p = 5000\left(1 - \dfrac{4}{4 + e^{-0.002x}}\right)$

(a) When $p = \$600$:

$$600 = 5000\left(1 - \frac{4}{4 + e^{-0.002x}}\right)$$
$$0.12 = 1 - \frac{4}{4 + e^{-0.002x}}$$
$$\frac{4}{4 + e^{-0.002x}} = 0.88$$
$$4 = 3.52 + 0.88e^{-0.002x}$$
$$0.48 = 0.88e^{-0.002x}$$
$$\frac{6}{11} = e^{-0.002x}$$
$$\ln\frac{6}{11} = \ln e^{-0.002x}$$
$$\ln\frac{6}{11} = -0.002x$$
$$x = -\frac{\ln(6/11)}{0.002} \approx 303 \text{ units}$$

(b) When $p = \$400$:

$$400 = 5000\left(1 - \frac{4}{4 + e^{-0.002x}}\right)$$
$$0.08 = 1 - \frac{4}{4 + e^{-0.002x}}$$
$$\frac{4}{4 + e^{-0.002x}} = 0.92$$
$$4 = 3.68 + 0.92e^{-0.002x}$$
$$0.32 = 0.92e^{-0.002x}$$
$$\frac{8}{23} = e^{-0.002x}$$
$$\ln\frac{8}{23} = \ln e^{-0.002x}$$
$$\ln\frac{8}{23} = -0.002x$$
$$x = -\frac{\ln(8/23)}{0.002} \approx 528 \text{ units}$$

116. $N = 68(10^{-0.04x})$

When $N = 21$:

$$21 = 68(10^{-0.04x})$$

$$\frac{21}{68} = 10^{-0.04x}$$

$$\log_{10}\frac{21}{68} = -0.04x$$

$$x = -\frac{\log_{10}(21/68)}{0.04} \approx 12.76 \text{ inches}$$

118. $P = \dfrac{0.83}{1 + e^{-0.2n}}$

(a)

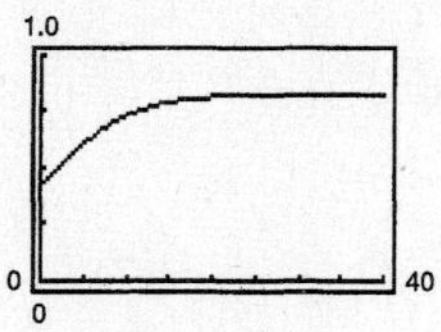

(b) Horizontal asymptotes: $P = 0, P = 0.83$
The upper asymptote, $P = 0.83$, indicates that the proportion of correct responses will approach 0.83 as the number of trials increases.

(c) When $P = 60\%$ or $P = 0.60$:

$$0.60 = \frac{0.83}{1 + e^{-0.2n}}$$

$$1 + e^{-0.2n} = \frac{0.83}{0.60}$$

$$e^{-0.2n} = \frac{0.83}{0.60} - 1$$

$$\ln e^{-0.2n} = \ln\left(\frac{0.83}{0.60} - 1\right)$$

$$-0.2n = \ln\left(\frac{0.83}{0.60} - 1\right)$$

$$n = -\frac{\ln\left(\frac{0.83}{0.60} - 1\right)}{0.2} \approx 5 \text{ trials}$$

120. $T = 20[1 + 7(2^{-h})]$

(a) From the graph in the textbook we see a horizontal asymptote at $T = 20$.
This represents the room temperature.

(b)

$$100 = 20[1 + 7(2^{-h})]$$

$$5 = 1 + 7(2^{-h})$$

$$4 = 7(2^{-h})$$

$$\frac{4}{7} = 2^{-h}$$

$$\ln\left(\frac{4}{7}\right) = \ln 2^{-h}$$

$$\ln\left(\frac{4}{7}\right) = -h \ln 2$$

$$\frac{\ln\left(\frac{4}{7}\right)}{-\ln 2} = h$$

$$h \approx 0.81 \text{ hour}$$

122. $\log_a(u + v) = (\log_a u)(\log_a v)$

False.

$2.04 \approx \log_{10}(10 + 100) \neq (\log_{10} 10)(\log_{10} 100) = 2$

124. $\log_a\left(\dfrac{u}{v}\right) = \log_a u - \log_a v$

True by Property 2 in Section 5.3.

126. $A = Pe^{rt}$

(a) $A = (2P)e^{rt} = 2(Pe^{rt})$ This doubles your money.

(b) $A = Pe^{(2r)t} = Pe^{rt}e^{rt} = e^{rt}(Pe^{rt})$

(c) $A = Pe^{r(2t)} = Pe^{rt}e^{rt} = e^{rt}(Pe^{rt})$

Doubling the interest rate yields the same result as doubling the number of years.

If $2 > e^{rt}$ (i.e., $rt < \ln 2$), then doubling your investment would yield the most money. If $rt > \ln 2$, then doubling either the interest rate or the number of years would yield more money.

128. (a) When solving an exponential equation, rewrite the original equation in a form that allows you to use the One-to-One Property $a^x = a^y$ if and only if $x = y$ or rewrite the original equation in logarithmic form and use the Inverse Property $\log_a a^x = x$.

(b) When solving a logarithmic equation, rewrite the original equation in a form that allows you to use the One-to-One Property $\log_a x = \log_a y$ if and only if $x = y$ or rewrite the original equation in exponential form and use the Inverse Property $a^{\log_a x} = x$.

130. $\sqrt{32} - 2\sqrt{25} = \sqrt{16 \cdot 2} - 2(5)$

$= 4\sqrt{2} - 10$

132.
$$\frac{3}{\sqrt{10} - 2} = \frac{3}{\sqrt{10} - 2} \cdot \frac{\sqrt{10} + 2}{\sqrt{10} + 2}$$
$$= \frac{3(\sqrt{10} + 2)}{10 - 4}$$
$$= \frac{3(\sqrt{10} + 2)}{6}$$
$$= \frac{\sqrt{10} + 2}{2}$$
$$= \frac{1}{2}\sqrt{10} + 1$$

134. $t = \dfrac{k}{s^3}$

136. $x = \dfrac{k}{b - 3}$

138. $\log_3 4 = \dfrac{\log_{10} 4}{\log_{10} 3} = \dfrac{\ln 4}{\ln 3} \approx 1.262$

140. $\log_8 22 = \dfrac{\log_{10} 22}{\log_{10} 8} = \dfrac{\ln 22}{\ln 8} \approx 1.486$

Section 5.5 Exponential and Logarithmic Models

Solutions to Even-Numbered Exercises

2. $y = 6e^{-x/4}$

This is an exponential decay model. Matches graph (e).

4. $y = 3e^{-(x-2)^2/5}$

This is a Gaussian model. Matches graph (a).

6. $y = \dfrac{4}{1 + e^{-2x}}$

This is a logistic growth model. Matches graph (f).

8. Since $A = 20{,}000e^{0.105t}$, the time to double is given by $40{,}000 = 20{,}000e^{0.105t}$, and we have

$$40{,}000 = 20{,}000e^{0.105t}$$
$$2 = e^{0.105t}$$
$$\ln 2 = \ln e^{0.105t}$$
$$\ln 2 = 0.105t$$
$$t = \frac{\ln 2}{0.105} \approx 6.60 \text{ years.}$$

Amount after 10 years: $A = 20{,}000e^{0.105(10)} \approx \$57{,}153.02$

10. Since $A = 10{,}000e^{rt}$ and $A = 20{,}000$ when $t = 12$, we have

$$20{,}000 = 10{,}000e^{12r}$$
$$2 = e^{12r}$$
$$\ln 2 = \ln e^{12r}$$
$$\ln 2 = 12r$$
$$r = \frac{\ln 2}{12} \approx 0.057762 = 5.7762\%.$$

Amount after 10 years:

$A = 10{,}000e^{0.057762(10)} \approx \$17{,}817.97$

12. Since $A = 600e^{rt}$ and $A = 19{,}205$ when $t = 10$, we have

$$19{,}205 = 600e^{10r}$$

$$\frac{19{,}205}{600} = e^{10r}$$

$$\ln\left(\frac{19{,}205}{600}\right) = \ln e^{10r}$$

$$\ln\left(\frac{19{,}205}{600}\right) = 10r$$

$$r = \frac{\ln(19{,}205/600)}{10} \approx 0.3466 \text{ or } 34.66\%.$$

The time to double is given by

$$1200 = 600e^{0.3466t}$$

$$t = \frac{\ln 2}{0.3466} \approx 2 \text{ years.}$$

14. Since $A = Pe^{0.08t}$ and $A = 20{,}000$ when $t = 10$, we have

$$20{,}000 = Pe^{0.08(10)}$$

$$P = \frac{20{,}000}{e^{0.08(10)}} \approx \$8986.58.$$

The time to double is given by

$$t = \frac{\ln 2}{0.08} \approx 8.66 \text{ years.}$$

16.

$$A = P\left(1 + \frac{r}{n}\right)^{nt}$$

$$500{,}000 = P\left(1 + \frac{0.12}{12}\right)^{12(40)}$$

$$P = \$4214.16$$

18. $P = 1000$, $r = 10.5\% = 0.105$

(a) $n = 1$

$$t = \frac{\ln 2}{\ln(1 + 0.105)} \approx 6.94 \text{ years}$$

(b) $n = 12$

$$t = \frac{\ln 2}{12\ln\left(1 + \frac{0.105}{12}\right)} \approx 6.63 \text{ years}$$

(c) $n = 365$

$$t = \frac{\ln 2}{365\ln\left(1 + \frac{0.105}{365}\right)} \approx 6.602 \text{ years}$$

(d) Compounded continuously

$$t = \frac{\ln 2}{0.105} \approx 6.601 \text{ years}$$

20.

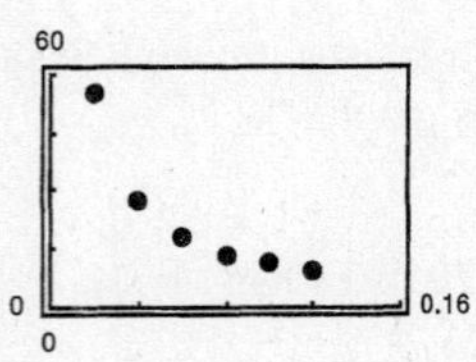

Using the power regression feature of a graphing utility,

$t = 1.099r^{-1}$.

22.

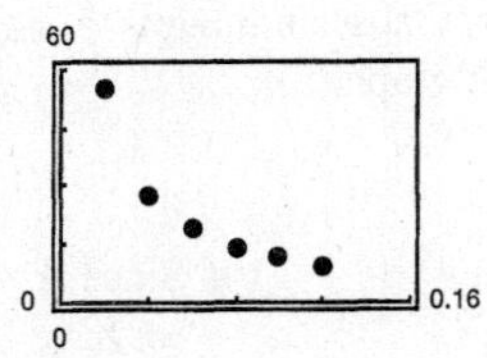

Using the power regression feature of a graphing utility,

$t = 1.222r^{-1}$.

24.

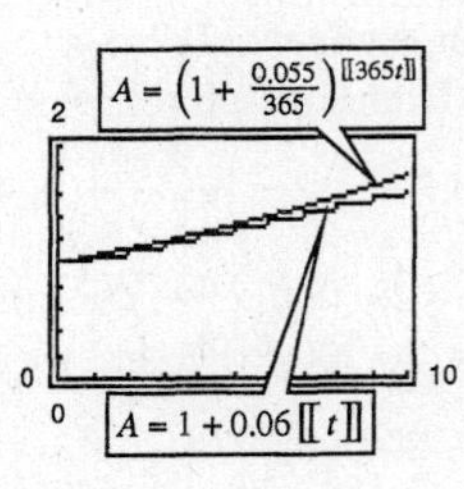

From the graph, $5\frac{1}{2}\%$ compounded daily grows faster than 6% simple interest.

26. $\frac{1}{2}C = Ce^{k(1620)}$

$\frac{1}{2} = e^{k(1620)}$

$\ln \frac{1}{2} = \ln e^{k(1620)}$

$\ln \frac{1}{2} = k(1620)$

$k = \frac{\ln \frac{1}{2}}{1620}$

Given $y = 1.5$ grams after 1000 years, we have

$1.5 = Ce^{[\ln(1/2)/1620](1000)}$

$C \approx 2.30$ grams.

28. $\frac{1}{2}C = Ce^{k(5730)}$

$\frac{1}{2} = e^{k(5730)}$

$\ln \frac{1}{2} = \ln e^{k(5730)}$

$\ln \frac{1}{2} = k(5730)$

$k = \frac{\ln \frac{1}{2}}{5730}$

Given $C = 3$ grams, after 1000 years we have

$y = 3e^{[(\ln 1/2)/5730](1000)}$

$y \approx 2.66$ grams.

30. $\frac{1}{2}C = Ce^{k(24,360)}$

$\frac{1}{2} = e^{k(24,360)}$

$\ln \frac{1}{2} = \ln e^{k(24,360)}$

$\ln \frac{1}{2} = k(24,360)$

$k = \frac{\ln \frac{1}{2}}{24,360}$

Given $y = 0.4$ grams after 1000 years, we have

$0.4 = Ce^{[(\ln 1/2)/24,360](1000)}$

$C \approx 0.41$ grams.

32. $y = ae^{bx}$

$\frac{1}{2} = ae^{b(0)} \Rightarrow a = \frac{1}{2}$

$5 = \frac{1}{2}e^{b(4)}$

$10 = e^{4b}$

$\ln 10 = \ln e^{4b}$

$\ln 10 = 4b$

$\frac{\ln 10}{4} = b \Rightarrow b \approx 0.5756$

Thus, $y = \frac{1}{2}e^{0.5756x}$.

34. $y = ae^{bx}$

$1 = ae^{b(0)} \Rightarrow 1 = a$

$\frac{1}{4} = e^{b(3)}$

$\ln\left(\frac{1}{4}\right) = \ln e^{3b}$

$\ln\left(\frac{1}{4}\right) = 3b$

$\frac{\ln\left(\frac{1}{4}\right)}{3} = b \quad \Rightarrow b \approx -0.4621$

Thus, $y = e^{-0.4621x}$.

36.

Country	2000	2010
Canada	31.3	34.3
China	1261.8	1359.1
Italy	57.6	57.4
United Kingdom	59.5	60.6
United States	275.6	300.1

(a) Canada:

$a = 31.3$

$34.3 = 31.3e^{b(10)}$

$\ln \frac{34.3}{31.3} = 10b \Rightarrow b \approx 0.00915$

For 2030, use $t = 30$.

$y = 31.3e^{0.00915(30)} \approx 41.2$ million

—CONTINUED—

36. —CONTINUED—

China:

$$a = 1261.8$$

$$1359.1 = 1261.8e^{b(10)}$$

$$\ln \frac{1359.1}{1261.8} = 10b \Rightarrow b \approx 0.00743$$

For 2030, use $t = 30$.

$$y = 1261.8e^{0.00743(30)} \approx 1576.9 \text{ million}$$

Italy:

$$a = 57.6$$

$$57.4 = 57.6e^{b(10)}$$

$$\ln \frac{57.4}{57.6} = 10b \Rightarrow b \approx -0.00035$$

For 2030, use $t = 30$.

$$y = 57.6e^{-0.00035(30)} \approx 57.0 \text{ million}$$

United Kingdom:

$$a = 59.5$$

$$60.6 = 59.5e^{b(10)}$$

$$\ln \frac{60.6}{59.5} = 10b \Rightarrow b \approx 0.00183$$

For 2030, use $t = 30$.

$$y = 59.5e^{0.00183(30)} \approx 62.9 \text{ million}$$

United States:

$$a = 275.6$$

$$300.1 = 275.6e^{b(10)}$$

$$\ln \frac{300.1}{275.6} = 10b \Rightarrow b \approx 0.00852$$

For 2030, use $t = 30$.

$$y = 275.6e^{0.00852(30)} \approx 355.9 \text{ million}$$

(b) The constant b determines the growth rates. The greater the rate of growth, the greater the value of b.

(c) The constant b determines whether the population is increasing ($b > 0$) or decreasing ($b < 0$).

38. $P = 224e^{kt}$

$$172 = 224e^{k(-20)}$$

$$\ln \frac{172}{224} = -20k \Rightarrow k \approx 0.0132$$

For 2020, use $t = 20$.

$$y = 224e^{0.0132(20)} \approx 292{,}000 \text{ million}$$

40. $N = 250e^{kt}$

$$280 = 250e^{k(10)}$$

$$1.12 = e^{10k}$$

$$k = \frac{\ln 1.12}{10}$$

$$N = 250e^{[(\ln 1.12)/10]t}$$

$$500 = 250e^{[(\ln 1.12)/10]t}$$

$$2 = e^{[(\ln 1.12)/10]t}$$

$$\ln 2 = [(\ln 1.12)/10]t$$

$$t = \frac{\ln 2}{(\ln 1.12)/10} \approx 61.16 \text{ hours}$$

42. $y = Ce^{kt}$

$$\frac{1}{2}C = Ce^{5730k}$$

$$\ln \frac{1}{2} = 5730k$$

$$k = \frac{\ln (1/2)}{5730}$$

The ancient charcoal has only 15% as much radioactive carbon.

$$0.15C = Ce^{[(\ln 0.5)/5730]t}$$

$$\ln 0.15 = \frac{\ln 0.5}{5730}t$$

$$t = \frac{5730 \ln 0.15}{\ln 0.5} \approx 15{,}683 \text{ years}$$

44. (0, 2000), (2, 500)

(a) $m = \dfrac{500 - 2000}{2 - 0} = -750$

$V = -750t + 2000$

(b) $500 = 2000e^{k(2)}$

$\ln\dfrac{1}{4} = 2k \Rightarrow k \approx -0.6931$

$V = 2000e^{-0.6931t}$

(c)

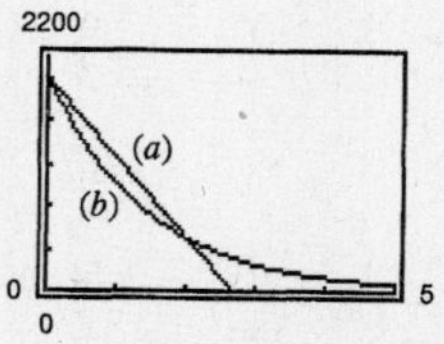

The exponential model depreciates faster in the first 2 years.

(d)

t	1	3
$V = -750t + 2000$	\$1250	−\$250
$V = 2000e^{-0.6931t}$	\$1000	\$250

(e) The slope of the linear model means that the computer depreciates \$750 per year.

46. $S = \dfrac{500{,}000}{1 + 0.6e^{kt}}$

(a)

$$300{,}000 = \frac{500{,}000}{1 + 0.6e^{2k}}$$

$$1 + 0.6e^{2k} = \frac{5}{3}$$

$$0.6e^{2k} = \frac{2}{3}$$

$$e^{2k} = \frac{10}{9}$$

$$2k = \ln\left(\frac{10}{9}\right)$$

$$k = \frac{1}{2}\ln\left(\frac{10}{9}\right) \approx 0.053$$

$$S = \frac{500{,}000}{1 + 0.6e^{0.053t}}$$

(b) When $t = 7$:

$$S = \frac{500{,}000}{1 + 0.6e^{0.053(7)}} \approx 267{,}450 \text{ units}$$

48.

$$y = ae^{bt}$$

$$632{,}000 = 742{,}000e^{b(2)}$$

$$\frac{632}{742} = e^{2b}$$

$$b = \frac{1}{2}\ln\left(\frac{632}{742}\right) \approx -0.0802$$

$$y = 742{,}000e^{-0.0802(5)} \approx \$496{,}880$$

50. $p(t) = \dfrac{1000}{1 + 9e^{-0.1656t}}$

(a) $p(5) = \dfrac{1000}{1 + 9e^{-0.1656(5)}} \approx 203$ animals

(b)

$$500 = \frac{1000}{1 + 9e^{-0.1656t}}$$

$$1 + 9e^{-0.1656t} = 2$$

$$9e^{-0.1656t} = 1$$

$$e^{-0.1656t} = \frac{1}{9}$$

$$t = -\frac{\ln(1/9)}{0.1656} \approx 13 \text{ months}$$

(c)

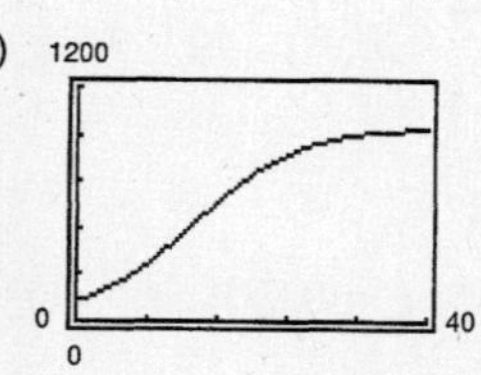

The horizontal asymptotes are $p = 0$ and $p = 1000$. The asymptote with the larger p-value, $p = 1000$, indicates that the population size will approach 1000 as time increases.

52. $R = \log_{10}\frac{I}{I_0} = \log_{10} I$ since $I_0 = 1$.

(a) $8.2 = \log_{10} I$

$10^{8.2} = I \approx 158{,}489{,}319$

(b) $6.7 = \log_{10} I$

$10^{6.7} = I \approx 5{,}011{,}872$

(c) $7.7 = \log_{10} I$

$10^{7.7} = I \approx 50{,}118{,}723$

54. $\beta(I) = 10 \log_{10}\frac{I}{I_0}$ where $I_0 = 10^{-12}$ watt/m^2

(a) $\beta(10^{-9}) = 10 \log_{10}\frac{10^{-9}}{10^{-12}} = 10 \log_{10} 10^3 = 30$ decibels

(b) $\beta(10^{-3.5}) = 10 \log_{10}\frac{10^{-3.5}}{10^{-12}} = 10 \log_{10} 10^{8.5} = 85$ decibels

(c) $\beta(10^{-3}) = 10 \log_{10}\frac{10^{-3}}{10^{-12}} = 10 \log_{10} 10^9 = 90$ decibels

(d) $\beta(10^{-0.5}) = 10 \log_{10}\frac{10^{-0.5}}{10^{-12}} = 10 \log_{10} 10^{11.5} = 115$ decibels

56. $\beta = 10 \log_{10}\frac{I}{I_0}$

$10^{\beta/10} = \frac{I}{I_0}$

$I = I_0\, 10^{\beta/10}$

$\%\text{ decrease} = \frac{I_0\, 10^{8.8} - I_0\, 10^{7.2}}{I_0\, 10^{8.8}} \times 100 \approx 97\%$

58. $\text{pH} = -\log_{10}[\text{H}^+] = -\log_{10}[11.3 \times 10^{-6}] \approx 4.95$

60. $3.2 = -\log_{10}[\text{H}^+]$

$10^{-3.2} = [\text{H}^+]$

$[\text{H}^+] \approx 6.3 \times 10^{-4}$ moles per liter

62. $\text{pH} - 1 = -\log_{10}[\text{H}^+]$

$-(\text{pH} - 1) = \log_{10}[\text{H}^+]$

$10^{-(\text{pH}-1)} = [\text{H}^+]$

$10^{-\text{pH}+1} = [\text{H}^+]$

$10^{-\text{pH}} \cdot 10 = [\text{H}^+]$

The hydrogen ion concentration is increased by a factor of 10.

64. Interest: $u = M - \left(M - \frac{Pr}{12}\right)\left(1 + \frac{r}{12}\right)^{12t}$

Principal: $v = \left(M - \frac{Pr}{12}\right)\left(1 + \frac{r}{12}\right)^{12t}$

(a) $P = 120{,}000$, $t = 35$, $r = 0.075$, $M = 809.39$

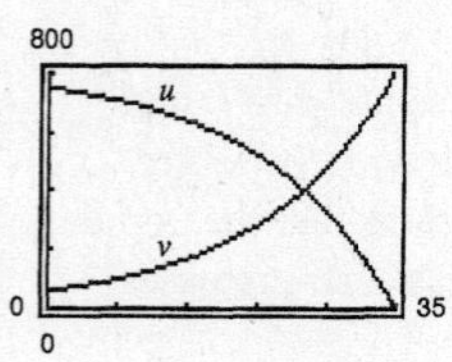

(b) In the early years of the mortgage, the majority of the monthly payment goes toward interest. The principal and interest are nearly equal when $t \approx 26$ years.

(c) $P = 120{,}000$, $t = 20$, $r = 0.075$, $M = 966.71$

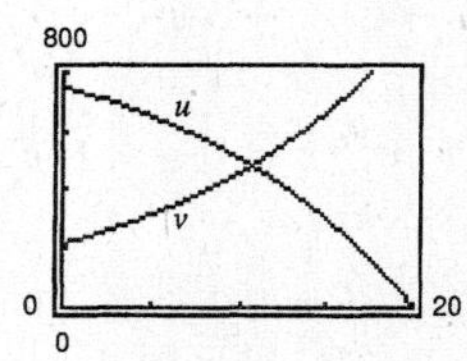

The interest is still the majority of the monthly payment in the early years. Now the principal and interest are nearly equal when $t \approx 10.729 \approx 11$ years.

66. $t_1 = 40.757 + 0.556s - 15.817 \ln s$

$t_2 = 1.2259 + 0.0023s^2$

(a) Linear model: $t_3 = 0.2729s - 6.0143$

Exponential model: $t_4 = 1.5385e^{0.02913s}$ or $t_4 = 1.5385(1.0296)^s$

(b)

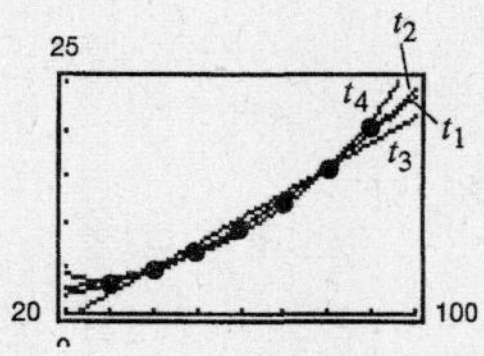

(c)

s	30	40	50	60	70	80	90
t_1	3.6	4.6	6.7	9.4	12.5	15.9	19.6
t_2	3.3	4.9	7.0	9.5	12.5	15.9	19.9
t_3	2.2	4.9	7.6	10.4	13.1	15.8	18.5
t_4	3.7	4.9	6.6	8.9	11.9	15.9	21.2

Note: Table values will vary slightly depending on the model used for t_4.

(d) Model t_1: $S_1 = |3.4 - 3.6| + |5 - 4.6| + |7 - 6.7| + |9.3 - 9.4| + |12 - 12.5| + |15.8 - 15.9| + |20 - 19.6| = 2.0$

Model t_2: $S_2 = |3.4 - 3.3| + |5 - 4.9| + |7 - 7| + |9.3 - 9.5| + |12 - 12.5| + |15.8 - 15.9| + |20 - 19.9| = 1.1$

Model t_3: $S_3 = |3.4 - 2.2| + |5 - 4.9| + |7 - 7.6| + |9.3 - 10.4| + |12 - 13.1| + |15.8 - 15.8| + |20 - 18.5| = 5.6$

Model t_4: $S_4 = |3.4 - 3.7| + |5 - 4.9| + |7 - 6.6| + |9.3 - 8.9| + |12 - 11.9| + |15.8 - 15.9| + |20 - 21.2| = 2.6$

The quadratic model, t_2, best fits the data.

68. False. A logistic growth function never has an x-intercept.

70. True. Powers of e are always positive, so if $a > 0$ a Gaussian model will always be greater than 0, and if $a < 0$ a Gaussian model will always be less than 0.

72. Answers will vary.

74. $y = -4x - 1$

Line

Slope: $m = -4$

y-intercept: $(0, -1)$

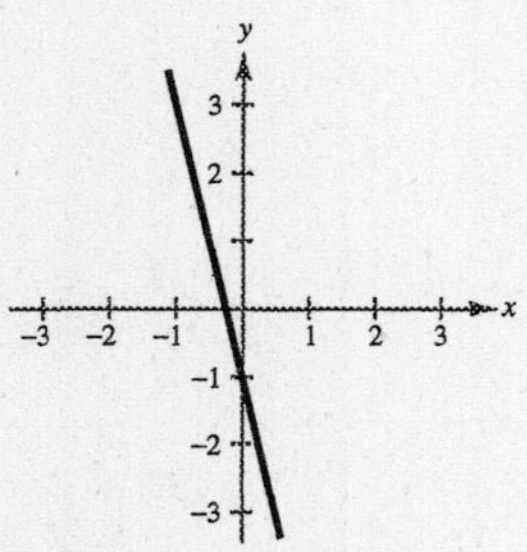

76. $y = 2x^2 - 7x - 30$

$= (2x + 5)(x - 6)$

$= 2\left(x - \frac{7}{4}\right)^2 - \frac{289}{8}$

Parabola

Vertex: $\left(\frac{7}{4}, -\frac{289}{8}\right)$

x-intercepts: $\left(-\frac{5}{2}, 0\right)$, $(6, 0)$

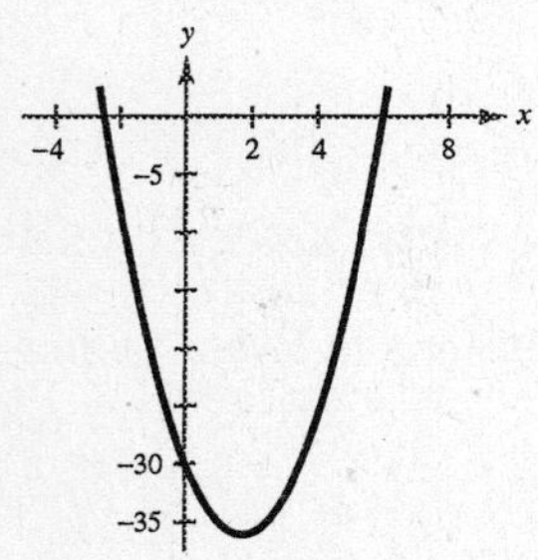

78. $-x^2 - 8y = 0$

$x^2 = -8y$

Parabola

Vertex: $(0, 0)$

Focus: $(0, -2)$

Directrix: $y = 2$

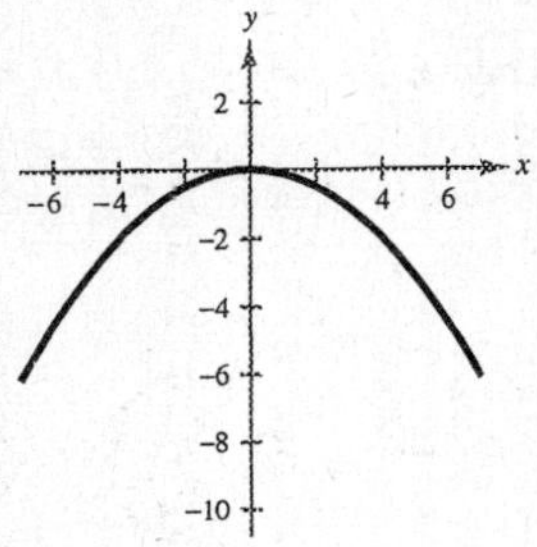

80. $y = \dfrac{x^2}{-x-2} = -x + 2 + \dfrac{4}{-x-2}$

Vertical asymptote: $x = -2$

Slant asymptote: $y = -x + 2$

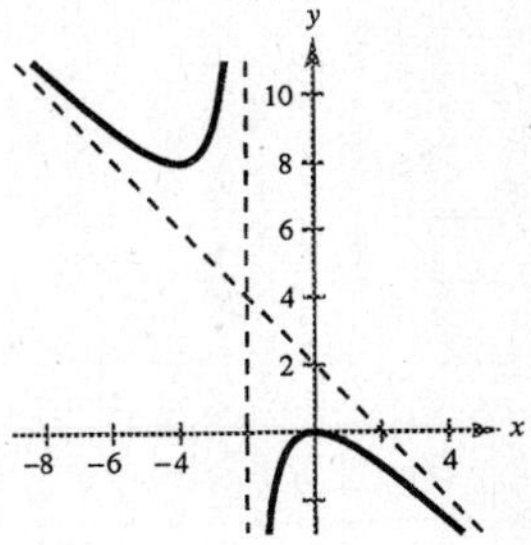

82. $(x - 4)^2 + (y + 7) = 4$

$(x - 4)^2 = -y - 7 + 4$

$(x - 4)^2 = -(y + 3)$

Parabola

Vertex: $(4, -3)$

$P = -\frac{1}{4}$

Focus: $(4, -3.25)$

Directrix: $y = -2.75$

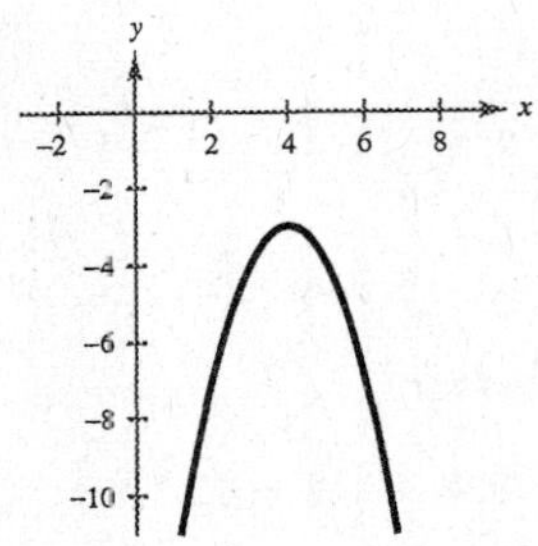

84. $f(x) = -2^{-x-1} - 1$

Horizontal asymptote: $y = -1$

x	-2	-1	0	1	2
$f(x)$	-3	-2	$-\frac{3}{2}$	$-\frac{5}{4}$	$-\frac{9}{8}$

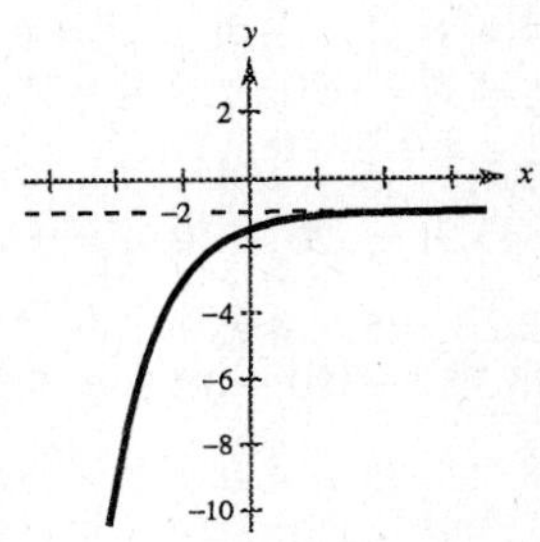

86. $f(x) = -3^x + 4$

Horizontal asymptote: $y = 4$

x	-2	-1	0	1	2
$f(x)$	$3\frac{8}{9}$	$3\frac{2}{3}$	3	1	-5

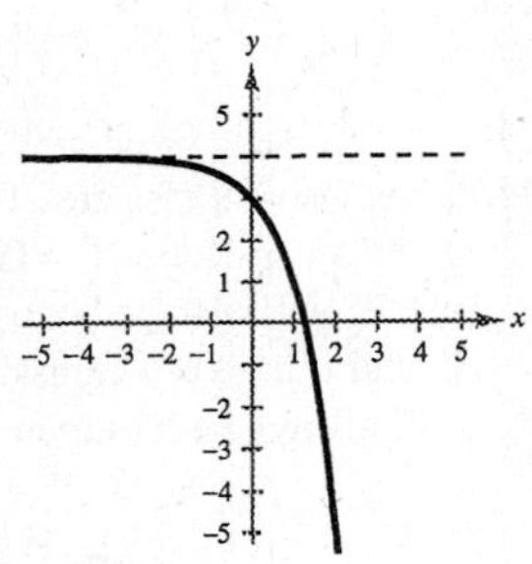

Review Exercises for Chapter 5

Solutions to Even-Numbered Exercises

2. $-14(5^{-0.8}) \approx -3.863$

4. $\sqrt[5]{1278} \approx 4.181$

6. $7^{-\sqrt{11}} \approx 0.002$

8. $f(x) = 4^{-x}$

Intercept: $(0,1)$
Horizontal asymptote: $y = 0$
Decreasing on: $(-\infty, \infty)$
Matches graph (d).

10. $f(x) = 4^x + 1$

Intercept: $(0, 2)$
Horizontal asymptote: $y = 1$
Increasing on: $(-\infty, \infty)$
Matches graph (b).

12. $f(x) = 4^x, (g)x = 4^{x-3}$

Because $g(x) = f(x) - 3$, the graph of g can be obtained by shifting the graph of f 3 units downward.

14. $f(x) = \left(\frac{2}{3}\right)^x, (g)x = 8 - \left(\frac{2}{3}\right)^x$

Because $g(x) = -f(x) + 8$, the graph of g can be obtained by reflecting the graph of f in the x-axis and shifting the graph of f 8 units upward.

16. $f(x) = -4^x - 3$

Horizontal Asymptote: $y = -3$

x	-2	-1	0	1	2
$f(x)$	-3.063	-3.25	-4	-7	-19

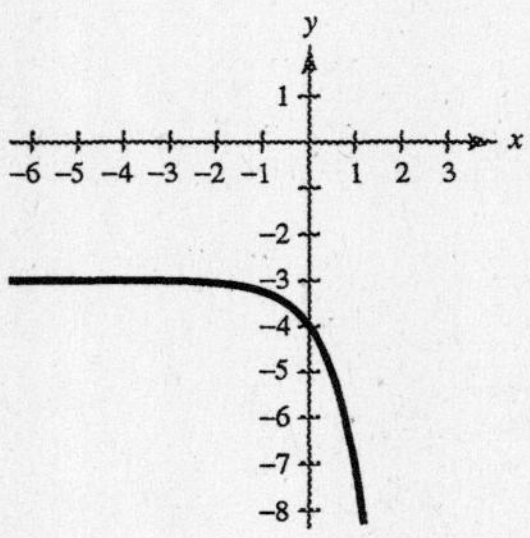

18. $f(x) = 2.65^{x-1}$

Horizontal Asymptote: $y = 0$

x	-3	-1	0	1	3
$f(x)$	0.020	0.142	0.377	1	7.023

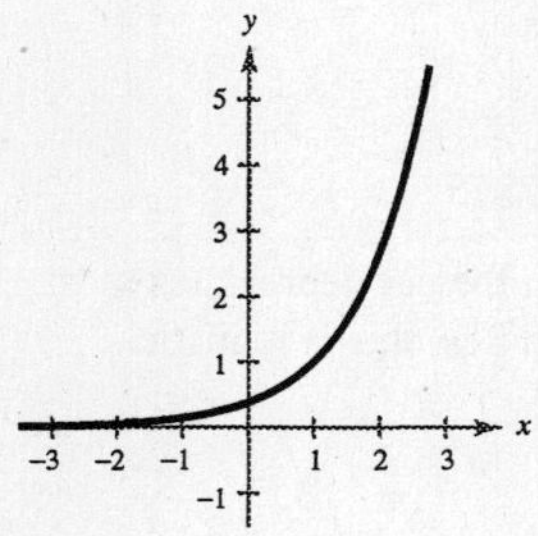

20. $f(x) = 2^{x-6} - 5$

Horizontal Asymptote: $y = -5$

x	0	5	6	7	8	9
$f(x)$	-4.984	-4.5	-4	-3	-1	3

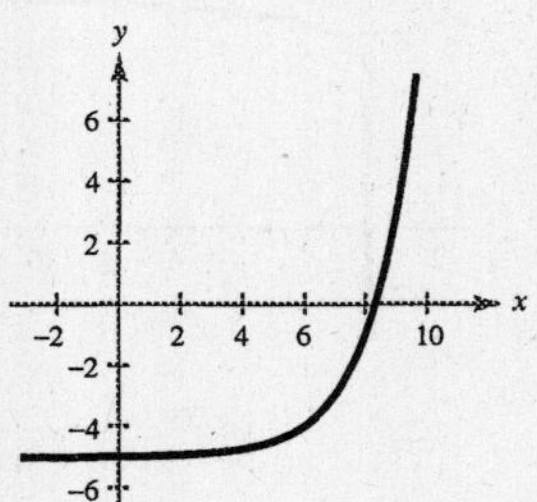

22. $f(x) = \left(\frac{1}{8}\right)^{x+2} - 5$

Horizontal Asymptote: $y = -5$

x	-3	-2	-1	0	2
$f(x)$	3	-4	-4.875	-4.984	-5

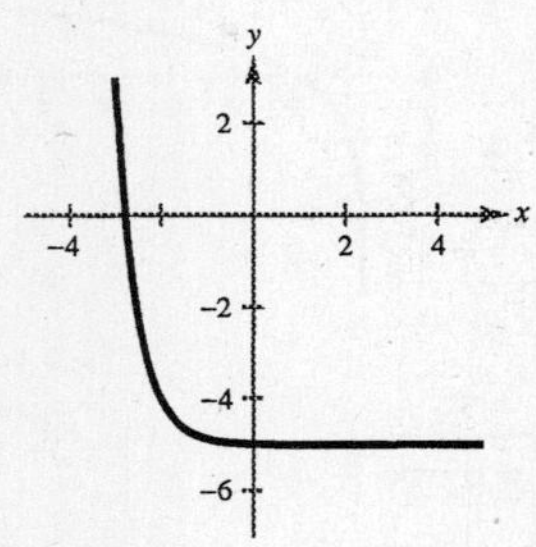

24. $e^{5/8} \approx 1.868$

26. $e^{0.278} = 1.320$

28. $h(x) = 2 - e^{-x/2}$

x	-2	-1	0	1	2
y	-0.72	0.35	1	1.39	1.63

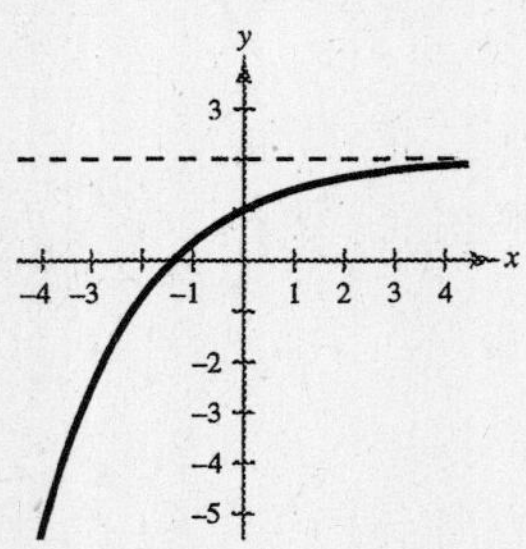

30. $s(t) = 4e^{-2/t}, t > 0$

t	$\frac{1}{2}$	1	2	3	4
y	0.07	0.54	1.47	2.05	2.43

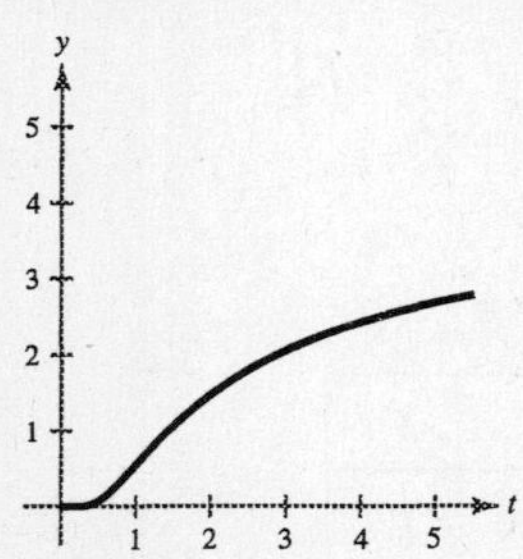

32. $A = 2000\left(1 + \frac{0.05}{n}\right)^{30n}$ or $A = 2000e^{(0.05)(30)}$

n	1	2	4	12	365	Continuous
A	\$8643.88	\$8799.58	\$3880.43	\$8935.49	\$8962.46	\$8963.38

34. $V(t) = 14{,}000\left(\frac{3}{4}\right)^t$

(a)

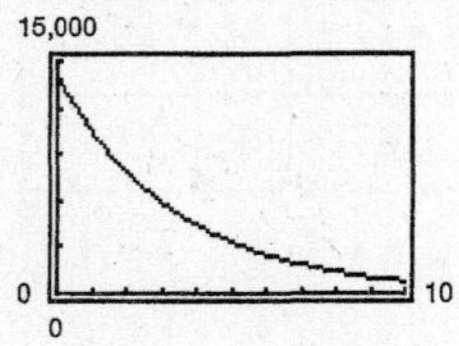

(b) $V(2) = 14{,}000\left(\frac{3}{4}\right)^2 = \$7875$

(c) According to the model, the car depreciates most rapidly at the beginning. Yes, this is realistic.

36. $Q = 100\left(\frac{1}{2}\right)^{t/13}$

(a) for $t = 0$: $Q = 100\left(\frac{1}{2}\right)^{0/13} = 100$ grams

(b) for $t = 10$: $Q = 100\left(\frac{1}{2}\right)^{10/13} \approx 58.7$ grams

(c)

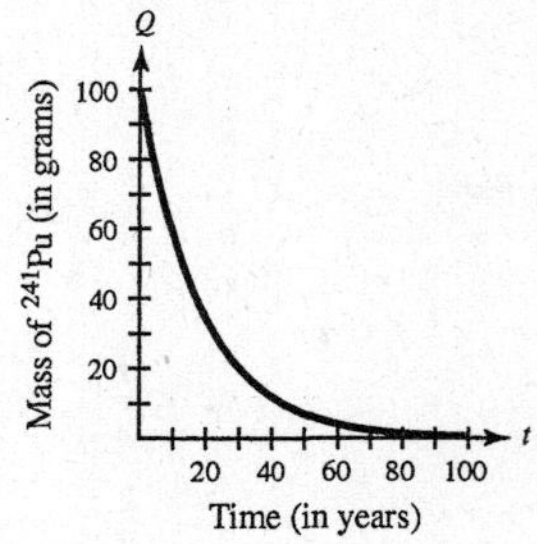

38. $25^{3/2} = 125$

$\log_{25} 125 = \frac{3}{2}$

40. $\log_9 3 = \log_9 9^{1/2} = \frac{1}{2}$

42. $f(x) = \log_4 x = \log_4 \frac{1}{4} = -1$

44. $g(x) = \log_5 x \implies 5^y = x$

Domain: $(0, \infty)$

$\log_5 x = 0$

$x = 5^0$

$x = 1$

x-intercept: $(1, 0)$

Vertical asymptote: $x = 0$

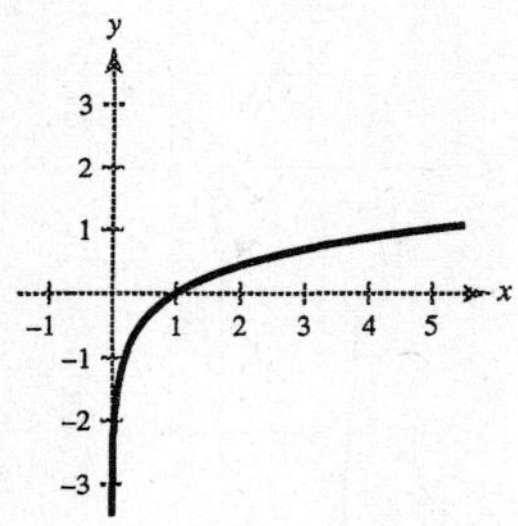

x	$\frac{1}{25}$	$\frac{1}{5}$	1	5	25
$g(x)$	-2	-1	0	1	2

46. $f(x) = 6 + \log_{10} x$

Domain: $(0, \infty)$

$6 + \log_{10} x = 0$

$\log_{10} x = -6$

$x = 10^{-6}$

$x = 0.000001$

x-intercept: $(0.000001, 0)$

Vertical asymptote: $x = 0$

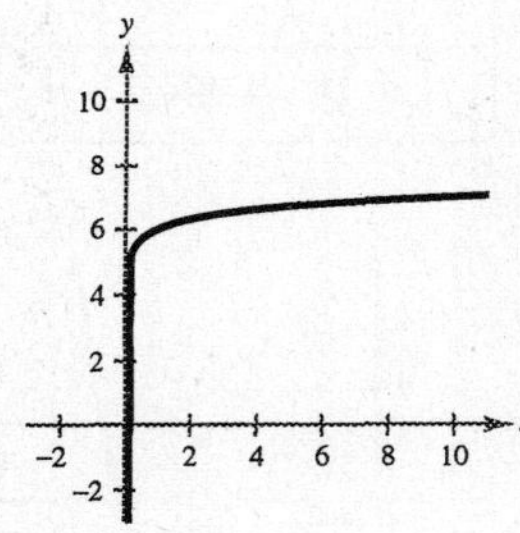

x	1	2	4	6	8	10
$f(x)$	6	6.3	6.6	6.8	6.9	7

48. $f(x) = \log_{10}(x - 3) + 1$

Domain: $(3, \infty)$

$\log_{10}(x - 3) + 1 = 0$

$\log_{10}(x - 3) = -1$

$x - 3 = 10^{-1}$

$x = 3.1$

x-intercept: $(3.1, 0)$

Vertical asymptote: $x = 3$

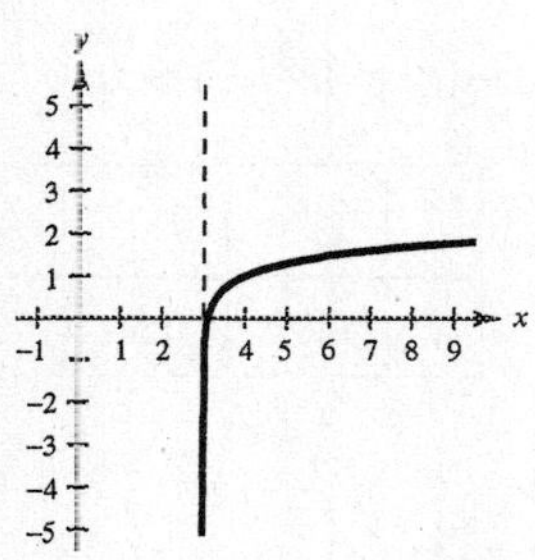

x	4	5	6	7	8
$f(x)$	1	1.3	1.5	1.6	1.7

50. $\ln 0.98 \approx -0.020$

52. $\ln e^7 = 7$

54. $\ln\left(\frac{\sqrt{3}}{8}\right) \approx -1.530$

56. $f(x) = \ln(x - 3)$

Domain: $(3, \infty)$

$\ln(x - 3) = 0$

$x - 3 = e^0$

$x = 4$

x-intercept: $(4, 0)$

Vertical asymptote: $x = 3$

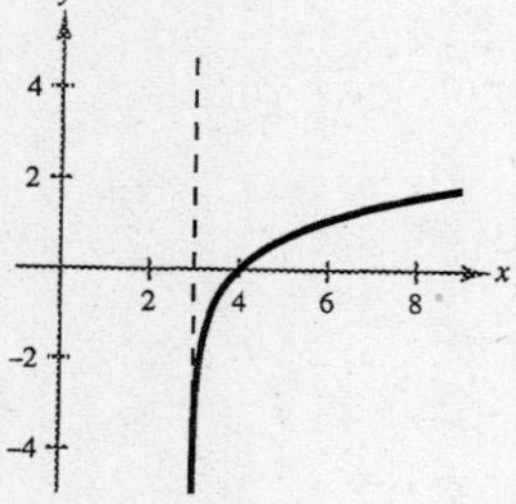

x	3.5	4	4.5	5	5.5
y	-0.69	0	0.41	0.69	0.92

58. $f(x) = \frac{1}{4}\ln x$

Domain: $(0, \infty)$

$\frac{1}{4}\ln x = 0$

$\ln x = 8$

$x = e$

$x = 1$

x-intercept: $(1, 0)$

Vertical asymptote: $x = 0$

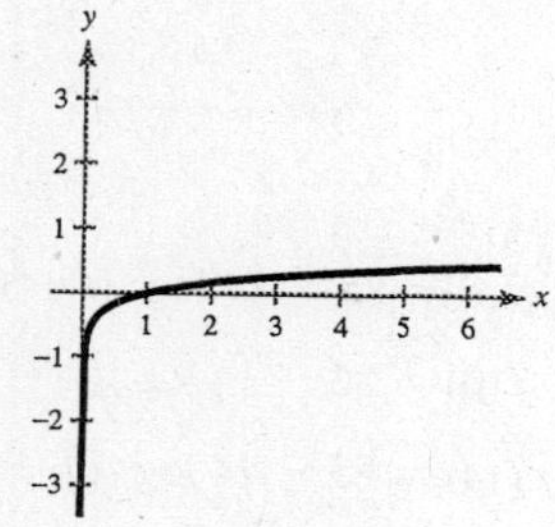

x	$\frac{1}{2}$	1	$\frac{3}{2}$	2	$\frac{5}{2}$	3
y	-0.17	0	0.10	0.17	0.23	0.27

60. $s = 25 - \dfrac{13\ln(10/12)}{\ln 3} \approx 27.16$ miles

62. $\log_{12} 200 = \dfrac{\log_{10} 200}{\log_{10} 12} \approx 2.132$

$\log_{12} 200 = \dfrac{\ln 200}{\ln 12} \approx 2.132$

64. $\log_3 0.28 = \dfrac{\log_{10} 0.28}{\log_{10} 3} \approx -1.159$

$\log_3 0.28 = \dfrac{\ln 0.28}{\ln 3} \approx -1.159$

66. $\log_{10} 7x^4 = \log_{10} 7 + \log_{10} x^4$

$= \log_{10} 7 + 4\log_{10}|x|$

68. $\log_7 \dfrac{\sqrt{x}}{4} = \log_7 \sqrt{x} - \log_7 4$

$= \log_7 x^{1/2} - \log_7 4$

$= \dfrac{1}{2}\log_7 x - \log_7 4$

70. $\ln 3xy^2 = \ln 3 + \ln x + \ln y^2 = \ln 3 + \ln x + 2\ln y$

72. $\ln\left(\dfrac{y-1}{4}\right)^2 = 2\ln\left(\dfrac{y-1}{4}\right)$

$= 2\ln(y - 1) - 2\ln 4$

$= 2\ln(y - 1) - \ln 16, y > 1$

74. $\log_6 y - 2\log_6 z = \log_6 y - \log_6 z^2$

$= \log_6 \dfrac{y}{z^2}$

76. $3\ln x + 2\ln(x + 1) = \ln x^3 + \ln(x + 1)^2$

$= \ln x^3(x + 1)^2$

78. $-2\log_{10} x - 5\log_{10}(x + 6) = \log_{10} x^{-2} - \log_{10}(x + 6)^5$

$= \log_{10} \dfrac{x^{-2}}{(x + 6)^5}$

$= \log_{10} \dfrac{1}{x^2(x + 6)^5}$

80. $$\begin{aligned} 5\ln(x-2) - \ln(x+2) - 3\ln x &= \ln(x-2)^5 - \ln(x+2) - \ln x^3 \\ &= \ln(x-2)^5 - [\ln(x+2) - \ln x^3] \\ &= \ln(x-2)^5 - \ln x^3(x+2) \\ &= \ln \frac{(x-2)^5}{x^3(x+2)} \end{aligned}$$

82. $f(t) = 85 - 14\log_{10}(t+1)$

$f(0) = 85 - 14\log_{10}(0+1) = 85 - 14\log_{10}1 = 85$

$f(4) = 85 - 14\log_{10}(4+1) = 85 - 14\log_{10}5 \approx 75.2$

Over the four-month period, the average score fell by nearly 10 points.

84. $3^x = 729$

$3^x = 3^6$

$x = 6$

86. $5^x = \frac{1}{25}$

$\log_5 5^x = \log_5 \frac{1}{25}$

$x = \log_5 \frac{1}{25}$

$x = -2$

88. $e^x = 6$

$\ln e^x = \ln 6$

$x = \ln 6 \approx 1.792$

90. $\log_6 x = -1$

$6^{\log_6 x} = 6^{-1}$

$6^{\log_6 x} = \frac{1}{6}$

92. $\ln x = -3$

$e^{\ln x} = e^{-3} \approx 0.498$

94. $e^{3x} = 25$

$\ln e^{3x} = \ln 25$

$3x = \ln 25$

$x = \dfrac{\ln 25}{3} \approx 1.073$

96. $14e^{3x+2} = 560$

$e^{3x+2} = 40$

$\ln e^{3x+2} = \ln 40$

$3x + 2 = \ln 40$

$x = \dfrac{(\ln 40) - 2}{3} \approx 0.563$

98. $6^x - 28 = -8$

$6^x = 20$

$\log_6 6^x = \log_6 20$

$x = \log_6 20$

$x = \dfrac{\ln 20}{\ln 6}$

$x \approx 1.672$

100. $2(12^x) = 190$

$12^x = 95$

$\ln 12^x = \ln 95$

$x \ln 12 = \ln 95$

$x = \dfrac{\ln 95}{\ln 12} \approx 1.833$

102. $e^{2x} - 6e^x + 8 = 0$

$(e^x - 2)(e^x - 4) = 0$

$e^x = 2$ or $e^x = 4$

$x = \ln 2$ $\quad$ $x = \ln 4$

$x \approx 0.693$ $\quad$ $x \approx 1.386$

104. $4^{-0.2x} + x = 0$

Graph $y_1 = 4^{-0.2x} + x$.

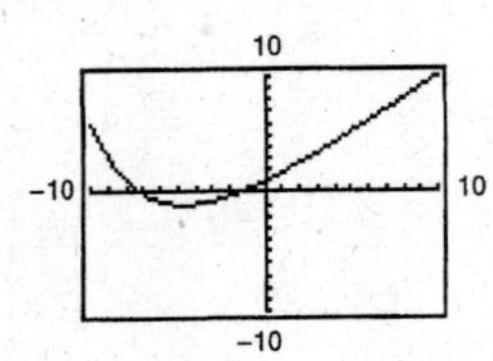

The x-intercepts are at $x \approx -7.04$ and $x \approx -1.53$.

106. $4e^{1.2x} = 9$

Graph $y_1 = 4e^{1.2x}$ and $y_2 = 9$.

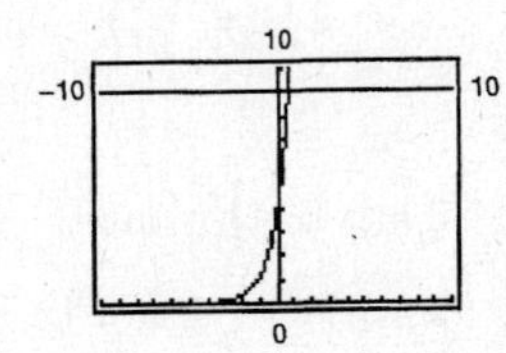

The intersection is at $x \approx 0.68$.

108. $\ln 5x = 7.2$

$$5x = e^{7.2}$$

$$x = \frac{e^{7.2}}{5} \approx 267.886$$

110. $4 \ln 3x = 15$

$$\ln 3x = \frac{15}{4}$$

$$3x = e^{15/4}$$

$$x = \frac{e^{15/4}}{3} \approx 14.174$$

112. $\ln\sqrt{x+8} = 3$

$$\tfrac{1}{2}\ln(x+8) = 3$$

$$\ln(x+8) = 6$$

$$x + 8 = e^6$$

$$x = e^6 - 8 \approx 395.429$$

114. $\ln x - \ln 5 = 4$

$$\ln\frac{x}{5} = 4$$

$$\frac{x}{5} = e^4$$

$$x = 5e^4 \approx 272.991$$

116. $\log_{10}(x+2) - \log_{10} x = \log_{10}(x+5)$

$$\log_{10}\left(\frac{x+2}{x}\right) = \log_{10}(x+5)$$

$$\frac{x+2}{x} = x + 5$$

$$x + 2 = x^2 + 5x$$

$$0 = x^2 + 4x - 2$$

$$x = -2 \pm \sqrt{6} \quad \text{Quadratic Formula}$$

Only $x = -2 + \sqrt{6} \approx 0.449$ is a valid solution.

118. $\log_{10}(-x - 4) = 2$

$$-x - 4 = 10^2$$

$$-x = 100 + 4$$

$$x = -104$$

120. $6 \log_{10}(x^2 + 1) - x = 0$

Graph

$y_1 = 6 \log_{10}(x^2 + 1) - x.$

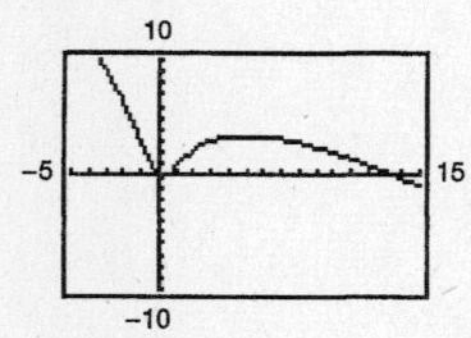

Zoom in to see the behavior near the origin.

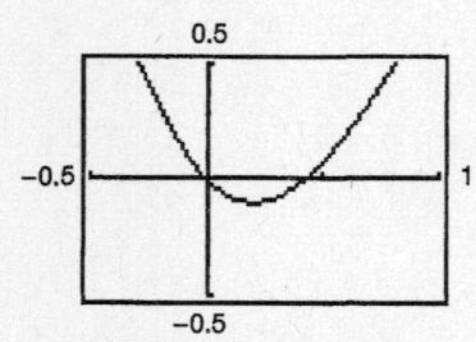

The solutions of the equation occur at the x-intercepts, which are at $x = 0$, $x \approx 0.42$, and $x \approx 13.63$.

122. $x - 2 \log_{10}(x + 4) = 0$

Graph

$y_1 = x - 2 \log_{10}(x + 4).$

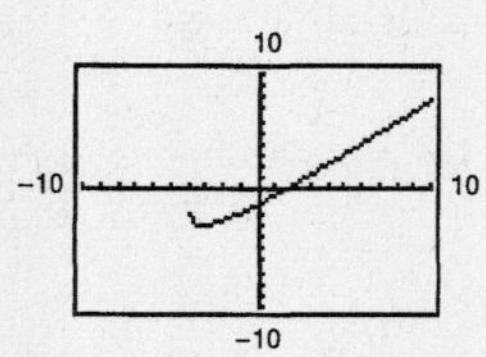

Note that $x = -4$ is a vertical asymptote, but the graph's behavior near $x = -4$ is not visible in most typical viewing windows. Zoom in to see the behavior near $x = -4$.

—CONTINUED—

122. —CONTINUED—

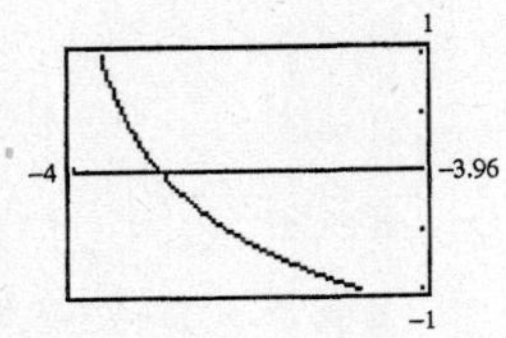

The solutions of the equation occur at the x-intercepts, which are at $x \approx -3.99$ and $x \approx 1.48$.

124. $p = 500 - 0.5e^{0.004x}$

(a) $p = 450$

$$450 = 500 - 0.5e^{0.004x}$$
$$0.5e^{0.004x} = 50$$
$$e^{0.004x} = 100$$
$$0.004x = \ln 100$$
$$x \approx 1151 \text{ units}$$

(b) $p = 400$

$$400 = 500 - 0.5e^{0.004x}$$
$$0.5e^{0.004x} = 100$$
$$e^{0.004x} = 200$$
$$0.004x = \ln 200$$
$$x \approx 1325 \text{ units}$$

126. $y = 4e^{2x/3}$

Exponential growth model

Matches graph (b).

128. $y = 7 - \log_{10}(x + 3)$

Logarithmic model

Vertical asymptote: $x = -3$

Matches graph (d).

130. $y = \dfrac{6}{1 + 2e^{-2x}}$

Logistics growth model

Matches graph (c).

132. $y = Ce^{kt}$

$$\frac{1}{2}C = Ce^{(250,000)k}$$
$$\ln\frac{1}{2} = \ln e^{(250,000)k}$$
$$\ln\frac{1}{2} = 250,000k$$
$$k = \frac{\ln(1/2)}{250,000}$$

When $t = 5000$, we have $y = Ce^{[\ln(1/2)/250,000](5000)} \approx 0.986C = 98.6\%C$.

After 5000 years, approximately 98.6% of the radioactive uranium II will remain.

134.

$$N = 200e^{kt}$$
$$350 = 200e^{k(5)}$$
$$1.75 = 200e^{5(k)}$$
$$\ln 1.75 = \ln e^{5k}$$
$$\frac{1}{5}\ln 1.75 = k$$
$$0.1119 \approx k$$

For the population to triple:

$$600 \approx 200e^{0.1119t}$$
$$3 \approx e^{0.1119t}$$
$$\ln 3 \approx \ln e^{0.1119t}$$
$$\ln 3 \approx 0.1119t$$
$$\frac{\ln 3}{0.1119} \approx t$$
$$9.82 \approx t$$

It will take about 9.8 hours for the population to triple.

136.

$$y = ae^{bx}$$
$$\frac{1}{2} = ae^{b(0)} \Rightarrow a = \frac{1}{2}$$
$$5 = \frac{1}{2}e^{b(5)}$$
$$10 = e^{5b}$$
$$\ln 10 = 5b$$
$$\frac{\ln 10}{5} = b$$
$$b \approx 0.4605$$
$$y = \frac{1}{2}e^{0.4605x}$$

138. $N = \dfrac{157}{1 + 5.4e^{-0.12t}}$

(a) When $N = 50$:

$$50 = \frac{157}{1 + 5.4e^{-0.12t}}$$
$$1 + 5.4e^{-0.12t} = \frac{157}{50}$$
$$5.4e^{-0.12t} = \frac{107}{50}$$
$$e^{-0.12t} = \frac{107}{270}$$
$$-0.12t = \ln\frac{107}{270}$$
$$t = \frac{\ln(107/270)}{-0.12} \approx 7.7 \text{ weeks}$$

(b) When $N = 75$:

$$75 = \frac{157}{1 + 5.4e^{-0.12t}}$$
$$1 + 5.4e^{-0.12t} = \frac{157}{75}$$
$$5.4e^{-0.12t} = \frac{82}{75}$$
$$e^{-0.12t} = \frac{82}{405}$$
$$-0.12t = \ln\frac{82}{405}$$
$$t = \frac{\ln(82/405)}{-0.12} \approx 13.3 \text{ weeks}$$

140. $R = \log_{10} I$ since $I_0 = 1$.

(a) $\log_{10} I = 8.4$

$$I = 10^{8.4} \approx 251{,}188{,}643$$

(b) $\log_{10} I = 6.85$

$$I = 10^{6.85} \approx 7{,}079{,}458$$

(c) $\log_{10} I = 9.1$

$$I = 10^{9.1} \approx 1{,}258{,}925{,}412$$

142. False. $\ln x + \ln y = \ln(xy) \neq \ln(x + y)$

Problem Solving for Chapter 5

Solutions to Even-Numbered Exercises

2.

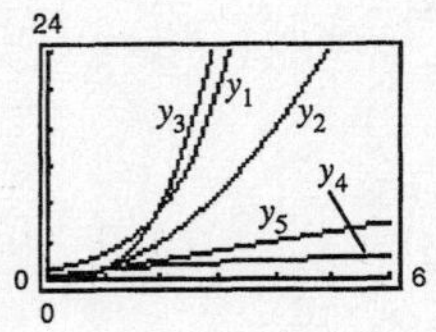

$y_1 = e^x$

$y_2 = x^2$

$y_3 = x^3$

$y_4 = \sqrt{x}$

$y_5 = |x|$

The function that increases at the fastest rate for "large" values of x is $y_1 = e^x$. (Note: One of the intersection points of $y = e^x$ and $y = x^3$ is approximately (4.536, 93) and past this point $e^x > x^3$. This is not shown on the graph above.)

4. It usually implies rapid growth.

6.
$$[f(x)]^2 - [g(x)]^2 = \left(\frac{e^x + e^{-x}}{2}\right)^2 - \left(\frac{e^x - e^{-x}}{2}\right)^2$$
$$= \left(\frac{e^{2x} + 2 + e^{-2x}}{4}\right) - \left(\frac{e^{2x} - 2 + e^{-2x}}{4}\right)$$
$$= \frac{4}{4}$$
$$= 1$$

8. $y_4 = 1 + \dfrac{x}{1!} + \dfrac{x^2}{2!} + \dfrac{x^3}{3!} + \dfrac{x^4}{4!}$

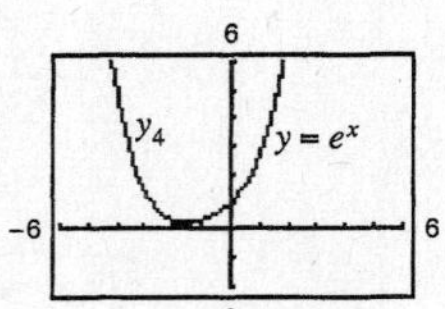

As more terms are added, the polynomial approaches e^x.

$$e^x = 1 + \frac{x}{1!} + \frac{x^2}{2!} + \frac{x^3}{3!} + \frac{x^4}{4!} + \frac{x^5}{5!} + \cdots$$

10. $f(x) = \dfrac{a^x + 1}{a^x - 1}, a > 0, a \neq 1$

$$x = \frac{a^y + 1}{a^y - 1}$$
$$x(a^y - 1) = a^y + 1$$
$$xa^y - a^y = x + 1$$
$$a^y(x - 1) = x + 1$$
$$a^y = \frac{x + 1}{x - 1}$$
$$y = \log_a\left(\frac{x + 1}{x - 1}\right) = \frac{\ln\left(\frac{x + 1}{x - 1}\right)}{\ln a} = f^{-1}(x)$$

12. (a) The steeper curve represents the investment earning compound interest, because compound interest earns more than simple interest. With simple interest there is no compounding so the growth is linear.

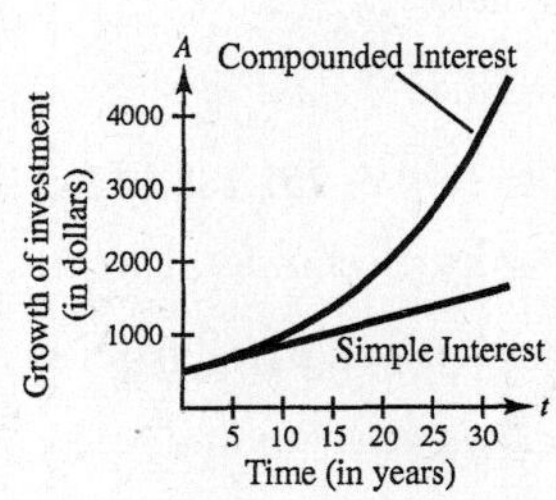

(b) Compound interest formula: $A = 500\left(1 + \dfrac{0.07}{1}\right)^{(1)t}$

$$= 500(1.07)^t$$

Simple interest formula: $A = Prt + P$

$$= 500(0.07)t + 500$$

14. $B = B_0 a^{kt}$ through (0, 500) and (2,200)

$B_0 = 500$

$200 = 500a^{k(2)}$

$\frac{2}{5} = a^{2k}$

$\log_a\left(\frac{2}{5}\right) = 2k$

$\frac{1}{2}\log_a\left(\frac{2}{5}\right) = k$

$B = 500a^{[1/2 \log_a(2/5)]t}$

$= 500[a^{\log_a(2/5)}]^{t/2}$

$= 500\left(\frac{2}{5}\right)^{t/2}$

16. Let $\log_a x = m$ and $\log_{a/b} x = n$. Then $x = a^m$ and $x = \left(\frac{a}{b}\right)^n$.

$a^m = \left(\frac{a}{b}\right)^n$

$a^{m/n} = \frac{a}{b}$

$a^{m/n-1} = \frac{1}{b}$

$\log_a \frac{1}{b} = \frac{m}{n} - 1$

$1 + \log_a \frac{1}{b} = \frac{m}{n}$

$1 + \log_a \frac{1}{b} = \frac{\log_a x}{\log_{a/b} x}$

18. $y = \ln x$

$y_1 = x - 1$

$y_2 = (x - 1) - \frac{1}{2}(x - 1)^2$

$y_3 = (x - 1) - \frac{1}{2}(x - 1)^2 + \frac{1}{3}(x - 1)^3$

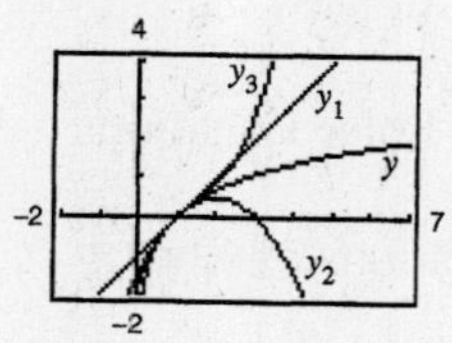

20. $y = ab^x$

$\ln y = \ln(ab^x)$

$\ln y = \ln a + \ln b^x$

$\ln y = \ln a + x \ln b$

$\ln y = (\ln b)x + \ln a$

Slope: $m = \ln b$

y-intercept: $(0, \ln a)$

$y = ax^b$

$\ln y = \ln(ax^b)$

$\ln y = \ln a + \ln x^b$

$\ln y = \ln a + b \ln x$

$\ln y = b \ln x + \ln a$

Slope: $m = b$

y-intercept: $(0, \ln a)$

22. (a) $\frac{450}{30} = 15$ cubic feet per minute

(b) $15 = 80.4 - 11 \ln x$

$11 \ln x = 65.4$

$\ln x = \frac{65.4}{11}$

$x = e^{65.4/11}$

$x \approx 382$ cubic feet of air space per child

(c) Total air space required: $382(30) = 11{,}460$ cubic feet

Let $x =$ floor space in square feet and $h = 30$ feet

$V = xh$

$11460 = x(30)$

$x = 382$

If the ceiling height is 30 feet, the minimum number of square feet of floor space required is 382 square feet.

CHAPTER 6
Topics in Analytic Geometry

CHAPTER 6
Topics in Analytic Geometry

Section 6.1 Lines

Solutions to Even-Numbered Exercises

2. $m = \tan \frac{\pi}{4} = 1$

4. $m = \tan \frac{2\pi}{3} = -\sqrt{3}$

6. $m = \tan \frac{5\pi}{6} = -\frac{\sqrt{3}}{3}$

8. $m = \tan 2.88 \approx -0.2677$

10. $-2 = \tan \theta$

$\theta = \tan^{-1}(-2) + \pi \approx 2.034 \text{ radians} \approx 116.6°$

12. $2 = \tan \theta$

$\theta = \tan^{-1} 2 \approx 1.107 \text{ radians} \approx 63.4°$

14. $-\frac{5}{2} = \tan \theta$

$\theta = \tan^{-1}\left(-\frac{5}{2}\right) + \pi \approx 1.9513 \text{ radians} \approx 111.8°$

16. $m = \frac{8-(-3)}{12-(-4)} = \frac{11}{16}$

$\frac{11}{16} = \tan \theta$

$\theta = \tan^{-1} \frac{11}{16} \approx 0.6023 \text{ radian} \approx 34.5°$

18. $m = \frac{100-0}{0-50} = -2$

$-2 = \tan \theta$

$\theta = \tan^{-1}(-2) + \pi \approx 2.0344 \text{ radians} \approx 116.6°$

20. $4x + 5y - 9 = 0$

$y = -\frac{4}{5}x + \frac{9}{5} \Rightarrow m = -\frac{4}{5}$

$-\frac{4}{5} = \tan \theta$

$\theta = \tan^{-1}\left(-\frac{4}{5}\right) + \pi$

$\approx 2.4669 \text{ radians} \approx 141.3°$

22. $x - y - 10 = 0$

$y = x - 10 \Rightarrow m = 1$

$1 = \tan \theta$

$\theta = \tan^{-1} 1 = 45° = \frac{\pi}{4} \text{ radian}$

24. $x + 3y = 2 \Rightarrow y = -\frac{1}{3}x + \frac{2}{3} \Rightarrow m_1 = -\frac{1}{3}$

$x - 2y = -3 \Rightarrow y = \frac{1}{2}x + \frac{3}{2} \Rightarrow m_2 = \frac{1}{2}$

$\tan \theta = \left|\frac{(1/2) - (-1/3)}{1 + (-1/3)(1/2)}\right| = 1$

$\theta = \tan^{-1} 1 = 45° = \frac{\pi}{4} \text{ radian}$

26. $2x - y = 2 \Rightarrow y = 2x - 2 \Rightarrow m_1 = 2$

$4x + 3y = 24 \Rightarrow y = -\frac{4}{3}x + 8 \Rightarrow m_2 = -\frac{4}{3}$

$\tan \theta = \left|\frac{(-4/3) - 2}{1 + (2)(-4/3)}\right| = 2$

$\theta = \tan^{-1} 2 \approx 63.4° \approx 1.1071 \text{ radians}$

28. $5x + 2y = 16 \Rightarrow y = -\frac{5}{2}x + 8 \Rightarrow m_1 = -\frac{5}{2}$

$3x - 5y = -1 \Rightarrow y = \frac{3}{5}x + \frac{1}{5} \Rightarrow m_2 = \frac{3}{5}$

$\tan\theta = \left|\frac{(-5/2) - (3/5)}{1 + (-5/2)(3/5)}\right| = \frac{31}{5}$

$\theta = \tan^{-1}\frac{31}{5} \approx 80.8° \approx 1.4109$ radians

30. $3x - 5y = 3 \Rightarrow y = \frac{3}{5}x - \frac{3}{5} \Rightarrow m_1 = \frac{3}{5}$

$3x + 5y = 12 \Rightarrow y = -\frac{3}{5}x + \frac{12}{5} \Rightarrow m_2 = -\frac{3}{5}$

$\tan\theta = \left|\frac{(3/5) - (-3/5)}{1 + (3/5)(-3/5)}\right| = \frac{15}{8}$

$\theta = \tan^{-1}\frac{15}{8} \approx 61.9° \approx 1.0808$ radians

32. $0.02x - 0.05y = -0.19 \Rightarrow y = \frac{2}{5}x + \frac{19}{5} \Rightarrow m_1 = \frac{2}{5}$

$0.03x + 0.04y = 0.52 \Rightarrow y = -\frac{3}{4}x + 13 \Rightarrow m_2 = -\frac{3}{4}$

$\tan\theta = \left|\frac{(-3/4) - (2/5)}{1 + (2/5)(-3/4)}\right| \approx \frac{23}{14}$

$\theta = \tan^{-1}\left(\frac{23}{14}\right) \approx 58.7° \approx 1.0240$ radians

34. Let $A = (-3, 2), B = (1, 3)$, and $C = (2, 0)$.

Slope of AB: $m_1 = \frac{2 - 3}{-3 - 1} = \frac{1}{4}$

Slope of BC: $m_2 = \frac{3 - 0}{1 - 2} = -3$

Slope of AC: $m_3 = \frac{2 - 0}{-3 - 2} = -\frac{2}{5}$

$\tan A = \left|\frac{(1/4) - (-2/5)}{1 + (-2/5)(1/4)}\right| = \frac{13/20}{18/20} = \frac{13}{18}$

$A = \tan^{-1}\left(\frac{13}{18}\right) \approx 35.8°$

$\tan C = \left|\frac{-3 - (-2/5)}{1 + (-3)(-2/5)}\right| \approx \frac{13/5}{11/5} = \frac{13}{11}$

$C = \tan^{-1}\left(\frac{13}{11}\right) \approx 49.8°$

$B = 180° - A - C \approx 180° - 35.8° - 49.8°$
$= 94.4°$

36. Let $A = (-3, 4), B = (2, 1)$, and $C = (-2, 2)$.

Slope of AB: $m_1 = \frac{4 - 1}{-3 - 2} = -\frac{3}{5}$

Slope of BC: $m_2 = \frac{1 - 2}{2 - (-2)} = -\frac{1}{4}$

Slope of AC: $m_3 = \frac{4 - 2}{-3 - (-2)} = -2$

$\tan A = \left|\frac{(-3/5) - (-2)}{1 + (-3/5)(-2)}\right| = \frac{7}{11}$

$A = \tan^{-1}\left(\frac{7}{11}\right) \approx 32.5°$

$\tan B = \left|\frac{(-3/5) - (-1/4)}{1 + (-3/5)(-1/4)}\right| = \frac{7}{23}$

$B = \tan^{-1}\left(\frac{7}{23}\right) \approx 16.9°$

$C = 180° - A - B \approx 180° - 32.5° - 16.9°$
$= 130.6°$

38. $(0, 0) \Rightarrow x_1 = 0$ and $y_1 = 0$

$2x - y - 4 = 0 \Rightarrow A = 2, B = -1$, and $C = -4$

$d = \frac{|2(0) + (-1)(0) + (-4)|}{\sqrt{2^2 + (-1)^2}}$

$= \frac{4}{\sqrt{5}} = \frac{4\sqrt{5}}{5} \approx 1.7889$

40. $(-2, 1) \Rightarrow x_1 = -2$ and $y_1 = 1$

$x - y - 2 = 0 \Rightarrow A = 1, B = -1$, and $C = -2$

$d = \frac{|1(-2) + (-1)(1) + (-2)|}{\sqrt{1^2 + (-1)^2}}$

$= \frac{5}{\sqrt{2}} = \frac{5\sqrt{2}}{2} \approx 3.5355$

42. $(10, 8) \Rightarrow x_1 = 10$ and $y_1 = 8$

$y - 4 = 0 \Rightarrow A = 0, B = 1$, and $C = -4$

$$d = \frac{|0(10) + 1(8) + (-4)|}{\sqrt{0^2 + 1^2}} = \frac{4}{1} = 4$$

44. $(4, 2) \Rightarrow x_1 = 4$ and $y_1 = 2$

$x - y - 20 = 0 \Rightarrow A = 1, B = -1$, and $C = -20$

$$d = \frac{|1(4) + (-1)(2) + (-20)|}{\sqrt{1^2 + (-1)^2}}$$

$$= \frac{18}{\sqrt{2}} = 9\sqrt{2} \approx 12.7279$$

46. (a) The slope of the line through AC is $m = \frac{0 + 2}{0 - 5} = -\frac{2}{5}$.

The equation of the line is $y - 0 = -\frac{2}{5}(x - 0) \Rightarrow 2x + 5y = 0$.

The altitude from vertex B to side AC is the distance between the line through AC and

$$B = (4, 5) \Rightarrow d = \frac{|2(4) + 5(5) + 0|}{\sqrt{2^2 + 5^2}} = \frac{33}{\sqrt{29}} = \frac{33\sqrt{29}}{29}.$$

(b) The distance between A and C is $d = \sqrt{(0 - 5)^2 + (0 + 2)^2} = \sqrt{29}$, which is the length of the base of the triangle.

So, $A = \frac{1}{2}\sqrt{29}\left(\frac{33\sqrt{29}}{29}\right) = \frac{33}{2}$ square units

48. (a) The slope of the line through AC is $m = \frac{12 - (-5)}{6 - (-4)} = \frac{17}{10}$.

The equation of the line through AC is $y - 12 = \frac{17}{10}(x - 6) \Rightarrow 17x - 10y + 18 = 0$.

The altitude from vertex B to side AC is the distance between the line through AC and

$$B = (3, 10) \Rightarrow d = \frac{|17(3) + (-10)(10) + 18|}{\sqrt{17^2 + (-10)^2}} = \frac{31}{\sqrt{389}} = \frac{31\sqrt{389}}{389}.$$

(b) The distance between A and C is $d = \sqrt{(6 + 4)^2 + (12 + 5)^2} = \sqrt{389}$, which is the length of the base of the triangle.

$$A = \frac{1}{2}\left(\sqrt{389}\right)\left(\frac{31\sqrt{389}}{389}\right) = \frac{31}{2}$$

50. $3x - 4y = 1$

$3x - 4y = 10$

A point on $3x - 4y = 10$ is $\left(0, -\frac{5}{2}\right)$. The distance between $\left(0, -\frac{5}{2}\right)$ and $3x - 4y = 1$ is:

$A = 3, B = -4, C = -1, x_1 = 0, y_1 = -\frac{5}{2}$

$$d = \frac{|3(0) + (-4)(-5/2) - 1|}{\sqrt{3^2 + (-4)^2}} = \frac{9}{5}$$

52. Slope: $m = \tan 0.2 \approx 0.2027$

Change in elevation:

$\sin 0.20 = \frac{x}{5280} \Rightarrow x = 5280 \sin 0.20 \approx 1049$ feet

54. Slope $= \frac{3}{5}$

Inclination $= \tan^{-1}\frac{3}{5} \approx 31.0°$

56. (a) $m = \tan\theta$

$0.709 = \tan\theta$

$\tan^{-1} 0.709 = \theta$

$\theta \approx 0.6167$ radian, or $35.34°$

(b) $\sin\theta = \dfrac{\text{elev } \Delta}{896.5}$

$896.5 \sin\theta = \text{elev } \Delta$

$\text{elev } \Delta = 896.5 \sin 0.6167 \approx 518.5$ ft

(c) $m = 0.709$ and y-intercept $= (0, 0)$, so $y = 0.709x$

(d)

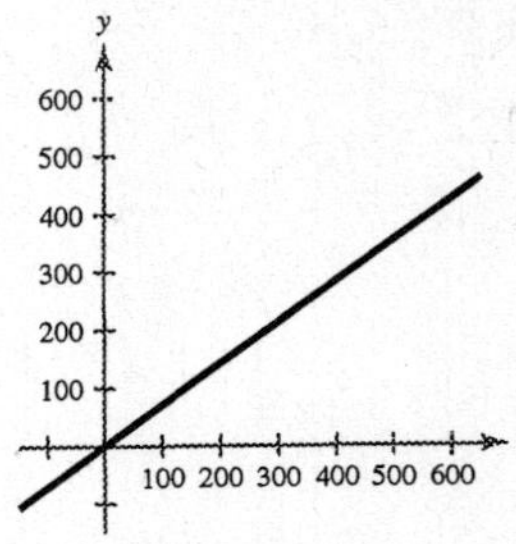

58. False. Substitute $m_1 = \tan\theta_1$ and $m_2 = \tan\theta_2$ into the formula for the angle between two lines.

60. Slope m and y-intercept $(0, 4)$

(a) $(x_1, y_1) = (3, 1)$ and line: $y = mx + 4$

$A = -m, B = 1, C = -4$

$$d = \frac{|(-m)(3) + (1)(1) + (-4)|}{\sqrt{(-m)^2 + 1^2}} = \frac{3|m + 1|}{\sqrt{m^2 + 1}}$$

(b)

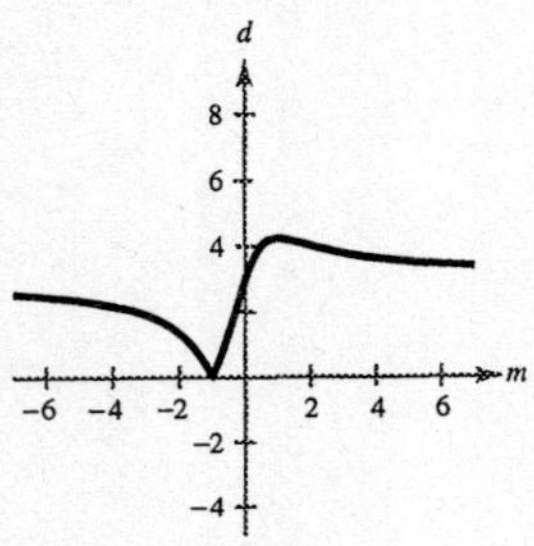

(c) From the graph it appears that the maximum distance is obtained when $m = 1$.

(d) From the graph it appears that the distance is 0 when $m = -1$.

(e) The asymptote of the graph in part (b) is $d = 3$. As the line approaches the vertical, the distance approaches 3.

62. $f(x) = (x + 9)^2$

$f(x) = (x + 9)^2 = 0 \Rightarrow x = -9$

x-intercept: $(-9, 0)$

$f(0) = (0 + 9)^2 = 81$

y-intercept: $(0, 81)$

64. $f(x) = (x + 11)^2 + 12$

$f(x) = (x + 11)^2 + 12 = 0$

$(x + 11)^2 = -12$

No solution

x-intercept: none

$f(0) = (0 + 11)^2 + 12 = 133$

y-intercept: $(0, 133)$

66. $f(x) = x^2 + 9x - 22$

$f(x) = x^2 + 9x - 22 = 0$

$(x + 11)(x - 2) = 0$

$x = -11, 2$

x-intercepts: $(-11, 0), (2, 0)$

$f(0) = -22$

y-intercept: $(0, -22)$

68. $f(x) = 2x^2 - x - 21$

$$= 2\left[x^2 - \tfrac{1}{2}x - \tfrac{21}{2}\right] = 2\left[x^2 - \tfrac{1}{2}x + \tfrac{1}{16} - \tfrac{1}{16} - \tfrac{21}{2}\right]$$

$$= 2\left[\left(x - \tfrac{1}{4}\right)^2 - \tfrac{169}{16}\right]$$

$$= 2\left(x - \tfrac{1}{4}\right)^2 - \tfrac{169}{8}$$

Vertex: $\left(\tfrac{1}{4}, -\tfrac{169}{8}\right)$

70. $f(x) = -x^2 - 8x - 15$

$= -[x^2 + 8x + 15] = -[x^2 + 8x + 16 - 16 + 15]$

$= -[(x + 4)^2 - 1] = -(x + 4)^2 + 1$

Vertex: $(-4, 1)$

72. $f(x) = -8x^2 - 34x - 21$

$= -8\left[x^2 + \frac{17}{4}x + \frac{21}{8}\right]$

$= -8\left[x^2 + \frac{17}{4}x + \frac{289}{64} - \frac{289}{64} + \frac{21}{8}\right]$

$= -8\left[\left(x + \frac{17}{8}\right)^2 - \frac{121}{64}\right]$

$= -8\left(x + \frac{17}{8}\right)^2 + \frac{121}{8}$

Vertex: $\left(-\frac{17}{8}, \frac{121}{8}\right)$

74. $f(x) = 6 - (x + 1)^2$

Vertex: $(-1, 6)$

x	-4	-3	-2	-1	0	1	2
$g(x)$	-3	2	5	6	5	2	-3

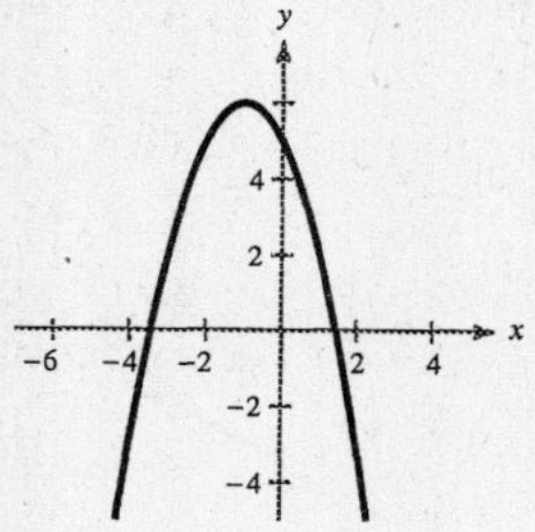

76. $g(x) = -x^2 + 6x - 8$

$\frac{-b}{2a} = \frac{-6}{2(-1)} = 3 \Rightarrow \text{Vertex} = (3, g(3)) = (3, 1)$

x	0	1	2	3	4	5	6
$g(x)$	-8	-3	0	1	0	-3	-8

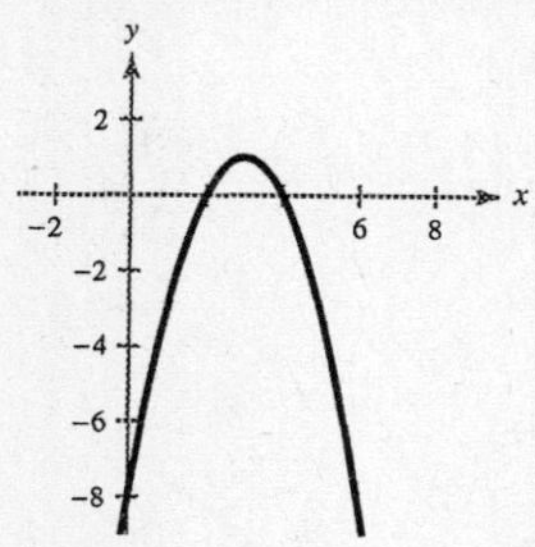

Section 6.2 Introduction to Conics: Parabolas

Solutions to Even-Numbered Exercises

2. An ellipse is formed when a plane intersects only the top or bottom half of a double-napped cone but is not perpendicular to the axis of the cone, not parallel to the side of the cone, and does not intersect the vertex.

4. A hyperbola is formed when a plane intersects both halves of a double-napped cone, is parallel to the axis of the cone, and does not intersect the vertex.

6. $x^2 = 2y$

Vertex: $(0, 0)$

$p = \frac{1}{2} > 0$

Opens upward.
Matches graph (b).

8. $y^2 = -12x$

Vertex: $(0, 0)$

$p = -3 < 0$

Opens to the left.
Matches graph (f).

10. $(x + 3)^2 = -2(y - 1)$

Vertex: $(-3, 1)$

$p = -\frac{1}{2} < 0$

Opens downward.
Matches graph (c).

12. $y = -2x^2 \Rightarrow x^2 = 4\left(-\frac{1}{8}\right)y$

Vertex: $(0, 0)$

Focus: $\left(0, -\frac{1}{8}\right)$

Directrix: $y = \frac{1}{8}$

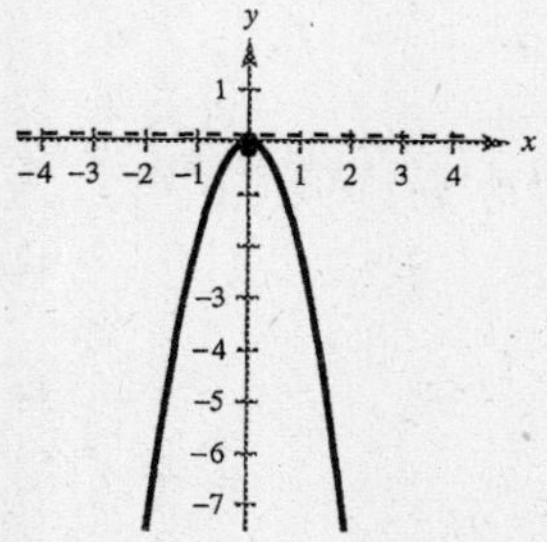

14. $y^2 = 3x \Rightarrow 4\left(\frac{3}{4}\right)x$

Vertex: $(0, 0)$

Focus: $\left(\frac{3}{4}, 0\right)$

Directrix: $x = -\frac{3}{4}$

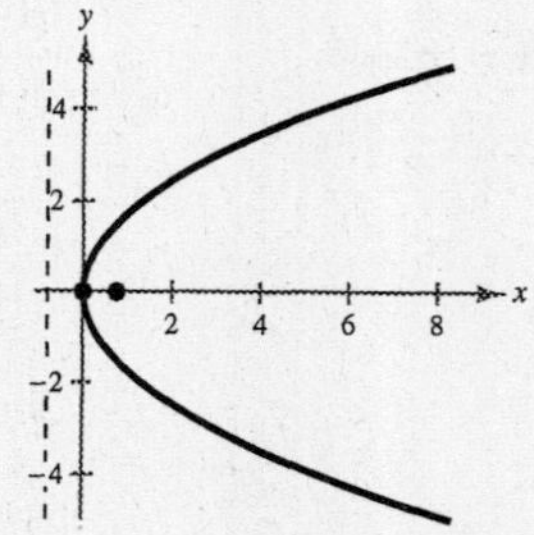

16. $x + y^2 = 0$

$y^2 = -x = 4\left(-\frac{1}{4}\right)x$

Vertex: $(0, 0)$

Focus: $\left(-\frac{1}{4}, 0\right)$

Directrix: $x = \frac{1}{4}$

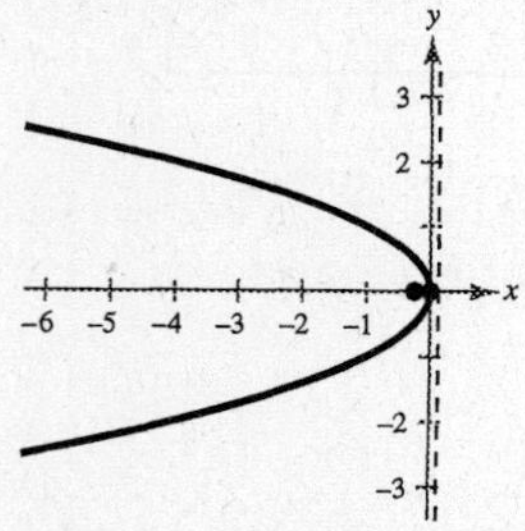

18. $(x + 5) + (y - 1)^2 = 0$

$(y - 1)^2 = 4\left(-\frac{1}{4}\right)(x + 5)$

Vertex: $(-5, 1)$

Focus: $\left(-5 + \left(-\frac{1}{4}\right), 1\right) \Rightarrow \left(-\frac{21}{4}, 1\right)$

Directrix: $x = -5 - \left(-\frac{1}{4}\right) = -\frac{19}{4}$

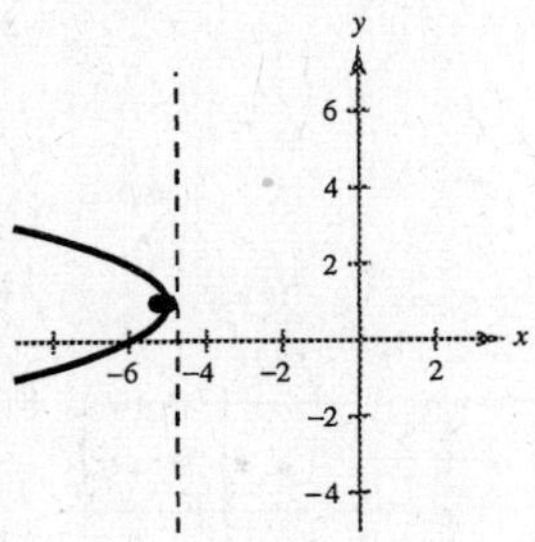

20. $\left(x + \frac{1}{2}\right)^2 = 4(y - 1) = 4(1)(y - 1)$

Vertex: $\left(-\frac{1}{2}, 1\right)$

Focus: $\left(-\frac{1}{2}, 1 + 1\right) \Rightarrow \left(-\frac{1}{2}, 2\right)$

Directrix: $y = 1 - 1 = 0$

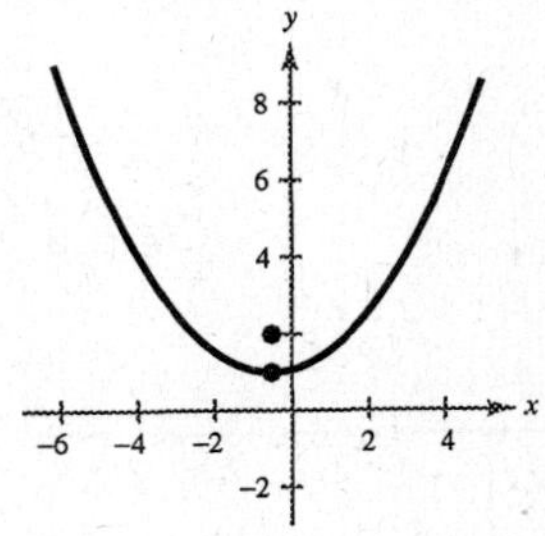

22. $x = \frac{1}{4}(y^2 + 2y + 33)$

$4x = y^2 + 2y + 1 - 1 + 33 = (y + 1)^2 + 32$

$(y + 1)^2 = 4(1)(x - 8)$

Vertex: $(8, -1)$

Focus: $(9, -1)$

Directrix: $x = 7$

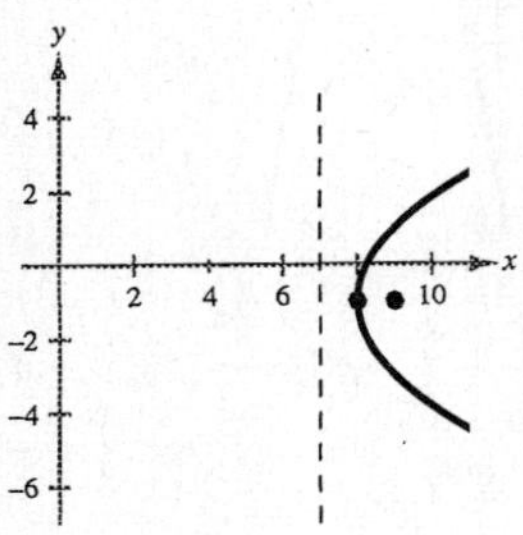

24. $y^2 - 4y - 4x = 0$

$y^2 - 4y + 4 = 4x + 4$

$(y - 2)^2 = 4(1)(x + 1)$

Vertex: $(-1, 2)$

Focus: $(0, 2)$

Directrix: $x = -2$

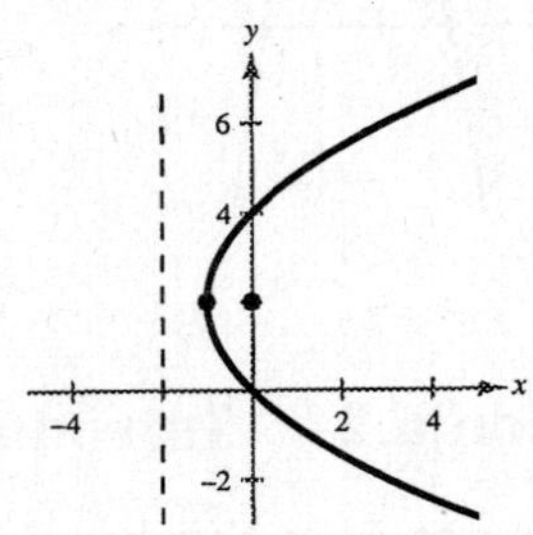

26. $x^2 - 2x + 8y + 9 = 0$

$x^2 - 2x + 1 = -8y - 9 + 1$

$(x - 1)^2 = -8(y + 1) = 4(-2)(y + 1)$

Vertex: $(1, -1)$

Focus: $(1, -3)$

Directrix: $y = 1$

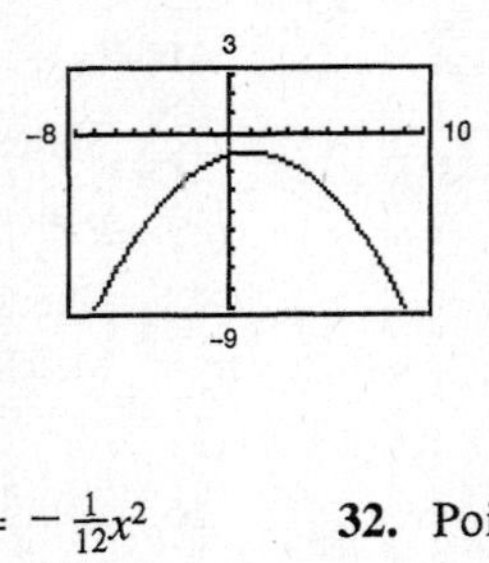

28. $y^2 - 4x - 4 = 0$

$y^2 = 4x + 4 = 4(1)(x + 1)$

Vertex: $(-1, 0)$

Focus: $(0, 0)$

Directrix: $x = -2$

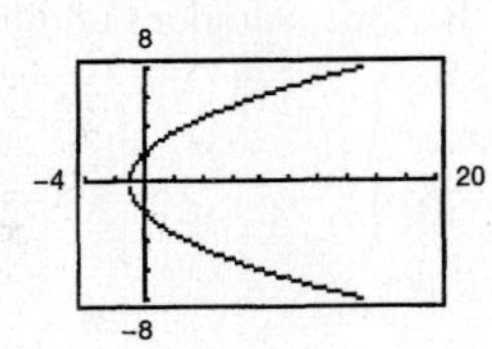

30. $x^2 + 12y = 0 \Rightarrow y_1 = -\frac{1}{12}x^2$

$x + y - 3 = 0 \Rightarrow y_2 = 3 - x$

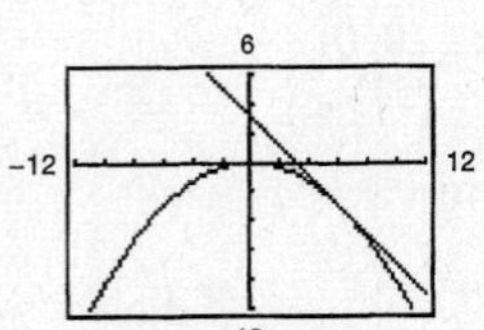

Using the trace or intersect feature, the point of tangency is $(6, -3)$.

32. Point: $(-2, 6)$

$x = ay^2$

$-2 = a(6)^2$

$-\frac{1}{18} = a$

$x = -\frac{1}{18}y^2$

$y^2 = -18x$

34. Focus: $(2, 0) \Rightarrow p = 2$

$y^2 = 4px$

$y^2 = 8x$

36. Focus: $(0, -2) \Rightarrow p = -2$

$x^2 = 4py$

$x^2 = -8y$

38. Directrix: $y = 3 \Rightarrow p = -3$

$x^2 = 4py$

$x^2 = -12y$

40. Directrix: $x = -3 \Rightarrow p = 3$

$y^2 = 4px$

$y^2 = 12x$

42. Vertical axis
Passes through: $(-3, -3)$

$$x^2 = 4py$$
$$(-3)^2 = 4p(-3)$$
$$9 = -12p$$
$$p = -\tfrac{3}{4}$$
$$x^2 = -3y$$

44. Vertex: $(5, 3) \Rightarrow h = 5,$ $k = 3$

Passes through: $(4.5, 4)$

$$(y - k)^2 = 4p(x - h)$$
$$(y - 3)^2 = 4p(x - 5)$$
$$1 = 4p(4.5 - 5)$$
$$p = -\tfrac{1}{2}$$
$$(y - 3)^2 = -2(x - 5)$$

46. Vertex: $(3, -3) \Rightarrow h = 3,$ $k = -3$

Passes through: $(0, 0)$

$$(x - h)^2 = 4p(y - k)$$
$$(x - 3)^2 = 4p(y + 3)$$
$$(0 - 3)^2 = 4p(0 + 3)$$
$$9 = 12p$$
$$p = \tfrac{3}{4}$$
$$(x - 3)^2 = 3(y + 3)$$

48. Vertex: $(-1, 2) \Rightarrow h = -1, \quad k = 2$
Focus: $(-1, 0) \Rightarrow p = -2$

$$(x - h)^2 = 4p(y - k)$$
$$(x + 1)^2 = 4(-2)(y - 2)$$
$$(x + 1)^2 = -8(y - 2)$$

50. Vertex: $(-2, 1) \Rightarrow h = -2, \quad k = 1$
Directrix: $x = 1 \Rightarrow p = -3$

$$(y - k)^2 = 4p(x - h)$$
$$(y - 1)^2 = 4(-3)(x + 2)$$
$$(y - 1)^2 = -12(x + 2)$$

52. Focus: $(0, 0)$

Directrix: $y = 8 \Rightarrow p = -4 \Rightarrow h = 0, k = 4$

$$(x - h)^2 = 4p(y - k)$$
$$x^2 = 4(-4)(y - 4)$$
$$x^2 = -16(y - 4)$$

54. $(y + 1)^2 = 2(x - 4)$

$$y + 1 = \pm\sqrt{2(x - 4)}$$
$$y = -1 \pm \sqrt{2(x - 4)}$$

Lower half of parabola: $y = -1 - \sqrt{2(x - 4)}$

56. $x^2 = 2y$

$$x^2 = 4\left(\frac{1}{2}\right)y$$
$$4\left(\frac{1}{2}\right)y = x^2$$
$$p = \frac{1}{2}$$

Focus: $\left(0, \dfrac{1}{2}\right)$

$$d_1 = \frac{1}{2} - b$$
$$d_2 = \sqrt{(-3 - 0)^2 + \left(\frac{9}{2} - \frac{1}{2}\right)^2} = 5$$
$$\frac{1}{2} - b = 5$$
$$b = -\frac{9}{2}$$
$$m = \frac{-(9/2) - (9/2)}{0 + 3} = -3$$

Tangent line: $y = -3x - \dfrac{9}{2} \Rightarrow 6x + 2y + 9 = 0$

x-intercept: $\left(-\dfrac{3}{2}, 0\right)$

58. $y = -2x^2$

$$-\frac{1}{2}y = x^2$$
$$4\left(-\frac{1}{8}\right)y = x^2$$
$$p = -\frac{1}{8}$$

Focus: $\left(0, -\dfrac{1}{8}\right)$

$$d_1 = \frac{1}{8} + b$$
$$d^2 = \sqrt{(2 - 0)^2 + \left(-8 - \left(-\frac{1}{8}\right)\right)^2} = \frac{65}{8}$$
$$\frac{1}{8} + b = \frac{65}{8}$$
$$b = \frac{64}{8} = 8$$
$$m = \frac{-8 - 8}{2 - 0} = -8$$

Tangent line: $y = -8x + 8 \Rightarrow 8x + y - 8 = 0$

x-intercept: $(1, 0)$

60. Maximum revenue occurs at $x = 135$

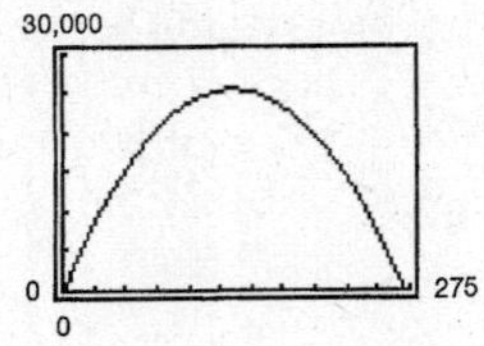

62. (a)

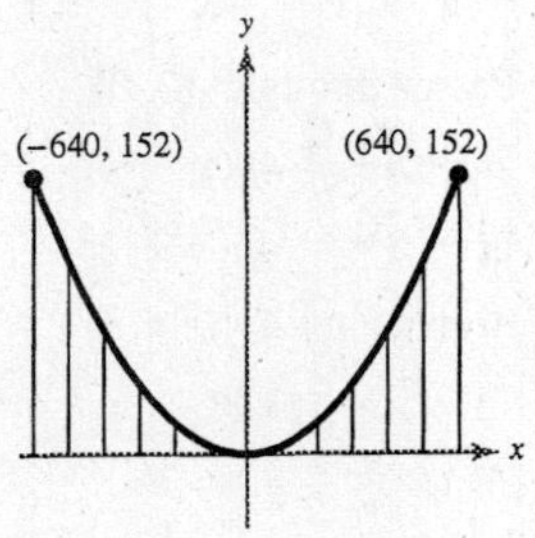

(b) vertex: (0, 0); opens upward

$$y - 0 = a(x - 0)^2$$
$$152 = a(640)^2$$
$$\frac{152}{640^2} = a$$
$$\frac{19}{51{,}200} = a$$

An equation of the cables is $y = \dfrac{19}{51{,}200}x^2$

(c)

Distance, x	Height, y
0	0
250	23.19
400	59.38
500	92.77
1000	371.09

64. Vertex: (0, 0)

$$(y - 0)^2 = 4p(x - 0)$$
$$y^2 = 4px$$

At (1000, 800): $800^2 = 4p(1000) \Rightarrow p = 160$

$$y^2 = 4(160)x$$
$$y^2 = 640x$$

66. (a)

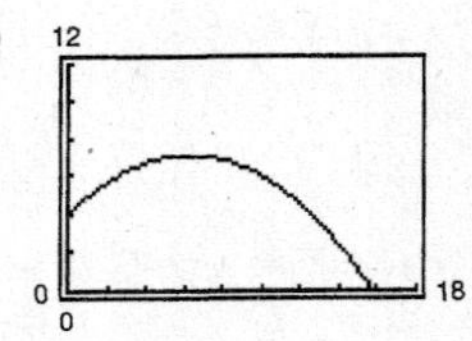

(b) Highest point: (6.25, 7.125)

Range: 15.69 feet

68. $v = \dfrac{550 \text{ mi}}{1 \text{ hr}} \cdot \dfrac{5280 \text{ ft}}{1 \text{ mi}} \cdot \dfrac{1 \text{ h}}{60 \text{ min}} \cdot \dfrac{1 \text{ min}}{60 \text{ s}} = 806\dfrac{2}{3} \text{ ft/s}$

$s = 42{,}000$

The bomb hits the ground when $y = 0$.

$$x^2 = -\frac{v^2}{16}(y - s)$$
$$x^2 = -\frac{\left(806\frac{2}{3}\right)^2}{16}(0 - 42{,}000)$$
$$x^2 \approx 1{,}708{,}120{,}000$$
$$x \approx 41{,}329$$

The distance is about 41,329 feet.

70. True. If the axis (line connecting the vertex and focus) is horizontal, then the directrix must be vertical.

72. (a) $A = \frac{8}{3}(2)^{1/2}(4)^{3/2} = \frac{8}{3}\left(\sqrt{2}\right)(8) = \frac{64\sqrt{2}}{3}$ square units

(b) As p approaches zero, the parabola becomes narrower and narrower, thus the area becomes smaller and smaller.

74. *Sample answer*: Any light ray (or other electromagnetic radiation) that enters a parabolic reflector (a surface for which any cross section containing the axis is a parabola) in a direction parallel to the axis of the surface will be reflected to the focus of the surface (the focus of any of the cross-sectional parabolas). Conversely, any ray projected from the focus in a direction that intersects the parabolic surface will be reflected in a direction parallel to the axis.

76. $f(x) = 2x^3 - 3x^2 + 50x - 75$

$$\begin{array}{r|rrrr} \frac{3}{2} & 2 & -3 & 50 & -75 \\ & & 3 & 0 & 75 \\ \hline & 2 & 0 & 50 & 0 \end{array}$$

$2x^2 + 50 = 0 \Rightarrow x^2 = -25 \Rightarrow x = \pm 5i$

Zeros: $x = \frac{3}{2}, \pm 5i$

78. $h(x) = 2x^4 + x^3 - 19x^2 - 9x + 9$

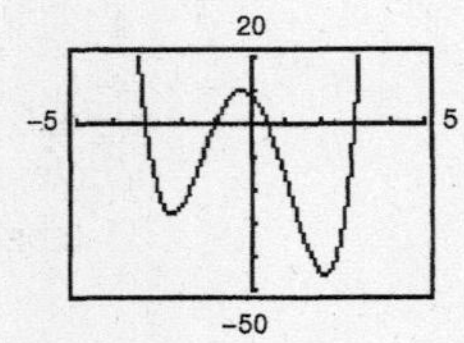

Zeros: $x = \pm 3, -1, \frac{1}{2}$

80. $B = 54°, b = 18, c = 11$

Because B is acute and $18 > 11$, one triangle is possible.

$$\sin C = \frac{c \sin B}{b} = \frac{11 \sin 54}{18} \approx 0.49440 \Rightarrow C \approx 29.63°$$

$$A = 180° - B - C \approx 180° - 54° - 29.63° = 96.37°$$

$$a = \frac{b}{\sin B}(\sin A) = \frac{18}{\sin 54}(\sin 96.37°) \approx 22.11$$

82. $B = 26°, C = 104°, a = 19$

$$A = 180° - B - C \approx 180° - 26° - 104° = 50°$$

$$b = \frac{a}{\sin A}(\sin B) = \frac{19}{\sin 50}(\sin 26°) \approx 10.87$$

$$c = \frac{a}{\sin A}(\sin C) = \frac{19}{\sin 50}(\sin 104°) \approx 24.07$$

84. $a = 58, b = 28, c = 75$

$$\cos A = \frac{b^2 + c^2 - a^2}{2bc} = \frac{784 + 5625 - 3364}{2(28)(75)} = 0.725 \Rightarrow A \approx 43.53°$$

$$\cos B = \frac{a^2 + c^2 - b^2}{2ac} = \frac{3364 + 5625 - 784}{2(58)(75)} = 0.943103 \Rightarrow B \approx 19.42°$$

$$C = 180° - A - B \approx 180° - 43.53° - 19.42° = 117.05°$$

86. $B = 71°, a = 21, c = 29$

$$b^2 = a^2 + c^2 - 2ac \cos B$$

$$b^2 = 441 + 841 - 2(21)(29)\cos 71° \approx 885.458$$

$$b \approx 29.76$$

$$\cos A = \frac{b^2 + c^2 - a^2}{2bc} \approx \frac{(29.76)^2 + 841 - 441}{2(29.76)(29)} \approx 0.74484 \Rightarrow A \approx 41.85°$$

$$C = 180° - A - B \approx 180° - 41.85° - 71° = 67.15°$$

Section 6.3 Ellipses

Solutions to Even-Numbered Exercises

2. $\frac{x^2}{9} + \frac{y^2}{4} = 1$

Center: $(0, 0)$

$a = 3, b = 2$

Horizontal major axis

Matches graph (c).

4. $\frac{y^2}{4} + \frac{x^2}{4} = 1$

Center: $(0, 0)$

Circle of radius: 2

Matches graph (f).

6. $\frac{(x + 2)^2}{9} + \frac{(y + 2)^2}{4} = 1$

Center: $(-2, -2)$

$a = 3, b = 2$

Horizontal major axis

Matches graph (e).

8. $\frac{x^2}{81} + \frac{y^2}{144} = 1$

$a^2 = 144, b^2 = 81, c^2 = 63$

Center: $(0, 0)$

Vertices: $(0, \pm 12)$

Foci: $(0, \pm 3\sqrt{7})$

Eccentricity: $e = \frac{c}{a} = \frac{\sqrt{7}}{4}$

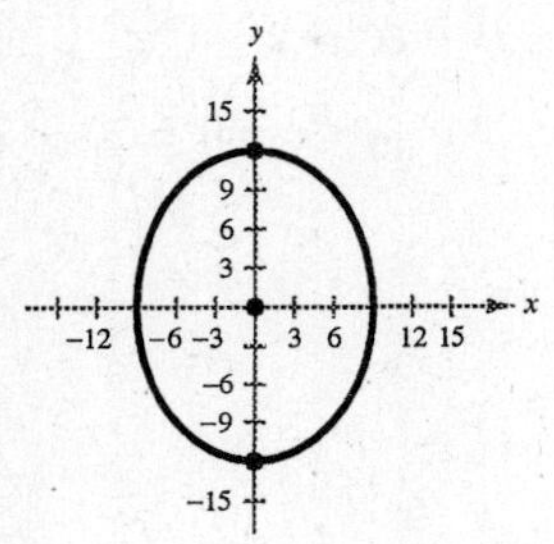

10. $\frac{x^2}{64} + \frac{y^2}{28} = 1$

$a^2 = 64, b^2 = 28, c^2 = 36$

Center: $(0, 0)$

Vertices: $(\pm 8, 0)$

Foci: $(\pm 6, 0)$

Eccentricity: $e = \frac{c}{a} = \frac{3}{4}$

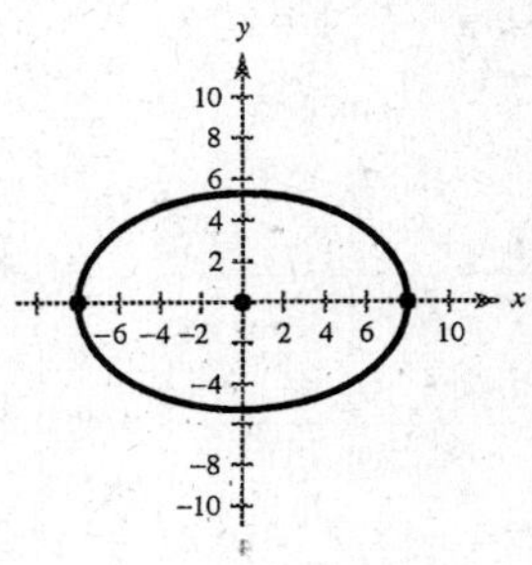

12. $\frac{(x - 4)^2}{12} + \frac{(y + 3)^2}{16} = 1$

$a^2 = 16, b^2 = 12, c^2 = 4$

Center: $(4, -3)$

Vertices: $(4, 1), (4, -7)$

Foci: $(4, -1), (4, -5)$

Eccentricity: $e = \frac{c}{a} = \frac{1}{2}$

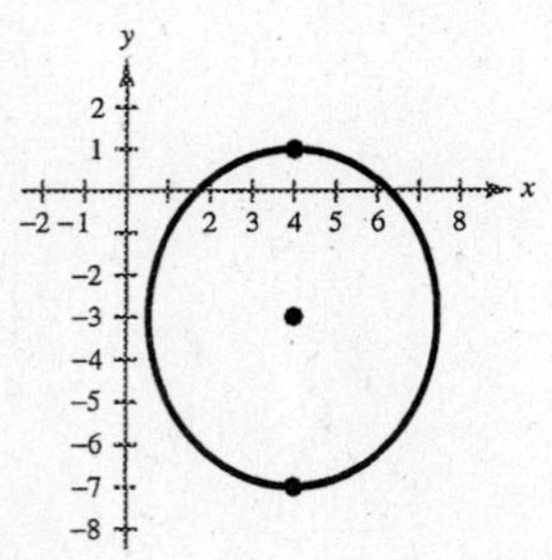

14. $\frac{(x + 2)^2}{1} + \frac{(y + 4)^2}{\frac{1}{4}} = 1$

$a^2 = 1, b^2 = \frac{1}{4}, c^2 = \frac{3}{4}$

Center: $(-2, -4)$

Vertices: $(-1, -4), (-3, -4)$

Foci: $\left(-2 \pm \frac{\sqrt{3}}{2}, -4\right) \Rightarrow \left(\frac{-4 \pm \sqrt{3}}{2}, -4\right)$

$e = \frac{\sqrt{3}}{2}$

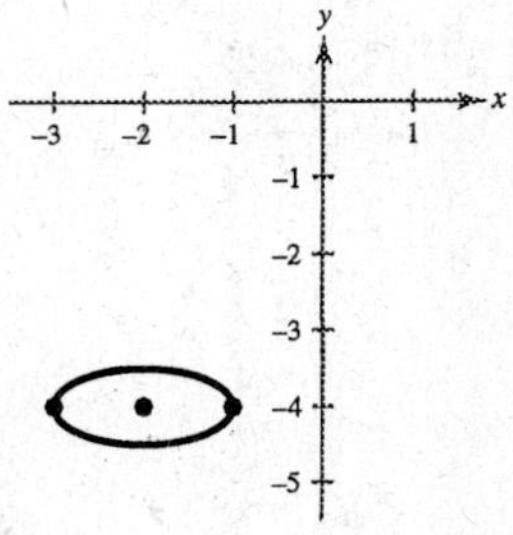

16. $9x^2 + 4y^2 - 54x + 40y + 37 = 0$

$9(x^2 - 6x + 9) + 4(y^2 + 10y + 25) = -37 + 81 + 100$

$\frac{(x-3)^2}{16} + \frac{(y+5)^2}{36} = 1$

$a^2 = 36, b^2 = 16, c^2 = 20$

Center: $(3, -5)$

Vertices: $(3, 1), (3, -11)$

Foci: $\left(3, -5 \pm 2\sqrt{5}\right)$

Eccentricity: $e = \frac{c}{a} = \frac{\sqrt{5}}{3}$

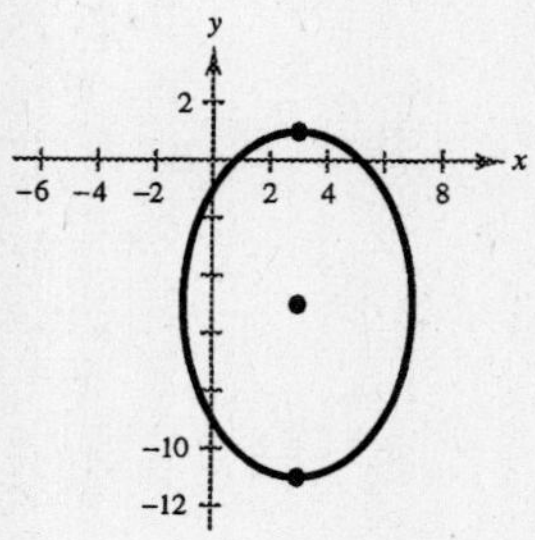

18. $3x^2 + y^2 + 18x - 2y - 8 = 0$

$3(x^2 + 6x + 9) + (y^2 - 2y + 1) = 8 + 27 + 1$

$\frac{(x+3)^2}{12} + \frac{(y-1)^2}{36} = 1$

$a^2 = 36, b^2 = 12, c^2 = 24$

Center: $(-3, 1)$

Vertices: $(-3, 7), (-3, -5)$

Foci: $\left(-3, 1 \pm 2\sqrt{6}\right)$

Eccentricity: $e = \frac{c}{a} = \frac{\sqrt{6}}{3}$

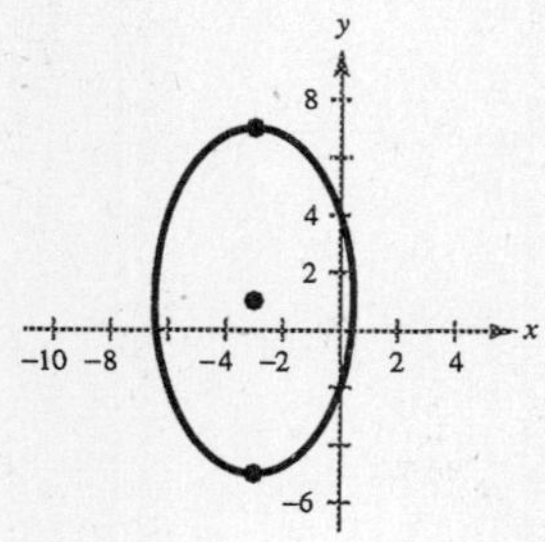

20. $x^2 + 4y^2 - 6x + 20y - 2 = 0$

$(x^2 - 6x + 9) + 4\left(y^2 + 5y + \frac{25}{4}\right) = 2 + 9 + 25$

$\frac{(x-3)^2}{36} + \frac{(y+\frac{5}{2})^2}{9} = 1$

$a^2 = 36, b^2 = 9, c^2 = 27$

Center: $\left(3, -\frac{5}{2}\right)$

Vertices: $\left(9, -\frac{5}{2}\right), \left(-3, -\frac{5}{2}\right)$

Foci: $\left(3 \pm 3\sqrt{3}, -\frac{5}{2}\right)$

Eccentricity: $e = \frac{c}{a} = \frac{\sqrt{3}}{2}$

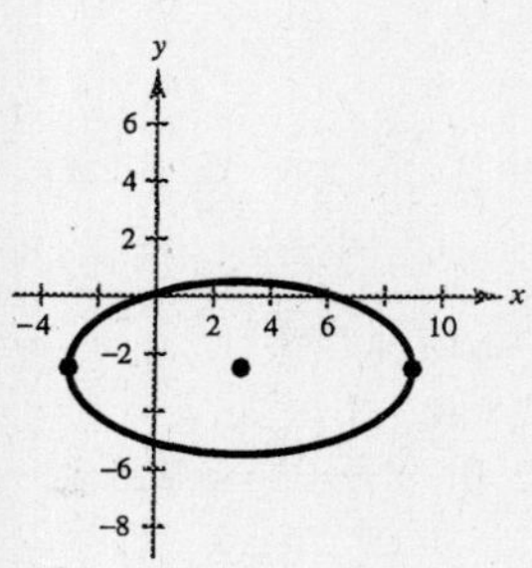

22. $9x^2 + 25y^2 - 36x - 50y + 60 = 0$

$9(x^2 - 4x + 4) + 25(y^2 - 2y + 1) = -60 + 36 + 25$

$\frac{(x-2)^2}{\frac{1}{9}} + \frac{(y-1)^2}{\frac{1}{25}} = 1$

$a^2 = \frac{1}{9}, b^2 = \frac{1}{25}, c^2 = \frac{16}{225}$

Center: $(2, 1)$

Vertices: $\left(\frac{5}{3}, 1\right), \left(\frac{7}{3}, 1\right)$

Foci: $\left(\frac{34}{15}, 1\right), \left(\frac{26}{15}, 1\right)$

Eccentricity: $e = \frac{c}{a} = \frac{4}{5}$

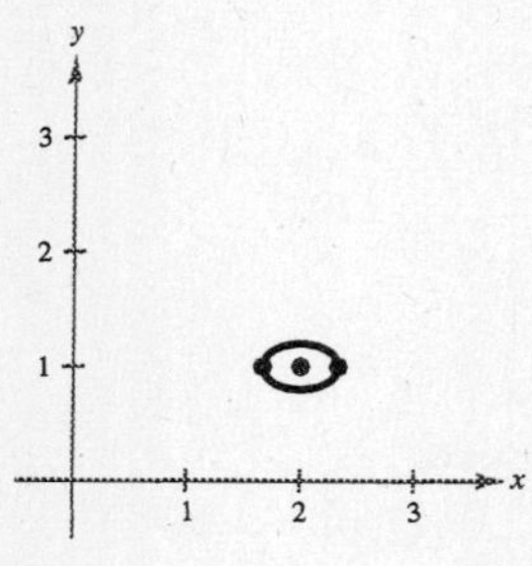

24. $3x^2 + 4y^2 = 12$

$$\frac{x^2}{4} + \frac{y^2}{3} = 1$$

$a^2 = 4, b^2 = 3, c^2 = 1$

Center: $(0, 0)$

Vertices: $(\pm 2, 0)$

Foci: $(\pm 1, 0)$

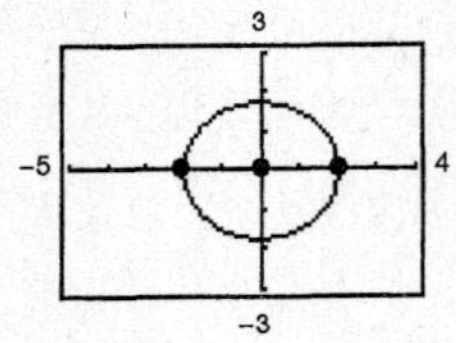

26. $36x^2 + 9y^2 + 48x - 36y - 72 = 0$

$$36\left(x^2 + \frac{4}{3}x + \frac{4}{9}\right) + 9(y^2 - 4y + 4) = 72 + 16 + 36$$

$$\frac{(x + \frac{2}{3})^2}{\frac{31}{9}} + \frac{(y - 2)^2}{\frac{124}{9}} = 1$$

$$a^2 = \frac{124}{9}, b^2 = \frac{31}{9}, c^2 = \frac{31}{3}$$

Center: $\left(-\frac{2}{3}, 2\right)$

Vertices: $\left(-\frac{2}{3}, 2 \pm \frac{2\sqrt{31}}{3}\right)$

Foci: $\left(-\frac{2}{3}, 2 \pm \frac{\sqrt{93}}{3}\right)$

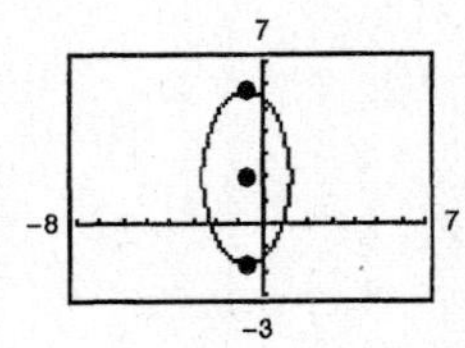

28. Vertices: $(\pm 2, 0) \implies a = 2$

Endpoints of minor axis: $\left(0, \pm\frac{3}{2}\right) \implies b = \frac{3}{2}$

$$\frac{x^2}{a^2} + \frac{y^2}{b^2} = 1$$

$$\frac{x^2}{2^2} + \frac{y^2}{(\frac{3}{2})^2} = 1$$

$$\frac{x^2}{4} + \frac{4y^2}{9} = 1$$

30. Vertices: $(0, \pm 8) \implies a = 8$

Foci: $(0, \pm 4) \implies c = 4$

$b^2 = a^2 - c^2 = 64 - 16 = 48$

$$\frac{x^2}{b^2} + \frac{y^2}{a^2} = 1$$

$$\frac{x^2}{48} + \frac{y^2}{64} = 1$$

32. Foci: $(\pm 2, 0) \implies c = 2$

Major axis length: $8 \implies a = 4$

$b^2 = a^2 - c^2 = 16 - 4 = 12$

$$\frac{x^2}{a^2} + \frac{y^2}{b^2} = 1$$

$$\frac{x^2}{16} + \frac{y^2}{12} = 1$$

34. Major axis vertical

Passes through: $(0, 4)$ and $(2, 0)$

$a = 4, b = 2$

$$\frac{x^2}{b^2} + \frac{y^2}{a^2} = 1$$

$$\frac{x^2}{4} + \frac{y^2}{16} = 1$$

36. Vertices: $(4, \pm 4) \implies a = 4$

Center: $(4, 0) \implies h = 4, k = 0$

Endpoints of minor axis: $(1, 0), (7, 0) \implies b = 3$

$$\frac{(x - h)^2}{b^2} + \frac{(y - k)^2}{a^2} = 1$$

$$\frac{(x - 4)^2}{9} + \frac{y^2}{16} = 1$$

38. Vertices: $(0, -1), (4, -1) \implies a = 2$

Center: $(2, -1) \implies h = 2, k = -1$

Endpoints of minor axis: $(2, 0), (2, -2) \implies b = 1$

$$\frac{(x - h)^2}{a^2} + \frac{(y - k)^2}{b^2} = 1$$

$$\frac{(x - 2)^2}{4} + \frac{(y + 1)^2}{1} = 1$$

40. Foci: $(0, 0), (4, 0) \implies c = 2, h = 2, k = 0$

Major axis length: $8 \implies a = 4$

$b^2 = a^2 - c^2 = 16 - 4 = 12$

$$\frac{(x-h)^2}{a^2} + \frac{(y-k)^2}{b^2} = 1$$

$$\frac{(x-2)^2}{16} + \frac{y^2}{12} = 1$$

42. Center: $(2, -1) \implies h = 2, k = -1$

Vertex: $\left(2, \frac{1}{2}\right) \implies a = \frac{3}{2}$

Minor axis length: $2 \implies b = 1$

$$\frac{(x-h)}{b^2} + \frac{(y-k)^2}{a^2} = 1$$

$$\frac{(x-2)^2}{1} + \frac{(y+1)^2}{\left(\frac{3}{2}\right)^2} = 1$$

$$\frac{(x-2)^2}{1} + \frac{4(y+1)^2}{9} = 1$$

44. Center: $(3, 2) \implies h = 3, k = 2$

$a = 3c$

Foci: $(1, 2), (5, 2) \implies c = 2, a = 6$

$b^2 = a^2 - c^2 = 36 - 4 = 32$

$$\frac{(x-h)^2}{a^2} + \frac{(y-k)^2}{b^2} = 1$$

$$\frac{(x-3)^2}{36} + \frac{(y-2)^2}{32} = 1$$

46. Vertices: $(5, 0), (5, 12) \implies a = 6$

Endpoints of the minor axis:

$(1, 6), (9, 6) \implies b = 4$

Center: $(5, 6) \implies h = 5, k = 6$

$$\frac{(x-h)^2}{b^2} + \frac{(y-k)^2}{a^2} = 1$$

$$\frac{(x-5)^2}{16} + \frac{(y-6)^2}{36} = 1$$

48. Vertices: $(0, \pm 8) \implies a = 8$

$e = \frac{1}{2} \implies \frac{c}{a} = \frac{1}{2}, c = 4$

$b^2 = a^2 - c^2 = 64 - 16 = 48$

Center: $(0, 0)$

$$\frac{x^2}{b^2} + \frac{y^2}{a^2} = 1$$

$$\frac{x^2}{48} + \frac{y^2}{64} = 1$$

50. The tacks should be placed at the foci and the length of the string is the length of the major axis, $2a$.

Center: $(0, 0)$

$a = 3, b = 2, c = \sqrt{5}$

Foci (Positions of the tacks): $(\pm\sqrt{5}, 0)$

Length of string: 6 feet

52. (a) $A = \frac{36.18}{2} = 18.09$

$e = \frac{c}{a} = 0.97$

$c = ea \approx 17.55$

$b^2 = a^2 - c^2 \approx 19.34$

$$\frac{x^2}{a^2} + \frac{y^2}{b^2} = 1$$

$$\frac{x^2}{327.25} + \frac{y^2}{19.34} = 1$$

(b)

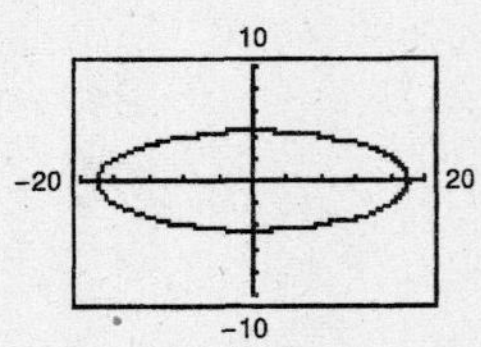

(c) The sun is at one focus. Because $c \approx 17.5473$ astronomical units, the apogee is about $18.09 + 17.5473 \approx 35.64$ astronomical units and the perigee is about $18.09 - 17.5473 \approx 0.54$ astronomical units.

54. For $\frac{x^2}{a^2} + \frac{y^2}{b^2} = 1$, we have $c^2 = a^2 - b^2$

When $x = c$: $\frac{c^2}{a^2} + \frac{y^2}{b^2} = 1$

$$y^2 = b^2\left(1 - \frac{a^2 - b^2}{a^2}\right) = \frac{b^4}{a^2}$$

$$y = \frac{b^2}{a}$$

Length of latus rectum: $2y = \frac{2b^2}{a}$

56. $\dfrac{x^2}{4} + \dfrac{y^2}{1} = 1$

$a = 2, b = 1, c = \sqrt{3}$

Points on the ellipse: $(\pm 2, 0), (0, \pm 1)$

Length of latera recta: $\dfrac{2b^2}{a} = \dfrac{2(1)^2}{2} = 1$

Additional points: $\left(-\sqrt{3}, \pm\dfrac{1}{2}\right), \left(\sqrt{3}, \pm\dfrac{1}{2}\right)$

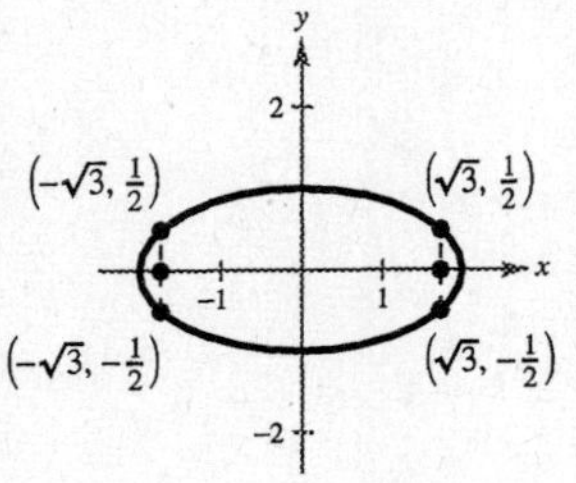

58. $5x^2 + 3y^2 = 15$

$\dfrac{x^2}{3} + \dfrac{y^2}{5} = 1$

$a = \sqrt{5}, b = \sqrt{3}, c = \sqrt{2}$

Points on the ellipse: $(\pm\sqrt{3}, 0), (0, \pm\sqrt{5})$

Length of latus recta: $\dfrac{2b^2}{a} = \dfrac{2 \cdot 3}{\sqrt{5}} = \dfrac{6\sqrt{5}}{5}$

Additional points: $\left(\pm\dfrac{3\sqrt{5}}{5}, -\sqrt{2}\right), \left(\pm\dfrac{3\sqrt{5}}{5}, \sqrt{2}\right)$

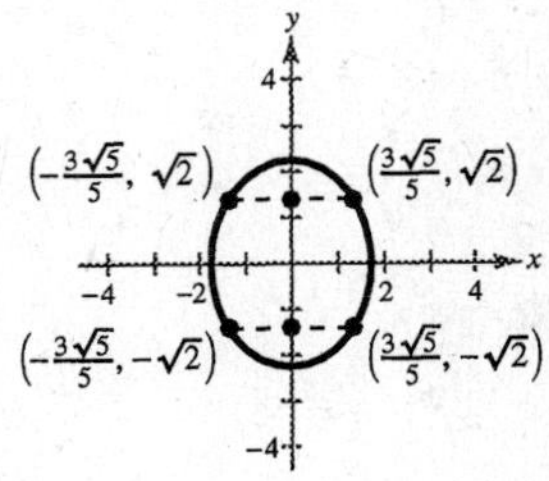

60. True. If e is close to 1, the ellipse is elongated and the foci are close to the vertices.

62. (a) Length of string $= 2a$

(b) By keeping the string taut, the sum of the distances from the two fixed points is constant (equal to the length of the string).

64. $100 - (x - 5)^2 = 0$

$(x - 5)^2 = 100$

$x - 5 = \pm 10$

$x - 5 = -10 \Rightarrow x = -5$

$x - 5 = 10 \Rightarrow x = 15$

66. $16x^2 - 40x + 25 = 0$

$(4x - 5)(4x - 5) = 0$

$4x - 5 = 0 \Rightarrow x = \frac{5}{4}$

68. $f(x) = x^2 - 10x$

(a) $f(4) = (4)^2 - 10(4) = 16 - 40 = -24$

(b) $f(-8) = (-8)^2 - 10(-8) = 64 + 80 = 144$

(c) $f(x - 4) = (x - 4)^2 - 10(x - 4) = x^2 - 8x + 16 - 10x + 40$

$= x^2 - 18x + 56$

70. $f(x) = \begin{cases} -3x, & x < -1 \\ 2x + 1, & x \geq -1 \end{cases}$

(a) $f(-1) = 2(-1) + 1 = -1$

(b) $f(7) = 2(7) + 1 = 15$

(c) $f(-6) = -3(-6) = 18$

Section 6.4 Hyperbolas

Solutions to Even-Numbered Exercises

2. $\dfrac{y^2}{25} - \dfrac{x^2}{9} = 1$

Center: $(0, 0)$

$a = 5, b = 3$

Vertical transverse axis

Matches graph (c).

4. $\dfrac{(x + 1)^2}{16} - \dfrac{(y - 2)^2}{9} = 1$

Center: $(-1, 2)$

$a = 4, b = 3$

Horizontal transverse axis

Matches graph (d).

6. $\dfrac{x^2}{9} - \dfrac{y^2}{25} = 1$

$a = 3, b = 5$

$c = \sqrt{3^2 + 5^2} = \sqrt{34}$

Center: $(0, 0)$

Vertices: $(\pm 3, 0)$

Foci: $\left(\pm\sqrt{34}, 0\right)$

Asymptotes: $y = \pm\dfrac{5}{3}x$

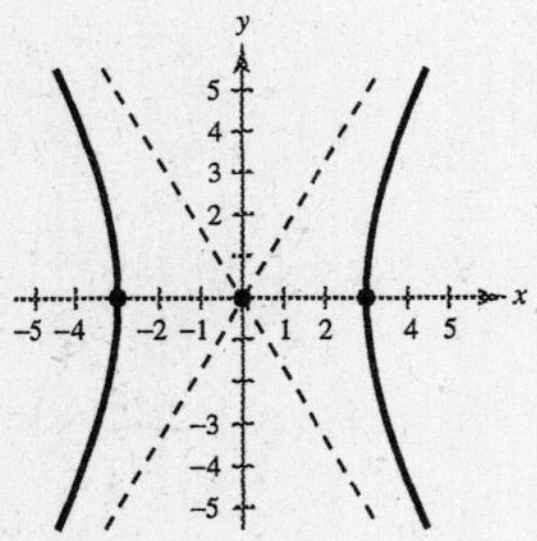

8. $\dfrac{x^2}{36} - \dfrac{y^2}{4} = 1$

$a = 6, b = 2,$

$c = \sqrt{36 + 4} = 2\sqrt{10}$

Center: $(0, 0)$

Vertices: $(\pm 6, 0)$

Foci: $\left(\pm 2\sqrt{10}, 0\right)$

Asymptotes: $y = \pm\dfrac{1}{3}x$

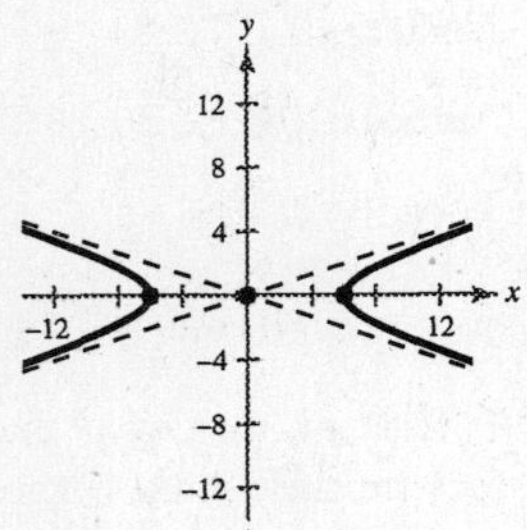

10. $\dfrac{(x + 3)^2}{144} - \dfrac{(y - 2)^2}{25} = 1$

$a = 12, b = 5$

$c = \sqrt{144 + 25} = 13$

Center: $(-3, 2)$

Vertices: $(9, 2), (-15, 2)$

Foci: $(10, 2), (-16, 2)$

Asymptotes: $y = 2 \pm \dfrac{5}{12}(x + 3)$

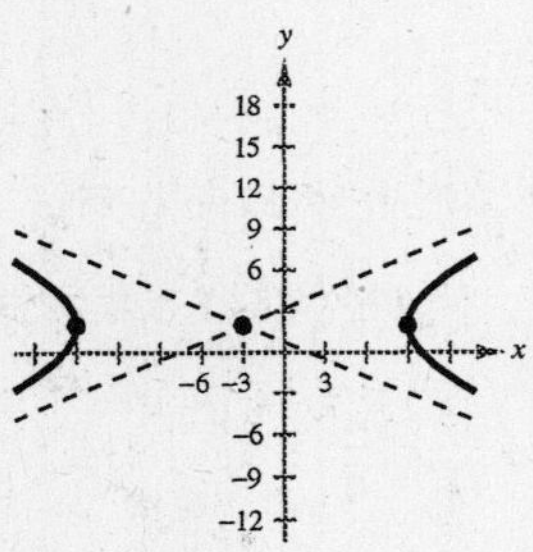

12. $\dfrac{(y - 1)^2}{\frac{1}{4}} - \dfrac{(x + 3)^2}{\frac{1}{16}} = 1$

$a = \dfrac{1}{2}, b = \dfrac{1}{4}$

$c = \sqrt{\dfrac{1}{4} + \dfrac{1}{16}} = \dfrac{\sqrt{5}}{4}$

Center: $(-3, 1)$

Vertices: $\left(-3, \dfrac{3}{2}\right), \left(-3, \dfrac{1}{2}\right)$

Foci: $\left(-3, 1 \pm \dfrac{\sqrt{5}}{4}\right)$

Asymptotes: $y = 1 \pm 2(x + 3)$

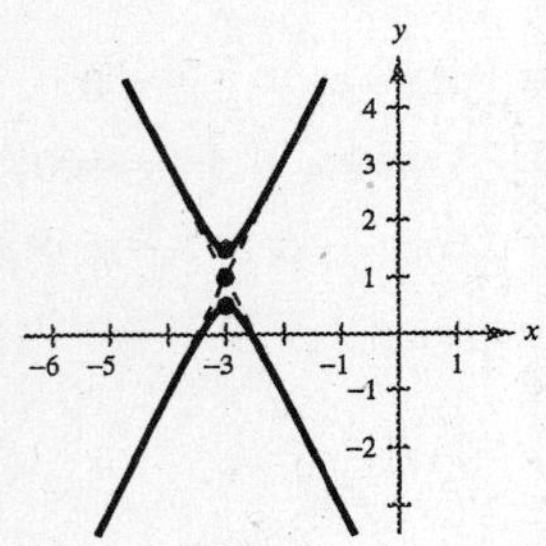

14. $x^2 - 9y^2 + 36y - 72 = 0$

$$x^2 - 9(y^2 - 4y + 4) = 72 - 36$$

$$x^2 - 9(y - 2)^2 = 36$$

$$\frac{x^2}{36} - \frac{(y - 2)^2}{4} = 1$$

$a = 6, b = 2, c = \sqrt{36 + 4} = 2\sqrt{10}$

Center: $(0, 2)$

Vertices: $(\pm 6, 2)$

Foci: $\left(\pm 2\sqrt{10}, 2\right)$

Asymptotes: $y = 2 \pm \dfrac{1}{3}x$

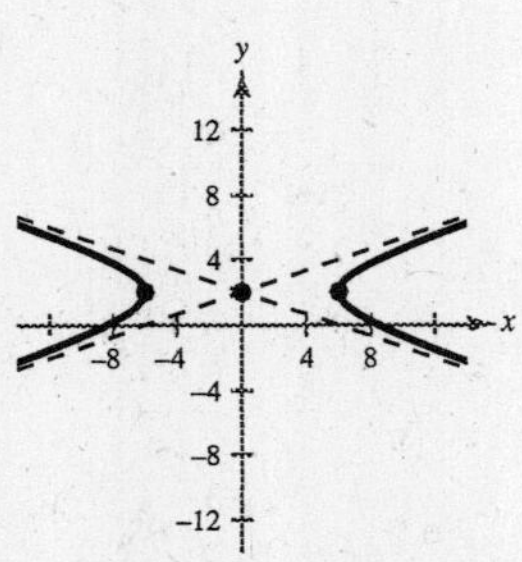

16. $16y^2 - x^2 + 2x + 64y + 63 = 0$

$$16(y^2 + 4y + 4) - (x^2 - 2x + 1) = -63 + 64 - 1$$
$$16(y + 2)^2 - (x - 1)^2 = 0$$
$$y + 2 = \pm\tfrac{1}{4}(x - 1)$$

Degenerate hyperbola: The graph is two lines intersecting at $(1, -2)$.

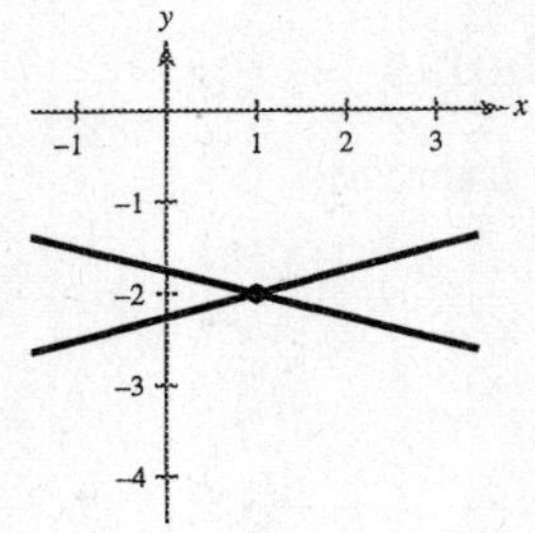

18. $6y^2 - 3x^2 = 18$

$$\frac{y^2}{3} - \frac{x^2}{6} = 1$$

$a = \sqrt{3}, b = \sqrt{6}, c = \sqrt{3 + 6} = 3$

Center: $(0, 0)$

Vertices: $(0, \pm\sqrt{3})$

Foci: $(0, \pm 3)$

Asymptotes: $y = \pm\dfrac{\sqrt{2}}{2}x$

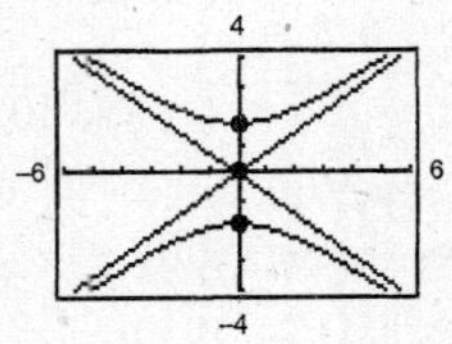

20. $9x^2 - y^2 + 54x + 10y + 55 = 0$

$$9(x^2 + 6x + 9) - (y^2 - 10y + 25) = -55 + 81 - 25$$
$$\frac{(x + 3)^2}{\frac{1}{9}} - \frac{(y - 5)^2}{1} = 1$$

$a = \dfrac{1}{3}, b = 1, c = \dfrac{\sqrt{10}}{3}$

Center: $(-3, 5)$

Vertices: $\left(-3 \pm \dfrac{1}{3}, 5\right) \Rightarrow \left(-\dfrac{10}{3}, 5\right), \left(-\dfrac{8}{3}, 5\right)$

Foci: $\left(-3 \pm \dfrac{\sqrt{10}}{3}, 5\right)$

Asymptotes:
$y = 5 \pm 3(x + 3)$

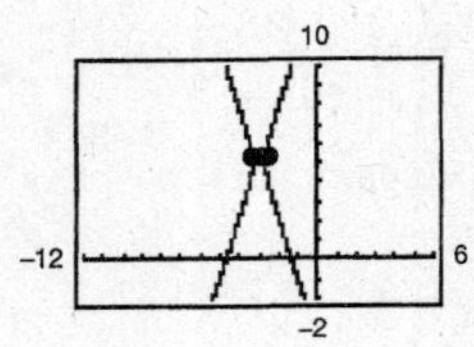

22. Vertices: $(\pm 4, 0) \Rightarrow a = 4$

Foci: $(\pm 6, 0) \Rightarrow c = 6$

$b^2 = c^2 - a^2 = 36 - 16 = 20 \Rightarrow b = 2\sqrt{5}$

$$\frac{x^2}{a^2} - \frac{y^2}{b^2} = 1$$

$$\frac{x^2}{16} - \frac{y^2}{20} = 1$$

24. Vertices: $(0, \pm 3) \Rightarrow a = 3$

Asymptotes: $y = \pm 3x \Rightarrow \dfrac{a}{b} = 3, b = 1$

$$\frac{y^2}{a^2} - \frac{x^2}{b^2} = 1$$

$$\frac{y^2}{9} - \frac{x^2}{1} = 1$$

26. Foci: $(\pm 10, 0) \Rightarrow c = 10$

Asymptotes: $y = \pm\dfrac{3}{4}x \Rightarrow \dfrac{b}{a} = \dfrac{3m}{4m}$

$$c^2 = a^2 + b^2 \Rightarrow 100 = (3m)^2 + (4m)^2$$
$$100 = 25m^2$$
$$2 = m$$

$a = 4(2) = 8$

$b = 3(2) = 6$

$$\frac{x^2}{a^2} - \frac{y^2}{b^2} = 1$$

$$\frac{x^2}{64} - \frac{y^2}{36} = 1$$

28. Vertices: $(2, 3), (2, -3) \Rightarrow a = 3$

Center: $(2, 0)$

Foci: $(2, 6), (2, -6) \Rightarrow c = 6$

$b^2 = c^2 - a^2 = 36 - 9 = 27$

$$\frac{(y-k)^2}{a^2} - \frac{(x-h)^2}{b^2} = 1$$

$$\frac{y^2}{9} - \frac{(x-2)^2}{27} = 1$$

30. Vertices: $(-2, 1),\ (2, 1) \Rightarrow a = 2$

Center: $(0, 1)$

Foci: $(-3, 1),\ (3, 1) \Rightarrow c = 3$

$b^2 = c^2 - a^2 = 9 - 4 = 5$

$$\frac{(x-h)^2}{a^2} - \frac{(y-k)^2}{b^2} = 1$$

$$\frac{x^2}{4} - \frac{(y-1)^2}{5} = 1$$

32. Vertices: $(-2, 1), (2, 1) \Rightarrow a = 2$

Center: $(0, 1)$

Point on curve: $(5, 4)$

$$\frac{(x-h)^2}{a^2} - \frac{(y-k)^2}{b^2} = 1$$

$$\frac{x^2}{4} - \frac{(y-1)^2}{b^2} = 1$$

$$\frac{25}{4} - \frac{9}{b^2} = 1$$

$$b^2 = \frac{12}{7}$$

$$\frac{x^2}{4} - \frac{(y-1)^2}{\frac{12}{7}} = 1$$

$$\frac{x^2}{4} - \frac{7(y-1)^2}{12} = 1$$

34. Vertices: $(1, \pm 2) \Rightarrow a = 2$

Center: $(1, 0)$

Point on curve: $\left(0, \sqrt{5}\right)$

$$\frac{(y-k)^2}{a^2} - \frac{(x-h)^2}{b^2} = 1$$

$$\frac{y^2}{4} - \frac{(x-1)^2}{b^2} = 1$$

$$\frac{5}{4} - \frac{1}{b^2} = 1$$

$$b^2 = 4$$

$$\frac{y^2}{4} - \frac{(x-1)^2}{4} = 1$$

36. Vertices: $(3, 0), (3, 6) \Rightarrow a = 3$

Center: $(3, 3)$

Asymptotes: $y = 6 - x, y = x$

$\frac{a}{b} = 1 \Rightarrow b = 3$

$$\frac{(y-k)^2}{a^2} - \frac{(x-h)^2}{b^2} = 1$$

$$\frac{(y-3)^2}{9} - \frac{(x-3)^2}{9} = 1$$

38. Vertices: $(3, 0), (3, 4) \Rightarrow a = 2$

Asymptotes: $y = \frac{2}{3}x,\ y = 4 - \frac{2}{3}x$

$\frac{a}{b} = \frac{2}{3} \Rightarrow b = 3$

Center: $(3, 2) = (h, k)$

$$\frac{(y-k)^2}{a^2} - \frac{(x-h)^2}{b^2} = 1$$

$$\frac{(y-2)^2}{4} - \frac{(x-3)^2}{9} = 1$$

40. Center: $(0, 0) = (h, k)$

Focus: $(24, 0) \Rightarrow c = 24$

Solution point: $(24, 24)$

$24^2 = a^2 + b^2 \Rightarrow b^2 = 24^2 - a^2$

$$\frac{(x-h)^2}{a^2} - \frac{(y-k)^2}{b^2} = 1$$

$$\frac{x^2}{a^2} - \frac{y^2}{24^2 - a^2} = 1 \Rightarrow \frac{24^2}{a^2} - \frac{24^2}{24^2 - a^2} = 1$$

Solving yields $a^2 = \frac{\left(3 - \sqrt{5}\right)24^2}{2} \approx 220.0124$ and

$$b^2 \approx 355.9876.$$

Thus, we have $\frac{x^2}{220.0124} - \frac{y^2}{355.9876} = 1.$

The right vertex is at $(a, 0) \approx (14.83, 0)$.

42. $x^2 + 4y^2 - 6x + 16y + 21 = 0$

$AC = (1)(4) > 0$ and $A \neq C$

The graph is an ellipse.

44. $y^2 - 6y - 4x + 21 = 0$

$AC = (0)(1) = 0$

The graph is a parabola.

46. $4y^2 - 2x^2 - 4y - 8x - 15 = 0$

$AC = (-2)(4) < 0$

The graph is a hyperbola.

48. $4y^2 + 4x^2 - 24x + 35 = 0$

$A = C$

The graph is a circle.

50. False. For the trivial solution of two intersecting lines to occur, the standard form of the equation of the hyperbola would be equal to zero.

$$\frac{(x-h)^2}{a^2} - \frac{(y-k)^2}{b^2} = 0 \quad \text{or} \quad \frac{(y-k)^2}{a^2} - \frac{(x-h)^2}{b^2} = 0$$

52. The extended diagonals of the central rectangle are the asymptotes of the hyperbola.

54. $x^2 + 14x + 49 = x^2 + 2(7)x + 7^2 = (x + 7)^2$

56.
$$\begin{aligned} 6x^3 - 11x^2 - 10x &= x(6x^2 - 11x - 10) \\ &= x(6x^2 - 15x + 4x - 10) \\ &= x[3x(2x - 5) + 2(2x - 5)] \\ &= x(3x + 2)(2x - 5) \end{aligned}$$

58. $y = \sin \pi x$

Period: $\dfrac{2\pi}{\pi} = 2$

Amplitude: 1

Key points: $(0, 0), \left(\frac{1}{2}, 1\right), (1, 0), \left(\frac{3}{2}, -1\right), (2, 0)$

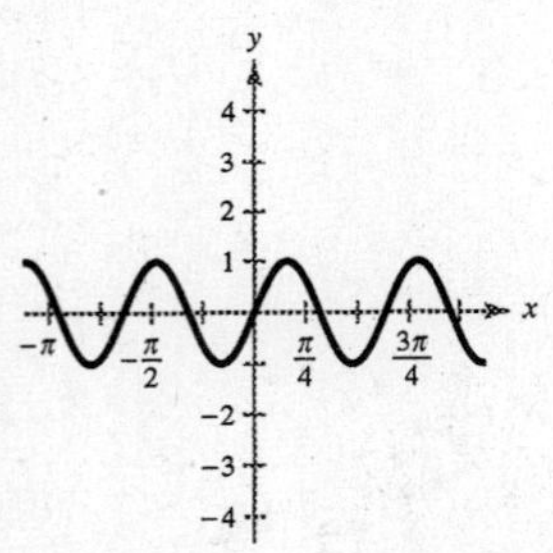

60. $y = -\dfrac{1}{2}\sec x$

Graph $y = -\dfrac{1}{2}\cos x$ first.

Period: 2π

One cycle: 0 to 2π

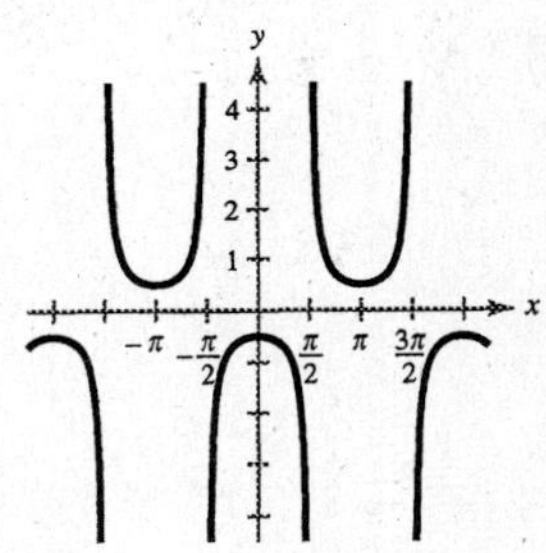

Section 6.5 Rotation of Conics

Solutions to Even-Numbered Exercises

2. $\theta = 45°$; Point: $(3, 3)$

$x' = x\cos\theta - y\sin\theta = 3\cos 45° - 3\sin 45° = 0$

$y' = x\sin\theta + y\cos\theta = 3\sin 45° + 3\cos 45° = 3\sqrt{2}$

Thus, $(x', y') = (0, 3\sqrt{2})$.

4. $\theta = 60°$; Point: $(3, 1)$

$x' = x\cos\theta - y\sin\theta = 3\cos 60° - 1\sin 60° = \dfrac{3}{2} - \dfrac{\sqrt{3}}{2}$

$y' = x\sin\theta + y\cos\theta = 3\sin 60° + 1\cos 60° = \dfrac{3\sqrt{3}}{2} + \dfrac{1}{2}$

Thus, $(x', y') = \left(\dfrac{1}{2}(3 - \sqrt{3}), \dfrac{1}{2}(3\sqrt{3} + 1)\right)$.

6. $\theta = 30°$; Point: $(2, 4)$

$x' = x\cos\theta - y\sin\theta = 2\cos 30° - 4\sin 30° = \sqrt{3} - 2$

$y' = x\sin\theta + y\cos\theta = 2\sin 30° + 4\cos 30° = 1 + 2\sqrt{3}$

Thus, $(x', y') = (\sqrt{3} - 2, 1 + 2\sqrt{3})$

8. $xy - 2 = 0$

$A = 0, B = 1, C = 0$

$\cot 2\theta = \dfrac{A - C}{B} = 0 \implies 2\theta = \dfrac{\pi}{2} \implies \theta = \dfrac{\pi}{4}$

$x = x'\cos\dfrac{\pi}{4} - y'\sin\dfrac{\pi}{4} = \dfrac{x' - y'}{\sqrt{2}}$ $\qquad$ $y = x'\sin\dfrac{\pi}{4} + y'\cos\dfrac{\pi}{4} = \dfrac{x' + y'}{\sqrt{2}}$

$$xy - 2 = 0$$

$$\left(\frac{x' - y'}{\sqrt{2}}\right)\left(\frac{x' + y'}{\sqrt{2}}\right) - 2 = 0$$

$$\frac{(x')^2 - (y')^2}{2} = 2$$

$$\frac{(x')^2}{4} - \frac{(y')^2}{4} = 1$$

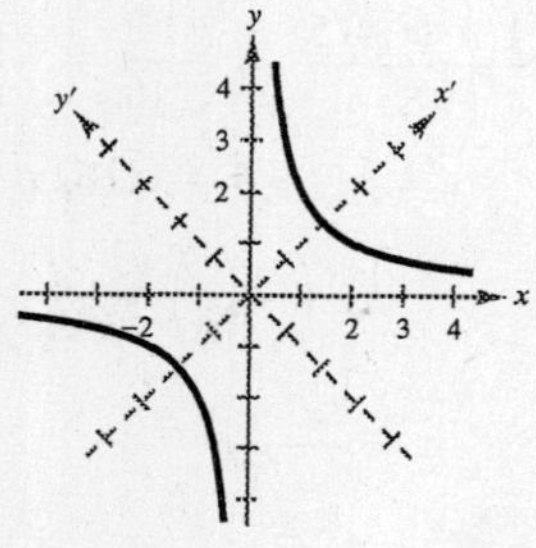

10. $xy + x - 2y + 3 = 0$

$A = 0, B = 1, C = 0$

$\cot 2\theta = \frac{A - C}{B} = 0 \Rightarrow 2\theta = \frac{\pi}{2} \Rightarrow \theta = \frac{\pi}{4}$

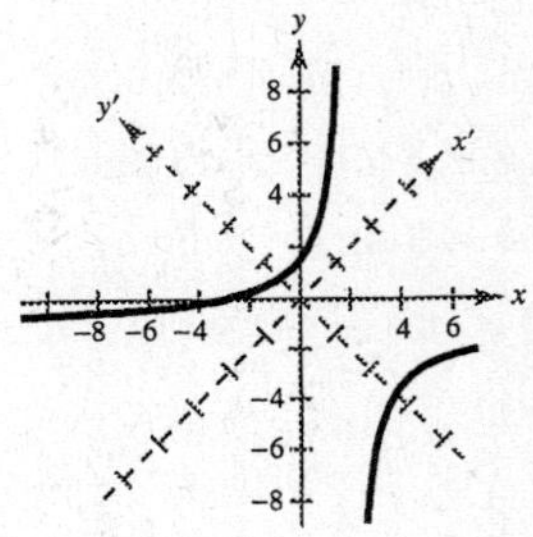

$$x = x' \cos\frac{\pi}{4} - y' \sin\frac{\pi}{4} = x'\left(\frac{\sqrt{2}}{2}\right) - y'\left(\frac{\sqrt{2}}{2}\right) = \frac{x' - y'}{\sqrt{2}}$$

$$y = x' \sin\frac{\pi}{4} + y' \cos\frac{\pi}{4} = x'\left(\frac{\sqrt{2}}{2}\right) + y'\left(\frac{\sqrt{2}}{2}\right) = \frac{x' + y'}{\sqrt{2}}$$

$$xy + x - 2y + 3 = 0$$

$$\left(\frac{x' - y'}{\sqrt{2}}\right)\left(\frac{x' + y'}{\sqrt{2}}\right) + \left(\frac{x' - y'}{\sqrt{2}}\right) - 2\left(\frac{x' + y'}{\sqrt{2}}\right) + 3 = 0$$

$$\frac{(x')^2}{2} - \frac{(y')^2}{2} + \frac{x'}{\sqrt{2}} - \frac{y'}{\sqrt{2}} - \frac{2x'}{\sqrt{2}} - \frac{2y'}{\sqrt{2}} + 3 = 0$$

$$\left[(x')^2 - \sqrt{2}x' + \left(\frac{\sqrt{2}}{2}\right)^2\right] - \left[(y')^2 + 3\sqrt{2}y' + \left(\frac{3\sqrt{2}}{2}\right)^2\right] = -6 + \left(\frac{\sqrt{2}}{2}\right)^2 - \left(\frac{3\sqrt{2}}{2}\right)^2$$

$$\left(x' - \frac{\sqrt{2}}{2}\right)^2 - \left(y' + \frac{3\sqrt{2}}{2}\right)^2 = -10$$

$$\frac{\left(y' + \frac{3\sqrt{2}}{2}\right)^2}{10} - \frac{\left(x' - \frac{\sqrt{2}}{2}\right)^2}{10} = 1$$

12. $2x^2 - 3xy - 2y^2 + 10 = 0$

$A = 2, B = -3, C = -2$

$\cot 2\theta = \frac{A - C}{B} = -\frac{4}{3} \Rightarrow \theta \approx 71.57°$

$\cos 2\theta = -\frac{4}{5}$

$\sin\theta = \sqrt{\frac{1 - \cos 2\theta}{2}} = \sqrt{\frac{1 - (-4/5)}{2}} = \frac{3}{\sqrt{10}}$

$\cos\theta = \sqrt{\frac{1 + \cos 2\theta}{2}} = \sqrt{\frac{1 + (-4/5)}{2}} = \frac{1}{\sqrt{10}}$

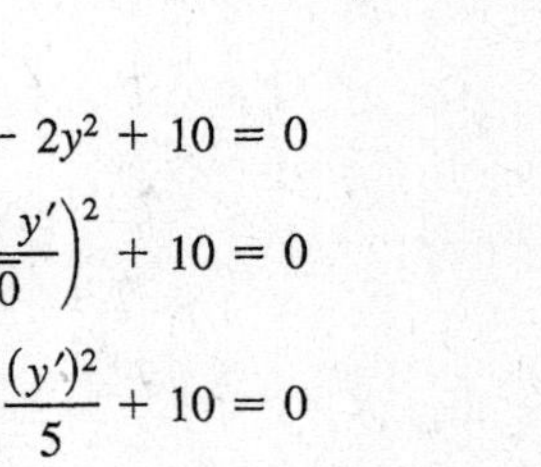

$$x = x' \cos\theta - y' \sin\theta = x'\left(\frac{1}{\sqrt{10}}\right) - y'\left(\frac{3}{\sqrt{10}}\right) = \frac{x' - 3y'}{\sqrt{10}}$$

$$y = x' \sin\theta + y' \cos\theta = x'\left(\frac{3}{\sqrt{10}}\right) + y'\left(\frac{1}{\sqrt{10}}\right) = \frac{3x' + y'}{\sqrt{10}}$$

$$2x^2 - 3xy - 2y^2 + 10 = 0$$

$$2\left(\frac{x' - 3y'}{\sqrt{10}}\right)^2 - 3\left(\frac{x' - 3y'}{\sqrt{10}}\right)\left(\frac{3x' + y'}{\sqrt{10}}\right) - 2\left(\frac{3x' + y'}{\sqrt{10}}\right)^2 + 10 = 0$$

$$\frac{(x')^2}{5} - \frac{6x'y'}{5} + \frac{9(y')^2}{5} - \frac{9(x')^2}{10} + \frac{24x'y'}{10} + \frac{9(y')^2}{10} - \frac{9(x')^2}{5} - \frac{6x'y'}{5} - \frac{(y')^2}{5} + 10 = 0$$

$$-\frac{5}{2}(x')^2 + \frac{5}{2}(y')^2 = -10$$

$$\frac{(x')^2}{4} - \frac{(y')^2}{4} = 1$$

14. $13x^2 + 6\sqrt{3}xy + 7y^2 - 16 = 0$

$A = 13, B = 6\sqrt{3}, C = 7$

$$\cot 2\theta = \frac{A - C}{B} = \frac{1}{\sqrt{3}} \implies 2\theta = \frac{\pi}{3} \implies \theta = \frac{\pi}{6}$$

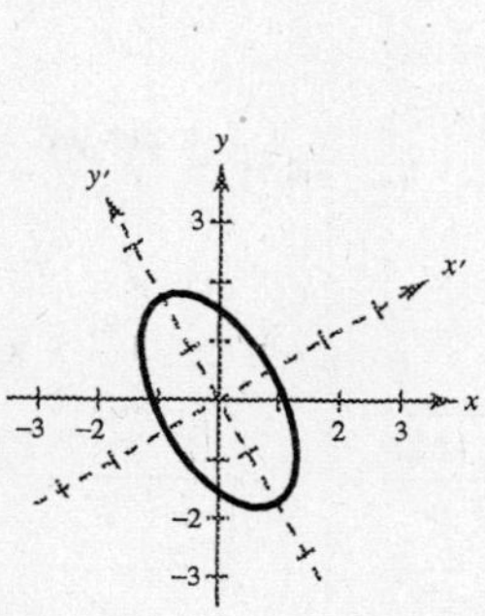

$$x = x'\cos\frac{\pi}{6} - y'\sin\frac{\pi}{6} \qquad y = x'\sin\frac{\pi}{6} + y'\cos\frac{\pi}{6}$$

$$= x'\left(\frac{\sqrt{3}}{2}\right) - y'\left(\frac{1}{2}\right) \qquad = x'\left(\frac{1}{2}\right) + y'\left(\frac{\sqrt{3}}{2}\right)$$

$$= \frac{\sqrt{3}x' - y'}{2} \qquad = \frac{x' + \sqrt{3}y'}{2}$$

$$13x^2 + 6\sqrt{3}xy + 7y^2 - 16 = 0$$

$$13\left(\frac{\sqrt{3}x' - y'}{2}\right)^2 + 6\sqrt{3}\left(\frac{\sqrt{3}x' - y'}{2}\right)\left(\frac{x' + \sqrt{3}y'}{2}\right) + 7\left(\frac{x' + \sqrt{3}y'}{2}\right)^2 - 16 = 0$$

$$\frac{39(x')^2}{4} - \frac{13\sqrt{3}x'y'}{2} + \frac{13(y')^2}{4} + \frac{18(x')^2}{4} + \frac{18\sqrt{3}x'y'}{4} - \frac{6\sqrt{3}x'y'}{4}$$

$$- \frac{18(y')^2}{4} + \frac{7(x')^2}{4} + \frac{7\sqrt{3}x'y'}{2} + \frac{21(y')^2}{4} - 16 = 0$$

$$16(x')^2 + 4(y')^2 = 16$$

$$\frac{(x')^2}{1} + \frac{(y')^2}{4} = 1$$

16. $16x^2 - 24xy + 9y^2 - 60x - 80y + 100 = 0$

$A = 16, B = -24, C = 9$

$$\cot 2\theta = \frac{A - C}{B} = -\frac{7}{24} \implies \theta \approx 53.13°$$

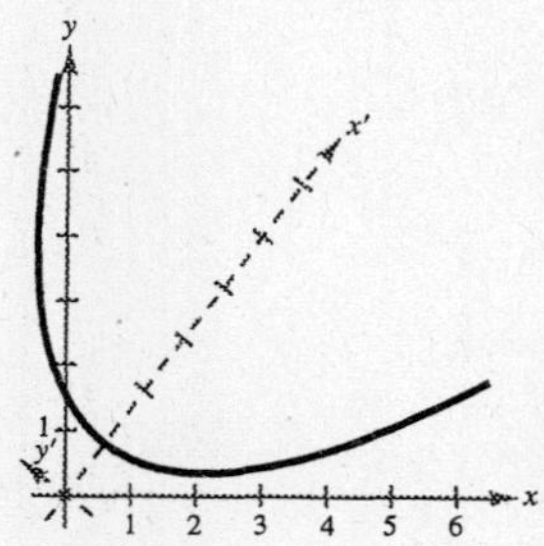

$$\cos 2\theta = -\frac{7}{25}$$

$$\sin\theta = \sqrt{\frac{1 - \cos 2\theta}{2}} = \sqrt{\frac{1 - (-7/25)}{2}} = \frac{4}{5}$$

$$\cos\theta = \sqrt{\frac{1 + \cos 2\theta}{2}} = \sqrt{\frac{1 + (-7/25)}{2}} = \frac{3}{5}$$

$$x = x'\cos\theta - y'\sin\theta \qquad y = x'\sin\theta + y'\cos\theta$$

$$= x'\left(\frac{3}{5}\right) - y'\left(\frac{4}{5}\right) \qquad = x'\left(\frac{4}{3}\right) + y'\left(\frac{3}{5}\right) = \frac{4x' + 3y'}{5}$$

$$16x^2 - 24xy + 9y^2 - 60x - 80y + 100 = 0$$

$$16\left(\frac{3x' - 4y'}{5}\right)^2 - 24\left(\frac{3x' - 4y'}{5}\right)\left(\frac{4x' + 3y'}{5}\right) + 9\left(\frac{4x' + 3y'}{5}\right)^2 - 60\left(\frac{3x' - 4y'}{5}\right)$$

$$- 80\left(\frac{4x' + 3y'}{5}\right) + 100 = 0$$

$$\frac{144(x')^2}{25} - \frac{384x'y'}{25} + \frac{256(y')^2}{25} - \frac{288(x')^2}{25} + \frac{168x'y'}{25} + \frac{288(y')^2}{25} + \frac{144(x')^2}{25} + \frac{216x'y'}{25}$$

$$+ \frac{81(y')^2}{25} - 36x' + 48y' - 64x' - 48y' + 100 = 0$$

$$25(y')^2 - 100x' + 100 = 0$$

$$(y')^2 = 4(x' - 1)$$

18. $9x^2 + 24xy + 16y^2 + 80x - 60y = 0$

$A = 9, B = 24, C = 16$

$$\cot 2\theta = \frac{A - C}{B} = -\frac{7}{24} \Rightarrow \theta \approx 53.13°$$

$$\cos 2\theta = -\frac{7}{25}$$

$$\sin \theta = \sqrt{\frac{1 - \cos 2\theta}{2}} = \sqrt{\frac{1 - (-7/25)}{2}} = \frac{4}{5}$$

$$\cos \theta = \sqrt{\frac{1 + \cos 2\theta}{2}} = \sqrt{\frac{1 + (-7/25)}{2}} = \frac{3}{5}$$

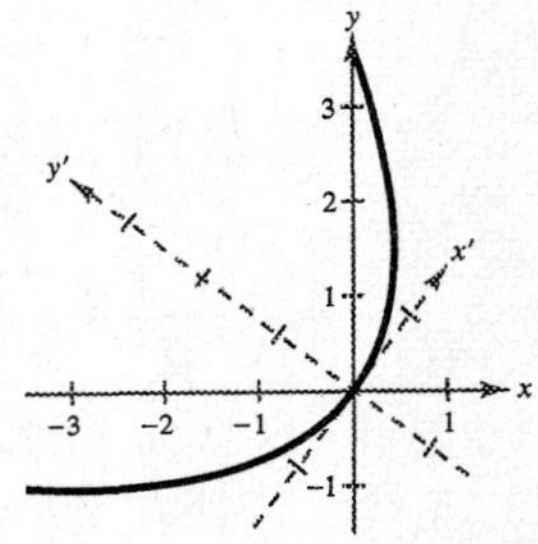

$$x = x' \cos \theta - y' \sin \theta \qquad y = x' \sin \theta + y' \cos \theta$$

$$= x'\left(\frac{3}{5}\right) - y'\left(\frac{4}{5}\right) \qquad = x'\left(\frac{4}{5}\right) + y'\left(\frac{3}{5}\right)$$

$$= \frac{3x' - 4y'}{5} \qquad = \frac{4x' + 3y}{5}$$

$$9x^2 + 24xy + 16y^2 + 80x - 60y = 0$$

$$9\left(\frac{3x' - 4y'}{5}\right)^2 + 24\left(\frac{3xy - 4y'}{5}\right)\left(\frac{4x' + 3y'}{5}\right) + 16\left(\frac{4x' + 3y'}{5}\right)^2 + 80\left(\frac{3x' - 4y'}{5}\right) - 60\left(\frac{4x' + 3y'}{5}\right) = 0$$

$$\frac{81(x')^2}{25} - \frac{216x'y'}{25} + \frac{144(y')^2}{25} + \frac{288(x')^2}{25} - \frac{168x'y'}{25} - \frac{288(y')^2}{25} + \frac{256(x')^2}{25} + \frac{384x'y'}{25}$$

$$+ \frac{144(y')^2}{25} + 48x' - 64x' - 48x' - 36x' = 0$$

$$25(x')^2 - 100y' = 0$$

$$(x')^2 = 4y'$$

$$\frac{1}{4}(x')^2 = y'$$

20. $x^2 - 4xy + 2y^2 = 6$

$A = 1, B = -4, C = 2$

$$\cot 2\theta = \frac{A - C}{B} = \frac{1 - 2}{-4} = \frac{1}{4}$$

$$\frac{1}{\tan 2\theta} = \frac{1}{4}$$

$$\tan 2\theta = 4$$

$$2\theta \approx 75.96$$

$$\theta \approx 37.98°$$

To graph conic with a graphing calculator, we need to solve for y in terms of x.

$$x^2 - 4xy + 2y^2 = 6$$

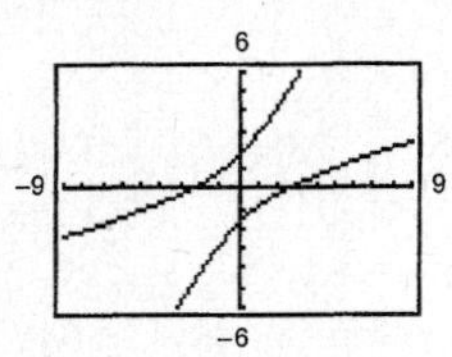

$$y^2 - 2xy + x^2 = 3 - \frac{x^2}{2} + x^2$$

$$(y - x)^2 = 3 + \frac{x^2}{2}$$

$$y - x = \pm\sqrt{3 + \frac{x^2}{2}}$$

$$y = x \pm \sqrt{3 + \frac{x^2}{2}}$$

Enter $y_1 = x + \sqrt{3 + \frac{x^2}{2}}$ and $y_2 = x - \sqrt{3 + \frac{x^2}{2}}$.

22. $40x^2 + 36xy + 25y^2 = 52$

$A = 40, B = 36, C = 25$

$$\cot 2\theta = \frac{A - C}{B} = \frac{40 - 25}{36} = \frac{5}{12}$$

$$\frac{1}{\tan 2\theta} = \frac{5}{12}$$

$$\tan 2\theta = \frac{12}{5}$$

$$2\theta \approx 67.38°$$

$$\theta \approx 33.69°$$

Solve for y in terms of x by completing the square:

$$25y^2 + 36xy = 52 - 40x^2$$

$$y^2 + \frac{36}{25}xy = \frac{52}{25} - \frac{40}{25}x^2$$

$$y^2 + \frac{36}{25}xy + \frac{324}{625}x^2 = \frac{52}{25} - \frac{40}{25}x^2 + \frac{324}{625}x^2$$

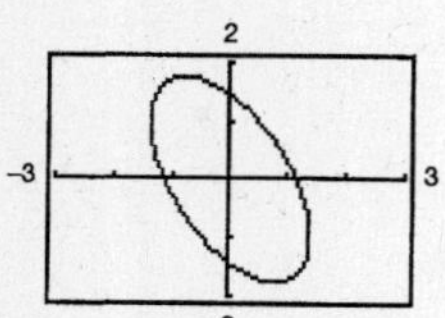

$$\left(y + \frac{18}{25}x\right)^2 = \frac{1300 - 676x^2}{625}$$

$$y + \frac{18}{25}x = \pm\sqrt{\frac{1300 - 676x^2}{625}}$$

$$y = \frac{-18x \pm \sqrt{1300 - 676x^2}}{25}$$

Enter $y_1 = \dfrac{-18x + \sqrt{1300 - 676x^2}}{25}$ and $y_2 = \dfrac{-18x - \sqrt{1300 - 676x^2}}{25}$.

24. $24x^2 + 18xy + 12y^2 = 34$

$A = 24, B = 18, C = 12$

$$\cot 2\theta = \frac{A - C}{B} = \frac{24 - 12}{18} = \frac{2}{3}$$

$$\tan 2\theta = \frac{3}{2}$$

$$2\theta \approx 56.31°$$

$$\theta \approx 28.15°$$

Solve for y in terms of x by completing the square:

$$12x^2 + 9xy + 6y^2 = 17$$

$$6\left(y^2 + \frac{3}{2}xy + \frac{9}{16}x^2\right) = 17 - 12x^2 + \frac{27}{8}x^2 = 17 - \frac{69}{8}x^2$$

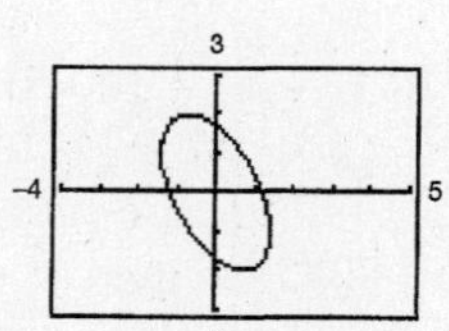

$$\left(y + \frac{3}{4}x\right)^2 = \frac{136 - 69x^2}{48}$$

$$y = -\frac{3}{4}x \pm \sqrt{\frac{136 - 69x^2}{48}} = \frac{-9x \pm \sqrt{3(136 - 69x^2)}}{12}$$

Enter $y_1 = \dfrac{-9x + \sqrt{3(136 - 69x^2)}}{12}$ and $y_2 = \dfrac{-9x - \sqrt{3(136 - 69x^2)}}{12}$.

26. $6x^2 - 4xy + 8y^2 + (5\sqrt{5} - 10)x - (7\sqrt{5} + 5)y = 80$

$A = 6, B = -4, C = 8$

$$\cot 2\theta = \frac{A - C}{B} = \frac{6 - 8}{-4} = \frac{1}{2}$$

$\tan 2\theta = 2$

$2\theta \approx 63.43°$

$\theta \approx 31.72°$

Solve for y in terms of x using the quadratic formula:

$8y^2 - (4x + 7\sqrt{5} + 5)y + 6x^2 + (5\sqrt{5} - 10)x - 80 = 0$

$$y = \frac{1}{16}\left[4x + 7\sqrt{5} + 5 \pm \sqrt{(4x + 7\sqrt{5} + 5)^2 - 32\left(6x^2 + (5\sqrt{5} - 10)x - 80\right)}\right]$$

Enter y_1 and y_2 from the above expression.

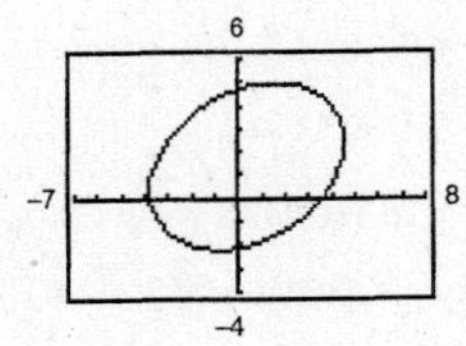

28. $x^2 + 2xy + y^2 = 0$

$(x + y)^2 = 0$

$x + y = 0$

$y = -x$

The graph is a line. Matches graph (f).

30. $x^2 - xy + 3y^2 - 5 = 0$

$A = 1, B = -1, C = 3$

$B^2 - 4AC = (-1)^2 - 4(1)(3) = -11$

The graph is an ellipse.

$$\cot 2\theta = \frac{A - C}{B} = \frac{1 - 3}{-1} = 2 \Rightarrow \theta \approx 13.28°$$

Matches graph (a).

32. $x^2 - 4xy + 4y^2 + 10x - 30 = 0$

$A = 1, B = -4, C = 4$

$B^2 - 4AC = (-4)^2 - 4(1)(4) = 0$

The graph is a parabola.

$$\cot 2\theta = \frac{A - C}{B} = \frac{1 - 4}{-4} = \frac{3}{4} \Rightarrow \theta \approx 26.57°$$

Matches graph (c).

34. $x^2 - 4xy - 2y^2 - 6 = 0$

$A = 1, B = -4, C = -2$

$B^2 - 4AC = (-4)^2 - 4(1)(-2) = 24 > 0$

Hyperbola

$-2y^2 - 4xy + x^2 - 6 = 0$

$$y = -\frac{1}{4}\left[4x \pm \sqrt{16x^2 + 8(x^2 - 6)}\right]$$

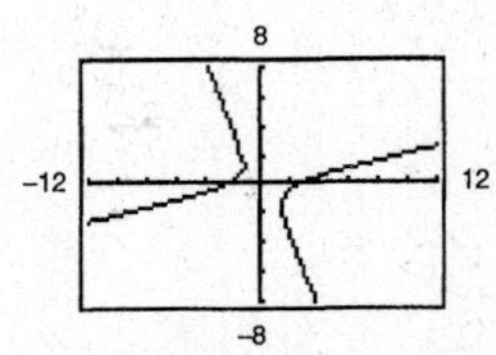

36. $2x^2 + 4xy + 5y^2 + 3x - 4y - 20 = 0$

$A = 2, B = 4, C = 5$

$B^2 - 4AC = 4^2 - 4(2)(5) = 16 - 40 = -24 < 0$

Ellipse

$5y^2 + (4x - 4)y + 2x^2 + 3x - 20 = 0$

$$y = \frac{1}{10}\left[-(4x - 4) \pm \sqrt{(4x - 4)^2 - 20(2x^2 + 3x - 20)}\right]$$

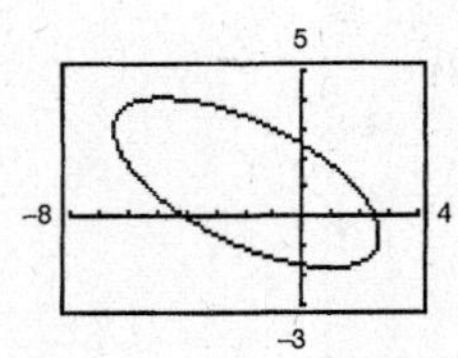

38. $36x^2 - 60xy + 25y^2 + 9y = 0$

$A = 36, B = -60, C = 25$

$B^2 - 4AC = (-60)^2 - 4(36)(25) = 0$

Parabola

$25y^2 - (60x - 9)y + 36x^2 = 0$

$$y = \frac{1}{50}\left[60x - 9 \pm \sqrt{(60x - 9)^2 - 3600x^2}\right]$$

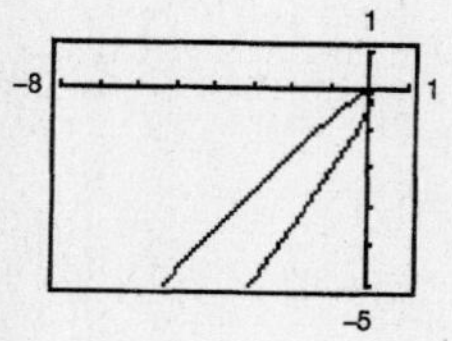

40. $x^2 + xy + 4y^2 + x + y - 4 = 0$

$A = 1, B = 1, C = 4$

$B^2 - 4AC = 1^2 - 4(1)(4) = -15$

Ellipse

$4y^2 + (x + 1)y + x^2 + x - 4 = 0$

$$y = \frac{1}{8}\left[-(x + 1) \pm \sqrt{(x + 1)^2 - 16(x^2 + x - 4)}\right]$$

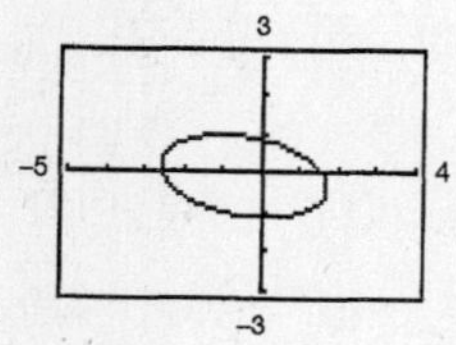

42.

$$\begin{aligned} x^2 + y^2 - 2x + 6y + 10 &= 0 \\ (x^2 - 2x + 1) + (y^2 + 6y + 9) &= -10 + 1 + 9 \\ (x - 1)^2 + (y + 3)^2 &= 0 \end{aligned}$$

Point at $(1, -3)$

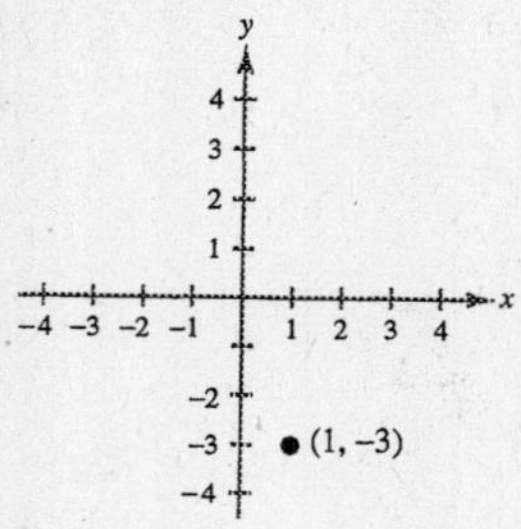

44.

$$\begin{aligned} x^2 - 10xy + y^2 &= 0 \\ y^2 - 10xy + 25x^2 &= 25x^2 - x^2 \\ (y - 5x)^2 &= 24x^2 \\ y - 5x &= \pm\sqrt{24x^2} \\ y &= 5x \pm 2\sqrt{6}x \\ y &= \left(5 \pm 2\sqrt{6}\right)x \end{aligned}$$

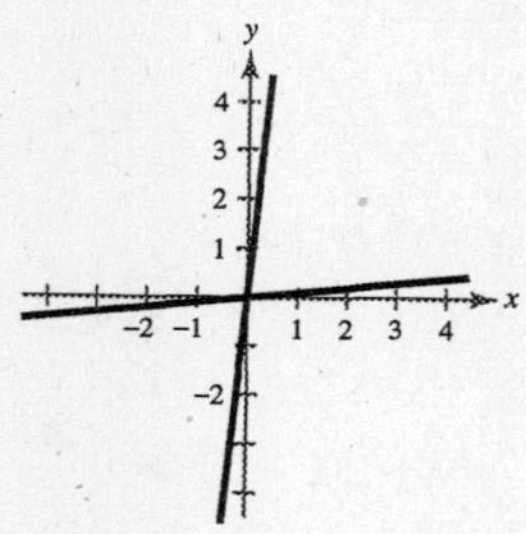

46.

$$\begin{aligned} -x^2 - y^2 - 8x + 20y - 7 = 0 &\Rightarrow (x + 4)^2 + (y - 10)^2 = 109 \\ x^2 + 9y^2 + 8x + 4y + 7 = 0 &\Rightarrow (x + 4)^2 + 9\left(y + \tfrac{2}{9}\right)^2 = \tfrac{85}{9} \\ \hline 8y^2 \quad + 24y \quad = 0 & \\ 8y(y + 3) = 0 & \\ y = 0 \text{ or } y = -3 & \end{aligned}$$

When $y = 0$: $x^2 + 9(0)^2 + 8x + 4(0) + 7 = 0$

$(x + 7)(x + 1) = 0$

$x = -7, -1$

When $y = -3$: $x^2 + 9(-3)^2 + 8x + 4(-3) + 7 = 0$

$x^2 + 8x + 76 = 0$

No real solution

Points of intersection: $(-7, 0), (-1, 0)$

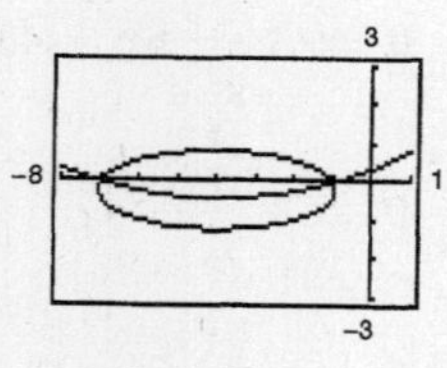

48.
$$\begin{aligned}
x^2 - 4y^2 - 20x - 64y - 172 &= 0 \Rightarrow (x - 10)^2 - 4(y + 8)^2 = 16 \\
\underline{16x^2 + 4y^2 - 320x + 64y + 1600} &= 0 \Rightarrow 16(x - 10)^2 + 4(y + 8)^2 = 256 \\
17x^2 - 340x \quad 1428 &= 0 \\
(17x - 238)(x - 6) &= 0 \\
x = 6 \text{ or } x &= 14
\end{aligned}$$

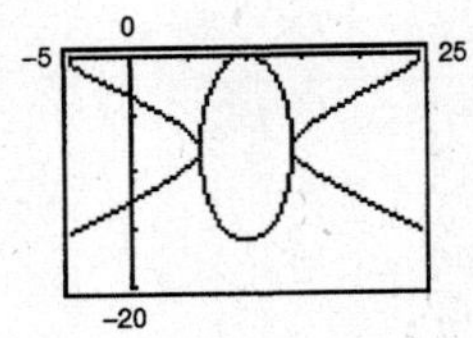

When $x = 6$:
$$\begin{aligned}
6^2 - 4y^2 - 20(6) - 64y - 172 &= 0 \\
-4y^2 - 64y - 256 &= 0 \\
y^2 + 16y + 64 &= 0 \\
(y + 8)^2 &= 0 \\
y &= -8
\end{aligned}$$

When $x = 14$:
$$\begin{aligned}
14^2 - 4y^2 - 20(14) - 64y - 172 &= 0 \\
-4y^2 - 64y - 256 &= 0 \\
y^2 + 16y + 64 &= 0 \\
(y + 8)^2 &= 0 \\
y &= -8
\end{aligned}$$

Points of intersection: $(6, -8), (14, -8)$

50.
$$\begin{aligned}
x^2 + 4y^2 - 2x - 8y + 1 &= 0 \Rightarrow (x - 1)^2 + 4(y - 1)^2 = 4 \\
\underline{-x^2 \qquad + 2x - 4y - 1} &= 0 \Rightarrow y = -\tfrac{1}{4}(x - 1)^2 \\
4y^2 \quad -12y &= 0 \\
4y(y - 3) &= 0 \\
y = 0 \text{ or } y &= 3
\end{aligned}$$

When $y = 0$:
$$\begin{aligned}
x^2 + 4(0)^2 - 2x - 8(0) + 1 &= 0 \\
x^2 - 2x + 1 &= 0 \\
(x - 1)^2 &= 0 \\
x &= 1
\end{aligned}$$

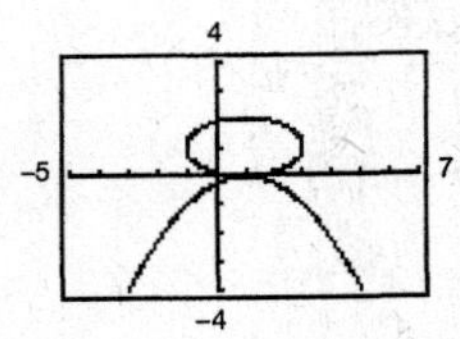

When $y = 3$:
$$\begin{aligned}
-x^2 + 2x - 4(3) - 1 &= 0 \\
x^2 - 2x + 13 &= 0
\end{aligned}$$

No real solution

Point of intersection: $(1, 0)$

52.
$$\begin{aligned}
16x^2 - y^2 \qquad + 16y - 128 &= 0 \Rightarrow 16x^2 - (y - 8)^2 = 64 \\
\underline{y^2 - 48x - 16y - 32} &= 0 \Rightarrow (y - 8)^2 - 48x = 96 \\
16x^2 - 48x \quad - 160 &= 0 \\
16(x^2 - 3x - 10) &= 0 \\
(x - 5)(x + 2) &= 0 \\
x = 5 \text{ or } x &= -2
\end{aligned}$$

When $x = 5$:
$$\begin{aligned}
y^2 - 48(5) - 16y - 32 &= 0 \\
y^2 - 16y - 272 &= 0 \\
y &= 8 \pm 4\sqrt{21}
\end{aligned}$$

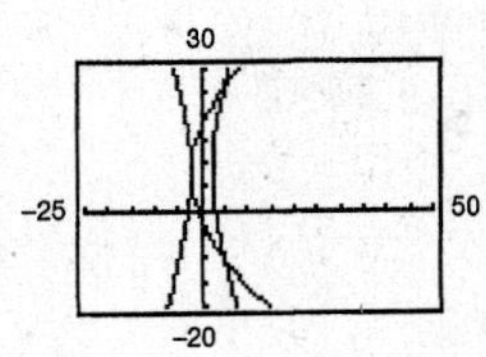

When $x = -2$:
$$\begin{aligned}
y^2 - 48(-2) - 16y - 32 &= 0 \\
y^2 - 16y + 64 &= 0 \\
(y - 8)^2 &= 0 \\
y &= 8
\end{aligned}$$

Points of intersection: $\left(5, 8 + 4\sqrt{21}\right), \left(5, 8 - 4\sqrt{21}\right), (-2, 8)$

54. $4x^2 + 9y^2 - 36y = 0 \Rightarrow 4x^2 + 9(y-2)^2 = 36$

$$x^2 + 9y - 27 = 0 \Rightarrow y = -\frac{x^2}{9} + 3$$

$$4(27 - 9y) + 9y^2 - 36y = 0$$
$$9y^2 - 72y + 108 = 0$$
$$9(y-6)(y-2) = 0$$
$$y = 6 \text{ or } y = 2$$

When $y = 6$: $x^2 = 27 - 9(6) = -27$

No real solution

When $y = 2$: $x^2 = 27 - 9(2) = 9$

$x = \pm 3$

Points of intersection: $(3, 2), (-3, 2)$

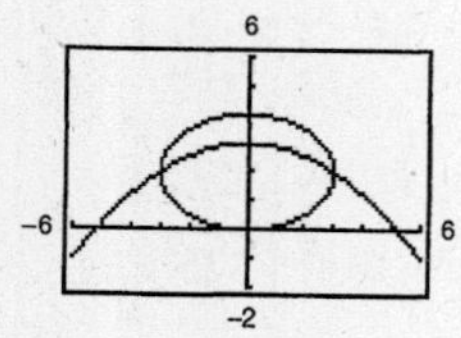

56. $x^2 + 2y^2 - 4x + 6y - 5 = 0 \Rightarrow 2(x-2)^2 + 4\left(y + \frac{3}{2}\right)^2 = 27$

$$x^2 - 4x - y + 4 = 0 \Rightarrow y = x^2 - 4x + 4$$

$$y - 4 + 2y^2 + 6y - 5 = 0$$
$$2y^2 + 7y - 9 = 0$$
$$(2y + 9)(y - 1) = 0$$
$$y = -\frac{9}{2} \text{ or } y = 1$$

When $y = 1$: $x^2 - 4x - 1 + 4 = 0$

$(x-3)(x-1) = 0$

$x = 1$ or $x = 3$

When $y = -\frac{9}{2}$: $x^2 - 4x - \left(-\frac{9}{2}\right) + 4 = 0$

$$x^2 - 4x + \frac{17}{2} = 0$$

No real solution

Points of intersection: $(1, 1), (3, 1)$

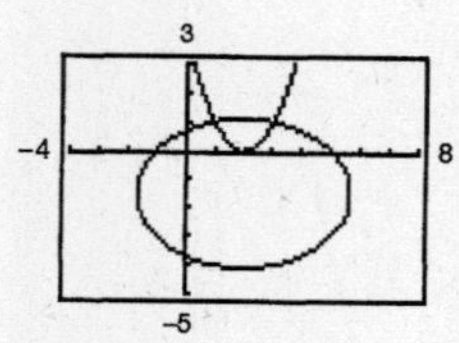

58. $5x^2 - 2xy + 5y^2 - 12 = 0$

$$x + y - 1 = 0 \Rightarrow y = 1 - x$$
$$5x^2 - 2x(1-x) + 5(1-x)^2 - 12 = 0$$
$$5x^2 - 2x + 2x^2 + 5(1 - 2x + x^2) - 12 = 0$$
$$5x^2 - 2x + 2x^2 + 5 - 10x + 5x^2 - 12 = 0$$
$$12x^2 - 12x - 7 = 0$$
$$x = \frac{3 \pm \sqrt{30}}{6}$$

When $x = \dfrac{3 + \sqrt{30}}{6}$: $y = 1 - \dfrac{3 + \sqrt{30}}{6} = \dfrac{3 - \sqrt{30}}{6}$

When $x = \dfrac{3 - \sqrt{30}}{6}$: $y = 1 - \dfrac{3 - \sqrt{30}}{6} = \dfrac{3 + \sqrt{30}}{6}$

Points of intersection: $\left(\frac{1}{6}(3 + \sqrt{30}), \frac{1}{6}(3 - \sqrt{30})\right), \left(\frac{1}{6}(3 - \sqrt{30}), \frac{1}{6}(3 + \sqrt{30})\right)$

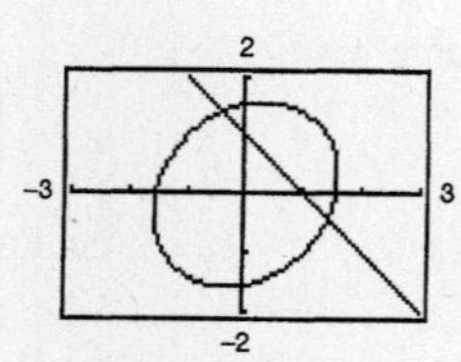

60. False. The coefficients of the new equation, after rotation of axes, are obtained by making the substitutions.

$x = x' \cos \theta - y' \sin \theta,$

$y = x' \sin \theta + y' \cos \theta$

62. In Exercise 14, the equation of the rotated ellipse is:

$$\frac{(x')^2}{1} + \frac{(y')^2}{4} = 1$$

$a^2 = 4 \Rightarrow a = 2$

$b^2 = 1 \Rightarrow b = 1$

Length of major axis is $2a = 2(2) = 4$.

Length of minor axis is $2b = 2(1) = 2$.

64. $f(x) = |x - 4| + 1$

Intercept: $(0, 4)$

x	-1	0	1	2	3
$f(x)$	6	5	4	3	2

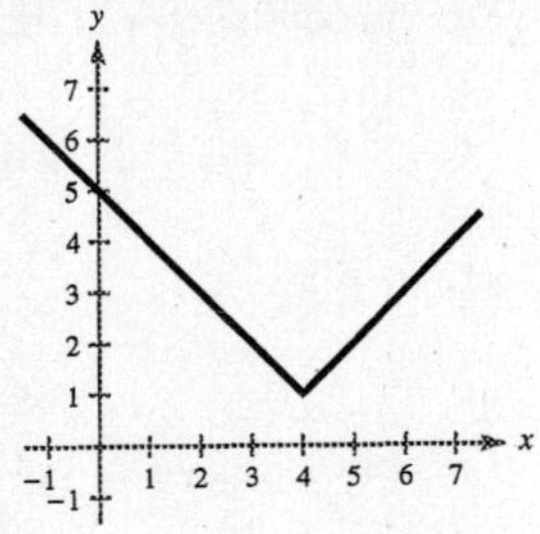

66. $g(x) = \sqrt{3x - 2}$

Domain: $x \geq \frac{2}{3}$

x	1	2	3	4
$g(x)$	1	2	2.65	3.16

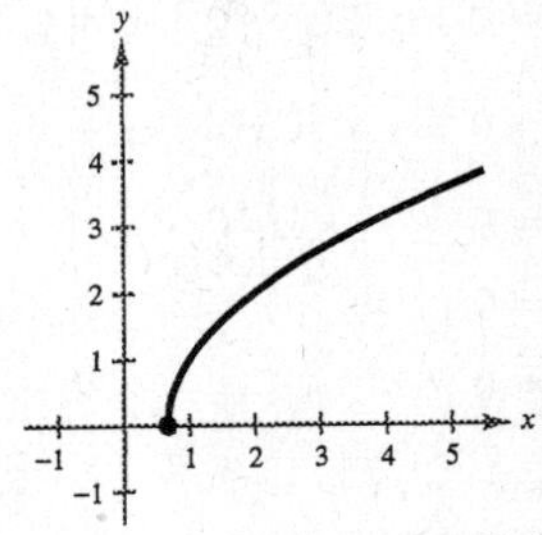

68. $h(x) = \frac{1}{2}(t + 4)^3$

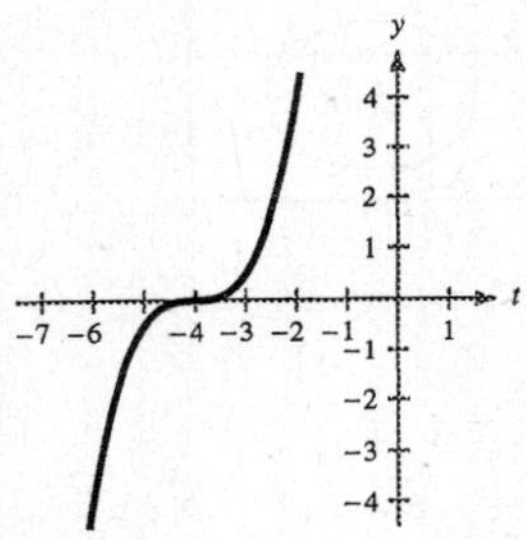

70. $f(t) = -2[\![t]\!] + 3$

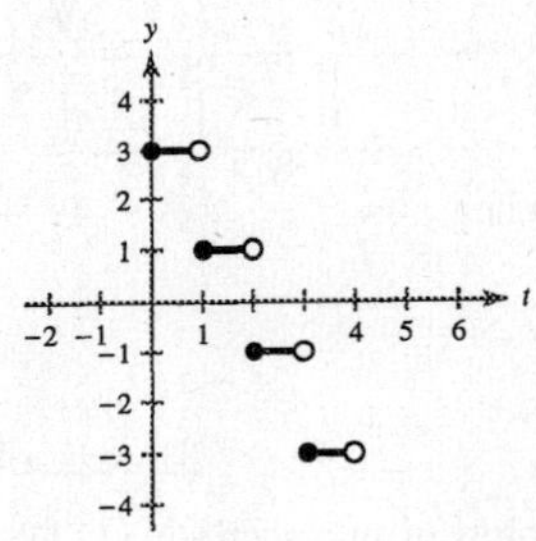

72. $B = 70°, a = 25, c = 16$

$$\text{Area} = \frac{1}{2}ac \sin B$$

$$= \frac{1}{2}(25)(16) \sin 70°$$

$$= 200 \sin 70° \approx 187.94 \text{ square units}$$

74. $a = 23, b = 35, c = 27$

$$s = \frac{a + b + c}{2} = \frac{23 + 35 + 27}{2} = \frac{85}{2} = 42.5$$

$$\text{Area} = \sqrt{s(s - a)(s - b)(s - c)}$$

$$= \sqrt{42.5(42.5 - 23)(42.5 - 35)(42.5 - 27)}$$

$$= \sqrt{42.5(19.5)(7.5)(15.5)} \approx 310.39 \text{ square units}$$

76. In 6 hours, the distance traveled is $6(20) = 120$ nautical miles.

Distance south: $\frac{x}{120} = \cos 27° \Rightarrow x = 120 \cos 27° \approx 107$ nautical miles south

Distance west: $\frac{x}{120} = \sin 27° \Rightarrow x = 120 \sin 27° \approx 54$ nautical miles west

Section 6.6 Parametric Equations

Solutions to Even-Numbered Exercises

2. $x = 4\cos^2\theta, y = 2\sin\theta$

(a)

θ	$-\pi/2$	$-\pi/4$	0	$\pi/4$	$\pi/2$
x	0	2	4	2	0
y	-2	$-\sqrt{2}$	0	$\sqrt{2}$	2

(b)

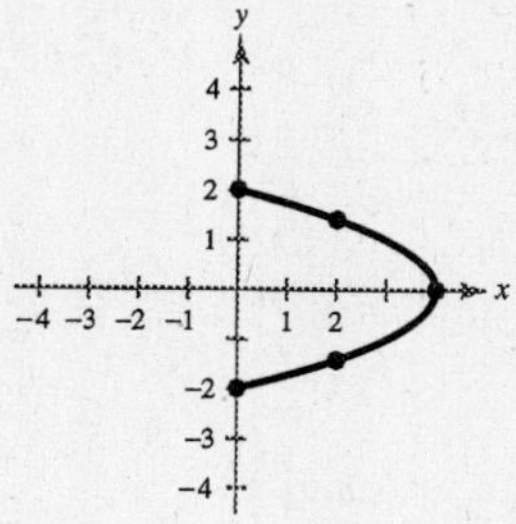

(c) $\dfrac{x}{4} = \cos^2\theta, \dfrac{y}{2} = \sin\theta$

$$\cos^2\theta + \sin^2\theta = 1$$

$$\frac{x}{4} + \left(\frac{y}{2}\right)^2 = 1$$

$$\frac{x}{4} + \frac{y^2}{4} = 1$$

$$x = -y^2 + 4$$

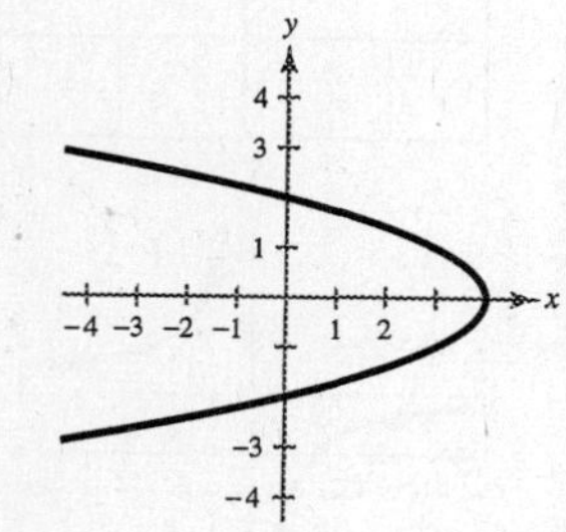

The rectangular version of the graph continues into the second and third quadrants.

4. $x = 3 - 2t \Rightarrow t = -\frac{1}{2}x + \frac{3}{2}$

$y = 2 + 3t$

t	-3	-2	-1	0	1	2	3
x	9	7	5	3	1	-1	-3
y	-7	-4	-1	2	5	8	11

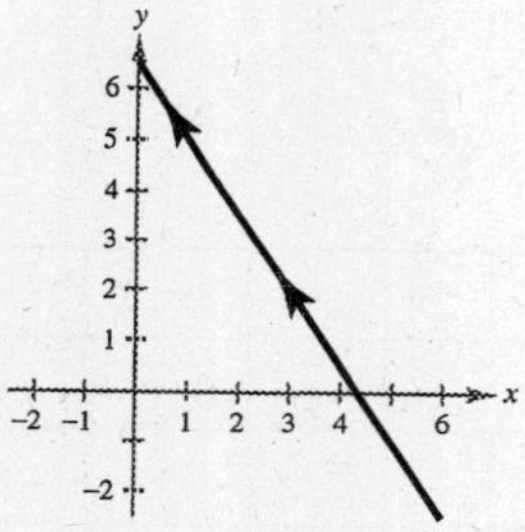

$$y = 2 + 3t$$

$$y = 2 + 3\left(-\tfrac{1}{2}x + \tfrac{3}{2}\right)$$

$$y = 2 - \tfrac{3}{2}x + \tfrac{9}{2}$$

$$2y = 4 - 3x + 9$$

$$3x + 2y - 13 = 0$$

6. $x = t$

$y = t^3$

t	-3	-2	-1	0	1	2	3
x	-3	-2	-1	0	1	2	3
y	-27	-8	-1	0	1	8	27

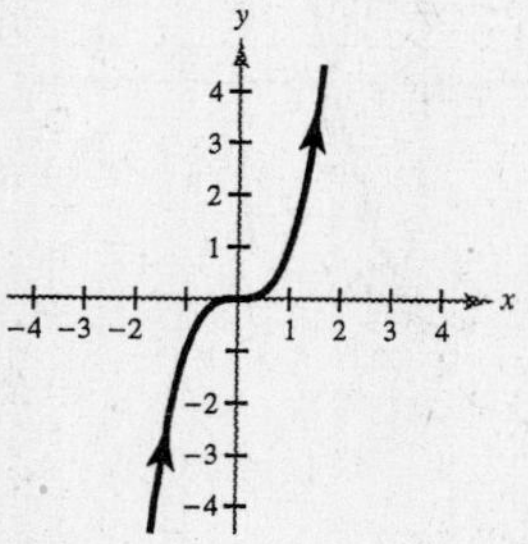

$y = t^3$

$y = x^3$

8. $x = \sqrt{t} \Rightarrow x^2 = t, t \geq 0$

$y = 1 - t$

t	0	1	2	3
x	0	1	$\sqrt{2}$	$\sqrt{3}$
y	1	0	-1	-2

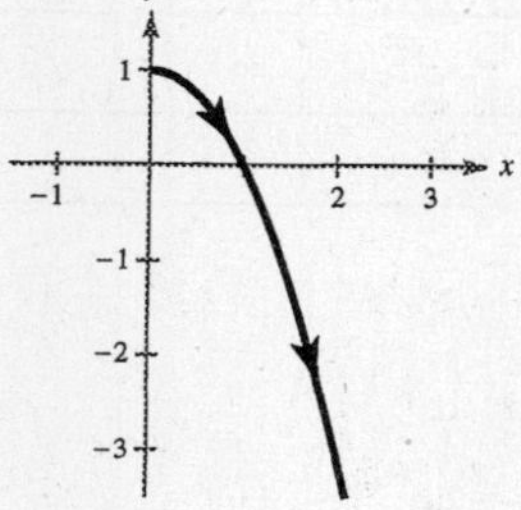

$y = 1 - t$

$y = 1 - x^2, x \geq 0$

10. $x = t - 1 \Rightarrow t = x + 1$

$y = \dfrac{t}{t-1}$

t	-3	-2	-1	0	2	3
x	-4	-3	-2	-1	1	2
y	$\frac{3}{4}$	$\frac{2}{3}$	$\frac{1}{2}$	0	2	$\frac{3}{2}$

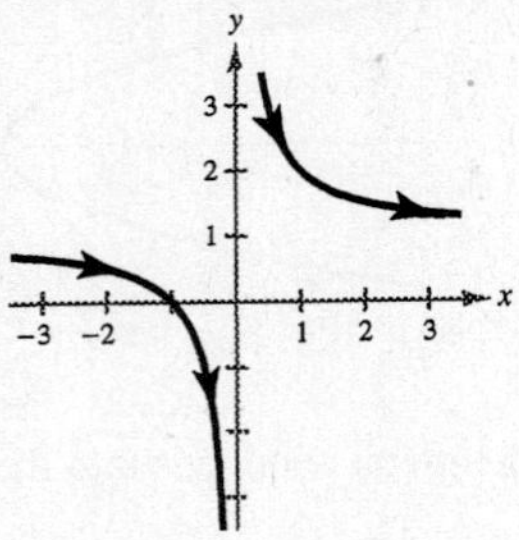

$$y = \frac{t}{t-1}$$

$$y = \frac{x+1}{x+1-1} = \frac{x+1}{x}$$

12. $x = |t - 1|$

$y = t + 2 \Rightarrow t = y - 2$

t	-3	-2	-1	0	1	2	3
x	4	3	2	1	0	1	2
y	-1	0	1	2	3	4	5

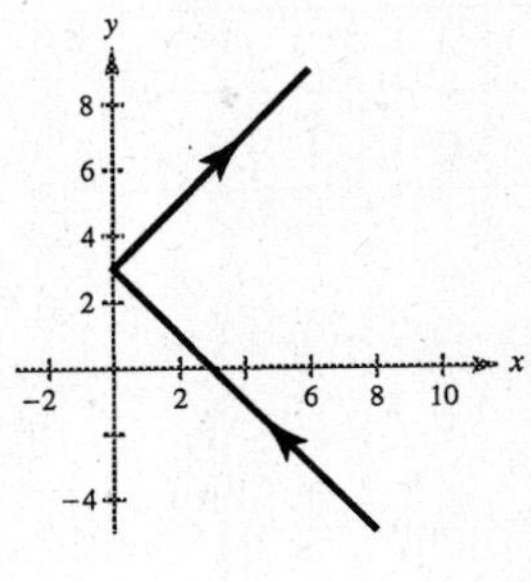

$x = |t - 1|$

$x = |y - 2 - 1| \Rightarrow x = |y - 3|$

14. $x = 2 \cos \theta$

$y = 3 \sin \theta$

θ	0	$\pi/4$	$\pi/2$	$3\pi/4$	π	$5\pi/4$	$3\pi/2$	$7\pi/4$	2π
x	2	$\sqrt{2}$	0	$-\sqrt{2}$	-2	$-\sqrt{2}$	0	$\sqrt{2}$	2
y	0	$3\sqrt{2}/2$	3	$3\sqrt{2}/2$	0	$-3\sqrt{2}/2$	-3	$-3\sqrt{2}/2$	0

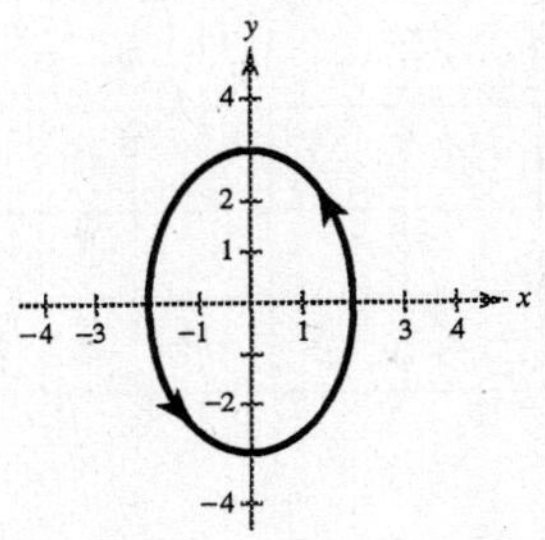

$$\cos^2\theta + \sin^2\theta = 1$$

$$\left(\frac{x}{2}\right)^2 + \left(\frac{y}{3}\right)^2 = 1$$

$$\frac{x^2}{4} + \frac{y^2}{9} = 1$$

16. $x = \cos \theta$

$y = 2 \sin 2\theta$

θ	0	$\pi/4$	$\pi/2$	$3\pi/4$	π	$5\pi/4$	$3\pi/2$	$7\pi/4$	2π
x	1	$\sqrt{2}/2$	0	$-\sqrt{2}/2$	-1	$-\sqrt{2}/2$	0	$\sqrt{2}/2$	1
y	0	2	0	-2	0	2	0	-2	0

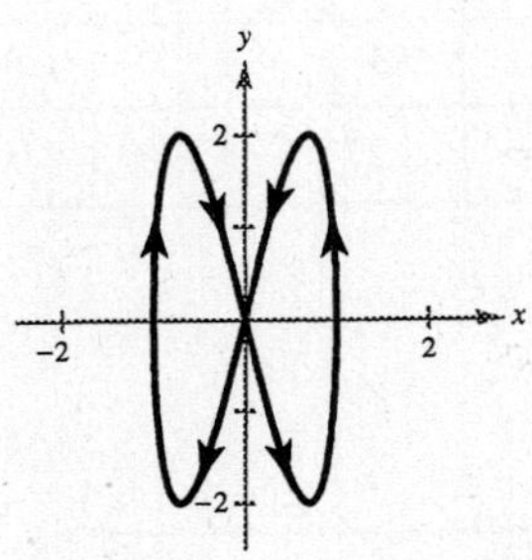

$$y = 2 \sin 2\theta$$

$$y = 2(2 \sin \theta \cos \theta)$$

$$y^2 = 16 \sin^2 \theta \cos^2 \theta$$

$$y^2 = 16(1 - x^2)x^2$$

$$y^2 = 16x^2(1 - x^2)$$

18. $x = 4 + 2\cos\theta$

$y = 2 + 3\sin\theta$

θ	0	$\pi/4$	$\pi/2$	$3\pi/4$	π	$5\pi/4$	$3\pi/2$	$7\pi/4$	2π
x	6	$4 + \sqrt{2}$	4	$4 - \sqrt{2}$	2	$4 - \sqrt{2}$	4	$4 + \sqrt{2}$	6
y	2	$2 + (3\sqrt{2}/2)$	5	$2 + (3\sqrt{2}/2)$	2	$2 - (3\sqrt{2}/2)$	-1	$2 - (3\sqrt{2}/2)$	2

$\cos^2\theta + \sin^2\theta = 1$

$$\frac{(x-4)^2}{4} + \frac{(y-2)^2}{9} = 1$$

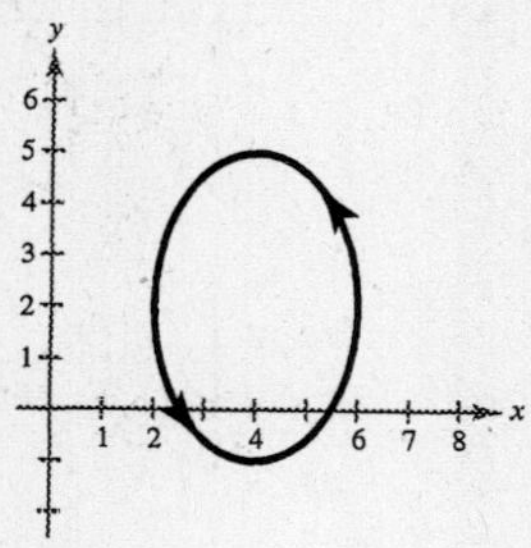

20. $x = e^{2t}$

$y = e^t \implies y^2 = e^{2t}$

t	-3	-2	-1	0	1	2
x	0.0025	0.0183	0.1353	1	7.3891	54.5982
y	0.0498	0.1353	0.3679	1	2.7183	7.3891

$x = e^{2t} = y^2$

$y^2 = x, y > 0$

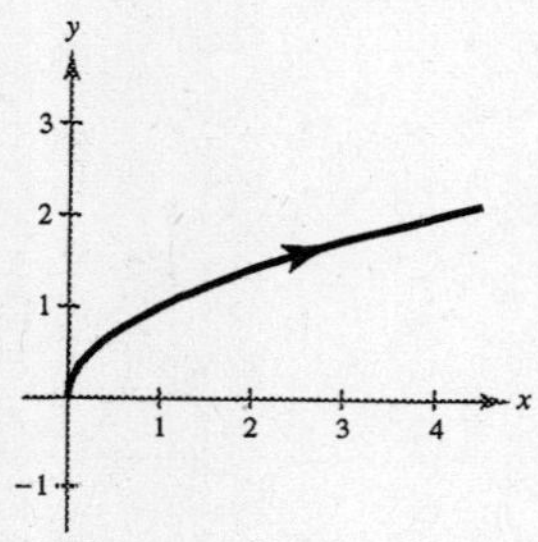

22. $x = \ln 2t \implies t = \frac{1}{2}e^x$

$y = 2t^2$

t	1	2	3	4
x	0.6931	1.3863	1.7918	2.0794
y	2	8	18	32

$y = 2t^2 = 2\left(\frac{1}{2}e^x\right)^2 = \frac{1}{2}e^{2x}$

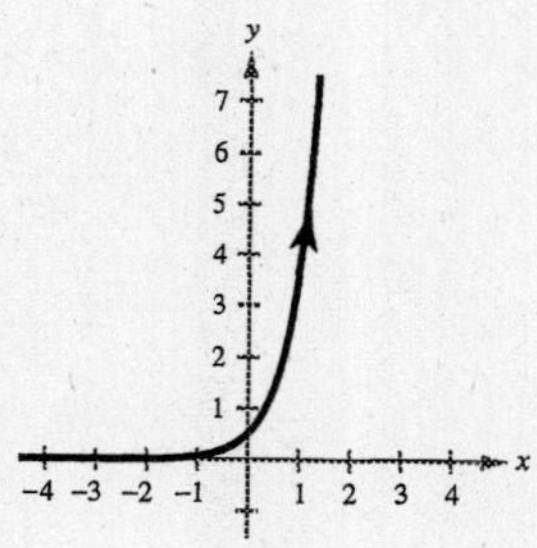

24. By eliminating the parameter, each curve represents a portion of $y = x^2 - 1$.

(a) $x = t$

$y = t^2 - 1$

There are no restrictions on x.

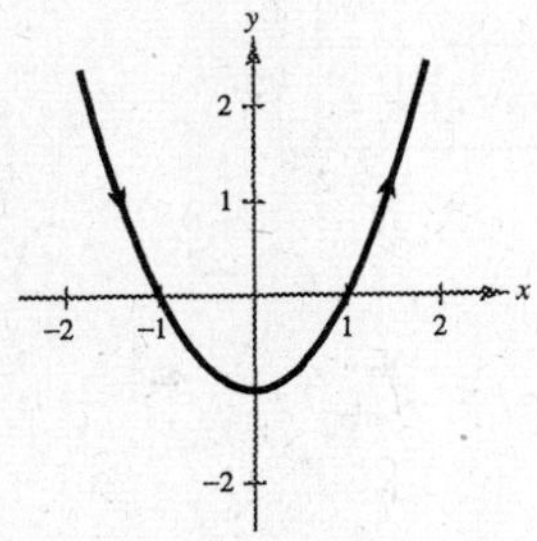

Domain: $(-\infty, \infty)$
Orientation: Left to right

(b) $x = t^2 \Rightarrow x \geq 0$

$y = t^4 - 1$

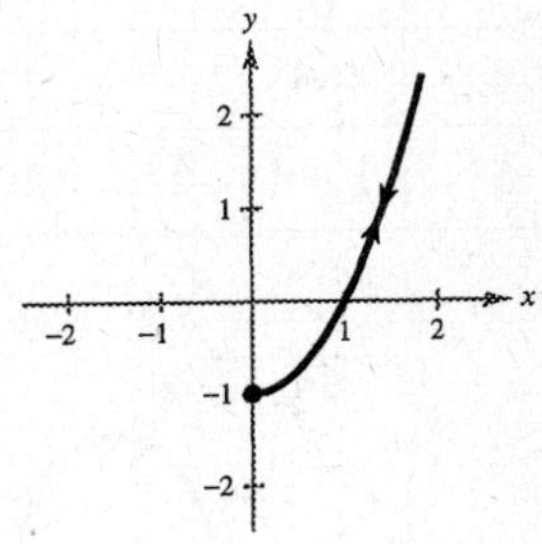

Domain: $[0, \infty)$
Orientation: Depends on t

(c) $x = \sin t \Rightarrow -1 \leq x \leq 1$

$y = \sin^2 t - 1$

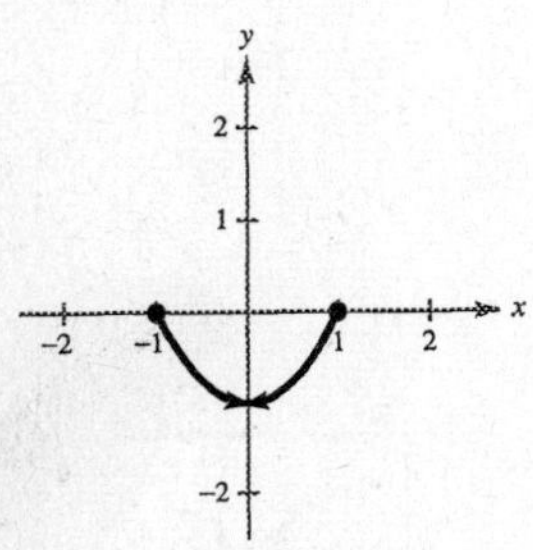

Domain: $[-1, 1]$
Orientation: Depends on t

(d) $x = e^t \Rightarrow x > 0$

$y = e^{2t} - 1$

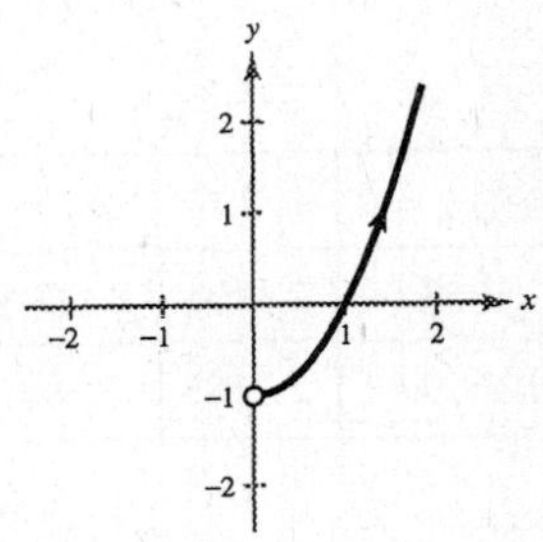

Domain: $(0, \infty)$
Orientation: Left to right

26. $x = h + r\cos\theta, y = k + r\sin\theta$

$$\cos\theta = \frac{x-h}{r}, \sin\theta = \frac{y-k}{r}$$

$$\cos^2\theta + \sin^2\theta = \frac{(x-h)^2}{r^2} + \frac{(y-k)^2}{r^2} = 1$$

$$(x-h)^2 + (y-k)^2 = r^2$$

28. $x = h + a\sec\theta, y = k + b\tan\theta$

$$\frac{x-h}{a} = \sec\theta, \frac{y-k}{b} = \tan\theta$$

$$\frac{(x-h)^2}{a^2} - \frac{(y-k)^2}{b^2} = 1$$

30. Line through $(2, 3)$ and $(6, -3)$

From Exercise 25 we have:

$$x = x_1 + t(x_2 - x_1) = 2 + t(6 - 2) = 2 + 4t$$

$$y = y_1 + t(y_2 - y_1) = 3 + t(-3 - 3) = 3 - 6t$$

32. Circle with center $(-3, 2)$; radius: 5

From Exercise 26 we have:

$$x = h + r\cos\theta = -3 + 5\cos\theta$$

$$y = k + r\sin\theta = 2 + 5\sin\theta$$

34. Ellipse

Vertices: $(4, 7), (4, -3) \Rightarrow (h, k) = (4, 2), a = 5$

Foci: $(4, 5), (4, -1) \Rightarrow c = 3$

$b^2 = a^2 - c^2 = 25 - 9 = 16 \Rightarrow b = 4$

From Exercise 27 we have:

$$x = h + a\cos\theta = 4 + 5\cos\theta$$

$$y = k + b\sin\theta = 2 + 4\sin\theta$$

36. Hyperbola

Vertices: $(\pm 2, 0) \Rightarrow (h, k) = (0, 0), a = 2$

Foci: $(\pm 4, 0) \Rightarrow c = 4$

$b^2 = c^2 - a^2 = 16 - 4 \Rightarrow b = 2\sqrt{3}$

From Exercise 28 we have:

$$x = h + a\sec\theta = 2\sec\theta$$

$$y = h + b\tan\theta = 2\sqrt{3}\tan\theta$$

38. $x = 3y - 2$

(a) $t = x$

$x = t$

$y = \frac{1}{3}(t + 2)$

(b) $t = 2 - x$

$x = 2 - t$

$y = \frac{1}{3}(x + 2) = \frac{1}{3}(4 - t)$

40. $y = x^3$

(a) $t = x$

$x = t$

$y = t^3$

(b) $t = 2 - x$

$x = 2 - t$

$y = (2 - t)^3$

42. $y = 2 - x$

(a) $t = x$

$x = t$

$y = 2 - t$

(b) $t = 2 - x$

$x = 2 - t$

$y = 2 - (2 - t) = t$

44. $y = \dfrac{1}{2x}$

(a) $t = x$

$x = t$

$y = \dfrac{1}{2t}$

(b) $t = 2 - x$

$x = 2 - t$

$y = \dfrac{1}{2(2 - t)} = \dfrac{1}{4 - 2t}$

46. $x = \theta + \sin\theta$

$y = 1 - \cos\theta$

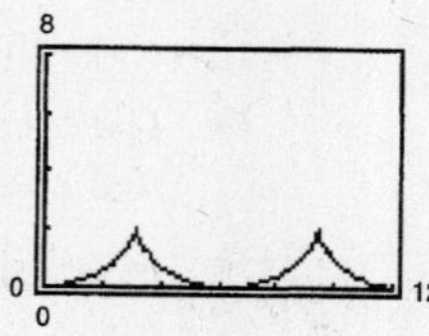

48. $x = 2\theta - 4\sin\theta$

$y = 2 - 4\cos\theta$

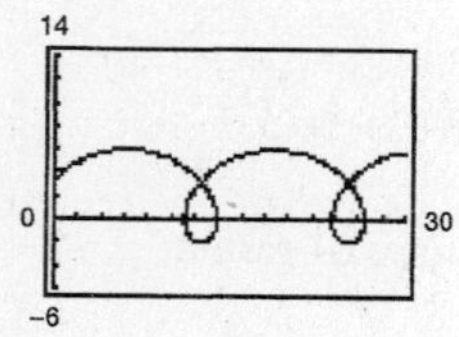

50. $x = 8\theta - 4\sin\theta$

$y = 8 - 4\cos\theta$

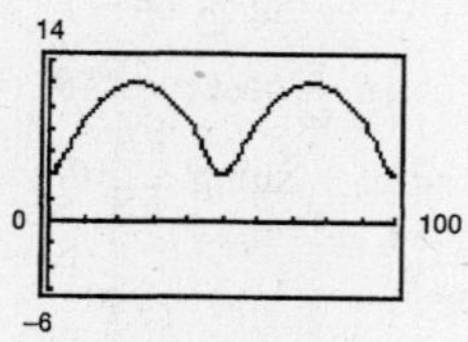

52. $x = \dfrac{3t}{1 + t^3}$

$y = \dfrac{3t^2}{1 + t^3}$

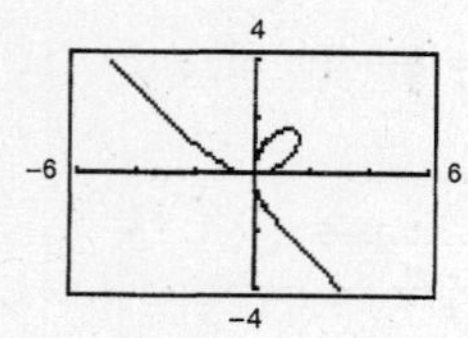

54. $x = 4\cos^3\theta \implies -4 \le x \le 4$

$y = 6\sin^3\theta \implies -6 \le y \le 6$

Matches graph (c).

Domain: $[-4, 4]$

Range: $[-6, 6]$

56. $x = \frac{1}{2}\cot\theta \implies -\infty < x < \infty$

$y = 4\sin\theta\cos\theta \implies -2 \le y \le 2$

Matches graph (a).

Domain: $(-\infty, \infty)$

Range: $[-2, 2]$

58. $x = (v_0 \cos\theta)t$

$y = h + (v_0 \sin\theta)t - 16t^2$

(a) $\theta = 15°$, $v_0 = 60$ ft/sec

Maximum height: 3.8 feet

Range: 56.3 feet

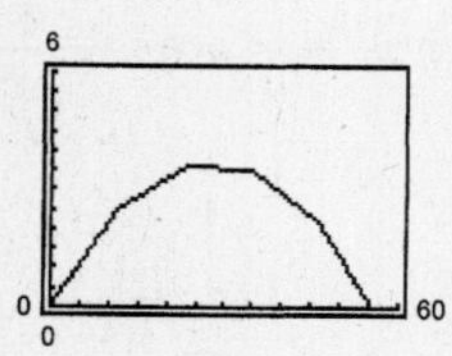

(b) $\theta = 15°$, $v_0 = 100$ ft/sec

Maximum height: 10.5 feet

Range: 156.3 feet

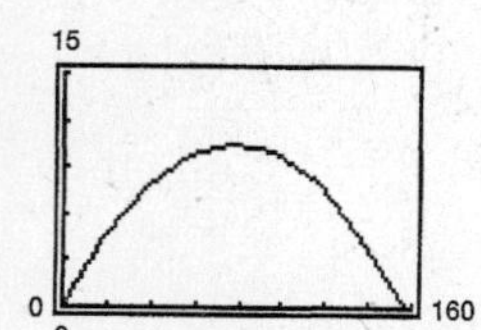

(c) $\theta = 30°$, $v_0 = 60$ ft/sec

Maximum height: 14.1 feet

Range: 97.4 feet

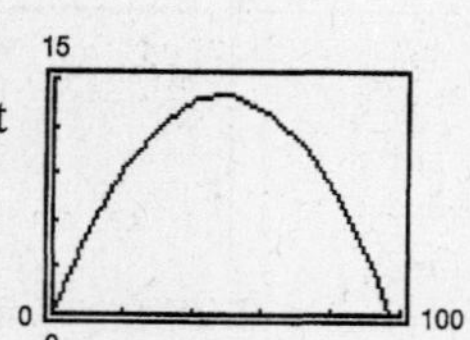

(d) $\theta = 30°$, $v_0 = 100$ ft/sec

Maximum height: 39.1 feet

Range: 270.6 feet

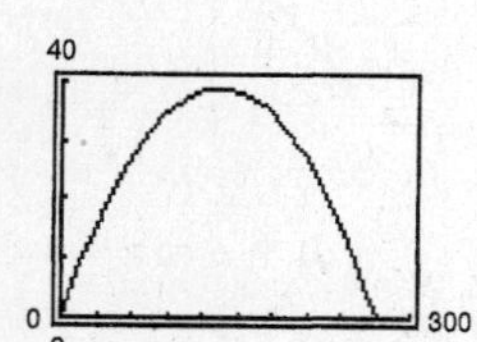

60. (a) $x = (v_0 \cos \theta)t$

$y = h + (v_0 \sin \theta)t - 16t^2$

$h = 5, v_0 = 240, \theta = 10°$

$x = (240 \cos 10°)t$

$y = 5 + (240 \sin 10°)t - 16t^2$

(b) $y = 5 + (240 \sin 10°)t - 16t^2 = 0$

$$t = \frac{-240 \sin 10° \pm \sqrt{(240 \sin 10°)^2 - 4(-16)5}}{2(-16)}$$

$t \approx -0.1149, 2.7196$

Distance traveled before arrow hits ground:

$(240 \cos 10°)(2.7196) \approx 643$ feet

(c) 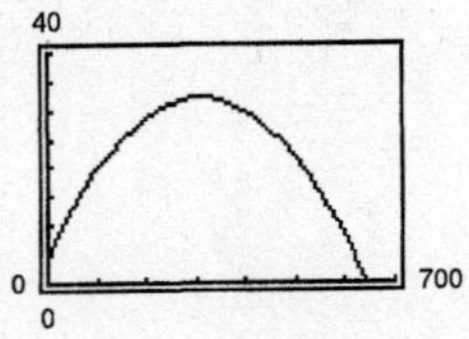

Maximum height: 32.1 feet

(d) Time arrow is in the air: approximately 2.72 seconds

(see part b)

62. $y = 7 + x - 0.02x^2$

(a) Exercise 61 result: $y = -\dfrac{16 \sec^2 \theta}{v_0^2}x^2 + (\tan \theta)x + h$

$h = 7$

$\tan \theta = 1 \implies \theta = 45°$

$\dfrac{16 \sec^2 45°}{v_0^2} = 0.02 \implies v_0 = 40$

$x = (v_0 \cos \theta)t = (40 \cos 45°)t$

$y = h + (v_0 \sin \theta)t - 16t^2 \approx 7 + (40 \sin 45°)t - 16t^2$

(b) 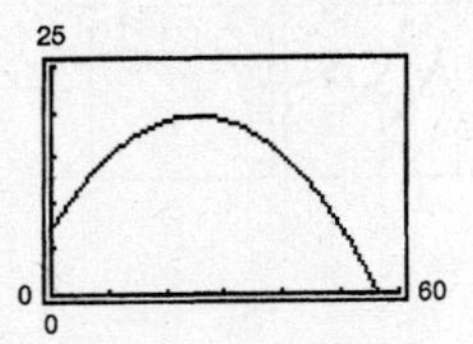

(c) Maximum height: 19.5 feet

Range: 56.2 feet

64. The coordinates of point (x, y) can be thought of as the sum of two vectors:

From origin to center of small circle:

$\langle 3 \cos \theta, 3 \sin \theta \rangle$

From center of small circle to point (x, y):

$\langle \cos \beta, \sin \beta \rangle$

Because the small circle rotates by 2θ when its center has rotated by θ, we have $\beta = \pi + 3\theta$.

$x = 3 \cos \theta + \cos(\pi + 3\theta) = 3 \cos \theta - \cos 3\theta$

$y = 3 \sin \theta + \sin(\pi + 3\theta) = 3 \sin \theta - \sin 3\theta$

66. False. Since $t^2 \geq 0$, the graph is that of $y = x$ for $x \geq 0$.

68. $\theta = 230°$

$\theta' = \theta - 180° = 230° - 180° = 50°$

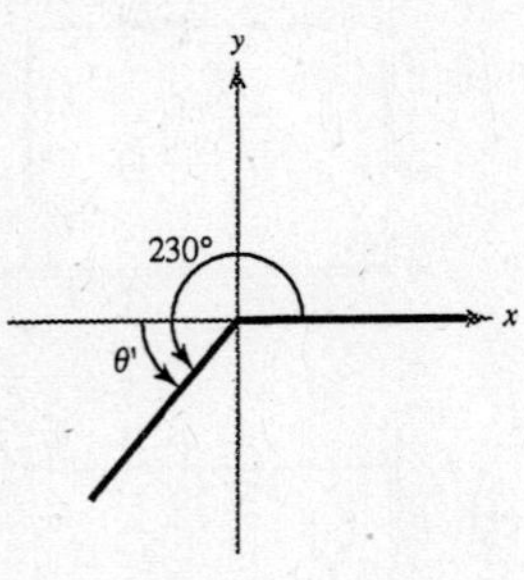

70. $\theta = \dfrac{5\pi}{6}$

$\theta' = \pi - \theta = \pi - \dfrac{5\pi}{6} = \dfrac{\pi}{6}$

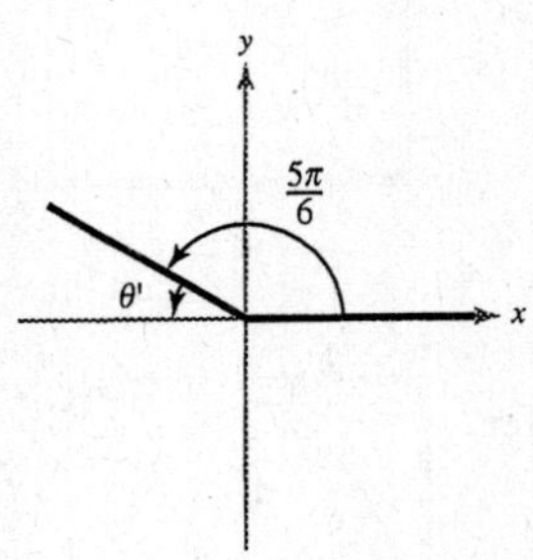

72. $y = \arccos(x - 1)$

Domain: $-1 \le x - 1 \le 1 \Rightarrow 0 \le x \le 2$

Range: $0 \le y \le \pi$

Horizontal translation to the right of $f(x) = \text{arcos } x$ by one unit.

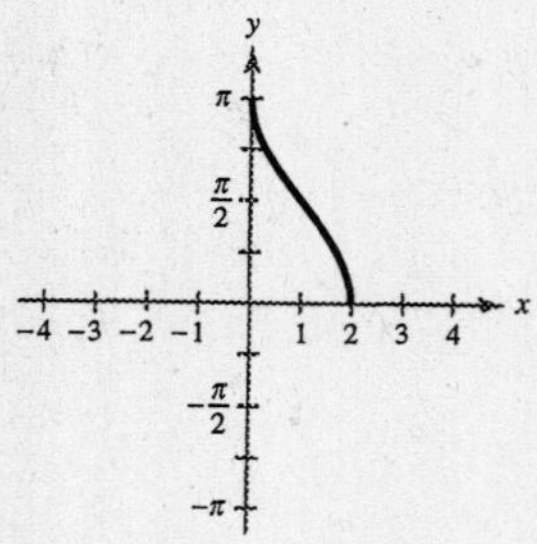

74. $f(x) = \arctan \dfrac{x}{2}$

Domain: all real numbers

Range: $-\dfrac{\pi}{2} < y < \dfrac{\pi}{2}$

Horizontal stretch of $f(x) = \arctan x$

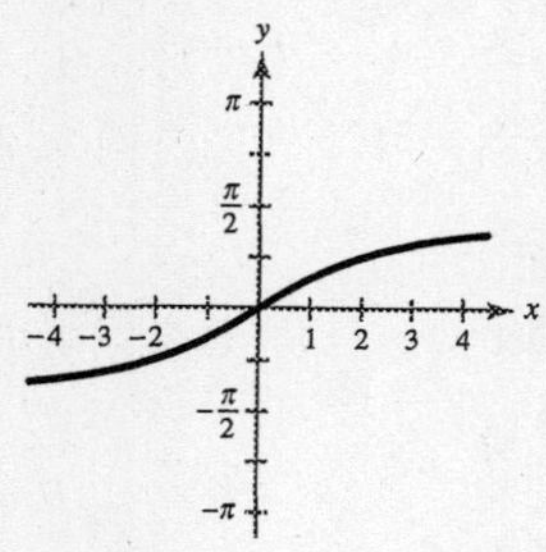

Section 6.7 Polar Coordinates

Solutions to Even-Numbered Exercises

2. $\left(-1, -\dfrac{3\pi}{4}\right)$

$\left(1, \dfrac{\pi}{4}\right)$

$\left(-1, \dfrac{5\pi}{4}\right)$

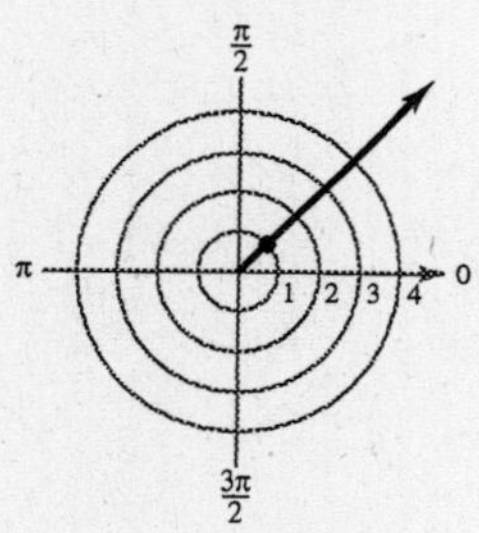

4. $\left(16, \dfrac{5\pi}{2}\right)$

$\left(16, \dfrac{\pi}{2}\right)$

$\left(-16, \dfrac{3\pi}{2}\right)$

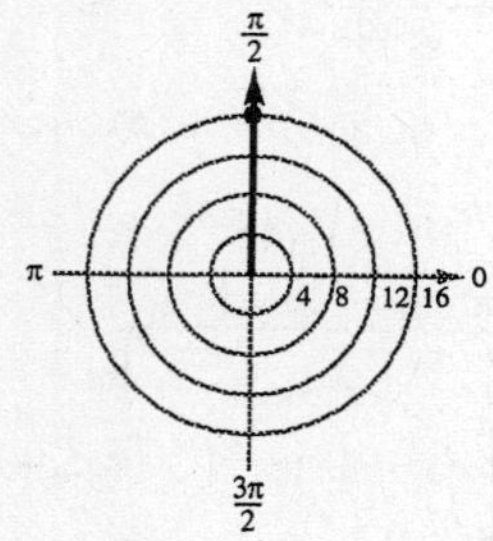

6. $(-3, -1.57)$

$(3, 1.5716)$

$(-3, 4.7132)$

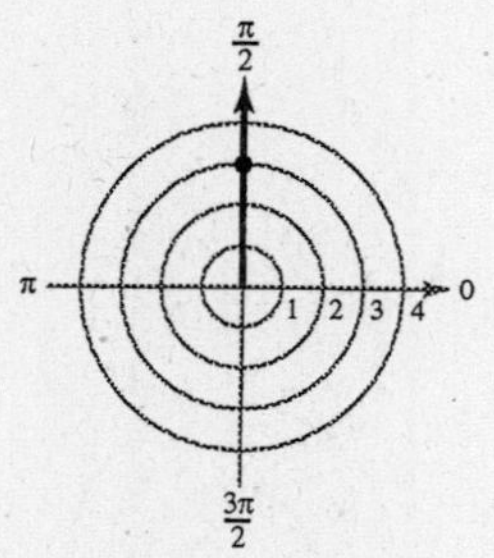

8. $(-5, -2.36)$

$(5, 0.7816)$

$(-5, 3.9232)$

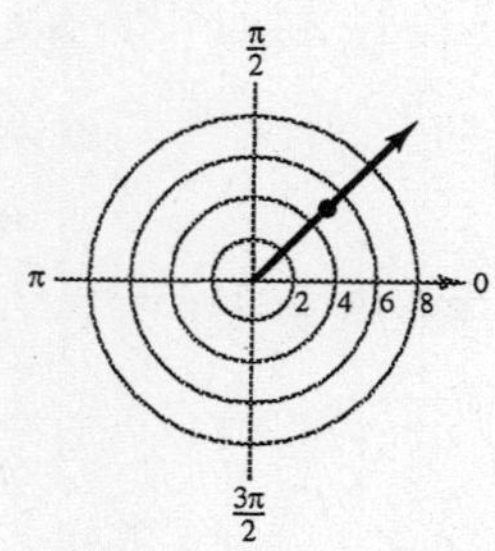

10. Polar coordinates: $\left(3, \dfrac{3\pi}{2}\right) = (r, \theta)$

$x = r\cos\theta = 3\cos\dfrac{3\pi}{2} = 0$

$y = r\sin\theta = 3\sin\dfrac{3\pi}{2} = -3$

Rectangular coordinates: $(0, -3)$

12. Polar coordinates: $(0, -\pi) = (r, \theta)$

$x = r\cos\theta = 0$

$y = r\sin\theta = 0$

Rectangular coordinates: $(0, 0)$

14. Polar coordinates: $\left(-2, \frac{7\pi}{6}\right) = (r, \theta)$

$x = r\cos\theta = -2\cos\frac{7\pi}{6} = \sqrt{3}$

$y = r\sin\theta = -2\sin\frac{7\pi}{6} = 1$

Rectangular coordinates: $\left(\sqrt{3}, 1\right)$

16. Polar coordinates: $(8.25, 3.5) = (r, \theta)$

$x = r\cos\theta = 8.25\cos 3.5 \approx -7.7258$

$y = r\sin\theta = 8.25\sin 3.5 \approx -2.8940$

Rectangular coordinates: $(-7.7258, -2.8940)$

18. Rectangular coordinates: $(-3, -3)$

$r = 3\sqrt{2}$, $\tan\theta = 1$, $\theta = \frac{5\pi}{4}$

Polar coordinates: $\left(3\sqrt{2}, \frac{5\pi}{4}\right)$

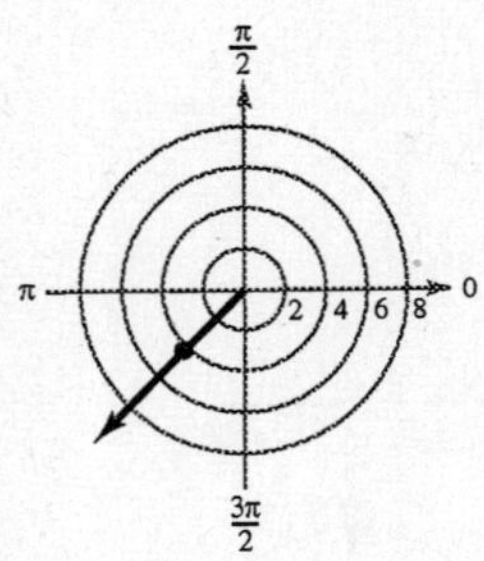

20. Rectangular coordinates: $(0, -5)$

$r = 5$, $\tan\theta$ undefined, $\theta = \frac{\pi}{2}$

Polar coordinates: $\left(5, \frac{3\pi}{2}\right)$

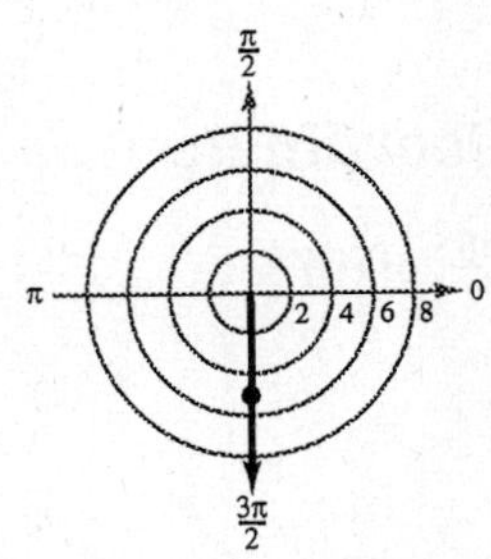

22. Rectangular coordinates: $(3, -1)$

$r = \sqrt{9+1} = \sqrt{10}$, $\tan\theta = -\frac{1}{3}$, $\theta \approx -0.322 \approx 5.9614$

Polar coordinates: $\left(\sqrt{10}, 5.961\right)$

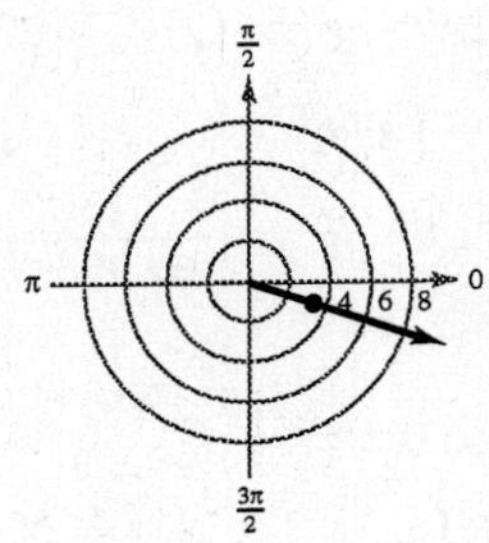

24. Rectangular coordinates: $\left(\sqrt{3}, -1\right)$

$r = \sqrt{3+1} = 2$, $\tan\theta = -\frac{1}{\sqrt{3}}$, $\theta = \frac{11\pi}{6}$

Polar coordinates: $\left(2, \frac{11\pi}{6}\right)$

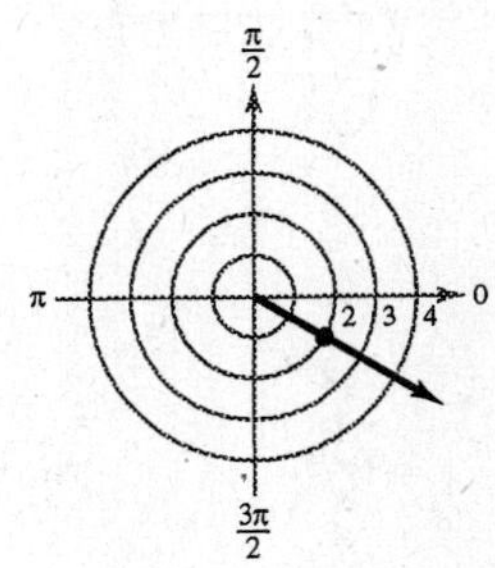

26. Rectangular coordinates: $(5, 12)$

$r = \sqrt{25+144} = 13$, $\tan\theta = \frac{12}{5}$, $\theta \approx 1.176$

Polar coordinates: $(13, 1.176)$

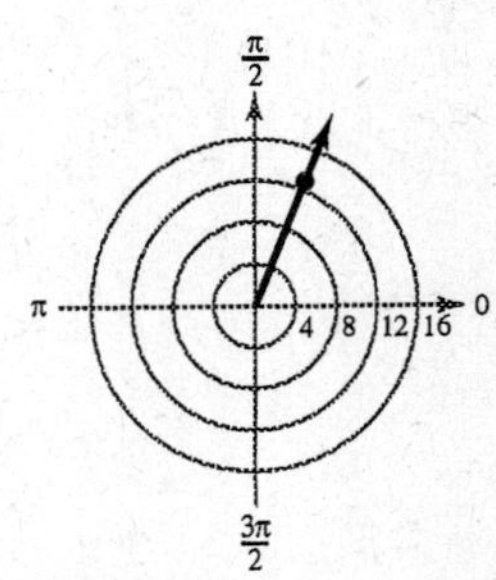

28. Rectangular coordinates: $(-5, 2)$

$R \blacktriangleright Pr(-5, 2) \approx 5.385$

$R \blacktriangleright P\theta(-5, 2) \approx 2.761$

$\approx (5.385, 2.761)$

30. Rectangular coordinates: $\left(3\sqrt{2}, 3\sqrt{2}\right)$

$R \blacktriangleright Pr\left(3\sqrt{2}, 3\sqrt{2}\right) = 6$

$R \blacktriangleright P\theta\left(3\sqrt{2}, 3\sqrt{2}\right) \approx 0.785$

$= \left(6, \frac{\pi}{4}\right)$

32. Rectangular coordinates: $\left(\frac{7}{4}, \frac{3}{2}\right)$

$R \blacktriangleright Pr\left(\frac{7}{4}, \frac{3}{2}\right) \approx 2.305$

$R \blacktriangleright P\theta\left(\frac{7}{4}, \frac{3}{2}\right) \approx 0.709$

$\approx (2.305, 0.709)$

34. $x^2 + y^2 = 16$

$r = 4$

36.

$$\begin{aligned} y &= x \\ r\cos\theta &= r\sin\theta \\ 1 &= \tan\theta \\ \theta &= \frac{\pi}{4} \end{aligned}$$

38.

$$\begin{aligned} x &= 4a \\ r\cos\theta &= 4a \\ r &= 4a\sec\theta \end{aligned}$$

40.

$$\begin{aligned} 3x + 5y - 2 &= 0 \\ 3r\cos\theta + 5r\sin\theta - 2 &= 0 \\ r(3\cos\theta + 5\sin\theta) &= 2 \\ r &= \frac{2}{3\cos\theta + 5\sin\theta} \end{aligned}$$

42. $2xy = 1$

$$\begin{aligned} 2(r\cos\theta)(r\sin\theta) &= 1 \\ 2r^2\cos\theta\sin\theta &= 1 \\ r^2 &= \frac{1}{2\cos\theta\sin\theta} \\ r^2 &= \frac{1}{2}\sec\theta\csc\theta \\ r^2 &= \csc 2\,\theta \end{aligned}$$

44. $(x^2 + y^2)^2 = 9(x^2 - y^2)$

$$\begin{aligned} (r^2)^2 &= 9(r^2\cos^2\theta - r^2\sin^2\theta) \\ &= 9r^2(\cos^2\theta - \sin^2\theta) \\ r^2 &= 9\cos 2\theta \end{aligned}$$

46. $x^2 + y^2 = 9a^2$

$r = 3a$

48. $x^2 + y^2 - 2ay = 0$

$$\begin{aligned} r^2 - 2ar\sin\theta &= 0 \\ r &= 2a\sin\theta \end{aligned}$$

50. Because $x = r\cos\theta$ and r is given as $2\cos\theta$, $x = 2\cos\theta\cos\theta = 2\cos^2\theta$.

$$\begin{aligned} r &= 2\cos\theta \\ r^2 &= 4\cos^2\theta \\ x^2 + y^2 &= 2\,(2\cos^2\theta) \\ x^2 + y^2 &= 2x \\ x^2 + y^2 - 2x &= 0 \end{aligned}$$

52.

$$\begin{aligned} \theta &= \frac{5\pi}{3} \\ \tan\theta &= -\sqrt{3} \\ \frac{y}{x} &= -\sqrt{3} \\ y &= -\sqrt{3}\,x \\ \sqrt{3}\,x + y &= 0 \end{aligned}$$

54.

$$\begin{aligned} r &= 10 \\ r^2 &= 100 \\ x^2 + y^2 &= 100 \end{aligned}$$

56.

$$\begin{aligned} r &= -3\sec\theta \\ \frac{r}{\sec\theta} &= -3 \\ r\sec\theta &= -3 \\ x &= -3 \end{aligned}$$

58.
$$r^2 = \sin 2\theta = 2 \sin \theta \cos \theta$$
$$r^2 = 2\left(\frac{y}{r}\right)\left(\frac{x}{r}\right) = \frac{2xy}{r^2}$$
$$r^4 = 2xy$$
$$(x^2 + y^2)^2 = 2xy$$

60.
$$r = 3 \cos 2\,\theta$$
$$r = 3(2 \cos^2 \theta - 1)$$
$$r = 3\left(\frac{2x^2}{r^2} - 1\right)$$
$$r = 3\left(\frac{2x^2 - r^2}{r^2}\right)$$
$$r = 3\left(\frac{2x^2 - x^2 - y^2}{x^2 + y^2}\right)$$
$$r = 3\left(\frac{x^2 - y^2}{x^2 + y^2}\right)$$
$$r^2 = 9\left(\frac{x^2 - y^2}{x^2 + y^2}\right)^2$$
$$x^2 + y^2 = 9\frac{(x^2 - y^2)^2}{(x^2 + y^2)^2}$$
$$(x^2 + y^2)^3 = 9(x^2 - y^2)^2$$

62.
$$r = \frac{1}{1 - \cos \theta}$$
$$r - r \cos \theta = 1$$
$$\sqrt{x^2 + y^2} - x = 1$$
$$x^2 + y^2 = 1 + 2x + x^2$$
$$y^2 = 2x + 1$$

64.
$$r = \frac{6}{2 \cos \theta - 3 \sin \theta}$$
$$r = \frac{6}{2(x/r) - 3(y/r)}$$
$$r = \frac{6r}{2x - 3y}$$
$$1 = \frac{6}{2x - 3y}$$
$$2x - 3y = 6$$

66.
$$r = 8$$
$$r^2 = 64$$
$$x^2 + y^2 = 64$$

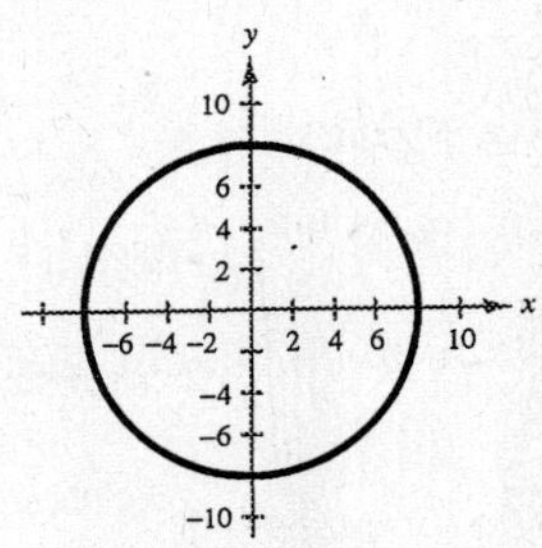

68.
$$\theta = \frac{3\pi}{4}$$
$$\tan \theta = \tan \frac{3\pi}{4}$$
$$\frac{y}{x} = -1$$
$$y = -x$$
$$x + y = 0$$

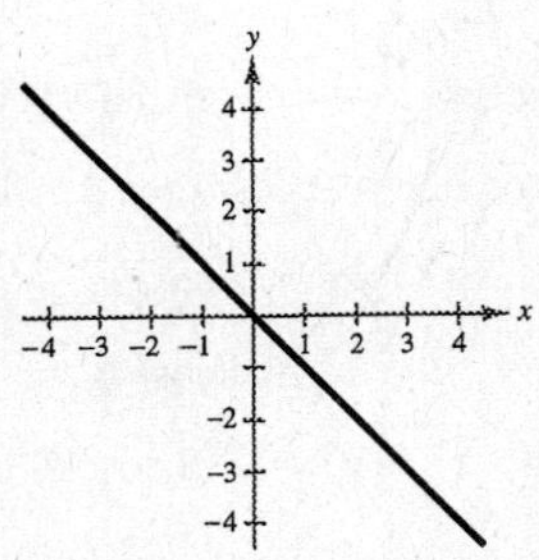

70.
$$r = 2 \csc \theta$$
$$r \sin \theta = 2$$
$$y = 2$$
$$y - 2 = 0$$

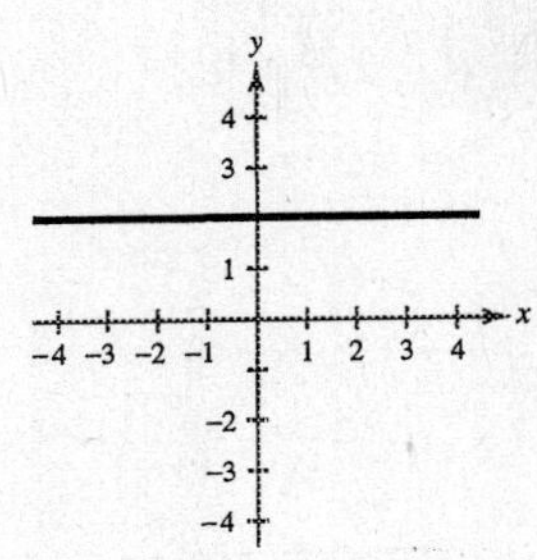

72. False. (r_1, θ) and (r_2, θ) represent the same point only if $r_1 = r_2$.

74.
$$r = \cos\theta + 3\sin\theta$$
$$r = \frac{x}{r} + \frac{3y}{r}$$
$$r^2 = x + 3y$$
$$x^2 + y^2 = x + 3y$$
$$x^2 - x + y^2 - 3y = 0$$
$$\left(x - \frac{1}{2}\right)^2 + \left(y - \frac{3}{2}\right)^2 = \frac{5}{2}$$

The graph is a circle.

76. (a) For horizontal moves, just the x-coordinate changes. For vertical moves, just the y-coordinate changes.

(b) For horizontal moves, both r and θ change. For vertical moves, both r and θ change.

(c) Unlike r and θ, x and y measure horizontal and vertical changes, respectively.

78. $\log_4 \frac{\sqrt{2x}}{y} = \log_4\sqrt{2x} - \log_4 y$

$$= \frac{1}{2}\log_4 2x - \log_4 y$$
$$= \frac{1}{2}\log_4 2 + \frac{1}{2}\log_4 x - \log_4 y$$
$$= \frac{1}{4} + \frac{1}{2}\log_4 x - \log_4 y$$

80. $\ln 5x^2(x^2 + 1) = \ln 5 + \ln x^2 + \ln(x^2 + 1)$

$$= \ln 5 + 2\ln x + \ln(x^2 + 1)$$

82. $\log_5 a + 8\log_5(x + 1) = \log_5 a + \log_5(x + 1)^8$

$$= \log_5 a(x + 1)^8$$

84. $\ln 6 + \ln y - \ln(x - 3) = \ln 6y - \ln(x - 3)$

$$= \ln\frac{6y}{x - 3}$$

Section 6.8 Graphs of Polar Equations

Solutions to Even-Numbered Exercises

2. $r = 5 - 5\sin\theta$

Cardioid

4. $r^2 = 16\cos 2\theta$

Lemniscate

6. $r = 3\cos\theta$

Circle

8. $r = 16\cos 3\theta$

$\theta = \frac{\pi}{2}$: $-r = 16\cos(3(-\theta))$

$-r = 16\cos(-3\theta)$

$-r = 16\cos 3\theta$

Not an equivalent equation

Polar axis: $r = 16\cos(3(-\theta))$

$r = 16\cos(-3\theta)$

$r = 16\cos 3\theta$

Equivalent equation

Pole: $-r = 16\cos 3\theta$

Not an equivalent equation

Answer: Symmetric with respect to polar axis

10. $r = \frac{3}{2 + \cos\theta}$

$\theta = \frac{\pi}{2}$: $-r = \frac{3}{2 + \cos(-\theta)}$

Not an equivalent equation

Polar axis: $r = \frac{3}{2 + \cos(-\theta)}$

Equivalent equation

Pole: $-r = \frac{3}{2 + \cos\theta}$

Not an equivalent equation

Answer: Symmetric with respect to polar axis

12. $r^2 = 36 \sin 2\theta$

$\theta = \frac{\pi}{2}$: $(-r)^2 = 36 \sin(-2\theta)$

Not an equivalent equation

Polar axis: $r^2 = 36 \sin(-2\theta)$

Not an equivalent equation

Pole: $(-r)^2 = 36 \sin 2\theta$

Equivalent equation

Answer: Symmetric with respect to pole

14. $|r| = |6 + 12 \cos \theta| \le |6| + |12 \cos \theta|$

$= 6 + 12|\cos \theta| \le 18$

$\cos \theta = 1$

$\theta = 0$

Maximum: $|r| = 18$ when $\theta = 0$

$0 = 6 + 12 \cos \theta$

$\cos \theta = -\frac{1}{2}$

$\theta = \frac{2\pi}{3}, \frac{4\pi}{3}$

Zero: $r = 0$ when $\theta = \frac{2\pi}{3}, \frac{4\pi}{3}$

16. $|r| = |3 \sin 2\theta| = 3|\sin 2\theta| \le 3$

$|\sin 2\theta| = 1$

$\sin 2\theta = \pm 1$

$\theta = \frac{\pi}{4}, \frac{3\pi}{4}, \frac{5\pi}{4}, \frac{7\pi}{4}$

Maximum: $|r| = 3$ when $\theta = \frac{\pi}{4}, \frac{3\pi}{4}, \frac{5\pi}{4}, \frac{7\pi}{4}$

$0 = 3 \sin 2\theta$

$\sin 2\theta = 0$

$\theta = 0, \frac{\pi}{2}, \pi, \frac{3\pi}{2}$

Zero: $r = 0$ when $\theta = 0, \frac{\pi}{2}, \pi, \frac{3\pi}{2}$

18. Circle: $r = 2$

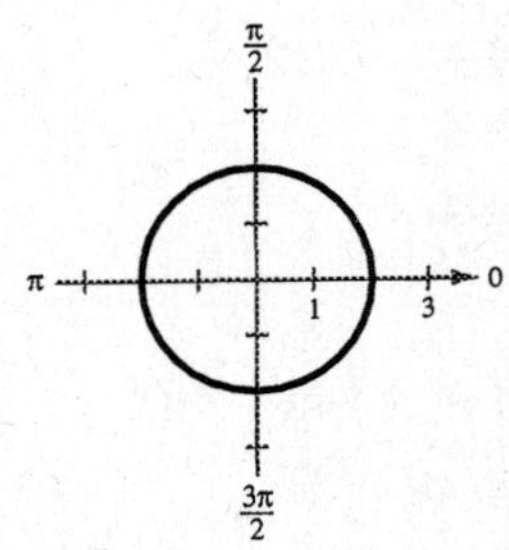

20. $r = -\frac{3\pi}{4}$

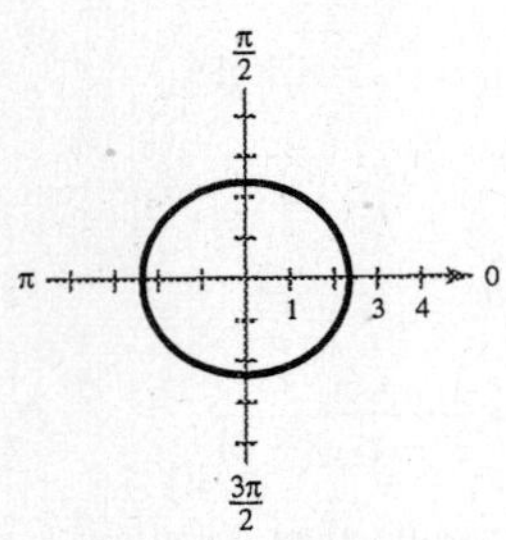

22. $r = 4 \cos \theta$

Symmetric with respect to polar axis.

Circle with radius 2.

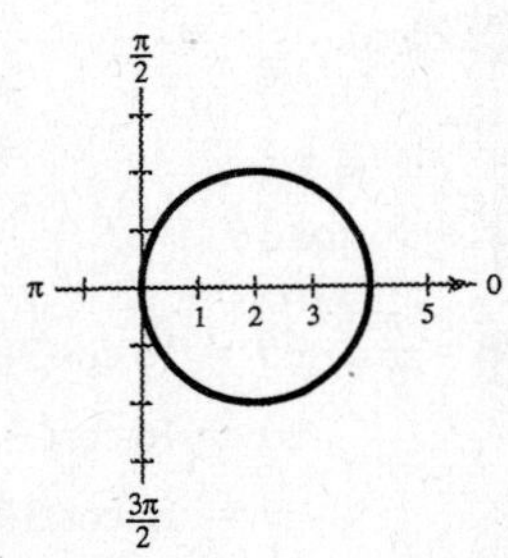

24. $r = 4(1 - \sin \theta)$

Symmetric with respect to $\pi/2$.

$\frac{a}{b} = \frac{4}{4} = 1 \Rightarrow$ Cardioid

$|r| = 8$ when $\theta = \frac{3\pi}{2}$.

$r = 0$ when $\theta = \frac{\pi}{2}$.

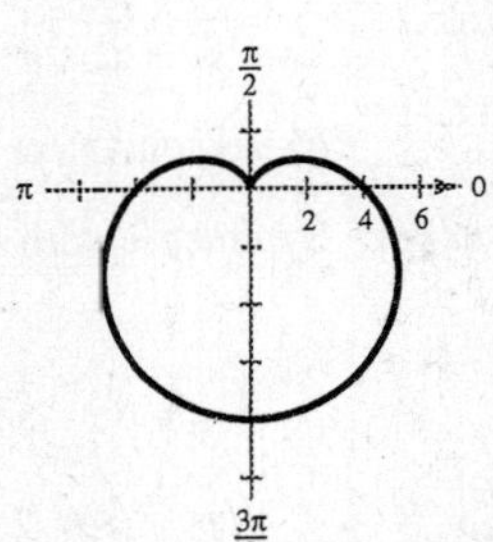

26. $r = 2(1 + \cos \theta)$

Symmetric with respect to polar axis.

$\frac{a}{b} = \frac{2}{2} = 1 \Rightarrow$ Cardioid

$|r| = 4$ when $\theta = 0$.

$r = 0$ when $\theta = \pi$.

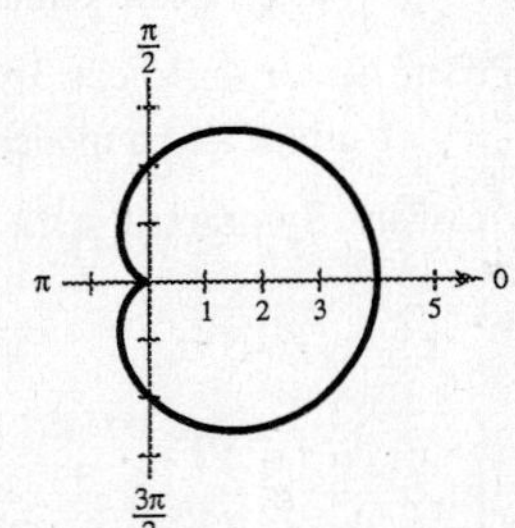

28. $r = 4 - 3\sin\theta$

Symmetric with respect to $\pi/2$.

$a = 4, b = 3$

$\frac{a}{b} = \frac{4}{3} \Rightarrow$ Dimpled limaçon

$|r| = 7$ when $\theta = \frac{3\pi}{2}$.

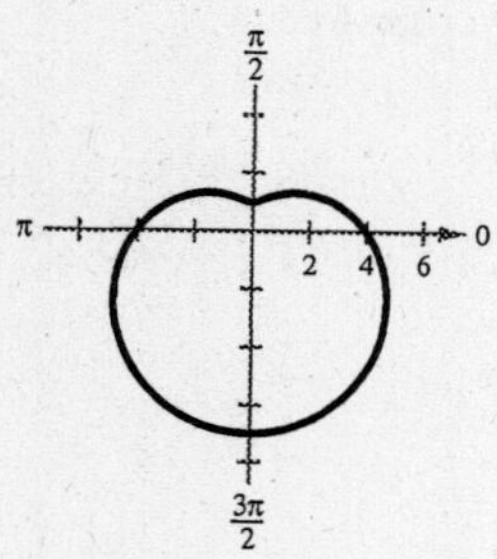

30. $r = 1 - 2\cos\theta$

Symmetric with respect to the polar axis.

$\frac{a}{b} = \frac{1}{2} \Rightarrow$ Limaçon with inner loop

$|r| = 3$ when $\theta = \pi$.

$r = 0$ when $\theta = \frac{\pi}{3}, \frac{5\pi}{3}$.

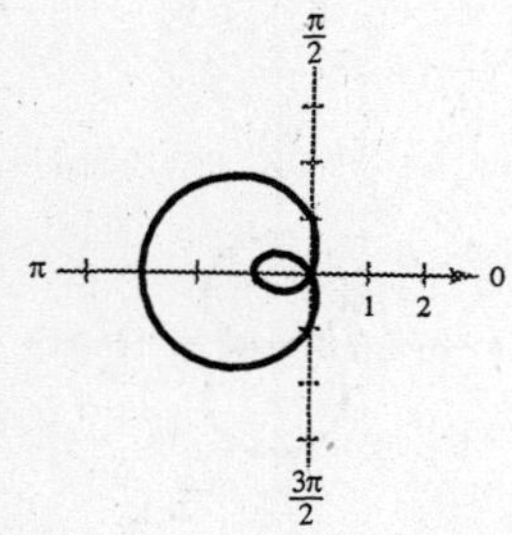

32. $r = 4 + 3\cos\theta$

Symmetric with respect to the polar axis.

$\frac{a}{b} = \frac{4}{3} > 1 \Rightarrow$ Dimpled limaçon

$|r| = 7$ when $\theta = 0$.

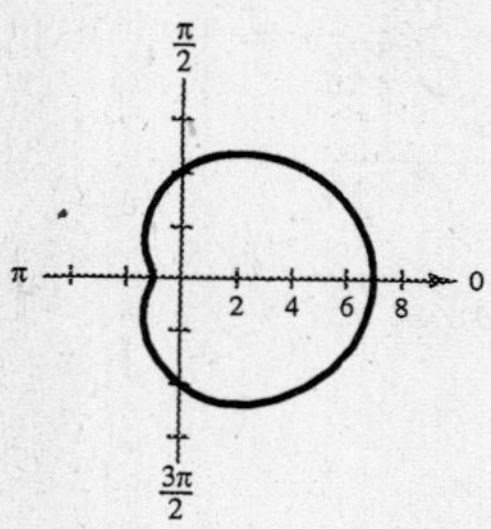

34. $r = 3\cos 2\theta$

Symmetric with respect to the polar axis.

Rose curve $(n = 2)$ with four petals

$|r| = 3$ when $\theta = 0, \frac{\pi}{2}, \pi, \frac{3\pi}{2}$.

$r = 0$ when $\theta = \frac{\pi}{4}, \frac{3\pi}{4}, \frac{5\pi}{4}, \frac{7\pi}{4}$.

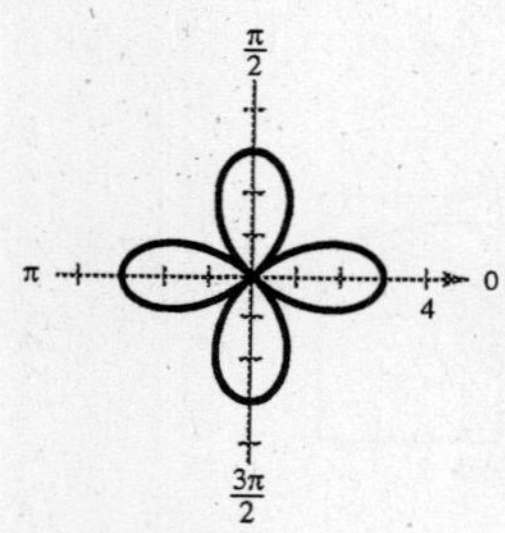

36. $r = 5\csc\theta$

$r\sin\theta = 5$

$y = 5 \Rightarrow$ Line

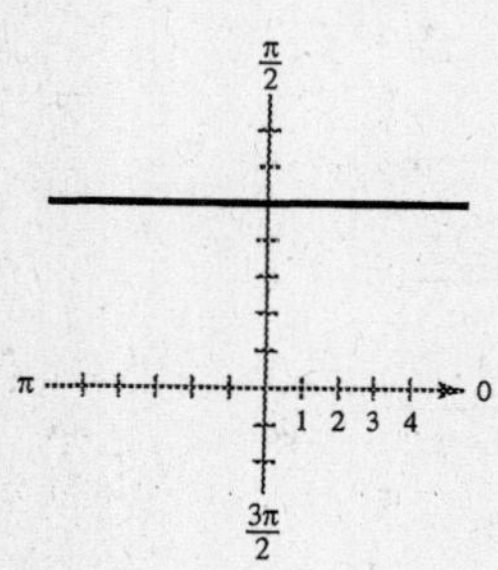

38. $r = \frac{6}{2\sin\theta - 3\cos\theta}$

$r(2\sin\theta - 3\cos\theta) = 6$

$2y - 3x = 6$

$y = \frac{3}{2}x + 3 \Rightarrow$ Line

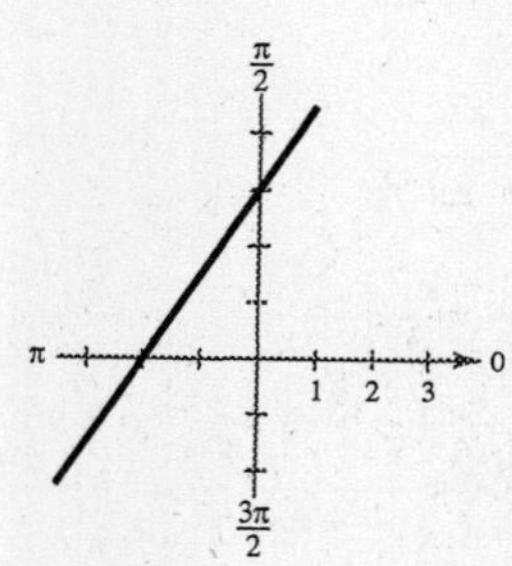

40. $r^2 = 4 \sin \theta$

$r = 2\sqrt{\sin \theta}$

$r = -2\sqrt{\sin \theta}$

$0 \le \theta \le \pi$

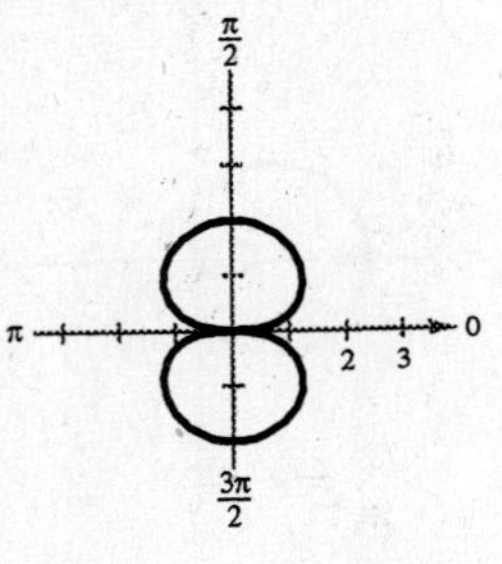

42. $r = \cos 2\theta$

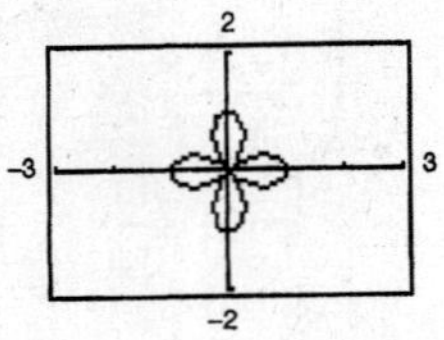

44. $r = 2\cos(3\theta - 2)$

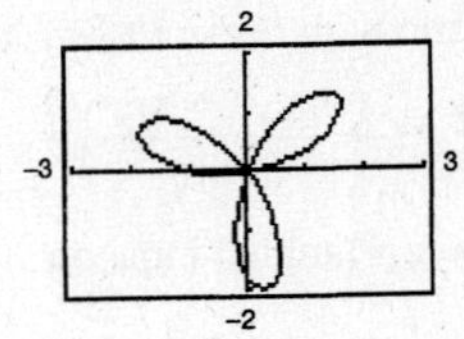

46. $r = 2 \csc \theta + 5 = \dfrac{2}{\sin \theta} + 5$

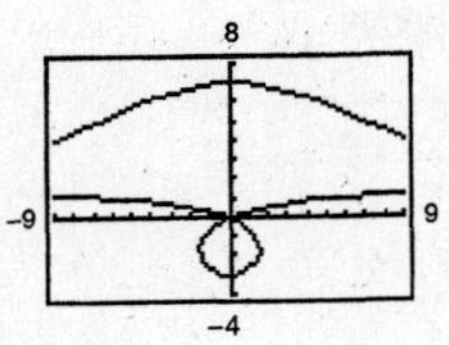

48. $r = 5 + 4 \cos \theta$

$0 \le \theta < 2\pi$

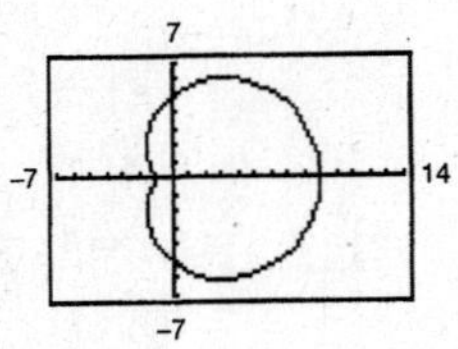

50. $r = 3 \sin\left(\dfrac{5\theta}{2}\right)$

$0 \le \theta < 4\pi$

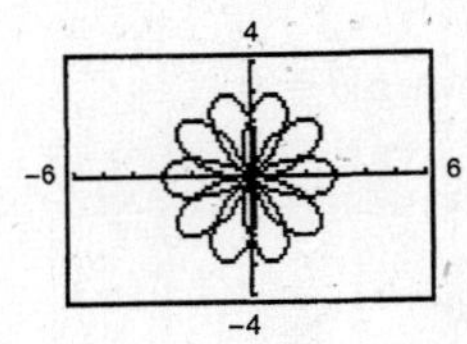

52. $r^2 = \dfrac{1}{\theta}$

$0 < \theta < \infty$

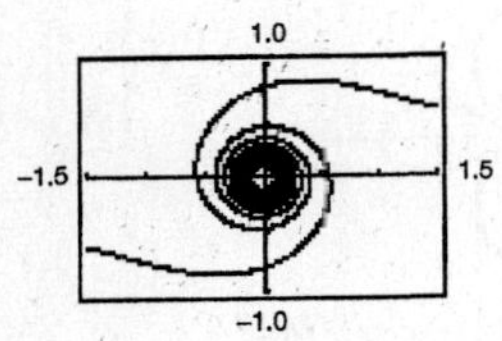

54.

$$r = 2 + \csc \theta = 2 + \frac{1}{\sin \theta}$$

$$r \sin \theta = 2 \sin \theta + 1$$

$$r(r \sin \theta) = 2r \sin \theta + r$$

$$\left(\pm\sqrt{x^2 + y^2}\right)(y) = 2y + \left(\pm\sqrt{x^2 + y^2}\right)$$

$$\left(\pm\sqrt{x^2 + y^2}\right)(y - 1) = 2y$$

$$\left(\pm\sqrt{x^2 + y^2}\right) = \frac{2y}{y - 1}$$

$$x^2 + y^2 = \frac{4y^2}{(y - 1)^2}$$

$$x^2 = \frac{y^2(3 + 2y - y^2)}{(y - 1)^2}$$

$$x = \sqrt{\frac{y^2(3 + 2y - y^2)}{(y - 1)^2}} = \pm\left|\frac{y}{y - 1}\right|\sqrt{3 + 2y - y^2}$$

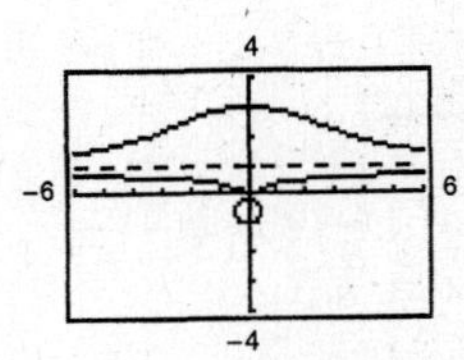

The graph has an asymptote at $y = 1$.

56. $r = 2\cos 2\theta \sec \theta = \dfrac{2\cos 2\theta}{\cos \theta}$

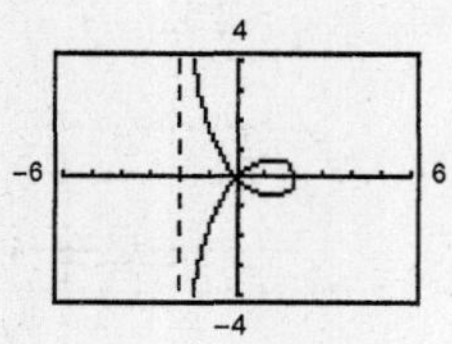

$r = \dfrac{2(\cos^2\theta - \sin^2\theta)}{\cos\theta}$

$r\cos\theta = 2(\cos^2\theta - \sin^2\theta)$

$x = 2(\cos^2\theta - \sin^2\theta)$

As $\theta \to \dfrac{\pi}{2}$, $x \to -2$.

58. False. For a graph to be symmetric about the pole, one portion of the graph coincides with the other portion when rotated π radians about the pole.

60. $r = 6[1 + \cos(\theta - \phi)]$

(a)

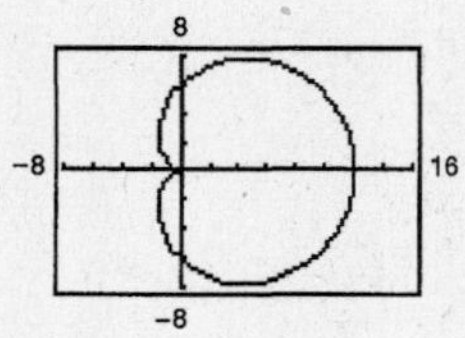

(b)

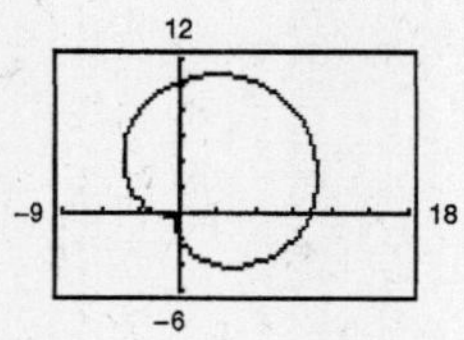

(c)

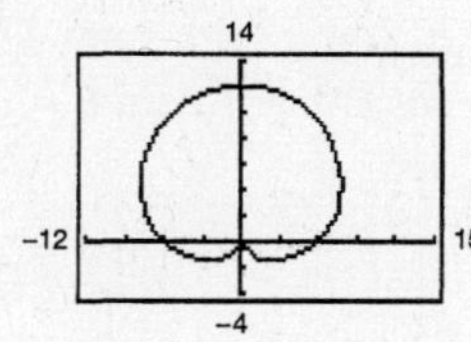

The angle ϕ has the effect of rotating the graph by the angle ϕ. For part (c),

$$r = 6\left[1 + \cos\left(\theta - \frac{\pi}{2}\right)\right]$$
$$= 6(1 + \sin\theta).$$

62. Use the result of Exercise 61.

(a) Rotation: $\phi = \dfrac{\pi}{2}$

Original graph: $r = f(\sin\theta)$

Rotated graph: $r = f\left(\sin\left(\theta - \dfrac{\pi}{2}\right)\right) = f(-\cos\theta)$

(b) Rotation: $\phi = \pi$

Original graph: $r = f(\sin\theta)$

Rotated graph: $r = f(\sin(\theta - \pi)) = f(-\sin\theta)$

(c) Rotation: $\phi = \dfrac{3\pi}{2}$

Original graph: $r = f(\sin\theta)$

Rotated graph: $r = f\left(\sin\left(\theta - \dfrac{3\pi}{2}\right)\right) = f(\cos\theta)$

64. $r = 2\sin 2\theta$

(a) $$r = 2\sin\left[2\left(\theta - \frac{\pi}{6}\right)\right]$$
$$= 2\left[2\sin\left(\theta - \frac{\pi}{6}\right)\cos\left(\theta - \frac{\pi}{6}\right)\right]$$
$$= 4\sin\left(\theta - \frac{\pi}{6}\right)\cos\left(\theta - \frac{\pi}{6}\right)$$

(b) $$r = 2\sin\left[2\left(\theta - \frac{\pi}{2}\right)\right]$$
$$= 2\sin(2\theta - \pi)$$
$$= -2\sin 2\theta$$
$$= -2(2\sin\theta\cos\theta)$$
$$= -4\sin\theta\cos\theta$$

(c) $$r = 2\sin\left[2\left(\theta - \frac{2\pi}{3}\right)\right]$$
$$= 2\left[2\sin\left(\theta - \frac{2\pi}{3}\right)\cos\left(\theta - \frac{2\pi}{3}\right)\right]$$
$$= 4\sin\left(\theta - \frac{2\pi}{3}\right)\cos\left(\theta - \frac{2\pi}{3}\right)$$

(d) $$r = 2\sin[2(\theta - \pi)]$$
$$= 2\sin(2\theta - 2\pi)$$
$$= 2\sin 2\theta$$
$$= 2[2\sin\theta\cos\theta]$$
$$= 4\sin\theta\cos\theta$$

66. (a)

$$r = 3 \sec \theta$$

$$r = \frac{3}{\cos \theta}$$

$$r \cos \theta = 3 \Rightarrow x = 3$$

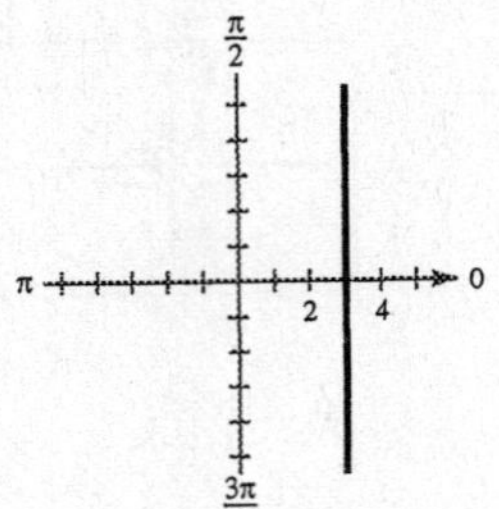

(b)

$$r = 3 \sec\left(\theta - \frac{\pi}{4}\right)$$

$$r = \frac{3}{\cos(\theta - (\pi/4))}$$

$$r = \frac{3}{\cos \theta \cos(\pi/4) + \sin \theta \sin(\pi/4)}$$

$$\frac{\sqrt{2}}{2} r \cos \theta + \frac{\sqrt{2}}{2} r \sin \theta = 3$$

$$\frac{\sqrt{2}}{2} x + \frac{\sqrt{2}}{2} y = 3$$

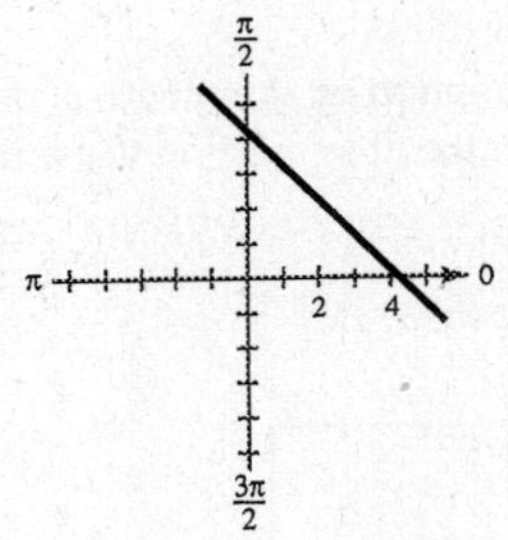

(c)

$$r = 3 \sec\left(\theta + \frac{\pi}{3}\right)$$

$$r = \frac{3}{\cos(\theta + (\pi/3))}$$

$$r = \frac{3}{\cos \theta \cos(\pi/3) - \sin \theta \sin(\pi/3)}$$

$$\frac{1}{2} r \cos \theta - \frac{\sqrt{3}}{2} r \sin \theta = 3$$

$$\frac{1}{2} x - \frac{\sqrt{3}}{2} y = 3$$

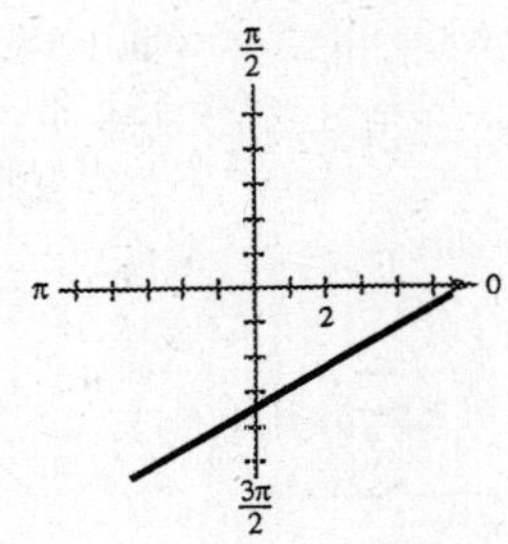

(d)

$$r = 3 \sec\left(\theta - \frac{\pi}{2}\right)$$

$$r = \frac{3}{\cos(\theta - (\pi/2))}$$

$$r = \frac{3}{\cos \theta \cos(\pi/2) + \sin \theta \sin(\pi/2)}$$

$$r \sin \theta = 3 \Rightarrow y = 3$$

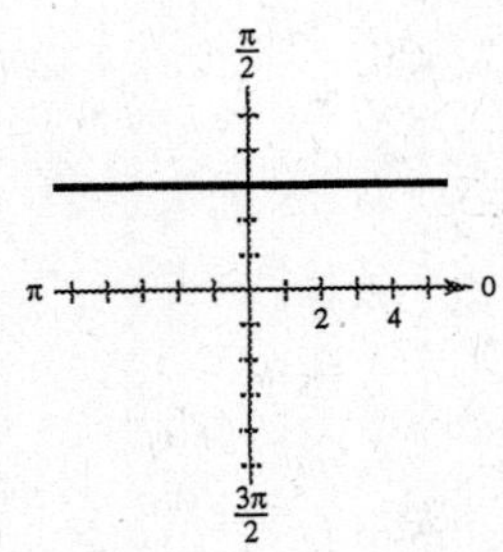

68. $r = 3 \sin k\theta$

(a) $r = 3 \sin 1.5\theta$

$0 \le \theta < 4\pi$

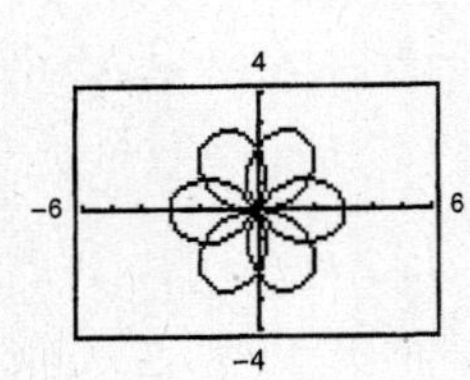

(b) $r = 3 \sin 2.5\theta$

$0 \le \theta < 4\pi$

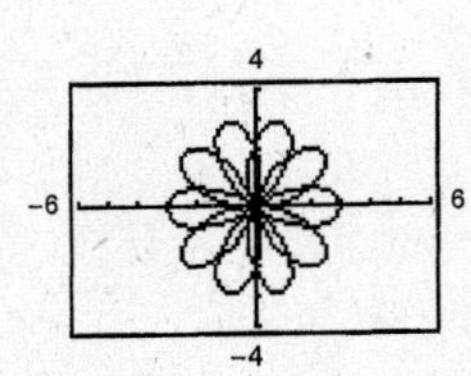

(c) Yes. $r = 3 \sin(k\theta)$.

Find the minimum value of $\theta (\theta > 0)$ that is a multiple of 2π that makes $k\theta$ a multiple of 2π.

70. Foci: $(3, 2), (3, -4)$; Major axis of length 8

Center: $(h, k) = (3, -1)$

Vertical major axis

$a = 4, c = 3, b^2 = a^2 - c^2 = 16 - 9 \Rightarrow b = \sqrt{7}$

$$\frac{(x-h)^2}{b^2} + \frac{(y-k)^2}{a^2} = 1$$

$$\frac{(x-3)^2}{7} + \frac{(y+1)^2}{16} = 1$$

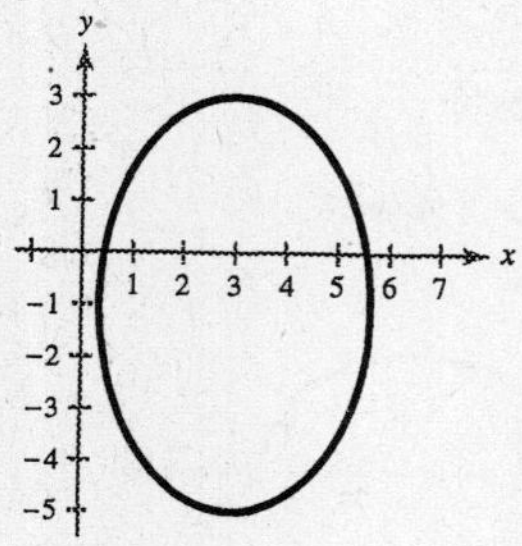

Section 6.9 Polar Equations of Conics

Solutions to Even-Numbered Exercises

2. $r = \dfrac{4e}{1 - e\cos\theta}$

$e = 1, r = \dfrac{4}{1 - \cos\theta}$, parabola

$e = 0.5, r = \dfrac{2}{1 - 0.5\cos\theta}$, ellipse

$e = 1.5, r = \dfrac{6}{1 - 1.5\cos\theta}$, hyperbola

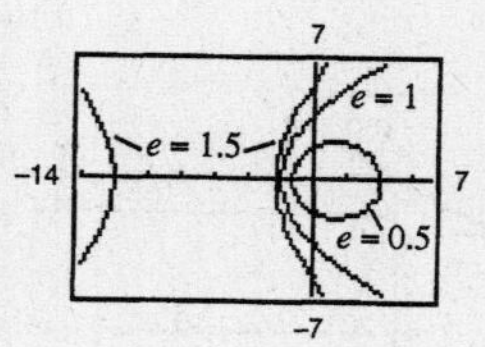

4. $r = \dfrac{4e}{1 + e\sin\theta}$

$e = 1, r = \dfrac{4}{1 + \sin\theta}$, parabola

$e = 0.5, r = \dfrac{2}{1 + 0.5\sin\theta}$, ellipse

$e = 1.5, r = \dfrac{6}{1 + 1.5\sin\theta}$, hyperbola

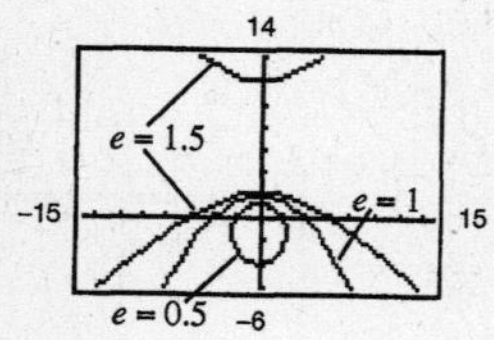

6. $r = \dfrac{3}{2 - \cos\theta}$

$e = \dfrac{1}{2} \Rightarrow$ Ellipse

Vertical directrix to the left of the pole
Matches graph (c).

8. $r = \dfrac{2}{1 - \sin\theta}$

$e = 1 \Rightarrow$ Parabola

Horizontal directrix below pole
Matches graph (e).

10. $r = \dfrac{4}{1 - 3\sin\theta}$

$e = 3 \Rightarrow$ Hyperbola

Horizontal directrix below pole
Matches graph (b).

12. $r = \dfrac{3}{1 + \sin\theta}$

$e = 1 \Rightarrow$ Parabola

Vertex: $\left(\dfrac{3}{2}, \dfrac{\pi}{2}\right)$

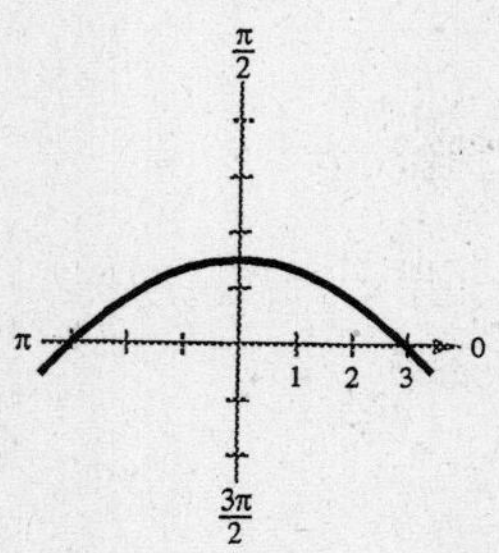

14. $r = \dfrac{6}{1 + \cos\theta}$

$e = 1 \Rightarrow$ Parabola
Vertex: $(3, 0)$

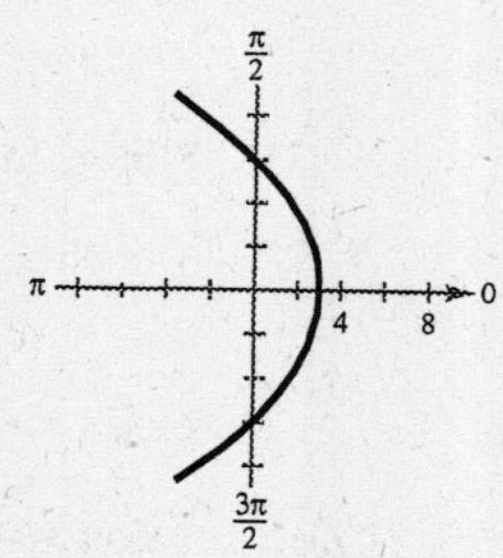

16. $r = \dfrac{3}{3 + \sin\theta} = \dfrac{1}{1 + (1/3)\sin\theta}$

$e = \dfrac{1}{3} < 1 \Rightarrow$ Ellipse

Vertices: $\left(\dfrac{3}{4}, \dfrac{\pi}{2}\right), \left(\dfrac{3}{2}, \dfrac{3\pi}{2}\right)$

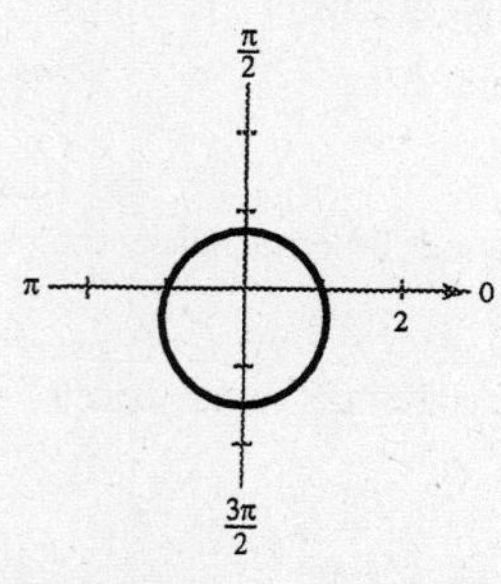

18. $r = \dfrac{9}{3 - 2\cos\theta} = \dfrac{3}{1 - (2/3)\cos\theta}$

$e = \dfrac{2}{3} < 1 \Rightarrow$ Ellipse

Vertices: $(9, 0), \left(\dfrac{9}{5}, \pi\right)$

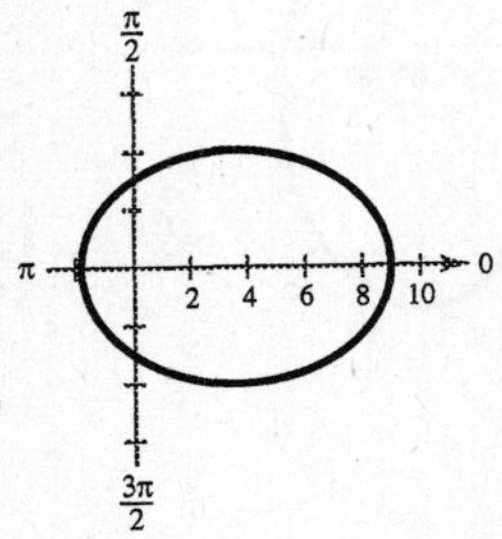

20. $r = \dfrac{5}{-1 + 2\cos\theta} = \dfrac{-5}{1 - 2\cos\theta}$

$e = 2 > 1 \Rightarrow$ Hyperbola

Vertices: $(5, 0), \left(-\dfrac{5}{3}, \pi\right)$

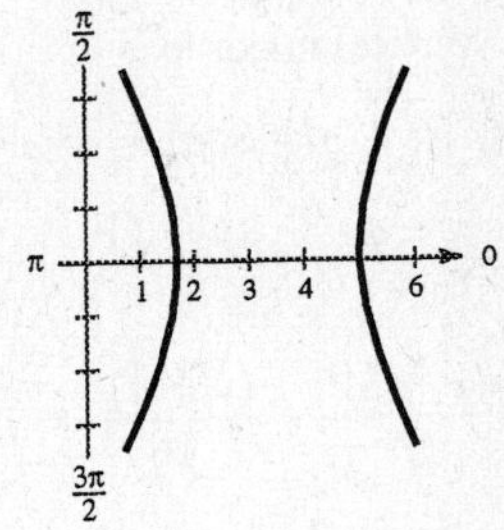

22. $r = \dfrac{3}{2 + 6\sin\theta} = \dfrac{3/2}{1 + 3\sin\theta}$

$e = 3 > 1 \Rightarrow$ Hyperbola

Vertices: $\left(\dfrac{3}{8}, \dfrac{\pi}{2}\right), \left(-\dfrac{3}{4}, \dfrac{3\pi}{2}\right)$

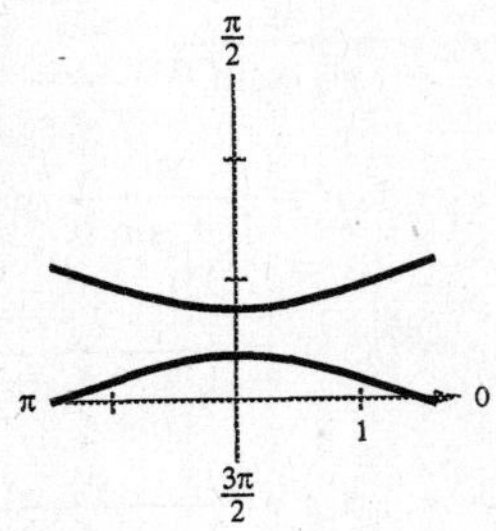

24. $r = \dfrac{2}{2 + 3\sin\theta} = \dfrac{1}{1 + (3/2)\sin\theta}$

$e = \dfrac{3}{2} > 1 \Rightarrow$ Hyperbola

Vertices: $\left(\dfrac{2}{5}, \dfrac{\pi}{2}\right), \left(-2, \dfrac{3\pi}{2}\right)$

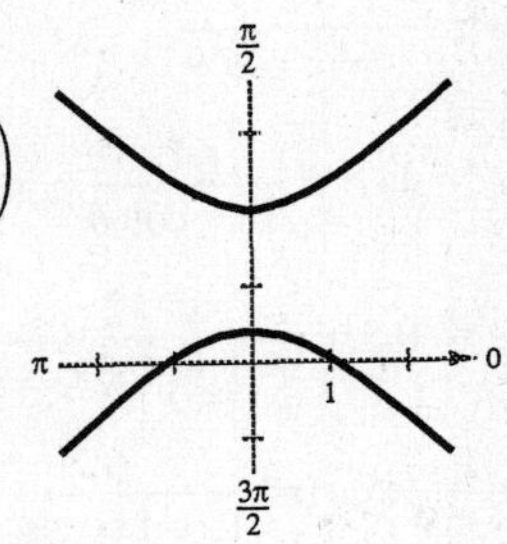

26. $r = \dfrac{-5}{2 + 4\sin\theta} = \dfrac{-(5/2)}{1 + 2\sin\theta}$

$e = 2 \Rightarrow$ Hyperbola

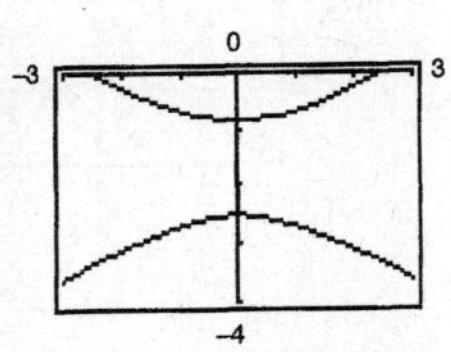

28. $r = \dfrac{4}{1 - 2\cos\theta}$

$e = 2 \Rightarrow$ Hyperbola.

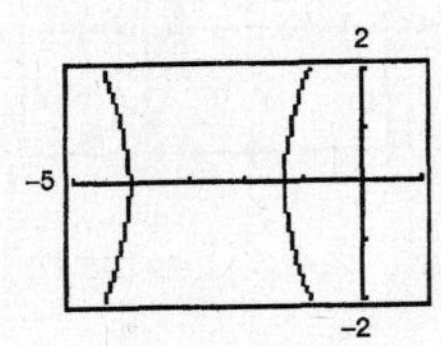

30. $r = \dfrac{3}{3 + \sin\left(\theta - \dfrac{\pi}{3}\right)}$

Rotate the graph in Exercise 16 through the angle $\pi/3$.

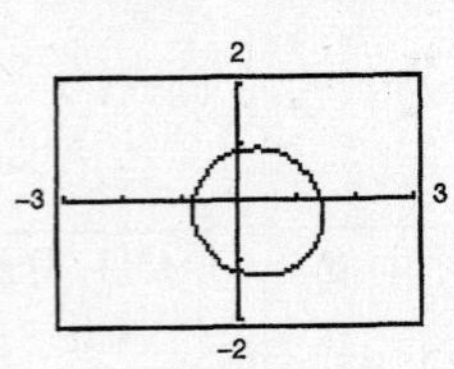

32. $r = \dfrac{5}{-1 + 2\cos\left(\theta + \dfrac{2\pi}{3}\right)}$

Rotate the graph in Exercise 20 through the angle $-2\pi/3$.

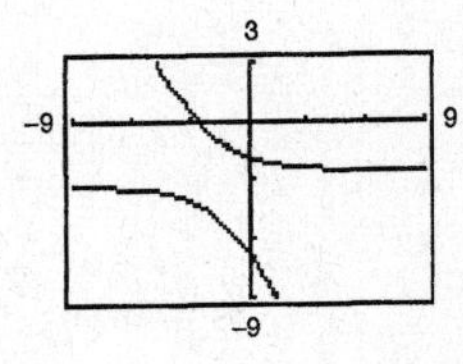

34. Parabola: $e = 1$

Directrix: $y = -2$

$p = 2$

Horizontal directrix below the pole

$r = \dfrac{1(2)}{1 - 1\sin\theta} = \dfrac{2}{1 - \sin\theta}$

36. Ellipse: $e = \dfrac{3}{4}$

Directrix: $y = -3$

$p = 3$

Horizontal directrix below the pole

$r = \dfrac{(3/4)(3)}{1 - (3/4)\sin\theta} = \dfrac{9}{4 - 3\sin\theta}$

38. Hyperbola: $e = \dfrac{3}{2}$

Directrix: $x = -1$

$p = 1$

Vertical directrix to the left of the pole

$$r = \frac{(3/2)(1)}{1 - (3/2)\cos\theta} = \frac{3}{2 - 3\cos\theta}$$

40. Parabola

Vertex: $(6, 0) \Rightarrow e = 1, p = 12$

Vertical directrix to the right of the pole

$$r = \frac{1(12)}{1 + 1\cos\theta} = \frac{12}{1 + \cos\theta}$$

42. Parabola

Vertex: $\left(10, \dfrac{\pi}{2}\right) \Rightarrow e = 1, p = 20$

Horizontal directrix above the pole

$$r = \frac{1(20)}{1 + 1\sin\theta} = \frac{20}{1 + \sin\theta}$$

44. Ellipse

Vertices: $\left(2, \dfrac{\pi}{2}\right), \left(4, \dfrac{3\pi}{2}\right)$

Center: $\left(1, \dfrac{3\pi}{2}\right)$; $c = 1, a = 3, e = \dfrac{1}{3}$

Horizontal directrix above the axis

$$r = \frac{1/3p}{1 + (1/3)\sin\theta} = \frac{p}{3 + \sin\theta}$$

$$2 = \frac{p}{3 + \sin(\pi/2)}$$

$$p = 8$$

$$r = \frac{8}{3 + \sin\theta}$$

46. Hyperbola

Vertices: $(2, 0), (8, 0)$

Center: $(5, 0)$; $c = 5, a = 3, e = \dfrac{5}{3}$

Vertical directrix to the right of the pole

$$r = \frac{(5/3)p}{1 + (5/3)\cos\theta} = \frac{5p}{3 + 5\cos\theta}$$

$$2 = \frac{5p}{3 + 5\cos 0}$$

$$p = \frac{16}{5}$$

$$r = \frac{5(16/5)}{3 + 5\cos\theta} = \frac{16}{3 + 5\cos\theta}$$

48. Hyperbola

Vertices: $\left(4, \dfrac{\pi}{2}\right), \left(1, \dfrac{\pi}{2}\right)$

Center: $\left(\dfrac{5}{2}, \dfrac{\pi}{2}\right)$; $c = \dfrac{5}{2}, a = \dfrac{3}{2}, e = \dfrac{5/2}{3/2} = \dfrac{5}{3}$

Horizontal directrix above the pole

$$r = \frac{(5/3)p}{1 + (5/3)\sin\theta} = \frac{5p}{3 + 5\sin\theta}$$

$$1 = \frac{5p}{3 + 5\sin\dfrac{\pi}{2}}$$

$$p = \frac{8}{5}$$

$$r = \frac{5(8/5)}{3 + 5\sin\theta} = \frac{8}{3 + 5\cos\theta}$$

50. Minimum distance occurs when $\theta = \pi$.

$$r = \frac{(1 - e^2)a}{1 - e\cos\pi} = \frac{(1 - e)(1 + e)a}{1 + e} = a(1 - e)$$

Maximum distance occurs when $\theta = 0$.

$$r = \frac{(1 - e^2)a}{1 - e\cos 0} = \frac{(1 - e)(1 + e)a}{1 - e} = a(1 + e)$$

52. $r = \dfrac{[1 - (0.0543)^2](1.429 \times 10^9)}{1 - 0.0543 \cos \theta} \approx \dfrac{1.4248 \times 10^9}{1 - 0.0543 \cos \theta}$

Perihelion distance: $r = 1.429 \times 10^9(1 - 0.0543) \approx 1.3514 \times 10^9$ kilometers

Aphelion distance: $r = 1.429 \times 10^9(1 + 0.0543) \approx 1.5066 \times 10^9$ kilometers

54. $r = \dfrac{[1 - (0.2056)^2](35.98 \times 10^6)}{1 - 0.2056 \cos \theta} \approx \dfrac{3.4459 \times 10^7}{1 - 0.2056 \cos \theta}$

Perihelion distance: $r = 35.98 \times 10^6(1 - 0.2056) \approx 2.8583 \times 10^7$ miles

Aphelion distance: $r = 35.98 \times 10^6(1 + 0.2056) \approx 4.3377 \times 10^7$ miles

56. $r = \dfrac{[1 - (0.0484)^2](778.40 \times 10^6)}{1 - 0.0484 \cos \theta} \approx \dfrac{7.7658 \times 10^8}{1 - 0.0484 \cos \theta}$

Perihelion distance: $r = 778.40 \times 10^6(1 - 0.0484) \approx 7.4073 \times 10^8$ kilometers

Aphelion distance: $r = 778.40 \times 10^6(1 + 0.0484) \approx 8.1607 \times 10^8$ kilometers

58. If e remains fixed and p changes, then the lengths of both the major axis and the minor axis change. For example, graph

$r = \dfrac{5}{1 - \frac{2}{3}\cos \theta}$, with $e = \dfrac{2}{3}$ and $p = \dfrac{15}{2}$, and graph

$r = \dfrac{6\frac{2}{3}}{1 - \frac{2}{3}\cos \theta}$, with $e = \dfrac{2}{3}$ and $p = 10$ on the same set of coordinate axes.

The first ellipse has a major axis of length 18 and a minor axis of length $6\sqrt{5}$, and the second ellipse has a major axis of length 24 and a minor axis of length $8\sqrt{5}$.

60. False. The graph has a horizontal directrix below the pole.

62.
$$\frac{x^2}{a^2} - \frac{y^2}{b^2} = 1$$
$$\frac{r^2\cos^2\theta}{a^2} - \frac{r^2\sin^2\theta}{b^2} = 1$$
$$\frac{r^2\cos^2\theta}{a^2} - \frac{r^2(1 - \cos^2\theta)}{b^2} = 1$$
$$r^2b^2\cos^2\theta - r^2a^2 + r^2a^2\cos^2\theta = a^2b^2$$
$$r^2(b^2 + a^2)\cos^2\theta - r^2a^2 = a^2b^2$$
$$a^2 + b^2 = c^2$$
$$r^2c^2\cos^2\theta - r^2a^2 = a^2b^2$$
$$r^2\left(\frac{c}{a}\right)^2\cos^2\theta - r^2 = b^2,\ e = \frac{c}{a}$$
$$r^2e^2\cos^2\theta - r^2 = b^2$$
$$r^2(e^2\cos^2\theta - 1) = b^2$$
$$r^2 = \frac{b^2}{e^2\cos^2\theta - 1}$$
$$= \frac{-b^2}{1 - e^2\cos^2\theta}$$

64. $\dfrac{x^2}{25} + \dfrac{y^2}{16} = 1$

$a = 5, b = 4, c = 3, e = \dfrac{3}{5}$

$r^2 = \dfrac{400}{25 - 9\cos^2\theta}$

66. $\frac{x^2}{36} - \frac{y^2}{4} = 1$

$a = 6, b = 2, c = 2\sqrt{10}, e = \frac{\sqrt{10}}{3}$

$$r^2 = \frac{-4}{1 - (10/9)\cos^2\theta} = \frac{-36}{9 - 10\cos^2\theta}$$

$$= \frac{36}{10\cos^2\theta - 9}$$

68. Ellipse

One focus: $(4, 0)$

Vertices: $(5, 0), (5, \pi)$

$a = 5, c = 4, b = 3, e = \frac{4}{5}$

$$r^2 = \frac{9}{1 - (16/25)\cos^2\theta} = \frac{225}{25 - 16\cos^2\theta}$$

70. $6\cos x - 2 = 1$

$$\cos x = \frac{1}{2}$$

$$x = \frac{\pi}{3} + 2\pi n, \frac{5\pi}{3} + 2\pi n$$

72. $9\csc^2 x - 10 = 2$

$$\csc^2 x = \frac{4}{3}$$

$$\sin x = \pm\frac{\sqrt{3}}{2}$$

$$x = \frac{\pi}{3} + \pi n, \frac{2\pi}{3} + \pi n$$

74. $\sqrt{2}\sec\theta = 2\csc\frac{\pi}{4}$

$$\cos\theta = \frac{1}{2}$$

$$\theta = \frac{\pi}{3} + 2\pi n, \frac{5\pi}{3} + 2\pi n$$

76. $\sin u = -\frac{3}{5}, \cos u = \frac{4}{5}$

$$\cos v = \frac{1}{\sqrt{2}} = \frac{\sqrt{2}}{2}, \sin v = -\frac{\sqrt{2}}{2}$$

$$\sin(u + v) = \sin u\cos v + \cos u\sin v$$

$$= \left(-\frac{3}{5}\right)\left(\frac{\sqrt{2}}{2}\right) + \left(\frac{4}{5}\right)\left(-\frac{\sqrt{2}}{2}\right)$$

$$= \frac{-7\sqrt{2}}{10}$$

78. $\sin u = -\frac{3}{5}, \cos u = \frac{4}{5}$

$$\cos v = \frac{1}{\sqrt{2}} = \frac{\sqrt{2}}{2}, \sin v = -\frac{\sqrt{2}}{2}$$

$$\sin(u - v) = \sin u\cos v - \cos u\sin v$$

$$= \left(-\frac{3}{5}\right)\left(\frac{\sqrt{2}}{2}\right) - \left(\frac{4}{5}\right)\left(-\frac{\sqrt{2}}{2}\right)$$

$$= \frac{\sqrt{2}}{10}$$

80. $\tan u = -\sqrt{3}, \frac{3\pi}{2} < u < 2\pi$

$$\sin u = -\frac{\sqrt{3}}{2}, \cos u = \frac{1}{2}$$

$$\sin 2u = 2\sin u\cos u = 2\left(-\frac{\sqrt{3}}{2}\right)\left(\frac{1}{2}\right) = -\frac{\sqrt{3}}{2}$$

$$\cos 2u = \cos^2 u - \sin^2 u = \left(\frac{1}{2}\right)^2 - \left(-\frac{\sqrt{3}}{2}\right)^2 = -\frac{1}{2}$$

$$\tan 2u = \frac{2\tan u}{1 - \tan^2 u} = \frac{2(-\sqrt{3})}{1 - (-\sqrt{3})^2} = \sqrt{3}$$

Review Exercises for Chapter 6

Solutions to Even-Numbered Exercises

2. $m = \frac{4 - 7}{3 - (-2)} = -\frac{3}{5} = \tan\theta$

$$\theta = \pi + \arctan\left(-\frac{3}{5}\right)$$

$$\approx 2.6012, \text{ or about } 149.04°$$

4. $6x - 7y - 5 = 0$

$$m = \tfrac{6}{7} = \tan\theta$$

$$\theta = \arctan\frac{6}{7}$$

$$\approx 0.7086, \text{ or about } 40.60°$$

6. $-5x + 3y = 3$

$-2x + 3y = 1$

$m_1 = \frac{5}{3}$

$m_2 = \frac{2}{3}$

$\tan\theta = \left|\frac{(5/3) - (2/3)}{1 + (5/3)(2/3)}\right| = \frac{9}{19}$

$\theta \approx 0.4424$, or about 25.35°

8. $0.02x + 0.07y = 0.18$

$0.09x - 0.04y = 0.17$

$m_1 = -\frac{2}{7}$

$m_2 = \frac{9}{4}$

$\tan\theta = \left|\frac{(9/4) - (-2/7)}{1 + (-2/7)(9/4)}\right| = \frac{71}{10}$

$\theta \approx 1.4309$, or about 81.98°

10. $(0, 4) \Rightarrow x_1 = 0, y_1 = 4$

$x + 2y - 2 = 0 \Rightarrow A = 1, B = 2, C = -2$

$d = \frac{|1(0) + (2)(4) + (-2)|}{\sqrt{1^2 + 2^2}} = \frac{6}{\sqrt{5}} = \frac{6\sqrt{5}}{5}$

12. A parabola is formed.

14. Vertex: $(2, 0) = (h, k)$
Focus: $(0, 0) \Rightarrow p = -2$

$(y - k)^2 = 4p(x - h)$

$y^2 = -8(x - 2)$

16. Vertex: $(2, 2) = (h, k)$
Directrix: $y = 0 \Rightarrow p = 2$

$(x - h)^2 = 4p(y - k)$

$(x - 2)^2 = 8(y - 2)$

18. $x^2 = -2y$

$p = -\frac{1}{2}$

Focus: $\left(0, -\frac{1}{2}\right)$

Tangent line through point $(-4, -8)$:

Slope: m

y-intercept: $(0, b)$

$d_1 = b + \frac{1}{2}$

$d_2 = \sqrt{(-4 - 0)^2 + \left(-8 + \frac{1}{2}\right)^2} = \frac{17}{2}$

$d_1 = d_2 \Rightarrow b = 8$

$m = \frac{-8 - 8}{-4 - 0} = 4$

$y = 4x + 8$

x-intercept of tangent line: $(-2, 0)$

20. $y^2 = 4px$

$p = 1.5$

$y^2 = 6x$

22. Vertices: $(2, 0), (2, 4) \Rightarrow a = 2, (h, k) = (2, 2)$
Foci: $(2, 1), (2, 3) \Rightarrow c = 1$

$b^2 = a^2 - c^2 = 4 - 1 = 3$

$\frac{(x - h)^2}{b^2} + \frac{(y - k)^2}{a^2} = 1$

$\frac{(x - 2)^2}{3} + \frac{(y - 2)^2}{4} = 1$

24. Vertices: $(-4, -1), (-4, 11) \Rightarrow a = 6, (h, k) = (-4, 5)$
Endpoints of the minor axis: $(-6, 5), (-2, 5) \Rightarrow b = 2$

$\frac{(x - h)^2}{b^2} + \frac{(y - k)^2}{a^2} = 1$

$\frac{(x + 4)^2}{4} + \frac{(y - 5)^2}{36} = 1$

26. $\dfrac{x^2}{324} + \dfrac{y^2}{196} = 1$

$a = \sqrt{324} = 18,\ b = \sqrt{196} = 14$

$c = \sqrt{a^2 - b^2} = \sqrt{128} = 8\sqrt{2}$

Longest distance: $2a = 36$ feet

Shortest distance: $2b = 28$ feet

Distance between foci: $2c = 16\sqrt{2}$ feet

28. $\dfrac{(x-5)^2}{1} + \dfrac{(y+3)^2}{36} = 1$

Center: $(5, -3)$

$a = 6,\ b = 1,\ c = \sqrt{a^2 - b^2} = \sqrt{35}$

Vertices: $(5, 3), (5, -9)$

Foci: $\left(5, -3 \pm \sqrt{35}\right)$

Eccentricity: $e = \dfrac{c}{a} = \dfrac{\sqrt{35}}{6}$

30. $4x^2 + 25y^2 + 16x - 150y + 141 = 0$

$4(x^2 + 4x + 4) + 25(y^2 - 6y + 9) = -141 + 16 + 225$

$\dfrac{(x+2)^2}{25} + \dfrac{(y-3)^2}{4} = 1$

Center: $(-2, 3)$

$a = 5,\ b = 2,\ c = \sqrt{a^2 - b^2} = \sqrt{21}$

Vertices: $(3, 3), (-7, 3)$

Foci: $\left(-2 \pm \sqrt{21}, 3\right)$

Eccentricity: $e = \dfrac{c}{a} = \dfrac{\sqrt{21}}{5}$

32. Vertices: $(2, 2), (-2, 2) \Rightarrow a = 2,\ (h, k) = (0, 2)$
Foci: $(4, 2), (-4, 2) \Rightarrow c = 4$

$b^2 = c^2 - a^2 = 16 - 4 = 12$

$\dfrac{(x-h)^2}{a^2} - \dfrac{(y-k)^2}{b^2} = 1$

$\dfrac{x^2}{4} - \dfrac{(y-2)^2}{12} = 1$

34. Foci: $(3, \pm 2) \Rightarrow c = 2,\ (h, k) = (3, 0)$

Asymptotes: $y = \pm 2(x - 3) \Rightarrow \dfrac{a}{b} = 2,\ a = 2b$

$b^2 = c^2 - a^2 = 4 - 4b^2 \Rightarrow b^2 = \dfrac{4}{5}, a^2 = \dfrac{16}{5}$

$\dfrac{(y-k)^2}{a^2} - \dfrac{(x-h)^2}{b^2} = 1$

$\dfrac{y^2}{16/5} - \dfrac{(x-3)^2}{4/5} = 1 \Rightarrow \dfrac{5y^2}{16} - \dfrac{5(x-3)^2}{4} = 1$

36. $\dfrac{(y-1)^2}{4} - x^2 = 1$

Center: $(0, 1)$

$a = 2,\ b = 1,\ c = \sqrt{a^2 + b^2} = \sqrt{5}$

Vertices: $(0, 3), (0, -1)$

Foci: $\left(0, 1 \pm \sqrt{5}\right)$

Asymptotes: $y = 1 \pm 2x$

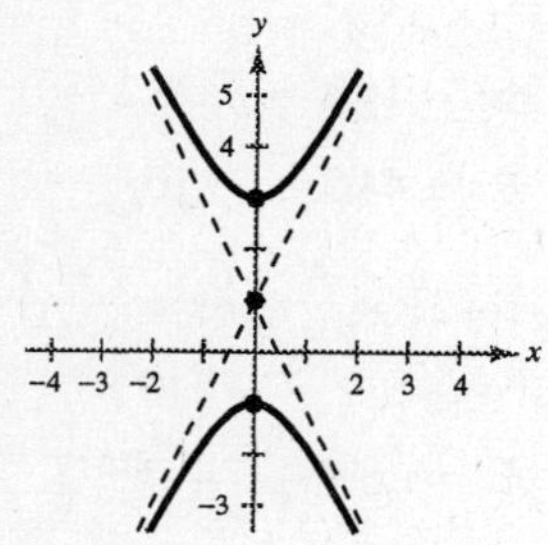

38. $-4x^2 + 25y^2 - 8x + 150y + 121 = 0$

$-4(x^2 + 2x + 1) + 25(y^2 + 6y + 9) = -121 - 4 + 225$

$\dfrac{(y+3)^2}{4} - \dfrac{(x+1)^2}{25} = 1$

Center: $(-1, -3)$

$a = 2,\ b = 5,\ c = \sqrt{a^2 + b^2} = \sqrt{29}$

Vertices: $(-1, -1), (-1, -5)$

Foci: $\left(-1, -3 \pm \sqrt{29}\right)$

Asymptotes: $y = -3 \pm \dfrac{2}{5}(x + 1)$

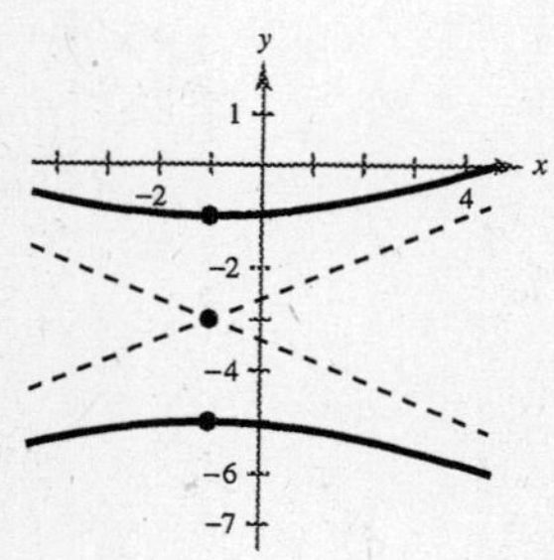

40. $BD = AD + 6\left(\frac{1100}{5280}\right)$

$CD = AD + 8\left(\frac{1100}{5280}\right)$

$2a = CD - BD = 2\left(\frac{1100}{5280}\right)$

$a = \frac{5}{24}, c = 2 \Rightarrow b^2 = \frac{2279}{576}$

Thus, we have $\frac{576x^2}{25} - \frac{576y^2}{2279} = 1$

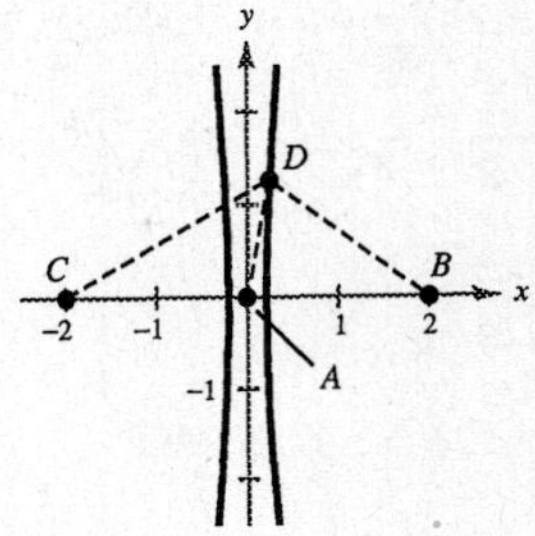

OR:

$CD = AD + 8\left(\frac{1100}{5280}\right)$

$BD = AD + 6\left(\frac{1100}{5280}\right)$

$2a = BD - AD = 6\left(\frac{1100}{5280}\right)$

$a = 3\left(\frac{5}{24}\right) = \frac{5}{8}, c = 1 \Rightarrow b^2 = \frac{39}{64}$

Center: $(1, 0)$

$\frac{64(x-1)^2}{25} - \frac{64y^2}{39} = 1$

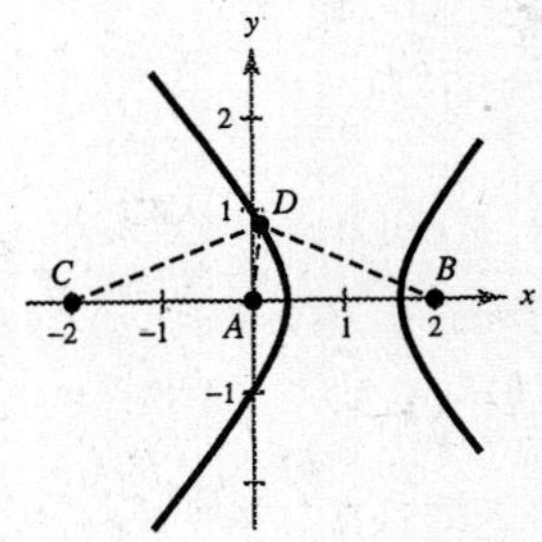

42. $-4y^2 + 5x + 3y + 7 = 0$

$AC = (0)(-4) = 0 \Rightarrow$ Parabola

44. $x^2 - 10xy + y^2 + 1 = 0$

$B^2 - 4AC = (-10)^2 - 4(1)(1) = 96 > 0 \Rightarrow$ Hyperbola

$\cot 2\theta = \frac{A - C}{B} = \frac{1-1}{-10} = 0 \Rightarrow 2\theta = \frac{\pi}{2} \Rightarrow \theta = \frac{\pi}{4}$

$x = x' \cos\frac{\pi}{4} - y' \sin\frac{\pi}{4} = \frac{1}{\sqrt{2}}(x' - y')$

$y = x' \sin\frac{\pi}{4} + y' \cos\frac{\pi}{4} = \frac{1}{\sqrt{2}}(x' + y')$

$\frac{1}{2}(x' - y')^2 - 5(x' - y')(x' + y') + \frac{1}{2}(x' + y')^2 + 1 = 0$

$6(y')^2 - 4(x')^2 + 1 = 0$

$\frac{(x')^2}{1/4} - \frac{(y')^2}{1/6} = 1$

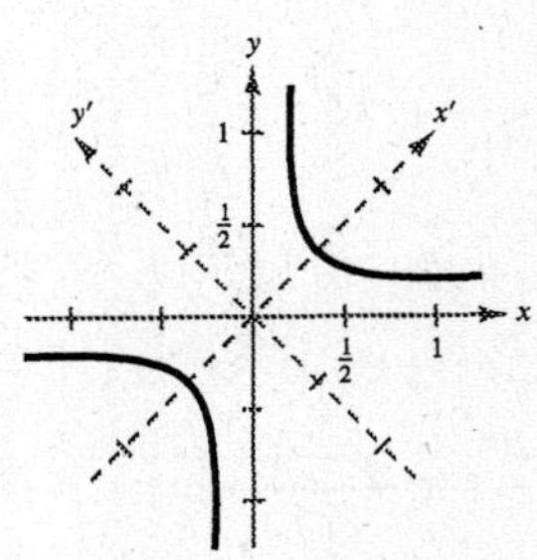

46. $4x^2 + 8xy + 4y^2 + 7\sqrt{2}x + 9\sqrt{2}y = 0$

$B^2 - 4AC = 8^2 - 4(4)(4) = 0 \implies$ Parabola

$\cot 2\theta = \frac{A - C}{B} = \frac{4 - 4}{8} = 0 \implies 2\theta = \frac{\pi}{2} \implies \theta = \frac{\pi}{4}$

$x = x'\cos\frac{\pi}{4} - y'\sin\frac{\pi}{4} = \frac{1}{\sqrt{2}}(x' - y')$

$y = x'\sin\frac{\pi}{4} + y'\cos\frac{\pi}{4} = \frac{1}{\sqrt{2}}(x' + y')$

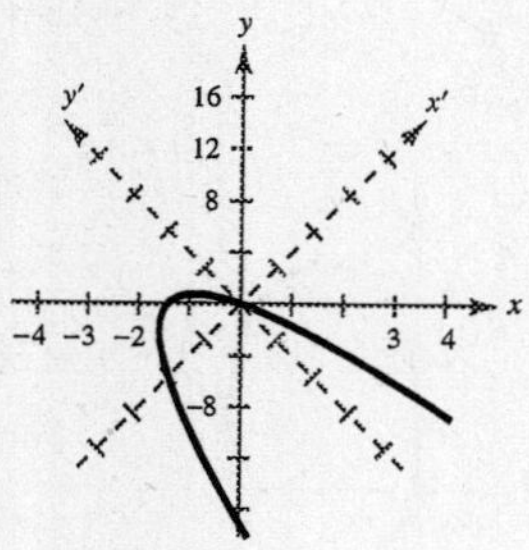

$2(x' - y')^2 + 4(x' - y')(x' + y') + 2(x' + y')^2 + 7(x' - y') + 9(x' + y') = 0$

$8(x')^2 + 16x' + 2y' = 0$

$y' = -4(x')^2 - 8x'$

48. $13x^2 - 8xy + 7y^2 - 45 = 0$

$B^2 - 4AC = (-8)^2 - 4(13)(7) = -300 < 0 \implies$ Ellipse

Use the Quadratic Formula to solve for y in terms of x:

$7y^2 - 8xy + 13x^2 - 45 = 0$

$y = \frac{1}{14}\left[8x \pm \sqrt{64x^2 - 28(13x^2 - 45)}\right]$

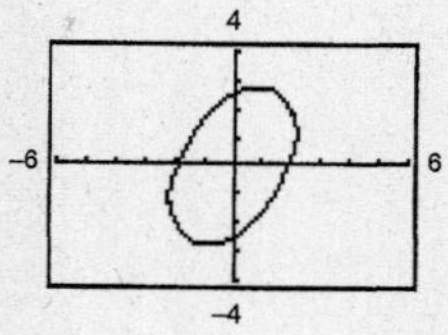

50. $x^2 - 10xy + y^2 + 1 = 0$

Since $B^2 - 4AC = (-10)^2 - 4(1)(1) > 0 \implies$ Hyperbola

Use the Quadratic Formula to solve for y in terms of x:

$y^2 - 10xy + x^2 + 1 = 0$

$y = \frac{1}{2}\left[10x \pm \sqrt{100x^2 - 4(x^2 + 1)}\right]$

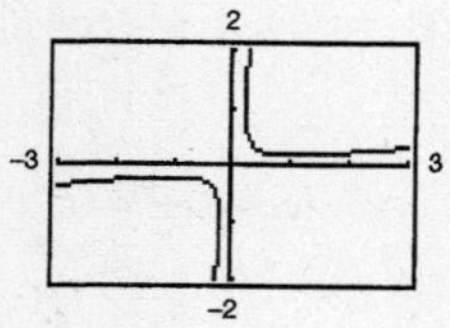

52. $x = 3\cos\theta, y = 2\sin^2\theta$

$\theta = \frac{\pi}{3}$

$x = 3\cos\frac{\pi}{3} = \frac{3}{2}, y = 2\sin^2\frac{\pi}{3} = \frac{3}{2}$

54. $x = 3\cos\theta, y = 2\sin^2\theta$

$\theta = -\frac{\pi}{4}$

$x = 3\cos\left(-\frac{\pi}{4}\right) = \frac{3\sqrt{2}}{2}, y = 2\sin^2\left(-\frac{\pi}{4}\right) = 1$

56. $x = 1 + 4t, y = 2 - 3t$

$t = \frac{x - 1}{4}$

$y = 2 - 3\left(\frac{x - 1}{4}\right)$

$3x + 4y = 11$

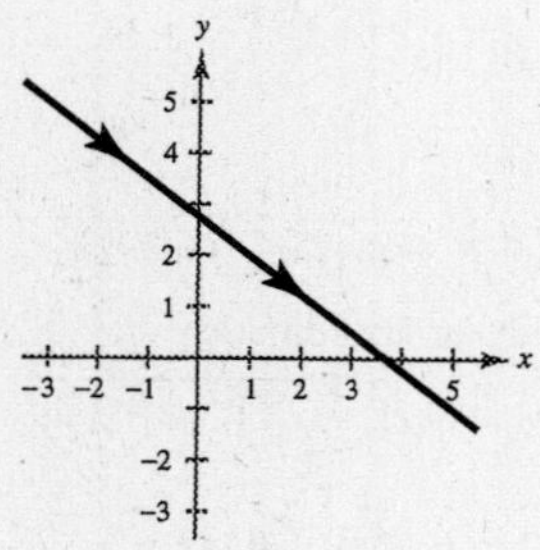

58. $x = t + 4, y = t^2$

$t = x - 4$

$y = (x - 4)^2$

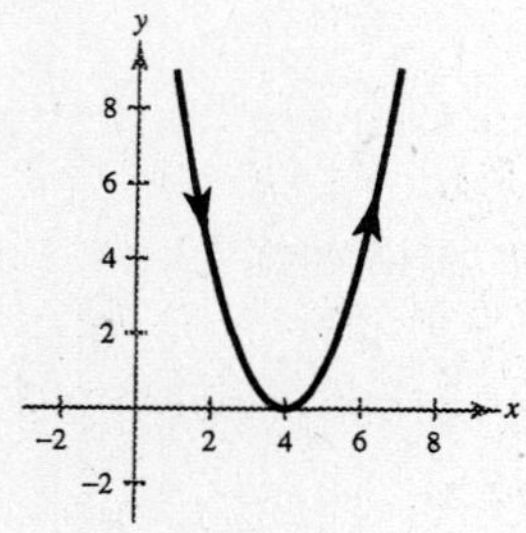

60. $x = 3 + 3\cos\theta,\ y = 2 + 5\sin\theta$

$$\cos\theta = \frac{x-3}{3},\ \sin\theta = \frac{y-2}{5}$$

$$\frac{(x-3)^2}{9} + \frac{(y-2)^2}{25} = 1$$

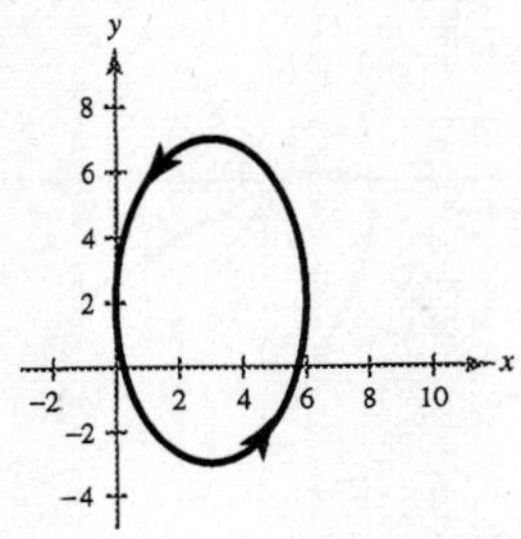

62. $(h, k) = (-3, 4)$

$2a = 8 \Rightarrow a = 4$

$2b = 6 \Rightarrow b = 3$

$$\frac{(x+3)^2}{16} + \frac{(y-4)^2}{9} = 1$$

$x = -3 + 4\cos\theta$

$y = 4 + 3\sin\theta$

This solution is not unique.

64. $y = \overline{QB} - \overline{QA}$

$\overline{QP} = \text{arc } QC = r\theta$

$\overline{QA} = r\theta\sin(90° - \theta)$

$= r\theta\cos\theta$

$\overline{QB} = r\sin\theta$

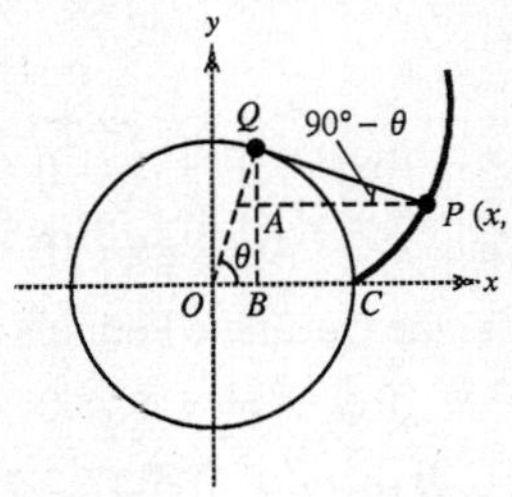

Therefore, $y = r\sin\theta - r\theta\cos\theta = r(\sin\theta - \theta\cos\theta)$.

Similarly, $x = \overline{OB} + \overline{AP}$.

Therefore, $x = r\cos\theta + r\theta\sin\theta = r(\cos\theta + \theta\sin\theta)$.

66. Polar coordinates: $\left(-5, -\frac{\pi}{3}\right) = \left(5, \frac{2\pi}{3}\right)$ or $\left(-5, \frac{5\pi}{3}\right)$

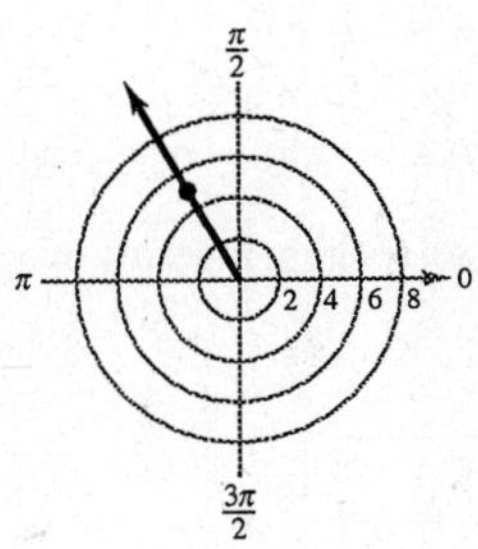

68. Polar coordinates:

$\left(\sqrt{3}, 2.62\right) \approx \left(\sqrt{3}, 8.90\right)$ or $\left(-\sqrt{3}, 5.76\right)$

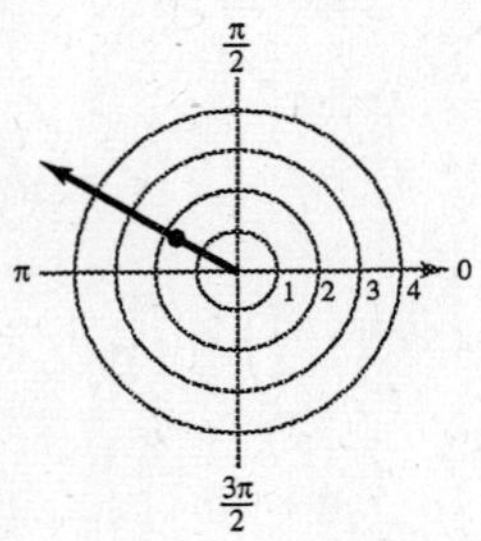

70. Polar coordinates: $\left(2, \frac{5\pi}{4}\right) = (r, \theta)$

$$x = r\cos\theta = 2\cos\frac{5\pi}{4} = -\sqrt{2}$$

$$y = r\sin\theta = 2\sin\frac{5\pi}{4} = -\sqrt{2}$$

Rectangular coordinates: $\left(-\sqrt{2}, -\sqrt{2}\right)$

72. Polar coordinates: $\left(0, \frac{\pi}{2}\right) = (r, \theta)$

$$x = r\cos\theta = 0\cos\frac{\pi}{2} = 0$$

$$y = r\sin\theta = 0\sin\frac{\pi}{2} = 0$$

Rectangular coordinates: $(0, 0)$

74. Rectangular coordinates: $\left(-\sqrt{5}, \sqrt{5}\right)$

Polar coordinates:

$$r = \sqrt{\left(-\sqrt{5}\right)^2 + \left(\sqrt{5}\right)^2} = \sqrt{10}$$

$$\tan\theta = -1,\ \theta = \frac{3\pi}{4}$$

$$\left(\sqrt{10}, \frac{3\pi}{4}\right)$$

76. Rectangular coordinates: $(3, -4)$

Polar coordinates:

$r = \sqrt{3^2 + (-4)^2} = 5$

$\tan\theta = -\frac{4}{3}, \theta \approx -0.9273$

$(5, 5.356)$

78. $x^2 + y^2 - 4x = 0$

$r^2 - 4r\cos\theta = 0$

$r = 4\cos\theta$

80.
$$xy = -2$$
$$r\cos\theta\, r\sin\theta = -2$$
$$r^2\cos\theta\,\sin\theta = -2$$
$$r^2 = \frac{-2}{\cos\theta\sin\theta}$$
$$r^2 = -2\sec\theta\,\csc\theta$$
$$r^2 = -4\csc 2\theta$$

82. Because $y = r\sin\theta$ and r is given as $8\sin\theta$,

$y = 8\sin\theta\sin\theta = 8\sin^2\theta.$

$$r = 8\sin\theta$$
$$r^2 = 64\sin^2\theta$$
$$r^2 = 8(8\sin^2\theta)$$
$$x^2 + y^2 = 8y$$
$$x^2 + y^2 - 8y = 0$$

84.
$$r^2 = \cos 2\theta$$
$$r^2 = \left(\frac{x}{r}\right)^2 - \left(\frac{y}{r}\right)^2$$
$$(x^2 + y^2)^2 = x^2 - y^2$$

86. $r = 11$

Symmetry: $\theta = \frac{\pi}{2}$, polar axis, pole

Maximum value of $|r|$: 11, for all values of θ

Zeros of r: none

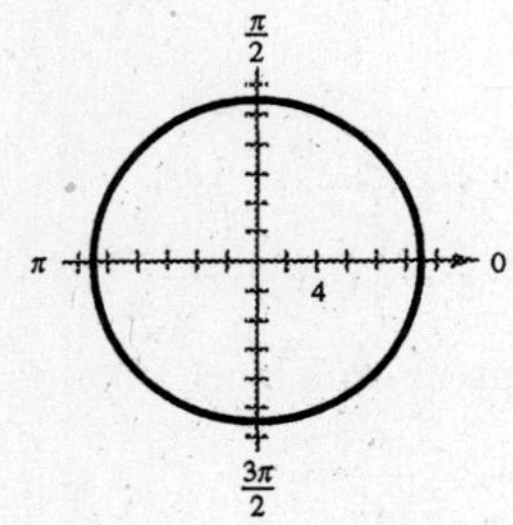

88. $r = \cos 5\theta$

Symmetry: polar axis

Maximum value of $|r|$: $|r| = 1$ when $\theta = 0, \frac{2\pi}{5}, \frac{4\pi}{5}, \frac{6\pi}{5}, \frac{8\pi}{5}$

Zeros of r: $r = 0$ when $\theta = \frac{\pi}{10}, \frac{3\pi}{10}, \frac{\pi}{2}, \frac{7\pi}{10}, \frac{9\pi}{10}$

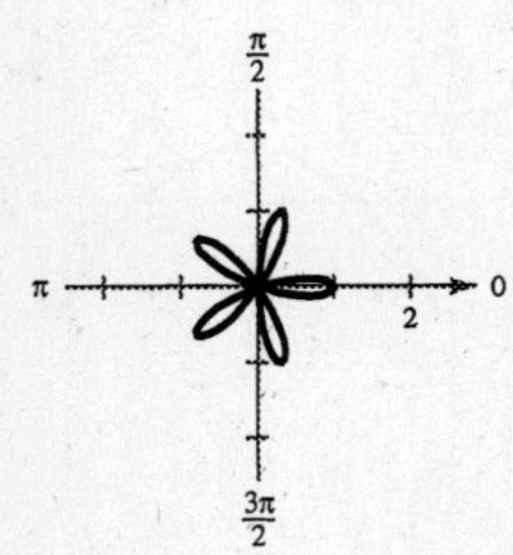

90. $r = 3 - 4\cos\theta$

Symmetry: polar axis

Maximum value of $|r|$: $|r| = 7$ when $\theta = \pi$

Zeros of r: $r = 0$ when $\theta = \arccos\frac{3}{4}, 2\pi - \arccos\frac{3}{4}$

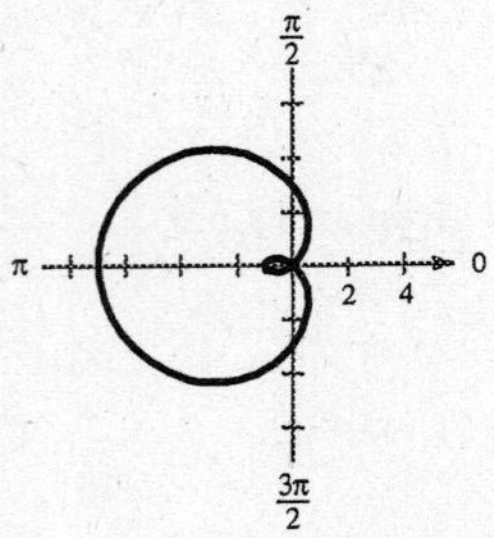

92. $r = 5 - 5\cos\theta$

$r = 5(1 - \cos\theta)$

Symmetry: polar axis

Maximum values of $|r|$: $|r| = 10$ when $\theta = \pi$.

Zeros of r: $r = 0$ when $\theta = 0, 2\pi$.

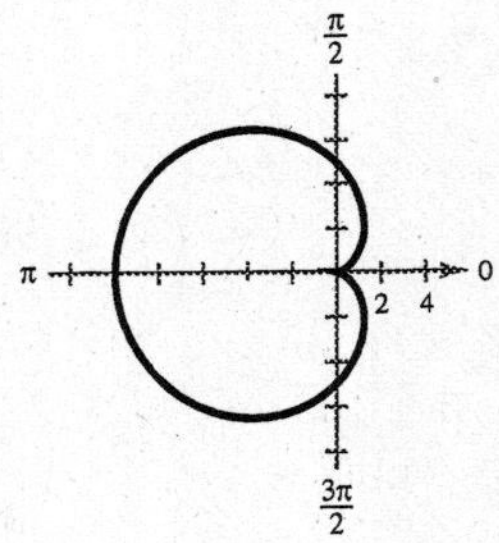

94. $r = \cos 2\theta$

Symmetry: polar axis

Maximum value of $|r|$: $|r| = 1$ when $\theta = 0, \frac{\pi}{2}, \pi, \frac{3\pi}{2}$

Zeros of r: $r = 0$ when $\theta = \frac{\pi}{4}, \frac{3\pi}{4}, \frac{5\pi}{4}, \frac{7\pi}{4}$

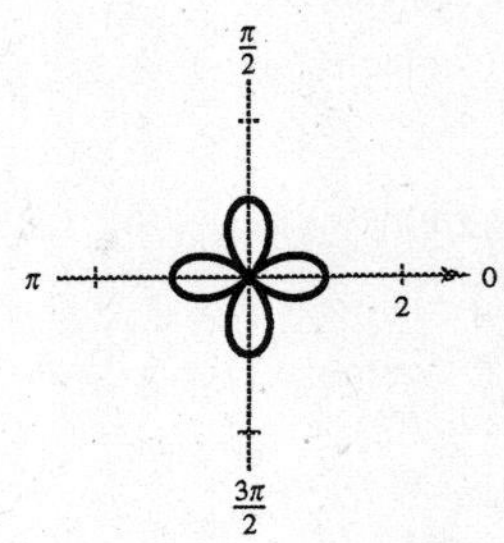

96. $r = 3(1 - 2\cos\theta)$

$r = 3 - 6\cos\theta$

Limaçon with inner loop.

98. $r^2 = 9\cos 2\theta$

Lemniscate

100. $r = \dfrac{2}{1 + \sin\theta}$

$e = 1 \Rightarrow$ parabola

Vertex: $\left(1, \frac{\pi}{2}\right)$

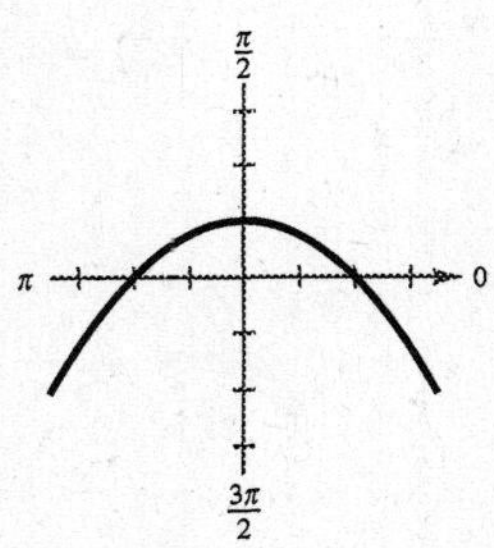

102. $r = \dfrac{16}{4 + 5\cos\theta}$

$r = \dfrac{4}{1 + (5/4)\cos\theta}$

$e = \frac{5}{4} > 1 \Rightarrow$ Hyperbola

Vertices: $\left(\frac{16}{9}, 0\right), (-16, \pi)$

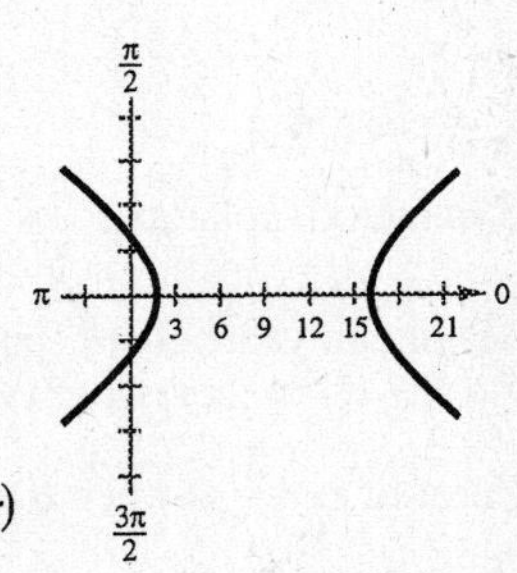

104. Parabola: $r = \dfrac{ep}{1 + e\sin\theta}$, $e = 1$

Vertex: $\left(2, \frac{\pi}{2}\right)$

Focus: $(0, 0) \Rightarrow p = 4$

$r = \dfrac{4}{1 + \sin\theta}$

106. Hyperbola: $r = \dfrac{ep}{1 + e\cos\theta}$

Vertices: $(1, 0), (7, 0) \Rightarrow a = 3$

One focus: $(0, 0) \Rightarrow c = 4$

$e = \frac{c}{a} = \frac{4}{3}, p = \frac{7}{4}$

$r = \dfrac{(4/3)(7/4)}{1 + (4/3)\cos\theta} = \dfrac{7/3}{1 + (4/3)\cos\theta} = \dfrac{7}{3 + 4\cos\theta}$

108. Parabola: $r = \dfrac{ep}{1 + e \sin \theta}, e = 1$

Vertex: $\left((6{,}000{,}000), \dfrac{\pi}{2}\right)$

Focus: $(0, 0) \implies p = 12{,}000{,}000$

$r = \dfrac{12{,}000{,}000}{1 + \sin \theta}$

$\theta = -\dfrac{\pi}{3}$

$r \approx 89{,}600{,}000$ miles

110. False.

$\dfrac{x^2}{4} - y^4 = 1$ is a fourth-degree equation.

The equation of a hyperbola is a second degree equation.

112. False.

$(r, \theta), (r, \theta + 2\pi), (-r, \theta + \pi)$, etc.

All represent the same point.

114. Yes. The orientation would be reversed.

116. (a) $\left(-4, \dfrac{\pi}{6}\right), \left(4, \dfrac{\pi}{6}\right)$: symmetric about the pole

(b) $\left(4, -\dfrac{\pi}{6}\right), \left(4, \dfrac{\pi}{6}\right)$: symmetric about the polar axis

(c) $\left(-4, -\dfrac{\pi}{6}\right), \left(4, \dfrac{\pi}{6}\right)$: symmetric about the $\theta = \dfrac{\pi}{2}$ axis.

Problem Solving for Chapter 6

Solutions to Even-Numbered Exercises

2.

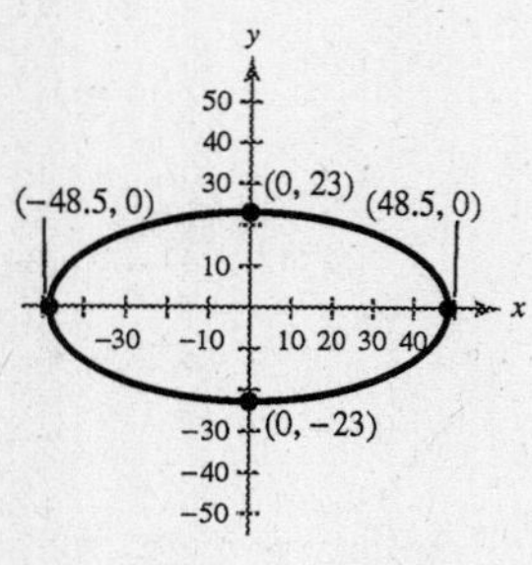

(a) Let $(0, 0)$ represent the center of the ellipse.
Then $2a = 97 \implies a = 48.5$ and $2b = 46 \implies b = 23$.

$$\frac{x^2}{(48.5)^2} + \frac{y^2}{23^2} = 1$$

$$\frac{x^2}{2352.25} + \frac{y^2}{529} = 1$$

(b) $c^2 = a^2 - b^2 = 2352.25 - 529 = 1823.25$

$c \approx 42.7$

The foci are $2c \approx 85.4$ feet apart

(c) $A = \pi ab = \pi(48.5)(23) = 1115.5\pi \approx 3504.45$ square feet

4. Let (x, x) be the corner of the square in Quadrant I.

$A = 4x^2$

$$\frac{x^2}{a^2} + \frac{x^2}{b^2} = 1 \implies x^2 = \frac{a^2b^2}{a^2 + b^2}$$

Thus, $A = \dfrac{4a^2b^2}{a^2 + b^2}$

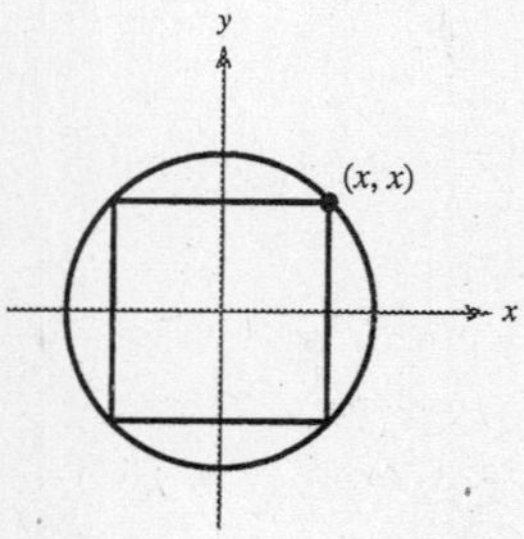

6. Foci: $(2, 2)$ and $(10, 2) \Rightarrow$ Center is $(6, 2)$ and $c = 4$

$|d_2 - d_1| = 2a = 6 \Rightarrow a = 3$

$c^2 = a^2 + b^2 \Rightarrow 16 = 9 + b^2 \Rightarrow b^2 = 7$

Horizontal Transverse axis

$$\frac{(x-6)^2}{9} - \frac{(y-2)^2}{7} = 1$$

8. (a) The first model describes linear motion, whereas the second model describes parabolic motion.

(b) $x = (v_0 \cos \theta)t \Rightarrow t = \dfrac{x}{v_0 \cos \theta}$

$y = (v_0 \sin \theta)t \Rightarrow t = \dfrac{y}{v_0 \sin \theta}$

$$\frac{x}{v_0 \cos \theta} = \frac{y}{v_0 \sin \theta}$$

$(v_0 \cos \theta)y = (v_0 \sin \theta)x$

$y = (\tan \theta)x$

$x = (v_0 \cos \theta)t \Rightarrow t = \dfrac{x}{v_0 \cos \theta}$

$y = h + (v_0 \sin \theta)t - 16t^2$

$$y = h + (v_0 \sin \theta)\left(\frac{x}{v_0 \cos \theta}\right) - 16\left(\frac{x}{v_0 \cos \theta}\right)^2$$

$$y = h + (\tan \theta)x - \left(\frac{16}{v_0 \cos^2 \theta}\right)x^2$$

(c) In the case $x = (v_0 \cos \theta)t$, $y = (v_0 \sin \theta)t$, the path of the projectile is not affected by changing the velocity v. When the parameter is eliminated, we just have $y = (\tan \theta)x$. The path is only affected by the angle θ.

10. $x = (a - b)\cos t + b \cos\left(\dfrac{a-b}{b}t\right)$

$y = (a - b)\sin t - b \sin\left(\dfrac{a-b}{b}t\right)$

(a) $a = 2, b = 1$

$x = \cos t + \cos t = 2 \cos t$

$y = \sin t - \sin t = 0$

The graph oscillates between -2 and 2 on the x-axis.

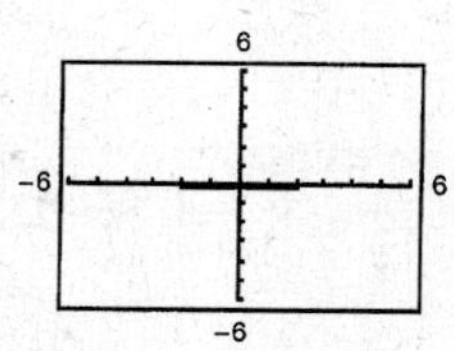

(b) $a = 3, b = 1$

$x = 2 \cos t + \cos 2t$

$y = 2 \sin t - \sin 2t$

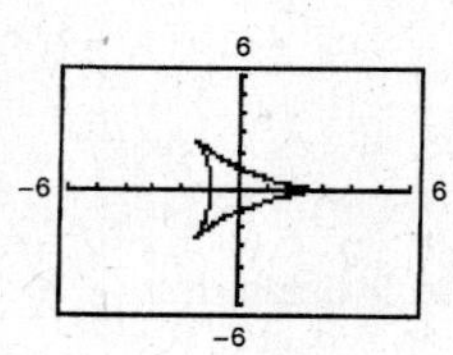

(c) $a = 4, b = 1$

$x = 3 \cos t + \cos 3t$

$y = 3 \sin t - \sin 3t$

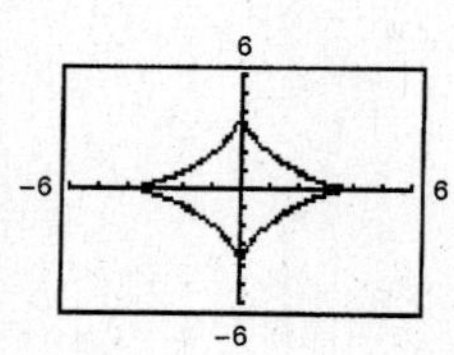

—CONTINUED—

—CONTINUED—

(d) $a = 10, b = 1$

$x = 9 \cos t + \cos 9t$

$y = 9 \sin t - \sin 9t$

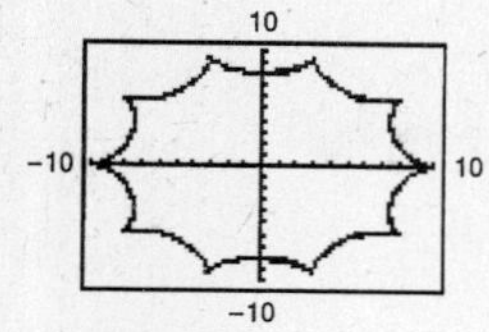

(e) $a = 3, b = 2$

$x = \cos t + 2 \cos \frac{t}{2}$

$y = \sin t - 2 \sin \frac{t}{2}$

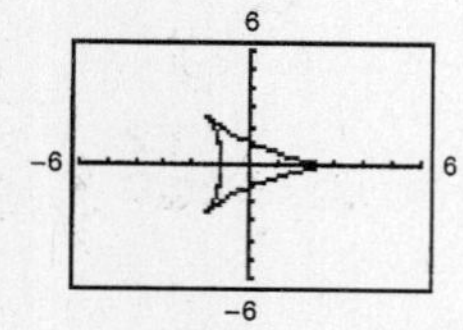

The graph looks the same as the graph in part (b), but is oriented clockwise instead of counterclockwise.

(f) $a = 4, b = 3$

$x = \cos t + 3 \cos \frac{t}{3}$

$y = \sin t - 3 \sin \frac{t}{3}$

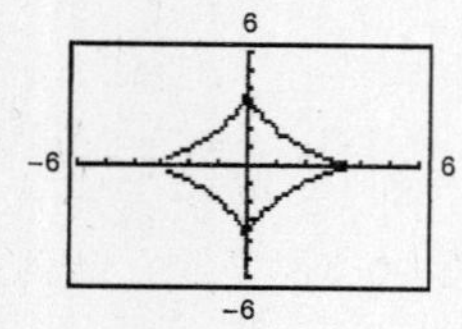

The graph is the same as the graph in part (c), but is oriented clockwise instead of counterclockwise.

12. $r = 2 \cos\left(\frac{1}{2}\theta\right)$

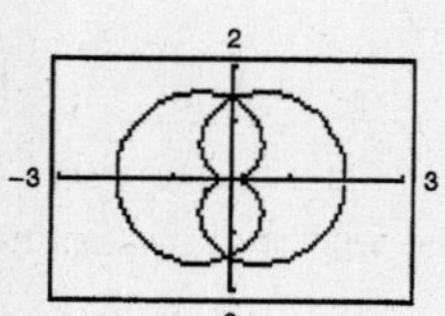

$r = 3 \sin\left(\frac{5\theta}{2}\right)$

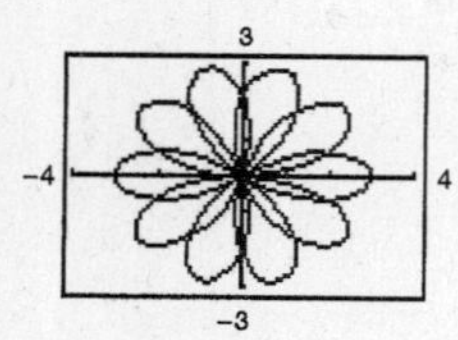

$r = -\cos\left(\sqrt{2}\,\theta\right)$

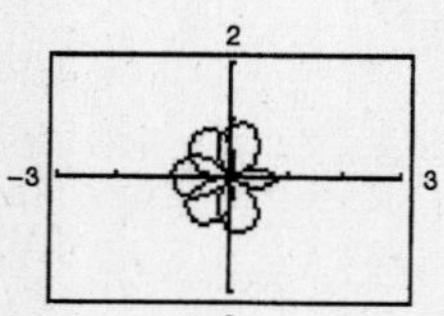

$r = -2 \sin\left(\frac{4\theta}{7}\right)$

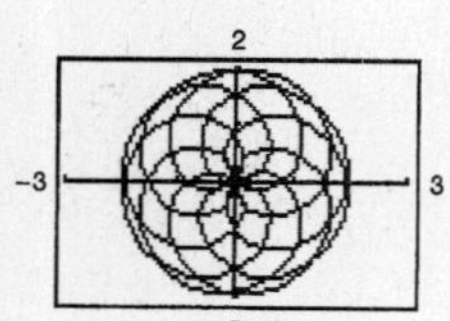

The graphs all contain overlapping loops or petals.

14. $r = e^{\cos \theta} - 2 \cos 4\theta + \sin^5\left(\frac{\theta}{12}\right)$

(a) No, the graph appears to have a period of 2π but does not. For example, $r(\pi) \neq r(3\pi)$.

(b) By using the table feature of the calculator we have $r \approx 4.077$ when $\theta \approx 5.54$ for $0 \leq \theta \leq 2\pi$ and $r \approx 4.46$ when $\theta \approx 11.83$ for $0 \leq \theta \leq 4\pi$. The graph is not periodic. As θ increases the value of r changes.

ISBN 0-618-31799-6
90000
9 780618 317998

HOUGHTON MIFFLIN
New Ways to Know ®

3-32764